Edited by
Sabu Thomas, Kuruvilla Joseph,
Sant Kumar Malhotra,
Koichi Goda, and Meyyarappallil
Sadasivan Sreekala

Polymer Composites

Related Titles

Nicolais, L., Carotenuto, G.

Nanocomposites

In Situ Synthesis of Polymer-Embedded Nanostructures

2012

ISBN: 978-0-470-10952-6

Thomas, S., Stephen, R.

Rubber Nanocomposites

Preparation, Properties and Applications

2010

ISBN: 978-0-470-82345-3

Nwabunma, D., Kyu, T. (eds.)

Polyolefin Blends and Composites

2008

ISBN: 978-0-470-19614-4

Nwabunma, D., Kyu, T. (eds.)

Polyolefin Composites

2008

ISBN: 978-0-471-79057-0

Klyosov, A. A.

Wood-Plastic Composites

2007

ISBN: 978-0-470-14891-4

Morgan, A. B., Wilkie, C. A. (eds.)

Flame Retardant Polymer Nanocomposites

2007

ISBN: 978-0-471-73426-0

Wereszczak, A., Lara-Curzio, E., Mizuno, M. (eds.)

Advances in Bioceramics and Biocomposites II, Ceramic Engineering and Science Proceedings

Volume 27, Issue 6

2007

ISBN: 978-0-470-08056-6

Kickelbick, G. (ed.)

Hybrid Materials

Synthesis, Characterization, and Applications

2007

ISBN: 978-3-527-31299-3

Edited by
Sabu Thomas, Kuruvilla Joseph,
Sant Kumar Malhotra, Koichi Goda,
and Meyyarappallil Sadasivan Sreekala

Polymer Composites

Volume 1

WILEY-VCH Verlag GmbH & Co. KGaA

The Editors

Prof. Dr. Sabu Thomas
Mahatma Gandhi University
School of Chemical Sciences
Priyadarshini Hills P.O.
Kottayam, Kerala 686560
India

Prof. Dr. Kuruvilla Joseph
Indian Institute of Space Science
and Technology
ISRO P.O., Veli
Thiruvananthapuram, Kerala 695022
India

Dr. Sant Kumar Malhotra
Flat-YA, Kings Mead
Srinagar Colony
14/3, South Mada Street
Saidafet, Chennai 60015
India

Prof. Koichi Goda
Yamaguchi University
Faculty of Engineering
Department of Mechanical Engineering
Tokiwadai 2-16-1
Ube, Yamaguchi 755-8611
Japan

Dr. Meyyarappallil Sadasivan Sreekala
Assistant Professor
Post Graduate Department of Chemistry
SreeSankara College
Kalady, Kerala 683574
India

■ All books published by **Wiley-VCH** are carefully produced. Nevertheless, authors, editors, and publisher do not warrant the information contained in these books, including this book, to be free of errors. Readers are advised to keep in mind that statements, data, illustrations, procedural details or other items may inadvertently be inaccurate.

Library of Congress Card No.: applied for

British Library Cataloguing-in-Publication Data
A catalogue record for this book is available from the British Library.

Bibliographic information published by the Deutsche Nationalbibliothek
The Deutsche Nationalbibliothek lists this publication in the Deutsche Nationalbibliografie; detailed bibliographic data are available on the Internet at http://dnb.d-nb.de.

© 2012 Wiley-VCH Verlag & Co. KGaA,
Boschstr. 12, 69469 Weinheim, Germany

All rights reserved (including those of translation into other languages). No part of this book may be reproduced in any form – by photoprinting, microfilm, or any other means – nor transmitted or translated into a machine language without written permission from the publishers. Registered names, trademarks, etc. used in this book, even when not specifically marked as such, are not to be considered unprotected by law.

Composition Thomson Digital, Noida, India

Printing and Binding Markono Print Media Pte Ltd, Singapore

Cover Design Adam Design, Weinheim

Printed in Singapore
Printed on acid-free paper

Print ISBN: 978-3-527-32624-2

Contents

The Editors *XXIII*
Preface *XXV*
List of Contributors *XXVII*

Part One Introduction to Polymer Composites *1*

1 **Advances in Polymer Composites: Macro- and Microcomposites – State of the Art, New Challenges, and Opportunities** *3*
Josmin P. Jose, Sant Kumar Malhotra, Sabu Thomas, Kuruvilla Joseph, Koichi Goda, and Meyyarappallil Sadasivan Sreekala
1.1 Introduction *3*
1.2 Classification of Composites *4*
1.2.1 Polymer Matrix Composites *4*
1.2.1.1 Factors Affecting Properties of PMCs *6*
1.2.1.2 Fabrication of Composites *7*
1.2.1.3 Applications *10*
1.2.1.4 Recent Advances in Polymer Composites *10*
1.3 Interface Characterization *14*
1.3.1 Micromechanical Technique *14*
1.3.2 Spectroscopic Tests *15*
1.3.3 Microscopic Techniques *15*
1.3.4 Thermodynamic Methods *15*
1.4 New Challenges and Opportunities *15*
References *16*

2 **Shock and Impact Response of Glass Fiber-Reinforced Polymer Composites** *17*
Vikas Prakash
2.1 Introduction *17*
2.2 Analytical Analysis *20*
2.2.1 Wave Propagation in Elastic–Viscoelastic Bilaminates *21*
2.2.2 Solution at Wave Front: Elastic Precursor Decay *23*
2.2.3 Late-Time Asymptotic Solution *24*

2.3	Plate-Impact Experiments on GRPs 33
2.3.1	Material: Glass Fiber-Reinforced Polymer 33
2.3.2	Plate-Impact Shock Compression Experiments: Experimental Configuration 36
2.3.2.1	t–X Diagram (Time versus Distance) and S–V Diagram (Stress versus Velocity) for Plate-Impact Shock Compression Experiments 37
2.3.3	Plate-Impact Spall Experiments: Experimental Configuration 38
2.3.3.1	t–X Diagram (Time versus Distance) and S–V Diagram (Stress versus Velocity) for Plate-Impact Spall Experiments 38
2.3.4	Shock–Reshock and Shock–Release Experiments: Experimental Configuration 40
2.3.4.1	t–X Diagram (Time versus Distance) for Shock–Reshock and Shock–Release Experiments 41
2.4	Target Assembly 42
2.5	Experimental Results and Discussion 42
2.5.1	Plate-Impact Shock Compression Experiments 42
2.5.1.1	Structure of Shock Waves in the GRP 44
2.5.1.2	Equation of State (Shock Velocity versus Particle Velocity) for S2-Glass GRP 49
2.5.1.3	Hugoniot Stress versus Hugoniot Strain (Hugoniot) 51
2.5.1.4	Hugoniot Stress versus Particle Velocity 54
2.5.2	Plate-Impact Spall Experiments 58
2.5.2.1	Determination of Spall Strength 59
2.5.2.2	Spall Strength of GRP Following Normal Shock Compression 61
2.5.2.3	Spall Strength of GRP Following Combined Shock Compression and Shear Loading 62
2.5.3	Shock–Reshock and Shock–Release Experiments 66
2.5.3.1	Self Consistent Method for the Determination of Dynamic Shear Yield Strength 67
2.5.3.2	Calculation of Initial Hugoniot Shocked State and Hugoniot Stress–Strain Curve 71
2.5.3.3	Calculation of Off-Hugoniot States for Reshock–Release Loading 72
2.5.3.4	Determination of the Critical Shear Strength in the Shocked State for S2-Glass GRP 74
2.6	Summary 76
	References 78
3	**Interfaces in Macro- and Microcomposites** 83
	Haeng-Ki Lee and Bong-Rae Kim
3.1	Introduction 83
3.2	Characterization of Interfaces in Macro- and Microcomposites 85
3.2.1	Surface Treatments of Reinforcements for Composite Materials 85
3.2.2	Microscale Tests 85
3.3	Micromechanics-Based Analysis 87

3.3.1	Micromechanical Homogenization Theory	87
3.3.1.1	Representative Volume Element	87
3.3.1.2	Eshelby's Equivalent Inclusion Method	88
3.3.2	Effective Elastic Modulus	89
3.3.2.1	Self-Consistent Method	89
3.3.2.2	Mori–Tanaka Method	90
3.3.2.3	Ensemble-Volume Average Method	90
3.3.3	Interface Model	92
3.3.3.1	Linear Spring Model	92
3.3.3.2	Interface Stress Model	93
3.3.3.3	Dislocation-Like Model	93
3.3.3.4	Free Sliding Model	94
3.4	Interfacial Damage Modeling	94
3.4.1	Conventional Weibull's Probabilistic Approach	94
3.4.2	Multilevel Damage Model	97
3.4.3	Cumulative Damage Model	98
3.5	Summary	100
	References	101
4	**Preparation and Manufacturing Techniques for Macro- and Microcomposites**	**111**
	Tibor Czigány and Tamás Deák	
4.1	Introduction	111
4.2	Thermoplastic Polymer Composites	111
4.2.1	Injection Molding	112
4.2.2	Extrusion	115
4.2.3	Compression Molding	115
4.2.4	Thermoplastic Prepreg Lay-up	117
4.2.5	Thermoplastic Tape Winding	118
4.2.6	Thermoplastic Pultrusion	119
4.2.7	Diaphragm Forming	119
4.2.8	Classification of Thermoplastic Composite Manufacturing Techniques	120
4.3	Thermosetting Polymer Composites	123
4.3.1	Hand Lamination	124
4.3.2	Spray-Up	124
4.3.3	Centrifugal Casting	124
4.3.4	Prepreg Lay-Up	125
4.3.5	Resin Transfer Molding and Vacuum-Assisted Resin Transfer Molding	126
4.3.6	Bulk Molding Compound and Sheet Molding Compound	127
4.3.7	Reaction Injection Molding and Structural Reaction Injection Molding	127
4.3.8	Thermosetting Injection Molding	128
4.3.9	Filament Winding	128

4.3.10	Pultrusion *129*	
4.3.11	Classification of Thermosetting Composite Manufacturing Techniques *130*	
4.4	Future Trends *133*	
	References *133*	

Part Two Macrosystems: Fiber-Reinforced Polymer Composites *135*

5	**Carbon Fiber-Reinforced Polymer Composites: Preparation, Properties, and Applications** *137*	
	Soo-Jin Park and Min-Kang Seo	
5.1	Introduction *137*	
5.2	Backgrounds *138*	
5.2.1	Manufacturing Processes *138*	
5.2.2	Surface Treatment *138*	
5.2.2.1	Surface Treatment of Carbon Fibers *139*	
5.2.3	Characterization of Polymeric Composites *141*	
5.2.4	Fiber Reinforcements *141*	
5.3	Experimental Part *143*	
5.3.1	Materials *143*	
5.3.2	Surface Treatment of Carbon Fibers *143*	
5.3.2.1	Electrochemical Oxidation *143*	
5.3.2.2	Electroplating *143*	
5.3.2.3	Oxyfluorination *144*	
5.3.2.4	Plasma Modification *145*	
5.3.3	Preparation of Carbon Fiber-Reinforced Polymer Composites *146*	
5.3.4	Characterization of Carbon Fibers *146*	
5.3.5	Theoretical Considerations of Dynamic Contact Angles *147*	
5.3.6	Characterization of Carbon Fiber-Reinforced Polymer Composites *148*	
5.3.6.1	Interlaminar Shear Strength *148*	
5.3.6.2	Critical Stress Intensity Factor (K_{IC}) *150*	
5.3.6.3	Mode I Interlaminar Fracture Toughness Factor *151*	
5.3.6.4	Fracture Behaviors *151*	
5.4	Results and Discussion *153*	
5.4.1	Effect of Electrochemical Oxidation *153*	
5.4.1.1	Surface Characteristics *153*	
5.4.1.2	Contact Angle and Surface Free Energy *154*	
5.4.1.3	Mechanical Interfacial Properties *155*	
5.4.2	Effect of Electroplating *156*	
5.4.2.1	Surface Characteristics *156*	
5.4.2.2	Contact Angle and Surface Free Energy *159*	
5.4.2.3	Mechanical Interfacial Properties *161*	
5.4.3	Effect of Oxyfluorination *163*	
5.4.3.1	Surface Characteristics *163*	
5.4.3.2	Contact Angle and Surface Free Energy *165*	

5.4.3.3	Mechanical Interfacial Properties	*167*
5.4.4	Effect of Plasma Treatment	*171*
5.4.4.1	Surface Characteristics	*171*
5.4.4.2	Contact Angle and Surface Free Energy	*172*
5.4.4.3	Mechanical Interfacial Properties	*173*
5.5	Applications	*176*
5.5.1	Automotives	*177*
5.5.2	Wind Energy	*177*
5.5.3	Deepwater Offshore Oil and Gas Production	*178*
5.5.4	Electricity Transmission	*178*
5.5.5	Commercial Aircraft	*178*
5.5.6	Civil Infrastructure	*178*
5.5.7	Other Applications	*179*
5.6	Conclusions	*179*
	References	*180*

6	**Glass Fiber-Reinforced Polymer Composites**	*185*
	Sebastian Heimbs and Björn Van Den Broucke	
6.1	Introduction	*185*
6.2	Chemical Composition and Types	*186*
6.2.1	Chemical Structure of Glass	*186*
6.2.2	Glass Fiber Types	*187*
6.3	Fabrication of Glass Fibers	*188*
6.3.1	Fiber Production	*188*
6.3.2	Sizing Application	*189*
6.4	Forms of Glass Fibers	*190*
6.4.1	Commercially Available Forms of Glass Fibers	*190*
6.4.2	Shaped Glass Fibers	*191*
6.5	Glass Fiber Properties	*192*
6.5.1	General Properties	*192*
6.5.2	Elastic Properties	*193*
6.5.3	Strength and Elongation Properties	*193*
6.5.4	Corrosion Properties	*195*
6.5.5	Thermal Properties	*195*
6.6	Glass Fibers in Polymer Composites	*196*
6.6.1	Polymers for Glass Fiber-Reinforced Composites	*196*
6.6.2	Determination of Properties	*197*
6.6.3	Manufacturing Processes and Related Composite Properties	*197*
6.6.4	Strength and Fatigue Properties	*199*
6.6.5	Strain Rate Effect in Glass Fiber-Reinforced Composites	*200*
6.6.6	Environmental Influences	*200*
6.6.7	Other Physical Properties of Glass Fiber-Reinforced Composites	*200*
6.7	Applications	*202*
6.8	Summary	*204*
	References	*205*

7	**Kevlar Fiber-Reinforced Polymer Composites** *209*
	Chapal K. Das, Ganesh C. Nayak, and Rathanasamy Rajasekar
7.1	Introduction *209*
7.2	Fiber-Reinforced Polymer Composites *210*
7.3	Constituents of Polymer Composites *210*
7.3.1	Synthetic Fibers *210*
7.4	Kevlar Fiber *211*
7.4.1	Development and Molecular Structure of Kevlar *211*
7.4.2	Properties of Kevlar Fibers *212*
7.5	Interface *212*
7.6	Factors Influencing the Composite Properties *214*
7.6.1	Strength, Modulus, and Chemical Stability of the Fiber and the Polymer Matrix *214*
7.6.2	Influence of Fiber Orientation and Volume Fraction *214*
7.6.2.1	Fiber Orientation in Injection Molded Fiber-Reinforced Composites *214*
7.6.3	Volume Fraction *216*
7.6.4	Influence of Fiber Length *217*
7.6.5	Influence of Voids *217*
7.6.6	Influence of Coupling Agents *218*
7.7	Surface Modification *218*
7.7.1	Surface Modification of Fibers *218*
7.7.2	Surface Modification of Matrix Polymers *219*
7.7.3	Fluorination and Oxyfluorination as Polymer Surface Modification Tool *219*
7.8	Synthetic Fiber-Reinforced Composites *220*
7.9	Effect of Fluorinated and Oxyfluorinated Short Kevlar Fiber on the Properties of Ethylene Propylene Matrix Composites *222*
7.9.1	Preparation of Composites *222*
7.9.2	FTIR Study *222*
7.9.3	X-Ray Study *223*
7.9.4	Thermal Properties *225*
7.9.5	Dynamic Mechanical Thermal Analysis (DMTA) *226*
7.9.6	Mechanical Properties *228*
7.9.7	SEM Study *228*
7.9.8	AFM Study *229*
7.9.9	Conclusion *229*
7.10	Compatibilizing Effect of MA-g-PP on the Properties of Fluorinated and Oxyfluorinated Kevlar Fiber-Reinforced Ethylene Polypropylene Composites *230*
7.10.1	Preparation of the Composites *230*
7.10.2	Thermal Properties *230*
7.10.3	X-Ray Study *233*
7.10.4	Dynamic Mechanical Thermal Analysis *233*
7.10.5	Flow Behavior *236*

7.10.6	SEM Study 237	
7.10.7	Conclusion 238	
7.11	Properties of Syndiotactic Polystyrene Composites with Surface-Modified Short Kevlar Fiber 238	
7.11.1	Preparation of s-PS/Kevlar Composites 238	
7.11.2	FTIR Study of the Composites 238	
7.11.3	Differential Scanning Calorimetric Study 239	
7.11.4	Thermal Properties 241	
7.11.5	X-Ray Study 242	
7.11.6	Dynamic Mechanical Thermal Analysis 243	
7.11.7	SEM Study 245	
7.11.8	AFM Study 245	
7.11.9	Conclusion 246	
7.12	Study on the Mechanical, Rheological, and Morphological Properties of Short Kevlar Fiber/s-PS Composites Effect of Oxyfluorination of Kevlar 246	
7.12.1	Rheological Properties 247	
7.12.2	Mechanical Properties 248	
7.12.3	Scanning Electron Microscopy Study 249	
7.12.4	Conclusion 249	
7.13	Effect of Fluorinated and Oxyfluorinated Short Kevlar Fiber Reinforcement on the Properties of PC/LCP Blends 250	
7.13.1	Preparation of Composites 250	
7.13.2	Differential Scanning Calorimetric Study 250	
7.13.3	Thermal Properties 252	
7.13.4	X-Ray Study 253	
7.13.5	Dynamic Mechanical Analysis (DMA) 253	
7.13.6	SEM Study 256	
7.13.7	Conclusion 256	
7.14	Simulation of Fiber Orientation by Mold Flow Technique 257	
7.14.1	Theoretical basis for Fiber Orientation Prediction 257	
7.14.2	Mold Flow's Fiber Orientation Model 259	
7.14.3	Simulation of Fiber Orientation by Mold Flow Technique on s-PS/Kevlar Composites 259	
7.14.4	Simulation of Fiber Orientation by Mold Flow Technique for PC/LCP/Kevlar Composites 265	
7.15	Kevlar-Reinforced Thermosetting Composites 270	
	References 272	
8	**Polyester Fiber-Reinforced Polymer Composites** 275	
	Dionysis E. Mouzakis	
8.1	Introduction 275	
8.2	Synthesis and Basic Properties of Polyester Fibers 277	
8.2.1	Fiber Manufacturing 278	
8.2.2	Basic Properties of Polyester Fibers 279	

8.2.3	Mechanical Response	279
8.2.4	Fiber Viscoelastic Properties	281
8.2.5	PEN Fibers	282
8.3	Polyester Fiber-Reinforced Polymer Composites	282
8.3.1	Elastomer Composites	282
8.3.2	Microfibrillar-Reinforced (MFR) PET Composites	283
8.3.3	Composites	285
8.3.4	PET Nanocomposites	286
8.4	Conclusions	287
	References	288

9 Nylon Fiber-Reinforced Polymer Composites 293
Valerio Causin

9.1	Introduction	293
9.2	Nylon Fibers Used as Reinforcements	294
9.3	Matrices and Applications	299
9.4	Manufacturing of Nylon-Reinforced Composites	305
9.5	Conclusions	311
	References	311

10 Polyolefin Fiber- and Tape-Reinforced Polymeric Composites 315
József Karger-Kocsis and Tamás Bárány

10.1	Introduction	315
10.2	Polyolefin Fibers and Tapes	315
10.2.1	Production	317
10.2.1.1	Hot Drawing	317
10.2.1.2	Gel Drawing	318
10.2.2	Properties and Applications	318
10.3	Polyolefin-Reinforced Thermoplastics	321
10.3.1	Self-Reinforced Version	321
10.3.1.1	Hot Compaction	321
10.3.1.2	Film Stacking	322
10.3.1.3	Wet Impregnation Prior to Hot Consolidation	324
10.3.2	Polyolefin Fiber-Reinforced Composites	324
10.3.2.1	Consolidation of Coextruded Tapes	324
10.3.2.2	Film Stacking	325
10.3.2.3	Solution Impregnation	325
10.3.2.4	Powder Impregnation (Wet and Dry)	326
10.3.2.5	*In Situ* Polymerization of the Matrix	326
10.3.3	Interphase	326
10.3.4	Hybrid Fiber-Reinforced Composites	327
10.4	Polyolefin Fiber-Reinforced Thermosets	327
10.4.1	Polyolefin Fiber-Reinforced Composites	327
10.4.2	Hybrid Fiber-Reinforced Composites	328
10.5	Polyolefin Fibers in Rubbers	329

10.5.1	Polyolefin Fiber-Reinforced Composites	329
10.5.2	Hybrid Fiber-Reinforced Composites	329
10.6	Others	330
10.7	Outlook and Future Trends	330
	References	331

11 Silica Fiber-Reinforced Polymer Composites 339
Sudip Ray

11.1	Introduction	339
11.2	Silica Fiber: General Features	339
11.2.1	Types	340
11.2.2	Characteristics	340
11.2.3	Surface Treatment of Silica Fiber	342
11.2.3.1	Surface Modification: Types and Methods	342
11.2.3.2	Characterization of Surface-Pretreated Silica Fiber	343
11.3	Silica Fiber-Filled Polymer Composites	347
11.3.1	Fabrication of Composite	347
11.3.2	Effect on Composite Properties	347
11.3.3	Surface-Modified Silica Fiber-Filled Polymer Composites	351
11.3.4	Reinforcement Mechanism	358
11.4	Applications	358
11.5	New Developments	360
11.6	Concluding Remarks	361
	References	361

Part Three Macrosystems: Textile Composites 363

12 2D Textile Composite Reinforcement Mechanical Behavior 365
Emmanuelle Vidal-Sallé and Philippe Boisse

12.1	Introduction	365
12.2	Mechanical Behavior of 2D Textile Composite Reinforcements and Specific Experimental Tests	366
12.2.1	Load Resultants on a Woven Unit Cell	366
12.2.2	Principle of Virtual Work	367
12.2.3	Biaxial Tensile Behavior	369
12.2.4	In-Plane Shear Behavior	370
12.2.5	Bending Behavior	372
12.3	Continuous Modeling of 2D Fabrics: Macroscopic Scale	373
12.3.1	Geometrical Approaches	373
12.3.2	Mechanical Approaches	374
12.3.2.1	Hypoelastic Model for Macroscopic Modeling of 2D Fabrics	375
12.3.2.2	Hyperelastic Model for Macroscopic Modeling of 2D Fabric	378
12.4	Discrete Modeling of 2D Fabrics: Mesoscopic Scale	382
12.4.1	Modeling the Global Preform	382
12.4.2	Modeling the Woven Cell	384

12.4.2.1	Longitudinal Behavior	385
12.4.2.2	Transverse Behavior	385
12.4.3	Use of the Mesoscale Modeling for Permeability Evaluation	386
12.5	Conclusions and Future Trend	388
	References	388

13 Three Dimensional Woven Fabric Composites 393
Wen-Shyong Kuo

13.1	Introduction	393
13.2	General Characteristics of 3D Composites	394
13.2.1	Multidirectional Structural Integrity	394
13.2.2	Near-Net-Shape Design	394
13.2.3	Greater Nonuniformity	394
13.2.4	Lower Fiber Volume Fraction	395
13.2.5	Higher Fiber Crimp	395
13.2.6	Lower Stress-to-Yield and Higher Strain-to-Failure	395
13.2.7	More Difficult in Material Testing	396
13.2.8	More Complex in Damage Mechanisms	396
13.3	Formation of 3D Woven Fabrics	396
13.3.1	Three-Axis Orthogonal Weaving	396
13.3.2	Design of Weaving Schemes	399
13.3.3	Yarn Distortion	403
13.3.4	Use of Solid Rods	405
13.4	Modeling of 3D Woven Composites	407
13.4.1	Fiber Volume Fractions	407
13.4.2	Elastic Properties of Yarns	409
13.4.3	Rule-of-Mixtures	409
13.4.4	Rotation of a Yarn	410
13.4.5	Equivalent Properties of 3D Composites	412
13.5	Failure Behavior of 3D Woven Composites	412
13.5.1	Tensile Fracture	413
13.5.2	Bending Fracture	413
13.5.2.1	Weaker Plane	414
13.5.2.2	Influence of Surface Loops	415
13.5.3	Compressive Damage	417
13.5.3.1	Microband	417
13.5.3.2	Miniband	420
13.5.3.3	Fracture Due to Compression-Induced Bending	423
13.5.4	Fracture Due to Transverse Shear	424
13.5.5	Impact Damage	427
13.6	Role of Interlacing Loops	428
13.6.1	Covering Weaker Planes	428
13.6.2	Holding Axial Yarns	429
13.7	Design of 3D Woven Composites	429

13.7.1	Modulus *429*	
13.7.2	Yield Point *430*	
13.7.3	Strain-to-Failure *430*	
13.7.4	Energy-Absorption and Damage Resistance *430*	
13.8	Conclusions *431*	
	References *431*	

14 Polymer Composites as Geotextiles *435*
Han-Yong Jeon

14.1	Introduction *435*	
14.1.1	Definition *435*	
14.1.2	Function of Composite Geotextiles *437*	
14.1.3	Application Fields of Composite Geotextiles *440*	
14.2	Developments of Composite Geotextiles *443*	
14.2.1	Raw Materials of Composite Geotextiles *443*	
14.2.1.1	Natural Fibers *443*	
14.2.1.2	Synthetic Fibers *444*	
14.2.1.3	Recycled Fibers *444*	
14.2.1.4	Advanced Functional Fibers *444*	
14.2.2	Advanced Trend of Composite Geotextiles *444*	
14.2.2.1	Geotextiles *444*	
14.2.2.2	Geosynthetic Clay Liners *445*	
14.2.2.3	Composite Geotextiles for Drainage and Filtration *445*	
14.2.2.4	Reinforced Concrete by Composite Geotextiles *446*	
14.2.2.5	Reinforced Geomembrane by Composite Geotextiles *446*	
14.3	Hybrid Composite Geotextiles *447*	
14.3.1	For Separation and Reinforcement *447*	
14.3.1.1	Manufacturing of Geotextiles/Geogrids Composites *447*	
14.3.1.2	Wide-Width Tensile Strength *448*	
14.3.1.3	Hydraulic Properties *448*	
14.3.2	For Drainage *449*	
14.3.2.1	Manufacturing of Geotextiles/Geonet Composites *449*	
14.3.2.2	Compressive Stress and Transmissivity *450*	
14.3.3	For Protection and Slope Stability *454*	
14.3.3.1	Manufacturing of Three-Layered Composite Geotextiles *454*	
14.3.3.2	Transmissivity *455*	
14.3.3.3	Thickness and Compressive Stress *457*	
14.3.3.4	Thickness and In-Plane Permeability *458*	
14.3.4	For Frictional Stability *461*	
14.3.4.1	Manufacturing of Geotextiles/Geomembranes Composites *461*	
14.3.4.2	Frictional Properties *461*	
14.4	Performance Evaluation of Composite Geotextiles *462*	
14.4.1	Performance Test Items *465*	
14.4.2	Required Evaluation Test Items *465*	
14.4.2.1	Composite Geotextiles *465*	

14.4.2.2	Geosynthetics Clay Liners 466	
	References 467	

15	**Hybrid Textile Polymer Composites** *469*	
	Palanisamy Sivasubramanian, Laly A. Pothan, M. Thiruchitrambalam, and Sabu Thomas	
15.1	Introduction 469	
15.2	Textile Composites 470	
15.2.1	Manufacture of Natural Fiber Textile-Reinforced Composites 475	
15.3	Hybrid Textile Composites 478	
15.4	Hybrid Textile Joints 479	
15.5	Conclusion 480	
	References 480	

Part Four Microsystems : Microparticle-Reinforced Polymer Composites *483*

16	**Characterization of Injection-Molded Parts with Carbon Black-Filled Polymers** *485*	
	Volker Piotter, Jürgen Prokop, and Xianping Liu	
16.1	Introduction 485	
16.2	Injection-Molded Carbon-Filled Polymers 486	
16.3	Processes and Characterization 488	
16.3.1	Rheological Characterization of the Compounds 488	
16.3.2	Molding and Electrical Characterization of Conductive Compounds 489	
16.3.3	Comparison of the Results Obtained with Simulation Calculations 490	
16.3.4	Electroplating of Injection-Molded Components 494	
16.3.5	Transfer of the Test Results to Standard Material Systems 495	
16.3.6	Summary of the Deposition Studies 500	
16.3.7	Demonstrator Production 501	
16.4	Mechanical Property Mapping of Carbon-Filled Polymer Composites by TPM *501*	
16.4.1	Introduction 501	
16.4.2	Multifunction Tribological Probe Microscope 502	
16.4.3	Specimen Preparation 505	
16.4.4	Multifunction Mapping 507	
16.5	Conclusions 512	
	References 512	

17	**Carbon Black-Filled Natural Rubber Composites: Physical Chemistry and Reinforcing Mechanism** *515*	
	Atsushi Kato, Yuko Ikeda, and Shinzo Kohjiya	
17.1	Introduction 515	
17.2	3D-TEM Observation of Nanofiller-Loaded Vulcanized Rubber 517	
17.3	Materials: CB-Filled Sulfur-Cured NR Vulcanizates 518	

17.4	Relationship Between the Properties of CB-Filled Sulfur-Cured NR Vulcanizates and CB Loading *519*	
17.4.1	Tensile Behavior *519*	
17.4.2	Coefficient of Thermal Expansion *520*	
17.4.3	Viscoelastic Properties *520*	
17.4.4	Electrical Properties: Volume Resistivity and Conductivity *522*	
17.4.5	Dielectric Properties: Time–Temperature Superposition of Dielectric Relaxation of CB-Filled NR Vulcanizates *525*	
17.5	CB Dispersion and Aggregate/Agglomerate Structure in CB-Filled NR Vulcanizates *529*	
17.5.1	3D-TEM Observation of CB-Filled NR Vulcanizates and Parameters of CB Aggregates/Agglomerates *529*	
17.5.2	Visualization of CB Network in the Rubber Matrix and the Network Parameters *532*	
17.5.3	CB Aggregate Network Structure and Dielectric Relaxation Characteristics *535*	
17.6	Conclusions *538*	
	References *540*	

18 Silica-Filled Polymer Microcomposites *545*
Sudip Ray

18.1	Introduction *545*
18.2	Silica as a Filler: General Features *545*
18.2.1	Types *546*
18.2.2	Characteristics *546*
18.2.3	Surface Treatment of Silica Filler *548*
18.2.4	Surface Modification: Types and Methods *549*
18.2.4.1	Physical Modification *549*
18.2.4.2	Chemical Modification *549*
18.2.4.3	Filler Surface Modification with Monofunctional Coupling Agents *550*
18.2.4.4	Filler Surface Modification with Bifunctional Coupling Agents *550*
18.2.5	Characterization of Surface Pretreated Silica *552*
18.3	Silica-Filled Rubbers *552*
18.3.1	Effect of Silica Filler on Processability of Unvulcanized Rubber Compounds *553*
18.3.2	Effect of Silica Filler on Vulcanizate Properties of Rubber Compounds *554*
18.3.2.1	Mechanical Properties *554*
18.3.2.2	Dynamic Mechanical Properties *555*
18.3.3	Surface-Modified Silica-Filled Rubbers *557*
18.3.4	Reinforcement Mechanism *558*
18.3.4.1	Rheological Properties *559*
18.3.4.2	Mechanical Properties *560*
18.3.4.3	Dynamic Mechanical Properties *564*
18.3.4.4	Frequency Dependence of Storage Modulus and Loss Tangent *565*

18.3.4.5	Equilibrium Swelling	566
18.3.5	Applications	568
18.4	Silica-Filled Thermoplastics and Thermosets	569
18.5	Concluding Remarks	571
	References	572

19 Metallic Particle-Filled Polymer Microcomposites 575
Bertrand Garnier, Boudjemaa Agoudjil, and Abderrahim Boudenne

19.1	Introduction	575
19.2	Metallic Filler and Production Methods	576
19.3	Achieved Properties of Metallic Filled Polymer	577
19.3.1	Electrical Conductivity	577
19.3.2	Thermal Conductivity	579
19.3.3	Mechanical Properties	583
19.4	Main Factors Influencing Properties	585
19.4.1	Effect of Volume Fraction	585
19.4.2	Effect of Particle Shape	587
19.4.3	Effect of Particle Size	589
19.4.4	Effect of Preparation Process	591
19.5	Models for Physical Property Prediction	593
19.5.1	Theory of Composite Transport Properties	593
19.5.1.1	Effective Medium Theory	593
19.5.1.2	Theory of Percolation	594
19.5.2	Models for Thermal Conductivity	595
19.5.3	Models for Electrical Conductivity	598
19.5.4	Models for Mechanical Properties	602
19.6	Conclusion	606
	References	606

20 Magnetic Particle-Filled Polymer Microcomposites 613
Natalie E. Kazantseva

20.1	Introduction	613
20.2	Basic Components of Polymer Magnetic Composites: Materials Selection	614
20.2.1	Introduction to Polymer Magnetic Composite Processing	614
20.2.2	Polymers	615
20.2.3	Magnetic Fillers	615
20.3	Overview of Methods for the Characterization of Materials in the Radiofrequency and Microwave Bands	621
20.4	Magnetization Processes in Bulk Magnetic Materials	628
20.4.1	Magnetostatic Magnetization Processes	628
20.4.2	Dynamic Magnetization Processes	632
20.4.2.1	Natural Resonance	633
20.4.2.2	Domain-Wall Resonance	634
20.4.3	Experimental Magnetic Spectra of Bulk Ferromagnets	635

20.4.3.1	Bulk Ferrites	635
20.4.3.2	Ferromagnetic Metals and Alloys	639
20.5	Magnetization Processes in Polymer Magnetic Composites	641
20.5.1	Effect of Composition	641
20.5.2	Effect of Particle Size, Shape, and Microstructure	646
20.5.3	Estimation of the Effective Permeability of a PMC by Mixing Rules	649
20.6	Polymer Magnetic Composites with High Value of Permeability in the Radiofrequency and Microwave Bands	651
20.6.1	Polymer Magnetic Composites with Multicomponent Filler	652
20.6.2	Polymer Magnetic Composites with Multicomponent Magnetic Particles	659
20.6.2.1	Preparation of Magnetic Particles with Corelike Structure	659
20.6.2.2	The Mechanism of PANI Film Formation	661
20.6.2.3	Electromagnetic Properties of Composites with Core-shell Structure Magnetic Particles	663
20.7	Conclusions	668
	References	669
21	**Mica-Reinforced Polymer Composites**	**673**
	John Verbeek and Mark Christopher	
21.1	Introduction	673
21.2	Structure and Properties of Mica	674
21.2.1	Chemical and Physical Properties	674
21.2.2	Structure	674
21.2.3	Applications	676
21.3	Mechanical Properties of Mica–Polymer Composites	677
21.3.1	Mechanism of Reinforcement	678
21.3.1.1	Shape and Orientation	682
21.3.1.2	Mica Concentration	684
21.3.1.3	Particle Size and Size Distribution	685
21.3.2	Interfacial Adhesion	687
21.3.2.1	Coupling Agents	689
21.3.2.2	Mechanisms of Interfacial Modification	690
21.3.3	The Hybrid Effect in Systems Using More Than One Filler	692
21.4	Thermal Properties	693
21.4.1	Crystallization	693
21.4.2	Thermal Stability	694
21.4.3	Flammability	695
21.5	Other Properties	696
21.5.1	Processability	696
21.5.2	Barrier Properties	698
21.5.3	Electrical Properties	698
21.6	Modeling of Mechanical Properties	700

21.6.1	Young's Modulus	*700*
21.6.2	Tensile Strength	*705*
21.6.3	Limitations to Existing Models	*707*
21.7	Conclusions	*709*
	References	*709*

22 Viscoelastically Prestressed Polymeric Matrix Composites *715*
Kevin S. Fancey

22.1	Introduction	*715*
22.1.1	Prestress in Composite Materials	*715*
22.1.2	Principles of Viscoelastic Prestressing	*716*
22.2	Preliminary Investigations: Evidence of Viscoelastically Generated Prestress	*716*
22.2.1	Objectives	*716*
22.2.2	Creep Conditions for Viscoelastic Recovery	*717*
22.2.3	Visual Evidence of Viscoelastically Induced Prestress	*717*
22.2.4	Initial Mechanical Evaluation: Impact Tests	*718*
22.3	Time–Temperature Aspects of VPPMC Technology	*719*
22.3.1	A Mechanical Model for Polymeric Deformation	*719*
22.3.2	Accelerated Aging: Viscoelastic Recovery	*722*
22.3.3	Force–Time Measurements	*724*
22.3.4	Accelerated Aging: Impact Tests	*726*
22.3.5	The Long-Term Performance of VPPMCs	*728*
22.4	VPPMCs with Higher Fiber Content: Mechanical Properties	*729*
22.4.1	Meeting the Objectives	*729*
22.4.2	Flexural Properties	*729*
22.4.3	Tensile Properties	*731*
22.5	Processing Aspects of VPPMCs	*733*
22.5.1	Background	*733*
22.5.2	Fiber Properties	*734*
22.5.3	Geometrical Aspects of Fibers in Composite Samples	*735*
22.6	Mechanisms for Improved Mechanical Properties in VPPMCs	*737*
22.6.1	Background	*737*
22.6.2	Flexural Stiffness	*738*
22.6.3	Tensile Strength	*738*
22.6.4	Impact Toughness	*739*
22.7	Potential Applications	*740*
22.7.1	High Velocity Impact Protection	*740*
22.7.2	Crashworthiness	*741*
22.7.3	Enhanced Crack Resistance	*742*
22.8	Summary and Conclusions	*742*
	References	*744*

Part Five Applications 747

23 Applications of Macro- and Microfiller-Reinforced Polymer Composites 749
Hajnalka Hargitai and Ilona Rácz

- 23.1 Introduction 749
- 23.2 Some Features of Polymer Composites 749
- 23.3 Transportation 750
- 23.3.1 Land Transportation 750
- 23.3.1.1 Shell/Body Parts 750
- 23.3.1.2 Compartment Parts 752
- 23.3.2 Marine Applications 754
- 23.3.3 Aviation 755
- 23.3.3.1 Military Aircrafts 755
- 23.3.3.2 Commercial Aircrafts 756
- 23.3.3.3 Business Aviation 756
- 23.3.3.4 Helicopter 757
- 23.3.3.5 Space 757
- 23.4 Biomedical Applications 757
- 23.4.1 External Fixation 758
- 23.4.2 Dental Composites 759
- 23.4.3 Bone Replacement 759
- 23.4.4 Orthopedic Applications 760
- 23.5 Civil Engineering, Construction 760
- 23.5.1 External Strengthening 761
- 23.5.1.1 Repair and Retrofitting 761
- 23.5.1.2 Composite Decks 762
- 23.5.1.3 Seismic Rehabilitation 764
- 23.5.1.4 Unique Applications 764
- 23.5.1.5 Structural Applications 764
- 23.5.1.6 Pipe Applications 765
- 23.5.1.7 Wind Energy Applications 766
- 23.5.1.8 Others 766
- 23.5.2 Internal Reinforcement for Concrete 767
- 23.6 Electric and Electronic Applications 767
- 23.7 Mechanical Engineering, Tribological Applications 769
- 23.7.1 Metal Forming Dies 770
- 23.7.2 Seals 770
- 23.7.3 Bearings 771
- 23.7.4 Brakes 771
- 23.7.5 Brake Pad 771
- 23.7.6 Brake Lining 771
- 23.7.7 Gears 772
- 23.8 Recreation, Sport Equipments 772
- 23.8.1 Summersports 772

23.8.1.1	Bicycle	772
23.8.1.2	Athletics	773
23.8.2	Wintersports	773
23.8.2.1	Skis, Snowboards	773
23.8.2.2	Boots	774
23.8.2.3	Poles	774
23.8.2.4	Hockey	774
23.8.3	Technical Sports	775
23.8.4	Racquets, Bats	777
23.8.5	Watersports	778
23.8.6	Leisure	779
23.9	Other Applications	780
23.9.1	Fire Retardancy, High Temperature Applications	780
23.9.2	Self-Healing Polymer Composites	781
23.9.3	Shape-Memory Polymers	782
23.9.4	Defense	783
23.10	Conclusion	784
	References	784

Index 791

The Editors

Sabu Thomas is a Professor of Polymer Science and Engineering at Mahatma Gandhi University (India). He is a Fellow of the Royal Society of Chemistry and a Fellow of the New York Academy of Sciences. Thomas has published over 430 papers in peer reviewed journals on polymer composites, membrane separation, polymer blend and alloy, and polymer recycling research and has edited 17 books. He has supervised 60 doctoral students.

Kuruvilla Joseph is a Professor of Chemistry at Indian Institute of Space Science and Technology (India). He has held a number of visiting research fellowships and has published over 50 papers on polymer composites and blends.

S. K. Malhotra is Chief Design Engineer and Head of the Composites Technology Centre at the Indian Institute of Technology, Madras. He has published over 100 journal and proceedings papers on polymer and alumina-zirconia composites.

Koichi Goda is a Professor of Mechanical Engineering at Yamaguchi University. His major scientific fields of interest are reliability and engineering analysis of composite materials and development and evaluation of environmentally friendly and other advanced composite materials.

M. S. Sreekala is an Assistant Professor of Chemistry at Post Graduate Department of Chemistry, SreeSankara College, Kalady (India). She has published over 40 paperson polymer composites (including biodegradable and green composites) in peer reviewed journals and has held a number of Scientific Positions and Research Fellowships including those from the Humboldt Foundation, Germany and Japan Society for Promotion of Science, Japan.

Preface

Composite materials, usually man-made, are a three-dimensional combination of at least two chemically distinct materials, with a distinct interface separating the components, created to obtain properties that cannot be achieved by any of the components acting alone. In composites, at least one of the components called the reinforcing phase is in the form of fibers, sheets, or particles and is embedded in the other materials called the matrix phase. The reinforcing material and the matrix material can be metal, ceramic, or polymer. Very often commercially produced composites make use of polymers as the matrix material. Typically, reinforcing materials are strong with low densities, while the matrix is usually a ductile, or tough, material. If the composite is designed and fabricated adequately, it combines the strength of the reinforcement with the toughness of the matrix to achieve a combination of desirable properties not available in any single conventional material.

The present book focuses on the preparation and characterization of polymer composites with macro- and microfillers. It examines the different types of fillers especially as the reinforcing agents. The text reviews the interfaces in macro- and microcomposites and their characterization. Advanced applications of macro- and micropolymer composites are discussed in detail. This book carefully analyses the effect of surface modification of fillers on properties and chemistry and reinforcing mechanism of composites. It also introduces recovery, recycling, and life cycle analysis of synthetic polymeric composites.

The book is organized into five parts. Part One contains four chapters. Chapter 1 is an introduction to composites, classification, and characteristic features of polymer composites, their applications in various fields, state of the art, and new challenges and opportunities.

Chapter 2 focuses on micro- and macromechanics of polymer composites. Knowledge of micro- and macromechanics is essential for understanding the behavior, analysis, and design of polymer composite products for engineering applications.

Chapter 3 deals with interfaces in macro- and microcomposites. Interface plays a big role in physical and mechanical behavior of polymer composites. It deals with the various techniques and analyses of the interfacial properties of various polymer composite materials.

Chapter 4 describes various preparation and manufacturing techniques for polymer composites starting with simplest hand lay-up (contact molding) to sophisticated autoclave molding and CNC filament winding methods.

Part Two deals with fiber-reinforced polymer composites and Part Three discusses textile composites.

Each of the seven chapters included in Part Two deals with a particular fiber as reinforcement for polymer matrices. These fibers are carbon, glass, Kevlar, polyester, nylon, polyolefin, and silica.

Each of the four chapters included in Part Three deals with a particular form of textiles as reinforcement. These textiles are 2D woven fabric, 3D woven fabric, geotextiles, and hybrid textiles.

The first five chapters included in Part Four deal with different microsized fillers reinforcing the polymer matrix. Different microparticulate fillers include carbon black, silica, metallic particles, magnetic particles, mica (flakes), and so on. The last chapter of this part deals with viscoelastically prestressed polymer composites.

Finally, Part Five studies applications of macro- and microfiller-reinforced polymer composites. Polymer composites find applications in all types of engineering industry, namely, aerospace, automobile, chemical, civil, mechanical, electrical, and so on. They also find applications in consumer durables, sports goods, biomedical, and many more areas.

Sabu Thomas, Kuruvilla Joseph,
Sant Kumar Malhotra, Koichi Goda,
and Meyyarappallil Sadasivan Sreekala

List of Contributors

Boudjemaa Agoudjil
Université El-Hadj-Lakhdar-Batna
LPEA
1, rue Boukhlouf Med El-Hadi
05000 Batna
Algeria

Tamás Bárány
Budapest University of Technology and Economics
Department of Polymer Engineering
Müegyetem rkp. 3
1111 Budapest
Hungary

Philippe Boisse
INSA Lyon
LaMCoS
Bat Jacquard
27 Avenue Jean Capelle
69621 Villeurbanne Cedex
France

Abderrahim Boudenne
Université Paris-Est Val de Marne
CERTES
61 Av. du Général de gaulle
94010 Créteil Cédex
France

Valerio Causin
Università di Padova
Dipartimento di Scienze Chimiche
Via Marzolo 1
35131 Padova
Italy

Mark Christopher
University of Waikato
Department of Engineering
Gate 1 Knighton Road
Private Bag 3105
Hamilton 3240
New Zealand

Tibor Czigány
Budapest University of Technology and Economics
Department of Polymer Engineering
Muegyetem rkp. 3
1111 Budapest
Hungary

Chapal K. Das
Indian Institute of Technology
Materials Science Centre
Kharagpur 721302
India

Tamás Deák
Budapest University of Technology and Economics
Department of Polymer Engineering
Muegyetem rkp. 3
1111 Budapest
Hungary

Kevin S. Fancey
University of Hull
Department of Engineering
Cottingham Road
Hull HU6 7RX
UK

Bertrand Garnier
Ecole Polytechnique de l'université de Nantes
LTN-UMR CNRS6607
Rue Christian Pauc, BP 50609
44306 Nantes Cdx 03
France

Koichi Goda
Yamaguchi University
Department of Mechanical Engineering
Tokiwadai
Ube 755-8611
Yamaguchi
Japan

Hajnalka Hargitai
Széchenyi István University
Department of Materials and Vehicle Manufacturing Engineering
9026 Győr Egyetem tér 1
Hungary

Sebastian Heimbs
European Aeronautic Defence and Space Company
Innovation Works
81663 Munich
Germany

Yuko Ikeda
Kyoto Institute of Technology
Graduate School of Science and Technology
Matsugasaki
Kyoto 606-8585
Japan

Han-Yong Jeon
Inha University
Division of Nano-Systems Engineering
253, Yonghyun-dong, Nam-gu
Incheon 402-751
South Korea

Josmin P. Jose
Mahatma Gandhi University
School of Chemical Sciences
Polymer Science & Technology
Priyadarshini Hills
Kottayam 686560
Kerala
India

Kuruvilla Joseph
Indian Institute of Space Science and Technology (IIST)
Department of Chemistry
Valiamala P.O.
Thiruvananthapuram 695547
Kerala
India

József Karger-Kocsis
Budapest University of Technology and Economics
Department of Polymer Engineering
Müegyetem rkp. 3
1111 Budapest
Hungary

and

Tshwane University of Technology
Faculty of Engineering and Built
Environment
Polymer Technology
P.O. X680
0001 Pretoria
Republic of South Africa

Atsushi Kato
NISSAN ARC, Ltd.
Research Department
Natsushima-cho 1
Yokosuka
Kanagawa 237-0061
Japan

Natalie E. Kazantseva
Tomas Bata University in Zlin
Faculty of Technology
Polymer Center
T.G. Masaryk Sq. 5555
760 01 Zlin
Czech Republic

Bong-Rae Kim
Korea Advanced Institute of Science
and Technology
Department of Civil and Environmental
Engineering
373-1 Guseong-dong
Yuseong-gu
Daejeon 305-701
South Korea

Shinzo Kohjiya
Kyoto Institute of Technology
Graduate School of Science and
Technology
Matsugasaki
Kyoto 606-8585
Japan

Wen-Shyong Kuo
Feng Chia University
Department of Aerospace and Systems
Engineering
No. 100 Wenhwa Road
Seatwen
Taichung 40724
Taiwan R.O.C.

Haeng-Ki Lee
Korea Advanced Institute of Science and
Technology
Department of Civil and Environmental
Engineering
373-1 Guseong-dong
Yuseong-gu
Daejeon 305-701
South Korea

Xianping Liu
University of Warwick
School of Engineering
Coventry CV4 7AL
UK

Sant Kumar Malhotra
Composites Technology Centre
IIT Madras
Chennai 600036
Tamil Nadu
India

Sreekala M. S.
Assistant Professor
Post Graduate Department of
Chemistry
SreeSankara College, Kalady
Kerala 683574
India

Dionysis E. Mouzakis
Technological Educational Institute of Larisa
School of Technological Applications
Department of Mechanical Engineering
T.E.I. of Larisa
411 10 Larisa
Greece

Ganesh C. Nayak
Indian Institute of Technology
Materials Science Centre
Kharagpur 721302
India

Soo-Jin Park
Inha University
Department of Chemistry
253, Yonghyun-dong, Nam-gu
Incheon 402-751
South Korea

Volker Piotter
Karlsruhe Institute of Technology (KIT)
Institute for Applied Materials
Hermann-von-Helmholtz-Platz 1
76344 Eggenstein-Leopoldshafen
Germany

Laly A. Pothan
Bishop Moore College
Department of Chemistry
Mavelikara 690101
Kerala
India

Vikas Prakash
Case Western Reserve University
Department of Mechanical and Aerospace Engineering
10900 Euclid Avenue
418 Glennan Building LC-7222
Cleveland, OH 44106-7222
USA

Jürgen Prokop
Karlsruhe Institute of Technology (KIT)
Institute for Applied Materials
Hermann-von-Helmholtz-Platz 1
76344 Eggenstein-Leopoldshafen
Germany

Ilona Rácz
Bay Zoltán Institute for Materials Science and Technology
Fehérvári u. 130
1116 Budapest
Hungary

Rathanasamy Rajasekar
Indian Institute of Technology
Materials Science Centre
Kharagpur 721302
India

Sudip Ray
University of Auckland
School of Chemical Sciences
Private Bag 92019
Auckland 1142
New Zealand

Min-Kang Seo
Jeonju Institute of Machinery and Carbon Composites
Aircraft Parts Division
750-1, Palbok-dong
Deokjin-gu
Jeonju 561-844
South Korea

Palanisamy Sivasubramanian
Department of Mechanical Engineering
SaintGITS College of Engineering
Pathamuttom
Kottayam-686532
Kerala
India

Meyyarappallil Sadasivan Sreekala
Department of Polymer Science and
Rubber Technology
Cochin University of science and
Technology
Cochin- 682022
Kerala
India

M. Thiruchitrambalam
Department of Mechanical Engineering
Tamilnadu College of Engineering
Coimbatore
Tamilnadu
India

Sabu Thomas
Mahatma Gandhi University
School of Chemical Sciences
Kottayam 686560
Kerala
India

Björn Van Den Broucke
European Aeronautic Defence and
Space Company
Innovation Works
81663 Munich
Germany

John Verbeek
University of Waikato
Department of Engineering
Gate 1 Knighton Road
Private Bag 3105
Hamilton 3240
New Zealand

Emmanuelle Vidal-Sallé
INSA Lyon
LaMCoS
Bat Jacquard
27 Avenue Jean Capelle
69621 Villeurbanne Cedex
France

Part One
Introduction to Polymer Composites

1
Advances in Polymer Composites: Macro- and Microcomposites – State of the Art, New Challenges, and Opportunities

Josmin P. Jose, Sant Kumar Malhotra, Sabu Thomas, Kuruvilla Joseph, Koichi Goda, and Meyyarappallil Sadasivan Sreekala

1.1
Introduction

Composites can be defined as materials that consist of two or more chemically and physically different phases separated by a distinct interface. The different systems are combined judiciously to achieve a system with more useful structural or functional properties nonattainable by any of the constituent alone. Composites, the wonder materials are becoming an essential part of today's materials due to the advantages such as low weight, corrosion resistance, high fatigue strength, and faster assembly. They are extensively used as materials in making aircraft structures, electronic packaging to medical equipment, and space vehicle to home building [1]. The basic difference between blends and composites is that the two main constituents in the composites remain recognizable while these may not be recognizable in blends. The predominant useful materials used in our day-to-day life are wood, concrete, ceramics, and so on. Surprisingly, the most important polymeric composites are found in nature and these are known as natural composites. The connective tissues in mammals belong to the most advanced polymer composites known to mankind where the fibrous protein, collagen is the reinforcement. It functions both as soft and hard connective tissue.

Composites are combinations of materials differing in composition, where the individual constituents retain their separate identities. These separate constituents act together to give the necessary mechanical strength or stiffness to the composite part. Composite material is a material composed of two or more distinct phases (matrix phase and dispersed phase) and having bulk properties significantly different from those of any of the constituents. Matrix phase is the primary phase having a continuous character. Matrix is usually more ductile and less hard phase. It holds the dispersed phase and shares a load with it. Dispersed (reinforcing) phase is embedded in the matrix in a discontinuous form. This secondary phase is called the dispersed phase. Dispersed phase is usually stronger than the matrix, therefore, it is sometimes called reinforcing phase.

Polymer Composites: Volume 1, First Edition. Edited by Sabu Thomas, Kuruvilla Joseph,
Sant Kumar Malhotra, Koichi Goda, and Meyyarappallil Sadasivan Sreekala
© 2012 Wiley-VCH Verlag GmbH & Co. KGaA. Published 2012 by Wiley-VCH Verlag GmbH & Co. KGaA.

Composites in structural applications have the following characteristics:

- They generally consist of two or more physically distinct and mechanically separable materials.
- They are made by mixing the separate materials in such a way as to achieve controlled and uniform dispersion of the constituents.
- They have superior mechanical properties and in some cases uniquely different from the properties of their constituents [2].

Wood is a natural composite of cellulose fibers in a matrix of lignin. Most primitive man-made composite materials were straw and mud combined to form bricks for building construction. Most visible applications pave our roadways in the form of either steel and aggregate reinforced Portland cement or asphalt concrete. Reinforced concrete is another example of composite material. The steel and concrete retain their individual identities in the finished structure. However, because they work together, the steel carries the tension loads and concrete carries the compression loads.

Most advanced examples perform routinely on spacecraft in demanding environments. Advanced composites have high-performance fiber reinforcements in a polymer matrix material such as epoxy. Examples are graphite/epoxy, Kevlar/epoxy, and boron/epoxy composites. Advanced composites are traditionally used in the aerospace industries, but these materials have now found applications in commercial industries as well.

1.2
Classification of Composites

On the basis of matrix phase, composites can be classified into metal matrix composites (MMCs), ceramic matrix composites (CMCs), and polymer matrix composites (PMCs) (Figure 1.1) [3]. The classifications according to types of reinforcement are particulate composites (composed of particles), fibrous composites (composed of fibers), and laminate composites (composed of laminates). Fibrous composites can be further subdivided on the basis of natural/biofiber or synthetic fiber. Biofiber encompassing composites are referred to as biofiber composites. They can be again divided on the basis of matrix, that is, nonbiodegradable matrix and biodegradable matrix [4]. Bio-based composites made from natural/biofiber and biodegradable polymers are referred to as green composites. These can be further subdivided as hybrid composites and textile composites. Hybrid composites comprise of a combination of two or more types of fibers.

1.2.1
Polymer Matrix Composites

Most commercially produced composites use a polymer matrix material often called a resin solution. There are many different polymers available depending upon the

1.2 Classification of Composites

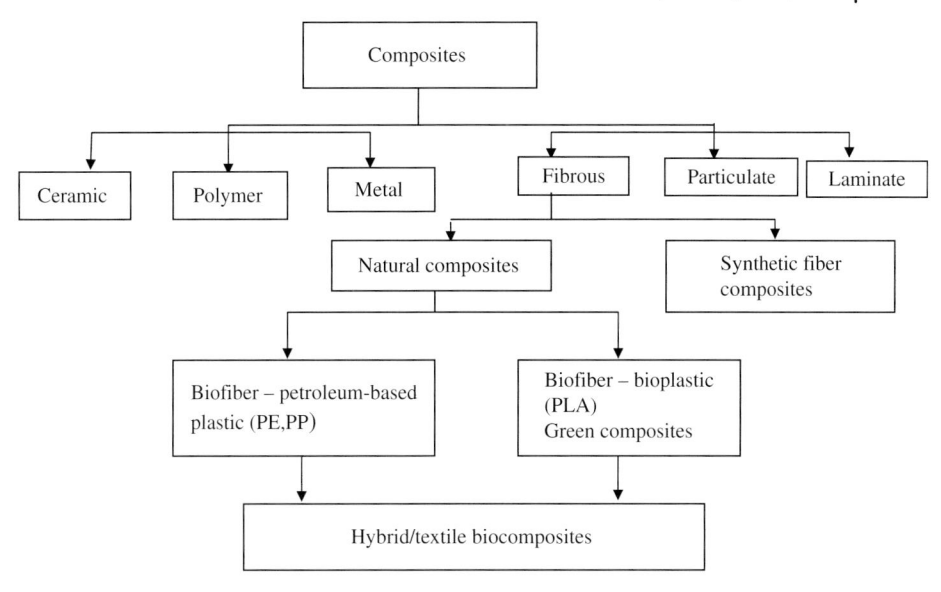

Figure 1.1 Classification of composites [5].

starting raw ingredients. There are several broad categories, each with numerous variations. The most common are known as polyester, vinyl ester, epoxy, phenolic, polyimide, polyamide, polypropylene, polyether ether ketone (PEEK), and others. The reinforcement materials are often fibers but can also be common ground minerals [6]. The various methods described below have been developed to reduce the resin content of the final product. As a rule of thumb, hand lay up results in a product containing 60% resin and 40% fiber, whereas vacuum infusion gives a final product with 40% resin and 60% fiber content. The strength of the product is greatly dependent on this ratio.

PMCs are very popular due to their low cost and simple fabrication methods. Use of nonreinforced polymers as structure materials is limited by low level of their mechanical properties, namely strength, modulus, and impact resistance. Reinforcement of polymers by strong fibrous network permits fabrication of PMCs, which is characterized by the following:

a) High specific strength
b) High specific stiffness
c) High fracture resistance
d) Good abrasion resistance
e) Good impact resistance
f) Good corrosion resistance
g) Good fatigue resistance
h) Low cost

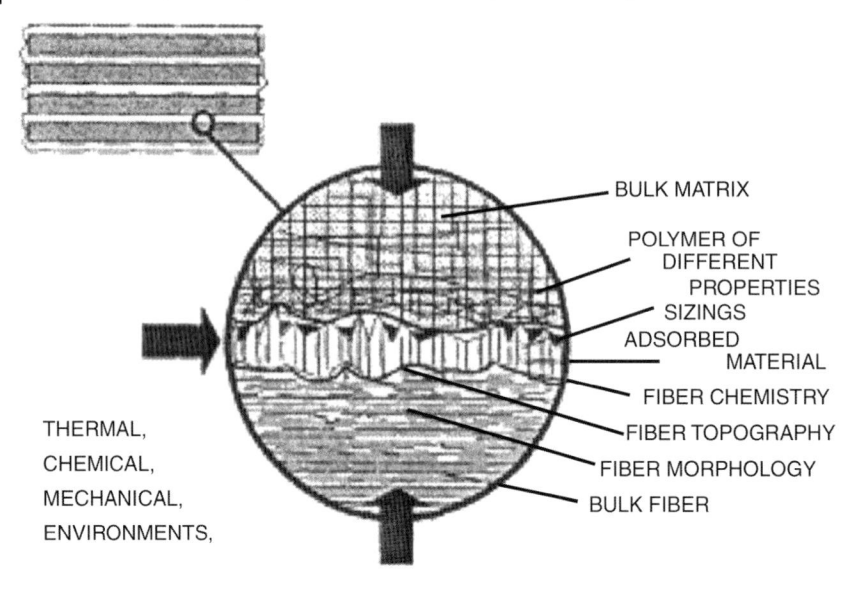

Figure 1.2 Schematic model of interphase [7].

The main disadvantages of PMCs are

a) low thermal resistance and
b) high coefficient of thermal expansion.

1.2.1.1 Factors Affecting Properties of PMCs

1.2.1.1.1 Interfacial Adhesion The behavior of a composite material is explained on the basis of the combined behavior of the reinforcing element, polymer matrix, and the fiber/matrix interface (Figure 1.2). To attain superior mechanical properties the interfacial adhesion should be strong. Matrix molecules can be anchored to the fiber surface by chemical reaction or adsorption, which determine the extent of interfacial adhesion. The developments in atomic force microscopy (AFM) and nano indentation devices have facilitated the investigation of the interface. The interface is also known as the mesophase.

1.2.1.1.2 Shape and Orientation of Dispersed Phase Inclusions (Particles, Flakes, Fibers, and Laminates) Particles have no preferred directions and are mainly used to improve properties or lower the cost of isotropic materials [8]. The shape of the reinforcing particles can be spherical, cubic, platelet, or regular or irregular geometry. Particulate reinforcements have dimensions that are approximately equal in all directions. Large particle and dispersion-strengthened composites are the two subclasses of particle-reinforced composites. A laminar composite is composed of two dimensional sheets or panels, which have a preferred high strength direction as found in wood. The layers are stacked and subsequently cemented together so that the orientation of the high strength direction varies with each successive layer [9].

1.2.1.1.3 Properties of the Matrix
Properties of different polymers will determine the application to which it is appropriate. The chief advantages of polymers as matrix are low cost, easy processability, good chemical resistance, and low specific gravity. On the other hand, low strength, low modulus, and low operating temperatures limit their use [10]. Varieties of polymers for composites are thermoplastic polymers, thermosetting polymers, elastomers, and their blends.

Thermoplastic polymers: Thermoplastics consists of linear or branched chain molecules having strong intramolecular bonds but weak intermolecular bonds. They can be reshaped by application of heat and pressure and are either semicrystalline or amorphous in structure. Examples include polyethylene, polypropylene, polystyrene, nylons, polycarbonate, polyacetals, polyamide-imides, polyether ether ketone, polysulfone, polyphenylene sulfide, polyether imide, and so on.

Thermosetting polymers: Thermosetts have cross-linked or network structures with covalent bonds with all molecules. They do not soften but decompose on heating. Once solidified by cross-linking process they cannot be reshaped. Common examples are epoxies, polyesters, phenolics, ureas, melamine, silicone, and polyimides.

Elastomers: An elastomer is a polymer with the property of viscoelasticity, generally having notably low Young's modulus and high yield strain compared with other materials. The term, which is derived from elastic polymer, is often used interchangeably with the term rubber, although the latter is preferred when referring to vulcanizates. Each of the monomers that link to form the polymer is usually made of carbon, hydrogen, oxygen, and silicon. Elastomers are amorphous polymers existing above their glass transition temperature, so that considerable segmental motion is possible. At ambient temperatures, rubbers are relatively soft ($E \sim 3$ MPa) and deformable; their primary uses are for seals, adhesives, and molded flexible parts. Natural rubber, synthetic polyisoprene, polybutadiene, chloroprene rubber, butyl rubber, ethylene propylene rubber, epichlorohydrin rubber, silicone rubber, fluoroelastomers, thermoplastic elastomers, polysulfide rubber, and so on are some of the examples of elastomers.

1.2.1.2 Fabrication of Composites
The fabrication and shaping of composites into finished products often combines the formation of the material itself during the fabrication process [11]. The important processing methods are hand lay-up, bag molding process, filament winding, pultrusion, bulk molding, sheet molding, resin transfer molding, injection molding, and so on.

1.2.1.2.1 Hand Lay-Up
The oldest, simplest, and the most commonly used method for the manufacture of both small and large reinforced products is the hand lay-up technique. A flat surface, a cavity or a positive-shaped mold, made from wood, metal, plastic, or a combination of these materials may be used for the hand lay-up method.

1.2.1.2.2 Bag Molding Process

It is one of the most versatile processes used in manufacturing composite parts. In bag molding process, the lamina is laid up in a mold and resin is spread or coated, covered with a flexible diaphragm or bag, and cured with heat and pressure. After the required curing cycle, the materials become an integrated molded part shaped to the desired configuration [12]. Three basic molding methods involved are pressure bag, vacuum bag, and autoclave.

1.2.1.2.3 Pultrusion

It is an automated process for manufacturing composite materials into continuous, constant cross-section profiles. In this technique, the product is pulled from the die rather than forced out by pressure. A large number of profiles such as rods, tubes, and various structural shapes can be produced using appropriate dies.

1.2.1.2.4 Filament Winding

Filament winding is a technique used for the manufacture of surfaces of revolution such as pipes, tubes, cylinders, and spheres and is frequently used for the construction of large tanks and pipe work for the chemical industry. High-speed precise lay down of continuous reinforcement in predescribed patterns is the basis of the filament winding method.

1.2.1.2.5 Preformed Molding Compounds

A large number of reinforced thermosetting resin products are made by matched die molding processes such as hot press compression molding, injection molding, and transfer molding. Matched die molding can be a wet process but it is most convenient to use a preformed molding compound or premix to which all necessary ingredients are added [13]. This enables the attainment of faster production rate. Molding compounds can be divided into three broad categories: dough molding, sheet molding, and prepregs.

1.2.1.2.6 Resin Transfer Molding

Resin transfer molding (RTM) has the potential of becoming a dominant low-cost process for the fabrication of large, integrated, high performance products. In this process, a dry reinforced material that has been cut and shaped into a preformed piece, generally called a perform, is placed in a prepared mold cavity. The resin is often injected at the lowest point and fills the mold upward to reduce the entrapping of air. When the resin starts to leak into the resin trap, the tube is clamped to minimize resin loss. When excess resin begins to flow from the vent areas of the mold, the resin flow is stopped and the mold component begins to cure. Once the composite develops sufficient green strength it can be removed from the tool and postcured (Figure 1.3).

1.2.1.2.7 Injection Molding

Injection molding is a manufacturing process for both thermoplastic and thermosetting plastic materials. Composites is fed into a heated barrel, mixed, and forced into a mold cavity where it cools and hardens to the configuration of the mold cavity. Injection molding is used to create many things such as wire spools, packaging, bottle caps, automotive dashboards, pocket combs, and most other plastic products available today. It is ideal for producing high volumes of the same object [15]. Some advantages of injection molding are high production rates, repeatable high tolerances, and the ability to use a wide range of

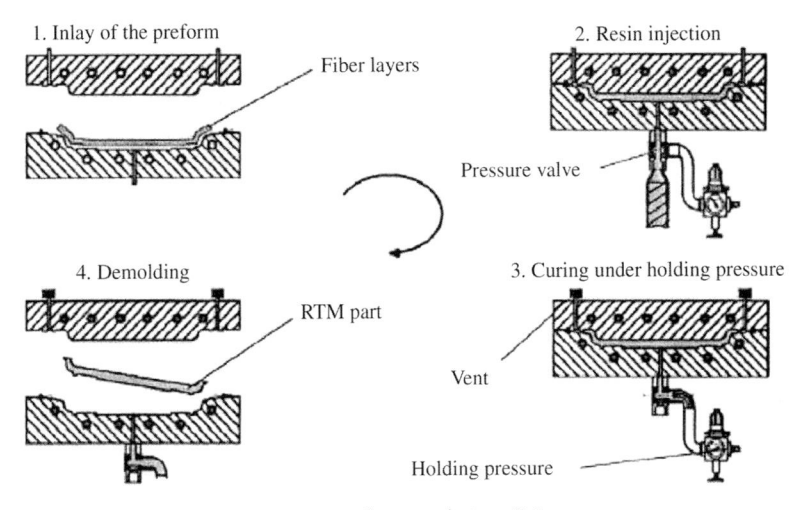

Figure 1.3 Schematic representation of RTM technique [14].

materials, low labor cost, minimal scrap losses, and little need to finish parts after molding. Some disadvantages of this process are expensive equipment investment, potentially high running costs, and the need to design moldable parts.

1.2.1.2.8 Reaction Injection Molding (RIM) RIM is similar to injection molding except that thermosetting polymers are used, which requires a curing reaction to occur within the mold. Common items made via RIM include automotive bumpers, air spoilers, and fenders. First, the two parts of the polymer are mixed together. The mixture is then injected into the mold under high pressure using an impinging mixer. The most common RIM processable material is polyurethane (generally known as PU-RIM), but others include polyureas, polyisocyanurates, polyesters, polyepoxides, and nylon 6. For polyurethane, one component of the mixture is polyisocyanate and the other component is a blend of polyol, surfactant, catalyst, and blowing agent. Automotive applications comprise the largest area of use for RIM-produced products. Polymers have been developed specifically for exterior body panels for the automotive industry. Non-E-coat polymers offer an excellent combination of stiffness, impact resistance, and thermal resistance for body panel applications. These provide excellent paintability and solvent resistance with the ability to achieve high distinction of image (DOI) when painted.

1.2.1.2.9 Reinforced Reaction Injection Molding If reinforcing agents are added to the mixture of RIM setting then the process is known as reinforced reaction injection molding (RRIM). Common reinforcing agents include glass fibers and mica. This process is usually used to produce rigid foam automotive panels. A subset of RRIM is structural reaction injection molding (SRIM), which uses fiber meshes for the reinforcing agent. The fiber mesh is first arranged in the mold and then the polymer mixture is injection molded over it.

1.2.1.2.10 Spray-Up

In spray-up process, liquid resin matrix and chopped reinforcing fibers are sprayed by two separate sprays onto the mold surface. The fibers are chopped into fibers of 1–2″ (25–50 mm) length and then sprayed by an air jet simultaneously with a resin spray at a predetermined ratio between the reinforcing and matrix phase. The spray-up method permits rapid formation of uniform composite coating, however, the mechanical properties of the material are moderate since the method is unable to use continuous reinforcing fibers.

1.2.1.3 Applications

PMCs are used for manufacturing

i) **Aerospace structures**: The military aircraft industry has mainly led the use of polymer composites. In commercial airlines, the use of composites is gradually increasing. Space shuttle and satellite systems use graphite/epoxy for many structural parts [16].
ii) **Marine**: Boat bodies, canoes, kayaks, and so on.
iii) **Automotive**: Body panels, leaf springs, drive shaft, bumpers, doors, racing car bodies, and so on.
iv) **Sports goods**: Golf clubs, skis, fishing rods, tennis rackets, and so on.
v) Bulletproof vests and other armor parts.
vi) Chemical storage tanks, pressure vessels, piping, pump body, valves, and so on.
vii) **Biomedical applications**: Medical implants, orthopedic devices, X-ray tables.
viii) Bridges made of polymer composite materials are gaining wide acceptance due to their lower weight, corrosion resistance, longer life cycle, and limited earthquake damage.
ix) **Electrical**: Panels, housing, switchgear, insulators, and connectors.
And many more.

1.2.1.4 Recent Advances in Polymer Composites

1.2.1.4.1 3-D FRP Composites

Fiber-reinforced polymer (FRP) composites are used in almost every type of advanced engineering structure, with their usage ranging from aircraft, helicopters, and spacecraft through to boats, ships, and offshore platforms and to automobiles, sports goods, chemical processing equipment, and civil infrastructure such as bridges and buildings. The usage of FRP composites continues to grow at an impressive rate as these materials are used more in their existing markets and become established in relatively new markets such as biomedical devices and civil structures. A key factor driving the increased applications of composites over the recent years is the development of new advanced forms of FRP materials. This includes developments in high performance resin systems and new styles of reinforcement, such as carbon nanotubes and nanoparticles [17].

Recent work on 3D FRP composites includes the following:

a) Manufacturing of 3D preforms by weaving, braiding, knitting, and stitching.
b) Fabrication of FRP composite products by preform consolidation followed by liquid molding.

Table 1.1 Mechanical properties of natural fibers compared with synthetic fibers [20].

Fiber	Density (10^3 kg/m^3)	Elongation (%)	Tensile strength (MPa)	Young's modulus (GPa)
Aramid	1.4	3.3–3.7	3000–3450	63–67
Carbon	1.4	1.4–1.8	4000	230–240
Kelvar 49	1.45	2.0	2800	124
Cotton	1.5	7.0–8.0	287–597	5.5–12.6
Jute	1.3	1.5–1.8	393–773	26.5
Flax	1.5	2.7–3.5	345–1035	27.6
Hemp	—	1.6	690	—
Ramie	—	3.6–3.8	400–938	61.1–128
Sisal	1.5	4–6	511–635	9.4–22
Coir	1.2	30	175	4.0–6.0
Banana	1.3	2–4	750	29–32
Pineapple	1.56	—	172	62
Oil palm	1.55	—	100–400	26.5
Soft wood craft	1.5	—	1000	40.0
E-glass	2.5	2.5	2000–3500	70.0
S-glass	2.5	2.8	4570	86.0
SiC	3.08	0.8	3440	400
Alumina	3.95	0.4	1900	379

c) Micromechanics model for mechanical properties of 3D woven/braided/knitted/stitched fabric polymer composites.

d) Designing microstructure of 3D FRP composite materials to obtain optimum performance (for both continuous and discontinuous fiber composites).

1.2.1.4.2 Natural Fiber Composites Glass, carbon, Kevlar, and boron fibers are being used as reinforcing materials in fiber-reinforced plastics, which have been widely accepted as materials for structural and nonstructural applications [18]. However, these materials are resistant to biodegradation and can pose environmental problems. Natural fibers from plants such as jute, bamboo, coir, sisal, and pineapple are known to have very high strength and hence can be utilized for many load-bearing applications. These fibers have special advantage in comparison to synthetic fibers in that they are abundantly available, from a renewable resource and are biodegradable. But all natural fibers are hydrophilic in nature and have high moisture content, which leads to poor interface between fiber and hydrophobic matrix. Several treatment methods are employed to improve the interface in natural fiber composite [19]. Automobile industry in Europe has started using natural fiber composites in a big way both for exterior and interior of car bodies because of stringent environmental requirements (Table 1.1).

Natural fibers are generally incompatible with the hydrophobic polymer matrix and have a tendency to form aggregates. Therefore, the surface of both (matrix and fibers) should be appropriately wetted to improve the interfacial adhesion and to remove any impurities. The surface of hydrophobic matrices should be modified by the introduction of polar groups by treating them with oxidative chemicals such as chromic acid/acetic acid or chromic acid/sulfuric acid [21]. Cold plasma chemistry opens up new avenues for the surface modifications of materials for composites and other applications. Various oxidative and nonoxidative chemical treatments are available for natural and synthetic fibers to improve the bonding at the interface. Alkali treatment has been proved to be an effective method for fiber modification from as early as 1935. It has been reported that on treatment with alkali, some of the wax components at the fiber surface are saponified and thereby removed from the fiber surface. Increased fiber/matrix adhesion as a result of improved surface area and increase in availability of the hydroxyl groups have also been reported as a result of alkali treatment.

Compared to unmodified composites, all chemically modified fiber composites show higher tensile properties and lower water uptake. As chemical treatment reduces hydrophobicity of the fiber it favors the strong interfacial adhesion between fiber and PP matrix. Tensile properties decrease with water uptake and time of immersion. Figure 1.4 shows the effect of chemical treatments on the tensile strength of the sisal/PP composites after immersion in water.

Compared to other natural fibers, banana and sisal have good mechanical properties. In general, the strength of a fiber increases with increasing cellulose

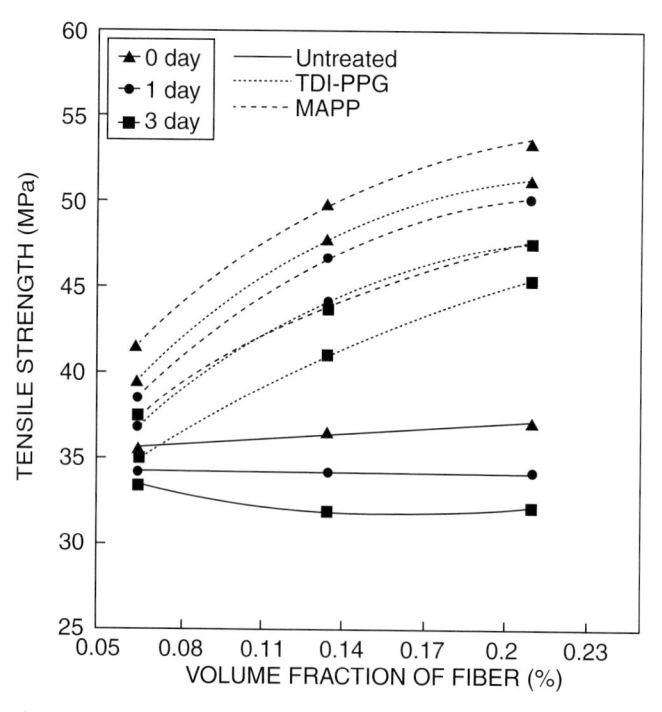

Figure 1.4 The effect of chemical treatments on the tensile strength of sisal/PP composites after immersion in water. Fiber loading 20%, temperature 20 °C [22].

Table 1.2 Properties of banana and sisal fiber [24].

	Banana	Sisal
Cellulose (%)	63–64	64–65
Hemicellulose (%)	19	12
Lignin (%)	5	9.9
Moisture content (%)	10–11	10
Microfibrillar angle (°)	11	20
Lumen size (μm)	5	11

content and decreasing spiral angle with respect to the fiber axis. The composition, microfibrillar angle, and lumen size of banana and sisal fibers are given in Table 1.2. The cellulose content of sisal and banana fibers is almost same, but the microfibrillar angle of banana fiber is much lower than sisal. Hence, the inherent tensile properties of banana fiber are higher than sisal fiber. The diameter of banana fiber is lower than sisal. As the surface area of banana fibers in unit area of the composite is higher, the stress transfer is increased in banana-reinforced composite compared to sisal-reinforced composites [23].

1.2.1.4.3 Fully Green Composites Research efforts are progressing in developing a new class of fully biodegradable green composites by combining fibers with biodegradable resins. The major attractions about green composites are that they are eco-friendly fully degradable and sustainable, that is, they are truly green in every way. The design and life cycle assessment of green composites have been exclusively dealt with by Baillie. Green composites may be used effectively in many applications such as mass-produced consumer products with short life cycles or products intended for one time or short time use before disposal. The important biodegradable matrices are polyamides, polyvinyl alcohol, polyvinyl acetate, polyglycolic acid, and polylactic acid, which are synthetic as well as polysaccharides, starch, chitin, cellulose, proteins, collagens/gelatin, lignin, and so on, which are natural [25]. Bio-based composites with their constituents developed from renewable resources are being developed and its application has extended to almost all fields. Natural fiber composites can be used as a substitute for timber and for a number of other applications. It can be molded into sheets, boards, gratings, pallets, frames, structural sections, and many other shapes. They can be used as a substitute for wood, metal, or masonry for partitions, false ceiling, facades, barricades, fences, railings, flooring, roofing, wall tiles, and so on [26]. It can also be used prefabricated housing, cubicles, kiosks, awnings, and sheds/shelters.

1.2.1.4.4 Other Emerging Areas

a) Five-axis weaving technology for the next generation of aircraft and mechanical performance of multiaxis weave structures.
b) Noncrimp fiber performs for helicopters composite parts.
c) Noncrimp braided carbon fiber-reinforced plastics for aeronautic applications.

d) Finite element modeling of textile-reinforced composites and comparison with real testing.
e) Textile composites in ballistics: modeling the material and failure response.
f) 3D textile composites: mechanical-progressive failure modeling and strength predictions.
g) Long-term durability of plain weaves polymer composites.

1.3
Interface Characterization

The characterization of interface gives relevant information on interactions between fiber and matrix. The mechanical properties of fiber-reinforced composites are dependent upon the stability of interfacial region. Thus, the characterization of interface is of great importance. The various methods that are available for characterization of the interface are as follows.

1.3.1
Micromechanical Technique

The extent of fiber/matrix interface bonding can be tested by different micromechanical tests such as fiber pull-out (Figure 1.5), micro-debond test, microindentation test, and fiber fragmentation test.

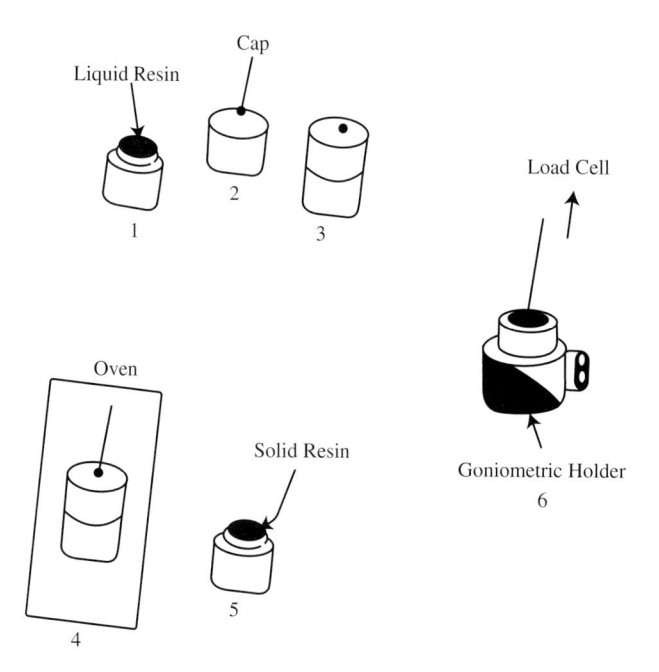

Figure 1.5 Schematic illustration of pull-out test preparation [27].

1.3.2
Spectroscopic Tests

Electron microscopy for chemical analysis/X-ray photoelectron spectroscopy, mass spectroscopy, X-ray diffraction studies, electron-induced vibration spectroscopy, and photoacoustic spectroscopy are successful in polymer surface and interfacial characterization.

1.3.3
Microscopic Techniques

Microscopic studies such as optical microscopy, scanning electron microscopy, transmission electron microscopy, and atomic force microscopy can be used to study the morphological changes on the surface and can predict the strength of mechanical bonding at the interface. The adhesive strength of fiber to various matrices can be determined by AFM studies.

1.3.4
Thermodynamic Methods

The frequently used thermodynamic methods for characterization in reinforced polymers are wettability study, inverse gas chromatography measurement, zeta potential measurement, and so on. Contact angle measurements have been used to characterize the thermodynamic work of adhesion between solids and liquids and surface of solids.

1.4
New Challenges and Opportunities

- In the context of eco-friendly materials, recyclability of the composites is one of the major problems. Recyclability of the composites will lead to the cost-effective products at the same time this is the remedy for the increased amount of waste materials. Green composites can replace all hazardous and waste-producing counterparts.
- Life cycle analysis should be done for all newly synthesized materials and thus the biodegradability can be measured. This will help us to select eco-friendly and acceptable materials.
- Microfibrillar composites, their properties and applications created a lot of interest in research because of its special properties and applications.
- Composite materials having long-term durability for continuous purposes are desirable and cost-effective.
- Since the interface has a significant role in property enhancement, new characterization techniques for interface will bring new opportunities.
- Online monitoring of morphology of composites during processing is another area, which requires a lot of attention of researchers.

References

1 Shaw, A., Sriramula, S., Gosling, P.D., and Chryssanthopoulo, M.K. (2010) *Composites Part B*, **41**, 446–453.
2 Mayer, C., Wang, X., and Neitzel, M. (1998) *Composites Part A*, **29**, 783–793.
3 Avila, A.F., Paulo, C.M., Santos, D.B., and Fari, C.A. (2003) *Materials Characterization*, **50**, 281–291.
4 Nicoleta, I. and Hickel, H. (2009) *Dental Materials*, **25**, 810–819.
5 Bunsell, A.R. and Harris, B. (1974) *Composites*, **5**, 157.
6 Mkaddem, A., Demirci, I., and Mansori, M.E. (2008) *Composites Science and Technology*, **68**, 3123–3127.
7 Downing, T.D., Kumar, R., Cross, V.M., Kjerengrtoen, L., and Keller, J.J.J. (2000) *Journal of Adhesion Science and Technology*, **14**, 1801.
8 Bednarcyk, B.A. (2003) *Composites Part B*, **34**, 175–197.
9 Tabiei, A. and Aminjikarai, S.B. (2009) *Composite Structures*, **88**, 65–82.
10 Huang, H. and Talreja, R. (2006) *Composites Science and Technology*, **66**, 2743–2757.
11 Sriramula, S. and Chryssanthopoulos, M.K. (2009) *Composites Part A*, **40**, 1673–1684.
12 Tay, T.E., Vincent, B.C., and Liu, G. (2006) *Materials Science and Engineering: B*, **132**, 138–142.
13 Friedrich, K., Zhang, Z., and Schlarb, A.K. (2005) *Composites Science and Technology*, **65**, 2329–2343.
14 Schmachtenberg, E. *et al.* (2005) *Polymer Testing*, **24**, 330.
15 Wakeman, M.D., Cain, C.D., Rudd, C.D., Brooks, R., and Long, A.C. (1999) *Composites Science and Technology*, **59**, 1153–1167.
16 Lekakou, C. and Bader, M.G. (1999) *Composites Part A*, **29**, 29–37.
17 Shokrieh, M.M. and Rafiee, R. (2010) *Computational Materials Science*, **50**, 437–446.
18 Liu, D., McDaid, A.D., and Xie, D.Q. (2011) *Mechatronics*, **21**, 315–328.
19 Geethamma, V.G., Thomas Mathew, K., Lakshminarayanan, R., and Thomas, S. (1998) *Polymer*, **39**, 1483–1491.
20 Cook, J.G. (1968) *Handbook of Textile Fibre and Natural Fibres*, 4th edn, Morrow Publishing, England.
21 Paul, S.A., Boudenne, A., Ibos, L., Candau, Y., Joseph, K., and Thomas, S. (2008) *Composites Part A*, **39**, 1582–1588.
22 Joseph, P.V. *et al.* (2002) *Computer Sciences Technology*, **62**, 1357.
23 John, M.A., Francis, B., Varughese, K.T., and Thomas, S. (2008) *Composites Part A*, **39**, 352–363.
24 Bledzki, A.K. and Gassan, J. (1999) *Progress in Polymer Science*, **24**, 221.
25 Jayanarayanan, K., Jose, T., Thomas, T., and Joseph, K. (2009) *European Polymer Journal*, **45**, 1738–1747.
26 Paul, S.A., Joseph, K., Mathew, J.D.G., Pothen, L.A., and Thomas, S. (2010) *Part A*, **41**, 1380–1387.
27 Bergeret, A. and Bozec, M.P. (2004) *Polymer Composites*, **25**, 12.

2
Shock and Impact Response of Glass Fiber-Reinforced Polymer Composites
Vikas Prakash

2.1
Introduction

A large body of knowledge currently exists in the literature on the propagation of acceleration waves and finite amplitude shock waves in heterogeneous materials. For such systems, scattering, dispersion, and attenuation play a critical role in determining the thermo-mechanical response of the media. In particular, the nonlinear behavior of the S2-glass fiber-reinforced polymer (GRP) composites can be attributed to the complex material architecture, that is, the impedance and geometric mismatch at the various length scales, and complex damage evolution in the form of extensive delamination, fiber shearing, tensile fiber failure, large fiber deflection, fiber microfracture, and local fiber buckling.

Even though some progress has been made in understanding the propagation of acceleration waves in model heterogeneous material systems, such as, bilaminates, the phenomenon of material and geometric dispersion in these materials continues to be poorly understood. For example, shock waves in the absence of phase transformations are understood to have a one-wave structure in most homogeneous materials. However, upon loading of a bilaminate, a two-wave structure is obtained – a leading shock front followed by a complex pattern that varies with time. This complex pattern is generated by a continuous interaction of compression and rarefaction waves due to the presence of interlaminar interfaces. Expressions for stress and particle velocity, based on the consideration of head–wave interaction with interfaces in linear elastic bilaminates under weak shock wave loading have been obtained by Laptev and Trishin [1]. It was shown that the attenuation of shock stress and particle velocity is primarily determined by the ratio of acoustic impedance of the layers and by the size of the periodic structure of the bilaminates (cell size). Smaller the cell size, the greater is the number of interfaces that interact with the propagating stress waves, and higher is the attenuation and dispersion. Analytical studies of wave dispersion relations for an infinite train of time–harmonic acceleration waves propagating in layered material systems have been conducted in a variety of elastic composites. Sun *et al.* [2] studied the case of waves in elastic bilaminates (i.e., composites consisting of alternating plane layers of different linear elastic materials) propagating in directions parallel or perpendicular to the laminates. These exact dispersion relations have been compared by Hegemier [3] to those obtained from various approximate theories. For viscoelastic bilaminates the understanding of dispersion relations is less complete. Stern et al. [4] considered wave propagation in a direction parallel to the laminates for alternating layers of elastic and viscoelastic materials. They simplified the analysis by neglecting the transverse displacement in the viscoelastic layers and the variation of the longitudinal displacement across the thickness of the elastic layers. Chen and Clifton [5] considered the exact theory of time-harmonic waves propagating in the direction of the normal to the laminates for general linear viscoelastic bilaminates. The dispersion relations obtained were similar to those obtained by Sve [6] for the closely related case of thermoelastic waves in laminates. Transient solutions for the case of step loading applied uniformly over the surface of a half space consisting of alternating plane layers of elastic materials have been obtained by Peck and Gurtman [7] and by Sve [8] who considered, respectively, waves propagating parallel and perpendicular to the layers. In both these cases late-time asymptotic solutions were obtained, which show the dispersive character of the main part of the wave. Sve [8] also considered, in an approximate way, the late-time solution for viscoelastic bilaminates in which the waves are propagating in the direction perpendicular to the laminates.

To date, only a limited number of experiments have been conducted that concern the finite amplitude wave-propagation in composite materials. Barker et al. [9] performed experiments on periodic laminates and found that below certain critical input amplitude, the stress wave amplitude decayed exponentially with distance and formed a structured shock wave above the critical amplitude. Lundergan and Drumheller [10] and Oved et al. [11] also conducted limited shock-wave experiments on layered stacks, which showed resonance phenomena due to layering. Nesterenko et al. [12] observed an anomaly in the precursor decay for the case of propagation of strong shock waves in periodic bilaminates with a relatively small cell size. They noted that for bilaminates with a relatively small cell size the jump in particle velocity at the wave front is essentially higher than one obtained with the larger cell size at the same distance of propagation. Similar observations were made for Ti–Al layered material systems under strong shock waves loading [13]. Comparison of the experimental results and computer simulations indicated that this effect is primarily due to the interactions of the secondary compression waves with the leading shock front. At early times, these secondary compression waves trail the shock front. However, with increasing distance of propagation these waves catch-up and eventually overtake the leading shock-wave front from behind. This increase in wave speed is facilitated by the propagation of the trailing secondary waves in a previously compressed material state. More recently, Zhuang [14] have conducted normal plate-impact experiments on layered stacks of polycarbonate and either glass, stainless steel, or aluminum systems to investigate dispersion versus dissipation characteristics due to heterogeneity of the layered material system during propagation of strong shock waves. They also reviewed existing models for propagation of shock waves and proposed new scaling laws for shock viscosity of heterogeneous layered solids.

Although glass reinforced polymers (GRP) were introduced in the 1930s, the dynamic failure of these material systems was not the focus until the 1970s when drop-weight testing machines were utilized to estimate their impact strength. Lifshitz [15] investigated the tensile strength and failure modes of unidirectional and angle-ply E-glass fiber-reinforced epoxy matrix composites at strain rates between 0.1 and 200 s^{-1}. The failure stresses under impact loading conditions were found to be considerably higher when compared to those obtained under quasi-static loading conditions. The dynamic response of GRPs has been investigated utilizing the split Hopkinson pressure bars (SHPBs) under relatively simple states of stress, for example, uniaxial compression, uniaxial tension, and pure shear [16–23]. In these studies the failure and ultimate strength of the GRP composites were found to increase with increasing strain rates. More recently, Zhuk et al. [24] studied the shock compressibility and sound wave velocity in commercial plain-weave fiberglass KAST-V (Soviet standard 102-92-74) composites using manganin gages in the range 5–22 GPa. They also utilized the VALYN™ VISAR to monitor the free surface particle velocity in experiments conducted at stress levels between 0.8 and 1.2 GPa. Hydrodynamic shock front attenuation was observed for the experiments impacted by thin (approximately 1.3 mm) aluminum flyer plates. Zaretsky et al. [25] also conducted plate-impact experiments on commercial KAST-V for stress range between 0.3 and 0.8 GPa. Spall signal was observed in the experiments with a shock stress level of about 0.3 GPa, and the spall strength of KAST-V was estimated to be about 0.1 GPa. The equation of state (EOS) of KAST-V was also determined, and the shear strength was found to be about 0.28 GPa. The authors proposed that the matrix–filler interface controlled the behavior of these materials in compression. Later Zaretsky et al. [26] performed plate-impact experiments on laminated glass fiber-reinforced epoxy 7781 composite. The free-surface velocities were recorded by the VISAR in the stress range of 0.5–2.4 GPa. The spall strength was calculated to be about 0.16 GPa. The dynamic viscosity was found to be much larger than of the epoxy matrix material. Oscillations in the free-surface particle velocity profile were observed; the frequency of the oscillations was found to increase with increasing impact stress.

Dandekar et al. [27] studied the elastic constants and spall strength of S2-glass GRP. They utilized ultrasonic wave velocity measurements along the six axes of the GRP to calculate the six independent elastic constants for its tetragonal stiffness matrix. The measured spall strength was between 0.007 and 0.06 GPa. On the same material, Boteler et al. [28] carried out a series of experiments using embedded polyvinylidene fluoride (PVDF) stress-rate gauges to study the shock-wave profiles in GRP as a function of propagation distance. The experimental stress histories displayed shock-wave attenuation with the increasing propagation distance. In the same year, Trott et al. [29] at Sandia National Laboratories applied a novel line-imaging velocity interferometer to simultaneously record the shock response of GRP at various points. The systematic difference in shock arrival time over a transverse distance of 2 mm and the relatively large amplitude fluctuations in the wave profiles reflected the complex periodic geometry of GRP. Later on, Dandekar et al. [30] in a joint research program with Sandia National Laboratory studied the shock response of GRP. The equation of state for the GRP was determined from a series of shock–reshock and plate-reverberation experiments. Tsou and Chou [31], using a combined analytical and numerical approach, studied the shock-wave propagation in unidirectional fiber-reinforced composite along the fiber direction. From these simulations the interface shear strength was estimated. Chen and Chandra [32], deBotton and coworkers [26], and Espinosa et al. [33] performed shock structure simulations on plain-woven glass fiber-reinforced composites. Fluctuations in particle velocity profiles due to stress wave reverberations between material interfaces were understood to play a critical role in controlling the overall behavior of the material.

Even though considerable progress has been made over the years in understanding the dynamic response of heterogeneous material systems under shock loading, the details of the shock structure including the phenomenon of material and geometric dispersion continues to be poorly understood. In view of this, in the present study asymptotic techniques have been employed to analyze propagation of acceleration waves in 2D layered material systems. Moreover, wave propagation in 2D elastic–viscoelastic bilaminates is analyzed to understand the effects of material inelasticity on both the wave-front and late-time solutions. The use of bilaminates provides a more tractable geometry from both analytical and experimental considerations. The analysis makes use of the Laplace transform and of the Floquet theory for ordinary differential equations (ODEs) with periodic coefficients [34]. Both wave-front and late-time solutions for step-pulse loading on layered half-space are presented. Moreover, a series of plate-impact shock-wave experiments were conducted by employing various different thicknesses of S2-GRP plates. The S2-glass GRP plates were impacted by Al 7075-T6 and D7 tool-steel flyer plates over a range of impact velocities. From this data, the details of the structure of the shock front, the Hugoniot elastic limit (HEL), EOS, and the Hugoniot states for the S2-glass GRP were determined. In the second series of experiments, both normal impact and combined compression-and-shear plate-impact experiments with skew angles ranging from 12° to 20°, were conducted to investigate the effects of normal compression and combined compression and shear loading on the spall strength of the two different architectures of GRP composites – S2-glass woven roving in Cycom 4102 polyester resin matrix and a 5-harness satin weave E-glass in a Ciba epoxy (LY564) matrix. In the third series of experiments, shock–reshock and shock–release plate-impact experiments were conducted to study the residual shear strength of the S2-glass GRP following normal shock-compression in the range 0.8–1.8 GPa.

In the following the results of the analytical analysis on elastic-elastic and elastic-viscoelastic bilaminates and the aforementioned plate-impact experiments on the GRP composites are presented.

2.2
Analytical Analysis

The objective of the proposed analytical analysis is to better understand wave scattering and dispersion at bimaterial interfaces and the role of material inelasticity in determining the structure of stress waves in heterogeneous material systems. To keep the problem analytically tractable the heterogeneous material systems are modeled as elastic–elastic and elastic–viscoelastic bilaminates. Both wave-front and late-time solutions for step-pulse loading on layered half-space are analyzed to understand the effects of layer thickness, impedance mismatch, and material inelasticity on the structure of acceleration waves in the bilaminates.

2.2.1
Wave Propagation in Elastic–Viscoelastic Bilaminates

Consider bilaminates consisting of elastic and viscoelastic layers of uniform thickness and infinite lateral extent. The elastic layers occupy odd-numbered layers, that is, $n = 1, 3, 5\ldots$, and the viscoelastic layers occupy even-numbered layers, that is, $n = 2, 4, 6\ldots$. Consider the individual layers to be homogeneous and isotropic and the layer thickness of both constituents to be the same, that is, $L_1 = L_2 = 0.5\,d$, where L_1 and L_2 are the thickness of the elastic and viscoelastic layers, respectively, and d is the total thickness of a typical bilaminate.

Let the laminates be subjected to a time-dependent normal stress loading, which is applied uniformly over the plane $x = 0$ (see Figure 2.1). Under these conditions longitudinal waves of one-dimensional strain propagate in the direction normal to the laminates. We consider the case in which the applied loading has a step function time dependence, that is, $\sigma = -\sigma_o H(t)$, and seek asymptotic solutions for the wave at the wave front and the waveform at late times.

For infinitesimal deformation, longitudinal waves propagating in the x-direction are governed by the balance of linear momentum and continuity. For the elastic layers, these equations can be written as

$$\varrho_1 \frac{\partial u_1}{\partial t}(x, t) - \frac{\partial \sigma_1}{\partial x}(x, t) = 0 \quad \text{and} \quad \frac{\partial \varepsilon_1}{\partial t}(x, t) = \frac{\partial u_1}{\partial x}(x, t) \qquad (2.1)$$

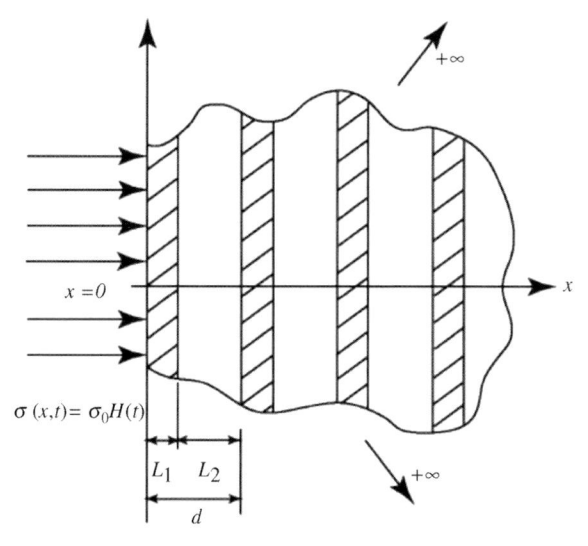

Figure 2.1 Schematic of the laminate used in the analytical analysis.

For viscoelastic layers, the balance of linear momentum and the continuity equations can be written as

$$\varrho_2 \frac{\partial u_2}{\partial t}(x,t) - \frac{\partial \sigma_2}{\partial x}(x,t) = 0 \quad \text{and} \quad \frac{\partial \varepsilon_2}{\partial t}(x,t) = \frac{\partial u_2}{\partial x}(x,t) \tag{2.2}$$

The constitutive equations for elastic and viscoelastic layers can be expressed as

$$\sigma_1(x,t) = E\,\varepsilon_1(x,t) \quad \text{and} \quad \sigma_2(x,t) = \int_{-\infty}^{t} G(t-\tau)d\varepsilon_2, \text{ respectively} \tag{2.3}$$

In Eqs. (2.1)–(2.3), σ_1 and σ_2 are the longitudinal components of the stress in the elastic and viscoelastic layers, u_1 and u_2 are the longitudinal components of the particle velocities in the elastic and viscoelastic layers, ϱ_1 and ϱ_2 are the mass density of the elastic and the viscoelastic layers, ε_1 and ε_2 are the longitudinal components of the strain in the elastic and viscoelastic layers, and E and $G(t)$ represent the elastic and the viscoelastic modulus, respectively.

The relaxation function for the viscoelastic material behavior is to be described by an exponential function of the following type:

$$G(t) = [G(0) - G(\infty)]e^{-t/\tau} + G(\infty) \tag{2.4}$$

where $G(0)$ denotes the "glassy" modulus at $t = 0$, $G(\infty)$ denotes the "rubbery" modulus at $t = \infty$, and τ denotes the characteristic relaxation time.

We seek solution to Eqs. (2.1)–(2.3) which satisfy zero stress and zero particle velocity initial conditions, and boundary conditions given by $\sigma(0, t) = -\sigma_0 H(t)$. Solutions to such problems are obtained most conveniently by means of Laplace transform methods in which the Laplace transform, $\hat{f}(x, s)$, of a function $f(x, t)$ is defined by

$$\hat{f}(x,s) = \int_0^\infty f(x,t)\,e^{-st}dt \tag{2.5}$$

Application of the Laplace transform to Eqs. (2.1)–(2.3) yields a system of four algebraic equations in the transformed plane. For a fixed s, these equations represent ODEs in which the coefficients are periodic functions of x with period $d = L_1 + L_2$. These equations contain four complex constants associated with the solution for the longitudinal component of stress. Two conditions on the four complex constants are obtained by requiring that the particle velocity and stress be continuous across the interface between the two adjacent layers comprising the bilaminate. The remaining conditions are obtained by the application of Floquet theory for periodic structures [34]. According to Floquet's theory, for such differential equations the solution at an arbitrary position x is related to the solution at $x-d$ by

$$\hat{w}(x,s) = e^{\mu(s)d}\hat{w}(x-d,s) \tag{2.6}$$

where $\hat{w}(x, s)$ represents the solution vector for the particle velocity and stress, and $\mu(s)$ is a characteristic parameter to be determined.

The characteristic parameter $\mu(s)$ in Eq. (2.6) can be obtained by solving the transcendental equation

$$\cosh \mu(s) d = \cosh \alpha_1(s) L_1 \cosh \alpha_2(s) L_2 + \frac{1}{2}\left(\frac{\varrho_1 \alpha_1(s)}{\varrho_2 \alpha_2(s)} + \frac{\varrho_2 \alpha_2(s)}{\varrho_1 \alpha_1(s)}\right) \sinh \alpha_1(s) L_1 \sinh \alpha_2(s) L_2 \quad (2.7)$$

where

$$\alpha_1(s) = \sqrt{\frac{s^2 \varrho_1}{E}}, \quad \text{and} \quad \alpha_2(s) = \sqrt{\frac{s \varrho_2}{\hat{G}(s)}} \quad (2.8)$$

Note that if μ is a solution of Eq. (2.7) then $-\mu$ is also a solution. By considering wave propagation in the direction of increasing x, we can restrict our attention to roots for which $\operatorname{Re}\mu(s) \leq 0$, so that the solution remains bounded as $x \to \infty$. This requirement uniquely determines μ, except for added integer multiples of $2\pi i$ that do not affect the solution.

2.2.2
Solution at Wave Front: Elastic Precursor Decay

Let the longitudinal wave fronts propagate with speeds c_1 and c_2 in the elastic and the viscoelastic layers, respectively. An average wave speed for the longitudinal wave fronts can be defined as

$$c_{\text{Wave}} = \frac{d}{(L_1/c_1 + L_2/c_2)} \quad (2.9)$$

At the arrival of the longitudinal wave at $x = x_n$, where $x_n = (n/2)d$ is the distance from $x = 0$ to the interface between the nth and the $(n+1)$th layers, the stress is given by

$$\sigma(x_n, x_n^+/c_{\text{Wave}}) = -\sigma_0 \left\{\exp\left[\frac{L_2 G'(0)}{2c_2 G(0)}\right]\right\}^{n/2} \theta^{-n/2} \quad (2.10)$$

where

$$\theta = \frac{1}{2} + \frac{1}{4}\left(\frac{\varrho_2 c_2}{\varrho_1 c_1} + \frac{\varrho_1 c_1}{\varrho_2 c_2}\right) \quad (2.11)$$

For the case of elastic–elastic bilaminates the argument of the exponential function is zero and the right-hand side of Eq. (2.10) can be interpreted as an average transmission coefficient for propagation of an elastic wave through a cell of length d. The attenuation of the amplitude at the wave front is the decay primarily due to successive elastic wave reflections. For the case of elastic–viscoelastic bilaminates, the argument of the exponential function gives rise to additional attenuation due to

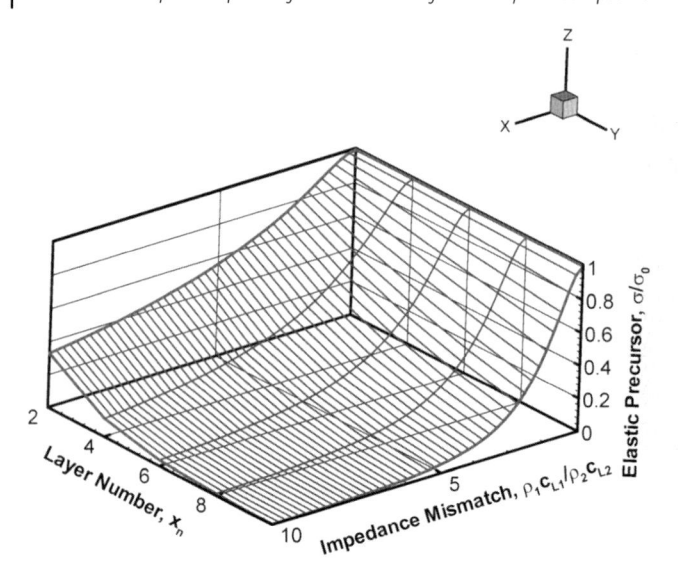

Figure 2.2 Effect of material mismatch and the number of layers on the elastic precursor decay for elastic–elastic bilaminates.

material inelasticity. The rate of decrease in stress is often so rapid that the stress at the wave front can become negligibly small at remote positions.

Figure 2.2 shows the magnitude of the elastic precursor as a function of number of layers and the impedance mismatch. A strong decay in the elastic precursor is observed with an increase in the number of layers and the increase in impedance mismatch.

Figure 2.3 shows the effect of viscoelasticity on the elastic precursor decay after wave propagation through 10 layers. The stress at the wave front is normalized by the amplitude of the corresponding elastic precursor for the case of elastic–elastic bilaminates. The x-axis represents the ratio of the time taken for the longitudinal wave to travel the thickness of a viscoelastic layer to the relaxation time constant for the material. It is to be noted that when the relaxation time is large and the viscoelastic layer thickness, that is, L_2 is small, the effect of material inelasticity on elastic precursor decay is small. Also, when the ratio between the instantaneous modulus and the rubbery modulus, that is, $\gamma^2 = G(0)/G(\infty)$, is close to one the effect of material inelasticity on the elastic precursor decay is negligible.

2.2.3
Late-Time Asymptotic Solution

At sufficiently late times after the arrival of the wave front, the stress at a remote position is expected to reach a level σ_o, which corresponds to the applied stress boundary condition at $x = 0$. The transition from the low-amplitude stress at the

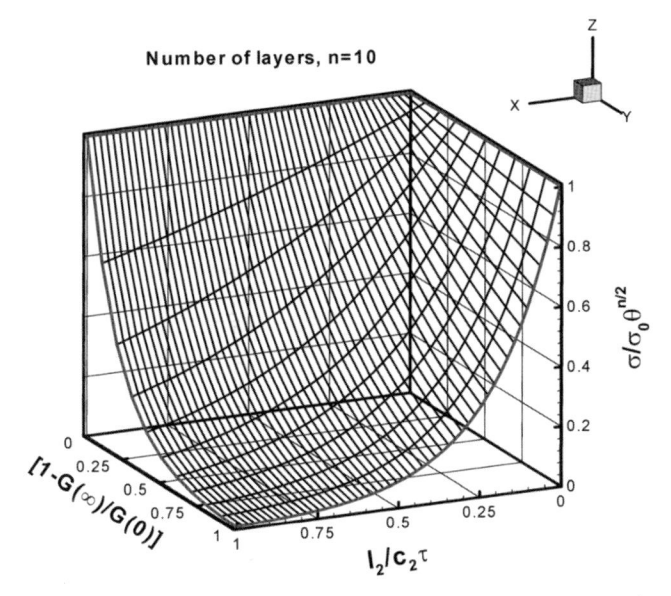

Figure 2.3 Effect of material inelasticity on the elastic precursor decay. The stress at the wave front (y-axis) is normalized by the amplitude of the elastic precursor for the case of elastic–elastic bilaminates.

wave front to this equilibrium state at late times can be characterized by obtaining the late-time asymptotic solution to the integral

$$\sigma(x_n, t) = \frac{1}{2\pi i} \int_{\gamma-i\infty}^{\gamma+i\infty} \hat{\sigma}(x_n, s) \, e^{st} ds \tag{2.12}$$

The integral in Eq. (2.12) can be evaluated asymptotically for large t by using the method of steepest descent. To this end, it is convenient to introduce the small time scale

$$\delta = t - x_n / c_{L_{ave}} \tag{2.13}$$

in which $c_{L_{ave}}$ denotes the average wave speed at which the main parts of the longitudinal disturbance propagates at late time and is given by

$$c_{L_{ave}} = \frac{d}{\left[(L_1/c_1)^2 + (L_2/c_2)^2 + ((\varrho_1 c_1 / \varrho_2 c_2) + (\varrho_2 c_2 / \varrho_1 c_1))(L_1/c_1)(L_2/c_2)\right]^{1/2}} \tag{2.14}$$

It should be noted that $c_{L_{ave}}$ is equivalent to defining the phase velocity of an infinite train of sinusoidal waves of zero frequency, that is, the long wavelength limit. It is interesting to note that the speed of longitudinal disturbance at late times depends on the impedance mismatch between the layers. For laminate architectures in which the

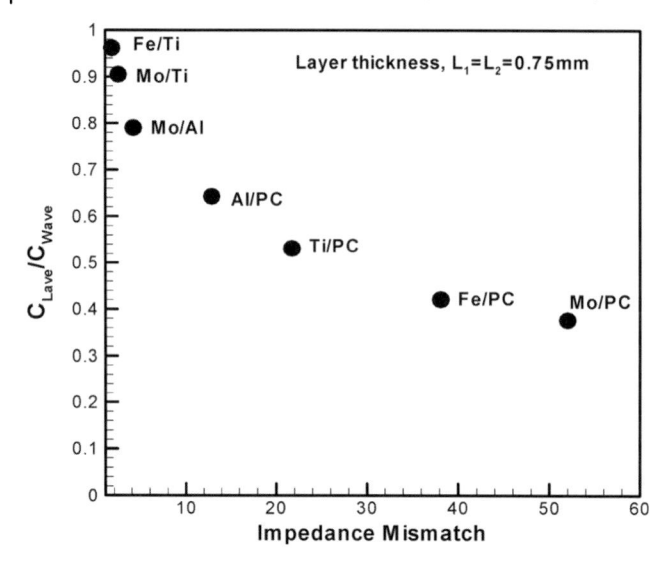

Figure 2.4 The effect of impedance mismatch on the average wave speed $c_{L_{ave}}$ at which the main parts of the longitudinal disturbance propagate at late times.

impedance mismatch is close to one, the late-time dispersion wave and the elastic precursor arrive at a particular location at the same time. However, for laminates in which the impedance mismatch is large, $c_{L_{ave}}$ is considerably less than $c_{W_{ave}}$. This effect is shown graphically in Figure 2.4 for a selected number of material pairs.

Substituting Eq. (2.13) in Eq. (2.12), we can obtain an alternate form of the inverse transform, that is,

$$\sigma(x_n, t) = \frac{\sigma_0}{2\pi i} \int_{\Gamma - i\infty}^{\Gamma + i\infty} \frac{e^{-\delta g(s)} e^{th(s)}}{s} ds \qquad (2.15)$$

where

$$g(s) = \mu(s) c_{L_{ave}} \qquad (2.16)$$

and

$$h(s) = \mu(s) c_{L_{ave}} + s \qquad (2.17)$$

In order to evaluate the integral in Eq. (2.15) for $t \to \infty$, we employ the method of steepest descent [35]. In view of this, it must be noted that the main contribution to the integral is expected to arise from $s = 0$. Expanding $h(s)$ about the saddle point $s = 0$ gives

$$g(s) \cong -s + \frac{h''(0)}{2!} s^2 + \frac{h'''(0)}{3!} s^3 \qquad (2.18)$$

and

$$h(s) \cong \frac{h''(0)}{2!}s^2 + \frac{h'''(0)}{3!}s^3 \tag{2.19}$$

In Eqs. (2.18) and (2.19), $h''(0)$ and $h'''(0)$ are given by

$$h''(0) = \frac{c_{L_{ave}}^2}{d^2}\left(\left[\frac{L_2^2}{c_2^2}\tau(\gamma^2-1)\right] - \frac{L_1 L_2}{c_1 c_2}\left[\tau(\gamma^2-1)\left(\frac{\varrho_1 c_1}{\varrho_2 c_2}\right)\right]\right) \tag{2.20}$$

and

$$h'''(0) = \frac{c_{L_{ave}}^4}{d^4}\frac{L_2(L_2\varrho_2 + L_1\varrho_1)}{4c_1^6\varrho_2^3\varrho_1^3 c_2^6}\{c_1^2 L_1^2 L_2\varrho_2^3\varrho_1(L_2\varrho_2-2L_1\varrho_1)c_2^4 + L_1^3 L_2\varrho_2^5 c_2^6$$

$$+ c_1^6 L_2\varrho_1^3(L_1^2 L_2\varrho_1^2 + 3(1+2\gamma^2-3\gamma^4)L_2\varrho_2^2\tau^2 c_2^2) + 3(1+2\gamma^2-3\gamma^4)L_1\varrho_2\varrho_1\tau^2 c_2^2$$

$$+ c_1^4 L_1\varrho_2\varrho_1^2 c_2^2(L_1^2 L_2\varrho_1^2 - 12\gamma^2(\gamma^2-1)L_2\varrho_2^2\tau^2 c_2^2 - 21\varrho_2\varrho_1(L_2^2 + 6\gamma^2(\gamma^2-1)\tau^2 c_2^2))\} \tag{2.21}$$

where, as before, γ^2 is the ratio between the instantaneous modulus and the rubbery modulus.

In view of Eqs. (2.15)–(2.19), the integral in Eq. (2.15) along the path of steepest descent Γ can be written as

$$\sigma(x,\delta) = \frac{\sigma_0}{2\pi i}\int_\Gamma \frac{1}{s} e^{-\left\{\frac{h'''(0)}{3!}s^3 + \frac{h''(0)}{2!}s^2 - s\right\}\delta + \left\{\frac{h'''(0)}{3!}s^3 + \frac{h''(0)}{2!}s^2\right\}t}ds$$

$$= = \frac{\sigma_0}{2\pi i}\int_\Gamma \frac{1}{s} e^{\delta s + (t-\delta)\frac{h''(0)}{2!}s^2 + (t-\delta)\frac{h'''(0)}{3!}s^3}ds \tag{2.22}$$

Substituting Eq. (2.13) in Eq. (2.22) and applying the transformation

$$s = z\left[\frac{6}{(x_n/c_{L_{ave}})h'''(0)}\right]^{1/3} - \frac{h''(0)}{h'''(0)} \tag{2.23}$$

yields

$$\sigma(x_n, t) = e^A \frac{\sigma_0}{2\pi i}\int_{\Gamma_z} \frac{e^{Bz+z^3}}{z}dz \tag{2.24}$$

In Eq. (2.24)

$$A = -(t-x_n/c_{L_{ave}})\frac{h''(0)}{h'''(0)} + \frac{1}{3}\frac{(h''(0))^3}{(h'''(0))^2}(x_n/c_{L_{ave}}) \tag{2.25}$$

and

$$B = \left[(t-x_n/c_{L_{ave}}) - \frac{1}{2}\frac{[h''(0)]^2}{h'''(0)}(x_n/c_{L_{ave}})\right]\left(\frac{6}{h'''(0)(x_n/c_{L_{ave}})}\right)^{1/3} \quad (2.26)$$

The path of steepest descent Γ approaches $s = 0$ along the directions $\arg(s) = -\pi/3$; it is indented to the right around the pole $s = 0$ and leaves the origin along the direction $\arg(s) = \pi/3$. The contribution to the integral from the one-third of a circle indentation around the origin is $\sigma_0/3$. Thus, the integral in Eq. (2.24) along the steepest descent path becomes [36]

$$\sigma(x_n, t) = e^A \sigma_0 \left[\frac{1}{3} + \int_0^B Ah(B) dB\right] \quad (2.27)$$

in which $Ah(B)$ is the Airy Hardy function

$$Ah(B) = \frac{1}{2\pi i} \int_\Gamma e^{Bz + z^3} dz \quad (2.28)$$

After certain algebraic manipulations it can be shown that Eq. (2.12) can be expressed as

$$\sigma(x_n, t) = \sigma_0 \, e^A \left[\frac{1}{3} + \sum_{m=0}^{\infty} \frac{\Gamma((m+1)/3)}{(m+1)!} B^{m+1} \sin\left(\frac{\pi}{3}(m+1)\right)\right] \quad (2.29)$$

where $\Gamma(m)$ is the gamma function and is defined as

$$\Gamma(m) = \int_0^\infty t^{m-1} e^{-t} dt \quad \text{for} \quad m > 0 \quad (2.30)$$

For infinitesimal deformation, solutions for elastic precursor decay and late-time dispersion, which satisfy zero stress and particle velocity initial conditions and boundary conditions given by a step loading function in time, are summarized in Figure 2.5. Upon impact of the laminate, a two-wave structure is obtained. The leading elastic precursor propagates at a speed dictated by the average wave speed in the two constituents given by c_W, while the late-time dispersed front arrives at a speed $c_{L_{ave}}$. The late-time stress wave oscillates about a mean level dictated by the amplitude of the input stress pulse.

Next, stress wave profiles obtained from the asymptotic solutions at the wave front and the waveform at late times are presented. These simulations are designed to illustrate the effect of impedance mismatch, distance of wave propagation, layer thickness, that is, cell size, density of interfaces, and material inelasticity on the amplitude of stress wave at the wave front and the dispersive characteristics of the waveform at late times. The simulations are carried out for elastic–elastic bilaminates comprising Ti–Fe and Mo–Fe material pairs, and elastic–viscoelastic bilaminates comprising Al–PC material pair. The acoustic impedance mismatch for Ti–Fe

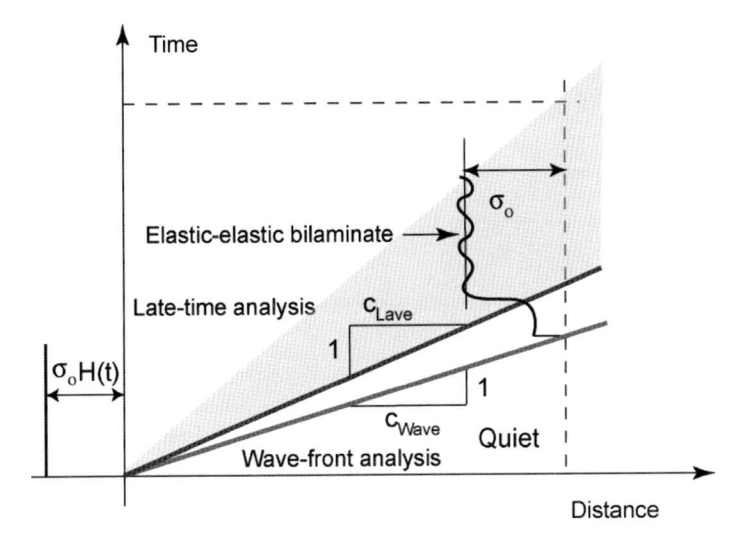

Figure 2.5 The two wave structure obtained during impact of a typical elastic bilaminate.

and Mo–Ti bilaminates is 1.75 and 2.45, respectively. For the Al–PC bilaminates, the acoustic impedance mismatch is 8.31. Three different layer thicknesses are evaluated for Ti–Fe and Mo–Fe laminates: 0.75, 1.5, and 2.25 mm for total laminate thickness of 9 mm. For the Al–PC laminates the simulations are presented for layer thickness of 0.125 mm with total laminate thicknesses of 0.5, 1.0, and 1.5 mm. The relaxation time in Eq. (2.4) for PC is taken to be $\tau = 2\,\mu s$. The ratio of the instantaneous modulus to the rubbery modulus, that is, $G(0)/G(\infty)$, is taken to be 1.01. Table 2.1 gives the physical properties of all the materials used in the simulations.

Figures 2.6 and 2.7 present the predictions of the wave profiles for elastic–elastic bilaminates illustrating the effects of impedance mismatch on the wave-front and late-time dispersion wave characteristics. Two different bilaminates are considered: Fe–Ti bilaminates with an impedance mismatch of 1.75 and Mo–Ti bilaminates with an impedance mismatch of 2.48. In each case, the thickness of the individual layers is

Table 2.1 Physical properties of the material layers employed in the simulations.

Layer material	Elastic modulus (GPa)	Density (g/cm³)	Longitudinal wave speed (m/s)	Acoustic impedance (GPa (mm/μs))
Titanium (Ti)	120.2	4.5	6716	30.22
Iron (Fe)	211.4	7.87	5950	46.83
Molybdenum (Mo)	324.8	10.22	6250	63.88
Aluminum (Al)	70.6	2.7	6420	17.33
Polycarbonate (PC)	2.3	1.2	1832	2.20

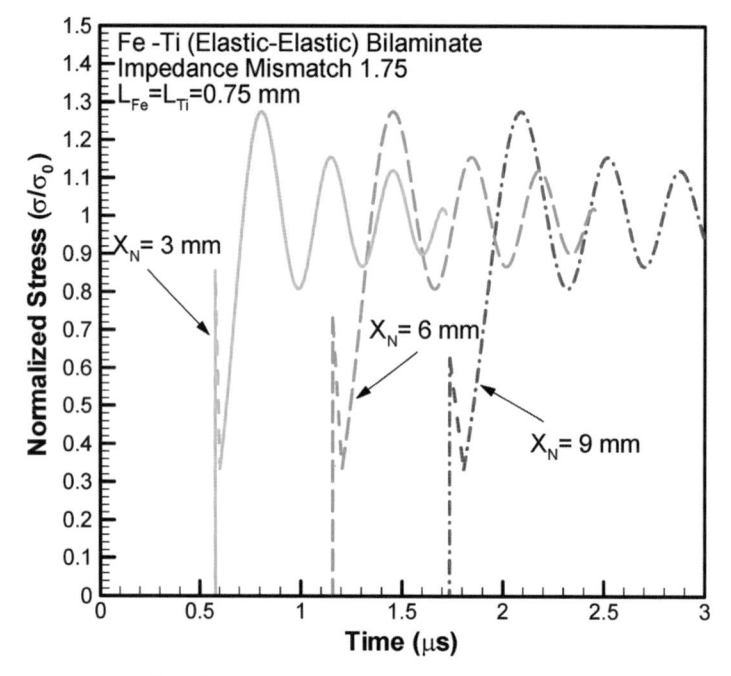

Figure 2.6 Effect of distance of propagation on the elastic precursor and late-time dispersion for Fe–Ti laminates.

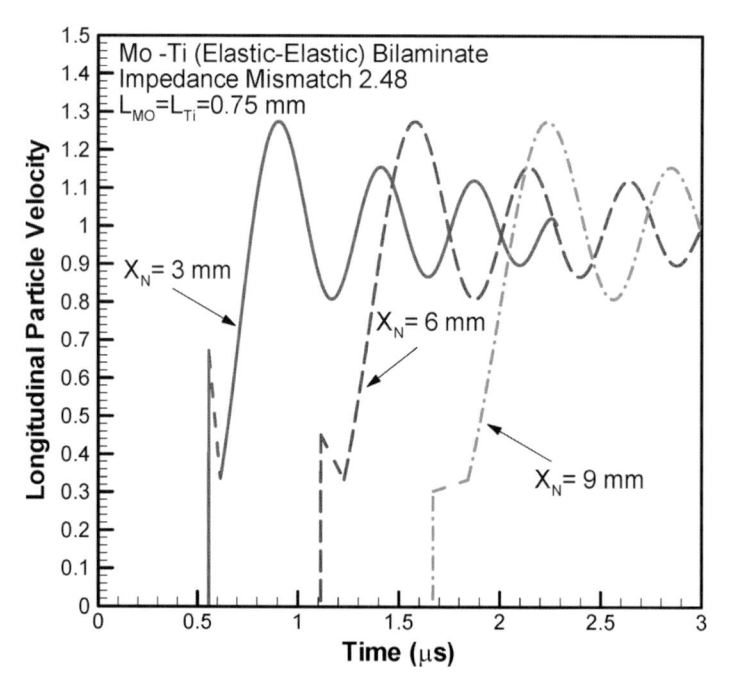

Figure 2.7 Effect of distance of propagation on the elastic precursor and late-time dispersion for Mo–Ti laminates.

0.75 mm. The abscissa represents the time after impact while the ordinate represents the theoretical normal stress normalized by the amplitude of the input stress. The stress profiles are shown at a propagation distance of 3, 6, and 9 mm. As discussed earlier, for each laminate a two-wave structure is obtained. A leading wave front (elastic precursor) that propagates at the speed $c_{W_{ave}}$ and a late-time dispersion wave propagating at the speed $c_{L_{ave}}$. For the case of Mo–Ti bilaminates a larger precursor decay and a lower frequency of the late-time dispersive waves is observed. Also, consistent with Eqs. (2.9) and (2.14), the time difference between the arrival of the leading wave-front and the late-time dispersive wave is much longer in the case of Mo–Ti laminates when compared with the Fe–Ti laminates. It is also interesting to note that for both cases the late-time dispersive waves show steady wave profiles with increasing distance of propagation into the bilaminates.

Figure 2.8 shows the effect of the layer thickness on the elastic precursor decay and late-time dispersion during propagation of stress waves in Ti–Fe laminates and the results for three different layer thicknesses are as follows: 0.75, 1.5, and 2.25 mm. As expected, the arrival of the elastic precursor at $x_n = 9$ mm occurs at the same time for the three different laminate architectures. However, the laminates with largest layer thickness, that is, 2.25 mm, shows the smallest elastic precursor decay, while the laminate with the smallest layer thickness, that is, 0.75 mm, shows the highest precursor decay. The late-time dispersive wave for the smallest layer thickness laminates contain the highest frequency oscillations while the largest layer thickness laminates contain the lowest frequency oscillations. Also, the rise-time associated

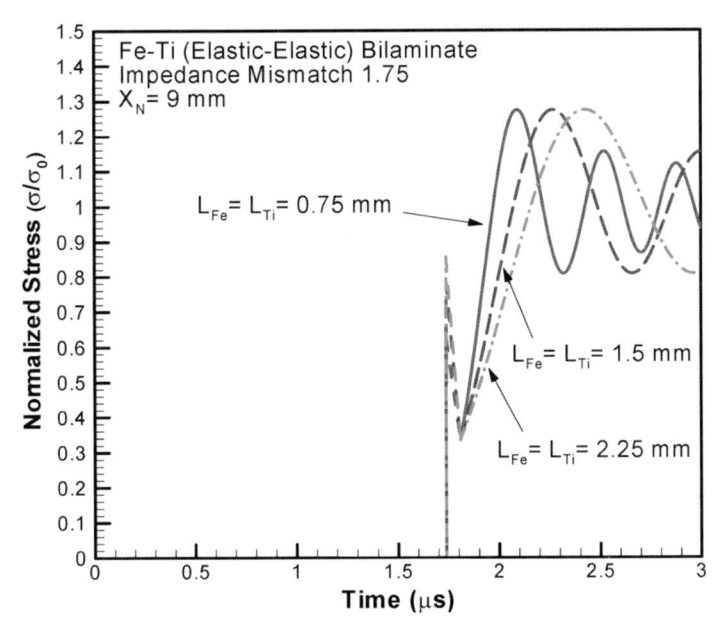

Figure 2.8 Effect of layer thickness on the elastic precursor and the late-time dispersion for Fe–Ti laminates.

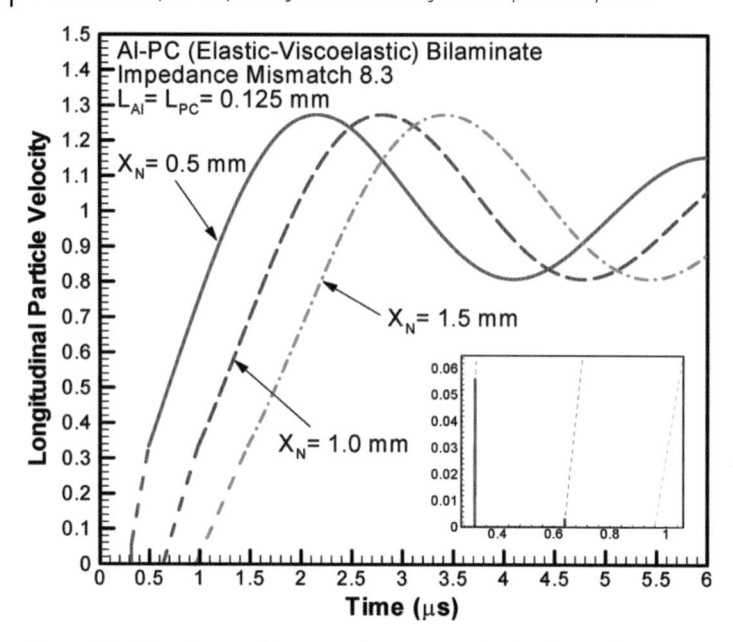

Figure 2.9 Wave-front and late-time dispersion results for Al–PC bilaminates.

with the late-time dispersive wave decreases with layer thickness and an increase in the density of interfaces (i.e., number of layers in a given laminate thickness).

Figure 2.9 shows the wave-front and the late-time solution for the Al–PC laminate for $x_n = 0.5$, 1, and 1.5 mm. The impedance mismatch for the Al–PC material pair is 8.3. The thickness of each Al and PC layer is 0.125 mm. Due to the relatively high impedance mismatch between the Al and PC layers, and also the viscoelasticity associated with PC, a very strong elastic precursor decay is observed; so-much-so that the elastic precursor is reduced to approximately zero as the stress wave propagates only 0.5 mm into the laminate. This is seen more clearly from the insert in Figure 2.9, which shows the early parts of the wave profiles for the three thicknesses. Also, it is interesting to observe that the frequency of the oscillations in the late-time dispersive wave solution is much smaller when compared to the lower impedance mismatch Mo–Ti and Ti–Fe material pair laminates.

Figure 2.10 compares the late-time dispersion characteristics for several select material pairs with different impedance mismatch laminates. Because of the dependence of $c_{L_{ave}}$ on the impedance mismatch, the dispersion profiles have been shifted in time so as to start at the same point in time. The late-time dispersion profiles can be characterized by the rise time and the frequency of the oscillations contained in the wave profiles. The rise times of the dispersion waves is observed to increase with an increase in impedance mismatch, while the frequency of the oscillations decreases with an increase in the impedance mismatch. Also, the late-time dispersive wave oscillates about the mean level corresponding to the input stress. The maximum amplitude of the late-time dispersion wave is $0.3\sigma_o$, and is

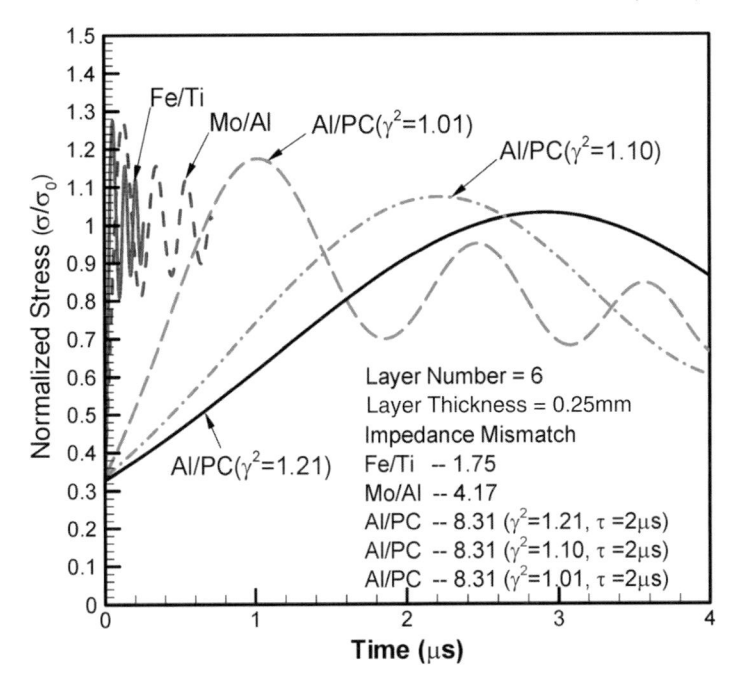

Figure 2.10 Characteristics of late-time dispersion for select material pairs with different impedance mismatch.

observed to be independent of the impedance mismatch of the elastic–elastic laminates. Also, it is interesting to note the effect of material inelasticity on the dispersion characteristics of the late-time wave profiles. With an increase in the ratio between the instantaneous modulus to the rubbery modulus the rise time associated with the late-time dispersive waves is observed to increase while the frequency of the oscillations decreases.

2.3
Plate-Impact Experiments on GRPs

2.3.1
Material: Glass Fiber-Reinforced Polymer

In the present investigation, two different GRPs were investigated: (a) S2-glass woven roving in Cycom 4102 polyester resin matrix and (b) a balanced 5-harness satin weave E-glass in a Ciba epoxy (LY564) matrix. The S2-glass GRP was fabricated at the Composites Development Branch, US Army Research Laboratory, Watertown, MA, while the E-glass GRP was fabricated by the DRA Land Systems, Great Britain. The S2-glass fibers (in which "S" stands for higher-strength glass fiber), are known to be stronger and stiffer than the E-glass fiber reinforcement – they have a 40% higher tensile strength, 10–20% higher compressive strength, and much greater abrasion resistance when compared to the E-glass fibers.

The S2-glass GRP specimens used in the present study were made from S2-glass woven roving in CYCOM 4102 polyester resin matrix with a resin content of $32\pm2\%$ by weight. The individual laminate plies were 0.68 mm in thickness. Composites of the desired thickness were manufactured by stacking an appropriate number of plies in a $\pm90°$ sequence. The desired number of laminates was stacked between two steel plates with release film. The stacked layers were then vacuum bagged and subjected to the following heat cycle:

1) Initially heated to 339 ± 4 K for 45 min.
2) Temperature raised to 353 ± 2 K for 2 h.
3) Temperature raised to 398 ± 4 K and held for 2 h.
4) Cooled to 312 ± 12 K at the rate of 7 K/min.

The curing cycle was initiated with a gradual temperature increase under vacuum conditions so that the volatile gases including the water vapor can be driven off. Next, the curing temperature was gradually increased to its maximum and held constant for a couple of hours to develop a high degree of cross-linking, followed by application of pressure to consolidate the laminate. The final density of S2-glass GRP was 1.959 ± 0.043 kg/m^3, while the longitudinal wave speed in the composite obtained from phase velocities of ultrasonic waves was 3.2 ± 0.1 km/s in the thickness direction [27]. The six independent elastic constants and elastic compliances of the tetragonal symmetry stiffness matrix are shown in Table 2.2.

The stiffness matrix [C] is given by

$$[C] = \begin{bmatrix} C_{11} & C_{12} & C_{13} & 0 & 0 & 0 \\ C_{12} & C_{22} & C_{23} & 0 & 0 & 0 \\ C_{13} & C_{23} & C_{33} & 0 & 0 & 0 \\ 0 & 0 & 0 & C_{44} & 0 & 0 \\ 0 & 0 & 0 & 0 & C_{55} & 0 \\ 0 & 0 & 0 & 0 & 0 & C_{66} \end{bmatrix}$$

where $C_{22} = C_{11}$, $C_{23} = C_{13}$, and $C_{55} = C_{44}$.

The E-glass laminates comprised of a balanced 5-harness satin weave E-glass with Ciba epoxy (LY564) as the matrix. The resin content was 50% by volume. The

Table 2.2 Values of elastic constants and elastic compliances of the GRP.

Elastic constants	GPa	Elastic compliances	10^{-2} GPa^{-1}
C_{11}	31.55	S_{11}	4.5039
C_{33}	20.12	S_{33}	6.2074
C_{44}	4.63	S_{44}	21.60
C_{66}	4.94	S_{66}	20.24
C_{12}	15.86	S_{12}	−1.8696
C_{13}	9.75	S_{13}	−1.2766

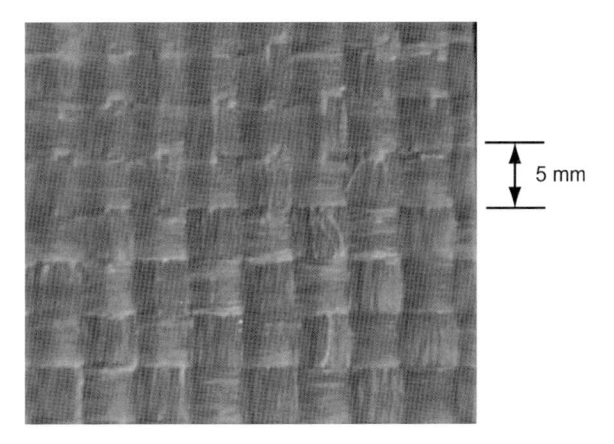

Figure 2.11 SEM micrograph of the S2-glass fiber woven roving layer.

individual laminate plies were 1.37 mm in thickness. The composite was manufactured by using the resin transfer molding process, in which an appropriate number of plies were stacked in $\pm 90°$ sequence to achieve the desired thickness. A low cure-time and temperature was used to produce a reasonably tough matrix. The final density of the E-glass GRP was 1.885 kg/m^3, while the longitudinal wave speed in the composite was 3.34 km/s in the thickness direction.

Figures 2.11 and 2.12 show SEM micrographs of the S2-glass and the E-glass fiber woven roving for the two GRPs, respectively. The E-glass GRP has a much smaller fiberglass bundle size when compared to the S2-glass GRP. Each fiberglass bundle is approximately 5 mm in width for the S2-glass GRP, while it was approximately 1.25 mm for the E-glass GRP.

Figure 2.12 SEM micrograph of the 5-harness satin weave E-glass fiber woven roving layer.

2.3.2
Plate-Impact Shock Compression Experiments: Experimental Configuration

The plate-impact shock compression experiments were conducted using the 82.5 mm single-stage gas-gun in the Department of Mechanical and Aerospace Engineering at Case Western Reserve University. The experiments involve the normal impact of a flyer plate with the GRP target. The shock-induced compression waves in the GRP are monitored at the free surface of the target by means of a multibeam VALYN VISAR system. A COHERENT VERDI 5W solid-state diode-pumped frequency doubled Nd:YVO$_4$ CW laser with wavelength of 532 nm is used to provide a coherent monochromatic light source. The schematic of the plate-impact experimental configuration is shown in Figure 2.13. A fiberglass projectile carrying the flyer plate is accelerated down the gun barrel by means of compressed helium gas. Rear-end of the projectile has sealing O-ring and a plastic (Teflon) key that slides in a key-way inside the gun barrel to prevent any rotation of the projectile. In order to reduce the possibility of an air cushion between the flyer and target plates, impact is made to occur in a target chamber that has been evacuated to 50 μm of Hg prior to impact. A laser-based optical system, utilizing a UNIPHASE helium–neon 5 mW laser (Model 1125p) and a high-frequency photodiode, is used to measure the velocity of the projectile. To ensure the generation of plane waves, with wave front sufficiently parallel to the impact face, the flyer and the target plates are aligned carefully to be parallel to within 2×10^{-5} rad by using an optical alignment scheme [5]. The actual tilt between the two plates is measured by recording the times at which four, isolated, voltage-biased pins, that are flush with the surface of the target plate, are shorted to ground. The acceptance level of the experiments is of the order of 0.5 milrad.

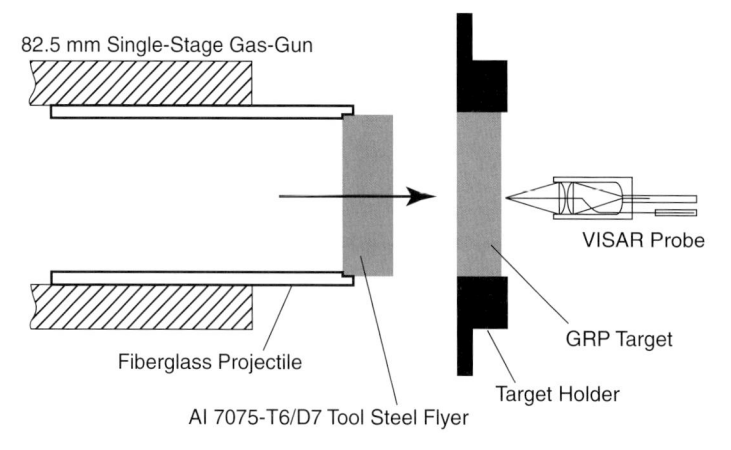

Figure 2.13 Schematic of the plate-impact shock compression experiments.

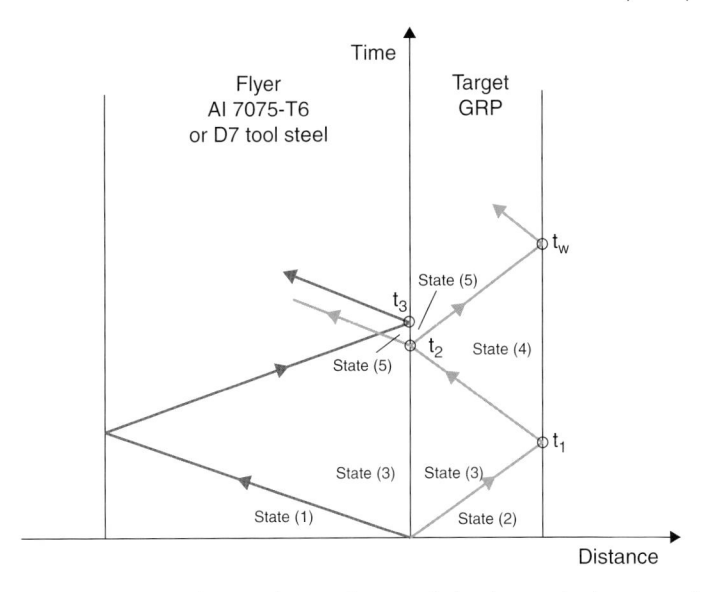

Figure 2.14 Time–distance diagram for normal plate-impact shock compression experiments.

2.3.2.1 t–X Diagram (Time versus Distance) and S–V Diagram (Stress versus Velocity) for Plate-Impact Shock Compression Experiments

The t–X diagram for normal plate-impact shock compression experiments on GRP is shown in Figure 2.14. The abscissa represents the distance from impact surface; while the ordinate represents the time after impact. Figure 2.15 shows the stress and particle velocity diagram for the same experiment. The S–V diagram provides the locus of all the stress and particle velocity states that can be attained during a typical experiment. The abscissa represents the particle velocity in the target and the flyer plates, while the ordinate represents the stress in the target and flyer. In order to avoid the possibility of spall (delamination) of the GRP during the experiment, the release

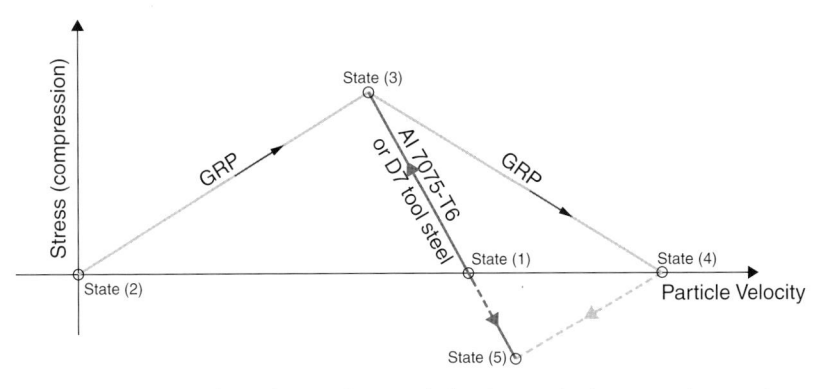

Figure 2.15 Stress–velocity diagram for normal plate-impact shock compression experiments.

waves from free surface of the flyer and the target plates should intersect within the GRP target during the time duration of the experiment. In order to achieve this, the thickness of the flyer plate is chosen such that the release wave from the free (back) surface of the flyer plate arrives later than the time of the arrival of the release wave from the free (back) surface of the target plate, that is, $t_3 > t_2$.

2.3.3
Plate-Impact Spall Experiments: Experimental Configuration

In order to conduct the plate-impact spall experiments, a thinner metallic flyer plate (Al 7075-T6) is impacted with a thicker GRP target plate at both normal and oblique incidence. Figure 2.16 shows the schematic of the experimental configuration used for the combined pressure-shear plate-impact spall experiments. For the case of the normal plate-impact spall experiments, the skew angle of the flyer plate is zero. The multibeam VALYN VISAR is also used to measure the history of the normal particle velocity at the rear surface of the target plate.

2.3.3.1 t–X Diagram (Time versus Distance) and S–V Diagram (Stress versus Velocity) for Plate-Impact Spall Experiments

A schematic of the time versus distance diagram (t–X diagram), which illustrates the propagation of compression waves and tensile waves through the target and flyer plates during the plate-impact spall experiments, is shown in Figure 2.17. The abscissa represents the distance in the flyer and the target plates from the impact surface while the ordinate represents the time after impact. The arrows indicate the direction of wave propagation. Upon impact of the flyer and the target plates, two compressive waves are generated. These waves propagate from the impact surface into the flyer and the target plates with wave speeds that are characteristic of the flyer and target plate materials. Since the flyer has a smaller thickness than the target and

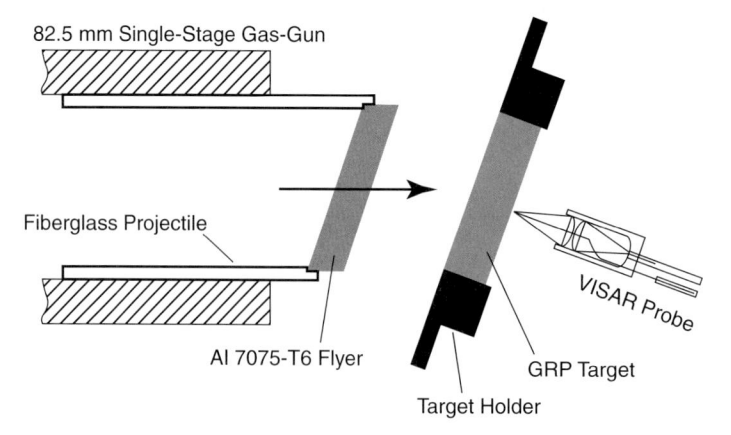

Figure 2.16 Schematic of the plate-impact spall experiments.

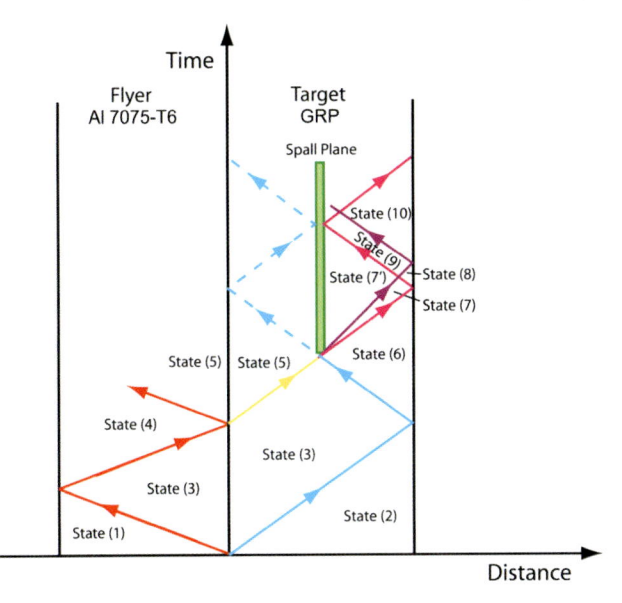

Figure 2.17 Time–distance diagram showing the wave propagation in the flyer and the target plates for plate-impact spall experiments. The spall plane occurs approximately in the middle of the target plate.

the Al alloy flyer has a higher longitudinal wave speed (6.23 km/s) than that of the GRP targets, the compressive wave in the flyer reflects as a release wave from its free surface, part of which is transmitted into the GRP target plate. Similarly, the compressive wave in the target reflects from its back surface as a rarefaction wave and interacts with the release wave from the flyer to generate a state of tensile stress at a predetermined plane in the target (represented as State (7) in the target). If the amplitude of the tensile wave is large enough, the GRP target will undergo spall failure. Since the spall failure is associated with the creation of a free surface, the tensile stress wave is reflected back from this surface as a compression wave, as shown in Figure 2.17.

Figure 2.18 shows the details of the locus of the stress and particle velocity states that can be attained during a typical plate-impact spall experiment. The abscissa represents the particle velocity in the target and the flyer plates, while the ordinate represents the stress in the target and flyer plates. For the case in which the spall strength is larger than the tensile strength, the stress and particle velocity in the GRP moves along the dashed lines from State (5) to the no-spall state denoted by State (7). However, if the tensile stress is greater than the spall strength of the GRP (σ_{spall} indicated by the short dashed lines), the GRP will spall and the tensile stress in State (7) will unload to the stress-free state denoted by State (7'). The compressive "*end of spall*" wave from State (7') arrives at the free surface of the GRP and brings the free surface particle velocity to State (10), which is the same as that in State (6) and also in State (7'). The free surface particle velocity in States (6), (7'), and (10), is

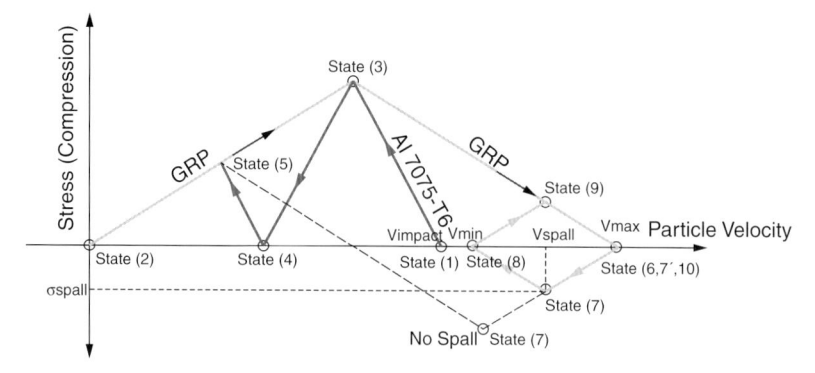

2 Shock and Impact Response of Glass Fiber-Reinforced Polymer Composites

Figure 2.18 Stress–velocity diagram showing the loci of all the stress and particle velocity states that can be achieved in a typical plate-impact spall experiment.

referred to as V_{max}, and the corresponding free surface particle velocity in State (8) is referred to as V_{min}. V_{max} and V_{min} can be used to determine the spall strength in the GRP.

2.3.4
Shock–Reshock and Shock–Release Experiments: Experimental Configuration

Figure 2.19 shows the schematic of the experimental configuration used for shock–reshock and shock–release experiments. In these experiments, a dual flyer plate assembly was used. The shock–reshock experiments were conducted by using a projectile faced with a GRP plate and backed by a relatively high shock impedance Al 6061-T6 plate; for the shock–release experiments, the GRP was backed by a

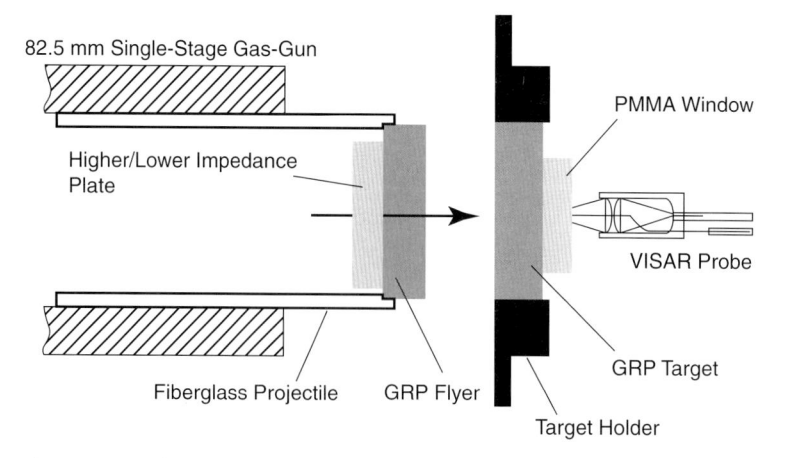

Figure 2.19 Schematic of the shock–reshock and shock–release experiments.

relatively lower impedance PMMA plate. The target comprises a disk machined from the GRP under investigation, which is backed by a PMMA window to prevent spall. The PMMA disk is of optical quality and is lapped and polished to a flat structure within a few bands of sodium light. Prior to gluing the PMMA window to the GRP disk, the bonding surface of the PMMA is lapped and coated with approximately 100 nm thick aluminum layer by vapor deposition to make it a diffusively reflecting surface. The VISAR probe is then focused on the diffuse interface and used for monitoring the shock-wave profile at the GRP/window interface.

2.3.4.1 t–X Diagram (Time versus Distance) for Shock–Reshock and Shock–Release Experiments

A schematic of the time versus distance diagram (*t–X* diagram), which illustrates the propagation of shock waves through the target and flyer plates during the shock–reshock and shock–release experiments, is shown in Figure 2.20. The abscissa represents the distance in the flyer and the target plates from the impact surface while the ordinate represents the time after impact. The arrows indicate the direction of wave propagation. The initial shock in the flyer reflects from the GRP/Al6061-T6 or GRP/PMMA interface as a reshock or release wave (which converts the material stress state from State (3) to (4)), as shown in Figure 2.20, and then propagates toward the GRP/PMMA window interface. As indicated in the *t–X* diagram, the GRP is shock loaded to State (3), reshocked or released to State (4), which are recorded as States (5) and (7) at the GRP/PMMA window interface.

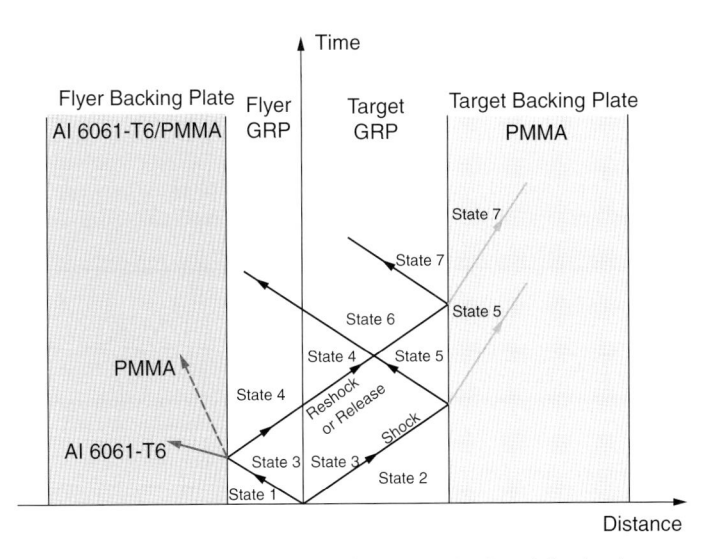

Figure 2.20 Time–distance diagram for shock–reshock and shock–release experiments.

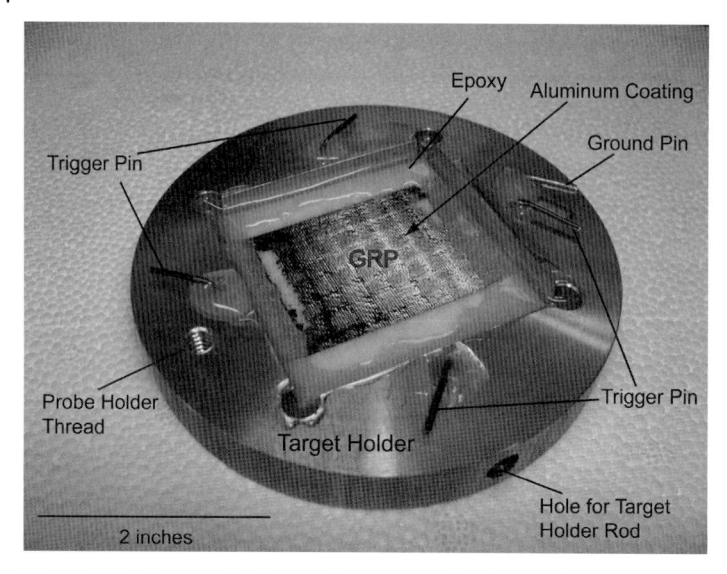

Figure 2.21 Photograph of a typical target assembly showing the GRP specimen, the aluminum ring holder, the tilt, and trigger wires.

2.4
Target Assembly

In all experiments, the Al 7075-T6 and D7 tool-steel flyer plates were 3 in. in diameter, while the GRP plates were $54 \times 54\,mm^2$. A typical target holder with the GRP specimen is shown in Figure 2.21. The target holder is made of aluminum. Besides being useful in holding and aligning the target, the target holder also provides the ground for the trigger and the tilt measurement systems. One ground pin and four trigger pins are mounted near the periphery of the GRP specimen. The GRP specimen, the ground, and the trigger pins are all glued in place by epoxy and lapped flush with the impact surface, shown facedown in Figure 2.21. In all the experiments conducted in the present study, a thin (60–125 nm) aluminum coating is applied to either the rear surface of the GRP specimen or the interface between the GRP target plate and the PMMA window plate so as to facilitate laser-based diagnostics using the multibeam VALYN VISAR.

2.5
Experimental Results and Discussion

2.5.1
Plate-Impact Shock Compression Experiments

In the present study, a series of plate-impact experiments were conducted to better understand the structure of shock waves in the S2-glass GRP under normal shock

Table 2.3 Summary of plate-impact shock compression experiments of the S2-glass GRP.

Experiment No.	Flyer material	Impact velocity (m/s)	Target thickness (mm)	Free surface particle velocity (m/s)	Shock wave arrival time (μs)
LT25	AL7075-T6	184.29	6.88	276.54	2.347
LT27	AL7075-T6	187.65	20.2	277.46	6.317
LT28	AL7075-T6	183.98	2.94	286.18	1.078
LT29	AL7075-T6	191.49	12.37	284.97	3.597
LT30	AL7075-T6	111.69	6.75	171.12	2.155
LT31	AL7075-T6	312.70	6.55	456.26	1.952
LT32	AL7075-T6	113.71	19.35	165.80	5.970
LT33	AL7075-T6	312.72	19.25	437.98	5.669
LT35	AL7075-T6	52.5	13.35	76.99	4.454
LT36	AL7075-T6	43.9	12.95	65.85	4.136
LT37	AL7075-T6	39.13	13.07	56.58	4.155
LT38	AL7075-T6	8.5	13.23	12.47	4.817
LT40	AL7075-T6	108.1	13.10	155.31	3.774
LT41	AL7075-T6	212.38	13.59	307.92	4.247
LT42	AL7075-T6	104.7	13.46	150.82	4.228
LT43	AL7075-T6	42.36	13.51	59.97	4.707
LT44	AL7075-T6	68.96	13.26	99.91	4.063
LT45	AL7075-T6	47.36	13.61	71.38	4.449
LT46	D7 tool-steel	329.13	6.95	575.34	2.060
LT47	D7 tool-steel	367.88	6.8	661.98	1.897
LT48	D7 tool-steel	417.96	6.76	807.16	1.978
LT49	D7 tool-steel	416.96	6.85	780.61	1.958
LT50	AL7075-T6	188.17	13.20	265.51	3.983
LT51	AL7075-T6	172.76	13.25	257.83	3.942
LT52	AL7075-T6	138.86	13.16	195.98	4.101
LT53	AL7075-T6	133.23	13.23	207.47	4.306
LT54	AL7075-T6	140.64	12.99	219.51	3.938
LT55	AL7075-T6	82.86	13.23	114.13	4.453
LT56	AL7075-T6	75.75	13.21	105.89	4.449
LT57	AL7075-T6	59.98	13.42	85.65	4.162
LT58	AL7075-T6	43.48	13.18	61.52	4.244
LT59	AL7075-T6	31.95	13.27	45.10	4.797
LT60	AL7075-T6	48.41	13.67	70.26	4.981
LT61	AL7075-T6	68.17	13.56	95.97	4.510

compression. Four different thicknesses of S2-glass GRP specimens (i.e., 3, 7, 13.5, and 20 mm) were utilized for characterization. In the experiments, the amplitude of the shock compression was varied from 0.03 to 2.6 GPa. The results of these experiments are analyzed to understand the structure of the shock waves in the GRP as a function of impact velocity and the distance of shock wave propagation. Moreover, the experimental shock data is analyzed to estimate the EOS (shock velocity versus particle velocity), the loci of Hugoniot stress versus Hugoniot strain and the Hugoniot elastic limit. Table 2.3 details the flyer material, impact velocity,

GRP target thickness, free surface particle velocity, and shock wave arrival time for each experiment.

2.5.1.1 Structure of Shock Waves in the GRP

From the results of the plate-impact shock compression experiments, the structure of the shock waves in the GRP are obtained at stress levels in the range 0.03–2.6 GPa. In order to establish the relationship between the structure of shock waves in the S2-glass GRP and distance of shock wave propagation, the experiments on GRP targets with nearly the same impact stress were employed. Figure 2.22 shows the free-surface particle velocity versus time profiles obtained from the four plate-impact experiments with different thickness target plates. In these experiments the amplitudes of shock compression were 865, 824, 874, and 842 MPa, respectively. A distinct knee in the velocity–time profile is observed during the rise-time of the particle velocity profiles in each of the four experiments. The slope of the velocity–time profiles after this knee decreases with the thickness of the GRP target, that is, with distance of wave propagation. The stress level at which this slope change occurs decreases with increasing GRP thickness. This phenomenon is similar to the elastic-precursor decay observed in elastic–viscoplastic materials, and in the case of the S2-glass-reinforced polymer composites is understood to be due to a result of both

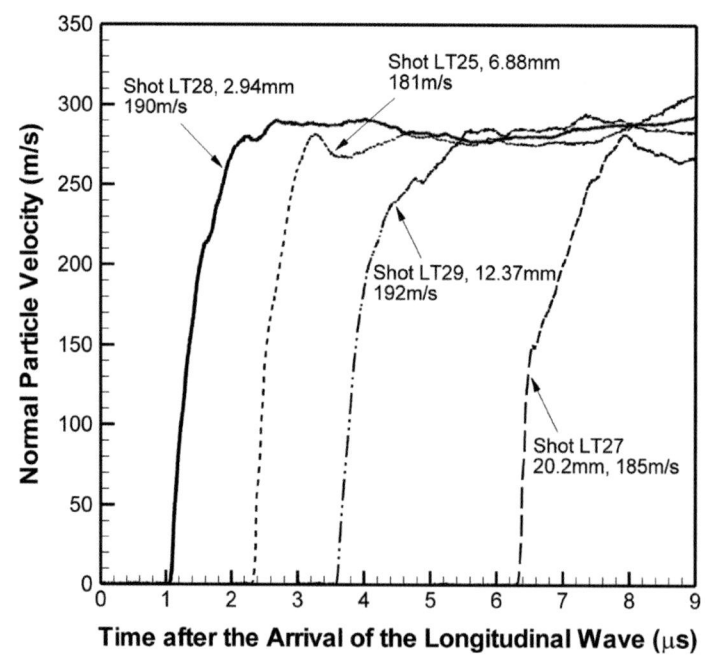

Figure 2.22 Free surface particle velocity profiles obtained from the four plate-impact shock compression experiments conducted on different thickness GRP target plates under nearly the same impact loading conditions.

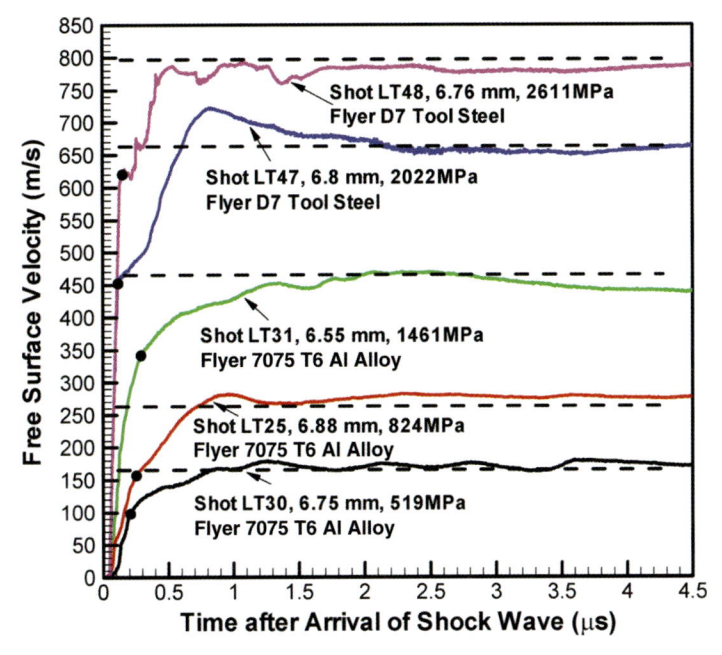

Figure 2.23 Free surface velocity profiles during five different shock loading on approximately 7 mm S2-glass GRP specimens and the dashed lines represent the elastic estimate level.

material and geometric dispersion of the shock wave. The effect becomes more prominent as the thickness of the GRP target increases. Following the knee, the shock wave is observed to rise to an equilibrium plateau level. It is interesting to note that the stress levels at equilibrium are nearly the same in all the four experiments. The equilibrium level is observed to drop for the thickest GRP specimen, that is, experiment LT27, at around 7.8 μs; however, the time corresponding to the drop in particle velocity coincides with the arrival of the release waves from the lateral boundary of the target plate.

The free-surface particle velocity profiles in GRP targets with nearly the same thickness were also used to establish the relationship between the shock stress and the structure of shock waves in the GRP. Figure 2.23 shows the free surface velocity profiles at five different levels of shock stress for approximately 7 mm thick GRP specimens. The abscissa represents the time after the arrival of the shock waves at the free surface of the target plate and the ordinate represents the free surface particle velocity. Experiments LT30, LT25, and LT31 were shock loaded to 519, 824, and 1461 MPa, respectively, using 7075-T6 aluminum flyer plates. Experiments LT47 and LT48 were shock loaded to 2022 and 2611 MPa, respectively, using D7 tool-steel flyer plates. It should be noted that in experiments with impact stresses less than 1.5 GPa the shock front is not observed, while in experiments with impact stresses greater than 2.0 GPa the presence of shock front is clearly evident. Moreover, the slope of wave front increases with the increasing impact stress. Barker et al. [9] proposed the idea of "critical amplitude," which represents the specific shock stress for a clear shock front to appear during shock loading for a variety of materials of interest. In this regards, the critical stress amplitude for the S2-glass GRP is estimated to be between 1.5 and 2.0 GPa.

In the results presented in Section 2.2, the late-time shock-wave profiles for the elastic–elastic bilaminates were determined to be oscillatory with the frequency of oscillations related to the density of interfaces. However, this oscillatory behavior is not observed in the shock experiments on S2-glass GRPs conducted in the present study, as seen in Figure 2.23. Some oscillations can be observed in the shock profile for shot LT30, but for all other experiments the late-time shock-wave profiles are relatively flat. The absence of the oscillatory behavior in the free surface velocity profiles is perhaps due to the development of a complex wave interference pattern within the composites due to the impedance mismatch between the S2-glass fiber reinforcement and the polymer matrix, and also the inelasticity of the polymer interlayer that tends to increase the wave dispersion and hence the rise time of the wave profiles. Also, in agreement with the analysis presented for the case of elastic–viscoelastic bilaminates, the elastic precursor is not observed in the shock profiles of the S2-glass GRP.

In Figure 2.23, the black circular markers indicate the position at which the slope of the particle velocity profiles changes during the rise-time. It should be noted that these markers do not indicate the Hugoniot elastic limit of the composites, and the origin of the change of slope at the black markers is likely due to the viscoelastic response of the polymer layers. Moreover, because of the layered architecture of the composite and the inelasticity of polymer matrix, the shock waves do not develop a sharp fronted wave. Rather, the wave profiles gradually approach the equilibrium level – like in experiments LT30, LT31, and LT48; or overshoot the equilibrium level and then settle down to the equilibrium level – like in experiments LT25 and LT47. According to Sve [8], whether the wave fronts approach the equilibrium level gradually or overshoot the equilibrium level depends on the relative importance of two competing attenuation mechanisms. When the inelasticity in polymer layers is the dominant factor the shock front approaches the equilibrium level gradually; otherwise, when the layered structure is the dominant factor the shock front overshoots the equilibrium level.

Figure 2.24 shows the free-surface particle velocity profiles from select experiments conducted at the different impact velocities on 13 mm thick S2-glass GRP specimens. The dashed lines indicate the elastic prediction for the experiments assuming the flyer plate and the GRP to remain elastic. The change in slope of wave front with increasing shock compression is quite evident. Moreover, the oscillatory nature of the shock-wave profile can be clearly observed in the experiments. Besides the experiment with the lowest compression stress, that is, shot LT38, the experiments shown in Figure 2.24 can be categorized into three main groups based on the level of the shock stress imparted to the composite: experiments with shock stresses below 350 MPa, shock stress between 350 and 700 MPa, and shock stresses above 700 MPa and less than 1 GPa. The particle velocity versus time profiles for these experiments are shown in Figures 2.25, 2.26 and 2.27, respectively. None of the

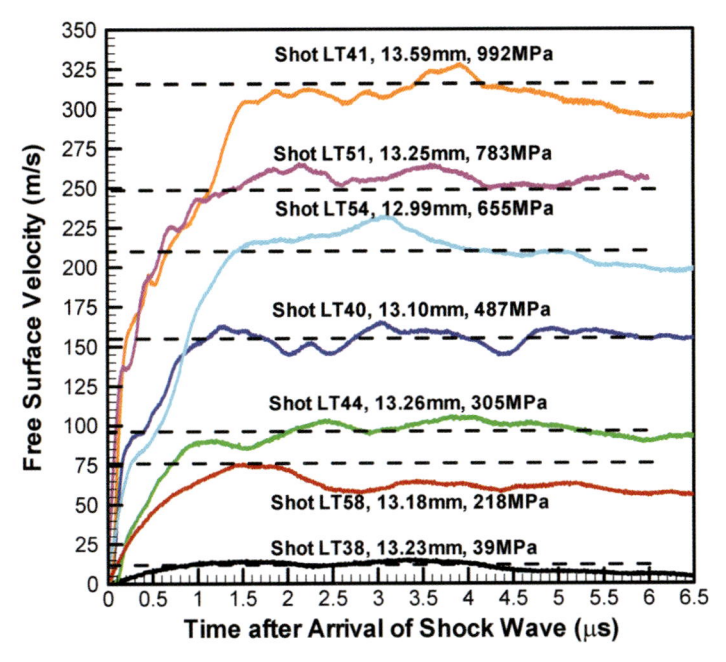

Figure 2.24 Selective free surface velocity profiles for approximately 13 mm S2-glass GRP specimens under different levels of shock compression loading. The dashed lines represent the elastic estimate levels.

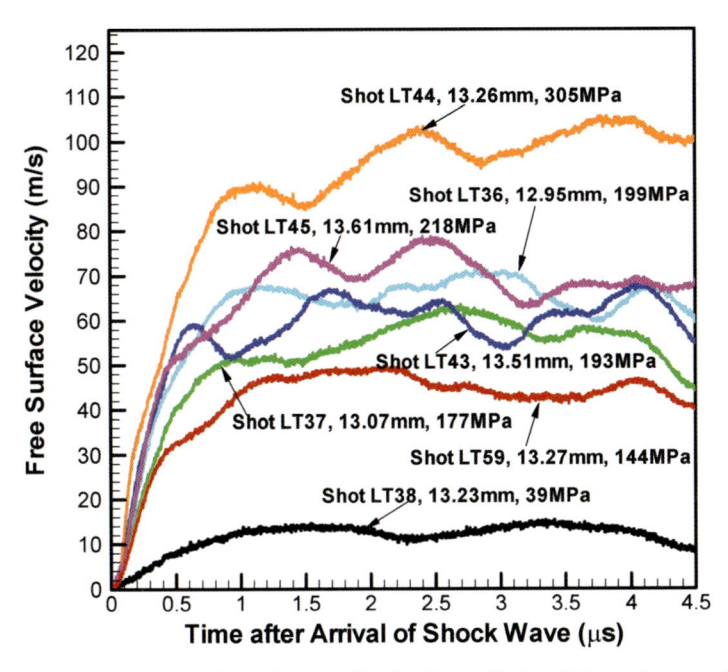

Figure 2.25 Free surface velocity profiles for 13 mm S2-glass GRP specimens at the low stress range.

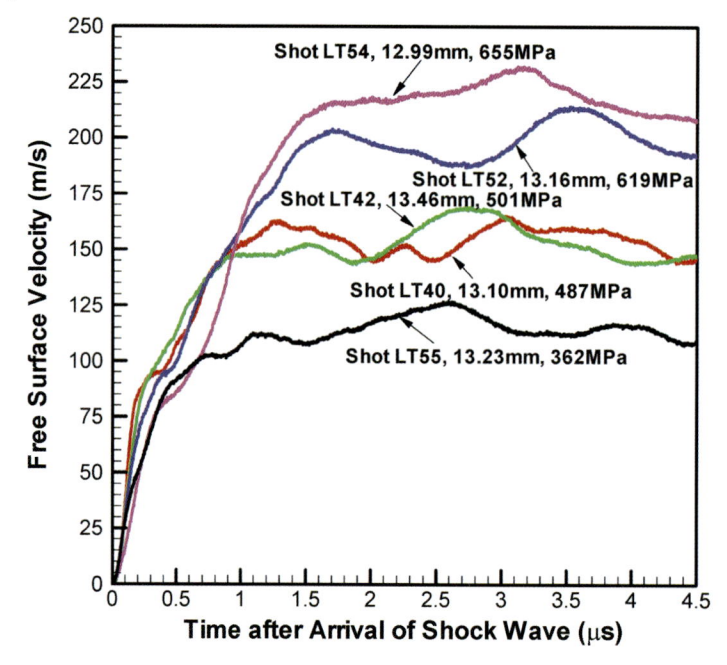

Figure 2.26 Free surface velocity profiles for 13 mm S2-glass GRP specimens at the medium stress range.

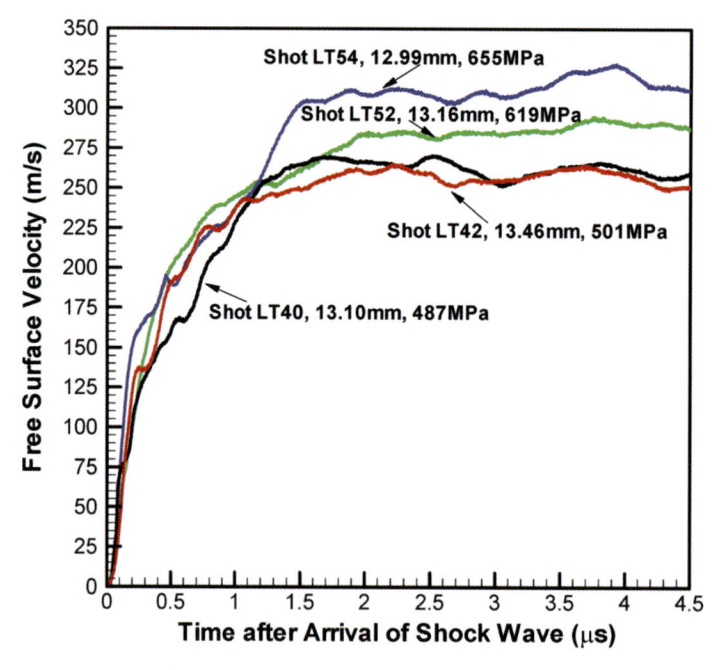

Figure 2.27 Surface velocity profiles for 13 mm thick S2-glass GRP specimens in the high stress range.

experiments show a clear shock front; however, the change in slope of the wave front with increasing impact stress is quite evident. Besides the increase in slope at the wave front, the difference in the number of oscillations in the late-time particle velocity versus time profiles for the three impact velocity regimes is quite evident. From Figure 2.25, it can be seen that the oscillations in the wave profiles have an amplitude between 7 and 10% of the equilibrium level. From Figure 2.26, it can be seen that the oscillations in the wave profiles were about 5–8% of the equilibrium level, while from Figure 2.27 the oscillations in the wave profiles are seen to be about 3% of the equilibrium level.

Thus, in summary, shock wave attenuation with increasing distance of shock wave propagation (target thickness) was not observed in the S2-glass GRP. The results of the present experiment indicate that with increasing levels of shock compression the slope of shock front increases continuously. Also, the amplitude of the oscillations in the wave profiles decreases with increasing levels of shock compression. In this regards, it could be argued that at the lower impact stresses the layered structure dominates the late-time wave profiles; however, at higher impact stresses the inelasticity of constituent materials dominate the S2-glass GRPs shock response.

2.5.1.2 Equation of State (Shock Velocity versus Particle Velocity) for S2-Glass GRP

The Rankine–Hugoniot conservation equations for shock waves in solids are derived under the assumption of a hydrodynamic state of stress within a solid. These equations, also referred to as the conservation of mass, momentum, and energy relations, represent three equations that relate the five variables: pressure (P), particle velocity (u_p), shock velocity (U_s), density (ϱ), and energy (E). Hence, an additional equation is needed to determine all parameters as a function of one of them [37]. This fourth equation, which can be conveniently expressed as the relationship between shock and particle velocities, has to be experimentally determined.

This relationship between shock velocity (U_s) and particle velocity (u_p) can be described by a polynomial equation of the form

$$U_s = C_0 + S_1 u_p + S_2 u_p^2 + \ldots \tag{2.31}$$

where S_i are experimental determined parameters and C_0 is the sound velocity in the material at zero pressure. For most materials, the equation of state can be approximated as a linear relationship between the shock velocity and the particle velocity (U_s versus u_p) given by

$$U_s = C_0 + S u_p \tag{2.32}$$

where S is an empirical constant determined experimentally. It is important to note that if there is porosity or phase transitions in the material, or if material undergoes large elastic–plastic deformations the linear EOS is no longer applicable and has to be modified [38].

Figure 2.28 shows the shock velocity versus particle velocity data obtained from the present plate-impact experiments on the GRP specimens. The shock velocity was estimated from the thickness of the GRP specimens and the shock wave arrival times.

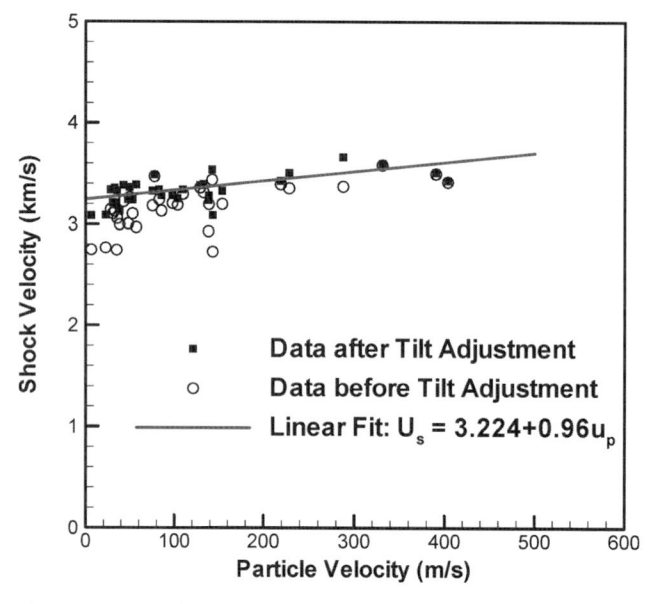

Figure 2.28 Shock velocity versus particle velocity of the S2-glass GRP for present work. The effect of tilt to shock velocity during impact is quite clear especially at lower impact velocity.

In Figure 2.28, the abscissa represents the particle velocity while the ordinate represents the shock velocity. The unfilled circles represent the data points before taking experiments' tilt data into consideration in the calculation of the shock velocity while the black squares represent the data points after the tilt adjustments have been applied. The effect of tilt on the shock velocity calculations is quite evident, especially at lower impact velocities where the tilt time measurements can be relatively larger. The linear fit for the EOS for the GRP under investigation in the present study is determined to be

$$U_s = 3.224 + 0.960 u_p \tag{2.33}$$

Figure 2.29 shows the EOS data obtained for the S2-glass GRP from the present experiments and from Ref. [30]. It also shows the EOS data for the GRP constituent materials, that is, the S2-glass and polyester [39, 40]. The combined data show that when the shock stresses are below 3 GPa, the EOS is essentially linear and lies between the EOS of S2-glass and the polyester materials. However, the slope of the shock velocity versus particle velocity line obtained in the present study is not as steep as that obtained by Dandekar et al. [30] at higher levels of shock compression. Moreover, the slope of the EOS for the S2-glass GRP is smaller than that of the EOS for the two constituents, that is, the S2-glass and polyester. This is probably because in monolithic materials, that is, in S2-glass and polyester, there are much fewer defects (e.g., voids, complex polymer/glass interfaces) when compared to those in the GRP. Hence, during shock compression of the S2-glass GRP, as the impact stress levels are

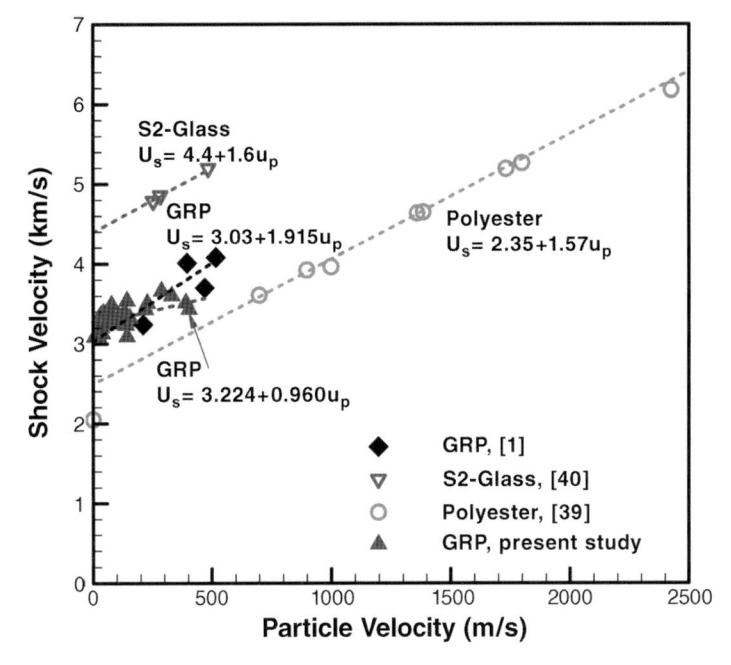

Figure 2.29 Shock velocity versus particle velocity for the S2-glass GRP and its component materials: S2-glass and polyester.

increased the composite is not able to carry the same level of shock stress when compared to its constituents – the S2-glass or polyester.

In order to estimate the EOS for the S2-glass GRP over a larger range of shock stress range, the data from Ref. [30] were combined with the data obtained in the present study. This combined data is shown in Figure 2.30. Linear fit of the three sets of experimental data show that the shock velocity versus the particle velocity relationship in S2-glass GRP in the shock compression range 0.04–20 GPa, and can be described by

$$U_s = 3.228 + 0.996 u_p \tag{2.34}$$

In Eq. (2.34), $C_0 = 3.228$ km/s, which is close to the ultrasonic wave-velocity measurement of 3.21 ± 0.012 km/s in the S2-glass GRP by Dandekar et al. [27] along the impact direction. The shock velocities, particle velocities, and tilt time data from the present study are provided in Table 2.4.

2.5.1.3 Hugoniot Stress versus Hugoniot Strain (Hugoniot)

From the Rankine–Hugoniot conservation relations, the relationship between stress and strain immediately behind wave front can be established. This stress versus strain relationship is generally referred to as the Rankine–Hugoniot equation, or simply as the "Hugoniot" [38]. For this reason, the stress and strain immediately

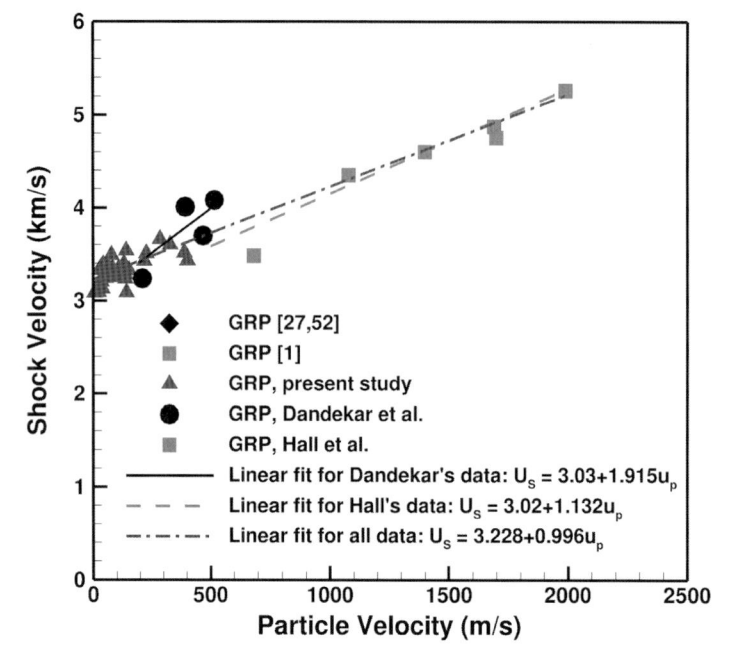

Figure 2.30 Shock velocity versus particle velocity for the S2-glass GRP composites.

behind shock wave front are also referred to as the Hugoniot stress and Hugoniot strain.

A "Hugoniot" is the locus of all the shock states in a material and essentially describes the shock response of a material. As mentioned in previous section, the Hugoniot of a material can be determined as long as its equation of state is known. From the Rankine–Hugoniot conservation relationships, it can be shown that the Hugoniot stress, σ_H, can be determined from shock velocity U_s and particle velocity u_p, as

$$\sigma_H = \varrho_0 U_s u_p \tag{2.35}$$

Also, the Hugoniot strain, ε_H, can be expressed as

$$\varepsilon_H = 1 - \frac{\varrho_0}{\varrho} \tag{2.36}$$

Using the Rankine–Hugoniot conservation of mass, the relationship between mass density, shock velocity, and particle velocity can be expressed as

$$\varrho_0/\varrho = (U_s - u_p)/U_s \tag{2.37}$$

Using Eqs. (2.36) and (2.37) the Hugoniot strain, ε_H, can be determined from shock velocity and particle velocity as

$$\varepsilon_H = \frac{u_p}{U_s} \tag{2.38}$$

2.5 Experimental Results and Discussion

Table 2.4 Shock velocity versus particle velocity data for S2-glass GRP from the present study.

Experiment No.	Free surface particle velocity (m/s)	Calculated shock velocity (km/s)	Impact tilt time (ns)	Adjusted shock velocity (km/s)
LT25	276.54	3.237	184.29	3.237
LT27	277.46	3.283	187.65	3.283
LT28	286.18	3.090	183.98	3.090
LT29	284.97	3.539	191.49	3.539
LT30	171.12	3.285	111.69	3.285
LT31	456.26	3.509	312.70	3.509
LT32	165.80	3.341	113.71	3.341
LT33	437.98	3.430	312.72	3.430
LT35	76.99	3.129	52.5	3.129
LT36	65.85	3.358	43.9	3.358
LT37	56.58	3.339	39.13	3.340
LT38	12.47	3.086	8.5	3.086
LT40	155.31	3.489	108.1	3.489
LT41	307.92	3.327	212.38	3.327
LT42	150.82	3.327	104.7	3.327
LT44	99.91	3.364	68.96	3.364
LT45	71.38	3.331	47.36	3.331
LT46	575.34	3.663	329.13	3.663
LT47	650.28	3.585	367.88	3.602
LT48	679.69	3.453	417.96	3.467
LT49	724.65	3.465	416.96	3.480
LT50	265.51	3.395	188.17	3.395
LT51	257.83	3.385	172.76	3.385
LT52	195.98	3.283	138.86	3.283
LT53	207.47	3.252	133.23	3.252
LT54	219.51	3.341	140.64	3.341
LT55	114.13	3.389	82.86	3.389
LT56	105.89	3.242	75.75	3.242
LT57	85.65	3.386	59.98	3.386
LT58	61.52	3.212	43.48	3.212
LT59	45.10	3.091	31.95	3.091
LT60	70.26	3.207	48.41	3.207
LT61	95.97	3.244	68.17	3.244

The Hugoniot stress and Hugoniot strain values, obtained from the measured shock and particle velocities and using Eqs. (2.35) and (2.38), are shown in Table 2.5 for the experimental data in Refs [27, 30] and in Table 2.6 for the data obtained from the experiments conducted in the present study.

Combining Eqs. (2.35), (2.38), and the EOS, the relationship between σ_H and ε_H can be expressed in terms of the sound velocity at zero pressure C_0, and the empirical constant S, as

$$\sigma_H = \frac{\varrho_0 C_0^2 \varepsilon_H}{(1-S\varepsilon_H)^2} \tag{2.39}$$

Table 2.5 Calculated Hugoniot stress and Hugoniot strain for experimental data on S2-glass GRP from Refs [27, 30].

Experiment No.	Particle velocity (km/s)	Shock velocity (km/s)	Hugoniot stress (GPa)	Hugoniot strain (%)
Dandekar et al. [30] at Sandia National Laboratories				
GRP-2	0.685	3.48	4.670	19.684
GRP-3	1.08	4.35	9.203	24.828
GRP-4	1.40	4.60	12.616	30.435
GRP-5	1.69	4.87	16.123	34.702
GRP-6	1.99	5.26	20.506	37.833
Dandekar et al. [27] at Army Research Laboratory				
448-1	0.211	3.24	1.403	6.821
438-1	0.395	4.01	3.103	9.850
503-1	0.5168	4.08	4.131	12.667
513	0.4695	3.70	3.403	12.690

The Hugoniot stress and strain states obtained from using data from the present experiments and Eqs. (2.35) and (2.38), are shown in Figure 2.31. The dashed line represents the linear fit to the Hugoniot stress versus Hugoniot strain data, while the solid line represents the relationship between Hugoniot stress and Hugoniot strain calculated using Eq. (2.39) and the EOS determined from Eq. (2.34) in the previous section. The concave-up shape of Hugoniot curve is more evident in Figure 2.32, which includes shock data in the highest stress range from Refs [1, 27, 30, 52]. Although the data shows good agreement with the linear fit at lower stress levels, the Hugoniot stress versus strain curve describes the experimental data more accurately over the entire range of stress.

2.5.1.4 Hugoniot Stress versus Particle Velocity

By utilizing the Rankine–Hugoniot relationship (2.35) and the EOS, the relationship between Hugoniot stress and particle velocity can be written as:

$$\sigma_H = \varrho_0(C_0 + Su_p)u_p \qquad (2.40)$$

This relationship can be determined as long as the sound velocity at zero stress C_0, and the empirical constant S, in the EOS are known. Figure 2.33 shows the Hugoniot stress versus particle velocity data from all the S2-glass GRP experiments conducted in the present study, and from Refs [27, 30]. The solid line represents the calculated Hugoniot stress versus particle velocity curve obtained using Eq. (2.40), while the dashed line represents the elastic prediction for the stress and particle velocity based on the elastic acoustic impedance of the GRP. The acoustic impedance was calculated from the sound velocity at zero stress multiplied the initial density of GRP [30]. As predicted, the Hugoniot stress versus particle velocity curve obtained by using Eq. (2.40) conforms well to the various data sets and has the regular concave-up profile. The dashed-line was drawn to identify the elastic limit in GRP under the shock loading conditions. Based on their experimental results, Dandekar et al. [30]

Table 2.6 Hugoniot stress and Hugoniot strain data for S2-glass GRP experiments conducted in the present study.

Experiment No.	Hugoniot stress (GPa)	Hugoniot strain (%)
LT25	0.877	4.272
LT27	0.892	4.226
LT28	0.866	4.631
LT29	0.988	4.026
LT30	0.551	2.260
LT31	1.568	6.501
LT32	0.543	2.481
LT33	1.471	6.385
LT35	0.236	1.230
LT36	0.217	0.981
LT37	0.185	0.847
LT38	0.038	0.202
LT40	0.531	2.226
LT41	1.003	4.628
LT42	0.491	2.267
LT44	0.329	1.485
LT45	0.233	1.072
LT46	2.064	7.854
LT47	2.294	9.027
LT48	2.308	9.802
LT49	2.470	10.412
LT50	0.883	3.911
LT51	0.855	3.808
LT52	0.630	2.984
LT53	0.661	3.190
LT54	0.718	3.285
LT55	0.379	1.684
LT56	0.336	1.633
LT57	0.284	1.265
LT58	0.194	0.958
LT59	0.137	0.730
LT60	0.221	1.095
LT61	0.305	1.479

estimated the HEL for S2-glass GRP to lie between 1.3 to 3.1 GPa. The reason for the estimate can be better understood from Figure 2.34. It shows the data points from zero to 1 km/s particle velocity, and thus provides a better look at the data points in the lower stress range. The dashed line represents the elastic relationship between stress and particle velocity based on the acoustic impedance of GRP before impact, while the solid line represents the calculated Hugoniot stress versus particle velocity curve. Because Dandekar et al. [30] show a data point at 1.3 GPa, which is consistent with the elastic estimate, and also a data point above 3.1 GPa that is considerably higher than the elastic estimate, it is reasonable to assume that HEL for the GRP lies somewhere

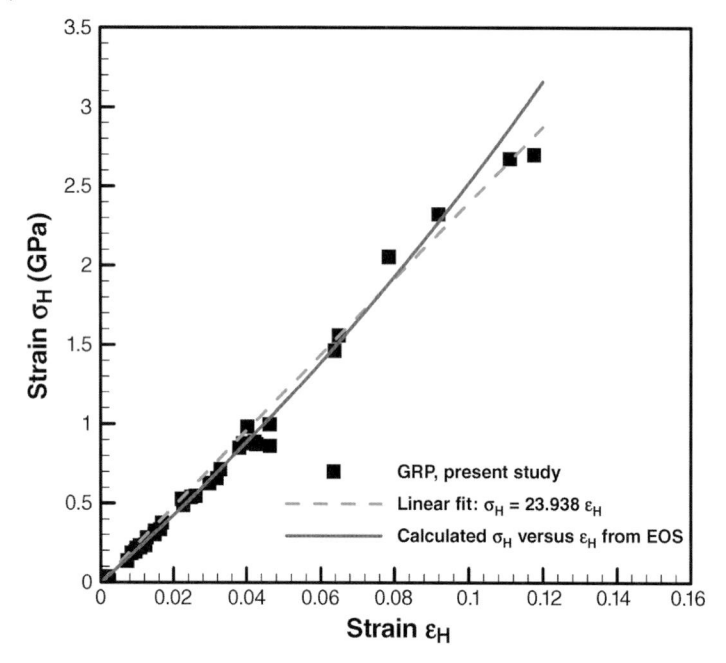

Figure 2.31 Hugoniot stress versus strain of S2-glass GRP in the present study. The linear fit indicates that the Hugoniot stress versus strain follows a linear relationship in the test range.

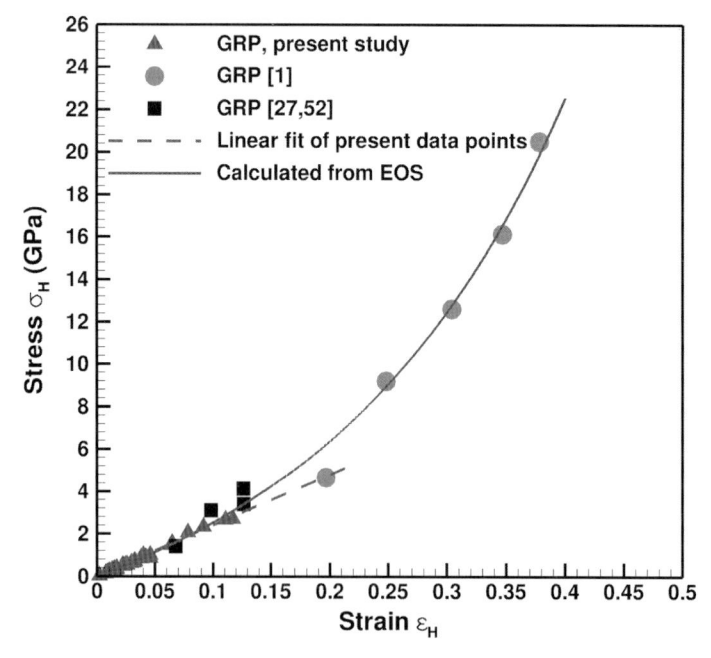

Figure 2.32 Hugoniot stress versus strain for S2-glass GRP composites with linear and third-order polynomial fit.

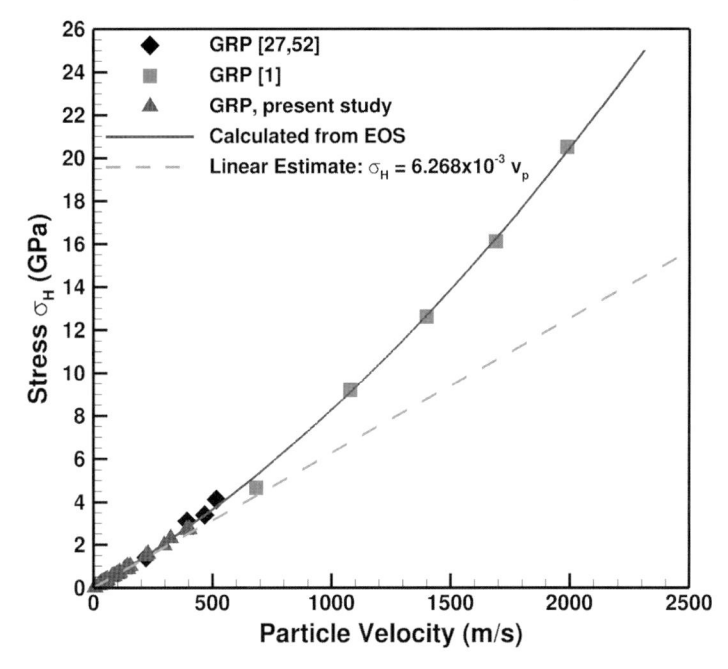

Figure 2.33 Hugoniot stress versus particle-velocity data of the S2-glass GRP.

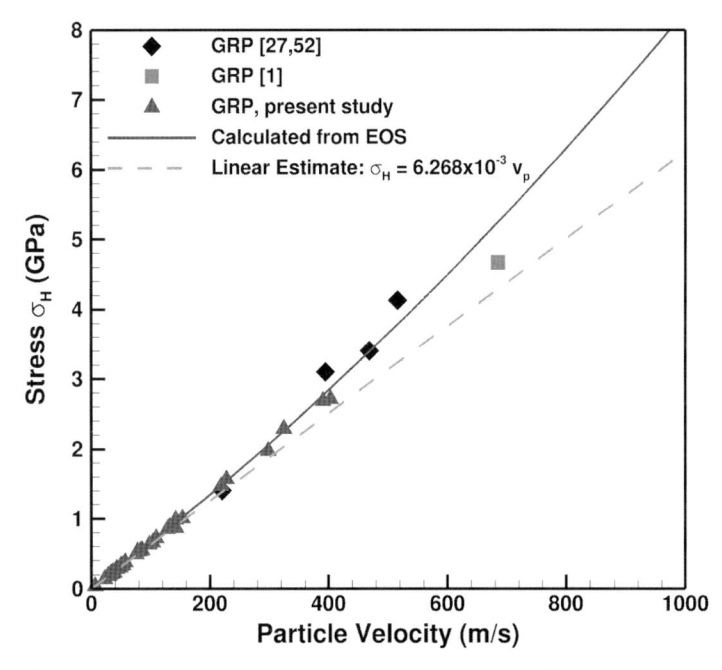

Figure 2.34 Hugoniot stress versus particle velocity data of the S2-glass GRP. Only particle velocity data below 1 km/s are shown here to have a better look of the data for the low stress range.

in between 1.3 and 3.1 GPa. Combining experimental data from the present research with that of Dandekar et al. [30], the HEL can be identified as the deviation of the elastic estimate (dashed line) with the Hugoniot stress versus particle velocity curve, and is estimated to be approximately 1.6 GPa for the S2-glass GRP.

2.5.2
Plate-Impact Spall Experiments

In the present study, results of a series of plate-impact experiments designed to study spall strength in glass fiber-reinforced polymer composites are presented. Two GRP architectures are investigated – S2-glass woven roving in Cycom 4102 polyester resin matrix and a balanced 5-harness satin weave E-glass in a Ciba epoxy (LY564) matrix. The spall strengths in these two composites were obtained as a function of the normal component of impact stress and the applied shear strain by subjecting the GRP specimens to shock-compression and combined shock compression and shear loading. The results were used to develop a failure surface for the two composites.

Table 2.7 provides a summary of all the experiments conducted on the S2-glass GRP in the present study. It shows the experiment no., the flyer and the target plate materials, the thickness of the flyer and target plates, the impact velocity, and the skew angle of impact. In this series of experiments, the impact velocity was varied from 8.5 to 138.8 m/s. In the case of the combined pressure and shear plate-impact experiments, skew angles of 12°, 15°, and 20° were utilized. Table 2.8 shows the

Table 2.7 Summary of all the normal plate-impact and the pressure–shear plate-impact experiments conducted to obtain the spall strength of S2-glass GRP.

Experiment No.	Flyer thickness: Al 7075-T6 (mm)	Target thickness: S2-glass GRP (mm)	Impact velocity (m/s)	Skew angle (°)
LT38	13.59	12.95	8.5	0
LT39	13.59	12.95	38.1	0
LT37	13.59	12.95	39.1	0
LT36	13.59	12.95	43.9	0
LT40	13.59	12.95	108.1	0
LT53	13.59	12.95	133.2	0
LT52	13.59	12.95	138.8	0
LT60	13.59	12.95	48.4	12
LT57	13.59	12.95	59.9	12
LT61	13.59	12.95	68.1	12
LT56	13.59	12.95	75.7	12
LT43	13.59	12.95	42.3	15
LT58	13.59	12.95	43.4	15
LT55	13.59	12.95	82.8	15
LT42	13.59	12.95	104.7	15
LT59	13.59	12.95	31.9	20
LT45	13.59	12.95	47.3	20
LT44	13.59	12.95	68.9	20

Table 2.8 Summary of all the normal plate-impact and the pressure–shear plate-impact experiments conducted to obtain the spall strength of E-glass GRP.

Experiment No.	Flyer thickness: Al 7075-T6 (mm)	Target thickness: E-glass GRP (mm)	Impact velocity (m/s)	Skew angle (°)
FY06001	12.5	10.34	71	0
FY06002	12.5	10.34	141	0
FY06003	12.5	10.34	199.8	0
FY06004	12.5	10.34	300.1	0
FY06005	12.5	10.34	448.8	0
FY06007	12.5	10.34	113.6	12
FY06006	12.5	10.34	213.3	12
FY06008	12.5	10.34	128.1	15
FY06009	12.5	10.34	177.2	15
FY06010	12.5	10.34	180.2	20

corresponding experiments on the E-glass GRP. In this series of experiments, the impact velocity was varied from 71 to 448.8 m/s. Moreover, as for the case of the S2-glass GRP, skew angles of 12°, 15°, and 20° were utilized.

2.5.2.1 Determination of Spall Strength

Figure 2.35 shows the measured free-surface particle velocity and the t–X diagram for a typical plate-impact spall experiment, FY06001, on the E-glass GRP. The abscissa represents the time after impact while the ordinate represents the free surface particle velocity measured at the rear surface of the GRP target plate. At time T1, when the compression wave arrives at the free surface of the GRP plate, the free surface particle velocity rises to the level V_{max}, which is consistent with the Hugoniot stress and particle velocity state corresponding to the impact velocity used in the experiment. At time $T2$, the release waves from the back of the target and the flyer plates intersect at the middle of the GRP plate; the corresponding "unloading tensile wave" and the "end of spall compressive wave" propagate and arrive at the free surface of the GRP plate at times $T3$ and $T4$, respectively. At time $T3$, the free surface particle velocity in the GRP plate starts to decrease and reaches a level V_{min}, at the time $T4$, before recovering to its Hugoniot state level of V_{max}. This initial decrease followed by a recovery in the free surface particle velocity, is also referred to as the "pull-back" characteristic of the spall signal, and is useful in the calculation of the material's spall strength, as detailed in the following.

The method applied for calculating the spall strength from the measured free surface particle velocity history is illustrated in Figure 2.36. The free surface particle velocity data for experiment FY06001 (shown in Figure 2.35) is used as an example. The abscissa represents the time after impact and the ordinate represents the free surface particle velocity measured by the VISAR. Due to the oscillatory nature of the measured free surface particle velocity profiles in GRP, V_{max} was taken to be the average free surface particle velocity during the shocked Hugoniot

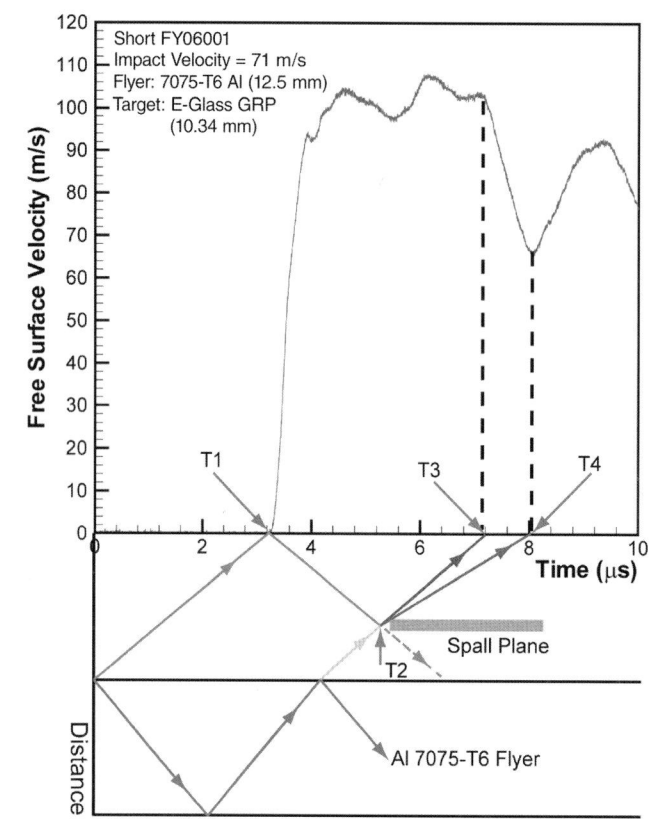

Figure 2.35 Time–distance diagram paired with the measured free surface particle velocity profile for experiment FY06001 to illustrate the "pull-back" phenomenon in the free surface particle velocity profile for a typical plate-impact spall experiment.

state. This level is also consistent with the prediction of the particle velocity in the Hugoniot state as obtained by using the EOS for the flyer and the target materials. After the spall event, the free surface particle velocity drops to V_{min}, followed by a pull back to V_o. In most spall experiments, V_o is expected to be equal to V_{max}; however, in experiments where V_o is observed to be smaller than V_{max}, the occurrence of a partial spall is indicated. $V_{no\ spall}$ corresponds to State (7) in Figure 2.18, when the tensile stress is not high enough to create spall.

The spall strength of the GRP is estimated to be approximately 119.5 MPa by using

$$\sigma_{spall} = Z_{GRP}(V_{max} - V_{min})/2 \tag{2.41}$$

In Eq. (2.41), Z_{GRP} is the acoustic impedance of the GRP in the zero stress condition and is calculated from the initial density and longitudinal wave speed of the GRP. The S2-glass GRP has an acoustic impedance of 6.288 MPa/(m s), and the E-glass GRP has an acoustic impedance of 6.296 MPa/(m s).

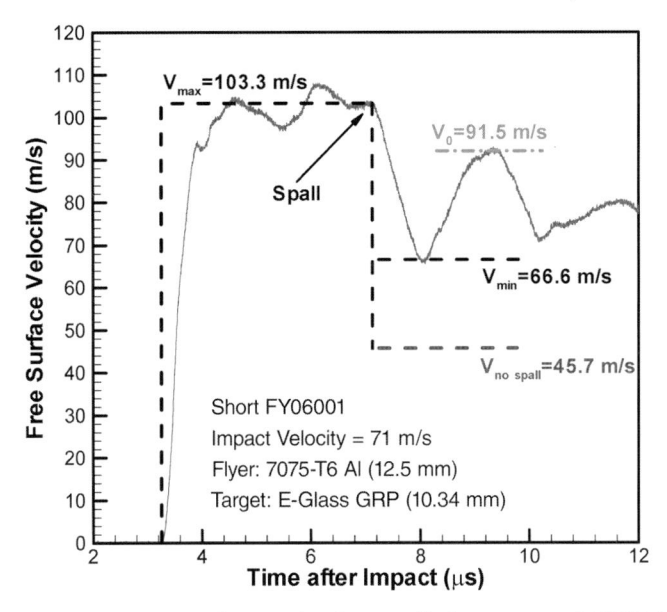

Figure 2.36 Free surface particle velocity profile for experiment FY06001 showing the calculation of the spall strength.

2.5.2.2 Spall Strength of GRP Following Normal Shock Compression

Figure 2.37 shows the spall strength data collected from all the normal plate-impact experiments on the E-glass and the S2-glass GRP composites conducted in the present work. The abscissa represents the impact stress while the ordinate shows the estimated spall strength obtained from the experiments using Eq. (2.41). Among the seven normal plate-impact experiments conducted on the S2-glass GRP composite, in experiments LT38 and LT39 (impact stresses lower than 180 MPa) the resultant tensile stress was not sufficient to cause spall in the specimens. In experiments LT36, LT37, and LT40 (i.e., with impact stresses in the range from 180 to 500 MPa), a finite spall strength was measured. In experiments LT52 and LT53 (with impact stresses greater than 600 MPa), no pull-back signal in the free surface particle velocity profile was observed, indicating that during shock compression the S2-glass GRP was damaged to such an extent that it could not support any tensile stress (i.e., delamination of the composite occurred with a negligible spall strength).

In all the five normal plate-impact spall experiments that were conducted on the E-glass GRP composite (impact stresses ranging from 330.7 to 2213.8 MPa), a finite spall strength was measured. Also, these levels of spall strength are significantly higher when compared to the spall strengths measured in S2-glass GRP composites. However, like in the case of the S2-glass GRP, the spall strengths in the E-glass GRP was observed to decrease with increasing levels of applied shock compression.

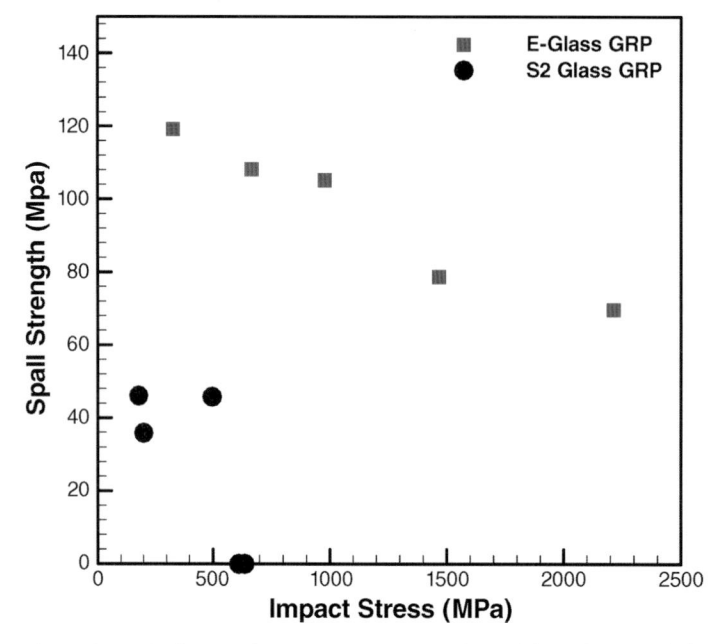

Figure 2.37 Spall strength versus impact stress obtained from the normal plate-impact experiments.

2.5.2.3 Spall Strength of GRP Following Combined Shock Compression and Shear Loading

In order to illustrate the effect of combined shock compression and shear loading on the spall strength, results of one normal impact and one oblique impact experiment on the E-glass GRP are presented in Figure 2.38. The figure shows the free surface particle velocity profiles for a normal plate-impact experiment (FY06003) and a 20° pressure-shear plate-impact experiment (FY06010). The normal component of the impact stress in the two experiments, FY06003 and FY06010, were 978.0 and 871.4 MPa, respectively. The magnitude of the shear strain, η_{13}, in the sample for experiment FY06010 was 1.465%. The shear strain was calculated using the analysis by Dandekar et al. [27].

$$\eta_{13} = \frac{\sigma'_{33}\sin\theta\cos\theta}{(\varrho_0/\varrho)\{C_{11}\sin^4\theta + C_{33}\cos^4\theta + ((1/2)C_{13} + C_{44})\sin^2 2\theta\}} \quad (2.42)$$

In Eq. (2.42), σ'_{33} is the impact stress along the gun barrel direction and is calculated from the impact velocity and the impedance of the flyer and the target materials; ϱ and ϱ_0 are the densities of the GRPs after and before impact, respectively; ϱ/ϱ_0 can be determined by shock velocity and particle velocity; C_{ij} are the elastic constants of GRP and are taken from Ref. [27]; and θ is the skew angle of the pressure–shear plate-impact experiments.

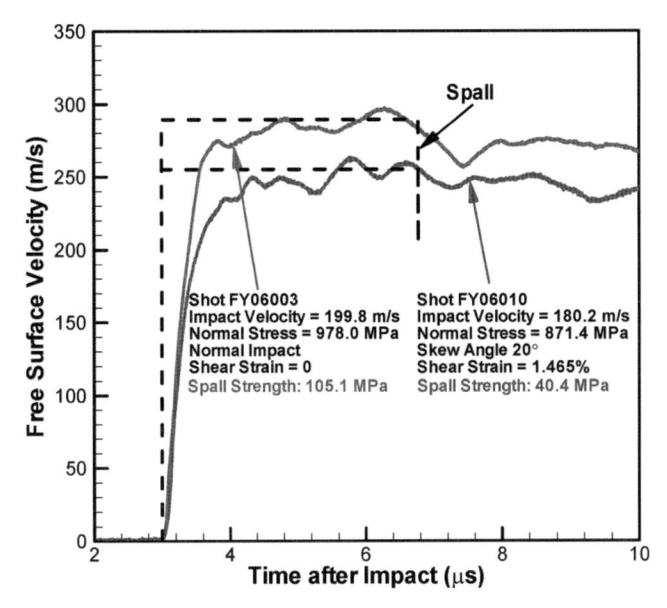

Figure 2.38 Free surface particle velocity profiles for experiments FY06003 and FY06010. The effect of the superimposed shear strain on the spall strength of the E-glass GRP is emphasized.

The spall strengths estimated in the two experiments with and without the presence of shear strain, that is, experiments FY06003 and FY06010, were 105.1 and 40.4 MPa, respectively. From these results, it is quite evident that the presence of shear strain decreases the spall strength of the E-glass GRP dramatically. For example, in experiment FY06006 on the E-glass GRP, the spall strength is reduced to essentially zero when the specimen is impacted at a normal stress of 1052.9 MPa and a shear strain of 1.056%.

To illustrate the effects of the shear stress on the spall strength of the S2-glass GRP, results of four pressure–shear plate-impact spall experiments (conducted at a normal impact stress of approximately 200 MPa), are shown in Figure 2.39. The abscissa represents the shear strain while the ordinate represents the spall strength. The normal components of the impact stresses in these experiments were 187.9, 204.4, 192.9, and 217.5 MPa, respectively. As seen from the figure, the spall strength in these experiments drops very rapidly, that is, from 39.4 MPa to essentially zero, as the shear strain is increased from 0.229 to 0.353%. These results indicate that for the E-glass GRP much higher levels of normal stress and shear strains are required to reduce its spall strength to essentially zero when compared to the S2-glass GRP.

Table 2.9 provides a summary of normal stress, shear strain, and the measured spall strength from all the experiments conducted in the present study on S2-glass GRP. In these experiments, the normal stress was varied from 39.0 to 637.9 MPa, while the shear strain was varied from 0 to 0.615%. Table 2.10 shows the corresponding data for the E-glass GRP. The normal stress was varied from 330.7 to 2213.8 MPa, and the shear strain varied from 0.549 to 1.465%.

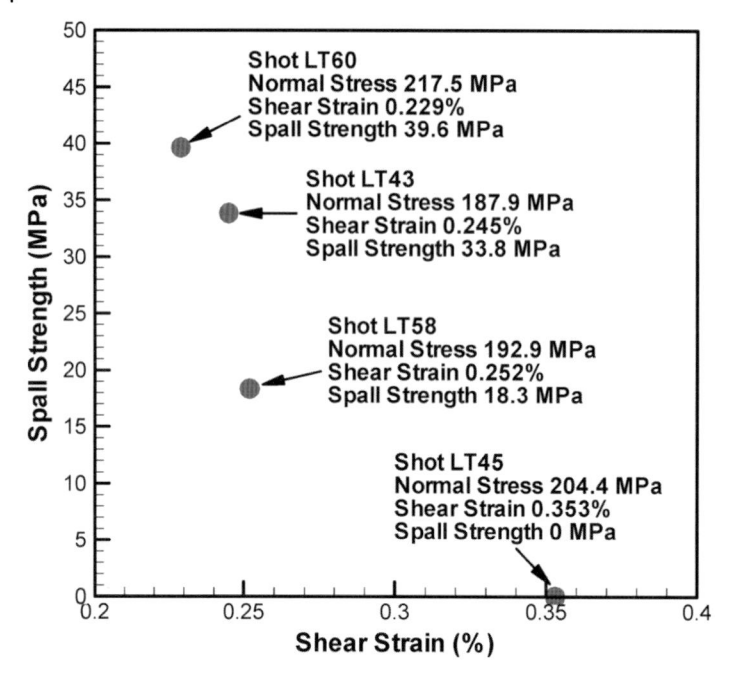

Figure 2.39 Spall strength as a function of the shear strain in the S2-glass GRP for selected experiments each having a normal component of the impact stress of about 200 MPa.

Data labels from figure:
- Shot LT60: Normal Stress 217.5 MPa, Shear Strain 0.229%, Spall Strength 39.6 MPa
- Shot LT43: Normal Stress 187.9 MPa, Shear Strain 0.245%, Spall Strength 33.8 MPa
- Shot LT58: Normal Stress 192.9 MPa, Shear Strain 0.252%, Spall Strength 18.3 MPa
- Shot LT45: Normal Stress 204.4 MPa, Shear Strain 0.353%, Spall Strength 0 MPa

Table 2.9 Summary of normal stress, shear–strain, and spall strength for S2-glass GRP.

Experiment No.	Normal stress (MPa)	Shear–strain (%)	Spall strength (MPa)
LT38	39.0	0	No spall
LT39	175.1	0	No spall
LT37	179.7	0	46.1
LT36	201.6	0	35.8
LT40	496.6	0	45.7
LT53	612.0	0	0
LT52	637.9	0	0
LT60	217.5	0.229	39.6
LT59	137.9	0.237	22.7
LT43	187.9	0.245	33.8
LT58	192.9	0.252	18.3
LT57	269.5	0.283	53.7
LT61	306.3	0.323	0
LT45	204.4	0.353	0
LT56	340.4	0.359	0
LT55	367.6	0.484	0
LT44	297.7	0.516	0
LT42	464.6	0.615	0

Table 2.10 Summary of normal stress, shear–strain, and spall strength for E-glass GRP.

Experiment No.	Normal stress (MPa)	Shear–strain (%)	Spall strength (MPa)
FY06001	330.7	0	119.5
FY06002	668.4	0	108.1
FY06003	978.0	0	105.1
FY06004	1467.8	0	78.7
FY06005	2213.8	0	69.7
FY06007	534.8	0.549	86.1
FY06006	1052.	1.056	0
FY06008	605.3	0.771	85.1
FY06009	855.3	1.094	73.9
FY06010	871.4	1.465	40.4

Figures 2.40 and 2.41 show the spall strengths as a function of the applied shear strain and the normal stress obtained from all the experiments conducted on S2-glass and the E-glass GRP composites. The abscissa represents the normal stress during impact while the ordinate represents the shear strain obtained in each experiment. The Z-axis represents the spall strength. The failure surface shows that the spall strength decreases with increasing shear strain and with increasing normal stress for the two GRP composites. As noted earlier, the E-glass GRP shows much larger levels for the spall strength when compared to the S2-glass GRP. The maximum spall strength measured for the E-glass GRP was 119.5 MPa, while the maximum measured spall strength for the S2-glass GRP was 53.7 MPa.

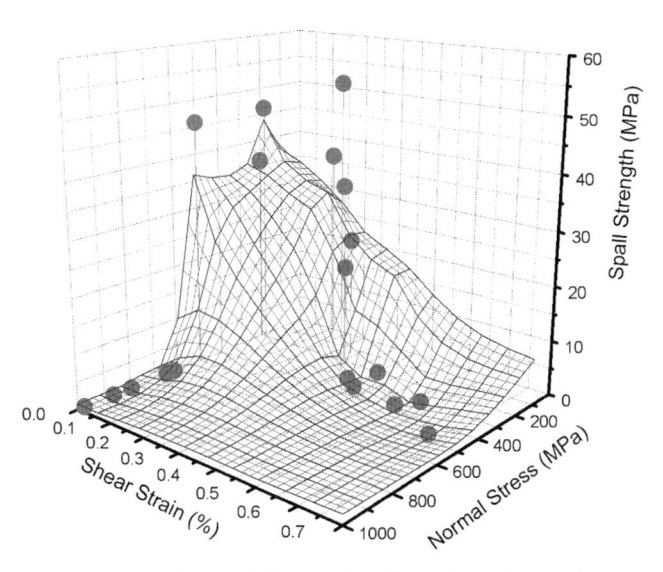

Figure 2.40 Spall strength illustrated in relationship with normal stress and shear strain for the S2-glass GRP.

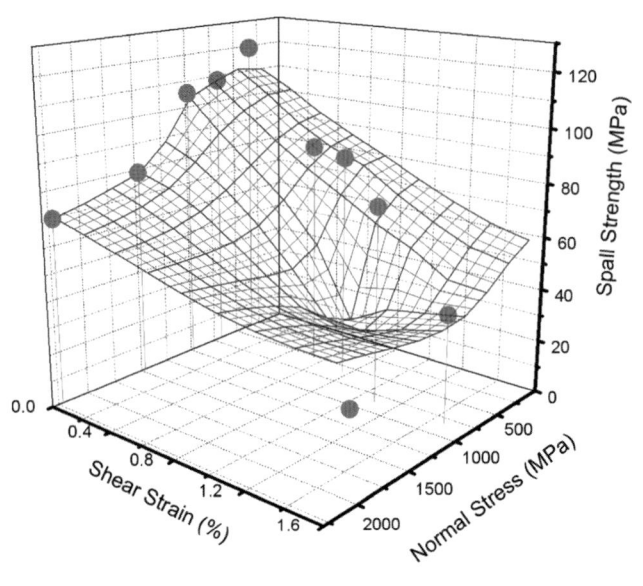

Figure 2.41 Spall strength illustrated in relationship with normal stress and shear strain for the E-glass GRP.

2.5.3
Shock–Reshock and Shock–Release Experiments

In the present study, two sets of shock–reshock and shock–release experiments on the S2-glass GRP were conducted. The shock–release experiment LT71, and the shock–reshock experiment LT73, were conducted at impact velocities of 252 and 264 m/s, respectively. The shock–release experiment LT76 and the shock–reshock experiment LT77 were performed at relatively higher impact speeds of 498 and 485 m/s, respectively. Table 2.11 provides a summary of all the four experiments conducted on the S2-glass GRP in the present study.

Figure 2.42 shows the particle velocity versus time profile at the GRP/PMMA window interface for experiments LT71 and LT73. The abscissa represents the time

Table 2.11 Summary of four shock–reshock and shock–release experiments conducted on the S2-glass GRP.

Experiment No.	Flyer (mm)	Flyer-backing plate (mm)	Target (mm)	PMMA window (mm)	Impact velocity (m/s)
LT 71	GRP, 4.50	PMMA, 12.08	GRP, 9.67	12.37	252
LT 73	GRP, 4.45	Al 6061-T6, 12.13	GRP, 9.53	12.35	264
LT 76	GRP, 4.33	PMMA, 12.16	GRP, 10.35	12.53	498
LT 77	GRP, 4.25	Al 6061-T6, 12.05	GRP, 9.43	12.28	485

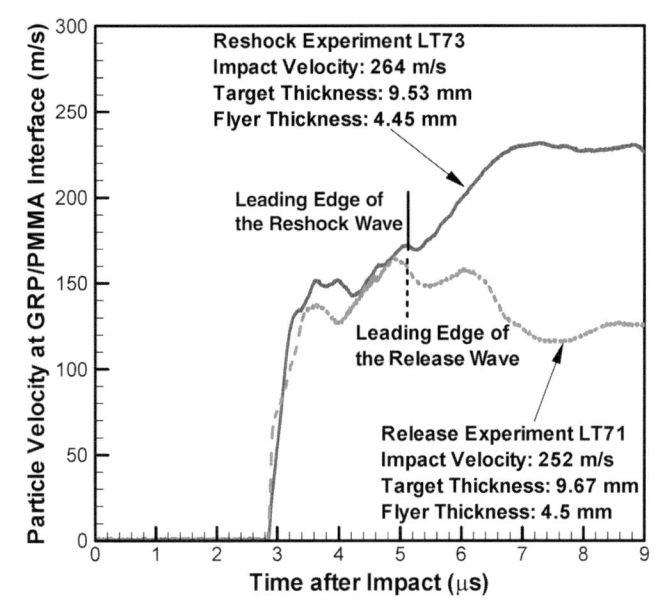

Figure 2.42 Particle velocity at GRP/PMMA window interface versus time profile for experiments LT71 and LT73. The shot LT71 is shifted to the left in order to match the arrival time of the first shock wave with the shot LT73.

after impact while the ordinate represents the measured particle velocity profile. The arrival times of shock and reshock/release waves do not exactly coincide due to small differences in the flyer and target thicknesses. The rise-time associated with the reshock waves and the fall-time associated with the release wave is much larger when compared to the rise of the first shock wave. Also, the oscillatory structure of the first shock Hugoniot state in the GRP is much less prominent in the reshock and release wave profiles. Figure 2.43 shows the particle velocity versus time profiles for experiments LT76 and LT77. The abscissa represents the time after impact while the ordinate represents the measured particle velocity at the window interface. Again, because of the slight difference in the flyer thicknesses, the arrival times of the reshock and release waves do not coincide. Also, unlike in experiments LT71 and LT73, the rise-time and the fall-times associated with the reshock and the release waves show small changes in slope compared with the rise time associated with the first shock wave. Moreover, the oscillatory characteristics of the first shock Hugoniot state in the GRP were not as prominent as that observed in experiments LT71 and LT73.

2.5.3.1 Self Consistent Method for the Determination of Dynamic Shear Yield Strength

For most materials (except most ceramics), the yield strength is typically less than 1 GPa; when shock compressed to above 10 GPa, the stress component normal to the shock front (which is an experimentally determined quantity) is often taken to be equal to the mean stress. This approximation is understood to be within experimental

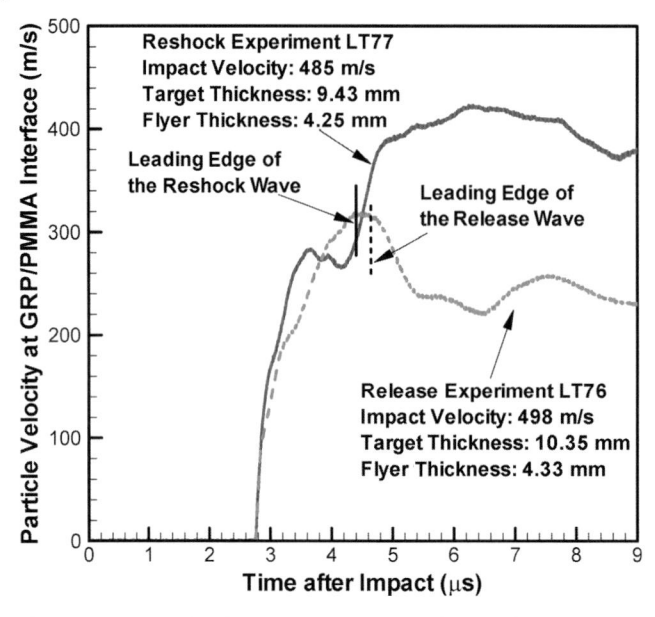

Figure 2.43 Particle velocity at GRP/PMMA window interface versus time profile for experiments LT76 and LT77. The shot LT76 is shifted to the left in order to match the arrival time of the first shock wave with the shot LT77.

error in shock wave experiments. However, recent results of shock wave studies suggest this assumption may not be valid; the dynamic shear yield strength of a solid under shock compression is given by the difference between the dynamic compressibility curve obtained under uniaxial strain conditions, also referred to as the Hugoniot curve, and the hydrostat, which is either measured directly or determined by extrapolating the pressure–volume behavior of the material determined at lower hydrostatic pressures. Based on von Mises yield criteria, the difference between the Hugoniot stress and the hydrostatic pressure curve is defined as two-thirds the dynamic shear yield strength.

Fowles [41] determined the shear strength of annealed Al 2024 when compressed to about 5 GPa by comparing the recorded shock impact stress with the predetermined hydrostat curve of annealed Al 2024. He successfully showed that at the same compression strain the stress normal to the shock front is larger than the hydrostatic pressure by an amount equal to its shear strength, that is, two-thirds of the dynamic yield strength. But in Fowles' [41] method, a predetermined hydrostat curve was required. Later, Asay and Lipkin [42] proposed a self-consistent technique for estimating the dynamic yield strength of a shock loaded material by utilizing shock–reshock and shock–release experimental data from a desired compression state. Asay and Chhabildas [43] utilized this technique to study the variation of shear strength in Al 6061-T6 under shock compression stress ranging from about 8 to 40 GPa. They found that the shear strength of Al 6061-T6 during shock compression increased with increasing shock stress. Reinhart and Chhabildas [44] also used this self-consistent technique to investigate the shear strength of AD995 alumina in the shock state over the stress range of 26–120 GPa; they found that the shear strength of AD995 alumina also increased with increasing shock stress.

Plate-impact experiments are often used to generate high strain rate under uniaxial strain conditions in materials. For uniaxial strain in the x-direction, the state of stress during shock compression is given by

$$\tau = \frac{\sigma_x - \sigma_y}{2} = \frac{\sigma_x - \sigma_z}{2} \tag{2.43}$$

where τ is the resolved shear stress; σ_x represents the longitudinal stress; and σ_y and σ_z represent the lateral stresses in the specimen. In accordance with the von Mises yield criterion, if the stress state is on the yield surface, the shear stress attains its maximum value, τ_c, which is the shear strength. Since in plate-impact experiments, under uniaxial strain conditions, the lateral stresses $\sigma_y = \sigma_z$, the mean stress in the shocked state can be expressed as

$$\bar{\sigma} = \frac{1}{3}(\sigma_x + 2\sigma_y) \tag{2.44}$$

From Eqs. (2.43) and (2.44), the longitudinal stress σ_x and the lateral stress σ_y can be expressed in terms of the mean stress and the shear stress as

$$\sigma_x = \bar{\sigma} + \frac{4}{3}\tau \tag{2.45}$$

$$\sigma_y = \bar{\sigma} - \frac{2}{3}\tau \tag{2.46}$$

In most typical normal plate-impact gas-gun experiments only the longitudinal stress is measured, and so it is necessary to infer the shear stress and/or material strength indirectly. For estimating the shear stress in the shocked state, it is often assumed that the mean stress can be approximated from low-pressure quasi-static measurements. This procedure is not accurate at high pressures due to uncertainties in extrapolating low-pressure response and also because thermal effects must be explicitly accounted for when estimating hydrostatic response. Conversely, if Hugoniot data are used to estimate the mean stress, $\bar{\sigma}$, the influence of a finite shear strength contribution to the Hugoniot requires consideration.

A more direct approach is to estimate shear stress in the shocked state. Taking the derivative of Eq. (2.45) with respect to engineering strain, ε, yields

$$\frac{d\sigma_x}{d\varepsilon} = \frac{d\bar{\sigma}}{d\varepsilon} + \frac{4}{3}\frac{d\tau}{d\varepsilon} \tag{2.47}$$

As will be shown later in this section, the relation allows for the determination of the shear stress from shock wave measurement without recourse to other data.

Figure 2.44 illustrates graphically the response expected if a material is shocked to an initial longitudinal stress, σ_H. Initial yielding is assumed to occur at the Hugoniot elastic limit, and steady shock compression proceeds along the Rayleigh line to a

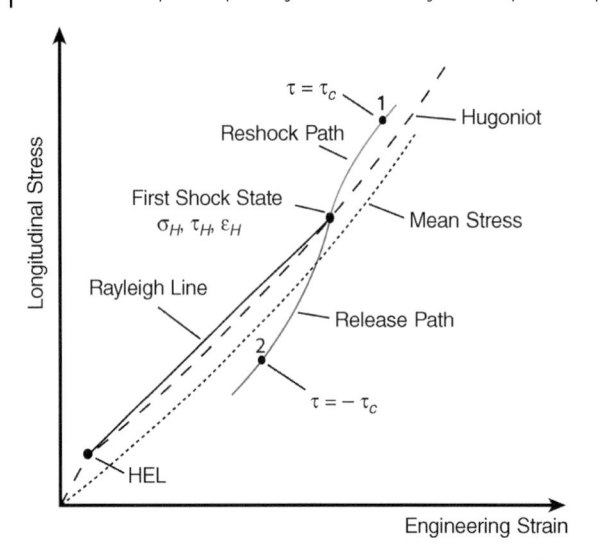

Figure 2.44 Stress versus strain states for a pair of reshock–release experiments loading or unloading from the same Hugoniot state.

point on the solid Hugoniot. At the Hugoniot stress σ_H, there is a deviatoric stress offset of $(4/3)\tau_H$ from the mean stress. In the idealized elastic–plastic theory, τ_H is equal to the maximum shear strength, τ_c. However, due to possible transient softening effects during the initial compression process or time-dependent hardening effects in the shocked state, the shear stress in the shocked state, τ_H, may differ from the maximum stress, τ_c, the material can support.

In view of Eq. (2.45), the relation between the Hugoniot stress σ_H, mean pressure $\bar{\sigma}$, and shear stress τ_H is given by

$$\tau_H = \frac{3}{4}(\sigma_H - \bar{\sigma}) \tag{2.48}$$

Equation (2.48) yields the following relations:

$$\sigma_H = \bar{\sigma} + \frac{4}{3}\tau_H \tag{2.49}$$

$$\sigma_{max} = \bar{\sigma} + \frac{4}{3}\tau_c \tag{2.50}$$

$$\sigma_{min} = \bar{\sigma} - \frac{4}{3}\tau_c \tag{2.51}$$

In Eqs. (2.50) and (2.51), σ_{max} and σ_{min} are the maximum and the minimum stresses at the common Hugoniot strain ε_H, and are estimated along with the shear stress in the shocked state τ_H, and the shear strength, τ_c, by employing a combination of shock–release and shock–reshock experiments from approximately the same "first

Hugoniot shocked state." If the material is released from the initial Hugoniot shocked state, σ_H, Eq. (2.45) should apply continuously as the shear stress varies from its initial value τ_H, to a final value, $-\tau_c$, corresponding to reverse yielding. For the sake of discussion, reverse yielding is assumed to occur at State (2) along the unloading path. Similarly, during reshock from σ_H, the shear stress increases from its initial value $\tau_H < \tau_c$, to the shear strength.

Combining Eqs. (2.49)–(2.51), the following equations can be obtained:

$$\tau_c + \tau_H = \frac{3}{4}(\sigma_H - \sigma_{min}) \tag{2.52}$$

$$\tau_c - \tau_H = \frac{3}{4}(\sigma_{max} - \sigma_H) \tag{2.53}$$

where the critical strength Y_C is defined as

$$Y_C = 2\tau_c = \frac{3}{4}(\sigma_{max} - \sigma_{min}) \tag{2.54}$$

2.5.3.2 Calculation of Initial Hugoniot Shocked State and Hugoniot Stress–Strain Curve

The equation of state for the S2-glass GRP is taken from Section 2.5.1:

$$U_s = 3.224 + 0.96 u_p \tag{2.55}$$

From the Rankine–Hugoniot conservation relationships, the initial Hugoniot shocked stress, σ_H, under plate-impact, can be determined by

$$\sigma_H = \varrho_0 U_s u_p = \varrho_0 (C_0 + S u_p) u_p \tag{2.56}$$

where ϱ_0 is the material density at zero pressure, u_p is the particle velocity, S and C_0 are material constant and taken from Eq. (2.55).

Because the first shock is GRP on GRP, which is a symmetry impact, the particle velocity can be estimated by the impact velocity u_I

$$u_p = \frac{1}{2} u_I \tag{2.57}$$

Combining Eqs. (2.56) and (2.57), the Hugoniot shocked stress, σ_H, under plate impact, can be determined by

$$\sigma_H = \frac{1}{2}\varrho_0 (C_0 + \frac{1}{2} S u_I) u_I \tag{2.58}$$

From the Rankine–Hugoniot conservation relations, the relationship between stress and strain immediately behind wave front can be established. This stress versus strain relationship is generally referred to as the Rankine–Hugoniot equation, or simply as the "Hugoniot" [38]. For this reason, the stress and strain immediately behind shock wave front are also referred to as the Hugoniot stress and Hugoniot strain.

The Hugoniot stress, σ_H, can be determined from shock velocity, U_s, and particle velocity, u_p, as

$$\sigma_H = \varrho_0 U_s u_p \tag{2.59}$$

While, the Hugoniot strain, ε_H, can be expressed as

$$\varepsilon_H = 1 - \frac{\varrho_0}{\varrho} \tag{2.60}$$

Using the Rankine–Hugoniot conservation of mass, the relationship between mass density, shock velocity, and particle velocity can be expressed as

$$\varrho_0/\varrho = (U_s - u_p)/U_s \tag{2.61}$$

Using Eqs. (2.60) and (2.61) the Hugoniot strain ε_H, can be determined from shock velocity and particle velocity as

$$\varepsilon_H = \frac{u_p}{U_s} \tag{2.62}$$

Combining Eqs. (2.59), (2.62), and the EOS (i.e., Eq. (2.55)), the relationship between σ_H and ε_H can be expressed in terms of the sound velocity at zero pressure C_0, and the empirical constant S, as

$$\sigma_H = \frac{\varrho_0 C_0^2 \varepsilon_H}{(1 - S\varepsilon_H)^2} \tag{2.63}$$

2.5.3.3 Calculation of Off-Hugoniot States for Reshock–Release Loading

An approach that employs the incremental form for stress and strain, both related to the corresponding Lagrangian velocity, was used for calculating the off-Hugoniot states of reshock–release loading

$$\sigma = \sum \varrho_0 C_L \Delta u_p$$
$$\varepsilon = \sum \frac{\Delta u_p}{C_L} \tag{2.64}$$

In Eq. (2.64), ϱ_o is the initial material density; u_p is the particle velocity; and C_L is the Lagrangian wave speed.

The particle velocity u_p was calculated from the measured particle velocity at the GRP/PMMA window interface by using the relation

$$u_p = \frac{u_{rs}(Z_{PMMA} + Z_{GRP})}{2 \quad Z_{GRP}} \tag{2.65}$$

where u_{rs} is the measured particle velocity at the GRP/PMMA window interface, Z_{PMMA} is the shock impedance for the PMMA window, and Z_{GRP} is the shock impedance for the GRP.

The Lagrangian velocity C_L was calculated using

$$C_L(u_{rs}) = \frac{\delta_T}{[T + \Delta t(u_{rs}) - t_{rise} - 2\delta_F/U_{eff}]} \quad (2.66)$$

In Eq. (2.66), T is the arrival time of the reshock–release front at the rear surface of the GRP target; t_{rise} is the rise time for the first shock state due to the layer structure of the GRP; δ_T and δ_F are the GRP target and the flyer thickness, respectively; and U_{eff} is the effective shock wave speed.

The effective shock velocity U_{eff} was introduced by Reinhart and Chhabildas [44] to take into account a single shock wave traversing at an effective shock velocity in the flyer. It is calculated by using the relation

$$U_{eff} = \frac{\sigma_H}{\varrho_0 u_H} \quad (2.67)$$

In Eq. (2.67), σ_H and u_H represent the stress and particle velocity of the first shock state, respectively.

The off-Hugoniot stress versus strain curves for experiments LT71, LT73, LT76, and LT77 from the "first Hugoniot shocked state" are presented in Figure 2.45. The abscissa represents the engineering strain and the ordinate represents the normal stress along impact direction. The four circles represent the first Hugoniot shock state calculated from Eq. (2.58), and the Hugoniot curve is calculated from Eq. (2.63).

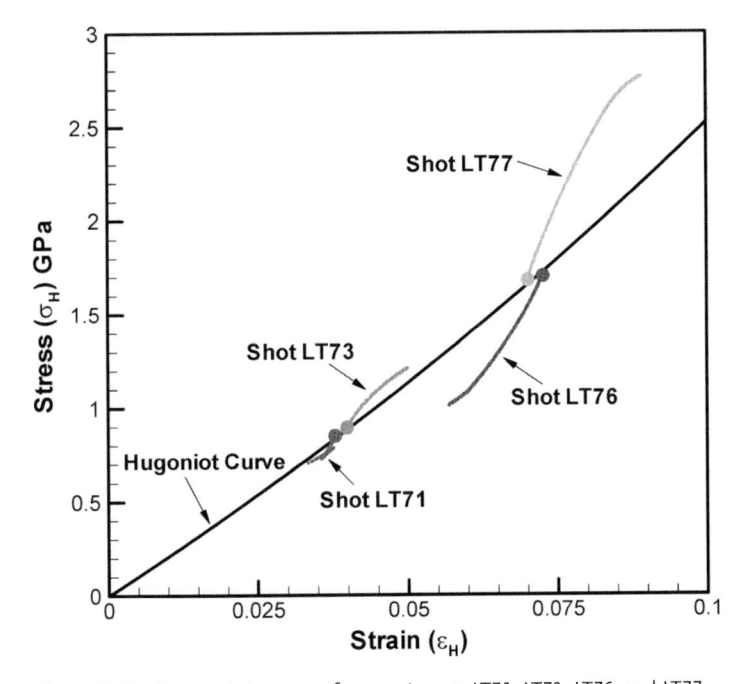

Figure 2.45 Stress–strain curves for experiments LT71, LT73, LT76, and LT77.

For higher impact stress experiments, LT76 and LT77, the path of the off-Hugoniot states was observed to deviate much more from the calculated Hugoniot curve when compared with the lower impact velocity experiments, LT71 and LT73. Some previous reshock–release tests on aluminum and alumina [43, 44], and silicon carbide [45] have shown elastic–reshock and elastic–release at the leading edge of the reshock–release waves. However, the present experiments show no indication of the elastic precursor in the reshock waves or elastic–release in the release waves at the leading front at the arrival of the reshock and the release waves in the GRP. The leading edge of reshock–release waves travels much faster than the first shock wave because of the increase in material density due to shock compression.

2.5.3.4 Determination of the Critical Shear Strength in the Shocked State for S2-Glass GRP

Figures 2.46 and 2.47 illustrate the calculation of the critical shear strength of the S2-glass GRP. In the figures, States (1) and (2) are defined as maximum or minimum shear stress states because the Lagrangian wave velocity reduces to the bulk wave speed at these states, that is, the reshock and release curves intersect with the $\pm \tau_c$ dashed Hugoniot lines (parallel to the Hugoniot curve). Once States (1) and (2) are decided, the maximum and the minimum stresses, σ_{max} and σ_{min}, can be calculated graphically [44]. Finally, the critical shear strength ($Y_C = 2\tau_c$) can be calculated from Eq. (2.54). Table 2.12 summarizes the results for stress and strain states obtained from this study.

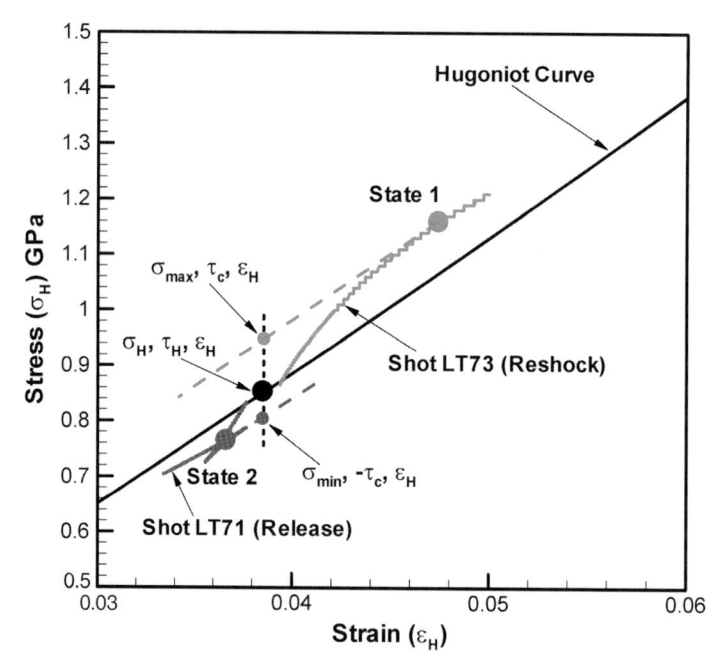

Figure 2.46 Stress–strain curves for experiments LT71 and LT73. The technique to estimate the maximum and the minimum stresses σ_{max}, σ_{min}, and the critical shear strength Y_C is illustrated.

2.5 Experimental Results and Discussion | 75

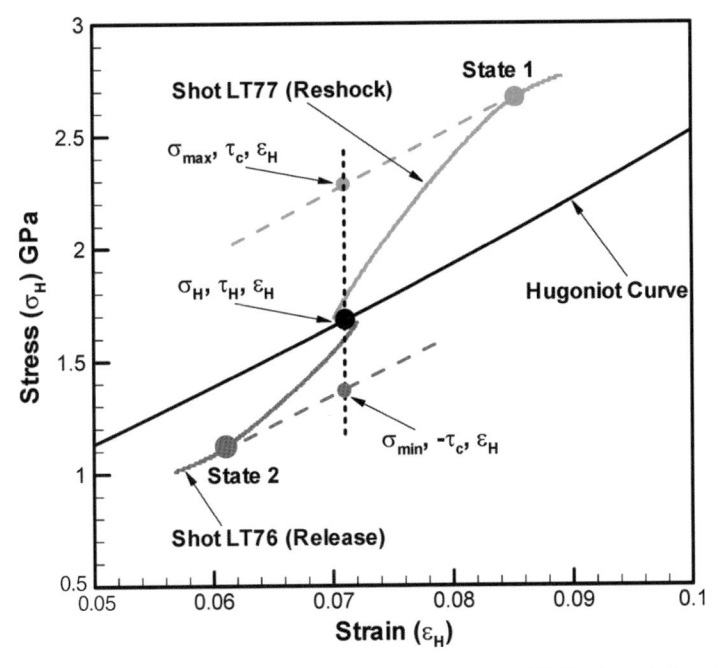

Figure 2.47 Stress–strain curves for experiments LT76 and LT77. The technique to estimate the maximum and the minimum stresses σ_{max}, σ_{min}, and the critical shear strength Y_C is illustrated.

The critical shear strength increased from 0.108 to 0.682 GPa when the first Hugoniot stress increased from 0.85 to 1.7 GPa. As noted earlier, the S2-glass GRP used in the present study has polyester resin matrix with a resin content of 32 ± 2% by weight; in general, polymeric systems exhibit high hydrostatic pressure and normal stress dependency of their flow and fracture behavior [46–49]. Moreover, damage in GRP materials is complicated by the presence of additional heterogeneities, and failure under impact loading is understood to proceed by various mechanisms – the incident energy is dissipated through the spread of failure laterally and through the thickness. It is envisioned that multiple factors such as rate and pressure dependence of the polyester resin combined with the complex damage modes in the GRP are collectively responsible for the increasing critical shear strength with increasing impact stress.

Table 2.12 Summary of stress and strain states for all shock–reshock and shock–release experiments.

Experiment No.	σ_H (GPa)	ε_H	σ_{max} or σ_{min} (GPa)	Y_C (GPa)
LT 71	0.833	0.0377	0.804	0.108
LT 73	0.868	0.0394	0.948	
LT 76	1.655	0.0702	1.37	0.682
LT 77	1.700	0.0719	2.28	

2.6
Summary

Synthetic heterogeneous systems, for example, layered composite materials with organic matrices reinforced by glass fibers, are attractive materials for a variety of lightweight armor applications. In an attempt to better understand the dynamic response of GRPs under shock wave loading. In the present study several series of plate-impact experiments were conducted on two different architectures of the GRP – S2-glass woven roving in Cycom 4102 polyester resin matrix and a balanced 5-harness satin weave E-glass in a Ciba epoxy (LY564) matrix.

In an attempt to better understand material and geometric dispersion of stress waves in the GRPs analytical techniques were used to investigate the structure of stress waves in elastic-elastic and elastic-viscoelastic bilaminates. The analysis makes use of the Laplace transform and Floquet theory for ODEs with periodic coefficients. Both wave-front and late-time solutions for step-pulse loading on layered half-space are compared with the experimental observations. The results of the study indicate that the structure of acceleration waves is strongly influenced by impedance mismatch of the layers constituting the laminates, density of interfaces, distance of wave propagation, and the material inelasticity. The speed of the elastic precursor is independent of the impedance mismatch of the individual laminae constituting the bilaminates and is equal to the average wave speed within the bilaminates. The speed of late-time dispersive wave is observed to decrease with an increase in impedance mismatch; however, it is found to be independent of the density of interfaces, that is, the number of layers in a given thickness laminate. The decay in elastic precursor is observed to increase with an increase in impedance mismatch, the density of interfaces, and the distance of wave propagation. Moreover, the rise-time of the late-time dispersion waves increases with an increase in impedance mismatch; however, it is observed to decrease with an increase in the density of interfaces. The frequency of oscillations of the late-time dispersive wave is observed to decrease with an increase in impedance mismatch; however, it is observed to increase with an increase in the density of interfaces.

In order to understand the shock response of the S2-glass GRP, plate-impact experiments were conducted to study its dynamic response under shock stresses ranging from 0.04 to 2.6 GPa. By varying the thickness of S2-glass GRP plates and the shock compression stress, the structure of shock waves and the effects of shock compression in S2-glass GRP were investigated. No elastic precursor was observed in the S2-glass GRP, and weak late-time oscillations were observed at low stress range. Moreover, no attenuation in shock wave was observed in the S2-glass GRP as a function of increasing distance of wave propagation. Moreover, the slope of shock-wave front was observed to increase continuously with increasing levels of shock compression. The amplitude of the oscillations in the wave profiles was observed to decrease with increasing levels of shock compression. The critical shock amplitude for GRP, which represents the specific shock stress for a clear shock front to appear during shock wave loading, was observed to be between 1.5 and 2.0 GPa. Although shock waves propagation in GRP was somewhat irregular, some important material shock parameters were determined through careful data analysis. Combining the results of the present experiments with data from Refs [27, 30], EOS of the S2-glass GRP was determined to be between 0.04 and 20 GPa. Besides EOS, the Hugoniot curve (Hugoniot stress versus Hugoniot strain) was calculated using Rankine–Hugoniot relationships; the departure of the Hugoniot stress versus particle velocity curve from linearity allowed the estimation of the Hugoniot elastic limit of the S2-glass GRP to be about 1.6 GPa.

In the second series of experiments, normal plate-impact and pressure–shear plate experiments were conducted to study the spall strength in two different glass fiber-reinforced polymer composites – S2-glass woven roving in Cycom 4102 polyester resin matrix and a balanced 5-harness satin weave E-glass in a Ciba epoxy (LY564) matrix. Based on the experimental results, the seven normal plate-impact experiments on the S2-glass GRP were placed in three different categories. Experiments in the first category were conducted at an impact stress between 0 and 175 MPa. In these experiments, the resultant tensile stress was too low to cause spall within the specimens, and the free surface particle velocity profiles were observed to unload completely to their no-spall predicted levels. Experiments in the second category were conducted at impact stresses in the range of 175 and 600 MPa; the resulting tensile stresses within the specimen were high enough to result in spall. In these experiments, a clear pull-back signal was observed in the measured free surface particle velocity profiles. In the third category of the experiments, the incident compression stress pulse amplitude was larger than 600 MPa. These relatively high levels of shock compression resulted in enough damage in the GRP specimens such that no resistance to spall (i.e., zero spall strength) was registered in the experiments. The corresponding free surface particle velocity profiles for these experiments show no signs of pull-back or unloading of the free surface particle velocity, and it remains at a level corresponding to the predicted Hugonoit state, V_{max}. On the other hand, experiments conducted on the E-glass GRP composites (at impact stresses ranging from 330.7 to 2213.8 MPa) showed a finite spall strength. However, like in the case of the S2-glass GRP, the spall strength of the E-glass GRP composite was observed to decrease with increasing levels of shock compression. Under the combined compression and shear loading (pressure–shear plate-impact experiments), the spall strengths in the two GRP composites were found to decrease with increasing levels of applied normal and the shear stress. A zero spall strength condition was found for the E-glass GRP when the specimen was impacted at a normal stress of 975 MPa and a shear strain of 1.056%, which is much higher than that obtained for the case of the S2-glass GRP composite. Based on these results, the spall strength for the two GRP composites is illustrated as a failure surface in the shear strain and the normal stress space. The measured spall strengths are much lower than those observed in monolithic metals, ceramics, polymer, and so on. In such homogeneous materials, the conventional spall process is thought to proceed from the coalescence/growth of inherent defects, such as impurities, micro-cracks, preexisting pores, and so on. However, damage in GRP materials is complicated by the presence of additional heterogeneities due to the composite material's microstructure, and failure under impact loading is understood to the proceed by various mechanisms – the incident energy is dissipated through the spread of failure laterally and through the thickness. Moreover, due to their inherent multimaterial heterogeneous composition of the GRPs, several distinctive modes of damage are observed which includes extensive delamination, fiber shearing, tensile fiber failure, large fiber deflection, fiber microfracture, and local fiber buckling. In particular, local fiber waviness is understood to led to interlaminar shear failure in such materials [50, 51]. Moreover, strong wave-reflection-effects, between components with different shock impedance, led to significant shock wave dispersion resulting in an overall loss of spall strength [24, 26, 52].

In the third series of experiments, plate-impact shock–reshock and shock–release experiments were conducted on a S2-glass GRP to estimate the critical shear strength of the GRP following shock compression by using the self-consistent technique described by Asay and Chhabildas [43]. The shear strength of the shocked GRP was determined for impact stresses in the range of 0.8–1.8 GPa. There is no indication of elastic behavior in the reshock waves and the release waves. The rise-time in the reshock waves and fall-time in the release waves decreases with increasing impact stress. The critical shear yield strength increases from 0.108 to 0.682 GPa when the Hugoniot stress was increased from 0.85 to 1.7 GPa – suggesting a rate-dependence and pressure-dependent yielding behavior.

Acknowledgments

The authors would like to acknowledge the financial support of the Case Prime Fellowship program at the Case Western Reserve University, TARDEC Armor/Structures Program (Dr. Doug Templeton) and the Army Research Office through grant ARO: DAAD 19-01-1-0782 (Program Manager – Dr. David Stepp) for conducting this research. Also, the authors would like to acknowledge the Major Research Instrumentation award by the National Science Foundation, MRI CMS: 0079458, for the acquisition of the multibeam VALYN VISAR used in the present experiments.

References

1 Dandekar, D.P., Hall, C.A., Chhabildas, L.C., and Reinhart, W.D. (2003) Shock response of a glass-fiber-reinforced polymer composite. *Composite Structures*, **61** (1–2), 51–59.

2 Laptev, V.I. and Trishin, Y.A. (1976) The increase of velocity and pressure under impact on inhomogeneous target. *Journal of Applied Mechanics and Technical Physics*, 837–841.

3 Sun, C.T., Achenbach, J.D., and Herrmann, G. (1968) Time-harmonic waves in a stratified medium propagating in the direction of the layering. *Journal of Applied Mechanics*, **35**, 408–411.

4 Hegemier, G.A. (1972) On a theory of interacting continua for wave propagation in composites, in *Dynamics of Composite Materials* (ed. E.H. Lee), ASME, pp. 70–121.

5 Stern, M., Bedford, A., and Yew, C.H. (1970) Wave propagation in viscoelastic laminates. *Journal of Applied Mechanics*, 70-WA/APM-40, 1–7.

6 Sve, C. (1971) Thermoelastic waves in a periodically laminated medium.

International Journal of Solids and Structures, **7**, 1363–1373.

7 Peck, J.C. and Gurtman, G.A. (1969) Dispersive pulse propagation parallel to the interfaces of a laminated composite. *Journal of Applied Mechanics*, **36**, 479–484.

8 Sve, C. (1972) Stress wave attenuation in composite materials. *ASME Transactions, Series E – Journal of Applied Mechanics*, **39**, 1151–1153.

9 Barker, L.M., Lundergan, C.D., Chen, P.J., and Gurtin, M.E. (1974) Nonlinear viscoelasticity and the evolution of stress waves in laminated composites: A comparison of theory and experiment. *Journal of Applied Mechanics*, **41**, 1025.

10 Lundergan, C.D. and Drumheller, D.S. (1971) *Dispersion of Shock Waves in Composite Materials* (eds J. Burke and V. Weiss), Syracuse University Press, New York, p. 141.

11 Oved, Y., Luttwak, G.E., and Rosenberg, Z. (1978) Shock wave propagation in layered composites. *Journal of Composite Materials*, **12**, 84.

12 Nesterenko, V.F., Fomin, V.M., and Cheskidov, P.A. (1984) Damping of strong waves in laminar materials. *Journal of Applied Mechanics and Technical Physics*, 567–575.

13 Benson, D.J. and Nesterenko, V.F. (2001) Anomalous decay of shock impulses in laminated composites. *Journal of Applied Physics*, **89** (7), 3622–3626.

14 Zhuang, S. (2002) Shock wave propagation in periodically layered composites, PhD dissertation, California Institute of Technology, Pasadena, CA.

15 Lifshitz, J.M. (1976) Impact strength of angle ply fiber reinforced materials. *Journal of Composite Materials*, **10**, 92–101.

16 Elhabak, A.M.A. (1991) Mechanical behavior of woven glass fiber reinforced composites under impact compression load. *Composites*, **22** (2), 129–134.

17 Agbossou, A., Cohen, I., and Muller, D. (1995) Effects of interphase and impact strain rates on tensile off-axis behavior of unidirectional glass-fiber composite – Experimental results. *Engineering Fracture Mechanics*, **52** (5), 923.

18 Tay, T.E., Ang, H.G., and Shim, V.P.W. (1995) An empirical strain rate-dependent constitutive relationship for glass-fibre reinforced epoxy and pure epoxy. *Composite Structures*, **33** (4), 201–210.

19 Barre, S., Chotard, T., and Benzeggagh, M.L. (1996) Comparative study of strain rate effects on mechanical properties of glass fibre-reinforced thermoset matrix composites. *Composites Part A – Applied Science and Manufacturing*, **27** (12), 1169–1181.

20 Sierakowski, R.L.C.S.K. (1997) *Dynamic Loading and Characterization of Fiber-Reinforced Composites*, John Wiley & Sons, Inc., New York.

21 Gama, B.A., Gillespie, J.W., Mahfuz, H., Raines, R.P., Haque, A., Jeelani, S., Bogetti, T.A., and Fink, B.K. (2001) High strain-rate behavior of plain-weave S-2 glass/vinyl ester composites. *Journal of Composite Materials*, **35** (13), 1201–1228.

22 Song, B., Chen, W., and Weerasooriya, T. (2002) Impact response and failure behavior of a glass/epoxy structural composite material, in *Proceedings of the 2nd International Conference on Structural Stability and Dynamics* (eds C.M. Wang, G.R. Liu, and K.K. Ang), World Scientific Publishing Co., Singapore, Singapore, pp. 949–954.

23 Vural, M. and Ravichandran, G. (2004) Failure mode transition and energy dissipation in naturally occurring composites. *Composites Part B – Engineering*, **35** (6–8), 639–646.

24 Zhuk, A.Z., Kanel, G.I., and Lash, A.A. (1994) Glass epoxy composite behavior under shock loading. *Journal De Physique IV*, **4** (C8), 403–407.

25 Zaretsky, E., Igra, O., Zhuk, A.Z., and Lash, A.A. (1997) Deformation modes in fiberglass under weak impact. *Journal of Reinforced Plastics and Composites*, **16**, 321–331.

26 Zaretsky, E., deBotton, G., and Perl, M. (2004) The response of a glass fibers reinforced epoxy composite to an impact loading. *International Journal of Solids and Structures*, **41** (2), 569–584.

27 Dandekar, D.P., Boteler, J.M., and Beaulieu, P.A. (1998) Elastic constants and

28 Boteler, J.M., Rajendran, A.M., and Grove, D. (2000) Shock wave profiles in polymer-matrix composites. AIP Conference Proceedings, Shock Compression of Condensed Matter-1999, Snowbird, UT. American Institute of Physics, Melville, NY (eds M.D. Furnish, L.C. Chhabildas, and R.S. Hixson), pp. 563–566.

29 Trott, W.M., Knudson, M.D., Chhabildas, L.C., and Asay, J.R. (2000) Measurements of spatially resolved velocity variations in shock compressed heterogeneous materials using a line-imaging velocity interferometer, in *Shock Compression of Condensed Matter-1999* (eds M.D. Furnish, L.C. Chhabildas, and R.S. Hixon), American Institute of Physics, Melville, NY, pp. 993–998.

30 Tsou, F.K. and Chou, P.C. (1969) Analytical study of Hugoniot in unidirectional fiber reinforced composites. *Journal of Composite Engineering*, **3**, 500–514.

31 Chen, X. and Chandra, N. (2004) The effect of heterogeneity on plane wave propagation through layered composites. *Composites Science and Technology*, **64** (10–11), 1477–1493.

32 Espinosa, H.D., Dwivedi, S., and Lu, H.C. (2000) Modeling impact induced delamination of woven fiber reinforced composites with contact/cohesive laws. *Computer Methods in Applied Mechanics and Engineering*, **183**, 259–290.

33 Chen, C.C. and Clifton, R.J. (1974) Asymptotic solutions for wave propagation in elastic and viscoelastic bilaminates. developments in mechanics. Proceedings of the 14th Mid-Eastern Mechanics Conference, pp. 399–417.

34 Achenbach, J.D. (1973) *Wave Propagation in Elastic Solids*, North-Holland, Amsterdam.

35 Cerrillo, M.V. (1950) *Technical Report No. 55, 2a*, MIT, Cambridge, MA.

36 Kumar, P. and Clifton, R.J. (1977) Optical alignment of impact faces for plate impact experiments. *Journal of Applied Physics*, **48**, 1366–1367.

37 Kinslow, R. (1970) *High-Velocity Impact Phenomena*, Academic Press, New York.

38 Meyers, M.A. (1994) *Dynamic Behavior of Materials*, John Wiley & Sons, Inc., New York.

39 Marsh, S.P. (1980) *LASL Shock Hugoniot Data*, University of California, LA.

40 Silling, S.A., Taylor, P.A., Wise, J.L., and Furnish, M.D. (1994) *Micromechanical Modeling of Advanced Materials*, Sandia National Laboratories.

41 Fowles, G.R. (1961) Shock wave compression of hardened and annealed 2024 aluminum. *Journal of Applied Physics*, **32** (8), 1475–1487.

42 Asay, J.R. and Lipkin, J. (1978) A self-consistent technique for estimating the dynamic yield strength of a shock-loaded material. *Journal of Applied Physics*, **49**, 4242–4247.

43 Asay, J.R. and Chhabildas, L.C. (1981) *Determination of Shear Strength of Shock-Compressed 6061-T6 Aluminum* (eds M.M. Meyers, and L.E. Murr), Plenum Publishing Corporation, New York.

44 Reinhart, W.D. and Chhabildas, L.C. (2003) Strength properties of Coors AD995 alumina in the shocked state. *International Journal of Impact Engineering*, **29** (1–10), 601–619.

45 Dandekar, D.P., Reinhart, W.D., and Chhabildas, L.C. (2003) Reshock behavior of silicon carbide. *Journal De Physique IV*, **110**, 827–831.

46 Bowden, L.M. and Jukes, J.A. (1972) Plastic flow of isotropic polymers. *Journal of Materials Science*, **7** (n1), 52–63.

47 Parry, T.V. and Wronski, A.S. (1985) The effect of hydrostatic pressure on the tensile properties of pultruded CFRP. *Journal of Materials Science*, **20** (6), 2141–2147.

48 Mehta, N. and Prakash, V. (2002) Pressure and strain-rate dependency of flow stress of glassy polymers, in *Modeling the Performance of Engineering Structural Materials III* (eds T.S. Srivatsan, D.R. Lesuer, and E.M. Taleff), The Minerals, Metals & Materials Society, Warrendale, PA, pp. 3–22.

49 Prakash, V. and Mehta, N. (2006) Uniaxial and multi-axial response of polycarbonate

at high strain rates, in *12th International Symposium on Plasticity and its Applications. Anisotropy, Texture, Dislocations and Multiscale Modeling in Finite Plasticity and Viscoplasticity and Metal Forming* (eds A.S. Khan and R. Kazmi), Neat, Inc., MD, USA, pp. 409–411.

50 Hsiao, H.M. and Daniel, I.M. (1996) Nonlinear elastic behavior of unidirectional composites with fiber waviness under compressive loading. *Journal of Engineering Materials and Technology-Transactions of the ASME*, **118** (4), 561–570.

51 Hsiao, H.M. and Daniel, I.M. (1996) Effect of fiber waviness on stiffness and strength reduction of unidirectional composites under compressive loading. *Composites Science and Technology*, **56** (5), 581–593.

52 Dandekar, D.P. and Beaulieu, P.A. (1995) Compressive and tensile strengths of glass reinforced polyester under shock wave propagation, in *High Strain-Rate Effects on Polymer, Metal, and Ceramic Matrix Composites and Other Advanced Materials* (eds Y.D.S. Rajapakse and J.R. Vinson), ASME, New York, pp. 63–70.

3
Interfaces in Macro- and Microcomposites
Haeng-Ki Lee and Bong-Rae Kim

3.1
Introduction

With continuing demand for materials offering improved performance in a wide variety of fields, the use of composite materials is increasing steadily due to their significant weight, cost, and performance advantages over conventional structural materials [1–6]. Composite materials, however, have a variety of imperfections caused by stress fluctuations due to their heterogeneity, and the damage that can be induced by such imperfections will potentially affect the performance of structural materials [6–8]. Therefore, an understanding of the complicated damage/fracture mechanisms in composite materials is of fundamental importance for adequate application.

An *interface* (or *interphase*) is a significant region that mainly determines the mechanical properties of all heterogeneous multiphase systems such as composite materials [9–17]. Composite materials are a microscopic or macroscopic combination of two or more chemically different materials with a recognizable interface (or interphase) between them [18]. The interface (or interphase) in composite materials is an important constituent in which the load stress will inevitably be transferred from the matrix to inclusions (or reinforcements) [19, 20]. Damage and fracture in composite materials are most likely to occur at the imperfect interface existing between inclusions and the matrix in the form of interfacial debonding, interfacial dislocations, interfacial cracks, and so on [8]. Damage at the interface leads to subsequent degradation of mechanical properties [8, 21]. Therefore, it is essential to fundamentally research the characteristics of imperfect interfaces in composite materials and their effects on structure–property relationships in order to achieve a comprehensive understanding of the science and technology of composite materials [1, 8, 22, 23].

Diverse studies on the imperfect interface of composite materials are being carried out in efforts to understand their physical characteristics, to determine the effects of the imperfect interface on the mechanical behavior of composite materials, and to characterize composite materials in terms of interfacial parameters for the prediction of interfacial debonding phenomenon [22]. To this end, experimental and analytical approaches have been employed. A variety of experimental methods have been developed for testing the interface quality using both macroscopic methods, where the macroscopic properties of composite materials are under nearly realistic loading conditions, and microscopic methods, which extend the understanding of damage/ failure mechanisms at the microscopic level by investigating single inclusion embedded in the matrix [24]. These experimental methods, however, are time-consuming and acquiring a thorough understanding of the mechanical behavior of composite materials remains challenging [25, 26]. Accordingly, in order to address these difficulties, the development of suitable analytical approaches is also imperative.

Generally, in the analysis of composite materials, there are four major levels corresponding to differentiated scale: nanolevel ($10^{-7} \sim 10^{-9}$ m), microlevel ($10^{-4} \sim 10^{-8}$ m), mesolevel ($10^{-3} \sim 10^{-8}$ m), and macrolevel ($10^{-2} \sim 10^{-5}$ m) [17, 27–31]. Among these, the interface of composite materials is mostly considered at the nano- and microlevels. In a *nanolevel analysis*, the interface is studied in terms of fundamental adhesion by chemistry and molecular physics [17, 32, 33]. A microlevel analysis entails modeling of the local constituents (fiber, particle, matrix, interface, etc.), which are treated as individual phases, and predicts the local damage between the constituents [26, 34, 35]. From the viewpoint of engineering, a microlevel analysis of the interface of composite materials plays the most important role with regard to the stress transfer efficiency and interfacial strength [17, 36]. Micro-level evaluations are generally conducted via the following procedures [17]: the interfacial parameters to be measured are first selected and micromechanical tests for the interface characterization are then properly chosen. Finally, adequate micromechanical models relating interfacial parameters from test results are developed.

This chapter is mainly focused on micromechanics-based analyses for interfacial characterization and understanding of the overall mechanical behavior of composite materials. General microscale tests for characterization of the interface and micromechanical models are reviewed. Details of the microscale tests and micromechanical models can be found in a variety of studies referenced in the following sections. In Section 3.2, the characterization of interfaces in composite materials is introduced. For the characterization of interfaces, the various surface treatments and microscale tests employed for interfacial parameters are summarized in this section. In Section 3.3, the micromechanics-based analysis is reviewed, and among the available micromechanical analysis techniques, a new micromechanical framework using ensemble-volume average method [37, 38] for a multiphase composite medium is explained in particular. Also, a variety of interface models are reviewed in this section. Finally, the interfacial damage modeling for composite materials having imperfect interfaces is presented in Section 3.4. In particular, in Section 3.4, the conventional Weibull probabilistic approach [39, 40], the multilevel damage model [41–45], and the cumulative damage model [46, 47] are reviewed and summarized.

3.2
Characterization of Interfaces in Macro- and Microcomposites

3.2.1
Surface Treatments of Reinforcements for Composite Materials

As mentioned in the previous section, the mechanical properties of composite materials are controlled by the interfacial region (interface or interphase) between the reinforcing inclusions (or fibers) and the matrix. It is worth noting that good interfacial bonds between inclusions and the matrix are a primary requirement for improvement of mechanical properties of composite materials and their effective use [48].

Surface treatments change the surface of reinforcing inclusions (or fibers) morphologically and chemically, and these techniques can improve the interfacial bond strengths between two distinct constituents [1, 18, 49–52]. In addition, surface treatments may also protect the reinforcing inclusions from degradation by moisture, chemicals, and adverse chemical reactions with the matrix at high temperatures [18]. The effectiveness of surface treatments for control of interfacial properties in composite materials has been widely researched [1, 50–62]. The numerous surface treatments developed to date can be classified according to the type of reinforcing inclusions: examples include dry (air, oxygen, oxygen-containing gases) and wet (liquid-phase oxidizing agent) oxidizing chemical treatments [63–68], nonoxidizing chemical treatments [69–71], plasma treatment (thermal plasma, cold plasma, corona discharge) [53, 72–79], and application of coupling agents [80, 81].

The effectiveness of surface treatments with respect to parameters such as interfacial bond strength and load-carrying capacity depends on the surface modification techniques. For selection of a suitable surface treatment, it is necessary to thoroughly understand the strengths or drawbacks of each treatment [49]. For an extensive and in-depth understanding of surface treatments and their effects on composite materials, details of surface treatments of reinforcing inclusions (e.g., glass, carbon, polymeric, and ceramic fibers) can be found in the relevant literature.

3.2.2
Microscale Tests

A large number of microscale tests have been developed to determine the interfacial parameters (e.g., interface fracture toughness, interface shear strength, and interface frictional strength) between reinforcing inclusions (or fibers) and the matrix [10, 12, 16, 17, 22, 24, 51, 82–84]. It is noted that *microscale tests* refer to experimental testing of specially constructed specimens containing single fiber and are aimed at understanding the interfacial properties and failure mechanisms at the microscopic level [17, 24]. The most common state-of-the-art tests are reviewed in this section.

The microscale tests for interfacial properties can be classified into two large classes based on the external loading points [17]. A classification for microscale tests is shown in Figure 3.1 and detailed descriptions of each technique can be found in the cited literature: single-fiber pull-out [17, 22, 51, 82, 85], microbond [12, 17, 51, 82, 84,

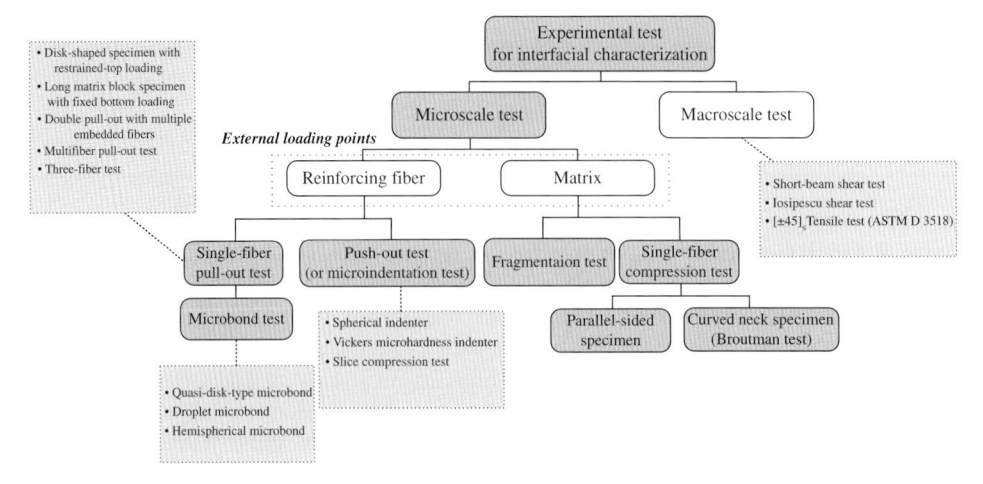

Figure 3.1 Classification of microscale tests for interfacial characterization.

86, 87], push-out (or microindentation test) [17, 51, 82, 88], fragmentation [17, 35, 51, 82, 89], and single-fiber compression tests [16, 17, 24, 51, 90] are summarized below.

The single-fiber pull-out test is a straightforward method that has been widely used, but it has some limitations associated with specimen quality issues and test scale [12, 51]. In order to alleviate the experimental difficulties encountered in single-fiber pull-out tests, the microbond test has been developed as a variation of the conventional pull-out test [51, 86, 91–94]. Although the microbond test can be used for almost any fiber/matrix composites, there are some inherent limitations (e.g., microdrop size, embedded fiber length, and variations in the location of points of contact between the blades and the microdrop) [82]. The push-out test (or microindentation test), meanwhile, is a fast, automated, and simple method capable of examining the interfaces of composites in the real environment [51, 82]. However, the push-out test also has some drawbacks: inability to observe the failure mode and location, problems related to crushing and splitting of fibers, limitation of the variety of fibers, and so on [51, 82, 95]. The single-fiber compression test is one of the earliest methods having two different types of specimen geometry: a parallel-sided specimen test (for shear debonding) and a curved neck specimen test (for tensile debonding in the transverse direction), which is commonly referred to as the "Broutman test" [16, 24, 51]. These test methods have not been as widely employed as other microscale tests due to problems related to specimen preparation and detection of interfacial debonding [51]. The fragmentation test is the most widely used method, and is based on tensile pulling of a totally embedded single-fiber specimen in the fiber direction [22, 51, 82, 96, 97]. However, this test method also has some shortcomings: stain limit and sufficient toughness of the matrix, higher interfacial shear stress due to Poisson's ratio, effects of penny-shaped cracks, and so on [82].

In addition to the aforementioned test methods (single-fiber pull-out test, microbond test, push-out test (or microindentaion test), single-fiber compression test, and fragmentation test), there are extensive tests available for determining the interfacial parameters. A number of factors, including specimen preparation, loading methods, and measurement and data reduction methods, contributing to the discrepancy of test results are also manifest with respect to each test method [51]. Therefore, a thorough understanding of each method including their strengths, weakness, and limitations is required for selection of an adequate test method for evaluating interfacial characteristics. Detailed microscale tests for the determination of interfacial characteristics, as well as the advantages and limitations of each test, can be found in the referenced literatures [51, 82].

3.3 Micromechanics-Based Analysis

3.3.1 Micromechanical Homogenization Theory

A micromechanics-based analysis aims to develop adequately related models with regard to microlevel investigation and to predict the effective (averaged) properties by understanding the fundamental concepts of composite materials and the relevant damage phenomena [26, 98–100]. In a micromechanics-based analysis, the composite materials are simplified in characterization models based on a widely accepted *homogenization process*. This is achieved by replacing a body of heterogeneous material by a constitutive equivalent body of a heterogeneous continuum [26, 37, 38, 101–104]. In order to obtain the effective behavior of a heterogeneous material, there are two basic approaches for homogenization [101, 105, 106]: *average-field homogenization theory* [107–112], in which the effective mechanical properties are determined as relations between the averaged microfields, and *asymptotic homogenization theory* [113–118], in which the effective properties naturally emerge as consequences of the relations between the microfield and macrofield using a multiscale perturbance method (cf. the terms "average-field homogenization theory" and "asymptotic homogenization" are expressed as "average-field theory" and "homogenization theory" in Hori and Nemat-Nasser [105]). Details of each approach can be found in the referenced literature. This chapter focuses on the micromechanical approach using average-field homogenization theory.

3.3.1.1 Representative Volume Element
The effective (averaged) properties of a heterogeneous material are obtained from the sample volume in a homogenization process using average-field homogenization theory [26]. The term effective (or averaged) is defined in a statistically homogeneous medium at a location independent of the sample volume [26, 103]. Therefore, the sample volume is considered to be both smaller than the macroscale characteristic volume and larger than the heterogeneities on the microscale in order to accurately represent the medium [26, 103, 106]. Such a sample volume is called the *representative volume element* (RVE) and some definitions of an RVE are given below.

RVE is a sample that (a) is structurally entirely typical of the whole mixture on average, and (b) contains a sufficient number of inclusions for the apparent overall moduli to be effectively independent of the surface values of traction and displacement, so long as these values are "macroscopically uniform." [109]

RVE is a "mesoscopic" length scale which is much larger than the characteristic length scale of inclusions (inhomogeneities) but smaller than the characteristic length scale of a macroscopic specimen. [37]

The size of RVE should be large enough with respect to the individual grain size in order to define overall quantities such as stresses and strains, but it should be small enough in order not to hide macroscopic heterogeneity. [119]

In a micromechanics-based analysis, based on the assumption of repeating characteristics of the microstructures, composite materials can be approximated by the RVE and the effective properties of composite materials are then determined [120]. It is worth noting that the RVE plays a key role in a micromechanics-based analysis. Therefore, an appropriate RVE size must be defined according to the statistical homogeneity assumption. The details of additional definitions of RVE, RVE existence, and RVE size determination can be found in the cited literatures [108, 121–135].

3.3.1.2 Eshelby's Equivalent Inclusion Method

Eshelby's equivalent inclusion method was proposed in a celebrated work carried out by Eshelby [107] involving the stress analysis of an infinite homogeneous elastic body that contains a subdomain undergoing a stress-free uniform transformation strain [136, 137]. Eshelby's inclusion method [107, 138, 139] is an indispensable part of the theoretical foundation of modern micromechanics and has been widely applied in evaluating the effective mechanical properties of heterogeneous media [111, 112, 140–142].

In Eshelby's inclusion problem, the elastic field for an infinite homogeneous medium **D** including an inclusion Ω and the matrix **D**–Ω is expressed as follows [26, 143]:

$$\boldsymbol{\varepsilon}(x) = \begin{cases} \boldsymbol{\varepsilon}_m (\neq 0) & \forall x \in D-\Omega \\ \boldsymbol{\varepsilon}_i (= \mathbf{S}:\boldsymbol{\varepsilon}^*) & \forall x \in \Omega \end{cases}, \quad \boldsymbol{\sigma}(x) = \begin{cases} \mathbf{C}:\boldsymbol{\varepsilon}_m & \forall x \in D-\Omega \\ \mathbf{C}:(\boldsymbol{\varepsilon}_i - \boldsymbol{\varepsilon}^*) & \forall x \in \Omega \end{cases} \quad (3.1)$$

where **C** denotes the elastic moduli and ε_m and ε_i are strains in the matrix and inclusion, respectively. Here, ε^* is *eigenstrain* (also called as stress-free transformation strains), a nonelastic strain resulting from thermal expansion, phase transformation, initial strains, plastic strains, and misfit strain [26, 111, 144, 145]. Note that *eigenstress* σ^* (=**C**: ε^*) denotes self-equilibrated internal stresses caused by these eigenstrains in the inclusion, which is free from any other external force and surface constraint [26, 145]. In addition, a fourth-order rank tensor **S** ($S_{ijkl} = S_{jikl} = S_{ijlk}$), which is referred to as Eshelby's tensor, relates the (constrained) strain ε_i inside the inclusion to its eigenstrain ε^* [26, 111, 145, 146]. The detailed expressions of Eshelby's tensor for various shapes (spheres, cylinders, ellipsoids, disks, and cuboids) are well documented in Refs [111, 112].

Based on Eshelby's equivalent inclusion method, the local strain and stress fields at any point **x** in the infinite medium under far-field strain ε^0 can be expressed as follows [26, 141, 143, 147]:

$$\varepsilon(x) = \begin{cases} \varepsilon^0 + \varepsilon'(x) & \forall x \in D-\Omega \\ \varepsilon^0 + \varepsilon'(x) & \forall x \in \Omega \end{cases}, \quad \sigma(x) = \begin{cases} C_m : (\varepsilon^0 + \varepsilon'(x)) & \forall x \in D-\Omega \\ C_i : (\varepsilon^0 + \varepsilon'(x)) & \forall x \in \Omega \end{cases} \quad (3.2)$$

where C_m and C_i denote the elastic modulus in the matrix and inclusion, respectively, and $\varepsilon'(x)$ is a perturbed strain field due to the presence of the inclusion [141]. From the eigenstrain $\varepsilon^*(x)$, the local strain and stress fields can be rewritten as follows [26, 143, 147]:

$$\varepsilon(x) = \begin{cases} \varepsilon^0 + \varepsilon'(x) & \forall x \in D-\Omega \\ \varepsilon^0 + \varepsilon'(x) - \varepsilon^*(x) & \forall x \in \Omega \end{cases}, \quad \sigma(x) = \begin{cases} C_m : (\varepsilon^0 + \varepsilon'(x)) & \forall x \in D-\Omega \\ C_m : (\varepsilon^0 + \varepsilon'(x) - \varepsilon^*(x)) & \forall x \in \Omega \end{cases} \quad (3.3)$$

Eshelby's equivalent inclusion method is an elegant method to mimic an inhomogeneity by an inclusion containing a fictitious transformation strain [146]. Here, an *inhomogeneity* is defined as an inclusion having material properties different from those of the surrounding matrix [146]. The result of Eshelby's equivalent inclusion method has a major impact on micromechanical modeling using a homogenization process [143]. Details of Eshelby's equivalent inclusion method can be found in the relevant literature.

3.3.2
Effective Elastic Modulus

In a micromechanics-based analysis, there are a variety of homogenization methods that can be used for determination of the effective elastic moduli for heterogeneous materials [26, 37, 38, 102, 103]: rule of mixture method [148, 149], variational principle method [108, 121, 122, 150–158], self-consistent method [110, 159–164], differential scheme [165–167], Mori–Tanak method [168–172], and other homogenization methods [173–184]. Among these approaches, the self-consistent and Mori–Tanaka methods are the most popularly used approaches for estimation of the effective elastic responses of heterogeneous materials [141, 185]. The self-consistent and Mori–Tanaka methods (see Refs [112, 143, 186] for details) and a new micromechanical framework using the ensemble-volume average method [37, 38] for a multiphase composite medium are described in the following section.

3.3.2.1 Self-Consistent Method

As means of estimating the average stress or strain in a typical inhomogeneity, the self-consistent method considers a single inclusion in a fictitious unbounded homogeneous matrix that has the yet-unknown overall properties of the composite materials [112, 143]. In the self-consistent method, the homogenized elastic operator $\bar{\mathbf{C}}$ of a two-phase composite is expressed as [143]

$$\bar{\mathbf{C}} = \mathbf{C}_0 + \phi_1(\mathbf{C}_1 - \mathbf{C}_0) : \mathbf{A} \tag{3.4}$$

where $\mathbf{C}_r$ and ϕ_r are the elastic tensor and volume fraction of r-phase, respectively. $\mathbf{A}$ denotes the strain concentration tensor, which relates the average strain of the inclusion $\langle \boldsymbol{\varepsilon}_1 \rangle$ to the average strain of a heterogeneous RVE $\langle \boldsymbol{\varepsilon} \rangle$, and $\langle \cdot \rangle$ means "averaged." Here, the strain concentration tensor for the self-consistent method is expressed as follows [143]:

$$\mathbf{A} = \left(\mathbf{I} + \mathbf{S} \cdot \left[\bar{\mathbf{C}}^{-1} \cdot \mathbf{C}_1 - \mathbf{I} \right] \right)^{-1} \tag{3.5}$$

where $\mathbf{I}$ denotes the fourth-rank identity tensor and $\mathbf{S}$ signifies Eshelby's tensor. The numerical solution can be obtained with the self-consistent method by using an iterative process, due to the unknown $\bar{\mathbf{C}}$. Generally, a physical estimate of the moduli should lie between the Voigt (uniform strain) [187] and Reuss (uniform stress) average bounds [189, 190]. Therefore, in an iterative process, Voigt's or Reuss' models can provide an initial estimate of $\bar{\mathbf{C}}$ [143]. Refer to Refs [143, 160, 161, 191] for more details on the self-consistent method.

3.3.2.2 Mori–Tanaka Method

The Mori–Tanaka method is also a widely used homogenization scheme and is mainly utilized for composite materials with moderate volume fractions of inclusions [143]. In particular, this method has been favored by many researchers due to its simplicity and universal applicability [192]. The basic concept of the Mori–Tanaka method is to relate the average stress in an inclusion to the average stress in the matrix [193]. In the Mori–Tanaka method, the strain concentration tensor $\mathbf{A}$ is expressed as follows [143, 194]:

$$\mathbf{A} = \left[\phi_1 \mathbf{I} + (1 - \phi_1)\left(\mathbf{I} + \mathbf{S} \cdot \mathbf{C}_0^{-1} \cdot [\mathbf{C}_1 - \mathbf{C}_0] \right) \right]^{-1} \tag{3.6}$$

Finally, the homogenized elastic operator $\bar{\mathbf{C}}$ of a two-phase composite for the Mori–Tanaka method can be obtained as an explicit formula from Eq. (3.4). Due to the explicit form of the Mori–Tanaka method, this method has received widespread attention and has been used with success to model the elastic behavior of composite materials [191]. Refer to Refs [143, 168–172, 191, 194, 195] for further details on the Mori–Tanaka method.

3.3.2.3 Ensemble-Volume Average Method

Eshelby's equivalent inclusion method plays a key role in homogenization models in the microfield. A number of studies have focused on the development of various micromechanical models using Eshelby's equivalent inclusion method to describe the overall mechanical properties of composite materials. In particular, Ju and Chen [37, 38] developed a new micromechanical framework using the ensemble-*volume average method*. For the ensemble-volume average method,

3.3 Micromechanics-Based Analysis

Eshelby's equivalent method and a renormalization procedure are employed to avoid the conditional convergence problem [26, 37, 38].

According to Ju and Chen [37, 38], upon loading, the strain at any point within an RVE is decomposed into two parts: the uniform far-field stress ε^0 and the perturbed strain $\varepsilon'(x)$ due to the presence of inclusions having eigenstrain $\varepsilon^*(x)$. A schematic description of RVE representation for a composite medium and decomposition of the strain field can be found in Refs [37, 38]. Here, the stresses of the inclusion in the original heterogeneous RVE can be expressed in the same stresses of the homogeneous equivalent medium using a *consistency condition* (or *mechanical equivalent*) in Eshleby's equivalent inclusion method [26]. The domain of the rth- phase inclusion with an elastic stiffness tensor C_r can be expressed as [26, 37, 38, 107, 139, 143]

$$C_r : [\varepsilon^0 + \varepsilon'(x)] = C_0 : [\varepsilon^0 + \varepsilon'(x) - \varepsilon^*(x)] \tag{3.7}$$

with

$$\varepsilon'(x) = \int_V G(x-x') : C_r : \varepsilon^*(x') dx' \tag{3.8}$$

where $x, x' \in V$, the subscript r signifies the rth phase, C_0 is the stiffness tensor of the matrix, and G is the (second derivative of the) Green's function in a linear elastic homogeneous matrix. Details of the Green's function can be found in Refs [37, 38, 111]. From Eqs. (3.7) and (3.8), the following equation can be obtained [37, 38].

$$-A_r : \varepsilon^*(x) = \varepsilon^0 + \int_V G(x-x') : \varepsilon^*(x') dx' \tag{3.9}$$

with

$$A_r \equiv (C_r - C_0)^{-1} \cdot C_0 \tag{3.10}$$

In addition, according to Ju and Chen [37, 38], the volume-averaged strain tensor is expressed as

$$\bar{\varepsilon} = \frac{1}{V} \int_V \varepsilon(x) dx = \varepsilon^0 + \frac{1}{V} \int_V \left[\int_V G(x-x') dx \right] : \varepsilon^*(x') dx' \tag{3.11}$$

where, since $[\cdot]$ is conditionally convergent, a renormalization procedure [196, 197] is applied. With the renormalization procedure [37, 38], the ensemble-volume-averaged strain field of Eq. (3.11) can be rephrased as

$$\bar{\varepsilon} = \varepsilon^0 + S : \left[\sum_{r=1}^n \phi_r \bar{\varepsilon}_r^* \right] \tag{3.12}$$

in which S is the (interior point) Eshleby's tensor and ϕ_r is the volume fraction of the rth phase. Eshleby's tensor S is fully documented in Refs [111, 112]. From a similar process, the ensemble-volume-averaged stress field can also be recast as [37, 38]

$$\bar{\boldsymbol{\sigma}} = \mathbf{C}_0 : \left[\bar{\boldsymbol{\varepsilon}} - \sum_{r=1}^{n} \phi_r \bar{\boldsymbol{\varepsilon}}_r^*\right] \quad (3.13)$$

Furthermore, assuming that the interinclusion interaction effects are neglected, the ensemble-volume average of the relation between $\bar{\boldsymbol{\varepsilon}}$ and $\bar{\boldsymbol{\varepsilon}}_r^*$ (or between $\boldsymbol{\varepsilon}^0$ and $\bar{\boldsymbol{\varepsilon}}_r^*$) over all rth-phase inclusions can be expressed as follows [37, 38]:

$$\bar{\boldsymbol{\varepsilon}}_r^* = -(\mathbf{A}_r + \mathbf{S}_r)^{-1} : \boldsymbol{\varepsilon}^0 \quad (3.14)$$

Finally, the effective elastic stiffness tensor $\mathbf{C}_*$ of the composite materials can be obtained from the strain–stress relationship as [37, 38]

$$\mathbf{C}_* = \mathbf{C}_0 \cdot \left[\mathbf{I} + \sum_{r=1}^{N} \left\{\phi_r (\mathbf{A}_r + \mathbf{S}_r)^{-1} \cdot \left[\mathbf{I} - \phi_r \mathbf{S}_r \cdot (\mathbf{A}_r + \mathbf{S}_r)^{-1}\right]^{-1}\right\}\right] \quad (3.15)$$

where "·" is the tensor multiplication and $\mathbf{I}$ denotes the fourth-rank identity tensor. Details of the expansion process for the ensemble-volume-averaged micromechanical equation in an elastic medium can be found in Refs [37, 38]. The ensemble-volume average method has been widely used for predicting the overall mechanical properties of composite materials. Refer to Refs [37, 38] for more detailed studies on ensemble-volume-averaged micromechanics–based analysis.

3.3.3
Interface Model

In the conventional analytical and numerical work, perfect bonding is assumed at the interface between an inclusion and the matrix; at the interface, however, material properties change discontinuously, while interfacial traction and displacements are continuous across it [8, 11, 23]. The interface is seldom perfect in actual cases and, therefore, it is of primary importance to formulate *imperfect interface* conditions mathematically [23]. Numerous studies on the interface have been conducted and many models have been proposed to simulate interface properties of composite materials. Among the various models, four interface models (linear spring model, interface stress model, dislocation-like model, and free sliding model) are reviewed in this section. Refer to Ref. [224] for details of these interface models.

3.3.3.1 Linear Spring Model
The linear spring model has been extensively used to model imperfect interfaces of composite materials [225, 226]. For the linear spring model, a thin interphase layer near the interface is introduced and its layer thickness is assumed to vanish [227]. In this case, although the interface tractions become continuous, the displacement at the inclusion and matrix sides can be discontinuous, and the jump in displacement is linearly proportional to the interface tractions, as given by the following interface conditions [224, 226, 227]:

$$[\boldsymbol{\sigma}] \cdot \mathbf{n} = 0, \quad \mathbf{P} \cdot \boldsymbol{\sigma} \cdot \mathbf{n} = \alpha \mathbf{P} \cdot [\mathbf{u}], \quad \mathbf{n} \cdot \boldsymbol{\sigma} \cdot \mathbf{n} = \beta [\mathbf{u}] \cdot \mathbf{n} \quad (3.16)$$

where $\mathbf{P} = \mathbf{I}^{(2)} - \mathbf{n} \otimes \mathbf{n}$, $\mathbf{I}^{(2)}$ is the second-order identity tensor, $\mathbf{n}$ is the unit normal vector to the interface between the inclusion and the matrix, $[\cdot] = $ (out)–(in), and α and β are the interface elastic parameters in the tangential and normal directions, respectively [224, 226]. In the linear spring model, the finite positive values of the interface parameters define an imperfect interface (e.g., perfectly bonded in case of α, $\beta \to \infty$ and completely debonded in case of $\alpha, \beta = 0$) [8, 226]. Details of linear spring models can be found in Refs [8, 224–226, 228–233].

3.3.3.2 Interface Stress Model

In case of large-scale composite materials with a small volume ratio of the interface to the bulk, the effect of the interface stress can be neglected. However, for nanocomposites with a large volume ratio of the interface to the bulk, the contribution of interface stress becomes significant [224, 234]. Therefore, an interface stress model is suitable for characterizing the elastic effects of interfaces/surfaces in nanocomposites. A general theoretical framework for surface/interface stress effects was proposed by Gurtin and Murdoch [235, 236] and Gurtin et al. [237]. The necessary basic equations for the interface stress model consist of the displacement continuity condition and the generalized Young–Laplace equations, as given by the following interface conditions [224, 226, 232, 233, 235, 238].

$$[\mathbf{u}] = 0, \quad \mathbf{n} \cdot [\boldsymbol{\sigma}] \cdot \mathbf{n} = -\boldsymbol{\tau} : \boldsymbol{\kappa}, \quad \mathbf{P} \cdot [\boldsymbol{\sigma}] \cdot \mathbf{n} = -\nabla_s \cdot \boldsymbol{\tau} \quad (3.17)$$

where $\boldsymbol{\tau}$ is the interface stress tensor, $\boldsymbol{\kappa}$ is the curvature tensor, and $\nabla_s \cdot \boldsymbol{\tau}$ denotes the interface divergence of $\boldsymbol{\tau}$. In addition, the last two equations signify the generalized Young–Laplace equations, where one is the interface condition in the normal direction and the other signifies that a nonuniform distribution of the interface stress will produce shear stress in the abutting bulk materials [224, 232, 233]. Details of interface stress models can be found in Refs [224, 226, 232, 233, 239].

3.3.3.3 Dislocation-Like Model

The dislocation-like model was proposed by Yu [240] to mathematically describe the effect of an imperfect interface on the load transfer. The interface condition is similar to the linear spring model in that the displacements can be discontinuous while the interface tractions become continuous at the limit of vanishing layer thickness. However, in the dislocation-like model, the displacement at one side of an interface is assumed to be proportional to the displacement at the interface of the constituent where the load is applied [224, 227, 240, 241]. The interface condition for the dislocation-like model can be expressed as [224, 240]

$$[\boldsymbol{\sigma}] \cdot \mathbf{n} = 0, \quad \mathbf{u}_1 = \boldsymbol{\gamma} \cdot \mathbf{u}_2 \quad (3.18)$$

where $\boldsymbol{\gamma} = \eta_T \mathbf{P} + \eta_N \mathbf{n} \otimes \mathbf{n}$ and η_T and η_N are two parameters that describe the bonding conditions in the tangential and normal directions of the interface, respectively. In the dislocation-like model, the two parameters take any values between 0 and 1, and define an imperfect interface (e.g., perfectly bonded in case of $\boldsymbol{\gamma} = \mathbf{I}^{(2)}$ and completely debonded in case of $\boldsymbol{\gamma} = 0$) [224, 240]. Details of the dislocation-like model can be found in Refs [224, 227, 240].

3.3.3.4 Free Sliding Model

The free sliding model allows free tangential slip at the interface, but prohibits relative normal displacement; that is, the interfacial shear stresses are assumed to be zero, and the normal traction and displacements are continuous across the interface according to the following interface conditions [224, 242, 243]:

$$\mathbf{n} \cdot [\boldsymbol{\sigma}] \cdot \mathbf{n} = 0, \quad \mathbf{n} \cdot [\mathbf{u}] = 0, \quad \mathbf{P} \cdot \boldsymbol{\sigma} \cdot \mathbf{n} = \mathbf{0} \tag{3.19}$$

Physically, the free sliding model may represent grain boundary sliding in polycrystals, particles in soils, or imperfectly bonded interfaces in composite materials [224, 242, 244]. Details of the free sliding model can be found in Refs [224, 242, 244–246].

3.4 Interfacial Damage Modeling

3.4.1 Conventional Weibull's Probabilistic Approach

Weibull's statistical approach [247] has been widely used to study survival/hazard problems (e.g., fatigue and fracture) [47, 248–250]. Progressive, interfacial debonding may occur under increasing deformations and influence the overall mechanical behavior of composite materials. After interfacial debonding between the reinforcing inclusions (or fibers) and the matrix, the debonded reinforcing inclusions lose their load-carrying capacity along the debonded directions [40, 204]. In this case, it is assumed that the debonding of reinforcing inclusions is controlled by their internal stress and the statistical behavior of interfacial strength [204]. The Weibull probability distribution function for fiber debonding can be expressed as [204]

$$P_d\left[(\boldsymbol{\sigma}_f)_m\right] = 1 - \exp\left\{-\left[\frac{(\boldsymbol{\sigma}_f)_m}{S_0}\right]^M\right\} \tag{3.20}$$

where $(\boldsymbol{\sigma}_f)_m$ denotes the internal stresses of the fibers, and S_0 and M are scale (Weibull modulus) and shape parameters, respectively. Refer to Refs [47, 204] for more detailed studies on the Weibull function for damage evolution.

Micromechanical damage models for effective elastoplastic behavior of composite materials considering complete particle debonding and partial particle debonding were developed by Ju and Lee [39, 40]. In this section, these models are summarized. Figure 3.2 shows a schematic diagram of complete particle debonding [39] and partial particle debonding [40] of composite materials. For the effective moduli of a composite material, Ju and Lee [39, 40] derived the ensemble-volume-averaged micromechanical equation (see Eq. (3.12)) with an evolutionary probabilistic interfacial debonding model. In addition, for the estimation of the overall elastoplastic behavior, the developed elastic framework was expanded by employing the J_2-type von Mises yield criterion with the isotropic hardening law. In particular, to meet the

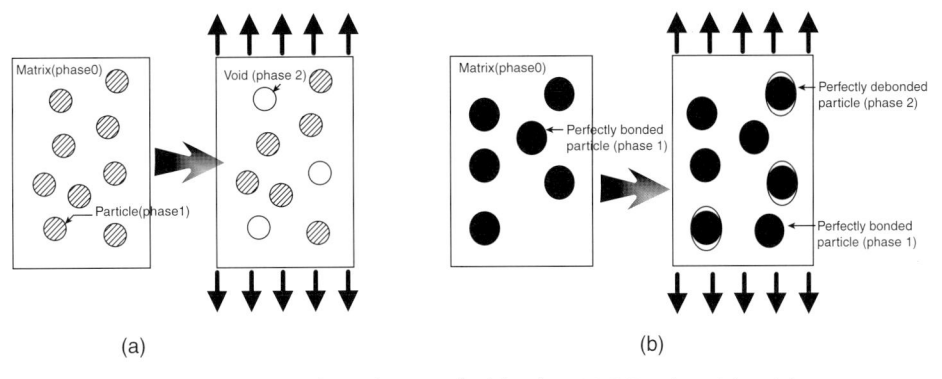

Figure 3.2 A schematic diagram of complete particle debonding (a) [39] and partial particle debonding (b) [40] of composite materials. Reproduced with permission from Elsevier from Ref. [39], Copyright 2000, and Ref. [40], Copyright 2001.

characteristics of partially debonded interfaces (see Figure 3.2b), a partially debonded isotropic spherical elastic particle was replaced by an equivalent, perfectly bonded spherical particle that possesses yet unknown transversely isotropic elastic moduli, as shown in Figure 3.3 [40, 251, 252]. It is worth noting that the effects of a random dispersion of inclusions and evolutionary interfacial debonding were considered in the micromechanical framework of Ju and Lee [39, 40].

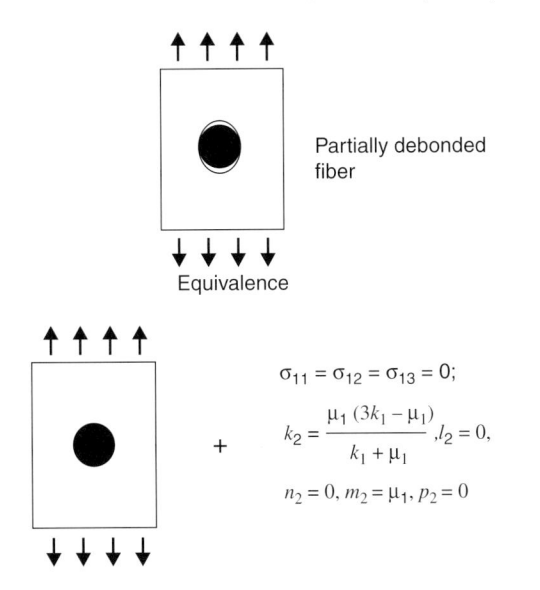

$$\sigma_{11} = \sigma_{12} = \sigma_{13} = 0;$$

$$k_2 = \frac{\mu_1 (3k_1 - \mu_1)}{k_1 + \mu_1}, l_2 = 0,$$

$$n_2 = 0, m_2 = \mu_1, p_2 = 0$$

Under Condition of plane stress with components in the 1-direction beign 0

Figure 3.3 A schematic representation of the equivalence between a partially debonded isotropic particle and an equivalent, perfectly bonded transversely isotropic particle. Reproduced from Ref. [40] with permission from Elsevier, Copyright 2001.

A variety of numerical simulations and a comparison between theoretical and experimental results were carried out. It was observed from the numerical simulations that the predicted overall stress–strain behavior of composite materials using the proposed micromechanical damage models [39, 40] is in good qualitative agreement with the theoretical and experimental results. In addition, from the numerical simulations, the following meaningful results were obtained by Ju and Lee [40]:

- The influence of partially debonded particles on the overall stress–strain responses is rather significant when the interfacial particle bonding strength S_0 is weak and the initial particle volume fraction ϕ is medium due to the relatively rapid damage evolution corresponding to weaker interfacial strength and higher volume fraction of debonded particles [40].
- On the contrary, if the interfacial strength is high and the particle volume fraction is low, the effects of partial particle debonding are not pronounced compared to that of the complete particle debonding model or even that of the perfect bonding model [40].

Figure 3.4a shows the predicted elastoplastic responses of the partially debonded damage model [40] and the completely debonded damage model [39]. In addition, comparisons with theoretical results [251] are given in Figure 3.4b. Details of the extended micromechanical damage model, and more detailed figures and results can be found in Refs [39, 40]. Also, refer to Refs [206, 207, 251–253] for more detailed studies on the conventional (one-step) Weibull's probabilistic approach.

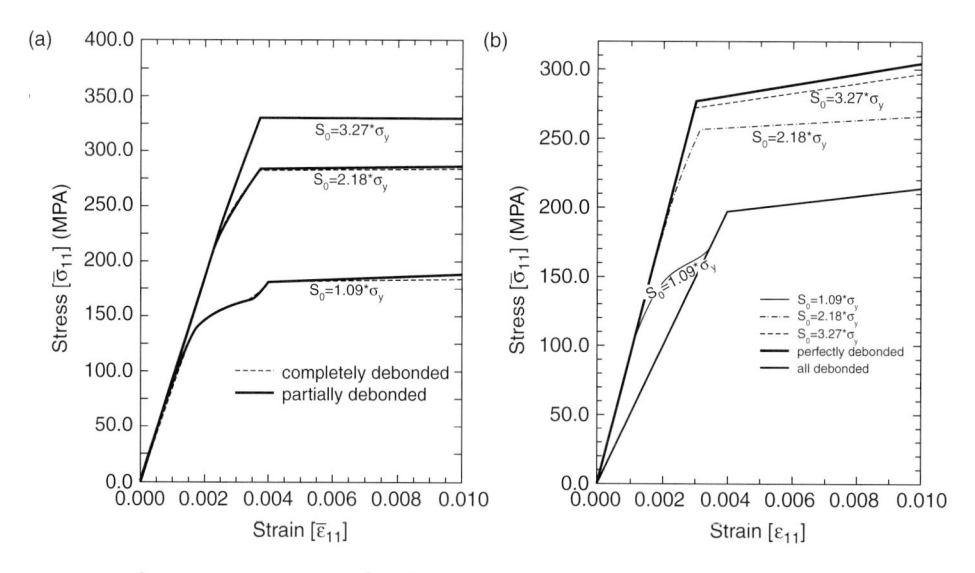

Figure 3.4 Comparisons of predicted elastoplastic responses of the partially debonded damage model [40] and completely debonded damage model [39] (a) and comparisons with theoretical results (b) [251]. Reproduced from Ref. [40] with permission from Elsevier, Copyright 2001.

3.4.2
Multilevel Damage Model

The conventional (one-step) Weibull's probabilistic approach has been used by many researchers to describe the interfacial debonding between the reinforcing inclusions (particles or fibers) and the matrix. However, a sequential probabilistic debonding analysis is necessary to realistically reflect the effect of loading history on the interfacial debonding [41]. Diverse studies on sequential probabilistic debonding analysis have been conducted by many researchers (e.g., Gurvich and Pipes [254], Hwang et al. [255], Liu et al. [214, 215], Ghosh et al. [256], Ju and Ko [220], and Lee and Pyo [41–45]). Recently, Lee and Pyo [219] proposed a micromechanical framework for predicting the effective elastic behavior and the weakened interface evolution of composite materials. In this section, the micromechanical framework proposed by Lee and Pyo [41, 42, 219] is summarized.

Lee and Pyo [41, 42, 219] developed a multilevel damage model in accordance with the Weibull's probabilistic function in order to realistically reflect the effect of loading history on the progression of a weakened interface, as shown in Figure 3.5. In order to model spherical particles having imperfect interfaces in composite materials, the modified Eshelby's tensor for slightly weakened interface proposed by Qu [8] was adopted in Refs [41, 42, 219]. In Ref. [8], the imperfect interface is modeled by a linear spring model and the final form of the modified Eshelby's tensor can be expressed as follows [8]:

$$S_{ijkl}^M = \frac{1}{\Omega}\int_\Omega S_{ijkl}(\mathbf{x}) dV(\mathbf{x}) = S_{ijkl} + (I_{ijpq} - S_{ijpq}) H_{pqrs} L_{rsmn}(I_{mnkl} - S_{mnkl}) \quad (3.21)$$

with

$$H_{pqrs} = \frac{1}{4\Omega} \int_S (\eta_{ik} n_j n_l + \eta_{jk} n_i n_l + \eta_{il} n_j n_k + \eta_{jl} n_i n_k) \quad (3.22)$$

where Ω signifies the inclusion domain and S and n_i denote the interface and its unit outward normal vector, respectively. In addition, $\mathbf{S}$ is the (original) Eshelby's tensor, $\mathbf{I}$ signifies the fourth-rank tensor, and $\mathbf{L}$ is the fourth-rank elasticity tensor. Here, the second-order tensor η_{ij} denotes the compliance of the interface spring layer and can be expressed as [8]

$$\eta_{ij} = \alpha \delta_{ij} + (\beta - \alpha) n_i n_j \quad (3.23)$$

in which δ_{ij} denotes the Kronecker delta and α and β represent the compliance in the tangential and normal directions of the interface, as previously mentioned in Section 3.3.3. Details of the modified Eshelby's tensor for ellipsoids, spheres, and cylinders can be found in Ref. [8].

With the help of the modified Eshelby's tensor [8] and the ensemble-volume-averaged micromechanical equation (Eq. (3.12)), micromechanics-based constitutive models for multilevel damage modeling were developed by Lee and Pyo [41, 42, 219]. A variety of numerical simulations were conducted for validation of the

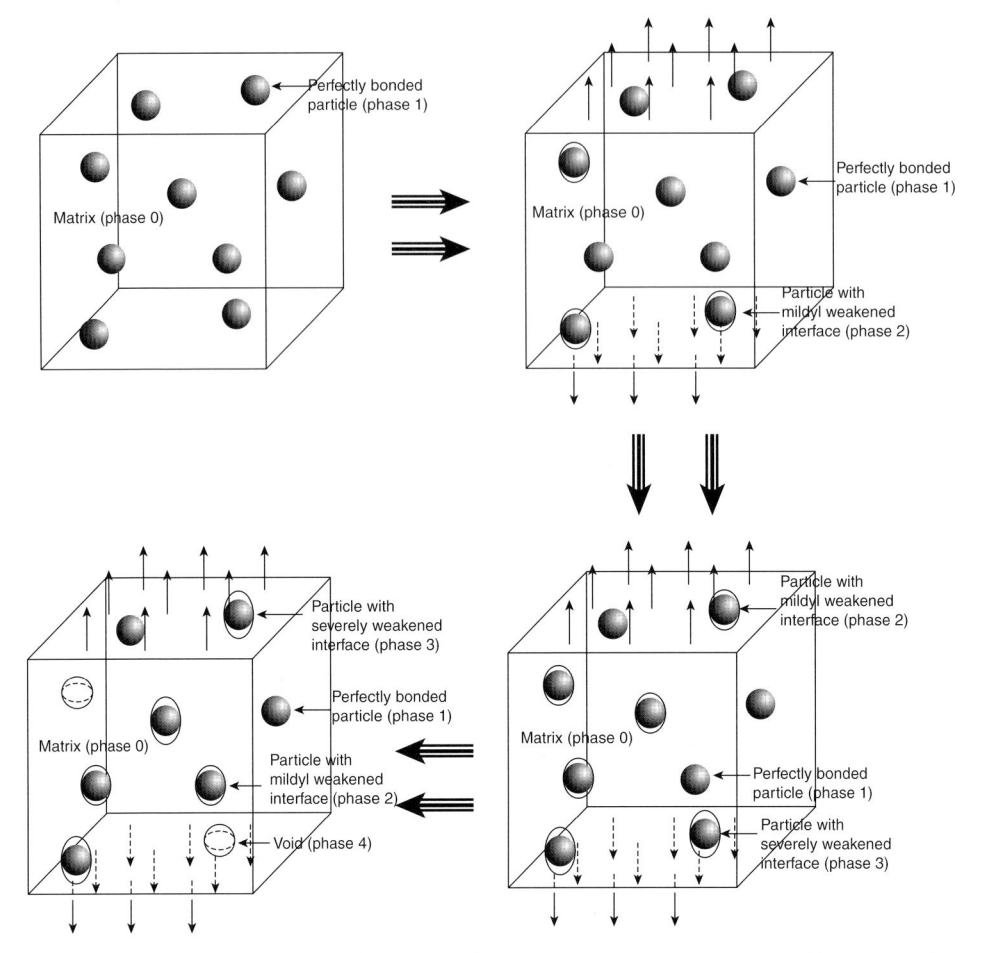

Figure 3.5 Schematics of a multilevel damage process in a composite material having imperfect interfaces. Reproduced from Ref. [41] with permission from Elsevier, Copyright 2007.

micromechanical damage model through comparison with theoretical and experimental results. Figure 3.6 shows the predicted stress–strain responses of composite materials and the corresponding damage evolution curves with regard to various interfacial types [219]. Details of the extended micromechanical damage model for multilevel interfacial damage and more detailed figures and results can be found in Refs [41–45].

3.4.3
Cumulative Damage Model

Cumulative damage models have been proposed with the assumption that microscale interfacial debonding damage is accumulated with increased loading or deformation and eventually leads to failure of composite materials [46, 47, 257]. Recently, Kim and

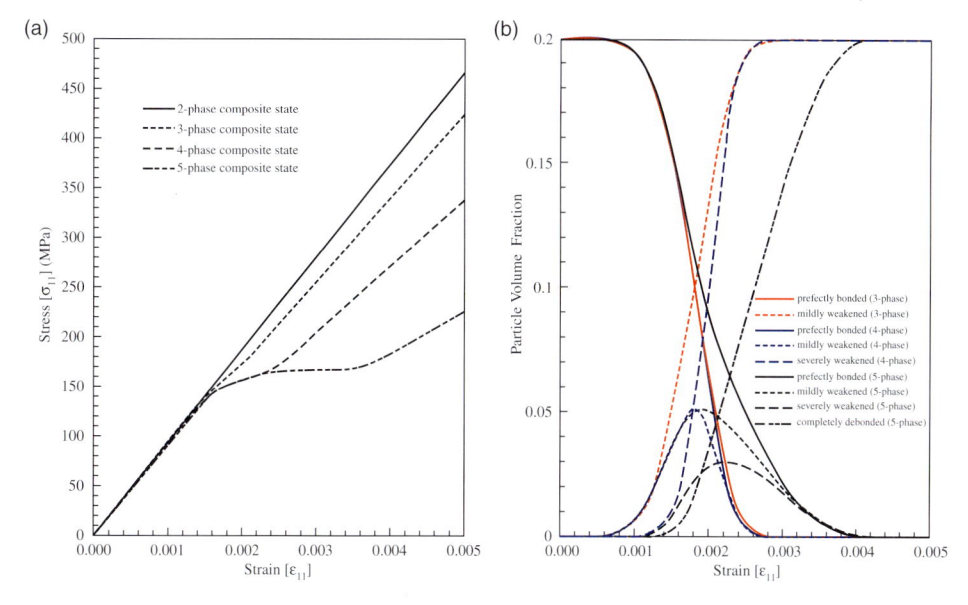

Figure 3.6 The predicted stress–strain responses of composite materials (a) and the corresponding damage evolution curves with regard to various interfacial types (b). Reproduced from Ref. [41] with permission from Elsevier, Copyright 2007.

Lee [46, 47] proposed a micromechanical damage model that incorporates a cumulative step-stress concept into the Weibull statistical function for a more realistic simulation of evolutionary interfacial debonding in composite materials. In this section, the micromechanical framework proposed by Kim and Lee [46, 47] is summarized. According to Kim and Lee [47], damage induced by interfacial debonding in composite materials may be accumulated as loading or deformations continue to increase. Therefore, the initial damage step needs to be modeled such that it is differentiated from the subsequent damage steps for more realistic prediction of progressive interfacial debonding; that is, the effects of previous damage steps have to be properly reflected in the prediction of interfacial debonding at the current damage step for accurate prediction of progressive interfacial debonding, since the subsequent damage steps are the following steps of the current damage step and are affected by the current and previous damage steps (see the figure in Refs [46, 47]). The Weibull probability distribution function at the nth damage step considering the cumulative damage steps can be expressed as [46, 47]

$$F_1\left[(\bar{\boldsymbol{\sigma}}_p)_1\right] = 1 - \exp\left\{-\left[\frac{(\bar{\boldsymbol{\sigma}}_p)_1}{(S_0)_1}\right]^M\right\} \quad (n = 1) \tag{3.24}$$

$$F_n\left[(\bar{\boldsymbol{\sigma}}_p)_n\right] = 1 - \exp\left\{-\left[\frac{[(\bar{\boldsymbol{\sigma}}_p)_n - (\bar{\boldsymbol{\sigma}}_p)_{n-1}] + (\boldsymbol{\sigma}_{\text{eq}})_{n-1}}{(S_0)_n}\right]^M\right\} \quad (n = 2, \ldots, N) \tag{3.25}$$

with

$$(\bar{\sigma}_{eq})_{n-1} = \left[(\bar{\sigma}_p)_{n-1} - (\bar{\sigma}_p)_{n-2}\right]\left[\frac{(S_0)_n}{(S_0)_{n-1}}\right] + (\bar{\sigma}_{eq})_{n-2} \quad (3.26)$$

From the proposed cumulative damage steps, a micromechanical damage model was presented by Kim and Lee [46, 47] and a variety of numerical simulations were conducted for validation of the micromechanical damage model through comparison with theoretical and experimental results. In particular, from the numerical simulation, the following observations and findings were investigated (see the figure in Ref. [46]):

- The four parts are observed from the stress–strain curve: (A) linear elastic part, (B) interface damage part, (C) plastic and interface damage part, and (D) plastic part. Part A reflects the elastic behavior of particulate ductile composites where no significant damage occurs [46]. Part B illustrates nonlinear behavior of the composites caused by particle interfacial debonding, as shown in the corresponding damage evolution curve [46]. This result accommodates the investigated phenomena [258–260], which show a clearly emerging view: there are both bonded and debonded regions that simultaneously present at the fiber–matrix interface. Distinguished from part B, part C indicates that the void growth initiated at the starting point of plastic behavior and the void nucleation by particle interfacial debonding influence the nonlinear behavior of the composites [46]. This result explains that for composite materials containing ductile matrices, the fiber–matrix interface region tends to yield in preference to clear-cut debonding [51]. Finally, part D corresponds to the plastic behavior following the isotropic hardening law/flow rule and ductile damage caused by the void evolution [46].
- It is worth noting that the proposed micromechanical damage model for the cumulative interfacial damage [46, 47] is a suitable micromechanics model for prediction of the interfacial damage of composite materials.

Details of the extended micromechanical damage model for the cumulative interfacial damage and more detailed figures and results can be found in Refs [46, 47].

3.5
Summary

Overall microscale tests utilized for characterization of the interface of composite materials and micromechanical models for composite materials were reviewed in this chapter. Imperfect interface existing in composite materials is the key region for understanding of damage and failure phenomenon. Therefore, it is essential to fundamentally research on imperfect interfaces and their effects through the analytical, numerical, and experimental studies. There exist many different microscale test methods to measure the interfacial parameters: single-fiber pull-out test, microbond test, push-out test, single-fiber compression test, and fragmentation test. Due to the diversity of experimental condition (e.g., specimen preparation, loading methods, and measurement and data reduction methods), they have the advantages and limitations for the determination of interfacial characteristics. Although the microscale tests for interfacial parameters were roughly introduced in this chapter, an in-depth understanding of each test methods has to be fully discussed for adequate investigation of the interface of composite materials. In addition, in order to efficiently understand the mechanical behavior of composite materials, a variety of micromechanics-based analyses have been developed: rule of mixture method, variational principle method, self-consistent method, differential scheme, and the Mori–Tanaka method. In particular, among various micromechanics-based analyses, a new micromechanical framework using the ensemble-volume average method [37, 38] for a multiphase composite medium and an interfacial damage modeling (e.g., the conventional Weibull probabilistic approach, the multilevel damage model, and the cumulative damage model) for composite materials having imperfect interfaces were mainly introduced in this chapter. The proposed methodologies for evaluating interfacial damage are suitable for more precise prediction of progressive interfacial debonding in composite materials.

However, since various damage mechanisms (e.g., fiber or particle breakage, void problems, microcrack problems, and interaction problems) exist in composite materials and they involve too many fitting parameters, micromechanics-based analyses are not always adequate for real material applications [6, 17, 26, 30]. Nevertheless, it is worth noting that the present micromechanics-based models open new perspectives for modeling composite materials; that is, these models can be significantly improved by further development and will yield good results for a range of technically important new composite materials. In addition, for a micromechanics-based analysis, the use of accurate model parameter values is essential to realistically predict the damage evolution behavior in composite materials. Therefore, it is important to point out that experimental studies for verification of the model parameters and further assessment of the model are also needed.

References

1. Drzal, L.T., Rich, M.J., and Lloyd, P.F. (1982) *The Journal of Adhesion*, **16** (1), 1–30.
2. Ashby, M.F. (1989) *Acta Metallurgica*, **37** (5), 1273–1293.
3. Matthews, F.L. and Rawlings, R.D. (1994) *Composite Materials: Engineering and Science*, Chapman & Hall.
4. Kaw, A.K. (1997) *Mechanics of Composite Materials*, CRC Press.
5. Berthelot, J.M. (1999) *Composite Materials: Mechanical Behavior and Structural Analysis*, Springer, New York.
6. Lee, H.K., Kim, B.R., and Na, S. (2011) *Advances in Materials Science Research*. Volume 1, pp. 235–258, Nova Science Publishers.
7. Griffith, A.A. (1920) *Philosophical Transactions of the Royal Society of London A*, **221**, 163–198.
8. Qu, J. (1993) *Mechanics of Materials*, **14** (4), 269–281.
9. Plueddemann, E.P. (1974) Composite materials, in *Interfaces in Polymer Matrix Composites*, vol. 6 (eds L.T. Broutman and R.H. Krock), Academic Press, New York.
10. Schultz, J., Lavielle, L., and Martin, C. (1987) *The Journal of Adhesion*, **23** (1), 45–60.

11 Achenbach, J.D. and Zhu, H. (1989) *Journal of the Mechanics and Physics of Solids*, **37** (3), 381–393.
12 Wu, H.F. and Claypool, C.M. (1991) *Journal of Materials Science Letters*, **10** (5), 260–262.
13 Hashin, Z. (1992) *Journal of the Mechanics and Physics of Solids*, **40** (4), 767–781.
14 Hoecker, F. and Karger-Kocsis, J. (1994) *Composites*, **25** (7), 729–738.
15 Torquato, S. and Rintoul, M.D. (1995) *Physical Review Letters*, **75** (22), 4067–4070.
16 Ageorges, C., Friedrich, K., Schüller, T., and Lauke, B. (1999) *Composites Part A*, **30** (12), 1423–1434.
17 Zhandarov, S. and Mäder, E. (2005) *Composites Science and Technology*, **65** (1), 149–160.
18 Mallick, P.K. (1997) *Composites Engineering Handbook*, Marcel Dekker, New York.
19 Lee, S.M. (1992) *Composites Science and Technology*, **43** (4), 317–327.
20 Majumdar, B.S. (1998) Chapter 4, in *Titanium Matrix Composites: Mechanical Behavior* (eds S. Mall and T. Nicholas), Technomic Publishing Company.
21 Ghorbel, E. (1997) *Composites Science and Technology*, **57** (8), 1045–1056.
22 Yue, C.Y. and Cheung, W.L. (1992) *Journal of Materials Science*, **27** (14), 3843–3855.
23 Hashin, Z. (2002) *Journal of the Mechanics and Physics of Solids*, **50** (12), 2509–2537.
24 Schüller, T., Beckert, W., Lauke, B., Ageorges, C., and Friedrich, K. (2000) *Composites Part A*, **31** (7), 661–670.
25 Greszczuk, L.B. (1969) *Interfaces in Composites*, STP 452-EB, ASTM International.
26 Kim, B.R. and Lee, H.K. (2010) Chapter 4, in *Composite Laminates: Properties, Performance and Applications* (eds A. Doughett and P. Asnarez), Nova Science Publishers.
27 Ghoniem, N.M. and Cho, K. (2002) *CMES: Computer Modeling in Engineering & Sciences*, **3** (2), 147–173.
28 Guz', I.A. and Rushchitskii, Y.Y. (2004) *Mechanics of Composite Materials*, **40** (3), 179–190.
29 Gates, T.S., Odegard, G.M., Frankland, S.J.V., and Clancy, T.C. (2005) *Composites Science and Technology*, **65** (15–16), 2416–2434.
30 Graham-Brady, L.L., Arwade, S.R., Corr, D.J., Gutiérrez, M.A., Breysse, D., Grigoriu, M., and Zabaras, N. (2006) *Probabilistic Engineering Mechanics*, **21** (3), 193–199.
31 Talreja, R. and Singh, C.V. (2008) *Multiscale Modeling and Simulation of Composite Materials and Structures* (eds Y.W. Kwon, D.H. Allen, and R. Talreja), Springer Science + Business Media, New York.
32 Mittal, K.L. (1995) *Adhesion Measurement of Films and Coatings*, VSP BV, The Netherlands.
33 Fowkes, F.M. (1987) *Journal of Adhesion Science and Technology*, **1** (1), 7–27.
34 Fish, J. (1998) *Mechanical Behavior of Advanced Materials*, American Society of Mechanical Engineers, Materials Division, vol. 84, 177–182.
35 Nishikawa, M., Okabe, T., Takeda, N., and Curtin, W.A. (2008) *Modelling and Simulation in Materials Science and Engineering*, **16**, 1–19.
36 Drzal, L.T., Sugiura, N., and Hook, D. (1997) *Composite Interfaces (Netherlands)*, **5** (4), 337–354.
37 Ju, J.W. and Chen, T.M. (1994) *Acta Mechanica*, **103** (1–4), 103–121.
38 Ju, J.W. and Chen, T.M. (1994) *Acta Mechanica*, **103** (1–4), 123–144.
39 Ju, J.W. and Lee, H.K. (2000) *Computer Methods in Applied Mechanics and Engineering*, **183** (3–4), 201–222.
40 Ju, J.W. and Lee, H.K. (2001) *International Journal of Solids and Structures*, **38** (36–37), 6307–6332.
41 Lee, H.K. and Pyo, S.H. (2008) *Composites Science and Technology*, **68** (2), 387–397.
42 Lee, H.K. and Pyo, S.H. (2008) *International Journal of Solids and Structures*, **45** (6), 1614–1631.
43 Lee, H.K. and Pyo, S.H. (2009) *ASCE Journal of Engineering Mechanics*, **135** (10), 1108–1118.
44 Pyo, S.H. and Lee, H.K. (2009) *CMES: Computer Modeling in Engineering & Sciences*, **40** (3), 271–305.

45. Pyo, S.H. and Lee, H.K. (2010) *International Journal of Plasticity*, **26** (1), 25–41.
46. Kim, B.R. and Lee, H.K. (2009) *CMES: Computer Modeling in Engineering & Sciences*, **47** (3), 253–281.
47. Kim, B.R. and Lee, H.K. (2010) *International Journal of Damage Mechanics*. doi: 10.1177/1056789509346688
48. Keusch, S. and Haessler, R. (1999) *Composites Part A*, **30** (8), 997–1002.
49. Bhushan, B. and Gupta, B.K. (1991) *Handbook of Tribology: Materials, Coatings, and Surface Treatments*, McGraw-Hill, New York.
50. Albertsen, H., Ivens, J., Peters, P., Wevers, M., and Verpoest, I. (1995) *Composites Science and Technology*, **54** (2), 133–145.
51. Kim, J.K. and Mai, Y.W. (1998) *Engineered Interfaces in Fiber Reinforced Composites*, Elsevier Science Ltd., The Netherlands.
52. Molitor, P., Barron, V., and Young, T. (2001) *International Journal of Adhesion & Adhesives*, **21** (2), 129–136.
53. Yuan, L.Y., Shyu, S.S., and Lai, J.Y. (1991) *Journal of Applied Polymer Science*, **42** (9), 2525–2534.
54. Li, Z.F. and Netravali, A.N. (1992) *Journal of Applied Polymer Science*, **44** (2), 333–346.
55. Silverstein, M.S. and Breuer, O. (1993) *Composites Science and Technology*, **48** (1–4), 151–157.
56. Roulet, J.F., Söderbolm, K.J.M., and Longmate, J. (1995) *Journal of Dental Research*, **74** (1), 381–387.
57. Mäder, E. (1997) *Composites Science and Technology*, **57** (8), 1077–1088.
58. Rashkovan, I.A. and Korabel'nikov, Yu.G. (1997) *Composites Science and Technology*, **57** (8), 1017–1022.
59. Valadez-Gonzalez, A., Cervantes-Uc, J.M., Olayo, R., and Herrera-Franco, P.J. (1999) *Composites Part B*, **30** (3), 309–320.
60. Rong, M.Z., Zhang, M.Q., Liu, Y., Yang, G.C., and Zeng, H.M. (2001) *Composites Science and Technology*, **61** (10), 1437–1447.
61. Joffe, R., Andersons, J., and Wallström, L. (2003) *Composites Part A*, **34** (7), 603–612.
62. Herrera-Franco, P.J. and Valadez-González, A. (2005) *Composites Part B*, **36** (8), 597–608.
63. Novak, R.C. (1969) *Composite Materials: Testing and Design, ASTM STP 460*, ASTM International, Philadelphia, PA.
64. McKee, D.W. (1970) *Carbon*, **8** (2), 131–136.
65. Scolar, D.A. (1974) *Composite Materials*, vol. 6 (ed. E.P. Plueddemann) Academic Press, New York.
66. Horie, K., Hiromichi, M., and Mita, I. (1976) *Fibre Science and Technology*, **9** (4), 253–264.
67. Donnet, J.B. and Ehrburger, P. (1977) *Carbon*, **15** (3), 143–152.
68. Fu, X., Lu, W., and Chung, D.D.L. (1996) *Cement and Concrete Research*, **26** (7), 1007–1012.
69. Goan, J.C. and Prosen, S.P. (1969) *Interfaces in Composites, STP452-EB*, ASTM International.
70. Reiss, G., Bourdeaux, M., Brie, M., and Jouquet, G. (1974) Proceedings of the 2nd Carbon Fibre Conference, The Plastics Institute, London.
71. Marks, B.S., Mauri, R.E., and Bradshaw, W.G. (1975) *Carbon*, **13** (6), 556.
72. Wertheimer, M.R. and Schreiber, H.P. (1981) *Journal of Applied Polymer Science*, **26** (6), 2087–2096.
73. Allred, R.E., Merrill, E.W., and Roylance, D.K. (1985) *Molecular Characterization of Composite Interfaces* (eds H. Ishida and G. Kumar), Plenum Press, New York.
74. Verpoest, I. and Springer, G.S. (1988) *Journal of Reinforced Plastics and Composites*, **7** (1), 2–22.
75. Bascom, W.D. and Chen, W.J. (1991) *The Journal of Adhesion*, **34** (1–4), 99–119.
76. Morra, M., Occhiello, E., Garbassi, F., and Nicolais, L. (1991) *Composites Science and Technology*, **42** (4), 361–372.
77. Brown, J.R., Ghappell, P.J.C., and Mathys, Z. (1991) *Journal of Materials Science*, **26** (15), 4172–4178.
78. Garbassi, F. and Occhiello, E. (1993) *Handbook of Composite Reinforcement* (eds S.M. Lee), Wiley-VCH Verlag GmbH, New York.

79 Pappas, D., Bujanda, A., Demaree, J.D., Hirvonen, J.K., Kosik, W., Jensen, R., and McKnight, S. (2006) *Surface and Coating Technology*, **201** (7), 4384–4388.

80 Vaughan, D.J. (1978) *Polymer Engineering and Science*, **18** (2), 167–169.

81 Inagaki, N., Tasaka, S., and Kawai, H. (1992) *Journal of Adhesion Science and Technology*, **6** (2), 279–291.

82 Herrera-Franco, P.J. and Drzal, L.T. (1992) *Composites*, **23** (1), 2–27.

83 Pitkethly, M.J. (1996) *Fiber, Matrix, and Interface Properties* (eds C.J. Spragg and L.T. Drzal), ASTM STP 1290, ASTM International.

84 Choi, N.S., Park, J.E., and Kang, S.K. (2009) *Journal of Composite Materials*, **43** (16), 1663–1677.

85 Kim, J.K., Baillie, C., and Mai, Y.W. (1992) *Journal of Materials Science*, **27** (12), 3143–3154.

86 Miller, B., Muri, P., and Rebenfeld, L. (1987) *Composites Science and Technology*, **28** (1), 17–32.

87 Wagner, H.D., Gallis, H.E., and Wiesel, E. (1993) *Journal of Materials Science*, **28** (8), 2238–2244.

88 Tandon, G.P. and Pagano, N.J. (1998) *Composites Science and Technology*, **58** (11), 1709–1725.

89 Nishikawa, M., Okabe, T., and Takeda, N. (2008) *Materials Science and Engineering A*, **480** (1–2), 549–557.

90 Broutman, L.J. (1969) *Interfaces in Composites*, STP452-EB, ASTM International.

91 McAlea, K.P. and Besio, G.J. (1988) *Polymer Composites*, **9** (4), 285–290.

92 Gaur, U. and Miller, B. (1989) *Composites Science and Technology*, **34** (1), 35–51.

93 Biro, D.A., McLean, P., and Deslandes, Y. (1991) *Polymer Engineering and Science*, **31** (17), 1250–1256.

94 Moon, C.K., Cho, H.H., Lee, J.O., and Park, T.W. (1992) *Journal of Applied Polymer Science*, **44** (3), 561–563.

95 Desaeger, M. and Verpoest, I. (1993) *Composites Science and Technology*, **48** (1–4), 215–226.

96 Kelly, A. and Davies, G.J. (1965) *International Materials Reviews*, **10** (1), 1–77.

97 Drzal, L.T., Rich, M.J., Koenig, M.F., and Lloyd, P.F. (1983) *The Journal of Adhesion*, **16** (2), 133–152.

98 Torquato, S. (2000) *International Journal of Solids and Structures*, **37** (1–2), 411–422.

99 Chen, J.S. and Mehraeen, S. (2005) *Modelling and Simulation in Materials Science and Engineering*, **13** (1), 95–121.

100 Ha, S.K., Jun, K.K., and Oh, J.H. (2006) *Journal of Korean Society of Mechanical Engineers.*, **A30** (3), 260–268 (in Korean).

101 Kouznetsova, V., Brekelmans, W.A.M., and Baaijens, F.P.T. (2001) *Computational Mechanics*, **27** (1), 37–48.

102 Kalamkarov, A.L. and Georgiades, A.V. (2002) *Smart Materials and Structures*, **11** (3), 423–434.

103 Yan, C. (2003) On homogenization and de-homogenization of composite materials. PhD dissertation. Drexel University.

104 Unnikrishnan, G.U., Unnikrishnan, V.U., and Reddy, J.N. (2007) *Journal of Biomechanical Engineering*, **129** (3), 315–323.

105 Hori, M. and Nemat-Nasser, S. (1999) *Mechanics of Materials*, **31** (10), 667–682.

106 Markovic, D. and Ibrahimbegovic, A. (2004) *Computer Methods in Applied Mechanics and Engineering*, **193** (48–51), 5503–5523.

107 Eshelby, J.D. (1957) *Proceedings of the Royal Society of London A*, **241** (1226), 376–396.

108 Hashin, Z. (1962) *ASME Journal of Applied Mechanics*, **29**, 143–150.

109 Hill, R. (1963) *Journal of the Mechanics and Physics of Solids*, **11** (5), 357–372.

110 Christensen, R.M. and Lo, K.H. (1979) *Journal of the Mechanics and Physics of Solids*, **27** (4), 315–330.

111 Mura, T. (1987) *Micromechanics of Defects in Solids*, 2nd revised edn, Martinus Nijhoff Publishers, Dordrecht.

112 Nemat-Nasser, S. and Hori, M. (1993) *Micromechanics: Overall Properties of Heterogeneous Materials*, Elsevier Science Publishers, The Netherlands.

113 Bensoussan, A., Lions, J.L., and Papanicolaou, G. (1978) *Asymptotic Analysis for Periodic Structures*, North Holland, The Netherlands.

114 Sanchez-Palencia, E. (1980) *Lecture Notes in Physics*, vol. 127, Springer, Berlin.
115 Bakhvalov, N. and Panasenko, G. (1984) *Homogenization: Averaging Processes in Periodic Media, Mathematical Problems of the Mechanics of Composite Materials*, Nauka, Moscow.
116 Suquet, P. (1987) *Homogenization Techniques for Composite Media* (eds E. Sanchez-Palencia and A. Zaoui), Springer, Berlin.
117 Michel, J.D. and Suquet, P. (2003) *International Journal of Solids and Structures*, **40** (25), 6937–6955.
118 Guinovart-Díaz, R., Rodríguez-Ramos, R., Bravo-Castillero, J., Sabina, F.J., Otero-Hernández, J.A., and Maugin, G.A. (2005) *Mechanics of Materials*, **37** (11), 1119–1131.
119 Evesque, P. (2000) *Poudres & Grains*, **11**, 6–17.
120 Tsai, J.L. and Chen, K.H. (2007) *Journal of Composite Materials*, **41** (10), 1253–1273.
121 Hashin, Z. and Shtrikman, S. (1962) *Journal of the Mechanics and Physics of Solids*, **10** (4), 335–342.
122 Hashin, Z. and Shtrikman, S. (1963) *Journal of the Mechanics and Physics of Solids*, **11** (2), 127–140.
123 Kröner, E. (1972) *Statistical Continuum Mechanics*, Springer, Berlin.
124 Hashin, Z. (1983) *Journal of Applied Mechanics*, **50** (3), 481–505.
125 Drugan, W.J. and Willis, J.R. (1996) *Journal of the Mechanics and Physics of Solids*, **44** (4), 497–524.
126 van Mier, J.G.M. (1997) *Fracture Processes of Concrete: Assessment of Material Parameters for Fracture Models*, CRC Press, Boca Raton, FL.
127 Hazanov, S. (1998) *Archive of Applied Mechanics*, **68** (6), 385–394.
128 Ostoja-Starzewski, M. (2002) *Journal of Applied Mechanics*, **69** (1), 25–35.
129 Ren, Z.Y. and Zheng, Q.S. (2002) *Journal of the Mechanics and Physics of Solids*, **50** (4), 881–893.
130 Ren, Z.Y. and Zheng, Q.S. (2004) *Mechanics of Materials*, **36** (12), 1217–1229.
131 Kanit, T., Forest, S., Galliet, I., Mounoury, V., and Jeulin, D. (2003) *International Journal of Solids and Structures*, **40** (13–14), 3647–3679.
132 Gitman, I.M., Gitman, M.B., and Askes, H. (2006) *Archive of Applied Mechanics*, **75** (2–3), 79–92.
133 Gitman, I.M., Askes, H., and Sluys, L.J. (2007) *Engineering Fracture Mechanics*, **74** (16), 2518–2534.
134 Swaminathan, S., Ghosh, S., and Pagano, N.J. (2006) *Journal of Composite Materials*, **40** (7), 583–604.
135 Trias, D., Costa, J., Turon, A., and Hurtado, J.E. (2006) *Acta Materialia*, **54** (13), 3471–3484.
136 Luo, H.A. and Weng, G.J. (1987) *Mechanics of Materials*, **6** (4), 347–361.
137 Ru, C.Q. (1999) *Journal of Applied Mechanics*, **66** (2), 315–322.
138 Eshelby, J.D. (1959) *Proceedings of the Royal Society of London A*, **252** (1271), 561–569.
139 Eshelby, J.D. (1961) *Progress in Solid Mechanics* (eds N.I. Sneddon and R. Hill), North Holland, Amsterdam.
140 Cristescu, N.D., Craciun, E.M., and Soós, E. (2003) *Mechanics of Elastic Composites, Modern Mathematics and Mechanics*, Chapman & Hall/CRC.
141 Yin, H.M., Sun, L.Z., and Paulino, G.H. (2004) *Acta Materialia*, **52** (12), 3535–3543.
142 Li, S., Sauer, R., and Wang, G. (2005) *Acta Mechanica*, **179** (1–2), 67–90.
143 Ouaar, A. (2006) Micromechanics of rate-independent multi-phase composites: application to steel fiber-reinforced concrete. PhD dissertation. Université Catholique de Louvain.
144 Collini, L. (2005) Micromechanical modeling of the elasto-plastic behavior of heterogeneous nodular cast iron. PhD dissertation. Universita' degli Studi di Parma.
145 Chen, J. (2006) Development and characterization of high strength NB$_3$SN superconductor. PhD dissertation. The Florida State University.
146 Maranganti, R. and Sharma, P. (2005) *Handbook of Theoretical and Computational Nanotechnology*, vol. 2, Nanotechnology Book Series (eds M. Rieth and W. Schommers), American Scientific Publishers.

147 Damiani, T.M. (2003) Mesomechanics of fabric reinforced composites. PhD dissertation. West Virginia University.
148 Jones, R.M. (1975) *Mechanics of Composite Materials*, McGraw-Hill, New York.
149 Tsai, S.W. and Hahn, H.T. (1980) *Introduction to Composite Materials*, Technomic Publishing Company, Lancaster, PA.
150 Hashin, Z. (1972) Theory of fiber reinforced materials, Technical report, NASA Report.
151 Hashin, Z. and Shtrikman, S. (1962) *Journal of the Mechanics and Physics of Solids*, **10** (4), 343–352.
152 Hashin, Z. and Rosen, B.W. (1964) *ASME Journal of Applied Mechanics*, **31**, 223–232.
153 Hill, R. (1964) *Journal of the Mechanics and Physics of Solids*, **12**, 199–212.
154 Walpole, L.J. (1966) *Journal of the Mechanics and Physics of Solids*, **14** (3), 151–162.
155 Walpole, L.J. (1966) *Journal of the Mechanics and Physics of Solids*, **14**, 289–301.
156 Walpole, L.J. (1969) *Journal of the Mechanics and Physics of Solids*, **17** (4), 235–251.
157 Ponte Castañeda, P. (1991) *Journal of the Mechanics and Physics of Solids*, **39** (1), 45–71.
158 Willis, J.R. (1991) *Journal of the Mechanics and Physics of Solids*, **39** (1), 73–86.
159 Hershey, A.V. (1954) *ASME Journal of Applied Mechanics*, **21**, 236–240.
160 Hill, R. (1965) *Journal of the Mechanics and Physics of Solids*, **13**, 213–222.
161 Budiansky, B. (1965) *Journal of the Mechanics and Physics of Solids*, **13**, 223–227.
162 Budiansky, B. and O'Connell, R.J. (1976) *International Journal of Solids and Structures*, **12** (2), 81–97.
163 Krajcinovic, D. and Sumarac, D. (1989) *Journal of the Mechanics and Physics of Solids*, **56** (1), 51–62.
164 Ju, J.W. (1991) *International Journal of Solids and Structures*, **27** (2), 227–258.
165 Roscoe, R. (1973) *Rheologica Acta*, **12** (3), 404–411.
166 McLaughlin, R. (1977) *International Journal of Engineering Science*, **15** (4), 237–244.
167 Hashin, Z. (1988) *Journal of the Mechanics and Physics of Solids*, **36** (6), 719–734.
168 Mori, T. and Tanaka, K. (1973) *Acta Metallurgica*, **21**, 571–574.
169 Taya, M. and Chou, T.W. (1981) *International Journal of Solids and Structures*, **17**, 553–563.
170 Benveniste, Y. (1986) *Mechanics Research Communications*, **13** (4), 193–201.
171 Weng, G.J. (1990) *International Journal of Engineering Science*, **28** (11), 1111–1120.
172 Qiu, Y.P. and Weng, G.J. (1990) *International Journal of Engineering Science*, **28** (11), 1121–1137.
173 Dewey, J.M. (1947) *ASME Journal of Applied Mechanics*, **18**, 578–581.
174 Kerner, E.H. (1956) *Proceedings of the Royal Society London B*, **69** (8), 807–808.
175 Willis, J.R. and Acton, J.R. (1976) *Quarterly Journal of Mechanics and Applied Mathematics*, **29**, 163–177.
176 Chen, H.S. and Acrivos, A. (1978) *International Journal of Solids and Structures*, **14** (5), 331–348.
177 Chen, H.S. and Acrivos, A. (1978) *International Journal of Solids and Structures*, **14** (5), 349–364.
178 Nemat-Nasser, S. and Taya, M. (1981) *Quarterly of Applied Mathematics*, **39**, 43–59.
179 Iwakuma, T. and Nemat-Nasser, S. (1983) *Composite Structures*, **16**, 13–19.
180 Nunan, K.C. and Keller, J.B. (1984) *Journal of the Mechanics and Physics of Solids*, **32** (4), 259–280.
181 Sangani, A.S. and Lu, W. (1987) *Journal of the Mechanics and Physics of Solids*, **35** (1), 1–21.
182 Christensen, R.M. (1990) *Journal of the Mechanics and Physics of Solids*, **38** (3), 379–404.
183 Kalamkarov, A.L. (1992) *Composite and Reinforced Elements of Construction*, John Wiley & Sons, Inc., New York.
184 Kalamkarov, A.L. and Kolpakov, A.G. (1997) *Analysis, Design and Optimization*

of *Composite Structures*, John Wiley & Sons, Inc., New York.
185 Kahler, B., Kotousov, A., and Swain, M.W. (2008) *Acta Biomaterialia*, **4** (1), 165–172.
186 Kanaun, S.K. and Levin, V.M. (2008) *Self-Consistent Methods for Composites, Vol. 1: Static Problems*, Springer, The Netherlands.
187 Voigt, W. (1928) *Lerrbuch der Kristallphysik*, Teubner-Verlag, Leipzig.
188 Reuss, A. (1929) *Zeitschrift für Angewandte Mathematik und Mechanik*, **9**, 49–58.
189 Mainprice, D. and Humbert, M. (1994) *Surveys in Geophysics*, **15** (5), 575–592.
190 Zaoui, A. (2002) *Journal of Engineering Mechanics*, **128** (8), 808–816.
191 Huysmans, G., Verpoest, I., and van Houtte, P. (1998) *Acta Materialia*, **46** (9), 3003–3013.
192 Zheng, Q.S. and Du, D.X. (2001) *Journal of the Mechanics and Physics of Solids*, **49** (11), 2765–2788.
193 Tan, H., Huang, Y., Liu, C., and Geubelle, P.H. (2005) *International Journal of Plasticity*, **21** (10), 1890–1918.
194 Schjødt-Thomsen, J. and Pyrz, R. (2001) *Mechanics of Materials*, **33** (10), 531–544.
195 Hu, G.K. and Weng, G.J. (2000) *Mechanics of Materials*, **32** (8), 495–503.
196 Willis, J.R. (1977) *Journal of the Mechanics and Physics of Solids*, **25** (3), 185–202.
197 Sen, A.K. and Torquato, S. (1989) *Physical Review B*, **39** (7), 4504–4515.
198 Ju, J.W. and Chen, T.M. (1994) *Journal of Engineering Materials and Technology*, **116** (3), 310–318.
199 Ju, J.W. and Tseng, K.H. (1996) *International Journal of Solids and Structures*, **33** (29), 4267–4291.
200 Ju, J.W. and Tseng, K.H. (1997) *ASCE Journal of Engineering Mechanics*, **123** (3), 260–266.
201 Ju, J.W. and Zhang, X.D. (1998) *International Journal of Solids and Structures*, **35** (9–10), 941–960.
202 Ju, J.W. and Zhang, X.D. (2001) *International Journal of Solids and Structures*, **38** (22–23), 4045–4069.
203 Lee, H.K. (1998) Three-dimensional micromechanical damage models for effective elastic and elastoplastic behavior of composite materials with inhomogeneities or microcracks. PhD dissertation. University of California, Los Angeles.
204 Lee, H.K. (2001) *Computational Mechanics*, **27** (6), 504–512.
205 Sun, L.Z. (1998) Micromechanics and overall elastoplasticity of discontinuously reinforced metal matrix composites. PhD dissertation. University of California, Los Angeles.
206 Lee, H.K. and Simunovic, S. (2000) *Composite Part B*, **31** (2), 77–86.
207 Lee, H.K. and Simunovic, S. (2001) *International Journal of Solids and Structures*, **38** (5), 875–895.
208 Ju, J.W. and Sun, L.Z. (2001) *International Journal of Solids and Structures*, **38** (2), 183–201.
209 Shen, L. and Yi, S. (2001) *International Journal of Solids and Structures*, **38** (32–33), 5789–5805.
210 Sarvestani, A.S. (2003) *International Journal of Solids and Structures*, **40** (26), 7553–7566.
211 Sarvestani, A.S. (2005) *Acta Mechanica*, **176** (3–4), 153–167.
212 Sun, L.Z., Ju, J.W., and Liu, H.T. (2003) *Mechanics of Materials*, **35** (3–6), 559–569.
213 Lee, H.K. and Liang, Z. (2004) *Computers and Structures*, **82** (7–8), 581–592.
214 Liu, H.T., Sun, L.Z., and Ju, J.W. (2004) *International Journal of Damage Mechanics*, **13** (2), 163–185.
215 Liu, H.T., Sun, L.Z., and Ju, J.W. (2006) *Acta Mechanica*, **181** (1–2), 1–17.
216 Sun, L.Z. and Ju, J.W. (2004) *ASME Journal of Applied Mechanics*, **71** (6), 774–785.
217 Ju, J.W., Ko, Y.F., and Ruan, H.N. (2006) *International Journal of Damage Mechanics*, **15** (3), 237–265.
218 Liang, Z., Lee, H.K., and Suaris, W. (2006) *International Journal of Solids and Structures*, **43** (18–19), 5674–5689.
219 Lee, H.K. and Pyo, S.H. (2007) *International Journal of Solids and Structures*, **44** (25–26), 8390–8406.
220 Ju, J.W. and Ko, Y.F. (2008) *International Journal of Damage Mechanics*, **17** (4), 307–356.

221 Skolnik, D.A., Liu, H.T., Wu, H.C., and Sun, L.Z. (2008) *International Journal of Damage Mechanics*, **17** (3), 247–272.

222 Kim, B.R. and Lee, H.K. (2009) *Composite Structures*, **90** (4), 418–427.

223 Kim, B.R. and Lee, H.K. (2010) *International Journal of Solids and Structures*, **47** (6), 827–836.

224 Duan, H.L., Wang, J., Huang, Z.P., and Luo, Z.Y. (2005) *Mechanics of Materials*, **37** (7), 723–736.

225 Hashin, Z. (1991) *Journal of the Mechanics and Physics of Solids*, **39** (6), 745–762.

226 Duan, H.L., Yi, X., Huang, Z.P., and Wang, J. (2007) *Mechanics of Materials*, **39** (1), 81–93.

227 Yu, H.Y., Wei, Y.N., and Chiang, F.P. (2002) *International Journal of Engineering Science*, **40** (14), 1647–1662.

228 Benveniste, Y. (1985) *Mechanics of Materials*, **4** (2), 197–208.

229 Aboudi, J. (1987) *Composites Science and Technology*, **28** (2), 103–128.

230 Zhong, Z. and Meguid, S.A. (1997) *Journal of Elasticity*, **46** (2), 91–113.

231 Shen, H., Schiavone, P., Ru, C.Q., and Mioduchowski, A. (2001) *Journal of Elasticity*, **62** (1), 25–46.

232 Duan, H.L., Wang, J., Huang, Z.P., and Karihaloo, B.L. (2005) *Journal of the Mechanics and Physics of Solids*, **53** (7), 1574–1596.

233 Duan, H.L. and Karihaloo, B.L. (2007) *Journal of the Mechanics and Physics of Solids*, **55** (5), 1036–1052.

234 Fang, Q.H. and Liu, Y.W. (2006) *Scripta Materialia*, **55** (1), 99–102.

235 Gurtin, M.E. and Murdoch, A.I. (1975) *Archive for Rational Mechanics and Analysis*, **57** (4), 291–323.

236 Gurtin, M.E. and Murdoch, A.I. (1978) *International Journal of Solids and Structures*, **14** (6), 431–440.

237 Gurtin, M.E., Weissmuller, J., and Larche, F. (1998) *Philosophical Magazine A*, **78** (5), 1093–1109.

238 Huang, Z.P. and Wang, J. (2006) *Acta Mechanica*, **182** (3–4), 195–210.

239 Yvonnet, J., He, Q.C., and Toulemonde, C. (2008) *Composites Science and Technology*, **68** (13), 2818–2825.

240 Yu, H.Y. (1998) *Composites Part A*, **29** (9–10), 1057–1062.

241 Pan, E. (2003) *ASME Journal of Applied Mechanics*, **70** (2), 180–190.

242 Jasiuk, I., Tsuchida, E., and Mura, T. (1987) *International Journal of Solids and Structures*, **23** (10), 1373–1385.

243 Valle, C., Qu, J., and Jacobs, L.J. (1999) *International Journal of Engineering Science*, **37** (11), 1369–1387.

244 Mura, T., Jasiuk, I., and Tsuchida, B. (1985) *International Journal of Solids and Structures*, **21** (12), 1165–1179.

245 Ghahremani, F. (1980) *International Journal of Solids and Structures*, **16** (9), 825–845.

246 Mura, T. and Furuhashi, R. (1984) *ASME Journal of Applied Mechanics*, **51** (2), 308–310.

247 Weibull, W. (1951) *Journal of Applied Mechanics*, **18**, 293–297.

248 Haldar, A. and Mahadevan, S. (2000) *Probability, Reliability and Statistical Methods in Engineering Design*, John Wiley & Sons, Inc., New York.

249 Blischke, W.R. and Murthy, D.N.P. (2003) *Case Studies in Reliability and Maintenance*, Wiley–Interscience, New York.

250 Chiou, M.J., Lin, S.F., and Lin, J.C. (2003) Proceedings of the Annual Reliability and Maintainability Symposium, pp. 295–302.

251 Zhao, Y.H. and Weng, G.J. (1996) *International Journal of Plasticity*, **12** (6), 781–804.

252 Zhao, Y.H. and Weng, G.J. (1997) *International Journal of Solids and Structures*, **34** (4), 493–507.

253 Liu, H.T., Sun, L.Z., and Wu, H.C. (2005) *ASME Journal of Engineering Materials and Technology*, **127** (3), 318–324.

254 Gurvich, M.R. and Pipes, R.B. (1996) *Composites Science and Technology*, **56** (6), 649–656.

255 Hwang, T.K., Hong, C.S., and Kim, C.G. (2003) *Composite Structures*, **59** (4), 489–498.

256 Ghosh, S., Bai, J., and Raghavan, P. (2007) *Mechanics of Materials*, **39** (3), 241–266.

257 Marotta, S.A., Kudiay, A., Ooi, T.K., Toutanji, H.A., and Gilbert, J.A. (2005) Proceedings of the 1st International

Forum on Integrated System Health Engineering and Management in Aerospace, Napa, California, November 7–10, 2005.

258 Favre, J.P., Sigety, P., and Jacques, D. (1991) *Journal of Materials Science*, **26**, 189–195.

259 Gulino, R. and Phoenix, L. (1991) *Journal of Materials Science*, **26** (11), 3107–3118.

260 Lacroix, Th., Tilmans, B., Keunings, R., Desaeger, M., and Verpoest, I. (1992) *Composites Science and Technology*, **43** (4), 379–387.

4
Preparation and Manufacturing Techniques for Macro- and Microcomposites

Tibor Czigány and Tamás Deák

4.1
Introduction

The manufacturing of polymer composites is a rather difficult task, since there are several requirements regarding the technology, design, function, and cost-effectiveness. The applicable manufacturing technique is determined by the balance of these requirements and strongly influenced by the chosen matrix material, reinforcement, fiber content, fiber length, architecture, and so on.

Matrix materials can be different thermoplastic and thermosetting polymers, which usually contain further modifying additives. The viscosity of resins varies in a very broad range. Reinforcements with different type, geometry (continuous or discontinuous), and structure (random or oriented) require different composite processing technologies. Reinforcements can be arranged to one-dimensional (roving, yarn), two-dimensional (mat, woven and knitted fabric), or three-dimensional (braid, fabric) systems. Geometry, holes, inserts, undercuts, and surface quality mean further constraints, similar to curing time, pressure, and temperature demands of the matrix.

The aim of this chapter is to give an overview of the characterization and categorization of manufacturing techniques of thermoplastic and thermosetting composites, including the applicable materials, tools, and applications.

4.2
Thermoplastic Polymer Composites

Nowadays the amount of thermoplastic matrix composites produced approaches to that of composites with thermosetting matrix. The most important character of thermoplastic composites – and their most important difference from thermosetting ones – is the fact that no chemical reaction occurs during processing. The thermoplastic matrix is heated over its softening or melting temperature, thus enabling the forming; subsequently, the part is cooled. This implies both advantages and disadvantages compared to thermosettings. Thermoplastics offer smaller health

Table 4.1 Feasible series lengths of thermoplastic composite manufacturing techniques.

	Prototype	Small series		Medium series		Large series		
Injection molding					■	■	■	■
Extrusion							■	■
Compression molding			■	■	■	■	■	■
Thermoplastic prepreg lay-up	■	■	■					
Thermoplastic tape winding	■	■	■					
Thermoplastic pultrusion					■	■	■	
Diaphragm forming	■	■						
Number of pieces	1	10	100	1000	10 000	100 000	1 000 000	

hazard, cleaner (and more environment-friendly) technologies, and short cycle times and techniques capable of producing large series with constant quality. On the other hand, the higher viscosity of thermoplastic melts impedes the impregnation, compaction, and consolidation of composites in many cases. The main goal of reinforcing thermoplastics with fibers – besides enhancing mechanical properties – is the increasing of heat deflection temperature, stiffness, creep resistance, wear resistance, and toughness, tailoring the electric properties and decreasing the thermal expansion coefficient. Thermoplastic matrix composites are mainly reinforced with glass, carbon, basalt, ceramic, and natural fibers. Table 4.1 shows the typical product series of thermoplastic composite manufacturing techniques to be introduced in this chapter. Injection molding, extrusion, and compression molding are typical technologies that can produce large quantities of thermoplastic composite parts with good dimensional accuracy and complex geometry. This explains the success of these techniques in the automotive industry; particularly, the compression molding has gained much development in the last few years.

4.2.1
Injection Molding

One of the most important processing technologies of thermoplastic matrix composites is injection molding, which is capable of producing products in large quantities with good dimensional accuracy and complex geometry. Injection molding applies a complex – at least two-sided – tempered mold, which is rapidly filled with the polymer melt with the aid of high pressures, often reaching 100 MPa [1]. This necessitates the closing of the mold with considerable force. The cycle times are very short, mostly around 1 min or less. The most common reinforcement in thermoplastic matrix composites is the glass fiber. One of the most significant problems of injection-molded composites is the fiber fragmentation. Mainly in the compounding extruder and in the feed zone of the injection molding screw, the shearing stresses reduce the length of fibers to the order of magnitude of a few tenths of millimeters, regardless of their original size [2]. Fibers fragment during compounding and plastification in the injection molding machine. During injection, further fragmentation occurs in the nozzle and the mold, meanwhile the average length decreases generally to 0.2–0.4 mm [3, 4]. The attainable fiber length depends on the geometry of the product and the injection molding speed. For example, in a product with 1.5 mm wall thickness, the attainable average fiber length is approximately 30% smaller than that in a product with 4 mm wall thickness [3]. The residual length is often under the critical fiber length; hence, short-fiber reinforcement is unable to utilize the full possibilities of fiber reinforcement. Since the enlargement of residual fiber length in the composite products can significantly enhance the mechanical and physical properties of injection-molded composites, numerous technologies have been developed in the industry for producing parts in which the residual fiber length exceeds the usual values of a few tenths of millimeters [5–7]. The manufacturing technologies of long-fiber-reinforced injection-molded thermoplastics can be divided into two groups: they are compounded on in-line blending units directly before molding (direct processing long-fiber-reinforced thermoplastics: D-LFT) or supplied as ready-to-use pellets (pelletized long-fiber-reinforced thermoplastics: P-LFT). P-LFT compounds, which can be processed by conventional injection molding machines, can be made by a variety of manufacturing processes, the principal difference between them being whether the fibers are cable coated (here specially treated glass fibers are enclosed within a plastic coat and not impregnated until processed) or fully impregnated (here the fibers are impregnated with the plastic matrix in the compounding process). The roving impregnated or coated with the matrix is cut into pieces with a length between 10 and 55 mm. If special cautious injection molding parameters are employed, the fragmentation of fibers can be decreased and fiber length can thus be largely retained during injection molding. D-LFT technologies employ complex machine lines that incorporate compounding and molding in one unit. The fiber orientation and formation of skin–core structure are basic attributes of fiber-reinforced injection-molded composites. During filling the mold, the melt forms a thin layer on the mold wall that has a thickness of 5–7% of the entire wall thickness of the product. The orientation of fibers is random in this thin layer. Inside this layer the growing shear stress creates a so-called skin layer where fibers strongly orientate to the direction of melt flow. In the core that composes the middle layer of the product, the fiber orientation is random or more or less perpendicular to the flow direction (Figure 4.1).

The average fiber orientation is determined by the proportion of the thickness of the skin and core layers. Generally, growing injection rate, melt temperature, and mold temperature decrease the thickness of the skin layer; therefore, the average fiber orientation will be smaller [8–11]. Fiber orientation is determined by the proportion of the skin and core layers inside the part. Larger injection speed, that is, larger resin flow rate, results in thinner skin layer and smaller average orientation (Figure 4.2).

The augmentation of the fiber content of the composite causes increased fiber fragmentation. The effect of average residual fiber length on the impact strength and modulus of elasticity is shown in Figure 4.3 [2, 12].

The wall thickness of the product also has considerable effect: larger thickness results in relatively thinner skin layer and smaller fiber orientation. This decreases

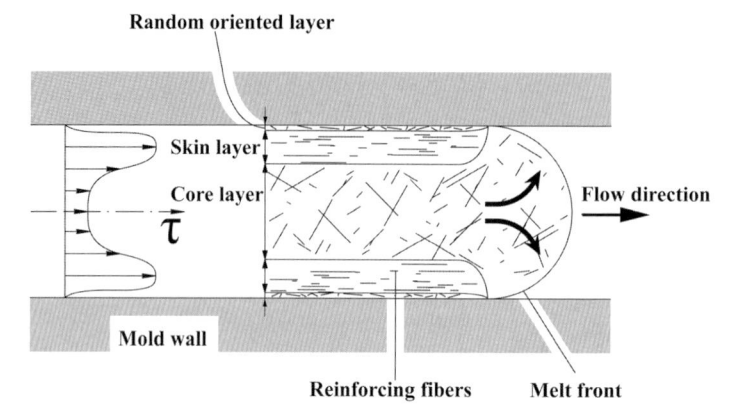

Figure 4.1 Distribution of shear stress (τ) along the cross section and formation of skin–core structure in a thin-wall injection-molded product.

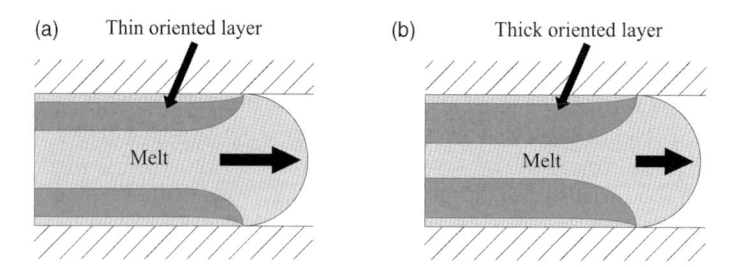

Figure 4.2 The thickness of the skin layer as the function of injection speed. (a) In case of high melt or mold temperature or high injection speed. (b) In case of low melt or mold temperature or low injection speed.

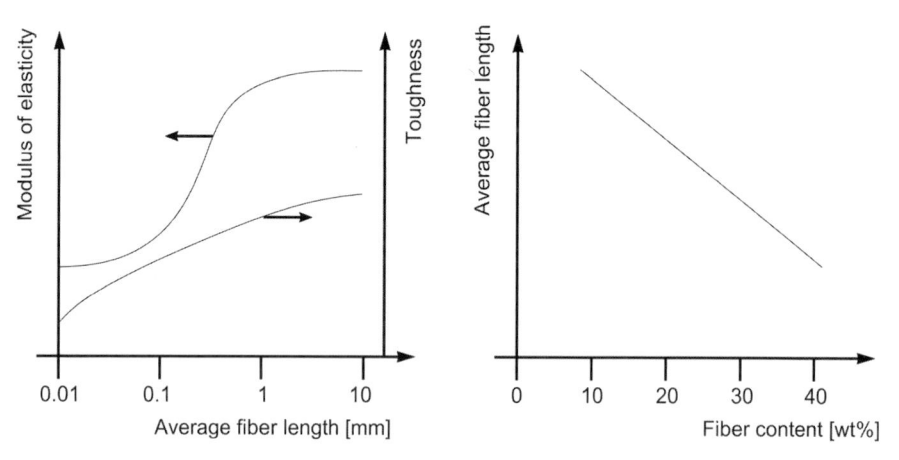

Figure 4.3 The effect of average fiber length on the impact strength and modulus of elasticity of injection-molded thermoplastic matrix composites and the effect of fiber content on average fiber length in injection-molded composites.

Table 4.2 The effect of different reinforcements and fillers on the basic properties of thermoplastic matrix composites (↑improve, ↓deteriorate).

Additive	Maximal content (wt%)	Modulus of elasticity	Elongation at break	Toughness	Warpage proneness	Fire resistance
Glass fiber	60	↑↑↑	↓↓	↓	↓	↑
Aramid fibers	20	↑	↓	↓	↓	↑
Carbon fibers	60	↑↑↑	↓↓↓	↓↓	↓	↑
Basalt fibers	60	↑↑↑	↓↓	↓	↓	↑
Natural fibers	60	↑	↓	↓	↓↓	↓↓
Antistatic agents	5	↓	↓↓	↓↓	—	—
Elastomers	15	↓	↑↑	↑↑↑	—	↓
Mineral fillers	40	↑	↓	↓	↑↑↑	↑
Wood fiber	80	↑	↓↓	↑	↓	↓

the modulus of elasticity and the tensile strength. Table 4.2 shows the most common reinforcements and fillers used in injection-molded composites and their effect on product properties.

4.2.2
Extrusion

Extruders play a dual role in composite technologies: first, they are used for making fiber-reinforced compounds that are further processed by injection molding or compression molding; second, typical extruded products (e.g., plates, pipes, and bars) are also manufactured with fiber reinforcement. The simplest method of adding fibers to the resin is dosing chopped fibers to the matrix granulate at the feeding hopper, but this leads to increased fiber fragmentation. Adding chopped fibers to the already melted resin stream in an ulterior segment of the extruder (side feeding) prevents major fragmentation, this solution is popular in extrusion–compression molding. Extruders are also capable of impregnating rovings by employing a method that is somewhat similar to pultrusion, this is how the pellets incorporating long strands (ready-to-use pellets, P-LFT) are made for injection molding and compression molding.

Reinforced extruded products involve glass and carbon fiber-reinforced semifinished products, which are often produced with large (up to 500 mm) diameters for making machine elements and parts by machining. The other major field is the production of natural fiber (mainly wood)-reinforced profiles and slats used in the building industry. The fiber content of these materials can reach 80 wt% [13, 14].

4.2.3
Compression Molding

Compression molding techniques are classified by whether there is notable resin flow during filling the mold or not. If there is a notable resin flow, we speak about

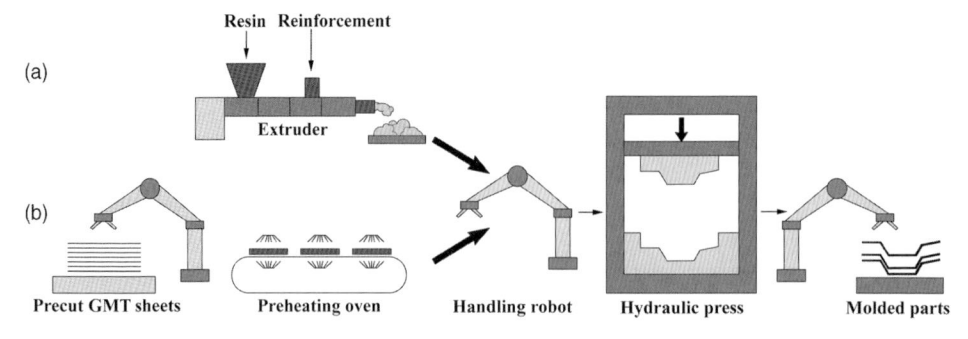

Figure 4.4 Scheme of automated extruder compression molding (a) and GMT compression molding (b) line.

glass mat-reinforced thermoplastics (GMT) or extrusion–compression molding. If there is little resin flow, the method is named hot stamping. For large-scale manufacturing of GMT compression molding, the cut to size raw material is conveyed through an oven (usually infrared or hot air) on a conveyor belt and then a robot places the completely melted material in the mold, which is rapidly closed. The material flows to fill the tempered mold, gets solidified, and then is ejected (Figure 4.4b). Extrusion–compression molding is used to produce reinforced parts principally for the automotive industry as lightweight parts for semistructural applications such as front ends, bumper beams, and underbody shields. Extrusion–compression molding uses an extruder to solidify the raw material for hot bulk molding compound. The low shear screw of the extruder melts the matrix material of the P-LFT with heat action and shear force of the screw without degrading the reinforcing fiber, therefore maintaining fiber length. The melted P-LFT is discharged from the extruder in the form of a slug. The melted charge is quickly transferred to a press where it is immediately molded, while still hot (Figure 4.4a). The part is removed after sufficient cooling. This process is generally referred to as LFT processing (this denomination is also used for describing injection-molded composites). In D-LFT processing, the fiber strands are introduced into the extruder in the form of rovings or chopped strands. The D-LFT process means that a wide variety of polymers, additives, and reinforcing fibers can be selected, enabling development of tailored material formulations for each application. It is also possible to use pellets incorporating long strands (P-LFT).

In case of continuous and aligned reinforcement, the material flow is largely restricted within the mold. This problem can be overcome by applying hot stamping. It means the employment of already consolidated prepregs that cover the entire mold surface and the final component thickness is the direct function of the thickness of the raw material at any given location. In this case, there is no considerable material flow during forming, only the shape of the prepreg is modified, similar to metal sheet stamping. The most common matrix material in compression molding is polypropylene (PP), although polybutylene terephthalate (PBT), polyamide (PA), and other resins are also available. The most common reinforcement is glass fiber. The molds are made of steel and are tempered, that is, can be cooled and also heated. The typical pressure within the mold during consolidation is 10–20 MPa, although in some cases it can be higher. Cycle times are usually between 20 and 60 s. The material flow is completed usually in 3 s, the remaining time is needed for the cooling. These methods enable the manufacturing of parts with large dimensions and up to 100 mm residual fiber length [15–19].

4.2.4
Thermoplastic Prepreg Lay-up

The most practical and prevalent method of laminating composites from thermoplastic prepreg is cutting and laying up by hand. At the same time, several automatic methods have been developed, based on technologies applied for thermosetting matrix composites. The instruments that are used for automatic lay up of thermoplastic matrix prepreg tapes are conceptually similar to the instruments used for thermosetting matrix prepreg tapes. The surface of the tape is melted and joined by means of pressure (Figure 4.5). Heat sources can be infrared beam, gas torch, laser beam, and the like.

In this case, if both surfaces (the previously laid layer and the tape) are completely melted, the component theoretically could be used without any postprocessing. However, residual stresses due to the inhomogeneous and localized heating make postprocessing necessary. Hence, usually full consolidation is not achieved during lay-up – permitting of the removal of the component from the mold without warpage – and a separate processing (autoclave curing) step is applied for full consolidation.

Thermoplastic prepreg lay-up is a costly technique, and as such mainly high-cost carbon fiber-reinforced composites are processed by this technique (glass fiber-reinforced composites can be processed as well, but it is not common.) The molds are

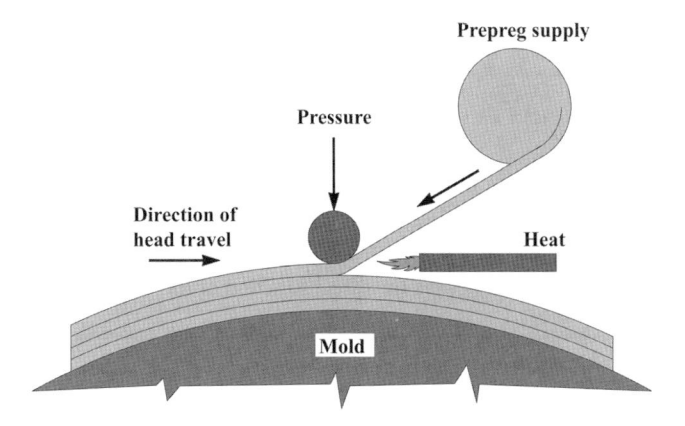

Figure 4.5 Scheme of automatic thermoplastic tape lay-up.

basically identical to those used in the corresponding thermosetting process, although heat tolerance usually must be higher in case of thermoplastics. Mold heating can be used for reducing residual stresses and increasing lay-up rate. Thermoplastic prepreg stacks can be consolidated by compression molding or diaphragm forming, which otherwise can be regarded as separate processing techniques. However, the autoclave consolidation is the most technically proven method, originating from the use of thermosetting prepregs in the aerospace industry. Vacuum bagging is theoretically feasible, but high processing temperatures make it nearly impossible [1, 19, 20].

4.2.5
Thermoplastic Tape Winding

The general scheme of thermoplastic tape winding instruments is shown in Figure 4.6. The prepreg tape is unwound from a spool and travels through a preheater. At the contact point, the already wound layer is heated and the incoming tape also gets a thermal boost. Heat sources can be infrared beam, gas torch, laser beam, and the like. The layers are pressed together by back tension; however, better compaction can be achieved by employing pressure rollers.

The raw material of tape winding must be in continuous tape form. Fully melt-impregnated prepregs are the most practical because the matrix is melted only for a very short time during lay-up. The consolidation is strongly influenced by the winding speed. Winding speed can be as low as 1 mm/s in case of small experimental devices or as high as 1 m/s in case of more sophisticated facilities. Better consolidation can be achieved with lower winding speeds. In case of helix winding, it is almost impossible to avoid the formation of voids at the crossing of prepregs, because (unlike thermosetting matrix composites) the matrix does not become entirely liquid. The void content can be decreased successfully by reconsolidation in an autoclave

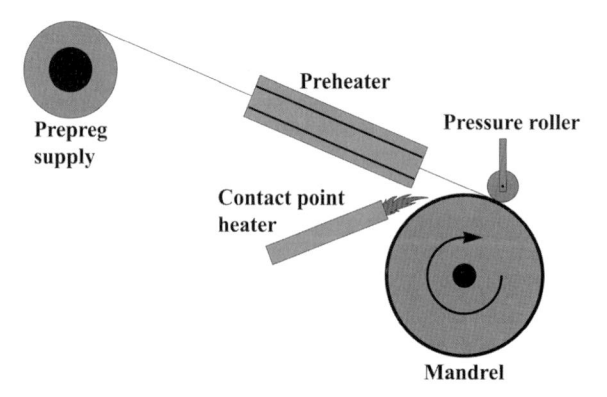

Figure 4.6 Scheme of thermoplastic tape winding.

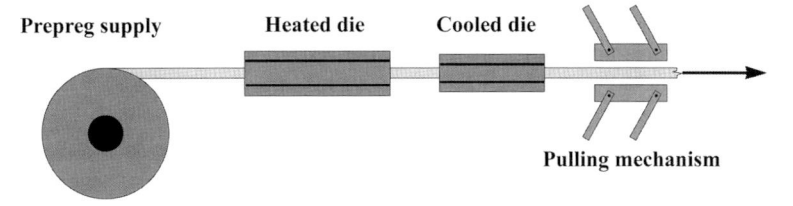

Figure 4.7 Scheme of thermoplastic pultrusion.

with vacuum bagging, although this process is definitely disadvantageous from economic point of view [1, 19].

4.2.6
Thermoplastic Pultrusion

Figure 4.7 shows the basic arrangement of thermoplastic pultrusion facilities. The raw material (prepreg) is unwound from a creel and pulled into a preheater that heats the prepreg to a temperature near to or over the melting point of the matrix. The material enters a heated die, where the final cross section of the product is gradually formed. The consolidation is taking place in a cooled die, which is followed by a pulling mechanism.

The process combines conventional glass fiber roving, aramid, or carbon fiber tows with thermoplastics, most commonly polyethylene terephthalate (PET) and polyamide. Other plastics that can be used include polyphenylene sulfide (PPS), styrene maleic anhydride (SMA), high-density polyethylene (HDPE), and polypropylene. Unlike thermosetting pultruded profiles, thermoplastic profiles can be postformed and reshaped. Higher continuous use temperatures are possible with some thermoplastic matrices, and line speeds are faster with raw materials usually costing less compared to thermosetting pultrusion. Outside coatings of most thermoplastics can be applied in-line while pultruding by using an extruder. Pultrusion process allows using recycled thermoplastics as well as virgin materials. Potential applications include round and flat profiles for tension cables and straps for transporting heavy construction materials. Production of towpreg, prepreg rods and ribbons, reinforcement bars for concrete, tool handles, and high-fiber-weight long-fiber pellets (P-LFT) is also possible [1, 19, 20].

4.2.7
Diaphragm Forming

The only technology that has been developed solely for thermoplastic composite manufacturing is diaphragm forming, which may be seen as a refined form of vacuum bag consolidation in an autoclave. Diaphragm forming permits the pro-

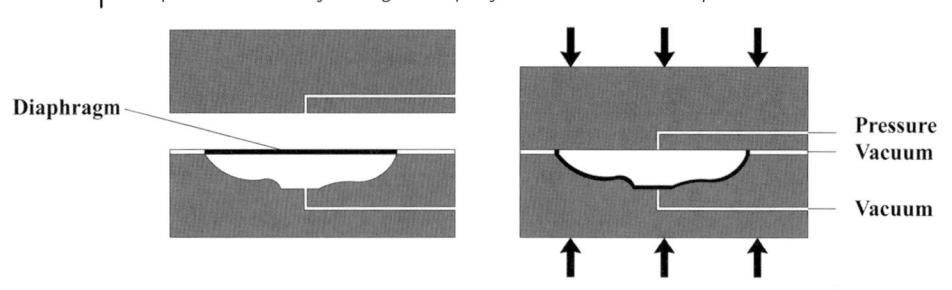

Figure 4.8 Scheme of diaphragm forming process.

duction of deeply drawn and geometrically complex components. The flat composite material to be formed is placed between two flexible diaphragms. The diaphragms (but not the composite) are clamped around in a clamping frame and the air is evacuated between the diaphragms. There are two basic versions of diaphragm forming: in the first case, the diaphragm–composite stack is placed in an oven and heated over the softening temperature of the matrix. The stack is rapidly placed on a female mold and vacuum is formed between the lower diaphragm and the mold, while pressure is applied over the stack. The mold is unheated, thus the composite solidifies as it comes in contact with the mold (Figure 4.8). In the other method, the stack is placed on the mold and the entire assembly is placed in an autoclave. After the evacuation of the air between the diaphragms, the internal atmosphere of the autoclave is heated over the softening temperature of the matrix. The air is evacuated under the lower diaphragm and this – combined with the pressure in the autoclave – shapes the material conforming the mold. When forming is completed, the heated gas surrounding the mold is replaced with cool air to solidify the product. Diaphragm forming is usually used for processing high-performance materials, for example, carbon fiber-reinforced polyether ether ketone (PEEK). The diaphragm can be made from superplastic aluminum, sheet rubber, or polymer film (mainly polyimide (PI)). Rubber diaphragms can be reused a few times, while the other types can be used only once because they suffer permanent deformation.

4.2.8
Classification of Thermoplastic Composite Manufacturing Techniques

The above-mentioned technologies can be categorized according to certain viewpoints. While in some cases (e.g., if we want to produce long slats with constant cross section) only one technology is available for preparing the required part, in most cases the manufacturer can choose from several different techniques. The deciding factors include capital investment, required series length, dimensional accuracy, part size, mechanical properties, raw material cost, and so on. Tables 4.3 and 4.4 summarize the techniques according to technological and economical features.

Table 4.3 Classification of thermoplastic composite manufacturing techniques according to their typical technological features.

	Product geometry	Production process	Fiber arrangement	Product size	Tool	Application
Injection molding	Complex 3D	Sequential	Randomly oriented chopped	Limited by machine size/clamping force	Two-sided metal (steel)	Wide range of products made in large serials
Extrusion	1D, 2D	Continuous	Randomly oriented chopped	Potentially infinite	Metal	Building industry
Compression molding	2D, 3D	Sequential	Randomly oriented chopped	Limited by press size	Two-sided metal	Automotive industry
Thermoplastic prepreg lay-up	2D, 3D	Sequential	Oriented anisotropic continuous	Limited by autoclave size	One-sided metal	Aircraft components
Thermoplastic tape winding	3D	Sequential	Oriented anisotropic continuous	Limited by mandrel length	Metal	Pressure vessels
Thermoplastic pultrusion	1D, 2D	Continuous	Unidirectional continuous	Potentially infinite	Metal	Elements for axial strain
Diaphragm forming	2D, 3D	Sequential	Planar, oriented and continuous	Limited by oven or autoclave size	One-sided metal, wood, plaster	Aircraft components

Table 4.4 Classification of thermoplastic composite manufacturing techniques according to their typical economical features.

	Raw material cost	Equipment cost	Mold cost	Advantages	Disadvantages
Injection molding	Low	High	High	Complex geometry, size accuracy, short cycle time	Costly tool
Extrusion	Low	High	High	Versatility, potentially infinite product length	Lack of fiber orientation
Compression molding	Intermediate	High	High	Complex geometry, large size, large residual fiber length	Poor appearance
Thermoplastic prepreg lay-up	High	High	Intermediate	Designable anisotropy, large product size	Costly machinery
Thermoplastic tape winding	High	High	High	Postprocessing unnecessary	Costly raw material
Thermoplastic pultrusion	High	High	High	Profiles can be postformed and reshaped	Costly machinery
Diaphragm forming	High	Low to intermediate	Low	Complex geometry, low mold cost	Costly raw material

4.3
Thermosetting Polymer Composites

The matrix materials of thermosetting composites are available as monomers or oligomers and they are polymerized/cross-linked during processing, that is, in the course of the formation of the product. This means that a chemical reaction occurs during processing. These techniques can be divided into two groups: in direct technologies, the components of the resin are mixed directly prior to or during processing. In this case, the components can be stored at room temperature for very long time. Indirect techniques employ mixed – usually oligomer – resins (e.g., prepregs, sheet molding compound, and bulk molding compound), where cross-linking is initiated by heat. These raw materials must be stored at low temperatures in order to avoid premature cross-linking and in most cases they have a limited shelf life. The duration of cross-linking can be as short as a few seconds (e.g., in reactive injection molding, sheet molding compound) or several hours (e.g., resin transfer molding (RTM) or hand lamination of very large components). Some thermosetting polymers can cross-link sufficiently at room temperature, while others require heating or autoclave treatment in order to reach full curing.

When we want to prepare a certain composite component with thermosetting matrix, there are several techniques to choose from. This is truer in case of thermosetting composites than in thermoplastics. Table 4.5 shows the typical series lengths of thermosetting composite manufacturing techniques to be introduced in this chapter.

Table 4.5 Feasible series lengths of thermosetting composite manufacturing techniques.

	Prototype	Small series	Medium series		Large series		
Hand lamination							
Spray-up							
Centrifugal casting							
Prepreg lay-up							
RTM and VARTM							
BMC and SMC							
RIM and SRIM							
Thermosetting injection molding							
Filament winding							
Pultrusion							
Number of pieces	1	10	100	1000	10 000	100 000	1 000 000

RTM: resin transfer molding, VARTM: vacuum-assisted resin transfer molding, BMC: bulk molding compound, SMC: sheet molding compound, RIM: reaction injection molding, SRIM: structural reaction injection molding.

4.3.1
Hand Lamination

Hand lamination is the simplest and most versatile thermosetting matrix composite preparing technique. Despite the continuous development of composite technologies, hand lamination is still popular because it requires small capital investment. Furthermore, it is highly economical for short production series and prototypes. Hand lamination requires a one-sided mold. At first this mold is treated with a mold release agent and then a gelcoat layer is sprayed on the surface. This layer determines the appearance, environmental resistance, and surface quality of the final product. The resin and the layers of reinforcing materials (usually mats and fabrics) are applied, impregnated, and compacted by brushes and handheld roller. The further layers are laid by repeating these steps. The impregnation and rolling must be carried out with great care because – due to the lack of postprocessing – its quality entirely determines the fiber content, void fraction, and thickness of the product. Hand lamination is usually applied with glass fibers and unsaturated polyester, although in some cases aramid, carbon fibers, epoxies, and vinylesters are also used. Molds do not have to endure high temperatures or pressure, thus they can be made from almost any material, for example, wood, but most often they are made from composites. Cross-linking usually takes place at room temperature, if necessary it can be accelerated with heating [1, 18, 21].

4.3.2
Spray-Up

In spray-up, a dedicated spray gun is used to disperse a mixture of chopped fibers and matrix onto the mold. Continuous rovings are fed into the gun, where they are chopped to a predetermined length. Resin components are mixed in the pistol by means of a static mixer or they are sprayed separately and mixing takes place on the way to the mold. In-gun mixing provides more thorough mixing, but it necessitates the cleaning of the gun. The gun can be handheld or placed on a robot, in the previous case the laminate quality greatly depends on the worker's skill. Laminate compaction is achieved by handheld rollers. Raw materials type and molds are similar to hand lamination. The rovings are chopped to lengths between 10 and 40 mm. Using thermoformed thermoplastic polymer films instead of gelcoat is more common in case of spray-up than for hand lamination. The removal of the product from the mold is facilitated by the inward shrinkage of the resin. Cross-linking usually takes place at room temperature [1, 19, 21].

4.3.3
Centrifugal Casting

Centrifugal casting is fundamentally similar to hand lamination and spray-up, the main difference is that the mold is rotating and the composite is built up on the inner wall of the mold. There are two options for achieving this goal: in the first case, the reinforcement is placed on the inner wall of the mold and resin is added during rotation; in the second case, an axially moving spray gun deposits the mixture of chopped fibers and the resin is in the mold. Cross-linking is initiated by heating the mold or injecting hot air into the mold. Large diameter tanks and pipes can be made by this method. External surfaces may be improved with gelcoat and resin-rich inner surface can be produced easily [19, 21].

4.3.4
Prepreg Lay-Up

Prepreg lay-up in some respects is similar to hand lamination. Using prepregs instead of impregnating the reinforcement during lay-up enables the production of higher performance components and reduces health hazards. Today most high-performance composites – particularly in the aerospace industry – are made with prepreg lay-up. The process can be enhanced by various degrees of automation in prepreg cutting and lay-up.

At first the prepreg is removed from the cold storage and allowed to thaw. Then, the prepreg is cut. For larger series, this can be automated, using ultrasonically vibrating knives, waterjet, or laser. Prepregs are mostly used for relatively short series, and are hence laid up by hand. After the removal of the backing paper, the plies are carefully placed on the one-sided mold treated with mold release agent. Automatic lay-up machines are large and costly instruments, which can handle only a limited mold complexity. However, there are applications (e.g., aircraft wing skins) where large components with relatively simple geometry must be manufactured precisely, and here the automatic lay-up machines can be used practically. Similar to hand lamination, molds for prepreg lay-up can be simple, although the high-tech components-made prepreg lay-up requires strict dimensional accuracy, making the molds more expensive. Laid-up prepregs must be consolidated using a vacuum bag, usually in an autoclave. A typical autoclave cure cycle is shown in Figure 4.9.

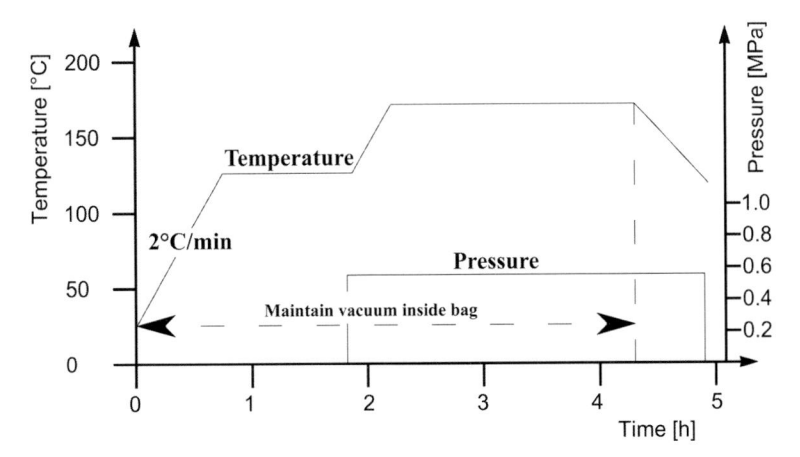

Figure 4.9 Typical autoclave consolidation cycle.

The vacuum bag and the internal pressure and temperature of the autoclave cooperate in consolidating the composite during cross-linking. The most common prepreg material combination is carbon fiber–epoxy, although glass, aramid and unsaturated polyester, and phenolic and bismaleimide resins are also available [1, 16, 18].

4.3.5
Resin Transfer Molding and Vacuum-Assisted Resin Transfer Molding

Resin transfer molding is the most popular composite manufacturing technique capable of producing structurally loadable components. Low mold costs and capital investments are further advantages of this method. Reinforcements (usually fabrics) are laid up as a dry stack of materials. These fabrics are sometimes prepressed to the mold shape and held together by a binder. These preforms are then laid into the lower half of the mold. It is possible to accommodate inserts, fasteners, and foam sandwich cores into the mold. Gelcoats may also be used. The mold is closed and resin is injected into the cavity. The mold can be kept closed with hydraulic or pneumatic presses or simply clamped together. Vacuum can also be applied at the vents of the mold cavity to assist resin in being drawn into the reinforcement. In this case, the seal between the mold halves must be very tight. This is known as vacuum-assisted resin transfer molding (VARTM) (see Figure 4.10).

Once all the reinforcement is impregnated, the resin inlets are closed and the laminate is allowed to cross-link. Both injection and cure can take place at either ambient or elevated temperatures. Resin injection is achieved with a pressure pot containing the liquid resin. This pot is connected to a pressurized air system. The pressures applied for RTM are usually between 0.1 and 1 MPa. Most RTM applica-

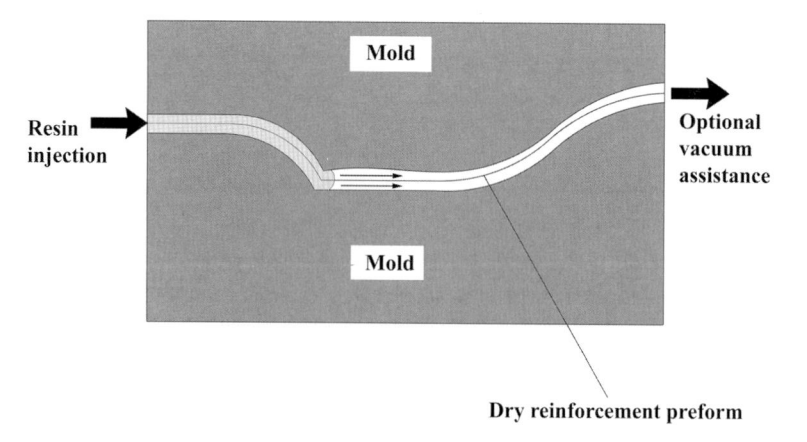

Figure 4.10 Scheme of RTM/VARTM process.

tions employ glass fibers and unsaturated polyester; more demanding products involve carbon fiber and epoxies, mainly in the aerospace industry. After filling the mold, the vents are closed and backpressure is applied until cross-linking in order to avoid bleeding out of the resin. High-temperature epoxy resins may require post-curing at elevated temperatures after removal from the mold. Vacuum infusion is a version of VARTM, where there is no pressure, the only driving force used to impregnate the reinforcement is vacuum. This process is feasible not only with closed mold (where the vacuum helps to keep together the mold halves) but also with a one-sided mold, covered by a vacuum bag [1, 15, 19, 20].

4.3.6
Bulk Molding Compound and Sheet Molding Compound

Bulk molding compound (BMC) and sheet molding compound (SMC)) are jointly named compression molding. This method is particularly popular for long production series. Vehicle body panels are the most prevalent products. BMC and SMC utilize two-sided molds. For series production, the molds are metal (mainly steel) dies. The stack of SMC sheets or bulk material (BMC) is placed on the lower mold half, and the mold is immediately closed with great force (usually hydraulic press is applied). The mold is heated, the heat at first decreases the viscosity of the resin making it possible for the resin to fill the mold. Subsequently, the cross-linking is initiated by the heat, and thus the component solidifies. Mold heating is usually carried out by hot oil or steam circulating with temperatures around 120 °C. The applied pressure usually varies between 3 and 20 MPa.

SMC semifinished products are manufactured the following way: a measured amount of specified resin is dispersed on a carrier polymer film. This carrier film passes underneath a chopper that cuts the glass rovings onto the surface. Once these have drifted through the depth of resin, another sheet is added on top that sandwiches the glass. The sheets are compacted and then enter onto a take-up roll, which is used to store the product while it matures. Prior to processing, the carrier film is removed and the material is cut into charges. Depending on what shape is required determines the shape of the charge and steel die to which it is then added [1, 20, 21].

4.3.7
Reaction Injection Molding and Structural Reaction Injection Molding

Reaction injection molding (RIM) and structural reaction injection molding (SRIM) are similar to resin transfer molding in that liquid resin (the mixture of two components) is injected into a closed mold where it cures spontaneously at ambient temperature. This process is an excellent choice for larger plastic parts produced in short run or low volume production quantities. The main difference is that RIM employs resins that harden very quickly (mostly within a few seconds after mixing), this is possible because RIM is a fast process and the liquid resin fills the mold very quickly. The most typical raw material of RIM is polyurethane. RIM is capable of producing relatively large parts (e.g., automotive bumpers and fenders) because – in spite of injection molding – the force required for closing the mold is not too large. It is also possible to produce foamed parts. Considerable design freedom is possible, including wall sections with large thickness differences that are not suitable for conventional injection molding, due to the uniform shrink characteristics. Foamed polyurethanes are natural thermal and acoustic insulators. Excellent flowability allows the encapsulation of a variety of inserts. Reaction injection molding is used in many industries for many types of parts. While bumpers for vehicles are produced in this process, most applications are for large, complex parts produced in quantities less than 5000 units. Examples include panels for electrical equipment, enclosures for medical devices, and housings for computer and telecommunications equipment. If fiber reinforcement is placed in the mold before injection (similar to RTM), we speak about SRIM. The equipment and mold for SRIM are much more expensive than that for RTM, but the SRIM enables the production of much longer series. Due to the high reactivity of the resins used for SRIM, the components cannot be as large as in case of RTM. The resins used for SRIM must have very low viscosity in order to impregnate the reinforcement with acceptably low void content and high fiber fractions [1, 19].

4.3.8
Thermosetting Injection Molding

Thermosetting injection molding is very similar to its thermoplastic counterpart in its machinery and tooling: the reciprocating screw, the steel mold, high pressures (up to 100 MPa), and high clamping forces are similar in the two processes. The thermosetting injection molding involves pelletized or powdered resin or BMC, which is liquefied in the screw due to friction and heat and then after injection it is cross-linked in the heated mold. It is possible to employ chopped fiber reinforcement, in this case significant fragmentation occurs in the screw, similar to thermoplastic injection molding. Unsaturated polyesters and phenolics are the most common raw materials. The cross-linking takes place within a few minutes after mold filling. Thermosetting injection molding is ideal for producing long series of parts made of thermosetting polymers or their composites [1, 20, 21].

4.3.9
Filament Winding

In this processing technique, continuous strands of fiber are used to achieve the maximum mechanical strength. Rovings of the reinforcing fiber are fed through a resin bath and wound on to a rotating mandrel. The resin must have low viscosity to

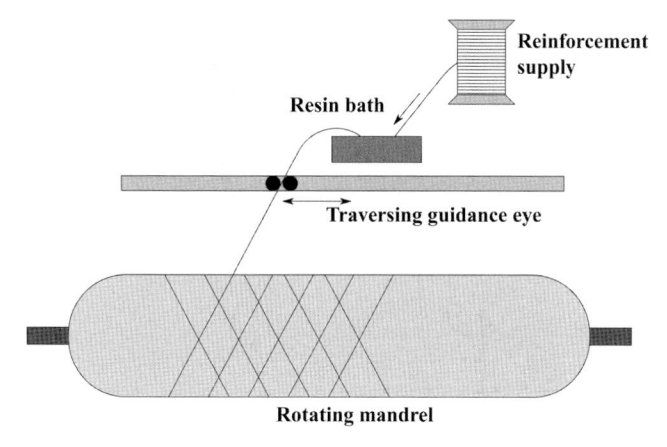

Figure 4.11 Scheme of filament winding process.

impregnate the fiber bundle easily. Preimpregnated rovings can also be used in the process. Some tension must be applied to the fiber to minimize voids and to have compact winding. When an external force is applied to a cured product, it is more or less equally shared by each strand wound under constant tension. This naturally imparts a superior strength to filament-wound products. Fibers can be arranged to traverse the mandrel at different angles with different winding frequencies in a programmed manner accomplished by automatic control or computer-aided control. Cylindrical objects can be easily manufactured by this process, as shown in Figure 4.11. However, items that have some degree of symmetry about a central axis such as conical shapes, isotensoids, cups, multisided boxlike structures, and even spheres can be produced by this technique. The shrinkage of the resin produces large compressive stresses between the composite and the core. In case of cores that remain inside the product (e.g., tanks with inside metal liners), this is an advantage; but in case of removable cores, the demolding must be facilitated by conical or disassemblable cores [18–20].

4.3.10
Pultrusion

Pultrusion is a process to produce composite materials in the form of continuous, constant cross-sectional profiles. Rovings are drawn through an impregnation tank and then through a die of desired geometry, as shown in Figure 4.12. The shaped rovings then pass through a tunnel oven for curing, and then pultruded composite is cut to proper lengths. The resin must have low viscosity for adequate impregnation of the rovings. Typical line speeds range from 15 to 40 mm/min. Among the thermosetting composites, this technique provides the highest attainable fiber

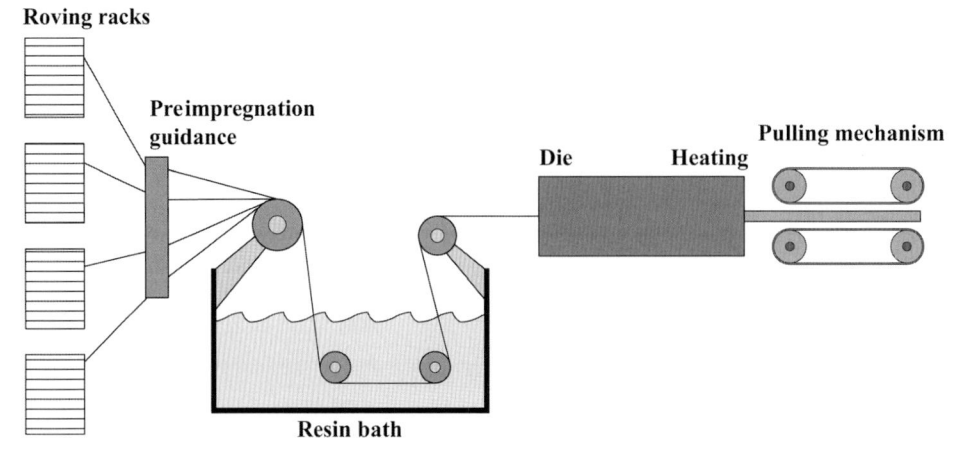

Figure 4.12 Scheme of pultrusion process.

percentage, up to 80 vol%. Capital costs are high, but labor costs can be minimal for long production runs. Very thin and quite thick profiles of varying geometries can be manufactured by this technique. More than 90% of all pultruded products are fiberglass-reinforced unsaturated polyesters. The products find applications in industry, transport, building materials, including roofing, awnings, canopies, domes, and sheeting, and also in electrical and sporting goods fields. Its ability to produce constant cross section of profiles with little waste materials makes pultrusion one of the most cost-effective processes. For equivalent strength, pultruded finished products can be 50% lighter than aluminum and 80% lighter than steel [9, 18, 19].

4.3.11
Classification of Thermosetting Composite Manufacturing Techniques

Thermosetting composite manufacturing techniques have some notable disadvantages compared to thermoplastics: higher health and environmental hazards and sensitivity to technological conditions (time, temperature, mixing ratio etc.). At the same time, most technologies capable of making components with very large dimensions, excellent mechanical properties, or heat resistance belong to thermosetting matrix composites. In many fields, composites gradually replace metals, including automotive and aerospace industry. Parts with large dimensions and complex geometry offer a further advantage, namely, the decrease of number of parts: it is possible to produce composite structures from one piece, while the construction of the same component from metal requires the welding, bolting, or riveting of several smaller, separate elements. Tables 4.6 and 4.7 summarize the techniques according to technological and economical features.

Table 4.6 Classification of thermosetting composite manufacturing techniques according to their typical technological features.

	Product geometry	Production process	Fiber arrangement	Product size	Tool	Application
Hand lamination	2D, 3D	Sequential	Planar, oriented and continuous or random chopped	Any	One-sided any material	Several kinds of other structures, for example, boats
Spray-up	2D, 3D	Sequential	Random chopped	Any	One-sided any material	Several kinds of other structures, for example, boats
Centrifugal casting	Cylindrical	Sequential	Random chopped or fabric	Large	Cylindrical metal	Tanks and pipes
Prepreg lay-up	2D, 3D	Sequential	Oriented anisotropic continuous	Limited by autoclave size	One-sided composite or metal	Aircraft components, sporting goods
RTM and VARTM	2D, 3D	Sequential	Planar, oriented and continuous or random chopped	Any	Two-sided composite or metal	Train, aircraft, and automotive components
BMC and SMC	2D, 3D	Sequential	Random chopped	Limited by press size	Two-sided metal	Automotive components
RIM and SRIM	2D, 3D	Sequential	Random chopped or fabric	Intermediate	Two-sided metal	Automotive components
Thermosetting injection molding	Complex 3D	Sequential	Random chopped	Limited by machine size/clamping force	Two-sided metal (steel)	Wide range of products made in large serials
Filament winding	3D	Sequential	Oriented anisotropic continuous	Limited by mandrel length	Metal	Pressure vessels, pipes
Pultrusion	1D, 2D	Continuous	Unidirectional continuous	Potentially infinite	Metal	Elements for axial strain

Table 4.7 Classification of thermosetting composite manufacturing techniques according to their typical economical features.

	Raw material cost	Equipment cost	Mold cost	Advantages	Disadvantages
Hand lamination	Low	Low	Low	Complex geometry, cheap tool and machinery	Long cycle time, not uniform quality, health hazard
Spray-up	Low	Low	Low	Complex geometry, cheap tool and machinery	Long cycle time, not uniform quality, health hazard
Centrifugal casting	Low	Intermediate	Low	Large dimensions possible	Weak mechanical properties
Prepreg lay-up	High	High	High	Precise, high fiber content, oriented	Long cycle time, high capital costs
RTM and VARTM	Low	Low	Low	Cheap tool, inserts, fasteners, and sandwich cores applicable	Long cycle time, low fiber content
BMC and SMC	Intermediate	High	High	Short cycle time, precise, few waste	Lack of fiber orientation
RIM and SRIM	Low	High	High	Short cycle time, complex geometry	Lack of fiber orientation
Thermosetting injection molding	Low	High	High	Complex geometry, size accuracy, short cycle time	Lack of fiber orientation, costly tool
Filament winding	Intermediate	High	High	High strength, high fiber volume fraction	Long cycle time
Pultrusion	Intermediate	High	High	High tensile strength	Costly machinery

4.4
Future Trends

A major breakthrough can be observed in the aircraft industry (both in the civilian and military sectors), where main structural parts are made from composite materials, although only less critical auxiliary parts were manufactured from composite materials in the last few decades. The most known examples are the Airbus 350 XWB, the Airbus 380, and the Boeing 787. Among others, the used techniques involve prepreg lay-up, resin transfer molding, and pultrusion. In the automotive industry, composites also approach new areas. Several components that were earlier made from metals are replaced by polymers and mostly by composites. A prominent example of this is the oil sump of truck engines that is now made from injection-molded glass fiber-reinforced PA. This trend is in line with the spreading of thermoplastic matrix composites. In the forthcoming years, the further growth of the usage of natural fibers can be expected, mainly in the automotive industry.

New reinforcements and matrix materials are constantly being developed, thanks to the new applications like the structurally load bearing cryogenic fuel tanks of space rockets, which must resist the diffusion of gases. A great success of self-reinforced composites can be expected, because they offer the advantage of reusability apart from also being lightweight and durable. Self-healing composites are intensively researched, their practical utilization can be expected mainly in areas where safety is critical (e.g., aircrafts). Another intensively researched area is the production of nano/macroscale reinforced hybrid composites.

Acknowledgment

This work is connected to the scientific program of the "Development of quality-oriented and harmonized R + D + I strategy and functional model at BME" project. This project is supported by the New Hungary Development Plan (TÁMOP-4.2.1/B-09/1/KMR-2010-0002).

References

1 Strong, A.B. (2000) *Plastics Materials and Processing*, Prentice Hall, Upper Saddle River.

2 Thomason, J.L. and Vlug, M.A. (1996) Influence of fiber length and concentration on the properties of glass fibre-reinforced polypropylene: 1. Tensile and flexural modulus. *Composites Part A*, **27**, 477–484.

3 Malloy, R.A. (1994) *Plastic Part Design for Injection Molding*, Carl Hanser Verlag, München.

4 Lafranche, E., Krawczak, P., Ciolczyk, J.P., and Maugey, J. (2005) Injection moulding of long glass fiber reinforced polyamide 66: processing conditions/microstructure/flexural properties relationship. *Advances in Polymer Technology*, **24**, 114–131.

5 Fourné, F. (1999) *Synthetic Fibers*, Carl Hanser Verlag, München.
6 McNally, D. (1977) Short fiber orientation and its effects on the properties of thermoplastic composite materials. *Polymer-Plastics Technology and Engineering*, **8**, 101–154.
7 Bright, P.F., Crowson, R.J., and Folkes, M.J. (1978) A study of the effect of injection speed on fibre orientation in simple mouldings of short glass fibre-reinforced polypropylene. *Journal of Materials Science*, **13**, 2497–2506.
8 Belofsky, H. (1995) *Plastics Product Design and Process Engineering*, Carl Hanser Verlag, München.
9 Chung, D.D.L. (2001) *Applied Materials Science*, CRC Press, Boca Raton.
10 Thomason, J.L. (2002) Influence of fiber length and concentration on the properties of glass fibre-reinforced polypropylene: 5. *Composites Part A*, **33**, 1641–1652.
11 Gu, H. (2006) Tensile and bending behaviours of laminates with various fabric orientations. *Materials and Design*, **27**, 1086–1089.
12 Mazumdar, S.K. (2002) *Composites Manufacturing: Materials, Product, and Process Engineering*, CRC Press, Boca Raton.
13 Gurvich, M.R. and Pipes, R.B. (1995) Strength size effect of laminated composite. *Composites Science and Technology*, **55**, 93–105.
14 Li, B. and He, J. (2004) Investigation of mechanical property, flame retardancy and thermal degradation of LLDPE-wood fibre composites. *Polymer Degradation and Stability*, **83**, 241–246.
15 Potter, K. (1997) *Resin Transfer Moulding*, Chapman & Hall, London.
16 Neitzel, M. and Mitschang, P. (2004) *Handbuch Verbundwerkstoffe: Werkstoffe, Verarbeitung, Anwendung*, Hanser, München.
17 Åström, B.T. (1997) *Manufacturing of Polymer Composites*, Chapman & Hall, London.
18 Cheremisinoff, N.P. (1998) *Advanced Polymer Processing Operations*, Noyes Publications, Westwood.
19 Chung, D.D.L. (1994) *Carbon Fiber Composites*, Butterworth-Heinemann, Boston.
20 Friedrich, K., Fakirov, S., and Zhang, Z. (2005) *Polymer Composites: From Nano- to-Macro Scale*, Springer, New York.
21 Peters, S.T. (1998) *Handbook of Composites*, Chapman & Hall, London.

Part Two
Macrosystems: Fiber-Reinforced Polymer Composites

5
Carbon Fiber-Reinforced Polymer Composites: Preparation, Properties, and Applications
Soo-Jin Park and Min-Kang Seo

5.1
Introduction

Carbon fibers have become one of the most important reinforcing materials in recent years, characterized by extremely high strength and modulus along with high-temperature resistance. A great deal of scientific effort has been directed toward improving and analyzing their performance in particular composite systems [1–5]. Despite these advantages and efforts, it is also well known that the interfacial adhesion between carbon fibers and matrix in a composite system is too poor to ensure good mechanical performance [6, 7].

In a composite system, the degree of adhesion at the interface between fiber and matrix plays an important role in improving the resulting mechanical behavior due to the existence of the physical or secondary interaction, that is, van der Waals attraction and hydrogen force at the interface. Furthermore, the degree of adhesion at interfaces is vital to the strength and durability of the composites formed from these materials [8, 9].

Polymeric composites such as carbon fibers/epoxy resins and carbon fibers/PEEK systems are now being used in numerous aerospace, marine, and recreational applications. The particular composites chosen depend on the application. Epoxy resins have proved to be the most versatile in this respect as the resin itself can be cross-linked with a number of different amines, anhydrides, and acids. They can also react with many other polymer substances [10–12].

The properties of these composites, however, are governed not only by the properties of individual components but also by the interface between the fibers and the matrix resins. Successful reinforcement of composite materials is achieved by only obtaining sufficient stress transfer between the fibers and the matrix resins. This can be realized by physical and chemical adhesion between the two. Epoxy resins do not bond strongly to untreated carbon fibers. To overcome this, pretreatments, usually oxidative in nature, of fiber surfaces have been developed that greatly improve the fiber/matrix adhesion.

This chapter focuses on the state of the art of the surface modification of polyacrylonitrile (PAN)-based carbon fibers using different methods such as electrochemical and plasma techniques, so as to control the increase in surface roughness in order to improve the fiber/matrix interface in carbon fiber-reinforced polymer composites (CFRCs).

5.2
Backgrounds

5.2.1
Manufacturing Processes

There are many manufacturing techniques in producing composite structural products, with many variations and patented processes such as lay-up, molding, winding, pultrusion, and so on. Among them, we address four basic manufacturing techniques: (1) the lay-up process engages a hand or machine buildup of mats of fibers that are held together permanently by a resin system. This method enables numerous layers of different fiber orientations to be built up to a desired sheet thickness and product shape. (2) Molding is a strip of material with various cross sections used to cover transitions between surfaces or for decoration. Molding is one of the most common methods of shaping plastic resins. (3) The filament winding process can be automated to wrap resin-wetted fibers around a mandrel to produce circular or polygonal shapes. (4) The pultrusion process involves a continuous pulling of the fiber rovings and mats through a resin bath and then into a heated die. The elevated temperature inside the die cures the composite matrix into a constant cross-sectional structural shape.

5.2.2
Surface Treatment

In the late 1960s and early 1970s, several different types of surface treatment were used:

a) **Wet methods:** Immersion of fibers in oxidizing agents such as nitric and chromic acid and electrochemical oxidation. Electrochemical oxidation is the most widely used industrial method for treating carbon fibers [13–16], but the details of such processes are mostly proprietary
b) **Dry methods:** Oxidation in air or oxygen. Plasma treatment is one of the convenient and effective methods to produce functional group on carbon fiber surfaces.

The physical and chemical properties of carbon fibers are usually modified to achieve good adhesion between the reinforcement and the matrix materials. To enlarge the surface polarity of nonpolar carbon fibers, various surface treatment techniques are applied, such as oxidation in acid solutions [17, 18], dry oxidation in oxygen, anodic oxidation, plasma treatments, mild fluorination [19–23], and metallic coating [24–26]. Interfacial adhesion between the fibers and the matrices cannot be achieved without intimate contact, that is, unless the fiber surface contacts the resin on an intermolecular equilibrium distance level [27–30].

A large interfacial area of intimate contact by the matrix resin is a prerequisite whenever the interfacial bond is primarily due to van der Waals physical adsorption force. In addition, complete wetting of the fibers by the liquid polymeric matrix is advantageous when the surface free energy of the fibers is well above that of the liquid polymer in view of the London dispersive–specific polar components of surface free energy.

Untreated commercial carbon fibers have a surface tension close to 40 mJ/m^2 and an extremely hydrophobic nature due to the extremely high temperature of the manufacturing process leading to carbonization ($\pm$700–1500 °C) or graphitization ($\pm$2500 °C). On the other hand, most polymeric matrix resins, such as phenolic resin, are slightly hydrophilic with a surface tension in the range of 35–45 mJ/m^2 [27–30].

5.2.2.1 Surface Treatment of Carbon Fibers

It is known that surface treatments improve fiber–resin bonding, but the mechanism by which the adhesion is improved is still controversial. There appear to be three major mechanisms contributing to the strength of the fiber/resin bond: (a) The adsorption (either physical or chemical) of the resin molecules onto surface complexes; acidic complexes have usually been considered to be of most importance. (b) Removal of contaminants providing a superior surface for adhesion. (c) A mechanical keying effect when the resin penetrates pits and channels on the roughened surface brought about by oxidation.

Figure 5.1 shows the functional groups that are thought to be present on oxidized carbon surfaces. It was hoped that by introducing one or more of these functional

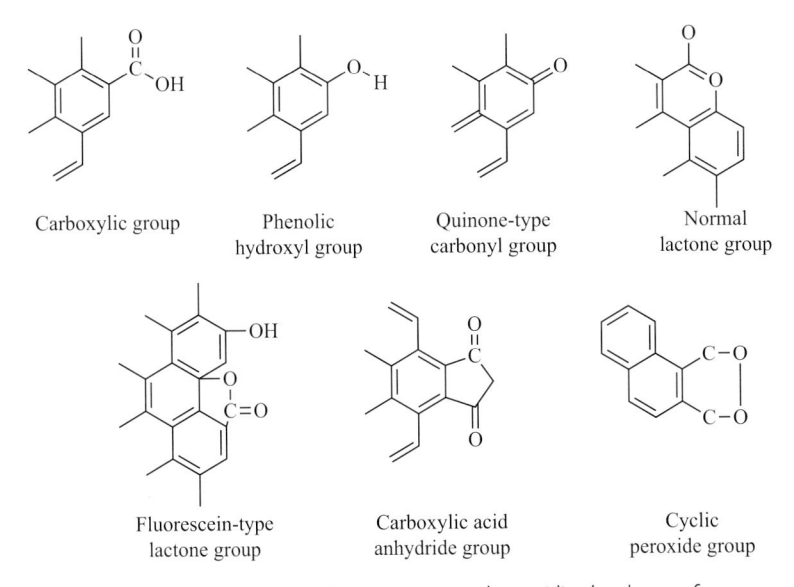

Figure 5.1 The possible functional groups presented on oxidized carbon surfaces.

Table 5.1 Description of carbon fibers.

Product name	Manufacturer	Precursor type	Properties
Thornel-25	Union carbide	Rayon	High strength
HMG-50	Hitco	Rayon	High modulus
AG	Le Carbone Lerraine	PAN	High modulus
AC	Le Carbone Lerraine	PAN	High strength
Type I	Not reported	PAN	High modulus
Type II	Not reported	PAN	High strength
HMU	Hercules Inc.	PAN	High modulus
HMS	Hercules Inc.	PAN	High modulus
T300	Amoco	PAN	High strength
P100	Amoco	Pitch	High modulus

groups, they would chemically interact with the epoxy resin, promoting bonding between fiber and resin. In most cases, an increase in the surface functionality is accompanied by an increase in surface area, and so the relative importance of these two factors is difficult to assess. Herrick [31] in 1968 reported that the treatment of Thornel-25 fibers in nitric acid increases their surface area and also the number of surface oxygen complexes. Similar results have been reported by Donner [32], Fitzer and Weis [33], and Rand and Robinson [34]. The interfacial shear strength (IFSS) of composites made from these treated fibers also increased.

Some fibers, after nitric acid treatment, were heated in hydrogen to reduce the surface oxygen complexes. The IFSS of the corresponding composites decreased significantly suggesting that the surface functionality was of most importance in enhancing fiber/resin bonding. Scola and Brooks [35], on the other hand, found no decrease in IFSS for composites produced from treated Hitco HMG-50 fibers (see Table 5.1) that had been heated in hydrogen prior to incorporation into the composite.

It has been found by many workers [36] that heating the fibers removes acidic surface oxidation almost completely. The treatment of AG (see Table 5.1) carbon fibers in Hummers reagent [32] produces a graphitic oxide layer. This lamellar graphitic oxide can be destroyed by heating. The formation of this lamellar coat greatly improves the IFSS of composites, but the IFSS made from treated fibers that had been heated decreased to a similar value of that of the untreated fiber/resin composites. For AC (see Table 5.1) fibers, heating did not produce such a marked effect. Although heating caused a decrease in the IFSS, the original value for the untreated fiber/resin composites was not regained.

Fitzer and Weis [33] studied the effect of electrochemically treating fibers in nitric acid on the shear strength of these composites. They used titration methods to determine the surface oxide concentration and blocking reagents such as diazomethane. Overoxidation of fibers was found to influence composite fracture behavior strongly, causing brittle fracture in the short-beam shear test (not a good indicator of the IFSS). With this in mind, only short oxidation times were used. They concluded that the amount of fiber–resin bonding solely depend upon the amount of acidic surface functionality. Surprisingly, Fitzer and Weis found that composites made from type I and II fibers (see Table 5.1) gave the same interfacial shear strength values, but type I had a surface oxide concentration one order of magnitude less than type II fibers. This surely indicates that surface treatments promote bonding via some other mechanism as well as the one they suggested.

Brelant [37] has a completely different view on the subject, suggesting the size and distribution of voids at the fiber–resin interface. The limiting strength is said to be the stress acquired to propagate cracks through the voids along the fiber/matrix boundary. Brelant proposed that the increase in shear strength noted in the experiments by Fitzer et al. [38] for the short treatment times could be attributed to the removal of contaminants or the reduction of protuberances that facilitate void formation. He suggested that fiber–resin bonding is primarily physical in nature.

This was later confirmed by Drzal and coworkers [39] who published evidence to show that the effect of commercial treatments was to remove weakly bound crystallites from the untreated fiber surface and hence provide a superior surface to which the resin could adhere. Kozlowski and Sherwood [40] found that these crystallites were removed by different mechanisms depending on the pH of the electrolyte used. In basic solutions, small fragments of fiber were detected in the working electrode solution; whereas in acidic solutions, they were oxidized to carbon dioxide. They then went on to study the effect of electrochemical oxidations on the IFSS of the resulting composites using treatment levels that were comparable to those of the industrially treated fibers.

5.2.3
Characterization of Polymeric Composites

The mechanical properties of composites depend on many variables such as fiber types, orientations, and architecture. The fiber architecture refers to the preformed textile configurations by braiding, knitting, or weaving. Polymer composites are anisotropic materials with their strength being different in any direction. Their stress–strain curves are linearly elastic to the point of failure by rupture. The polymeric resin in a composite material, which consists of viscous fluid and elastic solids, responds viscoelastically to applied loads. Although the viscoelastic materials will creep and relax under a sustained load, it can be designed to perform satisfactorily. Polymer composites have many excellent structural qualities such as high strength, material toughness, fatigue endurance, and lightweight. Other highly desirable qualities are high resistance to elevated temperature, abrasion, corrosion, and chemical attack.

5.2.4
Fiber Reinforcements

The fibers are an important constituent in polymer composites. A great deal of research and development has been done with the fibers on the effects in the types, volume fraction, architecture, and orientations. The fiber generally occupies 30–70% of the matrix volume in the composites. The fibers can be chopped, woven, stitched, and braided. They are usually treated with sizing such as starch, gelatin, oil, or wax to improve the bond as well as binders to improve the handling. The most common types of fibers used in advanced composites for structural applications are the fiber, glass, aramid, and carbon. Among them, carbon fibers are very expensive but they possess the highest specific physicochemical properties [41–44].

Carbon fiber-reinforced polymer (CFRP) composites are polymer matrix composite materials reinforced by carbon fibers. The reinforcing dispersed phase may be in the form of continuous or discontinuous carbon fibers of diameter about 7 μm commonly woven into a cloth. Carbon fibers are very expensive, but they possess the highest specific (divided by weight) mechanical properties such as elastic modulus and strength. Therefore, carbon fibers are used for reinforcing polymer matrix due to their following properties: (1) very high elastic modulus exceeding that of steel, (2) high tensile strength, which may reach 7 GPa, (3) low density of 1800 kg/m^3, (4) high chemical inertness, (5) good thermal stability in the absence of O_2, (6) high thermal conductivity, assisting good fatigue properties, and (7) excellent creep resistance. However, the main disadvantage of carbon fibers is their brittle mode of failure [45–50].

The graphites or carbon fibers are made from three types of polymer precursors such as PAN fibers, rayon fibers, and pitch. PAN-based carbon fibers are produced by conversion of PAN precursor through the following stages: (1) Stretching filaments from PAN precursor and their thermal oxidation at 400 °F (200 °C). The filaments are held in tension. (2) Carbonization is performed in nitrogen atmosphere at a temperature about 1200 °C for several hours. During this stage, noncarbon elements (O, N, and H) volatilize resulting in enrichment of the fibers with carbon. Graphitization is followed at about 2500 °C. Pitch-based carbon fibers are manufactured from pitch through the following stages: (1) Filaments are spun from coal tar or petroleum asphalt (pitch). (2) The fibers are cured at 315 °C. (3) Then, carbonization is performed in nitrogen atmosphere at a temperature about 1200 °C.

The tensile stress–strain curve is linear to the point of rupture. Although there are many carbon fibers available on the open market, they can be arbitrarily divided into three grades, as shown in Table 5.2. They have lower thermal expansion coefficients than both the glass and aramid fibers. The carbon fibers are anisotropic materials, and their transverse moduli are an order of magnitude less than its longitudinal modulus. The materials have a very high fatigue and creep resistance. Since its tensile

Table 5.2 Typical properties of three grades of carbon fibers.

Typical properties	High strength	Intermediate modulus	High modulus	Ultrahigh modulus
Density (g/cm^3)	1.8	1.8	1.9	2.0–2.1
Young's modulus (GPa)	230	290	370	520–620
Tensile strength (GPa)	3.5	5.5	2.2	3.4
Tensile elongation (%)	1.5	1.9	0.7	0.8

strength decreases with increasing modulus, its strain at rupture will be much lower. Because of the material brittleness at higher modulus, it becomes critical in joint and connection details, which can have high stress concentrations. As a result of this phenomenon, carbon composite laminates are more effective with adhesive bonding that eliminates mechanical fasteners.

5.3
Experimental Part

5.3.1
Materials

The carbon fibers used in this study were untreated and unsized PAN-based high-strength fibers, TZ-307 (12 K monofilaments), manufactured by Taekwang Company of Korea. The average diameter of these carbon fibers was approximately 7 μm, and typical tensile modulus and strength were about 245 and 3.5 GPa, respectively.

The epoxy resin used in this study was diglycidylether of bisphenol-A (DGEBA) YD-128 supplied by Kukdo Chemical Company of Korea). The epoxide equivalent weight was 185–190 g/equiv, and the density and viscosity were 1.16 g cm^3 and 5000 cPs, respectively, at $\pm 25\,°C$. Diaminodiphenylmethane (DDM) supplied by Aldrich Chemical Company was selected as a hardener for curing and methyl ethyl ketone (MEK) was used to reduce the high viscosity of DGEBA. Resol-type phenolic resin (CB-8057, supplied from Kangnam Chemical Company of Korea) was also used as the polymeric matrix.

5.3.2
Surface Treatment of Carbon Fibers

5.3.2.1 Electrochemical Oxidation
Electrochemical treatment of carbon fibers was conducted using the laboratory pilot-scale apparatus that we described earlier to introduce the functional groups and to improve the degree of adhesion at interfaces between fibers and matrix resins [51]. The electrolyte used was 10 wt% phosphoric acid solutions with a constant oxidation rate (1 m/min). The electric current densities used in this work were 0, 0.2, 0.4, 0.8, and 1.6 A/m^2. After anodic oxidation, the anodized carbon fibers were washed with freshly distilled water and then rinsed with acetone in a Soxhlet extractor for 2 h to remove surface impurities or residual oxides. The laboratory pilot plant is schematically illustrated in Figure 5.2.

5.3.2.2 Electroplating
Electroplating has been used to produce metal matrix composites reinforced with carbon fibers [52, 53]. Carbon fiber surfaces are metallized by electrolysis in molten salt solutions [54]. An electroplating device was constructed that can continuously plate nickel onto the fiber surfaces. The speed of carbon fibers was controlled by a

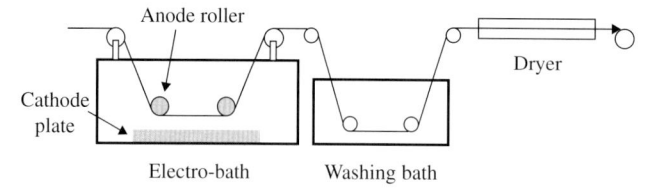

Figure 5.2 Schematic diagram of continuous electrolytic surface anodization process.

gearbox and the speed was normally about 0.72 m/min. Nickel sulfate was the main salt used in the electroplating solution, and electrolytic nickel plate was used as the anode. Carbon fibers are conductive, so they were used directly as the cathode to be plated. The compositions and operating conditions of the plating bath are given in Table 5.3. Before being plated, the carbon fibers were activated in nitric acid for 30 min in order to enhance the interfacial adhesion between the nickel coating and the carbon fibers. Prior to using fiber surface analysis or preparation of the composites, the residual chemicals were also removed by Soxhlet extraction with acetone at $\pm 70\,°C$ for 2 h. Finally, the carbon fibers were washed several times with distilled water and dried in a vacuum oven at $60\,°C$ for 12 h.

5.3.2.3 Oxyfluorination

For carbon fibers, the oxygen-containing fluorination treatment offers several advantages over other treatments. One primary advantage is that the mechanical properties of the fibers are not significantly degraded if optimum conditions are operated. Other important advantages are related to concerns about the development of more environment-friendly processes [55]. The carbon fibers were subjected to oxyfluorination under different conditions. An oxyfluorination reaction was performed in a batch reactor made of nickel with an outer electric furnace, as shown in Figure 5.3. After evacuation, the fluorine and oxygen mixtures (F_2/O_2 gases) were introduced to the reactor at room temperature, and then the reactor was heated to the treatment temperature. After the reaction, the specimens were cooled to room temperature, and then the reactive gases were purged from the reactor with nitrogen. In case of room temperature reaction, the reactor was cooled and evacuated in a cooling bath prior to charging of the fluorine. The reactor was removed from the cooling bath after the fluorine was purged with nitrogen. The total gas pressure was 0.2 MPa and the nominal reaction time was 10 min at the treatment temperature. Table 5.4 lists the experimental conditions of the carbon fibers studied.

Table 5.3 Composition and operating conditions of Ni-plating bath.

Compositions	$NiSO_4 \cdot 6H_2O$	280 g/L
	$NiCl_2 \cdot 6H_2O$	40 g/L
	H_3BO_3	30 g/L
Conditions	pH	4.5–5.0
	Temperature (°C)	45 ± 1
	Current density (A/m^2)	0–60

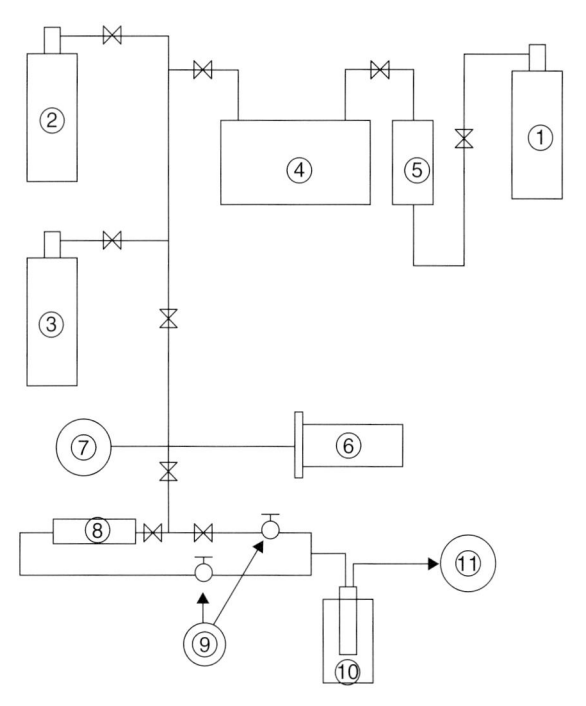

Figure 5.3 Schematic diagram of oxyfluorination reactor: (1) F_2 gas cylinder, (2) N_2 gas cylinder, (3) O_2 gas cylinder, (4) buffer tank, (5) HF absorber (NaF pellet), (6) reactor, (7) pressure gauge, (8) F_2 absorber (Al_2O_3), (9) glass cock, (10) liquid nitrogen, and (11) rotary vacuum pump.

5.3.2.4 Plasma Modification

One of the most practical surface modifications for commercial production of carbon fibers is plasma treatment. The carbon fibers were exposed to atmosphere argon plasma under the following conditions: 13.56 MHz radio frequency (RF), 150 W power, 10 mbar pressure at 50 sccm/min using the atmospheric pressure plasma devices equipped with mass flow controller (MFC) hardware, as shown in Figure 5.4.

Table 5.4 Experimental oxyfluorinated conditions of PAN-based carbon fibers used.

Specimens	F_2/O_2 mixtures (%)	Oxyfluorination temperature (°C)
No treatment	—	—
CF-RT	100/0	25
CFO-RT	50/50	25
CFO-100		100
CFO-300		300
CFO-400		400

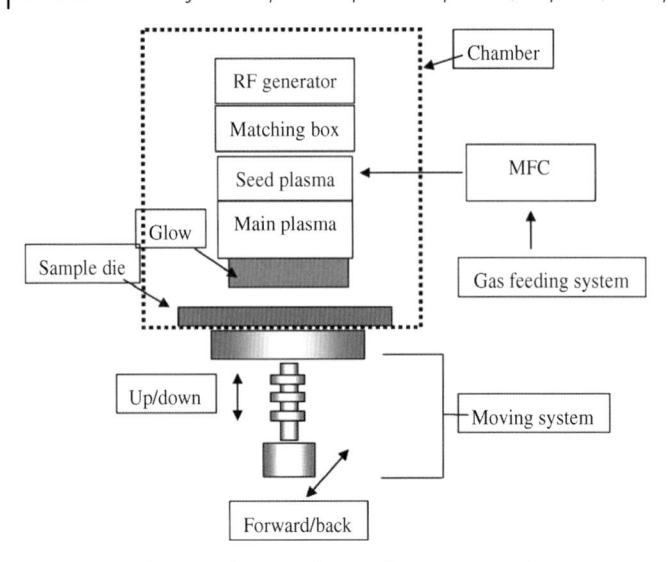

Figure 5.4 Schematic diagram of atmospheric pressure plasma apparatus.

The time of exposure was varied from 0 to 180 min and we designate them as CFs-0, (b) CFs-10, (c) CFs-20, (d) CFs-30, and (e) CFs-180, respectively.

5.3.3
Preparation of Carbon Fiber-Reinforced Polymer Composites

Unidirectional composite laminates were prepared by continuous impregnation of the fibers using a drum winding technique for manufacturing prepregs with subsequent hot pressing [56]. The laminates made with 32 plies of prepregs were fabricated in a hot press at 7.4 MPa at 150 °C for 150 min with a vacuum bagging method in a conventional composite processing [57]. To obtain an average fiber volume fraction V_f of the composites, a small-size rectangular specimen was cut from the laminate and the side section of the specimen was polished. The number N of fibers was counted using video images of the small area S of 1 mm (thickness direction) × 0.2 mm (width direction) and average diameter (d_f) of the fiber was measured at the cross section. Average fiber volume fraction V_f was then calculated from Eq. (5.1):

$$V_f = \frac{N \cdot d_f^2 \cdot \pi}{4S} \tag{5.1}$$

The average fiber volume fraction of bulk specimens was about 52% ($\pm 0.2\%$) for all composites.

5.3.4
Characterization of Carbon Fibers

In order to study the effect of surface treatment of carbon fibers, it is necessarily beneficial to determine the effect of surface treatments on the chemistry of the fiber surfaces. The X-ray photoelectron spectroscopy (XPS) measurement of fiber surfaces was performed using a VG Scientific ESCA LAB MK-II spectrometer equipped with Mg-Kα X-ray source. The base pressure in the sample chamber was controlled in the range of 10^{-8}–10^{-9} torr. In the survey spectrum (0–1100 eV), the XPS program package fixed the C1s peak at 284.6 eV considering the neutralization. For the high-resolution C1s spectra, the exact position of the C1s peak was not fixed. Quantification was carried out by applying a standard procedure, that is, linear background subtraction, fitting the measured lines by means of a set of Gaussian curves (except for the graphite line) and converting the intensities into atomic concentrations by using relative sensitivity factors.

However, carbon fibers are very difficult to work with as they do not lend themselves readily to the more common spectroscopic techniques owing to their color, size, and handling difficulties. Even with XPS, analysis of data has proved to be difficult owing to the extremely small concentration of chemical species present on fiber surfaces. Therefore, we performed contact angle method to examine the surface conditions of the fibers.

5.3.5
Theoretical Considerations of Dynamic Contact Angles

A precise contact angle measurement of fibrous materials is a difficult and complex process, although several methods have been proposed to measure their wettability [28, 58, 59]. In the early 1970s, Chwastiak [59] introduced the procedure for wicking rate measurements by enclosing the carbon fiber bundle in a glass tube so that the porosity is fixed for a given strand of carbon fibers. This measurement that was used successfully to evaluate the wettability of carbon fiber by water, glycol, heptane, and epoxy resins has been described. The wettability of carbon fibers is determined by measuring the wicking rates by the mass pickup technique or by the surface velocity method. The contact angle used in this investigation was then calculated using the Washburn's equation [59] defining the flow of a liquid through a capillary.

$$\frac{h^2}{t} = \frac{r \cdot \gamma_L \cdot \cos\theta}{2\eta} \tag{5.2}$$

where h is the rise height in a capillary of radius r, t is the flow time, θ is the contact angle of the liquid with the surface, and γ_L and η are the liquid surface tension and viscosity, respectively.

The fibers that have been packed in a column can be ascribed as a bundle of capillaries with mean radius $\bar{r}$. A modified Washburn's equation can be used for this system:

$$\frac{h^2}{t} = \frac{\bar{r} \cdot \gamma_L \cdot \cos\theta}{2\eta} \tag{5.3}$$

If the effects of gravity may be neglected, the height increase of the liquid is replaced with mass gain due to wicking of the liquid, and then the expression for the contact angle can be rewritten as

$$\frac{m^2}{t} = \frac{c \cdot \varrho^2 \cdot \gamma_L \cdot \cos\theta}{\eta} \tag{5.4}$$

where m is the weight of the penetrating liquid, ϱ is the density of the measuring liquid, and c is the packing factor.

The packing factor c is an empirical constant that depends on the size and degree of packing. In order to measure the packing factor for particular fibers, a liquid with low surface free energy is chosen. Hexane ($\gamma_L = 18.4\,\text{mJ/m}^2$) is generally the test liquid of choice, which wets completely. In this case, the contact angle against the fibers was assumed to be $0°$ and $\cos\theta = 1$. Therefore, the packing constant can be calculated by detecting the increase in weight per unit time and by using a modified Washburn's equation (5.4) with the viscosity, density, and surface free energy of hexane.

Also, the contact angle (θ) on carbon fibers can experimentally be determined in the same manner as the packing factor derived from n-hexane wetting method, after the values of packing factor c and liquid characteristics ϱ, η, γ_L of the testing liquids are known. The slope ($\mathrm{d}m^2/\mathrm{d}t$) was determined in the linear range for the measurements of c and $\cos\theta$. This is done by differentiating Eq (5.4) according to time and determining the value $\mathrm{d}m^2/\mathrm{d}t$. For use on a variety of fibrous materials, at least two unidentical liquids having known London dispersive and specific components of surface free energy are needed.

In this work, contact angle measurements of the carbon fibers were performed using the Krüss Processor Tensiometer K12 with fiber apparatus. The experimental setup is schematically illustrated in Figure 5.5. About 2 g of carbon fibers was packed into an apparatus, which was then mounted indirectly to the measuring arm of the microbalance. The packing factor of the fibers was measured for each continuous filament by measuring the increase in weight per unit time at zero depth of immersion of a test liquid (in this study, n-hexane), as shown schematically in Figure 5.6. The test liquids used for contact angle measurements were n-hexane, deionized water, diiodomethane, and ethylene glycol [60]. The surface free energy (or surface tension) and their London dispersive and specific (or polar) components for the test liquids are listed in Table 5.5.

5.3.6
Characterization of Carbon Fiber-Reinforced Polymer Composites

5.3.6.1 Interlaminar Shear Strength
Interlaminar shear strength (ILSS) was measured to estimate the interfacial adhesion strength of the composites. The test was conducted by three-point short-beam bending test method using an Instron model Lloyd LR-5K mechanical tester according to the ASTM D 2344. For a rectangular cross section of the composites, the ILSS determined from three-point bending tests is calculated as [61]

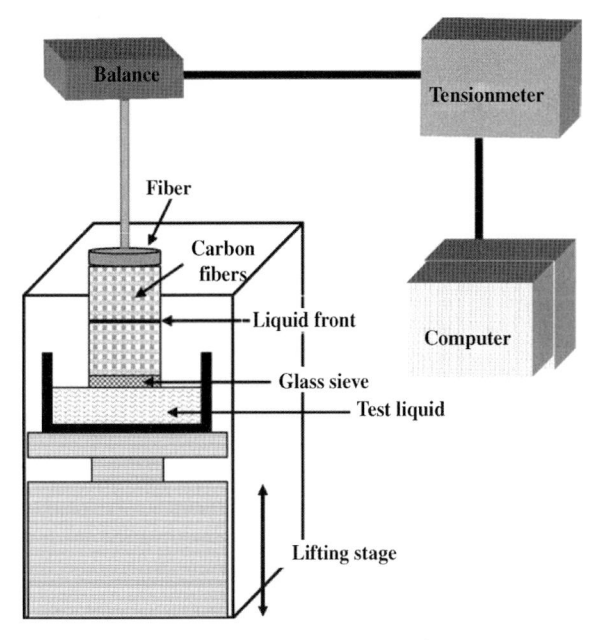

Figure 5.5 Schematic diagram of principles of the capillary rise method.

$$\text{ILSS} = \frac{3P}{4bd} \tag{5.5}$$

where P is the load at the moment of break, b is the width of the specimen, and d is the thickness of the specimen.

The accuracy and range of the load cell used were 0.5% grade and 5 kN, respectively. The composites were machined along the fiber direction into 30 mm × 6 mm

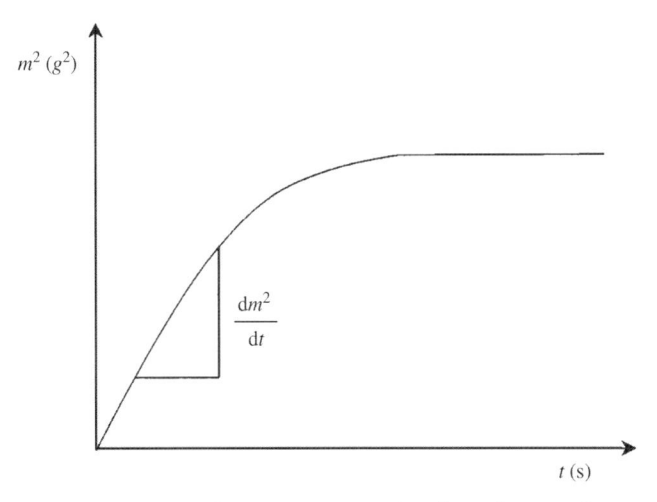

Figure 5.6 Typical wicking rate data in weight m^2 as a function of time t.

Table 5.5 The characteristics of wetting liquids used in this work [28].

Wetting liquids	γ_L^L (mJ/m²)	γ_L^{SP} [mJ/m²]	γ_L (mJ/m²)	η (MPa s)	ϱ (g/cm³)
n-Hexane	18.4	0	18.4	0.33	0.661
Water	21.8	51	72.8	1	0.998
Diiodo-methane	50.42	0.38	50.8	2.76	3.325
Ethylene glycol	31.0	16.7	47.7	17.3	1.100

γ_L^L: London dispersive component of surface free energy, γ_L^{SP}: specific component of surface free energy, γ_L: total surface free energy, η: viscosity, ϱ: density.

short-beam shear specimens with 5 mm thickness. The distance between supports divided by the thickness of specimens $L/d = 5$, and the crosshead speed was fixed at 2.0 mm/min.

5.3.6.2 Critical Stress Intensity Factor (K_{IC})

The critical stress intensity factor (K_C), which is one of the fracture toughness parameters, was described by the state of stress in the vicinity of the tip of a crack as a function of the specimen geometry, the crack geometry, and the applied load on the basis of linear elastic fracture mechanics [62]. The analytical expression for K_{IC} (K_C in mode I fracture) was characterized using a single-edge notched (SEN) beam fracture toughness test in 90° three-point bending flexure, as shown schematically in Figure 5.7. Notches were cut using a slow-speed diamond wire saw, approximately half the depth of the specimen. The SEN beam fracture toughness test was conducted using an Instron Model 1125 mechanical tester according to the ASTM E 399. A span-to-depth ratio of 4:1 and crosshead speed of 1 mm/min were used. For the SEN beam fracture toughness test, the value of K_{IC} is calculated using Eq (5.6).

$$K_{IC} = \frac{3PLa^{1/2}}{2b^2 d} Y \tag{5.6}$$

where P is the critical load for crack propagation (N), L is the length of the span (mm), a is the precrack length (mm), b is the specimen width (mm), d is the specimen thickness (mm), and Y is the geometrical factor given by Eq. (5.7):

$$Y = 1.98 - 3.07(a/b) + 14.53(a/b)^2 - 25.11(a/b)^3 + 25.80(a/b)^4 \tag{5.7}$$

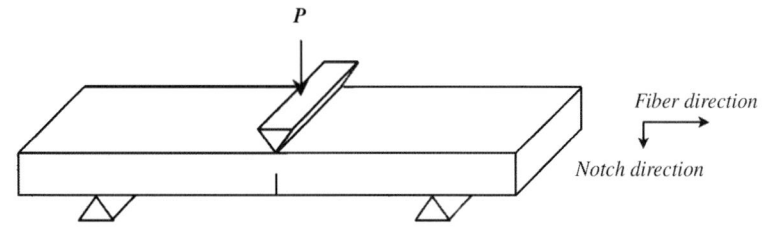

Figure 5.7 Schematic diagram of single-edge notched (SEN) beam fracture toughness test.

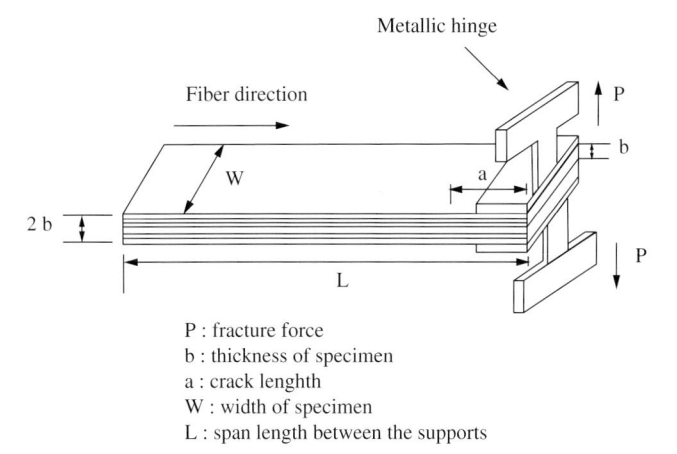

Figure 5.8 Schematic diagrams of double cantilever beam (DCB) specimen.

P : fracture force
b : thickness of specimen
a : crack lenghth
W : width of specimen
L : span length between the supports

5.3.6.3 Mode I Interlaminar Fracture Toughness Factor

Mode I interlaminar fracture toughness (G_{IC}) was determined according to ASTM D 5528-94 using the double cantilever beam (DCB) test, as shown in Figure 5.8. Specimens were cut to size using a diamond saw with the fiber direction parallel to the length of the sample, 20 mm in width, 3 mm in thickness, and 160 mm in length with a 30 mm crack starter film. Aluminum tabs were attached to the end of the specimens containing the insert using a high-strength epoxy adhesive and allowed to dry for 48 h. An Instron 1125 Universal Testing Machine was used to perform the tests using a crosshead speed of 1 mm/min and a load versus displacement curve was produced for each test. From these data, mode I critical strain energy release rate (G_{IC}) was calculated using corrected beam theory according to Eq. (5.8):

$$G_{IC} = \frac{3P \cdot \delta}{2W(a + |\Delta|)} \tag{5.8}$$

where P is the load, δ is the displacement, W is the specimen width, a is the crack length, and $|\Delta|$ is the crack length correction factor that is determined to be the x-axis intercept of the plot of the cube root of the compliance ($C = \delta/P$) versus the crack length a. Values of G_{IC} were plotted as a function of crack length a to produce a resistance R curve.

5.3.6.4 Fracture Behaviors

Ahearn and Rand [63] show the three general classes of fracture behavior result, depending on the relative magnitudes of the interfacial bond fracture energy: the fiber structural energy, the elastic mismatch between fiber and matrix, and the sliding resistance between fiber and debonded matrix. The general shapes of the load–deflection curves generated from these three types of behavior are shown in Figure 5.9.

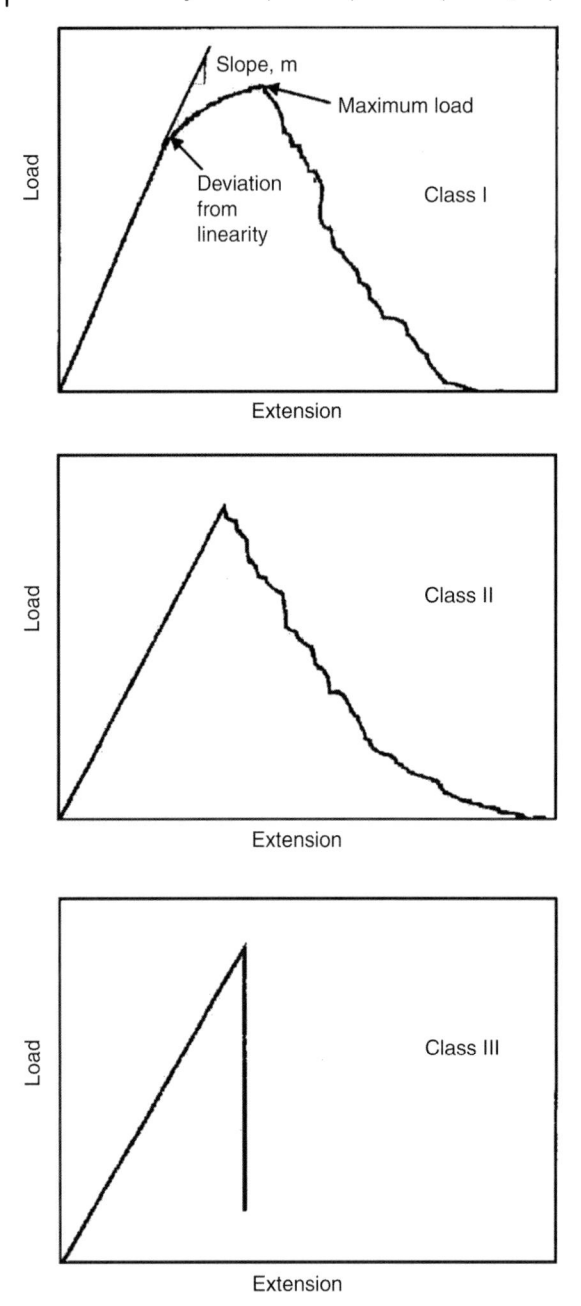

Figure 5.9 Load–extension curves generated by three main classes of fracture behaviors.

It is noted that debonding occurs leading to class I-type behavior especially in case of ductile matrix composites. Class I is a classical multiple fracture mode of failure, leading to the ultimate strength being largely controlled by the fiber properties, and arises when the interfacial sliding stress is low enough to allow relative motion of the

Table 5.6 Variation of surface chemical compositions of the carbon fibers with current density observed by XPS measurement (at%).

Current density (A/m^2)	XPS			
	C_{1s}	O_{1s}	N_{1s}	O_{1s}/C_{1s}
0	91.8	6.9	1.3	7.5
0.2	80.7	12.2	7.1	15.1
0.4	79.2	13.1	7.7	16.5
0.8	76.9	14.0	9.1	18.2
1.6	78.0	14.2	7.8	18.2

fibers. Conversely, class III-type behavior occurs with typical brittle matrix composites. Debonding takes place and the composite fails catastrophically.

5.4 Results and Discussion

5.4.1 Effect of Electrochemical Oxidation

5.4.1.1 Surface Characteristics

It is well known that XPS is a very useful apparatus in the determination of chemical composition and functional groups of the fiber surfaces [64–67]. As shown in Table 5.6, the amounts of both surface oxygen and nitrogen groups increased with an increase in the current density in electrolytic solution. The large variation of the oxygen group is due to the varying acidity of the surface functional groups. The O1s/C1s ratios as a function of the applied electric potential are also indicated in Table 5.6, and the ratios increase with the increasing electric current density up to about 0.4 A/m^2. When the O_{1s}/C_{1s} ratio is considered, it is clear that during electrochemical oxidation, the hydroxyl (−COOH) and carbonyl (C=O) groups increased on the fiber surfaces, as seen in Table 5.7.

Table 5.7 Variation of surface functional groups of the carbon fibers with current density observed from XPS measurement (at%).

Current density (A/m^2)	Functional groups			
	−C−C− (284.7 eV)	−C−OH− (286.1 eV)	−C=O− (287.8 eV)	−OH−C=O− (289.2 eV)
0	63.3	25.4	4.2	2.7
0.2	54.1	28.2	9.0	6.7
0.4	53.0	28.3	9.1	6.8
0.8	50.1	28.9	10.0	9.6
1.6	51.2	29.1	10.2	6.6

Table 5.8 Contact angle determinations (in degrees) on the carbon fibers as a function of current density.

Wetting liquids	0 (A/m^2)	0.2 (A/m^2)	0.4 (A/m^2)	0.8 (A/m^2)	1.6 (A/m^2)
Water	85	69	61	70	74
Diiodomethane	36	29	26	25	29

In this work, it is noted that the double bonds in the carbonyl and carboxyl functional groups on the fiber surfaces can largely enhance the physical attraction at interfaces between fibers and epoxy resins, and simultaneously the amide-type functional groups. Therefore, the increasing amounts of oxygen-containing functional groups on the fibers play an important role in improving the degree of adhesion at interfaces (hereby, Keesom's attraction of van der Waals force, hydrogen bonding, and the other small polar effects) between fibers and matrix. As seen in Tables 5.6 and 5.7, it is also noted that larger increases of oxygen than nitrogen groups of the fibers treated are related to enhancing the bulk degree of physical attraction as acidic functionality.

5.4.1.2 Contact Angle and Surface Free Energy

Table 5.8 shows contact angle data of carbon fibers with and without electrochemical oxidation. As a result, the angle of water largely decreases with the increasing current density up to 0.2–0.4 A/m^2. This suggests that electrochemical oxidation leads to a change in fiber surface nature, such as hydrophobic–hydrophilic properties. To obtain more detailed information about the surface energetic of the carbon fibers before and after electrochemical oxidation, an analysis of the surface free energy is evaluated in the surface energetic studies divided by two components: a London dispersive component of nonpolar interaction and a specific component describing all other types of interactions (Debye, Keesom, hydrogen bond, and other small polar effect). The London dispersive and specific (or polar) components of surface free energy of carbon fibers are determined by measuring the contact angles of a variety of testing liquids with known London dispersive and specific components and analyzing the results in accordance with the method proposed by Owens and Wendt [68] and Kaelble [69], using the geometric mean method.

$$\gamma_L \left(1 + \cos\theta\right) = 2\left(\sqrt{\gamma_S^L \gamma_L^L} + \sqrt{\gamma_S^{SP} \gamma_L^{SP}}\right) \tag{5.9}$$

where the subscripts L and S represent the liquid and solid states and γ^L and γ^{SP} are the London dispersive (superscript L) and specific (SP, Debye, Keesom of van der Waals, H-bonding, π-bonding, and other small polar effects) components of the surface free energy of the constitutive elements.

Using two wetting liquids (subscripts L_1 and L_2) with known γ_L^L and γ_L^{SP} for equilibrium contact angle (abbreviated θ here) measurements, one can easily determine γ_S^L and γ_S^{SP} and solve the following two equations [6, 70]:

$$\gamma_{L1}(1+\cos\theta_1) = 2\left(\sqrt{\gamma_S^L \gamma_{L1}^L} + \sqrt{\gamma_S^{SP} \gamma_{L1}^{SP}}\right) \tag{5.10}$$

$$\gamma_{L2}(1+\cos\theta_2) = 2\left(\sqrt{\gamma_S^L \gamma_{L2}^L} + \sqrt{\gamma_S^{SP} \gamma_{L2}^{SP}}\right) \tag{5.11}$$

In Table 5.9, the results of the surface free energy and its London dispersive and specific components of the carbon fibers are summarized. As a result, it can be seen that the specific component of the surface free energy, γ_S^{SP}, increases with the increasing electric current density up to 0.4 A/m²; then, a marginal decrease in the specific component is observed for the strongest current density used in this work. The London dispersive component γ_S^L remains almost constant as the current density increases. This result indicates that the moderate electrochemical oxidation of the carbon fibers leads to an increase in the surface free energy S, which is mainly influenced by its specific component. In previous studies [71], we have found that the electrochemical oxidation of the carbon fibers gives an increase in surface functionality, resulting in improved ILSS of the composites. Therefore, the present result seems to be partially due to the increase of surface oxygen functional groups created by surface oxidation on carbon fibers.

From a surface energetic point of view, it is expected that an increase in the specific component of the surface free energy of fibers will play an important role in improving the degree of adhesion at interfaces between the anodized fibers and the epoxy resin matrix containing oxide functional groups.

5.4.1.3 Mechanical Interfacial Properties

Fracture toughness is a critical property needed to resist crack propagation loaded from matrix to fiber, which ought to be considered in the evaluation of a composite material for a real application [72]. For a rectangular cross section of the composites, the fracture toughness of the composites can be measured by the three-point bending test for the critical stress intensity factor (K_{IC}).

Table 5.9 Results of surface free energy and their components of the carbon fibers as a function of current density.

Current density (A/m²)	γ_S^L [mJ/m²]	γ_S^{SP} [mJ/m²]	γ_S [mJ/m²]
0	40	2	42
0.2	42	7	49
0.4	42	11	53
0.8	44	6	50
1.6	42	5	47

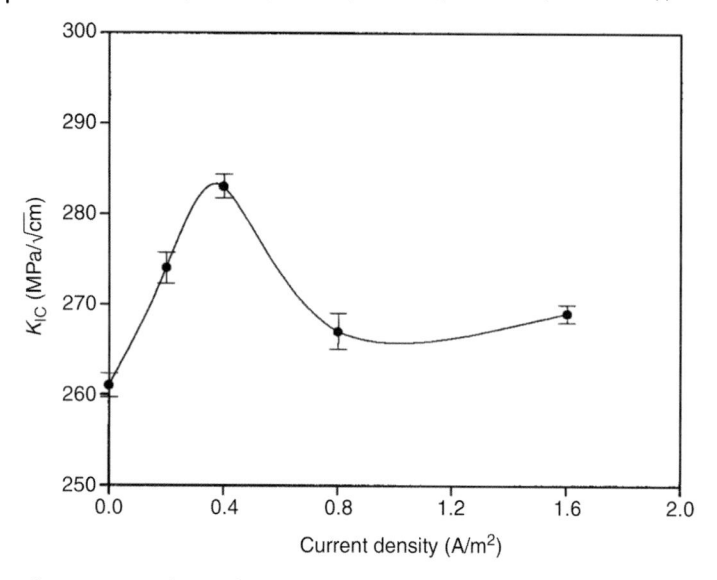

Figure 5.10 Evolution of K_{IC} as a function of the current density studied.

Figure 5.10 shows the evolution of K_{IC} with flexure of the composites as a function of electric current density. In the K_{IC} fracture toughness test, the K_{IC} of the composites continually increases with the increasing current densities of the treatments up to 0.4 A/m², and a maximum strength value is found about 283 MPa $\sqrt{cm}$ at the anodic treatment of 0.4 A/m². Additional energy needed to extend the interfacial crack under this condition is attributed to increased interfacial adhesion between fibers and matrix [73]. Also, a good linearity (regression coefficient $R = 0.974$; SD = 2.18) of the relationship between γ_S^{SP} S of the fibers and K_{IC} of the composites is shown in Figure 5.11. As mentioned above, this is a consequence of the increase of the specific component of fiber surface free energy, resulting in enhanced physical fiber–matrix adhesion of the composites.

5.4.2
Effect of Electroplating

5.4.2.1 Surface Characteristics

XPS spectra of untreated and nickel-plated carbon fibers are shown in Figure 5.12. As anticipated, the untreated carbon fibers (Figure 5.12a) show a C1s peak and a substantial O1s peak at 284.6 and 532.8 eV, respectively [74]. The O1s peak is probably due to the intrinsic surface carbonyl or carboxyl groups. Otherwise, for the nickel-plated carbon fibers (Figure 5.12b–e), carbon, oxygen, and nickel (BE = 857.6 eV) peaks are observed in XPS spectra [75]. The O1s peak of nickel-plated carbon fibers is probably due to the NiO, C=O, −OH, and O−C−O groups on the fiber surfaces.

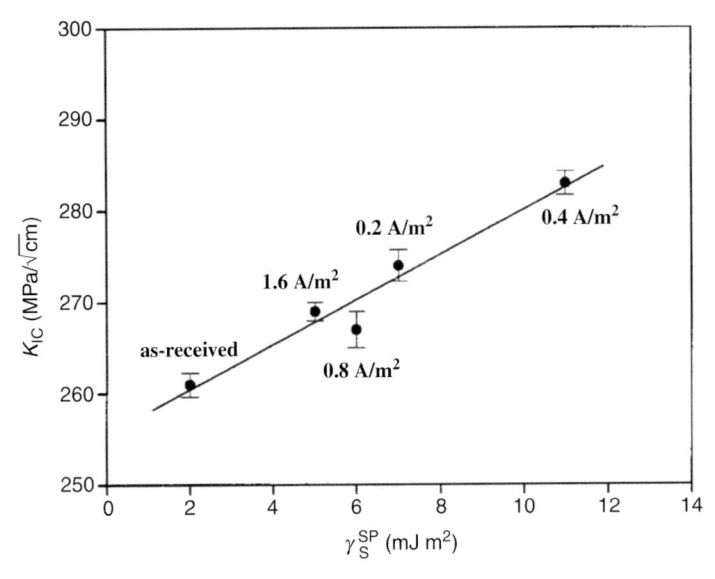

Figure 5.11 Dependence of K_{IC} of composites on the specific component of the fiber surface free energy studied.

Figure 5.13a shows expanded scale O1s XPS spectra for carbon fibers coated with nickel at 10 A/m² current density. The O1s spectra reveal the presence of three peaks corresponding to NiO groups (peak 1, BE = 529.6 eV), C=O, or —OH groups (peak 2, BE = 531.6 eV), and O—C—O groups (peak 3, BE = 532.6 eV) groups [76–80].

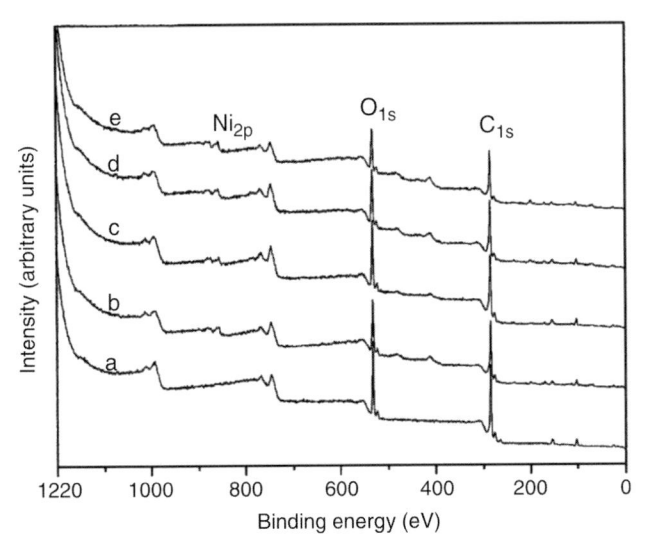

Figure 5.12 XPS spectra of the nontreated and nickel-plated carbon fibers: (a) 0, (b) 5, (c) 10, (d) 30, and (e) 60 A/m².

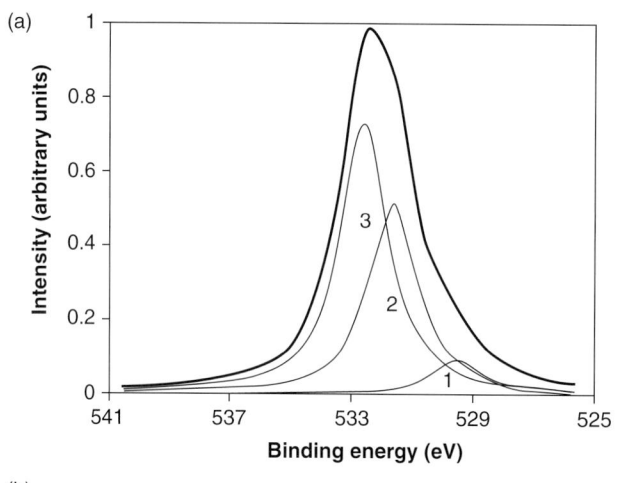

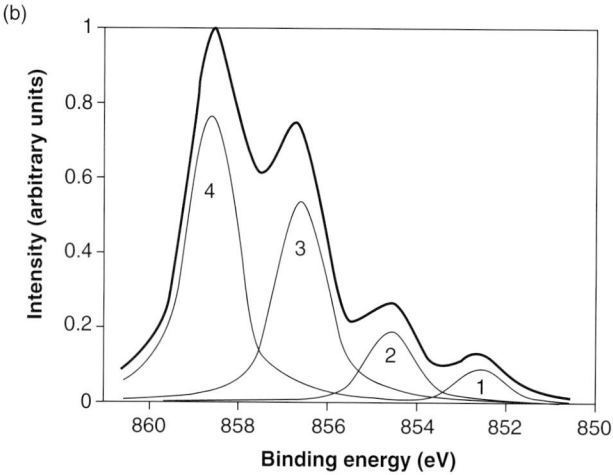

Figure 5.13 High-resolution O_{1s} and Ni_{2p} XPS spectra of nickel-plated carbon fibers (10 A/m² current density). (a) 1: NiO (BE = 529.6 eV), 2: C=O or –OH (BE = 531.6 eV), and 3: O–C–O (BE = 532.6 eV) (b) 1: Ni–metal (BE = 852.7 eV), 2: NiO (BE = 854.6 eV), 3: Ni(OH)₂ (BE = 856.4 eV), and 4: Ni–metal (BE = 858.5 eV).

The Ni_{2p} peak is shown on expanded scale in Figure 5.13b for nickel-plated carbon fibers. In case of nickel-plated carbon fibers in Figure 5.13b, several subpeaks are seen in addition to the main peak (BE = 858.5 eV). These subpeaks that are separated computationally in Figure 5.13b indicate that Ni metal (peak 1, BE = 852.7 eV, and peak 4, BE = 858.5 eV), NiO (peak 2, BE = 854.6 eV), and Ni(OH)₂ (peak 3, BE = 856.4 eV) are present as a result of electrolytic nickel plating [81–83]. However, these peaks are never seen for the untreated carbon fibers. For the untreated and nickel-plated carbon fibers, elemental composition and O_{1s}/C_{1s} ratio are listed in Table 5.10. The O_{1s}/C_{1s} composition ratios of nickel-plated carbon fibers are

Table 5.10 Elemental composition and O_{1s}/C_{1s} ratio of untreated and nickel-plated carbon fibers (at%).

Current density (A/m²)	XPS			
	C_{1s}	O_{1s}	N_{1s}	O_{1s}/C_{1s}
0	68.8	25.8	0.8	0.375
5	64.4	25.4	0.8	0.394
10	62.8	28.0	0.8	0.446
30	63.3	26.8	0.8	0.423
60	63.0	23.9	0.8	0.379

increased compared to that of untreated due to the deposition of more active forms, such as NiO, Ni(OH)$_2$, and Ni metal on the inactive carbon. However, nitrogen of carbon fiber surfaces has no significant changes as the concentration and distribution are varied.

From the XPS experimental results, it is found that the surface composition of carbon fibers changed substantially as a result of nickel plating. The carbon content of nickel-plated fibers decreased when the fibers were plated with metallic nickel, whereas the oxygen and nickel contents of plated fibers were higher than that for untreated fibers. The active groups on the carbon fiber surfaces after nickel plating may help to change the polarity and the functionality of the fiber surfaces.

5.4.2.2 Contact Angle and Surface Free Energy

The wetting of a solid surface by a liquid and the concept of contact angle (θ) was first formalized by Young [84]:

$$\gamma_S - \gamma_{SL} = \gamma_L \cos\theta \tag{5.12}$$

where γ_L is the surface energy of the liquid, γ_{SL} is the interfacial energy of solid/liquid interface, and γ_S is the surface energy of solid.

Fowkes [85] has suggested that the surface free energy of substances consists of two parts, the London dispersive and specific (or polar) components.

$$\gamma = \gamma^L + \gamma^{SP} \tag{5.13}$$

where the superscript L refers to the contribution due to London dispersive forces that are common to all substances, and superscript SP relates to the specific polar contribution.

In the context of surface energetic studies, Owens and Wendt [68] derived the following equation for the interfacial energy between liquid and solid assuming a geometric mean combination of the London dispersive and specific components:

$$\gamma_{SL} = \gamma_S + \gamma_L - 2\left(\sqrt{\gamma_L^L \gamma_S^L} - \sqrt{\gamma_L^{SP} \gamma_S^{SP}}\right) \tag{5.14}$$

Combining Eqs. (5.12) and (5.14) yields a linear equation:

$$\gamma_L(1+\cos\theta) = 2\left(\sqrt{\gamma_L^L \gamma_S^L} + \sqrt{\gamma_L^{SP} \gamma_S^{SP}}\right) \tag{5.15}$$

where γ_L, γ_L^L, and γ_L^{SP} are known for the testing liquids [86, 87] and γ_S, γ_S^L, and γ_S^{SP} can be calculated by the contact angle measurements.

Based on "harmonic" mean and force addition, Wu [88, 89] proposed the following equation:

$$\gamma_{SL} = \gamma_S + \gamma_L - \frac{4\gamma_S^L \gamma_L^L}{\gamma_S^L + \gamma_L^L} - \frac{4\gamma_S^{SP} \gamma_L^{SP}}{\gamma_S^{SP} + \gamma_L^{SP}} \tag{5.16}$$

Equation (5.16) can be written as follows with the aid of Eq. (5.12):

$$\gamma_L(1+\cos\theta) = \frac{4\gamma_S^L \gamma_L^L}{\gamma_S^L + \gamma_L^L} - \frac{4\gamma_S^{SP} \gamma_L^{SP}}{\gamma_S^{SP} + \gamma_L^{SP}} \tag{5.17}$$

Wu [88] claimed that this method applied accurately between polymers and between a polymer and an ordinary liquid.

The contact angles of nontreated and nickel-plated carbon fibers were measured for two testing liquids, deionized water and diiodomethane. The γ_S, γ_S^L, and γ_S^{SP} of the carbon fibers studied are given in Figure 5.14. The treatments at the current density range of 5–60 A/m² lead to a wetting level better than those for the sample without plating. The polar component γ_S^{SP} of the nickel-plated carbon fibers was significantly increased, but the dispersive component γ_S^L was barely changed. In this system, the surface polarity of carbon fibers is also increased compared to the nontreated one. These results reveal that the electroplating leads to an improvement of interaction between the polar liquid (deionized water) and the carbon fiber surfaces.

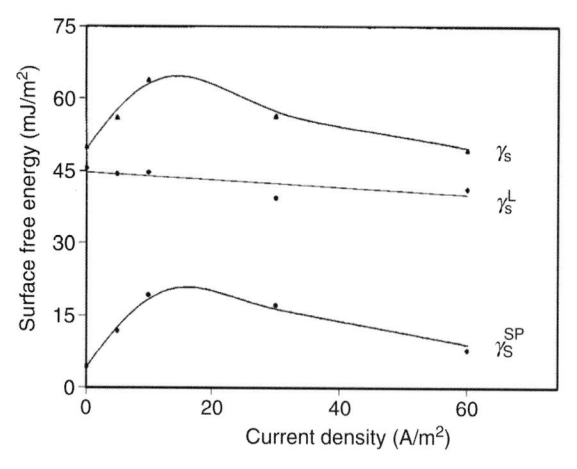

Figure 5.14 Surface free energy of the nontreated and nickel-plated carbon fibers. γ_S: surface free energy, γ_S^{SP}: specific (polar) component, and γ_S^L: London dispersive component.

To improve wettability, the surface energy of the fibers should be made larger than or equal to the surface energy of the matrix. In case of 10 A/m² current density, the surface free energy γ_S reaches the highest value of 67 mJ/m². Thus, it is expected that the interface between metallized carbon fiber surfaces and matrix resins may be good, since the metallized carbon fiber surface energy should allow extensive wetting by the matrix resins (35–45 mJ/m²) [28, 86, 87]. As mentioned above, these results are attributed to the introduction of van der Waals physical adsorption force and polar groups, such as $-CO$, $-COO$, NiO, $Ni(OH)_2$, and nickel metal of carbon fiber surfaces, resulting from the increasing specific component of the surface free energy, as already shown in Figure 5.13.

5.4.2.3 Mechanical Interfacial Properties

The effect of surface treatment can be expressed in terms of K_{IC} values. The critical stress intensity factor (K_{IC}; K_C in mode I fracture), which is one of the fracture toughness parameters, is described by the state of stress in the vicinity of the tip of a crack as a function of the specimen geometry, the crack geometry, and the applied load on the basis of linear elastic fracture mechanics (LEFM) [90]. K_{IC} can be characterized by a single-edge notched beam fracture toughness test in the three-point bending flexure [62].

Figure 5.15 shows K_{IC} of the Ni-plated carbon fiber-reinforced composites as a function of current density. The maximum strength is found at a current density of 10 A/m². Also, it is observed that K_{IC} of the composites for fibers treated at a relatively high current density (60 A/m²) is not significantly increased compared to that of the composites made with untreated fibers. Nevertheless, in this system, it appears that the nickel plating of carbon fiber surfaces affects K_{IC} of the composites, resulting

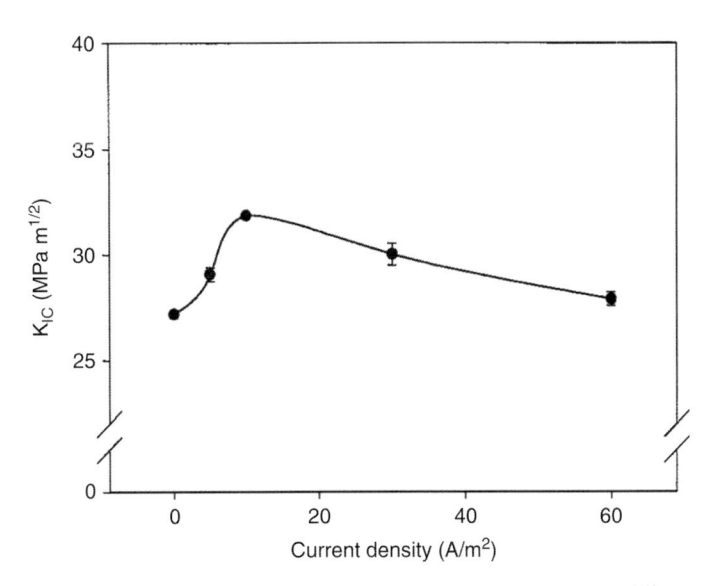

Figure 5.15 K_{IC} of Ni-plated carbon fiber-reinforced composites with different current densities.

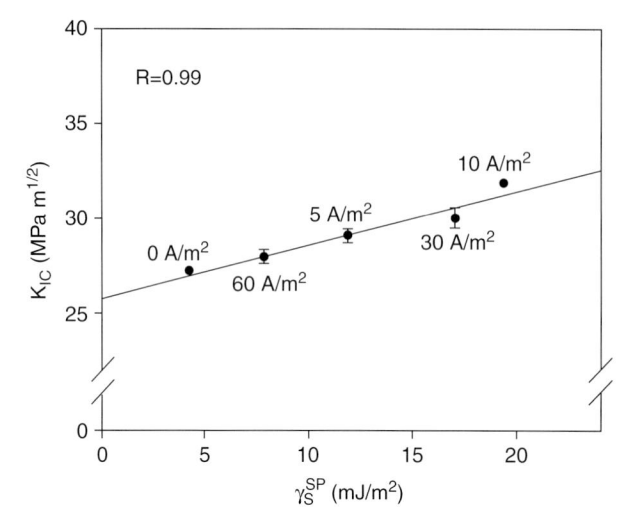

Figure 5.16 Dependence of K_{IC} on O_{1s}/C_{1s} ratio (R: coefficient of regression).

from the presence of both nickelized functional groups and increased oxide functional groups on the carbon fibers.

It is interesting to note that the K_{IC} results support the reliability of the data, since the trends in K_{IC} values seem to be very similar to trends in O_{1s}/C_{1s} or surface free energy. In fact, the relationships between K_{IC} and O_{1s}/C_{1s} and between K_{IC} and γ_S^{SP} are almost linear for all samples with different current densities, as shown in Figures 5.16 and 5.17, respectively. Therefore, the O_{1s}/C_{1s} ratio and γ_S^{SP} may prove

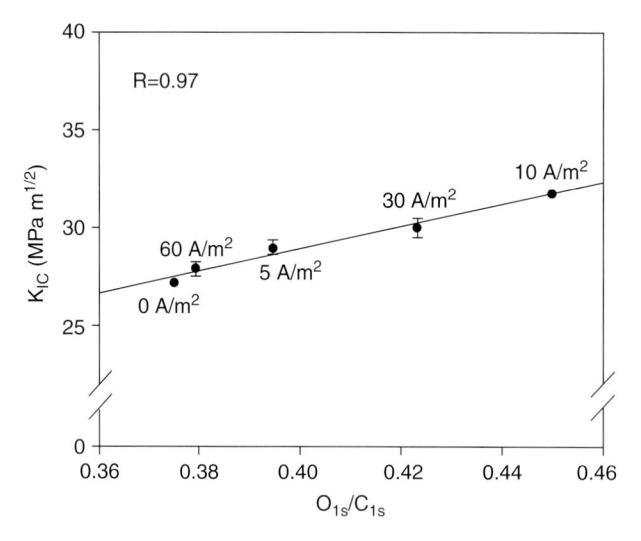

Figure 5.17 Dependence of K_{IC} on specific polar component γ_S^{SP} of surface free energy (R: coefficient of regression).

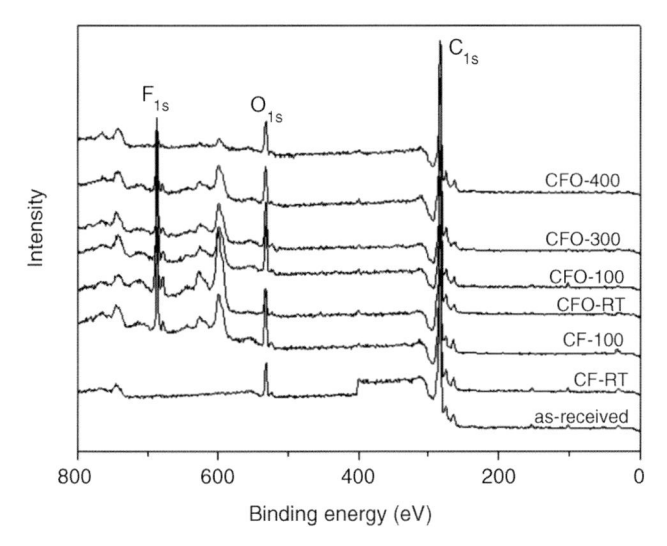

Figure 5.18 XPS wide scan spectra of the oxyfluorinated carbon fibers.

to be the governing factors in the adhesion between the fibers and the matrix resins in the present system studied.

5.4.3
Effect of Oxyfluorination

5.4.3.1 Surface Characteristics

XPS wide scan spectra of the oxyfluorinated carbon fiber specimens are shown in Figure 5.18. The intensity scale factor of the oxyfluorinated carbon fibers is higher than that of the as-received carbon fiber specimen. The XPS spectra show distinct carbon, oxygen, and fluorine peaks, representing the major constituents of the carbon fibers investigated. Relatively weak peaks of other major elements such as nitrogen are also observed. No other major elements are detected from the wide scan spectra on the surface of the carbon fibers. As expected, the fluorine peak intensity of the oxyfluorinated carbon fibers is also increased according to the fluorination temperature and oxygen content.

Figure 5.19a shows high-resolution spectra of the C_{1s} region of the oxyfluorinated carbon fibers. The binding energy (E_b) of the C_{1s} peak for the as-received carbon fibers is 285 eV, representing the most graphitic carbons (C–C), which undergoes a slight shift toward a lower E_b owing to a lowering of the Fermi level (E_F) and a subsequent decrease of the energy gap between the C_{1s} core level and E_F. The difference between the full-width half-maximum (FWHM) values of the as-received and oxyfluorinated carbon fibers, CFO-RT, is 0.65 eV. A higher FWHM value is found for the CFO-100 sample presumably due to the surface oxyfluorination. Figure 5.19b shows narrow scan spectra of the F_{1s} region of the oxyfluorinated carbon fibers. Due to the surface oxyfluorination, a higher FWHM value and a shoulder at a higher

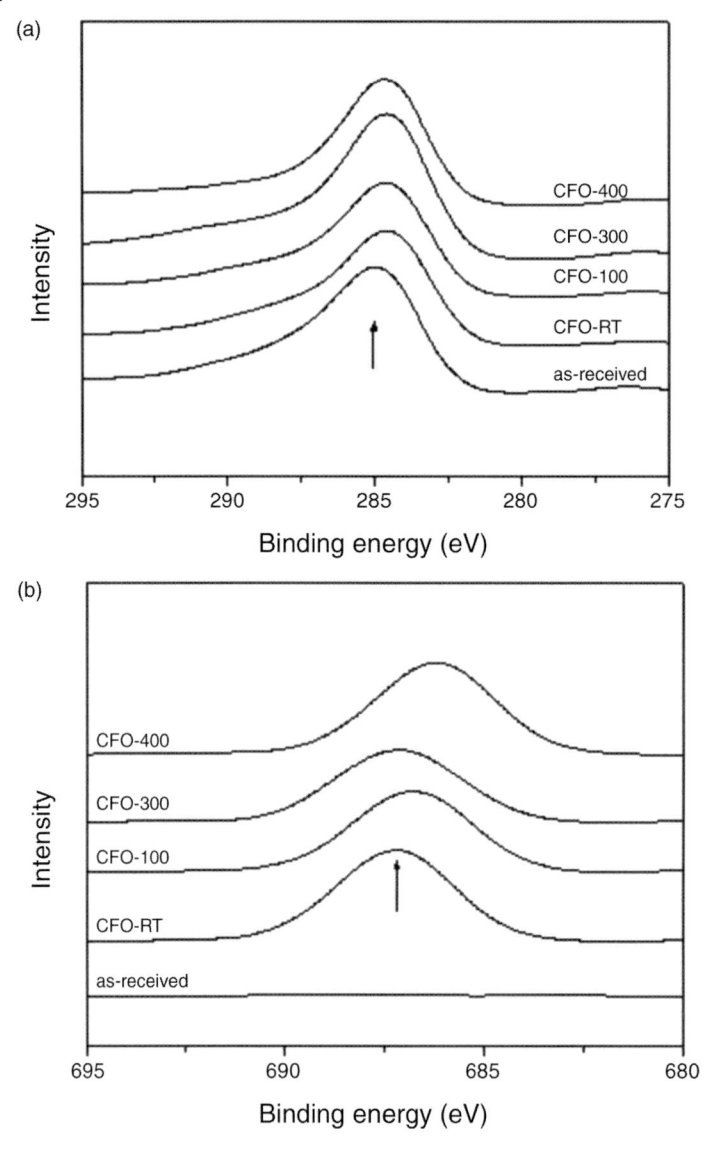

Figure 5.19 C_{1s} and F_{1s} narrow scan XPS spectra of oxyfluorinated carbon fibers. (a) C_{1s} spectra. (b) F_{1s} spectra.

binding energy range of 686.7~688 ± 0.1 eV are observed for the oxyfluorinated carbon fiber specimens.

A quantitative peak analysis is also carried out in order to determine the surface element concentrations that are listed in Table 5.11. It is found that the surface carbon concentrations of the as-received and CFO-100 carbon fiber specimens are 88.38 and 67.72 at%, respectively. A lower surface carbon concentration in the CFO-100 carbon

Table 5.11 Compositions of surface oxyfluorinated carbon fiber samples obtained by XPS measurement (at%).

Specimens	XPS			
	C_{1s}	F_{1s}	O_{1s}	$(F_{1s} + O_{1s})/C_{1s}$
No treatment	88.38	—	9.59	0.109
CF-RT	77.81	11.84	9.06	0.269
CFO-RT	74.21	13.56	10.93	0.330
CFO-100	67.72	19.50	9.27	0.425
CFO-300	79.96	10.64	8.06	0.234
CFO-400	76.93	8.91	10.13	0.247

fiber specimen compared to that in the as-received one is attributed to the bonding of oxygen or fluorine on the carbon fiber surfaces produced by the oxyfluorination. The surface fluorine concentration ranges for the oxyfluorinated carbon fiber specimens are 8.91~19.50 at%. The higher fluorine concentration on the CFO-100 specimen (19.50 at%) surfaces is also attributed to the surface oxyfluorination of the fibers.

Consequently, with increasing amounts of fluorination on the surface, the content of the graphite-type carbon decreases, whereas with increasing fluorination temperatures and oxygen contents, the relative amounts of C–F_x increases, as expected, probably due to kinetic reasons and possibly also due to the fluorination of the bulk phase of the fiber.

5.4.3.2 Contact Angle and Surface Free Energy

In our investigation, we next carry out contact angle measurements and combine them with the XPS results in order to obtain thermodynamic and chemical information on the outermost carbon layer. In contrast to the XPS data, the contact angles are sensitive to solid structure down to only about 1 nm, after which the van der Waals forces at the surface become negligible [91].

In this study, we are particularly interested in the influence of surface oxyfluorination on physical properties, such as wettability and surface polarity. We find that contact angle hysteresis, defined as the difference between the advancing θ_a and the receding θ_r contact angle, occurred for all the surface oxyfluorinated fibers. The hysteresis is not caused by swelling of the fibers due to the penetration of the test liquid, but is rather likely caused by the surface roughness and chemical inhomogeneity, resulting from the burn off at the fiber surface under the oxyfluorine atmosphere. This explanation is in good agreement with the Wenzel approach [7].

In Figure 5.20a, the contact angles of the oxyfluorinated carbon fibers, functions of oxyfluorination temperature and measured for water, indicate that a low temperature at the O_{1s}/C_{1s} ratio of 0.154 (the CFO-100 sample) increases the wettability of such fibers, resulting in a small decrease of the advancing θ_a contact angle. This behavior is caused by the polarization effect of the carbon fiber surfaces' F-atoms bonding to the neighboring C-atoms, which effect makes bonding to air oxygen easier, due to also

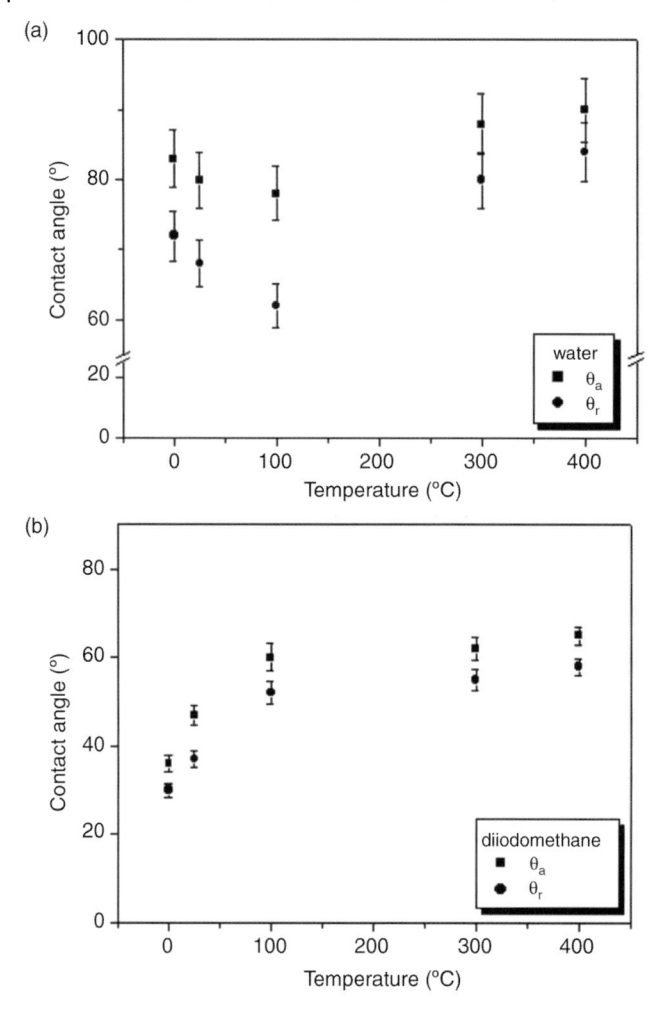

Figure 5.20 Dynamic contact angles of oxyfluorinated carbon fibers as functions of oxyfluorination temperature, measured (a) for water and (b) diiodomethane at 25 °C.

the van der Waals interaction of the slightly hydrophilic C–F bond. At higher oxyfluorination temperatures, the carbon fibers become, as expected, more hydrophobic, that is, θ_a increases. Oxyfluorination of fibers leads to a remarkable increase of the contact angle measured for diiodomethane, as shown in Figure 5.20b. θ_a is enlarged with increasing temperature and fluorine content, due to deteriorated interaction between the fiber surface and the nonpolar test liquid. In the present study, however, there are relatively small contact angle error limits owing to both the very high sensitivity of the Washburn's method to the measurement conditions and the impossibility of modifying all the fiber materials equally.

It is generally known that the specific component γ_S^P highly depends on the surface functional groups, which reflect the essential surface characteristics of

Table 5.12 Surface free energy (γ_S), London dispersive (γ_S^d), and specific (γ_S^p) component of oxyfluorinated carbon fibers as functions of oxyfluorination condition (mJ/m^2).

Specimens	γ_S^d	γ_S^p	γ_S	X_P
As-received	40.35	3.06	43.41	0.07
CFO-RT	30.80	8.61	39.41	0.22
CFO-100	29.70	11.68	41.38	0.28
CFO-300	28.62	9.51	38.13	0.25
CFO-400	28.45	9.10	37.55	0.24

carbon or the partial graphitized carbon framework of the fiber materials, whereas the dispersive component γ_S^d largely depends on the total electron density in the carbon fibers and thus does not vary much from one system to another. The surface free energies and their oxyfluorinated carbon fiber components are listed in Table 5.12. A decrease in the surface free energies is observed. The higher the fluorine content on the carbon fiber surfaces, the weaker the surface free energies. In other words, the polar component of the surface free energy seemingly increases at higher degrees of fluorination, and then decreases slightly again for lower degrees of fluorination and with increasing fluorination temperatures. All the experimental results listed in Table 5.12 are those usually observed for fluorinated carbon materials, considering that C—F bonding varies from chemical to physical bonding with increasing fluorination temperatures (for C—F$_x$ prepared at 100 °C or higher) [92].

Consequently, the nature of C—F bonding is affected mainly by the fluorination temperature and the fluorine content and therefore fluorination can cause a decrease in γ_S^d of the surface free energy. That is, when a certain amount of fluorine is present on carbon fiber surfaces, the amount of graphite may be decreased, which result is in good accordance with those of the present study's surface analysis of oxyfluorinated carbon fibers.

5.4.3.3 Mechanical Interfacial Properties

Fluorine directly reacts with graphite at temperatures greater than 300 °C, yielding physically bonded (C—F)$_n$ and (C$_2$—F)$_n$ with sp^3 hybridization [93]. At temperature less than 300 °C, the reaction rate of fluorine with graphite is drastically reduced. On the other hand, fluorine has been found to intercalate graphite at low temperatures (below 100 °C) in the presence of traces of fluorides such as hydrogen fluoride (HF) to form intercalation compounds C$_x$FH$_\delta$, in which carbon retains its sp^2 hybridization. In this work, thus, we are to investigate the effect of direct mild oxyfluorination on mechanical interfacial properties of oxyfluorinated carbon fiber-reinforced composites using pure F$_2$ and F$_2$/O$_2$ gas mixtures at different oxyfluorination temperatures.

It is generally accepted that good mechanical properties and long durability of the composites largely depend on fiber–matrix interfacial adhesion in case of composites, since load stress transfers from one matrix to the other via the fiber [94].

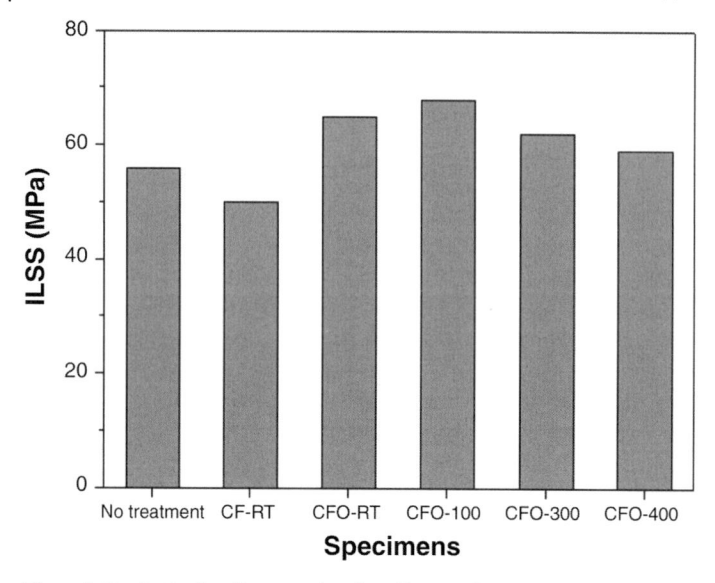

Figure 5.21 ILSS of oxyfluorinated carbon fiber-reinforced composites.

For example, the interlaminar shear strength (ILSS) is improved when the constitutive elements of composites are fabricated in modifications that increase interfacial surface areas and surface functional groups [61].

Figure 5.21 shows the results of ILSS for the carbon fiber-reinforced composites. A good relationship between the characters of oxyfluorinated carbon fiber surfaces and the resulting fiber–matrix adhesions on mechanical interfacial properties of the composites exists in this experimental condition. That is, ILSS value increases with the increasing wettability of the fibers for the degree of adhesion at interfaces due to the oxyfluorination, which can be attributed to the increasing polarity on the fiber surfaces. Also, the maximum strength value of ILSS is obtained at the oxyfluorinated carbon fiber sample condition at 100 °C (CFO-100).

Figure 5.22 shows the results of fracture toughness (K_{IC}) of the composites according to the fluorination conditions. Good relationships are shown between the fluorine content and the resulting fracture toughness of the composites, as listed in Table 5.11. That is, the value of K_{IC} increases with the increasing fluorine content on carbon fiber surfaces, which corresponds to the work of fracture for the degree of adhesion at interfaces. The maximum strength value of K_{IC} is obtained at the O_{1s}/C_{1s} ratio of 100 °C (CFO-100). Therefore, we suggest that additional energy is needed to extend the interfacial crack at this condition, which is attributed to the increasing interfacial adhesion between fibers and matrix [95]. Also, this is also in good agreement with the result of ILSS for the composites.

Generally, the fracture toughness (K_{IC}), elastic modulus (E), and Poisson ratio (ν) can be determined to obtain specific fracture energy of composite materials. Based on Griffith–Irwin equation [96], the resistance to crack propagation (G_{IC}) or specific

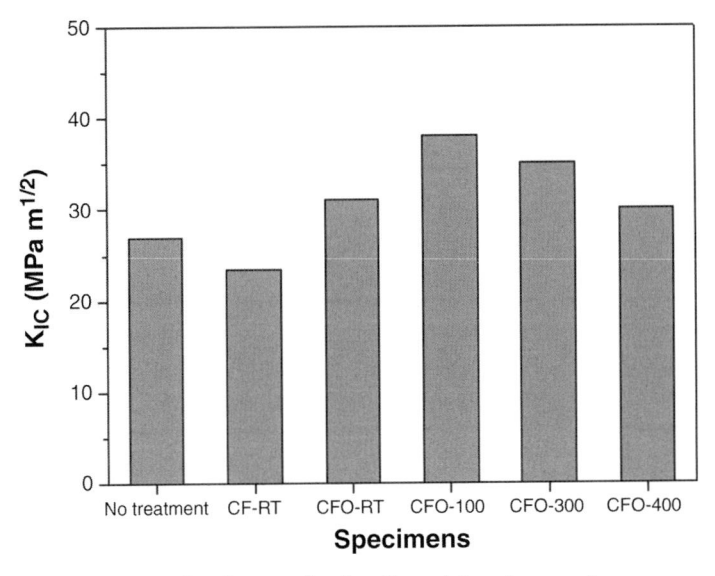

Figure 5.22 K_{IC} of oxyfluorinated carbon fiber-reinforced composites.

fracture energy increases as the fracture toughness and Poisson ratio increase or elastic modulus decreases, as seen in Eq. (5.18):

$$G_{IC} = \left(\frac{K_{IC}^2}{2E}\right) \cdot (1-\nu^2) \tag{5.18}$$

where E and ν are the elastic modulus (230 GPa) and the Poisson ratio (0.37) of carbon fibers, respectively, which are determined from the measurement of the speed of the longitudinal and transverse waves generated by an ultrasonic oscillator.

The specific fracture energy (G_{IC}) of the composites as a function of fluorine concentrations is shown in Figure 5.23. The composites made by fluorination at 100 °C show a significant improvement of G_{IC} studied in any experimental conditions, which probably results from the increase in fracture surface area through more tortuous path of crack growth or the increase in surface functional groups of the carbon fibers.

However, the effects of fiber pull-out, fiber bridging, and fiber fracture are not expected in the present composite system. This fracture mechanism can be due to the following reason: when bend stress (σ) is applied to the specimens with different loads, a shear (τ) or transverse tensile stress (σ_T) can be produced in the fiber–matrix interfaces. When the span length is low, the shear stress is produced in the fiber–matrix interfaces. Thus, the composites show a transition from fiber-dominant fracture to fiber–matrix interface-dominated fracture [97].

Figure 5.24 shows the relationship between surface properties of the carbon fibers and mechanical interfacial properties of the composites. It can be seen that both ILSS and K_{IC} are increased with the increasing ($F_{1s} + O_{1s}$)/C_{1s} ratio. From a good linearity of the results, it is found that there is strong correlation among the surface polarity,

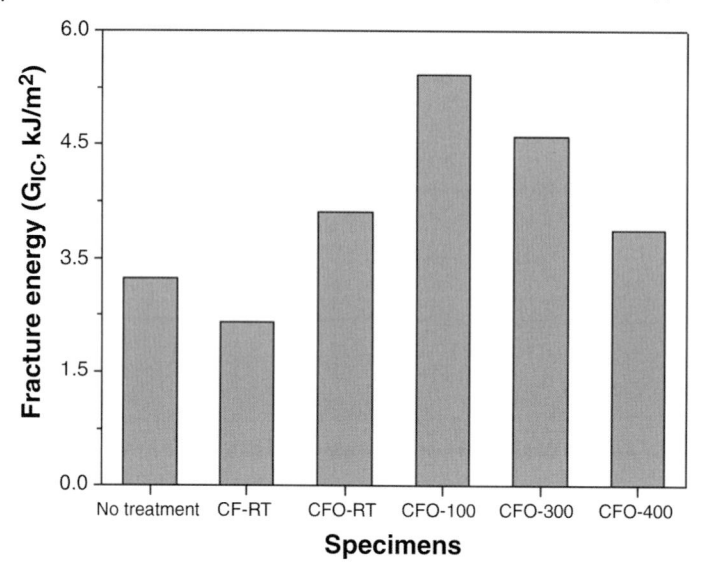

Figure 5.23 G_{IC} of oxyfluorinated carbon fiber-reinforced composites.

acidity, and mechanical interfacial properties of the composites. Consequently, this is a consequence of improving the surface functionality on fibers, resulting in growing fiber–matrix physical adhesion of the composites [98–100].

A good agreement on the fracture surface of the oxyfluorinated carbon fiber-reinforced composites is also shown in Figure 5.24. The morphology of the oxyfluorinated composites indicates the improvement of fiber impregnation into

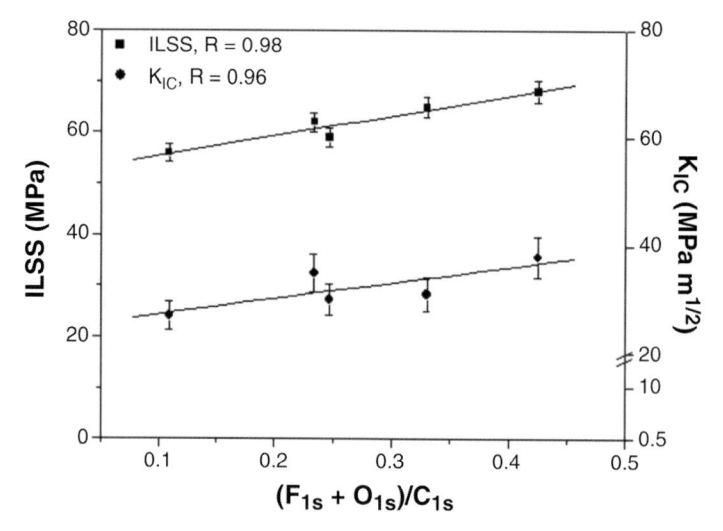

Figure 5.24 Dependence of ILSS and K_{IC} on the $(F_{1s} + O_{1s})/C_{1s}$ ratio of oxyfluorinated carbon fiber-reinforced composites.

matrix resins, which plays a role in increasing the mechanical properties of the composites.

Consequently, a good linear relationship between the surface characteristics and the physical properties, such as morphology, mechanical, and mechanical interfacial properties of surface oxyfluorinated carbon fiber-reinforced composites, is evaluated in the present study, which is probably due to the increasing fluorine/oxygen functional groups on carbon fiber surfaces, resulting in increasing van der Waals interaction between the fibers and the matrix in a composite system.

5.4.4
Effect of Plasma Treatment

5.4.4.1 Surface Characteristics

The broad scan XPS spectra of plasma-treated carbon fibers are shown in Figure 5.25 where the spectrum of the fibers after 10 min of plasma treatment now shows additional strong oxygen lines and a weak nitrogen line, resulting from the formation of oxygen functional groups such as hydroxyl, carbonyl, and carboxyl groups. According to the literature, the as-received PAN-based carbon fibers have about 1.7% N and 3.8% O on the fiber surfaces [101]. Therefore, in this work, original oxygen is found on the outer surface of carbon fibers after plasma modification and newly oxygen content is then added to the fiber surfaces due to the generation of oxygen groups.

Quantitative peak analysis is also carried out to determine the surface element concentrations. The resulting surface element concentrations of plasma-treated carbon fibers are listed in Table 5.13. It is found that the surface carbon concentrations of as-received (CFs-0) and CFs-30 specimens are 88.38 and

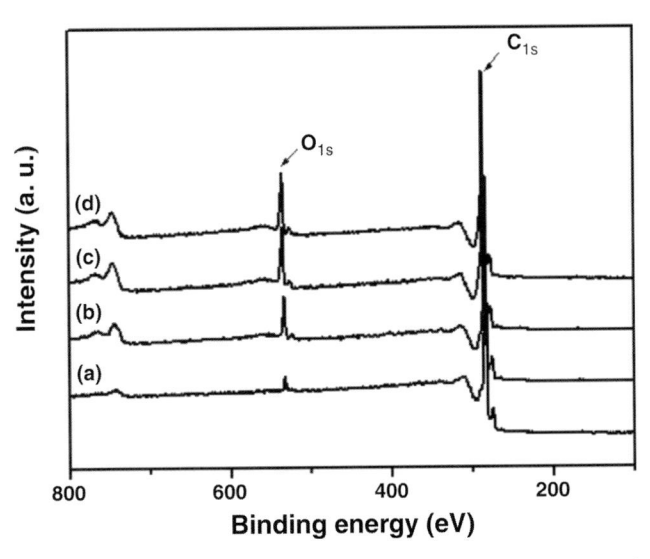

Figure 5.25 XPS wide scan spectra of the plasma-treated PAN-based carbon fibers. (a) CFs-0. (b) CFs-10. (c) CFs-20. (d) CFs-30.

Table 5.13 Atomic concentrations of plasma-treated carbon fibers obtained by XPS measurement (at%).

Specimens	C_{1s}					O_{1s}	N_{1s}	O_{1s}/C_{1s}
	Grap. (284.5)[a]	CHx (285.0)[a]	−C−OR (286.5)[a]	−C=O (287.8)[a]	−COOR (290.0)[a]			
CFs-0	88.38	1.6	0.4	0.4	0.2	9.59	2.12	0.1085
CFs-10	85.81	1.4	2.5	1.4	1.3	12.06	2.15	0.1405
CFs-20	82.40	1.3	3.4	1.6	1.4	16.68	2.19	0.2024
CFs-30	76.21	1.4	3.5	1.6	1.3	20.93	2.54	0.2746
CFs-180	86.72	1.5	1.2	0.7	0.6	11.27	2.46	0.1299

a) Binding energy (eV)

76.21 at%, respectively. A lower surface carbon concentration in the CFs-30 specimen compared to that in the as-received one can be attributed to the bonding of oxygen on the carbon fiber surfaces produced by the plasma modification.

Generally, it is also known that the C_{1s} lines are fitted by five or six component lines with different binding energies corresponding to different functional groups. The atomic concentration of these groups is given in Table 5.13 for the samples before and after plasma treatment. Now regarding the plasma-treated fibers for 30 min, the sample generates functional groups like the hydroxyl or ether group (C-OR), the carbonyl group (C=O), and the carboxyl group (COOR) [102]. The total oxygen concentration amounts to 20.93 at%, further plasma treatment does not enlarge this amount significantly and the concentration of superficial hydrocarbon (CH_x) is nearly fixed. Consequently, a plasma treatment time of 30 min is sufficient and longer times do not change the total oxygen concentration essentially.

5.4.4.2 Contact Angle and Surface Free Energy

Figure 5.26 shows contact angle data of carbon fibers as a function of plasma modification time. As a result, the angle of water is largely decreased with the increasing treatment time. As mentioned above, this suggests that a plasma modification leads to a change in nature of fiber surface from hydrophobic to hydrophilic properties.

It is generally known that the specific component γ_S^{SP} highly depends on surface functional groups, which reflect the surface characteristics of carbon or the partial graphitized carbon framework of the fiber materials, while the dispersive component γ_S^L largely depends on the total electron density in the carbon fibers and do not vary too much from one system to another.

In Table 5.14, the results of the surface free energy and its London dispersive and specific components of the carbon fibers are summarized. As a result, it can be seen that the specific component of surface free energy γ_S^{SP} increases with the increase in the plasma time at 30 min and then marginal decrease in its value, while the London dispersive component γ_S^L remains almost constant. This result indicates that the moderate plasma modification of carbon fibers leads to an increase in the surface free

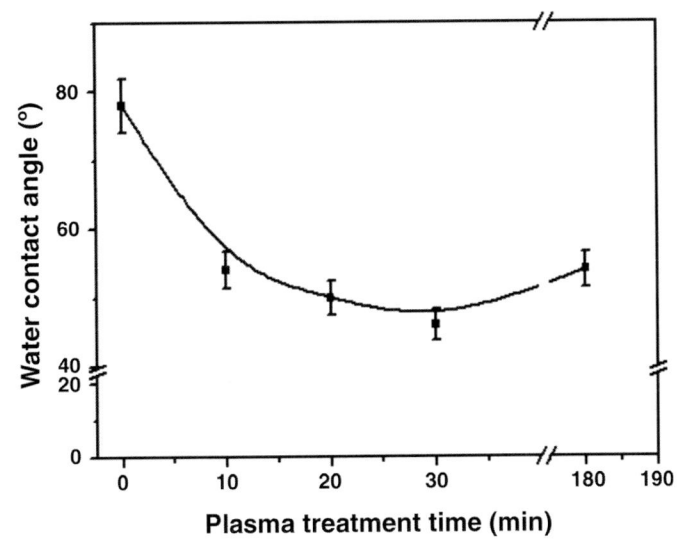

Figure 5.26 Water contact angle data of the carbon fibers as a function of plasma modification time.

energy (γ_S) that is mainly due to the increase in its specific component. Consequently, it is assumed that increase in the specific component of the surface free energy of the fibers plays an important role in improving the degree of adhesion at interfaces between the plasma-treated fibers and the epoxy matrix resin containing oxide functional groups.

5.4.4.3 Mechanical Interfacial Properties

The mode I interlaminar fracture toughness is characterized by the critical strain energy release rate (per unit area) G_{IC}. A variety of analytical approaches concerning the DCB test have been developed allowing the determination of G_{IC} from the experimental data. Typical load–displacement curves of the interlaminar fracture tests of plasma-treated carbon fiber-reinforced epoxy matrix composites are shown in Figure 5.27. In the load–displacement curves of the composites without plasma treatment, the load drops sharply at several points after the peak load, corresponding

Table 5.14 Surface free energy (γ_S), London dispersive (γ_S^L), and specific (γ_S^{SP}) components of plasma-treated carbon fibers (mJ/m^2).

Specimens	γ_S^L	γ_S^{SP}	γ_S
CFs-0	40.35	3.06	43.41
CFs-10	40.80	8.61	49.41
CFs-20	39.90	11.68	55.38
CFs-30	39.92	9.51	58.13
CFs-180	38.45	9.10	47.55

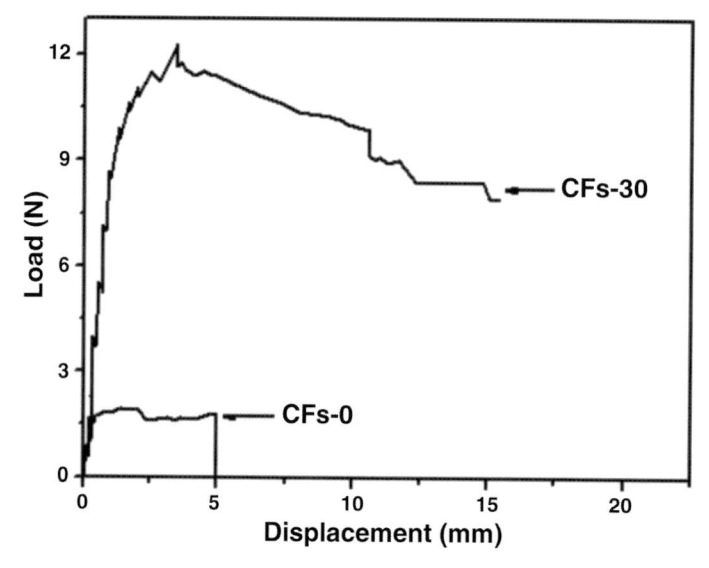

Figure 5.27 Typical load–displacement curves of mode I interlaminar fracture tests for plasma-treated carbon fiber-reinforced epoxy matrix composites: CFs-0 and CFs-30.

to unstable or fast crack propagation. The crack propagates in the main delamination plane without any side cracking and branching. Fiber bridging and breaking behind the crack tip are not observed macroscopically in crack propagation during the test, due to a relatively thick matrix interlaminar layer. The peak load is much higher for plasma-treated composites than for plasma-untreated ones, and the load–displacement curve shows typical stick-slip crack growth behaviors. The curve consists of repetitive drastic load drops followed by a gradual increase of the load. Most of time, delamination propagates unstably and quickly for a short distance, and then it is arrested. After sufficient continued loading, the crack reinitiates. Therefore, the crack propagates in a series of jumps accompanying the sharp load drop.

Figure 5.28a shows the typical relation for unidirectional mode I interlaminar fracture tests (DCB) specimen between load and crack opening displacement (COD) at the loading point. When the crack is propagated, the load is decreased abruptly, as shown in Figure 5.28a. Using the load–COD diagrams, the fracture toughness can be evaluated by compliance method [103]. Therefore, the maximum load was used as the critical load to determine critical energy release rate and fracture toughness. The results of fracture toughness and critical energy release rate by compliance method are shown in Figure 5.28b.

Figure 5.29 shows the relationship between surface properties of the carbon fibers and fracture toughness (K_{IC}) and critical energy release rate (G_{IC}) in mode I of the composites. It can be seen that both K_{IC} and G_{IC} are increased with the increasing O_{1s}/C_{1s} ratio. From a good linearity of the results, it is found that there is strong correlation among the surface characteristics and fracture toughness behaviors of the composites. Consequently, this is a consequence of improving the surface functionality on fibers, resulting in growing fiber–matrix physical adhesion and chemical reaction of the composites [104].

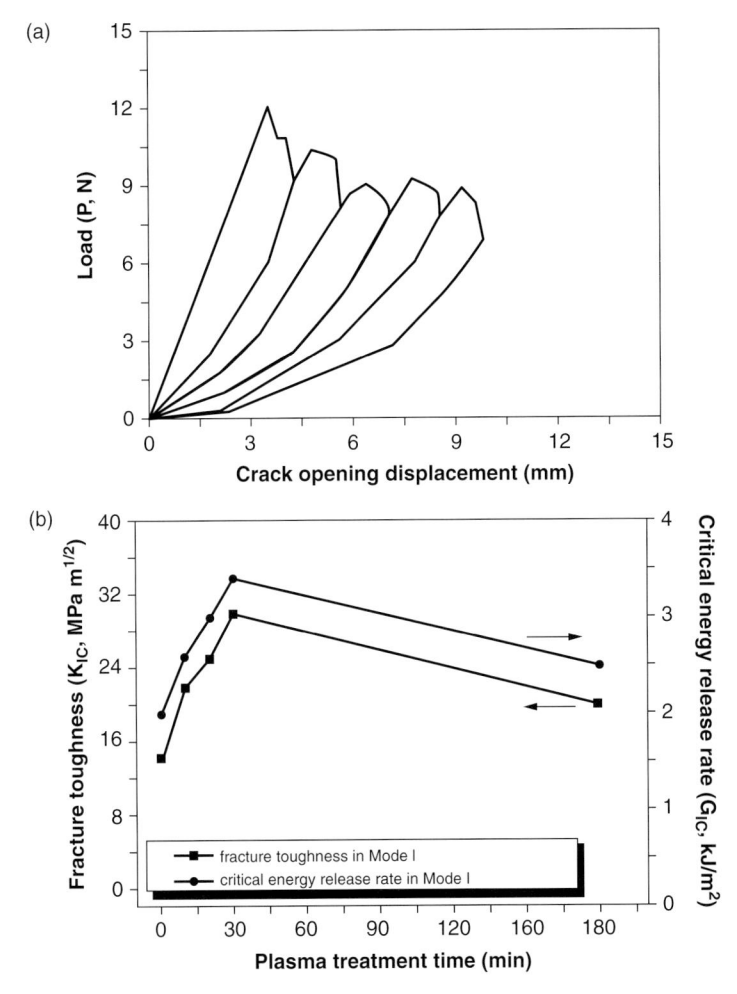

Figure 5.28 Typical load–displacement DCB curves (a), and fracture toughness and critical energy release rate in mode I (b) interlaminar fracture tests for carbon fiber-reinforced epoxy matrix composites in the fiber direction.

Carbon fibers are generally surface treated to maximize the degree of intimate molecular contact that is attained between the adhesive and the substrate during the bonding operation. Furthermore, some degree of surface roughening may often assist in attaining good interfacial contact. Therefore, the fibers are usually given a series of surface treatments to increase their surface reactivity and surface energy and to reduce the number of flaws by gaining new properties that are not found in pure carbon fiber materials. This study also shows that under reasonably surface treatment conditions, significant amounts of surface activity exist at the carbon fiber surfaces, which can activate the fiber surfaces.

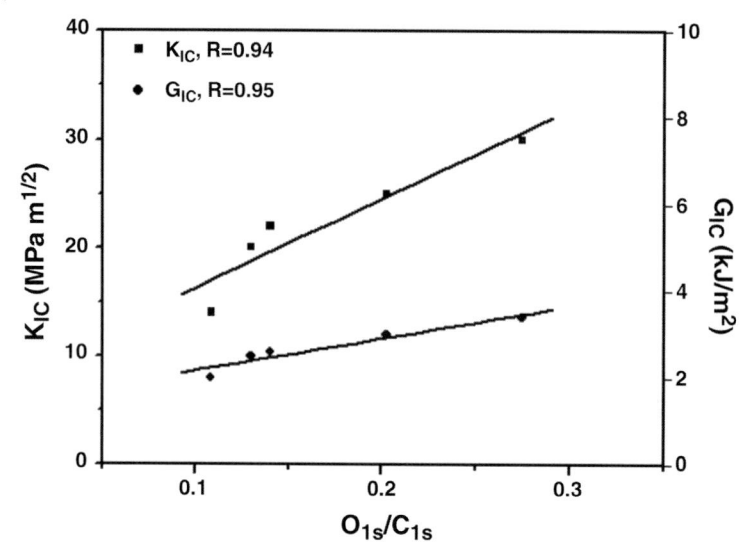

Figure 5.29 Dependence of K_{IC} and G_{IC} on O_{1s}/C_{1s} ratio of plasma-treated carbon fiber-reinforced epoxy matrix composites.

5.5
Applications

After World War II, US manufacturers began producing fiberglass and polyester resin composite boat hulls and radomes (radar cover). The automotive industry first introduced composites into vehicle bodies in the early 1950s. Because of the highly desirable lightweight, corrosion resistance, and high strength characteristics in composites, research emphasis went into improving the material science and manufacturing process. Such efforts led to the development of two new manufacturing techniques known as filament winding and pultrusion, which helped advance the composite technology into new markets. There was a great demand by the recreation industry for composite fishing rods, tennis rackets, ski equipment, and golf clubs. The aerospace industry began to use composites in pressure vessels, containers, and nonstructural aircraft components. The consumers also began installing composite bathtubs, covers, railings, ladders, and electrical equipment. The first civil application in composites was a dome structure built in Benghazi in 1968, and other structures followed slowly.

Carbon fibers are enabling materials for improved performance in many applications; however, it is too expensive for widespread utilization in most high-volume applications. In addition, the methods for manufacturing carbon fiber-reinforced composite structures tend to be slow, labor-intensive, and inconsistent in resulting product quality. As part of the US Department of Energy's (DOE's) FreedomCAR and Fuel Partnership, significant research is being conducted to develop lower cost, high-volume technologies for producing carbon fiber and composite materials containing carbon fiber, with specific emphasis on automotive applications that will reduce vehicle weight and thus fuel demand.

Through DOE sponsorship, Oak Ridge National Laboratory (ORNL) and its partners have been working with the Automotive Composites Consortium to develop technologies that would enable the production of commercial-grade carbon fiber at $5/lb–$7/lb. Achievement of this cost goal would allow the introduction of carbon fiber-based composites into a greater number of applications for future vehicles and accelerate its implantation in other energy-related applications. The goal of lower cost carbon fiber-reinforced composites has necessitated the development of alternative precursors and more efficient production methods, as well as new composites manufacturing approaches.

One of the ways in which transportation fuel efficiency can be improved is by reducing the weight of vehicles. In turn, this can be achieved by the use of lighter weight alloys (e.g., aluminum or magnesium) or polymer matrix composites in place of the more traditional steel. Carbon fiber-reinforced composites offer the greatest potential weight savings of all "nonexotic" materials. There are a number of other energy-related applications for carbon fibers.

5.5.1
Automotives

Carbon fiber-reinforced polymer is used extensively in high-end automobile racing. The high cost of carbon fibers is mitigated by the material's unsurpassed strength-to-weight ratio, and low weight is essential for high-performance automobile racing. Race car manufacturers have also developed methods to give carbon fiber pieces strength in a certain direction, making it strong in a load-bearing direction, but weak in directions where little or no load would be placed on the member. Conversely, manufacturers developed omnidirectional carbon fiber weaves that apply strength in all directions. This type of carbon fiber assembly is most widely used in the "safety cell" monocoque chassis assembly of high-performance race cars.

Many supercars over the past few decades have incorporated CFRP extensively in their manufacture, using it for their monocoque chassis as well as other components. Until recently, the materials had limited use in mass-produced cars because of the expense involved in terms of materials, equipment, and the relatively limited pool of individuals with expertise in working with it. Recently, several mainstream vehicle manufacturers have started to use CFRP in everyday road cars.

5.5.2
Wind Energy

Judicious use of carbon fibers reduces the weight of blades used in large utility scale wind turbines. Larger blades increase energy capture and decrease the cost of electricity. Thus, turbines have been growing larger in the push to reduce the cost of energy generated from the renewable wind resources to make the cost of wind-generated energy competitive with the cost of energy generated from conventional power sources.

5.5.3
Deepwater Offshore Oil and Gas Production

The cost of floating platforms can be significantly reduced by reducing their size. Reducing the weight of many components – especially riser strings, umbilicals, and other structures suspended from these platforms – can be accomplished using carbon fibers, thus enabling platform shrinkage and reducing the cost of deepwater oil and gas production.

5.5.4
Electricity Transmission

Carbon fiber composite cored transmission conductors can double the current-carrying capacity of transmission and distribution power cables. Carbon fiber's low thermal expansion also virtually eliminates the high-temperature line sag (the root cause of the northeastern US blackout in August 2003), thus increasing system reliability.

5.5.5
Commercial Aircraft

Carbon fiber composites have long been used in aircraft and aerospace applications, but the extensive use of carbon fibers in next generation commercial aircraft is reported to reduce fuel demand by about 20% as well as offer advantages that ultimately deliver improved passenger comfort. Much of the fuselage of the new Boeing 787 Dreamliner and Airbus A350 XWB will be composed of CFRP, making the aircraft lighter than a comparable aluminum fuselage, with the added benefit of less maintenance, thanks to CFRP's superior fatigue resistance. Due to its high ratio of strength to weight, CFRP is widely used in microair vehicles (MAVs). In MAVSTAR project, the CFRP structures reduce the weight of MAV significantly. In addition, the high stiffness of the CFRP blades overcomes the problem of collision between blades under strong wind.

5.5.6
Civil Infrastructure

Although composites have been extensively developed and used in the aerospace and marine industries, these technologies do not meet the specific needs of civil applications. Therefore, the composites are increasingly gaining acceptance in the construction and rehabilitation of civil infrastructure that were characterized by large scale and rapid construction. Compared to aircraft and marine component manufacture, civil applications deal in tons of product per project and thousands of projects per year, and are rapidly emplaced in composite bridge structures, thus reducing fuel waste incurred in construction-related traffic delays. Composites are also used for upgrading bridge columns to satisfy seismic standards, as well as patching cracks in concrete and masonry structures.

5.5.7
Other Applications

CFRP has found a lot of use in high-end sports equipment such as racing bicycles. With the same strength, a carbon fiber frame weighs less than a bicycle tubing of aluminum or steel. The choice of weave can be carefully selected to maximize stiffness. The variety of shapes it can be built into has further increased stiffness and also allowed aerodynamic considerations into tube profiles. Carbon fiber-reinforced polymer frames, forks, handlebars, seatposts, and crank arms are becoming more common on medium and higher priced bicycles. Carbon fiber-reinforced polymer forks are used on most new racing bicycles. Other sporting goods applications include rackets, fishing rods, longboards, and rowing shells.

CFRP has also found application in the construction of high-end audio components such as turntables and loudspeakers, again due to its stiffness. It is used for parts in a variety of musical instruments, including violin bows, guitar pickguards, and a durable ebony replacement for bagpipe chanters. It is also used to create entire musical instruments such as Blackbird Guitars carbon fiber rider models, Luis and Clark carbon fiber cellos, and Mix carbon fiber mandolins.

In firearms, it can substitute for metal, wood, and fiberglass in many areas of a firearm in order to reduce overall weight. However, while it is possible to make the receiver out of synthetic material such as carbon fiber, many of the internal parts are still limited to metal alloys as current synthetic materials are unable to function as replacements.

Shoes manufacturers may use carbon fiber as a shank plate in their basketball sneakers to keep the foot stable. It usually runs the length of the sneaker just above the sole and is left exposed in some areas, usually in the arch of the foot.

5.6
Conclusions

Carbon fiber-reinforced polymer composites are a type of fiber composite material in which carbon fibers constitute the fiber phase. Carbon fibers are a group of fibrous materials comprising essentially elemental carbon. This is prepared by pyrolysis of organic fibers. The main impetus for the development of carbon fibers has come from the aerospace industry with its need for a material with combination of high strength, high stiffness, and low weight. Recently, civil engineers and construction industry have begun to realize that carbon fibers have potential to provide remedies for many problems associated with the deterioration and strengthening of infrastructure. Effective use of CFRC could significantly increase the life of structures, minimizing the maintenance requirements.

There are different methods of manufacturing CFRC: continuous reinforcement process (filament winding, pultrusion, and hand lay-up), molding processes (matched die molding, autoclave molding, and vacuum bagging), and resin injection processes (resin transfer molding and reaction injection molding).

CFRCs are alkali resistant and resistant to corrosion; hence, they are used for corrosion control and rehabilitation of reinforced concrete structures. They also have low thermal conductivity, high strength to weight ratio eliminating requirements of heavy construction equipment, supporting structures, and a short curing time for taking application time shorter that reduces the project duration and down time of the structure to a great extent. CFRCs possess high ultimate strain; therefore, they offer ductility to the structure and they are suitable for earthquake-resistant applications. CFRCs have high fatigue resistance. So they do not degrade, which easily alleviates the requirement of frequent maintenance. Due to the lightweight of prefabricated components in CFRC, they can be easily transported. This thus encourages prefabricated construction and reduces site erection, labor cost, and capital investment requirements. In recent years, it has been found that the chemistry and manufacturing techniques for thermosetting plastics like epoxy are often poorly suited to mass production. One potentially cost-saving and performance-enhancing measure involves replacing the epoxy matrix with thermoplastic materials such as nylon or polyketone. Boeing's entry in the Joint Strike Fighter competition included a delta-shaped carbon fiber-reinforced thermoplastic wing, but difficulties in fabrication of this part contributed to Lockheed Martin winning the competition. Other materials can also be used as the matrix for carbon fibers. Due to the formation of metal carbides and corrosion considerations, carbon has seen limited success in metal matrix composite applications. However, the usage of composite materials like CFRC is still not widely recognized. The lack of knowledge of technology using CFRC and the simplicity of it will make some people hesitant to use it.

In conclusion, this chapter is focused on the state of the art of the surface modification of carbon fibers using different methods to improve the fiber/matrix interfacial adhesion of CFRC. The surface treatment proved efficient method, since it is easier to control the variables (time, concentration of the electrolytes, temperature, pressure, etc.) than by any other technique that can improve mechanical and mechanical interfacial properties of the polymer composites reinforced with surface-treated carbon fibers using many application fields.

References

1 Donnet, J.B. and Bansal, R.C. (1990) *Carbon Fibers*, 2nd edn, Marcel Dekker, New York.
2 Morgan, P.E. (2005) *Carbon Fibers and Their Composites*, CRC Press, Boca Raton, FL.
3 Park, S.J. (2006) *Carbon Materials*, Daeyoungsa, Seoul.
4 Paiva, M.C.R., Bernardo, C.A., and Nardin, M. (2000) *Carbon*, **38**, 1323.
5 Schwartz, M. (1992) *Composite Materials Handbook*, 2nd edn, McGraw-Hill, New York.
6 Wu, S. (1982) *Polymer Interface and Adhesion*, Marcel Dekker, New York.

7 Park, S.J., Seo, M.K., Ma, T.J., and Lee, D.R. (2002) *Journal of Colloid and Interface Science*, **252**, 249.

8 Ho, K.K., Lamoriniere, S., Kalinka, G., Schulz, E., and Bismarck, A., (2007) *Journal of Colloid and Interface Science*, **313**, 476.

9 Yosomiya, R., Morimoto, K., Nakajima, A., Ikada, Y., and Suzuki, T. (1990) *Adhesion and Bonding in Composites*, Marcel Dekker, New York.

10 May, C.A. (1988) *Epoxy Resins: Chemistry and Technology*, 2nd edn, Marcel Dekker, New York.

11 Reich, R. (1978) (Sep.: 1978) *Solid State Technology*, p. 82.

12 Melliar-Smith, C.M., Matsuoka, S., and Hubbauer, P. (1980) *Plastic and Rubber: Materials and Application*, **5**, 49.

13 Courtaulds Ltd. (1969) British Patent No. 144,341, March 19.

14 Ray, J.D., Steinsiger, S., and Cass, B.A. (1970) US Patent No. 16,251, March 3.

15 Morganite Research Development Ltd. (1969) British Patent No. 493,571, October 8.

16 Paul, J.T. (1974) Appl. No. 25208/74. Patent Specification 1433712.

17 Wang, S. and Garton, A. (1992) *Journal of Applied Polymer Science*, **45**, 1743.

18 Wu, Z., Pittman, C.U., Jr., and Gardner, S.D. (1995) *Carbon*, **33**, 597.

19 Fukunaga, A. and Ueda, S. (2000) *Composites Science and Technology*, **60**, 249.

20 Montes-Moran, M.A., Vanhattum, F.W.J., Nunes, J.P., Martinez-Alonso, A., Tascon, J.M.D., and Bernardo, C.A. (2005) *Carbon*, **43**, 1795.

21 Park, S.J. and Kim, M.H. (2000) *Journal of Materials Science*, **35**, 1901.

22 Oliver, S.R.J., Bowden, N., and Whitesides, G.M. (2000) *Journal of Colloid and Interface Science*, **224**, 425.

23 Chang, Y. and Ohara, H.J. (1992) *Journal of Fluorine Chemistry*, **57**, 169.

24 Beydon, R., Bernhart, G., and Segui, Y. (2000) *Surface and Coatings Technology*, **126**, 39.

25 Hage, E., Costa, S.F., and Pessan, L.A. (1997) *Journal of Adhesion Science and Technology*, **11**, 1491.

26 Li, H., Liang, H., He, F., Huang, Y., and Wan, Y. (2009) *Surface and Coatings Technology*, **203**, 1317.

27 Hoecker, F. and Kocsis, J.K. (1996) *Journal of Applied Polymer Science*, **59**, 139.

28 Park, S.J. (1999) *Interfacial Forces and Fields: Theory and Applications* (ed. J.P. Hsu), Marcel Dekker, New York, pp. 385–442.

29 Park, S.J., Jin, J.S., and Lee, J.R. (2000) *Journal of Adhesion Science and Technology*, **14**, 1677.

30 Park, S.J. and Lee, J.R. (1998) *Journal of Materials Science*, **33**, 647.

31 Herrick, J.W. (1968) Proceedings of the 23rd Annual Technical Conference, Reinforced Plastics Division, Section 16A, February.

32 Donnet, J.B. (1982) *Carbon*, **20**, 267.

33 Fitzer, E. and Weis, R. (1987) *Carbon*, **25**, 455.

34 Rand, B. and Robinson, R. (1977) *Carbon*, **15**, 2547.

35 Scola, D.A. and Brooks, S.C. (1970) 25th Annual Technical Conference of the Society of the Plastics Industry Meeting, Washington, DC.

36 Proctor, A. and Sherwood, P.M.A. (1982) *Journal of Electron Spectroscopy and Related Phenomena*, **27**, 39.

37 Brelant, S. (1981) *Carbon*, **19**, 329.

38 Fitzer, E., Geigl, K.H., Huttner, W., and Weiss, R. (1976) *Carbon*, **14**, 389.

39 Drzal, L.T., Rich, M.J., and Lloyd, P.F. (1982) *Journal of Adhesion*, **16**, 1.

40 Kozlowski, C. and Sherwood, P.M.A. (1985) *Journal of the Chemical Society*, **81**, 2745.

41 Neumann, A.W. and Spelt, J.K. (1996) *Applied Surface Thermodynamics*, Marcel Dekker, New York.

42 Kinloch, A.J. (1987) *Adhesion and Adhesives*, Chapman & Hall, London.

43 Oya, N. and Johnson, D.J. (2001) *Carbon*, **39**, 635.

44 Severini, F., Formaro, L., Pegoraro, M., and Posca, L. (2002) *Carbon*, **40**, 735.

45 Mallick, P.K. (1993) *Fiber-Reinforced Composites: Materials, Manufacturing, and Design*, 2nd edn, Marcel Dekker, New York.

46 Kumar, S., Rath, T., Mahaling, R.N., Reddy, C.S., Das, C.K., Pandey, K.N.,

Srivastava, R.B., and Yadaw, S.B. (2007) *Materials Science and Engineering B*, **141**, 61.
47. Jensen, C., Zhang, C., and Qiu, Y. (2003) *Composite Interfaces*, **10**, 277.
48. Kawai, M., Yajima, S., Hachinohe, A., and Kawase, Y. (2001) *Composites Science and Technology*, **61**, 1285.
49. Park, S.J. and Kim, B.J. (2005) *Materials Science and Engineering A*, **408**, 269.
50. Park, S.J., Seo, M.K., and Rhee, K.Y. (2003) *Materials Science and Engineering A*, **356**, 219.
51. Ryu, S.K., Park, B.J., and Park, S.J. (1999) *Journal of Colloid and Interface Science*, **215**, 167.
52. Lee, W.S., Sue, W.C., and Lin, C.F. (2000) *Composites Science and Technology*, **60**, 1975.
53. Lawton, R.A., Price, C.R., Runge, A.F., Doherty, W.J., and Saavedra, S.S. (2005) *Colloid Surface A*, **253**, 213.
54. Paiva, M.C., Bernardo, C.A., and Nardin, M. (2000) *Carbon*, **38**, 1323.
55. Nakajima, T. (2000) *Journal of Fluorine Chemistry*, **105**, 229.
56. Park, S.J., Seo, M.K., and Lee, J.R. (2003) *Journal of Colloid and Interface Science*, **268**, 127.
57. Schwartz, M. (1992) *Composite Materials Handbook*, 2nd edn, McGraw-Hill, New York.
58. Michalski, M.C. and Saramago, B.J.V. (2000) *Journal of Colloid and Interface Science*, **227**, 380.
59. Chwastiak, S. (1973) *Journal of Colloid and Interface Science*, **42**, 298.
60. Seo, M.K., Park, S.J., and Lee, S.K. (2005) *Journal of Colloid and Interface Science*, **285**, 306.
61. Park, S.J. and Kim, J.S. (2001) *Carbon*, **39**, 2011.
62. Gopal, P., Dharani, L.R., and Blum, F.D. (1997) *Polymer Composites*, **5**, 327.
63. Ahearn, C. and Rand, B. (1996) *Carbon*, **34**, 239.
64. Proctor, A. and Sherwood, P.M.A. (1982) *Surface and Interface Analysis*, **4**, 212.
65. Denison, P., Jones, F.R., and Watts, J.F. (1986) *Surface and Interface Analysis*, **9**, 431.
66. Denison, P., Jones, F.R., and Watts, J.F. (1987) *Journal of Physics D*, **20**, 306.
67. Kozlowski, C. and Sherwood, P.M.A. (1986) *Carbon*, **24**, 357.
68. Owens, D.K. and Wendt, R.C. (1969) *Journal of Applied Polymer Science*, **13**, 1741.
69. Kaelble, D.H. (1970) *Journal of Adhesion*, **2**, 66.
70. Park, S.J., Park, W.B., and Lee, J.R. (1999) *Polymer Journal*, **31**, 28.
71. Park, S.J. and Park, B.J. (1999) *Journal of Materials Science Letters*, **18**, 47.
72. Marieta, C., Schulz, E., Irusta, L., Gabilondo, N., Tercjak, A., and Mondragon, I. (2005) *Composites Science and Technology*, **65**, 2189.
73. Lindsay, B., Abel, M.L., and Watts, J.F. (2007) *Carbon*, **45**, 2433.
74. Yang, F., Zhang, X., Han, J., and Du, S. (2008) *Materials Letters*, **62**, 2925.
75. Costa, J., Turon, A., Trias, D., Blanco, N., and Mayugo, J.A. (2005) *Composites Science and Technology*, **65**, 2269.
76. Wagner, C.D., Riggs, W.M., Davis, L.E., and Moulder, J.F. (1979) *Handbook of X-Ray Photoelectron Spectroscopy*, Perkin-Elmer Corp., Norwalk.
77. Gardner, S.D., Singamsetty, C.S.K., Booth, G.L., He, G.R., and Pittman, C.U., Jr. (1995) *Carbon*, **33**, 587.
78. Yang, W.P., Costa, D., and Marcus, P. (1994) *Journal of the Electrochemical Society*, **141**, 2669.
79. Brown, N.M.D., Hewitt, J.A., and Meenan, B.J. (1992) *Surface and Interface Analysis*, **18**, 187.
80. Chung, D.D.L. (2001) *Carbon*, **39**, 279.
81. Podgoric, S., Jones, B.J., Bulpett, R., Troisi, G., and Franks, J. (2009) *Wear*, **267**, 996.
82. Tang, Y., Liu, H., Zhao, H., Liu, L., and Wu, Y. (2008) *Materials & Design*, **29**, 257.
83. Li, H., Wan, Y., Liang, H., Li, X., Huang, Y., and He, F. (2009) *Applied Surface Science*, **256**, 1614.
84. Kinloch, A.J. (1987) *Adhesion and Adhesives*, Chapman & Hall, London.
85. Fowkes, F.M. (1963) *The Journal of Physical Chemistry*, **67**, 2538.
86. Hoecker, F. and Kocsis, J.K. (1996) *Journal of Applied Polymer Science*, **59**, 139.
87. Park, J.M. and Bell, J.P. (1983) *Adhesion Aspects of Polymeric Coating*, Plenum, New York.

88 Wu, S. (1971) *Journal of Polymer Science Part C*, **34**, 19.
89 Wu, S. (1980) *Adhesion and Adsorption of Polymers, Polymer Science and Technology* (ed. L.H. Lee), vol. 12A, Plenum, New York, p. 53.
90 Griffith, A.A. (1921) *Philosophical Transactions of the Royal Society of London*, **221**, 163.
91 Lu, W., Fu, X., and Chung, D.D.L. (1998) *Cement and Concrete Research*, **28**, 783.
92 Bismark, A., Tahhan, R., Spriner, J., Schulz, A., Klapötke, T.M., Zell, H., and Michaeli, W. (1997) *Journal of Fluorine Chemistry*, **84**, 127.
93 Watanabe, N., Nakajima, T., and Ohsawa, N. (1982) *Bulletin of the Chemical Society of Japan*, **55**, 2029.
94 Varelidis, P.C., Papakostopoulos, D.G., Pandazis, C.I., and Papaspyrides, C.D. (2000) *Composites Part A*, **31**, 549.
95 Shim, J.S., Park, S.J., and Ryu, S.K. (2001) *Carbon*, **39**, 1635.
96 Kim, I.S. (1998) *Materials Research Bulletin*, **33**, 1069.
97 Park, S.J. and Donnet, J.B. (1998) *Journal of Colloid and Interface Science*, **206**, 29.
98 Park, S.J., and Jin, J.S. (2001) *Journal of Colloid and Interface Science*, **242**, 174.
99 Basova, Yu.V., Hatori, H., Yamada, Y., and Miyashita, K. (1999) *Electrochemistry Communications*, **1**, 540.
100 Fjeldly, A., Olsen, T., Rysjedal, J.H., and Berg, J.E. (2001) *Composites Part A*, **32**, 373.
101 Sarmeo, D., Blazewicz, S., Mermoux, M., and Touzain, Ph. (2001) *Carbon*, **39**, 2049.
102 Ham, H.T., Koo, C.M., Kim, S.O., Choi, Y.S., and Chung, I.J. (2004) *Macromolecular Research*, **12**, 384.
103 Wilkins, D.J., Eisenmann, J.R., Camin, R.A., Margolis, W.S., and Benson, R.A. (1980) *Characterizing Delamination Growth in Graphite–Epoxy, Damage in Composite Materials*, ASTM Special Technical Publication, vol. 775, p. 168.
104 Park, J.M. and Kim, J.W. (2002) *Macromolecular Research*, **10**, 24.

6
Glass Fiber-Reinforced Polymer Composites
Sebastian Heimbs and Björn Van Den Broucke

6.1
Introduction

The successful history of fiber-reinforced composites as lightweight materials in various engineering structures in the last decades was basically initiated by the mass production of glass fibers and glass fiber-reinforced plastics (GFRP). Besides the existence of other reinforcement fibers made from carbon or aramid, the vast majority of all fiber-reinforced composites today are still made from glass fibers. One of the reasons for this situation lies clearly in the relatively low price of glass fibers compared to other fibers. However, many other specific characteristics like the high tensile strength, high chemical resistance, or electrical insulation make glass fibers the ideal reinforcement for many applications, which are highlighted in this chapter. Although having been in use for more than 70 years, glass fiber composites are far away from going into retirement, especially as today's need for energy savings pushes the use of lightweight materials in all transportation industries and in particular in automotive manufacturing.

Although it is assumed that glass fibers are the first man-made reinforcement fibers that have already been produced in Egyptian and Roman times, the industrial mass production dates back to the 1930s. In 1938, two pioneering American companies combined and formed the Owens-Corning Fiberglas Corporation that for the first time continuously produced glass filaments. The breakthrough as a reinforcement fiber for composite materials then took place in the early 1940s in World War II resulting from electronic needs. Radomes of military aircraft were produced from GFRP, which is the first important application of this material. In general, most historical developments in the field of glass fibers and glass fiber composites go back to aeronautic and military applications. Since the start of the fiber-reinforced plastics industry in these days, GFRP found its way into numerous applications and markets, for example, in aerospace, automotive, marine, wind energy, construction, and leisure products [1].This chapter gives an overview on the

6.2
Chemical Composition and Types

6.2.1
Chemical Structure of Glass

Various glass fiber types are available for technical applications, each having a different chemical composition leading to unique properties for special purposes. All of those types have in common that they are based on silica, SiO_2, which exists in its pure form as a polymer, $(SiO_2)_n$. Silica has no true melting point, but at high temperatures around 2000 °C it softens and the molecules can move freely. If it is cooled very quickly from such temperatures, the molecules are unable to form a crystalline structure as in quartz (crystalline form of silica). Although a three-dimensional network is developed, the structure of glass is amorphous (noncrystalline) without orientation and with globally isotropic properties (Figure 6.1). The high working temperature of silica is a big disadvantage. Therefore, different other molecules, mostly oxides, are introduced into the glass in order to modify the network structure and to lower its working temperature and to provide other characteristics for special applications. Resulting from this choice of additional constituents, various different types of glass fibers have been developed during the last decades, which differ in the percentage of these additional molecules and the resulting global properties. These types, which are typically abbreviated with an individual letter, are shortly introduced as follows.

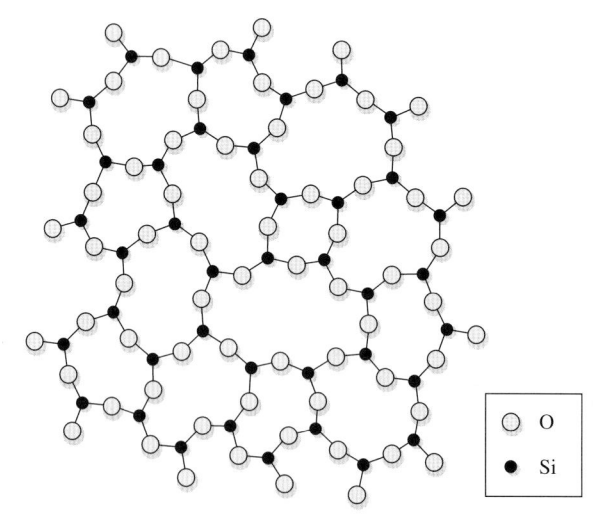

Figure 6.1 Illustration of chemical network of glassy silica.

6.2.2
Glass Fiber Types

A-glass (A = alkali): The first variant for the production of glass fibers was alkali-lime glass or A-glass, which is also found in windows or bottles. This glass is very cheap but shows poor resistance to alkali. Today it is typically not used for glass fibers anymore.

E-glass (E = electrical): E-glass was developed as a second form of glass fibers, named after its electrical insulation properties and radio-signal transparency. For these reasons, this aluminoborosilicate glass was the excellent fiber for the first applications in aircraft radomes, where this transparency of electromagnetic signals is of greatest importance. This type was the first to be produced in filaments and today still forms more than 90% of all produced glass fibers worldwide for glass fiber-reinforced composites [2].

C-glass (C = chemical): C-glass with high boron oxide content provides an increased corrosion resistance to acids compared to E-glass, which would be destroyed early in such an environment. Therefore, it is mostly used for chemical applications.

D-glass (D = dielectric): D-glass is a borosilicate glass that has a very low dielectric constant and was especially developed for modern aircraft radomes and covers of radar antennas.

L-glass (L = low-loss): L-glass is a recently introduced glass fiber by AGY showing a very low dielectric constant and low dissipation factor ideally suited for printed wiring board applications and designs requiring increased signal speeds and better signal integrity than with traditional E-glass.

M-glass (M = modulus): M-glass or YM-31A is a high modulus glass fiber, showing higher mechanical stiffness properties than E-glass. It was originally developed under an Air Force contract in the 1960s in order to increase the mechanical performance of glass fiber composites [3]. It is not produced anymore due to the toxicity of BeO.

R-glass (R = resistance), S-glass (S = strength): R-glass and S-glass are both aluminosilicate glass fibers with higher tensile strength, elastic modulus, impact resistance, and fatigue properties than E-glass with R-glass being the European version and S-glass the American version. These fibers are also more heat resistant as their strength reduces at 250 °C and not already at 200 °C. The development dates back to the 1960s for military applications like aircraft components and missile casings. These fibers have the highest tensile strength of all common reinforcement fibers, even higher than carbon and aramid fibers. The difference in chemical composition and higher manufacturing cost make them more expensive than conventional E-glass fibers. Therefore, a low-cost version of S-glass, so-called S2-glass, was developed. The main difference is that it is produced with less-stringent nonmilitary specifications. However, the tensile strength and elastic modulus are typically similar to S-glass [4].

T-glass (T = thermal): T-glass has the lowest thermal expansion coefficient of all glass fibers. It is a high performance fiber introduced by Nittobo Boseki Co Ltd and a comparable Japanese variant of the European R-glass and American S-glass [5].

Besides a 40% lower coefficient of thermal expansion (CTE) than E-glass, it has a 36% higher tensile strength and 16% higher modulus [6].

Z-glass (Z = zirconia): Z-glass has a high resistance to alkaline environments, comparable to the acid resistance of C-glass, both being developed for chemical applications [7].

An overview on the chemical composition of the different glass fiber types is given in Table 6.1, based on data published in Refs [2–4, 6–8]. These values typically differ slightly in percentage, but must be within a specific range.

6.3
Fabrication of Glass Fibers

6.3.1
Fiber Production

Glass fibers are produced by a viscous drawing process as illustrated in Figure 6.2. A furnace, which is continuously loaded with raw materials, produces molten glass at a temperature of 1400–1700 °C. The glass melt is constantly monitored and refined to ensure a proper composition. At a viscosity level between approx. 60 and 100 Pa s the glass flows into an electrically heated platinum–rhodium alloy bushing containing a large number of holes (200–400) in its base. Recent use of alloys containing traces of zirconia, yttria, or thoria enable the manufacturing of bushing plates with up to 6000 holes [9].

Table 6.1 Chemical composition in weight percent of different glass fiber types.

Chemical component	A	C	E	D	M	R	S	Z
SiO_2	72	64	54	74	53.7	60	65	68
Al_2O_3	1.5	4	14	0.2		25	25	0.7
CaO	9.5	14.5	19	0.5	12.9	9		
MgO	2.3	3	3	0.1	9	6	10	
Na_2O	13.5	9	0.4	1				11.8
K_2O	0.5	0.5	0.5	1				0.5
B_2O_3	0.2	5	8.5	23				
TiO_2			0.2	0.1	7.9			1.5
Fe_2O_3	0.2		0.2		0.5			
SO_3	0.3			0.1				
Li_2O					3			1
ZrO_2					2			16.5
BeO					8			
CeO_2					3			

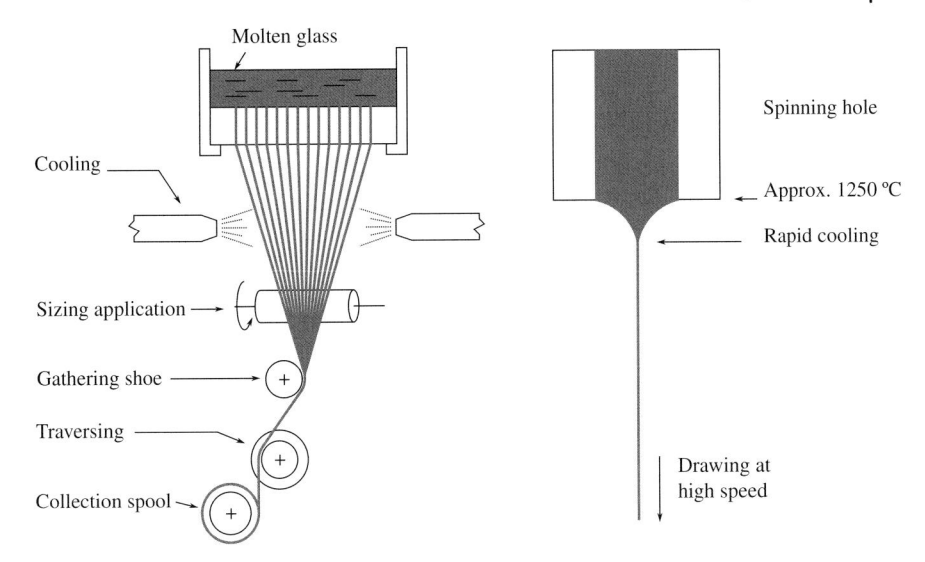

Figure 6.2 Schematic of the glass fiber production process.

The glass drops emerging from each hole are drawn into exactly one filament by pulling at speeds up to 50 m/s. The diameter of the resulting filament, typically around 5–20 μm, depends on the size of the bushing holes, the viscosity of the glass melt (i.e., the temperature) and the pulling speed of the filament [10].

Due to a significant reduction of diameter between the bushing hole and the filament a rapid cooling rate is achieved (>10 000 °C/s). This rapid cooling is required to freeze the internal molecular structure between 200 and 300 °C, which results in a significantly higher modulus and chemical resistance compared to bulk glass [10]. The filaments are then further cooled by sprayed water.

6.3.2
Sizing Application

The next step in the production of glass fibers is the coating of the filament with a sizing. The sizing agent is applied as an aqueous dispersion or emulsion in combination with a rubber roller. The size is a crucial component for glass fibers and consists of several components. The simplest component is a lubricant, often a mineral oil, which serves as protection agent against the highly abrasive nature between the individual filaments and aids in further processing steps such as filament winding or weaving [10]. A binder component is included in the sizing to bond or hold the filaments together into a glass fiber strand. Typical polymeric binders are polyvinylacetate, polyacrylate, or polystyrene. The third major component in a sizing is the coupling agent that works as an adhesion promoter between the filaments and the resin in the later composite. Common materials used as coupling agents are silane, polyvinylacetate, or polymers based on epoxy, polyester, or polyurethane [9].

After the application of the sizing agent, a bundle of filaments are gathered into a strand that typically contains 52, 102, or 204 fibers, depending on the number of holes in the bushing [10]. Finally, the glass fiber strands are collected on a spool and are next conveyed to a drying oven, where moisture is removed and the components of the sizing are cured.

6.4
Forms of Glass Fibers

6.4.1
Commercially Available Forms of Glass Fibers

Glass fibers are commercially available in different forms, necessary for the individual manufacturing processes of GFRP composite parts. Different terms are used for the several forms, which are standardized and, for example, defined in Ref. [11]. The basis are the individual *filaments* that are combined to *strands* during the manufacturing process with untwisted, parallel orientation. A group of untwisted, parallel strands – often 30–60 – can be combined to form a *roving*, which is typically wound on a cylindrical spool [9] (Figure 6.3a). Such rovings are used for pultrusion or filament winding manufacturing processes of composite parts. Furthermore, they are used for woven fabrics and prepreg manufacturing by impregnating the fibers with a polymeric matrix. In contrast to the untwisted roving, one or several strands can be twisted to form a *yarn*, with typically 28–40 twists/m [6, 12]. By this method, both integrity and workability are increased for later processing to textile fabrics. Woven fabrics with their bidirectional properties are perhaps the most versatile form that can be used from hand lay-up to autoclave processing [13] (Figure 6.3b). The properties depend on the weave pattern (e.g., plain weave, twill weave, satin weave, etc.), yarn type and yarn density in yarns per unit length. In addition to two-dimensional woven fabrics, braided textile preforms are increasingly used in engineering applications due to the high flexibility in complex geometries, possible reinforcement angles and changes in the cross-section shape and in the wall thickness [14–16]. Figure 6.3c shows a glass fiber tubular braid as a preform for a composite tube.

Another form of glass fibers are *chopped strands*, which are produced by cutting rovings or strands into short lengths, typically 1.5, 3, 6, 12, 25, or 50 mm [6, 9]. The shorter lengths are used for injection-molding operations, the longer strands are used to form *chopped strand mats* (CSM) [4]. Such nonwoven mats are produced by compacting the chopped strands with random orientation on a flat surface using a chemical binder, leading to nearly equal in-plane properties in all directions (Figure 6.3d). Their typical application are hand lay-up moldings. *Continuous strand mats* are also available, made from unchopped strands that are mechanically interlocked and require less chemical binder, resulting in better draping qualities suitable for matched-die molding [6, 12]. Short glass fiber or continuous glass fiber

Figure 6.3 (a–d) Photographs of different forms of glass fibers: (a) roving on coil, (b) woven fabric, (c) braided fabric, (d) chopped strand mat.

mats are additionally used for *glass mat thermoplastic* (GMT) composites, for example, for the industrial production of thermoplastic automobile parts [17].

Chopped strands in combination with thermoplastic or thermosetting resins are as well utilized for molding compounds. Typical examples based on a thermoset matrix are *sheet molding compound* (SMC) or *bulk molding compound* (BMC) that is, for example, used for high-volume low-cost automotive parts. One further type are *milled fibers* with a typical length of 0.08–6 mm, which are made by hammer-milling chopped strands or continuous strands [4, 6, 12]. The shorter the fibers, the lower the reinforcing effect, so that very short milled fibers mostly act as a filler material.

6.4.2
Shaped Glass Fibers

In addition to regular circular glass fibers described so far, extensive research on shaped glass fibers was conducted that are characterized by a cross-sectional shape other than a solid cylinder. Such studies resulted from early needs for composite materials with higher stiffness-to-weight ratios. Also many natural fibers have shapes other than regular cylinders. Theoretically, glass fibers can be drawn in a variety of different shapes by selecting an appropriate nozzle geometry in the manufacturing process.

One common form of shaped glass fibers are hollow fibers with a circular cross section, which may offer a 30% lower density with $1.8\,g/cm^3$ compared to conventional glass fibers with $2.5\,g/cm^3$. It is important that the hollow inside of the fiber is not filled with resin, otherwise the weight benefit would be lost. Their main potential is the increased weight-specific bending stiffness and buckling strength under compression loads due to the increase in moment of inertia compared to a solid circular fiber of the same mass per unit length [18]. The resistance of the fibers to microbuckling or kink band formation increases, which is the dominating failure initiation mode under compressive loads on the microscale [19]. An alternative to hollow round fibers are hollow hexagonal fibers.

Furthermore, rod-shaped fibers with triangular, square, or hexagonal cross sections have been investigated [20]. Properly shaped fibers can be packed together more efficiently resulting in composite materials with higher fiber volume fraction and therefore higher weight-specific stiffness and strength properties. A comparison of the single fiber tensile strength with circular and triangular geometry of similar cross-sectional area showed a 25% higher strength for the triangular shape, the compressive strength is even 60% higher [21]. This trend was also validated for the respective composite materials with similar fiber volume content, where the tensile strength was 20% higher and the compressive strength 40% higher with triangular fibers. This was attributed to a better fiber alignment within the composite and an increase in the moment of inertia [21]. Circular, peanut-shaped and oval glass fibers in composite laminates were compared in Refs [22, 23] with respect to their tensile, flexure, interlaminar shear, and impact behavior. The large contact area of the specially shaped fibers enabled longitudinal crack propagation and therefore reduced the delamination resistance, which was concluded to result in a better energy-absorbing capacity compared to conventional circular fibers.

All in all, although the weight-specific increase in mechanical properties often cannot compete with carbon fibers and there is very little commercial use of shaped glass fibers, a certain potential of such fibers is still seen today and current research is going on.

6.5
Glass Fiber Properties

6.5.1
General Properties

The density of glass is mainly depending on the composition of the glass, hence for different types of glass, a slightly different density is found. The density of the fiber is in general slightly lower than the value for the bulk material (e.g., 2.54–2.58 g/cm^3 for E-glass, 2.49–2.52 g/cm^3 for S-glass). The refraction index, which is in general determined for light with a wavelength of 550 nm, is directly proportional to the density of the glass and hence also varies depending on the glass composition. For all glass types used for fiber production a refraction index between 1.45 and 1.65 is measured [3].

According to Case *et al.* [24], E-glass fibers are approximately 50 times cheaper than HT carbon fibers, 250 times cheaper than HM carbon fibers, and 15 times cheaper than aramid fibers. They also indicate that S-glass is about eight times more expensive than E-glass.

6.5.2
Elastic Properties

The characterization of Young's modulus for glass fibers is not straightforward due to the difficulty in measuring the elastic properties of small fibers. Several experimental investigations have proven that there is little or no elastic anisotropy in glass fibers, which significantly reduces the experimental effort to determine their elastic properties. Young's modulus parallel to the fiber is determined either by mechanical or acoustic tests. Mechanical test methods are often tensile or bending tests, which require an accurate fiber diameter as input, but also allow for the simultaneous measurement of strength and elongation at failure. The acoustic method determines the velocity of an elastic wave traveling inside the fiber. The acoustic method can also be applied easily at elevated temperatures. The fiber's shear modulus and Poisson's ratio are determined by an additional torsion-resonance experiment, where a torsion pendulum is suspended on the fiber itself [3]. Typical values for glass fiber Young's modulus are between 70 and 75 GPa for A-, C- and E-glass whereas higher values are obtained for high strength (S- and R-) and high modulus (M-) glass fibers, 85 and 115 GPa, respectively (Table 6.2). The given values are valid at room temperature and in general slightly decrease at elevated temperatures. This difference is more pronounced for S-, R- and M-glasses as for E-glass. These stiffness values are significantly lower in comparison to other reinforcing fibers like carbon or aramid. Due to this difference, glass fibers and their composites are not well suited for ultralightweight structures.

6.5.3
Strength and Elongation Properties

The physical and mechanical properties of glass fibers and bulk glass are often very similar except for the strength. The significantly higher cooling rate during the production process of the glass fiber compared to bulk glass is the source of this difference. The strength of virgin glass fibers (i.e., fibers directly drawn from the bushing) is significantly higher than the strength of fibers tested at the end of the production process. This difference is caused by the subsequent steps since a glass fiber is very sensitive to any further mechanical contact as for instance during sizing

Table 6.2 Properties of commercial glass fibers.

Property	Unit	\ Glass fiber type						
		A	C	D	E	M	R	S
General properties								
Filament diameter	μm				5–20			8–14
Refraction index	—	1.512	1.541	1.47	1.548	1.635		1.523
Density	g/cm^3	2.47	2.49	2.16	2.54	2.89	2.49	2.49
Elastic properties								
Tensile modulus	GPa	73	74	55	71	115	86	85
Shear modulus	GPa				29.6			36
Poisson's ratio	—				0.20			0.22
Tensile strength (virgin filament)	GPa	3.1	3.1	2.5	3.4	3.4	4.4	4.58
Tensile strength (roving)	GPa	2.76	2.35	2.41	2.4	2.4	3.1	3.91
Stain at failure	%	3.6	3.5		2–4		4.2	5.4
Chemical resistance								
Weight loss after 1 h in boiling water	%	11.1	0.13		1.7			
Weight loss after 1 h in boiling H$_2$SO$_4$ (1.0 N)	%	6.2	0.10		48.2			
Weight loss after 1 h in boiling NaOH (0.1 N)	%	12–15	2.28		9.7			
Thermal properties								
Thermal conductivity	W/(K m)				1.1			
Specific heat capacity	J/(kg K)		836	732	800			735
CTE	10^{-6}/K	9	7–8	2–3	4.9–6		4	2.9–5.6
Maximum use temperature	°C				300		350	
Electrical properties								
Relative dielectric constant	—	6.9	6.2–6.3	3.5–3.8	5.9–6.6		6.2	4.5–5.6
Loss tangent at 10^6 Hz	10^{-6}			500	3900		1500	7200
Volume resistivity	Ω m		10^8		10^{13}			

application or in the winding process. Typical final strength values between 2.3 and 2.8 GPa can be found for A-, C-, E-, and M-glass, whereas high strength (R- and S-) glass fibers have a tensile strength between 3.1 and 3.9 GPa (Table 6.2). These high strength values are definitely one of the outstanding properties of glass fibers, exceeding other reinforcing fibers like HM carbon and aramid. Only HT carbon fibers can show comparable strength values.

The tensile strength of glass fibers is influenced by a number of environmental effects. A first effect is found in the conditioning of the glass melt during the manufacturing process. The conditioning temperature (i.e., temperature before drawing) significantly influences the final strength and its coefficient of variation; hence for each glass fiber type optimal processing conditions must be kept at all times. Glass fibers also suffer from corrosion when kept inside humid environments.

Often the glass fiber sizing agent will act as a protection layer and in this way try to avoid the corrosion process.

Another important issue is the strain rate dependency of the mechanical properties of glass fibers, which are especially important for applications or studies, where high loading rates are involved, for example, impact loads. Static and dynamic tests in the strain rate domain of 10^{-4} to $1100\,\text{s}^{-1}$ were conducted in Ref. [25] on pure E-glass fiber bundles and a rate dependency with an increase of strength and failure strain of almost 300% for the highest loading rates was reported. Also in Ref. [26], a considerable strain rate effect was obtained for pure E-glass fiber bundles.

The conditioning of glass fibers at high temperatures and for longer times has a reducing influence on the strength of glass fibers. The strength of glass fibers is also lower when tested at elevated temperatures due to the decomposition of the sizing where the resulting chemicals weaken the fiber. The strength of glass fibers is extremely sensitive to handling and care must be taken to prevent any contact of the fiber to any other solid at the test gage. Also the fiber diameter must be determined with care as it has a pronounced influence on the measurement result. The measurement of strengths at temperatures higher than regular room conditions may be performed only by local heating of the fiber inside the gage length.

The elongation of glass fibers has a relatively high value between 3 and 5%, which is significantly higher than for carbon or aramid fibers. Due to this property, a composite manufactured with glass fibers can be used for specific applications, where a large elastic deformability is required (e.g., leaf springs).

6.5.4
Corrosion Properties

As previously mentioned, the corrosion resistance of glass fibers is in general fully controlled by the fiber sizing. Noncoated fibers would significantly corrode in alkali or acid environments or even water would deteriorate the properties of the fiber. Specially developed glass compositions in C-glass and Z-glass have a higher corrosion resistance compared to E-glass in acidic and alkaline environments, respectively. The corrosion resistance of fibers is determined by the measurement of glass fiber weight loss when boiling in alkali or acid liquids or distilled water for 1 h [3].

6.5.5
Thermal Properties

The specific heat capacity of glass fibers is slightly lower compared to the value of its bulk glass equivalent, and mainly depends on the glass composition. A value between 700 and 800 J/(kg K) is found in literature [2]. The thermal conductivity of E-glass fibers is around 1.1 W/(m K). The coefficient of thermal expansion varies for the different glass compositions and is found between 2×10^{-6} and $9 \times 10^{-6}\,\text{K}^{-1}$ (Table 6.2).

Glass fibers have a very high thermal and electrical resistance and are hence often used in electronic and electrical components. D- and L-glass, which were especially developed for high signal transparency, have a significantly lower dielectric constant compared to other glass types.

A summary of the physical properties of different glass fiber types is given in Table 6.2. The values in this table are compiled from data published in Refs [3, 6, 8, 10].

6.6
Glass Fibers in Polymer Composites

6.6.1
Polymers for Glass Fiber-Reinforced Composites

Glass fibers are used as reinforcement material in combination with both thermoplastic and thermoset resins. The manufacturing processes used for thermoplastics are based on melting the polymer by introducing heat and solidifying the polymer–fiber blend after forming. Due to the high viscosity of the thermoplastic polymer, this matrix material is often premixed with the glass fiber reinforcement. For long glass fibers this is achieved in form of commingled yarns or as polymer-fabric layered preform whereas for short fiber-reinforced plastics, for example, for injection molding the fibers are premixed in the polymer pellets. Especially in case of short fiber-reinforced material, the local orientation of the fiber is strongly dependent on the polymer flow during the manufacturing process, which can significantly influence the physical properties of the composite material. Examples of thermoplastic materials that are commonly used for glass fiber-reinforced composites are polyamide (PA), polyethylene (PE), polypropylene (PP), polycarbonate (PC), polystyrene (PS), acrylonitrile butadiene styrene (ABS), polyoxymethylene (POM), polybutylene terephthalate (PBT), and polyether ether ketone (PEEK) [13].

Thermosetting resins are combined with long glass fibers within either infiltration processes (e.g., resin transfer molding or vacuum-assisted process) or are available as preimpregnated material. Spray-up, BMC, and SMC processes are commonly applied in combination with short fibers. After forming, the vitrification of the thermoset is started at higher temperatures of one-component resin systems or by mixing the components of resin system using two or more components. Polyester (UP), polyimide (PI), vinylester (VE), polyurethane (PUR), phenolic (PF) and epoxy (EP) resins are examples of thermosets that are reinforced with glass fibers [13].

The interfacial interaction between fibers and polymer matrix is a very important factor for composite materials, hence also for GFRP. Therefore, a coupling agent that works as an adhesion promoter between the filaments and the matrix is applied in form of a sizing during the production process of the fibers (see Section 6.3). The chemical composition of coupling agents can be varied for an optimum compatibility with different polymeric matrix materials. A detailed overview of different adhesion promoters is given in Ref. [9].

6.6.2
Determination of Properties

As in all composite materials, the physical properties of the resulting material are dependent on the physical behavior, the form and the concentration of all constituents as well as their compatibility. Exactly for this compatibility reason, several coupling agents in the fiber sizing are used to enhance the adhesion between resin and glass fiber, and hence improving the properties of the composite. The different forms of glass fibers as described earlier are mainly selected based on their performance and workability but also cost may be a selection criterion.

To determine the properties of the resulting composite both experimental and analytical methods can be used. For experimental investigations several testing standards are available and should be used to improve the quality and reproducibility of the obtained results. However, these standards are often only valid for a certain type of reinforcement and care should be taken when testing other reinforcement configurations. For instance the interlacing pattern of a woven glass reinforcement may result in local variations of the mechanical properties, and hence a wider test specimen is required than in case of a unidirectionally (UD) reinforced composite.

When it comes to the analytical determination of the physical properties, a variety of models are available in literature, which all can be grouped into three categories. The first category is the so-called micromechanical formulation, which determines the physical property as a simple function of the base constituent properties and the fiber volume fraction of the composite material. This method is very easy in use, but only provides limited detail and is often only accurate for a limited range of fiber volume fractions. Micromechanical formulations for different physical properties can for instance be found in Ref. [27]. The second category is based on mean-field Eshelby-based homogenization techniques as for instance shown in Ref. [28]. This technique is time inexpensive, easy to use as it is implemented in different software tools but only provides information on the averaged behavior and does not provide detail about the field variation (e.g., stress, strain, and temperature) inside the composite structure. The third category is based on multilevel modeling techniques and uses homogenization techniques to determine the average properties of the reinforced polymer [29]. This method does not only provide the average property of the composite but also provides detailed information on the internal physical phenomena that occur when loaded. The multilevel modeling approach is very computationally expensive but, especially for complex textile-reinforced composites, it can be the only possible approach.

6.6.3
Manufacturing Processes and Related Composite Properties

Short fiber-reinforced thermoplastic polymers are commonly used to manufacture injection molded parts due to the increased mechanical performance, improved toughness and creep resistance compared to unreinforced parts. The flow history during injection significantly influences the fiber orientation and tends to form a layer at the surface of the part, where the fiber orientation significantly deviates for the rest of the part, causing anisotropic properties. Temperature and rate dependency as well as moisture absorption and chemical resistance are mainly controlled by the thermoplastic matrix material.

Short fiber-reinforced composites based on chopped strands produced either by hand or spray lay-up are probably the most widespread reinforcement form of glass fibers due to the low cost of the tooling and process itself. Typically composites with fiber volume content between 10 and 25% are manufactured. Higher volume fractions are difficult to achieve due to the internal structure of the reinforcement [13]. The main disadvantage of chopped strand composite materials is the large variation in quality. As a direct consequence, the mechanical performance is strongly depending on this quality. When correctly designed, chopped strand composites can have good impact performance.

Another form of glass fiber-reinforced polymers is found in BMC and SMC materials. These materials contain next to glass fibers and the polymer matrix a large amount of filler material, often chalk, to obtain low shrinkage characteristics. The random nature of the reinforcement fiber results in relatively low physical properties often with high coefficient of variation [13].

Pultruded glass fiber profiles are manufactured in a variety of cross sections but always have a high amount of fibers oriented along the profile direction, hence having a highly anisotropic behavior. The parts manufactured with the pultrusion process are mainly intended to withstand axial tensile loads due to its high modulus and strength in this direction. Often these parts will not fail because of mechanical material failure but due to buckling of the flange or Web of the profile resulting from the thin-walled structural design.

The filament winding technique, which is regularly used for the production of parts in glass fiber-reinforced thermosetting polymers, provides high quality parts with high mechanical properties due to the good and controlled alignment of the fibers. A common problem during the filament winding process is the forming of voids, which can influence the performance of the material significantly. Typical applications for glass fiber filament winding are pipes, vessels, or torque tubes.

Glass fibers are also used as reinforcements in prepreg tapes, which are manufactured by preimpregnating the fibers with a thermosetting resin. In this configuration, the fibers are arranged in a unidirectional flat way and combined with the resin. Such prepregs require storage at low temperatures (typically $-18\,°C$) before use in order to prevent cross-linking of the resin. For the manufacturing of a composite part, the individual prepreg plies are stacked together in a so-called laminate until the required thickness is achieved and cured at elevated temperature and pressure. The highest anisotropic material behavior is achieved when all layers are stacked in the same direction. Laminates of unidirectional tapes have high in-plane stiffness and strength but lower performance in out-of-plane direction and under impact loads. Preimpregnated tapes must be stored at well-controlled conditions, resulting in additional expenses. Typical fiber volume fractions found for preimpregnated tapes are between 50 and 65%.

Woven and braided glass fibers are increasingly being used as a reinforcement material as the flexible and well-known textile production processes enable the tailoring of physical properties to a specific application. Woven glass reinforcements are also provided as preimpregnated tapes. The interlacing pattern makes the reinforcement architecture mechanically stable and hence these fabrics can easily be handled during the production process. As they are often combined with resin infusion techniques, they can be stored separately from the matrix material, hence reducing storage costs. The mechanical performance of woven glass fiber-reinforced composites depends on the interlacing pattern as well as the ratio between warp and weft yarns. For braided reinforcements also the braiding angle and an optional interlacing yarn will further influence the resulting physical properties. The fiber volume fraction for woven and braided reinforced polymers is found in the range of 25–60% [13].

More information on manufacturing processes and the resulting properties can be found for instance in Refs [3, 9, 13].

6.6.4
Strength and Fatigue Properties

The strength of glass fiber composites strongly depends on the form of the reinforcement and the quality of the adhesion between resin and fiber. The highest tensile strength is obtained for unidirectionally reinforced resin tested in reinforcement direction as the main fraction of the load is carried by the glass fiber. For all other reinforcement types or off-axis loading, the measured tensile strength will be lower. Due to the fact that GFRP parts have in general limited dimensions in thickness direction, the compressive strengths are often difficult to measure as these parts easily buckle under the applied load. Also the shear strength is not straightforward to determine even though several experimental methods are available. For composites with a high anisotropic nature, that is, for unidirectionally reinforced materials, a significant difference between tensile and compression strengths is found. This difference is less pronounced for materials having reinforcements in multiple directions, for example, woven or braided fabrics, SMC, BMC, or CSM.

GFRP have excellent fatigue resistance. For a unidirectional glass fiber-reinforced composite, the resulting decreased strength after cycling loading can be written according to

$$\sigma = \sigma_U(1 - k\log(N))$$

where σ_U is the ultimate tensile strength, N the number of load cycles, and k the fractional loss factor. Typical k values for GFRP at room temperature are between 0.08 and 0.1. At higher temperatures higher k values are found, that is, the material will have a lower fatigue resistance. High strength fibers (i.e., S- and R-glass fibers) do not only have a higher static strength but also higher fatigue strength compared to E-glass fibers [13].

An overview of the mechanical properties of GFRP can be found in Table 6.3. This table is compiled based on data published in Refs [2, 13].

6.6.5
Strain Rate Effect in Glass Fiber-Reinforced Composites

Characteristical for glass fiber composites is their strain rate effect, which is mainly attributed to the rate-dependent stiffness and strength behavior of the pure glass fibers, as mentioned earlier. Under high loading rates in a very short period of time, stiffness, strength, and failure strain of the composite can significantly increase, as reported in Refs [30, 31] for E-glass fiber/epoxy composites, in Ref. [32] for E-glass fiber/polyester composites, and in Ref. [33] for E-glass fiber/phenolic composites. Failure stress and failure strain can be two or three times higher for very high loading rates, which should be taken into account for applications or load cases involving high strain rates. This effect is much more significant for glass fiber composites than for carbon or aramid fiber composites, where such effects only become significant at much higher strain rates and despite some increase in strength, the failure strain rather tends to be unaffected or even decreases, respectively.

6.6.6
Environmental Influences

The properties of glass fiber-reinforced composites are influenced by moisture, temperature, chemical environment, and ultraviolet radiation. GFRP typically absorbs 0.3–1.5% of water, depending on the resin material, the temperature, the exposure time, and loading conditions (tensile loads will increase the absorption rate). The absorption of water will decrease the mechanical performance of the composite, may influence the dimensions of the part and can also result in chemical deterioration of the glass fibers. Temperature variations affect the physical properties that are highly dependent on the resin material like transverse and shear stiffness and strength. At higher temperatures the polymer matrix will start to degrade due to chemical decomposition. Typically the material will have a maximum operation temperature ranging from 90 °C for GMT up to 200 °C for continuously reinforced epoxy resin [13]. The chemical resistance against alkali or acid fluids is in general higher for GFRP than for metals, but strongly depends on the resistance of the resin polymer and to a smaller extent of the fiber. Ultraviolet light can also significantly damage the resin material, which is in general prevented by a gel-coating or protective paint.

6.6.7
Other Physical Properties of Glass Fiber-Reinforced Composites

The coefficient of thermal expansion of glass fiber-reinforced resins is, similar to the mechanical properties, depending on the thermoelastic properties of the base materials, the fiber volume fraction and the form of the reinforcement. For

Table 6.3 Properties of glass fiber-reinforced polymers.

Property Glass fiber type	Unit	E	S	E	E	E
Length		Cont.	Cont.	Cont.	Cont.	Short
Configuration		UD	UD	UD	Balanced plain weave	Random
Process		Prepreg	Prepreg	Hot pressed	Prepreg	Injection molding
Resin		EP	EP	PP	EP	PA
Fiber volume fraction	%	59	71	31	48	30
Density (ϱ)	g/cm^3	1.90	2.12	1.40	1.60	1.54
Longitudinal Young's modulus in tension (E_1^+)	GPa	47.0	66.0	21.5	25.2	10.6
Longitudinal Young's modulus in compression (E_1^-)	GPa	45.5		26.5		10.3
Transverse Young's modulus in tension (E_2^+)	GPa	16.4		3.6	25.2	7.9
Transverse Young's modulus in tension (E_2^-)	GPa	15.9		4.3		5.6
Shear modulus	GPa	6.0		1.34	4.41	1.6
Poisson's ratio	—	0.28		0.31	0.14	0.35
Longitudinal tensile strength (σ_1^+)	MPa	1140	1896	425	317	137
Longitudinal compressive strength (σ_1^-)	MPa	760	1380	272	303	180
Transverse tensile strength (σ_2^+)	MPa	63		11	317	112
Transverse compressive strength (σ_2^-)	MPa	213		53	303	148
Shear strength (σ_{12})	MPa	107		50	57	91

unidirectionally reinforced resins, there is a large difference between longitudinal and transverse CTEs, hence they are greatly anisotropic. This commonly introduces part deformations in manufacturing processes due to the difference between curing and room temperature. For GFRP with random-oriented fibers similar CTE values can be found as for metals like aluminum or steel. The low thermal conductivity of both polymeric matrix materials and glass fibers results in a low thermal conductivity of their composite. This property can be a problem during machining or high strain rate testing, where the generated heat is only slowly dissipated. The specific heat capacity of GFRP is very similar to the value for the unreinforced polymer and is found in a range of 10–13 kJ/(kg K). The specific heat capacity is dependent on the temperature and is higher at elevated temperatures [13].

Glass fiber-reinforced polymers have very good insulating properties and are hence often used in electronic and heavy electrical industry. Typical values for the electrical resistivity of GFRP are in the range between 10^{10} and $10^{15}\,\Omega\,m$, whereas their dielectric strengths are found between 6 and 22 MV/m [13].

Glass fibers and most polymers used as matrix material have a similar refraction index. As a result, the investigation of a GFRP with optical methods can be difficult and low contrast may be a problem. Therefore, methods like electron microscopy or computer tomography are better suited due to the difference in fiber and matrix density (Figure 6.4).

6.7
Applications

Today, in the majority of all fiber-reinforced composites the reinforcement fiber is made of glass. Therefore, the variety of technical applications is manifold. A breakdown of today's GFRP production into different industrial sectors shows that the construction sector is leading, followed by the transportation industry, sport and leisure products, electronic products and others [34]. A short overview of typical applications is given here, showing the potential of modern glass fiber composites.

Construction/civil infrastructure: Various examples of applications of GFRP in the construction sector exist, for example, panels, corrugated panels, light domes, lightwells, roof gutters, dormers, door surrounds, window canopies, chimneys,

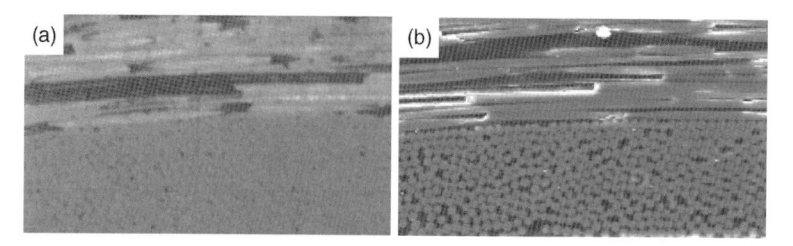

Figure 6.4 (a–b) Cross sections of a GFRP specimen: (a) optical microscopy (low contrast) and (b) computer tomography (high contrast).

coping systems, and so on. Such components not only fulfill durability requirements in corrosive media but also provide advantages in faster installation and handling due to the reduced weight. Presently, bridge decks and platforms have also received much attention because of their lightweight, high strength-to-weight ratio, and corrosion resistance [35].

Aeronautics: The development of GFRP started in the aeronautics industry and there are still numerous applications today. Although carbon fibers more and more find their way into primary structures of commercial airliners, there are classical applications for glass fibers mostly in sandwich structures like in the whole cabin interior (overhead bins, sidewall and ceiling panels, lavatories, etc.) [36]. Exterior applications can be found in the radar-signal-transparent radome of the aircraft and in fairing surfaces of the wings and empennage [37]. Applications in primary airframe structures are limited due to the relatively low stiffness of GFRP. Only in small aircraft and sailplanes the whole fuselage and wings are often made from GFRP [38]. Nevertheless, glass fibers are also found in the upper fuselage shells of the largest commercial aircraft, the Airbus A380. These shells are made from GLARE (glass-reinforced fiber metal laminate), which is a layered structure of GFRP and aluminum sheets with outstanding damage tolerance, fatigue, corrosion, and fire resistance properties [16]. Helicopter blades are also typically made from GFRP with some foam or honeycomb core. An overview on the use of GFRP in current aeronautic structures is given in Figure 6.5.

Marine: After the development of GFRP for military aircraft, the manufacturing of boats (rowing boats, sailing boats, motorboats, and lifeboats) was the first civilian application, which quickly gained acceptance since the 1950s. This was an ideal field for both hand lay-up and spray lay-up techniques for the manufacturing of complex

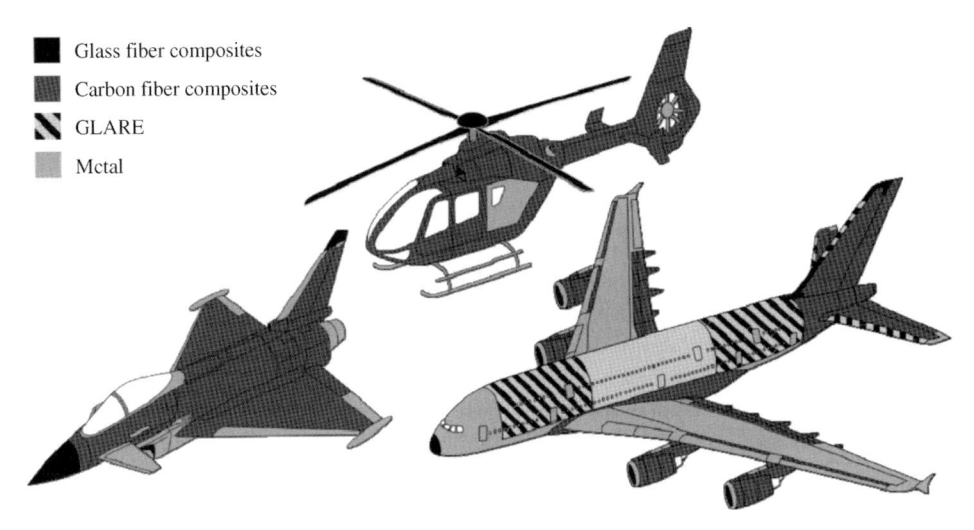

Figure 6.5 Use of composite materials in current aeronautic structures: Eurofighter jet, EC-135 helicopter, and Airbus A380 airliner.

shapes. Not only components but also large hulls and decks could be manufactured this way, revolutionizing the boat building. Today an ever-increasing number of boats and yachts are made from GFRP, both in the civil and military sector. One issue of recent studies is the impact performance of such structures to water slamming and projectile impact loads [39–41].

Automotive: GFRP was first used in the serial production of panels and fairings for commercial vehicles (wind deflectors, engine hood, etc.). Since the 1970s car bumpers were made from GFRP. After the introduction of SMC, further parts like trim panels, door components, trunk lids, hoods were made from such composite materials. Nonvisible parts for under-bonnet applications such as the front bulkhead supporting the radiator, cooling fan and front lights, are made from GMT due to the lower surface quality. The body of some small series sports cars is also made from GFRP. The use of composite materials in the automobile industry is significantly increasing in recent days following the trend of lighter cars for reduced energy consumption and glass fibers are a good candidate due to their price advantage.

Consumer and sporting goods: Many articles in the leisure and sporting industry are made from lightweight and cheap glass fiber composites like tent poles, pole vault poles, arrows, bows and crossbows, hockey sticks, surfboards, bowling balls, fishing rods, microwave cookware, and so on.

Tanks and pipes: Storage tanks, for example, for chemical storage are often made from GFRP with capacities up to 300 ton. While small and simple tanks are made from CSM, high-performance tanks and pressure vessels are typically made by filament winding. GFRP pipe systems are used for several different applications with liquid (e.g., drinking and waste water) or gaseous media of different temperatures.

Wind energy: Wind turbine blades are classical examples of large GFRP structures. The material has proven to be best suited for the aerodynamic and mechanical requirements of such lightweight components like stiffness, strength, durability, and shock resistance. It is also used for the nacelle, that is, the turbine housing.

6.8
Summary

Glass fibers are still the most common of all reinforcing fibers in polymer composites today, which is due to good reasons. This is mainly attributed to the lowest cost among all commercially available reinforcing fibers, which allow them to be used in numerous applications. In terms of mechanical properties, the high tensile strength is characteristic for glass fibers. Furthermore, a wide range of different glass fiber types with individual advantages make them favorable when, for example, high chemical resistance or electrical insulation qualities are required. On the other hand, disadvantages are the relatively low elastic modulus, poor abrasion resistance, and high density compared to other commercial reinforcing fibers, which ban them from use in ultralightweight structures. The glass fiber polymer composite market is dominated by E-glass fibers, which are available in various forms, for example, rovings, mats, fabrics, and so on. Combined with a wide choice of manufacturing technologies, for example, hand lay-up, filament winding, or resin infusion processes there are a number of different behaviors and properties that can be obtained with glass fiber-reinforced composites. Consequently, GFRP composites are used today in a product range from small electrical circuit boards to large ships. The status of glass fiber composites as an exotic material has long been replaced with the status of a generally accepted standard construction material. The range of applications in different industries is still increasing and academic research on glass fibers and glass fiber composites for improved properties and new applications is continuously conducted.

References

1 (2009) Global glass fiber market 2010–2015: Supply, demand and opportunity analysis. Report, Lucintel, Dallas, TX, December 2009.
2 Bunsell, A.R. (1988) *Fibre Reinforcements for Composite Materials*, Elsevier, Amsterdam.
3 Lowrie, R.E. (1967) Glass fibers for high-strength composites, in *Modern Composite Materials* (eds L.J. Broutman and R.H. Krock), Addison-Wesley, Reading, pp. 270–323.
4 Mallick, P.K. (2008) *Fiber-Reinforced Composites: Materials, Manufacturing and Design*, 3rd edn, CRC Press, Boca Raton.
5 (2008) *Aerospace Composites: A Design & Manufacturing Guide*, Gardner Publications Inc., Wheat Ridge.
6 Middleton, D.H. (1990) *Composite Materials in Aircraft Structures*, Longman, New York.
7 Bunsell, A.R. (2005) Oxide fibers, in *Handbook of Ceramic Composites* (ed. N.P. Bansal), Kluwer Academic Publishers, Boston, pp. 3–31.
8 Bunsell, A.R. and Renard, J. (2005) *Fundamentals of Fibre Reinforced Composite Materials*, Institute of Physics Publishing Ltd., Bristol.
9 Jones, F.R. (2001) Glass fibers, in *High-Performance Fibers* (ed. J.W.S. Hearle) Woodhead Publishing Ltd., Cambridge, pp. 191–238.
10 Baker, A.A., Dutton, S., and Kelly, D. (2004) *Composite Materials for Aircraft Structures*, 2nd edn, AIAA, Reston.
11 DIN 61850 (May 1976) *Textile Glass Products and Auxiliary Products, Terms and Definitions*, Beuth Verlag GmbH, Berlin, May 1976.
12 Agarwal, B.D., Broutman, L.J., and Chandrashekhara, K. (2006) *Analysis and Performance of Fiber Composites*, 3rd edn, Wiley, Hoboken.
13 Sims, G.D. and Broughton, W.R. (2000) Glass fibre reinforced plastics-properties, in *Comprehensive Composite Materials, Polymer Matrix Composites*, vol. 2 (eds A. Kelly and C. Zweben), Elsevier, Amsterdam, pp. 151–197.
14 Gessler, A., Maidl, F., and Schouten, M. (2007) Advancements in braiding technology for textile preforming. 28th SAMPE Europe International Conference, Paris, April 2–4.
15 Havar, T. and Stuible, E. (2009) Design and testing of advanced composite load introduction structure for aircraft high lift devices. ICAF 2009, Bridging the gap between theory and operational practice, Proceedings of the 25th Symposium of the International Committee on Aeronautical Fatigue, Rotterdam, The Netherlands, May 27–29 (ed. M. Bos), pp. 365–374.
16 Drechsler, K., Middendorf, P., Van Den Broucke, B., and Heimbs, S. (2008) Advanced composite materials – technologies, performance and modelling, in *Course on Emerging Techniques for Damage Prediction and Failure Analysis of Laminated Composite Structures* (eds D. Guedra Degeorges

and P. Ladeveze), Cepadues, Toulouse, pp. 147–197.

17 Li, Y., Lin, Z., Jiang, A., and Chen, G. (2004) Experimental study of glass-fiber mat thermoplastic material impact properties and lightweight automobile body analysis. *Materials and Design*, **25** (7), 579–585.

18 Hucker, M., Bond, I., Bleay, S., and Haq, S. (2003) Experimental evaluation of unidirectional hollow glass fibre/epoxy composites under compressive loading. *Composites Part A*, **34** (10), 927–932.

19 Heimbs, S., Strobl, F., Middendorf, P., and Guimard, J.M. (2010) Composite crash absorber for aircraft fuselage applications. 11th International Conference on Structures under Shock and Impact, Tallinn, Estonia, July 28–30.

20 Humphrey, R.A. (1967) Shaped glass fibers, in *Modern Composite Materials* (eds L.J. Broutman and R.H. Krock), Addison-Wesley, Reading, pp. 324–334.

21 Bond, I., Hucker, M., Weaver, P., Bleay, S., and Haq, S. (2002) Mechanical behaviour of circular and triangular glass fibers and their composites. *Composites Science and Technology*, **62** (7–8), 1051–1061.

22 Deng, S., Ye, L., and Mai, Y.W. (1999) Influence of fibre cross-sectional aspect ratio on mechanical properties of glass fibre/epoxy composites I. Tensile and flexure behaviour. *Composites Science and Technology*, **59** (9), 1331–1339.

23 Deng, S., Ye, L., and Mai, Y.W. (1999) Influence of fibre cross-sectional aspect ratio on mechanical properties of glass fibre/epoxy composites II. Interlaminar fracture and impact behaviour. *Composites Science and Technology*, **59** (11), 1725–1734.

24 Case, J., Chilver, L., and Ross, C.T.F. (1999) *Strength of Materials and Structures*, 4th edn, Arnold, London.

25 Xia, Y., Yuan, J., and Yang, B. (1994) A statistical model and experimental study of the strain-rate dependence of the strength of fibres. *Composites Science and Technology*, **52** (4), 499–504.

26 Rotem, A. and Lifshitz, J.M. (1971) Longitudinal strength of unidirectional fibrous composite under high rate of loading. Proceedings of the 26th Annual Technical Conference, Society of the Plastics Industry, Reinforced Plastics/Composites Division, Washington, D.C., Section 10-G, pp. 1–10.

27 Johnston, A.A. (1997) An integrated model of the development of process-induced deformation in autoclave processing of composite structures. PhD thesis, The University of British Colombia.

28 Pierard, O., Friebel, C., and Doghri, I. (2004) Mean-field homogenization of multi-phase thermo-elastic composites: a general framework and its validation. *Composites Science and Technology*, **64** (10–11), 1587–1603.

29 Kouznetsova, V., Geers, M., and Brekelmans, W. (2004) Multi-scale second-order computational homogenization of multi-phase materials: a nested finite element solution strategy. *Computer Methods in Applied Mechanics and Engineering*, **193** (48–51), 5525–5550.

30 Xia, Y. and Wang, X. (1996) Constitutive equation for unidirectional composites under tensile impact. *Composites Science and Technology*, **56** (2), 155–160.

31 Sierakowski, R.L. (1997) Strain rate effects in composites. *Applied Mechanics Reviews*, **50** (12), 741–761.

32 Welsh, L.M. and Harding, J. (1998) Effect of strain rate on the tensile failure of woven reinforced polyester resin composites. *Journal de Physique*, **46** (8), C5.405–C5.414.

33 Heimbs, S., Schmeer, S., Middendorf, P., and Maier, M. (2007) Strain rate effects in phenolic composites and phenolic-impregnated honeycomb structures. *Composites Science and Technology*, **67** (13), 2827–2837.

34 Witten, E. (2010) The European composites market 2008/2009. JEC Composites Magazine, No. 55, pp. 16–18, March.

35 Jeong, J., Lee, Y.H., Park, K.T., and Hwang, Y.K. (2007) Field and laboratory performance of a rectangular shaped glass fiber reinforced polymer deck. *Composite Structures*, **81** (4), 622–628.

36 Heimbs, S., Vogt, D., Hartnack, R., Schlattmann, J., and Maier, M. (2008) Numerical simulation of aircraft interior

components under crash loads. *International Journal of Crashworthiness*, **13** (5), 511–521.

37 Georgiadis, S., Gunnion, A.J., Thomson, R.S., and Cartwright, B.K. (2008) Bird-strike simulation for certification of the Boeing 787 composite moveable trailing edge. *Composite Structures*, **86** (1–3), 258–268.

38 Van Tooren, M.J.L. (1998) *Sandwich Fuselage Design*, Delft University Press, Delft.

39 Bull, P.H. (2004) Damage tolerance and residual strength of composite sandwich structures. PhD thesis, Royal Institute of Technology, Stockholm, Sweden.

40 Davies, P., Bigourdan, B., Chauchot, P., Choqueuse, D., Ferreira, A.J.M., Karjalainen, J.P., Hildebrand, M., Mustakangas, M., Gaarder, R., Carli, F., Van Straalen, I.J.J., Sargent, J.P., Adams, R.D., Broughton, J., and Beevers, A. (2000) Predicting the behaviour of a wide sandwich beam under pressure loading. Proceedings of the 5th International Conference on Sandwich Construction, Zurich, Switzerland, September 5–7, pp. 325–336.

41 Mäkinen, K. (1995) Dynamically loaded sandwich structures. Licentiate thesis, Royal Institute of Technology, Stockholm, Sweden, Report 95–20.

7
Kevlar Fiber-Reinforced Polymer Composites

Chapal K. Das, Ganesh C. Nayak, and Rathanasamy Rajasekar

7.1
Introduction

Thermoplastic-based composites are becoming more essential in several application fields due to the combination of toughness of thermoplastic polymers along with stiffness and strength of reinforcing fibers. Organic textile fibers can be used to prepare polymeric composites. Due to their low stiffness, organic textile fibers are used as reinforcing agent in polymer matrices for both rubbers and thermoplastics. It is well known that the behavior of the polymeric material strongly dependent on its structure, morphology, and relaxation processes. Furthermore, the properties of composite materials are determined by the characteristics of the polymer matrices themselves, together with reinforcements, and the adhesion of fiber/matrix interface, which mainly depends on the voids and the bonding strength at the interface.

In general, the Kevlar fibers are highly crystalline, their surface are chemically inert and smooth, thus its adhesion with matrix is very much poor. Therefore, surface modification of Kevlar fiber is essential to enhance its reinforcing efficiency. Direct fluorination is an effective approach for the surface modification of Kevlar fibers. This process does not need any initiation or catalyst. Elemental gases are used to modify the surface of the fibers. The fluorination can be carried out at room temperature. Surface modification of Kevlar fibers by fluorination and incorporation in the polymer matrix resulted in better mechanical stability, thermostability, and membrane properties.

The works carried out on the preparation of composites using both unmodified and modified Kevlar fibers, in various thermoplastics matrices are discussed in this chapter. In addition, the works carried out on thermosetting polymers containing Kevlar fibers were also presented. The various methods used for preparation of polymer composites and the effects of Kevlar fiber on their crystalline, thermal, dynamic mechanical, and morphological properties of polymer composites are focused. In addition the reports based on the effect of compatibilizers on the reinforcement of unmodified, fluorinated, and oxyfluorinated short Kevlar fibers incorporated in the polymer matrix had been surveyed.

Polymer Composites: Volume 1, First Edition. Edited by Sabu Thomas, Kuruvilla Joseph,
Sant Kumar Malhotra, Koichi Goda, and Meyyarappallil Sadasivan Sreekala
© 2012 Wiley-VCH Verlag GmbH & Co. KGaA. Published 2012 by Wiley-VCH Verlag GmbH & Co. KGaA.

The simulation reports analyzing the fiber orientation using mold flow technique under varying injection molding parameters of the Kevlar fiber-reinforced polymer composites were also centered in this chapter.

7.2
Fiber-Reinforced Polymer Composites

Fiber-reinforced composites comprise of fibers of high strength and modulus, embedded in a polymer matrix with distinct interfaces between them. In this composite both fiber and the polymer retains their physical and chemical identities. In general fibers act as main load bearing constituent, in turn surrounding matrix keeps them intact in their positions and desired orientations [1]. Fibrous composites are classified into two broad areas namely short fiber-reinforced composites and long fiber-reinforced composites, which in turn can be categorized into two segments:

- Fiber-reinforced thermoset plastics (polyester, epoxy, phenol, etc.)
- Fiber-reinforced thermoplastic (PPS, PEEK, PEI, etc.).

Other types of composites includes sandwich structures, fiber metal laminates, metal matrix composites, glass matrix composites, ceramic matrix composites, carbon carbon composites, and so on.

7.3
Constituents of Polymer Composites

The chief elements of composites include reinforcements and polymer matrix. Other additives such as catalysts, coupling agents, pigments and adhesives are also added to improve the properties of the composites. Reinforcements render stiffness, strength, dimensional stability, and thermal stability of the composite materials. They promote effective stress transfer and modify the failure mechanism in composite system. Various factors governing the properties of the composite material includes size, shape, distribution, concentration, and orientation of the reinforcing filler in the polymer matrix. Reinforcements are mainly categorized as fibrous particulate. Kevlar fibers come under the category of synthetic fibers.

7.3.1
Synthetic Fibers

Synthetic fibers are made entirely from chemicals. Synthetic fibers are usually stronger than either natural or regenerated fibers. Synthetic fibers and the regenerated acetate fibers are thermoplastic and can be softened by heat. Therefore manufacturers can shape these fibers at high temperatures, adding such features as pleats and creases. The most widely used synthetic fibers are mentioned as follows:

- Nylon fibers
- Glass fibers
- Kevlar fibers
- Metallic fibers

Among these synthetic reinforcing fibers, Kevlar had been discussed elaborately, which are as follows.

7.4 Kevlar Fiber

Kevlar is the DuPont trade name of poly(p-phenylene terephthalamide) (PPTA) and was first created in DuPont's labs in 1965 by Stephanie Kwolek and Herbert Blades [2]. It is an organic fiber in the aromatic polyamide family. It possesses unique combination of high strength, high modulus, toughness, and thermal stability. It can be spun into ropes or sheets of fabric that can be used in the construction of composite components. Kevlar is used in wide range of applications starting from bicycles to body armor, due to its high strength-to-weight ratio, and it is five times stronger than steel on an equal weight basis [2]. It is a member of the Aramid family of synthetic fibers and a competitor of Twaron manufactured by Teijin.

7.4.1 Development and Molecular Structure of Kevlar

In 1965, scientists at DuPont discovered a new method of producing perfect polymer chain extension. The polymer poly-p-benzamide was found to form liquid crystalline solutions due to the simple repetitiveness of its molecular backbone. The key structural requirement for the backbone is the *para* orientation on the benzene ring, which allows the formation of rodlike molecular structures. These developments led to the current formulation of Kevlar. DuPont developed the fiber of poly(p-phenylene terepthalamide), which was introduced as high strength Kevlar aramid fiber in 1971.

In aramid fiber, the fiber-forming substance is a long-chain synthetic polyamide in which at least 85% of the amide linkages are attached directly to two aromatic rings Thus, in an aramid, most of the amide groups are directly connected to two aromatic rings, with nothing else intervening [3].

Structure of kevlar

The properties of the Kevlar fibers can be explained on the basis of its physical and chemical microstructures. During the spinning process, the polymer chains were oriented in the draw direction producing chain crystallites having a very high aspect ratio. The chains inside the crystallites were interconnected by hydrogen bonding that makes these fibers extremely strong. The only way to break these fibers by tension is to break all the hydrogen bonds at once, which is very difficult to achieve because of their huge numbers.

Hydrogen bonding in Kevlar

7.4.2
Properties of Kevlar Fibers

- Due to their highly aromatic and ordered structure, aramids have very high thermal stability.
- These fibers are flame resistant but they can be ignited.
- Kevlar fiber possesses chemical inertness and low electrical conductivity in comparison to carbon dispersed glass fibers.
- Kevlar fiber composites have highest specific strengths among all composite materials. Although composites from newer fibers have taken over that position, aramids still offer outstanding combinations of properties, such as high specific strength, toughness, creep resistance, and moderate cost, for specific applications (Table 7.1).

7.5
Interface

The structure and properties of the fiber/matrix interface offers significant role in governing the mechanical and physical properties of the fiber-reinforced composites.

Table 7.1 Unique properties of Kevlar fiber over other fibers [4].

Properties	Nylon 66	Kevlar 29	Kevlar 49	E-Glass	Steel
Specific gravity	1.14	1.44	1.45	2.55	7.86
Tensile strength (MPa)	1000	2750	2760	1700	1960
Tensile modulus (GPa)	5.52	82.7	131	68.9	200
Elongation at break (%)	18	5.2	2.4	3.0	2.4

In fiber-reinforced thermoplastics, surface of the fiber side, surface of the matrix, and the phase between fiber and matrix are collectively called as interface [5]. The large difference between the properties of the fibers and the matrix are communicated through the interface or in other words the stress acting on the matrix polymer is being transferred to the fibers (the main load bearing constituents) through the interface. These interface effects are ascertained as a type of adhesion phenomenon and are often interpreted in terms of the surface structure of the bonded materials, that is, surface factors such as wettability, surface free energy, the presence of the polar groups on the surface, and surface roughness of the material to be bonded. A number of assumptions have to be made about the properties of the interface for the theoretical analysis of stress transfer from fiber to the matrix as given below:

- The matrix and the fibers behave as classic materials.
- The interfacial bond is infinitesimally tenuous.
- The bond between the fiber and the matrix is perfect.
- The fibers are arranged in a regular or repeating array.
- The materials close to the fiber has the same properties as material in bulk form.

Some mathematical models are necessary to correlate these assumptions in polymeric materials. Figure 7.1 shows the fracture surfaces of fiber-reinforced composite with different mode of interface.

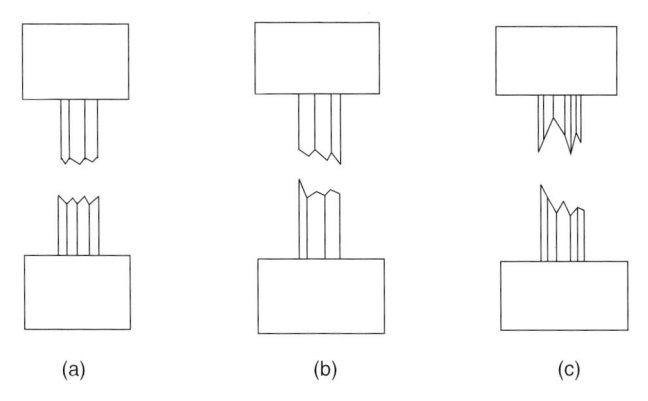

(a) (b) (c)

Figure 7.1 Diagrams of typical fracture surface of unidirectional composite loaded in tension along the fiber direction (a) composite with strong interfacial bond, (b) composite with intermediate interfacial bond, and (c) composite with poor interfacial bond [6].

7.6
Factors Influencing the Composite Properties

7.6.1
Strength, Modulus, and Chemical Stability of the Fiber and the Polymer Matrix

The mechanical properties of fiber-reinforced composites are widely regulated by the strength and modulus of the reinforcements [7]. As it is mentioned earlier that in fiber-reinforced composites fibers are the main load bearing constituents while the matrix keeps those fiber in their desired positions and orientations. Matrix should be chosen as per the requirements of the end product. There are many other factors such as cost, ease of fabrication and environmental conditions (e.g., temperature, humidity) in which the end product is going to be used should also be considered. The weathering and chemical resistance of the matrix, design stresses, and the required durability of the end product are also deserve supreme importance in regulating the composite properties. The main role of the matrix in a fiber-reinforced composite will alter depending on how the composite is stressed [11]. In case of compressing loading, the matrix must prevent the fibers from buckling, and thus offers a very critical part of the composite. If this fact is ignored then the reinforcing fibers could not carry any load. On the other hand, a bundle of continuous fibers could sustain high tensile loads in the direction of filaments without a matrix. The matrix also provides a transfer medium so that even if a single fiber breaks, it does not lose its load carrying capacity.

7.6.2
Influence of Fiber Orientation and Volume Fraction

Fiber orientation and volume fraction have significant role on the mechanical as well as other properties of the fiber-reinforced polymer composites [1]. With respect to fiber orientation, two extreme cases are possible-parallel alignment of the longitudinal axis of the fibers in a single direction and totally random orientation. These two extremes are depicted in Figure 7.2a and b, respectively. Normally continuous fibers are aligned and discontinuous ones are randomly oriented. Longitudinally aligned fiber-reinforced composites exhibit inherent anisotropic thermal, mechanical, and dynamic properties, where the maximum strength and reinforcements are achieved along the direction of fiber orientation [1]. On the other hand, transversely oriented fiber reinforcements are virtually nonexistent, fracture usually occurs at very low tensile stress, which may be even less that the pure matrix. In randomly oriented fibrous composites, strength lies between these two extremes. Hence, the prediction of fiber orientation in the fiber-reinforced composites is very much important in order to derive the properties of the same.

7.6.2.1 Fiber Orientation in Injection Molded Fiber-Reinforced Composites
In injection molding the fibers are considered to be suspended in the matrix and thereby orient themselves in response to the kinematics of the flow, mold cavity, and

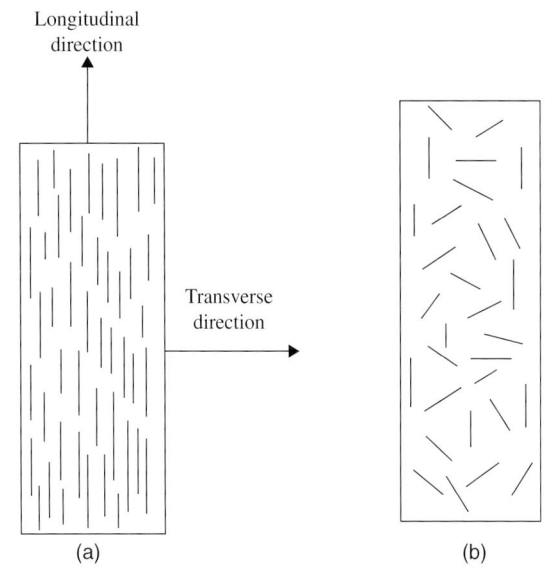

Figure 7.2 Schematic representation of (a) continuous and aligned, and (b) discontinuous and randomly oriented fiber-reinforced composites [6].

other neighboring fibers. However, fiber suspensions often demonstrate an anisotropic behavior due to a flow-induced fiber alignment in the flow direction. The properties of short fiber-reinforced thermoplastic composites, however, suffers from a problem associated with fiber orientation, which in turn depends on the processing conditions and the geometrical shape of the mold such as gating, inserts, and section thickness [8–14]. During the filling of an injection molding die, three flow regions normally exist.

These regions are

- a three-dimensional region near the gate,
- a lubrication region where no significant velocities out of the main flow plane exist and where the majority of the flow is contained,
- a fountain flow region at the flow front.

The effect of flow behavior on fiber orientation is complex but two rules of thumb have been demonstrated:

- Shearing flows tend to align fibers in the direction of flow.
- Stretching flows tend to align fibers in the direction of stretching. For a center gated disk, the stretching axis is perpendicular to the radial flow direction as can be seen from Figure 7.3.

The orientation of fibers in an injection molded parts consists of three different layers:

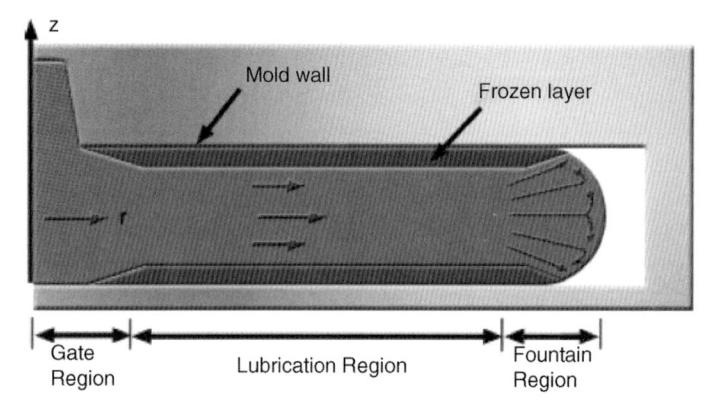

Figure 7.3 Flow regions during the injection molding process [6].

- A core created by in-plane fiber motion during mold filling.
- Shell layers on either side of the core, with a flow aligned orientation caused by gapwise shearing.
- Skin layers at the mold surface.

This difference in fiber orientation in different region affects the mechanical and thermal properties of the molded fiber-reinforced composite parts [15]. Hence it is of worth interest to predict the fiber orientation under different processing conditions. There are many reports available that demonstrates the numerical/analytical [16, 17] as well as computer aided simulation approach [18] to evaluate the fiber orientation in injection molded short fiber-reinforced composites under different molding conditions. Zhou and Lin [19] reported the behavior of fiber orientation probability distribution function in the planar flows. In their research work they have reported about the effect of fiber orientation on the planar flow behavior of the short fiber-reinforced composites. Chung et al. [20, 21] reported the polymer melt flow and weld line strength of injection moldings made by using the coinjection molding technique. A comprehensive study on the effect of processing variables on microstructure of injection molded short fiber-reinforced polypropylene composites have been reported by Singh and Kamal [11].

Imihezri et al. [22] investigated mold flow and component design analysis of polymeric-based composite automotive clutch pedals. Patcharaphun and Mennig [23] have reported the properties enhancement of short glass fiber-reinforced thermoplastics via sandwich injection molding. Although many research works have been reported leading to the theoretical or numerical approach for the prediction of fiber orientation, there are still a large scope remains for the computer aided simulation of the fiber orientation.

7.6.3
Volume Fraction

Fiber volume fraction plays a major role in predicting the properties of the fiber-reinforced composites, especially hybrid composites. Fiber volume fraction is represented in equation given below:

$$V_f = V_f/V_c \quad \text{(for normal composites)} \tag{7.1}$$

$$V_f = V_F/V_C \quad \text{(for hybrid composites)} \tag{7.2}$$

where V_f is the volume fraction of fiber, V_c is the total volume of composites, $V_F = V_{f1} + V_{f2}$. V_F is the total volume fraction of two types of fibers. Matrix volume fraction is given by the following equation:

$$V_m = V_m/V_c \tag{7.3}$$

where V_m is the volume fraction of the matrix.

For a fiber-reinforced composite an optimum spacing must be maintained to achieve the maximum properties, which is the minimum allowable spacing between the fibers, below which the structure will start to disintegrate under loading before the tensile failure. This minimum spacing is defined as the maximum volume fraction allowable for a composite.

7.6.4
Influence of Fiber Length

The strength of fiber-reinforced composites not only depends on the tensile strength but also on the extent to which an applied stress is transmitted to the fibers from matrix [1, 4]. The extent of this load transmittance is a function of fiber length and the strength of fiber–matrix interfacial bonds. Under an applied stress, this fiber–matrix bond ceases at the fiber ends resulting a matrix deformation. There is no stress transfer from the matrix at the fiber extremity. Hence a critical fiber length giving rise to critical aspect ratio, length to diameter ratio (L/D) should be maintained to effective stress transfer between the fiber and matrix at the interface.

7.6.5
Influence of Voids

During the incorporation of fibers into the matrix or during the manufacture of laminates, air or other volatiles are trapped in the material [7, 24]. The trapped air or volatiles exists in the cure laminate as microvoids, which may significantly affect some of its mechanical properties. Paul and Thompson [25] and Bascom [26] have investigated the origin of voids and described the various types of voids encountered in the composite and the means to reduce the void content. The most common cause of voids is the incapability of matrix to displace all the air that is entrained with in the roving or yarn as it passes through the resin impregnator. The rate at which the reinforcements pass through the matrix, the viscosity of the resin, the wettability or contact angle between the matrix and the fiber surface and the mechanical working of the fibers, and so on, will affect the removal of the entrapped air. A high void content (over 20%) usually leads to lower fatigue resistance, greater susceptibility to water diffusion, and increased variation (scatter) in mechanical properties.

7.6.6
Influence of Coupling Agents

Incorporation of coupling agents into the fiber-reinforced composites offers optimum physical properties of the same and retains these properties of the composites after environmental exposure. Good bonding at the interface can be achieved by modifying the interface with various coupling agents. An important technique for improving compatibility and dispersibility between the filler and the matrix is to develop a hydrophobic coating of a compatible polymer on the surface of the filler prior to mixing with the polymer matrix. Generally, coupling agents facilitate better adhesion between the filler and the matrix. The selection of coupling agents that can provide both strength and toughness to a considerable degree is important for a composite material.

7.7
Surface Modification

Surface modification of both matrix/or fiber is a key area of research at present to achieve optimum fiber/matrix properties.

7.7.1
Surface Modification of Fibers

Reinforcing fibers can be modified by physical as well as chemical methods.

i) **Physical methods of modification:** Physical treatments alter the structural and surface properties of the fibers thereby influence the mechanical bonding in matrix–fiber interface. These physical treatments include electric discharge method using corona or cold plasma. In the case of cold plasma treatment, depending on the type and nature of the gas used, a variety of surface modifications like cross-link could be introduced. In this process, surface free energy could be increased or decreased and reactive free radicals onto the fiber surface could be produced [27]. For Kevlar fiber where moisture absorption is known to have deteriorating effects, the plasma process is inherently an effective drying process providing further benefits [28]. Corona treatment is one of the most interesting techniques for surface oxidation activation. This process changes the surface free energy of the fibers [29].

ii) **Chemical methods of modification:** Chemical methods offer more convenient techniques to modify the fiber surface [30–32]. Chemical methods bring about the compatibility between the two polymeric materials either crating some functional groups on the fiber surface by means of chemical reactions or introducing a third material into the composites that has properties in between those other two.

Park *et al.* reported the chemical treatment of Kevlar surface by Phosphoric acid significantly affected the degree of adhesion at interfaces between Kevlar fibers and epoxy resin matrix [32]. Yue and Padmanabhan [33] improved the interfacial shear strength of Kevlar fiber/epoxy composites significantly, through chemical treatment of the fiber with organic solvents. Wu et al. [34] have proposed acid treatments as the modification tool for PBO, Kevlar, and carbon fiber. They have noticed a remarkable change in the surface free energy of those modified fibers. Mavrich et al. [35] have demonstrated the infrared mapping of epoxy reacted Kevlar/epoxy system. Different types of coupling agents are used to modify the fiber surface in order to enhance the fiber surface functionalities. Silanes have been the most frequently used coupling agents [36]. Silanes provide dual functionalities in one molecule, so that one part of the molecule forms a bond to the fiber surface while other part forms a bridge with the matrix molecule. Ai et al. [37] have used alkoxysilane as an effective coupling agent for Kevlar fiber in Kevlar/epoxy system. They have reported that with surface modification the adhesion between the fiber and the matrix increases resulting to the enhancement of interlaminar shear strength. Andreopoulos [38] has used acetic acid anhydride, sulfuric acid–acrylamide, and methacryloyl chloride as coupling agents to introduce the polar functional groups onto the Kevlar surface and among them methacryloyl chloride appeared to be the most effective coupling agents for Kevlar fiber. Menon et al. [39] have reported the effectiveness of the titanate coupling agents in Kevlar/phenolic composites. They have demonstrated an improved moisture resistant and flexural strength for modified systems. Copolymers can also be used as effective coupling agents for fibrous systems. MA-g-PP and MA-g-PE provide covalent bonds across the interface polypropylene–cellulose systems.

7.7.2
Surface Modification of Matrix Polymers

The wettability of the matrix polymer on the fiber surface depends on the viscosity of the matrix and the surface tension of both materials. For better wetting of the matrix on the fiber surface, surface tension of the matrix should be as low as possible; it should be at least lower than the reinforcing fiber materials. Hence the modification of matrix in a fiber-reinforced composites consist of the following methods:

a) **Chemical treatments:** The polymer surface can be modified by introducing polar groups onto the matrix surface. When a polymer is treated with highly oxidative chemicals such as chromic anhydride/tetracholoroethane, chromic acid/acetic acid, chromic acid/sulfuric acid under suitable conditions, polar groups are introduced on the polymer surface and the surface characteristics are improved [40, 41].
b) **Physical methods:** Physical method includes UV irradiation, corona discharge treatment, and plasma treatment.

7.7.3
Fluorination and Oxyfluorination as Polymer Surface Modification Tool

Direct fluorination and oxyfluorination plays a tremendous role as a surface modification technique for polymeric materials as it possess various advantages as given below:

- The elemental gas is used to modify polymeric materials. Due to very high energy release during the main elemental stages, the fluorination occurs at room temperature or below [42–44]. The process does not need any initiation or catalyst.
- Gaseous mixtures of fluorine, oxygen, nitrogen, helium, chlorine, and chlorine monofluoride may be used for direct fluorination.
- The reaction proceeds by the diffusion limited process.
- The fluorine-treated plastics consists of totally fluorinated, untreated virgin layer separated by a thin transition layer.
- The thickness of the modified layer polymer can be put under control over ~0.01–10 µm range.
- This technology is so called "dry" one (only gases are used) and polymeric.
- Article of any shape can be modified.
- One of the main advantages of direct fluorination and oxyfluorination is that the only thin surface layer of polymer is modified and hence the bulk properties of the polymer are not practically changed.

Fluorination and oxyfluorination offers improved properties in comparison to the unmodified polymeric materials as listed below.

7.8
Synthetic Fiber-Reinforced Composites

Fiber-reinforced plastics (FRP) have been widely accepted as materials for structural and nonstructural applications in recent years. The main reasons for the increased interest in FRP for structural applications, are high specific modulus, high strength of the reinforcing fibers. Various synthetic and natural fibers are being used as the reinforcing agents in FRP. Among natural fibers jute [45], sisal [46], henequen [47] fibers are used as successful reinforcements. Although natural fibers are inexpensive, easily available, biodegradable, they are not environmentally stable, chemically inert along with hydrophilic in nature, and of low strength.

Synthetic fibers are high-performance fibers possessing very high levels of at least one of the following properties: tensile strength, operating temperature, heat resistance, flame retardancy, and chemical resistance [48]. The resistance to heat and flame is one of the main properties of interest for determining the working conditions of the fibers [49].

Short fiber-reinforced thermoplastic-based polymer composites have drawn the attention of many engineering applications [50, 51] in recent years, particularly automobile and mechanical engineering industry have great interest for short fiber reinforcement in various polymer matrices for better mechanical, dynamic, and thermal properties of the concerned composites. Short fiber reinforcement is also favored over the continuous fiber reinforcement because of the combination of easier processability with low manufacturing cost. A large number of reports is available to analyze the short fiber reinforcement on the mechanical properties of the short fiber-reinforced composites [52, 53]. Broutman [54] had reported about the fibers made of phenol formaldehyde and also discussed about the properties of molded polypropylene thermoplastics as a function of phenolic fiber weight percent and surface treatment. Sudarisman et al. [55] have demonstrated the compressive failure of unidirectional hybrid fiber-reinforced epoxy composites containing carbon and silicon carbide fibers. Arikan et al. [56] have reported the fracture behavior of steel fiber-reinforced polymer composites. Fu and Mai [57] demonstrated the thermal conductivity of misaligned short fiber-reinforced polymer composites. They have used carbon fibers as reinforcements. Wang [58] had analyzed the toughness characteristics of synthetic fiber-reinforced cementitious composites. Kevlar, poly (p-phenylene terephthalamide), is an organic synthetic fiber with a distinct chemical composition of wholly aromatic polyamides (aramids). This fiber possesses a unique combination of high tensile strength and modulus, toughness, and thermal stability [59]. In air, PPTA demonstrates seven times the tensile strength of steel on an equal weight basis. In seawater, this advantage in tension increases by a factor of 20 [60]. Besides, Kevlar fiber possesses high thermal resistance and chemical inertness and low electrical conductivity with compared to metallic or carbon glass fibers. These superior properties of Kevlar fiber lead to the increasing applications of Kevlar fiber-reinforced composites in aircraft, missile, and space applications such as rocket motor casings and nozzles. Kevlar fiber composites are also used in conjunction with aluminum to give rise to superior hybrid composites. The fiber failure in the longitudinal splitting mode has led to the unique ballistic resistance of Kevlar when used with suitable combination of matrices. Thus, the development of Kevlar composites with high strength as well as stiffness, apart from their proven toughness, is highly relevant and significant in the current scenario where the thrust is on good damage tolerance, high strength and stiffness, good hot-wet properties, high fatigue life, and low density. The main drawback of the Kevlar fiber reinforcement is poor interfacial adhesion due to its chemical inertness and low surface energy, which affects the chemical and physical properties of the composites. In order to increase the surface adhesion of the fiber resulting good interaction with the matrix various physical and chemical surface modification techniques are being followed in present days.

Maalej [61] demonstrated the tensile properties of short fiber composites with fiber strength distribution. They have analyzed the influence of fiber rupture, fiber pull-out and fiber tensile strength distribution on the postcracking behavior of short randomly distributed fiber-reinforced brittle-matrix composites has been analyzed using an approach based on the Weibull weakest-link statistics. Yu et al. [62] have predicted the mechanical properties of short Kevlar fiber–nylon-6,6 composites. Sun et al. [63] have reported shear-induced interfacial structure of isotactic polypropylene (iPP) in iPP/Fiber composites, where they have studied the shear-induced interfacial structure of iPP in pulled iPP/fiber composites optical microscopy. Shaker et al. [64] have compared the low- and high-velocity impact response of Kevlar fiber-reinforced epoxy composites. Kutty and Nando [65] have reported the mechanical and dynamic properties of the short Kevlar fiber–thermoplastic polyurethane composites.

Murat İçten et al. [66] have studied the failure analysis of woven Kevlar fiber-reinforced epoxy composites pinned joints. Kim et al. [67] studied the graft copolymerization of the ε-caprolactam onto Kevlar-49 fiber surface and properties of grafted Kevlar-reinforced composite. Al-Bastaki [68] have designed the Kevlar fiber-reinforced epoxy tubes subjected to high strain rates using finite element analysis.

Kodama and Karino [69] demonstrated the polar–polar interaction between the reinforcement and matrix for Kevlar fiber-reinforced composite using the blend of polar polymers as matrix. They have used poly(hydroxy ether of bisphenol A) (I), with which poly(ethylene oxide) (II) or poly(ethylene adipate) (III) was blended as a part of matrix in Kevlar fiber-reinforced composites. It was shown by analyzing the storage modulus and loss modulus versus temperature curves that the reinforcement–matrix interaction is increased relatively to the primary transition temperature of matrix by blending II or III with I, and II is more efficient for increase of the interaction than III. Mahmoud [70] had discussed the tensile, impact, and fracture toughness behavior of unidirectional, chopped, and bidirectional fiber-reinforced glass and Kevlar/polyester composite in terms of fiber volume fraction and fiber arrangement. Wu and Cheng [71] demonstrated the interfacial studies on the surface-modified aramid fiber-reinforced epoxy composites. They have used solutions of rare earth modifier (RES) and epoxy chloropropane (ECP) grafting modification method were used for the surface treatment of aramid fiber. Kitagawa et al. [72] have evaluated the interfacial property in aramid fiber-reinforced epoxy composites. Haque et al. [73] have reported the moisture and temperature-induced degradation in tensile properties of Kevlar–graphite/epoxy hybrid composites.

7.9
Effect of Fluorinated and Oxyfluorinated Short Kevlar Fiber on the Properties of Ethylene Propylene Matrix Composites

Das et al. reported the surface modification of Kevlar fibers and their effect on the properties of ethylene propylene (EP) copolymer.

7.9.1
Preparation of Composites

Kevlar was fluorinated using a mixture of F_2 and He (5% : 95%) at 0.4 bar total pressure and 17 °C for 1 h. Oxyfluorination was carried out by 5% F_2, 5% air, 90% He (5% F_2, 1% O_2, 90% He, and 4% N_2) under 0.8 bar pressure at 17 °C for 1 h in a reaction chamber.

The unmodified, fluorinated, and oxyfluorinated Kevlar fibers were incorporated in ethylene propylene copolymer. The mixing operation was performed in Brabender mixer and the obtained composites were then compression molded into slabs. Table 7.2 shows the formulations of the composites.

7.9.2
FTIR Study

Figure 7.4 shows the FTIR spectra of the composite materials along with the pure EP. The peak associated with primary amine backbone appeared in the cases of unmodified, fluorinated, and oxyfluorinated Kevlar fiber-reinforced EP at

Table 7.2 Compounding formulations.

Sample Code	Ethylene propylene copolymer (wt%)	Kevlar (wt%)
EP	100	0
B	100	1.43 (original)
E	100	1.43 (fluorinated)
H	100	1.43 (oxyfluorinated)

$3340\,cm^{-1}$ [74] (absent in case of pure EP), which may be due to the incorporation of Kevlar fiber into the polymer matrix.

The peak at $1760\,cm^{-1}$, corresponding to the characteristics peak of C=O of Kevlar fiber, was observed for all the composites. The FTIR plot of fluorinated and oxyfluorinated Kevlar fiber composites showed a peak at $1200\,cm^{-1}$, which may be the characteristic peak of monofluorinate aliphatic groups (C–F bond), formed due to the fluorination and oxyfluorination. Oxyfluorinated Kevlar fiber-reinforced EP showed an additional peak at $3400\,cm^{-1}$ indicating the formation of OH group during the oxyfluorination process.

7.9.3
X-Ray Study

Figure 7.5 and Table 7.3 show the X-ray diffraction (XRD) diagram of various samples and their respective parameters. From Figure 7.5 it was revealed that the percent

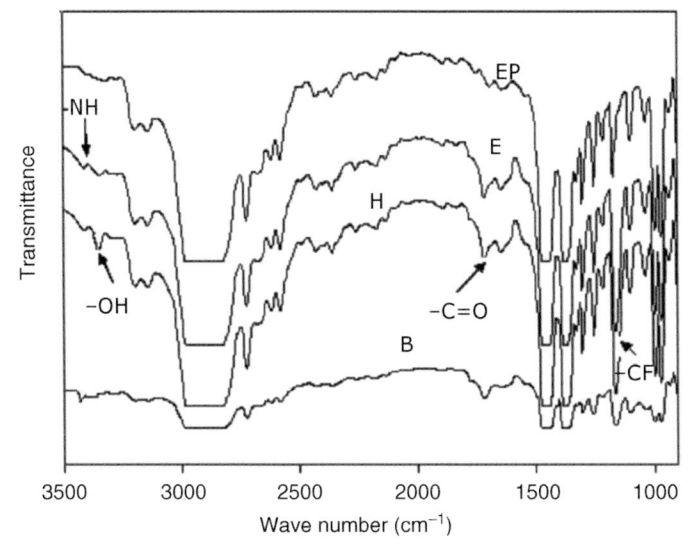

Figure 7.4 FTIR spectra of pure EP (EP), EP/unmodified Kevlar (B), EP/fluorinated Kevlar (E), and EP/oxyfluorinated (H) composites.

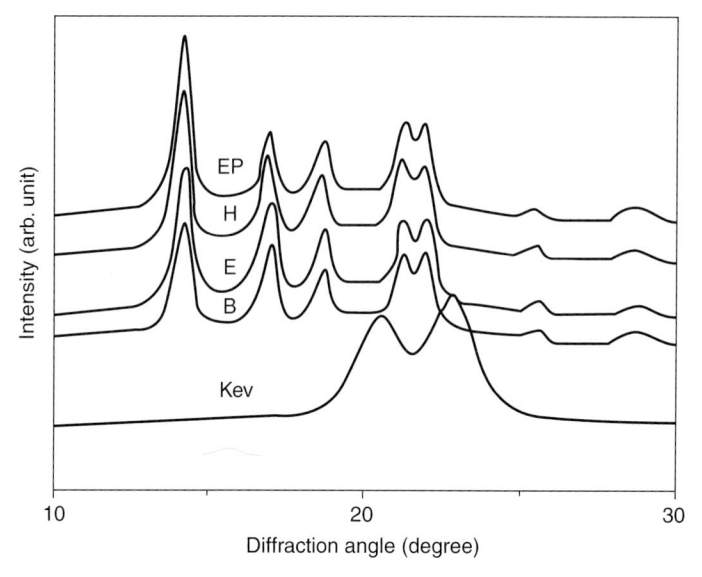

Figure 7.5 XRD pattern of Kevlar (Kev), pure EP (EP), EP/unmodified Kevlar (B), EP/fluorinated Kevlar (E), and EP/oxyfluorinated (H) composites.

crystallinity is highest for pure EP and the lowest for Kevlar fiber. The pure EP and Kevlar showed 76 and 30% crystallinity, respectively. The percentage of crystallinity significantly changes with the addition of Kevlar fiber into the EP matrix. The Kevlar fiber hinders the migration and diffusion of EP molecular chains to the surface of the

Table 7.3 X-ray parameters of EP and EP/Kevlar composites.

Sample code		Kevlar	EP	B	E	H
% Crystallinity		30.00	74.00	52.00	58.00	60.00
Peak angle (2θ) (°)	θ_1	—	—	14.20	14.20	14.20	14.10
	θ_2	—	—	16.95	17.0	17.05	16.90
	θ_3	—	—	18.70	18.70	18.68	18.66
	θ_4	20.58	21.30	21.32	21.34	21.20
	θ_5	22.80	21.90	21.88	21.95	21.88
Interplanar spacing (Å)	d_1	—	6.26	6.26	6.26	6.31
	d_2	—	5.25	5.23	5.22	5.23
	d_3	—	4.76	4.76	4.77	4.80
	d_4	4.45	4.12	4.18	4.18	4.02
	d_5	3.91	4.18	4.06	4.06	4.08
Crystallite size (Å)	p_1	—	116.60	144.70	144.70	155.80
	p_2	—	164.60	156.40	145.20	156.40
	p_3	—	165.30	145.60	146.70	145.50
	p_4	46.40	137.80	102.30	120.30	113.60
	p_5	39.50	137.90	102.40	120.40	114.10

growing polymer crystal in the composites, resulting in a decrease in the percent crystallinity.

In the case of composite E and H, the percent crystallinity of the composites increases with fluorination and oxyfluorination of Kevlar, compared to the composite B. This suggested that there are more interactions between the surface-modified Kevlar fiber and EP matrix. This may be due to the generation of more functional groups on the surface of Kevlar fiber during fluorination and oxyfluorination. It is also important to point out that the crystallite size of composites E and H, corresponding to all peak angles (2θ), are higher than composite B, which suggested that the nucleation and growth are also favored in case of E and H compared to B. The pure EP showed the reflection at 2θ of 21.6° and 24.0°, whereas Kevlar fiber showed the reflection at 20.6°, 22.8°, and 28.7°. In composites, the peak of EP at 24.0° and Kevlar at 28.7° were not visible, but in addition three broad peaks appeared. As the functional groups increased, the intensity and sharpness of the peaks also increased in the case of modified Kevlar fiber-reinforced composites. However, significant changes in the EP crystalline structure were observed in the presence of Kevlar fiber.

7.9.4
Thermal Properties

Figure 7.6a and b shows the differential scanning calorimetric (DSC) and thermogravimetric analysis (TGA) of pure polymer and the composites. The respective parameters are given in Table 7.4. Heat of fusion is proportional to the amount of crystallinity in the sample, that is, higher heat of fusion will correspond to higher crystallinity. It was observed from Table 7.4 that heat of fusion is highest for pure EP and then it decreases gradually in case of fiber-reinforced composites. This result supports the trend of crystallinity of the composites obtained from the XRD study. Pure EP shows a melting endotherm, T_m, at about 165 °C and the composites showed the values close to the T_m of EP (B = 162.3 °C, E = 162.4 °C,

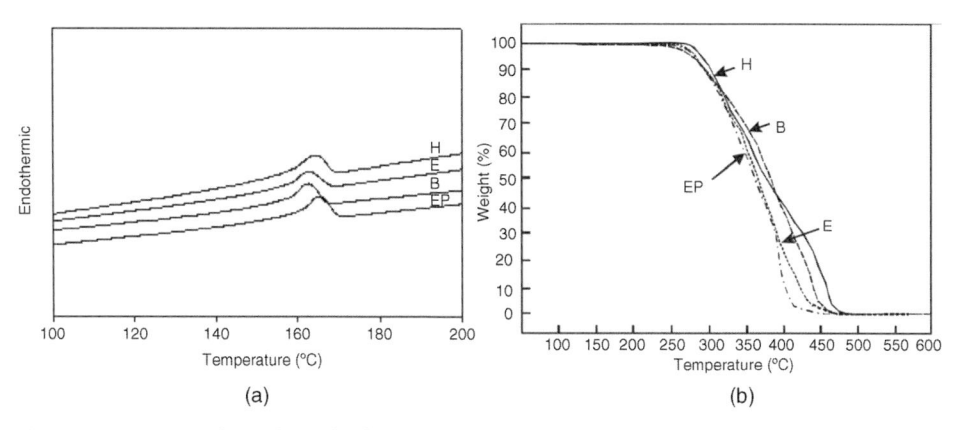

Figure 7.6 DSC (a) and TGA (b) study of pure EP (EP), EP/unmodified Kevlar (B), EP/fluorinated Kevlar (E), and EP/oxyfluorinated (H) composites.

Table 7.4 Thermal parameters of EP, EP/Kevlar composites.

Sample code	First decomposition temperature (°C)	Loss of wt% for first step	Second decomposition. temperature (°C)	Loss of wt% for second step	T_m (°C)	H_f (J/g)
EP	233.0	69.2	387	30.8	165.0	73.0
B	245.0	32	352	68	162.3	69.2
E	254.4	39	352	61	162.4	68.5
H	269.0	54	391	38	164.1	70.0

and H = 164.1 °C). The melting point of EP marginally decreases in case of composites. The depression of melting temperature of the composites may be due to the dilution effect. In case of pure EP and EP/Kevlar composites, degradation occurs in two steps. The initial degradation of pure EP starts at 233 °C and continues up to 387 °C at a faster rate accompanied by 69% weight loss. The second degradation step starts at 387 °C and proceed at a slower rate. For EP/Kevlar composite (sample B) degradation starts at 245 °C and continues up to 352 °C at a slowest rate and about 32% of weight loss occurred in this step. However, in case of EP/Kevlar (fluorinated) and EP/Kevlar (oxyfluorinated) composites, the initial degradation starts at a higher temperature (around 254.4 and 269 °C, respectively) compared to B. This first degradation process is slower than pure EP and associated with 39 and 54% sample degradation, respectively.

From the TGA study it proved that the incorporation of modified Kevlar fiber enhances the thermal stability of the composites. The presence of functional groups on oxyfluorinated Kevlar fiber may form a better interaction with the matrix polymer that may be responsible for higher thermal stability of the oxyfluorinated Kevlar fiber-reinforced (H) composite. Oxyfluorination generates controlled amount of long-living $RO_2^{\bullet}$ radical [75]. These radicals further may be responsible for graft polymerizations leading to higher thermal stability.

7.9.5
Dynamic Mechanical Thermal Analysis (DMTA)

Figure 7.7 represents the storage modulus E', for the EP, modified and unmodified Kevlar fiber containing EP composites. Kevlar-filled EP composites show higher storage modulus than pure EP due to the reinforcement imparted by Kevlar fibers that permits stress transfer from the EP matrix to Kevlar fiber.

The storage modulus increases especially in the glassy region of oxyfluorinated Kevlar fiber-reinforced EP composites. The increase in functional groups on Kevlar fiber resulted in improved interfacial adhesion between the fibers and matrix that restricts the molecular mobility of EP. The storage modulus graphs show a sharp decrease in the temperature range 0–10 °C, which correlates with the glass transition temperature.

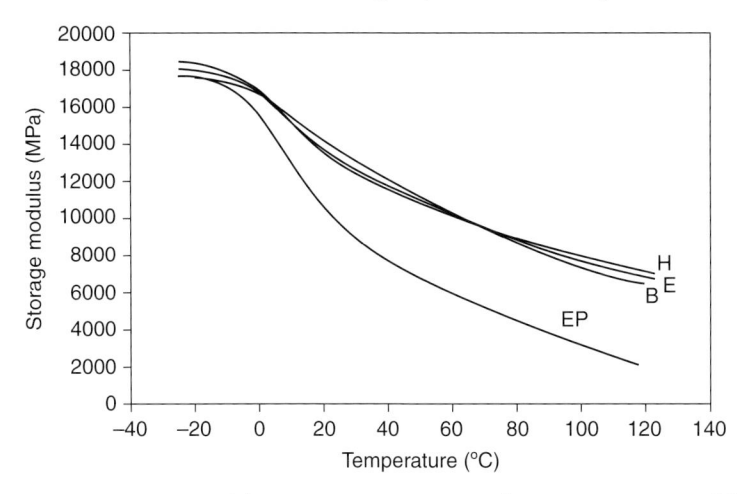

Figure 7.7 Storage modulus versus temperature curve of pure EP (EP), EP/unmodified Kevlar (B), EP/fluorinated Kevlar (E), and EP/oxyfluorinated (H) composites.

Figure 7.8 represents the tan δ versus temperature curve. It was that one relaxation peak appears around 10 °C that corresponds to the α-transition of EP polymer [76]. The α-transition shifted to higher temperatures for the composites due to the presence of Kevlar fiber in the EP matrix, which tends to decrease the mobility of the matrix chains. The loss peaks were indicative of the efficiency of the material to dissipate energy. The magnitude of the loss peak was lower in case of sample H. The lower magnitude of loss peak indicated lower mechanical energy dissipation capacity.

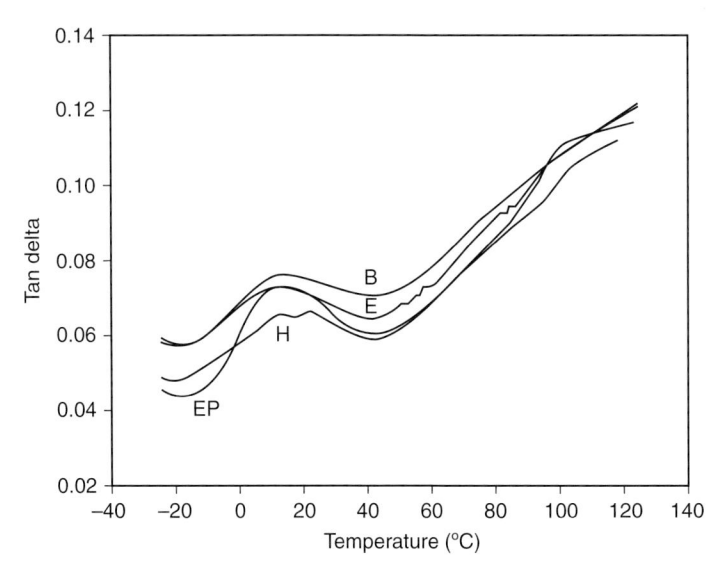

Figure 7.8 tan δ versus temperature curve of pure EP (EP), EP/unmodified Kevlar (B), EP/fluorinated Kevlar (E), and EP/oxyfluorinated (H) composites.

Hence, the molecular mobility of the composites decreases and the mechanical loss gets reduced after incorporation of oxyfluorinated Kevlar fibers in EP matrix. This can be also pointed out that the transition peak progressively broadens. The broadening of loss peak in the presence of fiber can be ascribed to matrix–fiber interaction. The matrix polymer in the adjacent portion of the fiber can be considered to be in different state, in comparison to the bulk matrix, which can disturb the relaxation of the matrix resulting in a broad tan δ peak.

7.9.6
Mechanical Properties

Table 7.5 shows the mechanical properties of the pure and composites. The matrix shows higher tensile strength and high elongation at break compared to EP/Kevlar composite. This arises from the very basic ordered structure of EP, which was also evident from % crystallinity value obtained from XRD study. The mechanical properties like modulus and tensile strength of polymeric materials strongly depend on their microstructure and crystallinity. So, crystallinity enhances the tensile properties of pure EP. The tensile strength decreases sharply on addition of Kevlar fibers (sample B) into the polymer matrix whereas it increases gradually in case of fluorinated Kevlar fiber/EP (sample E) and oxyfluorinated Kevlar fiber/EP composites (H). The resulted strength in the presence of fiber may be because of two opposing effects: dilution of the matrix and the reinforcement of the matrix by fibers. In case of sample B the matrix is not properly restrained by fibers because of poor adhesion between the fiber and matrix and concentration of the localized strains on the matrix, causing the pulling out of fibers and leaving the matrix diluted by nonreinforcing, debonded fibers. But in the case of E and H, where the matrix is sufficiently restrained and the stress is more evenly distributed that enhanced the mechanical properties. This can be ascribed to the surface characteristics of modified fibers, which increases the adhesion between the fiber and matrix.

7.9.7
SEM Study

Figure 7.9 shows the SEM images of the cross-sectional tensile fracture surface of EP and EP/Kevlar composites. In case of B, the fibers are randomly oriented and pulled

Table 7.5 Mechanical properties of EP, EP/Kevlar composites.

Sample code	Tensile strength (MPa)	Elongation at break (%)	Tensile modulus (GPa)
Pure EP	27	125	0.36 ± 0.03
B	20	6.85	0.40 ± 0.02
E	30	5.13	0.56 ± 0.02
H	33	4.02	0.68 ± 0.01

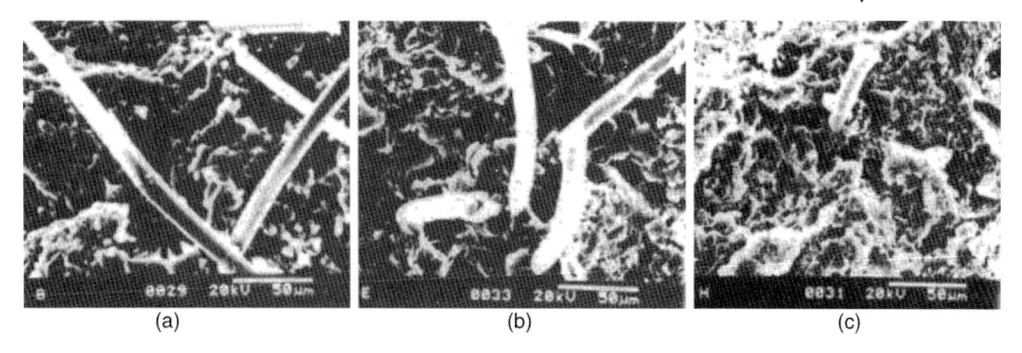

Figure 7.9 SEM photograph of (a) unmodified Kevlar/EP (B), (b) fluorinated Kevlar/EP (E), and (c) oxyfluorinated Kevlar/EP (H) composites.

out from the matrix indicating poor adhesion between the fiber and polymer matrix resulting in lower tensile strength as compared to pure EP. In case of E (Figure 7.9b), fibers are broken rather than pulled out due to the strong interfacial bonding between the fibers and the polymer matrix. However in case of sample H, fibers and the matrix forms network giving rise to the best mechanical and thermal properties as a result of good adhesion between fiber and the matrix.

7.9.8
AFM Study

Figure 7.10a–f represents the two-dimensional and three-dimensional AFM images of EP/Kevlar (B), EP/fluorinated Kevlar (E) and EP/oxyfluorinated Kevlar (H) composites. Comparison of the AFM topographies of composites with the surface roughness data suggested that the orientation of the Kevlar fiber during the preparation of composite samples might play more active roles in determining the surface morphology. The maximum surface roughness of B, E, and H composites were measured as 4, 17, and 26 nm, respectively.

The interface roughness is one of the most important parameter of interface adhesion strength as reported earlier [77]. So, the roughness data strongly suggested that the adhesion strength of oxyfluorinated Kevlar fiber/EP interface is considerably much higher than that of EP/fluorinated Kevlar and EP/Kevlar composites.

7.9.9
Conclusion

EP matrix containing modified Kevlar fibers show better crystalline, dynamic mechanical and thermal properties than the composites containing unmodified Kevlar fiber loading. This effect was more pronounced in case of oxyfluorinated Kevlar/EP composites. The polar functional groups on the Kevlar fibers surface played an important role in enhancing the dispersion and interfacial adhesion of the Kevlar fibers in the matrix polymer.

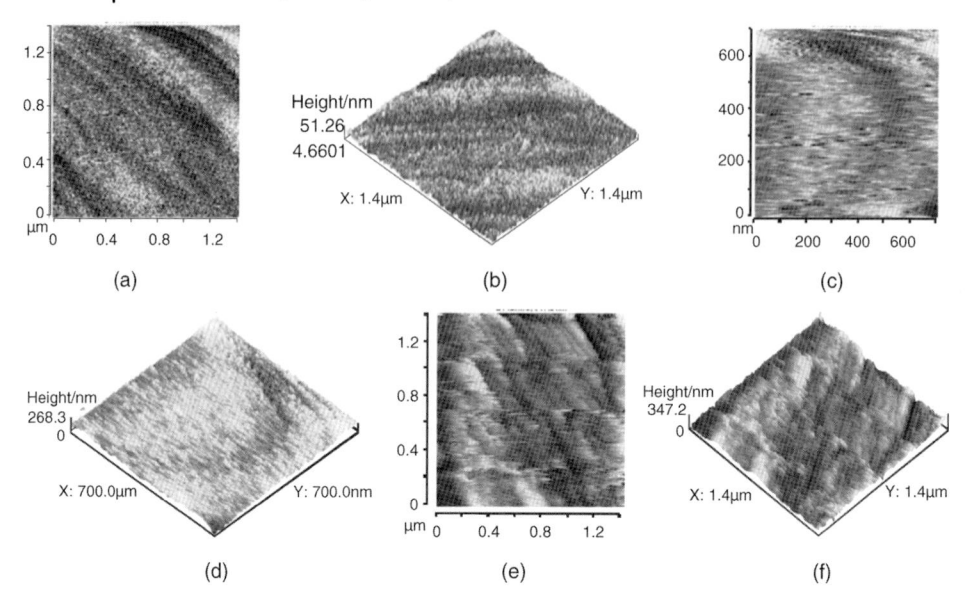

Figure 7.10 2D and 3D AFM image of (a) and (b) EP/unmodified Kevlar, (c) and (d) EP/fluorinated Kevlar, (e) and (f) EP/oxyfluorinated Kevlar composite.

7.10
Compatibilizing Effect of MA-g-PP on the Properties of Fluorinated and Oxyfluorinated Kevlar Fiber-Reinforced Ethylene Polypropylene Composites

Das et al. analyzed effect of compatibilizer on the properties of unmodified and modified Kevlar/EP composites. MA-g-PP was used as a compatibilizer for the composite preparation.

7.10.1
Preparation of the Composites

Ethylene polypropylene composites were prepared by them using both unmodified and modified Kevlar fiber in absence and presence of MA-g-PP (as compatibilizer) by melt blending technique. Table 7.6 shows the compounding formulation.

7.10.2
Thermal Properties

The thermal properties of the untreated, fluorinated, and oxyfluorinated Kevlar/EP composites were analyzed through TGA and DSC studies (Figure 7.11a–c and Table 7.7).

From Table 7.7 and Figure 7.11, it proves that in presence of compatibilizer, the variation in the magnitude of heat of fusion for the untreated, fluorinated and

Table 7.6 Compounding formulation.

Sample code	EP (wt%)	Kevlar (wt%)	Compatibilizer (wt%)
EP	100	0	—
B	100	1.43 (original)	0
B1	100	1.43 (original)	2
B2	100	1.43 (original)	5
E	100	1.43 (fluorinated)	0
E1	100	1.43 (fluorinated)	2
E2	100	1.43 (fluorinated)	5
H	100	1.43 (oxyfluorinated)	0
H1	100	1.43 (oxyfluorinated)	2
H2	100	1.43 (oxyfluorinated)	5

EP: ethylene propylene copolymer.

oxyfluorinated Kevlar fiber-reinforced EP composites were very small. Heat of fusion for the oxyfluorinated Kevlar/EP composites with compatibilizer was higher compared to other composites. This may be due to the better interaction between the oxyfluorinated Kevlar fibers and matrix as well as between MA-g-PP and EP at interfacial region, which results in a better adhesion. However, the melting point (T_m) of the composites shifts marginally toward lower value due to the imperfect or incomplete crystallization.

Figure 7.11b, d, f and Table 7.7 depict the TGA curves and the respective parameters of the composites. It was reported in their earlier work [78] that surface modification of the fibers by oxyfluorination leads to the enhanced thermal stability of EP/Kevlar composites.

It is evident from the TGA plots that degradation of all the composites along with pure EP follows two-step degradation process. As we have mentioned in our earlier

Table 7.7 Thermal parameters of EP, EP/Kevlar composites.

Sample code	First degradation temperature (°C)	50% degradation temperature (°C)	T_m (°C)	H_f (J/g)
EP	233.0	338.1	165.0	73.0
B	245.0	349.5	162.3	69.2
B1	260.5	365.3	162.9	70.2
B2	263.0	370.9	163.1	71.1
E	254.4	331.8	162.4	68.5
E1	266.6	346.2	163.1	69.8
E2	272.9	383.3	163.9	70.5
H	269.0	349.4	164.1	70.0
H1	292.0	374.0	164.8	72.3
H2	295.6	400.0	164.5	72.8

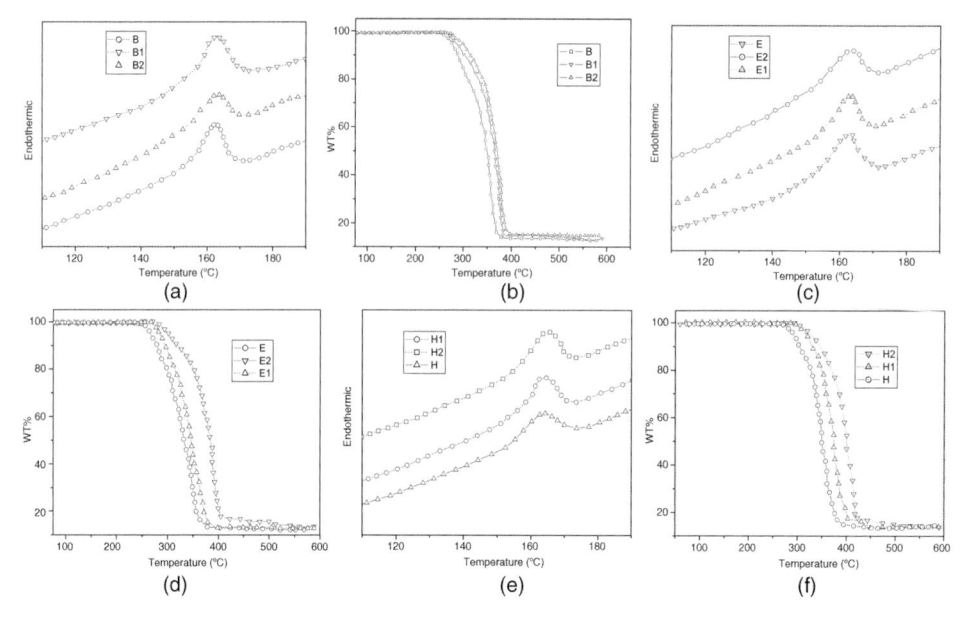

Figure 7.11 DSC and TGA study of (a) and (b) Kevlar/EP, (c) and (d) fluorinated Kevlar/EP, (e) and (f) oxyfluorinated Kevlar/EP composites with and without compatibilizer.

work that degradation of pure EP starts at 233 °C [78] and it gradually increases in case of all EP/Kevlar composites especially for modified Kevlar fiber-reinforced EP composites (B = 245 °C, E = 254.4 °C, and H = 269 °C).

In presence of compatibilizer, the thermal stability of aforementioned composites increases appreciably. From Table 7.7 it was evident that unmodified Kevlar/EP/MA-g-PP composites containing 2% (B1) and 5% (B2) of compatibilizer shows onset degradation temperature at about 260.5 and 263 °C, respectively. It was also observed that fluorinated Kevlar/EP and oxyfluorinated Kevlar/EP composites containing 2 and 5% of MA-g-PP show onset degradation temperature at about 266.6 and 272.9 °C, and 292.0 and 295.6 °C, respectively. The thermal stability of the composites increases in case of compatibilized systems in comparison to the uncompatibilized one. This may be due to the incorporation of the compatibilizer, which resulting better adhesion between the fiber and the matrix that increases the compatibility between them giving rise to enhanced thermal stability of the compatibilized systems. The enhancement of thermal stability is more distinct in case of MA-g-PP compatibilized oxyfluorinated Kevlar/EP composites (H1, H2) that can be ascribed to the better interaction between the fiber and the matrix polymer at the interface. This may be due to the generation of reactive functional group (peroxy radical) [75] on the Kevlar surface and as well as due to the presence of compatibilizer. At higher MA-g-PP contents thermal stability of the both unmodified, fluorinated, and oxyfluorinated Kevlar/EP composites increases very marginally.

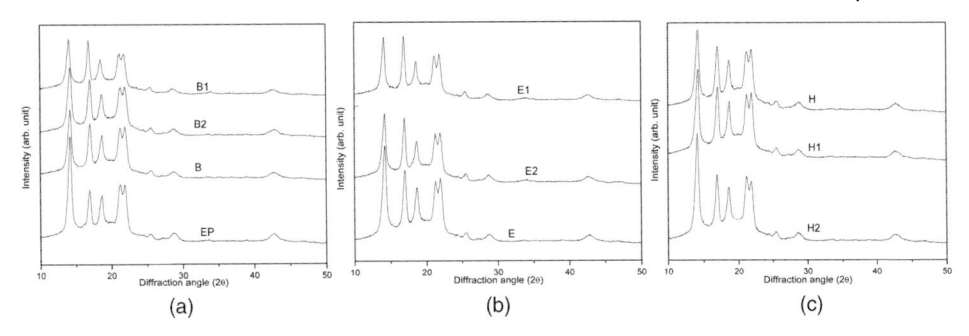

Figure 7.12 XRD pattern of (a) Kevlar/EP composite, (b) fluorinated Kevlar/EP composite, and (c) oxyfluorinated Kevlar/EP composite with and without compatibilizer.

7.10.3
X-Ray Study

Figure 7.12a–c and Table 7.8 show the XRD patterns and the respective parameters of uncompatibilized and compatibilized composites. From Table 7.8 it is evident that presence of compatibilizer appreciably affects the crystalline properties of Kevlar/EP composites.

This may be due to the better interaction between the fiber and matrix surface resulting improved adhesion between them at the interface, which in turn favors the crystal growth mechanism. In all cases the percent crystallinity of the composites are lower than that of the pure EP, which may be due to the hindrance offered by Kevlar fiber to the migration and diffusion of matrix polymer chains to surface of the growing polymer crystal in the composites. The increase in crystallinity is more prominent in case of MA-g-PP compatibilized oxyfluorinated Kevlar/EP composites (H1, H2). This can be ascribed to the effect of graft copolymerization, at the fiber/matrix interface, due to the generation of the peroxy radical ($RO_2^{\bullet}$) on the Kevlar surface and the reactive group on that of the matrix resulting better adhesion between the fiber and the matrix.

However, MA-g-PP content does not affect the crystallinity of the composites appreciably. It is also noteworthy that interplanar spacing corresponding to every peak position decreases in case of compatibilized unmodified and modified Kevlar/EP composites that supports the nucleating ability of MA-g-PP on Kevlar/EP systems and the results obtained from DSC study.

7.10.4
Dynamic Mechanical Thermal Analysis

Figure 7.13a–c shows the variation of storage modulus as a function of temperature of all composites. It is evident from the figure that incorporation of both untreated, fluorinated and oxyfluorinated Kevlar fiber leads to the increase in storage modulus of Kevlar/EP composites. Further enhancement of storage modulus is observed in case of compatibilized unmodified Kevlar/EP fluorinated and oxyfluorinated Kevlar/EP composites. In presence of MA-g-PP, the stiffness conferred upon the polymeric

Table 7.8 XRD parameters of EP/Kevlar composites with and with out MA-g-PP.

Sample code	% crystallinity	Peak angle (2θ) (°)					Interplanar spacing (Å)					Crystallite size (Å)				
		θ_1	θ_2	θ_3	θ_4	θ_5	d_1	d_2	d_3	d_4	d_5	p_1	p_2	p_3	p_4	p_5
EP	74	14.20	16.95	18.70	21.30	21.90	6.26	5.25	4.76	4.12	4.18	116.60	164.60	165.30	137.80	137.90
B	52	14.20	17.0	18.70	21.32	21.88	6.26	5.23	4.76	4.18	4.06	144.70	156.40	145.60	102.30	102.40
B1	55	14.10	16.93	18.57	21.16	21.87	6.30	5.28	4.79	4.21	4.08	146.23	158.11	146.70	104.10	103.91
B2	56	14.26	17.01	18.73	21.24	21.95	6.23	5.23	4.75	4.20	4.07	147.33	159.35	147.12	105.01	105.65
E	58	14.20	17.05	18.68	21.34	21.95	6.26	5.22	4.77	4.18	4.06	144.70	145.20	146.70	105.20	120.50
E1	59	14.10	16.93	18.57	21.08	21.79	6.30	5.25	4.79	4.23	4.09	145.32	146.63	147.80	121.57	121.87
E2	61	14.18	16.93	18.57	21.08	21.79	6.27	5.23	4.77	4.20	4.07	146.53	148.01	150.12	122.79	122.03
H	60	14.10	16.90	18.66	21.20	21.88	6.31	5.23	4.80	4.02	4.08	155.80	156.40	145.50	113.60	114.10
H1	66	14.25	17.00	18.72	21.16	22.05	6.24	5.23	4.76	4.21	4.07	158.03	159.21	147.32	116.65	115.13
H2	69	14.10	17.00	18.57	21.24	21.95	6.27	5.23	4.77	4.02	4.06	159.32	159.66	149.65	118.32	117.92

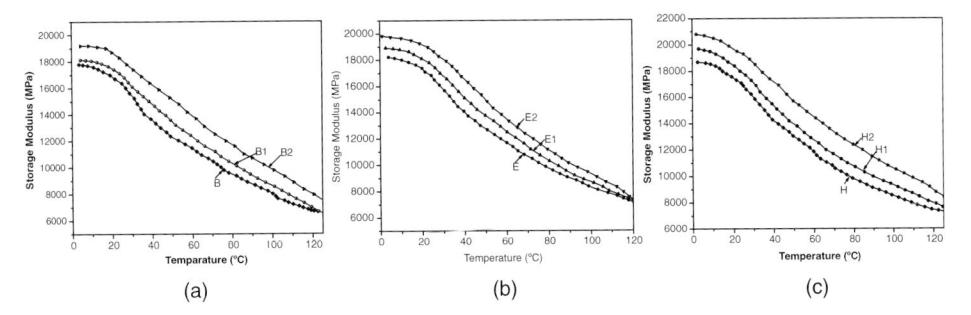

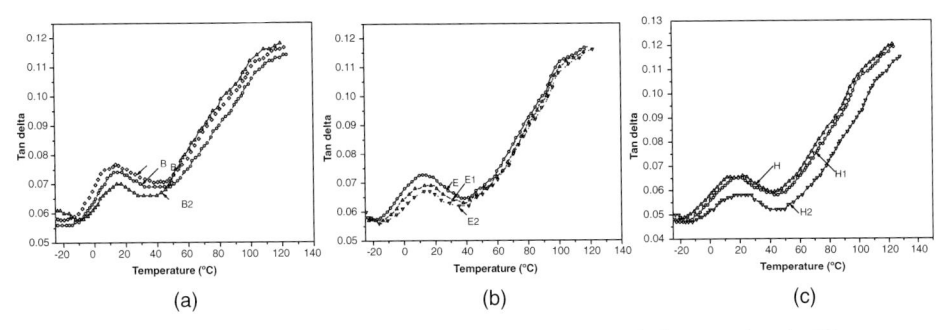

Figure 7.13 Storage modulus versus temperature curve of (a) Kevlar/EP composite, (b) fluorinated Kevlar/EP composite, and (c) oxyfluorinated Kevlar/EP composite with and without compatibilizer.

matrix by the fibers is more pronounced due to the efficient interfacial fiber matrix adhesion that increases the storage modulus of the composites.

The enhancement of the magnitude of storage modulus pronounced in case of oxyfluorinated Kevlar/EP composite (H1, H2) (Figure 7.13c) is due to the generation of reactive groups ($RO_2^\bullet$) on the oxyfluorinated Kevlar fiber surface and on the matrix surface (due to the incorporation of MA-g-PP) resulting stronger interfacial bonding between the fiber and the matrix. Storage moduli of the composites increase marginally with an increase in MA-g-PP content, indicating enhanced stiffness in case of composites with higher percentage of compatibilizer.

Figure 7.14a–c depicts the tan δ of the base polymer and their composites as a function of temperature. The loss peaks are indicative of the efficiency of a material to dissipate the mechanical energy. Form these figures it was clear that tan δ curve shows a distinct relaxation peak at around 10 °C, which corresponds to the α-transition of EP polymer. The magnitude of the loss peak is lower in case of all composites in comparison to the neat polymer. Addition of MA-g-PP further lowers the magnitude of loss peak indicating lower energy dissipation capacity as a result of decreased molecular mobility of the concerned composites giving rise to reduced mechanical loss for both unmodified and fluorinated oxyfluorinated Kevlar/EP composites. This indicated strong interfacial adhesion between fiber and the matrix in presence MA-g-PP.

Figure 7.14 tan δ versus temperature curve of (a) Kevlar/EP composite, (b) fluorinated Kevlar/EP composite, and (c) oxyfluorinated Kevlar/EP composite with and without compatibilizer.

7.10.5
Flow Behavior

Rheological parameters of the composites were studied with the help of a Dynamic analyzer RDA-II (Rheometrics Inc., USA) equipped with a parallel plate. All measurements were carried out at a temperature of 200 °C and a shear rate range of 1–100 s^{-1} using Bagley correction method.

Figure 7.15a–c shows the effect of MA-g-PP on the flow behavior of Kevlar/EP composites. Viscosity of the EP composites increases with the addition of Kevlar fiber due to the restriction to flow of the matrix imposed by the fibers. Viscosity of the treated Kevlar/EP composites increases due to the better fiber/matrix adhesion at the interface that is more prominent in case of oxyfluorinated Kevlar/EP (H, Figure 7.15c) composites because of the generation of more functional group on the fiber surface resulting better adhesion at the fiber–matrix interface.

From Figure 7.15a–c it observes that the addition of MA-g-PP further increases the viscosity of the composites and this compatibilizing effect was more pronounced in case of oxyfluorinated Kevlar/EP system (Figure 7.15c). The peroxy radicals (RO$_2^\bullet$) generated on the Kevlar surface during oxyfluorination may interact with the anhydride rings of MA-g-PP, thereby forming linkages at the interface. Furthermore, the PP moiety of the MA-g-PP adheres to the long hydrophobic chains of the matrix thus lowering the surface tension of the fibers and forms strong interfacial adhesion that is absent in case of unmodified and fluorinated Kevlar/EP composites thus the enhancement of viscosity is not so much for those particular cases. Additionally, the friction between the fiber and the matrix increases that further contributes to the enhancement of viscosity for the treated composites although MA-g-PP content slightly influences the viscosity of the treated composites.

In the high shear rate region, both the treated and untreated Kevlar-reinforced composites exhibit approximately same viscosities. Reduction in the interaction between the fibers with extensive alignment along the die entrance results the shear thinning of the polymer melt thereby leading to the flow curves to overlap at high shear rates, which in accordance with the results reported in the literature [79]. They

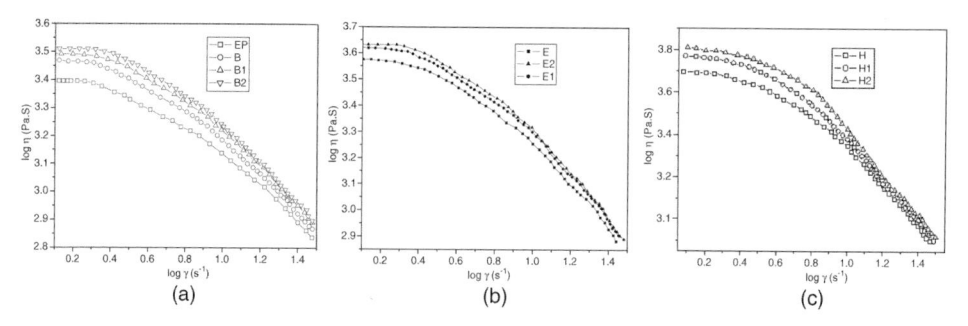

Figure 7.15 Viscosity versus shear rate curve (at 200 °C) for (a) Kevlar/EP composite, (b) fluorinated Kevlar/EP composite, and (c) oxyfluorinated Kevlar/EP composite with and without compatibilizer.

7.10.6
SEM Study

Figure 7.16a–f shows the SEM images of the cross-sectional tensile fracture surface of MA-g-PP compatibilized unmodified, fluorinated, and oxyfluorinated Kevlar/EP composites. The micrographs clearly indicate a significant difference in the interfacial characteristics of the composites. It was noticed in our previous work [78] that for untreated composites the fibers appeared to be pulled out from the matrix and for surface-treated fiber-reinforced Kevlar/EP composites it was evident that fibers were broken down. However, the addition of MA-g-PP-treated composites displayed better dispersion of the fibers over the continuous EP matrix. The SEM micrographs confirm that coupling agent facilitates the adhesion between the fibers and the matrix to a greater extent than the uncompatibilized systems.

There is a significant reduction in fiber pull-outs and barely any gaps between the fibers and the matrix is noticed (Figure 7.16e and f) that corresponds to a good adhesion between the fiber and matrix. This effect is much more pronounced in case of MA-g-PP compatibilized oxyfluorinated Kevlar/EP composites resulting best thermal, dynamic mechanical, crystalline, and rheological properties of that particular system among all the composites.

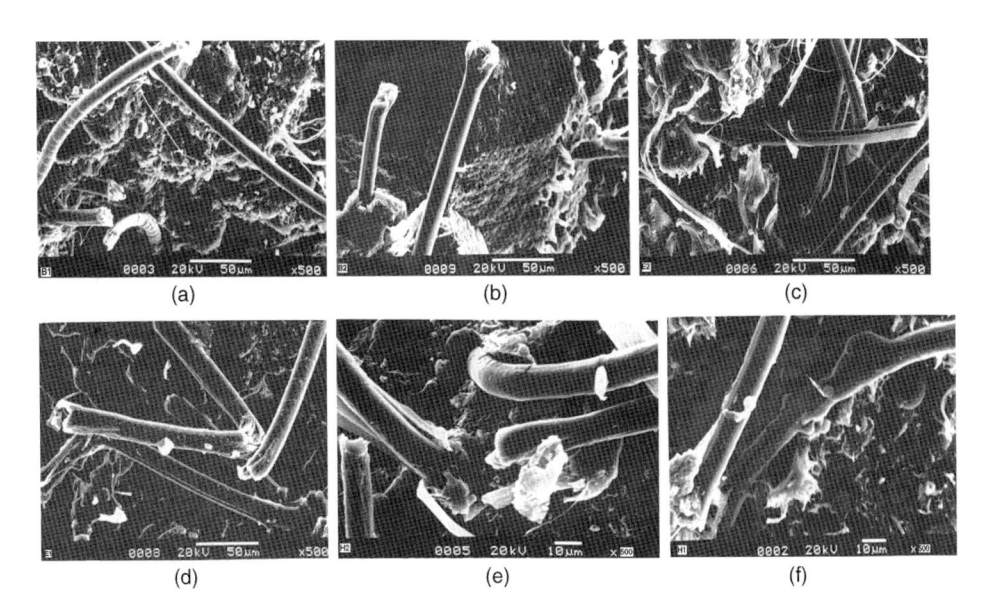

Figure 7.16 SEM photograph of (a) unmodified Kevlar/EP/2% MA-g-PP (B1), (b) unmodified Kevlar/EP/5% MA-g-PP (B2), (c) fluorinated Kevlar/EP/2% MA-g-PP (E1), (d) fluorinated Kevlar/EP/5% MA-g-PP (E2), (e) oxyfluorinated Kevlar/EP/2% MA-g-PP (H1), and (f) oxyfluorinated Kevlar/EP/5% MA-g-PP (H2).

7.10.7
Conclusion

The Kevlar fiber incorporated EP composites in presence of compatibilizer showed better improvement in thermal, dynamic mechanical, crystalline, and rheological properties. The utilization of compatibilizer forms a better adhesion between the fiber and matrix polymer. Among the compatibilized Kevlar fiber-reinforced EP composites, the system containing oxyfluorinated Kevlar fiber exhibited superior properties. The generation of the peroxy radical on to the Kevlar surface and the reactive groups on the EP matrix (due to MA-g-PP) resulted better interaction at interface.

7.11
Properties of Syndiotactic Polystyrene Composites with Surface-Modified Short Kevlar Fiber

In order to study the effect of surface modification of Kevlar fiber on the properties of syndiotactic polystyrene (s-PS)/Kevlar composites, Das et al. had prepared s-PS/Kevlar composites with and without surface modification of Kevlar fiber. The formulations of the composites are given in Table 7.9.

7.11.1
Preparation of s-PS/Kevlar Composites

Unmodified and modified Kevlar fibers were mixed with s-PS separately in a twin-screw extruder at 300 °C. Then the mixtures were injection molded in an injection molding machine (model no BOY22D equipped with a screw of L/D ratio of 17.5) at 290 °C, with a mold temperature of 60 °C and at a flow rate of 20 cm^3/s.

7.11.2
FTIR Study of the Composites

Figure 7.17 shows the FTIR spectrum of the composites along with pure s-PS. The peak associated with primary amine backbone appeared in the case of P, Q, and R at 3340 cm^{-1} is due to the incorporation of the Kevlar fiber into the matrix. The peak at 1760 cm^{-1} in case of P, Q, R is the characteristic peak of C=O stretching frequency of Kevlar fiber. The FTIR spectrum of Q and R show the peak at 1200 cm^{-1} that is due

Table 7.9 Compounding formulation.

Sample code	s-PS (wt%)	Kevlar (wt%)
B	100	(0)
P	100	0.60 (original)
Q	100	0.60 (fluorinated)
R	100	0.60 (oxyfluorinated)

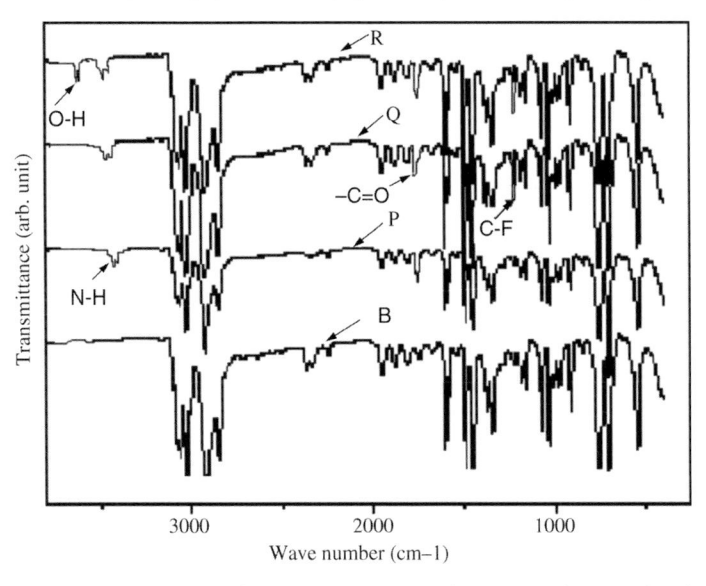

Figure 7.17 FTIR Study of pure s-PS (B), s-PS/Kevlar (P), s-PS/fluorinated Kevlar (Q), and s-PS/oxyfluorinated Kevlar (R) composites.

to the monofluorinate aliphatic groups (C—F bond) formed during fluorination. Sample R also shows a peak at 3400 cm^{-1} that corresponds to the —OH group that may be due to the carboxylic acid groups generated at the surface of the Kevlar fiber after oxyfluorination.

7.11.3
Differential Scanning Calorimetric Study

Figure 7.18 and Table 7.10 show the DSC plot and the corresponding parameters of the composites. All the composites are heated up to 300 °C to give the same thermal history. Figure 7.18 shows the glass transition temperature of pure s-PS at 100.8 °C. Incorporation of modified and unmodified Kevlar fibers slightly affects the T_g of s-PS phase. This shift of T_g to higher temperature may probably be due to the fiber/matrix interaction, which restricts the movement of the polymer chains in the

Table 7.10 Thermal parameters of s-PS/Kevlar composites.

Sample code	T_g (°C)		T_c (°C)	T_m (°C)	ΔH_f (J/g)	X_c (%)
	By DSC	By DMTA				
B	100.8	109.9	140.7	274.4	19.9	38.1
P	102.7	111.5	148.4	275.3	22.4	42.9
Q	103.8	112.3	148.9	275.8	23.4	44.2
R	104.1	113.8	146.5	274.8	21.0	40.3

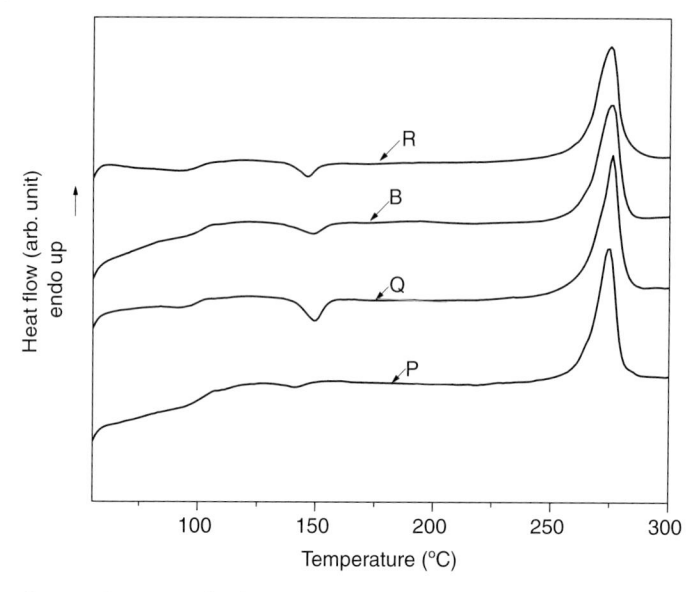

Figure 7.18 DSC study of pure s-PS (B), s-PS/Kevlar (P), s-PS/fluorinated Kevlar (Q), and s-PS/oxyfluorinated Kevlar (R) composites.

vicinity of the fiber and increases the T_g. The enhancement of T_g is more pronounced in case of oxyfluorinated Kevlar fibers composite. This may be due to the graft copolymers obtained from the free radicals generated in case of oxyfluorinated derivative. Formation of such products at the interface may enhance the miscibility of fiber and matrix (of the composite R) there by affecting the glass transition temperature of the respective component, which exceeds the effect of crystallization.

Upon increasing the temperature, the heating curve of s-PS shows an exothermic peak at 148.7 °C, which is the crystallization peak of s-PS [80]. Crystallization peaks shift to higher temperature in case of all Kevlar/s-PS composites. This may be due to the nucleating effect of the fibers on the s-PS matrix that supports the increasing crystallization trends in case of all s-PS/Kevlar composites obtained by both DSC and XRD studies. In case of oxyfluorinated Kevlar fiber-reinforced s-PS (composite R), T_c shifts to the low temperature, which may be due to the graft copolymerization taken place at the interface of the fiber and matrix leading to the cross-linking of the polymers resulting in lower percent crystallinity for this composite.

In this study, the percent of crystallinity of the s-PS/Kevlar composites, normalized for fractional content, increases with addition of Kevlar fiber into the s-PS matrix. This indicates that Kevlar fibers and its modification enhance the nucleation and formation of s-PS crystals. The percentage crystallinity is higher in case of fluorinated Kevlar fiber-reinforced s-PS, followed by unmodified Kevlar fiber/s-PS and oxyfluorinated/s-PS composites. This is due to the nucleating ability of the Kevlar fiber. But in case of oxyfluorinated derivative percent crystallinity decreases appreciably because of the graft polymerization, which leads to the increase in the degree of polymerization resulting to the cross-linking and branching at the interface for this composite. With increase in temperature the heating curve of s-PS shows an endotherm at about 274.36 °C, which is the melting endotherm of the neat s-PS. However, the melting temperatures of all s-PS/Kevlar composites does not vary appreciably in comparison to the pure component.

7.11.4
Thermal Properties

Figure 7.19 and Table 7.11 summarize the TG curves and their respective parameters of the composites obtained at a heating rate 10 °C/min in air. In order to avoid any ambiguity, the onset degradation temperature has been defined as the temperature at which polymer lost 1% of its weight. From the thermogram it is observed that degradation starts at higher temperature for all Kevlar/s-PS composites than neat s-PS. This enhancement in onset degradation temperature is more pronounced in case of modified Kevlar/s-PS composites.

The improvement in thermal stability of the modified Kevlar/s-PS composites may be due to the incorporation of functional groups on to the Kevlar surface that brings better compatibility between the two polymeric species. Moreover, it is known that crystalline polymer is thermally more stable than its amorphous counter part due to higher amount of energy required for overcoming both intermolecular and intramolecular forces. Oxy fluorination of Kevlar fiber produces more functional groups as compared to oxy fluorination that improves the fiber matrix interaction and hence enhanced the thermal stability of oxyfluorinated Kevlar/s-PS composite.

Oxyfluorination generates controlled amount of long-living ($RO_2^{\bullet}$) radicals that may be responsible for graft polymerizations and cross-linkings leading to the higher

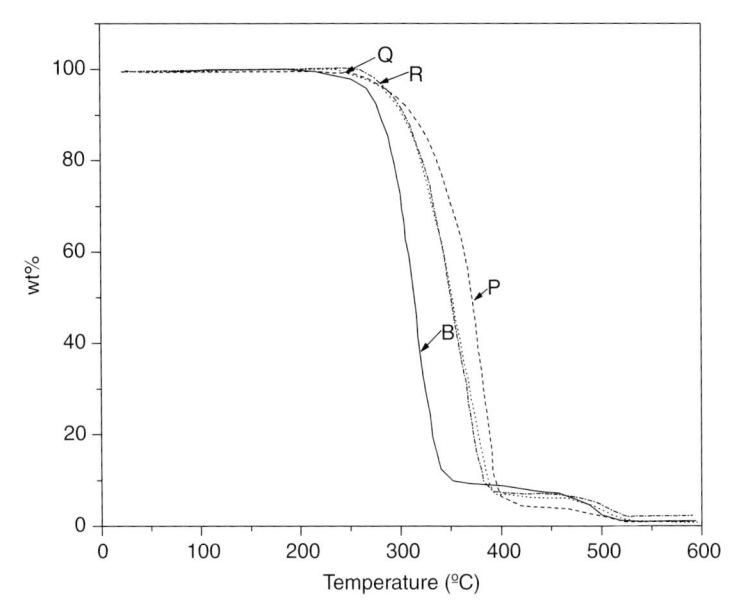

Figure 7.19 TG study of pure s-PS (B), s-PS/Kevlar (P), s-PS/fluorinated Kevlar (Q), and s-PS/oxyfluorinated Kevlar (R) composites.

7 Kevlar Fiber-Reinforced Polymer Composites

Table 7.11 Thermal parameters of s-PS/Kevlar composites.

Sample	Onset degradation temperature (°C)	5% degradation temperature (°C)	10% degradation temperature (°C)	50% degradation temperature (°C)
B	223.8	272.1	280.9	314.2
P	228.0	288.3	314.5	372.4
Q	241.4	289.7	303.4	351.4
R	249.9	292.4	304.8	350.4

thermal stability of R among all the composites resulting due to good adhesion between fiber and the matrix, which exceeds the crystallization effect.

7.11.5
X-Ray Study

Figure 7.20 shows the X-ray diffractograms of s-PS/Kevlar composites along with pure s-PS and the respective parameters are given in Table 7.12. From Figure 7.20, it is clear that pure s-PS shows only one broad peak at the 2θ value 20.05° and two small peaks at 11.75° and 13.35°. Upon addition of the unmodified and modified Kevlar fibers into s-PS matrix, the intensity of the peaks at 2θ value 11.75° and 13.35° increases. This may be due to the reinforcing ability of the Kevlar fiber into the s-PS matrix. The increasing intensity of these peaks is more pronounced in case of fluorinated Kevlar fiber-reinforced s-PS (Q).

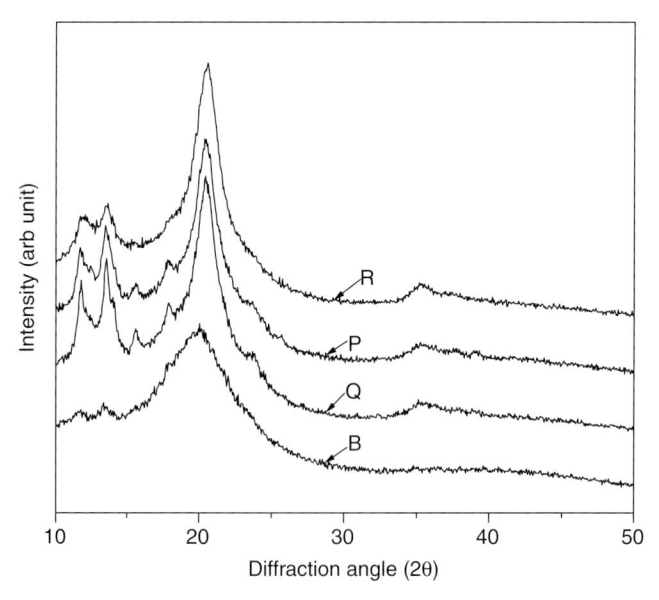

Figure 7.20 XRD pattern of pure s-PS (B), s-PS/Kevlar (P), s-PS/fluorinated Kevlar (Q), and s-PS/oxyfluorinated Kevlar (R) composites.

7.11 Properties of Syndiotactic Polystyrene Composites with Surface-Modified Short Kevlar Fiber

Table 7.12 XRD parameters of s-PS/Kevlar composites.

Sample code	Peak angle (θ)			Interplanar spacing (Å)			Crystallite size (Å)		
	θ_1	θ_2	θ_3	d_1	d_2	d_3	p_1	p_2	p_3
B	11.75	13.35	20.05	7.56	6.66	4.44	—	—	27.95
P	11.75	13.55	20.45	7.56	6.56	4.36	206.09	196.22	52.57
Q	11.7	13.5	20.4	7.59	6.58	4.37	186.75	146.34	48.59
R	12.05	13.65	20.6	7.37	6.51	4.33	116.47	146.65	44.85

It is also important to point out that the crystallite sizes of all the Kevlar/s-PS composites corresponding to all peak angles (2θ) are higher than pure s-PS, which suggests that the nucleation and growth of crystals is also favored in case of modified as well as unmodified s-PS/Kevlar composites.

7.11.6
Dynamic Mechanical Thermal Analysis

Figure 7.21 presents the storage modulus as a function of temperature of s-PS/Kevlar composites. The incorporation of modified and unmodified fibers gives rise to a considerable increase of s-PS stiffness. This may be due to the reinforcing nature of the Kevlar fiber. The storage modulus of the composites decreases with increasing temperature due to increase of segmental mobility of the s-PS. The storage modulus values increases especially in the glassy region for all Kevlar/s-PS composites. This

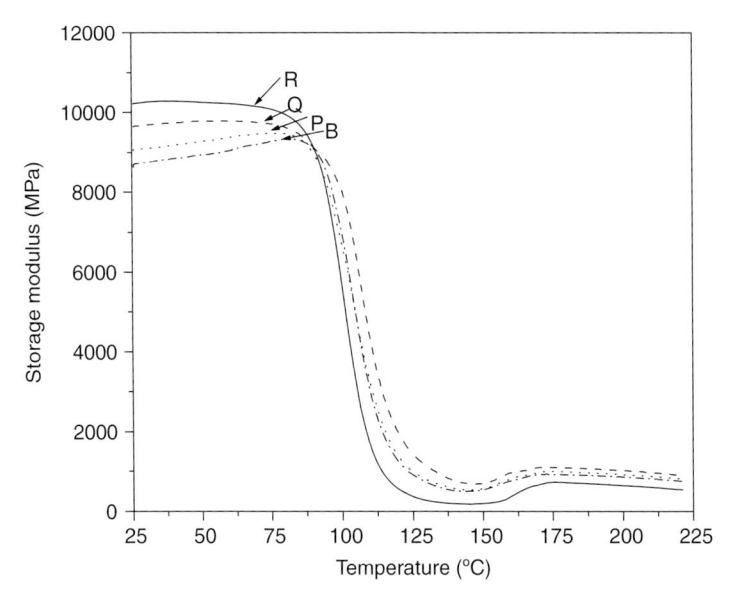

Figure 7.21 Storage modulus versus temperature curve of pure s-PS (B), s-PS/Kevlar (P), s-PS/fluorinated Kevlar (Q), and s-PS/oxyfluorinated Kevlar (R) composites.

effect is more pronounced in case of oxyfluorinated Kevlar fiber-reinforced s-PS composites due to more functional groups generated onto the Kevlar surface after the oxyfluorination which resulting in improved adhesion between the fiber and the matrix, and in turn reduces the mobility of the s-PS. The storage modulus graphs show a sharp decrease in the temperature range of 90–100 °C, which correlates the glass transition temperature of s-PS.

Figure 7.22 displays the tan δ as a function of temperature of the composites. The damping properties of the materials give the balance between the elastic phase and viscous phase in a polymeric structure. The damping peak in the treated fiber composites shows a decreased magnitude of tan δ in comparison to the virgin s-PS and untreated Kevlar fiber/s-PS composite. This is because of the fibers that carry a greater extent of stress and allows only a small part of it to strain the interface. Hence, energy dissipation will occur in the polymer matrix and at the interface, with a stronger interface characterized by less energy dissipation. Further, in comparison to neat s-PS, the tan δ peak of untreated fiber composite exhibit lower magnitude but higher than the treated Kevlar fiber/s-PS composites. This envisages that incorporation of the modified Kevlar fiber into the s-PS matrix leads to the better adhesion between the fiber and matrix at the interface.

The temperature corresponding to the tan δ peak is normally associated with the glass transition temperature (T_g) of a polymer. From Figure 7.22 and Table 7.10, it is observed that the α-peak of the composites are around 110–115 °C, corresponding to that of s-PS. The α-relaxation shifted toward the higher temperature in case of all the composites because of increasing crystallinity. But in case of oxyfluorinated Kevlar fiber-reinforced s-PS (R), the crystallization effect was over shadowed by the grafting mechanism leading to cross-linking, resulting the better fiber/matrix interaction

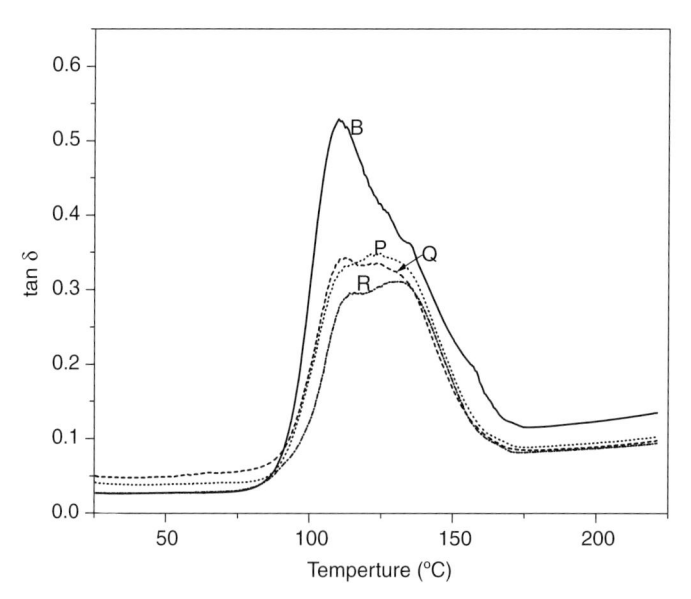

Figure 7.22 tan δ versus temperature curve of pure s-PS (B), s-PS/Kevlar (P), s-PS/fluorinated Kevlar (Q), and s-PS/oxyfluorinated Kevlar (R) composites.

7.11 Properties of Syndiotactic Polystyrene Composites with Surface-Modified Short Kevlar Fiber

among all the composite materials. The T_g values obtained from the two techniques (DSC and DMTA) are different because of the sensitivity difference between the two methods toward the glass transition temperature and also reported in literature [81]. In addition to the α-relaxation peak, all the composites show another peak at around 122–130 °C, which can be assigned as α-relaxation peak, which is highly pronounced in case of composite Q. This type of relaxation generally occurs in case of highly crystalline polymers between T_g and T_m (melting temperature), therefore it can be assigned to α-relaxation of s-PS. Sahoo et al. [82] has observed such type of relaxation in the PP/LCP system. This type of relaxation is attributed to the molecular motion of the polymeric materials within the crystalline phase.

7.11.7
SEM Study

Figure 7.23a–d shows SEM images of the fractured surfaces of pure s-PS and the composites. Figure 7.23a shows the fracture surface of the neat s-PS. Figure 7.23b is the micrograph of the Kevlar fiber-reinforced composite (P) where the fibers are randomly oriented and pulled out from the matrix indicating poor adhesion between the fiber and matrix resulting poor thermal as well dynamic mechanical stability of the composite. In case of functionalized Kevlar fiber-reinforced composite (Q, Figure 7.23c), fibers are broken rather pulled out which is the indication of good adhesion between the fiber and the matrix. The fractogram of oxyfluorinated Kevlar fiber-reinforced s-PS composite (R) (Figure 7.23d) shows that the fiber and the matrix form network, that is, fibers are well dispersed in the matrix phase giving rise to the best thermodynamical and thermal properties as a result of good adhesion between fiber and the matrix.

7.11.8
AFM Study

Figure 7.24a–h shows the two-dimensional and three-dimensional AFM images of neat s-PS (B), s-PS/Kevlar (P), s-PS/fluorinated Kevlar (Q), and s-PS/oxyfluorinated Kevlar (R) composites. Comparison of the AFM topographies of composites with the surface roughness data, however, suggest that the orientation of the Kevlar fiber during the preparation of composite samples might play more active roles in determining the surface morphology.

The maximum surface roughness of B, P, Q, and R composites were measured as 30, 46, 98, and 133 nm, respectively. Chen et al. [83] showed that the interface

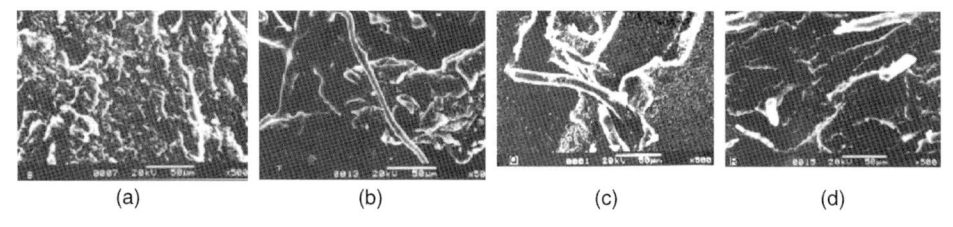

Figure 7.23 SEM picture of (a) pure s-PS (B), (b) unmodified Kevlar/s-PS (P), (c) fluorinated Kevlar/s-PS (Q), and (d) oxyfluorinated Kevlar/s-PS (R) composites.

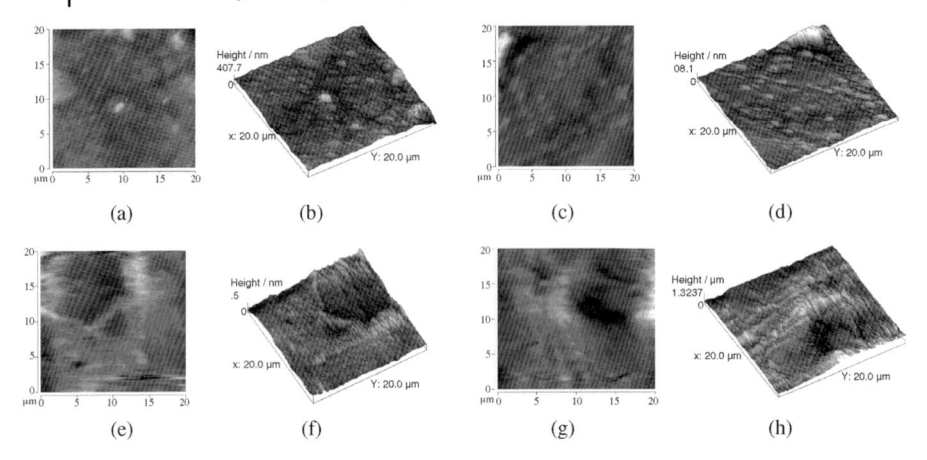

Figure 7.24 Two- and three-dimensional AFM image of (a) and (b) s-PS (B), (c) and (d) s-PS/Kevlar (P), (e) and (f) s-PS/fluorinated-Kevlar (Q), (g) and (h) s-PS/oxyfluorinated Kevlar (R) composites.

roughness could be considered as the index of interface adhesion strength. So, the roughness data of our samples strongly suggested that the adhesion strength of s-PS/oxyfluorinated Kevlar fiber interface is considerably much higher than of s-PS/fluorinated Kevlar and s-PS/Kevlar composites. This is due to the generation of more functional groups onto the surface of Kevlar fiber during oxyfluorination, which improved adhesion strength of concerned composite.

7.11.9
Conclusion

The incorporation of the short Kevlar fiber enhances the crystallization of the PS matrix due to heterogeneous nucleation. The nucleation effect was more evident for the system in presence of fluorinated short Kevlar fibers. The morphological studies showed homogenous dispersion and strong interfacial interaction for the s-PS composites containing oxyfluorinated Kevlar fibers. The same system showed superior enhancement in storage modulus and thermal stability. The presence of functional groups on the Kevlar fiber surface plays an important role in increasing its reinforcing efficiency in the matrix polymer

7.12
Study on the Mechanical, Rheological, and Morphological Properties of Short Kevlar Fiber/s-PS Composites Effect of Oxyfluorination of Kevlar

Das et al. had prepared s-PS composites with and without surface modification of Kevlar fiber and studied their rheological and mechanical properties. The formulations of the composites were given in Table 7.13.

Table 7.13 Compounding formulation.

Sample code	s-PS matrix (%)	Kevlar fiber (%)
S	100	0
KS	100	0.6 (untreated)
KSO	100	0.6 (oxyfluorinated)

7.12.1
Rheological Properties

Rheological parameters of the composites were studied with the help of a Dynamic analyzer RDA-II (Rheometrics Inc., USA) equipped with a parallel plate. All measurements were carried out at a temperature of 320 °C.

Figure 7.25 represents the variation of steady-state viscosity (η) as a function of shear rate (γ). From the figure it is evident that viscosity of the virgin polymer, s-PS (S), increases with the addition of Kevlar fiber in the composites (KS). In the case of fiber-filled systems, the fibers perturb the normal flow of the polymer and hinder the mobility of the chain segments in the direction of the flow [83]. The effect of oxyfluorinated Kevlar fiber on the viscosity of the s-PS/Kevlar composite (KSO) is also enumerated in Figure 7.25. It is observed from the figure that viscosity of KSO increases in comparison to the unmodified Kevlar/s-PS (KS) composite. Introduction of reactive functional groups on the surface of the Kevlar fiber increases the surface polarity of the fibers resulting in the formation of interfacial bonds between the fiber and matrix through the ester linkages, which in turn lowers the surface

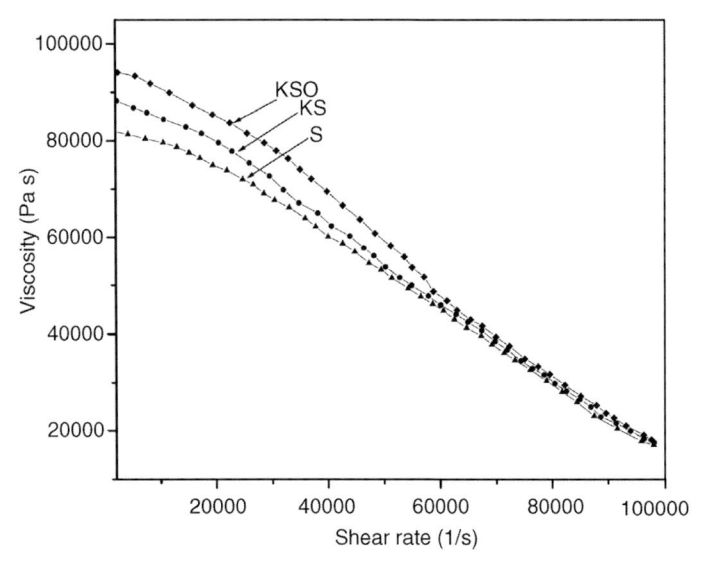

Figure 7.25 Viscosity versus shear rate curve of pure s-PS (S), s-PS/unmodified Kevlar (KS), and s-PS/oxyfluorinated Kevlar (KSO) composites.

tension of the fibers. Additionally, the friction between the fiber and the matrix is also increased, which leads to the further enhancement of the viscosity of the treated composites (KSO).

It is interesting to point out that composites as well as the virgin polymer, shows the typical pseudo plastic nature, showing a decrease in viscosity with shear rates. At low shear rates the fibers displayed larger reinforcing capabilities, which is attributed to the fiber–fiber interactions arising from weak structures made up by agglomerates of nonaligned fibers [84]. In the high range of shear rates, shear-thinning behavior of the composite melt persisted and all the composites show nearly the same viscosity. This can be attributed to the alignment of the fibers at high shear rates along the major axis thereby decreasing the fiber–fiber collision. However, the enhancement of viscosity at this temperature is attributed to the fine dispersion of the fibers into the matrix resulting from the oxyfluorination of the Kevlar fiber.

7.12.2
Mechanical Properties

Table 7.14 summarizes the tensile properties of the composites. Incorporation of unmodified Kevlar fiber in the s-PS matrix decrease the elastic modulus, tensile strength, and elongation at break as compared to pure s-PS. It may be due to the weak fiber–matrix interface derived from the divergent behavior in polarity between the hydrophilic Kevlar fibers and the hydrophobic s-PS matrix. This detrimental effect of Kevlar fibers on the mechanical properties of the resulting composite is a demarcation with the results reported in literature, for Kevlar fiber-reinforced polymers. Raj et al. [85] also reported the same observation in the high impact polystyrene and sisal fiber system. Surface treatment of Kevlar fiber by oxyfluorination improves the mechanical properties of the treated composite (KSO) appreciably due to the better fiber–matrix interaction.

From Table 7.14 it is clear that tensile strength of treated composite (KSO) increases about 12% compared to the untreated one (KS). Similarly, the magnitude of elongation at break (19%) and elastic modulus also increases in case of treated composite. This probably due to the improved bond strength between the fiber and matrix at interface arising from the oxyfluorination of Kevlar fiber that contributes to the efficient stress transfer from the matrix to the fiber. Impact strength of the composite KS decreases, with the addition of the fiber, indicating the immobilization of the matrix molecular chain by the fibers. This leads to the increased stress

Table 7.14 Mechanical properties of s-PS/Kevlar composites.

Sample code	Tensile strength (MPa)	Elastic modulus (MPa)	Elongation at break (%)	Impact strength (kJ/m^2)
S	48.08	3473	1.4	6.6
KS	46.87	3310	1.26	6.25
KSO	52.23	3559	1.51	7.45

concentration by limiting the composite's ability to adapt the deformation, which makes the concerned composite more brittle. This can be attributed as the weak fiber/matrix interaction at the interface. Oxyfluorinated Kevlar/s-PS composite (KSO) exhibits an appreciable enhanced magnitude of impact strength (at about 19.25%) in comparison to the untreated composite confirming the better fiber/matrix interaction at the interface.

7.12.3
Scanning Electron Microscopy Study

Figure 7.26a and b depicts the SEM micrographs of the tensile fractured surfaces of untreated and treated composites. The micrograph clearly indicates a significant difference in the interfacial characteristics of the composites.

In case of untreated fiber-reinforced composite (Figure 7.26a), the fibers appears to be free from the matrix material and a large number of fiber pull-outs are noticed. This indicates poor interfacial adhesion between the fibers and the matrix in case of untreated composite. Superior mechanical properties of treated Kevlar fiber-reinforced composites is due to the increasing functional groups on Kevlar fiber surfaces, resulting in the Van der Waals interaction [86] between the fiber and matrix at the interface-giving rise to effective stress transfer between the fiber and matrix resulting better mechanical property.

7.12.4
Conclusion

Untreated Kevlar fiber reinforcement in the s-PS matrix improves the dynamic mechanical and rheological properties of the composites but at the same time showed a drop in mechanical properties. Incorporation of oxyfluorinated Kevlar fiber in s-PS matrix significantly affects the surface morphology of that particular composite resulting better dynamic, mechanical, and rheological properties. This can be attributed to the improved interfacial adhesion between the fiber and the matrix, which may be due to the presence of functional group on to the Kevlar surface.

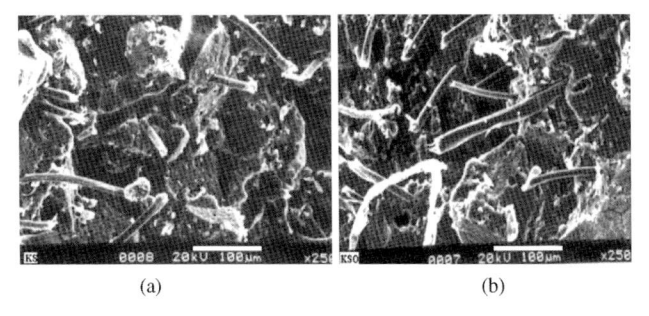

Figure 7.26 SEM picture of (a) unmodified Kevlar/s-PS (KS) and (b) oxyfluorinated Kevlar/s-PS (KSO).

7.13
Effect of Fluorinated and Oxyfluorinated Short Kevlar Fiber Reinforcement on the Properties of PC/LCP Blends

To evaluate the effect of fluorinated and oxyfluorinated short Kevlar fiber reinforcement on the properties of polycarbonate (PC)/LCP hybrid composites Das *et al.* had prepared PC/LCP/Kevlar composites with and without surface modification. Table 7.15 depicts the compounding formulations.

7.13.1
Preparation of Composites

The 0.5 wt% of original, fluorinated, and oxyfluorinated Kevlar fiber was mixed with a blend of PC/LCP using a twin-screw extruder at 320 °C, having screw of L/D ratio 17, at 20 rpm. Then the mixtures were injection molded in an injection molding machine (model no BOY22D equipped with a screw of L/D ratio of 17.5) at 320 °C, with a mold temperature of 40 °C and at a flow rate of 48 cm^3/s.

7.13.2
Differential Scanning Calorimetric Study

Figure 7.27 depicts the DSC heating profile of PC/LCP with unmodified and modified Kevlar composites along with pure PC. Table 7.16 shows the magnitude of glass transition temperature of the concerned composites. Figure 7.27 shows only one transition in the range of 150 °C for pure PC, which correlates with its glass transition temperature [87]. As can be seen in DSC curves with the addition of LCP the T_g of PC shifted to the lower temperature along with an additional peak near about 135 °C [88], which corresponds to the glass transition temperature of LCP. The double humped peak arises due to the partial miscibility of the PC with LCP phase, that is, the synergistic effect PC and LCP in the resulting blend. The shifting of the glass transition temperature of PC phase toward the lower temperature side may be due to three factors:

1) Partial miscibility of LCP with the PC matrix, that is, amorphous part of the LCP is miscible with amorphous PC phase.

Table 7.15 Compounding formulations.

Sample code	PC (wt%)	LCP (wt%)	Kevlar (wt%)
PC	100	0	0
H	100	20.8	0
I	100	20.8	0.5 (original)
J	100	20.8	0.5 (fluorinated)
K	100	20.8	0.5 (oxyfluorinated)

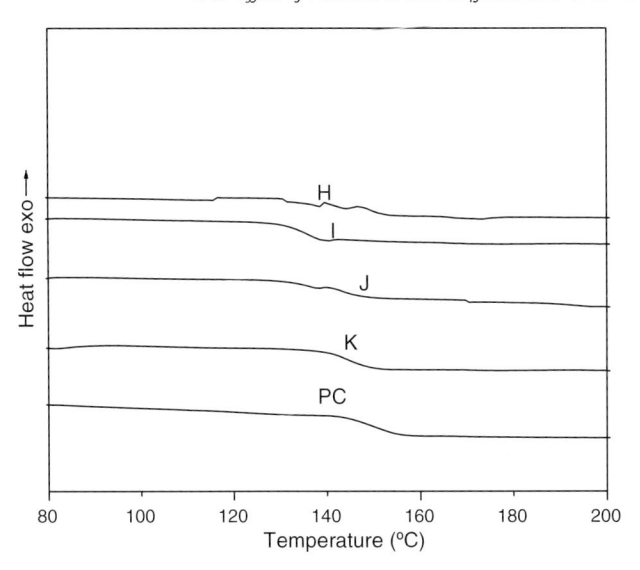

Figure 7.27 DSC profile of pure PC (PC), PC/LCP (H), PC/LCP/unmodified Kevlar (I), PC/LCP/fluorinated Kevlar (J), and PC/LCP/oxyfluorinated Kevlar (K).

2) Plasticization effect of low molecular weight fraction of the LCP.
3) Surface effect, that is, addition of LCP into PC increases the surface area (per unit volume) of the polycarbonate as polycarbonate molecules at the interface region have higher mobility than those in the bulk due to less constraint, in turn increased surface area should decrease the T_g of polycarbonate [87].

With addition of unmodified Kevlar fiber the DSC profile again exhibits a single T_g indicating the improved miscibility of PC/LCP in presence of Kevlar but it is lower than the pure PC that may be due to the poor fiber/matrix adhesion at the interface. Fluorination and oxyfluorination of Kevlar fiber further shift the T_g of the aforementioned blends to the higher temperature due to the better fiber/matrix adhesion at the interface, owing to the presence of reactive functional groups on the Kevlar

Table 7.16 Thermal properties of PC/LCP/Kevlar blends.

Sample code	Degradation temperature (T_d) (°C)	Mass change (%)	Residue at 649 °C (%)	Glass transition temperature (T_g) (°C)	
				By DSC	By DMTA
PC	427.7	71.1	25.4	149.3	168.0
H	434.3	69.4	26.0	141.3	156.6
I	442.4	68.1	26.5	138.9	153.6
J	446.6	67.8	27.5	144.5	161.0
K	450.5	67.0	27.9	148.3	171.2

surface. The improvement of T_g is more prominent in case of oxyfluorinated Kevlar fiber-reinforced composites.

7.13.3
Thermal Properties

Figure 7.28 and Table 7.16 show the TGA curves of the composites and the respective parameters. In order to avoid any ambiguity, the onset degradation temperature has been defined as the temperature at which polymer lost 1% of its weight. From the thermogram it is observed that degradation starts at higher temperature for all Kevlar/PC/LCP blends as compared to neat PC and pure PC/LCP blend. This enhancement in onset degradation temperature is more pronounced in case of modified Kevlar/PC/LCP composites.

Incorporation of LCP improves the thermal stability of PC/LCP blend as compared to pure PC that is ascribed to the better thermally stability of LCP. Addition of Kevlar fiber improves the thermal stability of the hybrid composites due to the introduction of aromatic content, of high thermal stability. Incorporation of surface-modified Kevlar fiber further increases the thermal stability of PC/LCP blends (J, K) that is due to the functional groups generated during the surface modification of Kevlar fiber, resulting in better fiber matrix compatibility and hence the higher thermal stability.

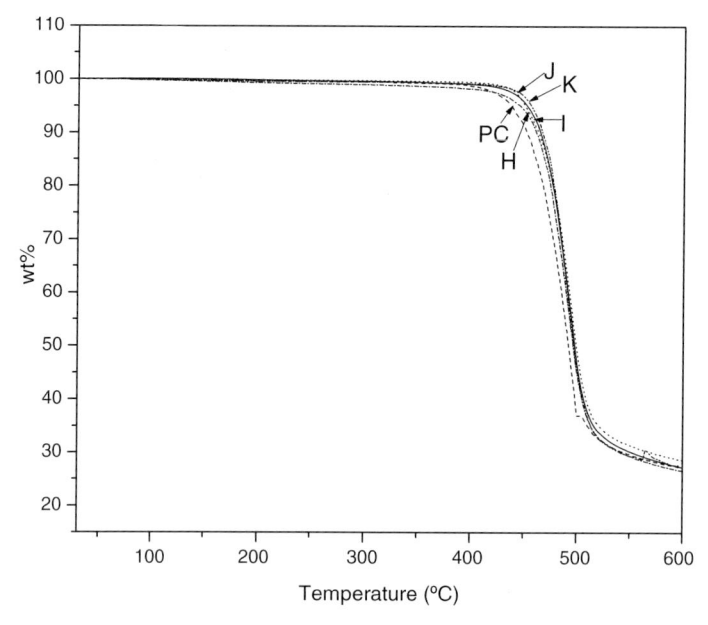

Figure 7.28 TG plot of pure PC (PC), PC/LCP (H), PC/LCP/unmodified Kevlar (I), PC/LCP/fluorinated Kevlar (J), and PC/LCP/oxyfluorinated Kevlar (K) composites.

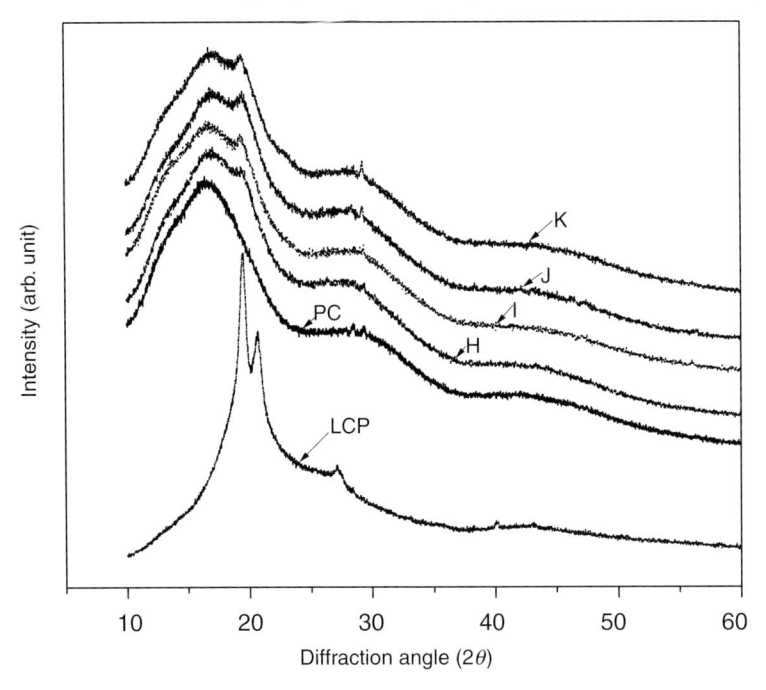

Figure 7.29 X-ray diffraction pattern of Pure PC (PC), PC/LCP blend (H), PC/LCP/unmodified Kevlar (I), PC/LCP/fluorinated Kevlar (J), and PC/LCP/oxyfluorinated Kevlar (K) composites.

7.13.4
X-Ray Study

Figure 7.29 shows the XRD curves of the blend systems. PC shows a broad spectrum near about $2\theta = 16.5°$ due to its amorphous nature. Addition of LCP generates a very minute change in the X-ray pattern near about $2\theta = 19.5°$, may be due to the incorporation of LCP ($2\theta \sim 20\,°C$). From this we can infer that LCP is partially miscible with the amorphous PC matrix. The intensity of that small peak increases with the introduction of Kevlar fiber into the PC/LCP blend matrix that can be ascribed as the nucleating effect of Kevlar in the PC/LCP matrix inducing very little crystallinity into PC. Surface modification of Kevlar further increases the intensity and shifts the peak to higher 2θ value for this small peak ($2\theta = 19.75°$) revealing better fiber matrix adhesion at the interface. This effect is more pronounced in case of oxyfluorinated derivative.

7.13.5
Dynamic Mechanical Analysis (DMA)

Figures 7.30 and 7.31 display the dynamic mechanical profile (storage modulus E' and tan δ) as a function of temperature for PC/LCP/Kevlar composites. The storage

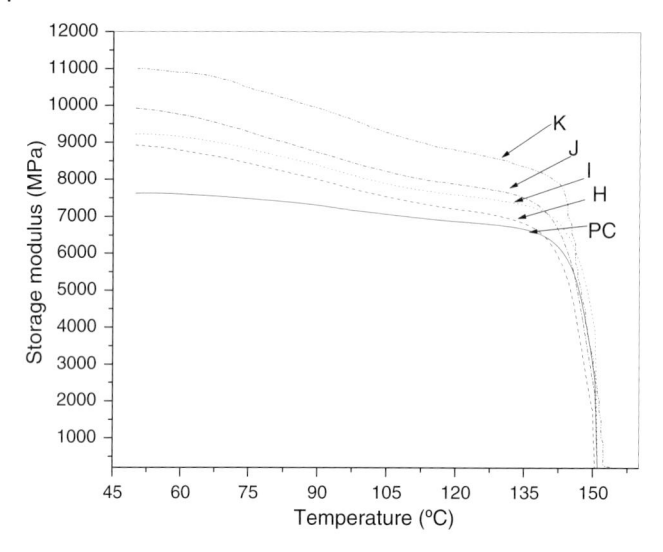

Figure 7.30 Storage modulus versus temperature curve of pure PC (PC), PC/LCP blend (H), PC/LCP/unmodified Kevlar (I), PC/LCP/fluorinated Kevlar (J), and PC/LCP/oxyfluorinated Kevlar (K) composites.

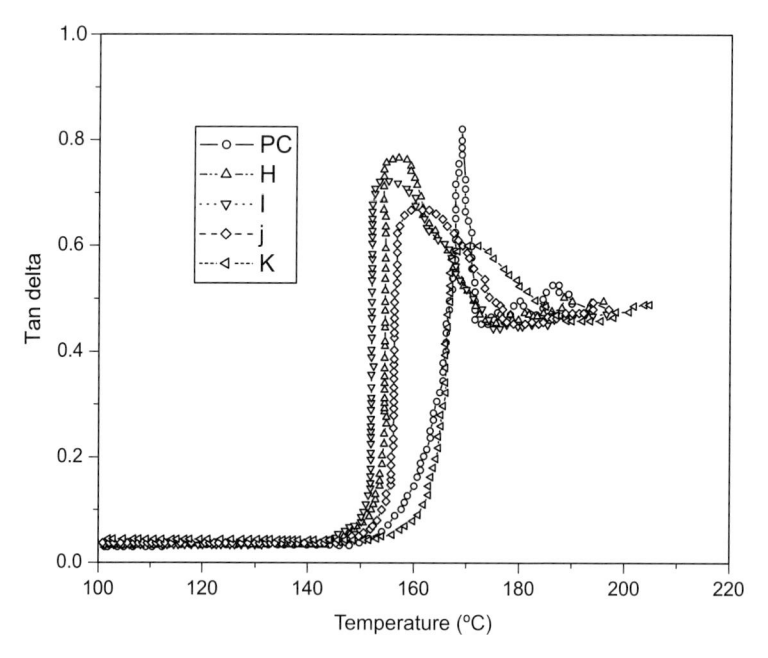

Figure 7.31 tan δ versus temperature curve of (i) pure PC (PC), (ii) PC/LCP blend (H), (iii) PC/LCP/unmodified Kevlar (I), (iv) PC/LCP/fluorinated Kevlar (J) and (v) PC/LCP/oxyfluorinated Kevlar (K) composites.

modulus is closely related to the capacity of a material to absorb or return energy that attributed to its elastic behavior [88]. From Figure 7.30 it is evident that addition of LCP leads to the appreciable enhancement of the magnitude of storage modulus. This is due to the high intrinsic modulus of LCP phase consisting rigid rodlike structure. Sahoo et al. [82] has reported the same observation in the PP/LCP blends.

Figure 7.30 shows a sharp drop in storage modulus of composite along with neat polymer that corresponds to the glass transition temperature (T_g). This modulus drop can be imputed to an energy dissipation phenomenon involving cooperative motions of the polymer chain. With the addition of Kevlar fiber into the PC/LCP blend further increases the storage modulus of the concerned blend, which may be due to the reinforcing nature of the Kevlar fiber. Surface modification leads to the further enhancement of the magnitude of storage modulus of the fluorinated and oxyfluorinated Kevlar fiber-reinforced PC/LCP blend due to the incorporation of the reactive functional groups onto the Kevlar surface leading to better fiber matrix adhesion resulting the improved stiffness of the modified composites. This phenomenon is also supported by the results obtained from the XRD study where surface modification induces some crystallinity into the PC/LCP blend.

The damping properties of the materials give the balance between the elastic phase and the viscous phase in a polymeric structure. The loss tangent (tan δ) of base polymer and their composites as a function of temperature is represented in Figure 7.31.

The T_g is selected as the peak position of the tan δ curve when plotted against temperature. From Figure 7.31 it is evident that all the composites along with the pure polymer show a single peak near about 168 °C, which is the glass transition temperature of PC. In case of PC/LCP/Kevlar composites the loss peak broadens, which may be due to the overlapping peak of PC and LCP, which suggests the partial miscibility of the LCP with PC matrix. The broadening of loss peak in the presence of fiber can also be ascribed to matrix–fiber (filler) interaction. The matrix polymer (PC/LCP) in the adjacent portion of the fiber can be considered to be in different state in comparison to the bulk matrix, which can disturb the relaxation of the matrix resulting in a broad tan δ peak T_g value shifted to the lower temperature in case of PC/LCP (H) in comparison to the pure matrix (PC) due to the partial miscibility of PC with LCP. In case of PC/LCP/unmodified Kevlar (I), the magnitude of T_g further decreases due to the poor fiber/matrix adhesion at the interface. Fluorination and oxyfluorination of Kevlar fiber enhances the glass transition temperature appreciably in case of the concerned composites (J, K). This can be ascribed to the better fiber/matrix adhesion at the interface due to the incorporation of reactive functional groups on the fiber surface. The relative decrease in the height of tan δ peak related to the increase in the extent of crystalline properties imposed by LCP and Kevlar fiber (since the transition behavior is associated with the local mobility of polymer chains in the amorphous region of the polymer), which is also reflected in the increase of storage modulus of the composites.

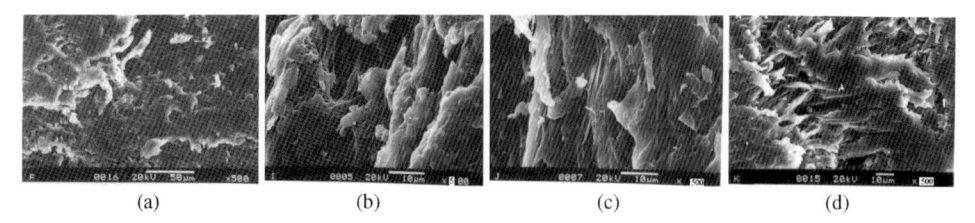

Figure 7.32 SEM picture of (a) PC/LCP blend (H), (b) PC/LCP/unmodified Kevlar (I), (c) PC/LCP/fluorinated Kevlar (J), and (d) PC/LCP/oxyfluorinated Kevlar (K).

7.13.6
SEM Study

From the micrograph (Figure 7.32a–d) of PC/LCP and PC/LCP/Kevlar blends, it is apparent that fibrillation of LCP occurs in the PC matrix (Figure 7.32a). Incorporation of Kevlar induces the fibrillation of the LCP in the PC phase resulting enhanced properties of the PC/LCP/Kevlar composites although the Kevlar fibers are covered by the matrix (Figure 7.32b).

In case fluorinated Kevlar-reinforced composites (J, Figure 7.32c), the LCP fibrils are more prominent and distributed throughout the matrix phase but appeared as bundle form. On the other hand oxyfluorinated derivative (K, Figure 7.32d) exhibits fine microfibrils and are homogeneously dispersed all over the PC matrix giving rises to best properties among all composites.

7.13.7
Conclusion

LCP enhances the thermal stability of the PC/LCP blend system. Incorporation of Kevlar fiber in the PC/LCP blend system showed higher thermal stability, which may be due to the presence aromatic content in the Kevlar fiber. Surface fluorination and oxyfluorination of Kevlar fibers further enhances the thermal stability of the PC/LCP blend systems due to the incorporation of functional groups on to the Kevlar fiber surface. Incorporation of LCP into the amorphous PC matrix induces some crystallinity of the matrix phase and the crystalline property was further enhanced by the incorporation of unmodified and modified Kevlar fiber in the blend system. DSC heating scan exhibits the double humped curve (glass transition temperature) in case of PC/LCP blend system, which indicates the partial miscibility of the blend partners. Incorporation of Kevlar fiber into the PC/LCP system shifted the glass transition temperature toward the lower temperature side, which may be due to the poor fiber/matrix adhesion at the interface. Surface modification (fluorination and oxyfluorination) increases the glass transition temperature in comparison to the blend system, which may be due to the better fiber/matrix bonding at the interface. The incorporation of functional groups onto the Kevlar surface may be responsible for the better interaction. Storage modulus of the PC/LCP system increases in comparison to the pure matrix. Further enhancement was achieved for the Kevlar fiber-reinforced systems and the superior properties were

7.14
Simulation of Fiber Orientation by Mold Flow Technique

7.14.1
Theoretical basis for Fiber Orientation Prediction

Motion of a single rigid ellipsoidal particle immersed in a viscous Newtonian liquid was considered by Jeffery [89]. Evolution equation for a single rigid ellipsoidal particle was developed, and it is the base of almost all the fiber orientation constitutive modeling. Fiber suspensions are characterized in terms of the number of fibers per unit volume n, fiber length L and diameter D as dilute ($nL^3 \ll 1$), semidilute ($nL^3 \gg 1$, $nL^2D \ll 1$) and concentrated ($nL^3 \gg 1$, $nL^2D \gg 1$) regime. nL^3 means the number of interacting fibers in a volume swept by a single fiber and nL^2D means an excluded volume of interacting fibers due to a line approximation of a fiber. Currently, many researchers have developed many numerical simulation programs with different methods for the description of fiber orientation including multiple fiber–fiber interaction. However, some of those simulation programs using direct calculation of fiber motion [90–93] require incredible computation time, which is the reason why such a numerical simulation method cannot be accommodated.

Calculation of fiber orientation state using the probability distribution function (DFC) is one of them. Instead, a tensor representation of orientation state [94], which is a preaveraging concept of DFC, is widely used for its efficiency, compactness and above all, manageable computation time. However, it is necessary to introduce a closure approximation to express the higher order tensor in terms of lower order tensors for a closed set of equation since the evolution equation for orientation tensor involves the next higher even order tensor. The second order orientation tensor a_{ij} provides an efficient description of fiber orientation in injection moldings. The predicted fiber orientation, which is a probability distribution in 3D space, can be represented by the Eq. (7.4), and graphically as the ellipsoid in Figure 7.33.

$$a_{ij} = \begin{pmatrix} a_{11} & a_{12} & a_{13} \\ a_{21} & a_{22} & a_{23} \\ a_{31} & a_{32} & a_{33} \end{pmatrix} \rightarrow \begin{pmatrix} \lambda_1 & 0 & 0 \\ 0 & \lambda_2 & 0 \\ 0 & 0 & \lambda_3 \end{pmatrix} ; (e_1 e_2 e_3) \qquad (7.4)$$

Second order orientation tensor — Eigen values — Eigen vector

Thus, the tensor has nine components, with the suffixes for the tensor terms being (1) in the flow direction, (2) transverse to the flow direction, and (3) in the thickness direction.

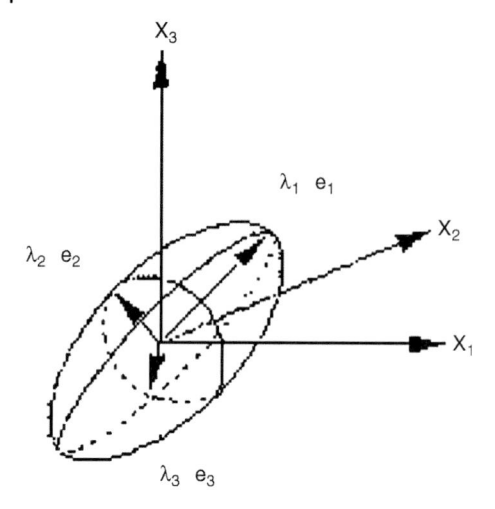

Figure 7.33 Orientation ellipsoid defined by general second-order orientation tensor [6].

The X–Y (or 1–2) axes are applied to flow plane and the Z-axis in the thickness direction, that is, out of the 1–2 flow plane. Due to tensor symmetry $a_{ij} = a_{ji}$, and a normalization condition $(a_{11} + a_{22} + a_{33}) = 1$, the original nine components reduce to five independent components. These three major orientation components have been included in the orientation considerations: a_{11} represents the fiber orientation in the flow direction, varying from 0 to 1.0, a_{22} represents the fiber orientation transverse to the flow, varying from 1 to 1.0, and a_{13} is the tilt of the fiber orientation in the 1–3 plane, varying from -0.5 to 0.5. However, the flow direction orientation (a_{11}) term possesses most of the quantitative information about microstructure and is most sensitive to flow, processing, and material changes. In modern concept the tensor components in synergy are named with a "T" rather than an "a", and show the specific axis names, x, y, and z, rather than the generic axis names 1, 2, and 3. For semiconcentrated suspension, Dinh and Armstrong proposed a model for prediction of fiber orientation [95]. In their proposed model, the fiber orientation follows the bulk deformation of the fluid with the exception that the particle cannot stretch. Whereas for concentrated suspensions a term, called "the interaction coefficient" (or C_I), has been incorporated in Folgar and Tucker [96] model of fiber orientation, where interactions among fibers tend to randomize the orientation and the term takes the same form as a diffusion term and since interactions only occur when the suspension is deforming, the effective diffusivity is proportional to the strain rate finally the dimensionless C_I term determines the strength of the diffusion term. The model is represented as given below:

$$\frac{\partial a_{ij}}{\partial t} + v k \frac{\partial a_{ij}}{\partial x_k} = -\frac{1}{2}(\omega_{ik} a_{kj} - a_{ik} \omega_{kj}) + \frac{1}{2}\lambda(\dot{\gamma}_{ik} a_{kj} + a_{ik} \dot{\gamma}_{kj} - 2\dot{\gamma}_{kl} a_{ijkl}) + 2 C_I \dot{\gamma}(\delta_{ij} - \alpha a_{ij})$$

where $\alpha = 3$ for 3D and 2 for 2D (planar) orientation, v_k is the velocity component, ω_{ij} is the velocity tensor, γ_{ij} is the deformation tensor, λ is the constant depending on the geometry of the particle, δ_{ij} is the unit tensor, and C_I is the interaction coefficient. As mentioned earlier that introduction of closure approximation is very much necessary to express the tensor in terms of lower order tensors for a closed set of equation because of the evolution equation for orientation tensor involves the next higher even order tensor. But closure approximation is too much challenging as it produces error in prediction of fiber orientation. In order to minimize these effects some proposal has been proposed in the Advani Tucker fiber orientation model – introduction of more accurate closure, introduction of a new orientation model that considers the closure error. The effect of the closure approximation is to predict too much out-of-plane orientation. This result has been addressed by the fiber orientation model form proposed by Mold flow.

7.14.2
Mold Flow's Fiber Orientation Model

In mold flow technique, the simulation of the orientation of fibers can be done using Jeffery's model [89] in conjunction with Tucker–Folgar term [96]. Mold flow model considers two assumptions:

- The Tucker–Folgar model gives acceptable accuracy for the prediction of fiber orientation in concentrated suspensions.
- Hybrid closure is used as its form is simple and has good dynamic behavior.

Into mold flow orientation model an extra term called "thickness moment of interaction coefficient (D_Z)" has been introduced.

In mold flow model while putting $C_I = 0.0$, the model takes the form of Jeffery's model, the magnitude of D_Z regulates the significance of the randomizing effect in the out-of-plane direction due to the fiber interaction, and when $D_Z = 1.0$, then this model gives the Folgar–Tucker orientation model. However, for injection molding situation, the flow hydrodynamics cause the fibers lie mainly in the flow plane. Their ability to flow out of plane is severely less. This mechanism predicts that the randomizing effect of fiber orientation is much smaller out of plane than in the in-plane directions, hence a small D_Z value. Thus, decreasing D_Z parameter out-of-plane orientation can be diminished and on the other hand thickness of the core layer can be increased. The simulation treats this problem as being symmetric about the mid plane.

7.14.3
Simulation of Fiber Orientation by Mold Flow Technique on s-PS/Kevlar Composites

Das *et al.* reported the fiber orientation of s-PS composites containing unmodified and modified Kevlar fibers using mold flow technique. The compounding formulations of s-PS matrix containing unmodified and modified Kevlar fibers are given in Table 7.17 and the variation in processing parameters of the composites are shown in Table 7.18.

Table 7.17 Compounding formulation.

Sample code	s-PS matrix (g)	Kevlar fiber (g)
Series I	600	0
Series II	600	3.5 (untreated)
Series III	600	3.5 (fluorinated)
Series IV	600	3.5 (oxyfluorinated)

Table 7.18 Processing parameters for injection molding.

Sample code	T_m (°C)	T_{mold} (°C)	Flow rate (cm³/s)
Series IA/IIA/IIIA/IVA	290	60	20
Series IB/IIB/IIIB/IVB	290	60	48
Series IC/IIC/IIIC/IVC	320	40	20
Series ID/IID/IIID/IVD	320	60	20

The variation of orientation tensor (T_{xx}) with normalized thickness for the unmodified and modified s-PS/Kevlar composites under varying processing conditions are displayed in Figure 7.34a–c. From Figure 7.34a–c it was clear that processing parameters has a significant effect on the orientation of fibers in the pure matrix, along the flow direction. Moreover, the fiber orientation seems to be different in skin and core region in different composites. Fiber orientation at the skin and core region for every composites under varying processing parameters are depicted in Figures 7.35–7.46.

In case of both unmodified and modified fiber-reinforced composites (Series IIA, IIIA, and IVA), at lower flow rate the fibers are more oriented in the core region than in the skin as evidenced from Figure 7.34a–c, 7.35, 7.40, and 7.45. But in case of higher flow rate the reverse trend had been observed, that is, the magnitude of the fiber orientation tensor (T_{xx}) was greater in the skin region than that of core region for all composites (Series IIB, IIIB, and IVB) (Figures 7.34a–c, 7.36, 7.40, and 7.44), which is in line with the observation of Bright et al. [88].

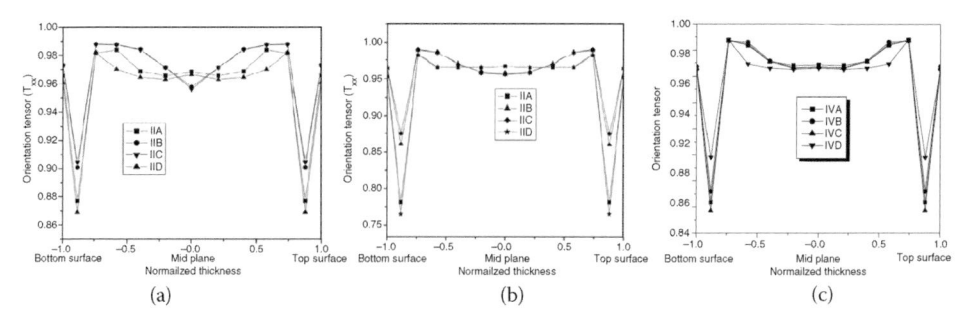

Figure 7.34 Fiber orientation tensor versus normalized thickness: (a) unmodified Kevlar/s-PS, (b) fluorinated Kevlar/s-PS, and (c) oxyfluorinated Kevlar/s-PS under different processing parameters.

7.14 Simulation of Fiber Orientation by Mold Flow Technique | 261

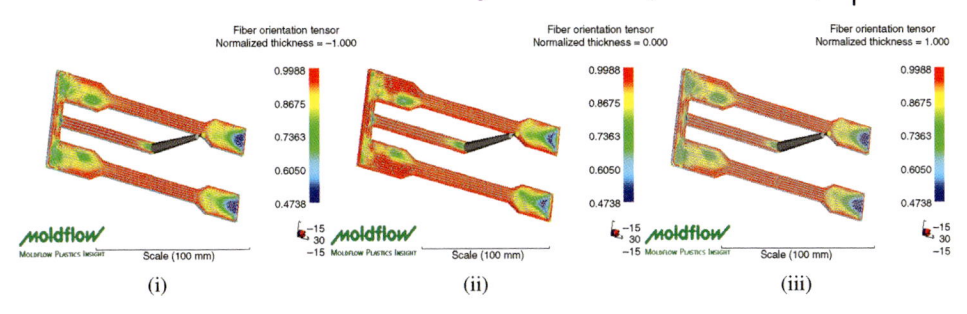

Figure 7.35 Fiber orientation at the (i) bottom surface (normalized thickness = −1, $T_{xx} = 0.965709$), (ii) core region (normalized thickness = 0, $T_{xx} = 0.968493$), (iii) top surface (normalized thickness = 1, $T_{xx} = 0.965709$) of unmodified Kevlar/s-PS composites (Series IIA).

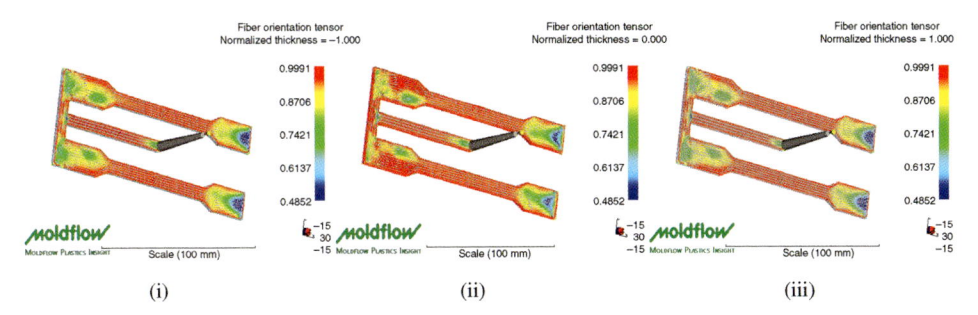

Figure 7.36 Fiber orientation at the (i) bottom surface (normalized thickness = −1, $T_{xx} = 0.972762$), (ii) core region (normalized thickness = 0, $T_{xx} = 0.957614$), (iii) top surface (normalized thickness = 1, $T_{xx} = 0.972762$) of fluorinated Kevlar/s-PS composites (Series IIB).

In the core region molding shearing flow is predominant and at higher flow rate the fibers flow almost without shearing thus orienting the fibers in the transverse to the flow direction in the concerned region. But in case of skin region the shear rate aligns more fibers in the flow direction resulting in the higher orientation tensor value.

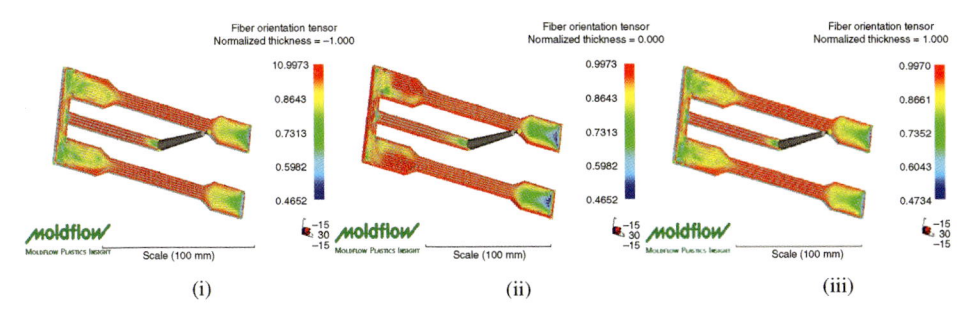

Figure 7.37 Fiber orientation at the (i) bottom surface (normalized thickness = −1, $T_{xx} = 0.972678$), (ii) core region (normalized thickness = 0, $T_{xx} = 0.956036$), (iii) top surface (normalized thickness = 1, $T_{xx} = 0.972678$) of fluorinated Kevlar/s-PS composites (Series IIC).

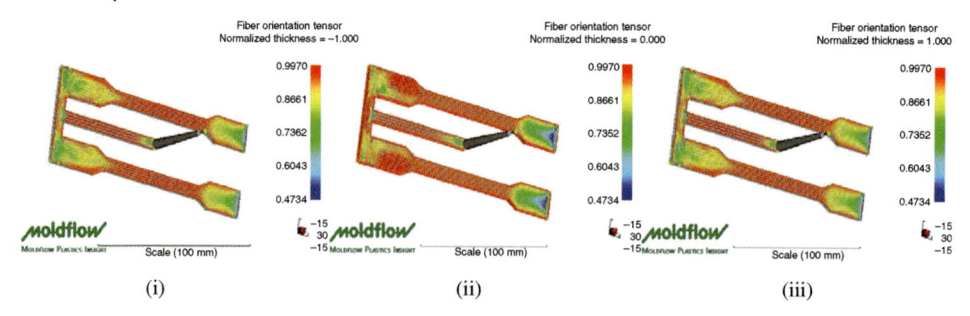

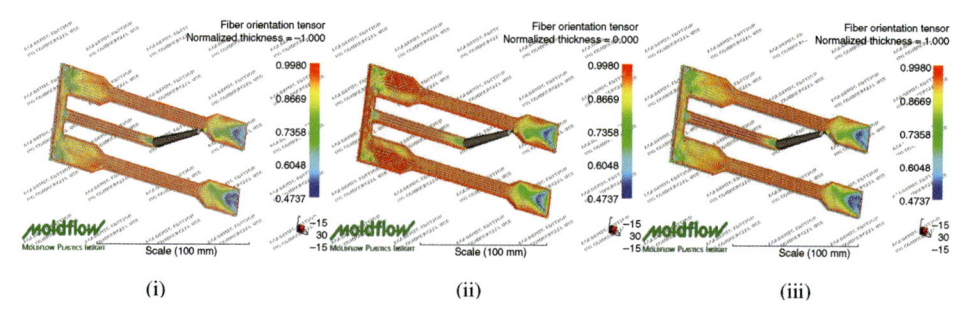

Figure 7.38 Fiber orientation at the (i) bottom surface (normalized thickness = −1, $T_{xx} = 0.96074$), (ii) core region (normalized thickness = 0, $T_{xx} = 0.966417$), (iii) top surface (normalized thickness = 1, $T_{xx} = 0.96074$) of fluorinated Kevlar/s-PS composites (Series IID).

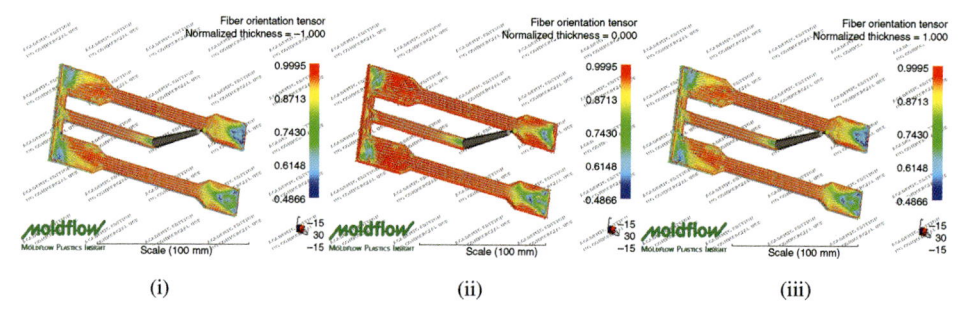

Figure 7.39 Fiber orientation at the (i) bottom surface (normalized thickness = −1, $T_{xx} = 0.952928$), (ii) core region (normalized thickness = 0, $T_{xx} = 0.966434$), (iii) top surface (normalized thickness = 1, $T_{xx} = 0.952928$) of fluorinated Kevlar/s-PS composites (Series IIIA).

An interesting trend of fiber orientation had been observed in case of varying mold temperature (Series IIC, IID, Figures 7.34a, 7.37, 7.38; IIIC, IIID, Figures 7.34b, 7.41, 7.42; and IVC, IVD, Figures 7.34c, 7.45, 7.46). At higher mold temperature more fibers are oriented in core region than that of the skin (greater orientation tensor value) and in case of lower mold temperature an opposite trend had been observed.

Figure 7.40 Fiber orientation at the (i) bottom surface (normalized thickness = −1, $T_{xx} = 0.963298$), (ii) core region (normalized thickness = 0, $T_{xx} = 0.956258$), (iii) top surface (normalized thickness = 1, $T_{xx} = 0.963298$) of fluorinated Kevlar/s-PS composites (Series IIIB).

7.14 Simulation of Fiber Orientation by Mold Flow Technique

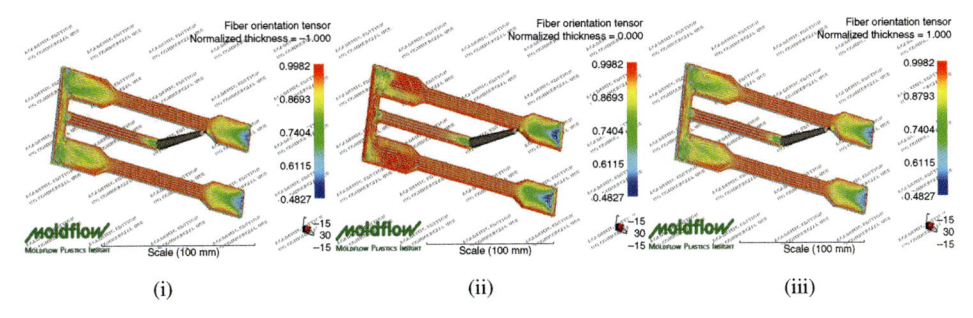

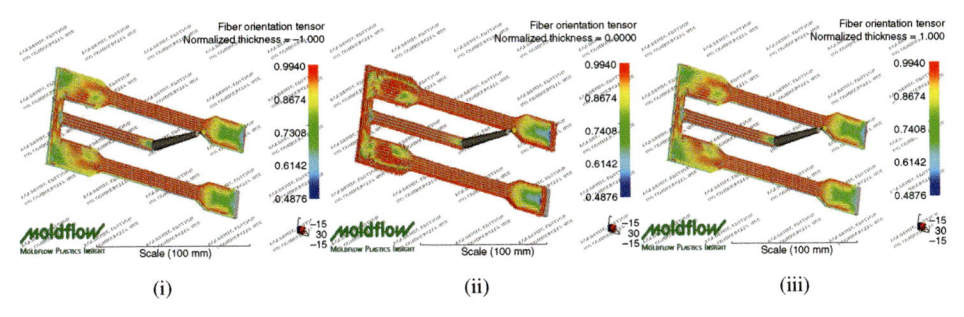

Figure 7.41 Fiber orientation at the (i) bottom surface (normalized thickness = −1, $T_{xx} = 0.964488$), (ii) core region (normalized thickness = 0, $T_{xx} = 0.956879$), (iii) top surface (normalized thickness = 1, $T_{xx} = 0.964488$) of fluorinated Kevlar/s-PS composites (Series IIIC).

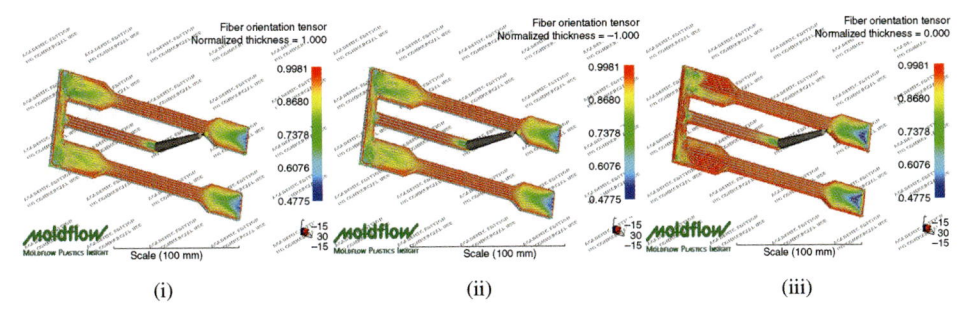

Figure 7.42 Fiber orientation at the (i) bottom surface (normalized thickness = −1, $T_{xx} = 0.952551$), (ii) core region (normalized thickness = 0, $T_{xx} = 0.965979$), (iii) top surface (normalized thickness = 1, $T_{xx} = 0.952551$) of fluorinated Kevlar/s-PS composites (Series IIID).

For high mold temperature, the temperature difference between the molten fluid and mold temperature leads to the thinner solidified skin leading to the lower flow field at the solid–melt interface and lower shear rates in the solid–melt interface, thus orienting lesser fiber in the flow direction in this concerned region.

However, melt temperature has also a pronounced effect on the fiber orientation as evidenced from Figures 7.34a–d, 7.35, 7.38, 7.39, 7.42, 7.43, and 7.46. Although the

Figure 7.43 Fiber orientation at the (i) bottom surface (normalized thickness = −1, $T_{xx} = 0.966716$), (ii) core region (normalized thickness = 0, $T_{xx} = 0.968575$), (iii) top surface (normalized thickness = 1, $T_{xx} = 0.966716$) of oxyfluorinated Kevlar/s-PS composites (Series IVA).

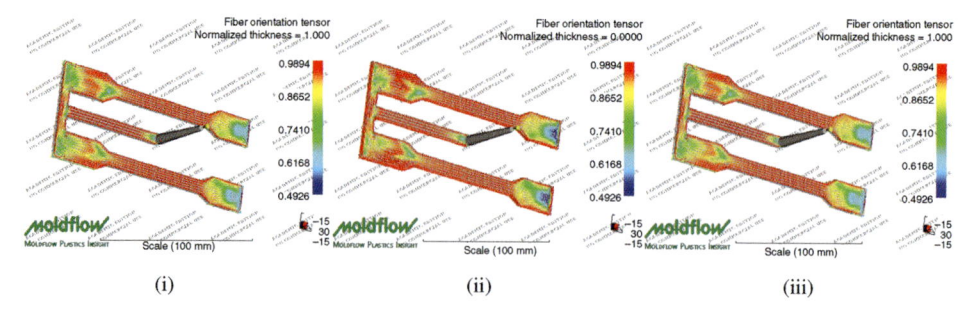

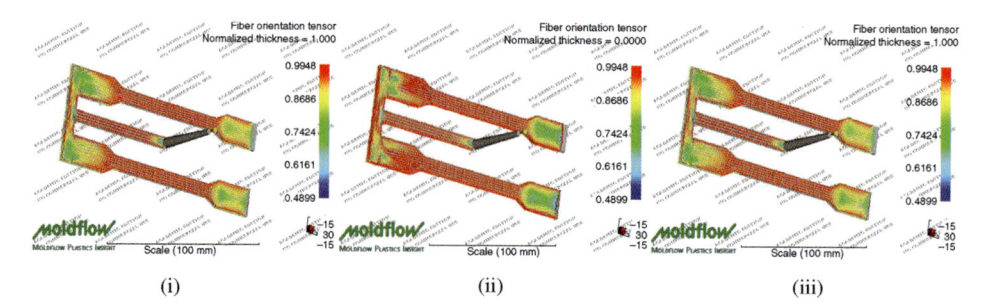

Figure 7.44 Fiber orientation at the (i) bottom surface (normalized thickness = −1, $T_{xx} = 0.96284$), (ii) core region (normalized thickness = 0, $T_{xx} = 0.960886$), (iii) top surface (normalized thickness = 1, $T_{xx} = 0.96284$) of oxyfluorinated Kevlar/s-PS composites (Series IVB).

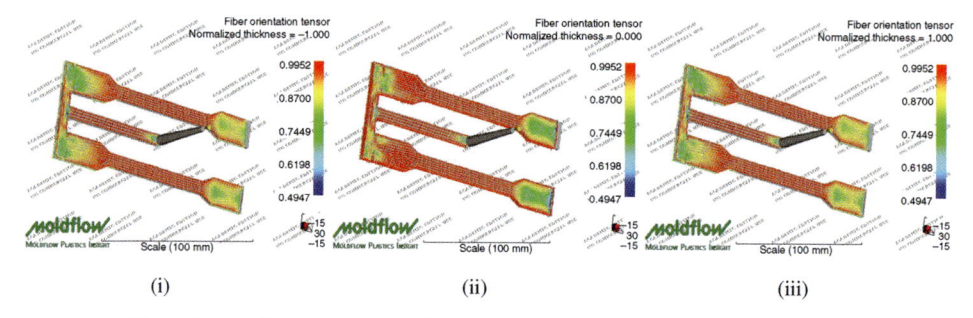

Figure 7.45 Fiber orientation at the (i) bottom surface (normalized thickness = −1, $T_{xx} = 0.967202$), (ii) core region (normalized thickness = 0, $T_{xx} = 0.966912$), (iii) top surface (normalized thickness = 1, $T_{xx} = 0.96284$) of oxyfluorinated Kevlar/s-PS composites (Series IVC).

T_{xx} shows the similar trend in the core as well as in skin region at both higher and lower melt temperature (Series IIA, 7.34a, 7.35, IID, 7.34a; IIIA, IIID; and IVA, IVD). The fibers are more oriented in the core region (higher magnitude of orientation tensor) than that of the skin. However, higher melt temperature possess somewhat less fiber orientation tensor than that of the lower variable.

Figure 7.46 Fiber orientation at the (i) bottom surface (normalized thickness = −1, $T_{xx} = 0.965483$), (ii) core region (normalized thickness = 0, $T_{xx} = 0.96677$), (iii) top surface (normalized thickness = 1, $T_{xx} = 0.965483$) of oxyfluorinated Kevlar/s-PS composites (Series IVD).

However, mold flow does not consider the adhesion between fiber and the matrix. As mentioned earlier that it uses the Folgar–Tucker model, which is a modification of the Jeffery model by adding a diffusive term to consider the fiber–fiber interaction. In the Jeffery model, inertia and Brownian motion of the fibers are neglected.

The fibers in the Jeffery model are considered rigid particles moving in a Newtonian fluid, which however do not disturb the motion of the fluid. Hence in the mold flow technique, the adhesion between the fiber and matrix does not play any significant role.

The orientation tensor value is the maximum in case of Series IIB and IIIB, both in the skin as well as in the core region manifesting the most fiber orientation in the flow direction, which is the optimum condition for the crystalline and thermal properties of the composites corroborating the results obtained from DSC and XRD (Figures 7.18 and 7.20).

Hence it proved that the fiber orientation was different in core and skin region. Moreover, processing parameters significantly affect the fiber orientation pattern in the skin and core region.

7.14.4
Simulation of Fiber Orientation by Mold Flow Technique for PC/LCP/Kevlar Composites

Simulation of fiber orientation of Kevlar fibers in PC/LCP blends by mold flow technique was also analyzed by Das et al. The compounding formulation and the variation in processing parameters of PC/LCP blends containing unmodified and modified Kevlar fibers are shown in Tables 7.19 and 7.20.

Table 7.19 Compounding formulations.

Sample code	PC (%)	LCP (%)	Kevlar (%)
I	100	20.8	0.5 (original)
J	100	20.8	0.5 (fluorinated)
K	100	20.8	0.5 (oxyfluorinated)

Table 7.20 Processing parameters for injection molding.

Sample code	T_m (°C)	T_{mold} (°C)	Flow rate (cm^3/s)
Series I1/Series J1/Series K1	300	60	20
Series I2/Series J2/Series K2	320	40	48
Series I3/Series J3/Series K3	320	40	20
Series I4/Series J4/Series K4	320	60	20

7 Kevlar Fiber-Reinforced Polymer Composites

In order to understand the fiber orientation in PC/LCP/unmodified, fluorinated, and oxyfluorinated Kevlar under different processing parameters, mold flow simulation technique had been used. The variation of orientation tensor with different normalized thickness under different processing parameters had been displayed in Figure 7.47a–c and corresponding orientation pattern under different normalized thickness have been depicted in Figures – (from Figures 7.47a–c, 7.48, 7.51, 7.52, 7.55, 7.56, and 7.59).

It is evident that melt temperature has pronounced effect on the fiber orientation of PC/LCP/Kevlar composites in both the skin and core region. Although the fiber orientation exhibited similar trend in the core as well as in skin region at both higher and lower melt temperature. The fibers are more oriented in the core region (higher magnitude of orientation tensor) than that of the skin. However, higher melt temperature possess somewhat less fiber orientation tensor than that of the lower variable.

Figures 7.47a–c, 7.48, 7.49, 7.52, 7.53, 7.56, and 7.57 exhibit the fiber orientation in the PC/LCP/Kevlar composites under varying flow rate. From those figures it was evident that flow rate has significant effect on the fiber orientation of the composites. At lower flow rate (Series I3, J3, K3), more fibers are oriented in the skin region in flow

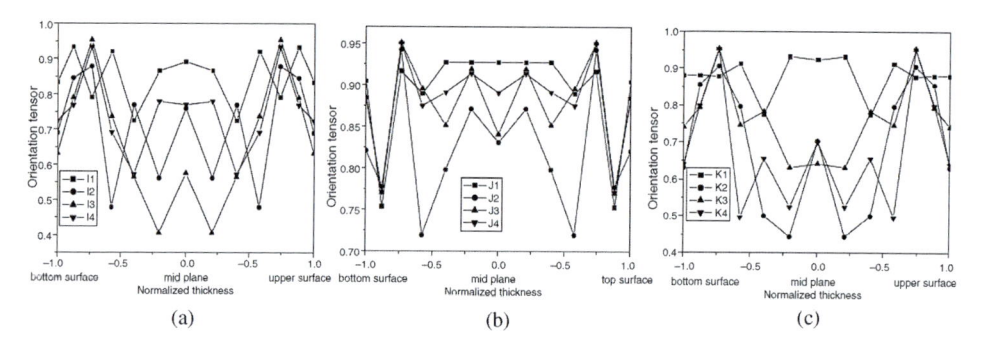

Figure 7.47 Fiber orientation tensor versus normalized thickness of (a) PC/LCP/unmodified Kevlar (I), (b) PC/LCP/fluorinated Kevlar (J), and (c) PC/LCP/oxyfluorinated Kevlar (K) under different processing parameters.

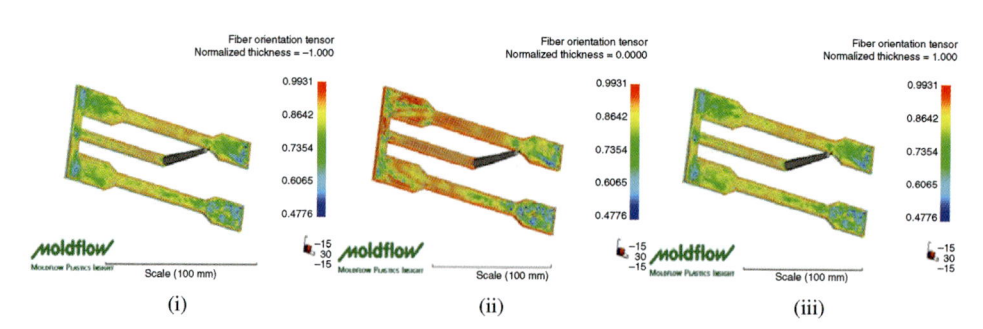

Figure 7.48 Fiber orientation at the (i) bottom surface (normalized thickness = −1, $T_{xx} = 0.834123$), (ii) core region (normalized thickness = 0, $T_{xx} = 0.891039$), (iii) top surface (normalized thickness = 1, $T_{xx} = 0.834123$) of unmodified Kevlar/PC/LCP composites (Series I1).

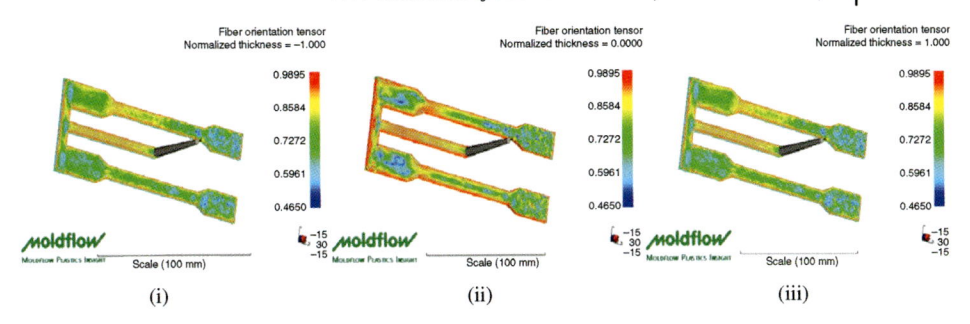

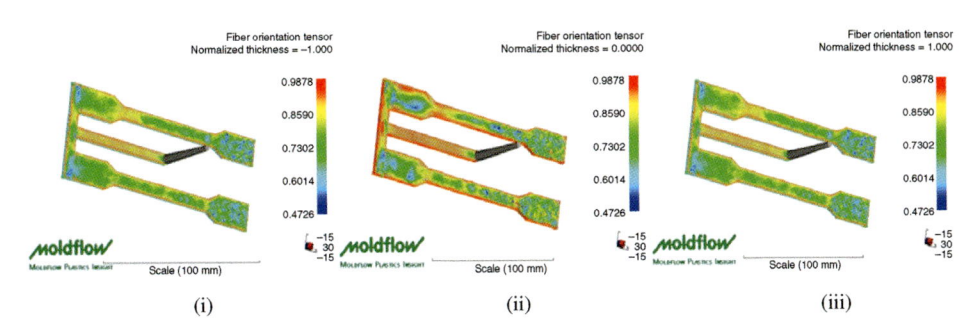

7.14 Simulation of Fiber Orientation by Mold Flow Technique | 267

Figure 7.49 Fiber orientation at the (i) bottom surface (normalized thickness = −1, $T_{xx} = 0.689984$), (ii) core region (normalized thickness = 0, $T_{xx} = 0.759418$), (iii) top surface (normalized thickness = 1, $T_{xx} = 0.689984$) of unmodified Kevlar/PC/LCP composites (Series I2).

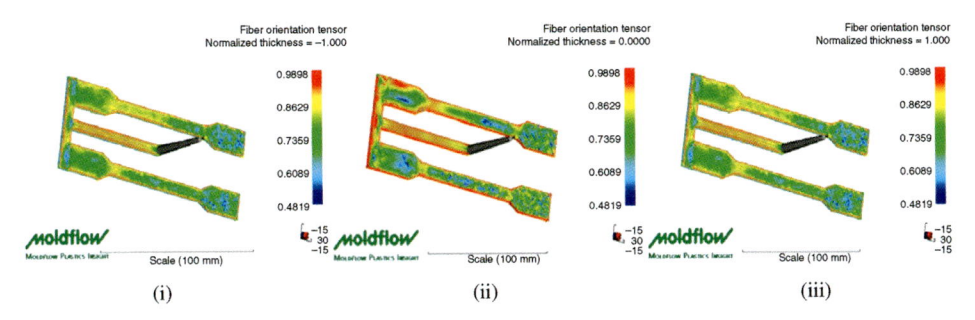

Figure 7.50 Fiber orientation at the (i) bottom surface (normalized thickness = −1, $T_{xx} = 0.632284$), (ii) core region (normalized thickness = 0, $T_{xx} = 0.575498$), (iii) top surface (normalized thickness = 1, $T_{xx} = 0.632284$) of unmodified Kevlar/PC/LCP composites (Series I3).

direction (evidenced from orientation tensor value), than that of higher counterpart (Series I2, J2, K2).

This can be ascribed as an increase in the mold wall–polymer contact time at lower flow rate leading to a thick solidified layer, that is, the skin structure. This solidified skin layer produce high shear flow field at the solid–melt interface. Such a shear flow orients the fibers in flow direction.

Figure 7.51 Fiber orientation at the (i) bottom surface (normalized thickness = −1, $T_{xx} = 0.724652$), (ii) core region (normalized thickness = 0, $T_{xx} = 0.76954$), (iii) top surface (normalized thickness = 1, $T_{xx} = 0.724652$) of unmodified Kevlar/PC/LCP composites (Series I4).

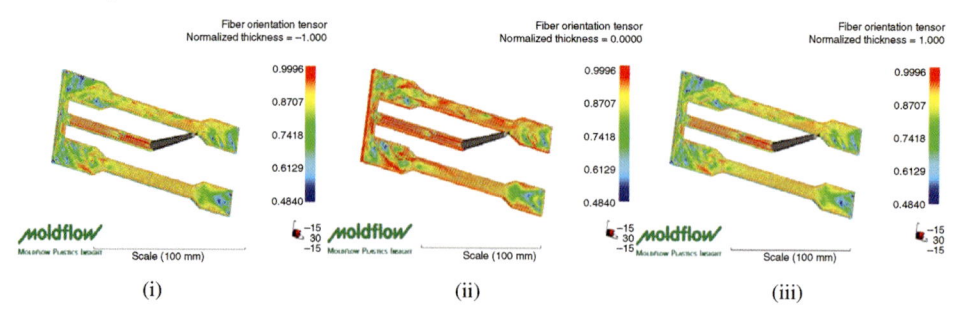

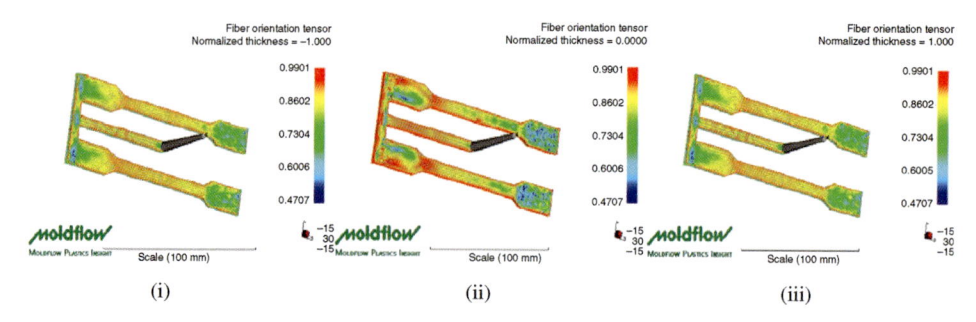

Figure 7.52 Fiber orientation at the (i) bottom surface (normalized thickness = −1, $T_{xx} = 0.904304$), (ii) core region (normalized thickness = 0, $T_{xx} = 0.927326$), (iii) top surface (normalized thickness = 1, $T_{xx} = 0.904304$) of fluorinated Kevlar/PC/LCP composites (Series J1).

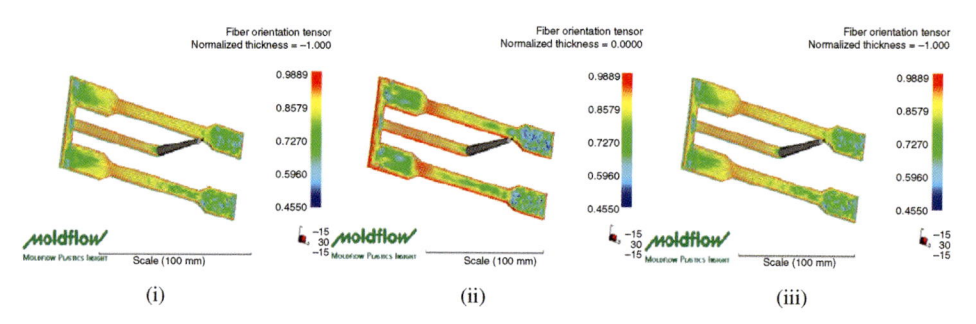

Figure 7.53 Fiber orientation at the (i) bottom surface (normalized thickness = −1, $T_{xx} = 0.820934$), (ii) core region (normalized thickness = 0, $T_{xx} = 0.830934$), (iii) top surface (normalized thickness = 1, $T_{xx} = 0.820934$) of fluorinated Kevlar/PC/LCP composites (Series J2).

Orientation in the flow direction increases with increasing shear rate. On the other hand, the long contact time provided by the low injection speed results in the thicker skin structure and in preservation of the fiber orientation patterns in the concerned zone (i.e., skin region).

Figure 7.54 Fiber orientation at the (i) bottom surface (normalized thickness = −1, $T_{xx} = 0.884709$), (ii) core region (normalized thickness = 0, $T_{xx} = 0.840919$), (iii) top surface (normalized thickness = 1, $T_{xx} = 0.884709$) of fluorinated Kevlar/PC/LCP composites (Series J3).

7.14 Simulation of Fiber Orientation by Mold Flow Technique | 269

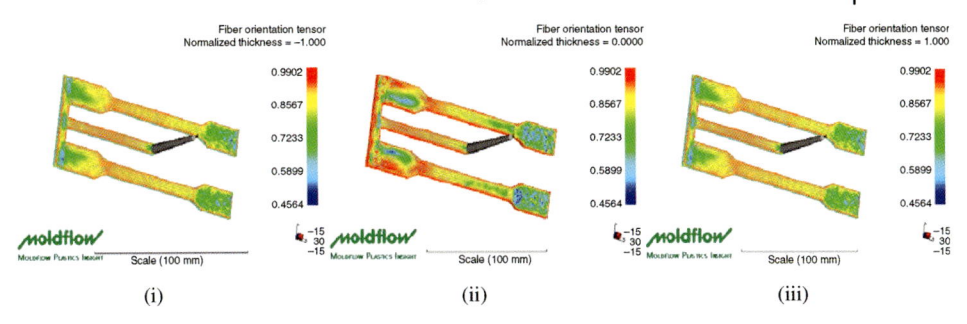

Figure 7.55 Fiber orientation at the (i) bottom surface (normalized thickness = −1, T_{xx} = 0.884479), (ii) core region (normalized thickness = 0, T_{xx} = 0.890948), (iii) top surface (normalized thickness = 1, T_{xx} = 0.884479) of fluorinated Kevlar/PC/LCP composites (Series J4).

An interesting trend of fiber orientation had been observed in case of varying mold temperature (I3, J3, K3 and I4, J4, K4) (Figures 7.47a–c, 7.50, 7.51, 7.54, 7.55, 7.58, and 7.59. At higher mold temperature more fibers are oriented in core region than that of the skin (greater orientation tensor value) and in case of lower mold temperature the opposite trend had been observed.

For high mold temperature, the temperature difference between the molten fluid and mold temperature leads to the thinner solidified skin leading to the lower flow field at the solid–melt interface and lower shear rates in the solid–melt interface thus orienting lesser fiber in the flow direction of the skin region.

However, mold flow does not consider fiber/matrix adhesion in this program. As mentioned earlier that it uses the Folgar–Tucker model, which is a modification of the Jeffery model by adding a diffusive term to consider the fiber–fiber interaction. In the Jeffery model, inertia and Brownian motion of the fibers are neglected. In the Jeffery model the fibers considers are rigid particles moving in a Newtonian fluid, which however do not disturb the motion of the fluid. So it is not important whether the fiber/matrix adhesion is good or not in mold flow technique.

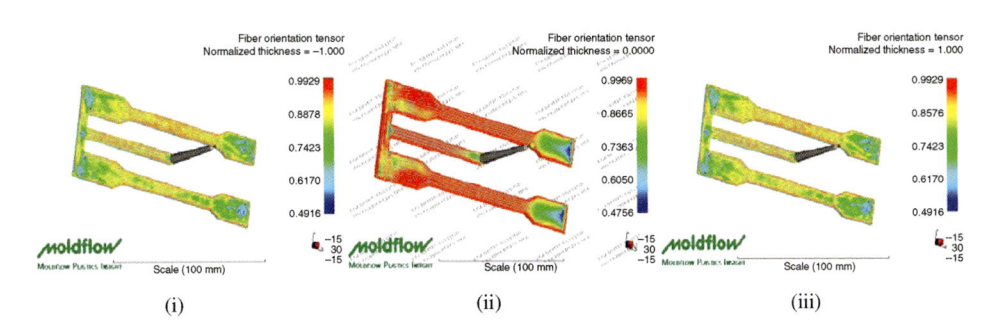

Figure 7.56 Fiber orientation at the (i) bottom surface (normalized thickness = −1, T_{xx} = 0.880099), (ii) core region (normalized thickness = 0, T_{xx} = 0.923133), (iii) top surface (normalized thickness = 1, T_{xx} = 0.880099) of oxyfluorinated Kevlar/PC/LCP composites (Series K1).

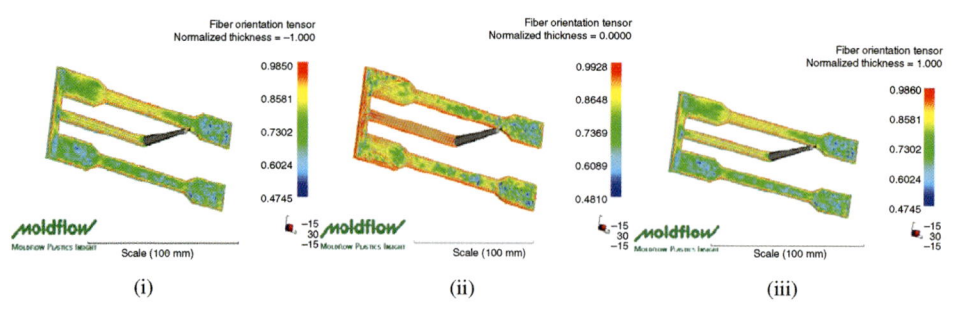

Figure 7.57 Fiber orientation at the (i) bottom surface (normalized thickness = −1, $T_{xx} = 0.630386$), (ii) core region (normalized thickness = 0, $T_{xx} = 0.702878$), (iii) top surface (normalized thickness = 1, $T_{xx} = 0.630386$) of oxyfluorinated Kevlar/PC/LCP composites (Series K2).

From the above observation, it proved that the fiber orientation was different in core and skin region. In particular, the processing parameters mainly influence the fiber orientation pattern in the skin and core region.

7.15
Kevlar-Reinforced Thermosetting Composites

Guo et al. studied the effect of plasma treatment of Kevlar fiber, on the tribological behavior of Kevlar fabric/phenolic resin composites [97]. The Kevlar fibers were exposed to cold air plasma that alters the surface characteristics without changing the bulk properties. Under the plasma environment the polymer chains at the surface were broken, which generates free radicals. These free radicals then combine with the other free radicals from the environment and forms the functional groups like −OH, −COOH, −NH$_2$ on the fabric surface (in air atmosphere). These functional groups strengthen the adhesion between the fabric and resin, which improves its wear resistance. Presence of the functional groups on the fabric, after plasma

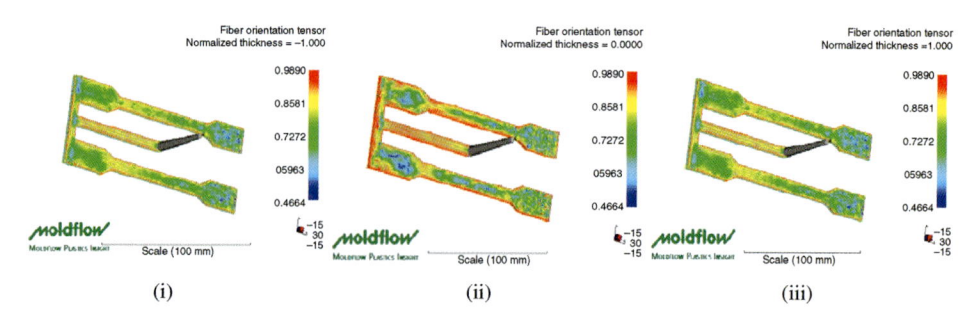

Figure 7.58 Fiber orientation at the (i) bottom surface (normalized thickness = −1, $T_{xx} = 0.740444$), (ii) core region (normalized thickness = 0, $T_{xx} = 0.641091$), (iii) top surface (normalized thickness = 1, $T_{xx} = 0.740444$) of oxyfluorinated Kevlar/PC/LCP composites (Series K3).

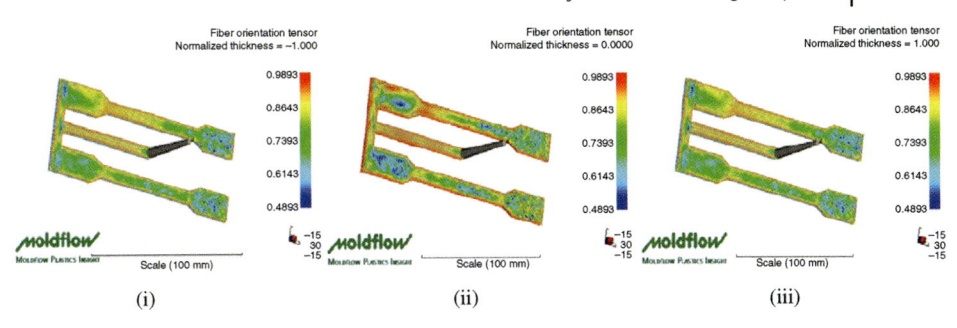

Figure 7.59 Fiber orientation at the (i) bottom surface (normalized thickness = −1, $T_{xx} = 0.639416$), (ii) core region (normalized thickness = 0, $T_{xx} = 0.699416$), (iii) top surface (normalized thickness = 1, $T_{xx} = 0.639416$) of oxyfluorinated Kevlar/PC/LCP composites (Series K4).

treatment, was confirmed by XPS and FTIR. The surface roughness of the treated fibers was found to be much more than the untreated ones. Due to the better adhesion between the fibers and matrix, the antiwear characteristics of the treated fiber/phenolic composites were found to be much better than that of the untreated fiber composites.

Guo et al. reported the tribological characteristics of spun Kevlar fabric composites filled with polyfluo wax (PFW) and lanthanum fluoride (LaF$_3$) under various load and rotating speed, in a pin-on-disc friction and wear tester [98]. The composites were prepared by mixing PFW and LaF$_3$ with phenolic resin and then used to impregnate the spun Kevlar fabric. For the wear test they fixed this fabric on the AISI-1045 steel (surface roughness $R_a = 0.45$ mm) with the phenolic resin and then cured at 180 °C for 2 h under pressure. They had found that the tribological characteristics of spun Kevlar fabric with 15 wt% of PFW and 5 wt% of LaF$_3$ were best among all the compositions studied. They suggested that, due to the low melting temperature and self-lubrication of the PFW filler, composite filled with 15 wt% PFW is compact, uniform, and smooth, corresponding to the best friction and wear abilities of the composite. PFW added composites shows better wear resistance compared to LaF$_3$ containing composites in different conditions. The increase of wear was found after the 15 and 5 wt% loading for PFW and LaF$_3$ (respectively) The reason may be that, excessive filler is prone to conglomerate and leads to the less uniformity of the system and thus the interfacial adhesion becomes poorer, which may lead to drawing out of the filler from the resin matrix during the friction process. Thus, abrasive wear occurred and the friction coefficient and wear rate increased.

Yue and Padmanabhan studied the alternation in interfacial properties of epoxy Kevlar composites by chemically modifying the Kevlar surface with different organic solvents [33]. In a typical experiment, they had prepared three sets of Kevlar fiber among which the first set was treated with acetic anhydride for 1 min followed by washing with distilled water. The second and third sets were prepared by treating the fibers with acetic anhydride for 1 min followed by washing with methanol for 3 and 10 min, respectively. Then all the samples were dried at 100 °C for 5 h under vacuum. To analyze the effect of surface treatment on the fiber matrix interfacial adhesion, the fiber pull-out test was carried out on Kevlar/epoxy samples (cured at 120 °C for 12 h) using an INSTRON universal testing machine. They found that interfacial shear strength of the sample, treated with methanol for 3 min, was highest followed by samples only treated with acetic anhydride. This improvement in the interfacial shear strength could be explained by the presence of an oxygen rich fiber surface whose blistered, striated, and undulated morphology adds to the mechanical interlocking of the adhering matrix. It was found that fiber samples treated with only acetic anhydride exhibited pimples and blisters on the skin. However, samples washed with methanol for 3 min showed striations and undulations along the fiber direction. Excessive methanol washing (10 min) was found to be detrimental to the skin layer thickness of the fiber by exposing the core in places as a result of excessively striated and undulated morphology.

References

1 Callistor, W.D., Jr. (1985) *Materials Science and Engineering*, John Wiley and Sons, Inc., New York.
2 Kadolph, Sara J. and Langford, Anna L. (2002) *Textiles*, 9th edn, Pearson Education, Inc., Upper Sadddle River, NJ.
3 Peters, S.T. (1997) *Handbook of Composites*, Chapman & Hall.
4 Clegg, D.W. and Collyer, A.A. (1986) *Mechanical Properties of Reinforced Thermoplastics*, Elsevier Applied Science Publishers, London and New York.
5 Sadov, F., Korchagin, M., and Matesky, A. (1978) *Chemical Technology of Fibrous Materials*, Mir Publishers, Moscow.
6 Mukherjee, M. (2007) Effect of fluorination and oxy-fluorination of Kevlar fiber on the properties of short fiber reinforced polymer composites and its mold flow simulation study. PhD thesis. IIT Kharagpur.
7 Broutman, L.J. and Crock, R.H. (1967) *Modern Composite Materials*, Addison Wesley Publishing Company, London.
8 Akay, M. and Barkley, D. (1991) *Journal of Materials Science*, **26**, 2731.
9 Fu, S.Y., Lauke, B., Mäder, E., Yue, C.Y., and Hu, X. (1999) *Journal of Materials Processing Technology*, **89**, 501.
10 Barbosa, S.E. and Kenny, J.M. (1999) *Journal of Reinforced Plastics and Composites*, **18**, 413.
11 Singh, P. and Kamal, M.R. (1989) *Polymer Composites*, **10**, 344.
12 Larsen, A. (2000) *Polymer Composites*, **21**, 51.
13 Malzahn, J.C. and Schultz, J.M. (1986) *Composites Science and Technology*, **25**, 187.
14 Gupta, M. and Wang, K.K. (1993) *Polymer Composites*, **14**, 367.
15 Bailey, R.S. (1994) Conventional thermoplastics, in *Handbook of Polymer–Fiber Composites, Polymer Science and Technology Series*, Longman Scientific & Technical, New York, p. 74.
16 Chung, D.H. and Kwon, T.H. (2002) *Journal of Non-Newtonian Fluid Mechanics*, **107**, 67.
17 Chung, D.H. and Kwon, T.H. (2002) *Korea-Australia Rheology Journal*, **14**, 175.
18 Senthilvelan, S. and Gnanamoorthy, R. (2006) *Applied Composite Materials*, **13**, 237.
19 Zhou, K. and Lin, J.Z. (2005) *Journal of Zhejiang University. Science*, **6A** (4), 257.
20 Chung, T.N. and Mennig, G. (2001) *Rheologica Acta*, **40**, 67.
21 Chung, T.N., Plichta, C., and Mennig, G. (1998) *Rheologica Acta*, **37**, 299.
22 Imihezri, S.S.S., Sapuan, S.M., Sulaiman, S., Hamdana, M.M., Zainuddin, E.S., Osman, M.R., and Rahman, M.Z.A. (2006) *Journal of Materials Processing Technology*, **171**, 358.
23 Patcharaphun, S. and Mennig, G. (2005) *Polymer Composites*, **26**, 823.
24 Mallick, P.K. (1998) *Fiber Reinforced Composites*, Marcel Dekker, Inc., New York.
25 Paul, J.T., Jr. and Thompson, J.B. (1965) Proceedings of the 20th Conference on

SPI Reinforced Plastics Division, Feb. 1965, Section 2-C.
26 Bascom, W.D. (1965) Proceedings of the 20th Conference on SPI Reinforced Plastics Division, Feb. 1965, Section 15-B.
27 Belgacem, M.N., Bataille, P., and Sapieha, S. (1994) *Journal of Applied Polymer Science*, **53**, 379.
28 Kaplan, S.L. and Wally, P., Hansen 4th State, Inc., Belmont, CA 94002. Gas Plasma Treatment of Kevlar® and Spectra® Fabrics for Advanced Composites. SAMPE International Conference, Orlando, FL, October 29, 1997.
29 Wakida, T. and Tokino, S. (1996) *Indian Journal of Fibre & Textile Research*, **21**, 69.
30 Lin, T.K., Wu, S.J., Lai, J.G., and Shyu, S.S. (2000) *Composites Science and Technology*, **60**, 1873.
31 Park, S.J. and Park, B.J. (1999) *Journal of Materials Science Letters*, **18**, 47.
32 Park, S.J., Seo, M.K., Ma, T.J., and Lee, D.R. (2002) *Journal of Colloid and Interface Science*, **252**, 249.
33 Yue, C.Y. and Padmanabhan, K. (1999) *Composites: Part B*, **30**, 205.
34 Wu, G.M., Hung, C.H., You, J.H., and Liu, S.J. (2004) *Journal of Polymer Research*, **11** (1), 31–36.
35 Mavrich, A.M., Ishida, H., and Koenig, J.L. (1995) *Applied Spectroscopy*, **49** (2), 149–155.
36 Broutman, L.J. and Crock, R.H. (1974) *Composite Materials*, vol. 6, Academic Press, New York.
37 Ai, T., Wang, R., and Zhou, W. (2007) *Polymer Composites*, **28** (3), 412–416.
38 Andreopoulos, A.G. (2003) *Journal of Applied Polymer Science*, **38** (6), 1053–1064.
39 Menon, N., Blum, F.D., and Dharani, L.R., Defence Technical Information Centre, Technical report, Jan 1, 1993–Apr 1, 1994.
40 Briggs, D., Brewis, D.M., and Konieczo, M.B. (1976) *Journal of Materials Science*, **11**, 1270.
41 Nakao, K. and Nishiuchi, M. (1966) *Journal of Adhesion Science and Technology*, **2**, 239.
42 Lagow, R.J. and Margrave, J.L. (1979) *Progress in Inorganic Chemistry*, **26**, 162.
43 Jagur-Grodzinski, J. (1992) *Progress in Polymer Science*, **17**, 361.
44 Anand, M., Hobbs, J.P., and Brass, I.J., (1994) *J. Organofluorine Chemistry* (eds RE Banks, BE Smart, and JC Tatlow) Plenum Press Div., New York, pp 469–481.
45 Mohanty, S., Verma, S.K., Tripathy, S.S., Nayak, S.K., and Reinf, J. (2004) *Plastics and Composites*, **23**, 625.
46 Joseph, K., Verghese, S., Kalaprasad, G., Thomas, S., Prasannakumary, L., Koshy, P., and Pavithran, C. (1996) *European Polymer Journal*, **32**, 1243.
47 Herrera-Franco, P.J. and Valadez-González, A. (2005) *Composites: Part B*, **36**, 597.
48 Hongu, T. and Philips, Go. (1997) *New Fibers*, Woodhead, Cambridge.
49 Pearce, E.M., Weil, E.D., and Barinov, V.Y. (2001) In *Fire and Polymers, ACS Symposium Series 797* (eds GL Nelson and CA Wilkie), American Chemical Society, Washington, DC, p. 37.
50 Chou, T.W. (1992) *Microstructural Design of Fiber Composites*, Cambridge University Press.
51 Du, S.Y. and Wang, B. (1998) *Meso-Mechanics of Composites*, China Science Press.
52 Varelidis, P.C., Papakostopoulos, D.G., Pandazis, C.I., and Papaspyrides, C.D. (2000) *Composites: Part A*, **31**, 549.
53 Saikrasuna, S., Amornskchaia, T., Sirisinhaa, C., Meesirib, W., and Bualek-Limcharoena, S. (1999) *Polymer*, **40**, 6437.
54 Broutman, L.J. (2004) *Polymer Engineering and Science*, **23**, 776.
55 Sudarisman, Davies, I.J., and Hamada, H. (2007) *Composites: Part A*, **38**, 1070.
56 Arikan, H., Avci, A., and Akdemir, A. (2004) *Polymer Testing*, **23**, 615.
57 Fu, S.Y. and Mai, Y.W. (2004) *Journal of Applied Polymer Science*, **88**, 1497.
58 Wang, Y. (1998) *Fatigue & Fracture of Engineering Materials & Structures*, **21**, 521.
59 DuPont Kevlar® Technical Guide DuPont Advanced Fiber Systems, 2000.
60 Tanner, D., Fitzgerald, J.A., and Phillips, B.R. (1989) *Advanced Materials*, **5**, 151.
61 Maalej, M. (2001) *Journal of Materials Science*, **36**, 2203.
62 Yu, Z., Brisson, J., and Ait-Kadi, A. (2004) *Polymer Composites*, **15**, 64.
63 Sun, X., Li, H., Wang, J., and Yan, S. (2006) *Macromolecules*, **39**, 8720.

64 Shaker, M., Ko, F., and Song, J. (1999) *Journal of Composites Technology & Research*, **21**, 224–229.
65 Kutty, S.K.N. and Nando, G.B. (2003) *Journal of Applied Polymer Science*, **43**, 1913.
66 Murat İçten, B., Karakuzu, R., and Evren Toygar, M. (2006) *Composite Structures*, **73**, 443.
67 Kim, E.-Y., An, S.-K., and Kim, H.-D. (1997) *Journal of Applied Polymer Science*, **65**, 99.
68 Al-Bastaki, N.M.S. (1998) *Applied Composite Materials*, **5**, 223.
69 Kodama, M. and Karino, I. (2003) *Journal of Applied Polymer Science*, **32**, 5345.
70 Kamel Mahmoud, M. (2003) *Polymer-Plastics Technology and Engineering*, **42**, 659.
71 Wu, J. and Cheng, X.H. (2006) *Journal of Applied Polymer Science*, **102**, 4165.
72 Kitagawa, K., Hamada, H., Maekawa, Z., Ikuta, N., Dobb, M.G., and Johnson, D.J. (1996) *Journal of Materials Science Letters*, **15**, 2091.
73 Haque, A., Mahmood, S., Walker, L., and Jeelani, S. (1991) *Journal of Reinforced Plastics and Composites*, **10**, 132.
74 Day, R.J., Hewson, K.D., and Lovell, P.A. (2002) *Composites Science and Technology*, **62**, 153.
75 Kolpakov, G.A., Kuzina, S.I., Kharitonov, A.P., Moskvin, Yu. L., and Mikhailov, A.I. (1992) *Soviet Journal of Chemical Physics*, **9**, 2283.
76 Lopez Manchado, M.A., Valentini, L., Biagiotti, J., and Kenny, J.M. (2005) *Carbon*, **43**, 1499.
77 Chen, X., Wu, L., Zhou, S., and You, B. (2003) *Polymer International*, **52**, 993.
78 Mukherjee, M., Das, C.K., and Kharitonov, A.P. (2006) *Polymer Composites*, **27**, 205.
79 Kalaprasad, G., Mathew, G., Pavitran, C., and Thomas, S. (2003) *Journal of Applied Polymer Science*, **89**, 432–442.
80 Wang, C., Lin, C.C., and Tseng, L.C. (2006) *Polymer*, **47**, 390.
81 Menard, K.P. (1999) *Dynamic Mechanical Analysis*, CRC Press, New York, p. 100.
82 Sahoo, N.G., Das, C.K., Jeong, H., and Ha, C.S. (2003) *Macromolecular Research*, **11**, 224.
83 Geethamma, V.G., Janardhhan, R., Ramamurthy, K., and Thomas, K.S. (1996) *International Journal of Polymeric Materials*, **32**, 147.
84 Carnerio, O.S. and Maia, J.M. (2000) *Polymer Composites*, **21**, 960.
85 Raj, R.G., Kokta, B.V., and Daneault, C. (1990) *International Journal of Polymeric Materials*, **14**, 223.
86 Park, S.J., Song, S.Y., Shin, J.S., and Rhee, J.M. (2005) *Journal of Colloid and Interface Science*, **283**, 190.
87 Engelsing, K. and Mennig, G. (2001) *Mechanics of Time-Dependent Materials*, **5**, 27.
88 Bright, P.F., Crowson, R.J., and Folks, M.J. (1978) *Journal of Materials Science*, **13**, 2497.
89 Jeffery, G.B. (1922) *Proceedings of the Royal Society*, **A102**, 161.
90 Yamamoto, S. and Matsuoka, T. (1993) *Journal of Chemical Physics*, **98** (1), 644–650.
91 Mackaplow, M.B. and Shaqfeh, E.S.G. (1996) *Journal of Fluid Mechanics*, **329**, 55–186.
92 Sundararajakumar, R.R. and Koch, D.L. (1997) *Journal of Non-Newtonian Fluid Mechanics*, **73**, 205–239.
93 Fan, X.J., Phan-Thien, N., and Zheng, R. (1998) *Journal of Non-Newtonian Fluid Mechanics*, **74**, 113–135.
94 Advani, S.G. and Tucker, C.L., III (1987) *Journal of Rheology*, **31**, 751–784.
95 Dinh, S.M. and Armstrong, R.C. (1984) *Journal of Rheology*, **28**, 207.
96 Folgar, F.P. and Tucker, C.L. (1984) *Journal of Reinforced Plastics and Composites*, **3**, 98.
97 Guo, F., Zhang, Z., Liu, W., Su, F., and Zhang, H. (2009) *Tribology International*, **42**, 243.
98 Guo, F., Zhang, Z., Zhang, H., Wang, K., and Jiang, W. (2010) *Tribology International*, **43**, 1466.

8
Polyester Fiber-Reinforced Polymer Composites
Dionysis E. Mouzakis

8.1
Introduction

The Federal Trade Commission (FTC) provides the following definition for the polyester fiber: "A manufactured fiber in which the fiber forming substance is any long-chain synthetic polymer composed of at least 85% by weight of an ester of a substituted aromatic carboxylic acid, including but not restricted to substituted terephthalic units, $p(\text{-R-O-CO-C}_6\text{H}_4\text{-CO-O-})x$ and parasubstituted hydroxy-benzoate units, $p(\text{-R-O-CO-C}_6\text{H}_4\text{-O-})x$." The most common polyester fibers are based on two polymers, namely poly(ethylene terephthalate) (PET) and poly(ethylene naphthalate) (PEN), respectively.

In 1996, 24.1 million metric tons of industrial fibers were produced worldwide. The main volume gain took place in production of PET fibers (PET filament 9% and PET staple 4%) [1]. Impetus for this growth was provided by the demand for fiber and container (e.g., beverages and water bottles) resin. Seventy-five percent of the entire PET production is usually directed toward fiber manufacturing. Hoechst, Dupont, and Eastman are the three world's largest polyester producers. Other polyester fiber producers especially in the United States are Acordis Industrial Fibers, Inc.; Allied-Signal Inc.; Cookson Fibers, Inc.; KoSa; Intercontinental Polymers, Inc.; Martin Color-Fi. Nan Ya Plastics Corp.; Wellman, Inc. [2]. Polyester fibers are hydrophobic, which is desirable for lightweight-facing fabrics used in the disposable industry. They provide a perceptible dry feel on the surface, even when the inner absorbent media is saturated. As new methods of processing and bonding of PET are developed, rayon is being replaced by polyester in the market. According to Harrison [3], 49% of the total nonwovens market share in the United States is dominated by polyester staple (291 million pounds back in 1996), and ranks number one among all kinds of fiber supplies.

Production of poly(ethylene terephthalate) fiber in the world is increasing every year and reached an amount of about 20 million tons a year in 2001. Dramatic growth in PET fiber production was foreseen in Asia about one decade ago [4]. The cost of polyester, with the combination of its superior strength and resilience, is lower than that of rayon, providing a success story.

Every 2 years the global leading plastic resin producers are gathered at the "K" plastic show held in Dusseldorf, Germany to present their product range and new developments. The 2007 event showed the topical stand and the market trends for thermoplastic resins and the basic raw material for the synthetic fiber production. According to Koslowski [5], world production of plastics in 2006 has reached 245 million tons and 15% of them went into the production of synthetic fibers. According to the same author, the global production of PET capacity has reached more than 45 million tons annually and polypropylene (PP) has gained increased importance with 6.6 million tons global production in 2007, while the global market for PA resins has been estimated with 6.6 million tons in 2006.

Developed PET market growth between 2002 and 2008 and after is shown in Table 8.1 [6].

The usual commercial names of PET products are Dacron®, Diolen®, Tergal®, Terylene®, and Trevira® fibers; Cleartuf®, Eastman®, and Polyclear® bottle resins; Hostaphan®, Melinex®, and Mylar® films; and Arnite®, Ertalyte®, Impet®, Rynite®, and Valox® injection-molding resins. The polyester industry makes up about 18% of world polymer production and is currently third only after polyethylene (PE) and polypropylene.

Currently a well-established industry's branch produces PET fibers from recycled products [7]. In the production of PET fiber, the wastes occur in the amount of about 3–5% of total production. The PET does not degrade in nature for a long time. Since PET is a derivative of petroleum, the wastes of PET are valuable and must be recycled due to economic reasons and ecological demands.

Recycling poly(ethylene terephthalate) involves two different technologies:

a) Chemical recycling, that is, back to the initial raw materials, for example, purified terephthalic acid (PTA) or dimethyl terephthalate (DMT) and ethylene glycol (EG) where the polymer structure is entirely destroyed (depolymerization), or in process intermediates, that is, bis-β-hydroxyterephthalate.
b) Mechanical recycling where many of the original polymer properties are being maintained (not molecular weight) or reconstituted. PET is converted to flakes by mechanical shredding. About 70% of the main portion, PET bottle flake, is converted to fibers and filaments by remelting. Various properties of the PET wastes are improved using the remelting method by employing an extruder. Degradation during remelting decreases the properties of the final product.

Table 8.1 PET market size per year.

Product type	2002 (million tons per annum)	2008 (million tons per annum)
Textile – PET	20	39
Resin, bottle/A – PET	9	16
Film – PET	1.2	1.5
Special polyester	1	2.5
Total	31.2	49

The results show that the properties of recycled PET wastes can be monitored by adjusting and carefully designing the parameters of the extruder. Thus, further degradation of the material can be prevented and recycling costs are minimized.

PET fibers can be incorporated into polymer matrices resulting in microscale polymer composites, or can be directly incorporated into the melt of bi- or triphasic (hybrid) melts leading to the formation of *in situ*-reinforced polymer composites, respectively.

Lately, many workers focus on the incorporation of nanofillers such as single- or multiwall nanotubes in PET resin during the melt and finally producing nanophase-reinforced PET fibers with interesting physical and mechanical properties [8]. The effort is directed into exploiting the superb mechanical properties of carbon nanotubes combined with the resilience and impact toughness of PET for the production of high modulus–high strength–high toughness fibers for further use as reinforcements in polymer composites or polymer composites with the aforementioned inherent properties. Another application regime of PET nanocomposites includes those applications where final material needs to possess electrical conductivity [9].

Some other advanced applications of PET fibers, due to their inherent biocompatibility, include manufacturing of prostheses for the human body, which are used in cardiovascular [10] and orthopedic surgical operations [11]. PET fibers have also been proposed for construction material applications, acting as Portland cement reinforcements with encouraging results [12]. It has been shown that the addition of synthetic fibers up to 2 wt% can improve the toughness of mortars and cements [13].

This work aspires to provide an insight into the latest advances in the technology of polyester fiber composites based on polymer matrices mentioned above and their spectrum of advanced applications.

8.2
Synthesis and Basic Properties of Polyester Fibers

PET is a typical linear polyester (see Figure 8.1), polymerized either by ester interchange between the dimethyl terephthalate and ethylene glycol monomers [14] or direct esterification reaction between terephthalic acid and ethylene glycol. Both ester interchange and direct esterification processes are combined with polycondensation steps either batch-wise or continuously. Batch-wise systems need two reactors, one for esterification or ester interchange and the other for polymerization.

Continuous systems need at least three reactors – the first for esterification or shear interchange, the second for reducing excess glycols, and the third for polymerization. An alternate way to produce PET is solid-phase polycondensation where a melt polycondensation is continued until the prepolymer has an intrinsic viscosity of 1.0–1.4 dl/g. At that specific point the polymer is cast into a solid firm. A precrystallization procedure is carried out by heating (above 200 °C) until the target molecular weight is obtained. Later the particulate polymer is melted for spinning (see Figure 8.2). This process is not extensively used for textile PET fibers but is used for some industrial fibers.

Figure 8.1 Chemical structure of PET.

8.2.1
Fiber Manufacturing

The degree of polymerization of PET is controlled, as a function of its final uses. PET for industrial fibers has a higher degree of polymerization, higher molecular weight, and higher viscosity. The normal molecular weight ranges between 15 000 and 20 000 g/mol. Under normal extrusion temperature (280–290 °C), it has a low shear

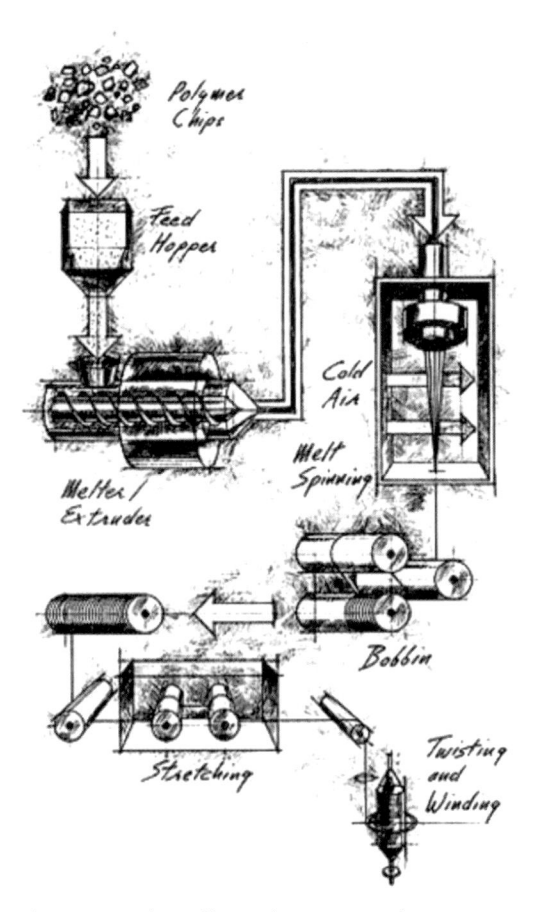

Figure 8.2 Polymer fiber melt spinning production process. http://www.fibersource.com.

viscosity of about 1000–3000 poise. Low molecular weight PET is spun at 265 °C, whereas ultrahigh molecular weight PET is spun at 300 °C or above. The degree of orientation is generally analog to the windup speeds in the spinning process as it controls the draw ratio [15]. A maximum orientation along with increase in productivity can be theoretically obtained at a wind-up speed of 10 000 m/min. However, under realistic production condition effects such as voided fiber skin may appear at wind-up speeds above 7000 m/min [16]. To produce uniform PET fiber surfaces, the drawing process should be performed at temperature above the usual unoriented polymer glass transition temperature (80–90 °C). Due to the fact that the drawing process induces additional orientation to products, the draw ratios (3 : 1–6 : 1) vary according to the final end uses. For higher tenacities, higher draw ratios must be applied. In addition to orientation, as expected, crystallinity can be developed during the drawing at the temperature range of 140–220 °C [17].

8.2.2
Basic Properties of Polyester Fibers

The wide commercial spectrum of uses and success of PET in the market is due to its following properties in the fiber form:

- Strong
- Resistant to stretching and shrinking
- Resistant to most chemicals
- Quick drying
- Crisp and resilient when wet or dry
- Wrinkle resistant
- Mildew resistant
- Abrasion resistant
- Retains heat-set pleats and crease
- Easily washed

8.2.3
Mechanical Response

Mechanical properties such as tensile strength and initial Young's modulus usually increase as a positive function of the degree of fiber stretch due to yielding higher crystallinity and molecular orientation [15]. Ultimate extensibility, that is, elongation at break is usually reduced. An increase of molecular weight further increases the tensile properties, modulus, and elongation. Typical physical and mechanical properties of PET fibers are shown in Table 8.2 and fiber-mechanical response in Figure 8.3. It can be seen that high tenacity filament and staple (curves A and B) exhibit very high breaking strengths and moduli, but relatively low elongations at break. The filament represented by curve C has a much higher initial modulus than the regular tenacity staple shown in curve D [16]. On the other hand, the latter exhibits a greater tenacity and elongation. Partially oriented yarn (POY) and spun filament

Table 8.2 Physical properties of polyester fibers [16].

Property	Filament yarn		Staple and tow	
	Regular tenacity[a]	High tenacity[b]	Regular tenacity[c]	High tenacity[d]
Breaking tenacity, N/tex[e]	0.35–0.5	0.62–0.85	0.35–0.47	0.48–0.61
Breaking elongation	24–50	10–20	35–60	17–40
Elastic recovery at 5% elongation, %	88–93	90	75–85	75–85
Initial modulus, N/tex[f]	6.6–8.8	10.2–10.6	2.2–3.5	4.0–4.9
Specific gravity	1.38	1.39	1.38	1.38
Moisture regain[g], %	0.4	0.4	0.4	0.4
Melting temperature, °C	258–263	258–263	258–263	258–263

a) Textile filament yarns for woven and knit fabrics.
b) Tire cord and high strength, high modulus industrial yarns.
c) Regular staple for 100% polyester fabrics, carpet yarn, fiberfill, and blends with cellulosic blends or wool.
d) High strength, high modulus staple for industrial applications, sewing thread, and cellulosic blends.
e) Standard measurements are conducted in air at 65% RH and 22 °C.
f) To convert N/tex to ge/den (multiply by 11.33).
g) The equilibrium moisture content of the fibers at 21 °C and 65% RH.

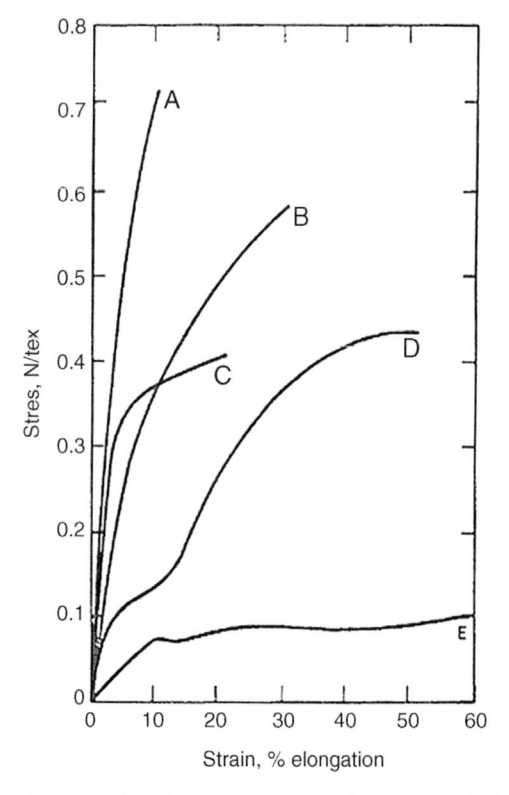

Figure 8.3 Tensile properties of PET fibers [16]. A – high tenacity filament, B – high tenacity staple, C – regular tenacity filament, D – regular tenacity staple, E – POY filament.

yarns, exhibit low strength but very high elongation (curve E). Fatigue testing, that is, exposing PET fiber to repeated compression (e.g., repeated bending), so-called kink bands start forming, and finally resulting in breakage of the kink band into a crack [16]. Slip (shear) bands during tensile testing have also been verified to appear in oriented PET [18]. It has been proven in that the compressive stability of PET is superior to that of nylons [19].

8.2.4
Fiber Viscoelastic Properties

Quite a few researchers have realized the high potential of polyester fibers for applications where high vibrational damping and damage tolerance is highly required. The capacity of such fibers to absorb and damp vibrations or shock loads (impact) due to their polymer nature in comparison to the much stiffer glass and carbon fiber systems has lead to studies regarding such loading cases. Of course usage of a polymer fiber is not bereft of problems regarding their viscoelastic nature; sensitivity to creep and stress relaxation hinters applications of polyester fibers where possible temperatures near to glass transition, high and long-term constant stress loading or constant strains for long periods of time, are likely to occur.

The damping capacities and dynamic moduli of polyester fiber-reinforced polymer composite (PFRPC) systems in naval applications were studied and modeled by House and Grant [20]. They showed that such systems could be incorporated in marine structures; however, their low stiffness must be taken into serious consideration. Davies *et al.* [21] have studied the creep and relaxation behavior of polyester fiber mooring lines and proposed a system for in-service strain measurements. The same author further proposed a nonlinear viscoelastic–viscoplastic model accurately describing the creep behavior of cables made from Diolen polyester [22]. The role of manufacturing conditions on the crystallinity and viscoelastic properties of different PET fibers has also been studied [23] and it was found that high fiber drawing rates lead to higher fiber crystallinity as confirmed by X-ray diffraction and dynamic mechanical analysis and fibers must be constrained during cooling to maintain high dynamic modulus data. Sewing polyester fiber properties have also been extensively studied with respect to their viscoelastic (stress and inverse relaxation) behavior [24]. These findings are important since polyester fibers are nowadays often used for stitching multiple-layers of noncrimp fabrics for polymer composites [25]. It has been shown, however, that appropriate mathematical modeling can be applied for the prediction of creep and relaxation responses of polyester fibers [26]. Even hybrid fiber systems (biological–organic) have been studied with terms of their dynamic viscoelastic response [27] and an important damping role was verified for the polyester constituent.

Finally, the viscoelastic response of polyester fibers in biomedical applications is analyzed by some workers. Their time-dependent behavior is very important due to the nature of their application field; that is, the human organism. Sutures for plastic surgery, for example, for repair of the flexor tendon made from braided Ticron®

polyester fibers exhibited a superior resistance against creep and stress relaxation compared to their usual polypropylene and nylon adversaries [28]. Woven and braided fabrics of high tenacity PET fibers are currently in focus as potential grafts to substitute the anterior cruciate ligament (ACL) in the human knee. Primary results show that the response of PET prostheses matches that of the natural ACL ligament [29], however, as the authors indicate, *in vivo* and *in vitro* long-term and fatigue studies are needed to justify the application.

8.2.5
PEN Fibers

Poly(ethylene 2,6-naphthalate) has excellent thermal and mechanical barrier and chemical resistance properties [30–32], but it possesses a severe problem; its viscosity is very high for melt processing such as fiber spinning, injection molding, and so on [33–35]. It has been proposed [36] that the processability in terms of viscosity can be improved by the addition of other polymers such as thermotropic liquid crystal polymers into polymer matrix. Also, blends of PET and PEN have been attracting increasing interest because they combine the superior properties of PEN with the low price of PET. Mixtures of these polyesters form random copolymers due to transesterification during melt processing. As a result, the glass transition temperature of PET/PEN increases linearly with volume fraction of PEN [37]. It is, therefore, possible to control properties connected with T_g by including PEN to PET. The addition of PEN improves gas-barrier properties as well [38]. On the other hand, due to the higher stiffness of PEN [38] low durability in repeated bending can be foreseen.

Lately, nanofibers of PEN have also been prepared by electrospinning technique [39] delivering very good results with respect to achieved fiber structure and diameter (1–2 μm). The smallest PEN nanofiber diameter (about 0.259 μm) ever reported was recently achieved by the exotic technique of CO_2 laser beam supersonic jet drawing carried under vacuum [40].

8.3
Polyester Fiber-Reinforced Polymer Composites

8.3.1
Elastomer Composites

In the past two decades considerable effort has been invested in reinforcing various types of elastomers with short polyester fibers (mainly PET). The rationale behind this is that PET fibers, yarns, and cords have been widely used as reinforcements for vehicle tires [41], especially the high tenacity–high modulus ones. Of course, their usage is plagued by PET degradation mechanisms such as hydrolysis [42] and aminolysis [43], which can lead to premature failure, sometimes due to fatigue initiation mechanisms. On the other hand, the application of short or longer polyester fibers for elastomer reinforcement would simplify things. This is because short fibers are easier to incorporate (impregnate) in the elastomer resin and resulting composites are amenable to the standard rubber processing steps: extrusion, calendering, and the various types of molding operations (compression, injection, and transfer) [44]. This way high-volume outputs become more cost-efficient, thus, feasible. This is in contrast and is much more economical than the slower processes required for incorporating and placing continuous fibers in the form of yarns or cords. The handicap is a sacrifice in reinforcing strength with discontinuous polymer fibers, although they considerably outperform simple particulate fillers such as carbon black [45]. Early studies showed that PET short fibers are prone to orient well with flow direction during rolling procedure, and provide good mechanical and viscoelastic responses in more than one different elastomer matrices [46]. The short polyester fibers are less effective in reinforcing low modulus materials than rigid ones, but when used properly, a positively sufficient level of reinforcement can be achieved by short fibers. Proper utilization comprises the following inelastic parameters: preservation of high aspect ratio in the fiber, control of fiber directionality, generation of a strong interface through physicochemical bonding, and establishment of a very good dispersion inside the matrix [44]. The effect of several bonding agents with interesting results has also been investigated both for styrene [47] and polyurethane rubber matrices [48, 49]. Natural rubber (NR) composites are also extensively studied; NR is an interesting matrix to reinforce with short polyester fibers [50–52]. It was found that electron-beam irradiation can effectively assist chemical graft polymerization to silica particles in the natural rubber matrix, resulting in superior tensile modulus and strength for silanes [50] and allyl methacrylate [51] coupling agents, respectively. Furthermore, it has been effectively proven that PET short fibers can provide superior price-to-mechanical performance characteristics as potential NR reinforcements to that of chopped aramide (Twaron®) and aramide fiber pulp [52].

8.3.2
Microfibrillar-Reinforced (MFR) PET Composites

Blending polyolefins with engineering plastics can prove as an useful solution to improve the mechanical properties of polymeric materials based on them. The constituents in many polymer blends are thermodynamically immiscible and incompatible with respect to processing. As a result, during processing a large variety of shapes of the dispersed phase can be formed, for example, spheres or ellipsoids, fibrils or plates [53]. It is well known that shape and size of the dispersed phase influences the properties of the final polymer blend [53, 54]. An incompatible pair of polymers can be processed in such a way that the dispersed phase forms *in situ* fibers. By this technique better mechanical properties can be achieved. Some researchers have proposed blending thermotropic liquid crystalline polymers (LCPs) into thermoplastics with substantial success [36, 55]. However, the LCPs are not cost-effective for many engineering applications. On the other hand, nowadays, there are considerable quantities of engineering plastics (e.g., PET from bottles and containers) in the form of recycled products such as flakes and pellets, which are an economic source of primary materials for the preparation of polymer blends.

A major issue when using common thermoplastics, contrary to LCPs, is that their molecular chains tend to relax during melt processing; therefore, a high molecular orientation is inevitable. In order to tackle this problem, a new type of processing methodology, the so-called microfibrillar-reinforced composite (MFC) concept was proposed in the early 1990s by the research group of Fakirov *et al.* [56–58]. According to them preparation of MFC includes the following three basic steps [59]:

i) melt blending with extrusion of two immiscible polymers having different melting temperatures T_m (mixing step);
ii) cold drawing of the extrudate with good orientation of the two phases (fibrillization step) (see Figure 8.4);
iii) thermal treatment at temperature between the T_m's of the two blend partners (isotropization step).

The above-mentioned research group has successfully applied the MFC concept in many thermoplastic polymer systems involving PET, PP, LCPs, poly(phenylene ether) (PPE), and studied their structure–property relationships [60]. Most of the MFC systems studied involve PP and PE matrices generally. Li *et al.* found that the viscosity ratio between PET and PE should remain less than 1, to obtain microfibrillation of PET in the PE matrix during slit-die extrusion [61]. The same research group [62] employed the essential work of fracture toughness approach [63] in order to validate the hot stretch ratio in PET/PE MFCs and found an optimum relationship between them. Of course, compatibilization of the polymer blend constituents also plays a key role in the microfibrillar morphology build-up. It is reported that by adding 7 wt% of ethylene/methacrylic acid copolymer in extruded 50/50 PET/HDPE blends, results in improved ductility and fracture toughness due to crack bridging by PET microfibers [64]. Pultrusion technique is lately reported also to work with

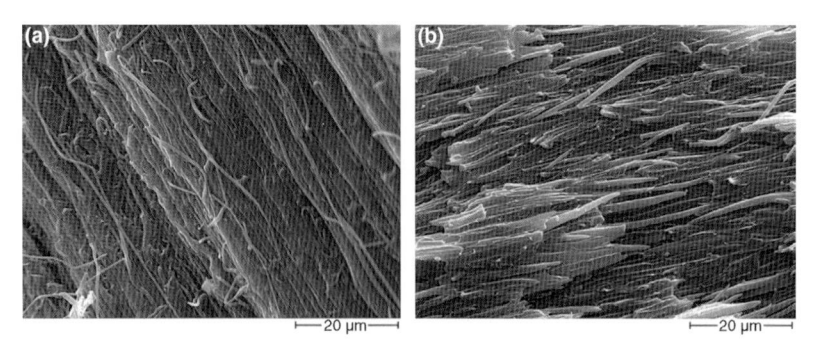

Figure 8.4 Surfaces of compression-molded PET/PP/E-GMA blends with an MFC structure in different ratios (by wt%). (a) 40/60/0 and (b) 40/54/6. PP was selectively dissolved leaving PET fibrils visible. After Friedrich *et al.* [59].

50/50 PET/LDPE blends inducing a composite structure of oriented bristles [65] with a mechanical performance compared to the compression molded (CM) corresponding microfibrillar blends.

Polypropylene/recycled PET MFCs have also been proved to exhibit fibrillation and reinforcement effects when compatibilized with ethylene-glycidyl methacrylate during extrusion and subsequently compression molded after an orientation-drawing procedure [61]. This study reported that CM procedure has resulted in specimens with mechanical performance better than injection-molded ones due to higher orientation degree of the PET fibrils.

Further studies have shown that tensile properties of the PET/PP MFC extrudates show a great improvement compared with the neat polymers when the processing window and weight ratio are carefully chosen. This is attributed to the reinforcement effect of the PET fibrils formed during extrusion [66]. Both tensile stiffness and strength increased with increasing content of the PET fibrils, as shown in Figure 8.5.

8.3.3
Composites

As already mentioned in Section 8.2.4, the superior performance of polyester fibers in polymer composites with respect to energy absorption especially under dynamic loading, that is, impact was early recognized. Effort was invested to incorporate these characteristics into high performance graphite fiber/epoxy matrix composites. Sometimes these composites are interleaved with one or more intermediate layers of PET fabrics [67] a technique called "hybridization" leading to improved impact resistance. Yuan et al. [68] reported that PET fiber mat when used as an interleaf material can improve the crash performance of carbon fiber epoxy tubes. Even PET films can be used as interleaves in order to produce the same toughening effect for

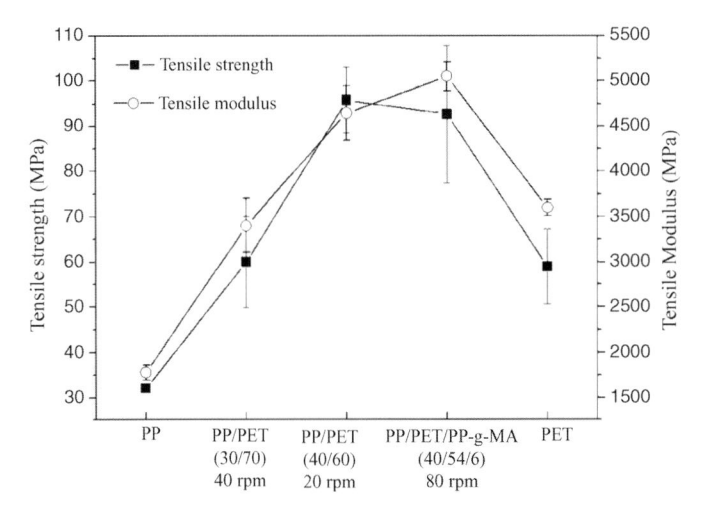

Figure 8.5 Tensile properties of PET/PP microfibrillar composites. After Li et al. [66].

high performance carbon epoxy composites [69]. Short PET fiber-reinforced epoxy composites (1 wt% fiber content) also exhibit high fracture toughness provided that the PET fibers are treated with NaOH solution to enhance adhesion [70]. In that case an almost 80% improvement was achieved in K_{IC} determination done by quasi-static three-point bending tests.

Thermoplastic matrices are also widely investigated with respect to polyester fiber reinforcements; the latter being added as fillers prior to compounding. Lopez-Manchado et al. [71] (cf. Figure 8.6) compared the effects of addition of nylon 66 and PET short fibers, both in sized and unsized states, on the thermal and dynamic properties of PP. They found that 20 wt% fiber content alters significantly the storage modulus E', and the PP matrix crystallization kinetics [71]. The latter effect on crystallinity was verified by Saujanya and Radhakrishnan [72] who also found that the addition of short PET fibers in a PP matrix can improve Young's modulus and Izod impact strength, having little impact on the tensile strength. Santos and Pezzin [73] also confirmed small effect of short-recycled PET fiber content on the tensile strength of PP matrix accompanied by a monotonous increase in Izod impact strength for contents up to 7 wt%. Finally, short PET fibers were even used to prepare hybrid composites of a poly(methylvinylsiloxane) matrix reinforced with wollastonite whiskers [74]. Positive results were obtained for the tensile strength and rupture energy for wollastonite volume fractions above 30% and short PET fibers volume fractions between 5 and 12% in that case.

8.3.4
PET Nanocomposites

The rise in nanotechnology as mentioned in the introduction has given impetus for exploiting the excellent properties of PET for a potential matrix for nanocomposites. The commercial availability of carbon nanofibers (CNFs) in the past decade led researchers to investigate the properties of systems of various polymers reinforced with CNFs. This is quite a challenging operation due to the inherent low dispersibility of the CNFs into the polymer matrix. Sophisticated compounding methods such as

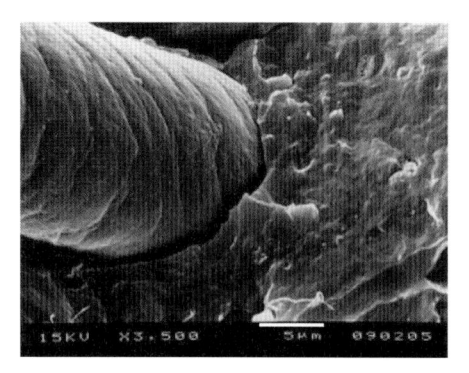

Figure 8.6 A polyester fiber pulled out from PP matrix showing characteristic kink bands [71].

ball-milling, high shear mixing in the melt, and extrusion using twin-screw extruders and melt spinning are nowadays common techniques to prepare nanocomposites with polymer matrices. PET has also been employed as a matrix for various types of CNFs. In a work published in 2003, PET/CNF nanocomposite fibers were studied [8]. It was stated that though tensile strength and modulus did not increase significantly by the addition of nanofibers, but compressive strength and torsional moduli of PET/CNF nanocomposite fibers were considerably higher than that for the control PET fiber. Conductive multilayer PET/4 wt% CNF nanocomposite fibers were manufactured by extruding and composite spinning, producing a textile cloth with excellent antistatic properties [75]. Poly(ethylene terephthalate) nanocomposites with single-walled carbon nanotubes (SWNTs) have been prepared by melt compounding. With increasing concentration (0–3 wt%) of SWNTs, mechanical (Young's modulus and tensile strength) and dynamic storage modulus (E') improved [76]. Electrical conductivity measurements on the PET/SWNT films showed that the melt-compounded SWNTs can result in electrical percolation albeit at concentrations exceeding 2 wt% [76]. Finally, multiwall carbon nanotubes (MWCNTs) PET/MWCNT nanocomposites that were prepared through melt compounding in a twin-screw extruder showed that the presence of MWCNT can act as nucleating agent and consequently enhances the crystallization of PET through heterogeneous nucleation [77]. The incorporation of a small quantity of MWCNTs (2 wt% at max) significantly improved the mechanical properties of the PET/MWCNT nanocomposites. A significant dependence of the rheological behavior of the PET/MWCNT nanocomposites as a function of the MWCNT content was also confirmed. The dynamic storage modulus and loss modulus of the PET/MWCNT nanocomposites increased with increasing frequency, and this increment effect was mostly pronounced at lower test frequencies.

8.4
Conclusions

A wide spectrum of applications is already verified for polyester fibers and composites ranging from vehicle tires and construction applications, to naval composites, and biomedical grafts and sutures. The field of biomaterials appears to be very friendly to this type of polymers. Microfibrillar *in situ*-reinforced polyester composites appear to have opened a window to a new category of composite materials with controllable properties. Nanotechnology applications such as PET/CNT conductive fibers and EMI shielding composites and films are also being investigated and there are still plenty of unexplored sides of polyester nanocomposite applications.

Moreover, recycled PET can be easily regained and further processed for useful purposes. In all, polyester materials are very practical, mostly cost-efficient, recyclable thermoplastics, and for the moment their applications are still increasing. It can be foreseen that in the near future also the expensive ones such as PEN shall become more economical or similar economic types shall evolve.

References

1 Froehlich, F.W. (1997) Restructuring, innovation see Akzo Nobel through difficult business environment. *International Fiber Journal*, **12** (3), 4–15.
2 http://www.fibersource.com/f-tutor/polyester.html (August 27, 2010).
3 Harrison, D. (1997) Synthetic fibers for nonwovens update. *Nonwovens Industry*, **28** (6), 32–39.
4 Harris, W.B. (1996) Is there a future for polyester investments outside Asia? *International Fiber Journal*, **11** (6), 4–9.
5 Koslowski, H.-J. (2007) Global market trends for synthetic fiber polymers. *Chemical Fibers International*, **57** (6), 287–293.
6 http://www.emergingtextiles.com (August 27, 2010).
7 Thiele, U.K. (2007) Polyester bottle resins production, in *Processing, Properties and Recycling*, PETplanet Publisher GmbH, Heidelberg, Germany, ISBN 978-3-9807497-4-9, p. 259 ff.
8 Ma, H., Zeng, J., Realff, M.L., Kumar, S., and Schiraldi, D.A. (2003) Processing, structure, and properties of fibers from polyester/carbon nanofiber composites. *Composites Science and Technology*, **63** (11), 1617–1628.
9 Kim, M.S., Kim, I.K., Byun, S.W., Jeong, S.H., Hong, Y.K., Jood, J.S., Song, K.T., Kim, J.K., Lee, C.J., and Lee, J.Y. (2002) PET fabric/polypyrrole composite with high electrical conductivity for EMI shielding. *Synthetic Metals*, **126** (2–3), 233–239.
10 Chaouch, W., Dieval, F., Chakfe, N., and Durand, B. (2009) Properties modification of PET vascular prostheses. *Journal of Physical Organic Chemistry*, **22** (5), 550–558.
11 Ambrosio, L., Netti, P.A., Iannace, S., Huang, S.J., and Nicolais, L. (1996) Composite hydrogels for intervertebral disc prostheses. *Journal of Materials Science: Materials in Medicine*, **7** (5), 251–254.
12 Silva, D.A., Betioli, A.M., Gleize, P.J.P., Roman, H.R., Gomez, L.A., and Ribeiro, J.L. (2005) Degradation of recycled PET fibers in Portland cement-based materials. *Cement and Concrete Research*, **35** (9), 1741–1746.
13 Wang, Y., Wu, H.C., and Li, V.C. (2000) Concrete reinforcement with recycled fibers. *Journal of Materials in Civil Engineering*, **12**, 314–319.
14 Fakirov, S. (2002) *Handbook of Thermoplastic Polyesters*, Wiley-VCH Verlag GmbH, Weinheim, ISBN 3-527-30113-5, p. 1223 ff.
15 Huisman, R. and Heuvel, H.M. (2003) The effect of spinning speed and drawing temperature on structure and properties of poly(ethylene terephthalate) yarns. *Journal of Applied Polymer Science*, **37** (3), 595–616.
16 Hegde, R.R., Dahiya, A., Kamath, M.G., and Kotra, R.,Gao (April 2004) http://web.utk.edu/~mse/Textiles/Polyester%20fiber.htm (27 August 2010).
17 Cook, J.G. (1968) *Handbook of Textile Fibers: Man-Made Fibres*, 4th edn, Woodhead Publishing Ltd., Cambridge, UK, pp. 358, 361.
18 Brown, N. and Ward, I.M. (1968) Deformation bands in oriented polyethylene terephthalate. *Philosophical Magazine*, **17** (149), 961–981.
19 Hearle, J.W.S. and Miraftab, M. (1995) The flex fatigue of polyamide and polyester fibers. Part II: The development of damage under standard conditions. *Journal of Materials Science*, **30** (4), 1661–1670.
20 House, J.R. and Grant, I.D. (1996) Viscoelastic composite materials for noise reduction and damage tolerance. *Advanced Performance Materials*, **3** (3–4), 295–307.
21 Davies, P., Huard, G., Grosjean, F., and Francois, M. (2000) Creep and relaxation of polyester mooring lines. Proceedings of the Offshore Technology Conference, Paper OTC 12176, pp. 1–12.
22 Chailleux, E. and Davies, P. (2005) A non-linear viscoelastic viscoplastic model for the behaviour of polyester fibres. *Mechanics of Time-Dependent Materials*, **9** (2–3), 147–160.
23 Diéval, F., Mathieu, D., and Durand, B. (2004) Comparison and characterization of polyester fiber and microfiber structure

24 Ajiki, I. and Postle, R. (2003) Viscoelastic properties of threads before and after sewing. *International Journal of Clothing Science and Technology*, **15** (1), 16–27.

25 Kong, H., Mouritz, A.P., and Paton, R. (2004) Tensile extension properties and deformation mechanisms of multiaxial non-crimp fabrics. *Composite Structures*, **66** (1–4), 249–259.

26 Demidov, A.V. and Makarov, A.G. (2007) System analysis of the viscoelasticity of polyester fibres. *Fibre Chemistry*, **39** (1), 83–86.

27 Jeddi, A.A.A., Nosraty, H. Taheri Otaghsara, M.R., and Karimi, M.A. (2007) Comparative study of the tensile fatigue behavior of cotton—polyester blended yarn by cyclic loading. *Journal of Elastomers and Plastics*, **39** (2), 165–179.

28 Vizesi, F., Jones, C., Lotz, N., Gianoutsos, M., and Walsh, W.R. (2008) Stress relaxation and creep: Viscoelastic properties of common suture materials used for flexor tendon repair. *The Journal of Hand Surgery*, **33** (2), 241–246.

29 Marzougui, S., Abdessalem, S.B., and Sakli, F. (2009) Viscoelastic behavior of textile artificial ligaments. *Journal of Applied Science*, **9** (15), 2794–2800.

30 Shin-Ichiro, I. and Masayoshi, I. (2008) Irreversible phase behavior of the ternary blends of poly(ethylene terephthalate)/poly(ethylene-2,6-naphthalate)/poly(ethylene terephthalate-*co*-ethylene-2,6-naphthalate). *Journal of Applied Polymer Science*, **110** (3), 1814–1821.

31 Schoukens, G. and Clerck, K.D. (2005) Thermal analysis and Raman spectroscopic studies of crystallization in poly(ethylene 2,6-naphthalate). *Polymer*, **46** (3), 845–857.

32 Karayannidis, G.P., Papachristos, N., Bikiaris, D.N., and Papageorgiou, G.Z. (2003) Synthesis, crystallization and tensile properties of poly(ethylene terephthalate-*co*-2,6-naphthalate)s with low naphthalate units content. *Polymer*, **44** (26), 7801–7808.

33 Cakmak, M., Wang, Y.D., and Simhambhatla, M. (1990) Processing characteristics, structure development, and properties of uni and biaxially stretched poly(ethylene 2,6 naphthalate) (PEN) films. *Polymer Engineering and Science*, **30** (12), 721–733.

34 Ülcer, Y. and Cakmak, M. (1994) Hierarchical structural gradients in injection moulded poly(ethylene naphthalene-2,6-dicarboxylate) parts. *Polymer*, **35** (26), 5651–5671.

35 Jager, J., Juijn, J.A., Van Den Heuvel, C.J.M., and Huijts, RA. (1995) Poly(ethylene-2,6 naphthalenedicarboxylate) fiber for industrial applications. *Journal of Applied Polymer Science*, **57** (12), 1429–1440.

36 Kim, S.Y., Kim, S.H., Lee, S.H., and Youn, J.R. (2009) Internal structure and physical properties of thermotropic liquid crystal polymer/poly(ethylene 2,6-naphthalate) composite fibers. *Composites: Part A*, **40** (5), 607–612.

37 Saleh, Y.S. and Jabarin, A. (2001) Glass transition and melting behavior of PET/PEN Blends. *Journal of Applied Polymer Science*, **81** (1), 11–22.

38 Higashioji, T. and Bhusdan, B. (2002) Creep and shrinkage behavior of improved ultra thin polymeric film. *Journal of Applied Polymer Science*, **84** (8), 1477–1498.

39 Nakayama, A., Takahashi, R., Hamano, T., Yoshioka, T., and Tsuji, M. (2008) Morphological study on electrospun nanofibers of aromatic polyesters. *Sen'i Gakkaishi*, **64** (1), 32–35.

40 Suzuki, A. and Yamada, Y. (2010) Poly(ethylene-2,6-naphthalate) nanofiber prepared by carbon dioxide laser supersonic drawing. *Journal of Applied Polymer Science*, **116** (4), 1913–1919.

41 Hendricks, N. (2008) Diolen high-tenacity polyester yarns provides solutions for many markets. *International Fiber Journal*, **23** (4), 44–45.

42 Sawada, S., Kamiyama, K. Ohcushi, S., and Yabuki, K. (1991) Degradation mechanisms of poly(ethylene terephthalate) tire yarn. *Journal of Applied Polymer Science*, **42** (4), 1041–1048.

43 Carduner, K.R., Paputa Peck, M.C., Carter, R.O. III, and Killgoar, P.C. Jr. (1989) An infrared spectroscopic study

of polyethylene terephthalate degradation in polyester fiber/nitrile rubber composites. *Polymer Degradation and Stability*, **26** (1), 1–10.
44 Ibarra Rueda, L., Chamorro Anton, C., and Tabernero Rodriguez, M.C. (1988) Mechanics of short fibers in filled styrene-butadiene rubber (SBR) composites. *Polymer Composites*, **9** (3), 198–203.
45 Kikuchi, N. (1996) Tires made of short fiber reinforced rubber. *Rubber World*, **214** (3), 31–32.
46 Ashida, M., Noguchi, T., and Mashimo, S. (1985) Effect of Matrix's type on the dynamic properties for short fiber-elastomer composite. *Journal of Applied Polymer Science*, **30** (3), 1011–1021.
47 Ibarra, L. (1993) The effect of a diazide as adhesion agent on materials consisting of an elastomeric matrix and short polyester fiber. *Journal of Applied Polymer Science*, **49** (9), 1595–1601.
48 Suhara, F., Kutty, S.K.N., and Nando, G.B. (1998) Mechanical properties of short polyester fiber-polyurethane elastomer composite with different interracial bonding agents. *Polymer-Plastics Technology and Engineering*, **37** (2), 241–252.
49 Suhara, F., Kutty, S.K.N., Nando, G.B., and Bhattacharya, A.K. (1998) Rheological properties of short polyester fiber-polyurethane elastomer composite with different interfacial bonding agents. *Polymer-Plastics Technology and Engineering*, **37** (1), 57–70.
50 Kondo, Y., Miyazaki, K., Takayanagi, K., and Sakurai, K. (2008) Surface treatment of PET fiber by EB-irradiation-induced graft polymerization and its effect on adhesion in natural rubber matrix. *European Polymer Journal*, **44** (5), 1567–1576.
51 Kondo, Y., Miyazaki, K., Takayanagi, K., and Sakurai, K. (2009) Mechanical properties of fiber-reinforced natural rubbers using surface-modified PET fibers under EB irradiation. *Journal of Applied Polymer Science*, **114** (5), 2584–2590.
52 Cataldo, F., Ursini, O., Lilla, E., and Angelini, G. (2009) A comparative study on the reinforcing effect of aramide and PET short fibers in a natural rubber-based composite. *Journal of Macromolecular Science, Part B*, **48** (6), 1241–1251.
53 Paul, D.R. and Bucknall, C.B. (2000) *Polymer Blends*, vol. 2, John Wiley and Sons, Inc., New York.
54 Cheremisinoff, N.P. (1999) *Handbook of Engineering Polymeric Materials*, Marcel Dekker, New York.
55 Robeson, L.M. (2007) *Liquid Crystalline Polymer Blends. Polymer Blends: A Comprehensive Review*, Hanser Verlag, Munich, pp. 163–164.
56 Evstatiev, M. and Fakirov, S. (1992) Microfibrillar reinforcement of polymer blends. *Polymer*, **33** (4), 877–880.
57 Fakirov, S., Evstatiev, M., and Petrovich, S. (1993) Microfibrillar reinforced composites from binary and ternary blends of polyesters and nylon 6. *Macromolecules*, **26** (19), 5219–5223.
58 Fakirov, S. and Evstatiev, M. (1994) Microfibrillar reinforced composites – New materials from polymer blends. *Advanced Materials*, **6** (5), 395–398.
59 Friedrich, K., Evstatiev, M., Fakirov, S., Evstatiev, O., Ishii, M., and Harrass, M. (2005) Microfibrillar reinforced composites from PET/PP blends: Processing, morphology and mechanical properties. *Composites Science and Technology*, **65** (1), 107–116.
60 Evstatiev, M., Fakirov, S., and Friedrich, K. (2005) Manufacturing and characterization of microfibrillar reinforced composites from polymer blends, in *Polymer Composites. From Nano- to Macro-Scale, Part II* (eds K. Friedrich, S. Fakirov, and Zhong Zhang), Springer, New York.
61 Li, Z.-M., Yang, M.-B., Feng, J.-M., Yang, W., and Huang, R. (2002) Morphology of *in situ* poly(ethylene terephthalate)/polyethylene microfiber reinforced composite formed via slit-die extrusion and hot-stretching. *Materials Research Bulletin*, **37** (13), 2185–2197.
62 Li, Z.-M., Xie, B.-H., Huang, R., Fang, X.-P., and Yang, M.-B. (2004) Influences of hot stretch ratio on essential work of fracture of *in-situ* microfibrillar poly(ethylene terephthalate)/polyethylene

blends. *Polymer Engineering & Science*, **44** (12), 2165–2173.

63 Mouzakis, D.E., Papke, N., Wu, J.S., and Karger-Kocsis, J. (2001) Fracture toughness assessment of poly(ethylene terephthalate) blends with glycidyl methacrylate modified polyolefin elastomer using essential work of fracture method. *Journal of Applied Polymer Science*, **79** (5), 842–852.

64 Fasce, L., Seltzer, R., Frontini, P., Rodriguez Pita, V.J., Pacheco, E.B.A.V., and Dias, M.L. (2005) Mechanical and fracture characterization of 50:50 HDPE/PET blends presenting different phase morphologies. *Polymer Engineering & Science*, **45** (3), 354–363.

65 Evstatiev, M., Angelov, I., and Friedrich, K. (2010) Structure and properties of microfibrillar-reinforced composites based on thermoplastic PET/LDPE blends after manufacturing by means of pultrusion. *Polymer Engineering & Science*, **50** (2), 402–410.

66 Li, W., Schlarb, A.K., and Evstatiev, M. (2009) Influence of processing window and weight ratio on the morphology of the extruded and drawn PET/PP blends. *Polymer Engineering & Science*, **49** (10), 1929–1936.

67 Jang, B.Z., Chen, L.C., Wang, C.Z., Lin, H.T., and Zee, R.H. (1989) Impact resistance and energy absorption mechanisms in hybrid composites. *Composites Science and Technology*, **34** (4), 305–335.

68 Yuan, Q., Kerth, S., Karger-Kocsis, J., and Friedrich, K. (1997) Crash and energy absorption behaviour of interleaved carbon-fibre reinforced epoxy tubes. *Journal of Materials Science Letters*, **16** (22), 1793–1796.

69 Pegoretti, A., Cristelli, I., and Migliaresi, C. (2008) Experimental optimization of the impact energy absorption of epoxy–carbon laminates through controlled delamination. *Composites Science and Technology*, **68** (13), 2653–2662.

70 Teh, S.F., Liu, T., Wang, L., and He, C. (2005) Fracture behaviour of poly(ethylene terephthalate) fiber toughened epoxy composites. *Composites: Part A*, **36** (8), 1167–1173.

71 Lopez-Manchado, M.A. and Arroyo, M. (2000) Thermal and dynamic mechanical properties of polypropylene and short organic fiber composites. *Polymer*, **41** (21), 7761–7767.

72 Saujanya, C. and Radhakrishnan, S. (2001) Structure development and properties of PET fibre filled PP composites. *Polymer*, **42** (10), 4537–4548.

73 Santos, P. and Pezzin, S.H. (2003) Mechanical properties of polypropylene reinforced with recycled-pet fibres. *Journal of Materials Processing Technology*, **143–144**, Proceedings of the International Conference on the Advanced Materials Processing Technology, 2001, pp. 517–520.

74 Fu, S., Wu, P., and Han, Z. (2002) Tensile strength and rupture energy of hybrid poly(methylvinylsiloxane) composites reinforced with short PET fibers and wollastonite whiskers. *Composites Science and Technology*, **62** (1), 3–8.

75 Li, Z., Luo, G., Wei, F., and Huang, Y. (2006) Microstructure of carbon nanotubes/PET conductive composites fibers and their properties. *Composites Science and Technology*, **66** (7–8), 1022–1029.

76 Anoop Anand, K., Agarwal, U.S., and Rani, J. (2007) Carbon nanotubes-reinforced PET nanocomposite by melt-compounding. *Journal of Applied Polymer Science*, **104** (5), 3090–3095.

77 Kim, J.Y., Park, H.S., and Kim, S.H. (2007) Multiwall-carbon-nanotube-reinforced poly(ethylene terephthalate) nanocomposites by melt compounding. *Journal of Applied Polymer Science*, **103** (3), 1450–1457.

9
Nylon Fiber-Reinforced Polymer Composites
Valerio Causin

9.1
Introduction

Composites may be generally defined as combinations of two or more materials deriving from the incorporation of some basic structural material into a second substance, that is, the matrix [1, 2]. On the basis of the concept of composite, there is the scope of exploiting the desirable properties of each component, obtaining a material with an increased performance with respect to the isolated constituents.

The incorporated material can appear in different morphologies, for example, particles, fibers, whiskers, or lamellae, but has always the same task, that is, to impart its own advantageous mechanical characteristics to the matrix material. The chemical nature and the morphology of the reinforcing agent are key additional variables that yield a competitive advantage to composites over homogeneous materials. A wise choice and optimization of these features, in fact, provides the opportunity to control physically uncorrelated parameters such as strength, density, electrical properties, and cost. Among the possible shapes of the reinforcement particles, fibers are particularly attractive for their high aspect ratio and very anisotropic nature [3]. This allows, if desired, to impart anisotropic physical properties to the whole composite. In this chapter, only fiber-reinforced composites will be discussed.

About the relative roles of the components in a composite material, three main instances may be encountered [1].

a) In the first one, the reinforcement has high strength and stiffness, whereas the matrix should transfer the stress from one reinforcement fiber to the other. This is the case of high performance composites, in which high strength reinforcement fibers are used in high volume fractions, calibrating their orientation, and dispersion for optimum physical mechanical property improvement.
b) In some other cases, the matrix already has desirable intrinsic physical, chemical, and processing properties, and addition of the reinforcement is intended to improve other important technological properties such as tensile strength, creep, or tear resistance. Preparation of this category of materials is driven by the necessity of improving the engineering properties of a matrix in order to

Polymer Composites: Volume 1, First Edition. Edited by Sabu Thomas, Kuruvilla Joseph, Sant Kumar Malhotra, Koichi Goda, and Meyyarappallil Sadasivan Sreekala.
© 2012 Wiley-VCH Verlag GmbH & Co. KGaA. Published 2012 by Wiley-VCH Verlag GmbH & Co. KGaA.

Table 9.1 Representative properties of different reinforcement fibers.

Material	Modulus (GPa)	Strength (GPa)	Specific gravity	Typical diameter (μm)
E-glass	70	3	2.6	10
S-glass	90	5	2.5	10
PAN-based carbon	200–500	2–3	1.7–2	7–10
Aramid	100–150	3	1.44	12–15
Polyolefin	1–5	2	0.97	5–500
Nylon	6	1	1.1	3–500
Rayon	7	1	1.5	3–500
Alumina	400–500	1–3	3.34	3–20
Silicon carbide	300–400	2–6	3.2	10–100

enlarge its range of applications. Moderate concentration of fibers, usually as discontinuous random fibers or flake are used for these composites.

c) In the third case, the matrix may be a high-performance material, albeit with a high cost, low processability, or insufficient aesthetical appeal. In this instance, addition of the fibers is aimed at maintaining adequate performance, while correcting the flaws of the matrix material, especially under a cost-effectiveness perspective.

The characteristics requested for a good reinforcing filler are that it is stiff, strong, and light, and it possibly modifies the failure mechanism in an advantageous way. Table 9.1 compares the strength, stiffness, and specific gravity of several different reinforcement fibers.

Nylon performs well among polymeric reinforcement fibers, and although its performance is inferior to that of inorganic and carbon fibers, but it shows extremely attractive cost-advantage balance.

9.2
Nylon Fibers Used as Reinforcements

The generic class of nylon comprises all linear polyamide polymers derived from aliphatic monomers. Linear, aliphatic polyamides can be subdivided into two groups: those synthesized from aminocarboxylic acids (either ω-aminocarboxylic acids or lactams, by ring-opening polymerization) and those made from the polycondensation reaction of diamines and dicarboxylic acids. The common nomenclature of nylon uses numbers for indicating the quantity of carbon atoms in the monomeric building blocks. For instance, nylon 6 is the polymer of caprolactone [$H_2N(CH_2)_5COOH$]. Nylon 6,10 is the polycondensate of hexamethylenediamine [$H_2N(CH_2)_6NH_2$] and sebacic acid [$HOOC(CH_2)_8COOH$].

The first nylons were developed in the 1930s and their commercialization started in 1939, taking particular advantage from the request of raw materials for military use during the Second World War. Since then, nylon fibers have provided a very versatile material for a large number of applications, because of their remarkable properties, primarily due to the extensive hydrogen bonding between the polymer chains. Nylon fibers are very attractive because of their high tenacity and elongation, abrasion resistance, and durability. The highly polar nature of the amide groups favors absorption of moisture by nylon fibers, which therefore suffer poor dimensional stability when exposed to water. Strict attention should be posed during processing of nylon to prevent it coming into contact with water. Purity is also an issue because reactive impurities could start uncontrolled chain growth, reducing the quality of the product, and causing yellowing. Nylon is prone to oxidation at high temperatures, so oxygen must be removed during polymerization and fiber spinning. Additives can be added to control the reactivity and sensitivity of nylon to certain conditions. Aging and light protection can be conferred by manganese (II) compounds or titanium dioxide. Heat aging resistance, for applications such as tire cords, is attained with antioxidants, such as substituted aromatic diamines.

A polyamide suitable for spinning contains mainly methylene groups between the amide groups. Cyclic segments in an aliphatic polyamide chain stiffen it and increase its melting point. Side carbon–carbon chains introduced in the polymer bring about a reduction of the melting temperature and an increase in the solubility in organic solvents. In low molecular mass nylon, melting point increases with increasing average molecular weight. This effect, though, fades and can be ignored in the case of high molecular weight polymers. An increase in molecular mass also brings about an increase in the strength of the fibers. A superior limit in the molecular mass that can be used is posed by the viscosity of the melt, which increases with increasing molecular weight, without allowing melt spinning.

Another factor that influences the physical, mechanical, and processing characteristics of nylon materials is the number of methylene groups in the chain. The melting point of nylon decreases as the ratio of methylene to carbonamide groups increases. An even–odd effect is observed in the trend of melting point as a function of methylene groups in the chain because the geometry of the even-numbered configuration allows a tighter packing of the chains, with a stronger hydrogen bonding, compared to odd-numbered configurations.

Nylons of the aminocarboxylic type suitable for fiber production are nylon 6, nylon 7, nylon 11, and nylon 12. Nylon 6 accounts for about half of the global production of nylon fibers, because its monomer, caprolactam, is easily obtainable by cheap petrochemicals. Nylon 6 is, though, not the ideal material for fiber production because of its quite high water absorption and the difficulty to convert the monomer into a continuously spinnable polymeric melt. Nylon 7 would be more suited, due to the high yield of its synthesis, easy polymerization–spinning process, very good tensile properties, and low water absorption. The high cost of monomer, though, does not allow an economical industrial production of nylon 7. Nylon 11 and nylon 12, although suitable for fiber production, are quite expensive and their use is restricted to a few peculiar applications such as ropes, transmission belts, or luggage.

Surely the most commonly employed nylons coming from the polycondensation of diamines and dicarboxylic acids are nylon 6,6 and nylon 6,10. Nylon 6,6 together with nylon 6 accounts for almost the total consumption of nylon for fiber manufacturing, due to its attractive cost and competitive performance. Due to its low water absorption and high elasticity, nylon 6,10 would be especially fit for usage in fiber preparation, but its higher cost with respect to nylon 6,6 prevented a wide commercial exploitation.

All nylon fibers of commercial interest are melt spun at temperatures up to about 300 °C, through spinnerets with holes 200–400 μm in diameter. The process can be carried out continuously from the monomer to the spun fiber, or it can be divided in different subsequent steps. Filaments are then combined into yarns. Obviously processing parameters such as draw ratio or spinning temperature are key in determining the structure attained by the polyamide molecules, and therefore, the physical and mechanical properties [4, 5]. Spinneret dimensions and draw ratio determine the final diameter of the fibers. About 1–20 dtex filaments are suitable for reinforcement purposes. High strength applications, such as tire cords, require quite coarse denier yarns, ranging from 400 to more than 35 000 dtex. The typical diameter of a nylon fiber used as reinforcement ranges from 3 to 500 μm and depends much on the application.

Also the length of fibers is important. Including short fibers in a polymeric matrix is easier and cheaper than to manufacture continuous fibers-containing composites. However, short fibers usually yield materials with lower strength, stiffness, and fracture toughness than continuous fibers.

The choice between long and short fibers can therefore only come from a trade-off between costs, advantages, and drawbacks on a case-by-case basis. Anyway, short fibers are more versatile materials than continuous fibers, and this is reflected by the fact that most of the scientific literature focuses on how to take the most advantage of short fibers. Long fibers, except some occasional works on tire cords, which are, however, already a mature and established product, are much less covered by recent scientific research.

A current trend in the research on fibers is the application of a nanotechnological approach for improving the physical and mechanical performance of the fibers. For example, nanosized fillers are added to the polymer matrix before the spinning process. An example of preparation of such fibrous composites is given by Mahfuz *et al.*, who enhanced the strength and stiffness of nylon 6 filaments by dispersing carbon nanotubes in a polyamide matrix by an extrusion process [6]. Carbon nanotubes are long and thin cylinders of covalently bonded carbon atoms with outstanding electronic, electrical, thermal, and mechanical properties which, since their discovery almost 20 years ago [7], have attracted large efforts of academic and industrial research. In this case, carbon nanotubes were included into nylon 6 by dry mixing and subsequent extruding in the form of continuous filaments by a single-screw extrusion method. Tensile tests on single filaments demonstrated that their Young's modulus and strength were increased by 220 and 164%, respectively, with the addition of only 1 wt% of carbon nanotubes. The authors also speculated, on the basis of SEM studies and micromechanics-based calculations, that the alignment of nanotubes in the filaments, and the high interfacial shear strength between the matrix and the nanotube reinforcement, were responsible for such a dramatic improvement in properties.

A further development of this idea was reported by Saeed et al. [8]. They prepared carbon nanotube/nylon 6 nanocomposites by *in situ* polymerization, obtaining a very good dispersion of the filler in the polyamide matrix. They subsequently electrospun these nanocomposites into nanofibers in which the carbon nanotubes were embedded and oriented along the nanofiber axis. Electrospinning is a spinning method capable of producing fibers of submicrometric diameter, and it will be further detailed in Section 9.3 of this chapter. Similarly to Mahfuz et al.'s work [6], the specific strength and modulus of the nanotube-reinforced nanofibers increased as compared to those of the pristine nylon 6 nanofibers. An interesting feature that was noted by the authors of this work was that the crystal structure of the nylon 6 in the composite fibers was mostly γ-phase, whereas that of the composite films prepared through compression molding of the mixture of the components was mostly α-phase. The difference in the polymorphism of such materials is due to the shear force exerted in the electrospinning process. The reinforcing effect of a filler is not only due to its presence but also to the modifications that it brings about in the structure and morphology of the matrix. Each time that a composite is prepared, it should be necessary to study the structure and morphology of the matrix in detail and on different length scales (especially that of polymer lamellae), in order to be able to identify structure–property relationships useful for a rational design of the materials.

Another example of preparation of nanocomposite nylon fibers is offered by the work of Francis et al. [9]. This paper offers an example of the possibilities offered by nanotechnology for creating functional materials. Silver nanoparticles (diameter 3 nm) were synthesized using silver nitrate as the starting precursor, ethylene glycol as solvent, and poly(*N*-vinylpyrrolidone) as a capping agent. These nano-Ag particles were then introduced in a nylon matrix by electrospinning of the nylon 6/Ag solution. Silver nanoparticles not only tripled the strength of the nylon fibers but they are also known to have an antibacterial activity, thereby conferring to the fibrous material novel properties which pristine nylon does not show.

Fibers can be "upgraded" also by coating them with a material that adds new properties to the whole fibrous material. If accurately designed, coating can moreover influence the surface properties of the fibers, which are crucial for a good reinforcement, because an optimal interfacial adhesion between the matrix and the filler is necessary for a proper transfer of stress from the former to the latter. Interfacial bonding must be strong enough for an efficient transfer of the applied load, but not excessive, since it could also promote crack propagation across the fibers and reduce the toughness of the composites [10]. On the opposite, a poor interfacial adhesion is not desirable because the weak interface would serve as preferential direction of crack propagation. In this case, a deformable interface that allows for fiber debonding and pull out can positively influence the toughness, although at the expense of the strength of the composite [10].

An example of fiber coating has been reported by Zhou et al. [11]. These authors coated nylon 66 fibers with a layer of carbon by physical vapor deposition in a vacuum chamber. By inclusion of these fibers in a concrete matrix, it was possible to obtain a smart material, which was not only mechanically reinforced by the fibrous material but that was also able to sense elastic and inelastic deformation, and fracture. The carbon coating, in fact, provided a conduction path for electrical signals, which was altered as a consequence of cracks or deformation. Changes in the resistance of the material could therefore be exploited for the early detection of failures or damage.

Conductivity was conferred to a nylon 6,6 fabric by covering it with a silver nanocrystalline thin film (from 220 to 2800 nm) by the radio-frequency sputtering technique, as recently reported by Wang et al. [12].

A further example that highlights the importance of surface interactions between polymer and filler is offered by Rangari et al. [13]. They were able to align silicon nitride nanorods in a nylon 6 matrix by a melt-extrusion process. Also in this case, they recorded better tensile properties, that is, increases in the strength and modulus by 273 and 610%, respectively, although at the expenses of elongation at break. The effect of the shape of the inorganic filler was evident, because substituting the silicon nitride nanorods with spherical nanoparticles it was possible to increase strength and modulus, with no decrease in elongation at break with respect to the matrix. This paper is particularly interesting because it shows how a positive interaction between polymer and filler is crucial for the functioning of the reinforcement mechanism. Figure 9.1 shows that nylon 6 polymer chains are entangled on the rod-shaped filler particle. This effect is similar to the formation of a transcrystalline layer, at the basis of the reinforcement of nylon by aramid fibers [14, 15]. When the matrix is able to crystallize, or otherwise cover with a layer, the filler particles, the reinforcement effect of such particles is exploited to its fullest extent [16, 17].

A quite original approach for controlling the mechanism of deformation of the material by adding a fibrous reinforcement was proposed by Lu et al. [10]. These researchers fabricated chain-shaped nylon 6 fibers (Figure 9.2) by compression molding, and they subsequently included them into a polypropylene/poly(propylene-

Figure 9.1 Rod-shaped Si_3N_4 particle entangled with a nylon 6 matrix. Reprinted from Ref. [13] with permission from Elsevier, Copyright 2009.

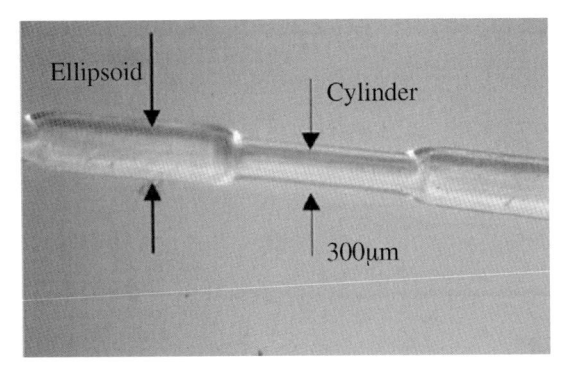

Figure 9.2 Optical micrograph of a chain-shaped fiber. Reprinted from Ref. [10] with permission from Elsevier, Copyright 2006.

co-octene) blend. The composites containing such chain-shaped short fibers displayed both higher strength and toughness than the materials prepared with straight short fibers of the corresponding chemical nature.

Single fiber pull-out tests for different embedded depths were carried out, and they showed a very relevant influence of the chain-shape geometry on the properties of the composites. The pulling-out load versus displacement curves of the chain-shaped-fiber-containing composites showed multiple peaks, which were not present in the material containing normal straight short fibers (Figure 9.3).

The number of peaks in the curve corresponded to the number of enlarged segments of the fiber, which were embedded in the matrix. When tension was applied, these enlarged sections were anchored in the matrix, and they required a plastic deformation of the matrix to be pulled out. The chain-shaped fibers therefore allowed an effective load transfer from matrix to fiber, regardless of the weakness of the interface between them.

9.3
Matrices and Applications

As can be seen in Table 9.1, the physical and mechanical properties of nylon are remarkable but not exceptional, compared, for example, to inorganic, carbon, or aramid fibers. Nylon is therefore most suited for the reinforcement of thermoplastic polymers and of elastomers.

Recent literature reports showed the potential in property improvement of nylon fibers especially in polymethylmethacrylate (PMMA) [18, 19], polyethylene terephthalate (PET) [20], polycarbonate (PC) [21], and rubber [22–25].

Nylon fibers are particularly attractive for the rubber industry. Reinforcement can be attained by thick and long fibers, such as tire cords [26–29], and also the use of short fibers has been extensively investigated. Reinforcement of rubber with short fibers combines the elasticity of rubber with the strength and stiffness of the fiber.

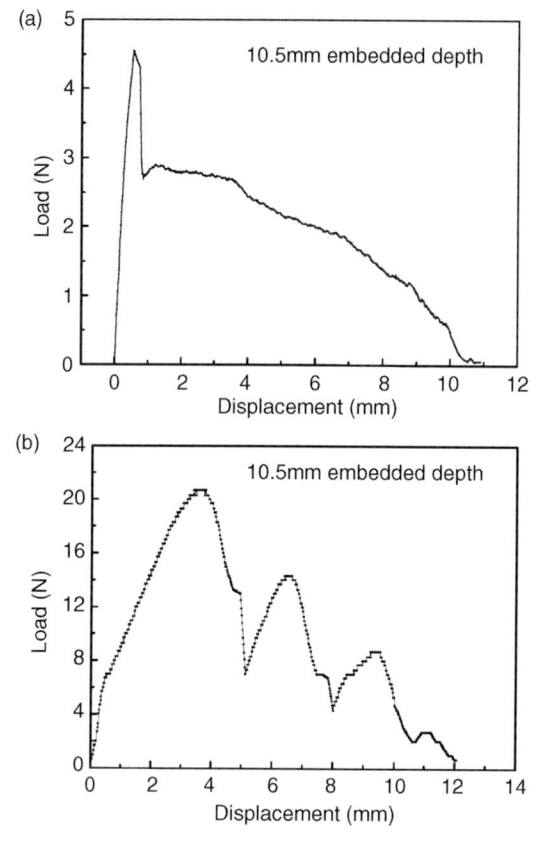

Figure 9.3 Pull-out load versus displacement curves: (a) straight fibers and (b) chain-shaped fibers. Reprinted from Ref. [10] with permission from Elsevier, Copyright 2006.

The additional benefit is that the fibers are incorporated into the compound as one of the ingredients of the recipe, and hence, they are amenable to the standard rubber-processing steps of extrusion, calendering, and various types of molding. Short fibers are also used to improve or modify certain thermodynamic properties of the matrix for specific applications or to reduce the cost of fabricated articles. The properties and performance of short fiber-reinforced rubber composites depend on several factors such as nature and concentration of the fiber, its aspect ratio, orientation, and the degree of adhesion of the fiber to the rubber matrix.

Among the many possible examples of investigations on the effect of short fibers on a rubber matrix, the work of Rajesh et al. can be cited [23]. These authors prepared acrylonitrile butadiene rubber (NBR)-based composites by incorporating short nylon fibers of different lengths (2, 6, or 10 mm) and concentration into the matrix using a two-roll mixing mill according to a base formulation. They studied the curing characteristics of the samples, and observed the influence of fiber length, loading, and rubber cross-linking systems on the properties of the composites. Addition of nylon fibers to NBR offered good reinforcement, causing improvement in mechanical properties. A fiber length of 6 mm was found to be optimal for the best balance of properties. At this critical fiber length, the load transmittance from the matrix to the fiber was maximum. If the critical fiber length was greater than the length of the fiber, the stressed fiber would debond from the matrix and the composite would fail at low load. If the critical fiber length was less than the length of the fiber, the stressed composites would lead to breaking of the fiber. Moreover, the elongation at break reduced with increase in fiber length because long fibers decrease the possibility and probability for rearrangement and deformation of the matrix under an applied load.

Since the presence of fibers causes an increase in viscosity and torque of the mixture, their quantity must be accurately chosen because, if excessive, it can lead to brittleness. The authors observed that, with an increase in fiber loading, the stiffness and brittleness of the composite increased gradually, with an associated decrease in elongation at break.

Interestingly, the presence of fibers in a rubber matrix not only modifies the mechanical properties but it also influences the response of the material to curing [23, 25]. The design of a process involving fiber-reinforced rubber should thus focus also on the optimization of the curing step.

In their study on short nylon fiber-reinforced NBR composites, Seema and Kutty explored the use of an epoxy-based bonding agent intended to improve the interfacial adhesion between fiber and matrix [25]. As said before, the fiber–matrix interfacial bond has a decisive effect on the mechanical properties of composites, so many attempts have been reported in the literature to improve it, mainly by addition of compatibilizing agents, either of polymeric, oligomeric, or of low-molecular-weight nature. The introduction of Seema and Kutty's work [25] contains several examples of such investigations, to which the reader is referred.

Other matrices that lend themselves to the use of nylon fibers as reinforcement are denture resins [30–33]. Conventionally, dental resins are reinforced with large quantities of inorganic fillers, as high as 75%, mostly consisting of ceramic (such as silica/glass) particles. Compared to dental amalgams, though, the strength and durability of resins is quite low (the strength of the dental composites reinforced by inorganic fillers is usually in the range from 80 to 120 MPa, and the average lifetime is 5 years or less, whereas dental amalgams have strength over 400 MPa and have a lifetime of more than 15 years [33]). Ironically, the inorganic fillers that are added for the purpose of strengthening the dental composites are actually responsible, at least in part, for their failures. Since the inorganic filler particles are much harder than the dental resin matrices, the stresses are transmitted through the filler to the resin. Sharp corners or edges of the filler particles may provoke stress concentration and thus generate cracks in the resin. So, interest has been posed in the use of fibers as reinforcement. The incorporation of various fibers, such as nylon, carbon, aramid, ultrahigh molecular weight polyethylene and glass fibers, into dental resins has provided substantial improvements on impact and flexural strength, and fatigue resistance. Several studies verified that inclusion of nylon fibers can substantially improve the mechanical properties, most notably the flexural properties and the work of fracture, while retaining good aesthetical characteristics, such as a brilliant white color [30, 32].

In this field, the work of Fong and coworkers [31, 33] is particularly innovative, since it makes use of electrospun nanofibers for the reinforcement. As said before, the diameter of fibrous fillers can exceed the tens of micrometers, therefore producing microcomposites, when dispersed in a polymeric matrix. The recent trend in the research and development of composites is that of going toward the preparation of nanocomposites. In such materials, the size of the dispersed filler is decreased to the nanometer length scale, thereby attaining a large increase in interfacial area, and thus largely changing the macroscopic properties of the material. As a consequence, it is of interest to be able to decrease the diameter of fibrous reinforcements, in order to further improve their effectiveness. One method of obtaining fibers of small size is by using electrospinning. The apparatus used for electrospinning basically consists of a high-voltage electric source, a syringe pump connected to a capillary spinnerette, and a conductive collector that in most of the cases is simply aluminum foil. The polymer, either in solution or in the melt, is carried by a syringe pump to the tip of the capillary, where it forms a pendant drop. The application of high-voltage potential (some tens of kilovolts) induces the formation of ions into the polymer solution, which move in the electric field toward the electrode of opposite polarity, thereby transferring tensile forces to the polymer liquid. A further effect of the application of the electrical field is that at the tip of the capillary, the pendant hemispherical polymer drop is deformed into a conelike projection and, after a threshold potential has been reached, necessary to overcome the surface tension of the liquid, a jet is ejected from the cone tip [34]. Exploitation of electrostatic forces for spinning fibers is not a new approach, since it has been known for more than 100 years. The recent interest in nanotechnology has brought to a rediscovery of the electrospinning technique, due to its capability of yielding fibers in the submicron range [34, 35] (Figure 9.4).

Mainly because of this reason, electrospun nanofibers attracted a huge interest in the research community, as testified by the ever-increasing scientific literature on the subject. Most of these reports are focused on the use of nanofibers in fields such as nanocatalysis, tissue scaffolds, protective clothing, filtration, and optical electronics. Surprisingly, very few reports exist on the use of electrospun nanofibers in the preparation of composites. Electrospinning is, though, a very attractive and versatile method for fabricating fillers for composite materials. The strength of this approach is that the filler properties can be tailored acting on its physical appearance (diameter, pore density of the mat, etc.) and chemical nature (type of polymer chosen for spinning). Recently, a fabrication approach was presented that allows to prepare composites based on polycaprolactone (PCL) filled with nylon 6 nanofibers by compression molding [36]. At very low filler contents (3%), the obtained composites exhibited remarkably improved stiffness with a simultaneous increase in ductility, differently from what is usually found in PCL nanocomposites with a variety of fillers, in which increases in modulus happen at the expense of elongation at break.

It is especially notable that a marked effect of the diameter of the fiber on the extent of interfacial adhesion was observed. Figure 9.5a, shows that the fibers with a larger

Figure 9.4 FESEM micrographs of (a) the cross-section of a nylon 6 electrospun fiber mat and (b) of a detail of the electrospun fibers Reprinted from Ref. [36] with permission from Elsevier, Copyright 2010.

diameter displayed a much poorer interfacial adhesion. The fibers oriented perpendicularly to the fracture surface did not undergo pulling out, whereas those oriented tangentially to the fracture surface were pulled out and hanged loose, especially when bundles were formed. Along with these large fibers, a number of very fine filaments were present. These appear firmly embedded in the matrix (Figure 9.5b and c), the exposed ends of the fibers protrude directly from within the matrix, without the presence of craters at their base due to debonding. The real advantage of using electrospinning is therefore the possibility to produce fibers with a very fine diameter, which are able to very efficiently interact with the matrix in which they are dispersed.

Electrospinning could also help to solve some other practical problems related to composite preparation. Enhancing the toughness of brittle composite laminates by introducing a thin thermoplastic layer is an attractive approach to mitigate the

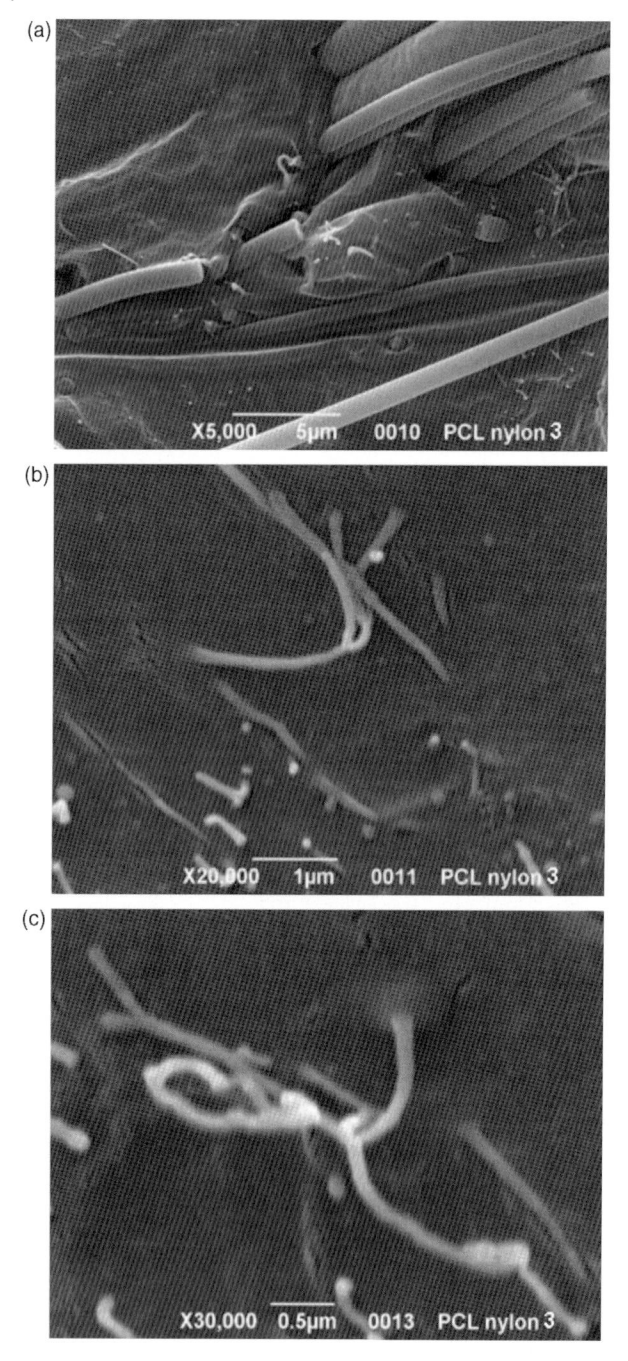

Figure 9.5 SEM micrographs at different magnification of the cryogenic fracture surface of a sample of PCL containing nylon 6 electrospun fibers Reprinted from Ref. [36] with permission from Elsevier, Copyright 2010.

delamination problem. Primary drawbacks of interleaved composites are increased laminate thickness (about 20%), decreased in-plane stiffness and strength (15–20%), and potential lowering of glass transition temperature [37].

Akangah et al. [37] prepared composite laminates of plain and interleaved epoxy resin, interleaved by electrospun nylon 6,6 nanofabric. Polymer nanofabric interleaving increased the threshold impact force by about 60%, reduced the rate of impact damage growth rate to one-half with impact height and reduced impact damage growth rate from 0.115 to 0.105 mm^2/N with impact force. More interestingly, polymer nanofabric interleaving marginally increased the laminate thickness, by about 2.0%.

Although not a polymeric matrix, it is worth noting that quite a vast amount of research has been carried out on the use of nylon fibers for the reinforcement of asphalt or concrete [38–42]. The use of fibers to reinforce architectural materials can be traced back to a 4000-year-old arch in China constructed with a clay earth mixed with fibers or the Great Wall built 2000 years ago [38]. Fibers known as monofilaments and fibrillated (from 13 to 38 mm long) are typically added into the concrete from 0.6 to 0.9 kg/m^3 [41], although higher quantities can be used for special applications. Fiber characteristics as content or length must be accurately chosen as a function of the intended application. For example, nylon monofilaments 19 mm long are more effective for the shrinkage crack reduction in lean mortars and concrete than in rich-cement mortar [41].

Macrofibers are distinguished by their typical long length (with 38 mm considered a minimum) and wide fibril cross-sections. They are added at higher dosage rates to the concrete, typically from 3 to 9 kg/m^3 [41], to limit the damaging effects of a seismic event or concussive energy. Moreover, the fibers are not affected by the alkaline environment and they demonstrate their long-term durability in the concrete. Properties such as compressive strength and dynamic elasticity modulus are mostly benefited by this procedure. Martínez-Barrera et al. reported the use of gamma irradiation to modify the surface of the fibers and thus to improve the interfacial adhesion between matrix and filler [39–41]. This approach is not limited to applications in the construction industry, but it is applicable to a number of other possible applications, since gamma radiation is a versatile tool to calibrate and modify the chemical structure and the morphology of the surface of the fiber, also by controlling the recrystallization process of the polymer. The ionizing energy can moreover promote cross-linking of the polymer, improving both the tensile stress and the tensile strain of the fibers.

9.4
Manufacturing of Nylon-Reinforced Composites

Nylon lends itself to most of the common procedures for fiber-reinforced composites. Regardless of the specific technology, the process of manufacture of such composites requires a first step in which the matrix, while in a fluid state, is forced to surround the fibers, and a subsequent step in which the matrix is solidified under controlled conditions, in order to maintain the dispersion of the fibers attained in the preceding phase.

In the case of thermoset matrices, the easiest approach consists in percolating a fluid prepolymer within a fiber mat or textile, followed by curing by application of heat to confer to the composite the desired shape and structure. "Prepreg" sheets are very commonly used as starting material. They are usually unidirectional tapes or fabrics made of fibers bonded by a thin film of partially cured resin. When dealing with thermoplastic matrices, the same concept is involved, except that the fluid form of the matrix is attained by melting. Processing temperature and pressure are in these instances crucial factors, since they must be high enough to afford a reasonable viscosity to the liquid, without triggering degradative side-reactions.

The basic ideas enumerated above have evolved into more sophisticated techniques where a dry fiber preform is placed into a closed mold after which the precursor is infused by applying vacuum or pressure, typically less than 10 bar. Following polymerization or cross-linking, the composite product is demolded. A further technique, important for its technological implications, is pultrusion, in which continuous filaments are impregnated drawing them through a matrix bath and subsequently through a die. The die serves to orient the reinforcement, to impart a shape to the laminate and to calibrate the quantity of matrix surrounding the fiber. In the case of a thermoset matrix, the cure can be completed in the die, or it can be carried out at a second time.

The choice of the most adequate processing method for composite preparation is very important for achieving the desired performances. Pang and Fancey, for example, recently reported a technique of sample preparation in which tension was applied to the reinforcing fibers, that was released just before inclusion of the fibers into the matrix [5]. During the solidification of the matrix, the compressive stresses imparted by the viscoelastically strained fibers improve the physical and mechanical properties. Preparing epoxy/nylon and polyester/nylon composites according to this procedure, the flexural modulus was increased by 50% with respect to control samples manufactured without prestressing the fibers [5]. This example shows the intrinsic potential of processing for obtaining, all the formulation conditions being equal, significant improvements in the final properties of the composite.

As previously said, if on one hand thermoplastic composites are advantageous over thermoset ones because of higher toughness, of faster manufacturing and of their recyclable nature, on the other hand, traditional melt processing limits thermoplastic composite parts in size and thickness. van Rijswijk and Bersee [43] recently showed the benefits of an alternative approach: reactive processing of textile fiber-reinforced thermoplastics. Somewhat in analogy with the thermoset composites, a low viscosity mono- or oligomeric precursor is used to impregnate the fibers, followed by *in situ* polymerization. Figure 9.6 shows a micrograph of a PMMA sample reinforced by nylon fibers obtained by *in situ* polymerization assisted by supercritical CO_2 [18]. This technique allows for the creation of kinetically trapped blends of immiscible polymers that are homogeneous down to the nanometer length scale so, as can be seen in Figure 9.6, the matrix completely fills the space between the reinforcing fibers.

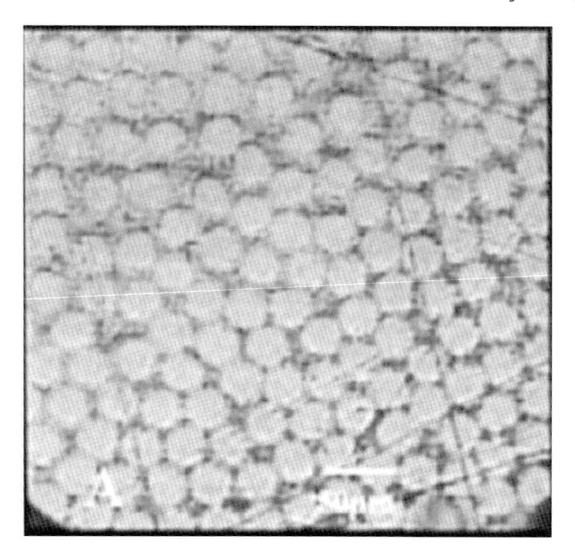

Figure 9.6 Optical micrograph of a PMMA/nylon composite prepared by *in situ* polymerization in supercritical CO_2 Reprinted from Ref. [18] with permission from John Wiley & Sons, Inc.

The fluidity of the matrix material in the impregnation step allows taking full advantage of the most efficient manufacturing processes that make use of injection molding or vacuum infusion of the mold. Polymerization can be initiated by heat or UV radiation and might require the addition of a catalyst system, which can be added to the precursor prior to impregnation. Due to their low molecular weight, precursors have low melt viscosities and proper fiber impregnation is therefore achieved without the need for high processing pressures [43]. Low-pressure infusion processes, therefore, become more accessible to thermoplastic matrices as well, allowing to obtain larger, thicker, and more integrated products than those currently achievable with melt processing. The flexibility of *in situ* polymerization allows moreover to create chemical bonds between the matrix and the filler, and to include further reinforcement concurrently with the fiber, such as, for example, nanoparticles.

When dealing with semicrystalline polymers, caution must be posed in choosing the processing temperature, in order to balance the rates of polymerization (favored by increasing temperature) and crystallization (slowed down at high temperature). When the temperature is too low, crystallization will be too fast and reactive chain ends and monomers can get trapped inside crystals before they can polymerize. On the other hand, when the temperature is too high, the final degree of crystallinity is reduced, which reduces the strength, stiffness, and chemical resistance of the polymer [44].

Caskey *et al.* showed the positive influence of supercritical CO_2 in composite synthesis, because it leads to complex morphologies, exhibiting long-range order, and orientation on multiple length scales from the nanometer to the centimeter scale [45]. It is beyond the scope of this chapter to review in detail such structural and morphological studies, but the reader will find a complete report of such aspects in

Caskey's papers [18, 45]. Suffice here to say that this morphology leads to improved flexural modulus and increased ultimate strength with only a small decrease in tensile modulus. These composites also exhibited significant improvements in stress distribution and load transfer without the use of fiber/matrix compatibilizers. The only drawback associated to the use of supercritical CO_2 is a drop in glass transition temperature and a reduction in the stiffness of the materials, in case some residual CO_2 remained trapped in the material. It is thus paramount that all CO_2 be removed from the samples after processing so as to regain the original unplasticized material properties.

Although it is clear, by the examples cited, that reactive processing is attractive and very promising, it has not found significant industrial applications yet, differently from the case of thermoset composites, in which reactive processing forms the mainstay of the composite industry worldwide.

Another alternative approach for the preparation of nylon fiber-reinforced composites was described by Hine and Ward [46], who reported a patented procedure, developed at the University of Leeds. In this process, an assembly of oriented elements, often in the form of a woven cloth, is held under pressure and taken to a critical temperature so that a small fraction of the surface of each oriented element is melted, which on cooling recrystallizes to form the matrix of a single-polymer composite. Same polymer composites, with a high volume fraction of reinforcement fibers can be thus prepared by this approach. The successful application of such hot compaction method to nylon 6,6 was described in Ref. [46]: the composites prepared in this way showed remarkable improvements in tensile properties.

A procedure that combines surface functionalization of the fibers, electrospinning, and hot compaction was proposed by Chen et al. [47]. They prepared core–shell composite nanofibers with nylon 6 as core and PMMA as shell by a coaxial electrospinning method. The fibers were afterwards transformed into a nanofiber-reinforced transparent composite through a hot press treatment at a temperature capable of melting the PMMA outer layer, which composed the matrix of the final material, but not high enough to melt the nylon 6 cores. The morphology was demonstrated by taking SEM pictures of the samples fractured in liquid nitrogen (Figure 9.7).

The fibers, firmly embedded in the PMMA matrix, can be clearly seen. Some nanofibers had diameters of about 1000 nm, though most had diameters within 200–500 nm. The embedded nanofibers absorbed energy when the composite was exposed to external forces, however, larger diameter fibers in the composite increased light scattering, resulting in a 10% loss of transmittance with respect to the neat PMMA matrix. The transparency of the composite would have been maintained if the fiber sizes were controlled to be significantly smaller than the wavelengths of visible light, which is feasible by a careful control of electrospinning conditions. The potential for preparing transparent nanocomposites based on PMMA by electrospinning is particularly attractive, since the other methods for reinforcing this polymer, such as the preparation of polymer alloys, the addition of nanoparticles, or of rubber-particles, or the traditional fiber reinforcement usually yield less transparent materials due to refractive index mismatches.

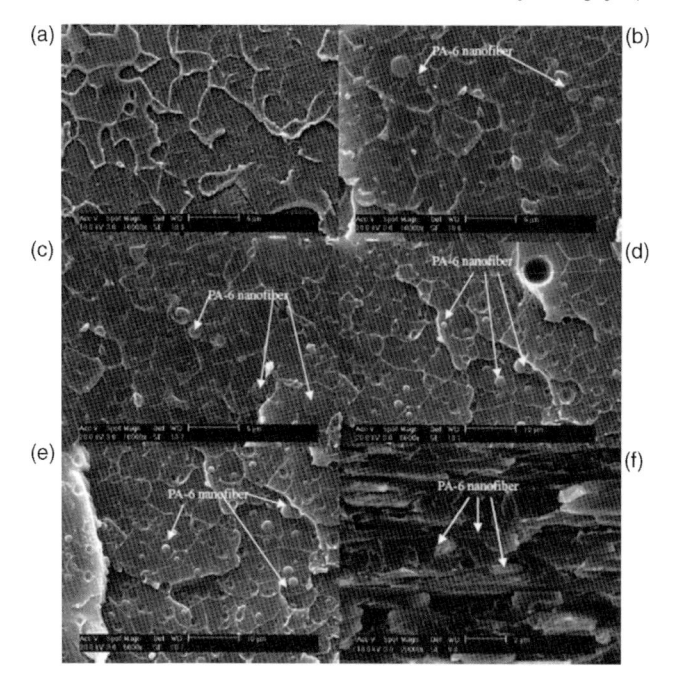

Figure 9.7 Fracture morphology of nylon 6 fiber-reinforced PMMA composites: (a) pure PMMA, (b) PMMA/0.5% nylon 6 composite, (c) PMMA/1.5% nylon 6 composite, (d) PMMA/2.5% nylon 6 composite, (e) PMMA/3.5% nylon 6 composite, and (f) PMMA/2.5% nylon 6 composite (tensile fracture surface) Reprinted from Ref. [47] with permission from John Wiley & Sons, Inc.

A very intriguing alternative way to produce fibrous reinforcements within a polymer matrix is by an *in situ* fibrillation technique [20, 21, 48–50]. The following three basic steps are required in order to obtain microfibrils within a polymer mixture [49]: (i) melt blending with extrusion of two immiscible polymers having different melting temperatures (mixing step); (ii) cold drawing of the extrudate with good orientation of the two components (fibrillation step); (iii) thermal treatment at a temperature between the melting points of the two blend partners (isotropization step) taking place during processing of the drawn blend via injection or compression molding.

The occurrence of fibrillation depends on parameters, such as interfacial tension, processing parameters, and the dispersed phase content [51, 52]. Although exceptions to the rule exist [52], it is usually accepted that fibers may be produced when the matrix is more viscous than the dispersed phase. Moreover, a factor that improves the stability of these morphologies in the molten state is the low interfacial tension between the components of the system.

Goitisolo *et al.* [20, 21, 50] recently reported the preparation of nylon nanofibrils within PET or PC matrices. In the case of PET, for example, they mixed a PET matrix with up to 40% nylon 6, reinforced with up to 7% of a fully dispersed organoclay.

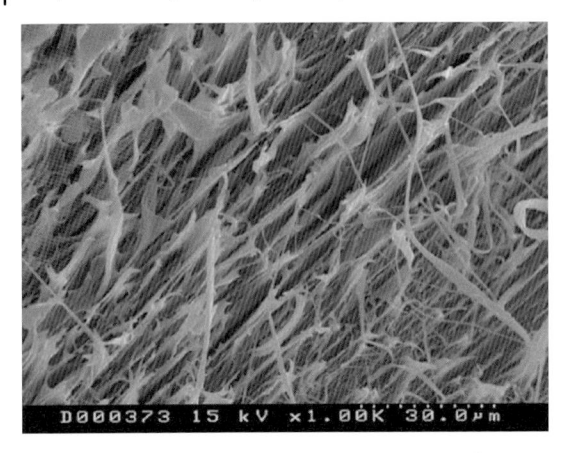

Figure 9.8 SEM micrograph of nylon fibrils obtained *in situ* in a PET matrix Reprinted from Ref. [20] with permission from Elsevier, Copyright 2010.

Figure 9.8 shows a micrograph of the so obtained nylon fibrils dispersed in the PET matrix.

Moreover, the authors were able to exploit the ability of nylon to exfoliate clay, and thus not only reinforce the material by the *in situ* formation of nanofibrils but also by preparing at the same time clay-based nanocomposites. The increase obtained in the elasticity modulus (modulus increases by 38% with respect to the PET matrix, with only 1.96% MMT in the whole nanocomposite) was higher than any previously observed PET-based composites containing nanoclay, due to the synergistic effect of the nanofibrils and of the very good dispersion of the nanoclay.

Very good results, also in this case due to the additive effect of nylon fibrils and of exfoliated nanoclay, were obtained in the case of the PC matrix, where very long fibrils, up to some millimeters long were observed [21]. It is noteworthy that in these samples the increases in modulus and yield stress were coupled to a retained ductility, since the elongation at break of the composites was similar to that of the neat PC matrix. This is unusually found in nanocomposites where, on the contrary, it is quite easy to increase the stiffness, but at the expense of ductility and toughness.

An approach like the one just described is useful when, in order to obtain the desired properties of the composite, too high fiber loadings are required. Very large fiber contents bring about brittleness and difficulties in processing and molding. Therefore, in these occasions, it is very attractive to substitute some of the fibrous filler with a lower loading of nanofiller. John *et al.* showed that, by adding just 1% nanosilica to a polypropylene/nylon-fiber composite, modulus and strength, both tensile and flexural, were significantly enhanced, obtaining a performance comparable to that of fiber-filled composites containing much higher fiber loadings [32].

In addition, Fakirov *et al.* [49] demonstrated, although for the case of PET microfibrils dispersed in a nylon matrix, that the reinforcing nanofibrils can be isolated as a separate material by means of a solvent extraction and used for a variety of purposes, such as, for example, gas and liquid nanofilters as nonwoven textiles, biomedical applications, as scaffold materials, or as carriers for controlled drug delivery.

9.5
Conclusions

Although with an inferior mechanical performance with respect to glass, carbon or aramid fibers, nylon fibers, due to their attractive cost and easy manufacture, pose themselves as a very convenient reinforcement for polymeric matrices. In this chapter, focus was posed on the most recent advances and trends in the quest for more performing fibers. Experimental approaches have been shown in fiber production, functionalization, and modification, and in composite manufacturing. Many of the examples shown have not lead to immediate commercialization, but the ideas proposed may spur further research and development, so that in the future, the field of application of nylon fibers can be enlarged and innovative materials can be brought into the market.

References

1 Kroschwitz, J.I. and Seidel, A. (eds) (2007) *Kirk-Othmer Encyclopedia of Chemical Technology*, Wiley–Interscience, Hoboken, NJ.
2 Bailey, J.E., Brinder, J., and Bohnet, M. (1998) *Ullmann's Encyclopedia of Industrial Chemistry*, 6th edn, Wiley-VCH, Weinheim.
3 Robinson, I.M. and Robinson, J.M. (1994) The influence of fibre aspect ratio on the deformation of discontinuous fibre-reinforced composites. *Journal of Materials Science*, **29**, 4663–4677.
4 Kim, S.J. and Shim, S.B. (2009) The effect of the processing factors on the physical properties of high shrinkable nylon composite yarns. *Fiber Polymers*, **10**, 813–821.
5 Pang, J.W.C. and Fancey, K.S. (2009) The flexural stiffness characteristics of viscoelastically prestressed polymeric matrix composites. *Composites A*, **40**, 784–790.
6 Mahfuz, H., Adnan, A., Rangari, V.K., Hasan, M.M., Jeelani, S., Wright, W.J. et al. (2006) Enhancement of strength and stiffness of Nylon 6 filaments through carbon nanotubes reinforcement. *Applied Physics Letters*, **88**, 083119.
7 Iijima, S. (1991) Helical microtubules of graphitic carbon. *Nature*, **354**, 56–58.
8 Saeed, K., Park, S.Y., Haider, S., and Baek, J.B. (2009) *In situ* polymerization of multi-walled carbon nanotube/Nylon-6 nanocomposites and their electrospun nanofibers. *Nanoscale Research Letters*, **4**, 39–46.
9 Francis, L., Giunco, F., Balakrishnan, A., and Marsano, E. (2010) Synthesis, characterization and mechanical properties of nylon–silver composite nanofibers prepared by electrospinning. *Current Applied Physics*, **10**, 1005–1008.
10 Lu, X., Zhang, Y., and Xu, J. (2006) Influence of fiber morphology in pull-out process of chain-shaped fiber reinforced polymer composites. *Scripta Materialia*, **54**, 1617–1621.
11 Zhou, Z., Xiao, Z., Pan, W., Xie, Z., Luo, X., and Jin, L. (2003) Carbon-coated-nylon-fiber-reinforced cement composites as an intrinsically smart concrete for damage assessment during dynamic loading. *Journal of Materials Science & Technology*, **19**, 583–586.
12 Wang, R.X., Tao, X.M., Wang, Y., Wang, G.F., and Shang, S.M. (2010)

Microstructures and electrical conductance of silver nanocrystalline thin films on flexible polymer substrates. *Surface & Coatings Technology*, **204**, 1206–1210.

13 Rangari, V.K., Shaik, M.Y., Mahfuz, H., and Jeelani, S. (2009) Fabrication and characterization of high strength Nylon-6/Si_3N_4 polymer nanocomposite fibers. *Materials Science and Engineering A – Structural Materials Properties Microstructure and Processing*, **500**, 92–97.

14 Shi, H.F., Zhao, Y., Dong, X., He, C.C., Wang, D.J., and Xu, D.F. (2004) Transcrystalline morphology of nylon 6 on the surface of aramid fibers. *Polymer International*, **53**, 1672–1676.

15 Feldman, A.Y., Wachtel, E., Zafeiropoulos, N.E., Schneider, K., Stamm, M., Davies, R.J. et al. (2006) In situ synchrotron microbeam analysis of the stiffness of transcrystallinity in aramid fiber reinforced nylon 66 composites. *Composites Science and Technology*, **66**, 2009–2015.

16 Causin, V., Yang, B.X., Marega, C., Goh, S.H., and Marigo, A. (2009) Nucleation, structure and lamellar morphology of isotactic polypropylene filled with polypropylene-grafted multiwalled carbon nanotubes. *European Polymer Journal*, **45**, 2155–2163.

17 Causin, V., Yang, B.X., Marega, C., Goh, S.H., and Marigo, A. (2008) Structure–property relationship in polyethylene reinforced by polyethylene-grafted multiwalled carbon nanotubes. *Journal of Nanoscience and Nanotechnology*, **8**, 1790–1796.

18 Caskey, T.C., Lesser, A.J., and McCarthy, T.J. (2003) Evaluating the mechanical performance of supercritical CO_2 fabricated polyamide 6,6/PMMA fiber reinforced composites. *Polymer Composites*, **24**, 545–554.

19 Ma, C.M. and Chen, C.H. (1992) Pultruded fiber-reinforced poly(methyl methacrylate) composites. II Static and dynamic mechanical properties, environmental effects, and postformability. *Journal of Applied Polymer Science*, **44**, 819–827.

20 Goitisolo, I., Eguiazábal, J.I., and Nazabal, J. (2010) Stiffening of poly(ethylene terephthalate) by means of polyamide 6 nanocomposite fibers produced during processing. *Composites Science and Technology*, **70**, 873–878.

21 Goitisolo, I., Eguiazabal, J.I., and Nazábal, J. (2010) Stiffening of PC by addition of a highly dispersed and fibrillated amorphous polyamide based nanocomposite. *Macromolecular Materials and Engineering*, **295**, 233–242.

22 Younan, A.F., Ismail, M.N., and Yehia, A.A. (1992) Reinforcement of natural rubber with nylon 6 short fibers. *Journal of Applied Polymer Science*, **45**, 1967–1971.

23 Rajesh, C., Unnikrishnan, G., Purushothaman, E., and Thomas, S. (2004) Cure characteristics and mechanical properties of short nylon fiber-reinforced nitrile rubber composites. *Journal of Applied Polymer Science*, **92**, 1023–1030.

24 Saad, A.L.G. and Younan, A.F. (1995) Rheological, mechanical and electrical properties of natural rubber-white filler mixtures reinforced with nylon 6 short fibers. *Polymer Degradation and Stability*, **50**, 133–140.

25 Seema, A. and Kutty, S.K.N. (2006) Effect of an epoxy-based bonding agent on the cure characteristics and mechanical properties of short-nylon-fiber-reinforced acrylonitrile–butadiene rubber composites. *Journal of Applied Polymer Science*, **99**, 532–539.

26 Naskara, A.K., Mukherjee, A.K., and Mukhopadhyay, R. (2004) Studies on tyre cords: Degradation of polyester due to fatigue. *Polymer Degradation and Stability*, **83**, 173–180.

27 Mark, J.B., Erman, B., and Eirich, F.R. (eds) (1994) *Science and Technology of Rubber*, Academic Press, New York.

28 De, S.K. and White, J.R. (eds) (2001) *Rubber Technologist's Handbook*, Rapra Technology, Shrewsbury, UK.

29 Cheremisinoff, N.P. (ed.) (1993) *Elastomer Technology Handbook*, CRC Press, Boca Raton.

30 Doğan, O.M., Bolayır, G., Keskin, S., Doğan, A., and Bülent, B., (2008) The evaluation of some flexural properties of a denture base resin reinforced with various aesthetic fibers. *Journal of Materials Science – Materials in Medicine*, **19**, 2343–2349.

31 Fong, H. (2004) Electrospun nylon 6 nanofiber reinforced BIS-GMA/TEGDMA dental restorative composite resins. *Polymer*, **45**, 2427–2432.

32 John, J., Gangadhar, S.A., and Shah, I. (2001) Flexural strength of heat-polymerized polymethylmethacrylate denture resin reinforced with glass, aramid, or nylon fibers. *The Journal of Prosthetic Dentistry*, **86**, 424–427.

33 Tian, M., Gao, Y., Liu, Y., Liao, Y., Xu, R., Hedin, N.E. et al. (2007) Bis-GMA/TEGDMA dental composites reinforced with electrospun nylon 6 nanocomposite nanofibers containing highly aligned fibrillar silicate single crystals. *Polymer*, **48**, 2720–2728.

34 Subbiah, T., Bhat, G.S., Tock, R.W., Parameswaran, S., and Ramkumar, S.S. (2005) Electrospinning of nanofibers. *Journal of Applied Polymer Science*, **96**, 557–569.

35 Huang, Z.M., Zhang, Y.Z., Kotakic, M., and Ramakrishna, S. (2003) A review on polymer nanofibers by electrospinning and their applications in nanocomposites. *Composites Science and Technology*, **63**, 2223–2253.

36 Neppalli, N., Marega, C., Marigo, A., Bajgai, M.P., Kim, H.Y., and Causin, V. (2010) Poly(epsilon-caprolactone) filled with electrospun nylon fibres: A model for a facile composite fabrication. *European Polymer Journal*, **46**, 968–976.

37 Akangah, P., Lingaiah, S., and Shivakumar, K. (2010) Effect of Nylon-66 nano-fiber interleaving on impact damage resistance of epoxy/carbon fiber composite laminates. *Composite Structures*, **92**, 1432–1439.

38 Hejazi, S.M., Abtahi, S.M., Sheikhzadeh, M., and Semnani, D. (2008) Introducing two simple models for predicting fiber-reinforced asphalt concrete behavior during longitudinal loads. *Journal of Applied Polymer Science*, **109**, 2872–2881.

39 Martínez-Barrera, G., Menchaca-Campos, C., Hernández-López, S., Vigueras-Santiago, E., and Brostow, W. (2006) Concrete reinforced with irradiated nylon fibers. *Journal of Materials Research*, **21**, 484–491.

40 Martínez-Barrera, G., Giraldo, L.F., Lòpez, B.L., and Brostow, W. (2008) Effects of γ radiation on fiber-reinforced polymer concrete. *Polymer Composites*, **29**, 1244–1251.

41 Martínez-Barrera, G., Menchaca-Campos, C., Vigueras-Santiago, E., and Brostow, W. (2010) Post-irradiation effects on Nylon-fibers reinforced concretes. *e-Polymers*, **42**, 1–13.

42 Sivaraja, M., Kandasamy, S., and Thirumurugan, A. (2010) Mechanical strength of fibrous concrete with waste rural materials. *Journal of Scientific & Industrial Research*, **69**, 308–313.

43 van Rijswijk, K. and Bersee, H.E.N. (2007) Reactive processing of textile fiber-reinforced thermoplastic composites – An overview. *Composites A*, **38**, 666–681.

44 van Rijswijk, K., Bersee, H.E.N., Beukers, A., Picken, S.J., and Van Geenen, A.A. (2005) Optimisation of anionic polyamide-6 for vacuum infusion of thermoplastic composites: influence of polymerisation temperature on matrix properties. *Polymer Testing*, **25**, 392–404.

45 Caskey, T.C., Lesser, A.J., and McCarthy, T.J. (2003) In situ polymerization and nano-templating phenomenon in nylon fiber/PMMA composite laminates. *Journal of Applied Polymer Science*, **88**, 1600–1607.

46 Hine, P.J. and Ward, I.M. (2006) Hot compaction of woven Nylon 6,6 multifilaments. *Journal of Applied Polymer Science*, **101**, 991–997.

47 Chen, L.S., Huang, Z.M., Dong, G.H., He, C.L., Liu, L., Hu, Y.Y. et al. (2009) Development of a transparent PMMA composite reinforced with nanofibers. *Polymer Composites*, **30**, 239–247.

48 Evstatiev, M., Schultz, J.M., Fakirov, S., and Friedrich, K.I. (2001) In situ fibrillar

reinforced PEWPA-6PA-66 blend. *Polymer Engineering and Science*, **41** (2), 192–204.

49 Fakirov, S., Bhattacharyya, D., and Shields, R.J. (2008) Nanofibril reinforced composites from polymer blends. *Colloids and Surfaces A*, **313–314**, 2–8.

50 Goitisolo, I., Eguiazábal, J.I., and Nazábal, J. (2008) Structure and properties of an hybrid system based on bisphenol a polycarbonate modified by a polyamide 6/organoclay nanocomposite. *European Polymer Journal*, **44** (7), 1978–1987.

51 Xu, X., Yan, X., Zhu, T., Zhang, C., and Sheng, J. (2007) Phase morphology development of polypropylene/ethylene-octene copolymer blends: Effects of blend composition and processing conditions. *Polymer Bulletin*, **58**, 465–478.

52 Zhang, L., Huang, R., Wang, G., Li, L., Ni, H., and Zhang, X. (2002) Fibrillar morphology of elastomer-modified polypropylene: Effect of interface adhesion and processing conditions. *Journal of Applied Polymer Science*, **86**, 2085–2092.

10
Polyolefin Fiber- and Tape-Reinforced Polymeric Composites
József Karger-Kocsis and Tamás Bárány

10.1
Introduction

Fibers, tapes, and their various textile assemblies (e.g., woven, knitted, braided fabrics) may act as efficient reinforcements in different thermoplastic, thermoset, and rubber matrices. This is due to their excellent mechanical characteristics, namely, high modulus (stiffness) and strength. The latter properties are achieved by various drawing procedures whereby strong uniaxial (in case of fibers and tapes) or biaxial (in films and sheets) orientations of macromolecules take place. In case of semicrystalline polymers, drawing is also associated with the formation of oriented supermolecular structural units (crystallites and crystalline lamellae). The "drawn" products are highly anisotropic, that is, their properties differ markedly from one another when measured in the orientation (machine) direction and transverse to it. Though the orientation of amorphous thermoplastics is accompanied with an upgrade in stiffness and strength, this is marginal compared to semicrystalline ones. It is noteworthy that semicrystalline polymers can be considered as composite materials themselves. Based on this analogy, the high-modulus and high-strength crystal units (reinforcing phase) are embedded in an amorphous matrix. The adhesion between the phases, which is a guarantee for the required stress transfer from the matrix toward the reinforcement, is given by chain entanglements, tie molecules, and absorption phenomena.

Although polymeric composites with reinforcing polyolefin films and sheets can also be produced (this is often the case with multilayer blown films that represent "composite laminates"), they are beyond the scope of this chapter. Besides polyethylene (PE) and polypropylene (PP), no further members of the polyolefin family will however be considered.

10.2
Polyolefin Fibers and Tapes

Polyolefins are flexible high molecular weight polymers that can be converted into fibers and tapes of excellent mechanical properties through drawing from the melt,

Polymer Composites: Volume 1, First Edition. Edited by Sabu Thomas, Kuruvilla Joseph,
Sant Kumar Malhotra, Koichi Goda, and Meyyarappallil Sadasivan Sreekala
© 2012 Wiley-VCH Verlag GmbH & Co. KGaA. Published 2012 by Wiley-VCH Verlag GmbH & Co. KGaA.

solid phase (using suitable preforms), and solution. The essential feature of these techniques is the transformation of coiled, entangled, amorphous or isotropic semicrystalline structure into a highly oriented and highly crystalline one. Note that the orientations of both the amorphous and crystalline phases are involved in this transformation.

In contrast to the modulus (theoretical value calculated for PE lies between 250 and 350 GPa), the tensile strength of the related products is far from the intrinsic strength of the C–C bond that is estimated to be at about 25 GPa [1]. Other sources suggest the range between 19 and 36 GPa for PE ([2] and references therein). Apart from this, the measured strength is at about 1/10 to 1/5 of the theoretical value [2, 3]. The theoretical modulus of linear PE may be at about 300 GPa (just for the sake of comparison, the value of steel is 200 GPa) [2]. This is attainable up to about 70% by suitable methods.

Different models were proposed to explain the relationships between the microstructure and mechanical performance of oriented polyolefins. Two of them are mostly favored (schematically depicted in Figure 10.1): (i) composite (crystalline fibril) model and (ii) intercrystalline bridge model ([4, 5] and references therein). The composite model is credited to Arridge, Barham and Keller, whereas the intercrystalline bridging concept can be traced to the activity of Gibson, Davis, Ward, and Peterlin [4, 5]. According to the composite model (Figure 10.1a), needlelike crystals act as reinforcements in the surrounding amorphous matrix. It is postulated that the mechanical improvement upon increasing draw ratio (DR) is due to an increase in the aspect ratio of the needlelike crystals. The intercrystalline bridging hypothesis (Figure 10.1b) assumes that the reinforcing action is given by taut tie molecules that bridge the adjacent crystal units. Interested reader may find further information on this topic in Refs [4–8].

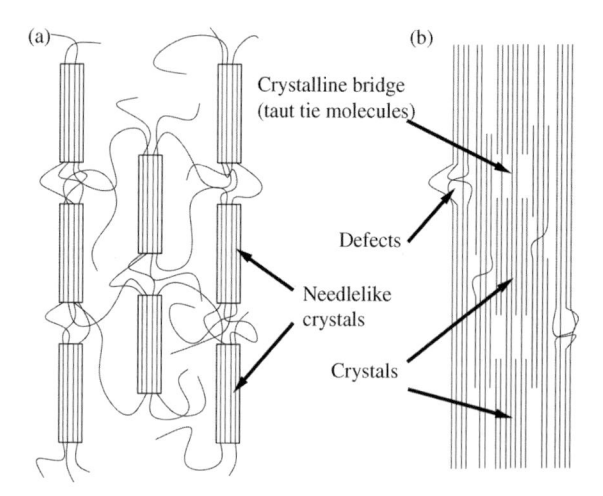

Figure 10.1 Structural schemes for ultrahigh modulus PE based on the "composite" (a) and "intercrystalline bridge" models (b), respectively.

Table 10.1 Basic properties of selected reinforcing fibers and tapes.

Material	Tensile modulus		Tensile strength		Elongation at break (%)	Density (g/cm³)	Heat resistance (°C)
	(cN/dtex)	(GPa)	(cN/dtex)	(GPa)			
Steel	290	200	3.5	2.8	1.0	7.80	—
Aramid (para)	<850	<110	<19.0	<2.8	2.4	1.45	<550
High-strength PE fiber	<1300	<120	<35.0	<3.5	<3.5	0.97	<140
High-strength PP tape	<170	<14	<6.0	0.5	6.0-15.0	<0.8	<170

Basic characteristics of polyolefin fibers and tapes are summarized in Table 10.1. For the sake of comparison, this table lists similar characteristics of some other reinforcing materials that are traditionally used to fabricate polymer composites.

10.2.1
Production

To produce (ultra) high-strength polyolefin fibers and tapes, two methods are mostly used: hot drawing (stretching and orientation drawing) and gel spinning (drawing).

10.2.1.1 Hot Drawing

For the drawing, a suitable preform (strand, strip) is used, which can be produced also on line (which is usually the case). A conventional drawing, whereby the polymer preform is extended between two sets of rollers rotating at different speeds, does not permit a DR substantially higher than 10 [4]. The DR range can, however, be expanded by using two- or multistage draw processes. During this process, the drawn polymer is further "thinned" in one or more steps. Between the steps, the drawing temperature is enhanced. Alternative solution is to make the drawing along a suitable temperature gradient. In multistage drawing, DR $\approx$ 40 can be easily reached [4]. With increasing DR, the modulus of the product also increases monotonously. On the other hand, this tendency does not hold for the strength, which either levels off at a given value (saturation) or goes through a maximum as a function of DR. Major reason for this behavior is that above a threshold DR, cavitations usually start. This can be recognized by bare eyes due to the concomitant stress whitening phenomenon (the microvoids generated scatter the light causing the "silverlike" appearance). This phenomenon is the reason why the density of PP tape is below that of the bulk PP in Table 10.1. Cavitations can be circumvented when the orientation thinning of the preform is done by rolling under side constraint. Under this condition, the prevailing compression stress hampers the onset of cavitations [9].

Drawing takes place usually between the glass transition (T_g) and the melting temperature (T_m) of the given semicrystalline polymer. During orientation, the folded chain crystal lamellae rotate, break up, defold, and finally form aligned chain crystals.

Fibers with very high DR can be produced in one or more drawing steps. In the latter case, the isothermal drawing temperature increases in the consecutive drawing steps. Elyashevich et al. [10, 11] manufactured in one-step orientation PE fibers having an E modulus and tensile strength of 35 and 1.2 GPa, respectively. Baranov and Prut [12] produced ultrahigh-modulus PP tapes by a two-step isothermal drawing process. The isothermal drawing of the parent film was done in a tensile testing machine equipped with a thermostatic chamber. The first drawing occurred at 163–164 °C, while the second one was at 165 °C. The tensile E modulus and strength of the tapes were 30–35 and 1.1 GPa, respectively.

Alcock et al. [13] produced highly oriented PP tapes by extrusion and drawing steps. The tensile deformation was achieved by pulling a tape from one set of rollers at 60 °C through a hot air oven to a second set of rollers working at the range of 160–190 °C. The results showed that the density was approximately constant with an increasing draw ratio up to DR $\approx$ 9, above which it sharply dropped. The decrease in density was associated with a change in opacity of the tape due to the onset of microvoiding within the tape. Karger-Kocsis et al. [14] noticed that microvoiding in stretched PP tapes takes place even at DR $\approx$ 8. This phenomenon, also termed "overdrawing," was studied in depth by Schimanski et al. [15]. The PP tapes, produced by Alcock et al. [13], exhibited tensile modulus and strength of $\sim$15 GPa and $\sim$450 MPa (see the related data in Table 10.1), respectively, at a DR $=$ 17. It is worth noting that tensile strength of PP tapes and strips, widely used for packaging purposes, are in the range of 220–350 MPa [16].

10.2.1.2 Gel Drawing

Via gel drawing (spinning), films and fibers can be produced from dilute polymer solutions. The principles of this solution technique were developed by Pennings and his colleagues Smith, Lemstra, and Kalb [3]. Their work represents the foundation of the production of ultrahigh molecular weight PE (UHMWPE) fibers under the trade name Dyneema® at DSM. A solution spinning process for UHMWPE was developed at Allied Signal by Kavesh and Prevorsek [3]. The gel drawing requires, however, a polymer with a high mean molecular weight ($M_w > 10^6$ Da) and suitable molecular weight distribution characteristics. If the molecules are less entangled in the gel, this guarantees drawing to high degrees [17–19]. Oriented synthetic fibers of UHMWPE, namely, *Dyneema* (www.dsm.com) and *Spectra* (www51.honeywell.com), produced by gel spinning may show tensile strengths as high as 3.5 GPa (cf. Table 10.1). The cross section of the related PE fibers is different.

10.2.2
Properties and Applications

As already mentioned, the basic mechanical properties of fibers and tapes are listed in Table 10.1. PP tapes, as already noted, are mostly used for packing whereby their easy weldability is of great practical importance.

UHMWPE fibers are mostly used to produce ballistic vest covers, safety helmets, cut-resistant gloves, bow strings, climbing ropes, fishing lines, spear lines for spear guns, high-performance sails, suspension lines in parachutes, and so on. The favorite ballistic application of UHMWPE fibers is linked with the fact that they become stiffer, stronger, and surprisingly even tougher with increasing deformation rate [3]. This behavior is similar to spider silk [20] that is exploited in protecting armors with suitable textile architectures of the fibers. The related products are either nonconsolidated or consolidated. The term "consolidated" refers to a polymer composite in which the reinforcements are embedded in a given matrix (usually PE or thermoplastic rubber). Considerable efforts are devoted to estimate and model the energy absorption of the fibers and their related composites ([21–25] and references therein). It was proven that unidirectional (UD) fiber layers when stacked in a cross-ply (CP) manner are far more resistant to high-speed transverse (out-of-plane) loading than the woven structures at the same layer number [26].

Recall that polyolefins are viscoelastic materials. This behavior is manifesting in both high (e.g., partial melting of the fibers under ballistic impact) and low frequency tests (e.g., creep performance) [27].

If a good adhesion between a polyolefin and another material is required, it is necessary to carry out some pretreatment of the polyolefin. Adhesion problems with polyolefins appear in many cases, like adhesive bonding, printing, metallizing, and heat sealing. The use of polyolefins as reinforcements in other polymers is a highly problematic issue. Brewis and Briggs [28] in their early paper pinpointed that the increase of the surface tension (for neat PE and PP between 30 and 33 mN/m), yielding substantially better wetting, is the key parameter of improved adhesion. The enhancement of the surface energy, resulting from the introduction of polar groups, can be achieved by different techniques (UV radiation, chemical etching, flame and corona treatments, fluorination, etc.). Accordingly, considerable research and development works were devoted to improve the interfacial adhesion between polyolefins (especially UHMWPE fibers) and different polymers covering thermoplastics, thermosets, and rubbers. The necessity of adhesion improvement was also urged by another aspect, namely, the unfavorable thermal expansion coefficient data of high-strength PE [29]. The axial coefficient of thermal expansion (CTE) of UHMWPE fibers (of about $-5 \times 10^{-6}\,\text{K}^{-1}$) is similar to those of carbon fibers (CFs) (CF $\approx -1 \times 10^{-6}\,\text{K}^{-1}$) and aramid fibers (AF $\approx -5 \times 10^{-6}\,\text{K}^{-1}$), all of them having negative values. This is due to their strongly oriented crystalline structure prone to shrinkage upon heating. Note that the CTE of usual epoxy resins (EP) is at about $7 \times 10^{-5}\,\text{K}^{-1}$. As a consequence, cooling from high curing temperature to ambient one is associated with compressive stresses acting axially on the fiber. As the ultimate compressive strain of UHMWPE fibers is low ($\approx 0.2\%$ – similar to aramid), they suffer compressive failure via formation of kink bands. Recall that this is caused by the mismatch of the thermal expansion coefficients of the fiber and EP matrix. The problem is even more grave in respect to the interfacial adhesion because the radial CTE of UHMWPE fiber is about two times higher (being at about $1.3 \times 10^{-5}\,\text{K}^{-1}$) than that of the EP. This will induce a high normal tensile stress rather than a beneficial compressive one at the interface on cooling from high curing temperature [29]. The normal tensile stress at the interface is detrimental to composites as their failure starts by fiber/matrix debonding in area where fibers are oriented transverse to the loading direction [30]. Therefore, it is imperative to enhance the adhesion of polyolefin fibers toward high-temperature curing thermosets. This is still a straightforward strategy when low-temperature curable resins are selected as matrices.

Accordingly, the large body of R&D activities addressed the surface modification of UHMWPE fibers. Chemical etching (chromic acid, potassium permanganate, and hydrogen peroxide etchants) was especially useful to detect the fibrillar structure hierarchy in the respective fibers. The etchant removed the outer "skin" layer consisting of process-related compounds but did not affect the UHMWPE itself [6, 31]. The modulus of the fiber remained unaffected, whereas the tensile strength and strain were reduced with duration of the etching. The observed changes also depend on the etchant type [31]. Bromination [32] and fluorination [33] were also adapted to modify the surface of UHMWPE fibers, even when the production of self-reinforced composites was targeted.

PE fibers have been surface modified in a two-stage wet grafting process in which acrylamide and acrylic acid were grafted onto the surface by redox initiation technique. Based on single-fiber pull-out test (cf. Figure 10.2), the grafted PE has shown markedly better adhesion to EP than the untreated PE fiber [34].

Interestingly, the majority of the surface modification works with polyolefins concentrated on plasma treatments. Biro et al. [35, 36] reported that the interfacial shear strength between UHMWPE and EP, quantified in microdebond tests (cf. Figure 10.2), could be doubled when the PE fiber is plasma treated in air. X-ray photoelectron spectra (XPS) revealed the presence of oxygen-containing groups on the fiber surface [35, 36]. Treatment in argon plasma caused micropittings at the fiber surface. This yielded some improvement in the adhesion via mechanical interlocking. The interfacial shear strength, measured by the short-beam shear test on UD fiber-reinforced vinyl ester composites, could be improved by ≈30%. The authors concluded that both the mechanical interlocking and the chemical modification of

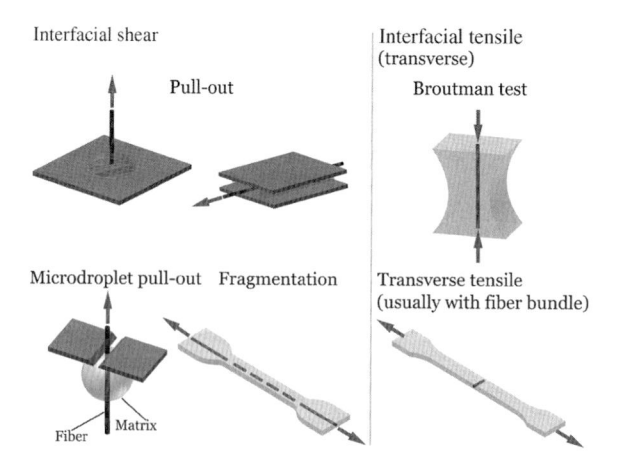

Figure 10.2 Schemes of the testing methods to determine the interfacial shear and transverse tensile debonding stresses in single-fiber composites.

the fiber surface should be targeted to improve the interfacial adhesion between high-strength PE fibers and thermosetting resins [37]. Intrater *et al.* [38] visualized the oxygen plasma-induced surface restructuring in PE fibers by means of atomic force microscopy (AFM). The observed restructuring was traced to preferential etching of the amorphous phase.

Plasma treatments (plasma activation, plasma-induced polymerization, and plasma-enhanced deposition) were also adapted for PP fibers. It has been concluded that the hydrophilicity of the PP fiber can be adjusted by using proper plasma treatments (method and conditions) [39].

10.3
Polyolefin-Reinforced Thermoplastics

The application of polyolefin fibers and tapes in thermoplastic polymer composites is rather limited by their temperature sensitivity (cf. Table 10.1) than by the problems with the interfacial adhesion between the composite's constituents.

10.3.1
Self-Reinforced Version

At present, research activities concentrate on the development of self-reinforced polymer composites (SRPCs). In these "all the same polymer composites," both the reinforcing and matrix phases are given by the same polymer. This material concept was introduced by Capiati and Porter in 1975 [40]. These composites are also referred to as single-phase or homocomposites. These materials may compete with traditional composites in various application fields based on their low density and easy recycling. Recall that the density of polymers, and especially those of PE and PP (cf. Table 10.1), is well below those of traditional reinforcements.

A commercial breakthrough with self-reinforced thermoplastic polymer composites occurred recently. Preforms for self-reinforced PE composites (e.g., FragLight® nonwoven) and PP composites (all-PP composites; e.g., Curv®) are now available on the market.

10.3.1.1 Hot Compaction
Ward *et al.* [41, 42] developed a new method to produce SRPCs that they called "hot compaction." The related research started with highly oriented PE fibers and tapes. When these preforms were put under pressure and the temperature was increased, their surface and core showed different melting behaviors. This finding was exploited to melt the outer layer of the fibers and tapes, which after solidification (crystallization) became the matrix. The residual part of the fibers and tapes (i.e., their core section) acted as the reinforcement in the resulting all-PE [43, 44] and all-PP composites [45]. It is intuitive that the processing window during the hot compaction of single-component polymeric systems is very narrow. It was also reported that in order to set optimum mechanical properties, a given amount of fiber should melt and

work later as the matrix. This was given by about 10% of the cross section (i.e., outer shell) of the fiber. The stress transfer between the residual fiber (reinforcement) and the newly formed matrix is guaranteed by a transcrystalline layer formed. The effect of transcrystallinity is controversially discussed from the point of view of the stress transfer from the matrix on the fiber. Though the development of the transcrystalline layer is necessary in all-polyolefin composites, its internal buildup may be of great relevance, as outlined by Karger-Kocsis [46]. Ratner et al. [47] found that the cross-linked interphase between the fiber and the matrix is more beneficial than the usual transcrystalline one, especially when long-term properties like fatigue are considered.

UHMWPE loses its stiffness and strength and becomes prone toward creep with increasing temperature. To overcome this problem, the UHMWPE fibers were exposed to γ-irradiation to trigger their cross-linking [48].

Due to the low-temperature resistance of PE, the hot compaction research shifted to PP. Jordan and coworkers [49–51] studied the effects of hot compaction on the performance of PE and PP tapes and fabrics. The latter differed in their mean molecular weights, which influenced the consolidation quality assessed by tear tests. Hine et al. [52] devoted a study to determine whether the insertion of film layers between the fabrics to be compacted results in improved consolidation quality, as well as whether this "interleaving concept" can widen the temperature window of the processing. This concept yielded the expected results: the consolidation quality was improved (well reflected in the mechanical property profile), the interlayer tear strength enhanced, and the processing temperature interval enlarged. This approach was also followed for PP fibers. Hine et al. [53] incorporated carbon nanofibers (CNF; up to 20 wt%) to improve the reinforcing activity of the PP preform after hot compaction. Introduction of CNF at 5 vol% increased the Young's modulus at room temperature by 60% and reduced the CTE by 35%. Attempts were also made to improve the bonding between CNF and PP via oxygen plasma treatment and also by using a maleic anhydride-grafted PP as compatibilizer.

One major goal of the hot compaction technology was to offer lightweight and easily recyclable thermoplastic composites for transportation. As further application fields, sporting goods, safety helmets, covers, and shells (also for luggage) were identified. Hot compacted PP sheets from woven PP fabrics are marketed under the trade name of Curv (www.curvonline.com).

10.3.1.2 Film Stacking

During film stacking, the reinforcing layers are sandwiched in-between the matrix-giving film layers before the whole "package" is subjected to hot pressing. Under heat and pressure, the matrix-giving material, which has a lower melting temperature than the reinforcement, becomes molten and infiltrates the reinforcing structure resulting in a "consolidated" composite. The necessary difference in the melting temperatures between the matrix and the reinforcement can be set by using different polymer grades (e.g., copolymers for the matrix and homopolymers for the reinforcement) or polymorphs (e.g., lower melting modification for the matrix and higher melting one for the reinforcement). It is of great importance to have a large enough difference between the melting temperatures of the composite constituents. In this way, the temperature-induced degradation in the stiffness and strength of the reinforcement can be kept on an acceptable level.

Bárány et al. [54–57] produced different all-PP composites. For reinforcement, highly oriented fibers in different textile architectures (carded mat, carded and needle-punched mat, and in-laid fibers in knitted fabrics) were used, whereas for matrices either PP fibers of lower orientation (the same textile assemblies as indicated above) or beta-nucleated PP films were selected. The matrix-giving phase in them was either a discontinuous fiber or a knitted fabric. Their consolidation occurred by hot pressing as in case of film stacking. It is noteworthy that the melting temperature of the beta-modification of isotactic PP is >20 °C lower than the usual alpha-form [58]. The beta-modification can be achieved by incorporating a selective beta-nucleating agent in the PP through melt compounding. The concept of this alpha(reinforcement)/beta(matrix) combination should be credited to Karger-Kocsis [59].

Bárány et al. [54, 57] also used PP fabric (woven-type from split yarns) as the reinforcement and beta-nucleated PP film as matrix-giving material. With increasing processing (pressing) temperature, the consolidation quality of the corresponding composites was improved. Parallel to this, the density, the tensile, and the flexural stiffness and strength increased, whereas the penetration impact resistance diminished.

Abraham et al. [60] produced all-PP composites with tape reinforcement by exploiting the difference in the melting behavior of alpha- and beta-polymorphs. The alpha-PP tapes were arranged in unidirectional (UD) and cross-ply (CP) manners by winding, putting beta-nucleated PP films in-between the related reinforcing tape layers. Inspection by polarized light microscopy proved the presence of transcrystalline layer between the PP reinforcement and PP matrix (cf. Figure 10.3).

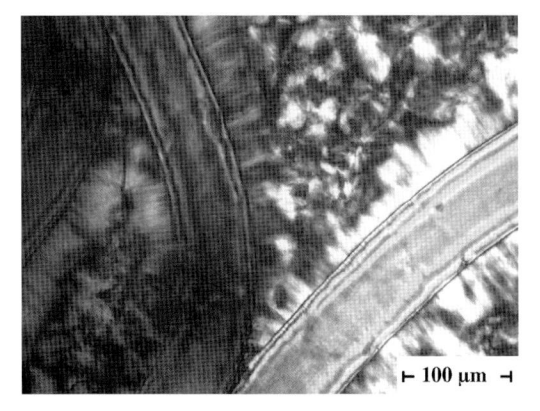

Figure 10.3 Formation of transcrystalline layer between the reinforcing and matrix phases in a tape-reinforced all-PP model composite.

10.3.1.3 Wet Impregnation Prior to Hot Consolidation

Cohen et al. [61] demonstrated that solution impregnation of UHMWPE fibers with a dilute tetralin solution of UHMWPE at $T = 132\,°C$ is helpful to achieve a good bonding between the reinforcement and the matrix without additional surface treatment of the former in hot compaction. This was traced to the appearance of an interphase with peculiar crystalline morphology. Surface swelling itself may also be a straightforward strategy to improve the adhesion between the same polymers acting as reinforcement and matrix, respectively.

10.3.2
Polyolefin Fiber-Reinforced Composites

In this section, those thermoplastic composites are surveyed whose matrix and reinforcement slightly (belonging to the same family of the corresponding polymer) or substantially (fully different polymers) differ from one another. Reinforcement/matrix combinations according to the former terminology cover, for example, UHMWPE/LDPE, UHMWPE/LLDPE, UHMWPE/HDPE (where LDPE, LLDPE, and HDPE designate low-density, linear low-density, and high-density PEs, respectively), PP homopolymer/PP copolymer systems. Note that the melting of the matrix-giving material is always below that of the reinforcement that is the guarantee of consolidation via hot pressing. The processing techniques of the related preforms and assemblies – except for the *in situ* polymerization of the matrix – are practically identical with those of the self-reinforced versions (cf. Section 10.3.1).

10.3.2.1 Consolidation of Coextruded Tapes

Peijs [62] developed a coextrusion technique for which the melting temperature difference between the composite constituents reached $20–30\,°C$. The invention was to "coat" a PP homopolymer tape from both sides by a copolymer through a continuous coextrusion process. Note that a copolymer always melts at lower temperatures than the corresponding homopolymer, owing to its less regular molecular structure. The coextruded tape was stretched additionally in two steps. This resulted in high-modulus, high-strength tapes (cf. Table 10.1). The primary tapes could be assembled in different ways, as in composite laminates (ply-by-ply structures with different tape orientations), or integrated in various textile structures (e.g., woven fabrics). The consolidation of these composite preforms occurred by hot pressing. The advantage of this method is that the reinforcement (core) content of the tape may be as high as 80%. This, along with the high draw ratio, yielded tapes of excellent mechanical properties (tensile E modulus $>6\,GPa$, tensile strength $>200\,MPa$; see also data in Table 10.1). Properties of the tape and related composites [63–68] as a function of the processing conditions have been reported in many publications. Among the beneficial properties of these all-PP composites, the resistance to perforation impact has to be additionally mentioned. Ballistic test results confirmed that the performance of composite sheets from Pure® tape is comparable to that of the state-of-the-art ballistic materials. Other groups were also involved in the characterization of this material [69–71]. Moreover, consolidated sheets of coextruded tapes in different lay-ups and textile structuring were used for the face covering of different sandwich structures with cores including honeycomb structures and foams. Recall that the coextruded PP tapes are known under the trade names of Pure and Armordon® (www.pure-composites.com; www.armordon.com).

10.3.2.2 Film Stacking

Shalom et al. [72] produced high-strength PE fiber- (Spectra®) reinforced HDPE composites by winding the fiber in a unidirectional manner and sandwiching the HDPE films in-between the wound fiber layers. The reinforcing fiber content in the UD assembly was 80 wt%. Its consolidation occurred by hot pressing ($T = 137\,°C$, $p = 16.5$ MPa).

Abraham et al. [60] produced high-strength alpha PP homopolymer tapes by a single-step hot stretching technique and used this as UD or CP reinforcement in alpha- and beta-phase random PP copolymer matrices. The interphase between the reinforcement and matrix was composed of a transcrystalline layer that was larger in the beta-phase than in the alpha-phase random PP copolymer matrix.

Houshyar and Shanks [73] used a mat from PP homopolymer fibers as the reinforcement and PP copolymer film as the matrix-giving material. The reinforcement content in the corresponding composites was 50 wt%. The difference between the melting temperatures of the PP homo- and copolymer used was about $16\,°C$ according to DSC results. Consolidation through hot pressing occurred between 155 and $160\,°C$. The surface of the homopolymer PP fiber acted as a heterogeneous nucleator and initiated transcrystalline growth. Objectives of further studies of the group of Shanks were to study effects of different textile architectures [74] and matrix modification [75] on the mechanical properties of the related all-PP composites. The mechanical results showed that the properties of the woven composites strongly depend on the woven geometry [74]. Blending of the matrix-giving PP with ethylene–propylene rubber was very straightforward to improve the energy absorption capability of the related composites [75]. It is noteworthy that the term "(hot) compaction" is frequently used in the literature, though this is reserved for those techniques in which a part of the reinforcing phase becomes molten and thus overtakes the role of the matrix after cooling. This is not the case in film stacking, where the melting temperature of the reinforcing fiber or tape is usually not surpassed.

Bárány et al. confirms that the best mechanical performance of all-PP composites by film stacking method can be achieved when the matrix is random copolymer whereby its beta-nucleation is even more beneficial [54, 56, 57, 76].

Basically we also have to do with "film stacking" when the reinforcing fibers and tapes are laid in between matrix-giving film layers by filament of tape winding techniques. Kazanci et al. [77] embedded UHMWPE fiber in between LLDPE films prior to composite consolidation via hot pressing.

10.3.2.3 Solution Impregnation

UHMWPE fiber bundles were coated with a xylene solution with dissolved LDPE. The material was dried, arranged in preforms prior to its hot pressing at different temperatures ($T = 120–140\,°C$). With increasing pressing temperature, the stiffness increased, whereas the strength remained practically unaffected. This impregnation technique has also been adapted for filament winding [78].

10.3.2.4 Powder Impregnation (Wet and Dry)

Impregnation of reinforcing fibers with polymer powders overtaking the role of matrix after melting was in the focus of preform-oriented research in the 1980s. Later this technique could not compete with the more economic "fiber commingling" technique. According to the latter, the reinforcing and matrix-giving fibers are commingled and the related bundles arranged in different preforms (e.g., woven, braided, and knitted fabrics) prior to their consolidation via hot pressing. Nevertheless, dry LDPE powder impregnation of UHMWPE fibers was explored by Chand et al. [79]. A wet powder coating method was developed by the group of Schulte [80]. In this procedure, HDPE powder slurry in propanol was used for the coating of UHMWPE fibers. Note that the final stage of composite production was always hot pressing.

10.3.2.5 *In Situ* Polymerization of the Matrix

The feasibility of this technique has been shown on the example of a composite consisting of braided PE fiber reinforcement and polymethyl methacrylate (PMMA) matrix [81]. The PE fiber was surface modified by chemical etching. Methyl methacrylate (MMA) was prepolymerized, introduced in the mold whose cavity was charged with the braided preform and polymerized afterward completely. Surface modification of the PE fibers improved the stiffness and strength of the composite but reduced its toughness. This technique is somewhat similar to that of resin transfer molding (RTM) that is widely used for thermoset matrix-based composites.

10.3.3
Interphase

As already mentioned, the interphase in self-reinforced PE and PP is similar to that in all-PE and all-PP composites: all of them are transcrystalline type. Note that the above material grouping makes a difference between "all-the-same-polymer" and "all-the-same-family polymer" composites. Essential prerequisite of transcrystallization is the presence of active nuclei on the surface of the reinforcing fiber and tape in high density. The closely spaced nuclei dictate that spherulitic growth occurs practically in one direction, namely, transverse to the nucleation surface. The resulting columnar structure, well resolvable by optical methods (cf. Figure 10.3), is usually considered to support the stress transfer from the matrix toward the reinforcing phase. Nevertheless, the reason and effects of transcrystallization are controversial issues in the literature [82]. It is, however, obvious that the inherent structure of the transcrystalline layer should influence the supposed stress transfer. This is likely the reason for the many works devoted to this topic ([83–85] and references therein). Effects of various surface treatments on the formation of the transcrystalline interphase were also investigated [86].

The interface/interphase characteristics in polyolefin reinforced other matrix systems (i.e., thermoplastics other than polyolefins, thermosetting resins, and rubbers) are far less studied and understood as in polyolefin/polyolefin combinations.

10.3.4
Hybrid Fiber-Reinforced Composites

The reader may find different definitions for hybrid fiber-reinforced composites. The term "hybrid reinforcement" usually means that the reinforcing phase is given by two or more materials. Accordingly, the combination of UHMWPE fibers with glass (GF) and carbon fiber represents hybrid reinforcement in a given composite. The other explanation for hybrid systems, more exactly for hybrid effects, is that the combined use of two (or more) reinforcing materials results in a synergistic effect in respect to the composite performance. Synergistic means a positive deviation from the linear rule of mixture (rule of additivity). The yet another and likely the original explanation is that the hybrid effect is linked with a residual stress (strain) field in the composite that is developed due to the difference in the CTEs of the reinforcing fibers. Composites containing both GF and CF (the original "hybrid" combination) exhibited enhanced failure strains compared to the solely GF- and CF-reinforced ones. Taketa *et al.* [87] demonstrated a similar effect on the example of hybrid composites composed of CF- and self-reinforced (Curv) PP plies. The failure strain of the resulting composite was enhanced via the incorporation of woven CF/PP plies that suffered compressive stresses due to the difference in the CTEs between the CF fabric and self-reinforced PP plies. The mechanical properties of hybrid composite laminates composed of self-reinforced PP (Curv), all-PP (Pure), and GF-PP (UD) layers have also been studied. The authors concluded the best lay-ups for the different tests, namely, for tensile, flexural, and perforation impact [88].

10.4
Polyolefin Fiber-Reinforced Thermosets

Because many thermosets are cured at ambient temperature or at temperatures well below the melting point of polyolefins, they may be favored matrices of composites with PE or PP reinforcements. However, to promote the adhesion of polyolefins to thermosets, the former have to be surface treated. Recall that this topic has already been addressed in Section 10.2.2.

10.4.1
Polyolefin Fiber-Reinforced Composites

Andreopoulos *et al.* [89] used chromosulfate and permanganate solutions as oxidative agents for the surface modification of UHMWPE fibers and ribbons. Ribbons were produced by hot calendaring in order to enhance the specific surface. For benchmarking of the chemical etching, corona treatment was adapted. Pull-out test (cf. Figure 10.2) results indicated that chemical etching is very efficient to improve the adhesion bonding; however, this treatment was associated with a drastic reduction of the strength. Calendered UHMWPE ribbons with corona treatment showed the optimum property profile. The effects of this treatment have also been studied after hygrothermal aging [90]. Chaoting et al. [91] used oxygen plasma treatment for UHMWPE fibers and assessed the changes in their surface energy and strength. With increasing surface energy, both the wetting and adhesion of the fiber/EP systems were improved. Pull-out tests indicated 10 times increase in the adhesion strength. Jana et al. [92] developed another method to improve the adhesion between UHMWPE and EP. They modified the EP matrix by reactive graphitic nanofibers that enhanced the wetting and adhesion to UHMWPE. Dutra et al. [93] demonstrated that the impact strength of EP can be prominently raised by the incorporation of UD aligned modified PP fibers. The PP fiber contained 5 wt% mercapto-modified ethylene/vinyl acetate copolymer. The T_g shift toward higher temperatures in the DMTA spectra reflected that the molecular mobility was hampered in the interphase being in a prestressed state owing to the mismatch of the CTEs of the components. It was also reported that the incorporation of short PP fibers may increase the resistance to thermal degradation of EP [94]. Vinyl ester-based thermoset composites with UD aligned UHMWPE fibers, produced by hand lay-up or vacuum-assisted RTM, exhibited excellent resistance to solid particle erosion [95]. The density-related specific strength of the related composites was the second highest after a polybisoxazole-type (PBO) fiber-reinforced composites. Ar^+ ion irradiation of UHMWPE fibers in oxygen atmosphere improved the quality of adhesion bonding to vinyl ester by about 20% [96]. Surface modification of chopped UHMWPE fibers in fluorine/oxygen reactive gas improved the stiffness and strength of the related thermoset polyurethane-based composites. Moreover, the toughness of the PU composites with surface-modified UHMWPE fibers was three times higher than that with unmodified ones at the same fiber content [97]. Surface-treated UHMWPE fibers (via chemical etching) in UD arrangement acted as efficient reinforcements in respect to stiffness and strength in phenolics of interpenetrating network (IPN) structure. The latter was achieved by combining a phenol/formaldehyde resol with vinyl-acetate/2-ethylhexyl acrylate copolymer [98]. UHMWPE fibers, with and without plasma treatments, were incorporated in PU/EP hybrids with grafted IPN structure [99]. "Grafted" here means that the continuous phases are linked to each other by chemical reactions [100]. The performance of UHMWPE was compared with the performance of that achieved by aramid. According to bulletproof tests, the composites UHMWPE/PU behaved similar to aramid/PU [99].

10.4.2
Hybrid Fiber-Reinforced Composites

PE due to its relative high hydrogen content is a promising shielding material against galactic cosmic irradiation. Structure integrity and safety requirements may force, however, the engineers to combine UHMWPE fibers with others, such as GF and CF, in the related EP-based composites. The feasibility of the strategy has been shown by Zhong et al. [101]. Park and Jang [102] studied the performance of woven CF and UHMWPE fiber fabrics in vinyl ester resins. The surface of the reinforcing fabrics was treated by low-temperature oxygen plasma followed also by silanization. In addition, the positioning of the reinforcing layers in the composites was assessed in flexural test. It was concluded that for optimum flexural loadability, woven CF-reinforced layers should be on the top and UHMWPE-based ones on the bottom experiencing compression and tensile loadings, respectively.

10.5
Polyolefin Fibers in Rubbers

Textile fibers (both discontinuous and continuous) and different fabrics are widely used (tyres, belts, air springs, and technical rubber goods) to reinforce rubbers as they impart the stiffness and strength without sacrificing the flexibility and resilience. Polyolefin fibers were already mentioned as potential rubber reinforcements in the 1970s [103]. It is worth mentioning that amorphous polymers, such as rubbers, usually show better wetting and adhesion toward different substrates than semicrystalline polymers [104, 105]. The simplest explanation for this is that no crystallization-induced shrinkage is at work in amorphous systems.

10.5.1
Polyolefin Fiber-Reinforced Composites

Shakar et al. [106] reported that bonding of chopped UHMWPE fiber to ethylene/propylene/diene rubber (EPDM) matrix could be markedly enhanced when the rubber was cured by electron beam irradiation. With increasing concentration of the fiber and increasing irradiation dose, the mechanical properties of the composites were also improved.

The excellent ballistic properties of UHMWPE fibers are exploited in protecting clothes (against fragments and debris of explosions and blasts), armors (architectural, vehicle) and so on. Some "consolidated" preforms with UHMWPE fibers, such as *Spectra Shield® Plus PCR* prepregs (www.customarmoring.com/SpectraShieldPlus.pdf), contain a proprietary matrix material [107]. This is likely a thermoplastic rubber. Earlier it was disclosed that as matrix a styrenic thermoplastic elastomer was selected [3]. The latter is usually a triblock copolymer with styrene/butadiene/styrene (SBS) or styrene/ethylene–butylene/styrene (SEBS) structure that becomes melted above the softening temperature of the polystyrene (PS) domains. These PS domains are formed by phase segregation during cooling and act as nodes of the physical network structure in this type of amorphous thermoplastic rubbers [108]. It was reported that UD reinforced Dyneema SB grades are thermoplastic elastomer matrix-based systems [109].

10.5.2
Hybrid Fiber-Reinforced Composites

Hybrid reinforcement in case of rubbers usually means the combination of traditional active fillers (carbon black and silica) and polyolefin fibers (mostly chopped UHMWPE fibers). Shakar *et al.* [106] reported that the tensile strength of carbon black containing EPDM decreased with increasing concentration of chopped UHMWPE fibers even though other properties (e.g., stiffness and resistance to swelling) were improved.

10.6
Others

As already mentioned, polyolefin fibers, especially PP fibers, have an important role in "commingled" composite preforms. Commingled PP/GF rovings and related textile assemblies (woven fabric) are marketed under the trade name *Twintex*® (www.twintex.com). Their processing structure–property relationships are well studied and disclosed in the open literature.

Polyolefins may also have some kind of reinforcing action in ceramic matrix-based composites. Polyolefin staple fibers, for example, are key additives in special concretes in which their role is to suppress microcracking and sedimentation phenomena [110]. UHMWPE fiber containing composites may have application in retrofitting of concrete structure, columns whereby their role is to guarantee the lateral stiffness and strength [111]. Polyolefin fibers with suitable additives may also be used as phase change materials (thermoregulation fibers) [112].

10.7
Outlook and Future Trends

Self-reinforced polyolefin composites will remain of interest due to their beneficial properties (very low density and easy recycling via reprocessing in the melt). To improve the reinforcing action of polyolefin fibers and tapes, they will be most probably modified by nanofillers. Among the latter, those having high aspect ratios (carbon nanotubes [113], carbon nanofibers, graphene layers, and layered silicates) and thus capable of increasing the stiffness, strength, and thermal stability of the related composites are most promising. Further exploratory work is expected in respect to electrospinning of polyolefins [114] and their use in composites. A similar note also holds for auxetic polyolefin fibers [115]. To elucidate the crystalline structure in ultradrawn polyolefins, high-resolution analytical techniques will be used, such as microbeam X-ray diffraction. The latter technique (during *in situ* loading of suitable specimens) along with laser Raman spectroscopy [116] will also be adapted to study the way of stress transfer in the interphase region.

Great efforts will be devoted to model the performance of polyolefin fiber-reinforced composites using finite element analysis. Major target of the modeling work will definitely be the ballistic behavior.

Acknowledgments

The authors want to thank the Hungarian Scientific Research Fund (OTKA K75117). T. Bárány is thankful for the János Bolyai Research Scholarship of the Hungarian

Academy of Sciences. This work is connected to the scientific program of the "Development of quality-oriented and harmonized R + D + I strategy and functional model at BME" project. This project is supported by the New Hungary Development Plan (TÁMOP-4.2.1/B-09/1/KMR-2010-0002).

References

1 Smook, J., Hamersma, W., and Pennings, A.J. (1984) The fracture process of ultrahigh strength polyethylene fibers. *Journal of Materials Science*, **19**, 1359–1373.

2 Wong, W.F. and Young, R.J. (1994) Analysis of the deformation of gel-spun polyethylene fibers using Raman-spectroscopy. *Journal of Materials Science*, **29**, 510–519.

3 Prevorsek, D.C. (1996) Structural aspects of the damage tolerance of spectra fibres and composites, in *Oriented Polymer Materials* (ed. S. Fakirov), Hüthig & Wepf Verlag Zug, Heidelberg, pp. 444–466.

4 Ward, I.M. and Sweeney, J. (2004) *The Mechanical Properties of Solid Polymers*, John Wiley & Sons, Inc., New York.

5 Wong, W.F. and Young, R.J. (1994) Molecular deformation processes in gel-spun polyethylene fibers. *Journal of Materials Science*, **29**, 520–526.

6 El-Maaty, M.I.A., Olley, R.H., and Bassett, D.C. (1999) On the internal morphologies of high-modulus polyethylene and polypropylene fibres. *Journal of Materials Science*, **34**, 1975–1989.

7 Fakirov, S. (1996) *Oriented Polymer Materials*, Hüthig & Wepf Verlag Zug, Heidelberg.

8 Kakiage, M., Tamura, T., Murakami, S., Takahashi, H., Yamanobe, T., and Uehara, H. (2010) Hierarchical constraint distribution of ultra-high molecular weight polyethylene fibers with different preparation methods. *Journal of Materials Science*, **45**, 2574–2579.

9 Galeski, A. and Regnier, G. (2009) Nano- and micromechanics of crystalline polymers, in *Nano- and Micro-Mechanics of Polymer Blends and Composites* (eds J. Karger Kocsis and S. Fakirov), Hanser, Munich, pp. 3–58.

10 Elyashevich, G.K., Karpov, E.A., Kudasheva, O.V., and Rosova, E.Y. (1999) Structure and time-dependent mechanical behavior of highly oriented polyethylene. *Mechanics of Time-Dependent Materials*, **3**, 319–334.

11 Elyashevich, K.G., Kapov, A.E., Rosova, Y.E., and Streltses, V.B. (1993) Orientational crystallization and orientational drawing as strengthening methods for polyethylene. *Polymer Engineering and Science*, **33**, 1341–1351.

12 Baranov, A.O. and Prut, E.V. (1992) Ultra-high modulus isotactic polypropylene. 1. The influence of orientation drawing and initial morphology on the structure and properties of oriented samples. *Journal of Applied Polymer Science*, **44**, 1557–1572.

13 Alcock, B., Cabrera, N.O., Barkoula, N.M., and Peijs, T. (2009) The effect of processing conditions on the mechanical properties and thermal stability of highly oriented PP tapes. *European Polymer Journal*, **45**, 2878–2894.

14 Karger-Kocsis, J., Wanjale, S.D., Abraham, T., Bárány, T., and Apostolov, A.A. (2010) Preparation and characterization of polypropylene homocomposites: exploiting polymorphism of PP homopolymer. *Journal of Applied Polymer Science*, **115**, 684–691.

15 Schimanski, T., Loos, J., Peijs, T., Alcock, B., and Lemstra, P.J. (2007) On the overdrawing of melt-spun isotactic polypropylene tapes. *Journal of Applied Polymer Science*, **103**, 2920–2931.

16 Kmetty, Á., Bárány, T., and Karger-Kocsis, J. (2010) Self-reinforced polymeric materials: a review. *Progress in Polymer Science*, **35**, 1288–1310.

17 Smith, P. and Lemstra, P.J. (1979) Ultrahigh-strength polyethylene filaments by solution spinning/drawing. 2. Influence of solvent on the drawability. *Makromolekulare Chemie*, **180**, 2983–2986.

18 Barham, P.J. and Keller, A. (1985) High-strength polyethylene fibers from solution and gel spinning. *Journal of Materials Science*, **20**, 2281–2302.

19 Pennings, A.J., Vanderhooft, R.J., Postema, A.R., Hoogsteen, W., and Tenbrinke, G. (1986) High-speed gel-spinning of ultrahigh molecular-weight polyethylene. *Polymer Bulletin*, **16**, 167–174.

20 Karger-Kocsis, J. and Bárány, T. (2007) Silk-containing polymeric systems, in *Handbook of Engineering Biopolymers* (eds S. Fakirov and D. Bhattacharyya), Hanser, Munich, pp. 485–505.

21 Morye, S.S., Hine, P.J., Duckett, R.A., Carr, D.J., and Ward, I.M. (2000) Modelling of the energy absorption by polymer composites upon ballistic impact. *Composites Science and Technology*, **60**, 2631–2642.

22 Jacobs, M.J.N. and Van Dingenen, J.L.J. (2001) Ballistic protection mechanisms in personal armour. *Journal of Materials Science*, **36**, 3137–3142.

23 Alves, A.L.S., Nascimento, L.F.C., and Suarez, J.C.M. (2005) Influence of weathering and gamma irradiation on the mechanical and ballistic behavior of UHMWPE composite armor. *Polymer Testing*, **24**, 104–113.

24 Holmes, G.A., Rice, K., and Snyder, C.R. (2006) Ballistic fibers: a review of the thermal, ultraviolet and hydrolytic stability of the benzoxazole ring structure. *Journal of Materials Science*, **41**, 4105–4116.

25 Chocron, S., Pintor, A., Galvez, F., Rosello, C., Cendon, D., and Sanchez-Galvez, V. (2008) Lightweight polyethylene non-woven felts for ballistic impact applications: material characterization. *Composites Part B*, **39**, 1240–1246.

26 Rodriguez, J., Chocron, I.S., Martinez, M.A., and Sanchez-Galvez, V. (1996) High strain rate properties of aramid and polyethylene woven fabric composites. *Composites Part B*, **27**, 147–154.

27 Kromm, F.X., Lorriot, T., Coutand, B., Harry, R., and Quenisset, J.M. (2003) Tensile and creep properties of ultra high molecular weight PE fibres. *Polymer Testing*, **22**, 463–470.

28 Brewis, D.M. and Briggs, D. (1981) Adhesion to polyethylene and polypropylene. *Polymer*, **22**, 7–16.

29 Grubb, D.T. and Li, Z.F. (1994) Single-fiber polymer composites. 2. Residual-stresses and their effects in high-modulus polyethylene fiber composites. *Journal of Materials Science*, **29**, 203–212.

30 Hoecker, F., Friedrich, K., Blumberg, H., and Karger-Kocsis, J. (1995) Effects of fiber/matrix adhesion on off-axis mechanical response in carbon-fiber/epoxy-resin composites. *Composites Science and Technology*, **54**, 317–327.

31 Silverstein, M.S. and Breuer, O. (1993) Mechanical-properties and failure of etched UHMW-PE fibers. *Journal of Materials Science*, **28**, 4153–4158.

32 Vaisman, L., Gonzalez, M.F., and Marom, G. (2003) Transcrystallinity in brominated UHMWPE fiber reinforced HDPE composites: morphology and dielectric properties. *Polymer*, **44**, 1229–1235.

33 Maity, J., Jacob, C., Das, C.K., Alam, S., and Singh, R.P. (2008) Direct fluorination of UHMWPE fiber and preparation of fluorinated and non-fluorinated fiber composites with LDPE matrix. *Polymer Testing*, **27**, 581–590.

34 Amornsakchai, T. and Doaddara, O. (2008) Grafting of acrylamide and acrylic acid onto polyethylene fiber for improved adhesion to epoxy resin. *Journal of Reinforced Plastics and Composites*, **27**, 671–682.

35 Biro, D.A., Pleizier, G., and Deslandes, Y. (1992) Application of the microbond technique. 3. Effects of plasma treatment on the ultra-high modulus polyethylene fiber epoxy interface. *Journal of Materials Science Letters*, **11**, 698–701.

36 Biro, D.A., Pleizier, G., and Deslandes, Y. (1993) Application of the microbond technique. 4. Improved fiber matrix adhesion by RF plasma treatment of

organic fibers. *Journal of Applied Polymer Science*, **47**, 883–894.

37 Moon, S.I. and Jang, J. (1999) The mechanical interlocking and wetting at the interface between argon plasma treated UHMPE fiber and vinylester resin. *Journal of Materials Science*, **34**, 4219–4224.

38 Intrater, R., Hoffman, A., Lempert, G., Gouzman, I., Cohen, Y., and Grossman, E. (2006) Visualization of polyethylene fibers surface restructuring induced by oxygen plasma. *Journal of Materials Science*, **41**, 1653–1657.

39 Wei, Q.F., Mather, R.R., Wang, X.Q., and Fotheringham, A.F. (2005) Functional nanostructures generated by plasma-enhanced modification of polypropylene fibre surfaces. *Journal of Materials Science*, **40**, 5387–5392.

40 Capiati, N.J. and Porter, R.S. (1975) The concept of one polymer composites modelled with high-density polyethylene. *Journal of Materials Science*, **10**, 1671–1677.

41 Ward, I.M. (2004) Developments in oriented polymers, 1970–2004. *Plastics Rubber and Composites*, **33**, 189–194.

42 Ward, I.M. and Hine, P.J. (2004) The science and technology of hot compaction. *Polymer*, **45**, 1413–1427.

43 Hine, P.J., Ward, I.M., El Matty, M.I.A., Olley, R.H., and Bassett, D.C. (2000) The hot compaction of 2-dimensional woven melt spun high modulus polyethylene fibres. *Journal of Materials Science*, **35**, 5091–5099.

44 Hine, P.J., Ward, I.M., Jordan, N.D., Olley, R.H., and Bassett, D.C. (2001) A comparison of the hot-compaction behavior of oriented, high-modulus, polyethylene fibers and tapes. *Journal of Macromolecular Science*, **B40**, 959–989.

45 El-Maaty, M.I.A., Bassett, D.C., Olley, R.H., Hine, P.J., and Ward, I.M. (1996) The hot compaction of polypropylene fibres. *Journal of Materials Science*, **31**, 1157–1163.

46 Karger-Kocsis, J. (2000) Interphase with lamellar interlocking and amorphous adherent: a model to explain effects of transcrystallinity. *Advanced Composites Letters*, **9**, 225–227.

47 Ratner, S., Pegoretti, A., Migliaresi, C., Weinberg, A., and Marom, G. (2005) Relaxation processes and fatigue behavior of crosslinked UHMWPE fiber compacts. *Composites Science and Technology*, **65**, 87–94.

48 Hine, P.J. and Ward, I.M. (2005) High stiffness and high impact strength polymer composites by hot compaction of oriented fibers and tapes, in *Mechanical Properties of Polymers Based on Nanostructure and Morphology* (eds F.J. Baltá-Calleja and G.H. Michler), CRC Press, Boca Raton, pp. 677–698.

49 Jordan, N.D., Olley, R.H., Bassett, D.C., Hine, P.J., and Ward, I.M. (2002) The development of morphology during hot compaction of Tensylon high-modulus polyethylene tapes and woven cloths. *Polymer*, **43**, 3397–3404.

50 Hine, P.J., Ward, I.M., Jordan, N.D., Olley, R., and Bassett, D.C. (2003) The hot compaction behaviour of woven oriented polypropylene fibres and tapes. I. Mechanical properties. *Polymer*, **44**, 1117–1131.

51 Jordan, N.D., Bassett, D.C., Olley, R.H., Hine, P.J., and Ward, I.M. (2003) The hot compaction behaviour of woven oriented polypropylene fibres and tapes. II. Morphology of cloths before and after compaction. *Polymer*, **44**, 1133–1143.

52 Hine, P.J., Olley, R.H., and Ward, I.M. (2008) The use of interleaved films for optimising the production and properties of hot compacted, self reinforced polymer composites. *Composites Science and Technology*, **68**, 1413–1421.

53 Hine, P., Broome, V., and Ward, I. (2005) The incorporation of carbon nanofibres to enhance the properties of self reinforced, single polymer composites. *Polymer*, **46**, 10936–10944.

54 Bárány, T., Izer, A., and Karger-Kocsis, J. (2009) Impact resistance of all-polypropylene composites composed of alpha and beta modifications. *Polymer Testing*, **28**, 176–182.

55 Bárány, T., Karger-Kocsis, J., and Czigány, T. (2006) Development and characterization of self-reinforced poly (propylene) composites: carded mat

reinforcement. *Polymers for Advanced Technologies*, **17**, 818–824.

56 Izer, A. and Bárány, T. (2010) Effect of consolidation on the flexural creep behaviour of all-polypropylene composite. *Express Polymer Letters*, **4**, 210–216.

57 Izer, A., Bárány, T., and Varga, J. (2009) Development of woven fabric reinforced all-polypropylene composites with beta nucleated homo- and copolymer matrices. *Composites Science and Technology*, **69**, 2185–2192.

58 Varga, J. (2002) Beta-modification of isotactic polypropylene: preparation, structure, processing, properties, and application. *Journal of Macromolecular Science Part B*, **41**, 1121–1171.

59 Karger-Kocsis, J. (2007) Composite composed of polypropylene reinforcement and polypropylene matrix and various production methods thereof. German Patent DE 102,37,803, Germany.

60 Abraham, T.N., Siengchin, S., and Karger-Kocsis, J. (2008) Dynamic mechanical thermal analysis of all-PP composites based on beta and alpha polymorphic forms. *Journal of Materials Science*, **43**, 3697–3703.

61 Cohen, Y., Rein, D.M., and Vaykhansky, L. (1997) A novel composite based on ultra-high-molecular-weight polyethylene. *Composites Science and Technology*, **57**, 1149–1154.

62 Peijs, T. (2003) Composites for recyclability. *Materials Today*, **6**, 30–35.

63 Alcock, B., Cabrera, N.O., Barkoula, N.M., Loos, J., and Peijs, T. (2006) The mechanical properties of unidirectional all-polypropylene composites. *Composites Part A*, **37**, 716–726.

64 Alcock, B., Cabrera, N.O., Barkoula, N.M., Loos, J., and Peijs, T. (2007) Interfacial properties of highly oriented coextruded polypropylene tapes for the creation of recyclable all-polypropylene composites. *Journal of Applied Polymer Science*, **104**, 118–129.

65 Alcock, B., Cabrera, N.O., Barkoula, N.M., and Peijs, T. (2006) Low velocity impact performance of recyclable all-polypropylene composites. *Composites Science and Technology*, **66**, 1724–1737.

66 Alcock, B., Cabrera, N.O., Barkoula, N.M., Reynolds, C.T., Govaert, L.E., and Peijs, T. (2007) The effect of temperature and strain rate on the mechanical properties of highly oriented polypropylene tapes and all-polypropylene composites. *Composites Science and Technology*, **67**, 2061–2070.

67 Alcock, B., Cabrera, N.O., Barkoula, N.M., Spoelstra, A.B., Loos, J., and Peijs, T. (2007) The mechanical properties of woven tape all-polypropylene composites. *Composites Part A*, **38**, 147–161.

68 Alcock, B., Cabrera, N.O., Barkoula, N.M., Wang, Z., and Peijs, T. (2008) The effect of temperature and strain rate on the impact performance of recyclable all-polypropylene composites. *Composites Part B*, **39**, 537–547.

69 Abraham, T., Banik, K., and Karger-Kocsis, J. (2007) All-PP composites (PURE®) with unidirectional and cross-ply lay-ups: dynamic mechanical thermal analysis. *Express Polymer Letters*, **1**, 519–526.

70 Banik, K., Abraham, T.N., and Karger-Kocsis, J. (2007) Flexural creep behavior of unidirectional and cross-ply all-poly (propylene) (PURE®) composites. *Macromolecular Materials and Engineering*, **292**, 1280–1288.

71 Kim, K.J., Yu, W.R., and Harrison, P. (2008) Optimum consolidation of self-reinforced polypropylene composite and its time-dependent deformation behavior. *Composites Part A*, **39**, 1597–1605.

72 Shalom, S., Harel, H., and Marom, G. (1997) Fatigue behaviour of flat filament-wound polyethylene composites. *Composites Science and Technology*, **57**, 1423–1427.

73 Houshyar, S. and Shanks, R.A. (2003) Morphology, thermal and mechanical properties of poly(propylene) fibre–matrix composites. *Macromolecular Materials and Engineering*, **288**, 599–606.

74 Houshyar, S. and Shanks, R.A. (2006) Mechanical and thermal properties of flexible poly(propylene) composites. *Macromolecular Materials and Engineering*, **291**, 59–67.

75 Houshyar, S. and Shanks, R.A. (2007) Mechanical and thermal properties of toughened polypropylene composites. *Journal of Applied Polymer Science*, **105**, 390–397.

76 Bárány, T., Izer, A., and Czigány, T. (2006) On consolidation of self-reinforced polypropylene composites. *Plastics Rubber and Composites*, **35**, 375–379.

77 Kazanci, M., Cohn, D., and Marom, G. (2001) Elastic and viscoelastic behavior of filament wound polyethylene fiber reinforced polyolefin composites. *Journal of Materials Science*, **36**, 2845–2850.

78 von Lacroix, F., Werwer, M., and Schulte, K. (1998) Solution impregnation of polyethylene fibre polyethylene matrix composites. *Composites Part A*, **29**, 371–376.

79 Chand, N., Kreuzberger, S., and Hinrichsen, G., (1994) Influence of processing conditions on the tensile properties of unidirectional UHMWPE fiber LDPE composites. *Composites*, **25**, 878–880.

80 von Lacroix, F., Lu, H.Q., and Schulte, K. (1999) Wet powder impregnation for polyethylene composites: preparation and mechanical properties. *Composites Part A*, **30**, 369–373.

81 Wan, Y.Z., Luo, H.L., Wang, Y.L., Raman, S., Huang, Y., Zhang, T.L., and Liu, H. (2006) Characterization of three-dimensional braided polyethylene fiber–PMMA composites and influence of fiber surface treatment. *Journal of Applied Polymer Science*, **99**, 949–956.

82 Karger-Kocsis, J. and Varga, J. (1999) Interfacial morphology and its effects in polypropylene composites, in *Polypropylene: An A–Z Reference* (ed. J. Karger-Kocsis), Kluwer Publishers, Dordrecht, pp. 348–356.

83 von Lacroix, F., Loos, J., and Schulte, K. (1999) Morphological investigations of polyethylene fibre reinforced polyethylene. *Polymer*, **40**, 843–847.

84 Kitayama, T., Utsumi, S., Hamada, H., Nishino, T., Kikutani, T., and Ito, H. (2003) Interfacial properties of PP/PP composites. *Journal of Applied Polymer Science*, **88**, 2875–2883.

85 Shavit-Hadar, L., Rein, D.M., Khalfin, R., Terry, A.E., Heunen, G., and Cohen, Y. (2007) Compacted UHMWPE fiber composites: morphology and X-ray microdiffraction experiments. *Journal of Polymer Science Part B*, **45**, 1535–1541.

86 Rochette, A., Bousmina, M., Lavoie, A., and Ajji, A. (2002) Effect of surface treatment on mechanical properties of polyethylene composite. *Journal of Composite Materials*, **36**, 925–940.

87 Taketa, I., Ustarroz, J., Gorbatikh, L., Lomov, S.V., and Verpoest, I. (2010) Interply hybrid composites with carbon fiber reinforced polypropylene and self-reinforced polypropylene. *Composites Part A*, **41**, 927–932.

88 Kuan, H.T., Cantwell, W.J., and Md Akil, H. (2009) The mechanical properties of hybrid composites based on self-reinforced polypropylene. *Malaysian Polymer Journal*, **4**, 71–80.

89 Andreopoulos, A.G., Liolios, K., and Patrikis, A. (1993) Treated polyethylene fibers as reinforcement for epoxy resins. *Journal of Materials Science*, **28**, 5002–5006.

90 Andreopoulos, A.G. and Tarantili, P.A. (1997) Effect of hygrothermal treatment on the flexural properties of UHMPE fibre/epoxy resin composites. *Advanced Composites Letters*, **6**, 103–108.

91 Chaoting, Y., Gao, S., and Mu, Q. (1993) Effect of low-temperature-plasma surface-treatment on the adhesion of ultra-high-molecular-weight-polyethylene fibers. *Journal of Materials Science*, **28**, 4883–4891.

92 Jana, S., Hinderliter, B.R., and Zhong, W.H. (2008) Analytical study of tensile behaviors of UHMWPE/nano-epoxy bundle composites. *Journal of Materials Science*, **43**, 4236–4246.

93 Dutra, R.C.L., Soares, B.G., Campos, E.A., De Melo, J.D.G., and Silva, J.L.G. (1999) Composite materials constituted by a modified polypropylene fiber and epoxy resin. *Journal of Applied Polymer Science*, **73**, 69–73.

94 Prabhu, T.N., Hemalatha, Y.J., Harish, V., Prashantha, K., and Iyengar, P. (2007) Thermal degradation of epoxy resin reinforced with polypropylene fibers.

Journal of Applied Polymer Science, **104**, 500–503.

95 Qian, D., Bao, L., Takatera, M., Kemmochi, K., and Yamanaka, A. (2010) Fiber-reinforced polymer composite materials with high specific strength and excellent solid particle erosion resistance. *Wear*, **268**, 637–642.

96 Rhee, K.Y., Oh, T.Y., Paik, Y.N., Park, H.J., and Kim, S.S. (2004) Tensile behavior of polyethylene fiber composites with polyethylene fiber surface-modified using ion irradiation. *Journal of Materials Science*, **39**, 1809–1811.

97 Williams, M.A., Bauman, B.D., and Thomas, D.A. (1991) Incorporation of surface-modified UHMWPE powders and fibers in tough polyurethane composites. *Polymer Engineering and Science*, **31**, 992–998.

98 Misra, R.K. and Datta, C. (2009) Mechanical behavior of polyethylene fibers reinforced Resol/VAC-EHA. *Journal of Macromolecular Science Part A*, **46**, 425–437.

99 Lin, S.P., Han, J.L., Yeh, J.T., Chang, F.C., and Hsieh, K.H. (2007) Composites of UHMWPE fiber reinforced PU/epoxy grafted interpenetrating polymer networks. *European Polymer Journal*, **43**, 996–1008.

100 Karger-Kocsis, J. (2006) Simultaneous interpenetrating network structured vinylester/epoxy hybrids and their use in composites, in *Micro- and Nanostructured Multiphase Polymer Blend Systems* (eds C. Harrats, S. Thomas, and G. Groeninckx), CRC Press, Boca Raton, pp. 273–293.

101 Zhong, W.H., Sui, G., Jana, S., and Miller, J. (2009) Cosmic radiation shielding tests for UHMWPE fiber/nano-epoxy composites. *Composites Science and Technology*, **69**, 2093–2097.

102 Park, R. and Jang, J. (1999) Performance improvement of carbon fiber polyethylene fiber hybrid composites. *Journal of Materials Science*, **34**, 2903–2910.

103 Goettler, L.A. and Shen, K.S. (1983) Short fiber reinforced elastomers. *Rubber Chemistry and Technology*, **56**, 619–638.

104 Hoecker, F. and Karger-Kocsis, J. (1993) Effects of crystallinity and supermolecular formations on the interfacial shear-strength and adhesion in GF PP composites. *Polymer Bulletin*, **31**, 707–714.

105 Hoecker, F. and Karger-Kocsis, J. (1995) On the effects of processing conditions and interphase of modification on the fiber/matrix load transfer in single fiber polypropylene composites. *Journal of Adhesion*, **52**, 81–100.

106 Shaker, M., Kamel, I., and Abdelbary, E.M. (1995) UHMW-PE fiber as reinforcing materials in EPDM rubber vulcanized by e-beam radiation. *Journal of Elastomers and Plastics*, **27**, 117–137.

107 Xu, T. and Farris, R.J. (2007) Comparative studies of ultra high molecular weight polyethylene fiber reinforced composites. *Polymer Engineering and Science*, **47**, 1544–1553.

108 Karger-Kocsis, J. (1999) Thermoplastic rubbers via dynamic vulcanization, in *Polymer Alloys and Blends* (eds G.O. Shonaike and G.P. Simon), Marcel Dekker, New York, pp. 125–153.

109 Chabba, S., van Es, M., van Klinken, E.J., Jongedijk, M.J., Vanek, D., Gijsman, P., and van der Waals, A. (2007) Accelerated aging study of ultra high molecular weight polyethylene yarn and unidirectional composites for ballistic applications. *Journal of Materials Science*, **42**, 2891–2893.

110 Chung, D.D.L. (2004) Use of polymers for cement-based structural materials. *Journal of Materials Science*, **39**, 2973–2978.

111 Fahmy, M.F.M. and Wu, Z. (2010) Evaluating and proposing models of circular concrete columns confined with different FRP composites. *Composites Part B*, **41**, 199–213.

112 Zhang, X.X., Wang, X.C., Tao, X.M., and Yick, K.L. (2005) Energy storage polymer/microPCMs blended chips and thermo-regulated fibers. *Journal of Materials Science*, **40**, 3729–3734.

113 Ciselli, P., Zhang, R., Wang, Z.J., Reynolds, C.T., Baxendale, M., and Peijs, T. (2009) Oriented UHMW-PE/CNT composite tapes by a solution casting-drawing process using mixed-solvents. *European Polymer Journal*, **45**, 2741–2748.

114 Yoshioka, T., Dersch, R., Tsuji, M., and Schaper, A.K. (2010) Orientation analysis of individual electrospun PE nanofibers by transmission electron microscopy. *Polymer*, **51**, 2383–2389.

115 Alderson, K.L., Alderson, A., Davies, P.J., Smart, G., Ravirala, N., and Simkins, G. (2007) The effect of processing parameters on the mechanical properties of auxetic polymeric fibers. *Journal of Materials Science*, **42**, 7991–8000.

116 Gonzalez-Chi, P.I. and Young, R.J. (2004) Deformation micromechanics of a thermoplastic–thermoset interphase of epoxy composites reinforced with polyethylene fiber. *Journal of Materials Science*, **39**, 7049–7059.

11
Silica Fiber-Reinforced Polymer Composites
Sudip Ray

11.1
Introduction

Fillers are added to polymers for a variety of purposes, of which the most important are reinforcement, increase in stiffness, reduction in material costs, and improvements in processing. Reinforcement is primarily the enhancement of strength and strength-related properties. In present days, varieties of fillers have been used to develop polymer composites, which could be particulate or fibrous in nature. Among the fibrous fillers glass fiber is the most widely used fiber in reinforced polymers. Several other fibers are also been commercially used, for example, cotton, sisal, and jute fiber obtained from natural resources; nylon, polyester, and rayon from synthetic process, or organic and inorganic high-performance fibers such as aramid, boron, and carbon/graphite fiber. Compared to other different fibers silica fiber is chemically very similar to glass fiber. The primary constituent of silica fiber and glass fiber is silicon dioxide. In contrast to glass fiber, silica fiber possesses very high silica content, usually more than 95% by weight, whereas a typical glass fiber used for polymer reinforcement contains only 52–56% by weight silica. Silica fibers can be fabricated synthetically and also available in nature whereas glass fiber can only be obtained via synthetic route. The surface chemistry of the silica-based filler plays a leading role toward processing and the final properties of the silica-filled polymer composites. The presence of active functional groups on silica fiber can contribute on polymer reinforcement. However, application of coupling agents can further enhance the ability of this silica-based fiber on the reinforcement process. A brief review of silica fibers used in various polymers and the role of silica fiber on determining the final properties of the composites have been documented in the following chapter.

11.2
Silica Fiber: General Features

The effect of fillers on properties of a composite depends on their concentration and their particle size and shape, and their interaction with the matrix [1, 2].

Polymer Composites: Volume 1, First Edition. Edited by Sabu Thomas, Kuruvilla Joseph,
Sant Kumar Malhotra, Koichi Goda, and Meyyarappallil Sadasivan Sreekala
© 2012 Wiley-VCH Verlag GmbH & Co. KGaA. Published 2012 by Wiley-VCH Verlag GmbH & Co. KGaA.

11.2.1
Types

Silica in its fiber form is available in nature and could be produced synthetically via chemical route. The synthetic variety of silica is commonly used in polymer composites, however, in recent years attempts have been made to explore the potential use of natural form of silica fiber to realize its suitability as reinforcing filler in polymer composites.

Natural silica fiber could be procured as highly pure as synthetic one. For example, natural inorganic silica short fiber, known as Silexil or Biogenic silica is obtained by multicellular animal fossilization derived from the skeleton of sponges (Demospongiae) and related organisms [3]. The mineral called "Spongolite" (or spongillite) is the source of these fibers. After extraction from land they are purified by conventional mineral treatment methods. These minerals are usually found in high amounts in sediments of lake beds together with clay, sand, and organic matter. In Brazil, it occurs in peat–bog ponds in the southwestern part of Minas Gerais, southern Goia's, northeastern Mato Grosso do Sul, Sao Paulo, and Bahia [4, 5].

Synthetic silica with highly elongated forms can be obtained during drying of thin films of silica sols, by either conversion from fibrous precursors or unidirectional crazing or cracking [6]. High purity silica fibers are manufactured usually by melt drawing soda glass through spinnerets of precise dimensions followed by leaching out the soda by acid, or by multistage drawing of fused silica rods through graphite bushes. Oxidation of silicon monoxide can also form fibrous silica. Vapor phase reaction of silica and silicon metal followed by condensation can lead to formation of hollow tubes and spiral fibers of amorphous silica with high aspect ratio, typically less than $0.04\,\mu m$ in diameter and many micrometers long [7]. Silica W, an unstable crystalline silica fiber can be converted to amorphous silica fibers by traces of moisture [8]. An electrically heated platinum surface exposed to nitrogen-diluted SiF_4, and water vapor at $1100\,°C$ can also produce amorphous anhydrous silica fiber [9].

11.2.2
Characteristics

The chief component of silica fiber is silicon dioxide. Silica fibers from natural sources or its synthetic variety consist of very high silica content, usually more than 95% by weight. Various other oxides such as oxides of titanium, aluminum, iron, calcium, magnesium, copper, potassium, sulfur at lower concentrations (below 1 wt%) may also be present. Production of this high purity silica fiber synthetically is quite expensive, but its natural form can be obtained from inexpensive mineral source at lower processing costs. These natural silica fibers mainly consist of pure amorphous silica (SiO_2) with a hollow cylindrical form and a sharp extremity. Silica consists of silicon and oxygen, tetrahedrally bound in an imperfect three dimensional structure and has strong polar surface groups, mostly hydroxyl groups bound to silicon known as silanol ($-Si-O-H$). The imperfections in its lattice structure provide free silanol

Figure 11.1 Scanning electron micrographs of natural amorphous silica fiber (a and c), and commercial glass fiber (b and d) at magnification of ×75 and ×1500 respectively. Reprinted from Properties of chemically treated natural amorphous silica fibers as polyurethane reinforcement [13], with permission from John Wiley & Sons, Inc., Copyright 2006 Society of Plastics Engineers.

groups on the surface. These functional groups are arbitrarily located on the fiber surface. Because of the siliceous nature with active functional surface silanol groups and morphological similarity with glass fiber, silica fiber also possesses potential reinforcing ability for polymeric composites.

Segatelli *et al.* examined the X-ray diffraction pattern for natural silica fiber [10]. A broad peak was obtained at a diffraction angle (2θ) ~22.5° corresponds to interlayer distance of 4 Å, which is characteristic of amorphous opaline silica [11, 12]. Surface area of the silica fiber determined by BET technique was found to be $1.14 \, \text{m}^2/\text{g}$ [10].

Scanning electron micrographs (Figure 11.1) [13] of the natural amorphous silica short fiber and a typical commercial glass fiber indicate that both fibers have similar fiber length and aspect ratio with smooth tubular needlelike shape. On the other hand, as compared to glass fiber, these natural fibers have lower density, for example, the density of natural amorphous silica short fiber is $1.64 \pm 0.41 \, \text{g/cm}^3$ and that of glass fiber is $2.54 \pm 0.71 \, \text{g/cm}^3$ [13]. As a result, natural amorphous silica short fiber could allow forming lighter weight polymer composites, while they were incorporated in polymer matrixes. Moreover, these fibers are harder than glass fibers, resulting in higher equipment wear. Since they are cheaper than glass fiber the compound cost can reduce. They are reported to be nontoxic [4, 13], which attracted further research to be done using this fiber.

In general, silica fiber can offer excellent low thermal conductivity and electrical insulation properties, high tensile strength and modulus, inert with respect to the majority of chemical reagents, resistant to organic and mineral acids even at the elevated temperature (except of hydrofluoric, phosphoric, and hydrochloric acid), and weak alkalies resists corrosion, have high chemical resistance to water and steam of high pressure, stable in vacuum and provide adequate flexibility.

11.2.3
Surface Treatment of Silica Fiber

Fibrous materials offer good reinforcement to polymers, depending on the strength, length, and aspect ratio of the fiber and the effectiveness of the fiber/matrix interaction. Silanol groups ($-SiOH$) present in silica fiber surface possess high affinity in their reactions with amines, alcohols, metal ions, and also water adsorbed on the fiber surface reduce reactivity of this functional group. Some of the reactions with silanols can have a significant effect on the processing and composite properties while used as filler for rubber compounds, especially where the chemical involved is an important part of the cure system. On the contrary, judicial utilization of these functional groups by modifying the surface chemistry and thus transformation from less-reactive filler to a more reactive ingredient for effective combination with polymer matrix could be possible. This relates to the compatibility of this fibrous filler with a specific polymer and the ability of the polymer to adhere to the filler surface. The compounded strength can be further improved if the matrix adheres to the mineral surface via chemical bonding.

11.2.3.1 Surface Modification: Types and Methods

Surface treatment of silica fiber can be done by both physical and chemical modification. In the physical modification, the interaction between the adsorbed surface modifier and the polymer matrix is weak, but interaction between the adsorbed surface modifier and the fiber surface is sufficient enough to modify the fiber surface polarity, such that it could match the polarity that of the polymer matrix. Chemical modification of silica fiber can be performed by the following methods: (a) grafts of chemical groups on the fiber surface to change the surface characteristics, that is, the case where interaction between the surface modifier and the fiber surface is strong but interaction between the adsorbed surface modifier and the polymer matrix is weak and (b) grafts that may react with fiber and polymer, that is, the case where both the interactions are strong. The former is referred to as monofunctional coupling agent where no chemical reaction with the polymer takes place with these grafts. The latter are called bifunctional coupling agents as they provide chemical linkages between the fiber surface and the polymer molecules.

In common practice, silane coupling agents are being employed to modify the silica fiber surface. Depending on the silane type and processing steps involved, an appropriate methodology is required to carry out the surface treatment of the silica fiber with the coupling agent. This could be either *in situ* method, that is, direct mixing of silane with silica filler during the compounding process or pretreatment method, that is, interaction of silane coupling agent with silica fiber prior to the compounding process. Both the processes have merits and demerits. Pretreatment of silica fiber with silanes require additional steps to modify it, whereas *in situ* process could affect efficiency of the silanization process and safety of the working atmosphere due to elimination of hazardous volatiles. Hence, selection of the silane treatment of the silica fiber primarily depends on the silane type, compounding formulation, processing type and conditions. Usually, pretreatment method involves interaction of silica fiber with silane coupling at elevated temperature to facilitate the coupling reaction between the surface functional groups at silica fiber and the coupling agent. Wang *et al.* [14, 15] investigated the effect of various silane treatment, namely γ-aminopropyl-triethoxysilane (γ-APS), γ-glycidoxypropyl-trimethoxysilane (γ-GPS), γ-methacryloxypropyl-trimethoxysilane (γ-MPS), and trimethylchlorosilane (TMCS), on silica fiber applying pretreatment method. Prior to the surface treatment, the silane coupling agents were hydrolyzed in aqueous ethanol (95%) for 1 h followed by dipping the desized silica fabrics in the silane solution (0.5%) for 30 min and dried in an oven at 110 °C for 30 min. However, there are few reports where silanization reaction was performed at room temperature. For example, Silva *et al.* [3] investigated the effect of silane-treated silica fiber as primary or secondary filler on the reinforcement of silicone rubber. The natural short silica fiber surface was modified with vinyltrimethoxysilane (VTMS) coupling agent. The silane coupling agent VTMS was hydrolyzed at a molar ratio of 1 : 1 (VTMS : H_2O). The solution of tetrahydrofuran (THF) containing VTMS and catalyst *n*-butyl titanate, 2% (wt/wt) with respect to the silane amount was mechanically stirred for 3 h at 25 °C. Natural short fiber was reacted with this coupling agent solution for 24 h at 25 °C under argon atmosphere with mechanical stirring. The modified silica short fiber was filtered and washed with THF for three times to eliminate the physically adsorbed silane on the fiber surface and dried at 60 °C for 48 h. Martins *et al.* [16] and Barra *et al.* [13], described pretreatment of amorphous silica fibers with aminosilane. Initially, the fibers were added at a 1 mol/L HCl solution and maintained under stirring for 2 h at a temperature of 60 °C, washed with distilled water, and dried under vacuum for 48 h. Then the fibers were added to a solution of 2% aminosilane in ethanol/water (90/10 v/v) and kept under stirring at room temperature for 3 h. Subsequently, fibers were washed with toluene and dried for 12 h, temperature of 60 °C.

11.2.3.2 Characterization of Surface-Pretreated Silica Fiber

The coupling agent can use several analytical techniques to estimate the extent of surface modification of silica fiber. This is quite essential to determine its loading in the compound formulation and for the quality control purposes prior to mixing of surface-pretreated silanized silica fibers with the polymer.

The evidence of the presence of coupling agent in the silane-pretreated silica fiber can be obtained by identifying the characteristics groups of silane by FTIR spectroscopy or Raman spectroscopy.

Wang et al. [14] applied diffuse reflectance infrared Fourier transform (DRIFT) spectroscopy technique to characterize the surface-treated silica fibers by investigating the possibility of adsorption of coupling agents on the silica fiber surface by the formation of hydrogen bonding (Figure 11.2). Free silanol groups observed at 3740 cm^{-1} for untreated silica fiber disappeared after the silane treatment specifying the involvement of the surface functional group in the silanization process. Moreover, appearance of new peaks at 2860–2980 cm^{-1} in the silane-treated silica fiber indicates the attachment of silane coupling agent on the silica fiber surface. The characteristics peaks due to C−O−C, C=O, and NH$_2$ stretching vibrations at 965,

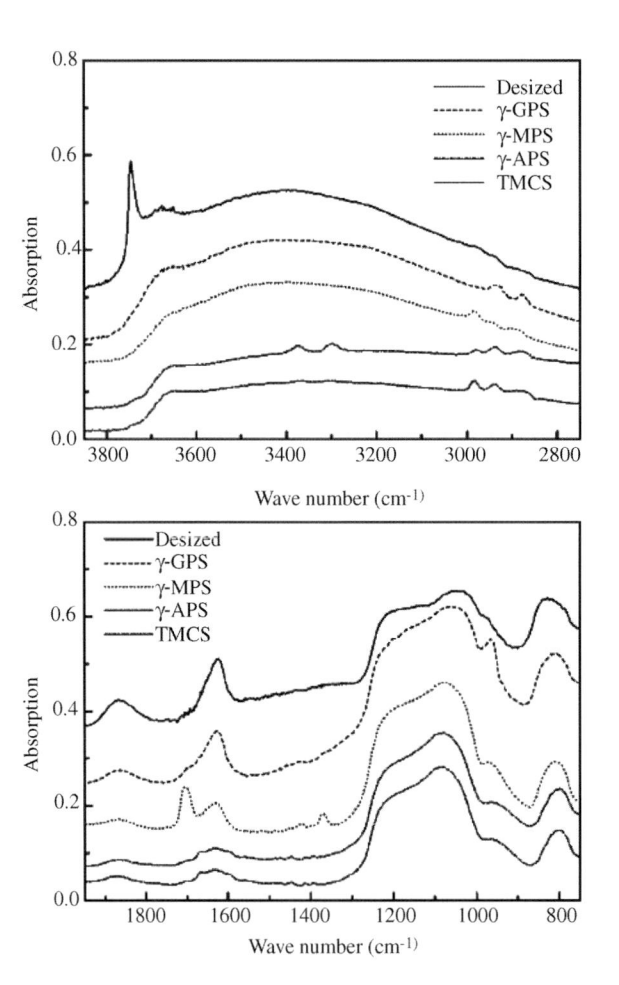

Figure 11.2 DRIFT spectra of desized silica fiber (a, without silane treated) and (b) treated with various coupling agents γ-GPS, γ-MPS, γ-APS, and TMCS, respectively. Reprinted from Effects of fibre surface silanisation on silica fibre/phenolics composites produced by resin transfer moulding process [14], with permission from Maney Publishing, Copyright 2006 Institute of Materials, Minerals and Mining.

1705, 3371, and 3301 cm^{-1} from the γ-GPS, γ-MPS, and γ-APS silanes were also identified in the respective silane-treated silica fiber.

Barra et al. [13], observed an enlargement of the band in the region of 1610 and 1200 cm^{-1}, which could be due to the occurrence of new bands related to the symmetric angular deformation of the N−H groups and the axial deformation of C−N, indicating the incorporation of the aminosilane coupling agent to the fibers.

Plausible interactions between a typical aminosilane (3-aminopropyltriethoxysilane) and silanol groups present in silica fiber surface is presented in Scheme 11.1. Silanol functional groups of silica fiber and methoxy functional groups from the silane coupling agent may form covalent linkages via condensation process. Additionally, the amine group from the silane coupling agent may also interact with the silanol groups of silica fiber via proton transfer and hydrogen bond formation.

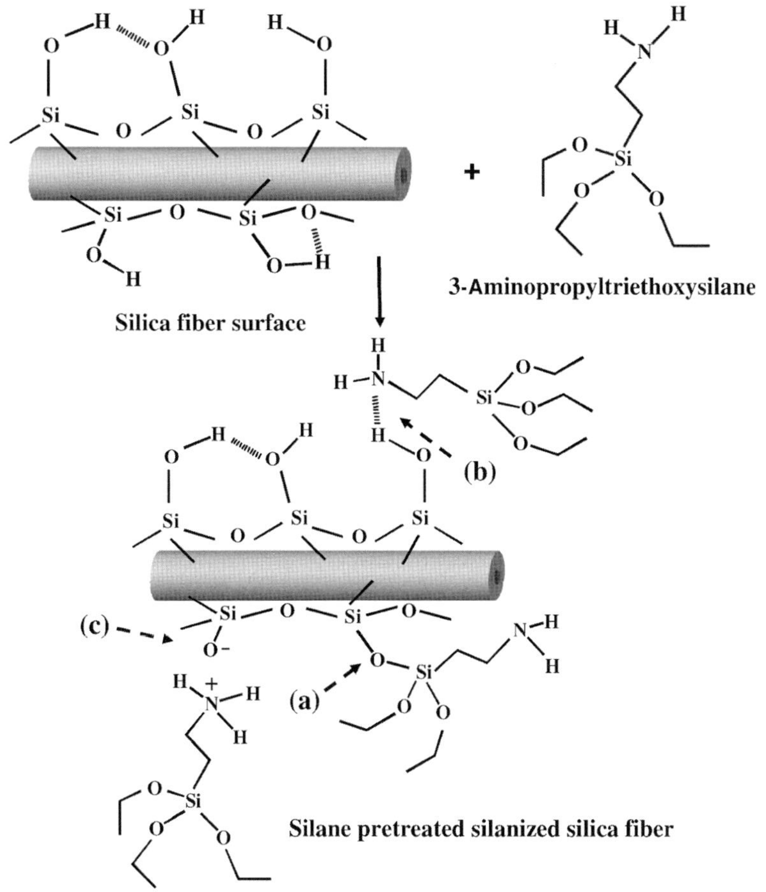

Scheme 11.1 Schematic presentation of plausible interactions between 3-aminopropyltriethoxysilane and silica fiber surface: (a) condensation with surface silanol groups (b) hydrogen bond formation, and (c) proton transfer.

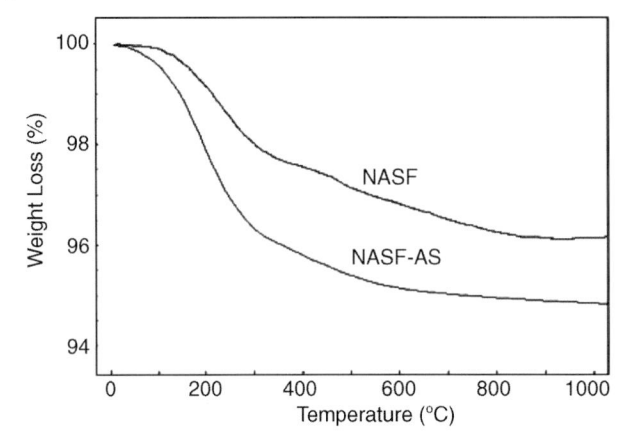

11 Silica Fiber-Reinforced Polymer Composites

Figure 11.3 TG thermograms for natural amorphous silica fiber (NASF) and aminosilane-treated silica fiber (NASF-AS). Reprinted from Properties of chemically treated natural amorphous silica fibers as polyurethane reinforcement [13], with permission from John Wiley & Sons, Inc., Copyright 2006 Society of Plastics Engineers.

While spectroscopic analysis can be used to detect the silane coupling agent in the silanized product, thermal analysis studies, for example, thermogravimetric analysis (TGA) could estimate the amount of silane coupling agent being incorporated in the treated samples from the additional weight loss due to the presence of silane coupling agent. Thus from TGA experiments Barra et al. [13], found three regions of thermal decomposition both for untreated and aminosilane-treated silica fiber, which can be accounted for the elimination of adsorbed water from the material at the different stages. However, additional weight loss in silane-treated silica fiber, between 100 and 400 °C, by 0.9 ± 0.2 wt% specifies the thermal decomposition of the silane agent and also quantifies its amount in the treated sample (Figure 11.3) [13].

In cases where a very small amount of coupling agent is attached with the silica fiber substrate, identification and quantitative estimation of the coupling agent in the pretreated filler is quite complicated and sometimes beyond the detection level of the experimental techniques, whereas tests on filler hydrophilicity/hydrophobicity can provide a simple way to assess the extent of surface modification. Generally the adsorption of silane coupling agents containing nonpolar groups, on the silica fiber surface induces decrease in the surface free energy of the modified silica fibers especially by reducing the surface energy of the polar component, which can be detected by wettability studies. Wang et al. [14] investigated the above study in details and found that silica fiber while treated with different silane coupling agent can reduce the surface free energy in the treated fiber. The effect was quite similar for γ-APS, γ-GPS, and γ-MPS treated silanes whereas in the case of TMCS-treated silica fiber the surface energy of the polar component was significantly dropped and hence the surface free energy of the modified silica fiber was reduced by about 40% from 50.9 to 29.2 mJ/m^2 as compared to the untreated silica fiber. Although TMCS is less reactive with respect to other silane coupling agents but the alkyl-terminated groups present in it, influenced the polarity of the silica fiber surface most effectively.

Some other techniques, for example, pyrolysis gas chromatography can be used for quantitative estimation of nature and amount of silane present in the treated fiber. Estimation of carbon content of the silane-treated silica fibers by ESCA, EDX, and so on can also be used for quantitative analysis. However, the difference between physically absorbed and chemically bound silane with the fiber surface can be confirmed by NMR spectroscopy.

11.3
Silica Fiber-Filled Polymer Composites

11.3.1
Fabrication of Composite

As compared to particulate fillers, compounding of fibrous fillers during the composite fabrication process require special care to obtain homogeneous dispersion in the polymer matrix. Dispersion of fibrous materials becomes more critical with increasing the fiber length and can seriously affect the polymer melt flow and the related processing. Hence, desizing the silica fiber prior to incorporation in the matrix polymer is a common practice. However, reducing the fiber length limits the effectiveness of its reinforcement. Long and continuous fibers can be used in compression molding, but these may require performing step.

11.3.2
Effect on Composite Properties

The reinforcing ability of the fiber in the polymer composites can be primarily determined by mechanical properties, for example, effect on modulus, tensile strength, elongation at break, dynamic mechanical properties, and so on. Apparently the reinforcing effect could be achieved by incorporating rigid fillers in the polymer matrix and from solvent swelling studies the effect on polymer–filler interaction could be estimated. Morphological studies are quite useful to comprehend the polymer–fiber adhesion by examining the extent of fiber pull-out in fractured composites. Several factors could affect the reinforcing ability of the fibrous fillers while they were incorporated in polymer matrix; for example, average length and diameter, aspect ratio, size distribution and orientation of fiber, amount of fiber loading. Furthermore, the polymer–fiber interactions play an important role in the reinforcement process depending on the type of polymer matrix and the fiber. In the following section, these are discussed with respect to several cases where silica fiber was used as the reinforcing filler in different polymer matrix and its reinforcing ability was inspected. A comparative study based on similar loadings of natural silica fiber in different polymers on composite properties was compiled in Table 11.1.

Table 11.1 Effect of reinforcement by natural silica fiber on the composite properties for different host polymer matrix.

Host polymer matrix	Fiber		Composite properties					Reference (adapted from)
	Type	Loading	Modulus (MPa)	Yield strength (MPa)	Tensile strength (MPa)	Elongation at break (%)	Notched Izod impact strength (J/m)	
Silicone rubber	Unfilled		0.07[a]	NA	0.29	620	NA	Silva et al. [3]
	Silica fiber	20 phr	0.2[a]	NA	0.55	639	NA	
	Silanized silica fiber	20 phr	0.18[a]	NA	0.67	579	NA	
Polyurethane	Unfilled		2.683[a]	NA	0.942	35.38	NA	Saliba et al. [5]
	Glass fiber	17 wt%	19.75[a]	NA	2.89	16.38	NA	
	Silica fiber	17 wt%	23.89[a]	NA	3.67	15.10	NA	
Polyurethane	Unfilled		470[c]	NA	0.94	81.6	NA	Barra et al. [13]
	Glass fiber	17 wt%	2330[c]	NA	2.58	38.7	NA	
	Silica fiber	17 wt%	3090[c]	NA	2.78	34.5	NA	
	Silanized silica fiber	17 wt%	NA	NA	3.1	38.8	NA	
Nylon 6	Unfilled		~2650[b]	77	NA	~33	27	Segatelli et al. [10]
	Silica fiber	20 wt%	~3500[b]	67	NA	~8	~27	
Nylon 6/EPDM-g-MA	Unfilled		~1600[b]	46	NA	~53	708	
	Silica fiber	20 wt%	~2400[b]	47	NA	~5	~125	
Epoxy	Unfilled		~0.92[c]	NA	~41.5	~2.8	NA	Martins et al. [16]
	Glass fiber	17.5 vol%	~1.03[c]	NA	~47.5	~0.8	NA	
	Silica fiber	17.5 vol%	~1.02[c]	NA	~47	~0.9	NA	
	Silanized silica fiber	17.5 vol%	~1.02[c]	NA	~49	~1.8	NA	

NA: data not available.
a) Young's modulus.
b) Flexural modulus.
c) Modulus of elasticity.

Saliba *et al.* [5] investigated the suitability of naturally occurring silica fiber (Silexil) as a reinforcing filler for plastics by incorporating this fiber to polyurethane (PU) and compared the composite properties with respect to commercial silane-treated glass fiber. Incorporation of only 17 wt% of silica fiber cause significant improvement in tensile strength and Young's modulus of unfilled polyurethane by about 290% (from 0.942 to 3.67 MPa) and 790% (from 2.683 to 23.89 MPa). In the case of commercial glass fiber as the reinforcing agent, the increment in tensile and Young's modulus properties were 207 and 636%, respectively. Elongation at break was dropped by 57 and 54%, respectively for Silexil and the silanized glass fiber as compared to the unfilled polyurethane. Thus, naturally occurring silica fiber even without any surface treatment showed potential usefulness as reinforcing agent for polyurethane and could provide superior reinforcing ability as compared to commercial silane-treated glass fiber.

Random dispersion of short silica fibers in the polyurethane matrix was evident from scanning electron micrographs of the fractured surface of composite material (Figure 11.4) [5]. The voids presents in the polymer matrix were described due to fiber pull-out during fracture, which could be due to inadequate level of polymer–fiber interaction.

Very recently Segatelli *et al.* [10] explored the use of short silica fiber as the reinforcing filler by incorporating different levels of this filler in nylon 6. They have also studied the effect of this filler in a rubber-toughened composition comprising 20 wt% ethylene–propylene–diene terpolymer (EPDM) grafted with maleic anhydride (EPDM-g-MA) in nylon 6. In EPDM-g-MA rubber, the anhydride maleic content was 0.5% and the ethylene–propylene ratio was 75/25 (wt/wt). The polymer/silica fiber composites were prepared in a twin-screw extruder and the test samples were prepared by injection molding.

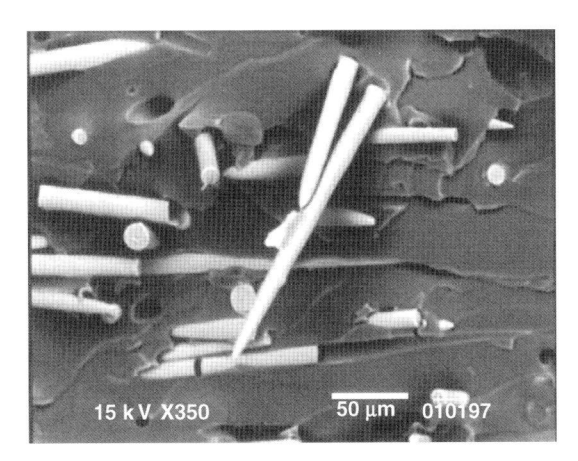

Figure 11.4 Scanning electron micrographs of the fractured surface of a polyurethane composite containing natural silica fibers. Reprinted from Effect of the incorporation of a novel natural inorganic short fiber on the properties of polyurethane composites [5], with permission from Elsevier, Copyright 2005 Elsevier Ltd.

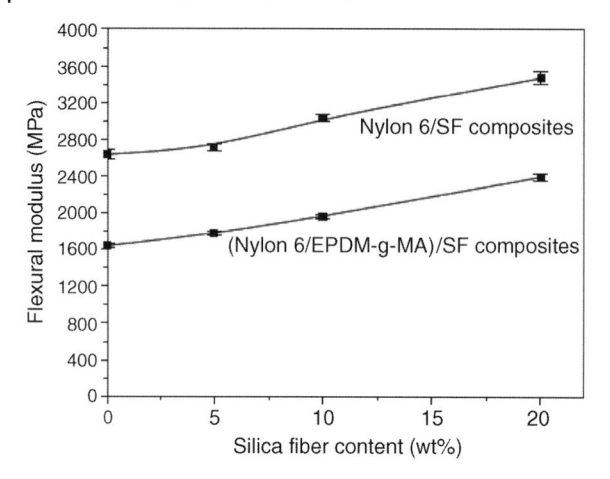

Figure 11.5 Effect of silica fiber (SF) content on flexural modulus property of nylon 6/SF and (nylon 6/EPDM-g-MA)/SF composites. Reprinted from Natural silica fiber as reinforcing filler of nylon 6 [10], with permission from Elsevier, Copyright 2009 Elsevier Ltd.

The torque value recorded for the unfilled nylon 6 and EPDM-g-MA toughened nylon 6 and their silica fiber filled composites did not show significant variation with increasing the fiber loading, even with a filler loading of 20 wt%. For both the cases of silica fiber filled nylon 6 and nylon 6/EPDM-g-MA composites, the flexural modulus was increased consistently with increasing the filler loading (Figure 11.5) [10]. Thus with a 20 wt% silica fiber loading, the flexural modulus values for nylon 6 and nylon 6/EPDM-g-MA composites were about 3500 and 2400 MPa with an increment by about 35 and 50%, respectively with respect to the unfilled polymers.

They noted a decreasing trend in yield strength of nylon 6/silica fiber composites with increasing the filler content. Thus with a 20 wt% silica fiber loading, the yield strength of unfilled nylon 6 was dropped from about 77 to 67 MPa. The effect was not so prominent at lower filler loading. Although the yield strength of unfilled nylon 6/EPDM-g-MA was 46 MPa but there was no noticeable change in this property while increasing the filler content even at 20 wt%. However, Laura *et al.* [17] observed that incorporation of glass fiber in nylon 6 and nylon 6/EPDM-g-MA matrices enhanced this property. According to Segatelli *et al.* [10] the anomalous behavior of the silica fiber-filled composites could be due to more random fiber orientation in the tested samples prepared by injection molding. The addition of silica fiber in nylon 6 and nylon 6/EPDM-g-MA matrices at different loadings resulted similar values in elongation at break property and the effect was comparable to those obtained by Laura *et al.* in the glass fiber-filled polymers [17].

The incorporation of silica fiber at various loadings in nylon 6 did not alter the Notched Izod impact strength property of the filled composites as compared to the unfilled polymer (Figure 11.6) [10].

Although blending of EPDM-g-MA with nylon 6 resulted in a substantial increase in this property from 27 J/m for nylon 6 to 708 J/m for nylon 6/EPDM-g-MA blend, addition of silica fiber considerably reduces the impact strength of the blend. Only at

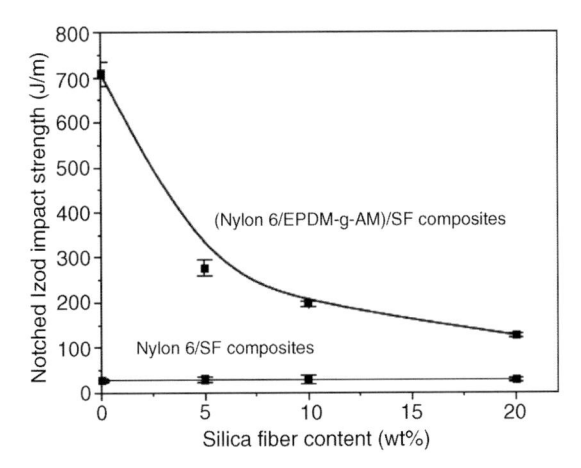

Figure 11.6 Effect of silica fiber (SF) content on Notched Izod impact strength property of nylon 6/SF and (nylon 6/EPDM-g-MA)/SF composites. Reprinted from Natural silica fiber as reinforcing filler of nylon 6 [10], with permission from Elsevier, Copyright 2009 Elsevier Ltd.

5 wt% silica fiber loading, impact strength value dropped by about 60% as compared to unfilled binary blend. Further drops in impact strength value were observed with increasing the filler loading, however, the impact strength of binary blend containing 20 wt% silica fiber was superior than that of the filled nylon 6 composites.

Segatelli et al. [10] found that scanning electron micrograph of the fracture surface of nylon 6/silica fiber composite possessed good adhesion between the polymer and the fiber, which could be due to interfacial interaction between polymer matrix and the filler surface via hydrogen bond formation between polar groups of the nylon 6 and silica fiber surface silanol groups. Similar morphological features were also observed in the case of nylon 6/EPDM-g-MA blend containing the above filler. Interestingly, the percentage crystallinity of the polymer matrix obtained by DSC technique showed increase in crystallinity from 30% to about 38% by incorporating only 5 wt% silica fiber in nylon 6, whereas a slight decrease in the crystallinity value was noted at 5 wt% silica fiber-filled nylon 6/EPDM-g-MA blend. At higher filler loadings, the crystallinity of the composites becomes similar to unfilled polymer. Such phenomena were explained by correlating the different characteristics of the melt rheology imposed by the presence of the silica fibers and their orientation, and the rubber phase dispersed in the nylon 6 matrices. In overall they concluded that the silica fiber can be used as a reinforcing filler for engineering thermoplastics such as nylon 6.

11.3.3
Surface-Modified Silica Fiber-Filled Polymer Composites

Although the silica fiber has the potential to function as a reinforcing filler for polymeric materials but their effectiveness could be further improved by using coupling agents. Coupling agents can help to interact between fiber surface and the polymer matrix by developing the surface adhesion. Hence, further enhancement in the physical properties could be achieved by improving fiber–polymer interaction.

Wang et al. [14, 15] studied the interaction between silica fiber and the polymer matrix in a silica fiber-filled phenolic resin-based polymer composite and the role of coupling agent by using different silanes in this process (Section 11.2.3). The polymer composites were obtained by the resin transfer molding (RTM) process, where the phenolic resin solution was injected into the RTM die in the presence of plies of fabric fiber followed by curing the resin infused under heat and pressure. Surface silanization of silica fiber significantly affect the dynamic adsorption process of phenolic resin solution. The presence of polar hydroxyl groups leads to high surface free energy in the untreated silica fiber. As a result, the untreated silica fiber favors preferential adsorption of highly polar ethanol solvent than phenolic resin during the RTM process. It was found that silane treatment decreased the initial wetting speed and adsorption amount of phenolic resin solution at equilibrium. The time to attain adsorption equilibrium for silane-treated silica fibers were longer compared to without silane-treated one. Hence, by reducing the adsorption strength of silica fiber, the modified surface could evenly adsorb the phenolic resin. Consequently, the surface-modified silica fiber-filled phenolic composites provided better distribution of the phenolic resin throughout the RTM part of which was realized from the TGA studies.

The degree of adhesion between the fiber and the matrix was determined by interlaminar shear strength (ILSS) measurements. All the silane-treated composites show less variation in ILSS values, measured at different positions of silica fiber/phenolics composites produced by RTM process as compared to untreated silica fiber-filled compound. This could be due to better distribution of the resin and porosity in the composites. Moreover, γ-APS, γ-GPS, and γ-MPS treated silica fiber-filled composites showed noticeable improvement in the ILSS values as compared to untreated silica fiber-filled composite, which indicates the improvement in the fiber–polymer interaction. However, ILSS value of the composite was reduced when the silica fiber was treated with TMCS (Table 11.2) [15]. Interestingly, from the wettability studies, it was observed that the surface treatment by alkyl-terminated silane TMCS

Table 11.2 Experimental data of the ILSS of silica fiber/phenolics composites.

Silica fiber	ILSS^a (MPa)	ILSS^b (MPa)	ILSS^a/ILSS^b
Desized	28.5	32.1	0.888
γ-APS	42.9	43.3	0.991
γ-GPS	39.2	40.1	0.978
γ-MPS	38.9	39.3	0.990
TMCS	26.4	25.8	1.023

The data of ILSS measured from inlet^a and outlet^b gate region of the RTM equipment. Reprinted from Effects of silica surface treatment on the impregnation process of silica fiber/phenolics composites [15], with permission from John Wiley & Sons, Inc., Copyright 2007 Wiley Periodicals, Inc.

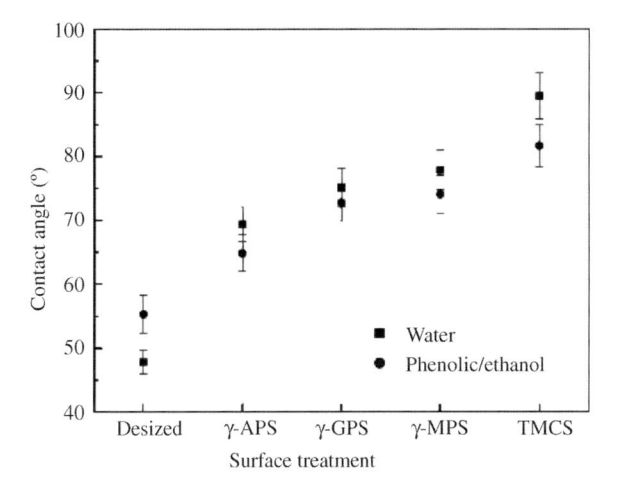

Figure 11.7 Effect of silanization on contact angle between silica fiber and water and phenolics/ethanol solution. Reprinted from Effects of silica surface treatment on the impregnation process of silica fiber/phenolics composites [15], with permission from John Wiley & Sons, Inc., Copyright 2007 Wiley Periodicals, Inc.

reduced the surface energy of the silica fiber most efficiently. Contact angle measurement between silanized silica fiber and phenolics/ethanol solution was also used to determine the effectiveness of the silane treatment and the degree of hydrophobicity of silica fiber. In all the cases, silane treatment increase the contact angle, however, it was most effective for TMCS-treated silica fiber (Figure 11.7) [15].

However, the inferior mechanical properties obtained for TMCS-treated silica fiber-filled composite even compared to untreated silica fiber-filled composite implies that it could improve the distribution of the resin in the composite but the lack of silanization coupling reaction due to the absence of functional group, it cannot develop the polymer–filler adhesion. This investigation led by Wang et al. manifests the importance of selecting coupling agents in order to achieve desired composite properties.

The superior interfacial adhesion between γ-APS, γ-GPS, and γ-MPS treated silica fiber and the phenolic composite was also noted from SEM studies. The photographs of fracture surface of these silica fiber/phenolics composites showed the improved quality (adhesion) and quantity (wetting) of interfacial contact resulting from coupling reaction and uniform adsorption of phenolic resin as compared to without silane-treated silica fiber-filled composites.

From this study, it was also found that among the different silane coupling agents, γ-APS was the most effective one. Wang and Huang [18] further studied the effects of different solvents, namely ethanol, acetone, and THF on the interface formation and properties of γ-APS modified silica fiber–phenolics composites produced by RTM process. It was observed that interaction occurring between silica fiber and phenolics solution had a significant effect on the competitive adsorption of phenolics onto silica fiber surface. Application of solvents did not alter chemically interfacial bonding

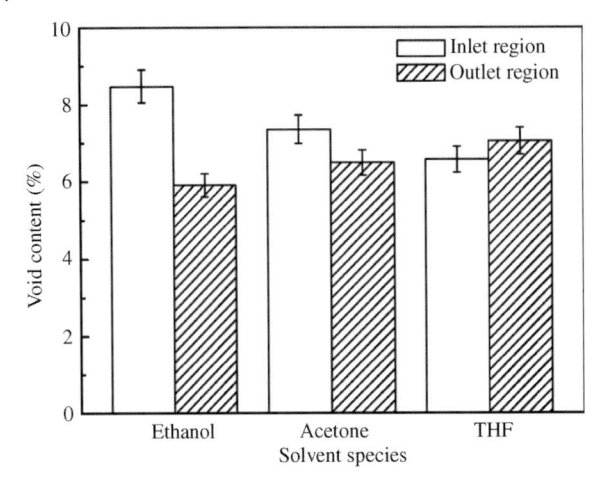

Figure 11.8 Effects of solvent species on void contents of the silica fiber–phenolics composites from inlet and outlet gate region of the RTM equipment. Reprinted from Interphase formation of a resin transfer molded silica–phenolics composites subjected to dynamic impregnation process [18], with permission from Elsevier, Copyright 2008 Elsevier B.V.

between fiber and resin matrix but affect the uniformity of resin distribution in mold cavity and thus affect those of void content and ILSS of a resulting resin transfer molded product (Figures 11.8 and 11.9).

The adsorption of phenolic resin onto γ-APS treated silica fiber was highly solvent-dependent. Microscopic interactions occurring among phenolics, solvent, and amino groups were responsible for the difference in the macroscopic distribution of resin

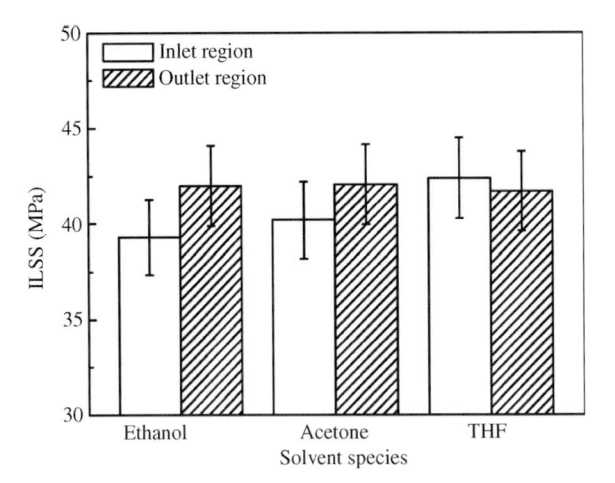

Figure 11.9 Effects of solvent species on ILSS of the silica fiber–phenolics composites from inlet and outlet gate region of the RTM equipment. Reprinted from Interphase formation of a resin transfer molded silica–phenolics composites subjected to dynamic impregnation process [18], with permission from Elsevier, Copyright 2008 Elsevier B.V.

throughout RTM apparatus. The competitive adsorption of solvent on γ-APS treated fiber surface suppressed the adsorption of phenolic resin and this suppression was unfavorable in solution impregnation route. Lower polarity and weaker hydrogen-bonding capacity of THF made it less effective on the adsorption of phenolics. Thus THF was found to be more suitable to be used in the production of silica fiber–phenolics composites than ethanol, which is more commonly used.

Silva et al. [3], investigated the effect of silane-treated silica fiber as a primary or secondary filler on the reinforcement of silicone rubber, poly(dimethylsiloxane) (PDMS). VTMS coupling agent was used to modify the natural short silica fiber surface (Section 11.2.3). To study the effect of silanized silica fiber as a primary reinforcing agent, the modified fiber at a loading of 20 phr was incorporated in silicone rubber and the results were compared with unfilled rubber and unmodified silica fiber (20 phr) filled rubber compounds. In another set of experiments, the effect of silica fiber as a secondary filler was inspected by incorporating the modified fiber at the same level of loading in a previously mixed silicone gum rubber containing about 27% (wt/wt) of nanometric fumed silica as the primary filler. The effect of modified silica fiber in fumed silica-filled rubber was assessed by comparing the results with fumed silica-filled silicone rubber with and without adding any unmodified silica fiber.

As primary filler, both unmodified and silane-modified silica short fibers improved the thermal stability of silicone rubber. The onset temperature of thermal degradation was raised by about 50 °C. Interestingly as secondary filler, thermal stability was dropped by about 25 °C. The authors suggested that the cross-linking of PDMS and the inclusion of silica fiber as primary filler in the rubber matrix could restrict the mobility of the polymer macromolecules by hindering inter- and intramolecular rearrangements of the PDMS chains and thereby improved the thermal stability. On the other hand, as a secondary filler in conjunction with fumed silica, the ionic contaminants particularly Al^{3+} and Ti^{4+} present in naturally occurring silica fiber, could catalyze nucleophilic attacks from the residual silanol groups of the fumed silica to the PDMS chains, creating volatile species that were eliminated from the material. The interfacial shear stress between silica short fibers and fumed silica could decrease the size of the fumed silica agglomerates during compounding process, which could result in new interactions of fumed silica particles with the PDMS chains, decreasing the thermal stability of the PDMS matrix.

Scanning electron micrographs of silica fiber-filled rubber showed uniform distribution of these fibers in the rubber matrix, with a tendency of fiber orientation, for both unmodified and silanized filler (Figure 11.10). Evidences of fiber pull-out regions due to inadequate polymer–fiber adhesion and good adhesion at the fiber–matrix interface were observed for both the above cases. However, the occurrence of fiber pull-out regions were more in the case of unmodified silica fiber-filled compound. Worthy of silane treatment on silica fiber was quite evident when silanized silica fiber was used as reinforcing filler showing superior fiber–matrix interface formation compared to its untreated counterpart.

From the swelling studies, it was found that as primary filler, incorporation of silane-modified silica short fibers increased the cross-linking density of this matrix from 3.0×10^{-4} to 3.7×10^{-4} mol/cm^3, whereas untreated silica fiber could not

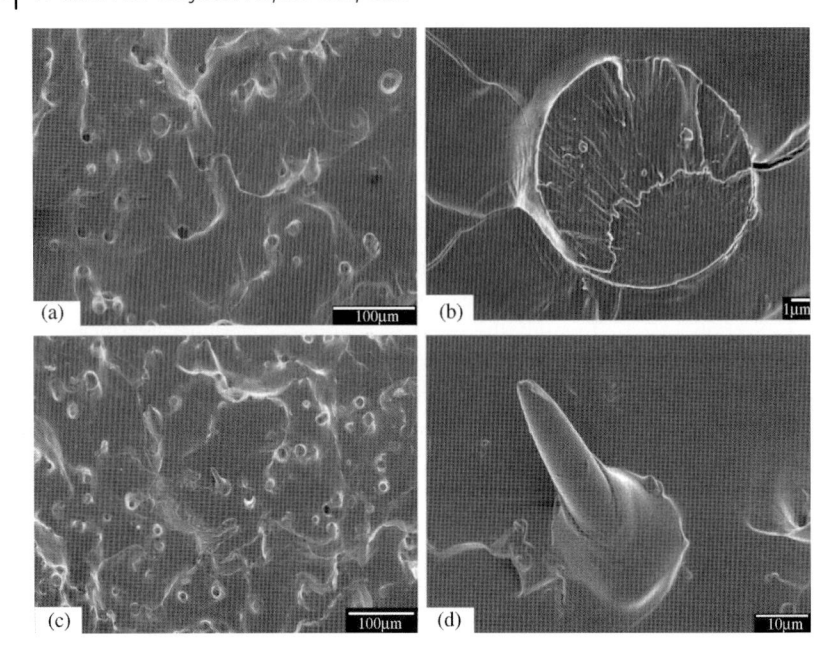

Figure 11.10 Scanning electron micrographs of silicone rubber composite containing (a, b) natural silica fiber and (c, d) silanized silica fiber as a primary reinforcing agent. Reprinted from Biogenic silica short fibers as alternative reinforcing fillers of silicone rubbers [3], with permission from John Wiley & Sons, Inc., Copyright 2006 Wiley Periodicals, Inc.

affect this property. Hence, the swelling studies corroborate the interaction of fiber and polymer matrix at the fiber–polymer interface whereas such interaction is absent in the case of untreated silica fiber-filled composite. Interestingly, swelling experiments could not identify the effect of silane treatment when the silica fiber was used as a secondary filler as both untreated and silane-treated silica fiber-filled composites showed increased cross-linking density of this matrix at the same level 6.4×10^{-4} mol/cm^3 as compared to 4.1×10^{-4} mol/cm^3 obtained for nanosilica filled rubber.

As a primary filler, both untreated and silane-treated silica fiber enhanced tensile strength and Young's modulus of the rubber compound. For the untreated silica-filled rubber, tensile strength was increased from 0.29 to 0.55 MPa and Young's modulus from 0.07 to 0.2 MPa. Those for silanized silica fiber-filled composite were 0.67 and 0.18 MPa, respectively. This additional improvement in tensile strength could be attributed due to the interaction of silica fiber and the polymer matrix occurred by using silane coupling agent. As a secondary filler, untreated silica fiber increased Young's modulus of the rubber compound from 0.11 to 0.36 MPa, which was further improved by silane-treated silica fiber to 0.47 MPa. However, a drop in tensile strength was incurred for both untreated and silane-treated silica fiber-filled composites from 8.3 to 6.0 MPa and 7.2 MPa, respectively. Elongation at break of all the silica fiber-filled composites was lower than the control compounds except untreated silica fiber-filled rubber when it was used as primary filler.

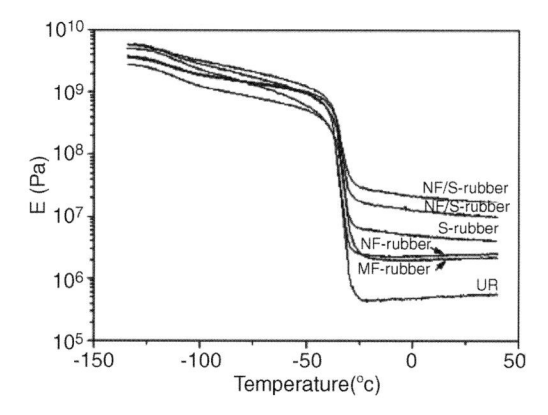

Figure 11.11 Variation of storage modulus (E′) with temperature for unfilled silicone rubber (UR); silicone rubber composite containing natural silica fiber (NF-rubber) and silanized silica fiber (MF-rubber) as a primary reinforcing agent; fumed silica-filled silicone rubber (S-rubber); silicone rubber composite containing natural silica fiber (NF/S-rubber); and silanized silica fiber (MF/S-rubber) as secondary reinforcing agent. Reprinted from Biogenic silica short fibers as alternative reinforcing fillers of silicone rubbers [3], with permission from John Wiley & Sons, Inc., Copyright 2006 Wiley Periodicals, Inc.

Storage modulus values of all the silica fiber-filled composites measured at room temperature by dynamic mechanical analyzer were higher than the control compounds in both occasions when used as primary and secondary filler and the additional reinforcing effect was realized in the case of silanized silica fiber-filled rubber (Figure 11.11) [3].

Likewise the case of silicone rubber as the matrix, Barra *et al.*, also observed the potential reinforcing ability of the pristine silica fiber as a reinforcing fiber for polyurethane [13]. Compared to glass fiber as the reinforcing filler, incorporation of naturally occurring silica fiber in PU can provide reinforcing effect superior or at equivalent level and even without any surface treatment. The effect was realized to its maximum extent at 17 wt% of fiber loading. Thus, silica fiber with an aspect ratio of 13.9 ± 1.2 and specific gravity of 1.68 ± 0.41 g/cm^3 at 17 wt% of loading, improved the tensile strength of unfilled PU from 0.94 ± 0.05 to 2.78 ± 0.15 MPa and modulus of elasticity from 0.47 ± 0.03 to 3.09 ± 0.23 GPa, respectively while the elongation at break was dropped from 81.6 ± 6.5 to $34.5 \pm 1.2\%$.

A further increase by 10% in the tensile strength and 12.5% in the strain were noted while using pretreated silica fiber using aminosilane coupling agent. Such enhancement in the reinforcing ability of the silica fiber could be due to better interfacial adhesion obtained by surface treatment. This was verified by morphological studies by SEM on fractured surfaces from different composites. Significant reduction in fiber pull-out regions in the case of silanized silica fiber-filled PU was noted as compared to without silane-treated silica fiber and glass fiber-filled compounds. This indicates that the application of aminosilane coupling agent could improve the polymer–fiber adhesion.

Fiber length and consequently aspect ratio of the silica fiber was also found to affect the reinforcing properties significantly. Thus, tensile strength and modulus of elasticity values for long fiber with higher aspect ratio (18.2 ± 0.8) were 3.99 ± 0.14 MPa and 3.94 ± 0.15 GPa, respectively whereas those properties were dropped to 2.58 ± 0.15 MPa and 2.03 ± 0.19 GPa, respectively by using short fiber with lower aspect ratio (11.6 ± 0.9). However, the elongation at break of short fiber-reinforced composites was superior than long fiber-filled polymer.

Later on in a similar study Martins *et al.* [16] investigated the ability of silica fiber as the reinforcing agent in epoxy resin. As noticed in other studies, incorporation of naturally occurring silica fiber in epoxy resin can also provide noticeable reinforcing effect even at low filler loading of 6% by volume without any surface treatment. Silanization of silica fiber by aminosilane further boosted its reinforcing ability. Moreover, the mechanical properties obtained by using silica fiber as the reinforcing agent could produce superior or at least similar results as compared to commercial short glass fiber-filled composites.

11.3.4
Reinforcement Mechanism

In principle, maximum reinforcement of filler can be achieved through good dispersion and better chemical or physical interaction with the matrix polymer. The case studies discussed in earlier sections showed enhancement in modulus of the composites while incorporating this fibrous filler in the polymer matrix, enabling its potential reinforcing ability. The increase in modulus of materials can be explained by the percolation theory, which considers the connectivity of the dispersed phase in the matrix [19]. According to this theory, random distribution of rigid fillers in the matrix establishes a variable interparticle distance, where the stress concentration is located in regions around these particles. Evidences of fiber pull-out under mechanical testing are also evident from the fractured surfaces indicating lack of polymer to fiber interaction. Adhesion between the fiber and the polymer can be improved by applying suitable coupling agents, which could occur via development of fiber–polymer interface. Especially, this effect can be realized while using bifunctional silane coupling agents, which facilitates the formation of permanent covalent bond between silanized silica fiber and host polymer matrix. Schematic presentation of reinforcement mechanism via interface formation, in a silanized silica fiber-reinforced polymer composite is shown in Scheme 11.2 with plausible interaction of silica fiber with bifunctional silane coupling agent and the interaction of this silanized silica fiber with matrix polymer.

11.4
Applications

Silica fiber possesses excellent electrical insulation properties and can offer low thermal conductivity, high tensile strength, and modulus properties. Additionally this inorganic fiber is inert to the majority of chemical reagents, resistant to organic and mineral acids even at the elevated temperature and weak alkalies. It can be used

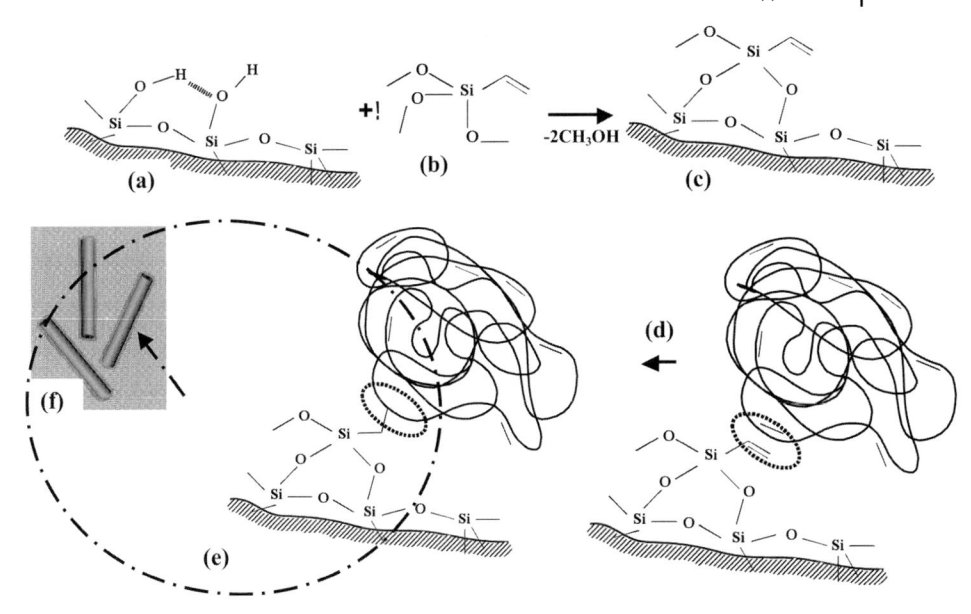

Scheme 11.2 Schematic presentation of plausible interaction of silica fiber with bifunctional coupling agent (a) silica fiber surface, (b) bifunctional coupling agent, (c) silane-pretreated silanized product, (d) interaction with matrix polymer, (e) interface formation, and (f) silanized silica fiber-polymer composite.

for corrosion protection. In addition to its outstanding chemical resistance against common chemical reagents this is also resistant to water and steam even at high-pressure conditions and stable in vacuum. As a result, it finds applications in the thermal protection systems as replacement of asbestos and used in refinery, space shuttle, aerospace vehicles, metallurgy, shipbuilding industries, and for the developments in fibrous refractory composite insulation.

More specifically it finds applications in thermal insulation rings for diffusion furnaces, furnace linings, heat-shielding curtains, thermal insulation seals and packing materials, thermal insulation coverings for thermocouple cables and wires, roller covering for tempered glass plate manufacturing, abrasives for plastic whetstones, insulators around generator and aircraft/rocket engines.

Silica fiber is also suitable for textile applications due to its stiffness and adequate flexibility. Silica fiber coated with EPDM, polyvinylchloride (PVC), silicone rubber are being used in bags, automobiles, hand gloves, hover crafts, and as a reinforcing material in different friction articles.

Other than conventional polymeric matrices, silica fiber can also be used as the reinforcing agent in silica-based composites. Mishra et al. [20] developed silica fiber-reinforced silica matrix composite foams with 84–90% porosity content involving random dispersion of 10 wt.% fibers with aspect ratios >1000. Fiber reinforcement was found to be beneficial for increase in stiffness of the composite foams. However, the presence of silica fiber could reduce the permeability in the porous silica foam by interrupting the pore interconnections in the foamed material.

Silicon nitride composite materials are suitable for high temperature electromagnetic window materials for communication, control, and thermal protection of spacecraft. However, to achieve adequate toughness and to overcome notch-sensitivity, long fiber-reinforced silicon nitride composites are being desired, where silica fiber can be used as the reinforcing agent [21].

11.5
New Developments

Polymer composites in its nanoscale dimension consisting of polymer matrix and inorganic fillers have recently gained attention as they could combine the advantages from both the components. While polymer materials can offer lightweight, flexibility, and good moldability, the inorganic materials can provide high strength, heat stability, and chemical resistance. These composite nanofibers can have enhanced mechanical, electrical, optical, thermal, and magnetic properties without losing transparency, and find uses in various multifunctional applications [22–24]. Weichold et al. [24] prepared flexible silica-nanofiber mats with filament diameters of 600–1000 nm by electrospinning of hydrolyzed tetraethyl orthosilicate (TEOS) gels. The silica nanofiber mats increased both tensile and bending strength of epoxy composites relative to unidirectional glass rovings. The random orientation of the fibers in the polymer matrix was found to be advantageous under complex loads and in bending tests.

Silica in the nanofiber form is used in mesoscopic research, nanoelectronic devices, and chromatographic supports with high adsorption capacities [25–30]. These materials can be fabricated by using different self-assembly techniques such as, chemical vapor deposition [31], thermal evaporation [32], carbothermal reduction synthesis [33], vapor–liquid–solid growth method [34], excimer laser ablation [35], stress-limited oxidation [36], and electrospinning process [26]. Shao et al. [26] demonstrated fabrication of silica nanofibers with diameters of 200–400 nm by using electrospun fibers of polyvinylalcohol (PVA)/silica composite as precursor. A silica gel was prepared by hydrolysis and polycondensation by addition of aqueous phosphoric acid to TEOS and reacted for 5 h. Then PVA solution was dropped slowly into the silica gels and the reaction proceeded in a water bath at 60 °C for another 12 h. Thus, a viscous gel of PVA/silica composite was obtained, which was electrospun to form PVA/silica composite nanofiber. Silica nanofibers were obtained by calcination of PVA/silica composite nanofiber. Huo et al. [37] prepared hexagonal phase mesoporous silica fibers by allowing to grow the fibers from an oil–water interface. Nakamura and Matsui [38] reported the formation of silica hollow tubules, 200–300 μm in length, by the hydrolysis of TEOS in a mixture of ethanol, ammonia, water, and DL-tartaric acid. Substituting DL-tartaric acid with citric acid wormlike structures of silica was obtained [39]. Sudheendra and Raju [40] reported the development of silica fibers by using long-chain carboxylic acids such as octanoic, decanoic, and lauric acids as templates. The presence of the acid was found to be necessary for the formation of the fibrous structure.

11.6
Concluding Remarks

As a reinforcing agent, silica fiber can be used as an integral part in thermoplastics, thermosets, and different elastomeric composites. In addition to its reinforcing ability by improving the strength of the composite materials its unique properties such as excellent thermal and chemical resistance find its usefulness in specialty applications. Although the use of silica fiber as the reinforcing filler for the polymeric materials is well known since several decades, the literature survey indicates that there is surprisingly little published work on this filler in spite of its obvious importance. High manufacture costs to produce silica fiber synthetically limits its use in general applications, but its natural form can be obtained from very cheap mineral source at low processing costs. Naturally occurring silica fiber also has potential ability to reinforce polymeric matrices. Further improvement could be achieved by using right combination of silane coupling agent. However, only a few instances, where attempts have been made to modify the fiber surface. Consequently, sufficient data and systematic studies are not available in this area. Compared to synthetic silica fiber or glass fiber they are cheaper and can be obtained in a high purity form. In various reports it was noted that the natural silica fiber is able to offer superior or at least similar level of reinforcement as compared to vastly used commercial glass fiber. Lower specific gravity of this natural fiber is also an advantage over the glass fiber. However, there are very limited reports on the use of this natural fiber whereas this could be successfully used as a reinforcing material to obtain composite materials with applications in the automotive industry, construction purposes. Nontoxicity of this material further widen up its scope to find application in food packaging sector. Hence, further exploration of using this natural fiber as reinforcing filler for polymeric composites could provide an economically viable alternative to synthetic silica fiber and glass fiber.

Recently the ideas of micron-sized fibrous materials have been expanded to fabricate a new and an emerging class of fibrous materials in its nanoscale dimension. Though there have been quite a few endeavors in producing the nanoscale silica fiber it is still on an exploratory level in the field of polymer composites based on nanosilica fiber and hence emphasis should be given on these relatively unexplored fields.

References

1 Wypych, G. (2010) *Handbook of Fillers*, 3rd edn, ChemTec Publishing, Toronto.
2 Leblanc, J.L. (2002) *Progress in Polymer Science*, **27**, 627.
3 Silva, V.P., Gon¸8calves, M.C., and Yoshida, I.V.P. (2006) *Journal of Applied Polymer Science*, **101**, 290.
4 Melnikov, P., Santagnelli, S.B., Santos, F.J., Delben, A., Delben, J.R., Teixeira, A.L.R., and Rommel, B.R. (2003) *Materials Chemistry and Physics*, **78**, 835.
5 Saliba, C.C., Oréfice, R.L., Carneiroa, J.R.G., Duarte, A.K., Schneider, W.T., and Fernandes, M.R.F. (2005) *Polymer Testing*, **24**, 819.

6 Iler, R.K. (1979) *The Chemistry of Silica*, John Wiley & Sons, Inc., New York.
7 Nemetschek, Th. and Hofmann, U. (1953) *Zeitschrift für Naturforschung*, **8b**, 410.
8 Weiss, A. (1954) *Die Naturwissenschaften*, **41**, 12.
9 Haller, W., (1961) *Nature*, **191**, 662.
10 Segatelli, M.G., Yoshida, I.V.P., and Gonçalves, M.D.C. (2010) *Composites: Part B*, **41**, 98.
11 Graetsch, H., Gies, H., and Topalović, I. (1994) *Physics and Chemistry of Minerals*, **21**, 166.
12 Graetsch, H., Mosset, A., and Gies, H. (1990) *Journal of Non-Crystalline Solids*, **119**, 173.
13 Barra, G.M.O., Fredel, M.C., Al-Qureshi, H.A., Taylor, A.W., and Clemenceau, C. Jr. (2006) *Polymer Composites*, **27**, 582.
14 Wang, B.C., Huang, Y.D., and Liu, L. (2006) *Materials Science and Technology*, **22**, 206.
15 Liu, L., Wang, H., Wang, B.C., and Huang, Y.D. (2008) *Journal of Applied Polymer Science*, **107**, 2274.
16 Martins, R.R., Pires, A.T.N., Al-Qureshi, H.A., and Barra, G.M.O. (2008) *Revista Matéria*, **13**, 605.
17 Laura, D.M., Keskkula, H., Barlow, J.W., and Paul, D.R. (2000) *Polymer*, **41**, 7165.
18 Wang, B.C. and Huang, Y.D. (2008) *Applied Surface Science*, **254**, 4471.
19 He, D. and Jiang, B. (1993) *Journal of Applied Polymer Science*, **49**, 617.
20 Mishra, S., Mitra, R., and Vijayakumar, M. (2009) *Ceramics International*, **35**, 3111.
21 Qi, G.J., Zhang, C.R., Hu, H.F., Cao, F., Wang, S.Q., Cao, Y.B., and Jiang, Y.G. (2005) *Materials Letters*, **59**, 3256.
22 Ji, L., Saquing, C., Khan, S.A., and Zhang, X. (2008) *Nanotechnology*, **19**, 085605.
23 Sawicka, K.M. and Gouma, P. (2006) *Journal of Nanoparticle Research*, **8**, 769.
24 Weichold, O., Tigges, B., Voigt, W., Adams, A., and Thomas, H. (2009) *Advanced Engineering Materials*, **11**, 417.
25 Gole, J.L., Wang, Z.L., Dai, Z.R., Stout, J., and White, M. (2003) *Colloid and Polymer Science*, **281**, 673.
26 Shao, C., Kim, H., Gong, J., and Lee, D. (2002) *Nanotechnology*, **13**, 635.
27 Katz, A. and Davis, M.E. (2000) *Nature*, **403**, 286.
28 Wang, Z.L., Gao, R.P., Gole, J.L., and Stout, J.D. (2000) *Advanced Materials*, **12**, 1938.
29 Kresge, C.T., Leonowicz, M.W., Roth, W.J., Vartuli, J.C., and Beck, J.S. (1999) *Nature*, **359**, 710.
30 Wang, N., Tang, Y.H., Zhang, Y.F., Lee, C.S., and Lee, S.T. (1998) *Physical Review B*, **58**, 16024.
31 Ozaki, N., Ohno, Y., and Takeda, S. (1998) *Applied Physics Letters*, **73**, 3700.
32 Niu, J., Sha, J., Liu, Z., Su, Z., Yu, J., and Yang, D. (2004) *Physica E*, **24**, 268.
33 Wu, X.C., Song, W.H., Wang, K.Y., Hu, T., Zhao, B., Sun, Y.P., and Du, J.J. (2001) *Chemical Physics Letters*, **336**, 53.
34 Givargizov, E.I. (1975) *Journal of Crystal Growth*, **32**, 20.
35 Morles, A.M. and Lieber, W.C. (1998) *Science*, **279**, 208.
36 Liu, H., Biegelsen, D.K., Johnson, N.M., Ponce, F.A., and Pease, R.F.W. (1994) *Applied Physics Letters*, **64**, 1385.
37 Huo, Q., Zhou, D., Feng, J., Weston, K., Buratto, S.K., Stucky, G.D., Schacht, S., and Schuth, F. (1997) *Advanced Materials*, **9**, 974.
38 Nakamura, H. and Matsui, Y. (1995) *Journal of the American Chemical Society*, **117**, 2651.
39 Nakamura, H. and Matsui, Y. (1995) *Advanced Materials*, **7**, 871.
40 Sudheendra, L. and Raju, A.R. (1999) *Bulletin of Materials Science*, **22**, 1025.

Part Three
Macrosystems: Textile Composites

12
2D Textile Composite Reinforcement Mechanical Behavior

Emmanuelle Vidal-Sallé and Philippe Boisse

12.1
Introduction

In liquid composite molding (LCM) processes, a resin is injected on a textile preform [1, 2]. The shape of this preform can be complex. It can be obtained by forming process from initially flat textile reinforcement. In case of double-curved shapes, the forming process may be tricky because it needs in-plane deformations of the textile reinforcement and especially in-plane shear strain. The analysis of this preforming process (e.g., its finite element simulation) needs to know and to model the mechanical behavior of the 2D composite reinforcement. It is dry (i.e., without matrix) during this stage. This chapter aims to present some experiments and some models that are dedicated to mechanical behavior of textile composite reinforcements. This behavior is linked to their internal structure. These are made up of continuous fibers. The diameter of these fibers is small (5–7 µm for carbon and 5–25 µm for glass). A yarn is made up of several thousands of fibers. These yarns are woven following standard weaves (plain, satin, and twill) or more complex structures such as ply-to-ply interlock weaves (Figure 12.1). An alternative consists in stitching a ply made of parallel fibers (noncrimp fabric (NCF)). The material resulting from this assembly of continuous fibers exhibits a very specific mechanical behavior since relative motions are possible between the yarns and the fibers (Figure 12.2). The textile reinforcement preforming process takes advantage of these possible motions.

Section 12.2 presents a simplified form of the principle of virtual works for textile reinforcements. The tensile, in-plane shear, and bending virtual works are separated. The related experiments are presented. They are specifically dedicated to 2D textile reinforcements.

Although fibrous reinforcements are not strictly continuous because of relative sliding between fibers, several mechanical constitutive equations have been proposed that consider the textile reinforcement as an anisotropic continuum. Some of them are presented in Section 12.3. Other approaches model the components at the meso- or microscopic scale, that is, the yarns or the fibers. Such modelings are presented in Section 12.4 for the simulation of both a full composite reinforcement forming and the deformation of a unit woven cell.

Polymer Composites: Volume 1, First Edition. Edited by Sabu Thomas, Kuruvilla Joseph,
Sant Kumar Malhotra, Koichi Goda, and Meyyarappallil Sadasivan Sreekala
© 2012 Wiley-VCH Verlag GmbH & Co. KGaA. Published 2012 by Wiley-VCH Verlag GmbH & Co. KGaA.

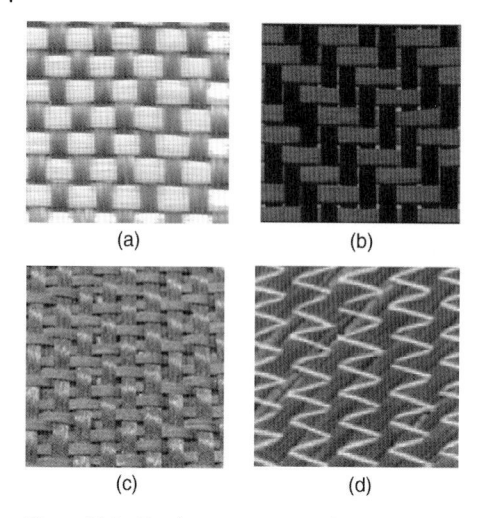

Figure 12.1 Textile composite reinforcements. (a) Plain weave. (b) Twill weave. (c) Interlock. (d) NCF.

12.2
Mechanical Behavior of 2D Textile Composite Reinforcements and Specific Experimental Tests

12.2.1
Load Resultants on a Woven Unit Cell

A textile composite reinforcement consists of woven unit cells. Its mechanical behavior is specific because of the possible relative displacements between yarns and between fibers within the yarns. Let us consider the loads on a unit woven cell

Figure 12.2 Woven yarns made of thousands of fibers.

12.2 Mechanical Behavior of 2D Textile Composite Reinforcements and Specific Experimental Tests

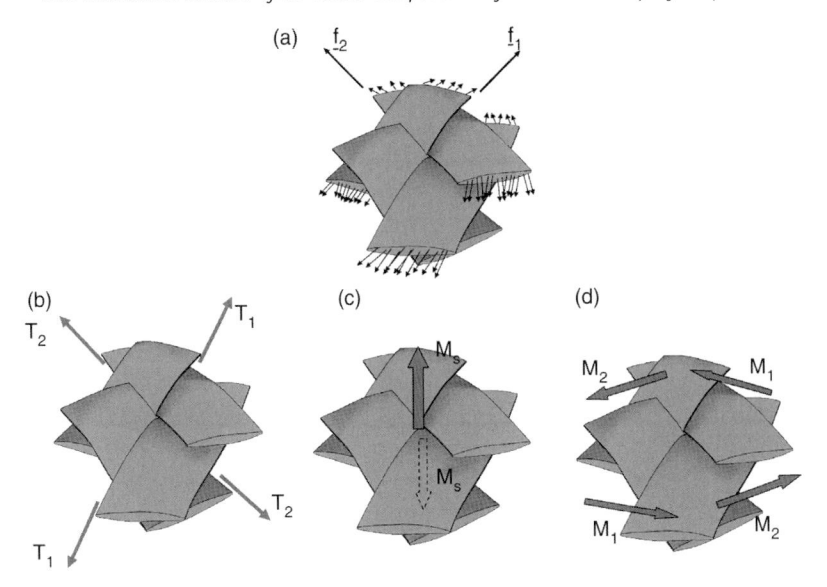

Figure 12.3 (a) Loads on a unit woven cell and resultants: (b) tensions, (c) in-plane shear moment, and (d) bending moment.

(representative unit cell (RUC)) (Figure 12.3a). The following resultants of these loads are considered:

- The tensions T_1 and T_2 are the resultants of the loads on warp and weft yarns, respectively, in the directions $\underline{f}_1$ and $\underline{f}_2$ of these yarns (see Figure 12.3b).
- The in-plane shear moment M_s is the moment at the center of the RUC in the direction normal to the fabric, resulting from the in-plane loads on the unit cell (Figure 12.3c).
- The bending moments M_1 and M_2 resulting on the warp and weft yarns, respectively (Figure 12.3d).

The loading on the unit woven cell is characterized by these load resultants: T_1, T_2, M_s, M_1, and M_2. This is a simplified modeling, and physical meanings of these quantities are clear. Furthermore, these loads are directly measured as functions of the deformation of the fabric in standard tests for composite reinforcements (see Sections 12.2.3–12.2.5). Finally these loads T_1, T_2, M_s, M_1, and M_2 are conjugated respectively to axial strains ε_{11} and ε_{22} in warp and weft directions, to in-plane shear angle γ, and to curvatures χ_{11} and χ_{22} of warp and weft yarns. Consequently, this will lead to a simple form of the internal virtual work presented in Section 12.2.2.

12.2.2
Principle of Virtual Work

A textile composite reinforcement is made of N_c unit woven cells, as shown in Figure 12.3. These figures have been drawn in case of a plain weave for simplicity, but the type of weaving can be different (twill, satin, etc.). Let us denote $W_{ext}(\underline{\eta})$ and $W_{acc}(\underline{\eta})$ as the virtual work of the exterior loads and the virtual work of the acceleration quantities, respectively, in the virtual displacement field $\underline{\eta}$ ($\underline{\eta}$ is equal to 0 on the boundary with prescribed displacements Γ_u). The principle of virtual work can be written as follows:

$$W_{ext}(\underline{\eta}) - W_{int}(\underline{\eta}) = W_{acc}(\underline{\eta}) \tag{12.1}$$

with

$$W_{int}(\underline{\eta}) = W^t_{int}(\underline{\eta}) + W^s_{int}(\underline{\eta}) + W^b_{int}(\underline{\eta}) \tag{12.2}$$

$W^t_{int}(\underline{\eta})$, $W^s_{int}(\underline{\eta})$, and $W^b_{int}(\underline{\eta})$ are the virtual internal works of tension, in-plane shear, and bending, respectively, with

$$W^t_{int}(\underline{\eta}) = \sum_{p=1}^{n\ cell} {}^p\varepsilon_{11}(\underline{\eta})\, {}^pT_1\, {}^pL_1 + {}^p\varepsilon_{22}(\underline{\eta})\, {}^pT_2\, {}^pL_2 \tag{12.3}$$

$$W^s_{int}(\underline{\eta}) = \sum_{p=1}^{n\ cell} {}^p\gamma(\underline{\eta})\, {}^pM_s \tag{12.4}$$

$$W^b_{int}(\underline{\eta}) = \sum_{p=1}^{n\ cell} {}^p\chi_{11}(\underline{\eta})\, {}^pM_1\, {}^pL_1 + {}^p\chi_{22}(\underline{\eta})\, {}^pM_2\, {}^pL_2 \tag{12.5}$$

The quantity A is denoted ${}^pA$ when it concerns the unit woven cell number p. $\varepsilon_{11}(\underline{\eta})$ and $\varepsilon_{22}(\underline{\eta})$ are the virtual axial strains in warp and weft directions. L_1 and L_2 are the lengths of the unit cell in warp and weft directions. $\gamma(\underline{\eta})$ is the virtual shear angle, that is, the virtual angle variation between the warp and weft directions in the displacement field $\underline{\eta}$. $\chi_{11}(\underline{\eta})$ and $\chi_{22}(\underline{\eta})$ are the virtual curvatures in warp and weft directions. The virtual strains $\varepsilon_{11}(\underline{\eta})$, $\varepsilon_{22}(\underline{\eta})$, $\gamma(\underline{\eta})$, $\chi_{11}(\underline{\eta})$, and $\chi_{22}(\underline{\eta})$ are functions of the virtual displacement field $\underline{\eta}$.

Remark 12.1 Concerning the In-Plane Shear Moment: In case of textile deformation, the shear angle γ is a significant and clearly defined quantity. It is used in all the studies concerning in-plane shear behavior of textile reinforcement (see Section 12.2.2). It is interesting to express the internal virtual work of in-plane shear as a function of this quantity. The load conjugated to this shear angle is the moment M_s of Eq. (12.4). As defined above, it is the component in the normal direction of moment at the center of the RUC due to the loads on the yarns (Figure 12.3c).

Remark 12.2 Concerning the Load Resultants: Defining the loading state on a woven unit cell by the load resultants T_1, T_2, M_s, M_1, and M_2 is a simplified modeling. Nevertheless, it is better fitted to describe the load state in a textile than a standard stress field. This stress tensor defines a load on an elementary surface according to its normal. In case of textile materials, the surfaces and, consequently, the normal vectors are not all well defined and the stress definition is questionable.

Remark 12.3 Concerning the Twisting Curvature: In Eq. (12.5), the bending virtual work should include $\chi_{12}(\underline{\eta})M_{12}$ coming from the twisting curvature. This term is neglected in the present approach mainly because of the lack of experimental data. Nevertheless, it is probable that this twisting term is small in case of woven material because of the structure of the fabric made of two sets of yarns.

The load resultants T_1, T_2, M_s, M_1, and M_2 depend on the strains in the textile material. Sections 12.2.3–12.2.5 describe the experiments used to characterize the mechanical behaviors, that is, the relations between the load resultants T_1, T_2, M_s, M_1, and M_2 and the strains in the material. These experiments are specific to textile composite reinforcements.

12.2.3
Biaxial Tensile Behavior

The tensile behavior of woven material is specific mainly because of the decrimping of tows when they are stretched. This leads to strong tensile behavior of nonlinearities. The fabric is very softer than the tow for small axial strains. When the yarns tend to become straight, the stiffness of the fabric increases and becomes close to that of the tow. Because of the weaving, the decrimping phenomena in warp and weft directions are interdependent and the tensile behavior is biaxial. Some biaxial tests have been developed in order to measure these properties [3–6]. Figure 12.4 shows a biaxial tensile device where the warp and weft strains are tested within a cross-shape specimen [4]. The measurement of tensions in warp and weft directions gives two tensile surfaces $T^{11}(\varepsilon_{11}, \varepsilon_{22})$ and $T^{22}(\varepsilon_{11}, \varepsilon_{22})$, as shown Figure 12.4b, in case of glass plain weave presented in Figure 12.5. There is only one surface presented because the properties in warp and weft directions are close. It has been experimentally shown

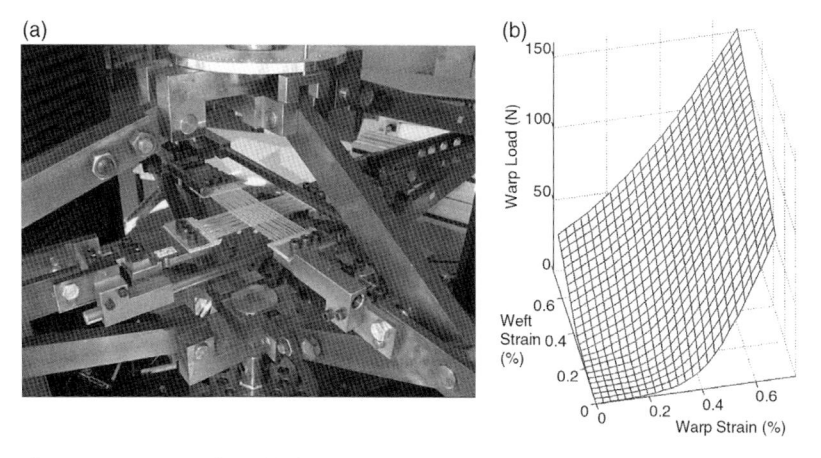

Figure 12.4 (a) Biaxial tensile device [7]. (b) Biaxial tensile surface.

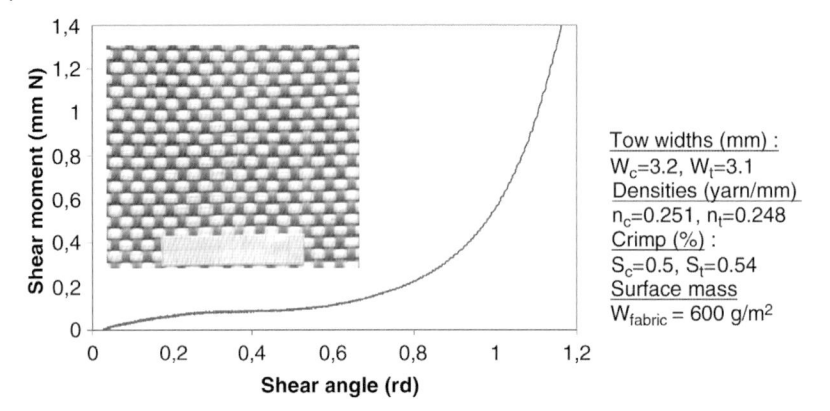

Figure 12.5 Shear curve of glass plain weave.

Tow widths (mm):
$W_c=3.2$, $W_t=3.1$
Densities (yarn/mm)
$n_c=0.251$, $n_t=0.248$
Crimp (%):
$S_c=0.5$, $S_t=0.54$
Surface mass
$W_{fabric} = 600$ g/m²

that the influence of the shear angle on the tensile behavior is usually weak and can be neglected [4].

12.2.4
In-Plane Shear Behavior

Two experimental tests are used to determine the in-plane shear behavior of textile composite reinforcements: the picture frame (Figure 12.6) and the bias extension tests (Figure 12.7). A great literature is dedicated to these tests [5, 8–16] mainly because the in-plane shear is the most dominant deformation mode in woven composite forming when the manufactured part is doubly curved. The shear angle (i.e., the angle variation between warp and weft yarns) can reach 50° (and sometimes more). For large values, wrinkling can occur depending on the process parameters and on the material properties. In addition, the two tests are difficult to perform from

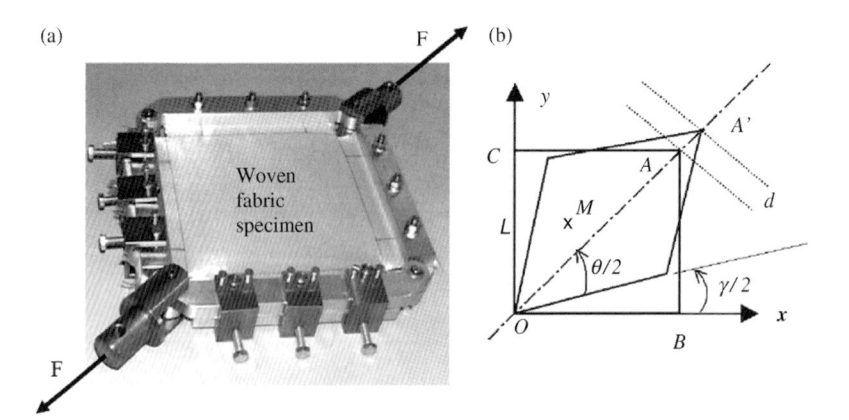

Figure 12.6 Picture frame test. (a) Experimental device. (b) Kinematics of the test.

12.2 Mechanical Behavior of 2D Textile Composite Reinforcements and Specific Experimental Tests

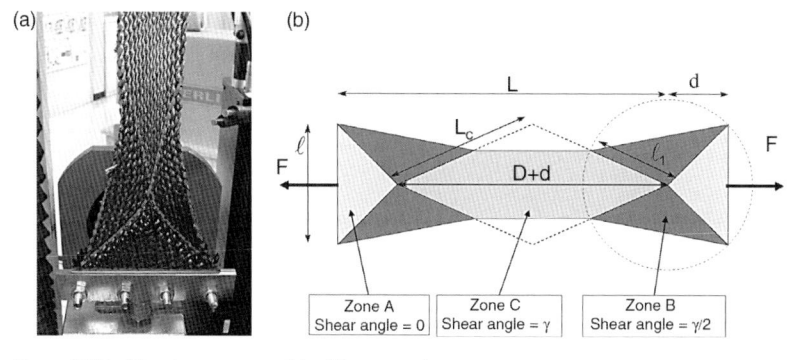

Figure 12.7 The three zones of the bias extension test.

both point of views: the experimental and the interpretation of the results. It has been confirmed by an international benchmark launched recently. Eight laboratories of six different countries have performed picture frame and bias extension tests on the same three textile composite reinforcements [16]. The results of these benchmarks have led to advances in understanding both the strengths and the limitations of the tests.

The picture frame (or trellis frame) is made of four hinged bars. It is subjected to a tensile force applied across diagonally opposing corners causing the picture frame to move from an initial square into a lozenge. The fabric specimen is initially square and the tows are parallel to the bars. Consequently, it is theoretically subjected to pure in-plane shear. The shear angle γ is function of the displacement of the frame corner d:

$$\gamma = \frac{\pi}{2} - 2 \arccos\left(\frac{1}{\sqrt{2}} + \frac{d}{2L}\right) \tag{12.6}$$

Neglecting the dissipation due to friction in the hinged bars of the frame, the equality of the power expressions obtained using the force on the frame F and the in-plane shear moment M_s leads to [17]

$$M_s(\gamma) = F \frac{S_c}{\sqrt{2}\,L}\left(\cos\frac{\gamma}{2} - \sin\frac{\gamma}{2}\right) \tag{12.7}$$

where S_c is the surface of the unit woven cell in the initial state. In Eq. (12.4), the shear moment M_s only depends on the shear angle γ. Some works have shown that stretching in the yarn directions can increase the shear stiffness [14, 18]. If material data giving M_s as function of angle γ and of ε_{11} and ε_{22} were available, they could be used in Eq. (12.4). Nevertheless, such data are usually not available and it will be considered that the picture frame (or the bias extension test) gives M_s depending on only γ for the fabric under consideration.

The bias extension test is an alternative to the picture frame test. It consists in clamping a rectangular specimen of woven fabric such that the warp and weft directions are initially oriented at 45° with respect to the tensile load applied by a tensile machine (Figure 12.7). The initial length L of the specimen must be larger than twice the width ℓ. The zone C in the center of the specimen is submitted to pure shear if the yarns are assumed to be nonstretchable. This is a correct assumption for the type of yarns used as composite reinforcements. This inextensivity imposes that the shear angle in the zone B is half the value in the central region C. The shear angle γ is function of the specimen elongation d:

$$\gamma = \frac{\pi}{2} - 2 \arccos\left(\frac{D+d}{\sqrt{2}D}\right) \quad (12.8)$$

D is the initial length of the central zone (Figure 12.7). The power made through the load F is turned into in-plane shear strain power in zones B and C:

$$F\dot{d} = M_s(\gamma)\dot{\gamma}N_{CC} + M_s\left(\frac{\gamma}{2}\right)\frac{\dot{\gamma}}{2}N_{CB} \quad (12.9)$$

N_{CC} and N_{CB} are respectively the numbers of woven unit cell in zones C and B. Hence,

$$M_s(\gamma) = \frac{FDS_c}{\ell(2D-\ell)}\left(\cos\frac{\gamma}{2} - \sin\frac{\gamma}{2}\right) - \frac{\ell}{2D-\ell}M_s\left(\frac{\gamma}{2}\right) \quad (12.10)$$

A shear curve $M_s(\gamma)$ measured in the case of glass plain weave is presented in Figure 12.5. In the first part of the curve (i.e., for small shear angles), the in-plane shear stiffness is small. For larger shear angles, this rigidity increases and becomes significant. It has been shown by optical full-field strain measurements at the mesoscopic level (i.e., within the tow) that during the first part of the loading, the tows rotate in a rigid body motion. When the shear angle becomes larger, lateral contacts between the yarns occur. The tows are progressively compressed and the shear rigidity increases significantly [15, 19]. The shear angle corresponding to the transition from low to higher stiffness is often called locking angle (about 40°–45° for textile composite reinforcements) [16]. Nevertheless, in practice, the transition is rather progressive (see, for instance, the large number of shear curves presented in Ref. [16]). As it will be shown in simulations presented below, the in-plane shear stiffness increase is important for the onset of wrinkling.

12.2.5
Bending Behavior

The bending stiffness of textile sheets is very weak in comparison to continuous materials such as sheet metal or composite plates with hardened matrix. This is due to the possible relative motions of fibers. Consequently, the relation given by the plate theory between the tensile and the bending stiffnesses is no longer valid. The tensile rigidities are so large in comparison to the bending ones that a membrane assumption is often made for fabric deformation simulations. This assumption is often correct in textile composite forming simulations.

Two main devices are used for the measurement of woven fabric bending properties. The KES-FB system [20] and the ASTM cantilever bending device developed from the work of Peirce (Figure 12.8a) [21–23]. In this test, the fabric

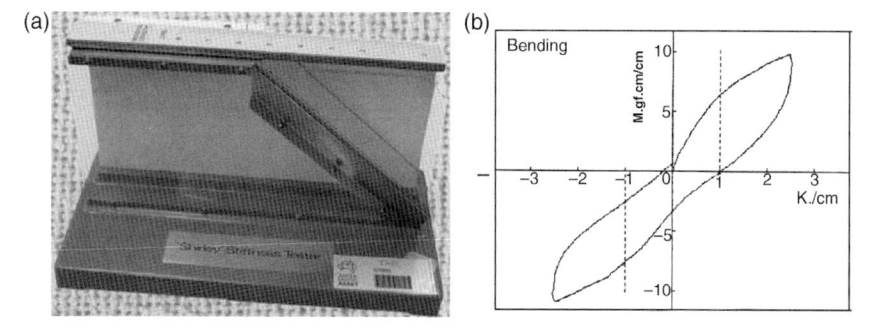

Figure 12.8 (a) ASTM, standard test method for bending stiffness of fabrics [22]. (b) A representative bending test result [23].

is cantilevered under gravity. The bending moment at the clamping section due to the weight of the fabric is related to the curvature calculated from the geometry of the specimen (Figure 12.8b).

In Eq. (12.5), only the bending moments in the warp and weft directions are taken into account and they are assumed to depend respectively on only warp and weft curvatures. Bending of textile material is a complex phenomenon. If more experimental data would be available, they could be introduced in Eq. (12.5) as mentioned in Section 12.2.2.

12.3
Continuous Modeling of 2D Fabrics: Macroscopic Scale

The multiscale structure of composite textile reinforcements has clear consequences on their mechanical behavior. The behavior specificities inherited by that nature can be observed at various scales. The softness of a woven fabric is largely due to the possible relative motions between warp and weft yarns. These fiber bundles exhibit themselves very low bending stiffness because of the possible sliding between fibers in the tow itself.

Modeling of this constitutive behavior can be made at both fabric and yarn scale. The present section deals with the former, which is the macroscopic scale, that is, the scale of the global structure as it allows using finite element codes with standard shell or membrane elements and does not need the description of the internal textile material structure.

Because of the possible relative motions between yarns, specific models have to be defined to correctly describe such particular behavior.

12.3.1
Geometrical Approaches

As the textile materials exhibit a tensile stiffness much higher than all the other rigidities, the first attempts to model the drapability of such materials have been based on geometrical approaches called fishnet (or fisherman's net) methods [24–26]. These approaches, based on the fishnet algorithm, state that the yarns are nonstretchable and that only the shear angles between warp and weft yarns are available. Each yarn is depicted by line segments representing the in-plane fiber bundle directions, connected at their crossover points by pivots. The method has the great advantage of simplicity and numerical efficiency for simulating the formability of a woven fabric as it is purely based on geometry.

However, geometrical approaches have strong limitations. In particular, since these methods do not take into account the mechanical characteristics of the fabric, they will give the same results whatever the weaving mode and whatever the nature of fibers used. Furthermore, the load boundary conditions (such as forces due to blank holders) cannot be taken into account. Moreover, because the shear and bending rigidities are not taken into account, the methods are unable to predict wrinkle appearance during the forming process.

Some authors have tried to extend the geometrical methods to estimate the mechanical characteristics of the final shapes of components [27].

12.3.2
Mechanical Approaches

Various approaches have been explored in the last decades. Some of them are based on the classical continuum mechanics approach, when another category uses a discrete modeling of the textile material. Chapter 3 presents continuous approaches.

At macroscopic level, a woven fabric can be seen as a continuous material with a very specific mechanical behavior, including high anisotropy and the ability to exhibit very large shearing and bending deformations. The first models have been proposed in the early 1990s [28]. They are based on the laminate theory with anisotropic viscoelastic behavior: the reinforcement is considered as an elastic material and the matrix is modeled as a viscous one. The composite material is seen as the superimposition of various layers, each one being representative of a yarn direction or of the matrix.

Layers representative of the textile material are anisotropic [29–31]. Such anisotropy has to be well monitored as it has been proved that a small error in the fiber orientation has large consequences on the mechanical response of a fabric [32]. During the deformation (during forming processes), the fabric can exhibit large inplane shear, with the consequence that the two fiber networks do not remain perpendicular to each other. In order to take into account this anisotropy, constitutive equations written in nonorthogonal frames have been proposed [33–36]. Unfortunately, some of these models [37] fail to give correct stress–strain responses for simple loadings, mostly because they are not able to strictly follow the fiber directions. This is the reason why some authors have given particular attention to that critical point [38–41]. All the above-mentioned models are said elastic or hypoelastic as they allow evaluating a stress increment (or the current stress state) as a function of a strain increment (or the current strain state).

Another solution for avoiding the difficulty linked to strong nonlinearities due to in-plane angle changes is to use hyperelastic models in the initial configuration. Various authors have proposed anisotropic hyperelastic models for modeling textiles [42]. Some models deal with dry fabrics [43], while others work on composites during their elaboration [44] or in use [45].

Most of these models neglect the bending stiffness of the textile material and only the membrane effects are taken into account.

12.3.2.1 Hypoelastic Model for Macroscopic Modeling of 2D Fabrics

Hypoelastic laws (or rate constitutive equations) are frequently used in finite element analyses at large strains [46, 47]. Truesdell has introduced these hypoelastic models in modern and general forms [48–50]. These rate constitutive equations are often used in finite element codes at finite strains [47, 51–53]. They link a strain rate tensor $\underline{\underline{D}}$ to a stress objective derivative $\underline{\underline{\sigma}}^{\nabla}$ of the Cauchy stress tensor:

$$\underline{\underline{\sigma}}^{\nabla} = \underline{\underline{\underline{C}}} : \underline{\underline{D}} \tag{12.11}$$

In Eq. (12.11), $\underline{\underline{\underline{C}}}$ is an Eulerian constitutive tensor oriented by the direction of the fiber. $\underline{\underline{\sigma}}^{\nabla}$ is an objective derivative of the Cauchy stress tensor $\underline{\underline{\sigma}}$ defined to avoid stress change due to rigid body rotations.

The objective derivative is the derivative for an observer fixed in a frame following the continuum in its deformation. As there are several possibilities for approaching this condition, there are several objective derivatives and several hypoelastic models [38, 46, 54–56]. It has been proved in case of textile reinforcements that the only consistent objective derivative is the one based on the fiber rotation tensor $\underline{\underline{\Phi}}$ [57]. This approach is different from the commonly used approaches in finite element codes of Jaumann corotational formulation (based on corotational frame rotation $\underline{\underline{Q}}$) [58] or Green–Naghdi approach (based on polar rotation $\underline{\underline{R}}$) [59]. The objective derivative of the Cauchy stress tensor with respect to fiber rotation tensor is

$$\underline{\underline{\sigma}}^{\nabla} = \underline{\underline{\Phi}} \cdot \left(\frac{d}{dt}(\underline{\underline{\Phi}}^T \cdot \underline{\underline{\sigma}} \cdot \underline{\underline{\Phi}}) \right) \cdot \underline{\underline{\Phi}}^T \tag{12.12}$$

where $\underline{\underline{\Phi}}$ is the rotation from the initial frame to the current fiber frame.

Since the constitutive tensor $\underline{\underline{\underline{C}}}$ is known along the fiber direction, it is mandatory to update the current fiber directions so that the constitutive law could be used properly.

If the fiber network is initially orthogonal, it loses that property rather quickly during the forming process.

There are two ways to follow the different fiber orientations: working on a nonorthogonal frame, each base direction corresponding to a fiber direction, or using as many orthogonal frames as there are fiber orientations. Contributions of each fiber orientation are superimposed in a common (working) frame (see Figure 12.9). The difficulties invoked above concerning models written on nonorthogonal frames lead to develop the presentation of the second approach [60]. For

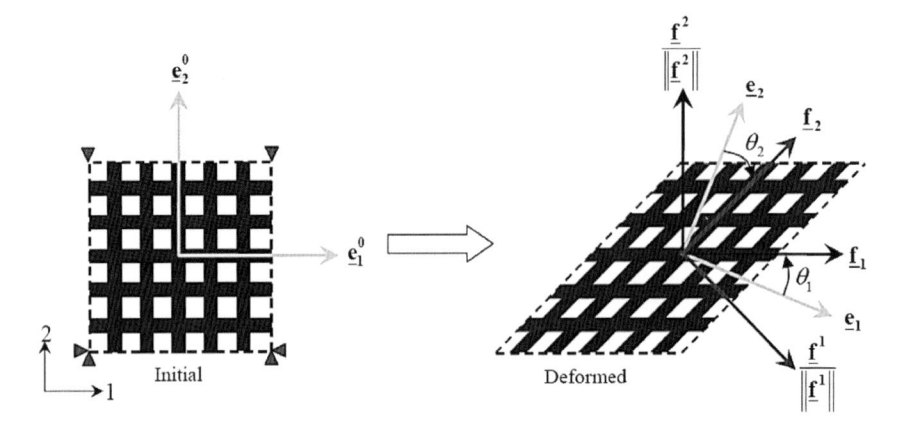

Figure 12.9 Orientation of Green–Naghdi axes $\underline{e}_\alpha$ and fibers axes, $\underline{f}_\alpha$ before and after deformation during a simple shear test. Initially both frames are superimposed.

commodity reasons, the working frame, that is, the frame in which the superimposition of stress contributions is made, is the classical Green Naghdi frame.

In Figure 12.9, vectors $\underline{e}_\alpha$ ($\alpha = 1, 2$) constitute the current Green–Naghdi basis, while $\underline{e}_\alpha^0$ is the initial frame. The current fiber directions are denoted $\underline{f}_\alpha$. Initially, when the fabric is undeformed, the initial fiber orientations $\underline{f}_\alpha^0$ and $\underline{e}_\alpha^0$ are superimposed.

As for many models, the bending stiffness of the material is not taken into account. Consequently, only the membrane effects are considered and the constitutive model is written in a 2D frame.

In the current configuration, it is possible to know the current directions $\underline{e}_\alpha$ and $\underline{f}_\alpha$ from the rotation tensor $\underline{\underline{R}}$ coming from the polar decomposition:

$$\underline{\underline{F}} = \underline{\underline{R}} \cdot \underline{\underline{U}} \tag{12.13}$$

where $\underline{\underline{U}}$ is the right stretch tensor and $\underline{\underline{F}}$ is the deformation gradient tensor.

The Green–Naghdi axes $\underline{e}_\alpha$ in the current configuration (the average rotation of the material axes) are updated using orthogonal rotation tensor $\underline{\underline{R}}$, and the initial orientation of Green–Naghdi axes $\underline{e}_\alpha^0$ by

$$\underline{e}_\alpha = \underline{\underline{R}} \cdot \underline{e}_\alpha^0, \quad \alpha = 1, 2 \tag{12.14}$$

whereas the current fiber directions $\underline{f}_\alpha$ are obtained using the deformation gradient tensor $\underline{\underline{F}}$ and the initial fiber orientations $\underline{f}_\alpha^0$:

$$\underline{f}_\alpha = \frac{\underline{\underline{F}} \cdot \underline{f}_\alpha^0}{\left\| \underline{\underline{F}} \cdot \underline{f}_\alpha^0 \right\|} = \frac{\underline{\underline{F}} \cdot \underline{e}_\alpha^0}{\left\| \underline{\underline{F}} \cdot \underline{e}_\alpha^0 \right\|}, \quad \alpha = 1, 2 \tag{12.15}$$

The two fiber axes ($\underline{f}_1$ and $\underline{f}_2$) usually do not remain orthogonal after deformation. Therefore, it is necessary to build two orthogonal frames for which one vector is a fiber direction. The first orthogonal frame is constructed from the first fiber axis $\underline{f}_1$ and the orthogonal unit vector $||\underline{f}^2/\underline{f}^2||$. The second one is built with $||\underline{f}^1/\underline{f}^1||$ and $\underline{f}_2$. Note that the covariant vectors (with subscripts) are parallel to their respective fiber or yarn directions, while the contravariant vectors (with superscripts) are orthogonal to the directions of the other fiber or yarn (Figure 12.9).

Let θ_α denote the angles between the current Green–Naghdi frame $\underline{e}_\alpha$ and vectors $\underline{f}_a$. Then, the transformation matrices $[T_\alpha]$ are defined by

$$[T_\alpha] = \begin{bmatrix} \cos\theta_\alpha & -\sin\theta_\alpha \\ \sin\theta_\alpha & \cos\theta_\alpha \end{bmatrix}, \quad \alpha = 1, 2 \tag{12.16}$$

The present approach has been implemented by mean of a user material VUMAT in ABAQUS/Explicit®. It calculates the strain increment tensor $[\Delta\varepsilon]_e$ in the Green–Naghdi frame at each increment. This strain increment is then to be projected in both fiber frames using the transformation matrices of Eq. (12.16):

$$[\Delta\varepsilon]_{f_\alpha} = [T_\alpha]^T [\Delta\varepsilon]_e [T_\alpha], \quad \alpha = 1, 2 \tag{12.17}$$

where, $[T_\alpha]^T$ is the transpose of $[T_\alpha]$

Then the constitutive equation is used in the fiber frames to evaluate the stress increments in both fiber orientations:

$$[\Delta\sigma]_{f_\alpha} = [C_\alpha]_{f_\alpha} [\Delta\varepsilon]_{f_\alpha} \tag{12.18}$$

Depending on the value of subscript α, the components of the constitutive tensor matrix are different:

$$[C_1]_{f_1} = \begin{bmatrix} E_1 & 0 & 0 \\ 0 & 0 & 0 \\ 0 & 0 & G_{12} \end{bmatrix} \quad \text{and} \quad [C_2]_{f_2} = \begin{bmatrix} 0 & 0 & 0 \\ 0 & E_2 & 0 \\ 0 & 0 & G_{12} \end{bmatrix} \tag{12.19}$$

The first component of the direct strain increment $(\Delta\varepsilon_{11})_{f_1}$ acts along the first fiber direction; the second direct strain increment component $(\Delta\varepsilon_{22})_{f_2}$ acts along the second fiber direction; and the in-plane shear strain increment $\Delta\gamma$ is defined to be the change in the angle between the warp and weft yarns and specifically here, it is equal to sum of the components of shear strain increments calculated from Eq. (12.17):

$$\Delta\gamma = (\Delta\varepsilon_{12})_{f_1} + (\Delta\varepsilon_{12})_{f_2} \tag{12.20}$$

The stress increments of Eq. (12.18) are cumulated in each fiber frame following the incremental formulation of Hughes and Winget [51]. This approach is commonly used in finite element codes at large strains. It is consistent with the objective derivative of Eq. (12.11) and uses the midpoint integration scheme to compute stress state at time t^{n+1} knowing at t^n.

$$\left[\sigma^{n+1}\right]_{f_\alpha^{n+1}} = \left[\sigma^n\right]_{f_\alpha^n} + [\Delta\sigma]_{f_\alpha}^{n+1/2} \tag{12.21}$$

The stresses computed in the two fiber frames using Eq. (12.21) are then expressed and superimposed in the Green–Naghdi frame as

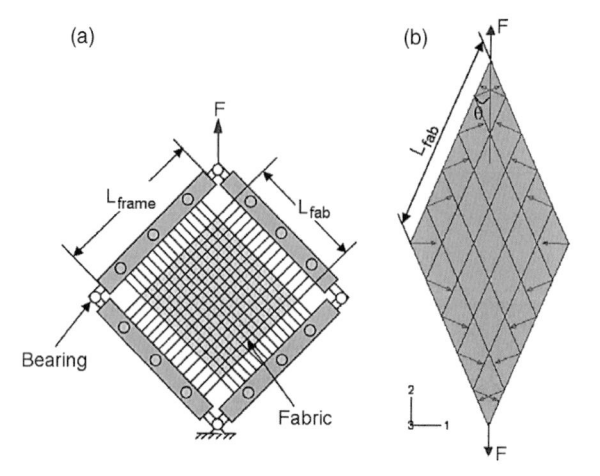

Figure 12.10 Picture frame test device (a) numerical simulation with nodal reaction forces perpendicular to the fiber directions (b).

$$[\sigma]_e = \sum_{a=1}^{2} [T_a][\sigma]_{f_a}[T_a]^T \qquad (12.22)$$

Such constitutive equation gives consistent results when applied to simple stress states. For a pure shear loading, reaction forces are orthogonal to the yarn directions exhibiting a zero tension (Figure 12.10) [60]. Mechanical characteristics are tensile moduli in both yarn directions E_1 and E_2 and the shear modulus G_{12}. These three quantities can vary with the current state of stress or strain. E_1 and E_2 are identified using biaxial tension tests [4] and G_{12} is known from in-plane shear test like picture frame or bias extension test [15, 16].

The model can be used for the modeling of forming processes as it is shown in Figure 12.11 where experimental and numerical deep drawings are compared [60, 61].

12.3.2.2 Hyperelastic Model for Macroscopic Modeling of 2D Fabric

Another approach for modeling the macroscopic mechanical behavior of 2D textile composite reinforcement consists in using hyperelastic formulations.

Some hyperelastic models for anisotropic materials analysis have been proposed. They mainly concern fiber–matrix composites and biomaterial analyses [62–68]. The

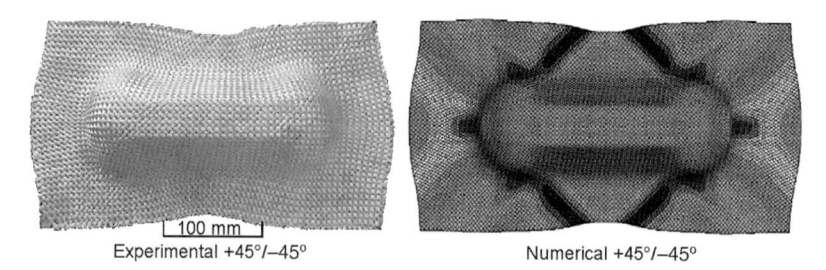

Figure 12.11 Deep drawing of a double dome for a commingled glass/polypropylene plain weave [61].

corresponding constitutive equations are related to the behavior of the fibers, the resin, and the interaction between both [69]. These models are not suitable for dry fabrics as their behavior is different due to the possible relative displacements between yarns and fibers. This is the reason why another model has been proposed [43] in order to take into account the specificities of dry textile reinforcements.

The formulation of a hyperelastic behavior law lies in the proposition of a potential energy from which derives the hyperelastic constitutive model. All the above-mentioned works are based on the existence of a potential energy based on the structural tensors introduced by Boehler [70].

The example presented here intends to take into account the macroscopic specificities of the woven reinforcement behavior such as small extensions in the warp and weft directions and large possible angular variations between these directions. Here again, the bending stiffness is neglected. In order to take into account these specificities, an energy potential based on the assumption that tensile and shear strain energies are uncoupled is proposed. The relation between the proposed strain energy potential and the stress components takes into account the material directions by means of structural tensors.

The two yarn networks (warp and weft) are characterized by two unit vectors $\underline{L}_1$ and $\underline{L}_2$, respectively, in the reference configuration (Figure 12.12a). These two vectors define locally privileged directions and a frame called initial orientation, from which the material directions change. In case of the woven fabric considered here, the initial anisotropy directions are supposed to be orthogonal in the reference (initial) configuration.

The constitutive law is formulated in the Lagrangian configuration. The elastic behavior at large strains is formulated as a function of the left Cauchy–Green strain tensor $\underline{\underline{C}}$ and the Piola–Kirchhoff second tensor $\underline{\underline{S}}$:

$$\underline{\underline{S}} = 2\frac{\partial W}{\partial \underline{\underline{C}}}, \quad \text{with} \quad W = W\left(\underline{\underline{C}}, \underline{L}_1, \underline{L}_2\right) \tag{12.23}$$

The Cauchy stress tensor can be calculated from $\underline{\underline{S}}$ and the deformation gradient tensor $\underline{\underline{F}}$:

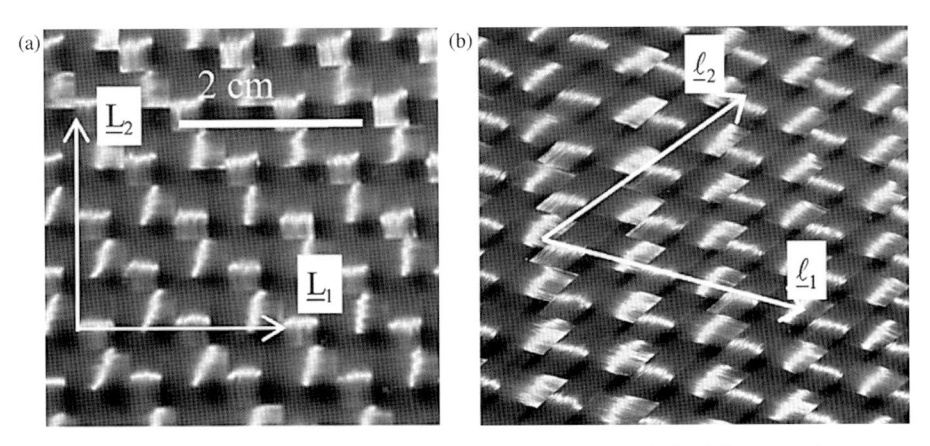

Figure 12.12 Woven fabric (glass plain weave) before deformation (a) and after deformation (b).

$$\underline{\underline{\sigma}} = \frac{1}{J} \underline{\underline{F}} \cdot \underline{\underline{S}} \cdot \underline{\underline{F}}^T \tag{12.24}$$

In order to model the woven reinforcement anisotropic mechanical behavior, the potential depends on invariants exhibiting a physical meaning:

$$W = \tilde{W}(I_4, I_6, I_{12}) \tag{12.25}$$

where I_4 and I_6 are the tensile invariants [69] and I_{12} is the shear invariant:

$$I_4 = \underline{L}_1 \cdot \underline{\underline{C}} \cdot \underline{L}_1 = \lambda_1^2 \tag{12.26}$$

$$I_6 = \underline{L}_2 \cdot \underline{\underline{C}} \cdot \underline{L}_2 = \lambda_2^2 \tag{12.27}$$

$$I_{12} = \frac{1}{I_4 I_6} \left[\underline{\underline{C}} : (\underline{L}_1 \otimes \underline{L}_1) \right] \cdot \left[\underline{\underline{C}} : (\underline{L}_2 \otimes \underline{L}_2) \right] = \cos^2 \theta \tag{12.28}$$

These invariants can also be defined using the fiber elongations in the direction $\alpha \lambda_\alpha$ and the current angle θ between yarns. I_4 and I_6 are called tensile invariants, while I_{12} is the shear invariant.

The second Piola–Kirchhoff stress tensor $\underline{\underline{S}}$ is derived from the strain energy by Eq. (12.23). With the present form of the potential, the stress tensor $\underline{\underline{S}}$ is given by

$$\underline{\underline{S}} = 2\left(\frac{\partial \tilde{W}}{\partial I_4} \frac{\partial I_4}{\partial \underline{\underline{C}}} + \frac{\partial \tilde{W}}{\partial I_6} \frac{\partial I_6}{\partial \underline{\underline{C}}} + \frac{\partial \tilde{W}}{\partial I_{12}} \frac{\partial I_{12}}{\partial \underline{\underline{C}}} \right) \tag{12.29}$$

The stress tensor $\underline{\underline{S}}$ requires the calculation of the first derivative of the considered invariants I_4, I_6, and I_{12} relatively to the right Cauchy–Green tensor $\underline{\underline{C}}$:

$$\frac{\partial I_4}{\partial \underline{\underline{C}}} = \underline{L}_1 \otimes \underline{L}_1 \tag{12.30}$$

$$\frac{\partial I_6}{\partial \underline{\underline{C}}} = \underline{L}_2 \otimes \underline{L}_2 \tag{12.31}$$

$$\frac{\partial I_{12}}{\partial \underline{\underline{C}}} = -\frac{I_{12}}{I_4}(\underline{L}_1 \otimes \underline{L}_1) - \frac{I_{12}}{I_6}(\underline{L}_2 \otimes \underline{L}_2) + \sqrt{\frac{I_{12}}{I_4 I_6}}(\underline{L}_1 \otimes \underline{L}_2 + \underline{L}_2 \otimes \underline{L}_1) \tag{12.32}$$

It can be noted that the derivative of the invariant I_{12} induces diagonal components of $\underline{\underline{S}}$. Finally, the second Piola–Kirchhoff stress tensor is calculated from Eqs. (12.29)–(12.32):

$$\begin{aligned}\underline{\underline{S}} &= 2\left(\frac{\partial \tilde{W}}{\partial I_4} - \frac{I_{12}}{I_4} \frac{\partial \tilde{W}}{\partial I_{12}} \right)(\underline{L}_1 \otimes \underline{L}_1) + 2\left(\frac{\partial \tilde{W}}{\partial I_6} - \frac{I_{12}}{I_6} \frac{\partial \tilde{W}}{\partial I_{12}} \right)(\underline{L}_2 \otimes \underline{L}_2) \\ &\quad + 2\sqrt{\frac{I_{12}}{I_4 I_6}} \frac{\partial \tilde{W}}{\partial I_{12}}(\underline{L}_1 \otimes \underline{L}_2 + \underline{L}_2 \otimes \underline{L}_1)\end{aligned} \tag{12.33}$$

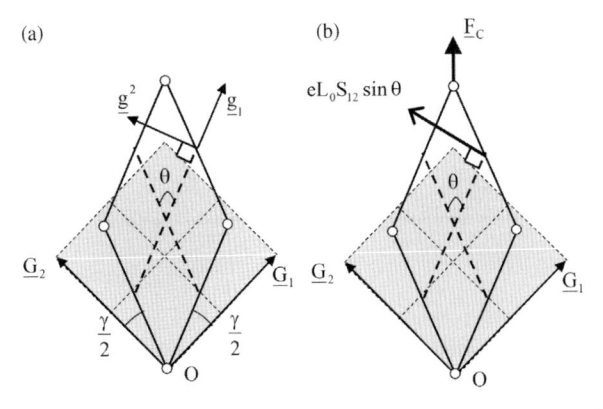

Figure 12.13 Reactions on the rigid edge of a picture frame. (a) Definition of the initial and actual frames. (b) Orientation of the reaction forces on the edges.

The Cauchy stress tensor in the current frame (Figure 12.12b) is then given by

$$\underline{\underline{\sigma}} = \sigma^{\alpha\beta} \underline{\ell}_\alpha \otimes \underline{\ell}_\beta = \frac{1}{J} S_{\alpha\beta} \underline{\ell}_\alpha \otimes \underline{\ell}_\beta, \quad \alpha, \beta = 1, 2 \qquad (12.34)$$

where $\underline{\ell}_\alpha = \underline{\underline{F}} \cdot \underline{L}_\alpha$ ($\alpha = 1, 2$).

The form of the potential is chosen in order to take the specificities of textile behavior into account. It has been checked that it gives a correct description of the reaction forces on the frame of a picture frame test, that is, the reaction forces within the picture frame are orthogonal to the fiber directions, which traduces that these fibers do not exhibit any tension (see Figure 12.13).

Two extra assumptions are made: yarn tensions and in-plane shear are independent; warp and weft tensions are uncoupled. These assumptions are questionable but are considered for simplicity reasons.

According to these hypotheses, the proposed potential energy is the addition of three terms: two tensile energies ($\tilde{W}_1$ and $\tilde{W}_2$) in the warp and weft directions and an in-plane shear energy $\tilde{W}_S$.

The potential is taken among the continuously differentiable functions of the considered invariants. It has to vanish in a stress-free configuration. Polynomial functions of the invariants are considered:

$$\tilde{W} = \sum_{i=0}^{r} \frac{1}{i+1} A_i (I_4^{i+1} - 1) + \sum_{j=0}^{s} \frac{1}{j+1} B_j (I_6^{j+1} - 1) + \sum_{k=0}^{t} \frac{1}{k} C_k I_{12}^k \qquad (12.35)$$

A_i, B_j, and C_k are constants determined in order to fit correctly experimental data. r, s, and t are the degrees of the considered polynomial functions.

For strain-free configuration, stresses have to vanish. This condition imposes

$$\sum_{i=0}^{r} A_i = 0, \quad \sum_{j=0}^{s} B_j = 0 \qquad (12.36)$$

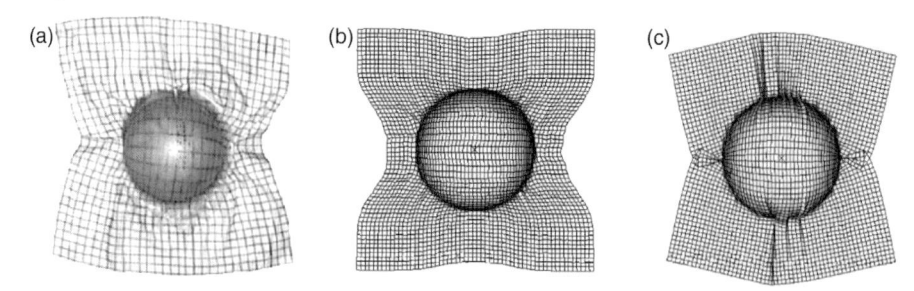

Figure 12.14 Deformed shape if an unbalanced fabric with an initial orientation of fibers 0°/90°). (a) Experimental shape [71]. (b) Simulation without shear rigidity. (c) Simulation with shear rigidity.

Because only initially orthogonal yarn network has been considered here, no specific condition is required for constants C_k. In case of initially nonorthogonal network, the same kind of conditions as for constants A_i and B_j must be fulfilled for constant C_k.

Three experimental tests are necessary to determine the potential: two uniaxial tensile tests in the warp and weft directions and one in-plane pure shear test (picture frame of bias extension test). The degrees of the polynomials are chosen in order to best fit the experimental curves. With these material parameters, it is possible to realize numerical simulations of dry reinforcement forming processes like for the hemispherical deep drawing of Figure 12.14. More details can be found in Ref. [43].

12.4
Discrete Modeling of 2D Fabrics: Mesoscopic Scale

Continuous modeling of mechanical behavior of 2D fabrics presents a weakness: it is not able to describe easily wrinkle forming and the possible sliding inside the fabric. Part of the mechanical behavior of the fabric is linked to these relative motions and the subsequent friction. Instead of modeling the fabric, if the scale of the modeling is the one of the fiber bundle, it is no more necessary to make assumptions in order to take into account these phenomena. The advantage of the continuous approach, that is, the not necessarily description of the fabric architecture, becomes a disadvantage. This is the reason why another approach exists: the discrete one.

12.4.1
Modeling the Global Preform

For the present approach, the continuity scale is no more the scale of the whole fabric but the scale of the yarns. The yarn material is modeled by means of the finite element tool. In such case, yarns are modeled and slippages are allowed between the warp and weft fiber networks. The main difficulty comes from the great number of yarns and of contacts that have to be taken into account. This one is modeled using elements simple enough to render the computation possible because it concerns the forming

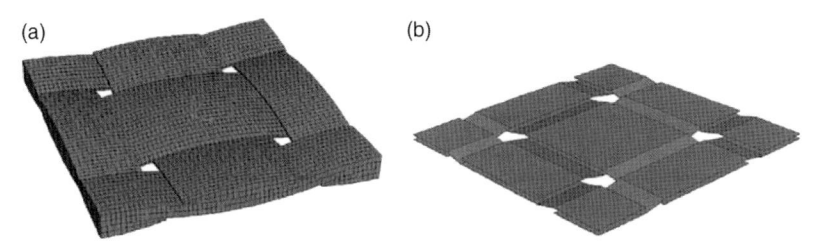

Figure 12.15 Mesomodeling of a unit cell of a plain weave. (a) FE model for the analysis of the behavior of the unit cell: 47 214 degrees of freedom. (b) FE model for simulations of the whole composite reinforcement forming: 648 degrees of freedom [79].

of the whole composite reinforcement and the number of yarns and contacts between these yarns is very large. The interactions between warp and weft directions are taken into account explicitly by considering contact behavior [72–74]. At the microscopic level, each fiber is satisfactorily described as a beam, but this approach is time consuming. The main difficulty is the great number of contacts with friction that have to be taken into account, especially for a woven fabric. For this reason, only very small elements of the fabric have been modeled to date [75–77]. Nevertheless, this approach is promising because it does not necessitate any assumptions regarding the continuity of the material, the specific mechanical properties resulting at the macroscopic level naturally follow the displacements and deformations of the yarns, and it provides an interesting way of taking the weaving operation into account. The fibers constituting the yarns can be modeled directly, but their very large number (3 to 48 K per yarn) requires that the computations are made for a number of fibers per yarn significantly smaller than in reality. An alternative possibility is to use a continuous behavior for each yarn.

This implies that the fibrous nature of the yarn is taken into account in this model especially in order to have rigidities in bending and transverse compression very small in comparison to the tensile stiffness. In any case, a compromise must be found between a fine description (which will be expensive from the computation time viewpoint) and a model simple enough to compute the entire forming process. Figure 12.15b shows the finite element model used for discrete simulations of forming processes (648 degrees of freedom). It is compared to another FE model of the unit cell used in Ref. [78] (Figure 12.15a) to analyze the local in plane shear of a plain weave unit cell (47 214 degrees of freedom). It cannot be considered (at least today) to use this last FE model to simulate the forming of a composite reinforcement that is made of several thousands of woven cells.

In the simplified unit cell (Figure 12.15b), each yarn is described by few shell elements and the contact friction and possible relative displacement of the yarns are considered. The in-plane mechanical behavior is the same as the one defined in [78] and is close of the one described in Section 12.3.2.1 dealing with the hypo-elastic continuous approach but for a single fiber direction.

The nature of the finite element used (thin shells), transverse compression of the yarn, cannot be modeled (only the "membrane" compression can be taken into

Figure 12.16 Pure shear on a square fabric blank. When the locking angle is reached, the wrinkles appear automatically [79].

account). The bending stiffness is independent of the tensile one and very much reduced in comparison to the one given by plate theories. In such approach, the fiber bundle direction is easily tracked and wrinkles appear naturally when the in-plane strain energy becomes greater than the bending one (see Figure 12.16), which corresponds slightly to the configuration for which the shear locking angle is reached. Moreover, such approach allows taking into account the possible sliding between yarns (see Figure 12.17). Such models are widely used for impact simulation [80, 81].

In order to chose convenient constitutive law for the simplified yarn behavior, precise finite element modeling, as shown in Figure 12.15a, is useful. The next section deals with mesoscopic scale modeling of textile behavior.

12.4.2
Modeling the Woven Cell

As previously mentioned, the constitutive behavior of the yarn material is not classical. During a preforming operation, the yarn shape can change dramatically.

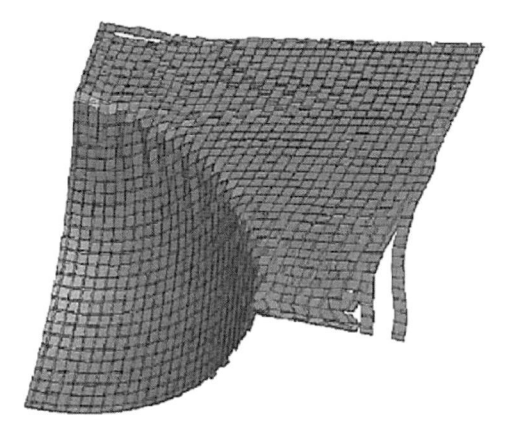

Figure 12.17 Deep drawing of a plain weave blank with a hemispherical punch: the unweaving is clearly pointed out [79].

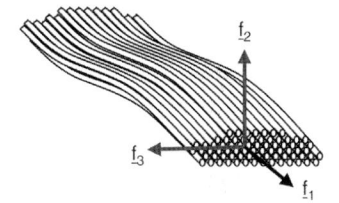

Figure 12.18 Local frame of the fiber bundle.

Shape and density changes have a great influence on the behavior of the global fabric. Being able to model the yarn behavior within the preform during characterization tests allows realizing virtual tests that make possible the estimation of the fabric mechanical characteristics before weaving in order to optimize the weaving pattern for a specific application.

Furthermore, during LCM processes, the final shape and density of the tows have a significant influence on the fabric permeability. Precise simulations of the elementary pattern evolution for simple loadings allow a more precise evaluation of the necessary parameters of the flow simulation [82–85].

The fiber bundle behavior as previously mentioned is as specific as the mechanical behavior of the whole fabric. Based on observations using X-ray tomography imaging [37], described in the previous paragraph, the fiber bundle behavior can be considered as transversely isotropic.

Based on this assumption, the constitutive relation used is decomposed into two parts: the longitudinal and the transverse behavior that can be split into a "spherical" and a "deviatoric" part. A local frame is associated with the fiber bundle with $\underline{f}_1$ being the fiber direction and $\underline{f}_2$ and $\underline{f}_3$ the in-plane section vectors (see Figure 12.18).

12.4.2.1 Longitudinal Behavior

The importance of a correct follow-up of the fiber direction has been already pointed out [40, 57]. Generally, the FE codes used give the strain increment in the Jaumann of Green–Naghdi frame; it is therefore necessary to use a specific procedure to correctly follow the fiber directions. The procedure is similar to the one described in the previous paragraph.

The longitudinal Young modulus E_1 is known from a tension test on a single yarn. To keep the yarn bending stiffness low enough, the longitudinal shear moduli G_{12} and G_{13} are significantly lower than the longitudinal Young modulus and assumed to have equal values: $G_{12} = G_{13} = G_1$.

12.4.2.2 Transverse Behavior

During the reinforcement deformation, the fiber bundle exhibits both surface and shape changes. These observations lead to use the same kind of decomposition of the strain tensor as used for metal plasticity. Consequently, the planar strain field is split into a "spherical" part, characterizing the surface change of the yarn section, and a "deviatoric" part, characterizing the shape change of the fiber bundle. Such decomposition leads to

$$[\tilde{\varepsilon}_T]_{f_1} = \begin{bmatrix} \varepsilon_s & 0 \\ 0 & \varepsilon_s \end{bmatrix} + \begin{bmatrix} \varepsilon_d & \varepsilon_{23} \\ \varepsilon_{23} & -\varepsilon_d \end{bmatrix} \tag{12.37}$$

with $\varepsilon_s = \frac{\varepsilon_{22} + \varepsilon_{33}}{2}$ and $\varepsilon_d = \frac{\varepsilon_{22} - \varepsilon_{33}}{2}$

This decomposition is assumed to be also valid for the strain increment and for the stress increment tensors, which leads to the constitutive Eq. (12.38):

$$\begin{aligned} \Delta\sigma_s &= A\Delta\varepsilon_s \\ \Delta\sigma_d &= B\Delta\varepsilon_d \\ \Delta\sigma_{23} &= C\Delta\varepsilon_{23} \end{aligned} \tag{12.38}$$

in which A, B, and C are material parameters depending on both the longitudinal tensile strain and the spherical strain ε_s and identified using an inverse method.

$\Delta\sigma_s$, $\Delta\sigma_d$, $\Delta\varepsilon_s$, and $\Delta\varepsilon_d$ are respectively spherical and deviatoric stress and strain increments.

Parameters A and B are given by Eq. (12.39).

$$\begin{aligned} A &= A_0 e^{-p\varepsilon_s} e^{n\varepsilon_{11}} \\ B &= B_0 e^{-p\varepsilon_s} \end{aligned} \tag{12.39}$$

Equation (12.39) shows that parameter A depends on both the spherical and longitudinal strain states of the yarn. Consequently, the identification of constant n requires a new experimental test of compression under controlled tension of the woven reinforcement.

The use of an in-plane shear test, ensuring a zero-tension state on the fiber tows gives a complementary test to identify constants A_0, B_0, and p.

Finally, the constitutive tensor matrix expressed in the current fiber frame (as defined in Figure 12.18) is

$$[C]_f = \begin{bmatrix} E_1 & 0 & 0 & 0 & 0 & 0 \\ 0 & (A+B)/2 & (A-B)/2 & 0 & 0 & 0 \\ 0 & (A-B)/2 & (A+B)/2 & 0 & 0 & 0 \\ 0 & 0 & 0 & G_1 & 0 & 0 \\ 0 & 0 & 0 & 0 & B & 0 \\ 0 & 0 & 0 & 0 & 0 & G_1 \end{bmatrix} \tag{12.40}$$

Such modeling allows predicting fine description of the fiber bundle deformation during large macroscopic in-plane deformations. It gives interesting information about the macroscopic behavior of a textile reinforcement knowing its architecture and about the solid skeleton of the reinforcement for the permeability calculations. Figure 12.19 shows the representative unit cell of a 2 × 2 carbon twill for the undeformed and deformed (sheared) configurations.

12.4.3
Use of the Mesoscale Modeling for Permeability Evaluation

The usefulness of the mesoscale simulations is particularly visible for the numerical determination of the permeability tensor of the reinforcement. Its experimental

(a)

(b)

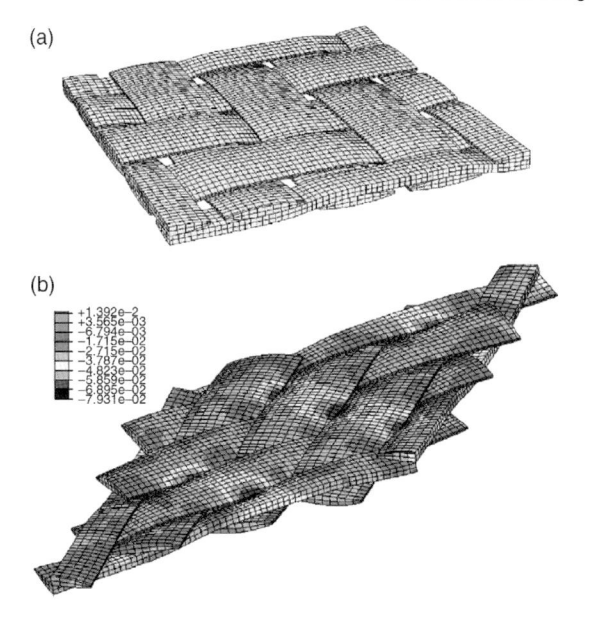

Figure 12.19 Simulation of shear of the 2 × 2 twill. (a) Initial unit cell and (b) 40° sheared cell with iso-values of local compaction of fibers [78].

determination is difficult and it is not really realistic to implement as many experiments as the possible deformed configurations. It has been proved that the reinforcement permeability highly depends on the actual shear angle between the warp and weft networks. After the simulation of the deformation of a unit woven cell, the complementary volume is extracted and meshed in order to realize a Stokes or Brinkman flow simulation (see Figure 12.20) [85, 86]. The same work has to be done for different deformations of the reinforcement in order to have a permeability tensor function of the actual strain state of the preform.

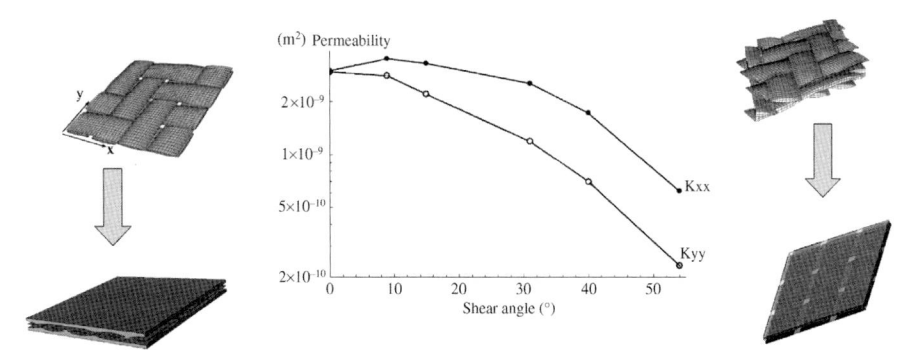

Figure 12.20 Influence of the shear angle on the permeability components for a 2 × 2 carbon twill [87].

12.5
Conclusions and Future Trend

The possible relative displacements between the fibers and the yarns that constitute a textile composite reinforcement lead to a very specific mechanical behavior. All the stiffnesses are small in comparison to the tensile rigidity in the fiber direction. The forming of these textile reinforcements on a double-curved surface takes advantage of these weak in-plane and bending stiffnesses. Specific experimental tests have been developed for textile composite reinforcements such as the picture frame and the bias extension test. They have been much studied, but they remain delicate mainly because of the very large difference between the tensile stiffness and the other rigidities.

The macroscopic continuous mechanical models are currently the most common approach in finite element analysis of fabric deformation, although there is no model that account for all the aspects of the behavior of fibrous fabrics. Some discrete analyses at mesoscale (the scale of the yarn) or microscale (the scale of the fiber) are also now developed. These are attractive and promising. The very specific mechanical behavior of the textile material due to the contacts and friction between the yarns and due to the change of direction is explicitly taken into account. If some sliding occurs between warp and weft yarns, they can be simulated. This is not possible by the continuous approaches that consider the textile material as a continuum. Nevertheless, it is necessary today to make a compromise between the accuracy of the model and the total number of degrees of freedom. The continuous increase of the computer power is a strong argument in favor of this approach.

References

1. Advani, S.G. (1994) *Flow and Rheology in Polymeric Composites Manufacturing*, Elsevier, Amsterdam.
2. Parnas, R.S. (2000) *Liquid Composite Molding*, Hanser Garner Publications.
3. Kawabata, S., Niwa, M., and Kawai, H. (1973) The finite deformation theory of plain weave fabrics. Part I: the biaxial deformation theory. *Journal of the Textile Institute*, **64** (1), 21–46.
4. Buet-Gautier, K. and Boisse, P. (2001) Experimental analysis and modeling of biaxial mechanical behavior of woven composite reinforcements. *Experimental Mechanics*, **41** (3), 260–269.
5. Willems, A., Lomov, S.V., Verpoest, I., and Vandepitte, D. (2008) Optical strain fields in shear and tensile testing of textile reinforcements. *Composites Science and Technology*, **68**, 807–819.
6. Carvelli, V., Corazza, C., and Poggi, C. (2008) Mechanical modelling of monofilament technical textiles. *Computational Materials Science*, **42**, 679–691.
7. Boisse, P., Gasser, A., and Hivet, G. (2001) Analyses of fabric tensile behaviour: determination of the biaxial tension–strain surfaces and their use in forming simulations. *Composites Part A*, **32**, 1395–1414.
8. Potter, K. (2002) Bias extension measurements on cross-plied unidirectional prepreg. *Composites Part A*, **33**, 63–73.
9. Lebrun, G., Bureau, M.N., and Denault, J. (2003) Evaluation of bias-extension and picture-frame test methods for the measurement of intraply shear properties of PP/glass commingled fabrics. *Composites Structures*, **61**, 341–352.

10. Peng, X.Q., Cao, J., Chen, J., Xue, P., Lussier, D.S., and Liu, L. (2004) Experimental and numerical analysis on normalization of picture frame tests for composite materials. *Composites Science and Technology*, **64**, 11–21.

11. Harrison, P., Clifford, M.J., and Long, A.C. (2004) Shear characterisation of viscous woven textile composites: a comparison between picture frame and bias extension experiments. *Composites Science and Technology*, **64**, 1453–1465.

12. Potluri, P., Perez Ciurezu, D.A., and Ramgulam, R.B. (2006) Measurement of meso-scale shear deformations for modelling textile composites. *Composites Part A*, **37**, 303–314.

13. Lomov, S.V. and Verpoest, I. (2006) Model of shear of woven fabric and parametric description of shear resistance of glass woven reinforcements. *Composites Science and Technology*, **66**, 919–933.

14. Launay, J., Hivet, G., Duong, A.V., and Boisse, P. (2008) Experimental analysis of the influence of tensions on in plane shear behaviour of woven composite reinforcements. *Composites Science and Technology*, **68**, 506–515.

15. Lomov, S.V., Boisse, P., Deluycker, E., Morestin, F. *et al.* (2008) Full-field strain measurements in textile deformability studies. *Composites Part A*, **39**, 1232–1244.

16. Cao, J., Akkerman, R., Boisse, P., Chen, J. *et al.* (2008) Characterization of mechanical behavior of woven fabrics: experimental methods and benchmark results. *Composites Part A*, **39**, 1037–1053.

17. Boisse, P., Zouari, B., and Daniel, J.L. (2006) Importance of in-plane shear rigidity in finite element analyses of woven fabric composite preforming. *Composites Part A*, **37** (12), 2201–2212.

18. Lomov, S.V., Willems, A., Verpoest, I., Zhu, Y., Barburski, M., and Stoilova, Tz. (2006) Picture frame test of woven composite reinforcements with a full-field strain registration. *Textile Research Journal*, **76** (3), 243–252.

19. Dumont, F., Hivet, G., Rotinat, R., Launay, J., Boisse, P., and Vacher, P. (2003) Field measurements for shear tests on woven reinforcements. *Mécanique et Industrie*, **4**, 627–635.

20. Kawabata, S. (1986) *The Standardization and Analysis of Hand Evaluation*, The Textile Machinery Society of Japan.

21. Peirce, F.T. (1937) The geometry of cloth structure. *The Journal of the Textile Institute*, **28**, 45–96.

22. ASTM (2002) D1388-96-ch. *Standard Test Method for Stiffness of Fabrics*, American Society for Testing and Materials.

23. de Bilbao, E., Soulat, D., Hivet, G., Launay, J., and Gasser, A. (2008) Bending test of composite reinforcements. *International Journal of Material Forming*, **1**, 835–838

24. Bergsma, O.K. and Huisman, J. (1988) Deep drawing of fabric reinforced thermoplastics. *Proceedings of the 2nd International Conference on Computer Aided Design in Composite Material Technology* (eds. C.A. Brebbia *et al.*), Springer, pp. 323–233.

25. Van Der Ween, F. (1991) Algorithms for draping fabrics on doubly curved surfaces. *International Journal of Numerical Methods in Engineering*, **31**, 1414–1426.

26. Cherouat, A., Borouchaki, H., and Billoët, J.L. (2005) Geometrical and mechanical draping of composite. *European Journal of Computational Mechanics*, **14**, 693–707.

27. Hofstee, J. and van Keulen, F. (2001) 3-D geometric modelling of a draped woven fabric. *Composite Structures*, **54**, 179–195.

28. Rogers, T.G. (1989) Rheological characterisation of anisotropic materials. *Composites*, **20**, 21–27.

29. Dong, L., Lekakou, C., and Bader, M.G. (2001) Processing of composites: simulations of the draping of fabrics with updated material behaviour law. *Journal of Composite Materials*, **35**, 138–163.

30. King, M.J., Jearanaisilawong, P., and Socrate, S. (2005) A continuum constitutive model for the mechanical behavior of woven fabrics. *International Journal of Solids and Structures*, **42**, 3867–3896.

31. Shahkarami, A. and Vaziri, R. (2007) A continuum shell finite element model for impact simulation of woven fabrics. *International Journal of Impact Engineering*, **34**, 104–119.

32. Breuer, U., Neitzel, M., Ketzer, V., and Reinicke, R. (1996) Deep drawing of

33 Yu, W.R., Pourboghrata, F., Chungb, K., Zampaloni, M., and Kang, T.J. (2002) Non-orthogonal constitutive equation for woven fabric reinforced thermoplastic composites. *Composites Part A*, **33**, 1095–1105.

fabric-reinforced thermoplastics: wrinkle formation and their reduction. *Polymer Composites*, **17** (4), 643–647.

34 Peng, X. and Cao, J. (2005) A continuum mechanics-based non orthogonal constitutive model for woven composite fabrics. *Composites Part A*, **36**, 859–874.

35 Xue, P., Cao, J., and Chen, J. (2005) Integrated micro/macro-mechanical model of woven fabric composites under large deformation. *Composite Structures*, **70**, 69–80.

36 Yu, W.R., Harrison, P., and Long, A. (2005) Finite element forming simulation for non-crimp fabrics using a non-orthogonal constitutive equation. *Composites Part A*, **36**, 1079–1093.

37 Badel, P., Vidal-Sallé, E., Maire, E., and Boisse, P. (2009) Simulation and tomography analysis of textile composite reinforcement deformation at the mesoscopic scale. *Composites Science and Technology*, **68**, 2433–2440.

38 Xiao, H., Bruhns, O.T., and Meyers, A. (1998) On objective corotational rates and their defining spin tensors. *International Journal of Solids and Structures*, **35** (30), 4001–4014.

39 Boisse, P., Gasser, A., Hagège, B., and Billoet, J.L. (2005) Analysis of the mechanical behavior of woven fibrous material using virtual tests at the unit cell level. *Journal of Materials Science*, **40**, 5955–5962.

40 Hagège, B., Boisse, P., and Billoët, J.L. (2005) Finite element analyses of knitted composite reinforcement at large strain. *European Journal of Computational Mechanics*, **14**, 767–776.

41 ten Thije, R.H.W., Akkerman, R., and Huétink, J. (2007) Large deformation simulation of anisotropic material using an updated Lagrangian finite element method. *Computer Methods in Applied Mechanics and Engineering*, **196**, 3141–3150.

42 Spencer, A.J.M. (2000) Theory of fabric-reinforced viscous fluids. *Composites Part A*, **31**, 1311–1321.

43 Aimène, Y., Vidal-Sallé, E., Hagège, B., Sidoroff, F., and Boisse, P. (2010) A hyperelastic approach for composite reinforcement large deformation analysis. *Journal of Composite Materials*, **44** (1), 5–26.

44 Wysocki, M., Toll, S., and Larsson, R. (2008) Hyperelastic constitutive models for consolidation of commingled yarn based composites. The 9th International Conference on Flow Processes in Composite Materials, Montreal (Québec), Canada – FPCM-9.

45 Holzapfel, G.A. and Gasser, T.C. (2001) A viscoelastic model for fiber-reinforced composites at finite strains: continuum basis, computational aspects and applications. *Computer Methods in Applied Mechanics and Engineering*, **190**, 4379–4430.

46 Xiao, H., Bruhns, O.T., and Meyers, A. (1997) Hypo-elasticity model based upon the logarithmic stress rate. *Journal of Elasticity*, **47**, 51–68.

47 Belytschko, T., Wing, K.L., and Moran, B. (2000) *Nonlinear Finite Elements for Continua and Structures*, John Wiley & Sons, Ltd, Chichester.

48 Truesdell, C. (1955) Hypo-elasticity. *Indiana University Mathematics Journal*, **4**, 83–133.

49 Truesdell, C. (1955) The simplest rate theory of pure elasticity. *Communications on Pure and Applied Mathematics*, **8**, 123–132.

50 Truesdell, C. (1956) Hypo-elastic shear. *Journal of Applied Physics*, **27**, 441–447.

51 Hughes, T.J.R. and Winget, J. (1980) Finite rotation effects in numerical integration of rate constitutive equations arising in large deformation analysis. *International Journal for Numerical Methods in Engineering*, **15**, 1862–1867.

52 Bathe, K.J. (1996) *Finite Element Procedures*, Prentice-Hall, Englewood Cliffs, NJ.

53 Criesfield, M.A. (1997) *Nonlinear Finite Element Analysis of Solids and Structure: Advanced Topics*, vol. 2, John Wiley & Sons, Ltd, Chichester.

54 Dienes, J.K. (1979) On the analysis of rotation and stress rate in deforming bodies. *Acta Mechanica*, **32**, 217–232.

55 Dafalias, Y.F. (1983) Corotational rates for kinematic hardening at large plastic deformations. *Journal of Applied Mechanics*, **50**, 561–565.
56 Liu, C.S. and Hong, H.K. (1999) Non-oscillation criteria for hypoelastic models under simple shear deformation. *Journal of Elasticity*, **57**, 201–241.
57 Badel, P., Vidal-Sallé, E., and Boisse, P. (2008) Large deformation analysis of fibrous materials using rate constitutive equations. *Computers & Structures*, **86**, 1164–1175.
58 Meyers, A., Xiao, H., and Bruhns, O. (2003) Elastic stress ratcheting and corotational stress rates. *Technische Mechanik*, **23** (2–4), 92–102.
59 Green, A.E. and Naghdi, P.M. (1965) A general theory of an elastic–plastic continuum. *Archive for Rational Mechanics and Analysis*, **18**, 251–281.
60 Khan, M.A., Mabrouki, T., Vidal-Sallé, E., and Boisse, P. (2010) Numerical and experimental analyses of woven composite reinforcement forming using a hypoelastic behaviour: application to the double dome benchmark. *Journal of Materials Processing Technology*, **210**, 378–388.
61 Woven Composites Benchmark Forum http://www.wovencomposites.org/.
62 Weiss, J.A. (1994) A constitutive model and finite element representation for transversely isotropic soft tissues. PhD thesis. Department of Bioengineering, University of Utah.
63 Weiss, J.A., Maker, B.N., and Govindjee, S. (1996) Finite element implementation of incompressible, transversely isotropic hyperelasticity. *Computer Methods in Applied Mechanics and Engineering*, **135**, 107–128.
64 Hirokawa, S. and Tsuruno, R. (2000) Three-dimensional deformation and stress distribution in an analytical/computational model of the anterior cruciate ligament. *Journal of Biomechanics*, **33**, 1069–1077.
65 Milani, A.S. and Nemes, J.A. (2004) An intelligent inverse method for characterization of textile reinforced thermoplastic composites using a hyperelastic constitutive model. *Composites Science and Technology*, **64**, 1565–1576.
66 Itskov, M. and Aksel, N. (2004) A class of orthotropic and transversely isotropic hyperelastic constitutive models based on a polyconvex strain energy function. *International Journal of Solids and Structures*, **41**, 3833–3848.
67 Diani, J., Brieu, M., Vacherand, J.M., and Rezgui, A. (2004) Directional model for isotropic and anisotropic hyperelastic rubber-like materials. *Mechanics of Materials*, **36**, 313–321.
68 Guo, Z.Y., Peng, X.Q., and Moran, B. (2006) A composite-based hyperelastic constitutive model for soft tissue with application to the human annulus fibrosus. *Journal of the Mechanics and Physics of Solids*, **54**, 1952–1971.
69 Spencer, A.J.M. (1984) *Continuum Theory of the Mechanics of Fibres-Reinforced Composites*, Springer, New York.
70 Boehler, J.P. (1978) Lois de comportement anisotrope des milieux continus. *Journal de Mécanique*, **17**, 153–170.
71 Daniel, J.L., Soulat, D., Dumont, F., Zouari, B., Boisse, P., and Long, A.C. (2003) Forming simulation of very unbalanced woven composite reinforcements. *International Journal of Forming Processes*, **6** (3–4), 465–480.
72 Ben Boukaber, B., Haussy, G., and Ganghoffer, J.F. (2007) Discrete models of woven structures: macroscopic approach. *Composites Part B*, **38**, 498–505.
73 Duhovic, M. and Bhattacharyya, D. (2006) Simulating the deformation mechanisms of knitted fabric composites. *Composites Part A*, **37**, 1897–1915.
74 Pickett, A.K., Creech, G., and de Luca, P. (2005) Simplified and advanced simulation methods for prediction of fabric draping. *European Journal of Computational Mechanics*, **14**, 677–691.
75 Zhou, G., Sun, X., and Wang, Y. (2004) Multi-chain digital element analysis in textile mechanics. *Composites Science and Technology*, **64**, 239–244.
76 Durville, D. (2010) Simulation of the mechanical behaviour of woven fabrics at the scale of fibers. *International Journal of Material Forming*. doi: 10.1007/s12289-009-0674-7.

77 Miao, Y., Zhou, E., Wang, Y., and Cheeseman, B.A. (2008) Mechanics of textile composites: micro-geometry. *Composites Science and Technology*, **68**, 1671–1678.

78 Badel, P., Gauthier, S., Vidal-Sallé, E., and Boisse, P. (2009) Rate constitutive equations for computational analyses of textile composite reinforcement mechanical behaviour during forming. *Composites Part A*, **40**, 997–1007.

79 Boisse, P., Aimène, Y., Dogui, A., Dridi, S. et al. (2010) Hypoelastic, hyperelastic, discrete and semi-discrete approaches for textile composite reinforcement forming. *International Journal of Material Forming*. doi: 10.1007/s12289-009-0664-9.

80 Nilakantan, G., Keefe, M., Gillespie, J.W., and Bogetti, T.A. (2009) Simulating the impact of multi-layer fabric targets using a multiscale model and the finite element method: recent advances in textile composites. *Proceedings of the 9th International Conference on Textile Composites: TEXCOMP9*, DEStech Publications, Inc., pp. 506–515.

81 Sapozhnikov, S.B., Forental, M.V., and Dolganina, N.Y. (2007) Improved methodology for ballistic limit and blunt trauma estimation for use with hybrid metal/textile body armor. Proceedings of the Conference Finite Element Modelling of Textiles and Textile Composites (CD ROM), St Petersburg.

82 Vandeurzen, P., Ivens, J., and Verpoest, I. (1996) A three-dimensional micromechanical analysis of woven-fabric composites: I. Geometric analysis. *Composites Science and Technology*, **56**, 1303–1315.

83 Bickerton, S., Simacek, P., Guglielmi, S.E., and Advani, S.G. (1997) Investigation of draping and its effects on the mold filling process during manufacturing of a compound curved composite part. *Composites Part A*, **28**, 801–816.

84 Belov, E.B., Lomov, S.V., Verpoest, I., Peters, T., Roose, D., Parnas, R.S., Hoes, K., and Sol, H. (2004) Modelling of permeability of textile reinforcements: lattice Boltzmann method. *Composites Science and Technology*, **64**, 1069–1080.

85 Laine, B., Boust, F., Boisse, P., Hivet, G., Lomov, S., and Fanget, A. (2005) Meso–macro optical experimental analysis of woven composite reinforcement in plane shear during their forming. *Revue des Composites et Matériaux Avancés*, **15-3**, 385–400.

86 Loix, F., Badel, P., Orgéas, L., Geindreau, C., and Boisse, P. (2008) Woven fabric permeability: from textile deformation to fluid flow mesoscale simulations. *Composites Science and Technology*, **68**, 1624–1630.

87 Laine, B. (2008) Influence des déformations d'un renfort fibreux sur sa perméabilité: modélisations et expériences, PhD thesis, pp. ENSAM 146.

13
Three Dimensional Woven Fabric Composites
Wen-Shyong Kuo

13.1
Introduction

Three dimensional (3D) textile composites are a unique class of materials, produced by impregnating matrix materials into the fabric to hold multidirectional yarns in position [1–5]. Based on the fabricating techniques, 3D preforms are usually categorized into the following groups: woven [6–9], braided [10–17], stitched [18–22], and knitted [23–26]. These techniques are fundamentally different in their formation principles and the resulting yarn structures. The needs in end use often dictate the choice of formation methods and the design of fabric structures. In conventional processing of 3D composites, dry fiber tows are often employed to construct 3D fabrics with desired external shapes and internal structures. Resin impregnation and consolidation are usually employed for the addition of the matrix material. In this chapter, the focus is on the 3D woven fabric composites.

In 3D weaving, yarns are commonly carried by shuttles to feed among axial yarns in the weaving plane, which is normal to the axial direction. Thus, this method is suitable for forming fabrics with an orthogonal yarn arrangement [27–29]. Compared with other methods, orthogonal weaving is able to form fabrics with much higher through–thickness reinforcements. Fiber distributions in three orthogonal directions can be designed according to the external load. This is in contrast to the many 3D fabrics having a majority of the fibers along the axial direction. By controlling the interlacing patterns, this weaving method also allows greater variability to the movement of the shuttles, making it possible to produce fabrics with intricate shapes such as hollow or rib-reinforced structures. This feature is nonexistent in other three groups of 3D textile composites.

For 3D fabrics, interlacing of yarns is a necessity that brings together the otherwise separated fiber tows. In 3D fabrics involving multiaxes reinforcements, the pattern of yarn interlacing can be more complicated. When making complex fabrics, designing the pattern is perhaps the most crucial part of fabric formation. The difficulty lies in the need that all yarns should be appropriately integrated in the near-net-shape fabrics. Thus, making complex 3D fabrics would call for a unique, and often ingenuous, design of the interlacing pattern and the techniques to accomplish it.

13.2
General Characteristics of 3D Composites

Some general characteristics of 3D composites are first discussed. These characteristics are the results of the internal yarn structure and yarn interlacing on surface.

13.2.1
Multidirectional Structural Integrity

This is the major reason that intensive research effort has been dedicated to 3D composites. 3D Composites are characterized by the 3D network of reinforcements and interlacing loops. The former prevents the materials from a large-scale, detrimental fracture, and the latter links yarns and provides an integrated structure for the composite [30–32]. The multidirectional reinforcing network and the distribution of fibers along each direction can be designed by taking into account the external loads. For this reason they are most suitable to be used in conditions where the loads are complex and multidirectional. Because of the 3D network, planar types of failure, such as delamination in 2D laminated composites, can be avoided. Due to the 3D network, weak planes are nonexistent, and cracks in the composite can be stopped or have to grow in a more tortuous manner. This often leads to a more damage-tolerant composite.

13.2.2
Near-Net-Shape Design

The various techniques developed for fabric formation allow more degree-of-freedom in the design of fiber architecture to obtain a fully integrated 3D fabric with a shape close to the final product. With this feature, the need for joining and further machining could be eliminated [33]. Fiber tows can remain unbroken in the composite, and this enhances load-carrying capability. Among the major techniques to form 3D textile fabrics, weaving is the most versatile in making fabrics with complex shapes. The reason is that each individual shuttle can move independently according to its specially designed path. To form a complex 3D fabric, the key lies in the selection of shuttle number and the design of the weaving path for each shuttle.

13.2.3
Greater Nonuniformity

The most marked property of 3D composites is perhaps the nonuniformity. The dimensional scale of interest to characterize nonuniformity can vary widely in different composites. In unidirectional composites, fibers are parallel, and the scale of interest is fiber diameter, which is of the order of $10\,\mu m$ (0.01 mm). In laminated composites, the thickness of the layers becomes the scale of interest, which is typically 100 μm (0.1 mm). For 3D textile composites, the scale of interest is the yarn size, which is usually greater than 1 mm. The pitch length, or spacing between yarns is typically 2–3 mm in 3D composites. For this reason, the stress and strain in 3D composites are highly nonuniform, even viewed in a macroscopic scale.

13.2.4
Lower Fiber Volume Fraction

While a 70% fiber volume fraction is not uncommon to unidirectional and laminated composites, the same level of fiber packing in 3D composites is virtually impossible. The main reason is that the interlacing of yarns inevitably creates interyarn spaces, which will become resin pockets or voids after resin impregnation [31]. The existence of interyarn spaces reduces the overall fiber content. In practice, it is difficult to make a 3D woven composite with a fiber volume fraction higher than 55% through the resin-transfer-molding approach. One simple way to increase the fiber content is to press the 3D fabric and squeeze out the excess resin before resin curing [34]. But this practice distorts the fabric and even causes fiber breakage.

13.2.5
Higher Fiber Crimp

Fiber crimp is inevitable in textile composites. In comparison with 2D woven fabrics, which are characterized by the wavelike undulation due to intertwining of the warp and weft yarns, 3D fabrics may develop more prominent and unpredictable yarn crimp due to their complex fabric architecture. Fiber crimp in 3D fabrics is primarily because of interlacing of yarns. At the point of interlacing, there is a resultant force to push the interlaced yarns. Yarns with flexible fibers can be distorted easily by the interlacing force. The distorted yarns not only create an irregular and possibly looser yarn packing but also alter the fabric from the originally designed architecture. This geometric problem in 3D fabrics should lower the stiffness and the strength of the resulting composites. If the composites are to benefit fully from the high stiffness of the fibers, fiber crimp in noninterlacing yarns must be minimized.

13.2.6
Lower Stress-to-Yield and Higher Strain-to-Failure

The stress-to-yield and strain-to-failure depend on the onset and growth of damage. Because of higher fiber crimp and lower fiber content, damage in 3D composite tends to trigger at a lower stress. Yet, while the stress-to-yield is low, 3D composites usually exhibit a higher strain-to-failure. Because of the 3D structure eliminates weak planes, large-scale disintegration can be prevented. Yarns damaged in the early stage of loading can be supported by undamaged yarns in other directions. Thus, these materials usually exhibit high load-resistance after initial damage has developed. With the loops and through-thickness yarns, the 3D composites are more capable in dissipating loads into various directions of fibers. These unique features, although unable to enhance the critical stress for initial damage, contribute to a high strain-to-failure, a long load-plateau and, consequently, a high capability of energy absorption.

13.2.7
More Difficult in Material Testing

In conducting material tests for 3D composites, some problems are usually encountered. First, in comparison with the existing standards designed for 2D laminated composites, the associated ones for 3D textile composites are less well defined. This is in part due to the wide varieties of 3D fabrics having dissimilar structures, and it is unlikely that a single testing standard is able to cover a broad range of 3D composites. For testing 3D composites, researchers in this area might have to define their specimen dimensions and testing specifications.

Second, 3D textile composites are inherently thick, and most important they are macroscopically inhomogeneous due to larger yarn spacing, as mentioned earlier. When testing thick specimens, the application of the load to fracture the specimen becomes difficult. For example, conducting tensile tests on short and thick composites is virtually impossible without the use of the dog-bone type, which should machine out the specimen surfaces. For end-tabbed specimens, the needed length of the end-tab, which must be proportional to the specimen thickness, can become unacceptably long.

13.2.8
More Complex in Damage Mechanisms

This characteristic is related to the greater nonuniformity discussed previously. 3D Composites are an assemblage of yarns oriented at various directions. After resin impregnation, each yarn becomes a unidirectional composite. When the yarn is loaded, damage can be initiated in a manner similar to that in unidirectional composites. Yet the growth of damage within a yarn can be interrupted at the yarn's boundary, at which the material properties are discontinuous. Intrayarn damage cannot grow across the boundary, and vice versa for interyarn damage. The discontinuity in material properties does not end the growth of damage, but it does diversify damage types and spread damage over a larger region – a mechanism that help to absorb energy by a larger but less damaged region.

13.3
Formation of 3D Woven Fabrics

13.3.1
Three-Axis Orthogonal Weaving

Weaving is perhaps the most important formation technique in traditional textile industry. Some 2D weaving techniques have been adopted for the formation of 3D

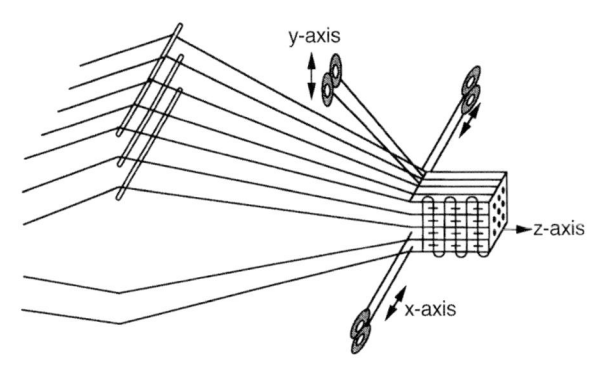

Figure 13.1 Schematic drawing of 3D orthogonal weaving.

fabrics. In 2D weaving, one set of axial yarns is arranged in a plane, and one shuttle is used to feed weft yarns in the transverse direction. To allow feeding of the weft yarn, selected axial yarns are lifted by the harnesses. Beat-up is usually followed to compact the fabric. In 3D weaving, multiple sets of axial yarns are arranged in a special pattern, according to the size and shape of the final composite. Multiple shuttles can be employed to feed transverse yarns among the axial yarns. Resembling the harness, a special mechanism is needed to displace some axial yarns and to allow shuttles to pass through.

Figure 13.1 is a schematic drawing of 3D orthogonal weaving. The axial yarns (denoted as z-axis) are arranged in an orthogonal pattern, and there are two sets of shuttles to feed yarns in the transverse directions (denoted as x-axis and y-axis). The result is so-called three-axis orthogonal fabric. Figure 13.2 is a schematic drawing of an idealized three-axis orthogonal fabric, assuming that the yarns have a circular cross-section. By adding two more sets of off-axis yarns to a three-axis fabric, we obtain a five-axis fabric. Figure 13.3 schematically shows a five-axis fabric with two off-axis yarn oriented along the $+45°$ and $-45°$ directions on the weaving (xy) plane. Note that in this fabric the off-axis yarns are smaller in bundle size than the on-axis

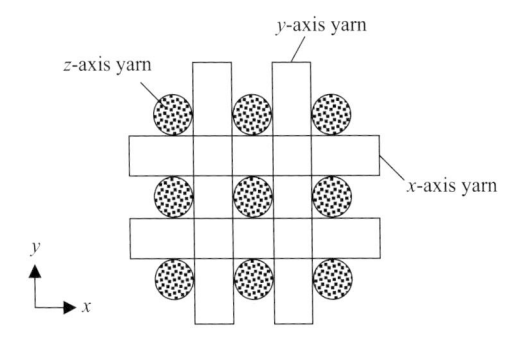

Figure 13.2 Schematic drawing of three-axis orthogonal fabric composed of circular yarns.

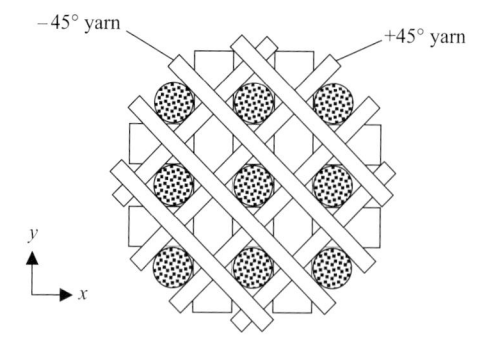

Figure 13.3 Schematic drawing of five-axis orthogonal fabric, adding +45° and −45° yarns to the three-axis orthogonal fabric.

yarns. If all five axes have the same bundle size, the spacing between the axial yarns must be increased, or the off-axis yarns must be in an undulated form.

Figure 13.4 shows a three-axis orthogonal composite. Solid Kevlar/vinyl ester rods with 1 mm in diameter are used in the y- and z-axis. Solid rods can be incorporated into the fabric to replace dry yarns [35, 36]. Figure 13.5 is a *xy*-section of a five-axis composite, showing the +45° yarns on the section. The off-axis yarns are slightly

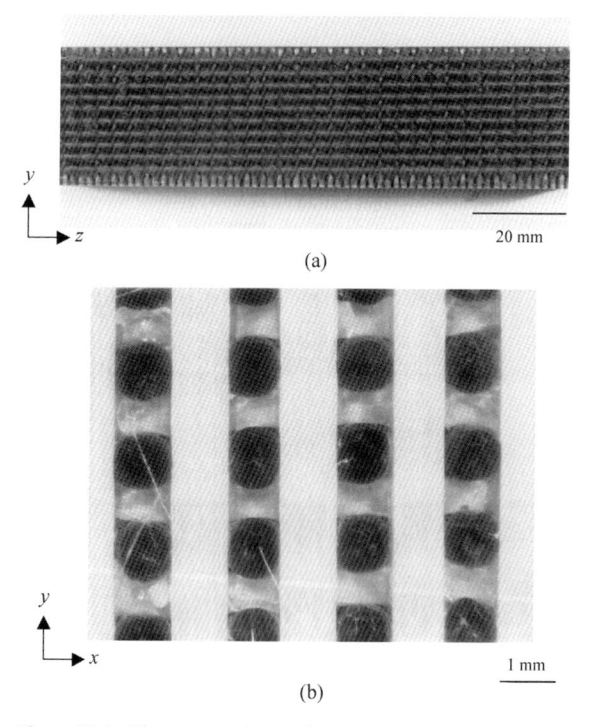

Figure 13.4 Three-axis orthogonal composite using solid rods in the axial (z) and transverse (y) directions. (a) Composite surface, (b) a cross-section [35].

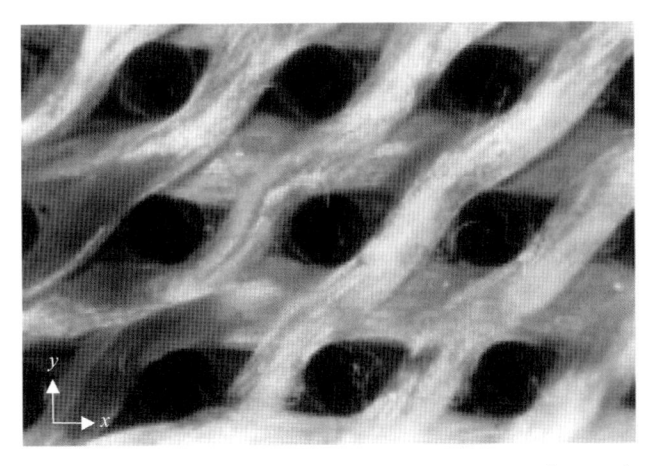

Figure 13.5 A xy-section of a five-axis orthogonal composite showing the wavy off-axis yarns [36].

undulated. The space for the off-axis yarns is smaller than for on-axis yarns, as illustrated in Figure 13.3.

13.3.2
Design of Weaving Schemes

To produce 3D fabrics with a prescribed internal yarn structure, usually there are several options in weaving schemes. Three examples are discussed in this section. The goal is a three-axis orthogonal fabric with axial yarns arranged in a 5×10 row-and-column pattern [31].

The first weaving scheme is termed *unidirectional weaving* (UW), meaning that in each step the weaving yarns move along one direction. As shown in Figure 13.6, the weaving shuttles are represented by arrows, and there are four steps in each cycle of weaving. The resulting interlacing pattern is shown in Figure 13.7. The second weaving scheme is termed *bidirectional weaving* (BW). The weaving yarns move in two opposite directions in each step, as shown in Figures 13.8 and 13.9. The third is *symmetric weaving* (SW), designed to achieve a complete and symmetric pattern of loops. This fabric needs only two steps to complete a cycle (Figure 13.10), and the resulting interlacing pattern is shown in Figure 13.11.

Figure 13.12 shows the fabrics and the composite cross-sections according to the three weaving schemes [31]. One advantage with the symmetric weaving is that the axial yarns, equally pushed by two weaving yarns in opposite directions, is not deformed in an undulated form, as is the case of the UW fabrics. The yarn deformation and the fiber misalignment are thus the minimum.

For axial yarns arranged in an m-by-n, row-and-column pattern, the number of shuttles used can vary from one to $m + n$. Figure 13.13 schematically shows how the one-shuttle scheme can be employed to make a three-axis orthogonal fabric. Note that there is a limitation for using the one-shuttle scheme for the m-by-n pattern; at least

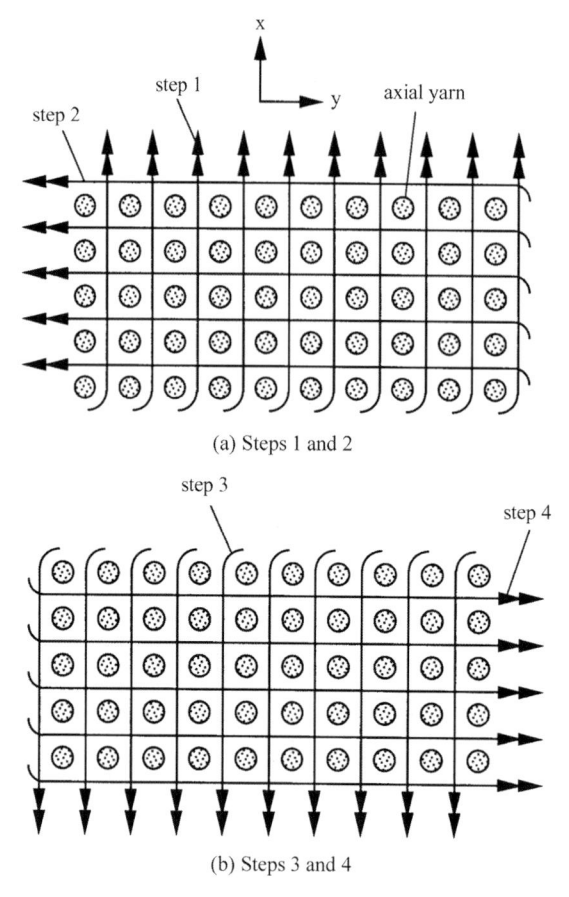

400 | *13 Three Dimensional Woven Fabric Composites*

(a) Steps 1 and 2

(b) Steps 3 and 4

Figure 13.6 Schematic drawing of unidirectional weaving.

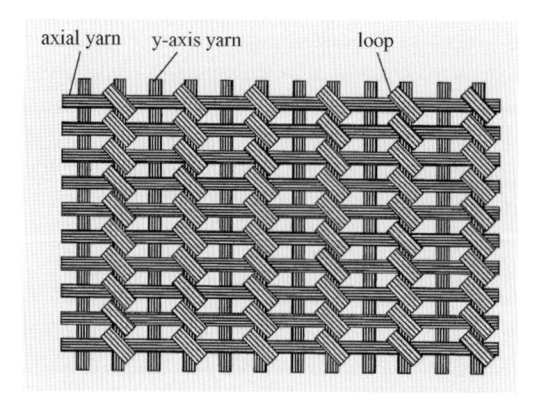

Figure 13.7 Schematic drawing of the pattern of the interlacing loops by unidirectional weaving.

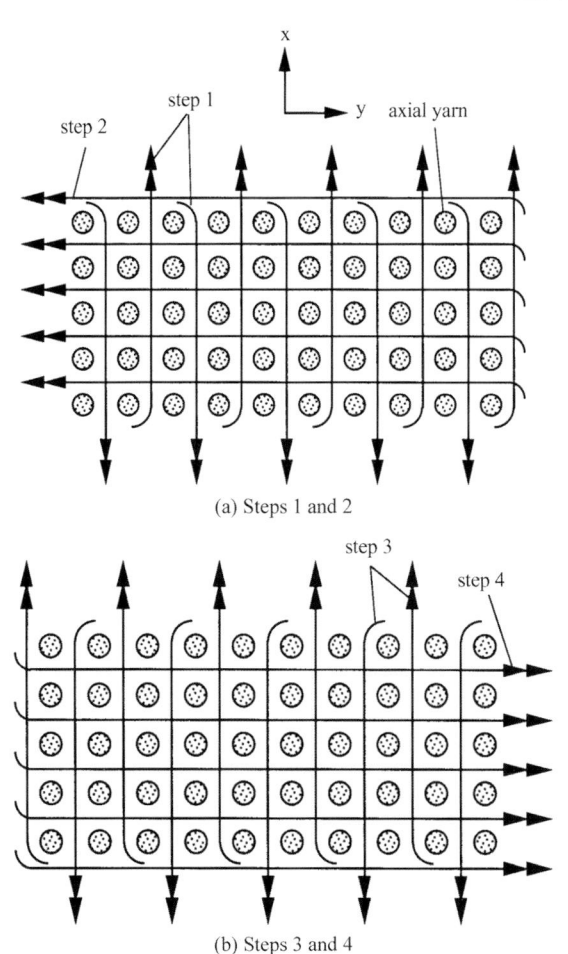

Figure 13.8 Schematic drawing of bidirectional weaving.

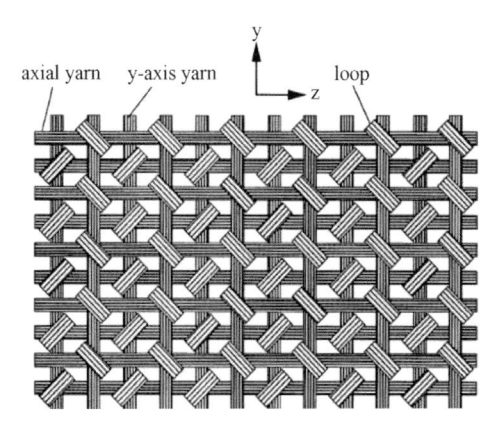

Figure 13.9 Schematic drawing of the pattern of the interlacing loops by bidirectional weaving.

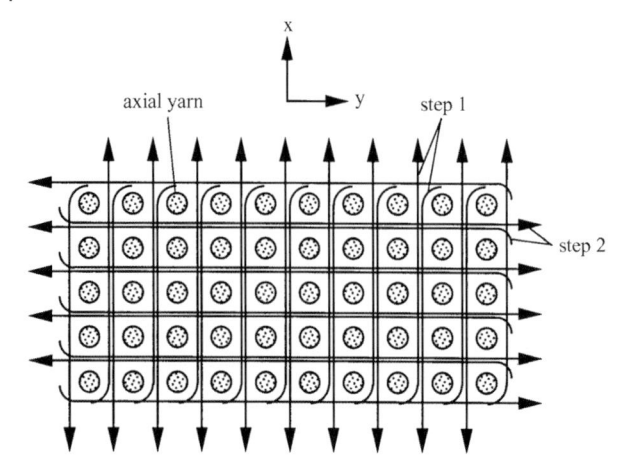

Figure 13.10 Schematic drawing of symmetric weaving.

one of the number m and the number n must be odd. If both are even, the shuttle returns to the starting point of step 1 after step 2. As a result, a half of the axial yarns located on the perimeter of the pattern cannot be interlaced. One unique feature of the one-shuttle scheme is that the interlacing loops are parallel to the weaving plane. In comparison, in the UW, BW, and SW fabrics, all interlacing loops are inclined to the weaving plane.

For more complex 3D fabrics such as hollow, I-shaped, T-shaped, ribbed fabrics, the weaving processes become proportionally complex. The keys of fabric formation lie in the selection of shuttle number and the design of the moving path for each shuttle. Two rules need to be met. First, all weaving paths must receive the equal amount of fibers on the same weaving plane. Otherwise, the axial yarns can be

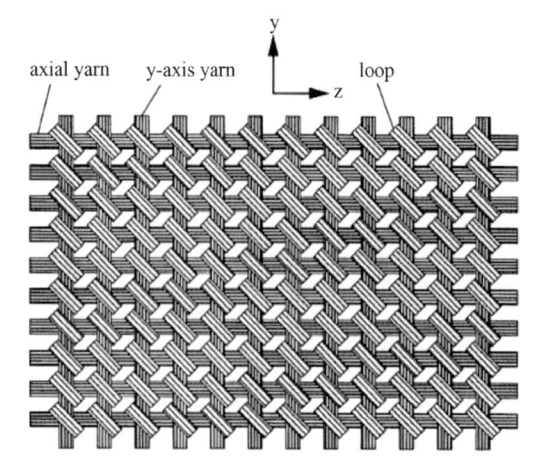

Figure 13.11 Schematic drawing of the pattern of the interlacing loops by symmetric weaving.

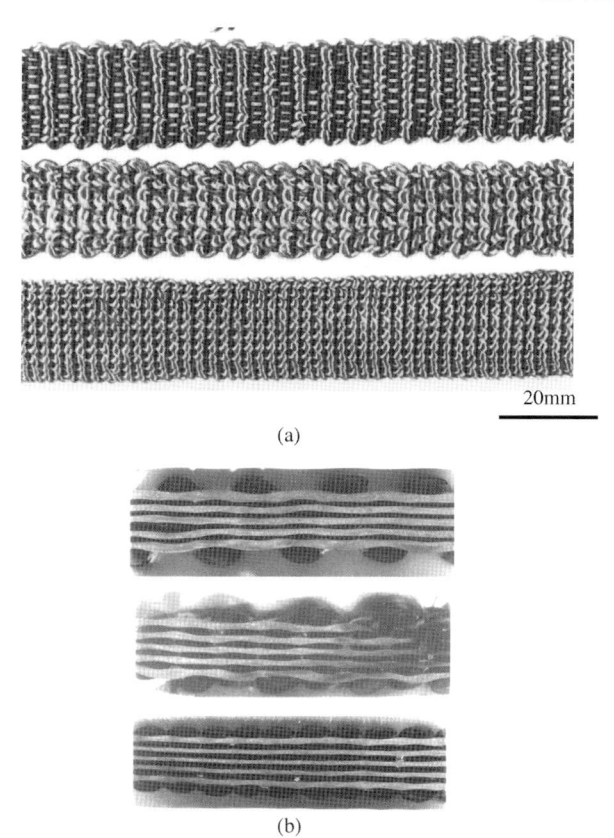

Figure 13.12 Fabrics from different weaving schemes. (a) Surface loops, (b) cross-sections (from top: unidirectional weaving, bidirectional weaving, and symmetric weaving) [31].

deflected and the fabric can be distorted due to yarn jamming. The second is that all axial yarns on fabric perimeter must be interlaced during a cycle of weaving. All axial yarns must be held in their designated positions. The one-shuttle scheme could be employed in fabrics with a simple shape. For more complex ones, however, multiple-shuttle schemes are generally needed. Figure 13.14 shows fabrics having the three-axis orthogonal yarn structure.

13.3.3
Yarn Distortion

Normally yarns are designed to be straight in fabric interior. However, because yarns are flexible before resin impregnation, yarn distortion is often inevitable. Two types of yarn distortion are discussed: yarn swelling and interlacing-induced deflection. The first is yarn swelling, as illustrated in Figure 13.15 [37]. In a three-axis orthogonal fabric, each yarn is periodically squeezed by the adjacent yarns that lie normal to it. At the region of contact, the yarn becomes thinner. At the region without the adjacent

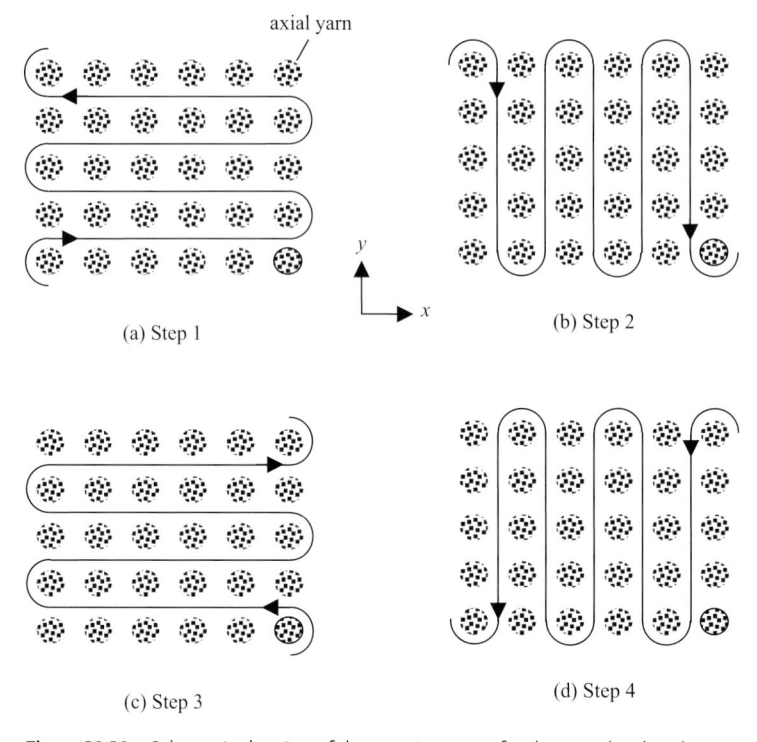

404 | *13 Three Dimensional Woven Fabric Composites*

Figure 13.13 Schematic drawing of the weaving steps for the one-shuttle scheme.

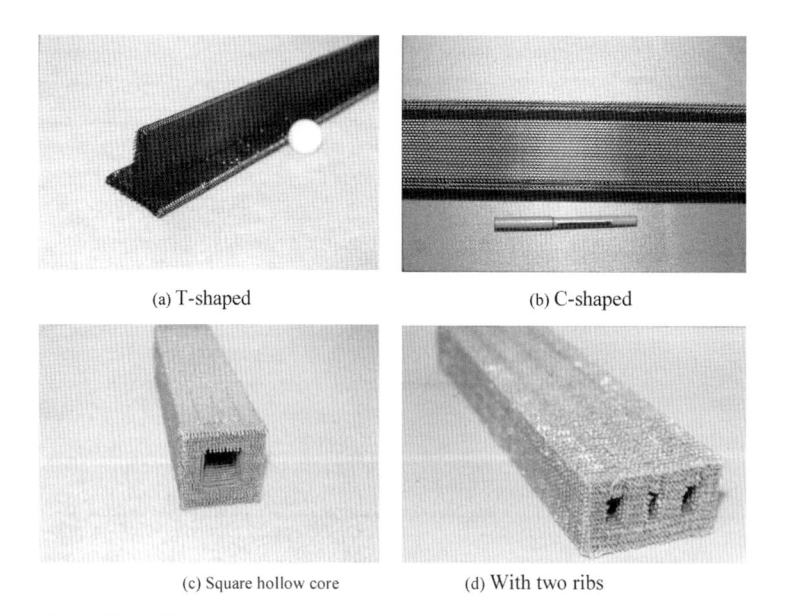

Figure 13.14 3D Fabrics with complex shapes.

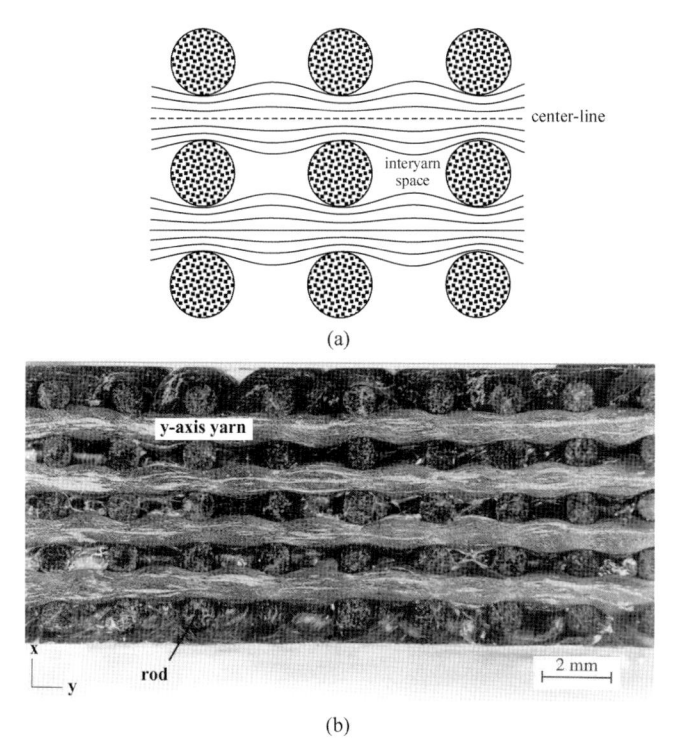

Figure 13.15 Yarn swelling. (a) Schematic illustration, (b) C/C composite [37].

yarns, the yarn tends to expand laterally. This results in a periodically swelling shape for the yarn. The second type is interlacing-induced deflection, as shown in Figure 13.16. To make the fabric compact, tension is usually applied to both axial and weaving yarns. The tension in weaving yarns can cause a force resultant that deflects the axial yarns.

13.3.4
Use of Solid Rods

Traditionally dry fiber yarns are used for making 3D fabrics. Before resin impregnation and curing, the fabrics are deformable and susceptible to external contacts. Various types of yarn deflection and deformation are also related to the flexible nature of dry yarns. For this reason, one possibility is the use of solid composite rods to replace some dry yarns in the formation of 3D fabrics [13, 14, 28–30]. In comparison with conventional 3D fabric composites, this approach is able to eliminate fiber crimp, enhance fabric consistency, and provide rigidity to the fabric. The major limitation is that solid rods cannot be used as weaving yarns to interlace others. For this reason, solid rods are often used in the axial direction.

To combine solid rods in a fabric, the size of the rods is the major concern. During fabric formation, bending of the rods is necessary in order to allow the weaving yarns

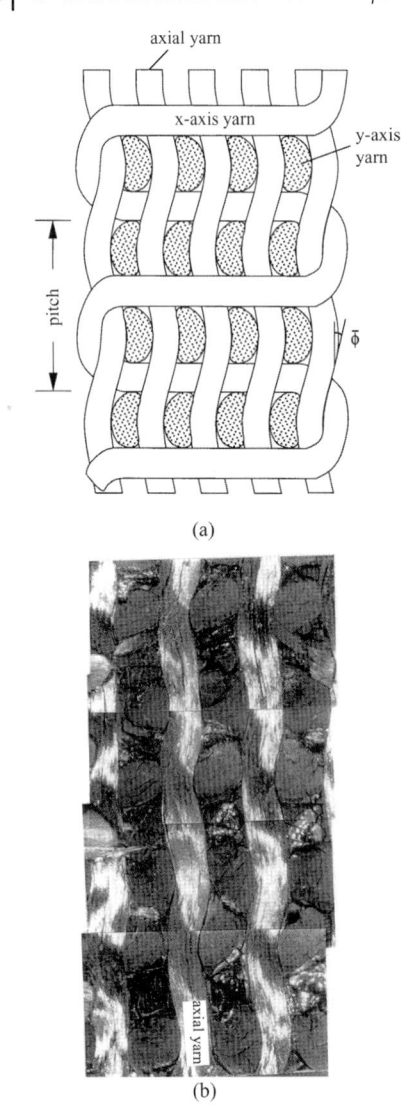

Figure 13.16 Interlacing-induced deflection in axial yarns. (a) Schematic illustration. (b) Three-axis C/C composite [29].

to travel among the rods. Thus, if the rods are too thick, bending of the rods will become difficult. On the other hand, if the rods are too small, fabrication of the rods becomes a problem. To combine the rods, the weaving setup must be modified. Since the rods are free from crimp, applying tension on the rods is unnecessary. In comparison with conventional 3D fabrics, using the rods can significantly reduce the crimp-induced problems and improve the geometry consistency. Some 3D composites composed of solid rods are shown in Figures 13.4, 13.5, 13.15 and 13.17.

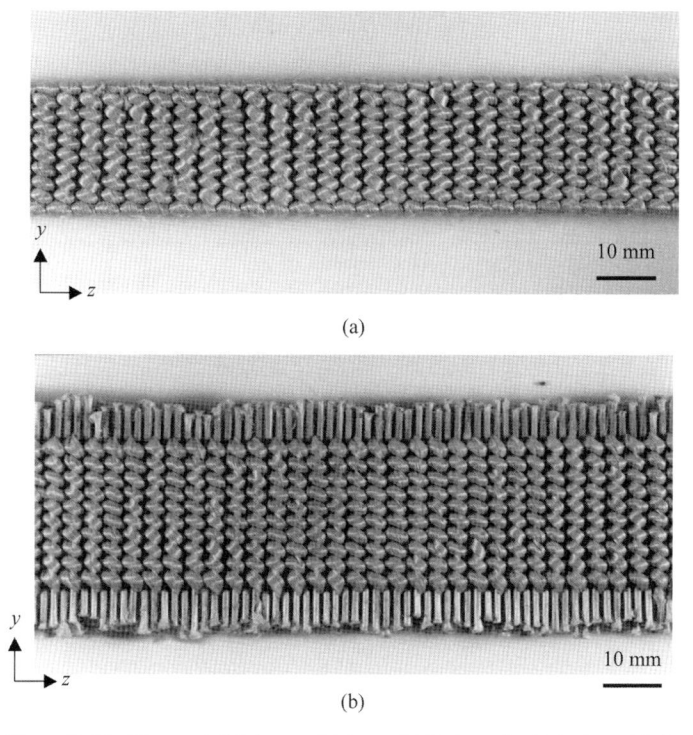

Figure 13.17 Three-axis fabrics combining solid Kevlar rods. (a) In the axial direction. (b) In both the axial and the transverse directions [36].

13.4
Modeling of 3D Woven Composites

13.4.1
Fiber Volume Fractions

For 3D composites, distribution of fibers in each component is more important than the total fiber content. For the present fabrics, fibers distribute in four components: x-axis, y-axis, z-axis, and interlacing loops. Fiber distribution in the internal composites is first discussed. A unit-cell is used to represent the composite interior. The unit-cell for a three-axis orthogonal fabric is schematically shown in Figure 13.18, assuming that the yarn's cross-sections are rectangular. Because the number of carbon filaments within a yarn is a constant, the total volume of fibers along each axis can be calculated. The fiber volume fraction along the i-axis is denoted as V_{f_i}, and can be calculated as

$$V_{f_i} = \frac{N_i A_f l_i}{L_x L_y L_z} \tag{13.1}$$

where N_i is the number of filaments within the i-bundle, A_f is the cross-sectional area of the fiber filament, and l_i is the length of the bundle within the unit cell. Assuming

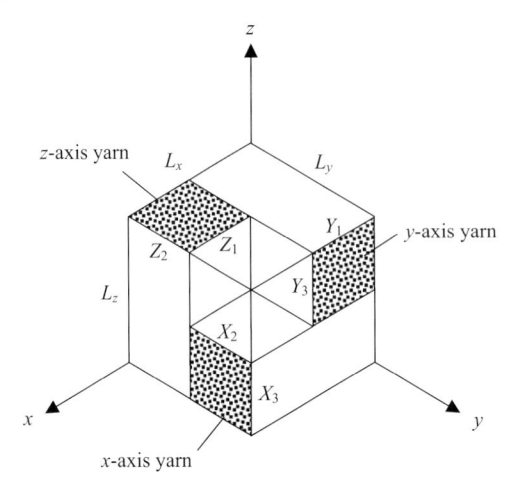

Figure 13.18 Schematic drawing of the unit-cell for the three-axis fabric composed of rectangle yarns.

straight fibers, l_i is equal to the unit cell dimension L_i. The overall fiber volume fraction (V_f) for the composite is thus the sum of the three components:

$$V_f = V_{f_x} + V_{f_y} + V_{f_z} \tag{13.2}$$

In the similar manner, when the fabric involves more axes or contains nonstraight yarns, we can calculate the fiber distributions in each component using Eq. (13.1). An example is the five-axis fabric. The associated unit cell must be redefined, as shown in Figure 13.19. The total fiber volume fraction is thus the sum of the five components.

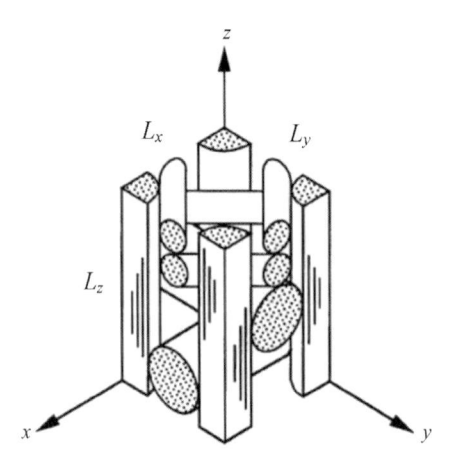

Figure 13.19 Schematic drawing of the unit-cell for the five-axis fabric composed of circular yarns.

13.4.2
Elastic Properties of Yarns

A fiber bundle is a unidirectional composite, and its properties are closely transversely isotropic with the axial direction being the symmetric axis. Therefore, if axis 3 is selected to be the fiber direction, the material is isotropic on the 12-plane. The stress–strain relation can be expressed as

$$\begin{Bmatrix} \sigma_1 \\ \sigma_2 \\ \sigma_3 \\ \tau_{23} \\ \tau_{13} \\ \tau_{12} \end{Bmatrix} = \begin{bmatrix} Q_{11} & Q_{12} & Q_{13} & 0 & 0 & 0 \\ Q_{12} & Q_{22} & Q_{23} & 0 & 0 & 0 \\ Q_{13} & Q_{23} & Q_{33} & 0 & 0 & 0 \\ 0 & 0 & 0 & Q_{44} & 0 & 0 \\ 0 & 0 & 0 & 0 & Q_{55} & 0 \\ 0 & 0 & 0 & 0 & 0 & Q_{66} \end{bmatrix} \begin{Bmatrix} \varepsilon_1 \\ \varepsilon_2 \\ \varepsilon_3 \\ \gamma_{23} \\ \gamma_{13} \\ \gamma_{12} \end{Bmatrix} \quad (13.3)$$

In the compliance form, the relation becomes

$$\begin{Bmatrix} \varepsilon_1 \\ \varepsilon_2 \\ \varepsilon_3 \\ \gamma_{23} \\ \gamma_{13} \\ \gamma_{12} \end{Bmatrix} = \begin{bmatrix} S_{11} & S_{12} & S_{13} & 0 & 0 & 0 \\ S_{12} & S_{22} & S_{23} & 0 & 0 & 0 \\ S_{13} & S_{23} & S_{33} & 0 & 0 & 0 \\ 0 & 0 & 0 & S_{44} & 0 & 0 \\ 0 & 0 & 0 & 0 & S_{55} & 0 \\ 0 & 0 & 0 & 0 & 0 & S_{66} \end{bmatrix} \begin{Bmatrix} \sigma_1 \\ \sigma_2 \\ \sigma_3 \\ \tau_{23} \\ \tau_{13} \\ \tau_{12} \end{Bmatrix} \quad (13.4)$$

Because the material is isotropic on the 12-plane, we have the following relations:

$$Q_{11} = Q_{22}, \quad Q_{13} = Q_{23}, \quad Q_{44} = Q_{55}, \quad Q_{66} = \frac{1}{2}(Q_{11} - Q_{12})$$
$$S_{11} = S_{22}, \quad S_{13} = S_{23}, \quad S_{44} = S_{55}, \quad S_{66} = 2(S_{11} - S_{12}) \quad (13.5)$$

The compliance elements are related to the engineering elastic constants as

$$S_{11} = \frac{1}{E_1}, \quad S_{33} = \frac{1}{E_3}$$
$$S_{12} = -\frac{\nu_{12}}{E_1}, \quad S_{13} = -\frac{\nu_{13}}{E_1} \quad (13.6)$$
$$S_{55} = \frac{1}{G_{13}},$$

Because of the relations in Eq. (13.5), the number of independent elements in the stiffness matrix (Eq. (13.3)) is five, and thus five engineering constants must be evaluated in order to obtain the stiffness matrix.

13.4.3
Rule-of-Mixtures

The engineering constants in Eq. (13.6) can be evaluated by either experimental testing methods or using the rule-of-mixtures. Here, the simple rule-of-mixtures is employed to find the elastic constants.

$$E_3 = E_{f1} V_f + E_m V_m$$

$$\frac{1}{E_1} = \frac{1}{E_2} = \frac{V_f}{E_{f2}} + \frac{V_m}{E_m}$$

$$\nu_{31} = \nu_{32} = \nu_{f1} V_f + \nu_m V_m \qquad (13.7)$$

$$\nu_{12} = \nu_{f2} V_f + \nu_m V_m$$

$$\frac{1}{G_{13}} = \frac{1}{G_{23}} = \frac{V_f}{G_f} + \frac{V_m}{E_m}$$

where the subscript f1 and f2 refers to the axial direction and transverse direction of the fiber, respectively. In general, the prediction on the axial modulus (E_3) is close to the experimental results. However, the others are less satisfactory because they involve deformation in the transverse direction.

13.4.4
Rotation of a Yarn

3D Composite can be viewed as a solid composed of spatially oriented yarns and matrix. The elastic constants of a rotated yarn are first analyzed. Two coordinates are needed. The 1–2–3 system is defined as the *material* coordinate as it relates the material directions. The x–y–z system is defined as the *geometrical* coordinate of the composite. To represent a spatial orientation of a solid body, three independent rotational vectors are required. For a yarn with transversely isotropic properties, the rotation with respect to yarn axis (axis 3) is irrelevant to the resulting properties. Thus, two angles (θ, ϕ) are sufficient to represent its orientation as shown in Figure 13.20.

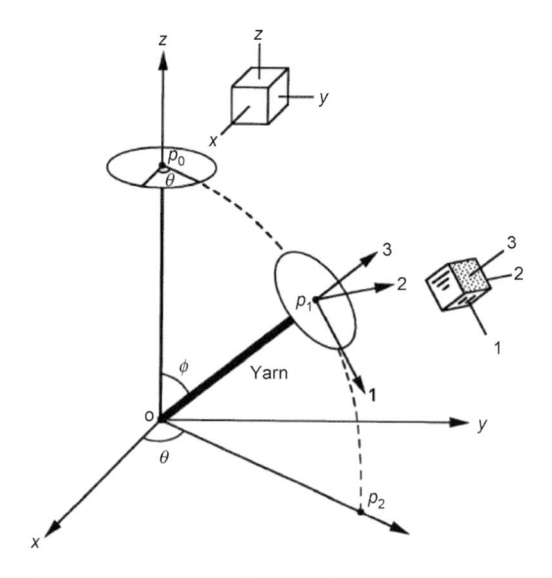

Figure 13.20 Rotation of a yarn and the two angles that define its spatial orientation.

13.4 Modeling of 3D Woven Composites

Let u_1, u_2, and u_3 be the unit vectors in the 1–2–3 coordinate and u_x, u_y, and u_z be the unit vectors in the x–y–z coordinate. The relations between these vectors can be expressed as

$$\begin{bmatrix} u_x \\ u_y \\ u_z \end{bmatrix} = \begin{bmatrix} mp & -n & mq \\ np & m & nq \\ -q & 0 & p \end{bmatrix} \begin{bmatrix} u_1 \\ u_2 \\ u_3 \end{bmatrix} \tag{13.8}$$

where $m = \cos(\theta)$, $n = \sin(\theta)$, $p = \cos(\phi)$, and $q = \sin(\phi)$.

The stress–strain relation of the yarn with respect to the x–y–z system can be written in the form:

$$\begin{Bmatrix} \sigma_x \\ \sigma_y \\ \sigma_z \\ \tau_{yz} \\ \tau_{xz} \\ \tau_{xy} \end{Bmatrix} = \begin{bmatrix} \bar{Q}_{11} & \bar{Q}_{12} & \bar{Q}_{13} & \bar{Q}_{14} & \bar{Q}_{15} & \bar{Q}_{16} \\ \bar{Q}_{12} & \bar{Q}_{22} & \bar{Q}_{23} & \bar{Q}_{24} & \bar{Q}_{25} & \bar{Q}_{26} \\ \bar{Q}_{13} & \bar{Q}_{23} & \bar{Q}_{33} & \bar{Q}_{34} & \bar{Q}_{35} & \bar{Q}_{36} \\ \bar{Q}_{14} & \bar{Q}_{24} & \bar{Q}_{34} & \bar{Q}_{44} & \bar{Q}_{45} & \bar{Q}_{46} \\ \bar{Q}_{15} & \bar{Q}_{25} & \bar{Q}_{35} & \bar{Q}_{45} & \bar{Q}_{55} & \bar{Q}_{56} \\ \bar{Q}_{16} & \bar{Q}_{26} & \bar{Q}_{36} & \bar{Q}_{46} & \bar{Q}_{56} & \bar{Q}_{66} \end{bmatrix} \begin{Bmatrix} \varepsilon_x \\ \varepsilon_y \\ \varepsilon_z \\ \gamma_{yz} \\ \gamma_{xz} \\ \gamma_{xy} \end{Bmatrix} \tag{13.9}$$

where the barred elements refer to the x–y–z system. Using the theory of elasticity, it can be proved that the stiffness elements $\bar{Q}_{ij}$ and Q_{ij} have the following relations [27, 39]:

$$\bar{Q}_{11} = Q_{11}(m^2p^2 + n^2) + Q_{33}m^2q^2 - K_0 m^2 q^2(m^2p^2 + n^2)$$
$$\bar{Q}_{12} = Q_{12}p^2 + Q_{13}q^2 + K_0 m^2 n^2 q^4$$
$$\bar{Q}_{13} = Q_{12}n^2q^2 + Q_{13}(n^2p^2 + m^2) + K_0 m^2 p^2 q^2$$
$$\bar{Q}_{14} = npq(Q_{13} - Q_{12} + K_0 m^2 q^2)$$
$$\bar{Q}_{15} = mpq(Q_{13} - Q_{11} + 2Q_{55} + K_0 m^2 q^2)$$
$$\bar{Q}_{16} = mnq^2(Q_{13} - Q_{11} + 2Q_{55} + K_0 m^2 q^2)$$
$$\bar{Q}_{22} = Q_{11}(n^2p^2 + m^2) + Q_{33}n^2q^2 - K_0 n^2 q^2(n^2p^2 + m^2)$$
$$\bar{Q}_{23} = Q_{12}m^2q^2 + Q_{13}(m^2p^2 + n^2) + K_0 n^2 p^2 q^2$$
$$\bar{Q}_{24} = npq(Q_{13} - Q_{11} + 2Q_{55} + K_0 n^2 q^2)$$
$$\bar{Q}_{25} = mpq(Q_{13} - Q_{12} + K_0 n^2 q^2)$$
$$\bar{Q}_{26} = mnq^2(Q_{13} - Q_{11} + 2Q_{55} + K_0 n^2 q^2) \tag{13.10}$$
$$\bar{Q}_{33} = Q_{11}q^2 + Q_{33}p^2 - K_0 p^2 q^2$$
$$\bar{Q}_{34} = npq(Q_{13} - Q_{11} + 2Q_{55} + K_0 p^2)$$
$$\bar{Q}_{35} = mpq(Q_{13} - Q_{11} + 2Q_{55} + K_0 p^2)$$
$$\bar{Q}_{36} = mnq^2(Q_{13} - Q_{12} + K_0 p^2)$$
$$\bar{Q}_{44} = Q_{55}(m^2p^2 + n^2) + Q_{66}m^2q^2 + K_0 n^2 p^2 q^2$$
$$\bar{Q}_{45} = mnq^2(Q_{55} - Q_{66} + K_0 p^2)$$
$$\bar{Q}_{45} = mpq(Q_{55} - Q_{66} + K_0 n^2 q^2)$$
$$\bar{Q}_{55} = Q_{55}(n^2p^2 + m^2) + Q_{66}n^2q^2 + K_0 m^2 p^2 q^2$$
$$\bar{Q}_{56} = npq(Q_{55} - Q_{66} + K_0 m^2 q^2)$$
$$\bar{Q}_{66} = Q_{55}q^2 + Q_{66}p^2 + K_0 m^2 n^2 q^4$$

where K_0 is a parameter for the transversely isotropic material

$$K_0 = Q_{11} + Q_{33} - 2Q_{13} - 4Q_{55} \tag{13.11}$$

It is noted that $K_0 = 0$ when the material is isotropic, and hence K_0 can be interpreted as a deviation from an isotropic medium.

13.4.5
Equivalent Properties of 3D Composites

Due to yarn architecture and material inhomogeneity, the stress and strain distributions in the composites are three dimensional. The complexity of 3D geometry precludes simple solutions satisfying all required conditions from the elasticity point of view. Therefore, simplification on the deformation must be made in order that analytical modeling of the composite can be performed. Some models have been proposed. Usually the concept of unit cell is adopted. These models are required to meet the displacement compatibility among different elements within the composite. The easiest and the most popular method is perhaps the so-called *iso-strain* or the *aggregate* model, assuming that the induced strains are constants throughout the composite.

The strains are assumed to be constants throughout the composite [27]. The stiffness matrix of the composite is assumed to be the sum of each individual yarns and matrix weighted by their volume fractions, as follows:

$$Q_{ij}^c = \sum_k \bar{Q}_{ij}^k V_k \tag{13.12}$$

where the superscript c refers to the composite, the index k runs for all yarns and the interyarn matrix. The V_k is the volume fraction of the kth material element. Once the composite stiffness matrix is calculated, the compliance matrix (S_{ij}^c) can then be obtained. The equivalent elastic moduli of the composite can be found as

$$\begin{aligned} E_x^c &= \frac{1}{S_{11}^c}, & E_y^c &= \frac{1}{S_{22}^c}, & E_z^c &= \frac{1}{S_{33}^c} \\ G_{yz}^c &= \frac{1}{S_{44}^c}, & G_{xz}^c &= \frac{1}{S_{55}^c}, & G_{xy}^c &= \frac{1}{S_{66}^c} \end{aligned} \tag{13.13}$$

Obviously, this method is unable to take into account strain concentrations in the 3D composite. For a more accurate analysis taking into account 3D variations of the stress and strains within the material, finite element methods must be used.

13.5
Failure Behavior of 3D Woven Composites

Textile composites are characterized by both intrayarn and interyarn types of failure. This section discusses these failure types based on experimental observations.

Figure 13.21 Tensile fracture of a three-axis orthogonal composite [27].

Because of the 3D structure, the induced damage mechanisms can be much more complicated than those in unidirectional and laminated composites.

13.5.1
Tensile Fracture

Tensile fracture of a three-axis carbon/epoxy composite is discussed, as shown in Figure 13.21 and 13.22 [27]. The specimens were machined from the composite. Because the specimens is thin and the surface loops are absent, the through-thickness (z-axis) yarns are susceptible to debonding, which is the first failure mode to appear. Yarns located on specimen edges were likely to debond earlier than those inside the specimen, leaving empty notches. During application of the load, some of these yarns were found to pop-out from specimen edges. For those inside the specimen, the yarns are constrained and are less likely to debond. As the applied load increases, debonding and splitting in widthwise (y-axis) yarns occur. The propagation of these cracks will be hindered by the strong fibers in the x-axis yarns. The creation of these transverse cracks depends upon the flaws on the interface or within the y-axis yarns. Once a transverse crack is formed, the stresses inside the yarn are changed due to the presence of the fractured surfaces. This stress redistribution reduces the possibility for another parallel crack to occur in the cracked yarn. Because the majority of the load is carried by the axial yarns, the final composite failure is determined by axial yarns. The results indicate that the axial yarns are brittle and the propagation of yarn fracture is unstable.

13.5.2
Bending Fracture

For unidirectional or 2D laminated composites under bending, failure typically appears on the compressive side in the form of fiber microbuckling. In 3D composites, transverse yarns and interlacing loops support the axial fibers. Their failure mechanisms can be different. Some important results of bending fracture in 3D composites are discussed.

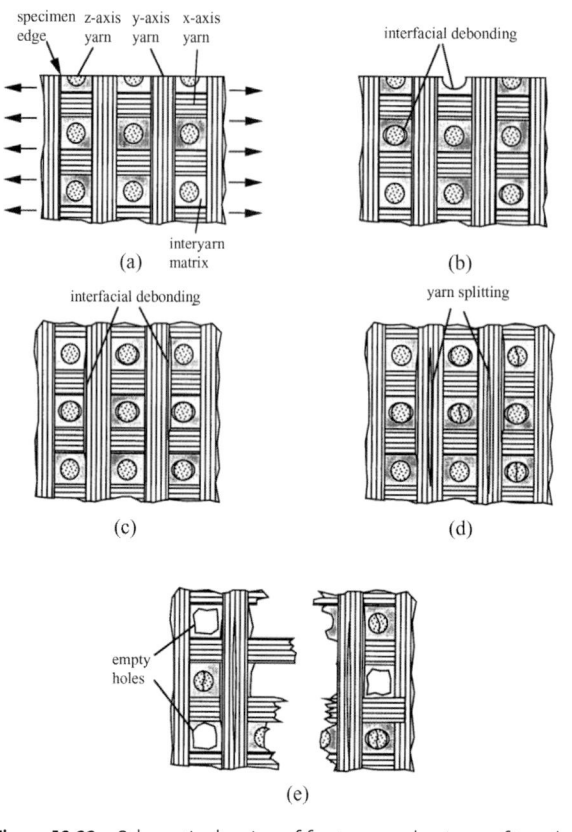

Figure 13.22 Schematic drawing of fracture mechanisms of tensile fracture [27].

13.5.2.1 Weaker Plane

Figure 13.23 shows a side surface of a 3D composite after the three-point bending [30]. Transverse matrix cracks are found to trigger from the tensile side and penetrate into the composite. The unique feature is that the cracks are regularly spaced, as marked by arrows. The induced crack patterns are related to the *weaker planes*. The concept of weaker planes is introduced.

Composites are nonuniform, and some planes are more vulnerable. In a 3D composite, no plane is absent of fibers and no plane is truly weak. However, some planes are inevitably weaker than others, and they are termed *weaker planes*. According to the experimental observations, the onset and the growth of cracks are most likely to follow the weaker planes.

The pattern of interlacing loops on fabric surface is critical to the location of the weaker plane. The locations of weaker planes are discussed, as schematically illustrated in Figure 13.24. There are *x*-axis loops on the left and right surfaces. If the tensile side is on the right, cracks tend to appear on this surface and grow toward left. As indicated, the weaker planes for this fabric are the regions without being covered by the loops. These sections are weaker than the others that contain loops.

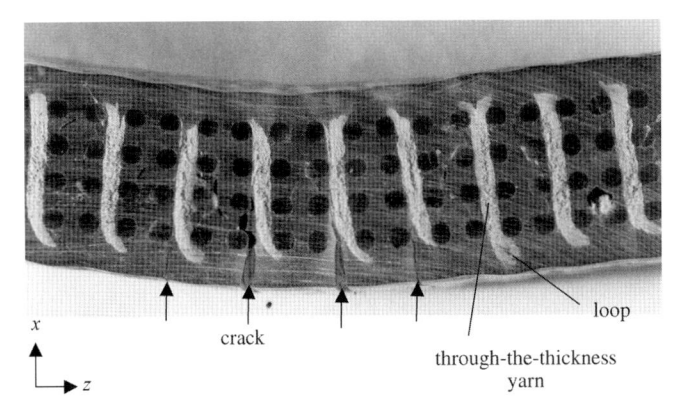

Figure 13.23 Transverse matrix cracks marked by arrows triggering from the tensile side and penetrating into the composite along the weaker planes [35].

13.5.2.2 Influence of Surface Loops

In order to elucidate the influence of interlacing loops on the induced damage mechanisms, some specimens were ground to remove all the loops on surfaces. Material characterizations based upon three-point bending tests were then carried out. The loop-retained and loop-removed specimens provide a comparison in the configuration of damage. Figure 13.25 shows loop-retained and loop-removed specimens after the three-point bending test. The dominant damage modes are rupture of the rods and transverse matrix cracking, both initiating from the middle of the bottom surface, where the tensile strain is the highest. Separation of the axial rods is common on the loop-removed specimens. When loaded in bending, the rods under tension tend to debond and slide. If this happens, the debonded rod can leave the surface in a mode termed as pop-out. For the loop-retained specimens, in

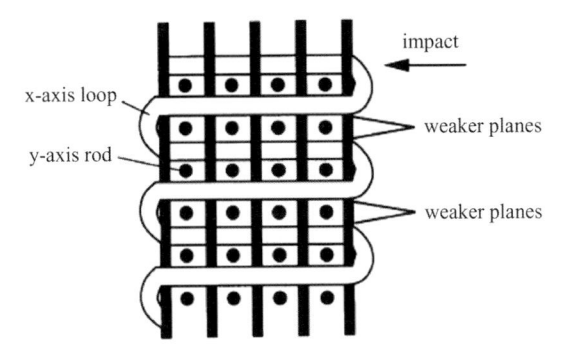

Figure 13.24 Weaker planes in a YRR composite. They exist in regions without the protection of interlacing loops.

(a) Tensile side

(b) Side view

Figure 13.25 Comparison of the bending-induced damage in loop-retained and loop-removed specimens [30].

comparison, pop-out of rods is impossible. Another mode of damage is transverse cracking. Two types of transverse cracking have been observed. One is the dominant transverse crack that originates at the middle of the tensile side and penetrates nearly throughout the thickness. The dominant transverse crack causes extensive ruptures of the rods. The crack propagates along a so-called weaker plane. The other type is the multiple matrix cracks that run across the specimen surface in a parallel manner. These cracks, usually stop when they meet the fiber bundles and are less harmful to the composite.

13.5.3
Compressive Damage

Compressive failure is perhaps the most complicated among the major composite failure modes. A compressed composite usually develops failure modes caused by noncompressive stresses. For instance, fiber kinking and matrix yielding, the two major modes of compressive failure, are caused by bending and shear stresses, respectively. This is because compressive failure is dominated by both the matrix and the interface. Many early works [40–42] have experimentally and theoretically examined the failure mechanisms in unidirectional composites and have established the underlying causes for the formation of kink bands. The knowledge learned from unidirectional composites is insufficient for 3D composites. Some of the differences are discussed.

The level of fiber misalignment is often much higher in 3D composites. For 3D composites, in contrast, fiber yarns are flexible and sensitive to lateral forces. Deformation of yarns is often inevitable before they are consolidated by matrix. Two types of yarn deformation have been discussed in Section 13.3.3. When conducting compressive tests for 3D composites, considerations should be given to the thickness problem. 3D Composites are typically thicker than 2D laminated composites. Some specimens are more than 10 mm in thickness. To test thick specimens, the key is how to apply the load without causing noncompressive types of failure. Conventional methods, such as end-tabbed types, are inappropriate for thick specimens, since failure can occur at the end tabs. Applying a compressive load directly on specimen ends is generally needed. To conduct compressive tests, it is important that global buckling must be avoided.

According to the observations, two types of material fracture are prevalent in the present 3D composites: the *microscopic fracture band* (termed as microband) and *miniscopic fracture band* (termed as miniband). The microband connects fiber kink bands in axial yarns and matrix cracks in transverse yarns. The miniband links shear failure in axial yarns, tensile rupture in through-thickness yarns, and large-scale shear sliding in transverse yarns. The width of the microband is about one order smaller than that of the miniband, which is of the size of yarn thickness. Both types of damage bands can go across several yarns in the composite. Both bands are mainly caused by the compression-induced shear along the through-thickness direction. This shear stress is responsible to all major modes of fracture.

13.5.3.1 Microband
A microband is composed of two parts: the microscopic fiber kink-bands in axial yarns and the small-scale matrix cracking in transverse yarns. Figure 13.26 shows growth of microbands. The zigzag multiple kink-bands in axial yarns and the matrix

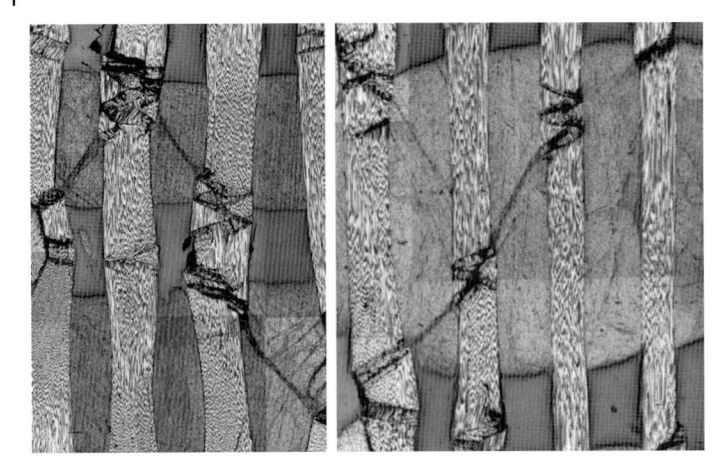

Figure 13.26 Typical microbands connecting fiber kink-bands in axial yarns and matrix cracks in transverse yarns [31].

cracks in transverse yarns are inseparable. The connection of these two failure modes forms a microband.

13.5.3.1.1 **Microscopic Kink-Band** This is the major mode in the microband. The focal point is the nucleation and growth of fiber buckling. The fiber kink-bands incline with an angle 20°–35° with respect to the horizontal direction. The unique feature is that there are multiple kink-bands connected in a zigzag form. Buckling of fibers is initiated at sites where fiber misalignment and material defects exist. A buckled fiber can push and trigger failure of the neighboring fibers, thus forming a band of fiber kinking. The kink band grows along an inclined direction until it meets the bundle surface. At this stage, the kink band stops if the energy supply is insufficient. Or, the kink band can grow in the opposite direction at the same inclining angle. Thus, the kink band rebounds from the boundary. This rebounding mechanism gives rise to multiple kink bands linked in a zigzag form, as shown in Figure 13.27. Note that the support of through-thickness yarns is essential for the forming of the zigzag bands. Without being suitably supported, the axial bundle can slide apart along the first kink band.

Figure 13.28 shows compressive damage in axial yarns of a three-axis, all-yarn composite. The most notable difference from Figure 13.27 is the number of bands – usually, there are only one or two bands in each kink-band group. This is in part because the yarns are thin and the misalignment is greater.

13.5.3.1.2 **Matrix Cracks in Transverse Yarns** This is the second mode in the microband and is an accompanied mode of damage resulting from the kinking-induced displacement. The matrix cracks form streamlinelike curves in contiguous transverse yarns, as shown in Figure 13.29. These curves are a combination of shear failure and sliding in the matrix of transverse yarns. When an axial bundle is damaged by forming microscopic kink-bands, the axial bundle tends to slide laterally and, as a

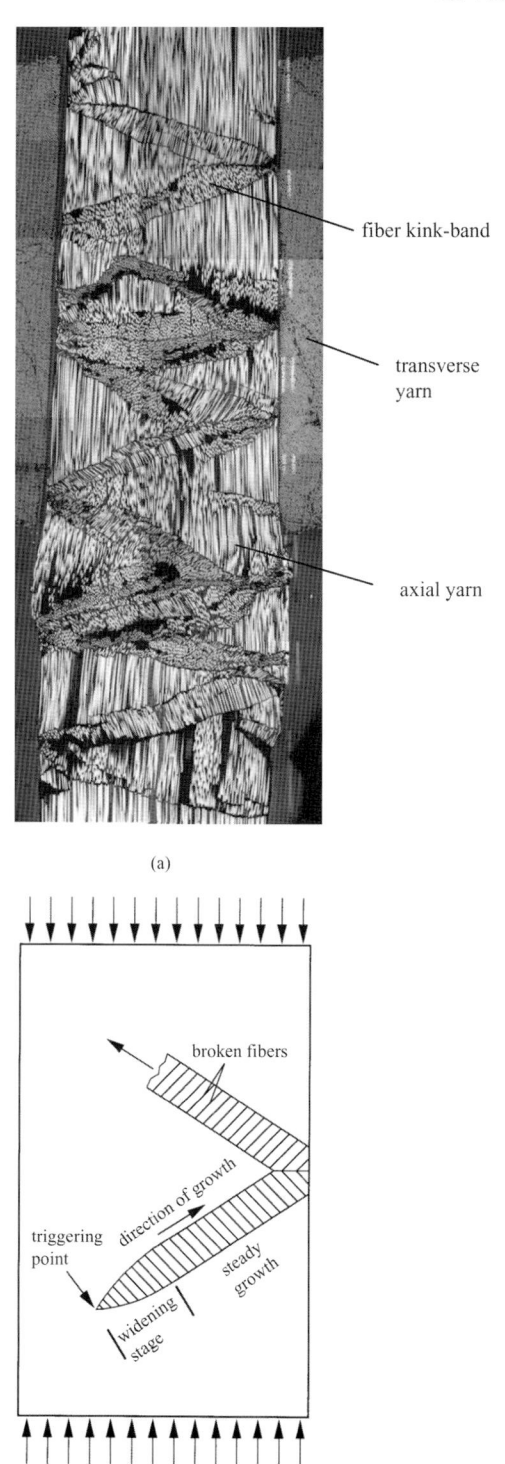

Figure 13.27 (a) Multiple kink-bands connected in a zigzag form. (b) Schematic illustration of the triggering and growth of the kink-bands [43].

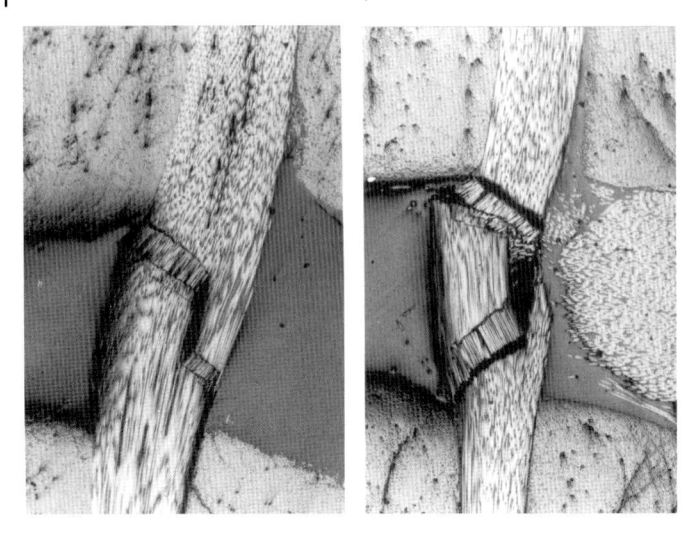

Figure 13.28 Micrographs showing induced kink-bands and yarn split in the axial yarns [43].

result, to push the neighboring transverse yarns. The pushed transverse yarns will result in shear deformation, which leads to matrix cracks, as shown in Figure 13.29. The streamlinelike curves are complex but not random. These streamlinelike curves are the flow of shear failure in the matrix and are related to the maximum shear stress resulting from the pushing force of the kink band.

13.5.3.2 Miniband

A miniband connects yarn-sized failure modes, including buckling in axial yarns, large-scale shear sliding in transverse yarns, and tensile rupture in through-thickness yarns. Figures 13.30–13.32 show typical examples of minibands. The minibands

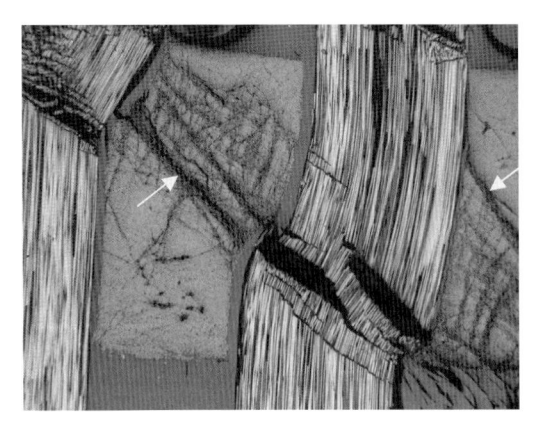

Figure 13.29 Matrix cracks in the transverse yarns (arrows) due to the lateral pushing by the fractured axial yarns.

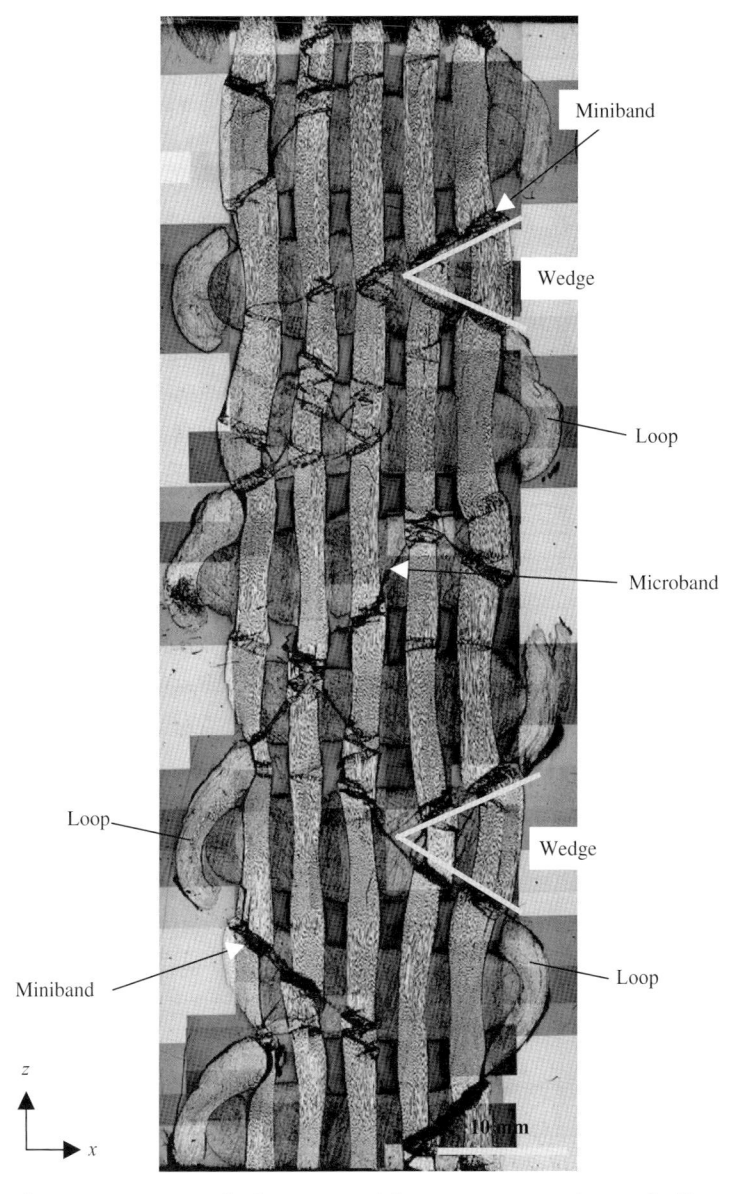

Figure 13.30 Micrograph of a compressed three-axis composite showing double-wave minibands [31].

undergo a much larger lateral displacement and cause more intensive shear failures in transverse yarns. The origin of the miniband is related to interlacing loops. The micrographs indicated that the miniband starts from the outside of a loop and grow around the loop into the interior along a closely 45° or −45° plane. The ±45° inclining bands of fracture indicate that the compression-induced shear is the cause of

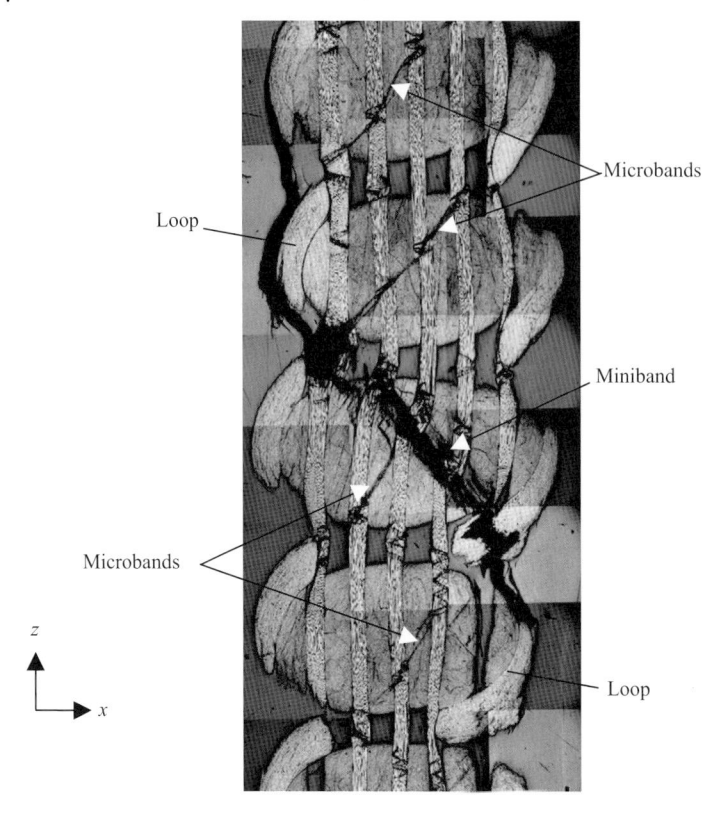

Figure 13.31 Compressed specimen showing one through-thickness miniband and several microbands [31].

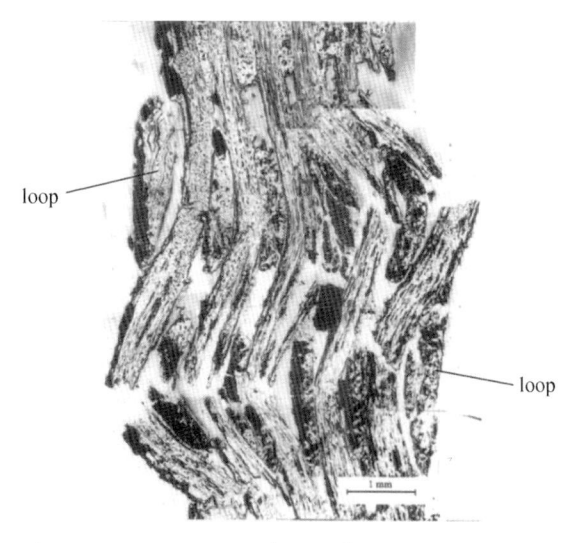

Figure 13.32 Fracture in axial yarns of a 3D C/C composite due to compression-induced bending [40].

fracture. Two types of the minibands have been observed: one in a *double-wave* form, and the other in a *through-thickness* form. The former comprises several shorter minibands, while the latter comes in a single but more dominating band.

13.5.3.2.1 **Double-Wave Miniband** The first type is the double-wave miniband shown in Figure 13.30. The minibands were found to originate from around the loops on the surface and grow into the composite. For each loop, two minibands grow around the top and bottom of the loop and penetrate into the material along 45° or −45° planes. In between two loops, the two neighboring minibands form a wedgelike fracture in the no-loop zone, as indicated. The regularly and repeatedly appearance of the fractured wedges form a long, wavelike band of fracture. Because two of such bands appear in the specimen, they are termed *double-wave* band. The double-wave minibands penetrate into the material only about two or three axial yarns, and they stop after the fractured wedge forms.

13.5.3.2.2 **Through-Thickness Miniband** The second type is the *through-thickness* miniband, as shown in Figure 13.31. The band is named because it runs across the entire thickness of the specimen. It is so dominating that each specimen shows only one band. The band was found to trigger from a loop on one side, grow with an angle close to either 45° or −45°, and leave out the material from another loop on the other side. The inclination angle of the band suggests that the fracture was due to the compression-induced shear stress. From energy point of view, the majority of the stored strain energy was released to create this single through-thickness band of fracture. With sufficient energy, this type of miniband can cut all yarns it encounters. Because of different orientations, the yarns fail in different modes. The axial yarns (z-axis) were fractured into long fragments, while the transverse (y-axis) yarns were sheared apart. The most interesting mode is the tensile rupture of the through-thickness (x-axis) yarns, even though the composite was loaded in compression. The sliding of the composite along the existing band results in the stretching of the through-thickness yarn, which is apt to fail in tensile rupture.

13.5.3.3 Fracture Due to Compression-Induced Bending

Interlacing of yarns can cause wavelike deformation in axial yarns. When the 3D composite is loaded in compression, bending is induced in the axial yarns due to the wavelike deformation. In some material systems, this compression-induced bending is critical in material failure.

Figure 13.32 is a 3D C/C composites fractured because of the compression-induced bending. The yarn imperfection was the major failure-determining factor. C/C composites are generally brittle and susceptible to bending. As illustrated in Figure 13.33, the yarns were found to break in the form of bending fracture at the points where imperfection is the greatest. The bending fracture grows horizontally almost across the section, until it meets the loop at the other side. The successive bending fracture creates a plane of fracture. Normally each specimen has two such fracture planes moving fractured yarns in opposite directions.

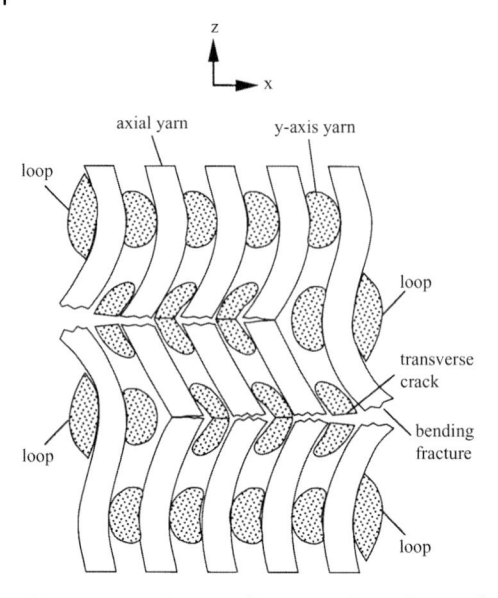

Figure 13.33 Schematic illustration of yarn fracture due to compression-induced bending.

13.5.4
Fracture Due to Transverse Shear

Among all the failure properties of composites, the one caused by transverse shear is perhaps the least well studied. This does not reflect the fact that fibers – and thus composites – are generally vulnerable when loaded by transverse shear. One possibility is that most composites are thin and designed to bear in-plane loads, and thus the induced transverse shear tends to be insignificant. Because of this, the related works aiming at the evolving nature of damage caused by transverse shear are very few in the literature. For 3D composites under 3D loads, however, the induced transverse shear is not always insignificant. Thus, the associated failure behavior and its influencing factors should deserve notice.

There are two types of shear stresses: in-plane and transverse. The transverse (or through-thickness) shear test is more difficult to conduct. The most popular method is the Iosipescu method [44–46], which results in a shear force in the test section. But to assure shear fracture along the test section, the Iosipescu method requires V-shaped notches at the top and bottom of the test section. The method is convenient but not suitable for 3D textile composites, because introducing notches could cut the loops on surface.

Figure 13.34 shows a fixture for applying the transverse shear. Figure 13.35 shows typical specimens after the transverse shear test. Due to the 3D yarn structure, the specimens are seriously damaged but they still remain connected. Figure 13.36 is a micrograph revealing some features of the shear failure. The axial yarns normal to the cutting section were fractured after the test. Another feature is that cracks usually

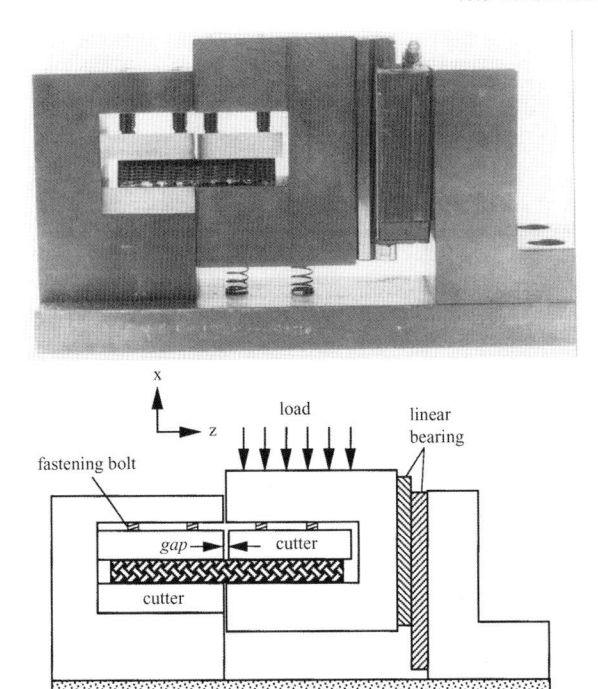

Figure 13.34 Fixture for applying the transverse shear [32].

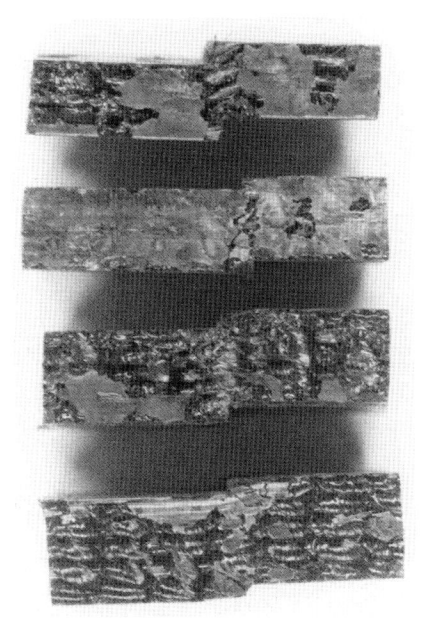

Figure 13.35 Typical damaged specimens after the transverse shear test [38].

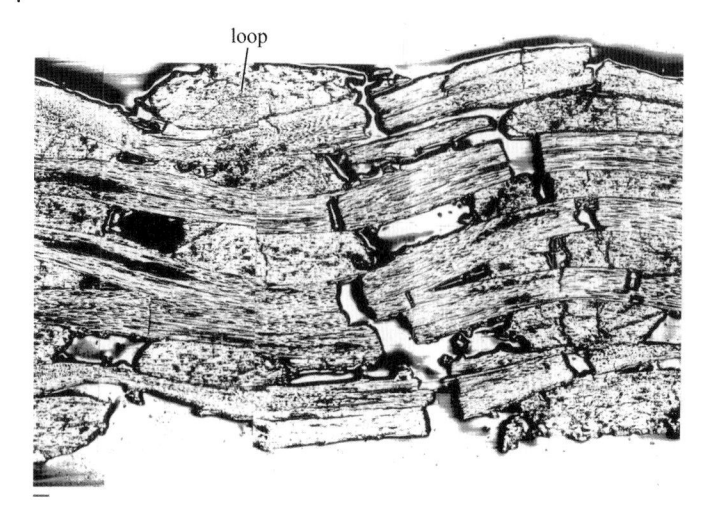

Figure 13.36 Fracture in a 3D C/C composite after the transverse shear test [38].

grow along the weaker planes, as illustrated in Figure 13.37. The weaker planes are located at the outer boundaries of each U-shaped portion of the x-axis yarns. Cracks most likely circumvent the loops and grow into the material along the weaker planes. By growing along the weaker plane, an incoming crack can avoid the loop, and thus it meets the least resistance along this path.

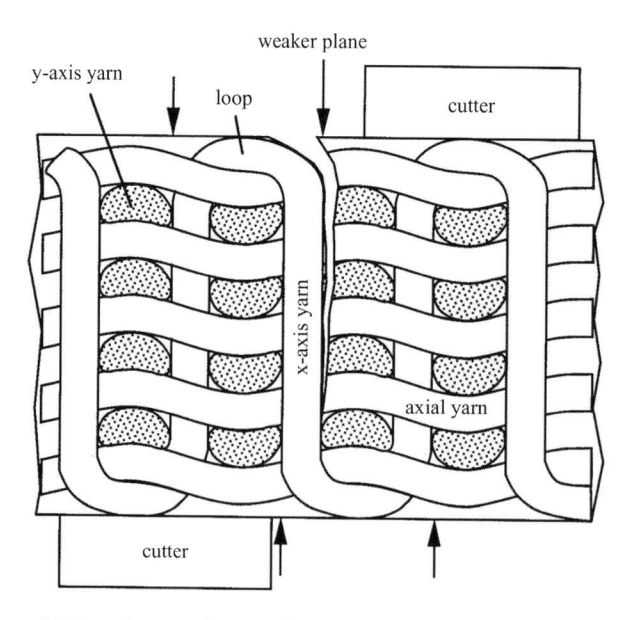

Figure 13.37 Schematic drawing of crack growth in fabric due to the transverse shear [32].

13.5.5
Impact Damage

Compared to the extensive research dedicated to impact behavior of 2D laminated structures, very little has been done on 3D woven composites. The major reason is the difficulty in making 3D woven specimens large enough for drop-weight impact tests [28, 47]. In this section, the impact behavior of the three-axis Kevlar/vinyl ester composite is discussed [28]. Solid rods were used along the axial direction. The specimen sizes are 100 × 100 × 10 mm. The impact tests were conducted using a Dynatup 8250 instrumented impact test machine. The mass of the impactor was 29.32 kg, and the tests were undertaken using a hemispherical impact tup 25.4 mm in diameter. The specimens were clamped by two parallel plates, each with a 76.2 mm diameter circular opening. The velocity of the drop–mass before impact was measured for calculating the incident impact energy.

All specimens under the impact loads are found to be subperforated. Typical damage configurations on the impacted surface are shown in Figure 13.38. At the center of the impacted surface, there is a depressed region at the impact location, known as *Hertzian cone*, which was subjected to a highest compressive stress due to direct impact. Most rods at this region were collapsed due to indentation. The micrographs indicate that the collapsed rods were damaged in the form of matrix crushing. The matrix crushing prevents the rod from resisting bending loads. The most apparent damage mode on the impacted surface is matrix cracking. The matrix cracks are formed in concentric ellipses with the major axis parallel to the rod direction. The concentric matrix cracks should result from the tensile stress wave moving out from the impact location.

On the nonimpact (bottom) surface, the damage modes are very different. As can be seen, the matrix cracks grow in radial directions. These radial cracks are due to the tangential tensile stress in the matrix. These radial cracks are similar to the bending cracks often seen in lower layers of unidirectional laminates due to transverse deflection.

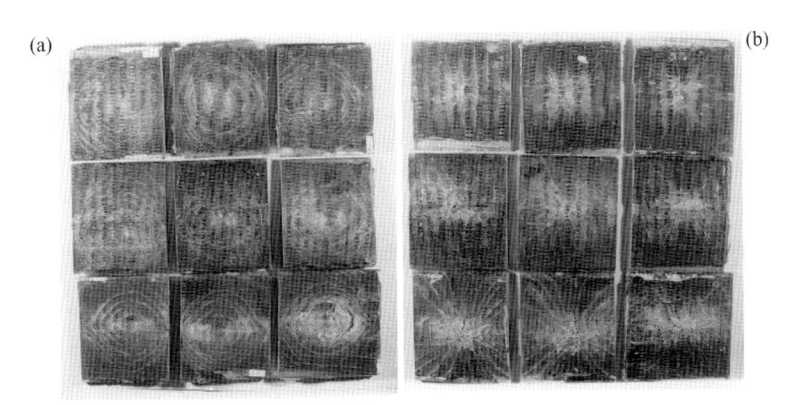

Figure 13.38 Impact damage on the top surface (a) and bottom surface (b) of the three-axis orthogonal composite [28].

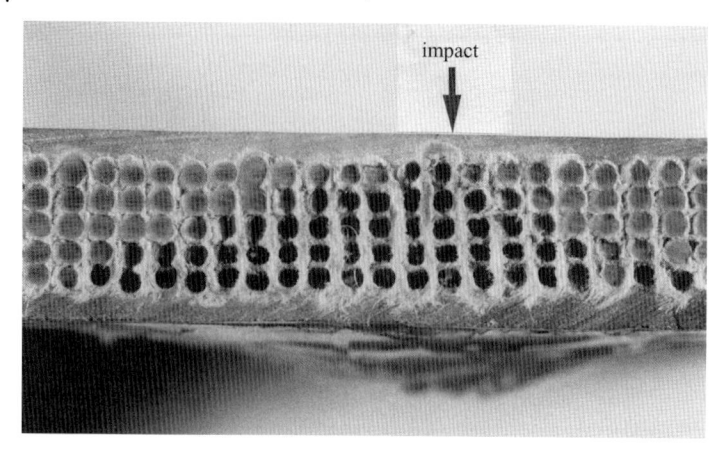

Figure 13.39 Pull-in of rods on the specimen side [28].

Another unique feature is rod pull-in, as shown in Figure 13.39. When a specimen is impacted, tensile stress is induced in the lower region. If the interfacial strength of the rod is weak, debonding of the rod can occur, causing the rod to be pulled into the interior. The pull-in leaves holes on the specimen edge. The depth of the hole is proportional to the deflection of the rod. It is interesting to see that fewer holes are created near the impact side while more near the nonimpact side, forming a triangular zone of rod pull-in.

13.6
Role of Interlacing Loops

Compared with the traditional 1D and 2D composites, 3D composites provide us with more tools for designing the materials. One of them is the interlacing loop pattern. The role of interlacing loops on the performance of 3D composites has long been overlooked. One possibility is that loops occupy only a small fraction of the volume, yet experiments revealed that their influence can be disproportionate in damage behavior. The key lies in the fact that loops not only build the 3D network of reinforcements by connecting through-thickness yarns, they also cover and protect the composites from external attack. In this section, two major functions of the loops are discussed.

13.6.1
Covering Weaker Planes

Loops act as a protective skin for 3D composites. They can defend the internal yarns from outer attacks through crack termination or deflection. The key lies in the fact that they cover the otherwise weaker planes, along which cracks can penetrate into the interior. With a proper design of the weaving process, it is possible to achieve a loop pattern that covers many weaker planes. This is a topic of interest to pursue. The experiments showed that when a crack meets a loop, the crack is either stopped or deflected. This suggests that loops on surface can cover the weaker planes. When covered by loops, the weaker planes are reinforced. Still, cracks can grow into the material from the uncovered weaker planes. To be more effective in resisting cracks, allowing fewer weaker planes to outer attacks can be helpful.

13.6.2
Holding Axial Yarns

In response to loads, 3D composites can develop complex deformation and stresses. The induced stresses could push apart the axial yarns on outer surfaces. Once an axial yarn moves apart, the transverse yarns must respond to prevent disintegration. In a loop-retained composite, the loops can firmly hold axial yarns. Loops are effective in holding axial yarns and in preventing disintegration. If loops are removed, the outer axial yarns are left unsupported. A cascade of succeeding damage mechanisms could otherwise be triggered. From the viewpoint of energy absorption, preventing the composite from disintegration is beneficial even when damage has heavily developed. Most important, the through-thickness yarns become fragmented if the loops are removed. The load-transfer efficiency and the 3D integrity will be adversely affected. This function makes it clear that loops should be retained whenever possible, and a near-net-shape design without post-machining is preferable.

13.7
Design of 3D Woven Composites

3D Composites can be designed to improve certain properties to meet the need in applications. However, trade-off effects often exist in the resulting properties when change a designing parameter. When one property is improved, others might be affected adversely.

13.7.1
Modulus

There are three Young's moduli and three shear moduli in a composite. These moduli dominate the stiffness and deformation of the composite. The three Young's moduli Ex, Ey, and Ez are directly proportional to the three components of fiber volume fractions, V_{f_x}, V_{f_y}, and V_{f_z}, respectively. The three-axis orthogonal composites are generally low in the shear moduli. To enhance shear rigidity and strength, off-axis yarns can be added to the associated plane, although actual adding of off-axis yarns relies on the technique of fabric formation. For the orthogonal weaving, adding off-axis yarn to the weaving plane (xy-plane) is easier than to the other two planes.

13.7.2
Yield Point

The yield point refers to the onset of fiber microbuckling in axial bundles. For applications where a damage-free state is needed, the yield point becomes most important. Methods that enhance critical strength of unidirectional composites – such as stiffer and stronger fibers, stronger interface, and lower fiber misalignment – are applicable to 3D composites. In addition, yarn undulation, which exists only in textile composites as a result of interlacing, must be reduced to a minimum by careful design of the material processing. A satisfactory resin impregnation and fiber wet-out is also important for enhancing the critical stress.

13.7.3
Strain-to-Failure

To extend the strain-to-failure, the yarn architecture plays a deciding role. The key is to prevent the composite from disintegration, after initial damage has developed. In this regard, 3D composites must make use of two unique features: interlacing loops and through-thickness yarns. Shorter pitch lengths, and thus denser loops, can better support axial yarns on surface. Surface loops become more crucial for thinner 3D composites, which are more likely to be sheared apart under compression. Surface loops cannot function properly without being linked to through-thickness yarns. When subjected to an axial compression, through-thickness yarns are most likely under tension. The stretched through-thickness yarns must be strong enough to maintain the structural integrity.

13.7.4
Energy-Absorption and Damage Resistance

The capability of energy-absorption is, in theory, the area under the stress–stain curve. From the viewpoint of damage, this capability depends how complex the damage can develop before final failure. Thus, the keys lie in achieving a high strain-to-failure, which has been discussed, retaining a high load after microscopic kink bands have developed, and developing intensive but nonfatal types of damage. To retain a high load, a proper support on damaged yarns is essential. Bundle thickness should be a critical factor. The width of microscopic kink bands does not diminish with the bundle thickness. If a bundle is too thin, the kink band-induced lateral displacement can disable the bundle from further resisting the load. To develop intensive but nonfatal damage, through-thickness yarns appear to play an important role. They tend to suppress composite splitting and favor branching out the growing cracks. They also surround and thus support the axial bundles in interior. The multiple kink bands in axial bundles would be impossible without their support.

The major modes of damage for 3D composites are all of miniscopic scales. To fully exploit the capability of energy absorption, however, the damage mechanisms must be of the microscopic scale. The major ways of dissipating energy are the shear deformation of the matrix and frictional sliding of fibers. The former has been well known in dominating the onset and growth of microbands in thermosetting composites. The latter has received less attention, in part because it does not greatly affect the microbands. However, frictional sliding of fibers must be taken into account when examining the capability of energy absorption.

13.8
Conclusions

Designing 3D woven composites involves additional parameters nonexistent in 1D and 2D composites. These parameters include selection of 3D yarn structure, yarn spatial orientation, through-thickness yarns, and 3D interlacing pattern. The first two dictate the resulting composite elastic properties and strengths. The last two shows importance in the ability of damage resistance and energy absorption. Using the 3D weaving, the distribution of fibers in spatial directions can be designed and optimized according to the external loads. The ability of forming complex fabrics is also the best among all major textile technologies. For 3D composites, the microstructure–behavior relationship and damage evolution are more complicated. Some inherent limitations deserve notice when applying these composites.

References

1 Chou, T.W. (1992) *Microstructural Design of Fiber Composites*, Cambridge University Press, Cambridge UK, pp. 374–422.
2 Ko, F.K. (1989) Three-dimensional fabrics for composites, in *Textile Structural Composites* (eds T.W. Chou and F.K. Ko), Elsevier Science Publisher, Amsterdam, pp. 129–171.
3 Chou, T.W. and Ko, F.K. (1989) *Textile Structural Composites*, Elsevier, Amsterdam.
4 Popper, P. (1990) Introduction to textile structural composites, in *International Encyclopedia of Composites*, vol. 1 (ed. S.M. Lee), Wiley-VCH Publishers, New York.
5 Tan, P., Tong, L., and Steven, G.P. (1997) Modelling for predicting the mechanical properties of textile composites—A review. *Composites Part A*, 28, 903–922.
6 Naik, N.K. (1994) *Woven Fabric Composites*, Technomic Publishing Company, Inc., Lancaster, Basel.
7 Du, G.W., Chou, T.W., and Popper, P. (1991) Analysis of three-dimensional textile preforms for multidirectional reinforcement of composites. *Journal of Materials Science*, 26, 3438–3448.
8 Kuo, W.S. (1996) Elastic behavior and damage of three-dimensional woven fabric composites. *Polymer & Polymer Composites*, 4, 369–375.
9 Whitney, T.J. and Chou, T.W. (1989) Modeling of 3-D angle-interlock textile structural composites. *Journal of Composite Materials*, 23, 890–911.
10 Byun, J.H., Whitney, T.J., Du, G.W., and Chou, T.W. (1991) Analytical characterization of two-step braided composites. *Journal of Composite Materials*, 25, 1599–1618.
11 Byun, J.H. and Chou, T.W. (1996) Process–microstructure relationships of 2-step and 4-step braided composites. *Composites Science and Technology*, 56, 235–251.
12 Ko, F.K., Chu, J.N., and Hua, C.T. (1991) Damage tolerance of composites: 3-D braided commingled PEEK/carbon.

Journal of Applied Polymer Science, **47**, 501–519.

13 Kuo, W.S. (1997) Topology of three-dimensionally braided fabrics using pultruded rods as axial reinforcements. *Textile Research Journal*, **67**, 623–634.

14 Kuo, W.S. and Chen, H.I. (1997) Fabrication and microgeometry of two-step braided composites incorporating pultruded rods. *Composites Science and Technology*, **57**, 1457–1467.

15 Kuo, W.S., Ko, T.H., and Chen, H.I. (1998) Elastic moduli and damage mechanisms in 3-D braided composites incorporating pultruded rods. *Composites Part A*, **29A**, 681–692.

16 Du, G.W. and Ko, F.K. (1993) Unit cell geometry of 3-D braided structures. *Journal of Reinforced Plastics and Composites*, **12**, 752–768.

17 Brookstein, D.S. and Tsiang, T.H. (1985) Load-deformation behavior of composite cylinders with integrally-formed braided and with machined circular holes. *Journal of Composite Materials*, **19**, 476–487.

18 Dransfield, K., Baillie, C., and Mai, Y.W. (1994) Improving the delamination resistance of CFRP by stitching—A review. *Composites Science and Technology*, **50**, 305–317.

19 Shu, D. and Mai, Y.W. (1993) Effect of stitching on interlaminar delamination extension in composite laminates. *Composites Science and Technology*, **49**, 165–171.

20 Mouritz, A.P., Leong, K.H., and Herszberg, I. (1997) A review of the effect of stitching on the in-plane mechanical properties of fibre-reinforced polymer composites. *Composites*, **28**, 979–991.

21 Lee, C. and Liu, D. (1990) Tensile strength of stitching joint in woven glass fabrics. *Journal of Engineering Materials and Technology*, **112**, 125–130.

22 Tong, L. and Jain, L.K. (1995) Analysis of adhesive bonded composite lap joints with transverse stitching. *Applied Composite Materials*, **2**, 343–395.

23 Ramakrishna, S. and Hull, D. (1994) Tensile behaviour of knitted carbon–fibre–fabric/epoxy laminates – Part I: Experimental. *Composites Science and Technology*, **50**, 237–247.

24 Leong, K.H., Falzon, P.J., Bannister, M.K., and Herszberg, I. (1998) An investigation of the mechanical performance of Milano rib weft-knit glass/epoxy composites. *Composite Science and Technology*, **58**, 239–251.

25 Khondker, O.A., Leong, K.H., and Herszberg, I. (2001) An investigation of the structure–property relationship of knitted composites. *Journal of Composite Materials*, **35**, 489–508.

26 Khondker, O.A., Leong, K.H., Herszberg, I., and Hamada, H. (2005) Impact and compression-after-impact performance of weft-knitted glass textile composites. *Composites, Part A*, **36**, 638–648.

27 Kuo, W.S. and Pon, B.J. (1997) Elastic moduli and damage evolution of three-axis woven fabric composites. *Journal of Materials Science*, **32**, 5445–5455.

28 Kuo, W.S. and Lee, L.C. (1998) Impact response of 3-D woven composites reinforced by consolidated rods. *Polymer Composites*, **19**, 156–165.

29 Kuo, W.S., Ko, T.H., and Chi, Y.C. (1999) Microstructures of 3D carbon/carbon composites using solid rods as reinforcements. *Polymer Composites*, **20**, 460–471.

30 Kuo, W.S. (2000) The role of loops in 3D fabric composites. *Composites Science and Technology*, **60**, 1835–1849.

31 Kuo, W.S., Ko, T.H., and Chen, C.P. (2007) Effect of weaving processes on compressive behavior of 3D woven composites. *Composites Part A*, **38**, 555–565.

32 Kuo, W.S., Fang, J., and Lin, H.W. (2003) Failure behavior of 3D woven composites under transverse shear. *Composites Part A*, **34**, 561–575.

33 Chang, L.W., Yau, S.S., and Chou, T.W. (1987) Notched strength of woven fabric composites with mould-in holds. *Composites*, **18**, 233–241.

34 Kuo, W.S., Ko, T.H., and Shiah, Y.C. (2006) Compressive damage in three-axis woven thermoplastic composites. *Journal of Thermoplastic Composite Materials*, **19**, 357–373.

35 Kuo, W.S. and Lee, L.C. (1999) Elastic and damage behavior of 3-D woven composites incorporating solid rods. *Composites Part A*, **30**, 1135–1148.

36 Kuo, W.S. and Cheng, K.B. (1999) Processing and microstructures of 3-D woven fabrics incorporating solid rods. *Composites Science and Technology*, **59**, 1833–1846.

37 Kuo, W.S., Ko, T.H., Cheng, K.B., and Hsieh, K.Y. (2001) Flexural behavior of three-axis woven carbon/carbon composites. *Journal of Materials Science*, **36**, 2743–2752.

38 Kuo, W.S., Ko, T.H., and Lo, T.S. (2002) Failure behavior of three-axis woven carbon/carbon composites under compressive and transverse shear loads. *Composites Science and Technology*, **62**, 989–999.

39 Kuo, W.S. and Huang, J.H. (1997) Stability and vibration of initially stressed plates composed of spatially distributed fiber composites. *Journal of Sound and Vibration*, **199**, 51–69.

40 Rosen, B.W. (1965) Mechanics of composite strengthening, in *Fibre Composite Materials*, American Society for Metals, pp. 37–75.

41 Argon, A.S. (1972) Fracture of composites, in *Treatise on Material Science and Technology*, vol. 1 (ed. H. Herman), Academic Press, New York, pp. 79–114.

42 Evans, A.G. and Adler, W.F. (1978) Kinking as a mode of structural degradation in carbon fiber composites. *Acta Metallurgica*, **26**, 725–738.

43 Kuo, W.S. and Ko, T.H. (2000) Compressive damage in 3-axis orthogonal fabric composites. *Composites Part A*, **31**, 1091–1105.

44 Pierron, F. and Vautrin, A. (1997) Measurement of the in-plane shear strengths of unidirectional composites with the Iosipescu test. *Composites Science and Technology*, **57**, 1653–1660.

45 Barnes, J.A., Kumosa, M., and Hull, D. (1987) Theoretical and experimental evaluation of the Iosipescu shear test. *Composites Science and Technology*, **28**, 251–268.

46 ASTM Standard (1993) D5379-93. *Standard Test Method for Shear Properties of Composite Materials by the V-Notched Beam Method*, American Society of Testing and Materials, Philadelphia.

47 Baucom, J.N. and Zikry, M.A. (2005) Low-velocity impact damage progression in woven E-glass composite systems. *Composites: Part A*, **36**, 658–664.

14
Polymer Composites as Geotextiles
Han-Yong Jeon

14.1
Introduction

14.1.1
Definition

Geosynthetics is a combination words "Geo (earth or ground)" and "Synthetics." ASTM D 4439 defines "Geosynthetics" as follows:

a planar product manufactured from polymeric material used with soil, rock, earth, or other geotechnical engineering related material as an integral part of a man-made project, structure, or system

Table 14.1 shows the polymeric raw materials that are used to make geosynthetics and various materials are employed for geosynthetics manufacturing.

Among geosynthetics, geotextiles is a kind of geosynthetics and their related materials are introduced by IGS (International Geosynthetics Society) educational resources leaflet, "Geosynthetics Classification," which are as follows:

1) *Geotextiles* are continuous sheets of woven, nonwoven, knitted, or stitch-bonded fibers or yarns. The sheets are flexible and permeable and generally have the appearance of a fabric. Geotextiles are used for separation, filtration, drainage, reinforcement, and erosion control applications (Figure 14.1).
2) *Composite geotextiles*
   a) *Geocomposites* are geosynthetics made from a combination of two or more geosynthetic types. Examples include geotextile-geonet; geotextile-geogrid; geonet-geomembrane; or a geosynthetic clay liner (GCL). Prefabricated geocomposite drains or prefabricated vertical drains (PVDs) are formed by a plastic drainage core surrounded by a geotextile filter (Figure 14.2).
   b) *Geosynthetic clay liners* are geocomposites that are prefabricated with a bentonite clay layer typically incorporated between a top and bottom geotextile layer or bonded to a geomembrane or single layer of geotextile.

Polymer Composites: Volume 1, First Edition. Edited by Sabu Thomas, Kuruvilla Joseph,
Sant Kumar Malhotra, Koichi Goda, and Meyyarappallil Sadasivan Sreekala
© 2012 Wiley-VCH Verlag GmbH & Co. KGaA. Published 2012 by Wiley-VCH Verlag GmbH & Co. KGaA.

Table 14.1 Geosynthetics with polymeric materials.

Polymeric material	Geosynthetics
Polyethylene	• Geotextiles • Geomembranes • Geogrids • Geopipes • Geonets • Composite geotextiles
Polypropylene	• Geotextiles • Geomembranes • Geogrids • Composite geotextiles
Polyvinyl chloride	• Geomembranes • Composite Geotextiles • Geopipes
Polyester	• Geotextiles • Geogrids
Polyamide	• Geotextiles • Composite geotextiles • Geogrids
Polystyrene	• EPS • Composite geotextiles

Geotextile-encased GCLs are often stitched or needle punched through the bentonite core to increase internal shear resistance. When hydrated, they are effective as a barrier for liquid or gas and are commonly used in landfill liner applications often in conjunction with a geomembrane (Figure 14.3).

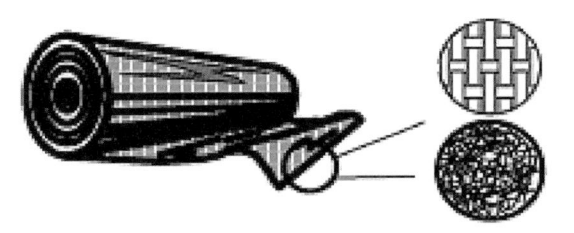

Figure 14.1 Schematic of geotextiles.

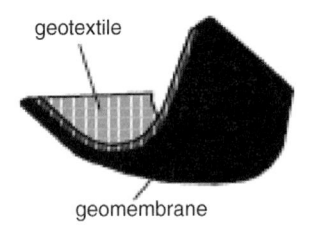

Figure 14.2 Schematic of geocomposites.

14.1 Introduction | 437

— bentonite
geotextile

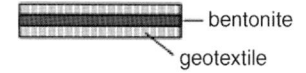

Figure 14.3 Schematic of GCLs.

Especially, advanced key technology of raw materials, manufacturing process, performance evaluation methods, design, and application for geotextiles and their composites are still introduced by USA, major European countries, for example, Germany, Italia, UK, and France, and so on. Figure 14.4 shows the movement of geotextiles-related technology in the world.

14.1.2
Function of Composite Geotextiles

Composite geotextiles include a variety of synthetic polymer materials that are specially fabricated to be used in geotechnical, geoenvironmental, hydraulic, and transportation engineering applications. It is convenient to identify the primary function of composite geotextiles as being one of separation, filtration, drainage, reinforcement, or erosion control. In some cases the geosynthetic may serve dual functions. Table 14.2 shows the functions of geosynthetics and main functions of geotextiles-related materials.

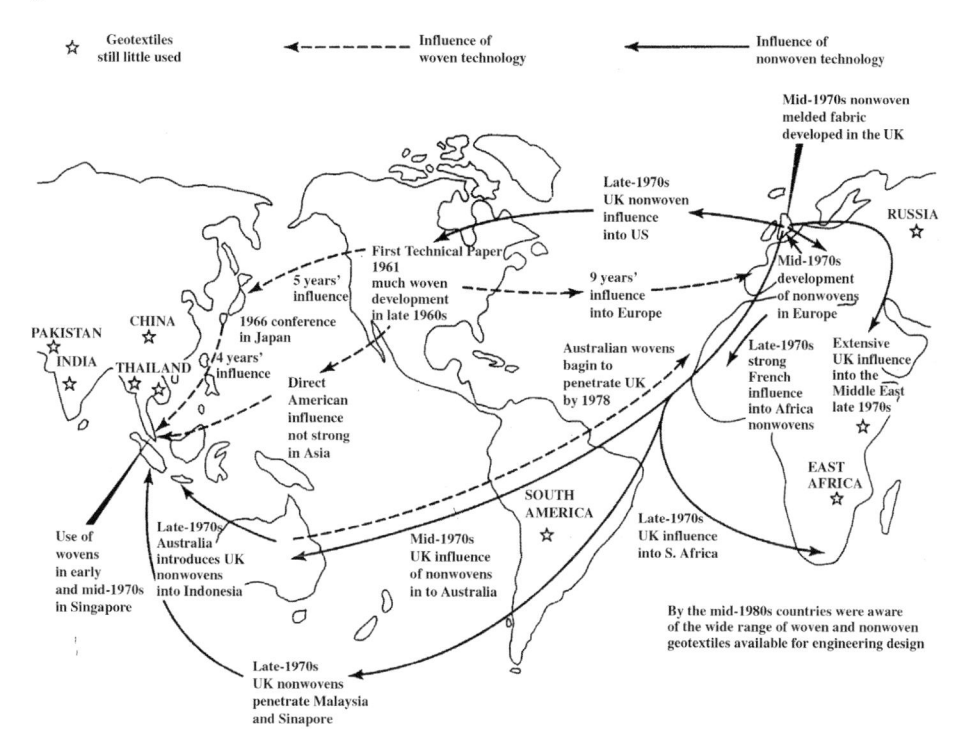

Figure 14.4 Technology transition route of geotextiles-related materials.

14 Polymer Composites as Geotextiles

Table 14.2 Functions of geosynthetics.

Function	Geosynthetics
Reinforcement	• Geotextiles • Geogrids • Composite geotextiles • EPS
Separation	• Geotextiles • Composite geotextiles
Filtration	• Geotextiles • PVD
Drainage	• Geotextiles • Composite geotextiles • Geonets • Geopipes
Water barrier and container	• Geomembranes • Composite geotextiles
Protection and energy absorber	• Geotextiles • EPS • Composite geotextiles

1) **Separation:** The geosynthetic acts to separate two layers of soil that have different particle size distributions. For example, geotextiles are used to prevent road base materials from penetrating into soft underlying soft subgrade soils, thus maintaining design thickness and roadway integrity. Separators also help to prevent fine-grained subgrade soils from being pumped into permeable granular road bases (Figure 14.5).

2) **Filtration:** The geosynthetic acts similar to a sand filter by allowing water to move through the soil while retaining all upstream soil particles. For example, geotextiles are used to prevent soils from migrating into drainage aggregate or pipes while maintaining flow through the system. Geotextiles are also used below riprap and other armor materials in coastal and river bank protection systems to prevent soil erosion (Figure 14.6).

3) **Drainage:** The geosynthetic acts as a drain to carry fluid flows through less permeable soils. For example, geotextiles are used to dissipate pore water

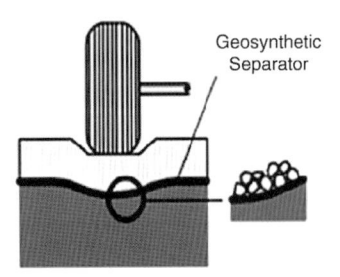

Figure 14.5 Separation function of geotextiles.

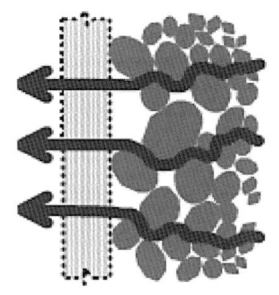

Figure 14.6 Filtration function of geotextiles.

pressures at the base of roadway embankments. For higher flows, geocomposite drains have been developed. These materials have been used as pavement edge drains, slope interceptor drains, and abutment and retaining wall drains. Prefabricated vertical drains have been used to accelerate consolidation of soft cohesive foundation soils below embankments and preload fills (Figure 14.7).

4) **Reinforcement:** The geosynthetic acts as a reinforcement element within a soil mass or in combination with the soil to produce a composite that has improved strength and deformation properties over the unreinforced soil. For example, geotextiles and geogrids are used to add tensile strength to a soil mass in order to create vertical or near-vertical changes in grade (reinforced soil walls). Reinforcement enables embankments to be constructed over very soft foundations and to build embankment side slopes at steeper angles than would be possible with unreinforced soil. Geosynthetics (usually geogrids) have also been used to bridge over voids that may develop below load bearing granular layers (roads and railways) or below cover systems in landfill applications (Figure 14.8).

5) **Erosion Control:** The geosynthetic acts to reduce soil erosion caused by rainfall impact and surface water runoff. For example, temporary geosynthetic blankets and permanent lightweight geosynthetic mats are placed over the otherwise exposed soil surface on slopes. Geotextile silt fences are used to remove suspended particles from sediment-laden runoff water. Some erosion control mats are manufactured using biodegradable wood fibers (Figure 14.9).

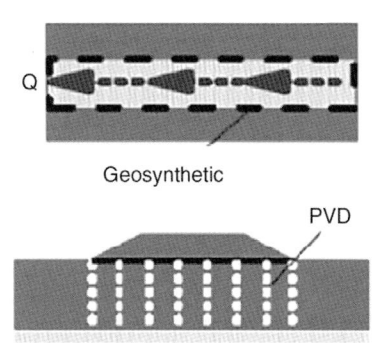

Figure 14.7 Drainage function of geotextiles.

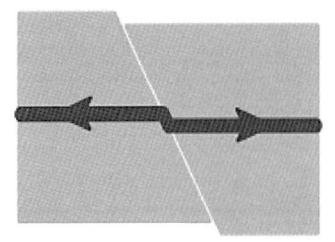

Figure 14.8 Reinforcement function of geotextiles.

Geotextiles are also used in other applications. For example, they are used for asphalt pavement reinforcement and as cushion layers to prevent puncture of geomembranes (by reducing point contact stresses) from stones in the adjacent soil, waste, or drainage aggregate during installation and while in service. Geotextiles have been used as daily covers to prevent dispersal of loose waste by wind or birds at the working surface of municipal solid waste landfills. Geotextiles have also been used for flexible concrete formworks and for sandbags. Cylindrical geotubes are manufactured from double layers of geotextiles that are filled with hydraulic fill to create shoreline embankments or to dewater sludge.

Figure 14.10 shows the main concept of composite geotextiles by role of function in the application soil structure.

14.1.3
Application Fields of Composite Geotextiles

The followings are the use areas of composite geotextiles with their functions and this is not an all-inclusive examples and it is gradually growing year by year and Figure 14.11 shows typical examples of composite geotextiles applications.

1) **Separation of dissimilar materials:**
   a) Between subgrade and stone base in unpaved/paved roads and airfields
   b) Between subgrade and ballast in railroads
   c) Between landfills and stone base courses
   d) Between geomembranes and sand drainage layers

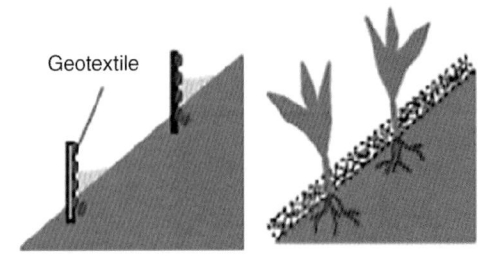

Figure 14.9 Erosion control function of geotextiles.

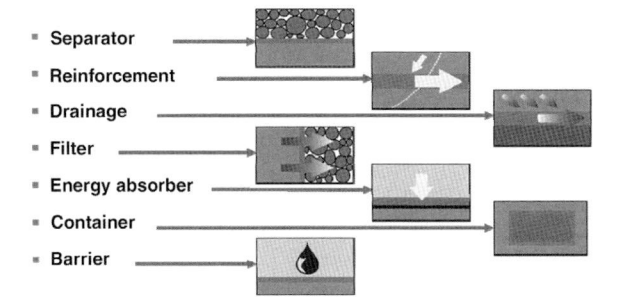

- Separator
- Reinforcement
- Drainage
- Filter
- Energy absorber
- Container
- Barrier

Figure 14.10 Concept of composite geotextiles by role of function.

e) Between foundation and embankment soils for surcharge loads, roadway fills, earth, and rock dams
f) Between foundation and soils and rigid retaining walls
g) Beneath parking lots
h) Beneath sport and athletic fields
i) Beneath precast blocks and panels for esthetic paving
j) Between various zones in earth dams
k) Between old and new asphalt layers

2) Reinforcement of Weak Soils and Other Materials
   a) Over soft soils for unpaved roads, airfields, railroads, landfills
   b) Over nonhomogeneous soils
   c) Over unstable landfills as closure systems
   d) To construct fabric-reinforced walls
   e) To reinforce embankments
   f) To reinforce earth and rock dams
   g) To stabilize slopes temporarily
   h) To halt or diminish creep in soil slopes
   i) To bridge over cracked or jointed rock
   j) To hold over graded-stone filter mattresses
   k) As substrate for articulated concrete blocks
   l) To prevent puncture of geomembranes by subsoils, landfill materials
   m) To contain soft soils in earth dam construction
   n) To bridge over uneven landfills during closure of the site

3) Filtration (Cross-Section Flow)
   a) In plane of granular soil filters
   b) Beneath stone base for unpaved/paved roads and airfields
   c) Beneath ballast under railroads
   d) Around crushed stone surrounding/without underdrains
   e) Around perforated underdrain pipe
   f) Beneath landfills that generate leachate

14 Polymer Composites as Geotextiles

(Railroad construction)	(Embankment reinforcement)	(Waste landfills)
(Road construction)	(Asphalt pavement)	(Levee reinforcement)
(Slope reinforcement)	(Erosion control)	(Chemical storage)
(Waste landfill liner)	(Ground reinforcement)	(Coast reinforcement)

Figure 14.11 Examples of composite geotextiles applications.

g) To filter hydraulic fills
h) As a silt fence, a silt curtain
i) Between backfill soil and voids in retaining walls
j) Between backfill soil and gabions
k) Against geonets and geocomposites to prevent soil intrusion
l) Around sand columns in sand drains
m) As a filter beneath stone riprap, precast block

4) Drainage (In-Plane Flow)
   a) As a chimney drain and drainage gallery in an earth dam
   b) As drainage blanket beneath a surcharge fill
   c) As a drain behind a retaining wall
   d) As a drain beneath railroad ballast
   e) As a drain beneath sport and athletic fields
   f) As a drain for roof gardens
   g) As a pore water dissipater in earth fills
   h) As a replacement for sand drains
   i) As a capillary break in frost-sensitive areas
   j) To dissipate seepage water from exposed soil or rock surfaces

5) Water Barrier
   a) Adjacent to geomembranes in vertical cutoff walls
   b) Above geomembranes as puncture protection against coarse gravel
   c) As liners for canals
   d) As a portion of a compacted clay liner in primary composite liners
   e) As a portion of a compacted clay liner in secondary composite liners
   f) As secondary liners for underground storage tanks
   g) As single liners for surface impoundments
   h) Beneath geomembranes as composite liners for surface impoundments
   i) Beneath geomembranes as composite liners for heap leach ponds
   j) Beneath geomembranes in the primary liners of landfills
   k) Beneath geomembranes in the secondary liners of landfills
   l) Beneath geomembranes and above clay liners of landfills
   m) Beneath geomembranes in the covers of landfills

14.2
Developments of Composite Geotextiles

14.2.1
Raw Materials of Composite Geotextiles

14.2.1.1 Natural Fibers

Natural fibers that are used to manufacture geotextiles are very restrictive and applied as shapes of fibers, yarns, fabrics, and knits for this purpose on the beginning stage of geotextile usage. Recently, nonwoven and woven fabrics are the main products for geotextiles and composite geotextiles and amounts of these materials are increasing gradually. Natural fibers such as cotton, jute, wool, coir, straw, basalt fiber, and natural waste fiber assembly have the advantage of eco-environmental property and are very popular as one of the degradable natural materials. Main application field of natural fibers used composite geotextiles are something special but slope stabilization, erosion control, drainage, vegetation matrix, and so on.

14.2.1.2 Synthetic Fibers

In general, typical synthetic fibers for composite geotextiles are those of polyolefin, polyester and these fibers have the strong merits of economy, low cost for manufacturing. Demand and amount used of these fibers will be increasing gradually by virtue of this merit. To add the special performance and apply special end-use to composite geotextiles, fibers of polyurethane, glass, carbon, and super fibers are used to the very restricted areas. Also, special technologies by additives and functional polymers to improve and supplement the disadvantages of typical fibers of polyolefin, polyester, and so on, are developing for specialty application fields.

14.2.1.3 Recycled Fibers

Lot of amounts of composite geotextiles are generally used for real installation and construction fields and therefore, high cost materials are insufficiently competitive from economic point of view. For this case, recycled fibers have a decided advantage especially from eco-environmental point of view. However, the weakest performance of recycled fibers used composite geotextiles is a falling-off in strength and to compensate and improve this problem is one of the hot issues for these composite geotextiles. To do this, development technology by special and high functional additives and polymeric materials should be progressed with advance of manufacturing process and equipment.

14.2.1.4 Advanced Functional Fibers

By increasing demand of smart composite geotextiles, advanced functional fibers, and hybrid polymeric materials of special performance should be needed and mechanical, chemical resistances, durability, weatherability, and so on, of composite geotextiles could be improved through this technology. Besides this, high absorbent, biodegradable, high tenacity, high modulus, high performance fibers could be used for this purpose but economic problem of high cost compared to typical polymeric materials is the most realistic barrier. To consider the introduction of future convergence technology, LCP (liquid crystal polymer), PBO (polybutylene oxide), PPS (polypropylene sulfide), *meta-* or *para* aramid fibers that can make advanced and hybrid materials should be applied to the more stability needed soil structure by composite geotextiles.

14.2.2
Advanced Trend of Composite Geotextiles

14.2.2.1 Geotextiles

Overall noticeable advanced technology of geotextiles is not appeared still now and advanced concepts of woven and nonwoven geotextiles are as follows.

14.2.2.1.1 Nonwoven Types

i) Development of high weight products over 5000 g/m^2.

ii) Development of smart geotextiles having separation, protection, optional drainage functions, and so on.
iii) Development of nanofibers-related products for environmental application.
iv) Development of eco-adaptative and biodegradable products for vegetation.
v) Development of multifunctional composites products, and so on.

14.2.2.1.2 Woven Types

i) Development of high strength products over design strength 80 ton/m.
ii) Development of advanced creep behavior products with low elongation by high performance fibers.
iii) Development of eco-adaptative and biodegradable products for vegetation.
iv) Development of multifunctional composites products, and so on.

14.2.2.2 Geosynthetic Clay Liners

GCLs is a kind of water barrier material as well as geomembrane as described in Sections 14.1.1 and 14.1.3 and bentonites of powder and granular type are filled between both geotextiles and final products are made by needle punching, stitch bonding, and chemical bonding. Here montmorilinite components have the high swelling property when contact water and structure of montmorilinite in GCLs should be changed from linear to house shape after swelling. And then, inner structures of GCLs are changed to very compact and GCLs have the powerful water barrier functions due to this structure change. However, some serious problems would be occurred in case of installation to seawater contact and slope areas. First, when GCLs are contacted with seawater, no welling occurs and GCLs has no water barrier effect. Second, when GCLs are installed to the slope of soil structure or waste landfills, bentonites may be lost because of powder state and some special agent should be added to avoid this phenomena. As a result, swelling is not occurred and GCLs has no water barrier effect. Therefore, the present solutions to solve these problems of GCLs are to make modified composition type GCLs not to be lost in the slope installation and to decrease the water barrier performance under freezing and thaw condition.

14.2.2.3 Composite Geotextiles for Drainage and Filtration

Hydraulic performances such as filtration and drainage are mainly dependent on EOS (effective opening size) of composite geotextiles and to maximize this is to keep the optimum EOS value by flow path procure and minimize the intrusion effect by confined loading. Plastic drain board (PDB) is one of the composite geotextiles for drainage and development trend of this material is to obtain the larger drainage capacity and filtration area. Also, PDB should have excellent resistance and durability to acidic, alkaline, and salt solutions. Especially, shrinkage and deformation by water component must not be occurred during service life period. Nonwoven filters of PDB should have higher strength and drainage capacity should be maximized and have almost no clogging phenomena. The following are some examples of advanced composite geotextiles for filtration and drainage.

- Composite geotextiles drains are formed of geotextile layers (often nonwoven) bonded either side of a discharge capacity core (5–25 mm thick).
- In-plane discharge capacities are in the range 0.0002–0.01 m^3/m width/s.
- A 20 mm thick geotextiles composite drain can have the same flow capacity as a 300 mm thick granular layer.

14.2.2.4 Reinforced Concrete by Composite Geotextiles

Composite geotextiles are specially used for the purpose of mechanical properties improvement and crack formation prevention from concrete structures. For this object, carbon, glass, metallic, aramid fibers are used as reinforcement materials of concretes as well as nylon, polyester, polypropylene, polyethylene, rayon, spandex fibers. With this reinforcement fibers application, we can obtained the following advantages: (1) crack resistance improvement, (2) share of cracked area, (3) resistance improvement by thermal shrinkage and expansion, (4) durability improvement, and so on. The detailed application fields of reinforced concrete by composite geotextiles should be carefully determined through examination and confirmation of tensile properties, fatigue properties, creep behaviors, acid and alkali resistance, thermal resistance, interface properties with concrete, and so on.

14.2.2.5 Reinforced Geomembrane by Composite Geotextiles

Geomembranes that are installed between valleys for waste landfills must have the ultimate resistance to partial cave-in and puncture conditions by external loading during construction and being used. For reinforcement of geomembranes, nonwoven geotextiles are bonded to one or both sides of geomembranes as a kind of composite geotextiles and this formation is a typical example of reinforced geomembranes manufacturing. Besides this formation, geomembrane/geonet and geomembrane/geogrid composites are generally used for improvement and supplementation of geomembranes disadvantages. Multilayered geotextiles/geogrids composites as shown in Figure 14.12 are the representative to be used in subground gasoline storage tank construction with the object of gas leakage blocking as liquid/gas barrier.

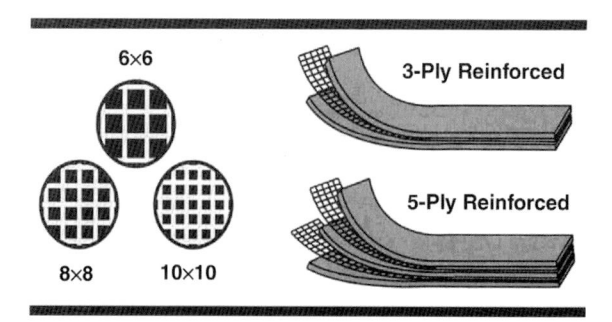

Figure 14.12 Composite geotextiles for liquid/gas barrier.

14.3
Hybrid Composite Geotextiles

14.3.1
For Separation and Reinforcement

Geotextiles/geogrids composites are generally used for slope and liner system of waste landfills, bedrock, soft ground, valleys, and so on, for permanent stabilization against erosion and ultimate tensile and tear strength, elastic modulus, bursting and puncture strength would be needed for this. Here it is introduced that hybrid composite geotextiles having both separation and reinforcement functions would be examined through evaluations of their main performance namely, separation/reinforcement performance.

14.3.1.1 Manufacturing of Geotextiles/Geogrids Composites

As raw materials, spun bonded geotextiles for separation SBNW-1, SBNW-2, and textiles type geogrids for reinforcement GG-1, GG-2, GG-3 are used to make geotextiles/geogrids composites, respectively. Tables 14.3 and 14.4 show the specifications of both raw materials.

For geotextiles/geogrids composites, GC-A is composite of GG-1 and SBNW-1, GC-B is that of GG-2 and SBNW-1, GC-C is that of GG-3 and SBNW-1, GC-D is that of GG-1 and SBNW-2, GC-E is that of GG-2 and SBNW-2, GC-F is that of GG-3 and

Table 14.3 Specifications of geotextiles used to manufacture goecomposites.

Specifications	SBNW-1	SBNW-2
Type of fiber	100% polypropylene	
Manufacturing method	Thermal bonding	
Weight (g/m^2)	136	220

Table 14.4 Specifications of geogrid to be used to manufacture goecomposite.

Specifications	GG-1		GG-2		GG-3	
	Woven type		Woven type		Warp-Knit type	
Aperture size (mm)	A × B[a]	C × D[b]	A × B[a]	C × D[b]	A × B[a]	C × D[b]
	24.0 × 24.6	17.2 × 20.2	26.4 × 26.7	19.1 × 23.2	27.0 × 26.6	19.2 × 20.2
Number of ribs (m^{-1})	Warp	Weft	Warp	Weft	Warp	Weft
	42 ± 1	41 ± 1	37 ± 1	37 ± 1	37 ± 1	39 ± 1

a) Aperture size from middle of ribs.
b) Inherent aperture size of geogrids.

SBNW-2, respectively. To improve the junction efficiency on junction point between geotextiles and geogrids, EVA (ethylene vinyl acetate) resin was used as bonding agent.

14.3.1.2 Wide-Width Tensile Strength

Figures 14.13 and 14.14 show tensile strength of geotextiles/geogrids composites and these values are larger than those of geogrids because tensile properties are developed by geogrids. However, it is seen that geogrids would be first broken before geotextiles because tensile strains of geogrids are smaller than those of geotextiles during tensile test. Consequently, it is seen that increase of tensile strength of geotextiles/geogrids composites is due to the strength hardening effect by junction bonding between geogrids and geotextiles and this is contributed to improvement of tensile.

14.3.1.3 Hydraulic Properties

Table 14.5 shows the permittivity, vertical water permeability of geotextiles/geogrids composites and this property is completely dependent on that of geotextiles. In Table 14.8, we can find the little decrease of permittivity and this is due to the decrease of water flow path area by junction area bonding between geotextiles and geogrids.

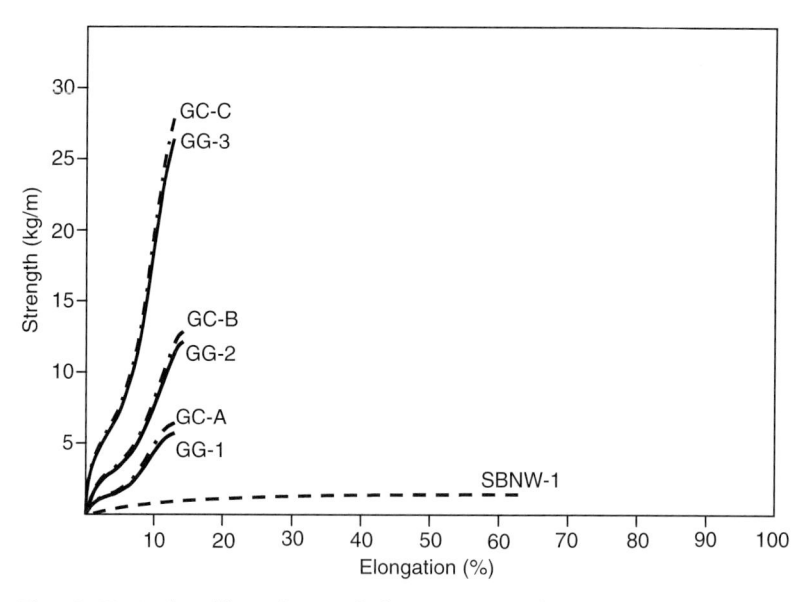

Figure 14.13 Wide-width tensile strength-elongation curves of composite geotextiles (GC-1, GC-2, GC-3) at machine direction.

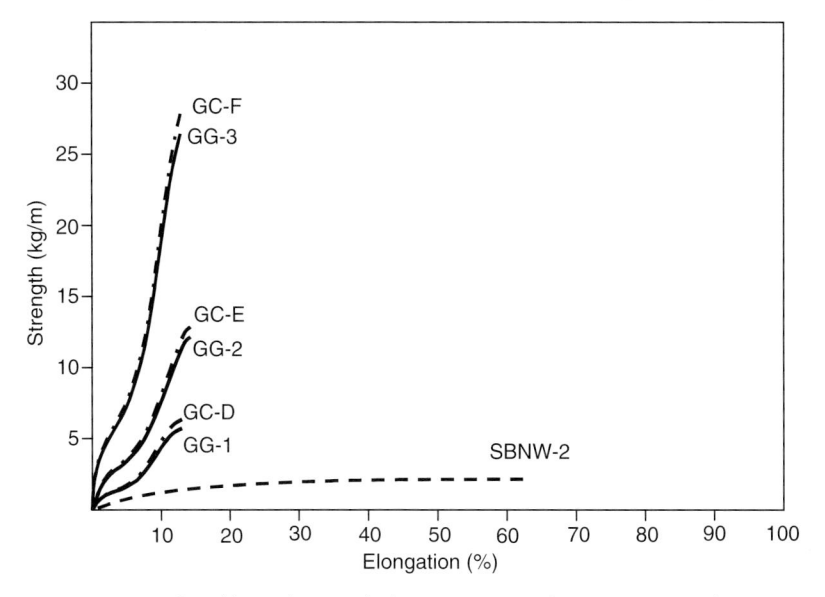

Figure 14.14 Wide-width tensile strength-elongation curves of composite geotextiles (GC-4, GC-5, GC-6) at machine direction.

Table 14.5 Hydraulic properties of geocomposites.

Hydraulic property	GC-A	GC-B	GC-C	GC-D	GC-E	GC-F
EOS, O_{95} (μm)	201	202	198	74	75	72
Permeability (cm/s) ($\times 10^{-2}$)	5.1	5.2	4.9	2.2	2.3	2.0
Permittivity (s^{-1})	1.18	1.19	1.16	0.43	0.44	0.42

14.3.2 For Drainage

14.3.2.1 Manufacturing of Geotextiles/Geonet Composites

A main function of geotextiles/geonet composites is drainage and these composites are always installed under confined loading system. When geotextiles/geonet composites are under confined loading system, transmissivity, horizontal water permeability is decreased by intrusion phenomena, especially needle punched nonwoven geotextiles.

14.3.2.1.1 **Nonwoven/Woven Composite Geotextiles** A 700–3000 g/m² needle punched nonwoven geotextiles of polyester staple fibers and woven geotextiles of polypropylene flat yarn are used to make nonwoven/woven composite geotextiles. Here nonwoven geotextiles were used for filtration and woven geotextiles were used for strength reinforcement of composite geotextiles and thermal calendar bonding were adapted to improve the bonding efficiency between nonwoven and woven geotextiles.

Table 14.6 Specification of geotextiles/geonet composites manufacturing-related materials.

Geosynthetics	Thickness (mm)	Weights (g/m²)
Nonwoven geotextiles (WD)	4.5	700
	6.0	1000
	8.0	1500
	12.0	2500
	14.0	3000
Geotextiles composites	5.0	1050
	5.0	1248
	5.0	1352
	5.0	1650
Geonet (GN)	55.0	3348

14.3.2.1.2 Nonwoven Geotextiles/Geonet Composites

These composite geotextiles were made to improve the geonet's drainage performance by bonding nonwoven geotextiles to geonet and nonwoven geotextiles can protect insertion of soil particles into inner side of geonet that is the serious cause of transmissivity decrease. However, this geotextiles composite has disadvantage of transmissivity decrease due to the intrusion of geotextiles under confined loading system. Here we introduced transmissivity change and dependence on compressive stress of nonwoven/woven composite geotextiles and nonwoven geotextiles/geonet composites. Table 14.6 shows the specification of geotextiles/geonet composites manufacturing-related materials.

14.3.2.2 Compressive Stress and Transmissivity

Figures 14.15–14.17 show transmissivity changes versus compressive stress of nonwoven geotextiles, nonwoven/woven composite geotextiles, nonwoven geotextiles/geonet composites, respectively. In Figure 14.15, nonwoven geotextiles show the rapid decrease of transmissivity till compressive stress $0.2\,\text{kg/cm}^2$ but shows the constant value 0.4–$0.7\,\text{cm}^2/\text{s}$ after $0.2\,\text{kg/cm}^2$. This tendency is clearly decreased for heavier weight nonwoven geotextiles and the reason of this tendency is due to pore size decrease of fiber assemblies in nonwoven geotextiles by increase of fiber density per constant area. Nonwoven/woven composite geotextiles shows slight decrease of transmissivity in Figure 14.16. Nonwoven geotextiles/geonet composites shows larger transmissivity without compressive stress but shows the rapid decrease with compressive stress and finally approached about $3.0\,\text{cm}^2/\text{s}$ under compressive stress $1.0\,\text{kg/cm}^2$ as shown in Figure 14.17.

Transmissivity decrease ratio against compressive stress of geotextiles composites could be determined by the following equation.

$$P_\text{r} = \frac{K_{\text{PE0}} - K_{\text{PE1}}}{K_{\text{PE0}}} \times 100 \tag{14.1}$$

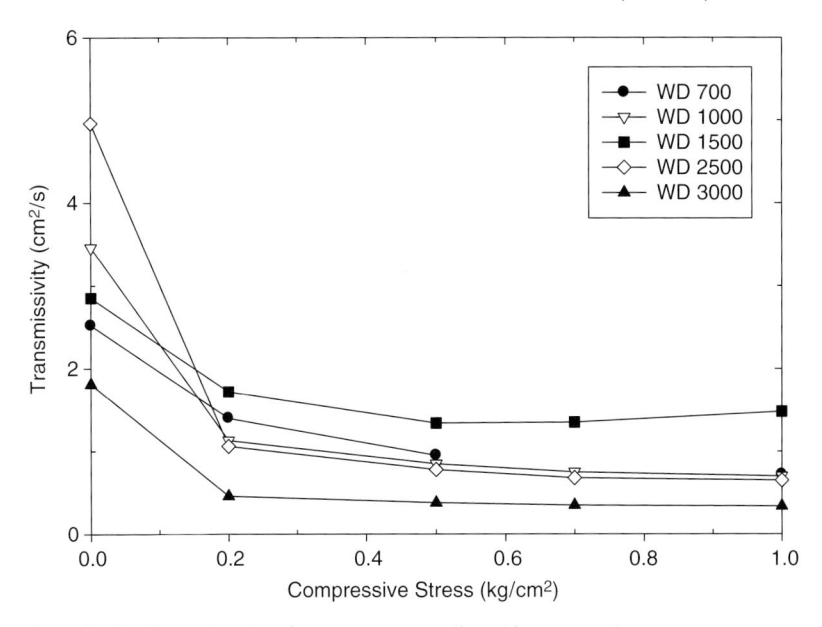

Figure 14.15 Transmissivity of nonwoven geotextiles with compressive stress.

where P_r is transmissivity (cm^2/s) decrease ratio (%) against compressive stress of composite geotextiles, K_{PE0} is transmissivity (cm^2/s) without compressive stress, K_{PE1} is transmissivity (cm^2/s) with compressive. Figures 14.18–14.20 show transmissivity decrease ratio against compressive stress, respectively.

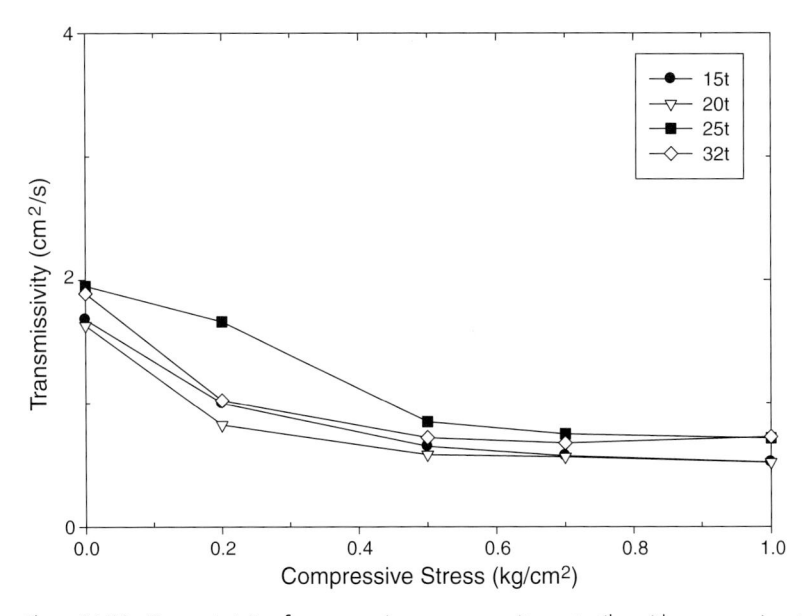

Figure 14.16 Transmissivity of nonwoven/woven composite geotextiles with compressive stress.

452 *14 Polymer Composites as Geotextiles*

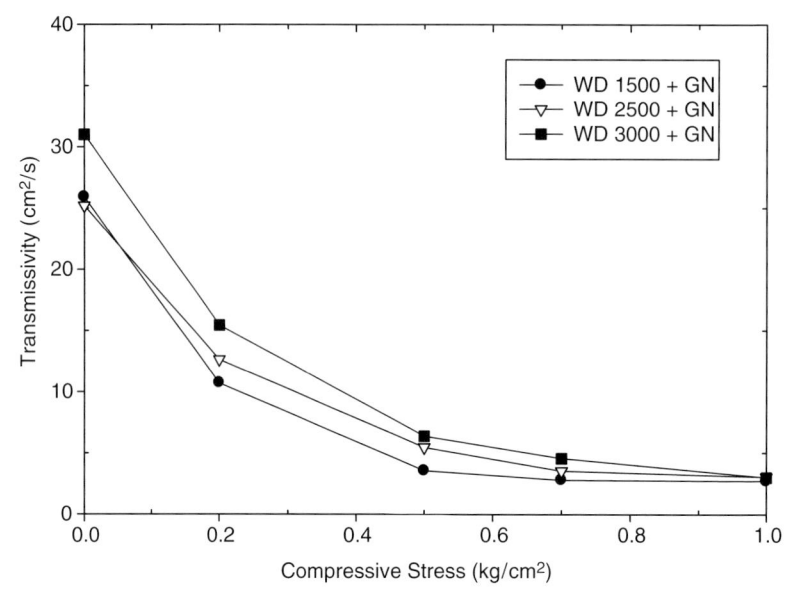

Figure 14.17 Transmissivity of nonwoven geotextiles/geonet composites with compressive stress.

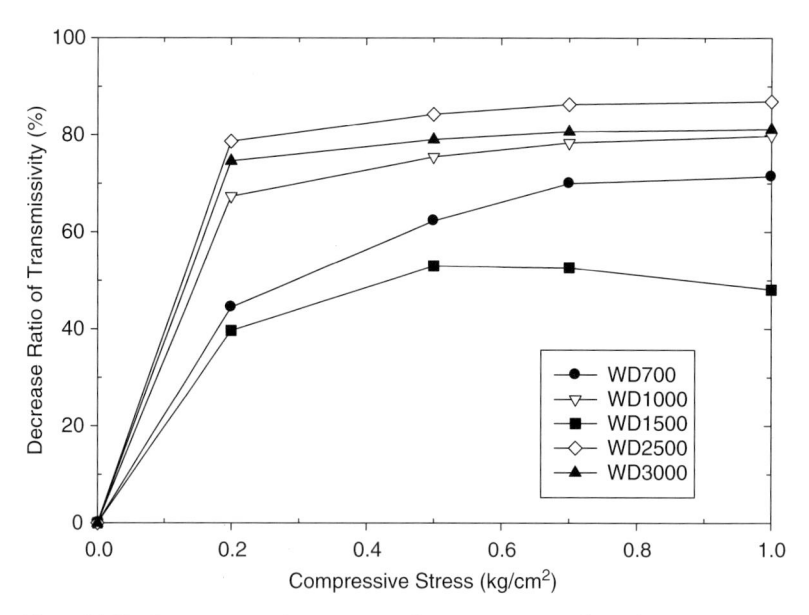

Figure 14.18 Transmissivity decrease ratio of nonwoven geotextiles with compressive stress.

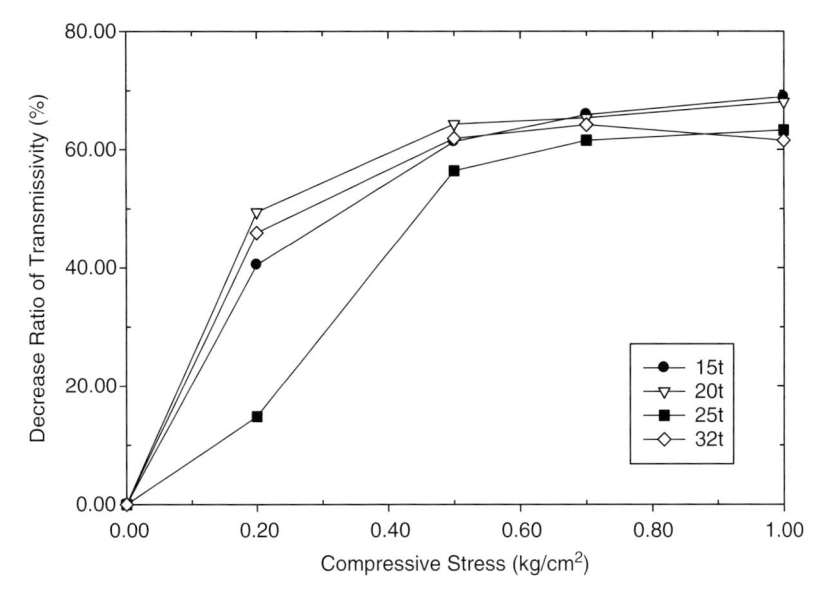

Figure 14.19 Transmissivity decrease ratio of nonwoven/woven composite geotextiles with compressive stress.

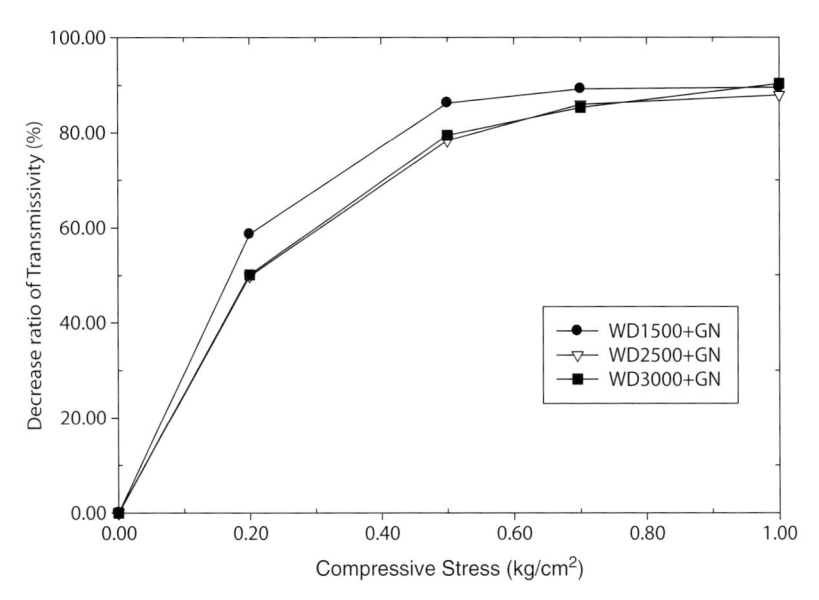

Figure 14.20 Transmissivity decrease ratio of nonwoven geotextiles/geonet composites with compressive stress.

In Figure 14.18, nonwoven geotextiles shows 45–85% transmissivity decrease ratio of initial transmissivity value under compressive stress 1.0 kg/cm² with thickness. Nonwoven/woven composite geotextiles shows 60–70% transmissivity decrease ratio of initial transmissivity value under compressive stress 1.0 kg/cm² with thickness in Figure 14.19. Nonwoven geotextiles/geonet composites show about 90% transmissivity decrease ratio of initial transmissivity value under compressive stress 1.0 kg/cm² with thickness in Figure 14.20. This tendency of transmissivity decrease is due to pore size decrease of fiber assemblies in nonwoven geotextiles by increase of fiber density per constant area as introduced before.

From the overall consideration, nonwoven geotextiles, nonwoven/woven composite geotextiles, and nonwoven geotextiles/geonet composites show the rapid transmissivity decrease ratio till the compressive stress 0.2, 0.5 kg/cm², respectively, but showed constant transmissivity after this compressive value. Nonwoven geotextiles/geonet composites show excellent transmissivity value compared to nonwoven geotextiles, nonwoven/woven composite geotextiles without compressive stress but showed transmissivity decrease in the less weight of nonwoven geotextiles. For nonwoven geotextiles, nonwoven/woven composite geotextiles, and nonwoven geotextiles/geonet composites showed thickness decrease with compressive stress and this tendency should be improved by special composition of composite geotextiles with optimum convergence composite manufacturing technology.

14.3.3
For Protection and Slope Stability

14.3.3.1 Manufacturing of Three-Layered Composite Geotextiles

Some composite types of geosynthetics that have the special functions are used in waste landfills and now geonet composites are adopted the proper materials that protect the geomembrane and have the drainage function. But for the waste landfills, marvel stones to be used as drainage materials of leachates over 50 mm diameter will cause to occur to the intrusion phenomena of geonet composites. These intrusion phenomena of geonet composites are the causes to decrease the drainage efficiencies in waste landfills. Therefore, it would be needed the smart geotextiles that have not only the excellent drainage function but also the distinguished protection function to the geomembranes for slope and liner system of waste landfills. Here it was designed and manufactured the smart nonwoven geotextiles that have the differential compositions by the special needle punching method. Changes of thickness, transmissivity, in-plane permeability with compressive stress were analyzed by the constitutive equations of drainage function.

The smart composite geotextiles of three-layer structure that have the adaptative drainage function under confined loading condition were manufactured by needle punching method. Three different punching patterns were applied to manufacture these geotextiles as ↑, ↑, and ↓ punching mechanism.

Table 14.7 showed the specifications of the smart composite geotextiles – SMGT 1, 2, 3 and three types of geotextiles/geonet composites – GNC-1, −2, −3, and so on –

Table 14.7 Specifications of smart geotextiles.

Geosynthetics for drainage		Thickness (mm)	Composition	Drainage layer
Smart geotextiles of three-layered	SMGT 1	1.2	Nonwoven/drainage layer/nonwoven	(Waste) PP or PET fibers; 20–1000 Deriers Accumulation of web Prepunched nonwovens
	SMGT 2	1.4		
	SMGT 3	1.7		
Geotextiles/Geonet composites	GNC-1	6.2	Nonwoven/drainage core/nonwoven	Two-layer HDPE core
	GNC-2	7.2		
	GNC-3	8.0		

having the same thickness as smart geotextiles were used as comparison materials for drainage function and Figure 14.21 shows the cross-sectional photograph of smart composite geotextiles and geotextiles/geonet composites.

14.3.3.2 Transmissivity

The principal transmissivity mechanism of smart composite geotextiles could be analyzed by Eqs. (14.2)–(14.5). If water flows along the surface of geotextiles horizontally and the amount of water-in should be equal to those of water-out, flow

(a)

(b)

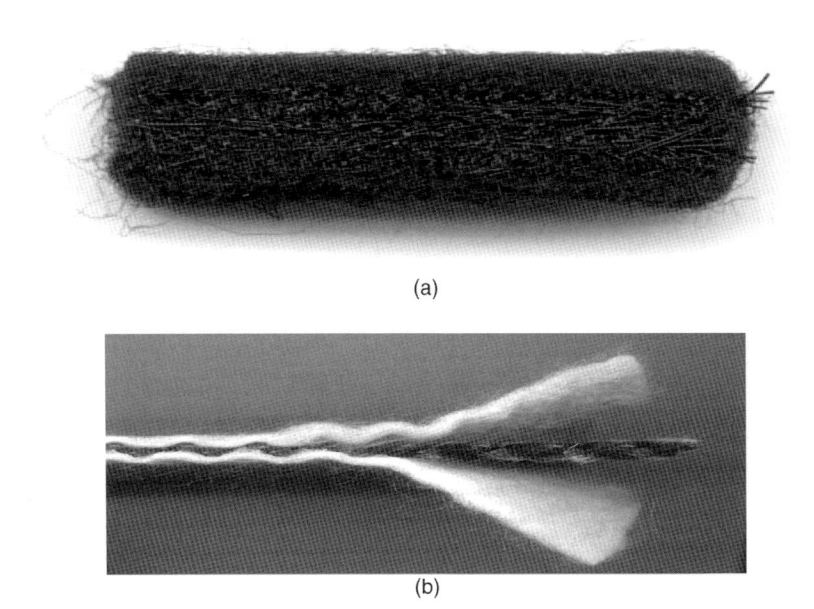

Figure 14.21 (a) Photographs of cross-sectional areas of smart composite geotextiles and (b) geotextiles/geonet composite.

rate of water, q, for drainage system could be written by Eq. (14.1) from Darcy's law,

$$q = K_p \times i \times A = K_p \frac{\Delta h}{L} \times w \times t \tag{14.2}$$

And transmissivity of geotextiles for drainage is as follows:

$$\theta = K_p \times t = q\frac{L}{\Delta h \times w} = \frac{q}{i \times w} \tag{14.3}$$

where θ is the transmissivity of the geotextile, i is the hydraulic gradient, K_p is the in-plane permeability, q is the flow rate, L, t are the length and thickness of the geotextile, respectively, Δh is the total water head lost, and w is the width of the geotextile.

If water flows radially through the geotextile and is collected around the outer perimeter of the device, the theory is adapted as follows:

$$q = K_p \times \frac{dh}{dr}(2\pi \times r \times t) \tag{14.4}$$

And the amounts of radial drainage are calculated by the following equations:

$$2\pi(K_p \times t)\int_{h_1}^{h_2} dh = q\int_{r_1}^{r_2} \frac{dr}{r}$$

$$\theta = \frac{q\ln(r_2/r_1)}{2\pi \Delta h} \tag{14.5}$$

where r_1 is the inner radius of the geotextile test specimen and r_2 is the outer radius of the geotextile test specimen.

From the above assumption, transmissivity of smart geotextile composites could be written as following as shown in Figure 14.22:

$$\theta_{SGT} = \sum \theta_i = \sum(\theta_U + \theta_I + \theta_L) = \sum k_{\theta_i} \times t_i \tag{14.6}$$

where θ_{SGT} is the transmissivity of the smart geotextile composites, θ_i is the transmissivity of i component of the smart geotextile composites, k_{θ_i} is the in-plane permeability of i component of the smart geotextile composites, and t_i is the thickness of i component of the smart geotextile composites.

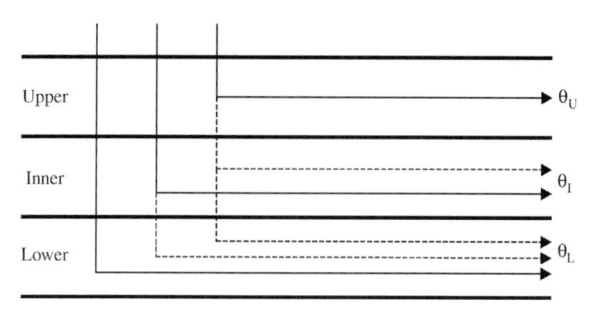

Figure 14.22 Schematic diagram for three-layer structure geotextiles.

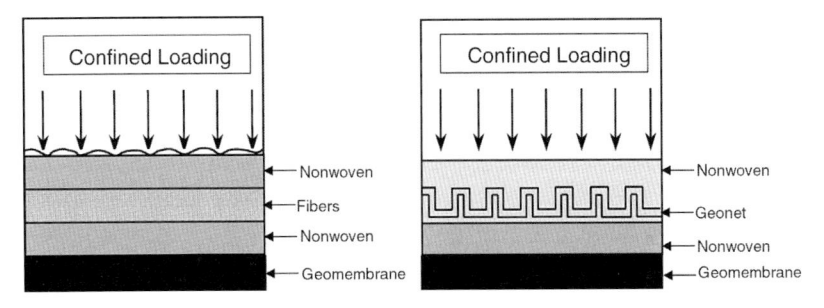

Figure 14.23 Schematic diagram of intrusion for smart composite geotextiles and geotextiles/geonet composites under confined loading.

14.3.3.3 Thickness and Compressive Stress

Figure 14.23 shows the schematic diagram of intrusion phenomena by the confined loading of smart geotextile composites and geotextiles/geonet composites. Here the upper nonwovens of geotextiles/geonet composites show the considerable intrusion by the confining load whereas smart geotextile composites show a bit of intrusion. This is closely related to the variations of thickness with compressive stress. In general, the thickness of geotextiles is decreased by the compressive stress for installation within the soil structure. For this case, transmissivity of the geotextiles would be the function of thickness and it is very important to evaluate the variation of thickness with the compressive stress. The relationship between thickness and compressive stress would be written as Eq. (14.6) by using the variation constant of the geotextiles,

$$T/T_0 = (\sigma/\sigma_0)^{-a} \tag{14.7}$$

where T_0, T are the thickness of the geotextiles with/without compressive stress, respectively, a is the variation constant of the geotextiles, σ_0, σ are the initial and compressive stress of the geotextiles, respectively ($\sigma_0 = 0.04\,\text{kg/cm}^2$).

From this equation, the variation constant, a, will be larger with the thickness of geotextiles and therefore, another variation constant, b, should be introduced to Eq. (14.7) to compensate the variation constant, a. Therefore, the variation of thickness with compressive stress could be written as follows:

$$T = T_0 - a\ln\frac{\sigma}{\sigma_0} = T_0\left(1 - \ln\frac{\sigma}{\sigma_0}\right) \tag{14.8}$$

$$b = a/T_0$$

Figure 14.24 shows the relative decrease of thickness with compressive stress of geotextiles by using Eq. (14.8). Geotextiles/geonet composites showed more significant decrease of thickness with confined loading due to the considerable intrusion of upper nonwovens than smart composite geotextiles.

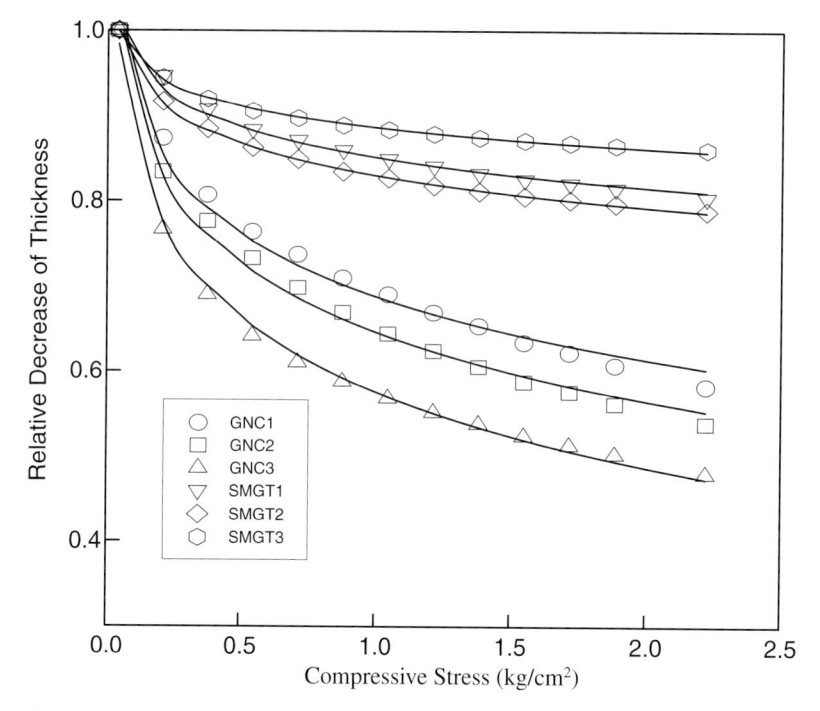

Figure 14.24 Relative decrease of thickness with compressive stress for smart composite geotextiles and geotextiles/geonet composites.

14.3.3.4 Thickness and In-Plane Permeability

The constants of Eq. (14.8), T_0, a_T, b_T, and correlation coefficient, R^2 for smart composite geotextiles and geotextiles/geonet composites were represented in Table 14.8 and Figure 14.25.

Table 14.8 Parameters to be related to thickness of smart composite geotextiles and geotextiles/geonet composites.

Geosynthetics for drainage	Coefficients to be related to thickness			
	T_0	a_T	b_T	R^2
GNC-1	3.4634	0.3585	0.1035	0.9853
GNC-2	4.3138	0.4946	0.1147	0.9923
GNC-3	2.7448	0.3564	0.1298	0.9964
SMGT 1	7.8463	0.3939	0.0502	0.9852
SMGT 2	9.8557	0.5285	0.0536	0.9992
SMGT 3	12.7900	0.4526	0.0354	0.9982

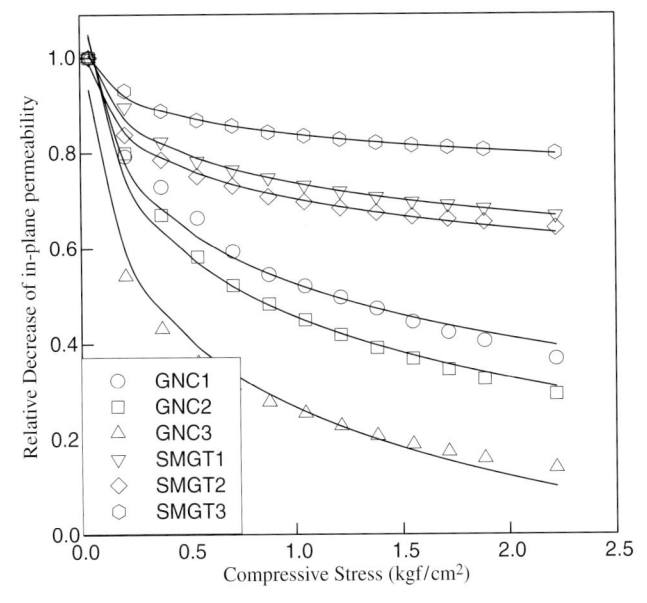

Figure 14.25 Relative decrease of in-plane permeability and compressive stress for smart composite geotextiles and geotextiles/geonet composites.

Transmissivity is a kind of parameter to determine the drainage properties of geotextiles and this is the function of the multiplication thickness by in-plane permeability of geotextiles. In-plane permeability of geotextiles to be derived from Eq. (14.8) is as follows:

$$K_p = K_0 - a_K \ln \frac{\sigma}{\sigma_0} = K_0 \left(1 - b_K \ln \frac{\sigma}{\sigma_0}\right) \quad (14.9)$$

$$b_K = a_K / K_0$$

where K_0 is the initial in-plane permeability, K_p is the in-plane permeability under confined loading, a_K and b_K are the variation constants of the geotextiles.

From Eqs. (14.2) and (14.9), transmissivity of geotextiles could be written as follows:

$$\begin{aligned} \theta &= T \times K_p \\ &= T_0 \left(1 - b \ln \frac{\sigma}{\sigma_0}\right) \times K_0 \left(1 - b_K \ln \frac{\sigma}{\sigma_0}\right) \\ &= (T_0 \times K_0) \cdot \left(1 - (b + b_K) \ln \frac{\sigma}{\sigma_0} + b \cdot b_K \ln^2 \frac{\sigma}{\sigma_0}\right) \\ &= \theta_0 \left(1 - (b + b_K) \ln \frac{\sigma}{\sigma_0} + b \cdot b_K \ln^2 \frac{\sigma}{\sigma_0}\right) \end{aligned} \quad (14.10)$$

where θ_0 and θ are the transmissivity with/without confined loading of the geotextiles, respectively.

For Eq. (14.10), the value of $(b \times b_K)$ is (0.02–0.03) and this value is smaller than $(b + b_K)$ (0.3–0.4). Therefore, the third term of Eq. (14.10) could be negligible to simplify this equation if the value of (σ/σ_0) is not larger than (0.02–0.03). Finally, transmissivity of geotextiles would be written as follows:

$$\theta = \theta_0 \left(1 - b_\theta \ln \frac{\sigma}{\sigma_0}\right) \tag{14.11}$$

where b_θ is the variation constant of the geotextile.

Figure 14.26 shows the relationship between transmissivity and compressive stress and solid line indicates the theoretical values of Eq. (14.10) in the condition of the initial compressive stress, $\sigma_0 = 0.04 \text{ kg/cm}^2$. Here the errors between experimental and theoretical values of transmissivities for GNC-1 were larger than those of the other materials. It means that the third term of Eq. (14.11) should not be negligible because of the larger (σ/σ_0) values. But the errors between experimental and theoretical values of transmissivities for GNC-1 will be smaller if the initial compressive stress is larger than 0.04 kg/cm^2. This means that the third term of Eq. (14.10) should be negligible and the initial compressive stress should be larger to be applied to Eq. (14.11) for the analysis of transmissivity of the geotextile. Table 14.9 shows the parameters to be related to in-plane permeability and transmissivity of smart composite geotextiles and geotextiles/geonet composites.

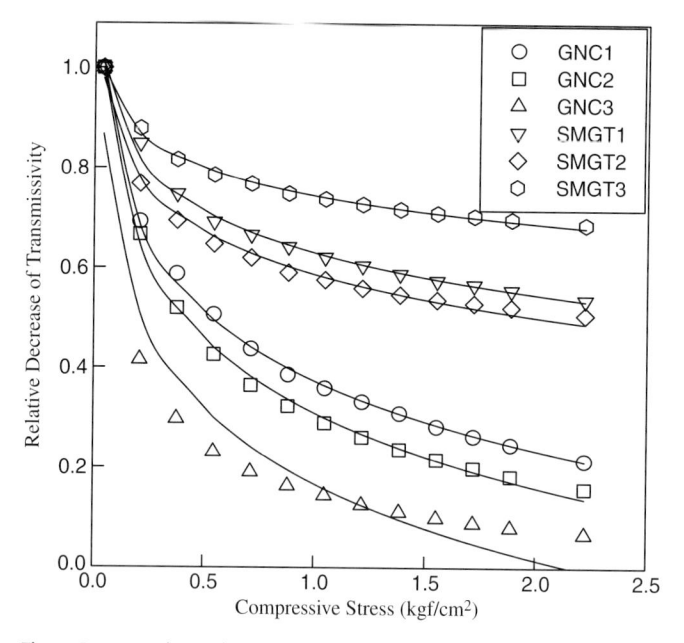

Figure 14.26 Relative decrease of transmissivity and compressive stress for smart composite geotextiles and geotextiles/geonet composites.

Table 14.9 Parameters to be related to in-plane permeability and transmissivity of composite geotextiles.

Samples	In-plane permeability				Transmissivity			
	K_0	a_K	b_K	R^2	θ_0	a_θ	b_θ	R^2
GNC-1	0.023	0.003	0.155	0.982	0.741	0.146	0.198	0.998
GNC-2	0.017	0.003	0.177	0.986	0.699	0.151	0.216	0.997
GNC-3	0.019	0.004	0.223	0.979	0.485	0.123	0.253	0.930
SMGT 1	0.026	0.002	0.085	0.993	2.023	0.239	0.118	0.994
SMGT 2	0.027	0.002	0.090	0.996	2.618	0.328	0.125	0.992
SMGT 3	0.028	0.001	0.052	0.996	3.581	0.286	0.080	0.997

The smart composite geotextiles that have the different structural compositions were manufactured by needle punching method. Variations of thickness and drainage properties such as in-plane permeability, transmissivity with compressive stress were evaluated between smart composite geotextiles and geotextiles/geonet composites by the constitutive equations. The variations of thickness with compressive stress of smart composite geotextiles were smaller than those of geotextiles/geonet composites. This is due to the difference of intrusion by compressive stress between smart composite geotextiles and geotextiles/geonet composites. The decrease of in-plane permeability and transmissivity with compressive stress of smart composite geotextiles showed the same tendency as the case of variations of thickness.

14.3.4
For Frictional Stability

Geomembranes that are used in waste landfills may be damaged by slippage between geosynthetics during installation period and this is the most serious cause of instability of waste landfills structure [3–5]. Here HDPE (high density polyethylene) geomembranes and PP (polypropylene) geotextiles were thermally bonded to make composites to improve frictional properties of geomembranes.

14.3.4.1 Manufacturing of Geotextiles/Geomembranes Composites
Figure 14.27 shows the photographs of surface and cross-sectional areas of geotextiles/geomembranes composites by thermal bonding.

14.3.4.2 Frictional Properties
Table 14.10 shows the interface friction angle between geomembranes and conducted soils and Table 14.9 shows the interface friction angle between geotextile/geomembrane composites and conducted soils, respectively. Figure 14.28 shows the shear stress versus horizontal displacement curves under relative density 40%, 80% of silt soils and Figure 14.29 shows the shear stress versus horizontal displacement

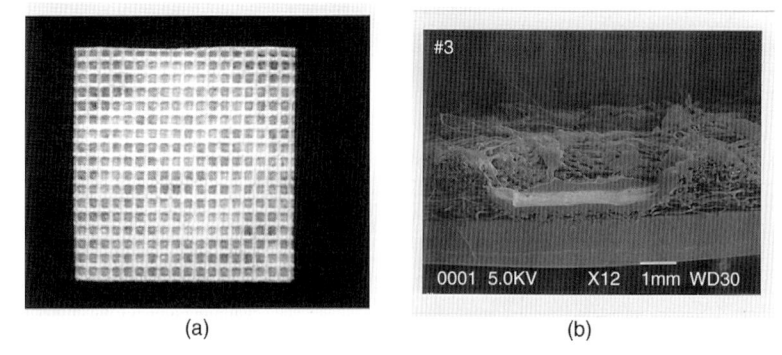

Figure 14.27 Photographs of geotextile/geomembrane composites: (a) surface area and (b) cross-sectional area.

curves under relative density 80%, 90% of weathered granite soils, respectively. As shown in these tables and figures, interface friction angle increased with shear stress and this is due to the convex and concave surface structures by spot shape thermal bonding between geotextiles and geomembranes. Also, interface friction angle increased with relative density for both silt and weathered granite soils. From Tables 14.10 and 14.11, it is seen that geomembranes and geotextile/geomembrane composites shows increase of interface friction angle with D_r in silt and weathered granite soils. Especially for geotextile/geomembrane composites, it is confirmed that outstanding improvement of frictional properties obtained through geotextiles complications.

14.4
Performance Evaluation of Composite Geotextiles

As introduced before, composite geotextiles act as important role in the soil-reinforced structure and therefore, evaluation of engineering performance is very important because durability and long-term behaviors of composite geotextiles are key factors for designing with composite geotextiles in soil reinforcement construction. ISO TC 221 and ASTM International D35 supply the standard test methods to specify and regulate for composite geotextiles. The following is the main list of required test items for composite geotextiles.

Table 14.10 Interface friction angles of geomembrane/conducted soils interface.

Conducted soils	Relative density	Interface friction angle, ψ (°)
Silt soil	$D_r = 40\%$	19.65
	$D_r = 80\%$	29.19
Weathered granite soil	90% of $Y_{d(max)}$	23.83

14.4 Performance Evaluation of Composite Geotextiles

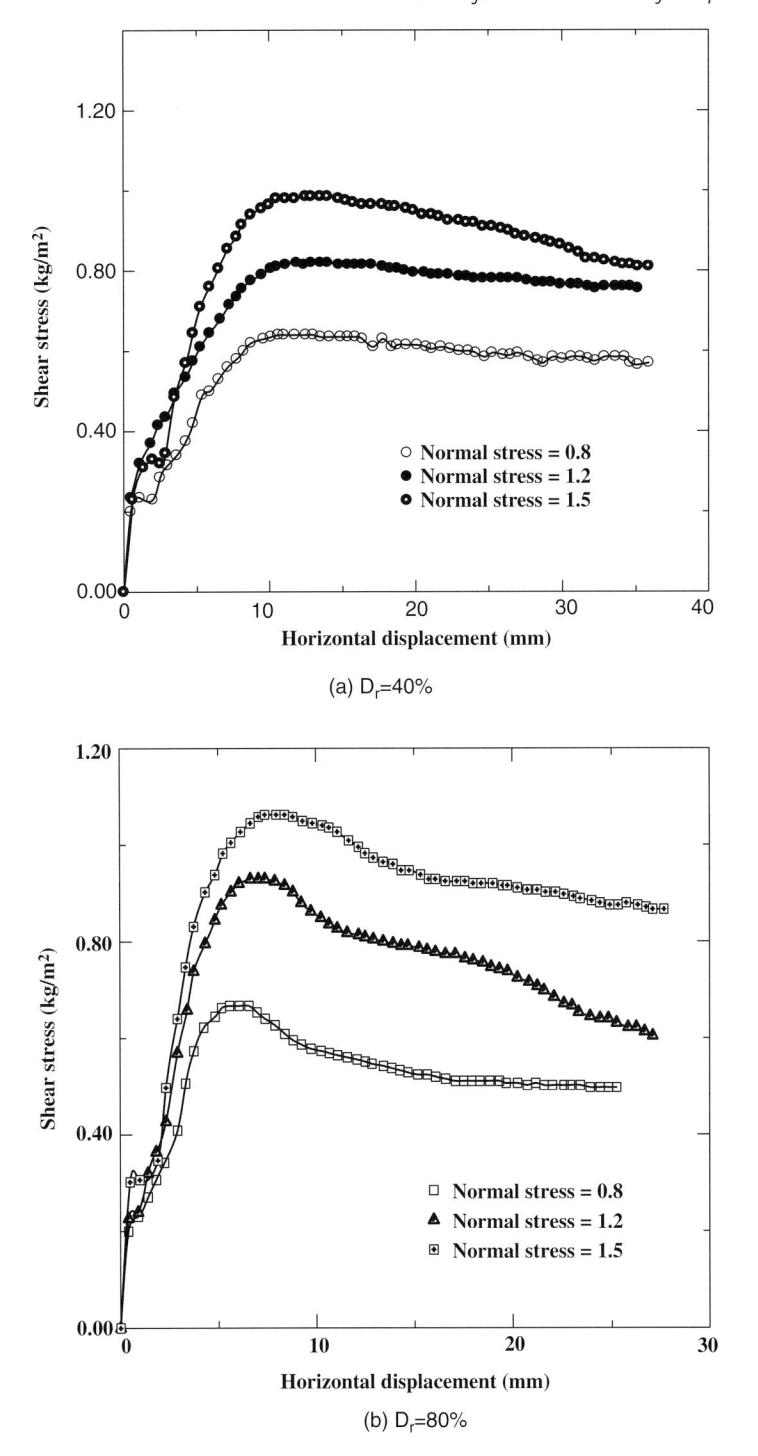

(a) D_r=40%

(b) D_r=80%

Figure 14.28 Results of direct shear test of geotextile/geomembrane composites with silt soil.

464 | *14 Polymer Composites as Geotextiles*

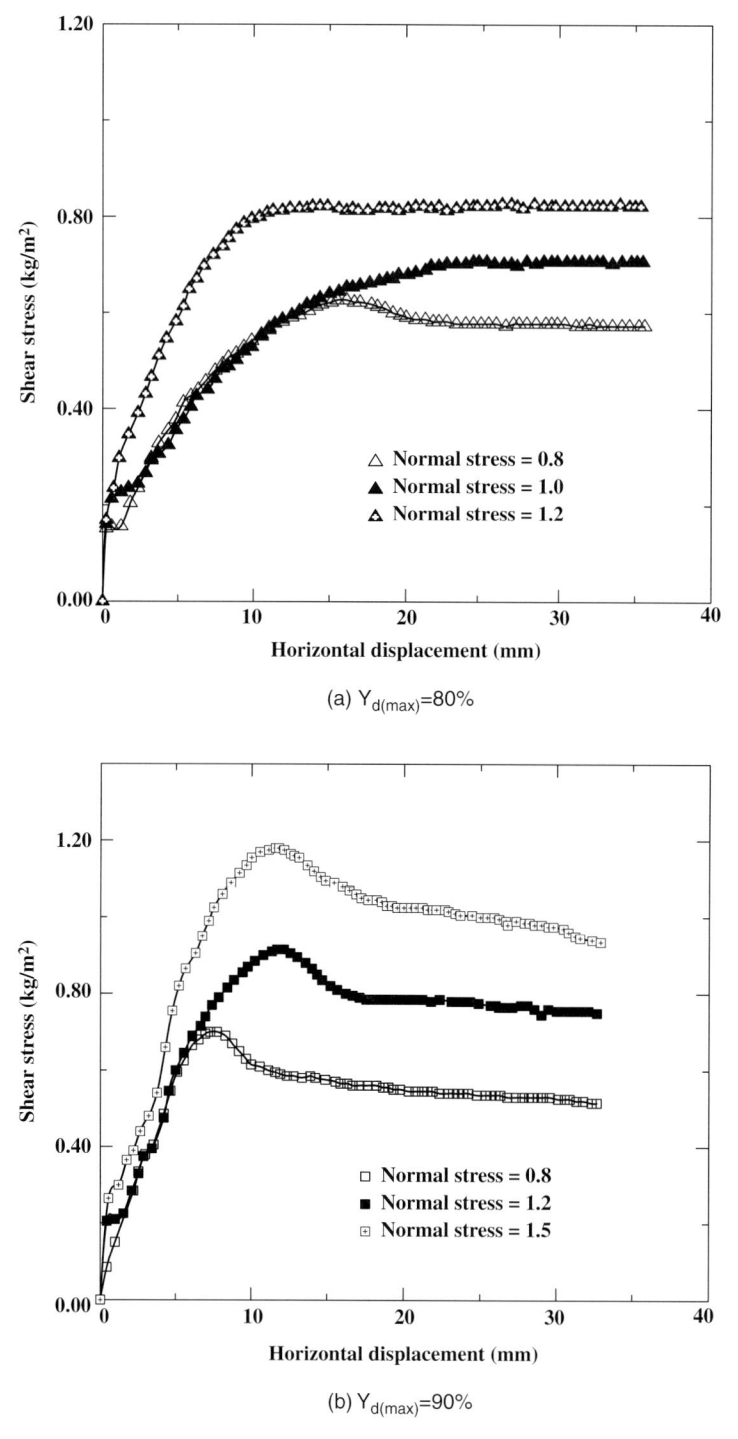

(a) $Y_{d(max)}=80\%$

(b) $Y_{d(max)}=90\%$

Figure 14.29 Results of direct shear test of geotextile/geomembrane composites with weathered granite soil.

Table 14.11 Interface friction angles of geotextile/geomembrane composites/conducted soils interface.

Conducted soils	Relative density	Interface friction angle, ψ (°)	Coherence, c (kg/cm²)
Silt soil	$D_r = 40\%$	29.536	0.257
	$D_r = 80\%$	31.728	0.241
Weathered granite soil	80% of $Y_{d(max)}$	27.480	0.254
	90% of $Y_{d(max)}$	36.793	0.183

14.4.1
Performance Test Items

1) Basic properties
   a) Weight and thickness
   b) Density
   c) Molecular weight, amounts of additives
   d) Melt flow index (MFI) and so on.

2) Mechanical properties
   a) Tensile, tear, bursting, puncture, impact, seam strength, and so on.
   b) Frictional properties – direct shear test, pull-out test, and so on.

3) Hydraulic properties
   a) Permittivity and transmissivity
   b) Porosity – AOS (apparent opening size), EOS (effective opening size)

4) Durability and endurance
   a) Creep and stress relaxation behaviors
   b) Failure property
   c) Clogging and so on.

5) Weatherability
   a) Sun light and ultraviolet resistance
   b) Chemical and biological resistance
   c) Thermal stability
   d) Environmental stress cracking resistance and so on.

14.4.2
Required Evaluation Test Items

14.4.2.1 Composite Geotextiles

1) Physical properties
   a) Weight and thickness
   b) Stiffness

2) Mechanical properties
   a) Wide-width tensile property
   b) Compressive property
   c) Tear, bursting, puncture, impact strength, and so on.
   d) Seam property
   e) Frictional properties – direct shear test, pull-out test, and so on.
   f) Failure property and so on.

3) Hydraulic properties
   a) Permittivity and transmissivity
   b) Porosity – AOS, EOS
   c) Soil retention and so on.

4) Durability
   a) Creep and stress relaxation behaviors
   b) Failure property
   c) Clogging and so on.

5) Weatherability
   a) Resistance to temperature, oxidation, hydrolysis, and so on.
   b) Chemical and biological resistance
   c) Resistance to radiation
   d) Sun light and ultraviolet resistance

14.4.2.2 Geosynthetics Clay Liners

1) Physical properties
   a) Clay type; bentonite property
   b) Moisture content
   c) Adhesives and cohesiveness, and so on.
   d) Moisture content
   e) Weight and thickness, and so on.

2) Mechanical properties
   a) Wide-width tensile property for geotextiles
   b) Seam property
   c) Puncture property
   d) Frictional properties – direct shear test, pull-out test, and so on.

3) Hydraulic properties
   a) Hydration (flux) in distilled water and field solution
   b) Free swell–swell index
   c) Moisture absorption
   d) Permeability – hydraulic conductivity, fluid loss, and so on.

4) Durability
   a) Freeze and thaw behavior

b) Shrink, swell behavior
c) Adsorption
d) Water breakout time and so on.

References

1. Koerner, R.M. (2005) *Designing with Geosynthetics*, 5th edn, Prentice-Hall Inc., New Jersey.
2. ASTM (1995) D-35: *ASTM Standards on Geosynthetics*, ASTM, West Conshohocken, PA.
3. IGS Educational Resource (2010) *Geosynthetics Functions*, International Geosynthetics Society.
4. Raymond, G.P. and Giroud, J.-P. (1993) *Geosynthetics Case Histories*, ISSMGE Technical Committee TC9 Geotextiles, and Geosynthetics, USA.
5. de Groot, M.B., den Hoedt, G., and Termatt, R.J. (1996) *Geosynthetics: Applications, Design and Construction*, A. A. BALKEMA, Nertherlands.
6. GRI (2007) *GRI Standard Test Methods on Geosynthetics*, Drexel University, Philadelphia, PA.
7. Baker, T.L. (1997) *Proceedings of '97 Geosynthetics Conference*, **3**, 829.
8. Koerner, G.R., Hsuan, G.Y., and Koerner, R.M. (1998) *Journal of Geotechnical and Geoenvironmental Engineering*, **124**, 1.
9. Artières, O., Gaunet, S., and Bloquet, C. (1997) *Geosynthetics International*, **4**, 393.
10. Salman, A. et al. (1997) *Proceedings of '97 Geosynthetics Conference*, **1**, 217.
11. IFAI (2008) *Geotechnical Fabrics Report – Specifier's Guide 2008*, Roseville, MN, USA.
12. Geosynthetic, Institute (2010) Proceedings of the GRI-21 Conference, Philadelphia.
13. IFAI (2009) *Geotechnical Fabrics Report – Specifier's Guide 2010*, Roseville, MN, USA.
14. Kent, K. (2001) Erosion control and geosynthetic and geosynthetic methods on venues for the 2002 Salt Lake City Olympic winter games. Geosynthetics Conference 2001, Portland, Oregon, pp. 305–315.
15. Leu, W. and Luane Tasa, P.E. (2001) Applications of geotextiles, geogrids, and geocells in northern Minnesota. Geosynthetics Conference 2001, Portland, Oregon, pp. 809–821.
16. Jenner, C.G. and Paul, J. (2000) Lessons learned from 20 years experience of geosynthetic reinforcement on pavement foundations. EuroGeo 2000 Conference Proceedings, Bologna, Italy.
17. Koerner, R.M. and Soong, T.-Y. (2000) Geosynthetic reinforced segmental retaining walls. 14th GRI Conference Proceedings, Las Vegas, Nevada, pp. 268–297.
18. Feodorov, V. (2000) Use of geosynthetics in landfills. Case studies from Romania. EuroGeo 2000 Conference Proceedings, Bologna, Italy, pp. 509–513.
19. Artieres, O. and Delmas, P. (2000) Examples of river band and coastal protection by new two-layer filtration system. EuroGeo 2000 Conference Proceedings, Bologna, Italy, pp. 613–618.
20. FHWA (1989) Geotextile Design and Construction Guidelines, U.S. Department of Transportation Federal Highway Administration, Publication No. FHWA HI-90-001, pp. 24–46.
21. Holtzs, R.D., Christopher, B.R., and Berg, R.R. (1995) Geosynthetic Design and Construction Guidelines, U.S. Department of Transportation Federal Highway Administration, Publication No. FHWA HI-95-038, pp. 27–105.
22. Koerner, R.M. (1990) *Geosynthetic Testing for Waste Containment Application*, STP 1081, ASTM, Philadelphia, pp. 257–272.
23. Inglod, T.S. (1994) *The Geotextiles and Geomembranes Manual*, Elsevier Advanced Technology, Oxford, pp. 1–66.
24. Frobel, R.K. and Taylor, R.T. (1991) *Geotextiles and Geomembranes*, **10**, 443.
25. Takasumi, D.L. et al. (1991) *Proceedings of Geosynthetics '91 Conference*, **1**, 87.

26 Koutsourais, M.M. *et al.* (1991) *Geotextiles and Geomembranes*, **10**, 531.
27 EPA Test Method 9090, "Compatibility Test For Waste and Membrane Liners", Technical Resource Document SW-846, Test Methods for Evaluating Solid Wastes, 3rd Ed., U.S. Environmental Protection Agency, Washington, D.C., 1986.
28 Koerner, R.M. (1990) *Geosynthetic Testing for Waste Containment Application*, ASTM, Philadelphia, pp. 25–54.

15
Hybrid Textile Polymer Composites
Palanisamy Sivasubramanian, Laly A. Pothan, M. Thiruchitrambalam, and Sabu Thomas

15.1
Introduction

The use of composite materials dates back to centuries while the use of natural fibers dates as far back as 3000 years ago to ancient Egypt, when clay was reinforced with straw to build walls [1, 2]. With time, the interest in natural fibers waned and other more durable construction materials such as metals were introduced. Recent developments in the use of natural fibers, such as hemp, flax, jute, sisal, coir, banana, and so on, in polypropylene matrix composites have shown that it is possible to obtain materials that perform well, using environment friendly materials [3]. In the past few years, natural fibers have been tested as an alternative to glass fibers since they satisfy essential requirements, such as, easy availability, low cost, and comparable properties with respect to traditional fibers. In addition, they have low environmental impact [4]. The main disadvantage in using natural fibers is related to their scattering of geometrical properties together with their water absorption and their low processing temperatures. As working temperatures cannot exceed 200 °C, the types of thermoplastic matrices that can be used to obtain composite materials are limited. Although surface treatments may have a negative impact on the cost thereof, they may improve compatibility and strengthen the interface in natural fiber composite materials [5].

The incorporation of several different types of fibers into a single matrix has led to the development of hybrid biocomposites. The behavior of hybrid composites is a weighed sum of the individual components in which there is a more favorable balance between the inherent advantages and disadvantages. While using a hybrid composite that contains two or more types of fiber, the advantages of one type of fiber could complement with what are lacking in the other. As a consequence, a balance in cost and performance can be achieved through proper material design [6]. The properties of a hybrid composite mainly depend upon the fiber content, length of individual fibers, orientation, extent of intermingling of fibers, fiber to matrix bonding, and arrangement of both the fibers. The strength of the hybrid composite is also dependent on the failure strain of individual fibers. Maximum properties of hybrid composites result when the fibers are highly strain compatible [7]. Textile

Polymer Composites: Volume 1, First Edition. Edited by Sabu Thomas, Kuruvilla Joseph,
Sant Kumar Malhotra, Koichi Goda, and Meyyarappallil Sadasivan Sreekala
© 2012 Wiley-VCH Verlag GmbH & Co. KGaA. Published 2012 by Wiley-VCH Verlag GmbH & Co. KGaA.

composites are used typically because of their high strength-to-weight and stiffness-to-weight ratios. In order to exploit these properties it is essential to have a good understanding of their behavior under load. It is normal to assume that composites (especially reinforced thermosets) are linearly elastic up to the point of failure, although there is some evidence that even thermoset matrix materials can yield in shear. Moreover, there is now increasing interest on the degradation in properties of a structure as damage progresses. However, in order to be able to reach that stage, an accurate understanding of the structure's elastic behavior must be established.

The mechanical behavior of textile reinforcements is studied, with the primary aim of understanding their behavior during formation and consolidation processes. A number of deformation mechanisms are available. However, during typical composites manufacturing processes, it is generally agreed that the most important of these are in-plane shear and tensile behavior and through-thickness compaction. Of these mechanisms, the ability of fabrics to shear in-plane is their most important feature during formation, although their low shear stiffness, in-plane tensile behavior, all represent largest source of energy dissipation. Compaction behavior defines the fiber volume fraction that can be obtained after manufacturing. Other properties such as fabric bending and ply/tool friction are not considered here, primarily because these have received relatively little attention elsewhere and little data are available. These methods have been developed within research studies, usually to obtain material data for manufacturing simulation. The geometry of textile reinforcements can be described at a number of length scales. Individual fibers represent the microscopic scale, with large numbers of fibers (typically several thousand) making up the tow or yarn. The scale of the yarns and of the fabric repeating unit cell is the mesoscopic scale. Finally, the fabric structure constitutes the macroscopic scale. The macroscopic mechanical behavior of fabrics depends on phenomena at smaller scales, and in particular it is dependent on geometric and contact nonlinearities.

15.2
Textile Composites

Fibers in textile forms are used in composites to derive advantages from textile structures such as better dimensional stability, subtle conformability, and deep draw moldability/shapeability. Although considerable insight into the thermomechanical behavior of unidirectional composites (UD) has been gained, such studies on textile composites are limited [8].

Woven fabric composite material represents a type of textile composite where the process of weaving forms strands; these strands are interlaced in two mutually orthogonal directions to one another and impregnated with a resin material. A growing interest in textile composites has been observed in recent years. Woven fabric-reinforced composites are the most widely used form of textile structural reinforcement. Their use in the fabrication of structures with high mechanical performance is increasing in the field of aeronautics, naval construction, and automobile engineering [9]. Woven fabrics are attractive as reinforcement since they provide excellent integrity and conformability for advanced structural composite applications. The major driving force for the increased use of woven fabrics, compared to their nonwoven counterparts, are excellent drapeability (allowing complex shapes to be formed), reduced manufacturing costs (e.g., a single two dimensional biaxial fabric replaces two nonwoven plies) [10] and increased resistance to impact damage (improved compressive strengths after impact follow from a reduction in the area of impact damage) [11]. These woven fabric composite materials have easier handling in production quality. The ultimate properties of these composites depend on many parameters such as fiber yarn and matrix properties, weaving architecture, yarn spacing and thickness, overall fiber volume fraction, and so on. The varieties of manufacturing methods have made textile composites cost-competitive with unidirectional properties. Woven fabric composites provide more balanced properties in the fabric plane than unidirectional composites. As a consequence of the complex architecture of the textile fabrics, many parameters such as fiber yarn and matrix properties, weaving/braiding architectures, yarn spacing, and so on influence the mechanical performance of fabric composites [12].

Shin and Jang [13] reported on the important role in delamination resistance of woven fabric composites. Gommers et al. [14] investigated the application of Mori–Tanaka method for the calculation of the elastic properties of textile composite materials. Attempts have been made by various researchers to make use of natural fibers in different forms (short fibers, long fiber, and woven form) to make composite materials [15–19].

Textile composites are composed of textile reinforcements combined with a binding matrix (usually polymeric). This describes a large family of materials used for load-bearing applications within a number of industrial sectors. The term textile is used here to describe an interlaced structure consisting of yarns, although it also applies to fibers, filaments, and yarns, and most products derived from them. Textile manufacturing processes have been developed over hundreds or even thousands of years. Modern machinery for processes such as weaving, knitting, and braiding operates under automated control, and is capable of delivering high-quality materials at production rates of up to several hundreds of kilograms per hour. Some of these processes (notably braiding) can produce reinforcements directly in the shape of the final component. Hence such materials can provide an extremely attractive reinforcement medium for polymer composites. Textile composites are attracting growing interest from both the academic community and the industry. This family of materials, at the center of the cost and performance spectra, offers significant opportunities for new applications of polymer composites. Although the reasons for adopting a particular material can be various and complex, the primary driver for the use of textile reinforcements is undoubtedly cost. Textiles can be produced in large quantities at reasonable cost using modern, automated manufacturing techniques. While direct use of fibers or yarns might be cheaper in terms of materials costs, such materials are difficult to handle and to form into complex component shapes. Textile-based materials offer a good balance in terms of the cost of raw materials and ease of manufacture. Target application areas for textile composites are primarily within the aerospace, marine, and defense; land transportation, construction, and power generation sectors. As an example, thermoset composites based on 2D braided preforms have been used by Dowty Propellers in the United Kingdom since 1987 [20]. Here a polyurethane foam core is combined with glass and carbon fiber fabrics, with the whole assembly over-braided with carbon and glass tows. The resulting preform is then impregnated with a liquid thermosetting polymer via resin transfer molding (RTM). Compared with conventional materials, the use of textile composites in this application results in reduced weight, cost savings (both initial cost and cost of ownership), damage tolerance, and improved performance via the ability to optimize component shape. A number of structures for the Airbus A380 passenger aircraft rely on textile composites, including the 6 m diameter dome-shaped pressure bulkhead and wing trailing edge panels, both manufactured by resin film infusion (RFI) with carbon noncrimp fabrics, wing stiffeners and spars made by RTM, the vertical tail plane spar by vacuum infusion (VI), and thermoplastic composite (glass/poly(phenylene sulfide)) wing leading edges. Probably the largest components produced are for off-shore wind power generation, with turbine blades of up to 60 m in length being produced using (typically) noncrimp glass or carbon fabric reinforcements impregnated via vacuum infusion. Other application areas include construction, for example, in composite bridges that offer significant cost savings for installation due to their low weight. Membrane structures, such as that used for the critically acclaimed (in architectural terms) Millennium Dome at Greenwich, UK, are also a form of textile composite. Numerous automotive applications exist, primarily for niche or high-performance vehicles and also in impact structures such as woven glass/polypropylene bumper beams. The term "textile composites" is used often to describe a rather narrow range of materials, based on three dimensional reinforcements produced using specialist equipment. Such materials are extremely interesting to researchers and manufacturers of very high-performance components (e.g., space transportation); an excellent overview is provided by Miravette [21].

Pothan et al. [22] studied the static and dynamic mechanical properties of banana and glass woven fabric-reinforced polyester composites with special reference to the effect of fiber volume fraction, layering pattern, and the weaving architecture. Composites with high tensile strength could be obtained using banana and glass in the fabric form using two layers of the fabric. The impact strength of the composites increased with the number of layers and the fiber volume fraction. By the incorporation of fabric, the G' curve showed an improved rubbery plateau indicating that the incorporation of the fabric induces reinforcing effects in the polyester matrix. The storage modulus was found the highest for composites with four layers of the fabric. In all the composite samples, the relaxation peak of polyester was visible as peak in G'' at about 98 °C. In the case of composites with four layers of the fabric, two peaks and one shoulder were seen. Increase in the number of layers made a second relaxation peak visible. The damping peaks were found to be lowered by the incorporation of more number of layers.

Researchers have looked into tensile strength of ramie–cotton hybrid fiber textile-reinforced polyester composites [23]. The plain woven fabrics had ramie strings on the warp and cotton strings on the weft. Four kinds of fabrics with varying amounts of fibers were manufactured. The fibers used in this work were threads classified as 2.70

Table 15.1 Configuration of ramie–cotton fabrics.

Fabrics	Ramie thread	Volume fraction of ramie fibers (%)
I	2.70	52
II	2.70	56
III	3.10	72
IV	3.10	83

and 3.10. This classification stems from the textile industry, and the first figures refer to the number of filaments per thread. Therefore, the designation 2.70 means that two filaments of ramie are twisted along the thread length, and so on. The practical result for this work is that thread 3.10 is slightly thicker than thread 2.70. The details for these fabrics are shown in Table 15.1. Composites with varying volume fraction and orientation of ramie fibers were made by compression molding. The results obtained are summarized. It is observed that tensile behavior was dominated by volume fraction of ramie fibers aligned in the test direction. The fabric and diameter of the thread did not play any role in tensile characteristics. Cotton fabric was found to have minor reinforcement effect due to weak cotton/polyester interface. Similar studies were performed by Mwaikambo and Bisanda [24] on kapok–cotton fiber-reinforced polyester composites. Novolac-type phenolic composites reinforced with jute/cotton hybrid woven fabrics were fabricated and their properties were investigated as a function of fiber orientation and roving/fabric characteristics [25]. Results showed that the composite properties were strongly influenced by test direction and roving/fabric characteristics. The anisotropy degree was shown to increase with test angle and strongly depend on the type of architecture of fabric used, that is, jute roving diameter, relative fiber content, and so on. The best overall mechanical properties were obtained for the composites tested along the jute roving direction. Composites tested at 45° and 90° with respect to the jute roving direction exhibited a controlled brittle failure combined with a successive fiber pull-out, while those tested in the longitudinal direction (0°) exhibited a catastrophic failure mode. The researchers are of the opinion that jute fiber promotes a higher reinforcing effect and cotton fiber avoids catastrophic failure. Therefore, this combination of natural fibers is suitable to produce composites for lightweight structural applications. The thermal diffusivity, thermal conductivity, and specific heat of jute/cotton, sisal/cotton, and ramie/cotton hybrid fabric-reinforced unsaturated polyester composites were investigated by Alsina *et al.* [26]. These properties were measured both parallel and perpendicular to the plane of the fabrics. Thermal properties of hybrid fabrics, composites with as-received fabrics and composites with predried fabrics were studied. The results obtained are shown in Table 15.2. It can be seen that higher values were obtained parallel to the plane of the fibers. Sisal/cotton composites showed a particular behavior, with thermal properties very close to those of the resin matrix. The thermal properties of the fabrics, that is, without any resin, were also evaluated and were used to predict the properties of the composites from the theoretical series and parallel model equations. The effect of fabric predrying on the thermal properties of the composites was also evaluated. The

Table 15.2 Thermal properties of lignocellulosic fabrics and their composites.

Materials	Direction of the heat flux	Thermal properties		
		Specific heat (J/(mol × K))	Thermal diffusivity (mm^{2}s^{-1})	Thermal conductivity (Wm^{-1}°C^{-1})
Resin	—	0.987 (0.002)*	0.153 (0.0004)	0.15
Sisal/cotton				
fabrics	Parallel	1.037 (0.003)	0.178 (0.001)	0.185 (0.005)
	Perpendicular	0.94 (0.02)	0.20	0.19
Composites^1	Parallel	1.236 (0.007)	0.200 (0.0006)	0.25
	Perpendicular	1.065 (0.021)	0.194 (0.002)	0.213 (0.006)
Composites^2	Parallel	1.194 (0.006)	0.203 (0.001)	0.24
	Perpendicular	1.553 (0.071)	0.132 (0.004)	0.205 (0.007)
Ramie/cotton				
Fabrics	Parallel	1.128 (0.008)	0.510 (0.004)	0.575 (0.005)
	Perpendicular	0.640 (0.01)	0.648 (0.002)	0.415 (0.006)
Composites^1	Parallel	0.894 (0.005)	0.251 (0.002)	0.22
	Perpendicular	0.839 (0.01)	0.220 (0.0007)	0.19
Composites^2	Parallel	1.467 (0.019)	0.164 (0.002)	0.24
	Perpendicular	0.861 (0.014)	0.218 (0.002)	0.19
Jute/cotton				
Fabrics	Parallel	1.068 (0.076)	0.524 (0.037)	0.555 (0.006)
	Perpendicular	0.536 (0.04)	0.677 (0.006)	0.36 (0.02)
Composites^1	Parallel	1.017 (0.017)	0.231 (0.003)	0.237 (0.006)
	Perpendicular	0.869 (0.015)	0.218 (0.003)	0.19
Composites^2	Parallel	0.793 (0.027)	0.252 (0.0007)	0.20
	Perpendicular	1.032 (0.005)	0.192 (0.0007)	0.20

* Standard deviations
Coimposites^1 is made by as- received fabrics
Composite^2 is made by pre-dried fabrics

results showed that the drying procedure used did not bring any relevant change in the properties evaluated.

Developments in textile technologies such as weaving, knitting, and braiding have resulted in the formation of textile composites that have superior mechanical properties. Woven fabrics are attractive as reinforcements since they provide excellent integrity and conformability for advanced structural applications. The driving force for the increased use of woven fabrics compared to their nonwoven counterparts are excellent drapeability, reduced manufacturing costs, and increased mechanical properties, especially the interlaminar or interfacial strength. The interconnectivity between adjacent fibers in the textile reinforcement provides additional interfacial strength to supplement the relatively weak fiber–resin interface. The nondelamination characteristics of three dimensional braided composites under ballistic impact also make them possess considerable potential in ballistic protection applications. Formation of different textile preforms is an important stage in composite technology. Plain, twill, satin, basket, leno, and mock leno are some

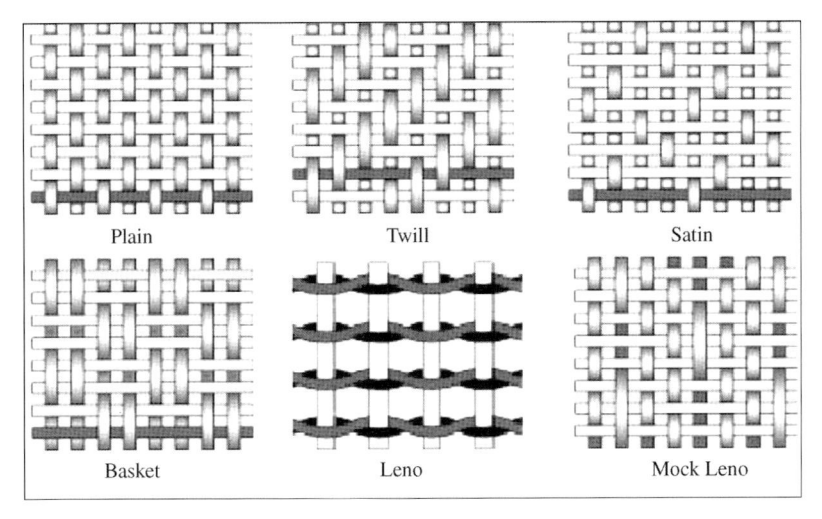

Figure 15.1 Some typical woven styles used as reinforcements in making composites.

commonly used woven patterns for the textiles, which are employed as reinforcements in making composites. The woven characteristics are shown in Figure 15.1 and their respective properties are summarized in Table 15.3. Textile structural composites are finding use in various high-performance applications recently. Among them, three dimensional fiber-reinforced polymer composites made by the textile processes of weaving, braiding, stitching, and knitting were found to have tremendous potential for improving the performance of composite structures and reducing their cost of manufacture. The current applications of three dimensional composites, including examples in aerospace, maritime, automotive, civil infrastructure, and biomedical fields are also enumerated.

15.2.1
Manufacture of Natural Fiber Textile-Reinforced Composites

Compression molding is a conventional and simple method to make fiber-reinforced composites. The prepreg is laid inside the mold by hand or robot and then hot pressed

Table 15.3 Properties comparison of the fabrics with different woven styles.

Property	Plain	Twill	Satin	Basket	Leno	Mock Leno
Stability	****	***	**	**	*****	***
Drape	**	****	*****	***	*	**
Porosity	***	****	*****	**	*	***
Smoothness	**	***	*****	**	*	**
Balance	****	****	**	****	**	****
Symmetrical	*****	***	*	***	*	****
Crimp	**	***	*****	**	**/*****	**

***** = excellent, **** = good, *** = acceptable, ** = poor, * = very poor.

at a certain pressure by a compression molding machine. With the aid of pressure and heat, the resin immerses into the reinforcement and cures inside the mold. Because the whole process involves a lot of human's efforts and controlling parameters, the quality of the final product is scattered a lot and the working efficiency is quite low. But it is still a popular method to make fiber-reinforced polymer composites due to its extremely flexible nature, capable of making a wide variety of shapes.

RTM technology refers to a group of processes that inject resin into a fiber preform captured in a closed tool. Recent advances in textile processes have produced a variety of options for making very complex fiber preform by automatic means. Research interest has focused intently on the mold filling problem that is critical to the success of RTM. RTM appears to be best suited for medium volume, small-to-medium sized complex parts. One of the major attractions of using natural fibers as reinforcements in making composites is their low cost. Finding an economic processing method to manufacture natural fiber-reinforced composites is a key factor for the successful application of this kind of material. Based on the above discussion, it seems that both compression molding and RTM could be suitable for making natural textile-reinforced polymer composites due to their low cost. Indeed, some research groups have reported making natural textile composites by these two processing techniques [27, 28]. Li also compared the mechanical properties of plain woven sisal textile-reinforced vinyl ester composites made by these two methods. The quality of the final products was examined with the aid of optimal microscopy [29]. Figure 15.2 shows the tensile strength, flexural strength, and impact energy of permanganate-treated sisal textile-reinforced vinyl ester composites made by RTM and compression molding, respectively. It can be seen that composites made by RTM possessed higher tensile strength, flexural strength, and Charpy impact energy than those of the composites made by compression molding. Sisal textile used in the study shows a relatively loose woven pattern and larger fiber diameter compared to man-made fibers, such as glass or carbon fibers. Therefore, much air can be trapped inside the sisal fiber bundles. As we know, the void content of the composites is a factor that affects their mechanical performances. High void content could reduce the mechanical properties of the composites. During RTM process both the injection pressure and the vacuum could work together to draw the resin penetrating the reinforcements. With the flowing of

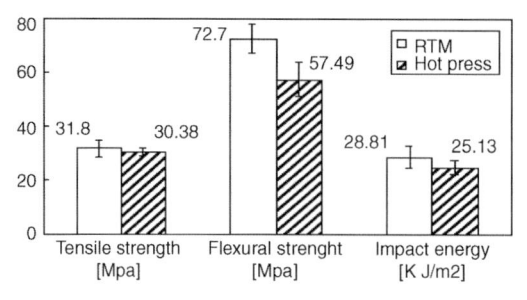

Figure 15.2 Comparisons of mechanical properties of sisal textile-reinforced composites made by compression molding and RTM, respectively.

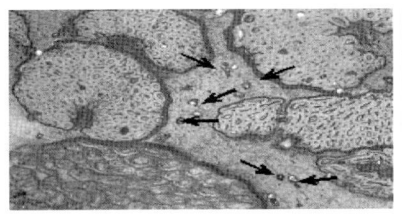

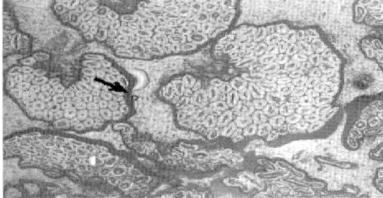

Figure 15.3 Cross-sections of sisal fiber-reinforced vinyl ester composites made by (a) compression molding, (b) RTM. (The arrows indicate the voids/bubbles.)

the resin through the reinforcement, the air which was trapped inside sisal textile could be driven out through the vent. So the void content of the composites can be reduced and good mechanical properties can be expected. Compression molding, however, drives the air out by compressive pressure between the two heating plates. The air bubbles caused by the matrix polymerization could be quenched by the applied pressure. But the small air bubbles inside the loose sisal bundles are hard to be driven out. The relatively higher void content led to the lower mechanical properties of sisal textile-reinforced vinyl ester made by compression molding. Microstructure analysis did indicate the presence of higher void in composites made by compression molding than those made by RTM (Figure 15.3). The above conclusions drawn from the study of sisal textile-reinforced composites would be different if other kinds of natural fibers and woven styles were used.

Swelling index and solubility parameter, S was higher for textile composite than neat Ecoflex. This was due to the more interaction of neat Ecoflex and solvent. High penetrating power of hydrocarbon contributed to the higher sorption of neat Ecoflex. But in Ecoflex/ramie mat composite, the hydrocarbon in diesel diffused into the interfaces and voids, if any, in the composite and the fibers restricted the movement of diesel. Hence, percentage swelling index and solubility parameter decreases in this case. The decreased amount of matrix due to fabric content and the good fiber matrix interaction and effective binding of fiber in the composite contributed to the lower diesel uptake in the textile composite sample. Studies on the diesel uptake of natural fibers (isora) reinforced natural rubber composite were also reported elsewhere. The study reported the effect of natural fiber content upon the sorption behavior. They reported that the percentage-swelling index and swelling coefficient of the composite were found to decrease with increase in fiber content. This was due to the increased hindrance exerted by the fibers at higher fiber content and also due to the good interactions between fibers and rubber [30].

Silva et al. studied the fracture toughness of natural fibers reinforced castor oil polyurethane composites by conducting compact tension test [31]. Short sisal fiber, coconut fiber, and sisal fabrics were used to make the composites. Alkaline was selected to treat natural fiber surfaces. Note that the volume fraction of the treated fabric composites was higher than the corresponding untreated ones. This was due to the fabric shrinkage during the drying stage of the alkaline treatment that promoted a more closed woven and consequently larger volume fraction. From the results, it can be seen that the best fracture toughness performance was displayed by the sisal fabric composite, which clearly indicated that natural fiber-reinforced composites in textile form would possess better fracture properties than the composites made by short fibers. The alkaline treatment showed to be harmful for fracture toughness of sisal fiber composites since the improved interfacial adhesion impaired the main energy absorption mechanisms. On the other hand, an enhancement on the fracture toughness of coconut fiber composites was observed, which was caused by the fibrillation process occurring under the severest condition of the alkaline treatment, which created additional fracture mechanisms. We can take banana fiber textile manufacturing as an example. Sapuan et al. summarized the manufacturing of banana fiber textile in the following three steps [32]. It is easy to understand the large variations of properties of natural fiber nonwoven mats reinforced composites due to the normally nonuniform distribution of natural fibers in the composites. Therefore, it is difficult to describe the properties of natural fiber nonwoven mats reinforced composites accurately. Kaveline et al. proposed a method that could quantitatively estimate the properties of this kind of composites quite well [33].

We have considered here very simple textile reinforcement. It is a periodic glass plain weave, which is balanced since the warp and the weft yarns have identical geometrical and mechanical properties. Its geometry is based on circle arcs and tangent segments. It is simple but ensures consistency of the model, which means that yarns do not penetrate each other [34].

15.3
Hybrid Textile Composites

Novolac-type phenolic composites reinforced with jute/cotton hybrid woven fabrics were fabricated and its properties were investigated as a function of fiber orientation and roving/fabric characteristics [35]. Results showed that the composite properties were strongly influenced by test direction and roving/fabric characteristics. The anisotropy degree was shown to increase with test angle and strongly depend on the type of architecture of fabric used, that is, jute rovings diameter, relative fiber content, and so on. The best overall mechanical properties were obtained for the composites tested along the jute rovings direction. Composites tested at 45° and 90° with respect to the jute roving direction exhibited a controlled brittle failure combined with a successive fiber pull-out, while those tested in the longitudinal direction (0°) exhibited a catastrophic failure mode. The researchers are of the opinion that jute fiber promotes a higher reinforcing effect and cotton fiber avoids catastrophic failure. Therefore, this combination of natural fibers is suitable to produce composites for lightweight structural applications.

Advanced three dimensional textile-reinforced polymers are predestined for the use in impact and crash relevant structures due to their high-specific mechanical properties and the adjustable energy absorption capacity. For the reliable design of highly dynamic loaded components, a deep knowledge of the strain rate dependent deformation and failure behavior and realistic simulation models are necessary, but absent for the new group of textile-reinforced materials [36–39].

Researchers [40] developed a new low dielectric constant material suited to electronic materials applications using hollow keratin fibers and chemically modified soybean oil. The unusual low value of dielectric constant was due to the air present in the hollow microcrystalline keratin fibers and the triglyceride molecules. The authors are of the opinion that the low cost composite made from avian sources and plant oil has the potential to replace the dielectrics in microchips and circuit boards in the ever-growing electronics materials field. In an extension of the above study the authors have also observed that the coefficient of thermal expansion (CTE) of the composite was low enough for electronic applications and similar to the value of silicon materials or polyimides used in printed circuit boards [41].

In the automotive industry [42], cotton fibers embedded in polyester matrix was used in the body of the East German "Trabant" car. The use of flax fibers in car disk brakes to replace asbestos fibers is also another example. Daimler–Benz has been exploring the idea of replacing glass fibers with natural fibers in automotive components since 1991. A subsidiary of the company, Mercedes Benz pioneered this concept with the "Beleem project" based in Sao Paolo, Brazil. In this case, coconut fibers were used in the commercial vehicles over a 9 year period. Mercedes also used jute-based door panels in its E-class vehicles in 1996. In September 2000, Daimler Chrysler began using natural fibers for their vehicle production. The bast fibers are primarily used in automotive applications because they exhibit greatest strength. The other advantages of using bast fibers in the automotive industry include weight savings of between 10 and 30% and corresponding cost savings. Recent studies have also indicated that hemp-based natural fiber mat thermoplastics are promising candidates in automotive applications where high specific stiffness is required [43].

15.4
Hybrid Textile Joints

Particularly with regard to the on-site installation and handling of prefabricated components made of textile-reinforced concrete, hybrid joints are tested in addition to bonded plane joints. For the test specimens, the same basic geometry as already established for the adhesive joint tests is used. Hybrid joints combine the advantages of an easy assembly and disassembly on the one hand, and the two dimensional transmission of force on the other hand [44, 45]. This joining geometry involves so-called "collar jackets" that are inserted into subsequently drilled holes as shown in Figure 15.4. The collar flange surfaces are bonded to the concrete surfaces by rigid epoxy resin adhesives.

Compared to punctiform bolts or rivet joints, the application of collar jackets leads to a further increase of the tensile capacity. The collar of the jacket reduces the tension stresses concentrated in the region next to the hole, perpendicular to the loading direction. Due to its high stiffness, the jacket distributes these stresses two

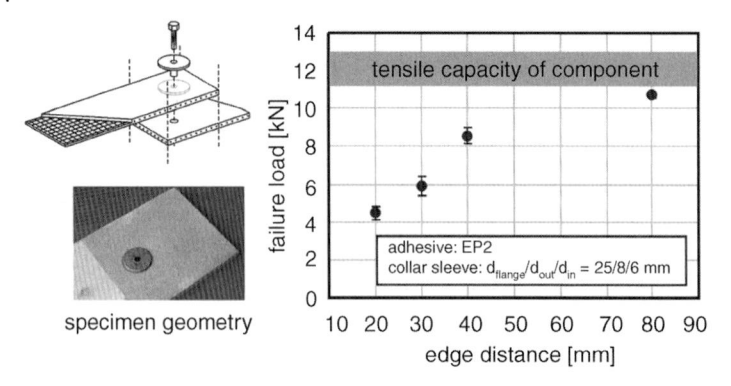

Figure 15.4 Tensile capacity of hybrid joints with pasted-in "collar jackets."

dimensionally in a homogeneous way through the high-modulus adhesive to a larger area of concrete.

15.5
Conclusion

Textile-reinforced composites based on natural fibers have been studied by many research groups in recent years due to their good mechanical performances, easy to handle, excellent integrity, and reduced manufacturing cost. It can be concluded that properties of the composites made by fiber textiles are better than the composites made by short fibers. Permeability, mechanical, and fracture properties are all affected by the weaving architectures of the reinforcing fabrics. Keeping up with the developments in technology, it is compulsory that we stop developing materials, which are environmentally hazardous. This concept of textile composite materials has now become of key importance due to the need to preserve our environment. The field of hybrid textile composites research has experienced an explosion of interest, particularly with regard to its comparable properties to glass fibers within composites materials. The main area of increasing usage of these composites materials is the automotive industry, predominantly in interior applications.

References

1 Brouwer, W.D. (2000) *SAMPE Journal*, **36**, 6.
2 Taha, I. and Ziegmann, G. (2006) *Journal of Composites Technology & Research*, **40**, 1933.
3 Lopez-Manchado, M.A., Biagiotti, J., and Kenny, J.M. (2002) *Journal of Thermal Composite Materials*, **15**, 337.
4 Sanadi, A.R., Caufield, D.F., and Rowell, R.M. (1994) *Plastics Engineering*, **4**, 27.
5 Joseph, K., Vardese, S., Kalaprasad, G., Thomas, S., Prasannakumari, L., Koshy, P., and Pavithran, C. (1996) *European Polymer Journal*, **32**, 1243.
6 Thwe, M.M. and Liao, K. (2003) *Composites Science and Technology*, **63**, 375.

7 Sreekala, M.S., George, J., Kumaran, M.G., and Thomas, S. (2002) *Composites Science and Technology*, **62**, 339.
8 Naik, N.K. (1994) *Woven Fabric Composites*, Technomic Publishing.
9 Scida, D., Aboura, Z., Benzeggagh, M.L., and Bocherens, E. (1999) *Composites Science and Technology*, **59**, 505.
10 Gao, F., Boniface, L., Ogin, S.L., Smith, P.A., and Greaves, R.P. (1999) *Composites Science and Technology*, **59**, 123.
11 Bishop, S.M. and Curtis, P.T. (1983) RAE Technical Report 83010, HMSO, London.
12 Huang, Z.M. (2000) *Composites Science and Technology*, **60**, 479.
13 Shin, S.G. and Jang, J.S. (2000) *Journal of Materials Science*, **35** (8), 2047.
14 Gommers, B., Verpoest, I., and VanHoutte, P. (1998) *Acta Materialia*, **46** (6), 2223.
15 Satyanarayana, K.G., Sukumaran, K., Kulkarni, A.G., Pillai, S.G.K., and Rohatgi, P.K. (1986) *Composites*, **17**, 4.
16 Gowda, T.M., Naidu, A.C.B., and Chhaya, R. (1999) *Composites Part A*, **30**, 277.
17 Ghosh, P. and Das, D. (2000) *European Polymer Journal*, **36**, 10.
18 Pothan, L.A., Thomas, S., and Neelakantan, N.R. (1997) *Journal of Reinforced Plastics and Composites*, **16**, 744.
19 Pothan, L.A., Thomas, S., Oommen, Z., and George, J. (1999) *Polimery Nr*, 11–12.
20 Gu, B. and Ding., X. (2005) *Journal of Composite Materials*, **39**, 685–710.
21 Miravette, A. (ed.) (2004) *3-D Textile Reinforcements in Composite Materials*, Woodhead Publishing Ltd., Cambridge.
22 Pothan, L.A., Potschke, P., Habler, R., and Thomas, S. (2005) *Journal of Composite Materials*, **39** (11), 1007–1025.
23 Paiva Junior, C.Z., de Carvalho, L.H., Fonseca, V.M., Monteiro, S.N., and d'Almeida, J.R.M. (2004) *Polymer Testing*, **23**, 131–135.
24 Mwaikambo, L.Y. and Bisanda, E.T.N. (1999) *Polymer Testing*, **18**, 181–189.
25 De Medeiros, E.S., Agnelli, J.A.M., De Carvalho, L.H., and Mattoso, L.H.C. (2004) *Polymer Composites*, **26**, 1–11.
26 Alsina, O.L.S., de Carvalho, L.H, Filho, F.G.R., and d'Almeida, J.R.M. (2005) *Polymer Testing*, **24**, pp. 81–85.
27 Bullions, T.A., Hoflman, D., Gillespie, R.A., Price-O'Brien, J., and Loos, A.C. (2006) *Composites Science and Technology*, **66**, 102–114.
28 Gassan, J. (2002) *Composites Part A*, **33**, 369–374.
29 Li, Y. (2002) Eco-composite: Structural, processing, mechanical and fracture properties of sisal fiber reinforced composites. PhD dissertation, School of Aerospace, Mechanical and Mechatronic Engineering, The University of Sydney, Sydney, NSW, Australia.
30 Mathew, L., Joseph, K.U., and Joseph, R. (2006) *Bulletin of Materials Science*, **29**, 91–99.
31 Silva, R.V., Spinelli, D., Bose Filho, W.W., Claro Neto, S., Chierice, G.O., and Tarpani, J.R. (2006) *Composites Science and Technology*, **66**, 1328–1335.
32 Sapuan, S.M., Leenie, A., Harimi, M., and Beng, Y.K. (2006) *Materials & Design*, **27**, 689–693.
33 Kaveline, K.G., Ermolaeva, N.S., and Kandachar, P.V. (2006) *Composites Science and Technology*, **66**, 160–165.
34 Hivet, G. and Boisse, P. (2005) *Finite Elements in Analysis and Design*, **42**, 25–49.
35 De Medeiros, E.S., Agnelli, J.A.M., De Carvalho, L.H., and Mattoso, L.H.C. (2004) *Polymer Composites*, **26**, 1–11.
36 Hufenbach, W., Kroll, L., Gude, M., and Langkamp, A. (2002) 11th International Scientific Conference Achievements in Mechanical & Materials Engineering, Zakopane, Poland.
37 Hufenbach, W., Petrinic, N., Gude, M., Langkamp, A., and Andrich, M. (2003) Experimentelle Versagensanalyse von textilverstärkten Verbundwerkstoffen bei hochdynamischer Belastung, in *Verbundwerkstoffe und Werkstoffverbunde* (ed. H.P. Degischer), Springer, Berlin, pp. 423–4238.
38 Hufenbach, W., Kroll, L., Langkamp, A., and Böhm, R. (2002) International Symposium on Mechanics of Composites, Prague, Czech Republic, pp. 61–68.
39 Langkamp, A. (2002) Bruchmodebezogene Versagensmodelle für faser- und textilverstärkte Basisverbunde mit polymeren,

metallischen sowie keramischen Matrices. PhD thesis, TU Dresden.

40 Hong, C.K. and Wool, R.P. (2004) *Journal of Natural Fibers*, **2**, 83.

41 Hong, C.K. and Wool, R.P. (2005) *Journal of Applied Polymer Science*, **95** (6), 1524–1538.

42 Suddell, B.C. and Evans, W.J. (2005) Natural fiber composites in automotive applications: Natural fibers, in *Biopolymers and Biocomposites* (eds A.K. Mohanty, M. Misra, and L.T. Drzal), CRC Press, p. 23.

43 Pervaiz, M. and Sain, M.M. (2003) *Macromolecular Materials and Engineering*, **288** (7), 553–557.

44 Sedlacek, G., Trumpf, H., and Kammel, C. (2003) Proceedings 21 Internationales Klebtechnik-Symposium Verbindungstechnik im Bauwesen, Munich, pp. 65–73.

45 Sedlacek, G., Feldmann, M., Völling, B., Trumpf, H., and Wellershoff, F. (2005) 5th Colloquium Research in Adhesive Technology, Düsseldorf.

Part Four
Microsystems: Microparticle-Reinforced Polymer Composites

16
Characterization of Injection-Molded Parts with Carbon Black-Filled Polymers

Volker Piotter, Jürgen Prokop, and Xianping Liu

16.1
Introduction

Today's urgent necessity of saving resources and reduce energy consumption is pushing the interest in small components and microparts. This worldwide trend is accompanied by increasing demands on the precision and dimensional tolerance of components. In addition, increased integration entails a growing demand for complex-shaped microprecision components of metal materials. With this in view, new approaches for microfabrication processes have to be developed.

A famous method for manufacturing high-quality microcomponents is the LIGA process that – roughly spoken – consists of lithographical structuring of a negative resist, electroplating of the degradable sections, and replication. If the last step is omitted, LIGA in principle enables the fabrication of metal microcomponents. Owing to the high costs of mask manufacturing and synchrotron irradiation, however, such devices are quite expensive. A much more economical option is obtained if the original LIGA parts are used as tool inserts for a two-component injection molding process to produce partial conductive preforms. The latter has to consist of an electrically conductive substrate with insulating microstructures mounted on the surface. These 2C parts can be used further as preforms in an electroplating process by which the interspaces between the insulating microstructures are filled successively from the bottom, that is, from the conductive substrate. As injection molding and electroplating are typical mass production processes, cost-effective microcomponents can be fabricated in large or medium series. The whole process chain is illustrated in Figure 16.1.

It is easily understandable that for the fabrication of uniform microparts, a constantly conductive substrate plate is a prerequisite. This demand, of course, directly leads to the necessity of driving the injection molding step in such a way that a uniform distribution of the conductive filler is achieved. In the following sections, innovative solutions for this challenge are described.

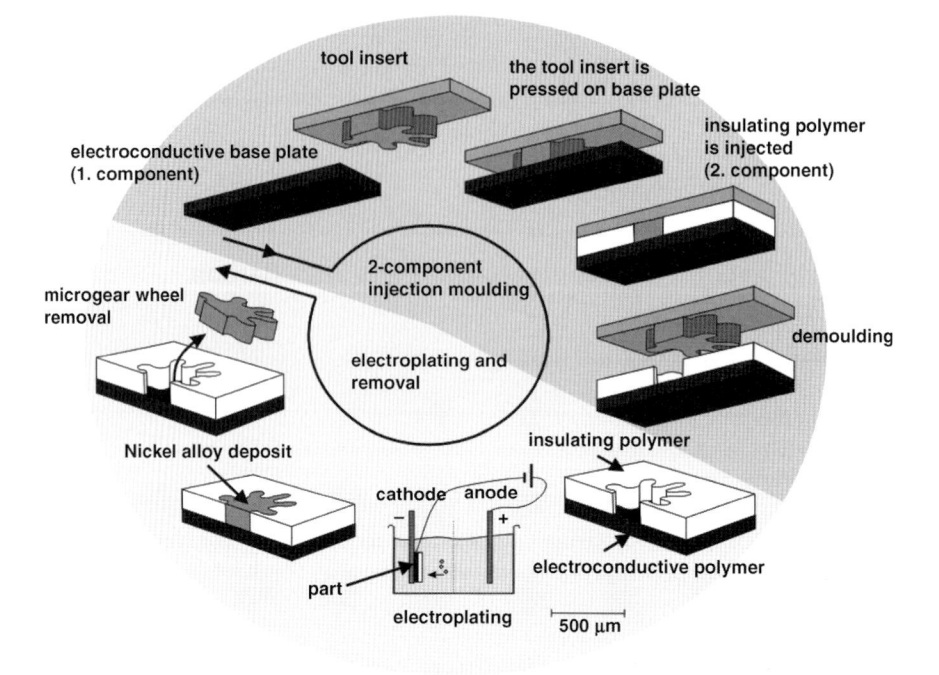

Figure 16.1 Scheme of the microfabrication process based on 2C injection molding and electroplating.

16.2
Injection-Molded Carbon-Filled Polymers

Injection molding of filled polymers with the aim of obtaining components with electrical properties is of interest to diverse applications. Recent studies, for example, have been reporting on direct molding of pcb tracks on manufacturing of surfaces for electroplated coatings. As a matter of fact, however, the electrical properties of such molding components are mainly attractive with regard to the discharge properties.

The different properties occurring during processing of electrically conductive materials become apparent as one regards the electroplating processes that are required for microtechnical gas-shielded metal arc welding applications.

The very good electrical conductivities that are achieved using carbon fibers (CFs) in plastics are described in the literature [1]. These carbon fibers, however, are not suited for electroplating purposes because the local surface conductivity resulting from the several micrometer-sized fibers causes a low galvanic start point density [2] (see Figure 16.1). The different shrinkage properties of polymers compared to carbon fibers, moreover, lead to inferior surface qualities that will show on the metal layer surfaces after the process of electroforming.

In plastics that are filled only with conductive carbon black, the electrical resistance mostly varies considerably due to discrepancies in filler concentration ([1–4]; [5]) that are caused by inhomogeneities in filler distribution on account of the specific processing conditions.

The specific processing characteristics of carbon black compounds are discussed extensively by Gilg [3] who points out in his studies that the conductivity of an injection-molded part depends on the flow of melt, and hence is largely influenced by the degree of orientation of the carbon black agglomerates. The poor conductivity that is detected around the gate is explained by the orientation taken by the agglomerates in the presence of the high flow gradient during the injection process. In brochures published by the SIMONA [6] and CABOT [7] companies and by Evonik Industries SE, one finds comparable studies and observations. The publication issued by CABOT outlines that there is a relationship between shear forces and poor conductivity. CABOT points out that a shear-induced separation of the carbon black agglomerates occurs during processing and that separation of the carbon black agglomerates interrupts the conductivity network link causing poor conductivity in areas with high shear forces. Extremely poor conductivities in the area of the gate are shown to occur during injection molding.

According to extensive studies performed on the above phenomena by Knothe [1], the differences in the electrical resistances of molded parts of carbon black-filled polystyrene may amount to more than two orders of magnitude. Knothe attributes these decisive differences both to the orientations and to the decomposition of the carbon black structure. Knothe, too, speaks of a poor electrical conductivity of the molded parts in the area of the gate.

Holstein et al. [2] have carried out injection molding and electroplating tests on carbon black-filled materials. Their tests of carbon black-filled polyoxymethylenes and polyamide 12 proved that the best results regarding processing and metal deposition are achieved using carbon black-filled polyamide 12 available as Vestamid LR1-MHI. Both Holstein and Knothe find inhomogeneous electrically conductive surfaces on the injection-molded components. According to statistical test planning procedures performed by Finnah [5], a screw feed rate of 31 mm/s is optimal for reaching surface resistances in the range of 80–100 Ω.

Finnah uses the carbon black-filled Vestamid LR1-MHI also for insert injection molding of microcomponents. Upon inspection, the molded insert plates revealed deficiencies caused by the inhomogeneity of deposition rates during injection molding. During electroforming, the deposition of nickel was observed to be decreasing with increasing distance from the contacting point and partial differences were found to occur regarding the nickel layer [5]. According to component tests by means of MOLDFLOW Plastics Insight[1], this can be attributed to the effect caused by the shear rate 0.2–0.5 mm below the surface (see Figure 16.2).

---

[1] MOLDFLOW Plastics Insight is a standard program for simulation of injection molding processes. Since June 2008, this software belongs to Autodesk. Hereinafter, the program will be referred to as MOLDFLOW software.

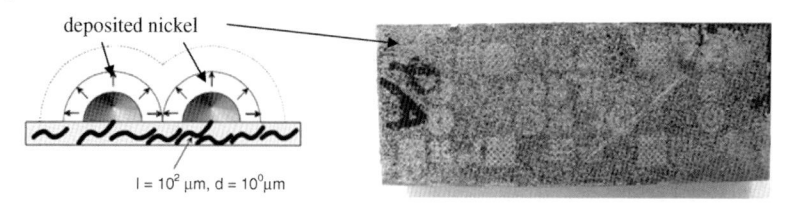

Figure 16.2 Galvanic start on carbon fiber-filled polyamide 6.6 carbon fiber of a mean fiber length of 25 μm [2].

In that very area, the conductive link between the surface and the center of the component is destroyed by the shear [5]. By widening the cavity from 1 to 3 mm, Finnah achieved surface resistances ranging from 40 to 100 Ω in further tests.

A test series allows taking a closer look at all these facts and shows how to select injection-molding parameters that ensure homogeneous electrical resistances around the gate. To achieve this, a material is filled with carbon black of different vol.%. For testing, the thermoplastic Grilamid L16A by EMS Chemie is chosen due to its flexibility, its hydrolytic resistance, and its chemical resistance, while Ketjenblack® EC-600 JD by Akzo Nobel is selected for being conductive. To introduce the very fine-scaled, highly conductive carbon black [8] homogeneously into the material while causing only little compound shear, a screw configuration is selected that subjects the material to comparatively low shear forces. In case of high shear rates, particles of carbon black may be coated by the polymer. Such isolated particles are not available anymore to the conductivity network.

16.3
Processes and Characterization

16.3.1
Rheological Characterization of the Compounds

To ensure processability, the material is tested rheologically[2)] before injection. Mainly the highly filled systems are critical as regards that criterion. Figure 16.3 shows shear viscosities of the material when filled with 7.1 and 9.0 vol.% of carbon black. The increase in the volume percentage of carbon black also considerably increases viscosity (see Figure 16.3).

The shear viscosity of the commercially available Vestamid LR1-MI is low. Hence, Vestamid LR1-MI is better suited for injection molding processing than the two-compound materials. Both of the latter, however, still are appropriate for injection molding processing procedures.

---

2) Rheology is the study of the flow of matter. It places emphasis on the analysis of the mechanical behavior of continuous deformable matter (elasticity, plasticity, and viscosity).

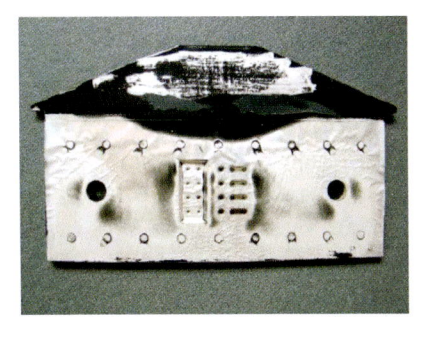

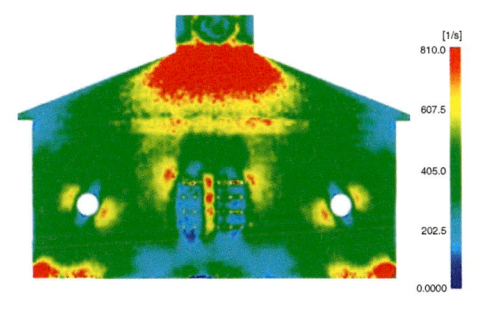

Figure 16.3 (a) Nickel electroplating starting layer on a Vestamid LR1-MHI substrate plate. (b) Mold flow simulation of the same component showing the maximum shear rate 0.35 mm below the surface [5].

16.3.2
Molding and Electrical Characterization of Conductive Compounds

Varying the injection velocities, components of 60×40 mm^2 in size and 1 and 2 mm in thickness were manufactured from the different compounds (see Figure 16.4).

For the purpose of testing, the screw feed rate is adjusted to 10, 20, 30, 40, 50, and 55 mm/s. The electrical resistance of the injection-molded parts is measured based on standard VDE DIN 0303 provided by the German Electrical Engineering Association and a percolation curve is plotted.

Figure 16.5 shows the percolation threshold of the samples to be at approximately 4 vol.% of carbon black.

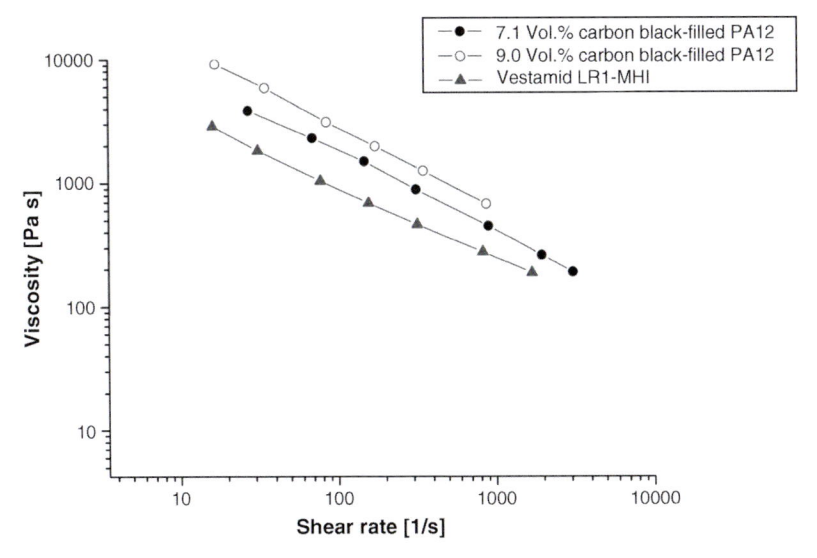

Figure 16.4 Shear viscosity of molding compounds filled with 7.1 and 9.0 vol.% of carbon black compared to the shear viscosity of LR1-MHI.

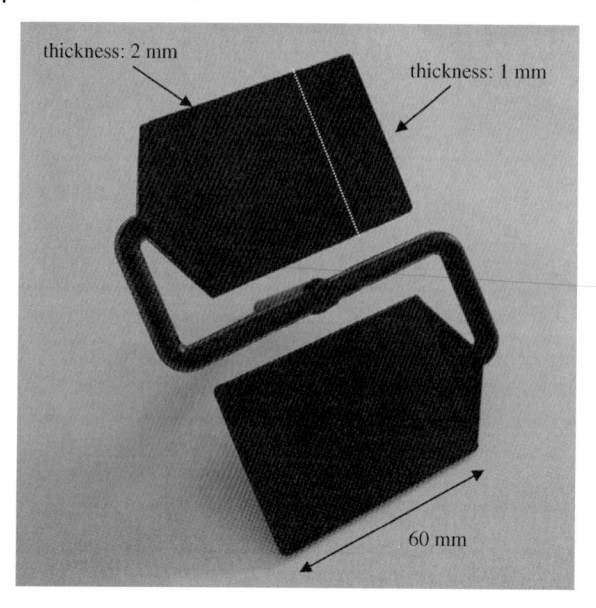

Figure 16.5 Test specimen geometry for determination of the percolation threshold[3] and tests of the dependence between injection rate and surface resistance.

For surface electroplating processes, the percolation threshold, however, is less important than the lowest possible electrical surface resistance. The latter parameter decreases as the volume percentage of the carbon black increases (Figure 16.6).

A more detailed view at the results of the components filled with 7.1 and 9.0 vol.% elucidates that increase (Figure 16.7).

A more detailed view at the results of the components filled with 7.1 and 9.0 vol.% elucidates that increase.

16.3.3
Comparison of the Results Obtained with Simulation Calculations

The influence of shear on the surface resistance of components is discussed in the literature [5, 7]. To examine and evaluate the respective assumptions, 9.0 vol.% carbon black component samples were manufactured. The surface resistances of the specimens were determined and comparative simulations of the shear rates were performed by means of MOLDFLOW. Figure 16.8 shows the sample geometry dimensions for tapered and stepped cross sections.

Molded filling simulations and calculations of the maximum shear rates in the case of the tapered sample show that the shear increases each time the cross section changes (see Figure 16.9).

---

3) The critical filler concentration that causes a sudden increase in the system's conductivity is referred to as the percolation threshold.

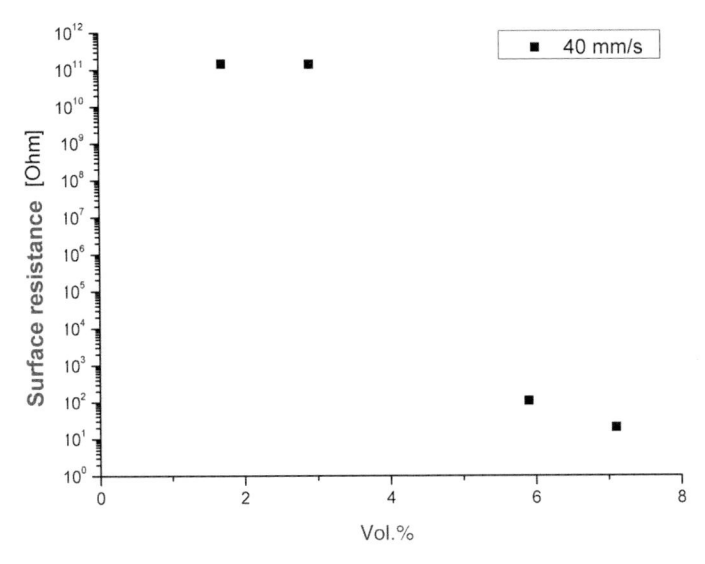

Figure 16.6 Surface resistances of injection molding components measured based on standard DIN VDE 0303 after compounding of Grilamid L16A supplied by EMS Chemie and Akzo Nobel's Ketjenblack® EC-600 JD at a screw feed rate of 40 mm/s.

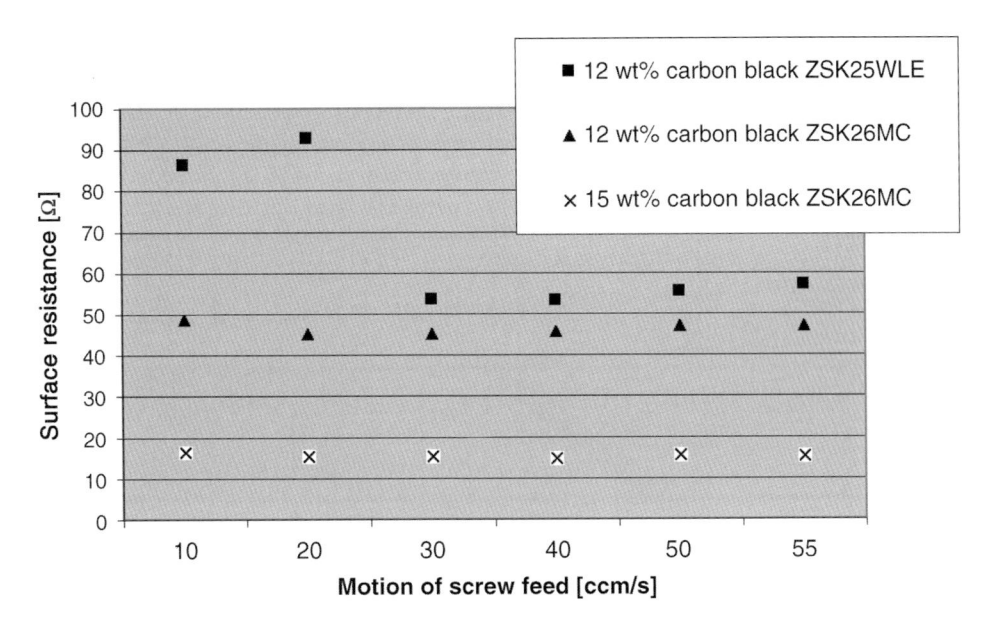

Figure 16.7 Influence of the injection rate on the surface resistances of thermoplastics filled with different quantities of carbon black [0–85 Ω].

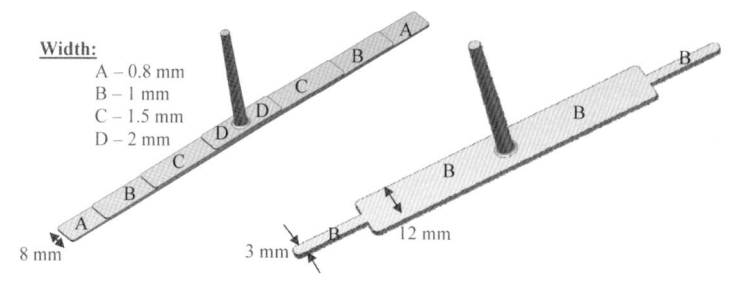

Figure 16.8 Samples with tapered and stepped cross sections.

Four-point surface resistance measurements yield the results shown in Figure 16.10.

In view of the above, the surface resistance increases with each move further along the sample's stages. The component resistance is observed to increase as shear increases during molding. Comparing the simulated MOLDFLOW shear rate with the measured surface resistance, a linear dependence becomes evident. As the shear rate increases, there is a linear increase in the surface resistance (see Figure 16.11).

Similarly, the relationship between shear rate and electrical surface resistance is evident from the sample with stepped cross section (see Figure 16.12).

In the narrowed cross section, shear rates of 600/s are indicative of an almost double as high value as in the wider areas where they are calculated to be amounting to 300/s. By comparing these values with the relationship between shear rate and surface resistance illustrated in Figure 16.10, a value of approximately 95 Ω is expected for the wide cross section, while a value of about 200 Ω is assumed for the narrowed area. The actually measured values are, hence, within the range of the previously calculated ones (see Figure 16.12).

CT analyses of powder injection-molded components performed by Heldele [9, 10] proved that shear acting during the process of injection molding causes segregation of fillers from the edge areas. These segregation effects are considered to be the cause of higher surface resistances in the case of high shear effects. As particles of carbon

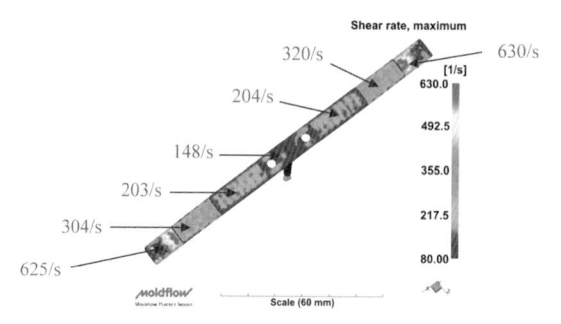

Figure 16.9 Results of the simulation of the maximum shear rate occurring as one fills the tapered sample geometry.

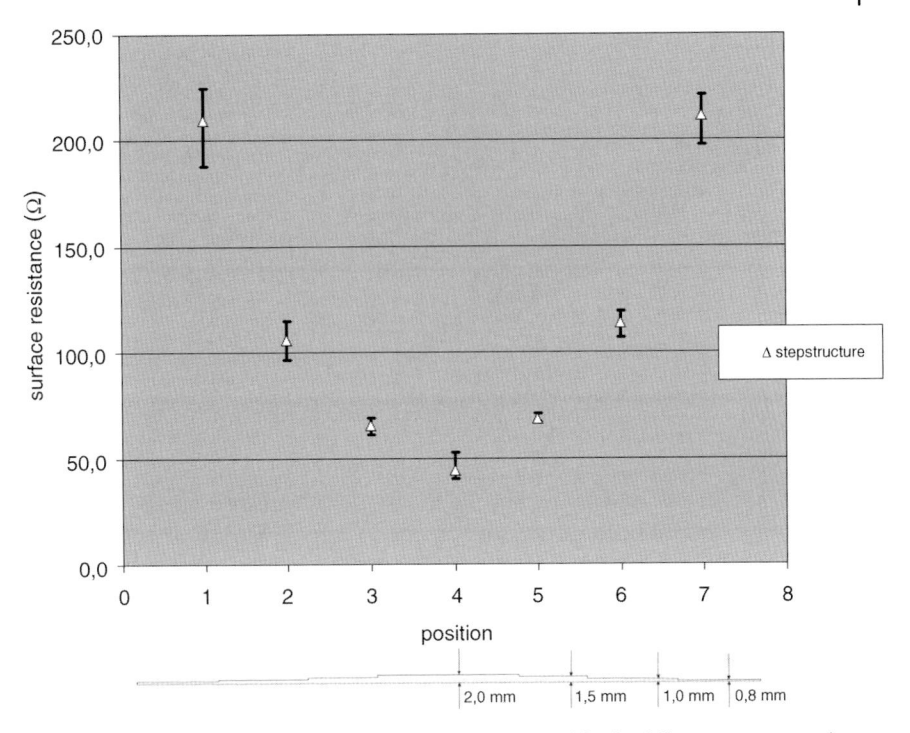

Figure 16.10 Tapered sample and surface resistances measured for the different stages using the four-point measuring method.

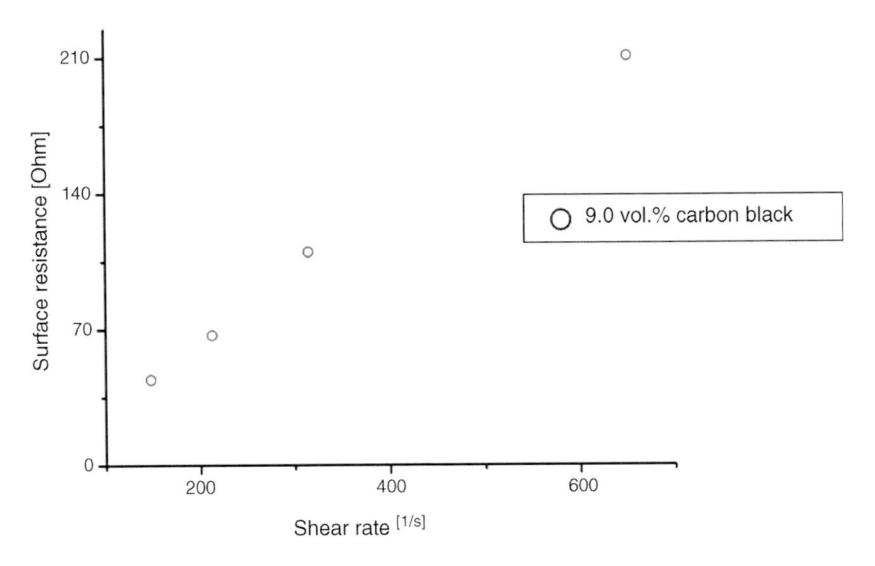

Figure 16.11 Measured surface resistances in correlation with the calculated shear rate.

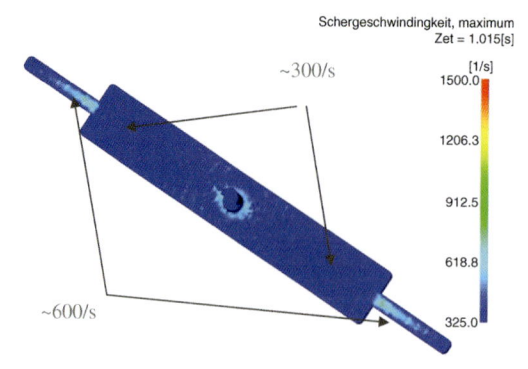

Figure 16.12 Simulation of the maximum shear rate that occurs when filling the sample.

black move from the surface to the interior of the component, one finds surface areas with less carbon black. Hence, the surface resistance increases (Figure 16.13).

16.3.4
Electroplating of Injection-Molded Components

Figure 16.14 shows the results of nickel deposition on injection-molded samples made of the material filled with 9.0 vol.% of carbon black.

After 10 min of deposition on the molded part, a homogeneous nickel layer is observed to have developed even though only one side of the part has been contacted.

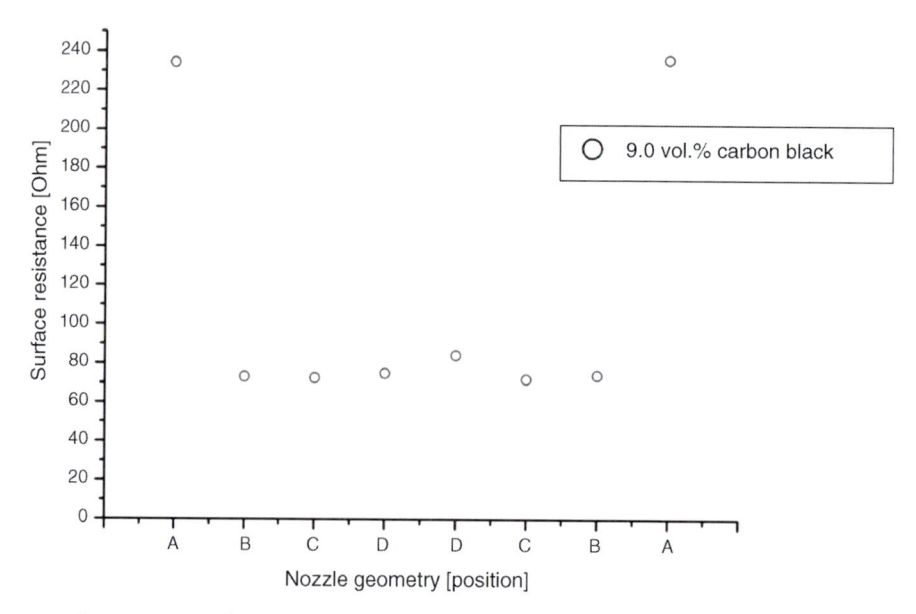

Figure 16.13 Surface resistance of the nozzle geometry showing a defined increase in resistance in the area of high shear rates during simulation.

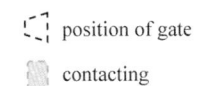

Figure 16.14 Sample made of the material filled with 9.0 vol.% of carbon black after 10 min of electrode position at a voltage of 2.55 V.

This proves that a homogeneous deposition on injection-molded parts can be achieved without the need for posttreatment of the respective surfaces. Using fillers that reduce viscosity, the injection molding capability of the compound filled with 9.0 vol.% of carbon black can be improved. On the basis of these results, existing processes can also be improved and commercially available materials can be manufactured with more homogeneous electrical conductivity properties.

16.3.5
Transfer of the Test Results to Standard Material Systems

Inhomogeneous resistance profiles were found on the surfaces of the injection-molded Vestamid LR1-MHI parts. Tests of plate-type hot runner parts identified an area of conductivity that was insufficient for electroforming purposes (see Figure 16.15).

Since the area around the gate has hardly been considered in the past, findings such as the above neither have yet been discussed quite clearly nor have been

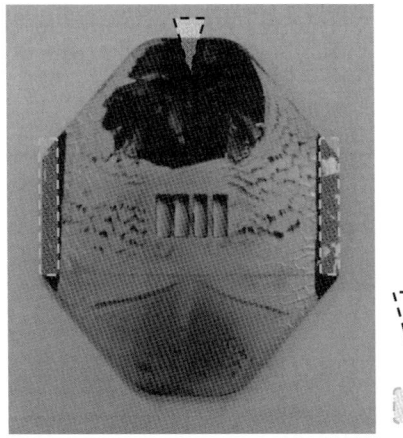

Figure 16.15 Vestamid LR1-MHI specimen after 10 min of electrodeposition at a voltage of 2.55 V. The specimen was contacted on two sides. Deposition fails to occur in the area of the hot runner connection.

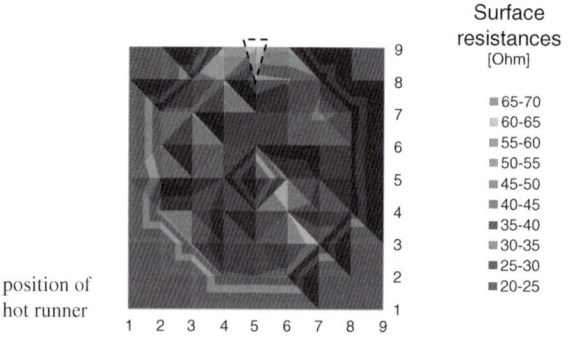

Figure 16.16 Surface resistance distributions on Vestamid LR1-MHI components molded at a screw feed rate of 1 mm/s.

analyzed in detail. So far, only the influence of the injection rate on the degree of surface resistance has been assessed [5].

To gain an overview of the effect of the injection rate on the electrical resistance, components are molded applying varied screw feed rates and the respective conductivities are determined using the four-point measuring method. To analyze additionally and more comprehensively the resistance distribution, the components are subdivided into 52 quadrants by which the conductivity is measured more precisely (see Figures 16.16 and 16.17).

The test results reveal that as the resistance decreases in parts of the component with increasing injection rate, the inhomogeneity of the electrical resistance on the surface increases. Comparing, for example, the resistance distributions of components manufactured at 1 mm/s (Figure 16.15) with those molded at 50 mm/s, a very pronounced increase is detected in the area around the gate, while the resistances on the component surfaces are found to be decreased at the same time.

Figure 16.17 Surface resistance distributions on Vestamid LR1-MHI components molded at a screw feed rate of 50 mm/s.

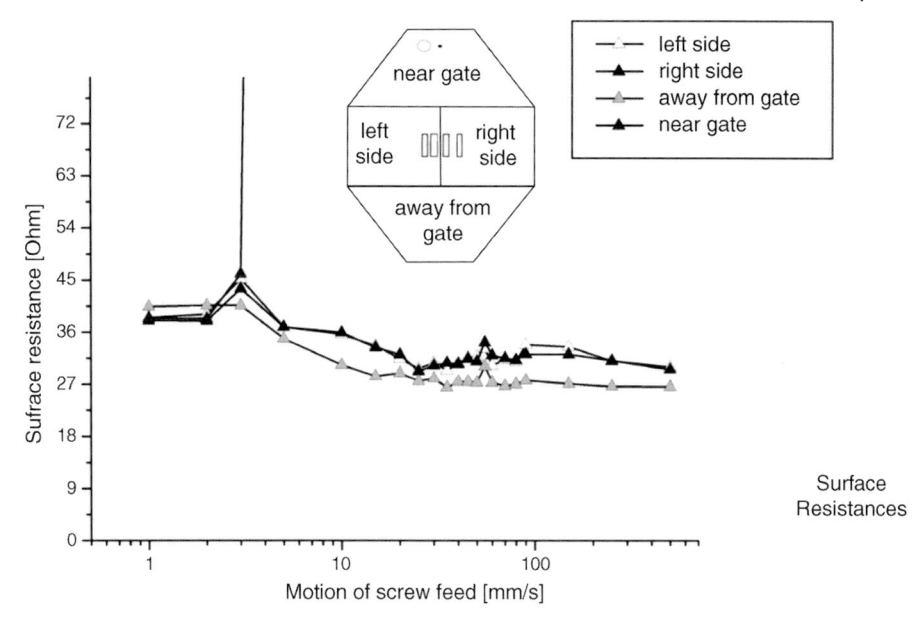

Figure 16.18 Surface resistance distributions on Vestamid LR1-MHI components as a function of screw feed rates at defined component positions.

The above effect becomes still clearer as one combines the measured values within four defined sections on the surfaces of the components. To do so, each component is subdivided into segments referred to as "around the gate," "away from the gate," and in the area of the component center, as "right side" and "left side" (see Figure 16.17). While an area with moderate surface resistance is first detected "around the gate," the resistance increases as the screw feed rate reaches 2 mm/s. At 10 mm/s and above, that is, at resistances exceeding 500 Ω, it is not possible anymore to carry out measurements using the above four-point measuring method. Likewise, a moderate surface resistance is obtained initially on the "right side," "left side," and "away from the gate." That value, however, improves as the injection rate increases until reaching a screw feed rate of 22 mm/s. Characteristically low resistances are not obtained anymore when the feed rate is increased further (see Figure 16.17).

In line with the findings obtained using the proprietary compound, it is assumed that an area of high shear develops around the gate on account of the increasing injection rate and that carbon black is not dispersed perfectly in the commercially available material. To illustrate this increase in the shear rate, mold filling was simulated at an injection rate of 2 and 55 mm/s. At both these rates, pronounced differences in resistance show on the component surfaces.

Considering the resulting values, the maximum shear rate occurring around the gate at an injection rate of 2 mm/s amounts to approximately 576/s (see Figure 16.18).

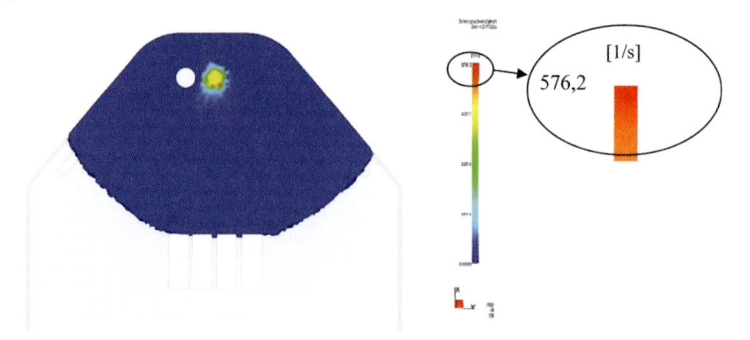

Figure 16.19 Simulation of the shear profile in the area of the gate at an injection rate of 2 mm/s.

Simulating mold filling at an injection rate of 55 mm/s, one observes a pronounced increase in the calculated shear rate (see Figure 16.19).

At that injection rate, the maximum shear rate reaches a calculated value of approximately 8916/s that corresponds to a value that is about 15 times higher than the one observed for molding at a lower injection rate. It suggests itself that the high-resistance area around the gate develops due to this increase in the shear rate.

Optical micrograph analysis (see Figure 16.20) reveals a polished area near the gate in the case of the high-shear components. While the surface resistance measured in that area is $\gg 500\,\Omega$, values of $\sim 30\,\Omega$ are measured for the area appearing dark.

This micrograph, too, supports the hypothesis that high shear in the area of the gate causes the carbon black particles to segregate and move from the surface to the interior of the component.

To apply the results obtained to the case of component electroforming, a filling study using variable injection rates is carried out. The galvanic starting tests on the components corroborate the simulation result by showing that the problems of conductivity in the area around the gate already occur after injection of small quantities of material (see Figure 16.21).

The analyses performed show that, on the one hand, increased injection rates cause segregations when using hot runner moulds. On the other hand, the surface

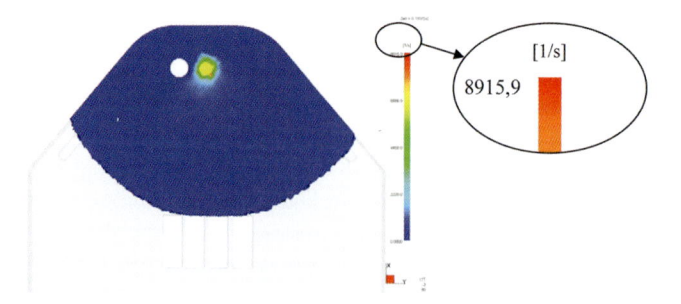

Figure 16.20 Simulation of the shear profile in the area of the gate at an injection rate of 55 mm/s.

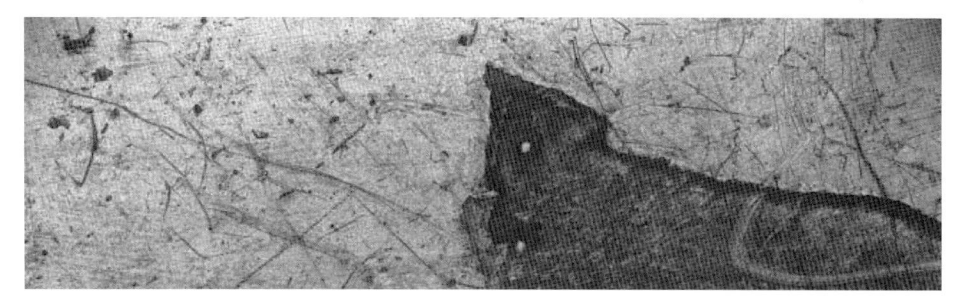

Figure 16.21 Micrograph of the area at the interface between conductive and nonconductive surface of a Vestamid LR1-MHI component molded at a constant screw feed rate of 50 mm/s.

resistance increases if materials such as Vestamid LR1-MHI are molded at insufficient injection rates. To implement these two divergent requirements in one process cycle a staged injection rate during the filling of the mold reduced the shear effect in the area of the gate. A staged injection profile like that reduces the shear effect in the area of the gate. Injection rates that during the process of injection range from 3 (material quantity: 3 mm^3) to 8 mm/s ensure very homogeneous electrical surface resistances and, hence, homogeneous deposition rates during electroforming (see Figure 16.22).

Completely filled specimens, too, prove that the staged injection rates affect homogeneous component resistances.

A staged injection rate of 1 mm/s (3 mm^3) → 30 mm/s was determined through further optimization of the parameter values within the process. Figure 16.23 shows excellent electroforming results for components that have been molded applying these specifically assessed values.

The above analyses and tests illustrate that shear has a decisive effect on components and that it is basically possible to calculate the local shear profiles through mold filling simulation to predict the conductivity distributions on component surfaces. A staged injection profile is required to avoid poor conductivities on the surface in the area of the gate (Figure 16.24).

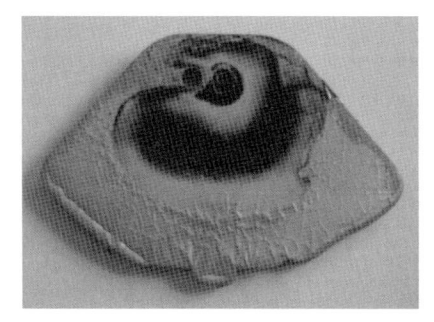

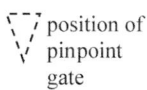

position of pinpoint gate

Figure 16.22 Galvanic starting tests on components filled by 1/5. Constant injection rate (55 mm/s).

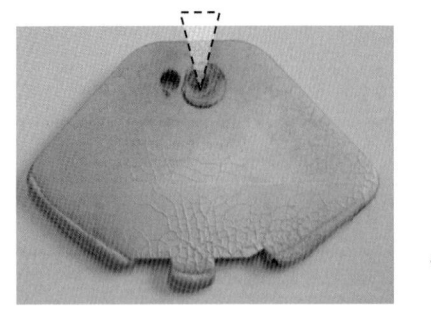

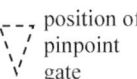

Figure 16.23 Galvanic starting tests on components filled by 1/5 at staged injection rates (3 – 8 mm/s).

16.3.6
Summary of the Deposition Studies

Poor conductivity is often said to be caused by the inappropriate orientation of the fillers [3, 4]. This seemingly state-of-the-art assumption is not corroborated by the studies introduced. Moreover, the results that have been discussed disprove the reason given in the literature, that is, the idea that the conductivity is interrupted by a layer that is located below the surface [5]. Instead, the areas of insufficient conductivity have been identified hereunder as being due to shear-induced segregation directly on the surface [20].

In line with the above, the following conclusions can be drawn from the studies carried out so far [21]:

- Injection rates have a decisive effect on the surface resistances of carbon black-filled materials during processing. The selective adjustment of injection rates can reduce the electrical surface resistance of components.
- The shear rate occurring on the surface during mold filling has a major influence on the surface percolation and, thus, on the surface resistance. The resistance increases along with the shear rate.

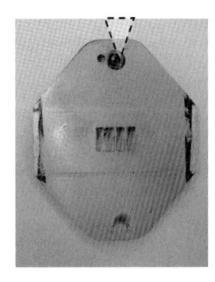

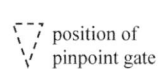

Figure 16.24 Galvanic starting tests on completely filled components with different geometries.

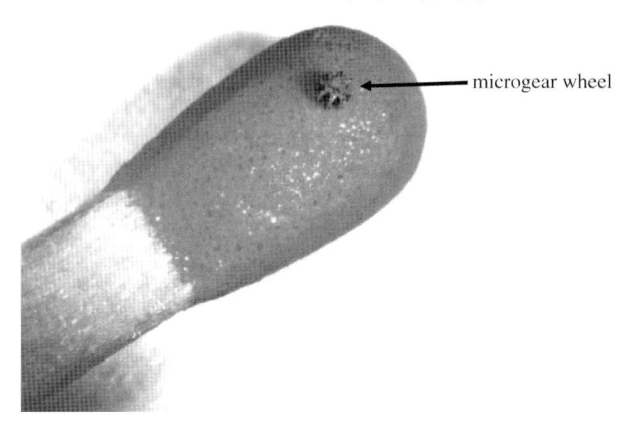

Figure 16.25 Microgear wheel made of nickel in comparison to a match.

16.3.7
Demonstrator Production

To demonstrate the capabilities of the new process with a realistic microcomponent, gearwheels with an outside diameter of 550 μm were produced (Figure 16.25). As mold insert for the second injection molding step, a Ni-artifact made by UV-lithography had been used. Analyses of the microcomponents with respect to hardness and structure revealed values comparable to that of nickel electroplated from a sulfamate electrolyte.

16.4
Mechanical Property Mapping of Carbon-Filled Polymer Composites by TPM

16.4.1
Introduction

This section presents a new approach to the characterization of carbon-filled polymer composites by using a novel multifunction tribological probe microscope (TPM). The TPM is capable of measuring four functions in a single scan to provide area mappings of topography, friction, Young's modulus, and nanohardness. The measurement is based on point-by-point scanning so values of the four measured functions are linked in space and in time. The specimens are PA6.6 and PA12 filled with carbon black or carbon fiber, especially prepared at the Institute for Materials Research III of Karlsruhe Research Centre. The four-in-one measurement of TPM enables us to identify the material difference on the surface in order to estimate the distribution of a particular material within the composite. For each specimen, mappings of topography, hardness, and Young's modulus were obtained, and from the last two, it is easy to see the existence of two different materials. Comparing the topography and hardness mappings, we are able to pick up the areas where carbons are located.

In manufacturing microstructure parts by injection molding of polymers, the polymers are often filled with certain amount of carbons to make it conductive. These parts can then be further treated by electroplating in order to replicate them into metallic forms. The conductive polymer parts have to serve both as substrates for electroplating step and as templates. Unfortunately, such kind of electroplating (undirected growth) is limited to an aspect ratio not higher than 5 because the galvanic deposit will overgrow the apertures of the template before they are complete, and partially filled cavities are left in the product. To overcome this problem, an insert molding process was developed [11, 12]. First, an electrically conductive base plate is made by injection molding of thermoplastics filled with carbon black or carbon fibers. Then by a second injection molding step, microstructures made of insulating plastics are mounted on the plate. The micromolds thus produced make the electroplating starting at the base plate only. However, the conductivity of the injection-molded parts is quite anisotropic and depends obviously on process parameters such as temperature, pressure, melt velocity, conductive filler type, and polymer combination. In general, it is very difficult to detect the distribution of the fillers in the surface region. Surface profilers and SEM/TEM have been used but without any sufficient results, in identifying the filler material from the polymer substrate. In the next section, we introduce a new method of mapping the mechanical properties of such formed polymeric composites with a novel instrument system, the multifunction TPM.

16.4.2
Multifunction Tribological Probe Microscope

In the characterization of surface and surface-related properties, the available instruments can be categorized into two types, atomic force microscopy (AFM)-based instruments and hardness measurement-based instruments. The AFM-based instruments are capable of measuring multifunctions of a surface at nanometer/nanonewton level and have already made a great impact on the study of surface topography, tip–sample interaction, and mechanical/physical properties [13–15]. However AFM-based instruments have a limited force range due to the fact that force is provided by bending a cantilever that has a typical stiffness of 0.03–3 N/m. This can apply a force in a range from nN up to μN, which is not large enough for most engineering surfaces. The hardness-based instruments are primarily derived from conventional hardness measuring devices. These instruments can deliver a large range of loading force from N/mN down to μN but usually measure hardness/elastic modulus only. Although some nanoindentation instruments have been equipped with a separate AFM/STM to image the surface before and after the indentation, measuring individual functions separately will not give the direct correlated measurements.

The TPM has been developed at the University of Warwick over the past few years, in order to bridge the gap between AFM and hardness-based instruments by bringing the multifunction measurement to the middle force range [16–18]. The novel features of the TPM are its capability of measuring multifunctions in a single scan and the fact that contact force is independently controlled to a range up to 30 mN. The four

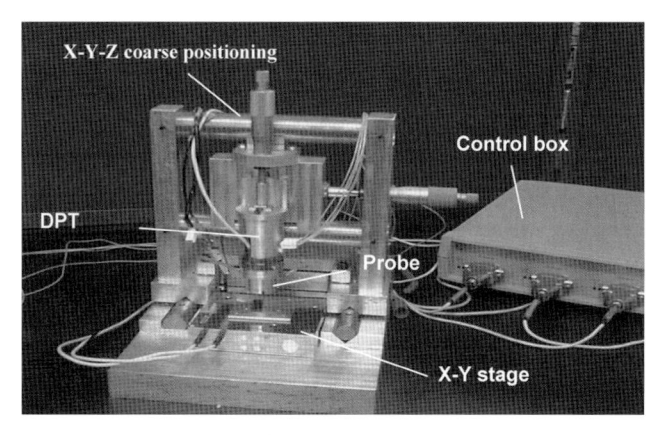

Figure 16.26 Photograph of the TPM system.

functions of topography, friction, hardness, and elastic modulus of a surface are mapped point by point in a short time period; thus, the measured four function mappings can be linked in space and time.

Figure 16.26 shows a photograph of the mechanical system of the TPM. It consists of a sensor probe attached to a DPT (Digital Piezoelectric Translator) from Queensgate Instruments Ltd., which can be coarsely adjusted by 3 μm, a precision X–Y stage, and a drive control unit. The essential part of the TPM is the sensor probe, which has a magnet/coil force actuator and two precision capacitive sensors for measuring surface height/deformation and frictional force between the probe tip and the surface being scanned. The scan area is $100 \times 100\,\mu m^2$ with the position feedback giving a linearity of 0.004%. The contact force at the probe tip can be electromagnetically controlled in a range of 0.01–30 mN. The tip is a Berkovich diamond tip with a radius of 0.1 μm. Other specifications and details of the instrumentation and calibration can be found in the references above.

As shown in Figure 16.27, the system is centrally controlled by a personal computer with a 14-bit data acquisition card and Labview software. The X–Y movements are

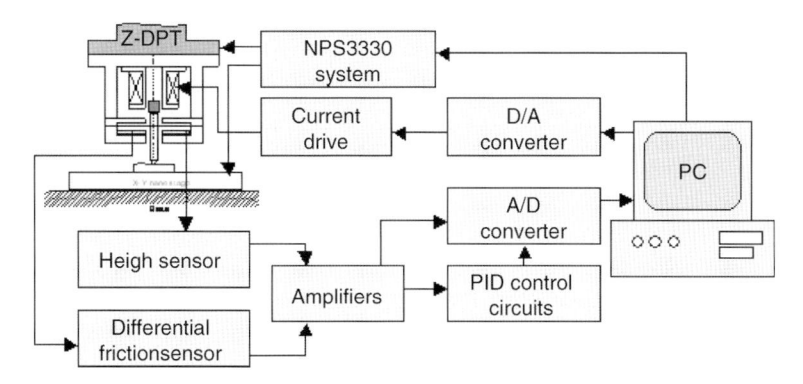

Figure 16.27 The schematic diagram of the TPM.

controlled by the computer via a high-voltage drive control unit, NPS3330 from Queensgate Instruments. The Z-DPT is a feedback-loop controlled by the setting of the height sensor. The contact force at the probe tip is controlled via the current drive unit by the combination of the magnet/coil force actuator and the cantilever beam. The cantilever beam is of a cross shape made of beryllium/Copper (Be/Cu) foil with a thickness of 25 μm. It acts as a flexible spring and a common electrode to the other four electrodes at the base plate of the probe. The four capacitance sensors thus formed were configured into two pairs: one pair acts as the height sensor for measuring the vertical displacement of the tip, which represents the geometrical features of a surface during scanning or the deformation of a surface through a force ramping, and the other pair of sensors is operated in a differential mode, for measuring the torsional motion of the beam and hence the frictional force at the tip along the scanning axis. The general dimension for the probe is 36 mm in diameter and 22 mm in total height.

The measurement of the four functions of a surface, that is, topography, frictional force, Young's modulus, and nanohardness, is achieved by operating the TPM in two scanning modes: the normal scanning mode and the force ramping mode. At each surface point, the TPM measures the surface height first in the normal scanning mode and then switches to the ramping mode to increase the contact force to a preset value and decrease it again, while the deformation/penetration is measured. Then, the TPM moves to next surface point and at the same time measures the frictional force. The process then repeats itself. At the end of scanning, four sets of data representing surface topography, friction, Young's modulus, and nanohardness are available for 2D and 3D displays. The calculation for topography and friction is straightforward, basically using the calibrated coefficients to convert the voltage signals into the corresponding height and frictional force.

The hardness and Young's modulus are obtained from the force–deformation curve at each point. As the load and deformation are continuously monitored, the hardness H is determined by the maximum load divided by the projected area that is related to the contact depth. The Young's modulus is derived from the initial slope of the unloading curve called "stiffness" because the initial unloading is elastic in nature [13–15]. In our case, the force and deformation curve is obtained at each scanning point in the force ramping mode. The indenter is Berkovich diamond tip with a nominal radius of 100 nm and an included angle of 142°. Table 16.1 lists the formulas used in calculation of the hardness and the reduced Young's modulus E_r.

Table 16.1 Details of formulas used in the calculation of hardness and elastic modulus (ISO 14577-1: 2002).

Hardness	$H = \dfrac{P_{max}}{A_C}$		where $A_C = 24.5 h_C^2$
Young's modulus	$E_r = \dfrac{\sqrt{\pi}}{2}\dfrac{S}{\sqrt{A_C}}$ and $\dfrac{1-v_{sample}^2}{E_{sample}} = \dfrac{1}{E_r} - \dfrac{1-v_{tip}^2}{E_{tip}}$		$S = \dfrac{dP}{dh}\bigg|_{P_{max}}$
			$h_C = h_{max} - 0.75\dfrac{P_{max}}{S}$

Table 16.2 General specifications of the TPM.

X–Y stage	$100 \times 100\,\mu m^2$ with a position feedback loop to give a linearity of 0.004% and a resolution of 1 nm
Height sensor	Up to $20\,\mu m$ with a linearity of 0.02% and a resolution of 0.1 nm
Contact force	Up to 30 mN with a linearity of better than 1% and a resolution of $5\,\mu N$
Friction sensor	Up to 10 mN with a resolution of better than $5\,\mu N$
Tip	A Berkovich diamond tip with a radius of $0.1\,\mu m$

Due to the fact that $E_{tip} = 1140\,GPa$, $\nu_{tip} = 0.07$, the difference between E_r and E_{sample} will be very small for most materials, thus the reduced Young's modulus is presented here.

The characteristics of displacement, contact force, and frictional force of the sensing probe have been tested and calibrated. Displacement calibration was carried out by loading the probe tip against a digital piezoelectric translator. The DPT provides direct reading and repeatability to 1 nm referenced to its internal capacitance gage. The DPT was ramped to give specific ranges, and the response of the probe was recorded. The linearity of the height sensor is $0.0298\,V/\mu m$ over a range of $20\,\mu m$. The force actuator was tested by loading the probe onto a load cell (D2000, Thames Side-Maywood Limited), which has a range of 10 g (0.0982 N) with a resolution of 1.0 mg (about $10\,\mu N$). The load cell was used here for calibrating the sensitivity of the force actuator, as it has a fairly large electronic noise on itself. The sensitivity of the force actuator is 0.217 mN/mA. The friction sensor of the TPM was calibrated against a set of dead weights, which were cut from a piece of silicon wafer and calibrated by a precision balance with an accuracy of 1 mg. The dynamic resonance of the sensing probe was measured in a typical "hammer test," free from contact but lifted to its neutral position. The response of the probe to a gentle tap was recorded at 175 Hz by a digital spectrum analyzer, AD-3525 FFT Analyzer. The general specifications of the TPM are summarized in Table 16.2. Details of the calibration procedures and performance evaluation can be found in Ref. [18].

16.4.3
Specimen Preparation

Many different thermoplastics such as PMMA, POM, PS, PA12, PA6.6, and their conductive versions have been applied for the injection molding tests at the Institute for Materials Research III of Karlsruhe Research Centre. As best suited combinations, conductively filled polyamides with insulating microstructures on top were chosen for this investigation. There are four specimens: two are polyamides PA12 mixed with carbon black and two are PA6.6 mixed with carbon fibers. There is about 15–20% of carbon black in PA12 and 20–30% of carbon fibers in PA6.6. The conductivity of the filled polymers is in general lower than that of metallic and metal-coated substrates. The measured specific surface resistance for PA12 composite is less than $1\,k\Omega$, while the PA6.6 is less than $10\,k\Omega$, although it has higher

Figure 16.28 SEM micrographs of (a) PA12 filled with carbon black and (b) PA6.6 filled with carbon fibers.

percentage of filler. Figure 16.28 shows the SEM micrographs of two types of specimens. The specimens were also measured for their topography by an optical profiler, WYKO NT2000. We have used vertical scanning interferometry mode for the measurement. It has a range of 500 μm with a vertical resolution of 3 nm. The topography results for each type of specimens are shown in Figure 16.29; here, a scan area of 121 μm × 93 μm was selected for each sample in order to match the scan area of 100 × 100 μm^2 made by TPM. The carbon black-filled polymer surface has fine and randomly distributed features, while the carbon fiber-filled polymer has distinguished lines or scratches on the surface, which may indicate the existence of carbon fiber features. For the carbon black-filled polymer surface, it is very difficult to tell where the carbon powder is. There was a general problem during the measurement by WYKO NT2000 as only about 70–80% data was received and thus data restoration was used in the data processing. We have tried a different scanning range and found it getting worse with a larger scanning area. The filtered red light source was also tried but there was no further improvement. This is possibly due to the scattered nature and the dark color associated with the specimens. The roughness

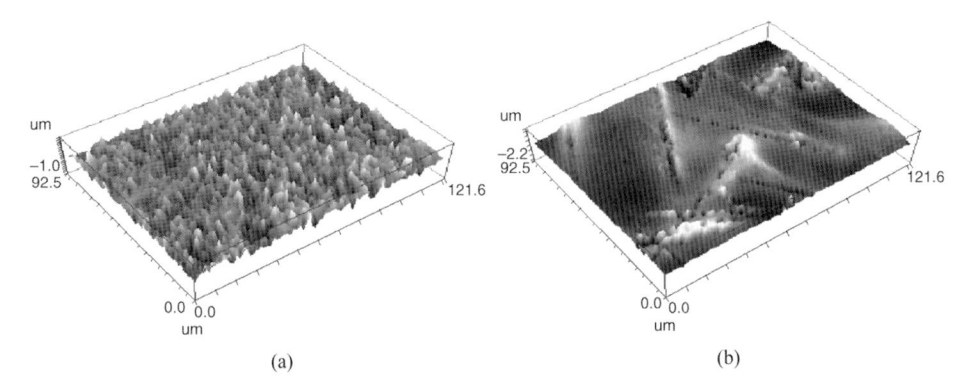

Figure 16.29 Surface topography of (a) PA12 filled with carbon black and (b) PA6.6 filled with carbon fibers, measured by WYKO NT2000.

16.4 Mechanical Property Mapping of Carbon-Filled Polymer Composites by TPM

Table 16.3 The averaged function parameters over the scanned area by TPM.

Sample No.	R_a (μm)	R_q (μm)	R_t (μm)	Hardness, H (GPa)			Young's modulus, E (GPa)			Comments
				Max.	Mean	Min.	Max.	Mean	Min.	
1 (PA12/C-black)	0.120	0.160	1.88							WYKO
	0.070	0.089	1.36	2.00	0.17	0.04	7.21	1.49	0.11	TPM
2 (PA12/C-black)	0.119	0.157	1.94							WYKO
	0.102	0.133	1.59	0.78	0.33	0.06	4.79	2.74	1.11	TPM
3 (PA6.6/C-fiber)	0.111	0.154	3.43							WYKO
	0.096	0.127	2.32	6.51	0.22	0.03	42.05	4.41	0.11	TPM
4 (PA6.6/C-fiber)	1.41	2.04	14.96							WYKO
	1.228	1.480	15.04	11.99	0.32	0.01	60.06	4.46	0.05	TPM

parameters, R_a and R_q, and the peak-to-valley parameter R_t for each specimen are listed in Table 16.3.

In general, surfaces measured by optical methods tend to give higher roughness values than the contact methods [19]. This is mainly due to the fact that the optical-based surface instruments are unable to cope with changes in optical surface properties, feature with wavelengths smaller than the laser focus spot and rapid slope changes in the surface. With the contact methods, the contact force, the size, and shape of the stylus tip will affect the measured result. It is not our intention in this chapter to justify which method is more suitable. We simply list the results obtained by two methods here for cross checking purpose.

16.4.4
Multifunction Mapping

The multifunction mapping of mechanical properties by TPM was then carried out on each of the four specimens. All the measurements were taken in a metrology laboratory with temperature and humidity controlled at 20 ± 1 °C and 40 ± 5% RH.

First, a series of indentation tests were carried out to see the behaviors of both hardness and Young's modulus under a sequential load. Then, the maximum force of 1 mN in the ramping mode was determined for all specimens. As mentioned before, the four functions were measured in two modes, the scanning mode and the force ramping mode. The topography and the frictional force were measured in the scanning mode with a contact force set at 0.2 mN, while the hardness and Young's modulus were obtained in the force ramping mode with the preselected force range

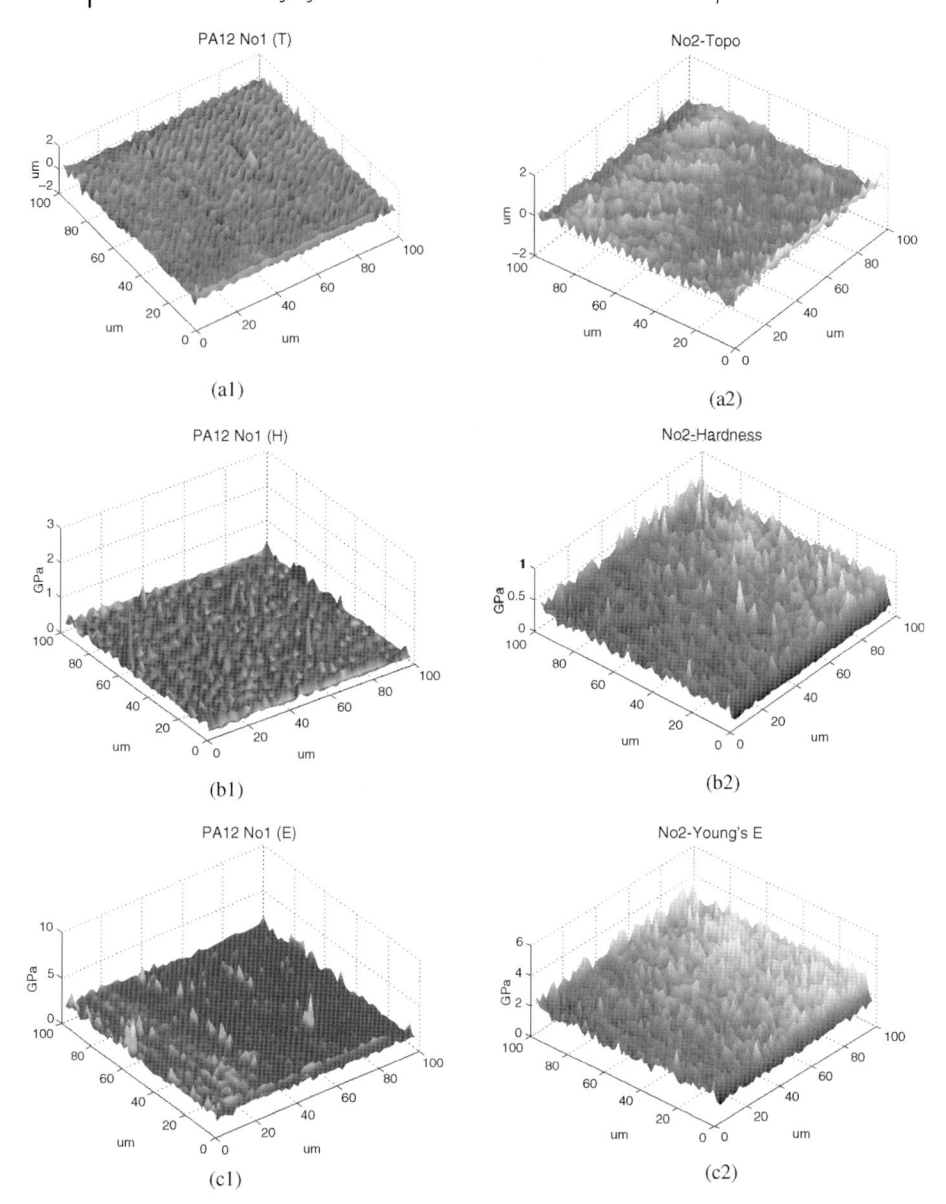

Figure 16.30 Mappings of topography (a1, a2), hardness (b1, b2), and Young's modulus (c1, c2) for PA12 filled with carbon black. The x–y axes are in micrometer.

of 1 mN. Here, the function maps of topography, hardness, and Young's modulus of four specimens are shown in Figures 16.30 and 16.31. The averaged function parameters over the scanned area for each specimen are listed in Table 16.3. For this investigation, the friction function was not concerned.

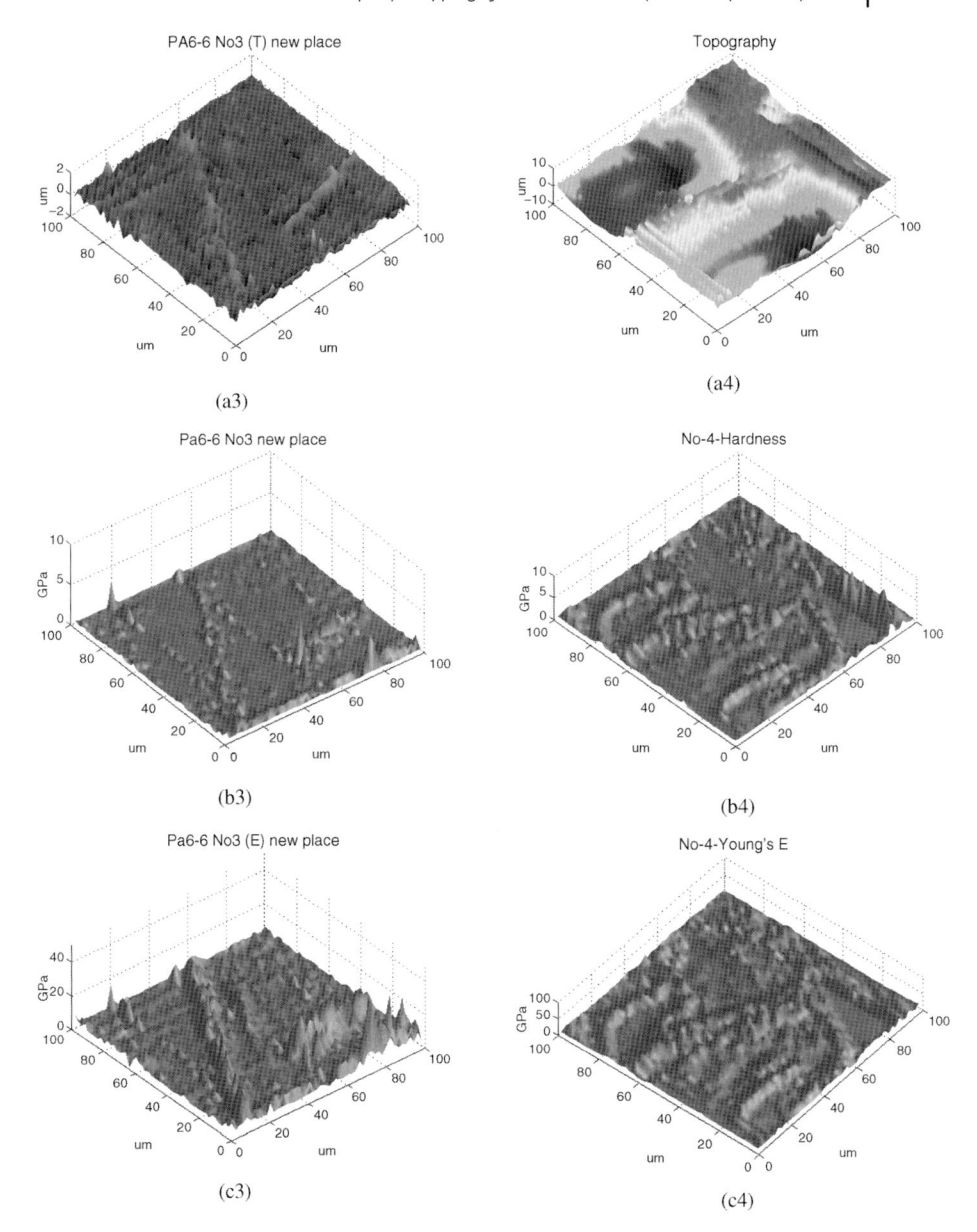

16.4 Mechanical Property Mapping of Carbon-Filled Polymer Composites by TPM

Figure 16.31 Mappings of topography (a3, a4), hardness (b3, b4), and Young's modulus (c3, c4) for PA6.6 filled with carbon fiber. The x–y axes are in micrometer.

For each specimen, the multifunction measurement was taken over an area of $100 \times 100\,\mu m^2$ with 50×50 data points; this gives a spacing of $2\,\mu m$. The choice of scanning points in the measurement is based upon the lateral resolution of the tip and the interaction/deformation between the tip and the material being measured.

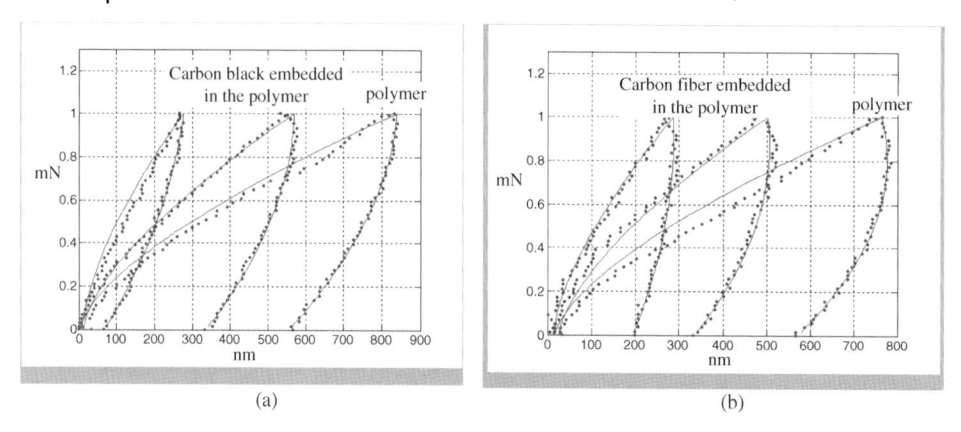

Figure 16.32 Typical indentation curves for (a) PA12 filled with carbon black and (b) PA6.6 filled with carbon fiber. The x-axis is the penetration depth in nm and the y-axis is the loading force in mN.

If only topography is to be measured, the scanning points can be made as many as 1000×1000 for the tip of 0.1 μm in radius when a very light contact force is used. As for multifunction mapping, the choice has to depend on the deformation made in the force ramping mode and the criterion is to make sure that there is no interference between two adjacent points being investigated. For a typical contact depth $h_c = 0.5$ μm, the affected area can be estimated to be within a radius of 1.4 μm from the projected area calculation. This typical indentation depth was aimed in the sequential loading mentioned above for determining the maximum ramping force. Due to the composite nature of the specimens in this case, the affected area could be bigger than the estimated one on some soft polymer parts, as shown in Figure 16.32. In the worst-case scenario, taking a contact depth up to 0.8 μm, the calculated project area would be within a radius of 2.2 μm. We had tried 5 μm spacing mapping, that is, 20×20 data points over 100×100 μm^2, but found that some features were simply lost due to the low lateral resolution. Therefore, a spacing of 2 μm was chosen, as a compromise, for this investigation, which was mainly to map the distribution of a harder material embedded in the soft. It is worth to note that the guideline (ISO 14577-1: 2002) for minimizing the interference between adjacent indents should be applied otherwise.

A slow scanning rate was used for all multifunction measurements. This is because at each of the surface points, a full indentation penetration was conducted to obtain a force/deformation curve. This involves 500 steps on each force loading and unloading and this process takes most of the measurement time. But still it takes less than the recommended time for conventional indentation tests. To save time, there was no dwelling time at the peak loading to allow a full penetration at each data point. For a typical 4 s for each indentation process, the total measurement time for each specimen was about 3 h for 50×50 data points. Obviously, the more data points (or higher lateral resolution) there are, the longer time it takes.

For hardness and Young's modulus mappings, the maximal values, average (mean) values, and minimal values are listed separately in Table 16.3, in order to see the difference between the base material (polymer) and the filler material

16.4 Mechanical Property Mapping of Carbon-Filled Polymer Composites by TPM

Table 16.4 Material distributions on the specimen surfaces over the scanned $100 \times 100\,\mu m^2$.

Sample No.	Depth < 400 nm	400 nm < Depth < 600 nm	600 nm < Depth
1 (PA12/C-black)	$0.88 < H < 1.68$ 0.2%	$0.23 < H < 0.9$ 12.7%	$0.08 < H < 0.2$ 87.1%
2 (PA12/C-black)	$0.6 < H < 0.9$ 0.7%	$0.21 < H < 0.62$ 84.6%	$0.06 < H < 0.27$ 14.7%
3 (PA6.6/C-fiber)	$0.35 < H < 9.23$ 8.8%	$0.15 < H < 0.54$ 21%	$0.02 < H < 0.21$ 70.2%
4 (PA6.6/C-fiber)	$0.35 < H < 8.3$ 18.8%	$0.13 < H < 0.68$ 29.6%	$0.01 < H < 0.49$ 51.6%

(carbon). From the hardness and Young's modulus mappings, it is easy to see the existence of two different materials on the surface of interest. The areas show high hardness values indicating existence of the carbon filler, where the substrate of polymer gives much low value in both hardness and Young's modulus. The carbon fiber-filled polymer specimens have a hardness range of 6–12 GPa, while the carbon black-filled polymer specimens are in a region of 0.8–2 GPa. The corresponding Young's modulus is 5–7 GPa for carbon black-filled polymer and 42–60 GPa for carbon fiber-filled specimens. The substrate polymer has a hardness of less than 0.2 GPa and a Young's modulus of less than 4.0 GPa.

In general, these carbons are embedded in the polymer, and this cushion effect makes the measured hardness/Young's modulus values much lower than those expected from pure carbons. Comparing the topography and hardness mappings, we are able to pick up the areas where carbons are located. For carbon fiber-filled polymer, the carbons seem to be hidden in the valleys. With carbon black-filled polymer, it is difficult to tell the whereabouts of carbons from topography only, but the hardness/Young's modulus mappings clearly show the difference.

For each specimen, all the scanned indentation curves were analyzed and there are three typical curves as shown in Figure 16.32. These curves represent characteristics of hard, intermediate, and soft nature of materials, and they can be used to distinguish the filler from the polymer substrate. The left curves in both graphs are for carbons and the right curves are for polyamides, while the middle ones give something in between, indicating the cushion effect as mentioned above. As the composite surface was scanned under the same ramping force, the maximum penetration depth is used to work out the distribution of the filler on the surface. Table 16.4 shows the results of such an analysis. Here, all indentation curves are sorted into three groups according to their penetration depths. The threshold for each category is determined after looking at the overall distribution. Within each category, the hardness range and its percentage are given. For instance with samples No. 1 and No. 3, the results show that percentage of the hardness values with penetration depth lower than 600 nm is 12.9% for No. 1 and 29.8% for No. 3, which match very well with the real percentages of fillers, 15–20% for carbon black in PA12 and 20–30% for carbon fiber in PA6.6. However, for the samples No. 2 and No. 4, the theme has changed: the percentage of harder material has increased to 85.3% for sample No. 2 and 48.4% for sample No. 4. This seems to be an effect of the percolation of carbon particles. It is particularly true for PA12 as we have found out that the measured electrical resistance on the surface is usually less than 1 kΩ, while the resistance deep below surface area is much higher, typically >10 kΩ. In return, we could tune the manufacturing process to enhance this carbon percolation on the surface region to increase the product conductivity.

16.5
Conclusions

We have demonstrated a new approach to characterizing the mechanical properties of polymer composites by using a newly developed multifunction tribological probe microscope. The TPM is capable of mapping surface topography, hardness, Young's modulus, and friction in a single set. It has shown that using the difference in hardness or elastic modulus, the TPM can be used to identify a certain material from its embedded substrate. The measured results show that the carbon fiber-filled polymer specimens have a hardness range of 6–12 GPa, while the carbon black-filled polymer specimens are in a region of 0.8–2 GPa. The corresponding Young's modulus is 5–7 GPa for carbon black-filled polymer and 42–60 GPa for carbon fiber-filled specimens. The substrate polymer has a hardness of less than 0.2 GPa and a Young's modulus of less than 4.0 GPa. In general, these carbons are embedded in the polymer and this cushion effect makes the measured hardness/Young's modulus values much lower than those expected from pure carbons. The results also show that carbon particles were mostly percolated to surface areas making the surface favorable to the electroforming processes.

References

1 Knothe, J. (1996) *Elektrische Eigenschaften von Spritzgegossenen Kunststoffformteilen aus Leitfähigen Compounds*, Institut für Kunststoffverarbeitung, RWTH Aachen.
2 Holstein, N. et al. (2005) Metallic microstructures by electroforming from conducting polymer templates. *Microsystem Technologies*, 11, 179–185.
3 Gilg, R. (1979) Ruß für leitfähige kunststoffe, in *Schriftenreihe Pigmente. s.l*, Degussa.
4 Gilg, R.G. (2002) Ruß und andere pigmente für leitfähige kunststoffe regensburg. [Hrsg.] otti Technik Kolleg. *Elftes Fachforum Elektrisch leitfähige Kunststoffe: Eigenschaften, Prüfung, Anwendungen*, OTTI.
5 Finnah, G. (2005) Injection molding processes for integration and assembly in MST, in *The World of Electronic Packaging and System Integration* (eds B. Michel and R. Aschenbrenner), ddp Goldenbogen, pp. 314–319.
6 Simona (2006) *Produktinformation Elektrisch leitfähige Kunststoffe*, Simona, Kirn.
7 Cabot (2006) *Verarbeitungsrichtlinien für Cabelec Compounds. s.l*, Cabot Corporation.
8 Clingerman, M.L. (1998) Development and modelling of electrically conductive

composite materials, Michigan Technological University, Dissertation.

9. Heldele, R. *et al.* (2006) Micro powder injection molding: process characterization and modeling. *Microsystem Technologies*, **12** (10–11), 941–946.

10. Heldele, R. *et al.* (2006) X-ray tomography of powder injection moulded parts using synchroton radiation. *Nuclear Instruments and Methods in Physics Research Section B: Beam Interactions with Materials and Atoms*, **246** (1), 211–216.

11. Piotter, V., Holstein, N., Merz, L., Ruprecht, R., and Hausselt, J. (2002) Methods for large scale manufacturing of high performance micro parts. Proceedings of 3rd Euspen International Conference, Eindhoven, The Netherlands, vol. 1, pp. 337–340.

12. Piotter, V., Holstein, N., Oskotski, E., Ruprecht, R., and Hausselt, J. (2003) Metal micro parts made by electroforming on two-component lost polymer moulds. International Topical Conference on Precision Engineering, Micro Technology, Measurement Techniques and Equipment, Aachen, Germany, Vol. 2, pp. 367–370.

13. Doerner, M.F. and Nix, W.D. (1986) A method for interpreting the data from depth-sensing indentation instruments. *Journal of Materials Research*, **1** (4), 601–609.

14. Oliver, W.C. and Pharr, G.M. (1992) An improved technique for determining hardness and elastic modulus using load and displacement sensing indentation experiment. *Journal of Materials Research*, **7** (6), 1564–1583.

15. Herrmann, K., Jennett, N.M., Wegener, W., Meneve, J., Hasche, K., and Seemann, R. (2000) Progress in determination of the area function of indenters used for nanoindentation. *Thin Solid Films*, **377–378**, 394–400.

16. Liu, X. (2002) A four-in-one tribological probe microscope for characterising surface properties at the micro and nanometre level. *Proceedings of the Royal Microscopical Society*, **37** (4), 215–223.

17. Liu, X., Bell, T., Chetwynd, D.G., and Li, X.Y. (2003) Characterisation of engineered surfaces by a novel four-in-one tribological probe microscope. *Wear*, **255**, 385–394.

18. Liu, X. and Gao, F. (2004) A novel multi-function tribological probe microscope for mapping surface properties. *Measurement Science & Technology*, **15**, 91–102.

19. Mattsson, L. and Wågberg, P. (1993) Assessment of surface finish on bulk scattering materials: a comparison between optical laser stylus and mechanical stylus profilometers. *Precision Engineering: Journal of the American Society for Precision Engineering*, **15** (3), 141–149.

20. Piotter, V. *et al.* (2010) Multi-component microinjection moulding: trends and developments. *International Journal of Advanced Manufacturing Technology*, **47**, 63–71.

21. Prokop, J. *et al.* (2008) Manufacturing process for high aspect ratio metallic micro parts made by electroplating on partially conductive templates. *Microsystem Technologies*, **14**, 1669–1674.

17
Carbon Black-Filled Natural Rubber Composites: Physical Chemistry and Reinforcing Mechanism

Atsushi Kato, Yuko Ikeda, and Shinzo Kohjiya

17.1
Introduction

The higher-order structures governing the properties of polymers range in size from the nanometer to the micrometer order in term of the length. Activities have been under way in the field of polymer nanotechnology to create high-performance materials by intentionally manipulating the higher-order structures, microphase separation, and the state of nanoparticle dispersion [1, 2]. These efforts are beginning to result in the creation of several new nanomaterials that possess novel functionality and structures [3–6]. A traditional yet very good example of a nanocomposite is the carbon black-filled rubber used to make the tires of automobiles and airplanes that we see in everyday life [7, 8]. Such a nanocomposite can be regarded as a typical "soft material." This material is obtained by vulcanizing rubber compounded with carbon black (CB) nanofiller particles of several tens of nanometers in size. The structure of nanofiller aggregates/agglomerates in vulcanized rubber is said to govern the properties of the resultant product. However, the compounding technique and the manufacturing process have often been carried out on the basis of an engineer's many years of experience and intuition. Here, an aggregate stands for more or less fundamental units of carbon black that have been formed during their manufacturing process of primary particles, and have not been broken down by a mechanical mixing [1]. An agglomerate is the product of further associations of aggregates, which are the case in the rubbery matrix. So far, the agglomerates have been called "structures," and the influence of the structures on filled rubber properties has been considered much. However, the concrete structures are not elucidated much. The present authors are proposing a CB network structure in the rubbery matrix as a general structure of CB agglomerates [1, 2].

Recent years have also seen the development of high-performance carbon black and silica fillers. For example, silica has come to be used frequently in recent years especially in "green" tires [9–11]. However, because there are no methods currently available for observing the structure of such fillers in three dimensions, it is virtually impossible to execute a rational material design. For that reason, there have been strong demands for the development of a device or method for visualizing quantitatively three dimensional structures at the nanometer level. In recent years, an analysis method called three dimensional transmission electron microscopy (3D-TEM) has been developed as a technique for observing 3D morphologies, based on the combined use of TEM slice imaging and electron tomography. Because 3D-TEM images provide information in the thickness direction of a sample, it is possible to observe and analyze the 3D structure of nanoparticle aggregates/agglomerates [1, 2, 5–16].

Carbon black is blended with polymers for the purpose of reinforcement and for improving electrical properties. It is well known that adding electrically conductive fillers such as carbon black or a metal powder to plastics or rubber lowers the electrical resistance of the material, a phenomenon that is referred to as percolation. It has been reported that percolation behavior varies depending on the type of carbon black used or the surface treatment applied [17–19]. In the percolation process, volume resistivity generally does not decrease linearly in relation to the loading level of the electrically conductive filler. Until a certain loading level (i.e., percolation threshold) is reached, volume resistivity hardly changes. Once that percolation threshold is exceeded, volume resistivity declines sharply and becomes nearly constant.

This percolation behavior has traditionally been used to impart electrical conductivity to polymer materials in order to eliminate electrostatic charges in the material for the sake of safety. Moreover, there have been cases in recent years where the plastic or rubber parts used in electrical and electronic equipment required a balance between the imparted conductivity and elimination of static electricity. For example, the transfer methods used in laser printers include the corona transfer technique, whereby a corona charger imparts to the paper surface an electrical charge having the opposite polarization of the toner, and the roller transfer technique that passes the paper through narrow guides between the belt and the photoconductive drum [20–23].

The transfer belt consists of a urethane rubber base layer and a fluororubber release layer. It is reported that volume resistivity of approximately $100\,G\Omega/cm$ is needed to enable the belt to exert sufficient electrostatic adsorption force even if resistance decreases due to high humidity. A roller made of flexible conductive rubber is used in the roller transfer method. Like the transfer belt, the rubber base material of the roller must have precise electrical properties. For that reason, resistance must be controlled precisely in the vicinity of the percolation threshold. It is believed that the existence of the percolation threshold suggests the occurrence of one type of transition phenomenon, although no basis for this assumption has yet been found. Additionally, it is said that the conductive filler forms an electric network in the region where volume resistivity becomes constant, which accounts for this tendency.

The following sections describe the changes that occur in the mechanical, thermal, electrical, and dielectric properties of carbon black-filled sulfur-cured natural rubber (NR) vulcanizates near the percolation threshold and the relationship with the dispersed structure of the CB aggregates/agglomerates, based on the use of 3D-TEM.

17.2
3D-TEM Observation of Nanofiller-Loaded Vulcanized Rubber

As Coran [24, 25] has pointed out, the activator, zinc oxide (ZnO), used in the sulfur vulcanization process reacts with sulfur and with the vulcanization accelerator such as N-cyclohexyl-2-benzothiazole sulfenamide (CBS) to produce rubber-soluble Zn compounds. It was found previously that these Zn compounds scattered the electron beam in 3D-TEM observations, making it impossible to obtain distinct 3D-TEM images of the nanofiller (CB, silica, etc.) in thin specimens that were merely cut from the vulcanized rubber as mentioned above. To overcome that problem, two types of original pretreatment processes were applied to the vulcanized rubber in this study to obtain samples for 3D-TEM observation. One process made use of the NISSAN ARC-AK method for removing the rubber-soluble Zn compounds that cause the electron beam to scatter. Following that pretreatment process, an ultramicrotome was used to cut thin specimens from the vulcanized rubber that had been frozen at the temperature of liquid nitrogen (approximately $-198\,°C$). The other original pretreatment process (NISSAN ARC-SG method) was then applied to the thin specimens to make their thickness uniform and simultaneously make the surface smooth [1, 2, 26, 27].

The instrument used in the 3D-TEM observations was a Tecnai G2 F20 TEM (FEI Company). The accelerating voltage of the electron beam was set at 200 kV. Samples were tilted over a range of angles from $-70°$ to $+70°$, and image data (tilted images) were continuously obtained in $2°$ increments as shown in Figure 17.1. This meant that 71 consecutive tilted images altogether were automatically loaded into the computer. The positions of the tilted images were aligned at that time and a search was made for their axis of rotation. These tilted images were not simply 2D image slices, but rather 2D projected images of the mass–density distribution of the samples. Using the IMOD software (a program created at Colorado University [28])

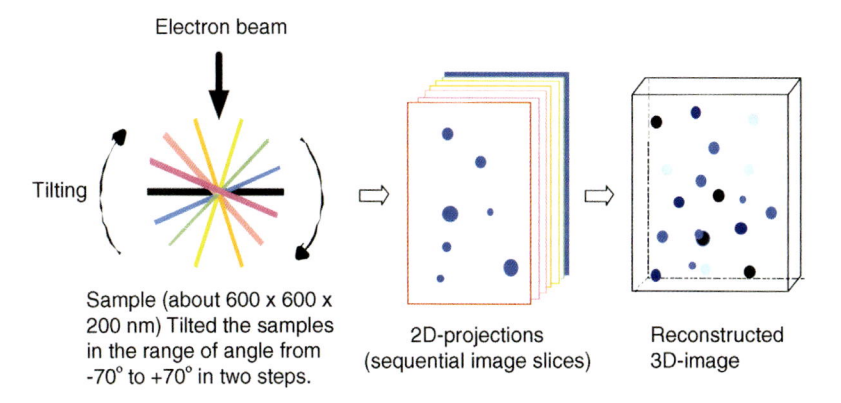

Figure 17.1 3D-TEM observation principle and conditions. Image slices are obtained while tilting the sample in one axial direction. Radon transform operation on the image slices to construct 3D-images.

installed on the TEM, the consecutive tilted images thus obtained were converted to image slices showing the mass–density distribution at each angle. Then, 3D images were reconstructed from the image slices by applying a Radon transform using the Amira software developed by TGS, Inc. [29].

Simultaneously, volume rendering or surface rendering was performed to construct the 3D images on a nanometer scale [30]. The former method reconstructs a 3D object by stacking cross-sectional images vertically and displaying the data semitransparently. The latter method reconstructs a 3D object by extracting the contours of the object depicted in a cross-sectional image in terms of curves and inserting surfaces between many of the curves thus obtained.

The foregoing discussion has explained the principle of 3D-TEM, which is shown in Figure 17.1. This principle is virtually the same as that of X-ray computed tomography. Because an electron beam is used as a probe, this technique is also called electron tomography. The use of 3D-TEM not only facilitates 3D visualization of nanoparticles but also makes it possible to calculate their number, density, 3D shape, volume, and other details [1, 2, 10–16, 26, 27, 31–33]. Since it is well known that organic polymers are apt to be damaged by an electron beam, this is an issue that must always be kept in mind when making 3D-TEM measurements [34].

17.3
Materials: CB-Filled Sulfur-Cured NR Vulcanizates

The compounding recipes of the CB-filled sulfur-cured NR vulcanizates investigated in this study are given in Table 17.1. The only ingredient changed in this series of NR vulcanizate samples was the CB loading, which was varied in a range of 5 to 80 phr (grams of additive per 100 g of rubber). The quantities of the other compounding agents were all the same in every vulcanizate sample: 2 phr of stearic acid (LUNAC S-25, Kao Co.), 1 phr each of ZnO (average diameter: 0.29 μm, Sakai Chemical Industry Co., Ltd.) and CBS (Sanceler CM-G, Sanshin Chemical Industry Co., Ltd.),

Table 17.1 Recipes of carbon black-filled NR compounds[a].

Samples	CB-0	CB-5	CB-10	CB-20	CB-30	CB-40	CB-50	CB-60	CB-80
Ingredient (phr)[b]: NR (RSS # 1)	100	100	100	100	100	100	100	100	100
Stearic acid (ST)	2	2	2	2	2	2	2	2	2
ZnO	1	1	1	1	1	1	1	1	1
CBS[c]	1	1	1	1	1	1	1	1	1
Sulfur	1.5	1.5	1.5	1.5	1.5	1.5	1.5	1.5	1.5
Carbon black[d]	0	5	10	20	30	40	50	60	80

a) Curing conditions: 140 °C, 15 min under pressure.
b) Per 100 g of rubbers.
c) N-cyclohexyl-2-benzothiazole sulfenamide.
d) HAF, dry at 120 °C for 2 h.

17.4
Relationship Between the Properties of CB-Filled Sulfur-Cured NR Vulcanizates and CB Loading

and 1.5 phr of sulfur (powder, 150 mesh, Hosoi Chemical Industry Co., Ltd.). Vulcanization was performed under pressure at 140 °C for 15 min in a mold.

17.4.1
Tensile Behavior

The tensile stress (σ_t) versus elongation ratio (α) behavior of the CB-filled sulfur-cured NR vulcanizates at room temperature is plotted in Figure 17.2. It will be noted here that α represents the ratio (L/L_0) of the sample length before (L_0) and after elongation (L). Samples with a CB loading of 20 phr or lower showed a low elastic modulus, high breaking elongation, and nearly the same rupture strength. In contrast, samples with a CB loading of 40 phr or higher displayed a high elastic modulus, low breaking elongation, and noticeably high rupture strength. These results indicate that the tensile behavior of the samples changed markedly near a CB loading of about 30 phr. This sort of discontinuous change in the elastic modulus cannot be explained by the theory of rubber elasticity even if we take into account a slight increase in the network density of the samples accompanying an increase in the CB loading. Omnèsa et al. [35] have also reported an improvement in the elastic modulus, tensile strength, and other mechanical properties of CB-filled NR as a result of increasing the CB loading. They suggested that this improvement can be understood in terms of the interactions

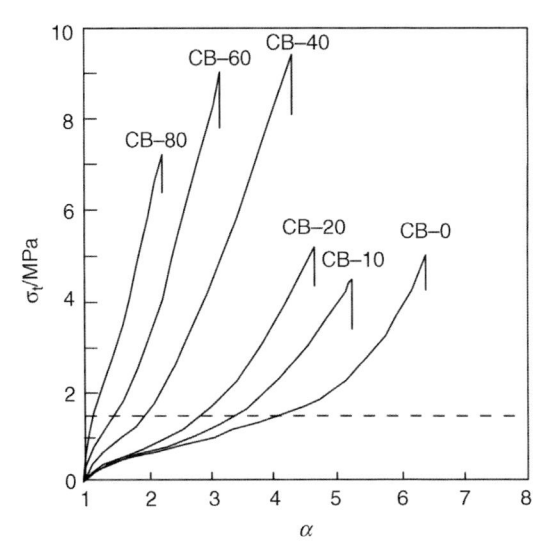

Figure 17.2 Tensile stress (σ_t) versus elongation ratio (α) behavior of CB-filled NR vulcanizates at room temperature. $\alpha = L/L_0$ (L, L_0: elongated and original length).

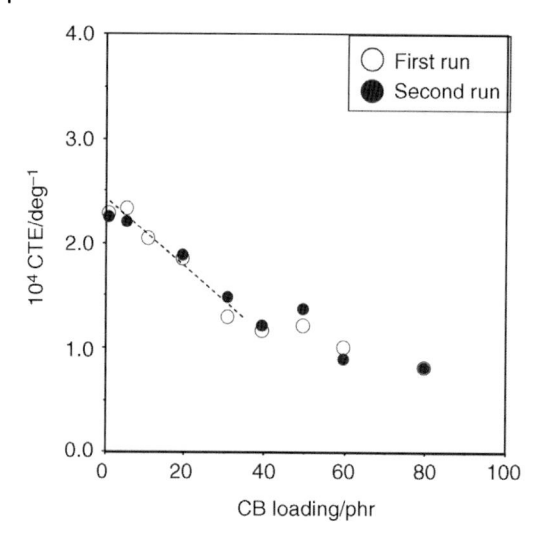

Figure 17.3 Relationship between coefficient of thermal expansion (CTE) and CB loading of CB-filled sulfur-cured NR in a temperature region from 296 to 363 K under N_2 gas atmosphere.

between the fillers and between the fillers and the rubber, based on a consideration of the occluded rubber, bound rubber, and the percolating network. However, these are all so-called circumstantial evidences, not a conclusive one.

17.4.2
Coefficient of Thermal Expansion

The coefficient of thermal expansion (CTE) found for the CB-filled sulfur-cured NR vulcanizate samples in a temperature range from room temperature to 90 °C is plotted in Figure 17.3 as a function of the CB loading. The results measured for the same sample during the first temperature increase (first run) and the second temperature increase (second run) are indicated by the open (o) and closed (•) circles, respectively. For CB loadings of 0–30 phr, the CTE values decreased almost linearly with an increase of CB loading. From around a CB loading of 40 phr, the CTE values tended to increase relatively gradually or converge to a nearly constant level. These results suggest that a transition occurred in the CB dispersion and agglomeration in a CB loading range around 40 phr and that the CB network that formed at higher CB loadings reduced the CTE. In this regard, Hong et al. [36] have proposed that the CB network interferes with the thermal motion of the rubber molecule chains, based on the fact that the CTE of CB-filled rubber is a decreasing function of the CB loading.

17.4.3
Viscoelastic Properties

Dynamic viscoelastic tests were conducted on CB-filled sulfur-cured NR vulcanizate samples in a tensile mode at a frequency of 10 Hz in a nitrogen atmosphere in a

17.4 Relationship Between the Properties of CB-Filled Sulfur-Cured NR Vulcanizates and CB Loading

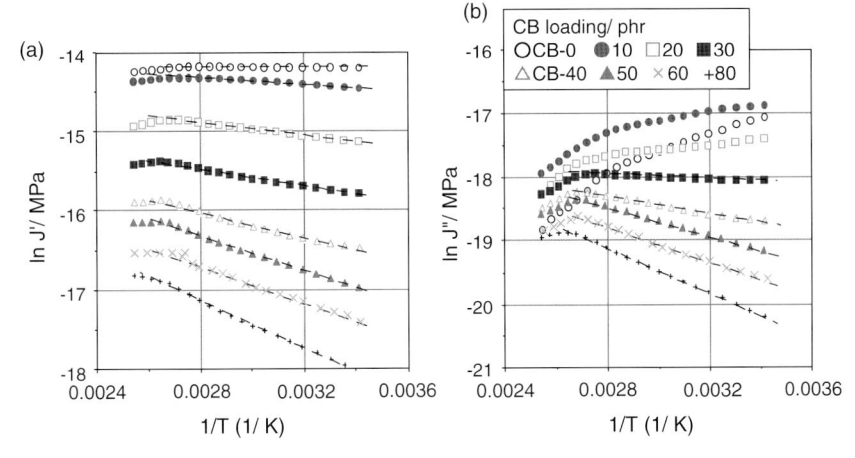

Figure 17.4 Dependence of (a) storage compliance (J') and (b) loss compliance (J'') on temperature (T). Conditions of dynamic mechanical thermal analysis: mode, tensile; temperature, 293–393 K; frequency, 10 Hz; atmosphere, N_2 gas.

temperature range of 293–393 K. Figure 17.4 presents the Arrhenius plots showing the relationship between the logarithms of the storage compliance (J') and loss compliance (J'') and the reciprocal of temperature ($1/T$). All of the tested samples showed a negative linear relationship for the $1/T$ dependence of J'. Additionally, the samples with a CB loading of 30 phr or higher showed a negative linear relationship for the $1/T$ dependence of J''.

The activation energies ($\Delta E_{J'}$ and $\Delta E_{J''}$) calculated from these two negative linear relationships are plotted in Figure 17.5 as a function of the CB loading. The

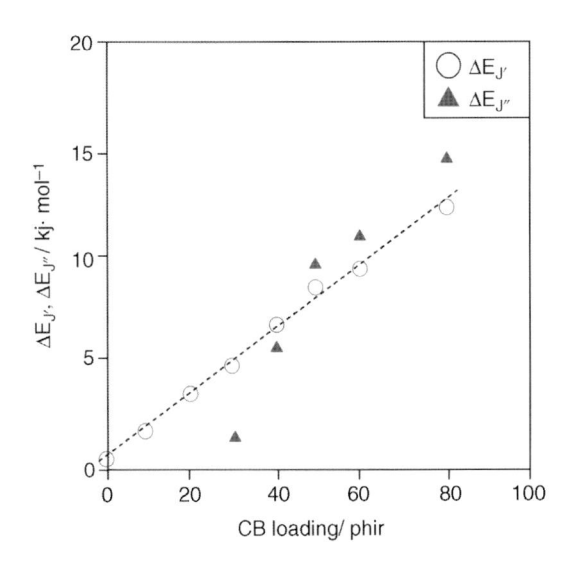

Figure 17.5 Relationship between activation energy $\Delta E_{J'}$, $\Delta E_{J''}$, and CB loading.

relationship between $\Delta E_{J'}$ and the CB loading was nearly linear for all the samples. This linearity is therefore assumed to be attributable to the rubber matrix. The other activation energy $\Delta E_{J''}$ appeared near a CB loading of 30 phr and tended to increase almost linearly with a higher CB loading. It is inferred from these results that this tendency is ascribable to the detailed structure of the CB aggregates/agglomerates. In this connection, Caoa et al. [37] have pointed out that the activation energy of thermally induced percolation in CB-filled high-density polyethylene (HDPE) is markedly higher than the viscous flow or molecular relaxation in the polymer matrix.

17.4.4
Electrical Properties: Volume Resistivity and Conductivity

Three principal mechanisms have usually been advanced to explain electrical conduction in CB-filled polymer materials. One mechanism involves the theory of conduction channels. This theory posits that CB particles form chain structures through which π electrons migrate to achieve conduction. Frenkel [38] has proposed a particle gap of 1 nm as the barrier over which electrons can easily cross. However, there are cases where conduction is manifested even though the polymers are filled with CB particles that do not readily form chain structures or a much larger particle gap separates CB particles that easily form such structures. Moreover, conduction also takes place in polymers filled with a metal powder filler, which does not form any chain structures at all. Considering these cases, it is difficult to explain the conduction mechanism of CB-filled polymer materials solely on the basis of the theory of conduction channels.

The second conduction mechanism concerns the tunneling effect or the hopping of electrons. Polley and Boonstra [39] proposed that electrical conduction is achieved by means of electrons hopping across the gap between CB particles dispersed in a polymer matrix, rather than simply through contact between CB particles. Based on TEM studies, they reported that conduction occurred even in cases where no CB chain structures were observed in the rubber matrix and in cases where the distance between CB particles was much greater than 1 nm. Based on the relationship between the specific volume and resistance of compressed CB-filled vulcanized rubber, Voet [40] found that electrical conductivity is related to the size of the gap between the CB particle chain structures rather than to the length of the chains. Bahder and Garcia [41] also showed that the conduction mechanism of CB-filled polyethylene is related to both the theory of conduction channels and the hopping of electrons. El-Tantawy et al. [42] investigated the electrical and thermal stability of epoxy composites at high CB concentration. They thought that the negative resistance phenomena in CB-filled epoxy composites was attributed to the negative temperature coefficient of conductivity (NTCC). Konishi and Cakmak [43] reported that organoclay can be used as a dispersion control agent in the polyamide 6-carbon based nanoparticle hybrids to induce self-assembly of CB network at low CB content, simultaneously, partial blocking the electron hopping pathways to level the slope of percolation curves.

The third conduction mechanism is based on the field emission theory. Van Beek and van Pul [44] reported that sulfur-cured NR vulcanizates filled with two types of CB

17.4 Relationship Between the Properties of CB-Filled Sulfur-Cured NR Vulcanizates and CB Loading

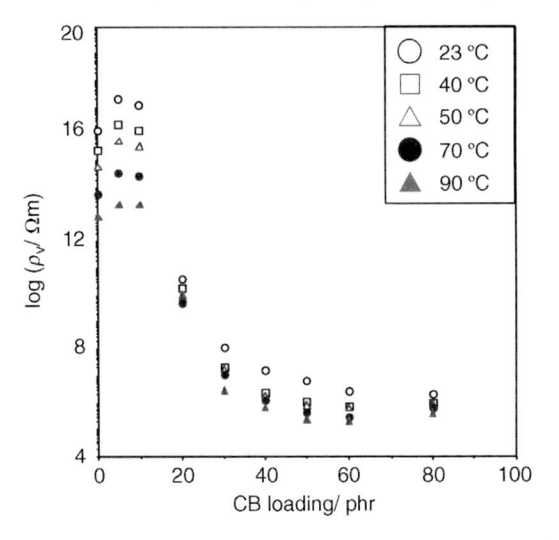

Figure 17.6 Relationship between volume resistivity (ϱ_v) and CB loading of CB-filled NR vulcanizates at various temperatures.

(high abrasion furnace (HAF) and medium thermal (MT)) displayed non-Ohmic current–voltage characteristics. Accordingly, he suggested that an insulator was present between the CB particles and that the high field intensity generated between the particles produced field emissions, causing a current to flow.

The relationship found in the present study is shown in Figure 17.6 between the volume resistivity (ϱ_v) and CB loading of CB-filled sulfur-cured NR vulcanizates at a temperature range of 23–90 °C. Let us first look at the results for 23 °C. For CB loadings of 10 phr or less, the ϱ_v values are nearly the same, but once the CB loading is increased above 10 phr, ϱ_v decreases sharply and then tends to remain nearly constant at CB loadings of 40 phr or higher. This phenomenon involving a sudden change at a certain point from an insulator to a conductor is referred to as electrical percolation. It is said that a conduction circuit or an electrical network is completed at the point where ϱ_v becomes constant [45]. Meanwhile, Blythe [46] suggested on the basis of their investigation of the temperature dependence of conductivity that it is possible to distinguish between the hopping mechanism whereby electrons hop over barriers and the tunneling mechanism whereby electrons pass through barriers. Therefore, let us focus here on the relationship between ϱ_v and the CB loading at each temperature. The tendency seen in Figure 17.6 for ϱ_v to decrease with increasing temperature suggests that the conduction mechanism of CB-filled NR vulcanizates is dependent on an activation process.

Figure 17.7 presents the Arrhenius plots showing the temperature dependence of the conductivity ($\sigma = 1/\varrho_v$) of the CB-filled sulfur-cured NR vulcanizates (Figure 17.7b) and the activation energy (ΔE_a) calculated from the results, which is shown as a function of the CB loading. The Arrhenius plots in Figure 17.7a show that the negative slopes of the linear approximation curves change profoundly near a

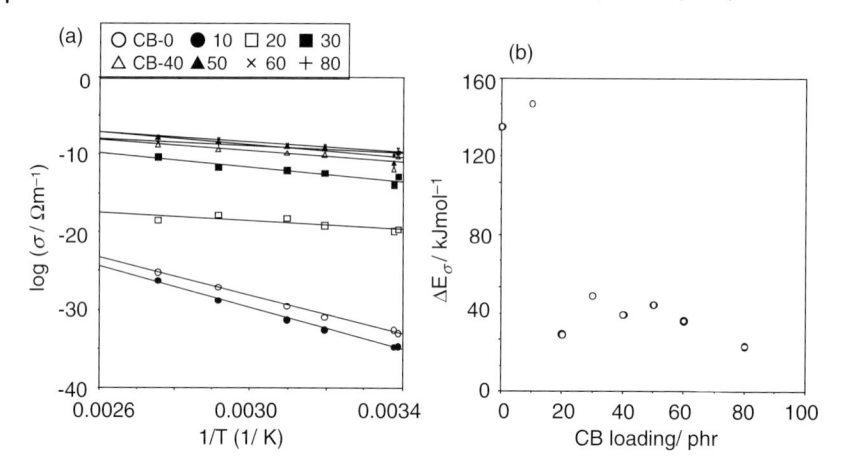

Figure 17.7 Electric conductivity of CB-filled sulfur-cured NR. (a) Dependence of conductivity ($\sigma = 1/\varrho_v$, ϱ_v: volume resistivity) on temperature. (b) Relationship between activation energy (ΔE_σ) of conductivity and CB loading.

CB loading of 20 phr. Specifically, while exceptionally large ΔE_σ values are seen at CB loadings of 10 phr or less, ΔE_σ is only about one-fifth as much (20–40 kJ/mol) for CB loadings of 20 phr or higher. These values are nearly equal to the activation energy levels reported by Jaward and Alanajjar [47], which they calculated from the dielectric relaxation characteristics of rubber compounded with 45 and 60 phr of graphitized CB. This suggests that an electrical network is formed between the CB aggregates in the high CB loading region through which electrons migrate by means of electron hopping or tunneling. In the low CB loading region, electron hopping between CB aggregates or between conductive impurities located in close proximity due to the thermal motion of the rubber molecule chains has to occur in order for current to flow in the samples. That is presumably the reason why relatively high activation energy was observed in the low loading region.

In other words, electrical conduction is achieved in both the higher and lower ΔE_σ mechanisms as a result of electrons hopping over energy barriers. At CB loadings of 10 phr or less, the conduction mechanism is governed by the rubber matrix, whereas in the CB loading region above 20 phr, ΔE_σ converges to a nearly constant value, implying that the conduction mechanism is attributable to the formation of a CB network, that is, the linkage of the CB structures consisting of CB aggregates/ agglomerates. It is assumed that the formation and destruction of this CB network is dependent on the compounding, vulcanizing, and molding conditions and that these conditions influence the mechanical properties of vulcanized rubber [45, 48–58]. Reffaee et al. [45] have suggested that there is a correlation between the electrical properties (conductivity and dielectric relaxation) and mechanical properties of HAF CB-filled styrene butadiene rubber (SBR), linear low-density polyethylene (LLDPE), and nitrile butadiene rubber (NBR)/LLDPE blends. Moreover, Satoh et al. [59] examined the viscoelastic properties of CB-filled SBR and separated the components contributing to the elasticity of the CB network. In addition, based on the strain

17.4 Relationship Between the Properties of CB-Filled Sulfur-Cured NR Vulcanizates and CB Loading

dependence of the viscoelastic properties and electrical conductivity, they suggested that the CB network in rubber is destroyed above a certain strain level.

17.4.5
Dielectric Properties: Time–Temperature Superposition of Dielectric Relaxation of CB-Filled NR Vulcanizates

Figure 17.8 shows the relationship between the loss tangent (tan δ) and the reciprocal of frequency (f) for the CB-0, CB-10, CB-20, and CB-40 NR vulcanizate samples in a temperature range of 203–373 K. A marked increase in tan δ is observed for all the samples on the long time (low frequency) side in the region of 363–373 K. At other temperatures the samples show complex relaxation behavior depending on the CB loading. Specifically, only one tan δ relaxation peak is seen for the CB-0 sample outside the low-frequency region. This peak is attributed solely to the vulcanized rubber. In contrast, in the CB loading range of 5–10 phr, two or three relaxation peaks

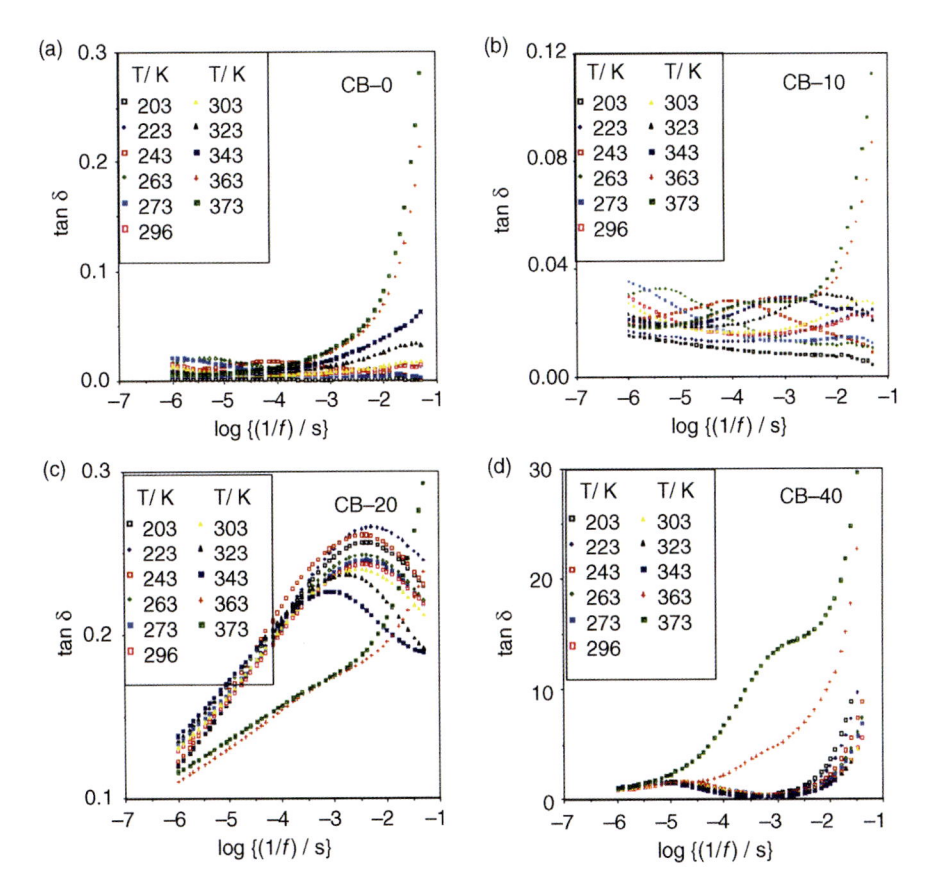

Figure 17.8 Relationship between loss tangent (tan δ) and frequency of dielectric measurements for (a) CB-0, (b) CB-10, (c) CB-20, and (d) CB-40.

are observed at frequencies lower than that of the relaxation peak ascribable to the rubber matrix. It is reasonable to assume that these peaks are related to the dispersion state of the CB aggregates/agglomerates, as will be described later. In the CB loading range of 20–40 phr, on the other hand, only one relaxation peak is observed and it has shifted to the high frequency side compared with the peak attributed to the vulcanized rubber. Although not shown in Figure 17.8, the results for the CB-80 sample displayed a monotonic curve without any relaxation peaks. This implies that the CB aggregates/agglomerates form a more strongly linked network in the relatively high CB loading region.

A master curve can sometimes be composed at a standard temperature with respect to the dielectric relaxation phenomena attributable to the constituent units of polymers and the orientation polarization of the functional groups [60–64]. This is done by applying the time–temperature superposition principle to shift the time- or frequency-dependent curves of the dielectric characteristics at various temperatures. Figures 17.9–17.11 show the master curves obtained for the CB-0, CB-5, CB-10, CB-20, CB-40, and CB-80 samples by superposing the frequency (20 Hz–1 MHz) dependence of their dielectric loss tangent (tan δ) at various temperatures by setting a standard temperature at 23 °C. Excluding the low-frequency region on the right-hand side of the horizontal axis in Figures 17.9–17.11, one relaxation peak (indicated by the arrow) is seen for the CB-0 sample in Figure 17.9a, whereas three or two peaks

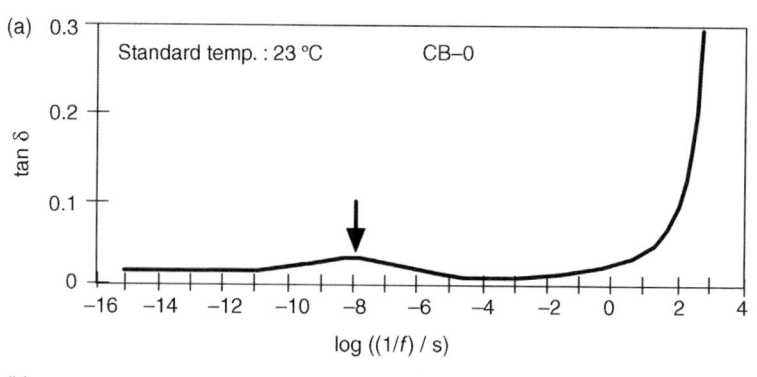

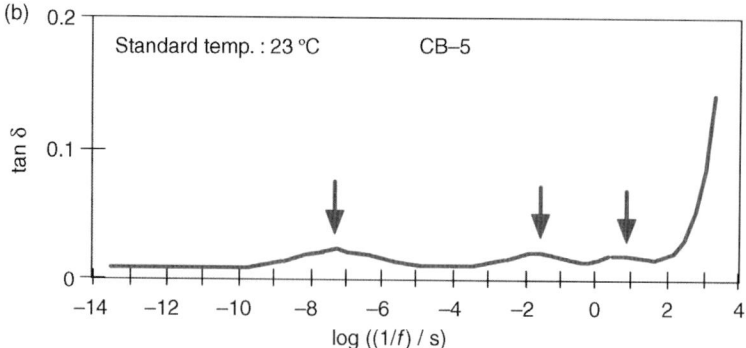

Figure 17.9 Time–temperature superposition results for (a) CB-0 and (b) CB-5.

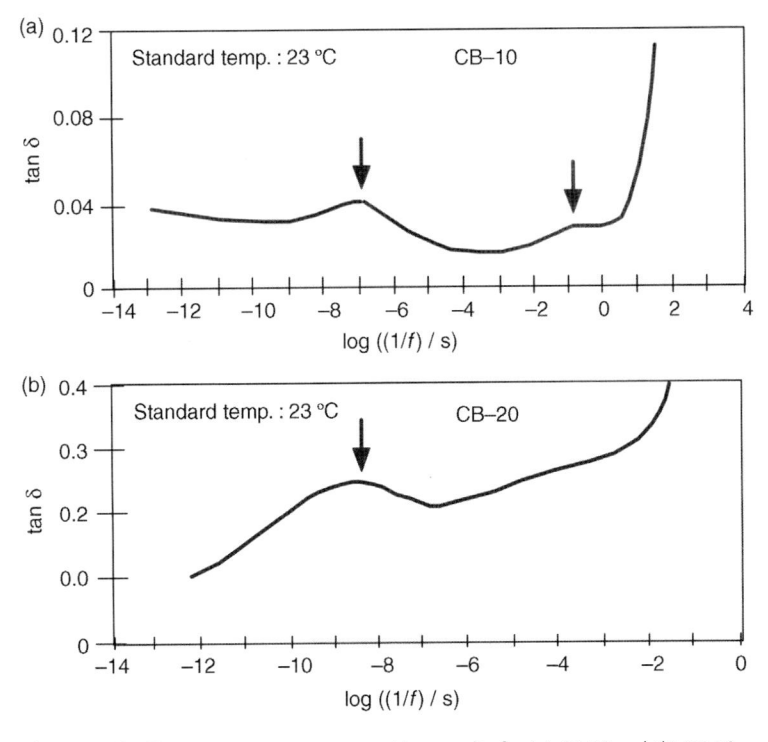

17.4 Relationship Between the Properties of CB-Filled Sulfur-Cured NR Vulcanizates and CB Loading

Figure 17.10 Time–temperature superposition results for (a) CB-10 and (b) CB-20.

are observed for the CB-5 and CB-10 samples in Figures 17.9b and 17.10a. The master curves of the CB-20 and CB-40 samples in Figures 17.10b and 17.11a show only one relaxation peak. However, no relaxation peak was observed for the CB-80 sample in Figure 17.11b. Among these relaxation peaks, the peak on the short time (high frequency) side is presumably attributable to the rubber matrix, as was true for the CB-0 sample. The other relaxation peaks observed for the CB-5, CB-10, and other samples are attributed to the state of CB dispersion and the structure of the CB aggregates/agglomerates.

Figure 17.12 shows the temperature (T) dependence of the horizontal (time) axis shift factor ($\log a_T$) and the vertical ($\tan \delta$) axis shift factor ($\log b_T$) used to obtain the $\tan \delta$ master curves (Figures 17.9–17.11) of the CB-filled NR vulcanizates in a CB loading range of 0–80 phr. Surprisingly, the temperature dependence of $\log a_T$ in Figure 17.12a shows a dramatic difference between the CB loading ranges of 0–10 and 20–80 phr. In the low loading range, the temperature dependence of $\log a_T$ shows behavior typical of polymers, that is, $\log a_T$ decreased in a downward curve with increasing temperature. This implies that the lateral shift is attributable to viscoelasticity. In contrast to that tendency, the results for the higher loading region indicate that $\log a_T$ became nearly zero at temperatures below 70 °C. At higher temperatures above 70 °C, some type of thermal factor induced a structural change in the samples and that change tended to increase with a higher CB loading.

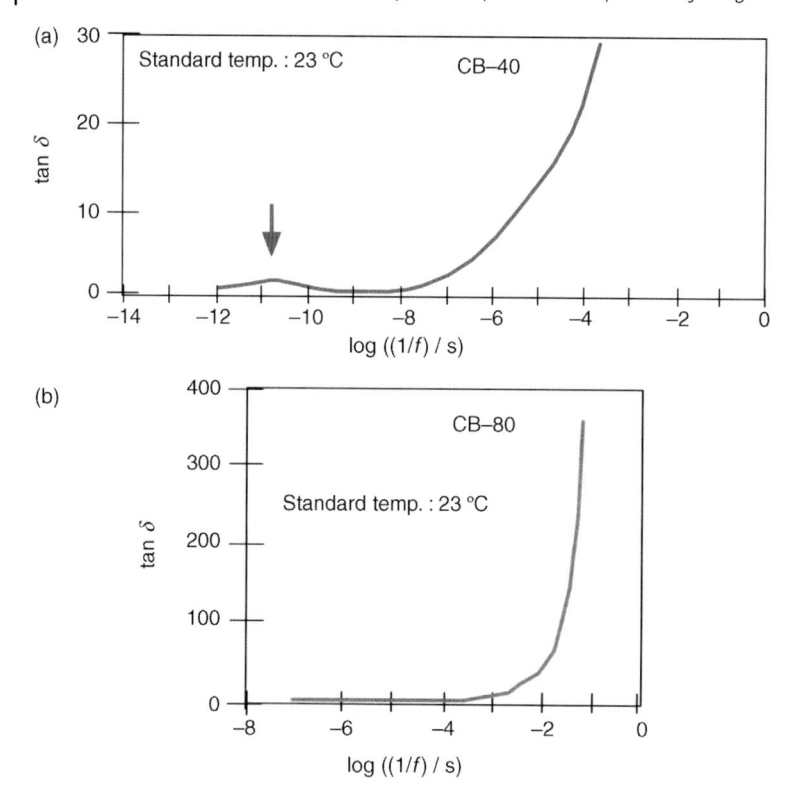

528 | *17 Carbon Black-Filled Natural Rubber Composites: Physical Chemistry and Reinforcing Mechanism*

Figure 17.11 Time–temperature superposition results for (a) CB-40 and (b) CB-80.

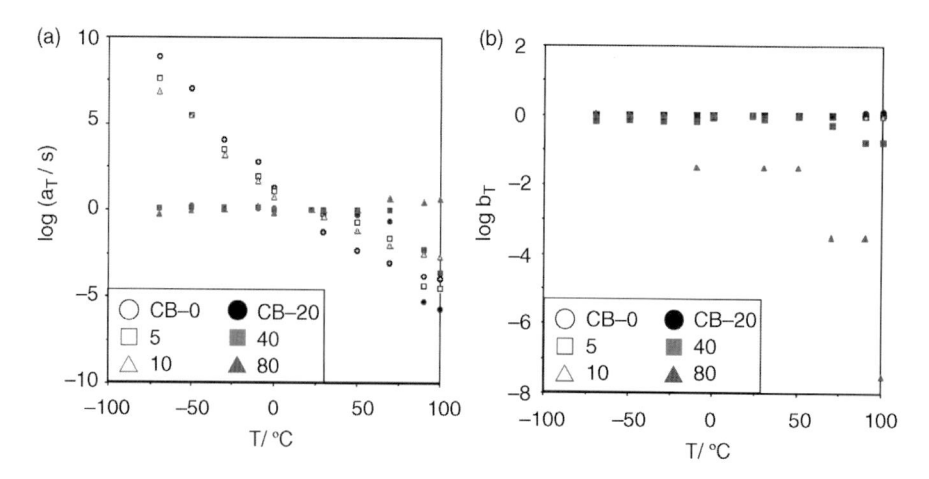

Figure 17.12 Temperature (T) dependence of shift factors (a) $\log a_T$ and (b) b_T.

The results in Figure 17.12b for the temperature dependence of $\log b_T$ are also very interesting. For the CB-0, CB-5, CB-10, and CB-20 samples with a CB loading in a range of 0–20 phr and for the CB-40 sample in the low temperature region below 70 °C, $\log b_T$ was almost zero, which is consistent with the $\log b_T$ behavior generally seen for polymers. In contrast to that result, for the CB-40 sample in the high temperature range above 70 °C and the CB-80 sample, $\log b_T$ tended to decrease with increasing temperature. This tendency was more pronounced with a lower CB loading. Presumably, this tendency for $\log b_T$ to decline can also be attributed to the state of CB dispersion and the structure of the CB aggregates/agglomerates.

Based on the foregoing discussion of the changes observed in the physical properties of the CB-filled sulfur-cured NR vulcanizates, the following section will describe the visualization and analysis of the CB dispersion in the rubber matrix and the structure of the CB aggregates/agglomerates, based on the use of 3D-TEM.

17.5
CB Dispersion and Aggregate/Agglomerate Structure in CB-Filled NR Vulcanizates

17.5.1
3D-TEM Observation of CB-Filled NR Vulcanizates and Parameters of CB Aggregates/Agglomerates

Figure 17.13 presents 3D-TEM images of the CB-filled sulfur-cured NR vulcanizates from which the Zn compounds had been removed. The CB loading of the samples labeled as CB-10, CB-20, CB-40, and CB-80 was 10, 20, 40, and 80 phr, respectively. The white-colored particles are CB and the black portions are the rubber matrix. Even for the CB-10 sample with a low CB loading, a number of aggregates consisting of several CB particles are seen. The images indicate that the CB aggregates grew larger in size with a higher CB loading. For the CB-80 sample having the highest CB loading, many large aggregates positioned closely together are observed.

The distance between the centers of gravity (d_g) of the closest CB aggregates and the distance between the CB aggregates (d_p) can be found from the image processing results [1, 2, 12–16, 31–33] in Figure 17.13. First, we provisionally define a CB aggregate as a collection of adjacent primary CB particles within a distance of approximately 1 nm, which is the resolution of the 3D-TEM equipment. The definitions given for d_g and d_p of the two closest CB aggregates are shown schematically in Figure 17.14. The CB aggregates are represented as circles (spheres) for convenience. The distance between the centers of gravity of the two closest CB aggregates is defined as d_g, and the distance between the two aggregates along a line connecting their centers of gravity is defined as d_p.

Figure 17.15 shows the average values of d_g and d_p and their standard deviations (STD) as a function of the CB loading. The results in Figure 17.15a indicate that d_g and d_p decreased sharply with increasing CB loading and tended to become constant at a CB loading of around 30 phr or higher. As shown in Figure 17.15b, the standard

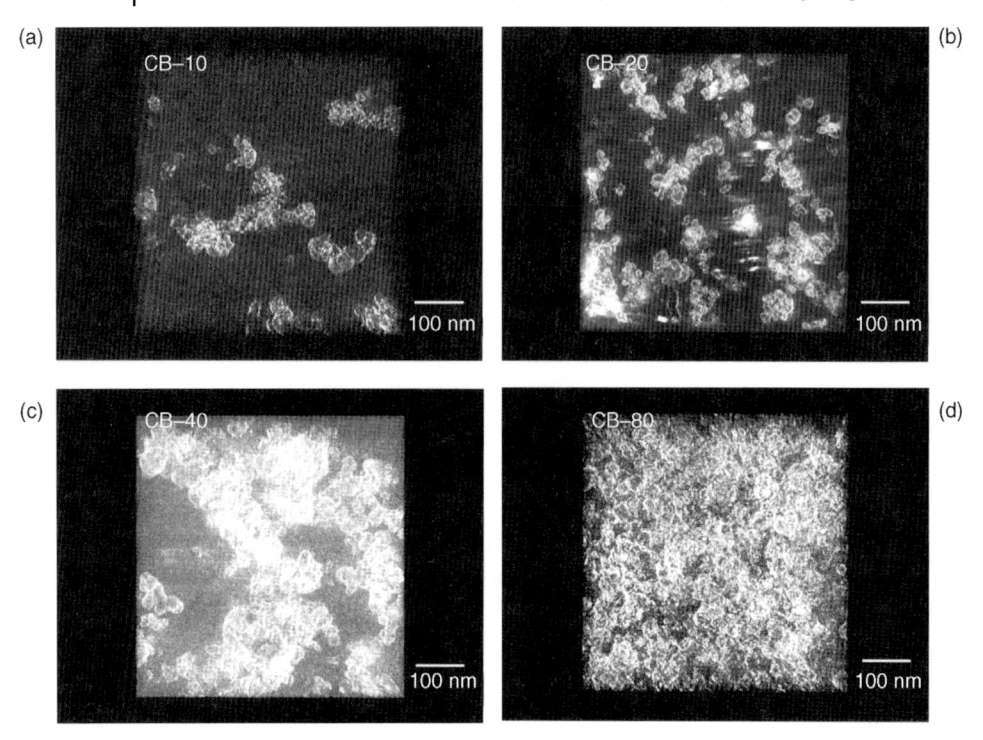

17 Carbon Black-Filled Natural Rubber Composites: Physical Chemistry and Reinforcing Mechanism

Figure 17.13 3D-TEM images of (a) CB-10, (b) CB-20, (c) CB-40, and (d) CB-80.

deviations of d_g and d_p also exhibited the same tendency of decreasing sharply with a higher CB loading and becoming constant at a CB loading of around 30 phr or more. These results suggest that the CB aggregates were nonuniformly dispersed in the rubber matrix in the low CB loading range below 30 phr and that they became closely

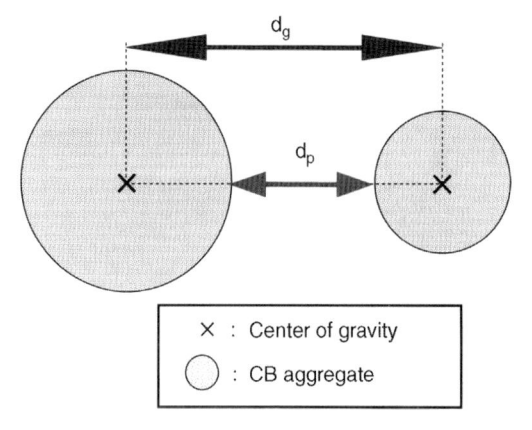

Figure 17.14 Definitions of the closest distance (d_p) between two neighboring aggregates and the closest distance (d_g) between the center of gravity of two neighboring aggregates as 3D parameters.

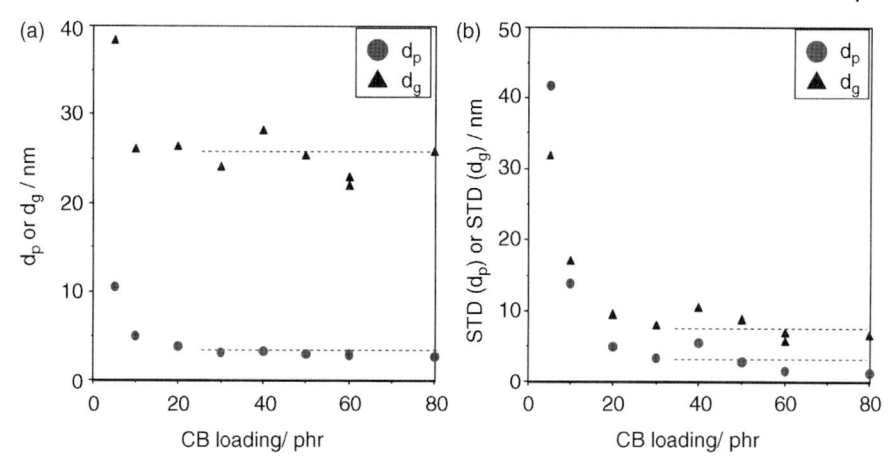

17.5 CB Dispersion and Aggregate/Agglomerate Structure in CB-Filled NR Vulcanizates

Figure 17.15 Dependence of average minimum distance (d_g) between centers of gravity of CB aggregates and average minimum distance (d_p) between CB aggregates and their standard deviations (STD(d_p) and STD(d_g)) on CB loading.

and uniformly distributed as the CB loading was increased. The results also imply that the CB aggregates became linked and formed a CB network at certain critical values of d_g and d_p in a CB loading range of about 30 phr or higher. The critical value of d_p was approximately 3 nm. This tendency for d_p to become constant at a CB loading of 30 phr or higher corresponds well with the tendency seen in Figure 17.16 for volume resistivity to become nearly constant at CB loadings of around 30 phr or more. This correspondence is further evidence of the formation of a CB network.

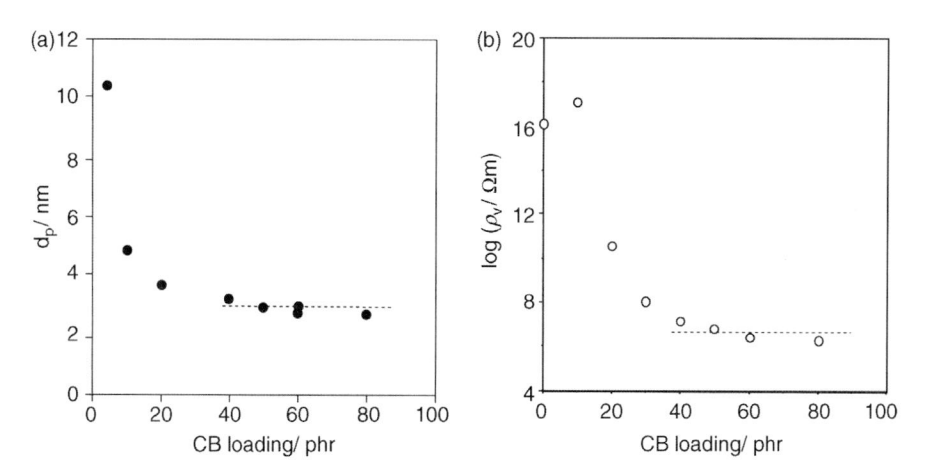

Figure 17.16 Dependence of average minimum distance between CB aggregates (d_p) and volume resistivity (ϱ_v) on CB loading.

17.5.2
Visualization of CB Network in the Rubber Matrix and the Network Parameters

It was shown in the previous section that the distance between two CB aggregates (d_p) tended to converge to a constant value from a CB loading of approximately 30 phr. It was also noted that the critical value of d_p as the closest distance two CB aggregates can approach one another at CB loadings of 30 phr or higher was approximately 3 nm. Curiously, this value nearly coincides with the thickness of the bound rubber around CB aggregates, which Nishi [65] and O'Brien et al. [66] estimated by nuclear magnetic resonance (NMR) to be on the order of several nanometers. It is inferred from this result that CB aggregates are linked together by means of bound rubber to form a CB network.

In order to visualize the 3D network structures of CB aggregates in the CB-10, CB-20, CB-40, and CB-80 vulcanizate samples, diagrams were made of the connections between the centers of gravity of adjacent CB aggregates at the closest distance of $d_p = 3$ nm [1]. The diagrams of the 3D network structures are shown in Figure 17.17. The boldfaced lines at the corners of the rectangular parallelepipeds of the observed samples indicate a length of 100 nm. It will be noted that a value of 3 nm was used for the purpose of linking the closest CB aggregates [1, 2, 12–16, 31–33]. This distance was the value at which volume resistivity and the closest interaggregate distance tended to saturate. For a low CB loading of 20 phr or less, isolated networks exist locally, whereas for CB loadings of 40 phr or higher, it is judged that the CB network structure is linked and extends throughout the entire sample. Therefore, the high conductivity seen for the samples with a CB loading of 40 phr or higher can presumably be attributed to the movement of electrons through such a network. It is estimated that this network structure linked by relatively strong interactions is also closely related to various other properties besides the electrical characteristics. It will be noted that branched chains of CB aggregates were observed in all of the samples examined, but virtually no isolated chains unconnected to the network were observed.

Two parameters of the CB aggregate network structures were then defined as shown schematically in Figure 17.18. They are referred to here as simply CB network parameters. The circles in the figure represent CB aggregates, and the short arrows indicate the linkages of the aggregates to the surrounding network structures. The long arrow and the thick arrow represent the cross-linked point and the branched point, respectively. The cross-linked points are connected by cross-linked chains (NdNd) and branched chains (NdTm) extend outwardlike branches from the branched points. N_{NdNd} and N_{NdTm}, respectively represent the number of cross-linked chains and the number of branched chains contained in the targeted observation total volume (TV), the densities of the cross-linked chains, and the branched chains per unit volume can be expressed as N_{NdNd}/TV and N_{NdTm}/TV. Assuming that each CB aggregate is analogous to a monomer functional group (f_{av}) in the gelation theory [67, 68], Eq. (17.1) below was introduced to express the relationship between N_{NdTm}/TV and N_{NdNd}/TV. This equation signifies that a network structure is generated when f_{av} has a value of 2.0 or more.

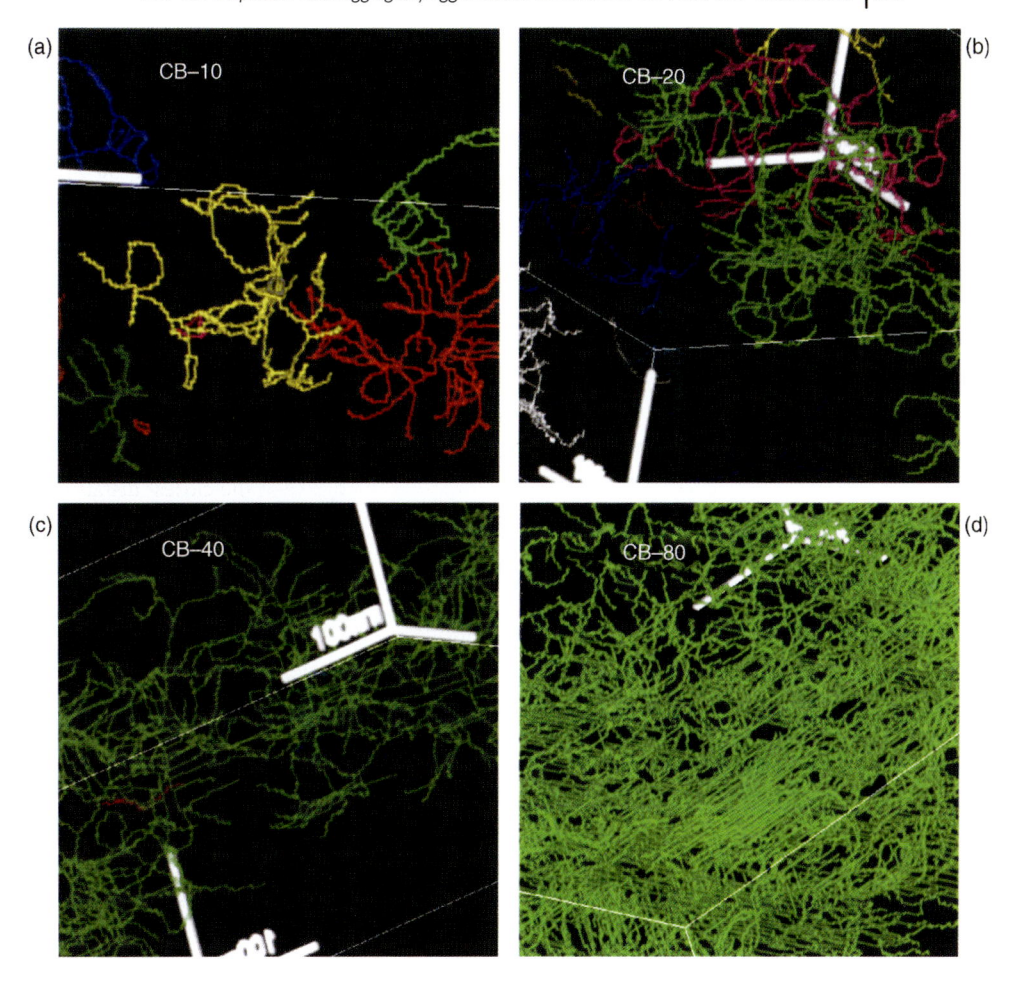

Figure 17.17 Visualization of CB aggregate network structures and their characteristics for (a) CB-10, (b) CB-20, (c) CB-40, and (d) CB-80. The centers of gravity of CB aggregates are connected by a line. The aggregates are linked by a d_p of 3 nm. The unconnected patterns are shown in different colors.

$$N_{\text{NdTm}}/TV = (f_{\text{av}} - 2) N_{\text{NdNd}}/TM \tag{17.1}$$

$f_{\text{av}} (= \sum N_i f_i / \sum N_i)$: functionality of CB aggregates (f_i: functionality of i; N_i: number of monomers with i functionality).

The dependence of N_{NdNd}/TV and N_{NdTm}/TV on the CB loading is shown in Figure 17.19a and the relationship between N_{NdNd}/TV and N_{NdTm}/TV is plotted in Figure 17.19b. The results in Figure 17.19a indicate that N_{NdNd}/TV and N_{NdTm}/TV increased nonlinearly with an increase in the CB loading. Compared with N_{NdTm}/TV, N_{NdNd}/TV tended to show a more pronounced increase. In Figure 17.19b, a linear relationship passing through the origin is seen between N_{NdNd}/TV and N_{NdTm}/TV in

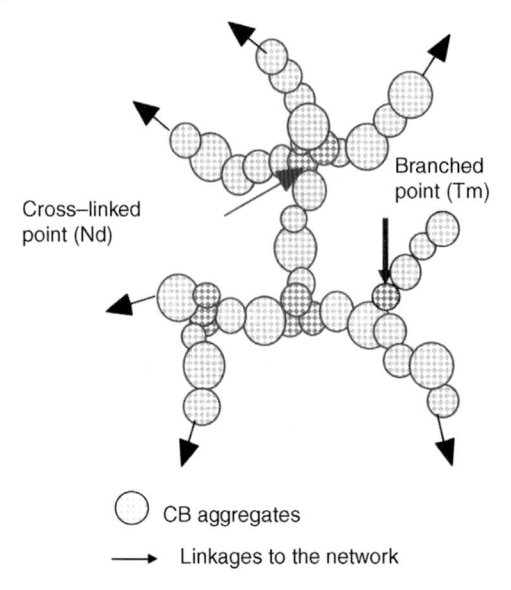

Figure 17.18 Schematic diagram of CB aggregate network parameters. The chains connecting the cross-linked points are the cross-linked ones and the chains extending outwardlike branches are the branched ones.

the CB loading region of 20 phr or lower. Moreover, because the f_{av} value obtained in this region is greater than 2.0, it is assumed that a CB network formed in this CB loading range through a process resembling gelation. In the region of higher CB loadings above 20 phr, N_{NdNd}/TV and N_{NdTm}/TV exhibit a nonlinear relationship that

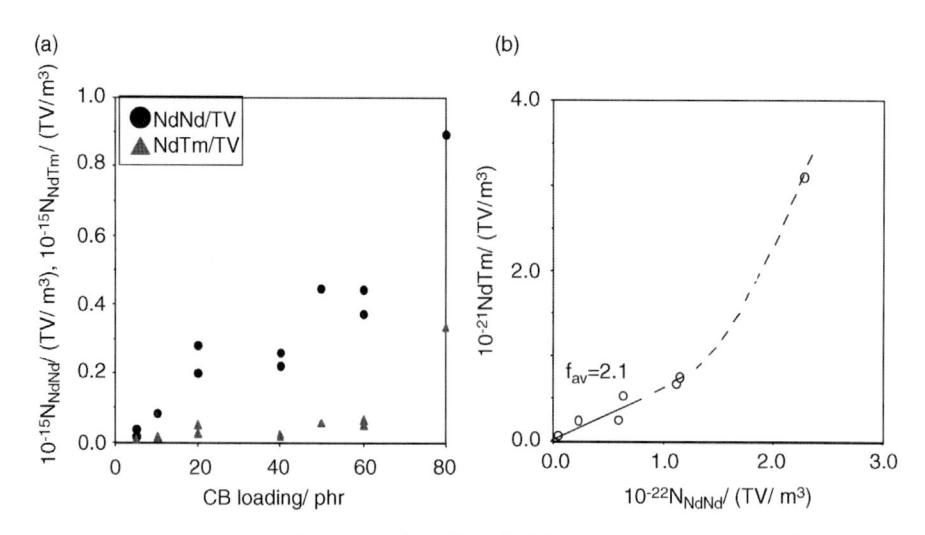

Figure 17.19 (a) Dependence of number of branched chains (N_{NdTm}/TV) and cross-linked chains per unit volume (N_{NdNd}/TV) on CB loading and (b) relationship between N_{NdTm}/TV and N_{NdNd}.

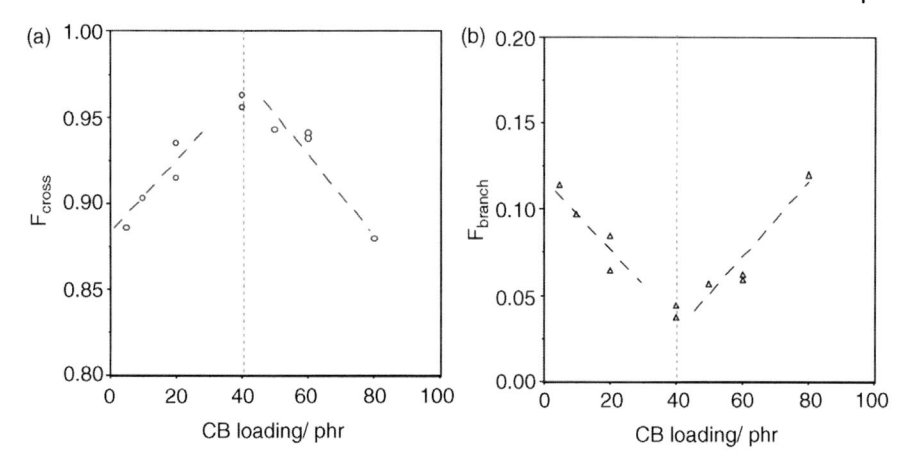

Figure 17.20 Fractions (F_{cross} and F_{branch}) of cross-linked and branched chains of CB aggregates as a function of CB loading. At CB loadings larger than 40 phr, (a) the fraction of cross-linked chains decreases linearly and (b) the fraction of branched chains increases linearly with increasing CB loading.

does not pass through the origin. Accordingly, it is difficult to interpret their relationship at this time [2, 32].

The fractions of cross-linked chains (F_{cross}) and branched chains (F_{branch}) forming the network structure can be defined as indicated in the following equations.

$$F_{cross} = N_{NdNd}/(N_{NdNd} + N_{NdTm}) \tag{17.2}$$

$$F_{branch} = N_{NdTm}/(N_{NdNd} + N_{NdTm}) \tag{17.3}$$

The dependence of these fractions on the CB loading is shown in Figure 17.20. The fraction of cross-linked chains increased almost linearly with increasing CB loading up to 40 phr, above which it then decreased nearly linearly. In contrast, the fraction of branched chains showed exactly the opposite tendency. These results imply that branched chains are linked to form cross-linked chains in the low CB loading region of 40 phr or less, whereas the generation of cross-linked chains in the CB loading range above 40 phr is obstructed by the three dimensional network structure, resulting in an increase in branched chains [1, 2, 31–33].

17.5.3
CB Aggregate Network Structure and Dielectric Relaxation Characteristics

The dielectric characteristics (permittivity ε' and dielectric loss ε'') of the CB-filled NR vulcanizates were measured at room temperature in a frequency range of 20 Hz to 1 MHz. The Cole–Cole plots of ε'' versus ε' are shown in Figure 17.21. In Figure 17.21a, no notable pattern is observed for the samples with a CB loading of 10 phr or less, though an incomplete pattern of a circular arc is seen for a CB

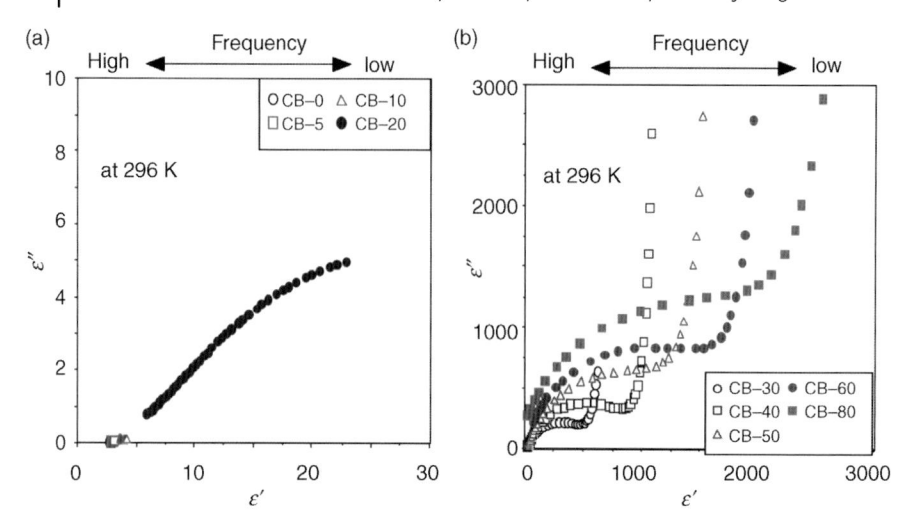

Figure 17.21 Relationship between dielectric properties and CB aggregate network structure. Cole–Cole plots for CB-filled NR vulcanizates showing dielectric loss (ε'') versus permittivity (ε').

loading of 20 phr. In contrast, a circular arc-shaped pattern is clearly seen in the high-frequency region for the samples with a CB loading of 30 phr or higher in Figure 17.21b. Moreover, this circular arc-shaped pattern becomes larger with a higher CB loading. Because this circular arc-shaped pattern is probably attributable to some type of relaxation phenomenon [46, 69], it is inferred that CB aggregates form a particular structure at a CB loading of 30 phr or higher and also that this structure grows as the CB loading is further increased. This value of 30 phr correlates well with the CB loading threshold mentioned earlier for the formation of the CB network structure. The CB-20 sample displayed only a portion of the circular arc-shaped pattern, and the pattern was markedly incomplete compared with that seen for the samples with higher CB loadings. Therefore, the CB-20 sample was excluded from the subsequent analysis described below.

An original graphic analysis method was then applied to the Cole–Cole plots, albeit a detailed explanation of the procedure [32] will be omitted here. An experimental equation was used to remove the relaxation component on the low frequency side from the circular arc-shaped pattern observed for CB loadings of 30 phr or higher in Figure 17.21b. Then, a geometrical analysis was applied to separate the semicircular relaxation component and the remaining residual component [32]. These two relaxation components (semicircular and residual) are shown in Figure 17.22. The vertical axis shows $\Delta\varepsilon''$, representing the remaining value after removing the low-frequency relaxation component from the measured value of ε'' in Figure 17.21. For the samples with a CB loading of 30 phr or higher, it is possible to separate the two relaxation components in this way. The results indicate that both relaxation components increased with a higher CB loading and that they also tended to extend into the high ε' region (low-frequency region). It will be noted that virtually no relaxation components were observed for the samples with a CB loading of 20 phr or less.

17.5 CB Dispersion and Aggregate/Agglomerate Structure in CB-Filled NR Vulcanizates

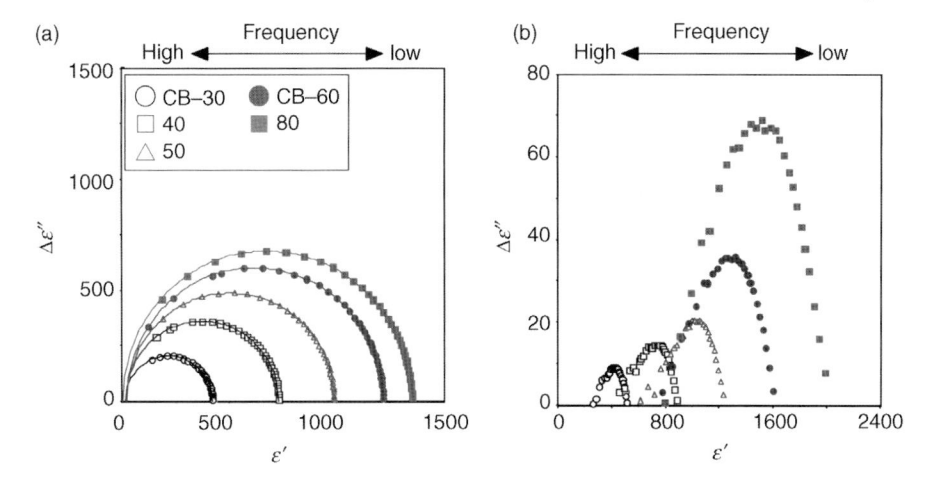

Figure 17.22 Two relaxation components ((a) semicircular and (b) residual) of CB-30, 40, 50, 60, and 80. $\Delta\varepsilon''$ is defined as the value obtained by subtracting an extrapolated curve of a Cole–Cole plot on the high-frequency side from the circular arc-shaped relaxation. The differences ($\Delta\varepsilon'_{cir}$ and $\Delta\varepsilon'_{res}$) are the relaxation strength of the two relaxation components.

The relaxation strength of the semicircular component ($\Delta\varepsilon'_{cir}$) and that of the residual component ($\Delta\varepsilon'_{res}$) can be defined with respect to the difference from the value of ε' when the frequency is ∞ (ε'_∞) and the value of ε' when the frequency is 0 (ε'_s). Then, the fractions (F_{cir} and F_{res}) of the relaxation strength of each component can be expressed as

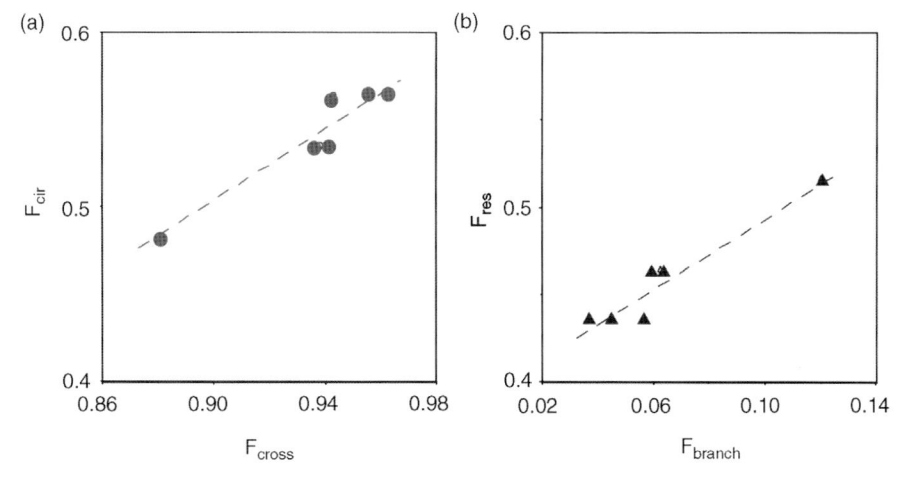

$$F_{cir} = \Delta\varepsilon'_{cir}/(\Delta\varepsilon'_{cir} + \Delta\varepsilon'_{res}) \tag{17.4}$$

Figure 17.23 Dependence of two fractions ((a) F_{cir} and (b) F_{res}) of relaxation strength on F_{cross} and F_{branch} in the CB loading region of 40 phr and higher.

$$F_{\text{res}} = \Delta\varepsilon'_{\text{res}}/(\Delta\varepsilon'_{\text{cir}} + \Delta\varepsilon'_{\text{res}}) \qquad (17.5)$$

The relationship between F_{cir} and F_{cross} and that between F_{res} and F_{branch} are presented in Figure 17.23. A good positive linear correlation is seen between the fraction of relaxation strength of the semicircular component and F_{cross} and between the fraction of relaxation strength of the residual component and F_{branch}. This result suggests that the CB aggregate network structure is closely related to the dielectric relaxation characteristics. In other words, the semicircular relaxation component and the residual relaxation component of dielectric relaxation are probably attributable to the polarization of the rubber around the cross-linked chains and to the interaction between the branched chains and the CB aggregate network.

17.6
Conclusions

The tensile properties and coefficient of thermal expansion of CB-filled sulfur-cured NR vulcanizates changed markedly around a CB loading of 30 phr. The elastic modulus of the NR vulcanizate samples changed profoundly with the CB loading. It is assumed that the state of CB dispersion and the CB aggregate/agglomerate structure change in the vicinity of 30 phr. The dependence of the coefficient of thermal expansion on the CB loading reversed direction near 30 phr, which implies that a transition related to CB dispersion and the aggregate/agglomerate structure occurs around this loading level. Moreover, activation energy related to loss compliance was manifested near 30 phr and tended to increase with a higher CB loading. This tendency also supports the hypothesis that the state of CB dispersion and aggregate/agglomerate structure change in the neighborhood of 30 phr.

The percolation occurrence of the CB-filled sulfur-cured NR vulcanizates was found to be near a CB loading of 10 phr. Beginning from that loading level, the volume resistivity (ϱ_v) of the samples decreased sharply and became nearly constant at a CB loading of 30 phr or higher. From Arrhenius plots of conductivity, which is the reciprocal of ϱ_v, it was found that the activation energy differed markedly between a CB loading of 10 phr or less and a CB loading of 20 phr or more. Since the distribution of CB aggregates is locally nonuniform in the low loading region, electrons jump between CB aggregates and also between conductive impurities in the rubber matrix by means of thermal energy. For that reason, the activation energy is rather high and is presumed to be approximately the same as that of the vulcanized rubber matrix. In the higher CB loading region, on the other hand, electrons move by hopping through a network that links the CB aggregates. The activation energy is relatively low and presumably becomes nearly constant independent of the CB loading.

In this study, master curves were obtained for the loss tangent of the dielectric relaxation of CB-filled sulfur-cured NR vulcanizates by applying a time–temperature superposition principle. In the master curves for the vulcanizate samples with a CB loading of 5–10 phr, a relaxation peak ascribable to the rubber matrix (also seen in the curve for CB-0) was observed on the short time (high frequency) side and a peak was also seen on the relatively long time (low frequency) side that was attributed to the nonuniform dispersion of the CB aggregates/agglomerates. The peak on the long time side disappeared from the master curves of the samples having a CB loading of 20 phr or higher because the distribution of the CB aggregates/agglomerates become uniform. In addition, the temperature dependence of the horizontal axis shift factor ($\log a_T$) used in the time–temperature superposition method differed dramatically between the samples with a CB loading of 10 phr or less and the samples with a higher CB loading. For the former samples, $\log a_T$ decreased nonlinearly with increasing temperature, as is the case with polymers in general. For the latter samples, $\log a_T$ was approximately zero and did not show any temperature dependence. The vertical axis shift factor ($\log b_T$) was virtually zero for the samples with a CB loading of 20 phr or less, which is consistent with the general behavior of polymers. For the samples with a CB loading greater than 20 phr, $\log b_T$ tended to decrease with a higher CB loading. These results imply that the viscoelastic behavior of the samples with a CB loading of 10 phr or less was governed by the rubber matrix, whereas the CB aggregate/agglomerate structure strongly influenced the viscoelastic behavior of the samples with a higher CB loading.

The results of 3D-TEM observations revealed that the average distance (d_p) between the closest CB aggregates and the standard deviation tended to decrease sharply with increasing CB loading in the loading range of 10 phr or less, whereas the values were nearly constant for the samples with a CB loading of 30–40 phr or higher. This tendency corresponds remarkably well with electrical percolation behavior. The results suggest that the CB aggregates/agglomerates were distributed rather nonuniformly with a relatively large distance between them in the CB loading range of 10 phr or less, whereas the CB aggregates were linked at the same short distance in the samples with a CB loading of 30–40 phr or higher. The critical distance of the latter samples was found to be approximately 3 nm, which probably corresponds to the thickness of the bound rubber.

The structure of the CB network was visualized by means of lines connecting the centers of gravity of the closest CB aggregates located within this 3 nm distance. A gelation theory was applied to the relationship between the density of cross-linked chains and that of branched chains. The results indicated that the formation of a CB network was feasible in the CB loading range of 20 phr or less. It is assumed that the branched chains are linked together to form cross-linked chains in the region of 20 phr or lower; in contrast, the three dimensional network structure obstructs the formation of cross-linked chains in the loading range above 30 phr, promoting the generation of branched chains instead.

Moreover, a circular arc-shaped pattern was observed in the Cole–Cole plots of dielectric relaxation for the vulcanizate samples with a CB loading of 30 phr or higher. An analysis of this relaxation pattern indicated that it could be divided into a semicircular component and the remaining residual component. A positive linear correlation was found between the fraction of relaxation strength of the semicircular component and the fraction of cross-linked chains of the CB network and between the fraction of relaxation strength of the residual component and the fraction of branched chains of the network. This made it clear that the relaxation characteristics are closely related to the three dimensional structure of the CB network.

Acknowledgments

The authors are grateful to Messrs. M. Hashimoto, J. Shimanuki, H. Sawabe, T. Suda, and T. Hasegawa, and Mss. M. Gonda, A. Isoda, and M. Nishioka (NISSAN ARC, LTD.) and Messrs. Y. Kasahara and Y. Morita (KIT) for their experimental assistance and stimulating discussions. Professor M. Tsuji and Dr. M. Tosaka (Institute for Chemical Research, Kyoto University) are also thanked for their excellent advice on electron microscopy.

References

1 Kohjiya, S., Kato, A., and Ikeda, Y. (2008) Visualization of nanostructure of soft matter by 3D-TEM: Nanoparticles in a natural rubber matrix. *Progress in Polymer Science*, **33** (10), 979–997.
2 Kohjiya, S., Ikeda, Y., and Kato, A. (2008) Current topics in elastomers research, in *Visualization of Nano-Filler Dispersion and Morphology in Rubbery Matrix by 3D-TEM* (ed. A.K. Bhowmick), CRC Press, Taylor & Francis Group, Boca Raton, pp. 543–551.
3 Friedrich, K., Fakirov, S., and Zhang, Z. (eds) (2005) *Polymer Composites: From Nano- to Macro-Scale*, Springer, New York.
4 Michler, G.H. and Batta-Calleja, F.J. (eds) (2005) *Mechanical Properties of Polymers Based on Nano-Structure and Morphology*, CRC Press, New York.
5 Shahinpoor, M., Kim, K.J., and Mojarrad, M. (2007) *Artificial Muscles: Applications of Advanced Polymeric Nano-Composites*, Taylor & Francis Group, London.
6 Karger-Kocsis, J. and Fakirov, S. (2009) *Nano- and Micro-Mechanics of Polymer Blends and Composites*, Hanser Gardner Pubs, New York.
7 Roberts, A.D. (ed.) (1988) *Natural Rubber Science and Technology*, Oxford University Press, Oxford.
8 Gent, A.N. (ed.) (1992) *Engineering with Rubber: How to Design Rubber Composites*, Hanser, Munich.
9 Rauline, R. (1992) Copolymer rubber composition with silica filler, tires having a base of said composition and method of preparing same, EP 0501227, US Patent No. 5,227,425 (1992/1993) (Compagnie Gdnrale des Establissements Michelin).
10 Ikeda, Y., Kato, A., Shimanuki, J., and Kohjiya, S. (2004) Nano-structural observation of *in situ* silica in natural rubber matrix by three dimensional transmission electron microscopy. *Macromolecular Rapid Communications*, **25** (12), 1186–1190.
11 Ikeda, Y. (2005) Characterization of rubbery nanocomposites by three-dimensional transmission electron microscopy/electron tomography. *Sen'i Gakkai-Shi*, **61** (2), 34–38.
12 Kato, A., Ikeda, Y., and Kohjiya, S. (2005) Three-dimensional observation of nano filler-filled natural rubber vulcanizates by 3D-transmission electron microscopy (3D-TEM). *Nippon Gomu Kyokaishi*, **78** (5), 180–186.
13 Kohjiya, S., Kato, A., Shimanuki, J., Hasegawa, T., and Ikeda, Y. (2005) Three-dimensional nano-structure of in situ silica in natural rubber as revealed by 3D-TEM/electron tomography. *Polymer*, **46** (12), 4440–4446.
14 Kohjiya, S., Kato, A., Shimanuki, J., Hasegawa, T., and Ikeda, Y. (2005) Nano-structural observation of carbon black dispersion in natural rubber matrix by three-dimensional transmission electron microscopy. *Journal of Materials Science*, **40** (9, 10), 2553–2555.
15 Kohjiya, S. and Kato, A. (2005) Visualization of nanostructure in soft materials by 3D-TEM. *Kobunshi Ronbunshu*, **62** (10), 467–475.
16 Kato, A., Kohjiya, S., and Ikeda, Y. (2006) Three-dimensional electric transmission microscopy. *Koubunshi*, **55** (8), 616–619.

17 Carmona, F. and Ravier, J. (2002) Electrical properties and mesostructure of carbon black-filled polymers. *Carbon*, **40** (2), 151–156.

18 Bauhofer, W. and Kovacs, J.Z. (2009) A review and analysis of electrical percolation in carbon nanotube polymer composites. *Composites Science and Technology*, **69** (10), 1486–1498.

19 Morozov, I., Lauke, B., and Heinrich, G. (2010) A new structural model of carbon black framework in rubbers. *Computational Materials Science*, **47** (3), 817–825.

20 Spruth, W.G. and Bahr, G. (1983) Printing technologies. *Computer & Graphics*, **7** (1), 51–57.

21 Nakamura, T., Kisu, H., Araya, J., and Okuda, K. (1991) The mechanism of charging roller. *Denshi Shashin Gakkaishi (Electrophotography)*, **30** (3), 302–305.

22 Takagi, K., Castleb, G.S.P., and Takeuchi, M. (2003) Tribocharging mechanism of mono-component irregular and spherical toners in an electrophotograph development system. *Powder Technology*, **135** (1), 35–42.

23 Lovelace, R.W. and Thom, K. (2008) *Paper and Print Technology, Encyclopedia of Materials: Science and Technology*, Elsevier, pp. 1–7.

24 Coran, AY. (1994) Vulcanization, in *Science and Technology of Rubber*, 2nd edn (eds J.E. Mark, B. Erman, and F.R. Eirich), Academic Press, San Diego.

25 Coran, A.Y. (2003) Chemistry of the vulcanization and protection of elastomers: A review of the achievements. *Journal of Applied Polymer Science*, **87** (1), 24–30.

26 Kato, A., Kohjiya, S., and Ikeda, Y. (2007) Nanostructural in traditional composites of natural rubber and reinforcing silica. *Rubber Chemistry and Technology*, **80** (4), 690–700.

27 Kato, A., Ikeda, Y., Kasahara, Y., Shimanuki, J., Suda, T., Hasegawa, T., Sawabe, H., and Kohjiya, S. (2008) Optical transparency and silica network structure in cross-linked natural rubber as revealed by spectroscopic and three-dimensional transmission electron microscopy techniques. *Journal of the Optical Society of America B*, **25** (10), 1602–1615.

28 The IMOD Home page, Bolder Lab for 3D Electron Microscopy of Cells. Available at http://bio3d.colorado.edu/imod/index.html.

29 TGS – MAXNET-amira documentation. Available at http://www.maxnt.co.jp/support/amira_doc.

30 The Japanese Society of Microscopy. (2004) Feature articles: Electron tomography. *Microscopy*, **39** (1), 2–33.

31 Kohjiya, S., Kato, A., Suda, T., Shimanuki, J., and Ikeda, Y. (2006) Visualisation of carbon black networks in rubbery matrix by 3D-TEM image. *Polymer*, **47** (10), 3298–3301.

32 Kato, A., Shimanuki, J., Kohjiya, S., and Ikeda, Y. (2006) Three-dimensional morphology of carbon black in NR vulcanizates as revealed by 3D-TEM and dielectric measurements. *Rubber Chemistry and Technology*, **79** (4), 653–673.

33 Ikeda, Y., Kato, A., Shimanuki, J., Kohjiya, S., Tosaka, M., Poompradub, S., Toki, S., and Hsiao, B.S. (2007) Nano-structural elucidation in carbon black loaded NR vulcanizate by 3D-TEM and *in situ* WAXD measurements. *Rubber Chemistry and Technology*, **80** (2), 251–264.

34 Tsuji, M., Fujita, M., and Kohjiya, S. (1997) On the correlationship between modulus of polymer crystals and resistance against electron-beam irradiation. *Nihon Reorogi Gakkaishi*, **25**, 193–194.

35 Omnèsa, B., Thuilliera, S., Pilvina, P., Grohensa, Y., and Gilletb, S. (2008) Effective properties of carbon black filled natural rubber: Experiments and modeling. *Composites Part A*, **39** (7), 1141B–1149B.

36 Hong, G.C., Nikiel, L., and Gerspacher, M. (2004) Electrical resistivity and thermal expansion coefficient of carbon-black-filled compounds around T_g. *Journal of the Korean Physical Society*, **44** (4), 962–966.

37 Caoa, Q., Songa, Y., Tana, Y., and Zhenga, Q. (2009) Thermal-induced percolation in high-density polyethylene/carbon black composites polymer. *Polymer*, **50** (26), 6350–6356.

38 Frenkel, J. (1930) On the electrical resistance of contacts between solid

conductors. *Physical Review*, **36** (11), 1604–1618.
39 Polley, M.H. and Boonstra, B.B.S.T. (1957) Carbon blacks for highly conductive rubber. *Rubber Chemistry and Technology*, **30** (1), 170–179.
40 Voet, A. (1964) Electric conductance of carbon black. *Rubber Age*, **95** (5), 746–753.
41 Bahder, G. and Garcia, F.G. (1971) Electrical characteristics and requirements of extruded semi-conducting shields in power cables, power apparatus and systems. *IEEE Transactions on Power Apparatus and Systems*, **PAS-90** (3), 917–925.
42 El-Tantawy, F., Kamada, K., and Ohnabe, H. (2002) *In situ* network structure, electrical and thermal properties of conductive epoxy resin–carbon black composites for electrical heater applications. *Materials Letters*, **56** (1–2), 112–126.
43 Konishi, Y. and Cakmak, M. (2006) Nanoparticle induced network self-assembly in polymer–carbon black composites. *Polymer*, **47** (15), 5371–5391.
44 Van Beek, L.K.H. and van Pul, B.I.C.F. (1962) Internal field emission in carbon black-loaded natural rubber vulcanizates. *Journal of Applied Polymer Science*, **6** (24), 651–655.
45 Reffaee, A.S.A., Nashar, D.E.E.I., Abd-El-Messieh, S.L., and Abd-El Nour, K.N. (2009) Electrical and mechanical properties of acrylonitrile rubber and linear low density polyethylene composites in the vicinity of the percolation threshold. *Materials & Design*, **30** (9), 3760–3769.
46 Blythe, A.R. (1979) *Electrical Properties of Polymers*, Cambridge University Press, Cambridge.
47 Abdul Jaward, S. and Alnajjar, A. (1997) Frequency and temperature dependence of ac electrical properties of graphitized carbon-black filled rubbers. *Polymer International*, **44** (2), 208–212.
48 Tang, H., Chen, X., and Luo, Y. (1996) Electrical and dynamic mechanical behavior of carbon black filled polymer composites. *European Polymer Journal*, **32** (8), 963–966.
49 Ali, M.H. and Abo-Hashem, A. (1997) Percolation concept and the electrical conductivity of carbon black-polymer composites 2: Noncrystallisable chloroprene rubber mixed with HAF carbon black. *Journal of Materials Processing Technology*, **68** (2), 163–167.
50 Ali, M.H. and Abo-Hashem, A. (1997) Percolation concept and the electrical conductivity of carbon black-polymer composites 3: Crystallisable chloroprene rubber mixed with FEF carbon black. *Journal of Materials Processing Technology*, **68** (2), 168–171.
51 Flandin, L., Hiltner, A., and Baer, E. (2001) Interrelationships between electrical and mechanical properties of a carbon black-filled ethylene–octene elastomer. *Polymer*, **42** (2), 827–838.
52 Drozdov, A.D. and Dorfmannb, A. (2002) The nonlinear viscoelastic response of carbon black-filled natural rubbers. *International. Journal of Solids and Structures*, **39** (23), 5699–5717.
53 Yamaguchi, K., Busfield, J.J.C., and Thomas, A.G. (2003) Electrical and mechanical behavior of filled elastomers. I. The effect of strain. *Journal of Polymer Science Part B*, **41** (17), 2079–2089.
54 Zaborski, M. and Donnet, J.B. (2003) Activity of fillers in elastomer networks of different structure. *Macromolecular Symposia*, **194** (1), 87–100.
55 Sajjayanukul, T., Saeoui, P., and Sirisinha, C. (2005) Experimental analysis of viscoelastic properties in carbon black-filled natural rubber compounds. *Journal of Applied Polymer Science*, **97** (6), 2197–2203.
56 Podhradská, S., Prokeš, J., Omastová, M., and Chodák, I. (2009) Stability of electrical properties of carbon black-filled rubbers. *Journal of Applied Polymer Science*, **112** (5), 2918–2924.
57 Mickinney, J.E. and Roth, F.L. (1952) Carbon black differentiation by electrical resistance of vulcanizates. *Industrial & Engineering Chemistry*, **44** (1), 159–163.
58 Boonstra, B.B.S.T. and Medalia, A.I. (1963) Effect of carbon black dispersion on the mechanical properties of rubber vulcanizates. *Rubber Age*, **92** (6), 892–902.

59 Satoh, Y., Suda, K., Fujii, S., Kawahara, S., Isono, Y., and Kagami, S. (2007) Differential dynamic modulus of carbon black filled, uncured SBR in single-step large shearing deformations. *e-Journal of Soft Materials*, **3**, 29–40.

60 IIan, B. and Loring, R.F. (1999) Local vitrification model for melt dynamics. *Macromolecules*, **32** (3), 949–951.

61 Watanabe, H., Matsumiya, Y., and Inoue, T. (2002) Dielectric and viscoelastic relaxation of highly entangled star polyisoprene: Quantitative test of tube dilation model. *Macromolecules*, **35** (6), 2339–2357.

62 Bedrov, D. and Smith, G.D. (2006) A molecular dynamics simulation study of segmental relaxation processes in miscible polymer blends. *Macromolecules*, **39** (24), 8526–8535.

63 Cerveny, S., Zinck, P., Terrier, M., Arrese-Igor, S., Alegla, A., and Colmenero, J. (2008) Dynamics of amorphous and semicrystalline 1,4-trans-poly(isoprene) by dielectric spectroscopy. *Macromolecules*, **41** (22), 8669–8676.

64 Chen, Q., Matsumiya, Y., Masubuchi, Y., Watanabe, H., and Inoue, T. (2008) Component dynamics in polyisoprene/poly(4-*tert*-butylstyrene) miscible blends. *Macromolecules*, **41** (22), 8694–8711.

65 Nishi, T. (1974) Effect of solvent and carbon black species on the rubber–carbon black interactions studied by pulsed NMR. *Journal of Polymer Science Polymer Physics*, **12** (4), 685–693.

66 O'Brien, J., Cashell, E., Wardell, G.E., and McBriety, V.J. (1976) An NMR investigation of the interaction between carbon black and *cis*-polybutadiene. *Macromolecules*, **9** (4), 653–660.

67 Flory, P.J. (1946) Fundamental principles of condensation polymerization. *Chemical Reviews*, **39** (1), 137–197.

68 Flory, P.J. (1953) *Principles of Polymer Chemistry*, Cornell University Press, Ithaca, NY.

69 Diogo, A.C., Marin, G., and Monge., Ph. (1987) Decomposition of mechanical spectra into elementary cole–cole domains. Application to the viscoelastic behaviour of a styrene–butadiene–styrene block copolymer. *Journal of Non-Newtonian Fluid Mechanics*, **23**, 435–447.

18
Silica-Filled Polymer Microcomposites
Sudip Ray

18.1
Introduction

Commercial importance of polymers has been the driving force behind the intense investigation of polymeric composites reinforced by particulates fillers. Their performance depends on the right combination of polymers, filler systems, and other compounding ingredients and the processing technique. Among the reinforcing fillers, as carbon blacks could only be used in black objects, the search for alternative nonblack active fillers, which can partially or completely replace the carbon blacks and permit the production of highly durable colored products led to the introduction of synthetic silicas in early 1940s. Since then enormous developments have taken place in this nonblack particulates filler resulting in a number of grades which find specific applications in various polymer industries both in thermoplastics and elastomeric sectors. The advent of silane coupling agent further enhanced the silica market. Subsequently, it has also been found that, in addition to its reinforcing properties, appropriate use of silica filler can also offer advanced crucial properties over carbon black, which leads to its wider use in the tire industry and several other applications in elastomeric nontire industries. Synthetic silicas are also extensively used in various thermoplastics and thermosets as a thixotropic or thickening agent and matting agent in paints and coatings; antiblock agent in packaging films. The surface chemistry of the silica filler plays a dominant role toward processing and the final properties of the silica-filled polymer composites. A summary of different kinds of silica fillers used in various polymers and the role of silica filler on determining the final properties of the composites have been documented in the following chapter.

18.2
Silica as a Filler: General Features

Fillers have important roles in modifying various properties of polymers. The effect of fillers on properties of a composite depends on their concentration and their particle size and shape, as well as on the interaction with the matrix [1–5].

Polymer Composites: Volume 1, First Edition. Edited by Sabu Thomas, Kuruvilla Joseph,
Sant Kumar Malhotra, Koichi Goda, and Meyyarappallil Sadasivan Sreekala
© 2012 Wiley-VCH Verlag GmbH & Co. KGaA. Published 2012 by Wiley-VCH Verlag GmbH & Co. KGaA.

18.2.1
Types

Silica, or silicon dioxide, SiO_2, in its pure natural form ranges from colorless to white. It occurs in several forms and is insoluble in water, slightly soluble in alkali, and soluble in dilute hydrofluoric acid [6]. Two different forms of silica are mostly commercially available for various industrial applications, namely crystalline (natural) silica and amorphous (synthetic) silica. Crystalline silica is extensively and richly distributed throughout the earth, both in the pure state and in silicates. The synthetic variety of silica is commonly used in polymer composites. Depending on the method of preparation, the amorphous silica, or the specialty silica is classified mainly into three forms: precipitated silica, fumed silica or pyrogenic silica, and silica gel. Examples of diverse commercially available amorphous silica, their leading producers in the global market, manufacturing processes, key filler properties and applications are summarized in Table 18.1.

18.2.2
Characteristics

Silica consists of silicon and oxygen, tetrahedrally bound in an imperfect three-dimensional structure and has strong polar surface groups, mostly hydroxyl groups bound to silicon known as silanol ($-Si-O-H$). The imperfections in its lattice structure provide free silanol groups on the surface. The number of silanols and their distribution depends on the method of preparation. The type and concentration of these surface silanol groups significantly influence the processing and the final properties of the composites. Three different types of silanol groups occur namely – isolated ($-OH$ groups on separated silicon atom), germinal (two $-OH$ groups on the same silicon atom), and vicinal ($-OH$ on adjacent silicon atoms) [6]. These functional groups are arbitrarily located on the filler surface. Isolated hydroxyls exist predominantly on dehydrated silica, fumed silica, and to a lesser extent on precipitated silica. Vicinal hydroxyls are stronger adsorption sites and hence, have stronger reinforcement effect than the isolated hydroxyls. The occurrence of these functional groups can be identified by FTIR spectroscopy.

The properties of different types of synthetic silica differ primarily based on particle size, specific surface area, and oil absorption values (Table 18.1). Due to the presence of highly populated surface silanol groups, there is a tendency of primary particles to conglomerate to form a chainlike structure, generally termed as a secondary structure or filler network. These secondary structures of silica fillers further associate through strong hydrogen bond to form agglomerates, known as tertiary structures. Usually the high shear forces generated during the compounding process can break the relatively stable tertiary structures but may not disrupt the aggregates of the secondary particles to further disperse the primary filler particles in the polymer matrix (Scheme 18.1) [7].

18.2 Silica as a Filler: General Features

Table 18.1 Comparison of different amorphous silica filler: types, examples, preparation methodologies, key properties, and applications.

Silica type	Example (product name/manufacturer)	Preparation	Properties[a] Average particle size (μm)	Surface area (m²/g)	Oil absorption, (cm³/100 g)	Key application area
Precipitated silica	HK 125/Degussa, Hi-Sil 233/PPG	Reaction of sodium silicate solution with sulfuric acid	2–10	50 (MS), 200 (HS), 450 (VHS)	150 (MS), 200 (HS), 230 (VHS)	Reinforcing filler for natural rubber and several synthetic rubber like SBR, butyl, neoprene, nitrile, and so on
Fumed silica	Cab-O-Sil M-5/Cabot, Aerosil 200/Degussa	Hydrolysis of silicone tetrachloride in a flame of hydrogen and oxygen at ≥1000 °C. The molten spheres of fumed silica (primary particle) fused together to form three-dimensional aggregates (secondary particles)	0.8	50–380	150–250	Reinforcing filler for silicone rubber, thixotropic agent, and matting agent in paints and plastics
Silica gel	Syloid 244/Davison, Silcron G-100/Glidden	Reaction of sodium silicate solution with sulfuric acid to form hydrosol, which under aging forms hydrogel. An aerogel can be produced by quickly removing water from hydrogel without shrinkage of silica structure, whereas xerogel can be produced by slow removing water from hydrogel	4–10 (HG), 2–3 (AG)	250–675 (HG), 250–300 (AG)	60–275 (HG), 275–300 (AG)	Matting agent in paints and plastics, thixotropic and viscosity control agents in thermoset resins such as epoxy and unsaturated polyester resins, antiblocking agent in polyethylene film

a) HG: hydrogel; AG: aerogel; MS: medium structure; HS: high structure; VHS: very high structure.

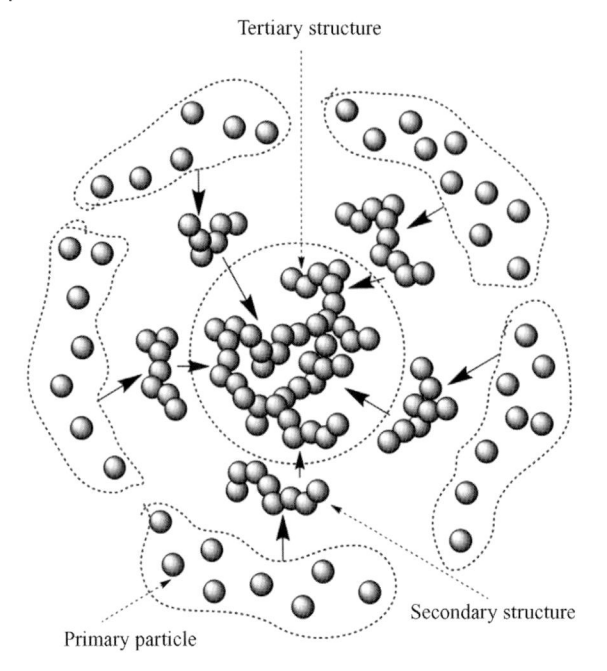

Scheme 18.1 Schematic presentation of structure formation from primary silica filler particles.

18.2.3
Surface Treatment of Silica Filler

A filler that offers small particle size, high surface area, could still provide relatively poor reinforcement if it has low specific surface activity. This relates to the compatibility of the filler with a specific polymer and the ability of the polymer to adhere to the filler surface. The compounded strength can be further improved if the matrix adheres to the mineral surface via chemical bonding.

Compared to carbon black fillers, silica generally offers less affinity and less surface activity to common polymers. The silanol group (−SiOH) present in silica surface behave as acid, however, due to high chemical activity of these surface silanol groups the primary filler particles react with each other to form aggregates and agglomerates. Silanol groups show high affinity in their reactions with amines, alcohols, and metal ions and also water adsorbed on the surface of filler particles reduce silanol reactivity. Some of the reactions with silanols can have a significant effect on the properties of rubber compounds, especially where the chemical involved is an important part of the cure system. Most of the accelerators used in sulfur cure systems contain an amine group. Strong adsorption or reaction with silica filler particles can decrease the amount of accelerator available for vulcanization reactions. This might cause slower cure rates and a reduced state of cure. Moreover, use of ZnO in the rubber formulation may cause reaction of zinc ions with silica filler particles and hence significantly reduce the curing efficiency.

Surface modification of particulate silica filler opens up new prospects for the composite materials into high-value applications, resulting in significant economic and product improvement benefits to industry. This approach is by modification of the surface chemistry and thus transformation from less reactive filler to a more reactive ingredient for effective combination with polymer matrix.

18.2.4
Surface Modification: Types and Methods

Diverse methods can be used for filler surface modification. However, the two most frequently practiced methods on the filler surface are physical modification and chemical modification.

18.2.4.1 Physical Modification

If the interaction between the adsorbed surface modifier and the polymer matrix is weak, but interaction between the adsorbed surface modifier and the filler surface is sufficient enough to modify the filler surface polarity, such that it matches that of the polymer matrix, a noninteracting type treatment is defined. It happens when certain chemicals are added to filler, particularly silica, they may strongly adsorb on the surface via dispersive interaction, polar interaction, hydrogen bonding, and acid–base interaction. Examples include treatment of glycols, glycerol, triethanolamine, secondary amines, as well as diphenyl guanidine (DPG), or di-o-tolylguanidine (DOTG) with silica. Generally, the polar or basic groups of these materials are directed toward the silica surface and the less polar or the alkylene groups toward the polymer matrix thereby increasing affinity with the hydrocarbon polymer. Consequently, the filler networking of silica can be substantially depressed resulting in better dispersion in the polymer matrix, lower viscosity of the compound, and lower hardness of the vulcanizate. Nevertheless, surface modification by physical adsorption of chemicals may not be preferred as they can be extracted by solvent or evaporated at high temperature. In addition, such an approach has rarely been applied in highly reinforced compounds because of the relatively poor polymer–filler interaction.

18.2.4.2 Chemical Modification

In case of chemical modification, the filler surface could be changed in such a way that it is tailored to its application. Generally, two types of chemicals have been used for surface modification: (a) grafts of chemical groups on the filler surface to change the surface characteristics, that is, the case where interaction between the surface modifier and the filler surface is strong but interaction between the adsorbed surface modifier and the polymer matrix is weak and (b) grafts that may react with filler as well as polymer, that is, the case where both the interactions are strong. The former is referred to as monofunctional coupling agent even though no chemical reaction with the polymer takes place with these grafts. The latter are called bifunctional coupling agents as they provide chemical linkages between the filler surface and the polymer molecules.

18.2.4.3 Filler Surface Modification with Monofunctional Coupling Agents

Grafted silica has been found to have many applications in different fields. Surface chemical modification to change the surface chemistry of silica fillers with monofunctional coupling agents has also been investigated widely. According to the thermodynamic analysis of filler networking, the grafting of oligomers or polymer chains would expectedly result in significant changes in the nature and intensity of the surface energy of silica, providing a surface similar to that of the polymer thus eliminating the driving force of filler networking. For example, Vidal et al. [8] have shown a drastic reduction in filler surface energy, by esterification of silica with methanol and hexadecanol. Donnet et al. [9] have described surface modification of silica filler by esterification method for the improvement on the reinforcement properties of elastomers. It was found that grafting monofunctional silanes with a long alkyl chain, for example, octadecyltrimethoxy silane (ODTMS) or hexadecyltrimethoxy silane (HDTMS), to silica also provides a nonpolar and low-energetic surface to increase compatibility and affinity toward the hydrocarbon polymer [10]. When such modified silica is incorporated in NR, a low-filler network compound can be formed [11]. Although the filler surface modification by monofunctional chemicals may greatly improve the microdispersion of the filler in the polymer matrix but still it suffers from the lack of polymer–filler interactions.

18.2.4.4 Filler Surface Modification with Bifunctional Coupling Agents

The bifunctional coupling agents are a group of chemicals, which are able to establish molecular bridges at the interface between the polymer matrix and filler surface. These chemicals generally enhance the degree of polymer–filler interaction; hence, impart improved performance properties to the filled materials. The most important coupling agents for surface modification of silica filler is the group of bifunctional organosilanes with the general formula as $X_{3-p}R_pSi(CH_2)_qY$, where X is a hydrolysable group, such as halogen, alkoxyl, or acetoxyl groups and Y is a functional group that itself is able to chemically react with the polymer either directly or through other chemicals. It may also be a chemical group able to develop a strong physical interaction with polymer chains. The Y groups in important silane coupling agents include amino, epoxy, acrylate, vinyl, and sulfur-containing groups, such as mercapto, thiocyanate, and polysulfide [12, 13]. The bifunctional silane coupling agents most often contain three ($p=0$) X groups and the functional group Y is generally in the γ position ($q=3$).

The demand for precipitated silica and silane increased continuously after the recognition in the late 1960s, that, by the addition of a coupling agent like 3-mercaptopropyltrimethoxysilane in sulfur-cured compounds and 3-methacryloxypropyltrimethoxysilane in peroxide cured system significant improvement in reinforcing properties could be achieved [14]. The introduction of the "*Green-Tire*" technology by *Michelin* in the early 1990s further boosted it up [15].

The sulfur-functional silanes and the vinylsilanes are very widely used coupling agents. The tetrasulfide silane, bis-(3-triethoxysilylpropyl)tetrasulfide (TESPT) known as Si 69, is reportedly the most commonly used coupling agent for sulfur-cured rubber compound, for example, in tire tread application [16]. Presence of copious amount of silanol groups on the silica surface causes the silanization reaction rates to occur very fast at elevated temperatures. However, because of the steric hindrance around the silyl propyl group in TESPT, this processing is favored at higher temperatures. The reaction is thus carried out *in situ* between 150 and 160 °C in an internal mixer. While sulfur-functional silanes are used in sulfur-cured compounds, vinylsilanes are best suited for peroxide cross-linking. The use of vinylsilanes in rubber applications started in the late 1960s in peroxide-cured ethylene propylene rubber (EPR) compounds [17]. The main applications for vinylsilanes are in wire and cable, hose, profile, and so on. The other types of silanes, for example, halogen containing silanes are suitable for the reinforcement of chloroprene rubber and require a metal oxide cross-linking (Scheme 18.2) [18].

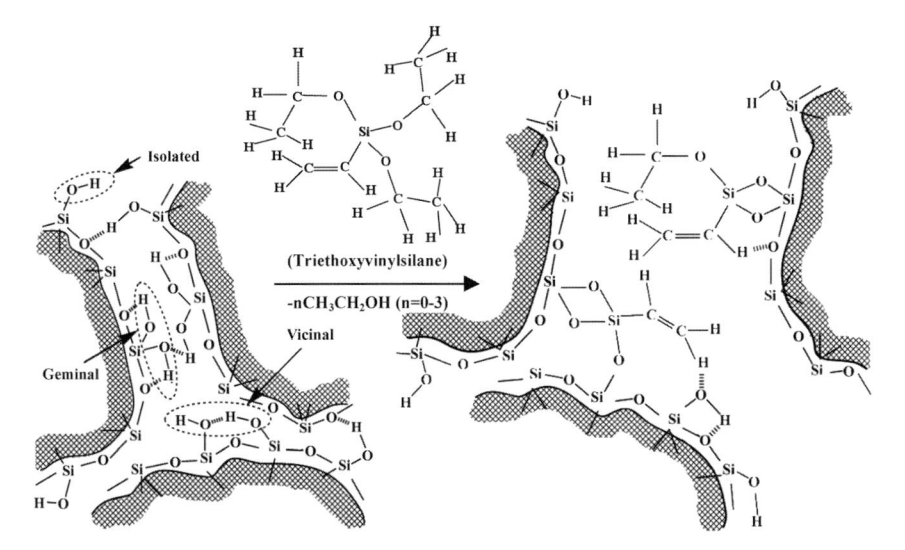

Scheme 18.2 Schematic presentation of different types of silanol groups on silica filler surface and the silane coupling reaction at the silica filler surface using silane coupling agent, triethoxyvinylsilane.

Depending on the silane type and processing steps involved, an appropriate methodology is required to carry out the surface treatment of the silica filler with the coupling agent. For example, mercapto silane possesses strong obnoxious odor that is quite unhealthy in factory environment. A special attention however, is required to control the compounding temperature to avoid scorching as a minimum high temperature is required for the formation of the active mercapto functional group. Such processing intricacies can be avoided by pretreatment of silica filler with the coupling agent. The following methods are commonly used for the silica surface treatment with the silane coupling agents [19]:

a) *In situ* method: direct mixing of silane with silica filler during the compounding process. The main advantage of this method is in its simple one step process by avoiding pretreatment steps.

b) **Pretreatment methods**:
- *Dry method*: control spraying of silane on silica filler under high shear mixing followed by heat treatment to assist the silica–silane chemical reaction and to remove the reaction by products.
- *Slurry method*: control spraying or dipping of dilute solution or emulsion of silane on silica filler under high shear mixing followed by setting, isolation, and drying of silanized silica.
- *Masterbatch method*: predispersion of silane in a host matrix at high concentration, for example, wax and then blending of this silane masterbatch with polymer and silica during compounding.

18.2.5
Characterization of Surface Pretreated Silica

Prior to mixing of surface pretreated silanized silica fillers with the polymer, it is quite crucial to estimate the extent of surface modification of silica filler by the coupling agent as a measure of quality control and to determine its loading in the compound formulation. Several analytical techniques can be used for this purpose [19]. For example, the evidence of the presence of coupling agent in the silane pretreated silica can be obtained by identifying the characteristics groups of silane by FTIR or Raman spectroscopy. Pyrolysis gas chromatography can be used for quantitative estimation of nature and amount of silane present in the treated filler. However, the difference between physically absorbed and chemically bound silane with the filler surface can be confirmed by NMR spectroscopy. Estimation of carbon content of the silane-treated silica fillers by ESCA, EDX, and so on, has been found to be quite useful method for such characterization. Chemical interaction of the silane coupling agent with the silica filler can cause significant reduction in free surface silanol groups and hence reduce the hydrophilicity of the silica filler. In some cases, identification and quantitative estimation of the coupling agent in the pretreated filler is quite complicated whereas tests on filler hydrophilicity/hydrophobicity can provide a simple way to assess the extent of surface modification.

Figure 18.1 depicts TEM images of precipitated silica fillers. The untreated filler (V) consists of very irregular aggregates of silica particles with branched chainlike structure. The aggregates are again composed of firmly fused nodular subunits and the particle size is in the range of 20–40 nm. The filler surface treatment by using bifunctional organosilane, triethoxyvinylsilane (3 wt%) followed by electron-beam treatment (100 kGy) reduces the aggregated nature of the filler. Water contact angle of the hydrophilic precipitated silica filler (V) measured by dynamic wicking method increases from 0 to 58° by pretreatment of vinyl silane in VV103 [20].

18.3
Silica-Filled Rubbers

Silica in its amorphous form, especially precipitated silica has been extensively used in the majority of the common elastomers in diverse applications [21]. Among the

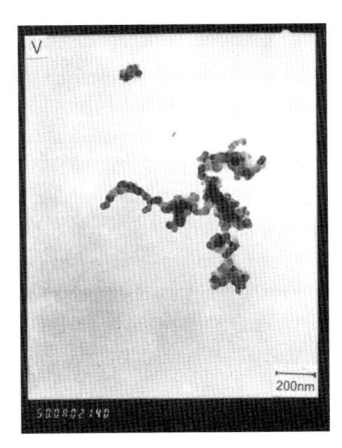

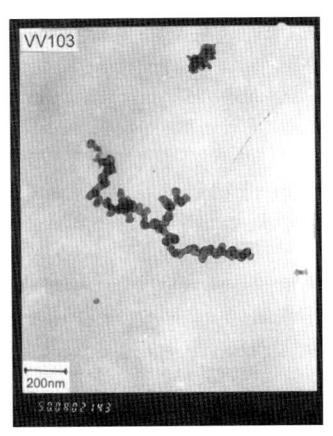

Figure 18.1 TEM photographs of silica fillers: (V) precipitated silica filler VULKASIL S and (VV103) surface-treated silica filler using silane coupling agent, triethoxyvinylsilane followed by electron-beam irradiation [20].

nonblack fillers, silica is the most effective reinforcing filler for rubber compounds. Nonblack fillers like clay, calcium carbonate are primarily used to reduce the compound cost. As reinforcing filler silica possesses several special attributes as compared to carbon black. Although both of these particulate fillers have similar morphology but due to their dissimilar surface functionality and hence different surface activities, they perform differently while incorporated in various elastomers. In general, addition of active filler to a polymer results in a change in properties, such as viscoelastic response and stress–strain behavior [22–28].

18.3.1
Effect of Silica Filler on Processability of Unvulcanized Rubber Compounds

Rheological and curing characteristics of unvulcanized filled rubber compounds are of considerable importance in both processing and forming operations. Various steps of rubber processing generally involve a flow state and the viscosity of rubber mixes increases with increasing structure and the surface area and decreasing particle size of silica. The presence of large amounts of surface silanol and siloxane functionality cause strong filler–filler interaction by hydrogen bonding and hence increases filler aggregation and subsequently increases the Mooney viscosity of the unvulcanized filled rubber compounds. It becomes further critical at higher silica filler loadings.

The surface activity and the surface energy of the filler as well as the polarity of the host matrix play important roles in the reinforcement process and consequently influence the processing and the vulcanizate properties of the filled composites. Silica in its pristine form is hydrophilic substance. Hence it has been found that polar elastomers, for example, silicone rubber, NBR, and so on, provide comparatively better rubber–silica interaction than the nonpolar hydrocarbon elastomers, for example, natural rubber (NR) and styrene–butadiene rubber (SBR). Also the occurrence of filler aggregation is more pronounced in nonpolar elastomers than polar elastomers. Hence, the compound viscosity is higher while silica is incorporated in nonpolar rubber than polar rubber.

Moreover, at low shear rates, strong filler–filler interaction exists quite profoundly and hence the silica-filled rubber compounds exhibit higher Mooney viscosity. At higher shear rates, disruption of silica filler aggregates relatively reduces the effect, which has been discussed in the subsequent section.

It is also probable that active surface groups present in silica filler also critically affect the curing characteristics of unvulcanized filled rubber compounds. Amine groups present in the accelerators commonly used in sulfur cure systems are prone to interact with silanol groups present in silica surface and hence reduce their availability for vulcanization reactions. Likewise, use of ZnO in the rubber formulation may cause reaction of zinc ions with silica filler particles and hence significantly reduce the curing efficiency.

As compared to carbon black, the adverse effect on processability and cross-linking reactions by increasing the compound viscosity and slowing down the cure rates especially in the cases of ZnO and sulfur based cure systems restrict the use of silica filler [29].

Therefore it is quite essential to reduce the hydrophilicity, that is, increasing the hydrophobicity of silica while using this as the reinforcing filler to design the compound formulation. Hence, for nonpolar elastomers, it is a common practice to use surface modifiers along with silica filler to improve the filler dispersion in the polymer matrix by reducing filler–filler interactions and enhancing polymer–filler interactions. Surface pretreatment of the silica filler also contributes significantly in controlling compound viscosity during processing stage and enhancing the vulcanizate properties. To diminish the silica–silica interaction silane coupling agents are commonly employed in the silica-based formulations. Use of softeners, for example, hydrogenated rosin or aromatic resins are also helpful in reducing viscosity. Activators such as glycols, for example, diethylene glycol, polyethylene glycol, amines, for example, triethanolamine and so on, are used as well in compounding, which interact with silanol groups of silica and forms a protective layer on silica particles and disrupts the silica–silica interaction and hence reduces such undesirable effects and consequently improves the dispersion quality and improves the cure efficiency.

18.3.2
Effect of Silica Filler on Vulcanizate Properties of Rubber Compounds

18.3.2.1 Mechanical Properties

As a reinforcing filler, silica can improve modulus at high elongation, tensile strength, tear strength, heat resistance of a rubber compound. Addition of silica filler helps dissipating the strain energy and hence provides its ability to function as reinforcing filler. Due to strong filler–filler interaction in silica, a portion of the rubber gets trapped within the silica aggregates, known as occluded rubber. With increasing the filler loading, the occluded rubber formation increases that contribute in the process of modulus increments of the vulcanizates. A comparison of primary physical properties of the vulcanizate based on equivalent loadings of carbon black and amorphous silica in a rubber formulation is mentioned in Table 18.2. Compared to carbon black with a similar surface area and filler loading, precipitated silica can provide improved tear strength, heat aging, and elongation at break properties with similar hardness, tensile strength, and compression set, while modulus and abrasion resistance remain inferior. This lower modulus and abrasion resistance is due to poor polymer–filler interaction. By the use of coupling agent with silica, performance enhancement closer to that with carbon black could be achieved. The ability of silica to offer better physical properties over carbon black on tear strength, cut growth resistance that are highly desirable in conveyor belts, off-road and heavy service tire applications. Moreover, improved heat resistance and adhesion to fabrics and metals obtained using silica as compared to its black counterpart allows for partial replacement of the carbon black in the compound formulation. Additionally, for reinforcing nonblack products precipitated silica is an obvious choice and can replace carbon black completely.

18.3.2.2 Dynamic Mechanical Properties

The incorporation of polar silica into the rubber especially hydrocarbon rubber results in the formation of low polymer–filler and high filler–filler interaction, which leads to its low reinforcing ability as compared to carbon blacks. At low strains, the silica–silica network results in a high shear modulus, but with increasing strain, part of the network breaks down. The breakdown of the filler–filler networks, that is, stress softening at small deformation, also known as *Payne effect* plays an important role in the reinforcement process. At higher filler loading and using high surface area silica the interaggregate distance decreases that facilitate the filler network formation [30–32]. Compared to carbon black with a similar surface area and filler loading, silica forms a much stronger network, due to the strong hydrogen bonding between the aggregates and agglomerates [33–35]. Silica-filled rubber exhibits larger "Payne effect" or strain dependent drop in modulus than the carbon black-filled rubber. However, using silane coupling agents, the "Payne effect" can be reduced significantly [36, 37]. Alternatively, by modifying the host rubber matrix the "Payne effect" can be reduced. Sahakaro and Beraheng demonstrated that in the case of maleic anhydride-modified NR with appropriate maleic anhydride content can reduce filler–filler interaction and hence improve silica dispersion, resulting in an enhancement of the mechanical and dynamical properties [38].

In the case of nonpolar rubbers, carbon black, being less hydrophilic in nature as compared to silica, has been found to be more suitable reinforcing filler. Hence carbon black can be dispersed more effectively in the polymer matrix and can form better polymer–filler interaction, which is especially important for the dynamic applications that require better abrasion resistance.

On the contrary, silica can offer several superior dynamic mechanical properties over carbon black. Dynamic mechanical properties of rubber compound are very useful for predicting the performance of final products. For example, tan δ value at 60 °C measures the rolling resistance property of tire tread. As compared to carbon black with a similar surface area and filler loading, precipitated silica can provide

Table 18.2 Comparison of primary physical properties of the vulcanizate based on equivalent loadings of carbon black and amorphous silica in a rubber formulation.

Physical property	Carbon black	Silica	Comment
Hardness	Lower	Higher	Higher level of filler aggregation in silica, comparable results can be obtained by using silane coupling agent
Tensile strength	Comparable		Using silane coupling agent it may increase
Elongation at break	Lower (superior)	Higher (inferior)	Lower level of polymer-to-filler bonding in silica-filled compound. Comparable results can be obtained by using silane coupling agent
Modulus at 300% elongation	Higher (superior)	Lower (inferior)	
Compression set	Lower (superior)	Higher (inferior)	
Tear strength	Lower (inferior)	Higher (superior)	Using silane coupling agent it may reduce
Cut growth resistance	Lower (inferior)	Higher (superior)	
Abrasion resistance	Lower (superior)	Higher (inferior)	Comparable results can be obtained by using silane coupling agent or "zinc-free" cure system
Loss modulus	Higher (inferior)	Lower (superior)	Less deformation energy requires to disrupt the silica structure than the permanent structure of carbon black
tan δ	Higher (inferior)	Lower (superior)	
Heat buildup	Higher (inferior)	Lower (superior)	
Solvent swelling	Lower (superior)	Higher (inferior)	Comparable results can be obtained by using silane coupling agent
Aging resistance	Lower (inferior)	Higher (superior)	Using silane coupling agent it further improves
Color of final product	Dark (inferior)	Transparent (superior)	Any desired color can be obtained using pigments and dyes

lower tan δ value at 60 °C resulting lower rolling resistance lead to increased fuel economy.

Similarly tan δ values above 60 °C are the measure of heat buildup in tire. Lower tan δ values at this high temperature ensure the longer tread life and hence the superiority of silica over carbon black.

18.3.3
Surface-Modified Silica-Filled Rubbers

The synthesis of surface-modified precipitated silica for use as a reinforcing filler in polymeric matrixes has been described in several literatures. *Waddell et al.* [39, 40] reported the surface modification of silica by the *in situ* polymerization of organic monomers, for example, copolymers of isoprene or 1,3-butadiene with vinyl acetate, acrylonitrile, and so on. They found that cure times were decreased, and breaking strength, tear energy, elongation to break, and cut growth resistance were increased. Okel and Wagner [41], developed high-performance precipitated silica equivalent to fumed silica in silicone reinforcement. Vidal *et al.* [42] studied the characterization and the kinetics of the grafting reaction of isocyanate-terminated liquid polybutadiene with silica, where the elastomer was grafted onto the solid surface through a urethane bond. Ou *et al.* [43, 44] reported effects of different surface alkylation of silica filler on rubber reinforcement. The extent of reinforcement in NR and SBR vulcanizates is reduced when the silicas are alkylated, especially in the case of the hexadecyl group. Choi *et al.* [45] scrutinized the influence of silane coupling agent on polymer–filler interactions of styrene and butadiene units in silica-filled SBR compounds. Salimi *et al.* [46] found that in the case of silica-filled NR/SBR blends compounded with optimum levels of TESPT silane coupling agent, the cure time decreases with increasing NR content while tensile properties of the NR/SBR blends decrease with increasing SBR content. In nitrile rubber vulcanizates, the alkylation seems to have less influence on the reinforcement characteristics of the silica. Thomas [47] described the reinforcement modification achieved with the use of silane coupling agent in blends of plasticized poly(vinyl chloride) and copolyester thermoplastic elastomer. Addition of silica filler increased hardness, modulus and tear strength of the blends whereas tensile strength, elongation at break, impact strength, and tensile set were decreased. However, use of coupling agent reduced the viscosity of the uncured system and improved tear resistance, tensile strength, elongation at break and impact strength of the filled compounds.

Investigations on the interactions between the silica filler and the silane coupling agents and their use for the improvements in the vulcanizate properties have also been described in many reports. In one early report, Wagner [48] found that silica modified with silane coupling agents could be used in tire treads, carcass, and steel-belt skim compounds for good mileage and improved wet road traction as compared to carbon blacks. He has also noted that chemical bonding of silica fillers to natural and styrene–butadiene rubbers, achieved by modifying the filler with γ-mercaptopropyltrimethoxysilane and γ-methacryloxypropyltrimethoxysilane, resulted in restricting solvent swelling, improved modulus and increased abrasion resistance [49]. Doran et al. [50] patented a process, where silica-filled SBR and EPDM (ethylene–propylene–diene monomer) rubber with improved wear resistance and other properties were prepared by incorporating unsaturated triethoxysilane into the vulcanization mixture. For example, SBR tire tread comprising of this silane coupling agent and silica filler showed faster 90% cure time, lower viscosity, lower heat buildup and compression set, and higher Pico abrasion index and road wear index than the rubber prepared without silane coupling agent. Bhowmick et al. [51] reported that adhesion between NR and BR (butadiene rubber) could be substantially improved using silica filler and coupling agent.

Surface modification of silica filler by different silane coupling agents is an ongoing research area [31, 52–58]. Ziegler and Schuster [59] studied the influence of silanization of precipitated silica on dynamic mechanical properties and the filler partition in acrylonitrile–butadiene rubber (NBR)/butadiene rubber blends. Reuvekamp et al. [60] investigated the effects of time and temperature on the reaction of TESPT silane coupling agent during mixing with silica filler and rubber used in tire. Burhin [61] demonstrated an improved mixing procedure for ensuring optimum silica compound properties for both silica "silanization" and low silane degradation. Wideman and Zanzig [62], patented a process for enhancing the electrical conductivity of a silica-reinforced rubber composition with a nonsilane coupling agent as N-3-(1,2-dihydroxypropyl)-N-oleylammonium bromide and/or N-3-(1,2-dihydroxypropyl)-N-methyl-2-mercaptoimidazolium bromide. Advanced studies on this topic including investigations on the polymer–filler interaction by use of new silane coupling agents or tailoring the surface energy of the filler without using any coupling agent is in progress [63–65].

18.3.4
Reinforcement Mechanism

As discussed in earlier sections, incorporation of silica filler to rubber has a strong impact on its static and dynamic behavior. Maximum reinforcement of filler can be achieved through good dispersion and better chemical or physical interaction with the rubber. The improvement in properties due to the presence of filler in rubber compounds can be attributed to the additive effects of the following factors [1]:

a) *Polymer network*: enhancement of apparent cross-link density of the filled rubber due to strong rubber–filler interaction.
b) *Hydrodynamic effect*: an increase in viscosity of the unvulcanized filled rubber and enhancement of modulus of the vulcanized filled rubber.
c) *In-rubber structure*: development of rubber to filler interaction.
d) Filler–filler interaction.

A model case study by Ray et al. [20, 37, 66–69] on silica filler and the role of silane coupling agent on rubber reinforcement related to the above factors is demonstrated in the following section by investigating the rheological, mechanical, dynamic mechanical, and swelling properties of the filled rubber. Silica filler pretreated with

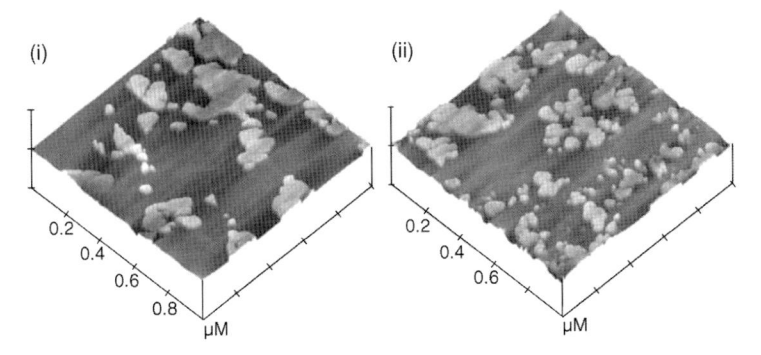

Figure 18.2 Tapping mode AFM three-dimensional surface plots (phase image, z-scale = 60°) showing the microdispersion of 50 phr loaded silica filler in the ethylene–octene copolymer rubber: (i) unmodified silica-filled rubber and (ii) silanized and electron-beam irradiated silica-filled rubber [37].

an organosilane, triethoxyvinylsilane followed by irradiation of the coated filler via electron-beam technique has been mentioned in Section 2.3. The surface treatment resulted in considerable improvement in the hydrophobicity of the pristine silica filler [20]. Atomic force microscopic studies showed that incorporation of this modified filler in ethylene–octene copolymer rubber successfully reduced the filler–filler interaction, which in turn reduced the formation of big agglomerates and improved the filler dispersion in the rubber matrix (Figure 18.2) [37, 66].

18.3.4.1 Rheological Properties

The plots of storage modulus (G') versus frequency (ω) for the unfilled and uncured ethylene–octene copolymer (UE) and its 15 phr loaded, untreated (UEV00015) and surface-modified silica-filled composites (UEVM10315 and UEVV10315, respectively) at 90 and 150 °C are presented in Figure 18.3 [67]. It was observed that with an increase in frequency, G' increased for all the compounds, that is, G' was strongly frequency dependent as the materials were stiffer at higher frequencies. The unfilled ethylene–octene copolymer rubber showed the lower storage modulus values compared to its filled composites. Incorporation of the silica filler in the rubber increased the storage modulus due to filler action and the effect was more pronounced at the lower temperatures and at the lower frequencies. It was interesting to note that G' was also found to be dependent on the filler type. At a particular frequency, reduction in storage modulus values were noted in the case of modified silica-filled rubber. The above finding was presumably governed by the aggregate–aggregate interaction. In addition to this, rubber trapped within the filler agglomerates or secondary structure might considerably increase modulus value in the untreated silica-filled compound.

Figure 18.4 displays log–log plots of G' against G'' (Han plot) [70] for the unfilled and uncured ethylene–octene copolymer and its untreated and surface-modified silica-filled composites at 15 phr loading at different temperatures [67]. The unfilled rubber (UE) exhibited temperature independence in the *Han plot*, which may be ascribed to "thermo-rheological simplicity" of the melt. This implies that all the

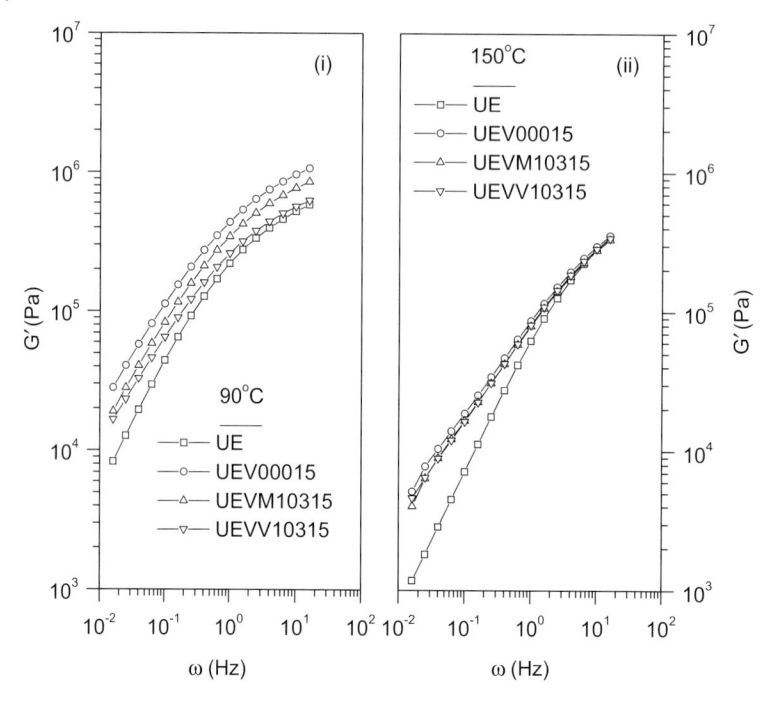

Figure 18.3 Storage modulus (G′) versus frequency (ω) plots for uncured and unfilled (UE) and silica-filled [15phr (UEV00015: untreated silica); (UEVM10315: silica filler coated with an acrylate monomer, trimethylolpropanetriacrylate and electron-beam treated); and (UEVV10315: silica filler coated with triethoxyvinylsilane and electron-beam treated)] ethylene–octene copolymer rubber at (i) 90 °C and (ii) 150 °C [67].

relaxation processes that determine the observed rheological behavior of the polymer had the same dependency with temperature (the same flow activation energy), that is, the melt was homogeneous at four different temperatures (90, 110, 130, and 150 °C). For the melt containing untreated silica (UEV00015), the initial values were slightly scattered followed by a downward tailing, which may be attributed to the formation of heterogeneous melt structure. The acrylated and electron-beam-treated silica-filled compound (UEVM10315) behaved similar to the untreated silica-filled compound, whereas the silanized and electron-beam-treated silica-filled compound (UEVV10315) showed a proximity toward the behavior of unfilled rubber. Thus, it could be stated that the melt rheology of the untreated silica-filled compound was highly influenced by the filler aggregation. The silane treatment was found to be more successful over the acrylate treatment to reduce the filler aggregation as a result of which the silanized and electron-beam-treated silica-filled compound behaved similar to the unfilled rubber.

18.3.4.2 Mechanical Properties

Engineering stress–strain curves of untreated and 100 kGy irradiated silanized silica-filled composites are shown in Figure 18.5 [68, 69]. An empirical curve fit was

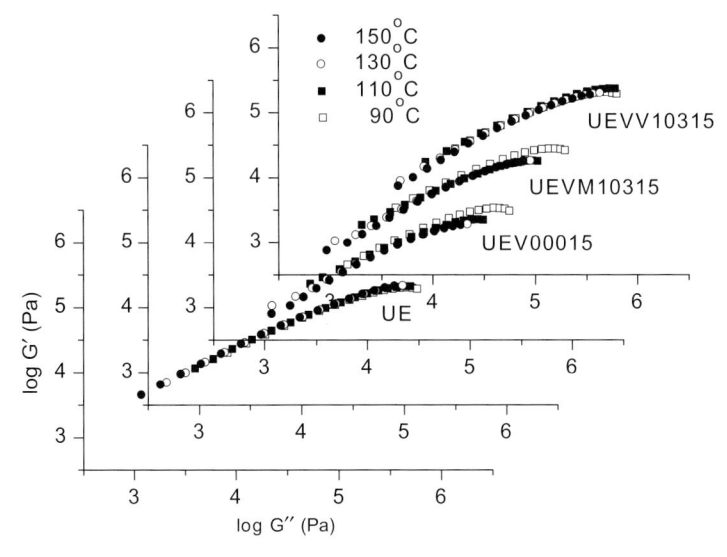

Figure 18.4 Log–log plot of storage modulus (G') versus loss modulus (G'') for unfilled and silica-filled ethylene–octene copolymer rubber at different temperatures [67].

proposed in order to further envisage the effect of electron-beam irradiation and the silanization process. Attempts had been made in the past by several researchers to correlate the modulus with volume concentration of fillers [71, 72]. In this study, it was observed that the ratio of the modulus of the filled to that of the gum vulcanizate followed an exponential nature with the volume concentration of filler. Hence, the filler–filler and the polymer–filler interaction were explained considering a generalized exponential equation correlating the modulus and the volume concentration of filler as follows:

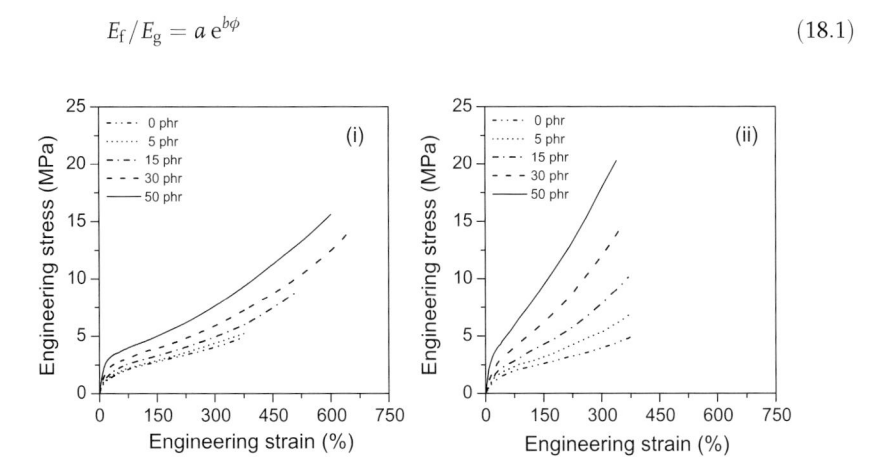

$$E_f/E_g = a\, e^{b\phi} \tag{18.1}$$

Figure 18.5 Stress–strain curves of (i) untreated (EV) and (ii) electron-beam irradiated silanized silica (EVV103)-filled ethylene–octene copolymer rubber [68, 69].

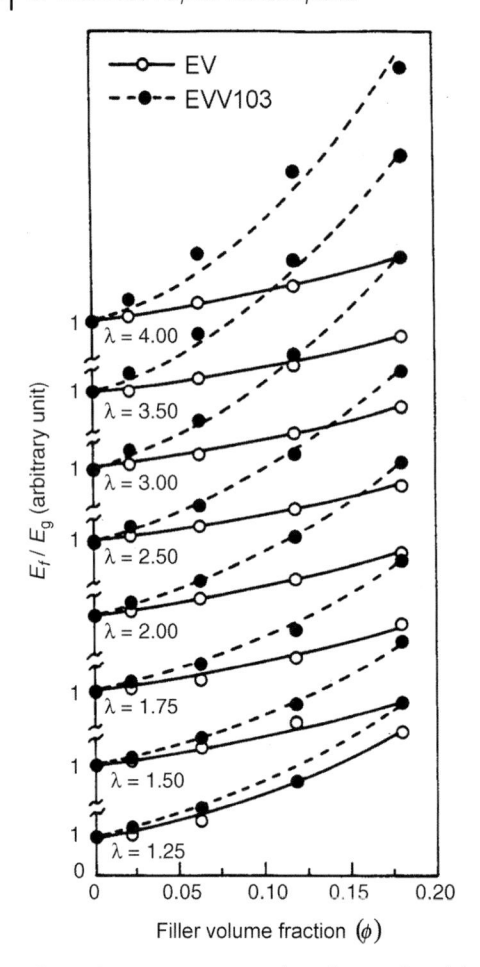

Figure 18.6 Representative plots of ratios of modulus of filled to that of the gum compound with respect to the volume fraction of filler for untreated (EV) and electron-beam irradiated silanized silica (EVV103), in the composites at different strain levels [68, 69].

where E_g is the modulus of the gum vulcanizate, E_f is the modulus of the filled vulcanizate, ϕ is the volume fraction of filler in the filled composite, a and b are the constants at a particular strain level.

The ratio of modulus of the filled to that of the gum compounds at different strain levels against the volume fraction of filler in the composites was curve fitted following Eq. (18.1) (Figure 18.6) [68, 69]. It was observed that for all the cases of the filled vulcanizates, the experimental data nicely fit the above empirical equation. The value of "b" varied with strain levels and with the type of the filler.

By incorporating the value of "a" ($a = 1$), Eq. (18.1) reduces to

$$E_f/E_g = e^{b\phi} \tag{18.2}$$

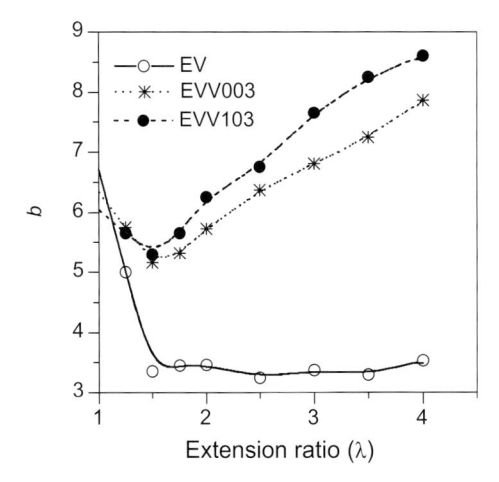

Figure 18.7 Plots of "b" against extension ratio (λ) for untreated (EV), unirradiated (EVV003), and electron-beam irradiated silanized silica (EVV103)-filled composites [68, 69].

On expanding the above equation, one obtains,

$$E_f/E_g = (1 + b\phi + b^2\phi^2 + \cdots) \tag{18.3}$$

The above equation closely resembled that of the earlier models [71–73]. Here the factor "b" was responsible for the filler–filler and the polymer–filler interaction.

In Figure 18.7 [68, 69], the "b" values so obtained were plotted against the extension ratio. An entirely different nature of the curves for the unmodified and the modified silica-filled composites was observed. The initial part of the curves (below $\lambda = 1.25$) was obtained by interpolation technique. In all the cases, a drop in the above curves was observed while increasing the extension ratio (up to $\lambda = 1.5$). This part of the curves resembled that of the *Payne effect*, which might be attributed to the breakdown of aggregated structure of filler upon straining. The above effect was most significant in the case of untreated silica-filled compounds (EV), which was noticeably reduced when the filler was replaced by the unirradiated silanized silica filler (EVV003). However, a further reduction was noted in the case of irradiated silanized silica-filled compounds. The "b" value obtained for untreated silica-filled composites at $\lambda = 1.5$ was 3.35. Interestingly, at higher extension ratios (beyond $\lambda = 1.5$), there was no noticeable change in the "b" values. On the other hand, in the case of irradiated silanized silica-filled composites, the "b" value at $\lambda = 1.5$ was 5.3. With $\lambda > 1.5$, there was a steady increase in the "b" values and at $\lambda = 4$, $b = 8.6$, which was about 2.4 times higher than that of the untreated one. Although, the unirradiated silanized silica-filled composite showed a similar trend, as that of its irradiated counterpart, but the values were noticeably lower (e.g., about 9.3% at $\lambda = 4$). From the atomic force microscopy studies a noticeable reduction in the aggregates size and the occurrence of the primary and secondary silica filler particles were clearly visualized in the case of electron-beam irradiated silanized silica-filled compound. This clearly emphasized

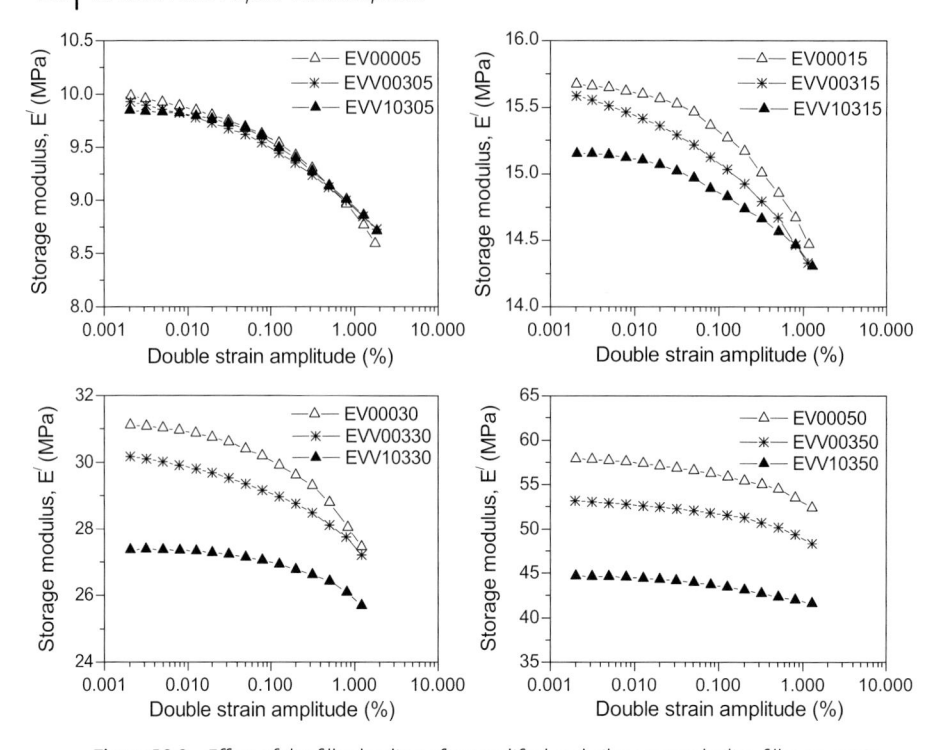

Figure 18.8 Effect of the filler loading of unmodified and silane-treated silica filler on strain dependence of storage modulus [37].

that the pretreatment of silanized silica filler by electron-beam irradiation successfully helped the silanization process.

18.3.4.3 Dynamic Mechanical Properties

Figure 18.8 depicts the dependence of storage modulus on strain amplitude for the vulcanizates with 5, 15, 30, and 50 phr of unmodified and silanized silica fillers [37]. The decrease of storage modulus upon increasing strain amplitude showed a typical nonlinear behavior of *Payne effect* and the effect was more pronounced with increasing the filler loading. The difference between the unmodified and the modified silica-filled compounds was a high storage modulus at very low strain followed by a faster decrease with increasing strain for the unmodified silica.

The results were further inspected from the interaction parameter "b'" calculated from a similar generalized exponential Eq. (18.1) correlating the storage modulus and the volume concentration of filler. Figure 18.9 [37] shows the "b'" values plotted against double strain amplitude (DSA). Interestingly, in the case of the control silica-filled rubber, "b'" was found to be significantly higher than those of the silanized silica-filled compounds. Such phenomena could be explained by *Payne effect* similarly as noted earlier at this low strain level on the mechanical properties of silica-filled ethylene–octene copolymer.

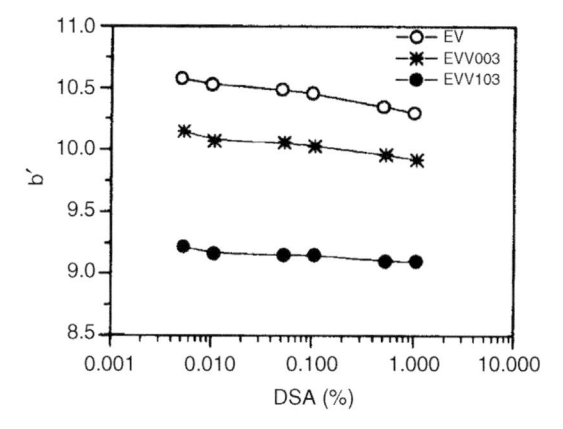

Figure 18.9 Plots of "b'" against DSA for untreated silica (EV), unirradiated silanized silica (EVV003), and electron-beam irradiated silanized silica (EVV103)-filled composites [37].

18.3.4.4 Frequency Dependence of Storage Modulus and Loss Tangent

The response of polymers to mechanical oscillations was found to be dependent upon a change of frequency of oscillation or a change in temperature of test [74]. The temperature–frequency dependence of the dynamic properties of vulcanizates cured by peroxide and filled with electron-beam-treated silanized silica filler (EVV10315) was investigated by applying time–temperature superposition principle in comparison to that filled with untreated silica filler (EV00015). The observed phenomena are explained in the light of polymer–filler and filler–filler interactions. Master curves of storage modulus and loss tangent are presented in Figure 18.10i and ii, respectively [37].

The master curves can be divided into three regions: high temperature region (low frequency region), medium temperature region (medium frequency region), and low temperature region (high frequency region). At high temperature (i.e., low

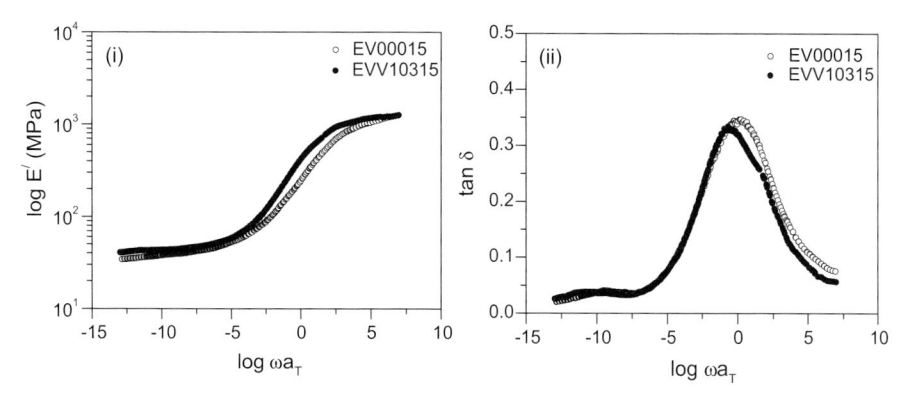

Figure 18.10 Master curves of ethylene–octene copolymer filled with control and electron-beam irradiated silane-modified silica filler: (i) storage modulus and (ii) loss tangent [37].

frequency) there was a pseudoelastic plateau region [1] and the storage modulus was found to be slightly high in the case of silanized silica-filled rubber. Depending on the strength of polymer–filler interaction, physical and/or chemisorption of rubber molecules might take place on the filler surface [75]. This interaction leads to an effective immobilization of the elastomer segments. Depending on the intensity of the polymer–filler interaction and the distance from the filler surface, the mobility of the polymer segments near the interface was lower than in the matrix. Based on the above assumption, Smit [76] proposed a rubber shell model of a definite thickness around the filler. It was also assumed that the modulus of the inner shell was very high and decreased gradually with increasing distance from the filler surface. The amount of the rubber in this quasi-glass state or volume of the shell obviously depends on the strength of the polymer–filler interaction and the surface area of the filler. At this high temperature filler aggregates formed a thin layer of rubber shell and the polymer–filler interaction predominates over filler–filler interaction. The higher storage modulus of silanized silica-filled composite ensures the improvement in the polymer–filler interaction due to silanization of silica filler.

At higher frequencies there was a sharp rise in the storage modulus as the rubber changes from a rubberlike material to a rigid glass. There was a noticeable shifting of this transition region toward low frequency, that is, high temperature was observed both in the case of storage modulus and loss tangent peak. Shifting of the transition zone toward lower frequency clearly indicates the formation of rubber shell at the early stage due to better polymer–filler interaction in the silanized silica-filled compound. With increasing frequency, the influence of thickness of rubber shell increased. It was supposed that the polymer in rubber shell was closer to or inside the transition zone, while the polymer matrix was still in its rubbery state, as a result the modulus of the rubber shell predominates over that of the polymer matrix.

At even higher frequencies, after this transition region, there was a glassy-amorphous plateau region, where the effect of thickness and the modulus of the rubber shell increased rapidly. At this stage, the deformation of joint shell was so hard that the rubber trapped inside the filler network loses its identity as polymer and behaving as filler. Thus at this stage, the storage modulus of silanized silica-filled compound matches with that of the control silica-filled compound.

The shift factors obtained from the above vulcanizates were used to derive the WLF coefficients C_1 and C_2 in order to quantify the temperature–frequency dependence of the dynamic properties of the above vulcanizates. It was found that silanization of silica filler causes increase in C_1 value from 19.6 to 25.3 and C_2 value from 102.1 to 143.0, respectively.

18.3.4.5 Equilibrium Swelling

For cases where the cross-link density is not affected by the addition of filler, Kraus developed an equation to describe the relationship between equilibrium swell and filler loading [77]:

$$\frac{v_{r_o}}{v_{r_f}} = 1 - m\left(\frac{\phi}{1-\phi}\right) \tag{18.4}$$

18.3 Silica-Filled Rubbers

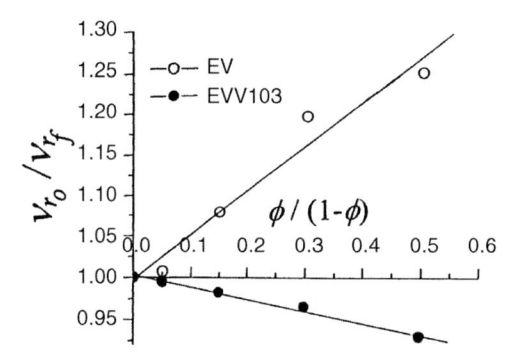

Figure 18.11 Kraus plots of swelling in toluene for untreated (EV) and electron-beam irradiated silanized silica (EVV103)-filled ethylene–octene copolymer [68].

where v_{r_o} is the volume fraction of rubber in the gum vulcanizate, v_{r_f} is the volume fraction of rubber in the filled vulcanizate, ϕ is the volume fraction of filler in the filled vulcanizate.

Polymer–filler interaction parameter "C" can be calculated using the Kraus equation:

$$C = \frac{m - v_{r_o} + 1}{3(1 - v_{r_o}^{1/3})} \tag{18.5}$$

where "m" is the slope obtained from the linear plot of Eq. (18.4).

The volume fraction of rubber in the swollen gel for the vulcanizates containing the electron-beam irradiated silane-modified fillers (EVV103) was found to be much higher than that corresponding to the untreated compound (EV). The plots in Figure 18.11 corroborate this fact quantitatively and show a trend of v_{r_o}/v_{r_f} against $\phi/(1-\phi)$. v_{r_f} values being lower for unmodified filler loaded rubber, the ratio of v_{r_o}/v_{r_f} increases with $\phi/(1-\phi)$, whereas the same decreases with $\phi/(1-\phi)$ for the modified filler loaded vulcanizate. From the slope of the plots, the "m" value obtained for the former was −0.55 and that of the modified filler loaded system was 0.135. By substituting this exact value of "m" in the Kraus equation, the C value for the unmodified system was obtained as 0.168, which was increased to 0.818 in the case of the modified system. Although the value of C is mainly dependent on filler–rubber interaction, it was found that both the secondary and tertiary structures of the filler have significant effect on C [78]. For reinforcing fillers such as carbon black and silica, the C values can directly be used to characterize the restriction of swelling caused by the filler. Here, it was assumed that swelling was completely restricted at the particle surface and that the rubber matrix at some distance from the particle was swollen isotropically in a manner characteristic of the unfilled rubber. The restriction of swelling suggested firm bond formation between the rubber and the filler. A negative value of "m" and a very low value of "C" in the case of untreated silica-filled composites indicate that the solvent desorbs the rubber, opening up vacuoles and the restriction of swelling is less. The above process was rather enhanced with the addition of higher doses of filler. On the other hand, a fivefold increase in the "C" value was observed after modifying the silica surface by the silanization process, which manifests the higher swelling restriction in the silanized silica-filled composites. This is also indicative of a coupling reaction, which must have taken place between the filler surface and the polymer and which indirectly would have led to an increase in the apparent cross-link density. Hence it can be concluded that the surface modification of silica improves the adhesion between polymer and filler.

18.3.5
Applications

In elastomeric applications, tire industries lead the market by occupying the major share. The key properties that determine the performance of tire tread compounds are abrasion resistance, wet traction, rolling resistance, heat buildup and low temperature flexibility. Though using silica alone in the compound formulation cannot meet the desired requirements, however, addition of silane coupling agents can enormously improve the above properties. The ability of silane coupling agents to reduce the silica filler–filler interaction helps in improving filler dispersion in the rubber matrix and hence reduce the heat buildup. Such effect also reduces the hysteresis loss and hence reduces the rolling resistance and increases low temperature flexibility. These improved filler dispersion and better polymer–filler interaction achieved by silane coupling agent further helps in improving abrasion resistance and wet grip properties. However, it is noteworthy to mention that the selection of right grade of silica filler, coupling agent, elastomer, curing system and so on and also the tire tread design contribute to obtain the optimized properties. The excellent combination of above properties obtained by silica-silane coupling system over carbon black filler helps to enhance road safety as well as reduce the fuel consumption and hence replacing carbon black filler partially in the Passenger Car, Winter Tire Treads as well as motorcycle tires. Low heat buildup, rolling resistance, and ample reinforcing ability of silica–silane filler system permit application in other components of the tire, for example, undertread, carcass, belt, fabric, and metal adhesion compound. Silica filler also finds inclusion in compound formulation for off-road, truck and heavy service tire applications for its excellent tear strength and cut growth resistance and thus partly replaces carbon black. In addition to these distinctive properties, silica filler is the obvious choice in white or colored bicycle tires.

Silica–silane filler system is also used in several other dynamic applications such as conveyor belts, V-belts, power transmission belts, and so on, as it can offer high tear strength, reduce the heat buildup with considerable abrasion resistance.

In sealing compounds, the above filler system can help to meet the requirements of low compression set, high tear strength, high flexibility, and low solvent swelling. With improved polymer–filler interaction using silane coupling agent, the ability for the polymer to flow under stress will restrict and hence reduces compression set. The enhanced polymer–filler interaction increases the apparent cross-link density of the compound and hence reduces solvent swelling.

In situ or pretreatment of silane coupling agent makes this filler more valuable in several other applications. In the case of rice hulling rollers, silane coupling agent can effectively reduce the viscosity of the unvulcanized rubber compound and helps during processability. This also allows high level of silica filler incorporation in the formulation to achieve the desired abrasion resistance and compression set properties. Similarly in the case of soft rollers, silane coupling agent helps to balance the strength properties and the flexibility and also ease the processability.

Due to its ability to provide good abrasion resistance and hardness properties, silica fillers are also used in shoe soles, tennis balls, golf balls, and elastomeric floor coverings.

In the fuel hose application, enhanced heat aging properties can be achieved by using silica as the main filler as compared to carbon black.

In the above applications, use of silica filler cannot only provide reinforcement effect but also offer the possibility to fabricate nonstaining and white or colored products.

18.4
Silica-Filled Thermoplastics and Thermosets

As with elastomers, the synthetic silica are being commonly used in various thermoplastic and thermoset composites compared to its natural variety. The major market for silica in thermoplastics and thermosets is as a thixotropic or thickening agent in inks, paints and coatings; matting agent in paints and coatings; antiblock agent in packaging films, and so on.

It has been pointed out earlier that silica has an inherent property to form aggregated structure by hydrogen bonding between the surface silanol groups. Hence, adding silica in liquid-based formulations can increase the viscosity of the system and find application such as thixotropic and viscosity control agent. For example, in the fiberglass-reinforced plastic parts based on thermoset resins, for example, unsaturated polyester for applications like marine and sanitary ware industries, synthetic silicas are commonly used. Due to the thixotropic nature of silica even adding about 1 phr or at a concentration of 2% of this material in the polyester resin can increase the viscosity of the resin fairly effectively and hence reduce the sagging of resin while applied to vertical surfaces or draining from the vertical mold surface prior to gelation [79]. Using hydroxyl containing polar compounds like ethylene glycol by 5–10% based on silica weight can further improve the thixotropic property of the silica and the aging property, that is, long-term viscosity stability and nonsagging property of the coatings by acting as a bridging agent between the surface silica silanol groups.

In epoxy based thermoset resins for the applications of high performance coatings, adhesives, reinforced plastics, and so on, synthetic silicas, especially fumed silica and silica gel are widely used as viscosity control agent to avoid undesirable sagging.

Synthetic silicas are also been used in PVC plastisols as a rheology control agent for automobile undercoats and synthetic leather applications.

Synthetic silicas are commonly used in coating formulations requiring matte or flat finish and reduced gloss. For example, the refractive index of silica gels allows producing clear coating formulations in which silica remains invisible. Thus, silica gels can reduce gloss without changing the film clarity.

There is a tendency of silica fillers to absorb the chemical ingredients used in the compound formulation due to their strong surface polarity. While this phenomenon is quite detrimental for elastomeric applications but due to this property silica finds main applications in thermoplastics as an antiblock and antislip agent. Films, sheets, and tapes based on polyethylene, polypropylene, PVC, PET, and so on, can stick together during the formation stage, while slip can occur between the surfaces of packaging foils. Addition of synthetic silica even at levels of about 0.1 wt% in the resin formulation can significantly reduce such adverse effects without degrading other film properties by absorbing the plasticizers that primarily cause this difficulty. Moreover, fine particulate silica may also assist as a nucleating agent for crystallite formation without major loss to color and transparency of the film.

Plastic formulations based on high levels of plasticizers, coloring agents, for example, in the case of plasticized PVC, such ingredients may leach out from the mixed compound and tend to deposit on the processing equipment. Use of high-structured silica in the formulation can overcome this plate-out problem by absorbing these plating ingredients.

Due to the strong surface absorption property of high structured precipitated silica, silica gel, they are used as a carrier by absorbing liquid ingredients in the compound formulation. Thus in PVC formulations, silica are used as a carrier to convert liquid PVC stabilizers to free-flowing dry powders without altering their heat stability.

In the PVC-based wire and cable applications, silica are been used to further improve the insulating property due to their low heat and electrical conductivity.

Precipitated silicas are used in the preparation of microporous thermoplastic, for example, polyethylene battery separators in a high proportion of about by 50% by weight with respect to the thermoplastic resin. Precipitated silica is also used in other porous thermoplastic applications like in semipermeable membrane and filter applications.

Recently, the ideas of microcomposites have expanded to a new and an emerging class of filled polymers, called nanocomposites. Similar to the case of microscale silica-filled polymer composites, the preformed nanosilica can be incorporated into various polymers by melt or solution blending [80]. Alternatively, the nanosilica can be introduced as a precursor via sol–gel process in the polymer matrix where silica nano particles can be formed *in situ* in the presence of a polymer, simultaneous formation of nanosilica during the polymerization of the monomer(s) [81] or nanosilica filler as the template for polymer synthesis [82, 83]. The decrease in size of this inorganic component into the nano dimension, and the increase of the interfacial area, results in extraordinary materials properties and molecular modeling approach applied to understand the interfacial interactions [84]. In addition to polymer reinforcement other applications occur in coatings, optical devices, flame-retardant materials, electronics and packaging materials, sensors, membranes, and so on [85].

18.5
Concluding Remarks

Ideally the primary aim of incorporation of filler in a polymer matrix is to improve the ultimate properties and to reduce the overall materials cost of the composite compared to the unfilled polymer. Addition of silica in elastomers can improve several essential final properties of the base polymer and find primary application as nonblack reinforcing filler. In tire industries the demand of silica filler is growing constantly with the realization of its superiority on rolling resistance and wet grip properties over carbon blacks, which could be effective in reduction of fuel consumption and environmental pollution. As a reinforcing filler, silica is also well accepted in various other applications especially in transparent or light colored polymeric products.

On the shady side, the applications where wear or abrasion resistance is of prime importance, silica filler even in presence of coupling agents may not always meet the requirements and would not be able to entirely replace carbon black. In addition, use of silanes further increases the material cost and has limited its uses. Recently, Cabot Corporation has introduced carbon–silica dual phase filler combining the unique features of conventional carbon black and silica fillers. Modifying surface characteristics and morphology of the silica and carbon black domains in this hybrid filler the key properties required for passenger tire applications like wet traction, rolling resistance and abrasion resistance were well balanced while reducing the requirement of doses of silane coupling agent. On the contrary, higher cost and dark color as compared to conventional silica filler limit the use of this new filler. Processing difficulties associated with poor filler dispersion, undesirable interactions with curing agents, and so on, need special care while using silica in rubber formulation. However, development of highly dispersible silica may again widen its use.

Amorphous silicas are extensively used in various thermoplastics and thermosets as a thixotropic or thickening agent and matting agent in paints and coatings; antiblock agent in packaging films. Incorporation of silicas in polyethylene and polypropylene, the two highest volume and comparatively least expensive polymers can certainly increase modulus and hardness of these composites due to their higher modulus and specific gravity as compared to these polymers. However, polyolefins being hydrophobic and silica being hydrophilic in nature causes poor polymer–filler interactions and some of the mechanical properties of the composite deteriorates, while use of silanes further increases the material cost and hence limits its use as reinforcing filler.

Materials or process development for rising above the problems associated with the silica reinforcement of polymers is always a significant research topic. Although nano silicas provide some spectacular property improvement but due to high cost and intricate processing their use are still very much in research and development stage and could not replace microsilicas in major commercial applications.

References

1 Kraus, G. (1965) *Reinforcement of Elastomers*, Interscience, New York.
2 Leblanc, J.L. (2002) *Progress in Polymer Science*, **27**, 627.
3 Rothon, R.N. and Hancock, M. (2003) General principles guiding selection and use of particulate materials, in *Particulate-Filled Polymer Composites*, 2nd edn (ed. R.N. Rothon), Rapra Technology Limited, Shawbury.
4 Donnet, J.-.B. and Custodero, E. (2005) Reinforcement of elastomers by particulate fillers, in *Science and Technology of Rubber*, 3rd edn (eds J.E. Mark, B. Erman, and F.R. Eirich), Elsevier, Burlington.
5 Wypych, G. (2010) *Handbook of Fillers*, 3rd edn, ChemTec Publishing, Toronto.
6 Iler, R.K. (1979) *The Chemistry of Silica*, John Wiley & Sons, Inc., New York.
7 Wagner, M.P. (1976) *Rubber Chemistry and Technology*, **49**, 703.
8 Vidal, A., Papirer, E., Wang, M.-J., and Donnet, J.-B. (1987) *Chromatographia*, **23**, 121.
9 Donnet, J.B., Wang, M.-J., Papirer, E., and Vidal, A. (1986) *Kautschuk Gummi Kunststoffe*, **39**, 510.
10 Wang, M.-J. and Wolff, S. (1992) *Rubber Chemistry and Technology*, **65**, 715.
11 Wolff, S., Wang, M.-J., and Tan, E.-H. (1994) *Kautschuk Gummi Kunststoffe*, **47**, 102.
12 Grillo, T.A. (1971) *Rubber Age*, **103**, 37.
13 Ranney, M.W., Pageno, C.A., and Ziemiansky, L.P. (1970) *Rubber World*, **163**, 54.
14 Wagner, M.P. (1971) *Rubber World*, **165**, 46.
15 Rauline, R. (1992) EP 501227, Michelin et Cie., invs.
16 Wolff, S. (1982) *Rubber Chemistry and Technology*, **55**, 976.
17 Fusco, J.V. (1966) *Rubber World*, **48**, 147.
18 Wolff, S. (1980) *Kautschuk Gummi Kunststoffe*, **33**, 1000.
19 Borup, B. and Weissenbach, K. (2010) Silane coupling agents, in *Functional Fillers for Plastics*, 2nd edn (ed. M. Xanthos), Wiley-VCH, Weinheim.
20 Ray, S. and Bhowmick, A.K. (2002) *Journal of Applied Polymer Science*, **83**, 225.
21 Ciullo, P.A. and Hewitt, N. (1999) *The Rubber Formulatory*, Noyes Publications, Norwich.
22 Payne, A.R. (1961) *Rubber and Plastic Age*, **42**, 96.
23 Wagner, M.P., Wartmann, H.J., and Sellers, J.W. (1967) *Kautschuk Gummi Kunststoffe*, **20**, 407.
24 Medalia, A.I. (1974) *Rubber Chemistry and Technology*, **47**, 411.
25 Bhowmick, A.K., Gent, A.N., and Pulford, C.T.R. (1983) *Rubber Chemistry and Technology*, **56**, 226.
26 Maiti, S., De, S.K., and Bhowmick, A.K. (1992) *Rubber Chemistry and Technology*, **65**, 293.
27 Waddell, W.H. and Evans, L.R. (1996) *Rubber Chemistry and Technology*, **69**, 377.
28 Shim, S.E. and Isayev, A.I. (2001) *Rubber Chemistry and Technology*, **74**, 303.
29 Reuvekamp, L.A.E.M., Debnath, S.C., ten Brinke, J.W., van Swaaij, P.J., and Noordermeer, J.W.M. (2004) *Rubber Chemistry and Technology*, **77**, 34.
30 Payne, A.R. and Whittaker, R.E. (1992) *Rubber Chemistry and Technology*, **44**, 440.
31 Luginsland, H.-D., Frohlich, J., and Wehmeier, A. (2002) *Rubber Chemistry and Technology*, **75**, 563.
32 Wang, M.-J., Wolff, S., and Tan, E.-H. (1993) *Rubber Chemistry and Technology*, **66**, 178.
33 Evans, L.R., Hope, J.C., and Waddell, W.H. (1995) *Rubber World*, **212**, 21.
34 Wolff, S., Wang, M.-J., and Tan, E.H. (1994) *Kautschuk Gummi Kunststoffe*, **47**, 873.
35 Voet, A., Morawski, J.C., and Donnet, J.B. (1977) *Rubber Chemistry and Technology*, **50**, 342.
36 Ramier, L., Chazeau, L., Gauthier, C., Guy, L., and Bouchereau, M.N. (2007) *Rubber Chemistry and Technology*, **80**, 183.
37 Ray, S. and Bhowmick, A.K. (2004) *Polymer Engineering & Science*, **44**, 163.
38 Sahakaro, K. and Beraheng, S. (2008) *Journal of Applied Polymer Science*, **109**, 3839.

39 Waddell, W.H., O'Haver, J.H., Evans, L.R., and Harwell, J.H. (1995) *Journal of Applied Polymer Science*, **55**, 1627.
40 O'Haver, J.H., Harwell, J.H., Evans, L.R., and Waddell, W.H. (1996) *Journal of Applied Polymer Science*, **59**, 1427.
41 Okel, T.A. and Wagner, M.P. (1992) *Rubber World*, **206**, 30.
42 Vidal, A., Papirer, E., and Donnet, J.B. (1979) *European Polymer Journal*, **15**, 895.
43 Ou, Y.-C., Yu, Z.-Z., Vidal, A., and Donnet, J.B. (1994) *Rubber Chemistry and Technology*, **67**, 834.
44 Ou, Y.-C., Yu, Z.-Z., Vidal, A., and Donnet, J.B. (1996) *Journal of Applied Polymer Science*, **59**, 1321.
45 Choi, S.-S., Kim, I.-S., Lee, S.G., and Joo, C.W. (2004) *Journal of Polymer Science Part B: Polymer Physics*, **42**, 577.
46 Salimi, D., Khorasani, S.N., Abadchi, M.R., and Veshare, S.J. (2009) *Advances in Polymer Technology*, **28**, 224.
47 Thomas, S. (1987) *International Journal of Polymeric Materials*, **12**, 1.
48 Wagner, M.P. (1976) *Rubber Chemistry and Technology*, **49**, 703.
49 Wagner, M.P. (1977) *Rubber Chemistry and Technology*, **50**, 356.
50 Doran, T.J., Wagner, M.P., and Stevens, H.C. (1973) US Patent 3737334, PPG Industries, Inc., invs.
51 Bhowmick, A.K., Loha, P., and Chakravarty, S.N. (1989) *International Journal of Adhesion and Adhesives*, **9**, 95.
52 Nakamura, Y., Honda, H., Harada, A., Fujii, S., and Nagata, K. (2009) *Journal of Applied Polymer Science*, **113**, 1507.
53 Phewphong, P., Saeoui, P., and Sirisinha, C. (2008) *Journal of Applied Polymer Science*, **107**, 2638.
54 Krysztafkiewicz, A., Jesionowski, T., and Binkowski, S. (2000) *Colloids and Surfaces, A: Physicochemical and Engineering Aspects*, **173**, 73.
55 Jesionowski, T. and Krysztafkiewicz, A. (2001) *Applied Surface Science*, **172**, 18.
56 Krysztafkiewicz, A. and Binkowski, S. (1999) *Pigment & Resin Technology*, **28**, 270.
57 Göerl, U., Hunsche, A., Mueller, A., and Koban, H.G. (1997) *Rubber Chemistry and Technology*, **70**, 608.
58 Thammathadanukul, V., O'Haver, J.H., Harwell, J.H., Osuwan, S., Na-Ranong, N., and Waddell, W.H. (1996) *Journal of Applied Polymer Science*, **59**, 1741.
59 Ziegler, J. and Schuster, R.H. (2003) *Kautschuk Gummi Kunststoffe*, **56**, 159.
60 Reuvekamp, L.A.E.M., ten Brinke, J.W., Van Swaaij, P.J., and Noordermeer, J.W.M. (2002) *Rubber Chemistry and Technology*, **75**, 187.
61 Burhin, H.G. (2002) *Kautschuk Gummi Kunststoffe*, **55**, 175.
62 Wideman, L.G. and Zanzig, D.J. (2002) US Patent 6,476,115, The Goodyear Tire & Rubber Company, invs.
63 Peng, H., Liu, L., Luo, Y., Hong, H., and Jia, D. (2009) *Journal of Applied Polymer Science*, **112**, 1967.
64 Tiwari, M., Datta, R.N., Talma, A.G., Noordermeer, J.W.M., Dierkes, W.K., and Van Ooij, W.J. (2009) *Rubber Chemistry and Technology*, **82**, 473.
65 Yan, H., Tian, G., Sun, K., Zhang, Y., and Zhang, Y. (2005) *Journal of Polymer Science Part B: Polymer Physics*, **43**, 573.
66 Ray, S., Bhowmick, A.K., and Bandyopadhyay, S. (2003) *Rubber Chemistry and Technology*, **76**, 1091.
67 Ray, S., Bhowmick, A.K., and Swayajith, S. (2003) *Journal of Applied Polymer Science*, **90**, 2453.
68 Ray, S. and Bhowmick, A.K. (2003) *Journal of Materials Science*, **38**, 3199.
69 Ray, S., Shanmugharaj, A.M., and Bhowmick, A.K. (2002) *Journal of Materials Science Letters*, **21**, 1097.
70 Han, C.D. and Jhon, M. (1986) *Journal of Applied Polymer Science*, **32**, 3809.
71 Einstein, A. (1906) *Annalen der Physik*, **19**, 289.
72 Guth, E. and Gold, O. (1938) *Physical Review*, **53**, 322.
73 Medalia, A.I. (1972) *Rubber Chemistry and Technology*, **45**, 1171.
74 Roland, C.M. and Ngai, K.L. (1992) *Macromolecules*, **25**, 363.
75 Wang, M.-J. (1998) *Rubber Chemistry and Technology*, **71**, 520.
76 Smit, P.P.A. (1970) *Kautschuk Gummi Kunststoffe*, **23**, 4.
77 Kraus, G. (1963) *Journal of Applied Polymer Science*, **7**, 861.

78 Wolff, S. and Wang, M.-J. (1992) *Rubber Chemistry and Technology*, **65**, 329.
79 Wason, S.K. (1987) Synthetic silicas, in *Handbook of Fillers for Plastics* (eds H.S. Katz and J.V. Milewski), Van Nostrand Reinhold, New York.
80 Boisvert, J.-P., Persello, J. and Guyard, A. (2003) *Journal of Polymer Science Part B: Polymer Physics*, **41**, 3127.
81 Reynaud, E., Jouen, T., Gauthier, C., Vigier, G., and Varlet, J. (2001) *Polymer*, **42**, 8759.
82 Perruchot, C., Khan, M.A., Kamitsi, A., Armes, S.P., Watts, J.F., von Werne, T., and Patten, T.E. (2004) *European Polymer Journal*, **40**, 2129.
83 Yilmaz, E., Ramström, O., Möller, P., Sanchezc, D., and Mosbach, K. (2002) *Journal of Materials Chemistry*, **12**, 1577.
84 Odegard, G.M., Clancy, T.C., and Gates, T.S. (2005) *Polymer*, **46**, 553.
85 Zou, H., Wu, S., and Shen, J. (2008) *Chemical Reviews*, **108**, 3893.

19
Metallic Particle-Filled Polymer Microcomposites
Bertrand Garnier, Boudjemaa Agoudjil, and Abderrahim Boudenne

19.1
Introduction

Polymers are widely used due to their ease in realizing complex parts and interesting mechanical or physical properties as impact strength, lightness, and so on. However, their insulating features prevent any application when higher electrical or thermal conductivity is required. To overcome this drawback, a lot of investigations were performed by incorporating metallic particles in polymers. Indeed, the electrical conductivity of metallic particles is 8–10 orders of magnitude higher than those of other fillers, except carbon fiber. The contrast in thermal conductivity is less pronounced with only 2–3 orders of magnitude. The final properties of metallic particle-filled polymer composite will be a mixture of properties of the phases and will depend on the concentration, size, and shape of filler, on the microstructure of the composite, and on the processing and interfacial conditions. Very often, the great challenge in preparing metal-filled polymer is to achieve the highest electrical and thermal conductivities while minimizing the expense on mechanical properties, density, or cost. In a more general manner, compared to metals, metal-filled polymers are less expensive and lighter, have better corrosion resistance, and usually require less processing steps. In addition, not only the electrical and thermal conductivities but also the thermal expansion coefficient, density, and so on of the composites can be tailored to the end user wishes to suit various applications. However, it is difficult to change one property without changing the others. Therefore, very often a compromise between the requirements has to be found.

Metallic particles have not been used much to strength polymers but are very useful for increasing electrical and thermal conductivities. Consequently, there are a lot of applications for metal-filled polymers: heat conduction, electromagnetic shielding, microwave absorbers, magnets, antistatic devices, magnetic recording, electrical heating, as well as thermistor or chemical sensor [1, 2]. Heat conduction applications are mainly in electronic, electrotechnic, or car industries (electronic cases, car hood, thermoplastic hose for hot air circulation, heat exchanger for electric car batteries, etc.).

Polymer Composites: Volume 1, First Edition. Edited by Sabu Thomas, Kuruvilla Joseph,
Sant Kumar Malhotra, Koichi Goda, and Meyyarappallil Sadasivan Sreekala
© 2012 Wiley-VCH Verlag GmbH & Co. KGaA. Published 2012 by Wiley-VCH Verlag GmbH & Co. KGaA.

In this chapter, the various types of metallic filler and production methods used in filled polymers will be studied. Microsized particles and also metalized microparticles will be considered. Then, we will focus on achievements on the properties of metal-filled polymers with emphasis on electrical and thermal ones. As dispersion in the final properties is usually obtained, the effect of various factors on the effective electrical and thermal conductivities will be discussed. Finally, this chapter will deal with the effective property prediction that is helpful to tailor composites by reducing the amount of experimental runs. It will be shown that the complex interaction between the phases of the composites leads to numerous models in the literature.

19.2
Metallic Filler and Production Methods

The powder technology allows almost all metals or alloys to be reduced to powder. So aluminum, copper, iron, nickel, lead, silver, tungsten, zinc, and diverse metallic alloys (brass, bronze, and stainless steel) are available as powders with different shapes like grains, fibers, flakes, and whiskers. Many of the fillers that induce electrical conductivity also improve the thermal conductivity of polymers. However, the most popular electrical conductive filler is silver, while for thermal conductivity it is aluminum due to its superior conductivity and lower cost.

There are three main production methods to manufacture metallic particle fillers [3]:

- **Atomization**: The raw material is melted and the molten metal is disintegrated into small droplets and quenched using a water, air, or inert gas stream (vertical atomization) or a centrifugation chamber (centrifugal atomization).
- **Mechanical methods**: Billets are comminuted using lathe turning or chipping and large particles are reduced in size by milling devices such as ball, hammer, vibratory, attrition, and tumbler mills.
- **Chemical methods**: Metal powders can be manufactured by the reduction of metallic oxides, precipitation from solution, and thermal decomposition.

The choice of the process is linked to the nature of the metal. For example, mechanical methods are valid only for brittle metals. The characteristics (size, shape, morphology, internal porosity, oxidation, etc.) of the obtained particles will depend on the manufacturing method.

One can found various suppliers for metallic particle by looking for companies working in the metal production or in the powder metallurgy. Suppliers mostly still valid and the application of each type of metallic particles used as filler in polymers are well described by Sussman [2].

Increasing conductive properties of composites usually require a high amount of filler resulting in a much higher density of metal-filled polymers. To overcome this drawback, conductive filler with a lower density is realized by metallization of nonmetallic fillerlike polymeric grains, mica, carbon fibers, glass spheres, or fibers. The coating, typically of thickness less than 1 μm, is applied as a continuous layer with

19.3
Achieved Properties of Metallic Filled Polymer

19.3.1
Electrical Conductivity

The incorporation of metallic filler in polymers is known to modify the characteristics of polymers in a different way. Electrical of such composites is similar to that of fillers, whereas their mechanical properties and processing techniques are typical of these of polymers.

When the filler amount is increased to some range, the change is smooth when particles are surrounded by polymer, but when the filler amount is such that there is contact between particles, the characteristics of polymer composites change very rapidly. This is especially the case for the electrical conductivity. Figure 19.1 shows the typical dependence of the logarithm of the electrical conductivity on the filler amount. Experimental values that show such a behavior for both the electrical and thermal conductivities can be found in many works [4–9]. So, composites become electrically conductive at a critical volume generally called percolation threshold, φ_c, corresponding to the formation of an infinite cluster of particles within the polymer matrix. Table 19.1 shows various electrical characteristics of a wide variety of thermoplastics and thermosets filled with metallic particles.

The percolation threshold can be reduced by using fibrous conductive particles especially with a high aspect ratio (AR). This can be noticed in the measurements obtained by Bigg and Bradbury [12] where the percolation threshold decreases from

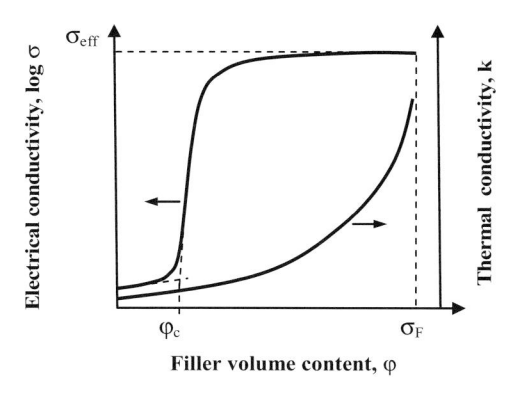

Figure 19.1 Typical dependence of the log of electrical and the thermal conductivities of polymer composites filled with metallic particles.

Table 19.1 Electrical properties of polymers filled with metallic particles.

Matrix	Filler	Shape	Size (μm)	Percolation threshold (vol.%)	σ_{eff} $(\Omega^{-1}\,m^{-1})$[a]	Reference
ER	Cu	Irregular	100	5	181	[9]
ER	Cu	Dendritic	5	25	3560	[10]
ER	Ni	Powder	10	8.5	121	[9]
LPDE	Cu	Powder	<38	19	11	[8]
LLDPE	Cu	Powder	<38	19	12	[8]
PBT	Al	Fiber	90/2000	30	0.1	[11]
PE	Al	Fiber	100/1250	9	52	[12]
PE	Cu	Powder	3.3	15	2.5	[13]
PP	Al	Fiber	100/1250	10	60	[12]
PP	Al	Fiber	127/3050	6	60	[12]
PP	Cu	Powder	30	34	—	[5]
PP	Cu	Powder	280	43	—	[5]
PVC	Cu	Irregular	100	5	330	[9]
PVC	Ni	Powder	10	4	90	[9]
PVC	Al	Powder	<75	20	—	[14]

ER: epoxy resin, LDPE: low-density polyethylene, LLDPE: linear low-density polyethylene, PBT: polybutylene terephtalate, PE: polyethylene, PP: polypropylene, PVC: poly(vinyl chloride).
a) After the peak.

10 to 6 vol.% for a fiber aspect ratio increased from 12 to 24. The percolation threshold can be slightly decreased by using smaller particles as shown by Boudenne et al. [5] where the copper particle size reduced by a factor of 9 decreases the filler content at the percolation threshold from 6 to 11 vol%, as shown in Figure 19.2. This comes from the highest probability to form a conductive chain with smaller particles. Another way to lower the percolation threshold is to provide a segregative distribution

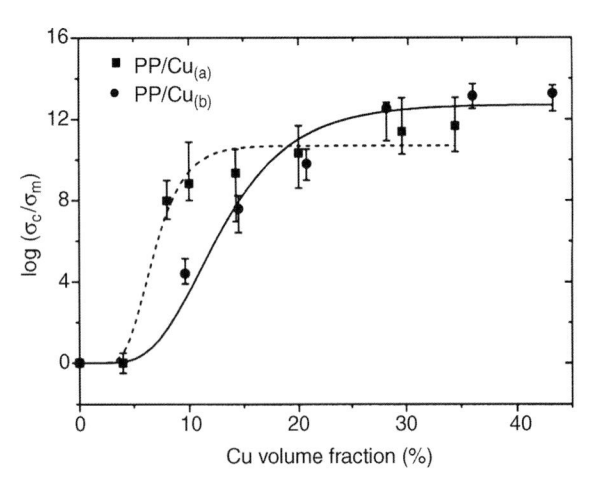

Figure 19.2 Electrical conductivity of PP/Cu composites versus filler content (measured at 120 °C, $Cu_{(a)}$ and $Cu_{(b)}$: powders with average particle size of respectively 30 and 280 μm [5].

of metal particles. Such a structure of particles can be obtained by using nonmiscible polymers with one already filled with conductive material, by the use of both conductive and nonconductive fillers, or by compaction of both plastic and metal powders. Experimental realizations were performed by various authors using PVC and copper or nickel particles [9, 15, 16]. Finally, except for composites with a segregative distribution or with a high aspect ratio of fiber, the canceling of insulating property of polymers requires volumetric filler contents higher than 20% (Table 19.1).

In Table 19.1, the obtained values after the peak, σ_{eff}, are widely scattered. They do not depend much on filler ratio or shape, but they are affected by the electrical properties of the phases and the microstructure of the composite. One should notice that aluminum, copper, and zinc powders are susceptible to oxidation that can alter the electrical conductivity of composite since these oxides are nonelectrical conductors [2, 17].

Composites can be exposed to large temperature fluctuations and during the heating of metal-filled composites, an abrupt decrease in the electrical conductivity for filler concentration above the percolation threshold may occur. Very often, it comes from the fact that the thermal expansion of polymer is 2–8 times higher than the ones of metal that causes the metallic particles to separate as temperature is increased [18, 19].

As the filler has a density 2–7 times higher than that of the polymer, the composite will show a higher density than the pure polymer that is not desired especially for transport applications. In addition, the high rate of sedimentation could make it difficult to obtain homogeneous composites during their processing. This can be overcome by metal coatings of nonmetallic particles. Usually silver coating is chosen since its oxide is conductive unlike aluminum and copper. Table 19.2 shows various realizations of composites filled with metal-coated particles. Composites become conductive for very small metal content, lower than 15 vol.%. The composites with high filler content or high aspect ratio particles content are highly conductive. The specific density of the filler is reduced by a factor of 2–5 compared to the one for silver that is of 10.5.

19.3.2
Thermal Conductivity

As illustrated in Figure 19.3, the dependence of the thermal conductivity on the filler volume content does not show a percolation behavior. Indeed, heat conduction increases gradually and then shows an exponential increase up to its highest value corresponding to the maximum packing fraction. Table 19.3 for thermoplastics and Table 19.4 for thermosets and elastomers show realizations of thermally conductive composites with metallic particles and various polymers. Compared to electrical conductivity, the highest value of the effective thermal conductivity is obtained for larger filler concentrations. For example, with aluminum fibers in polypropylene, Bigg [4] has found increased values of the electrical and thermal conductivities for volumetric filler content of about 10 and 30%, respectively. Increasing the thermal conductivity is not easy, for isotropic molded composite, the highest value does not

Table 19.2 Electrical conductivity of polymers filled with metal-coated particles.

Matrix	Filler coating/core	Filler density	Shape	Size (μm)	e (μm)[a]	φ_c (vol.%)[b]	φ_{mc} (vol.%)[c]	$\varphi_{\sigma eff}$ (vol.%)[d]	σ_{eff} (Ω^{-1} m^{-1})	Reference
ER	Ag/basalt	2.3	Irregular	5–15	1	28	14	50.6	10^4	[7]
ER	Ag/basalt	2.3	Fiber	2/10–20	0.1	29	3	50.6	$3.2\ 10^4$	[7]
EVA	Ag/CaSiO$_3$	5.7	Fiber	4/44	0.5	8	3	29	$1.8\ 10^5$	[20]
EVA	Ag/glass	2.7	Sphere	14	0.2	15	0.48	49.6	27	[21]
PU	Ag/basalt	2.3	Irregular	5–15	1	29	14	50.6	$7.9\ 10^4$	[7]
PU	Ag/basalt	2.3	Fiber	2/10–20	0.1	29	3	50.6	$2.5\ 10^5$	[7]
PC	Ni/glass	—	Fiber	14/500	—	—	—	43	100	[12]
HPDE	Ag/PA	4.4	Powder	6 and 12	<1	4	1.4	32.9	680	[22]

ER: epoxy resin, EVA: ethylene vinyl acetate, PA: polyamide, PC: polycarbonate, PU: polyurethane.
a) Thickness of the metal coating.
b) Filler content at the percolation threshold.
c) Metal coating content at the percolation threshold.
d) Filler contents for which σ_{eff} values are provided.

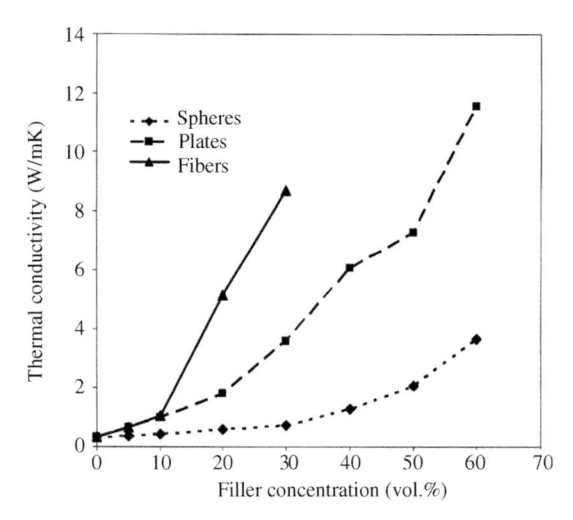

Figure 19.3 Thermal conductivity of polyamide/Cu composites as a function of filler content based on Ref. [24] – the average Cu particle size is 45, 50, and 600 μm for spheres, plates, and fibers, respectively.

exceed 10–20 times of that for bulk polymer and is still much lower than that for metal. With the same filler content and as shown in Figure 19.3, much higher value can be reached for composites with a high aspect ratio of particles and especially if a preferential orientation of the particles is realized [23, 24]. In fact, the insulating matrix is short-circuited by the highly conductive particles. A specific orientation of the particles can be induced by the processing method. So, with injected PBT thermoplastic and aluminum fibers, Dupuis [11] has obtained a composite with an in-plane thermal conductivity 3–4 times higher than that of the transverse one as fibers are partially aligned in the flow direction. Such thermally conductive polymers are useful as they act as a heat sink or reduce hot spots and, therefore, increase lifetime of the polymer.

In terms of conductivity range, metals are the adequate material to improve heat conduction in composites since the influence of increasing the thermal conductivity of the filler is negligible if it is 100 times more than that of the matrix as shown experimentally [23] or using numerical models [40].

To optimize properties, a maximum loading is generally sought. However, the volumetric loading of metallic particles is limited to some amount according to the mixing and processing technique that corresponds to a marked increase in viscosity. This maximal amount corresponds to the maximum packing fraction φ_F that depends on the shape, size, and spatial distribution of conductive filler particles. Many properties change quickly on reaching this maximum packing fraction, and therefore it is incorporated into many mathematical treatments to describe the concentration dependence of several properties [9]. To achieve high packing density composites, the use of large size particles with multimodal particle size distribution and low aspect ratio is recommended [3].

Table 19.3 Transverse thermal conductivity of thermoplastics filled with metallic particles.

Matrix k_m (W/(m K))	Filler k_f (W/(m K))	Shape	Size (μm)	Filler (vol.%)	k_{eff} (W/(m K))	Reference
HDPE (0.50)	Al (220)	Powder	40–80	33	3.6	[25]
HDPE (0.50)	Cu (386)	Powder	15–40	10	0.7	[26]
HDPE	Cu	Powder	<38	24	1.7	[27]
HPDE (0.50)	Bronze (64)	Powder	<100	23	1.9	[28]
HPDE (0.50)	Cu (384)	Powder	<60	24	1.1	[28]
HPDE (0.50)	Fe (80)	Powder	<100	24	1.3	[28]
HPDE (0.55)	Sn (64)	Spherical	20–40	16	1.1	[29]
HPDE (0.50)	Zn (116)	Powder	<5	20	0.88	[28]
i-PP (0.12)	Ni (90)	Powder	40	3.4	0.15	[30]
LDPE (0.31)	Cu	Powder	<38	24	0.72	[8]
LLDPE (0.36)	Cu	Powder	<38	24	0.76	[8]
PA (0.32)	Cu (384)	Plate	50	60	11.6	[24]
PA (0.32)	Cu (384)	Sphere	45	60	3.6	[24]
PA (0.32)	Cu (384)	Fiber	50/600	30	8.6	[24]
PE (0.26)	Cu (390)	Powder	3.3	30	1.8	[13]
PP (0.26)	Al (220)	Fiber	100/1250	15	0.72	[23]
PP (0.26)	Al (220)	Fiber	100/1250	28	2.1	[4]
PP (0.24)	Al (237)	Powder	8	59	2.7	[17]
PP (0.24)	Al (237)	Powder	44	58	4.2	[17]
PP (0.24)	Cu (389)	Powder	30	34	2.1	[5]
PP (0.24)	Cu (389)	Powder	280	43	2.3	[5]
PP (0.24)	Cu (401)	Irregular	15	36	2.3	[31]
EVA (0.16)	Ni (91)	Powder	1–50	26	0.68	[32]
PVC (0.46)	Cu	Irregular	100	46	1.6	[9]
PVC (0.46)	Ni	Powder	10	32	1.0	[9]
PBT (0.24)	Al (160)	Fiber	90/2000	45	2.2	[6]

HDPE: high-density polyethylene, i-PP: isotactic polypropylene, LDPE: low-density polyethylene, LLDPE: linear low-density polyethylene, PA: polyamide, PBT: polybutylene terephtalate, PE: polyethylene, PP: polypropylene, PVC: poly(vinyl chloride).

With increasing temperature, the effective thermal conductivity of metal-filled polymers usually follows that of the polymer. So, the effective thermal conductivity increases under the glass transition temperature of amorphous polymers and decreases until the melting temperature [29, 37]. One should note that for metal-filled polymers, the uniformity of the filler distribution has to be checked. Danes et al. [41] report a strong nonuniformity of the transversal electrical and thermal conductivities in injection-molded parts with PBT and aluminum fibers. They show that this nonuniformity comes from the nonuniformity of the filler distribution and can be reduced by decreasing the size of the fibers.

Metal-coated particles were developed mainly for electrical purpose and just a few works were performed for heat conduction application. As shown in Table 19.5, the effective thermal properties are similar to that of compact metallic particle, but they are obtained for a much smaller amount of metal. Higher thermal conductivity values

Table 19.4 Transverse thermal conductivity of thermosets filled with metallic particles.

Matrix k_m (W/(m K))	Filler k_f (W/(m K))	Shape	Size (μm)	Filler (vol.%)	$k_{,eff}$ (W/(m K))	Reference
ER (0.24)	Ag (420)	Spherical	27	55.7	1.33	[33]
ER (0.15)	Al (220)	Spherical	2	37.7	0.66	[34]
ER (0.15)	Al (220)	Fiber	—	29.8	0.50	[35]
ER (0.23)	Al	Powder	44	40	1.43	[36]
ER (0.23)	Al	Fiber	5000	40	1.82	[36]
ER (0.24)	Au (315)	Spherical	37	50	1.36	[33]
ER (0.24)	SS (16.3)	Spherical	92	56.7	1.27	[37]
ER (0.23)	CI	Flake	<150	40	0.91	[36]
ER (0.24)	Cu (383)	Spherical	100	40.8	0.82	[37]
ER (0.24)	Cu (383)	Spherical	100	57.7	1.37	[37]
ER (0.24)	Cu (383)	Spherical	46	39.1	0.86	[33]
ER (0.26)	Cu (383)	Wire	76//3040	12.0	2.52	[38]
ER (0.24)	Cu (383)	Spherical	46	55.2	1.36	[33]
ER (0.24)	Sn (61.7)	Spherical	46	54.2	1.33	[37]
SR (0.19)	Al (220)	Spherical	1200	27	0.41	[39]

CI: cast iron, ER: epoxy resin, SS: stainless steel, SR: silicone rubber.

might be obtained by increasing the thickness of the metal coating up to 20% of the radius of the particle as shown by numerical simulation [42].

19.3.3
Mechanical Properties

Generally, the composites present reduced properties due to incorporation of filler compared to the unfilled polymer.

For the low filler content, the thermal expansion coefficient of composite is mainly influenced by the polymers. However, for the high amount of particle, the filler will have a dominating influence and can be used to reduce the thermal expansion coefficient.

Table 19.5 Transverse thermal conductivity of polymers filled with metal-coated particles.

Matrix	Filler coating/core	d [a]	Shape	Size (μm)	e (μm)[b]	Filler (vol.%)	Metal (vol.%)	$k_{,eff}$ (W/(m K))	Reference
VA	Ag/glass	2.7	Sphere	14	0.2	49.6	1.2	1.29	[21]
EVA	Ag/glass	2.6	Sphere	47	0.3	49.2	0.61	1.38	[21]
EVA	Ag/CaSiO$_3$	5.7	Fiber	4/44	0.5	41	14	1.32	[43]
HPDE	Ag/PA	4.4	Sphere	6 and 12	<1	33.4	11.1	1.93	[44]

EVA: ethylene vinyl acetate, HDPE: high-density polyethylene.
a) Filler density.
b) Thickness of the metallic coating.

For most applications, the main mechanical properties are the modulus of elasticity, tensile strength, and impact strength. From many observations on short metallic particles, the tensile strength and impact strength decrease with an increase in filler content [4, 14, 25, 45–47]. This can be explained by the weak adhesion between filler and matrix and, therefore, in this situation, metallic particles raise locally the stress in the composite.

The bond between matrix and particles can be enhanced by surface treatment or interfacial reactive agents. So, much lower loss of tensile strength can be obtained by using additives as reported by Bigg [4]. In addition, Tan et al. [48] have reported an increase of impact strength with the increase of the filler amount as fiber and surface treatment are used.

The modulus of elasticity depends not only on Young's modulus of phases but also on the particles dispersion and interfacial adhesion and shows various behaviors. The modulus of elasticity usually increases in metallic particles that are stiffer than the polymer matrix [45, 49, 50], and it can decrease for high filler amount, this coming from particles aggregation [25] or weak bond between matrix and filler [50], or can stay unchanged [46].

Elongation to break is very often considered in mechanical tests dealing with filled polymers. Elongation to break usually drops rapidly with the amount of metallic particle typically between 5 and 20 vol.% [12, 25, 45, 46].

Figure 19.4 presents a typical stress–strain curve for polymer filled with metallic particle illustrating the decrease of elongation to break and tensile strength and an unchanged modulus of elasticity with the increase of the filler content.

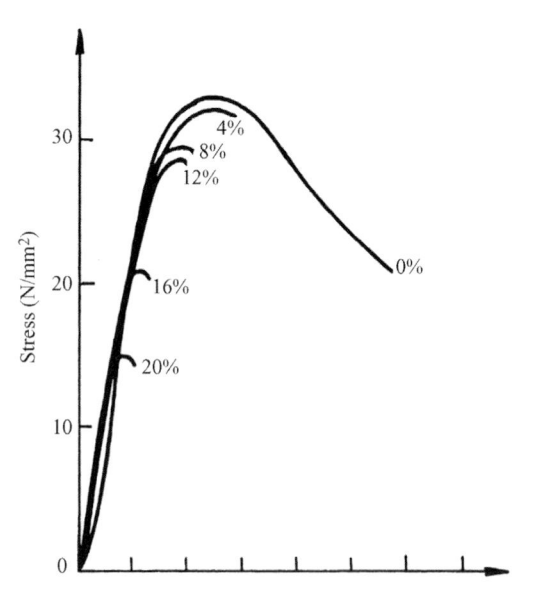

Figure 19.4 Stress–strain curves for HDPZE/zinc composites with various contents of zinc particle (vol.%.) based on Ref. [46].

Finally, the mechanical properties of metal-filled polymers are strongly influenced by type, concentration, size, shape, orientation, type of dispersion of the particles, and strength of the bond between the two phases.

19.4
Main Factors Influencing Properties

When the metallic conductive fillers are incorporated into a polymer matrix, thermal, electrical, mechanical, and other properties of the pure material may change depending on the properties of each component as well as on the shape, size, filler concentration, morphology of the system, and the preparation of the test samples [5, 51–55]. There are numerous examples where metallic fillers have been added to plastics to produce conductive composites, including aluminum [56], steel [57], silver [58], iron [59], nickel [9], copper [5, 9, 26, 60], and metal-coated glass [21, 61, 62].

Clarifying the role of the factors influencing the properties of the composites enables us to choose the suitable processing method for obtaining the composites and to improve the different properties of these systems. The effect of each of these parameters needs to be addressed separately, and will be done so in the following sections.

19.4.1
Effect of Volume Fraction

The particle loading is an important factor that affects mechanical properties of metal-filled polymers. Xia *et al.* [60] investigated the mechanical properties of the Cu/LDPE composites. They showed that the Young's modulus, tensile strength, and elongation at break of the Cu/LDPE microcomposite decrease with the increasing copper particles loading. An opposite effect was reported by Zhang *et al.* [63] when using $Mg(OH)_2$-filled EPDM (see Table 19.6). The authors of Ref. [63] explained that with increasing content, the particles form agglomerates. The extension deformation of these agglomerates under the stress and the slide of interfacial molecules will improve the mechanical properties of the composites.

Some of the important results obtained from literature are listed in Table 19.6. It can be clearly seen that the variation of the mechanical properties versus filler volume loading must not be considered as a general rule for all metal-filled polymer composites as reported by some authors [8, 26, 46, 58, 63]. As Rusu *et al.* [46] described, the occurrence of particle agglomeration is responsible for the appearance of metal particle–metal particle contacts instead of polymer–metal particle contacts, characterized by a total lack of adhesion. Under such circumstances, a 3D network-type structure made of metal particles is formed. Thus, the nature of this network influences the evolution of the mechanical properties of the composites. The poor dispersion of microparticles in matrix and poor interfacial adhesion between particles and matrix resulting from the microsize of particles lead to Young's modulus, tensile strength, and elongation at break of the microcomposite decrease with the increasing of the microparticles loading [60]. The strong interfacial adhesion

Table 19.6 Effect of metallic particles loading on composites properties: literature data.

Composite	k	σ_e	η	E	σ	ε	Reference
LDPE/Cu	—	—	—	Decrease	Decrease	Decrease	[60]
EPDM/Mg(OH)$_2$	—	—	—	—	Increase	Increase	[63]
HDPE/Zn	Increase	—	—	Increase	Increase	Decrease	[46]
LDPE/Cu	Increase	Increase	—	Increase	Decrease	Decrease	[8]
HDPE/Cu	Increase	—	—	Increase	Decrease	Decrease	[26]
PA/Ag	—	—	—	Constant	Decrease	Decrease	[58]
PP/Cu	Increase	Increase	—	—	—	—	[5]
EVA/silver-coated glass	Increase	Increase	Increase	—	—	—	[21]
Epoxy/Cu Epoxy/Ni	Increase	Increase	—	—	—	—	[9]
PVC/Cu PVC/Ni	Increase	Increase	—	—	—	—	[9]
PVDF/Al	Increase	—	—	—	—	—	[64]
PVC/Al	—	Increase	—	—	Decrease	Decrease	[14]
ABS/Cu	Increase	—	—	—	—	—	[65]
ABS/Fe	Increase	—	—	—	—	—	[65]
LDPE/Cu	—	—	—	Constant	Decrease	Decrease	[66]

σ: tensile strength, ε: elongation at break, σ_e: electrical conductivity, ABS: acrylonitrile butadiene styrene.

between particles and matrix results in a considerable decrease of the mechanical properties of the composites [46].

It is well known that the filling of a polymer with metallic particles results in an increase of both electrical and thermal conductivities of the composites obtained. This behavior is due to the fact that the metallic fillers have higher thermal and electrical conductivities than the polymeric matrix. This result (see Table 19.6) was reported by some authors [5, 8, 9, 21, 26, 46].

Concerning the electrical conductivity σ, as explained by McCullough [67] and Mamunya *et al.* [9], when the volume filler fraction φ reaches a critical value φ_c (the so-called percolation threshold), an infinite conductive cluster (IC) is formed and, consequently, the composite becomes electrically conductive. As the filler concentration increases from φ_c to the filling limit F, the value of σ increases rapidly over several orders of magnitude, from the value σ_c at the percolation threshold to the maximal value σ_m.

Below the percolation threshold, the electrical conductivity change is negligible and the conductivity of the composite is equal to the polymer conductivity σ_p or slightly higher. The typical dependence of the logarithm of electrical conductivity on the filler volume fraction is shown in Figure 19.5 [9, 67].

For the thermal conductivity, several investigations on polymer composites with dispersed fillers show the absence of percolation behavior of the thermal conductivity k with increasing dispersed filler concentration [5, 8, 9, 21, 26, 46, 55, 68, 69]. The reason for the percolation threshold absence is explained in Ref. [9]. This is due to the

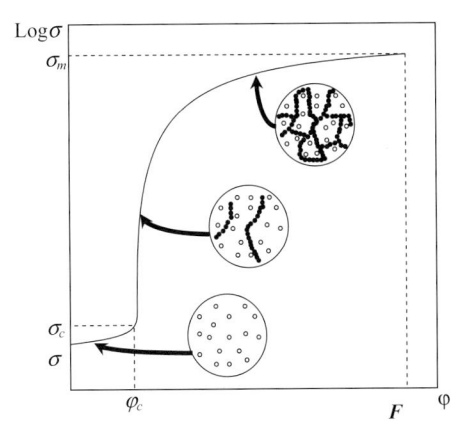

Figure 19.5 Illustration of chain formation in a particulate-filled composite. Open circles indicate isolated particles; filled circles indicate contacting particles participating in chain formation based on Refs [9, 67].

fact that the thermal conductivities of the dispersed filler k_f and of the polymer matrix k_p are comparable to each other, their ratio not being more than 10^3, whereas the filler electrical conductivity σ_f is 10^{10}–10^{20} times larger than the polymer conductivity σ_p.

As a conclusion, the incorporation of metallic particles in polymer results in an increase of both electrical and thermal conductivities of the composites obtained, while the nature of particle–particle network influences the evolution of the mechanical properties of the composites.

19.4.2
Effect of Particle Shape

The reinforcement shape has important effects on the mechanical properties of composites. The failure mode of composites can be changed by modifying the particle morphology [70]. The possible scenarios include spherical fillers, irregularly shaped fillers, flakes (3D random and random in plane), short fibers (3D random and random in plane), long fibers (unidirectional and random in plane), and continuous fibers (unidirectional and cross-ply laminated) [55]. The composite reinforced with angular particles tends to fail through particle fracture, whereas the composite reinforced with spherical particles tends to fail through void nucleation, growth, and coalescence in the matrix regions near particles [10, 19].

Experimental and numerical results have shown that the composite reinforced with spherical particles exhibits a slightly lower yield strength and work hardening rate, but a considerably higher ductility than the composite reinforced with angular particles [70–72]. Qualitatively, it is well known that deleting particle sharp corners will improve the mechanical properties of composites, and the angular particle is more likely to fail than the spherical particle [70]. Besides, the spherical particle gives better stress distribution than the cylindrical [73]. This behavior can be explained by value of the aspect ratio of particle that plays an important role in the overall response of composites [74]. According to Bigg [4], further increases in tensile strength can only be achieved by increasing the aspect ratio of the inclusions.

The shape and the orientation of filler particles also have a significant influence on the thermal conductivity of a composite material. Many theoretical models are valid for only specific types of filler particles and composite constructions [55]. Bigg [23] reported that for spherical and dimensionally isotropic irregularly shaped filler particles, the influence of increasing filler conductivity is negligible when the ratio of filler conductivity to matrix exceeds 100 : 1. This indicates that the effect of filler shape on the thermal conductivity of composites is linked to the ratio of filler conductivity to matrix.

The effective electric conductivity and the percolation threshold of metal–polymer composite are investigated theoretically by Xue [75] as a function of the electric conductivities of the constituents, of the particle shape and size, and of the volume of loading. They consider the metal–polymer composite in which an assembly of randomly oriented ellipsoidal metal particles (with axial ratio M) are embedded in a homogeneous matrix. Using Maxwell–Garnett theory and the relation between two distinct topological structures, several simulations were carried out. The shape dependence of the percolation threshold and the effective electrical conductivity are shown in Figures 19.6 and 19.7. The results indicate that the percolation threshold value decreases significantly with increasing metal particle axial ratio M. Many more metal particles with smaller axial ratio are needed to form a conductive network through the entire polymer matrix.

As shown in Figure 19.7, the effective electric conductivity increases rapidly with the increase of the metal particle axial ratio before it reaches *the axial ratio threshold*. It should be kept in mind that the theoretical results on the effective electric conductivity of tin–polypropylene composites are in good agreement with the experimental data [75].

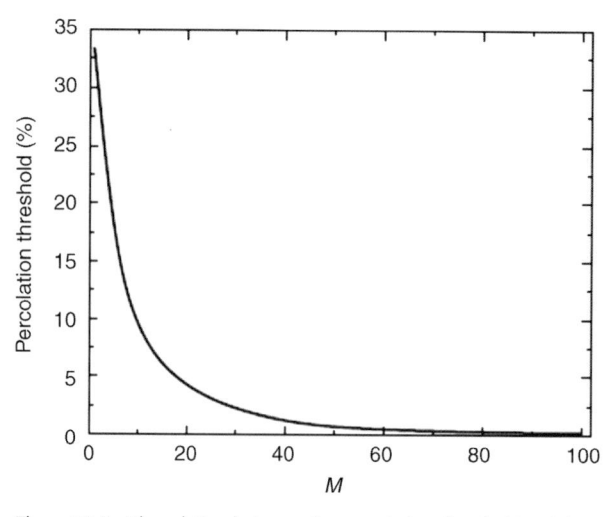

Figure 19.6 The relation between the percolation threshold and the metal particle axial ratio [75].

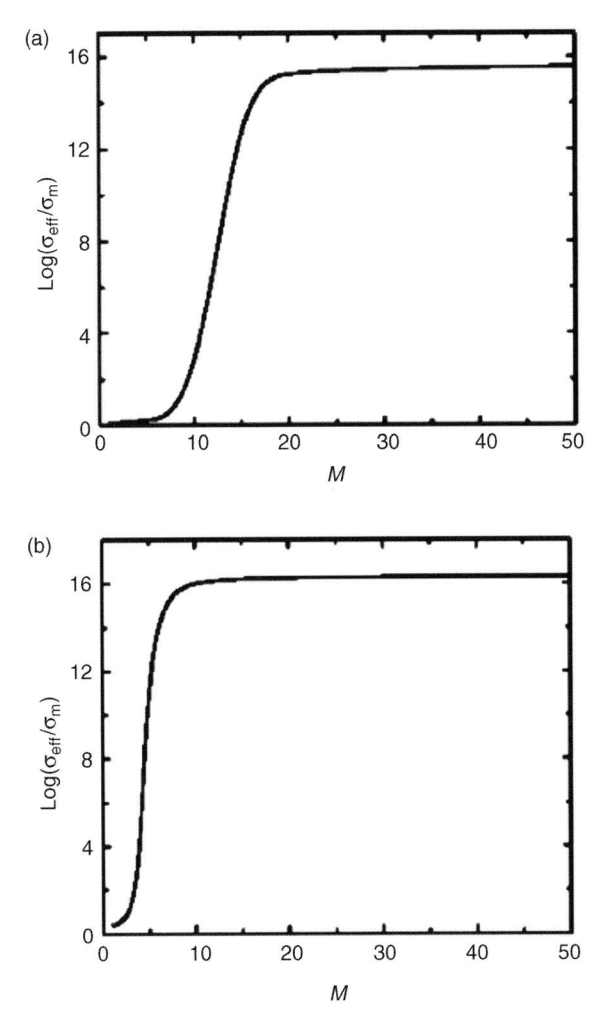

Figure 19.7 The relation between the logarithm of the effective electric conductivity and the axial ratio M. The volume fraction (φ) of the whole metal particles is selected as 5% (a) and 20% (b), respectively [75].

The above results indicate that the axial ratio of the metal particles is a principal factor that should be carefully examined when making conductive polymer-based composites containing metal particles.

19.4.3
Effect of Particle Size

The effect of particle size on physical performance of composites has been investigated by many authors [5, 21, 56, 63, 76–78]. Particle size greatly influences the mechanical, thermal, and electrical properties of polymeric composites, such as

Young's modulus, tensile strength, thermal conductivity, and electrical conductivity. Particle size effect mainly results from the synergism mechanism between particles and the host matrix. Due to its great importance, large amount of researches have been performed, including experiments, theories, and numerical computations [79].

Cho et al. [76] investigated the effect of particle size of alumina on the mechanical properties of polymeric composites with spherical particles ranging from 3 to 70 μm in diameter. They found that particle sizes at microscale had little influence on Young's modulus and the tensile strength increases as the particle size decreases. The effect of particle size (average diameter: 1, 2, 5, 8, and 12 μm) on the elastic modulus of epoxy/alumina trihydrate composites was studied by Radford [77]. It is seen that the modulus is not very much affected by particle size in the range studied. Zhang et al. [63] investigated the effect of $Mg(OH)_2$ loading with different particle sizes on tensile strength of $Mg(OH)_2$/EPDM composites. They concluded that the particle size clearly has a significant effect on the strength of particulate-filled polymer composites, which generally increases with decreasing size.

Agoudjil et al. [21] demonstrated that the thermal and electrical conductivities of EVA/silver-coated glass do not depend on the size of particles. This behavior was reported in several studies (see Table 19.7) reviewed by Bigg [23]. Nevertheless, this result must not be considered as a general rule as some authors [5, 67, 80] reported that the size of fillers influences the thermal conductivity of composites. Thus, it seems that the effect of filler size on the thermal conductivity of composites is not the same for all systems. This behavior can be linked to the thermal contact resistance. According to Nogales and Böhm [81], the thermal contact resistances between

Table 19.7 Effect of metallic particle sizes on composites properties: literature data.

Composite	Property						Reference
	k	φ_c or σ_e	η	E	σ	ε	
Silicone/Al_2O_3	$k_L > k_S$	—	—	—	$\sigma_L < \sigma_S$	$\varepsilon_L < \varepsilon_S$	[56]
EVA/silver-coated glass	$k_L \approx k_S$	$\varphi_{cL} > \varphi_{cS}$	$\eta_L \approx \eta_S$	—	—	—	[21]
PP/Cu	$k_L < k_S$	$\varphi_{cL} > \varphi_{cS}$	—	—	—	—	[5]
Epoxy/Cu	$k_L \approx k_S$	—	—	—	—	—	[23]
Epoxy/Ag							
Epoxy/Al							
Epoxy/SS							
Epoxy/Au							
Vinyl ester/Al_2O_3	—	—	—	$E_L \approx E_S$	$\sigma_L < \sigma_S$	—	[76]
Epoxy/Al_2O_3	—	—	—	—	$\sigma_L < \sigma_S$	—	[77]
EPDM/$Mg(OH)_2$	—	—	—	—	$\sigma_L < \sigma_S$	—	[63]
Epoxy/Cu	—	$\sigma_{eL} > \sigma_{eS}$ $\varphi_{cL} = \varphi_{cS}$	—	—	—	—	[78]
LDPE/Cu	—	—	—	$E_L < E_S$	$\sigma_L > \sigma_S$	$\varepsilon_L > \varepsilon_S$	[66]

Subscript L: composite filled with large particles, subscript S: composite filled with small particles, φ_c: electrical percolation threshold, SS: stainless steel.

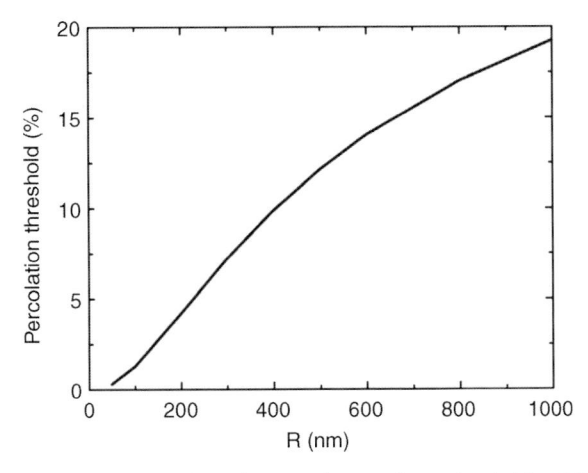

Figure 19.8 The relation between the percolation threshold and the metal particle size. The metal particles are regarded as spheres with radius R [75].

reinforcements and matrix give rise to a size effect in the overall conductivity of composites, so even highly conductive particles fail to increase the overall conductivity once their size falls bellow some critical value.

Mamunya et al. [78] investigated the dependence of electrical conductivity of copper powders used as dispersed conductive fillers in polymer composite materials. It was shown that powder with small particle size dispersed within a polymer matrix has lower electrical conductivity value compared to the composite with large particles. The percolation threshold of metal–polymer composite is investigated theoretically by Xue [75] as a function of the particle size (see Section 3.2). As shown in Figure 19.8, the percolation threshold of metal–polymer composites increases rapidly with the increasing metal particle size. Thus, the size of the metal particles is the factor that should be carefully examined when making conductive polymer-based composites containing metal particles. The use of metal particles with smaller size helps to reduce the minimum metal content required to reach a certain conductivity value [22].

To conclude, as discussed above and shown in Table 19.7, there is no good general rule about the effect of the particle size on all properties of composites for all systems. We think that the influence of the particles size on the properties of the composites is linked to the shape of particles and to the degree of adhesion between filler and matrix. Thus, to clarify the effect of this parameter, a study of large number of systems filled with several particle sizes and different particle shapes should be performed.

19.4.4
Effect of Preparation Process

Another important item for consideration is the method by which the composites are made and subsequently molded into parts. Different types of process methods have been used to prepare polymer composites [51, 61, 82–87].

Melt mixing is the most widely used process for producing polymer composites [87]. This technique facilitates the random dispersion of fillers within matrix with strong damage to filler aspect ratio compared to some others process methods. Al-Saleh and Sundararaj [87] reported that for melt mixing technique, electrical conductivity decreased with increasing mixing time and mixing speed due to filler breakage and due to the uniform distribution of fillers without segregated network formation. However, in case of poor adhesion between the polymer matrix and the filler, changing processing conditions did not affect the tensile properties of composites. This indicates that adhesion is one of the major factors that influences the tensile properties of polymer composites [87]. Therefore, optimizing mixing conditions and understanding the relationship between final properties and filler dispersion, distribution and aspect ratio is essential for the preparation of composite at the lowest filler loading.

An alternative technique to melt mixing is polymerization filling method by which the polymer is synthesized in the presence of conductive particles. The electrical properties of polymer-filled composites are enhanced by improving the distribution of conductive filler in the polymer matrix. This method providing intimate contacts improves the interfacial interaction between conductive particles and polymer molecules by decreasing average agglomerate size [51].

Extrusion and injection molding of a composite can align fillers that have an aspect ratio greater than 1 in a specific direction due to the flow through the nozzle of the different machines and the mold. This alignment will create an anisotropic thermal and electrical conductivity within the samples, meaning that conductivity will be greater in one direction over another [61]. Compression-molded samples will also display an effect of filler orientation on some composite properties. Weber and Kamal [85] found that injection-molded samples had a 33.6° angle of orientation, while the compression-molded samples with the fillers had a 45.2° angle. Deviation from a 45° angle is evidence that there has been alignment of the fillers. It has been shown by Bayer et al. [82, 83] that the injection-molded composite material exhibits not only a lower percolation threshold φ_c than the conventionally pressure-molded isotropic sample but also electrical conductivities two to three orders of magnitude larger than the latter.

Feller [84] investigated the influence of extrusion temperature and screw speed on electrical properties of extruded composites. The authors found that for extruded tapes of composites, an increase of both processing temperature and screw speed leads to a significant decrease of resistivity. This behavior was attributed to viscosity variations of polymers.

As a conclusion, it seems that to optimize the electrical properties of composites, the use of melt mixing or polymerization is convenient to reduce the electrical percolation threshold concentration. This is due to that nonuniform distribution of fillers is highly desired to reduce the electrical percolation threshold concentration. Extrusion and injection molding can be used to produce homogeneous distribution of fillers that is essential to enhance the mechanical properties of the composites [86]. However, when the thermophysical properties are concerned, melt mixing or polymerization produces composites with homogeneous thermal conductivity, while extrusion and injection molding allow obtaining composites with heterogeneous thermal conductivity. For all these physical properties, a high aspect ratio is highly desirable. For electrical applications, the electrical percolation threshold decreases with increase in the conductive filler aspect ratio, and for tensile properties, a larger aspect ratio of filler was reported to maximize the stress transfer to the filler [86, 87].

19.5 Models for Physical Property Prediction

19.5.1 Theory of Composite Transport Properties

Several approaches have been used to estimate the effective properties of composite materials. According to Refs [88, 89], Maxwell obtained for the first time the electrical conductivity of a heterogeneous material using the *effective medium* (EM) theory. Bruggeman [90] in 1935 introduced an extension of this theory using the theory of *effective medium approximation* (EMA) to study the electrical conductivity of composites. Then, Brown [91] showed in 1955 that there is a link between macroscopic properties and microscopic structure of a composite. Other studies have later defined the parameters affecting the evolution of properties of heterogeneous materials, including the volume fraction of fillers in the matrix that has been studied for the first time in 1957 by Broadbent and Hammersley [92]. The latter has been used for the percolation theory to study the evolution of fluids through porous media. Basic concepts of percolation were also used to explain various physical phenomena. For example, Kirkpatrick [93] used the concepts of percolation to explain the electrical behavior of composites. The combination of the basic concepts of percolation theory and effective medium approximation allowed the introduction of another concept of *generalized effective medium* (GEM) to explain the electrical and dielectric behavior of composite materials [89, 94–97]. According to McCullough [67], models for predicting transport properties of composites can be classified into two classes: models based on the principle of diffusion and effective medium and models based on the percolation theory.

19.5.1.1 Effective Medium Theory

Effective medium theory is based on assumptions of Maxwell and Bruggeman [90], where Maxwell assumed that the composite medium can be substituted by an effective medium consisting of particles of ellipsoidal (or spherical) isolated and dispersed in a continuous medium (Figure 19.9). The term isolated means that there is no contact between particles, which represents the limit of this theory. Indeed, the model proposed by Maxwell (see Figure 19.9) is valid only for low filler concentrations. This limit has been confirmed by Meredith and Tobias [98], where they demonstrated that this model is valid only for low filler concentrations.

To solve this problem, another theory was proposed by Bruggeman in 1935 [90]. In the Bruggeman model (or model of effective medium approximation), the principle of effective medium is adopted, but the hypothesis is excluded from isolated inclusions (see Figure 19.10). According to Bruggeman, this theory takes into account the geometry of the inclusions. As will be detailed in the following section, this concept has been modified by other authors where they take into account other factors like the size of loads.

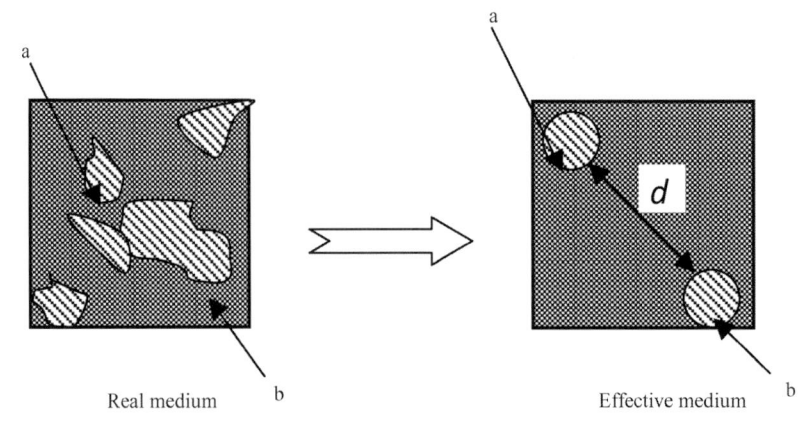

Figure 19.9 The effective medium according to Maxwell. Inclusions (modeled by spheres) are no interacting because *d* is bigger.

19.5.1.2 Theory of Percolation

Percolation means the passage of information between two points of a system. In case of electrical transport in a composite, the percolation can be defined by monitoring changes in electrical conductivity of a composite material made of a metallic phase randomly in dispersed insulating phase. For low metallic filler fraction, the composite behaves as an electrical insulator. When the volume of filler increase, electrons can travel between two opposite sides of the material through the metal phase and the composite becomes an electrical conductor. The volume fraction at which this change in behavior is observed is the percolation threshold.

The mathematical model of percolation was introduced for the first time by Broadbent and Hammersley in 1957 [92, 93]. Initially, the author focuses on characteristics of random media and specifically porous media. This approach was used for several physical models and particularly for electrical behavior of composites [85, 99].

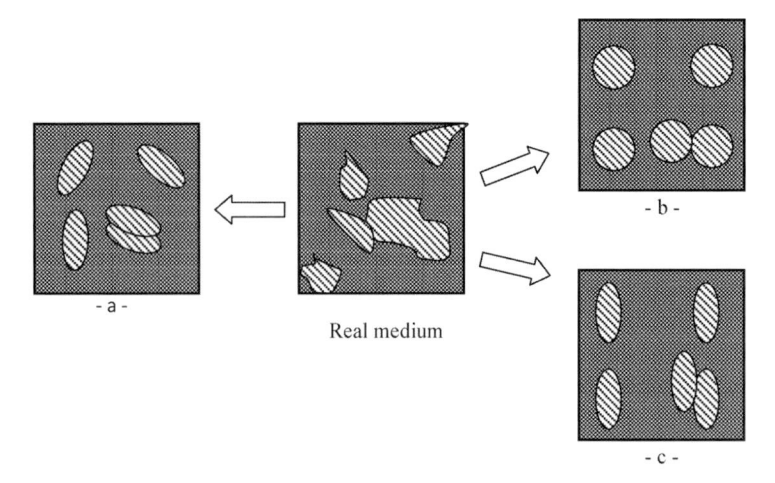

Figure 19.10 The Bruggeman effective medium approximation (the inclusions are interacting).

19.5.2
Models for Thermal Conductivity

Model	Reference	Equation	Remarks
Models of first order			
Parallel	[100]	$k_{sup} = k_m(1-\varphi) + k_f \varphi$	$k_f/k_m < 10$
Series	[100]	$\dfrac{1}{k_{low}} = \dfrac{(1-\varphi)}{k_m} + \dfrac{\varphi}{k_f}$	$k_{low} < k_{eff} < k_{sup}$
Models of second order			
Maxwell	[101]	$k = k_m \dfrac{2k_m + k_f - 2\varphi(k_m - k_f)}{2k_m + k_f + \varphi(k_m - k_f)}$	1. Valid for spheres dispersed in a matrix 2. Not valid at finite concentration of particles 3. Interaction between the particles is not considered 1. $d = 3$ for spherical fillers 2. $d = 2$ for cylindrical fillers
Bruggeman	[90, 102]	$1 - \varphi = \dfrac{k_f - k}{k_f - k_m}\left(\dfrac{k_m}{k}\right)^{1/d}$	
Hatta and Taya	[102, 103]	$k = k_m \left[1 - \varphi \dfrac{(k_m - k_f)[(k_f - k_m)(2S_{33} + S_{11}) + 3k_m]}{3(k_f - k_m)^2(1-\varphi)S_{11}S_{33} + k_m(k_f - k_m)R + 3k_m^2}\right]$ where $R = 3(S11 + S_{33}) - \varphi(2S_{11} + S_{33})$	1. $S_{11} + S_{22} + S_{33} = 1$ 2. Disks: $S_{11} = S_{22} = 0$, $S_3 = 1$ 3. Spheres: $S_{11} = S_{22} = S_{33} = 1/3$ 4. Fibers, rods, or long cylinders: $S_{11} = S_{22} = 1/2$, $S_{33} = 0$

(Continued)

Model	Reference	Equation	Remarks
			5. Randomly oriented short fibers (length L and diameter D) $$\begin{cases} S_{11} = \dfrac{\alpha}{2(\alpha^2-1)^{3/2}} \\ \quad \times \left[\alpha(\alpha^2-1)^{1/2} - \cosh^{-1}\alpha\right] \\ S_{33} = 1 - 2S_{11} \\ \alpha = \dfrac{L}{D} \end{cases}$$ 6. Random dispersion of spheres in a continuous matrix, $R = 2 - \varphi$
Hashin and Shtrikman	[102,104, 105]	$k = k_m \left(1 + \dfrac{\varphi}{((1-\varphi)/3) + (k_m/(k_f - k_m))}\right)$	For spherical fillers
Models of third and fourth order			
Torquato	[106]	$\dfrac{k_c}{k_m} = \dfrac{1 + \varphi\beta - (1-\varphi)\varsigma\beta^2}{1 - \varphi\beta - (1-\varphi)\varsigma\beta^2}$, where $\beta = \dfrac{k_f - k_m}{k_f + (d-1)k_m}$	1. $d = 3$ for spheres and 2 for cylinders 2. ξ: Microstructural three-point parameter range $0 < \xi < 1$
Semitheoretical predicted models			
Mean geometric	[69]	$k = k_f^{\varphi} \cdot k_m^{(1-\varphi_i)}$	Nevertheless, gives generally a good estimate of k

Agari and Uno	[13]	$\log k = \varphi C_2 \log k_f + (1-\varphi)\log(C_1 k_m)$	1. For all types of inclusions in a matrix 2. For multiphase systems 3. For both low and high φ 4. C_1 and C_2 are obtained by fitting experimental data 5. C_1 represents the effect of particles on the polymer structure 6. C_2 represents the ability of filler particles to create continuous chains
Lewis and Nielsen	[107]	$\begin{cases} k = k_m \dfrac{1+AB\varphi}{1-B\psi\varphi} \\ \psi = 1 + \dfrac{(1-\varphi_{max})\varphi_f}{\varphi_{max}^2} \\ B = \dfrac{k_f/k_m - 1}{k_f/k_m + A} \end{cases}$	1. Taking into account: • maximum packing fraction of the filler (φ_{max}) • shape and orientation of the fillers $A = \lambda_E - 1$ 2. λ_E: Einstein coefficient

Numerical predicted models

There are numerous studies relating to the numerical predicted models on thermal conductivity of filler polymer composites such as two-dimensional numerical model of Kumlutas et al. [88], 3D model of Kumlutas and Tavman [36], Cai et al. [37], and some other numerical models. Recently, Sanada et al. [108] studied the thermal conductivity of polymer composites with microfillers of alumina numerically and experimentally. The authors used finite element analyses (ANSYS[R] soft) to predict the potential of fillers to enhance thermal conductivity of the composites and to analyze the effect.

k_m, k_f: thermal conductivities of the polymeric matrix and the fillers, respectively; φ: filler volume fraction, k_{sup}, k_{low}: upper and lower bounds of effective thermal conductivity.

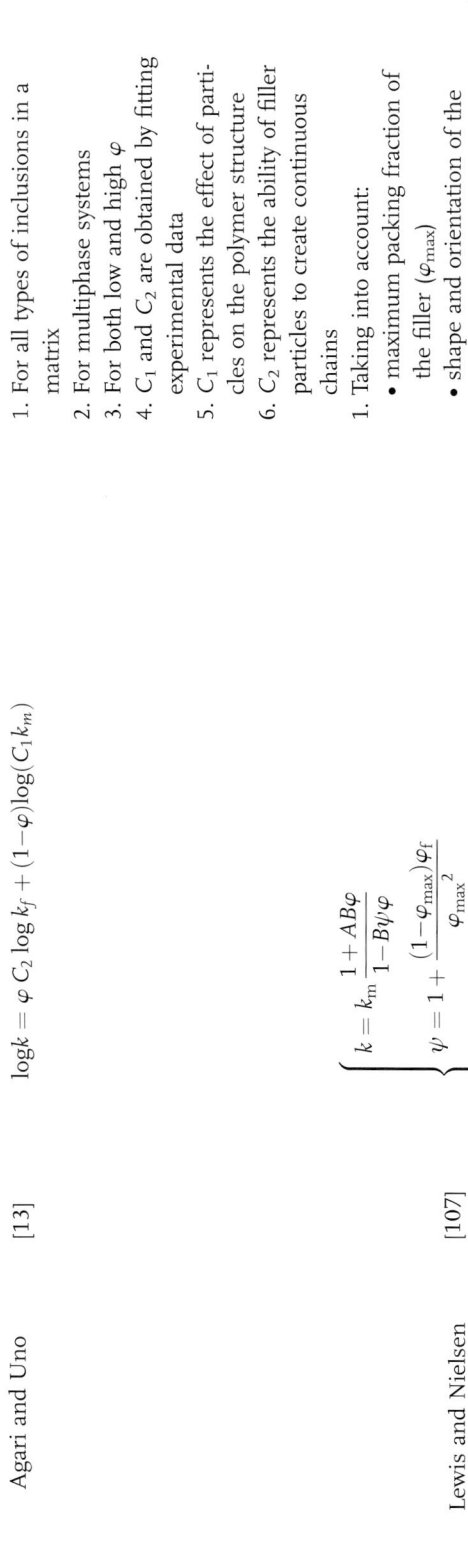

19.5.3
Models for Electrical Conductivity

Model	Reference	Equation	Remarks
Statistical percolation models			
Kirkpatrick and Zallen	[93, 109]	$\sigma \propto (\varphi-\varphi_c)^t$	1. φ_c: Volume percolation concentration
			2. t determines the power of the conductivity increase above φ_c and ranges from 1.6 to 1.9
McLachlan	[89]	$\dfrac{(1-\varphi)\left(\sigma_m^{1/t}-\sigma^{1/t}\right)}{\sigma_m^{1/t}+((1-\varphi_c)/\varphi_c)\sigma^{1/t}} + \dfrac{\varphi\left(\sigma_f^{1/t}-\sigma^{1/t}\right)}{\sigma_{ch}^{1/t}+((1-\varphi_c)/\varphi_c)\sigma^{1/t}} = 0$	This equation is an empirical generalization of effective media and percolation theories By changing the constants (t, φ_c), many different shapes can be obtained and almost every situation predicted
Bueche	[110]	$\dfrac{\varrho}{\varrho_m} = \dfrac{\varrho_f}{(1-\varphi)\varrho_c + \varphi\omega_g\varrho_m}$ with $\omega_g = 1 - \dfrac{(1-\alpha)^2 \gamma}{(1-\gamma)^2 \alpha}$	1. γ is the smallest root of the equation: $a(1-\alpha)^{f-2} = \gamma(1-\gamma)^{f-2}$
			2. f is the maximum number of contacts per particle

		3. $a = \dfrac{\varphi}{\varphi_{max}}$ where φ_{max} is the maximum packing fraction of the filler
Thermodynamic percolation models		
Mamunya	[109]	Before the percolation threshold φ_c: $$\log \sigma = \log \sigma_m + (\log \sigma_c - \log \sigma_m)\dfrac{\varphi}{\varphi_c}$$ After the percolation threshold φ_c: $$\log \sigma = \log \sigma_c + (\log \sigma_{max} - \log \sigma_c)\left(\dfrac{\varphi - \varphi_c}{F - \varphi_c}\right)^k$$ with $k = \dfrac{(A - B\gamma_{pf})\varphi_c}{(\varphi - \varphi_c)^{0.75}}$ and $$F = \dfrac{5}{(75/(10+k)) + k}$$	1. σ_c is the electrical conductivity of the system at percolation threshold 2. σ_{max} is the conductivity at the maximum packing fraction F 3. γ_{pf} is the interfacial tension: $\gamma_{pf} = \gamma_m + \gamma_f - 2\sqrt{\gamma_m \gamma_f}$ where γ_m is the surface energy of the matrix and γ_f is the surface energy of the filler 4. A and B are constants 5. k is the aspect ratio

(*Continued*)

Model	Reference	Equation	Remarks
Sumita	[9]	$\dfrac{1-\varphi_c}{\varphi_c} = \dfrac{3}{g^{*}R}$ $\times \left[(\gamma_c + \gamma_m - 2\sqrt{\gamma_c \gamma_m})\left(1 - e^{-(ct/\eta)}\right) + K_0 e^{-(ct/\eta)} \right]$	1. γ_c is the surface energy of carbon black particles 2. g^{*} is the interfacial free energy 3. K_0 is the interfacial energy at $t=0$ (before the mixing phase) 4. c is a constant taking into account the speed of evolution of the interfacial energy 5. t is the duration of mixing the two components 6. η is the viscosity of the matrix during preparation 7. R is the particle diameter
Geometric percolation models			
Slupkowski	[111]	$\sigma = 2\pi\sigma_{ch} \dfrac{d([x] + p_n)}{D\ln[1 + (1/([x]+1)\alpha)]}$ with $[x] = \left[\left(\dfrac{1}{1-\varphi}\right)^{1/3} - 1\right]\dfrac{D}{2d}$	1. This model takes into account the spherical particles of polymer and conductive fillers

2. p_n is the probability of forming a conductive network
3. $[x]$ is the number of layers (structures) that are completely filled with conductive particles
4. D is the diameter of the polymer particle
5. d is the diameter of the conductive particle

Malliaris and Turner proposed a model that yields the evolution of electrical conductivity and percolation threshold values according to diameter ratios

This model provides a better estimate of the percolation threshold than that by the Malliaris and Turner model

r is the radius of the contact zone between adjacent loads

1. Taking into account: maximum packing fraction of the filler (φ_{max})

Malliaris and Turner	[16]	$\varphi_c = 50 p_n \left[1 + \left(\dfrac{\varphi D}{4 d} \right) \right]^{-1}$
Bhattacharya and Chaklader	[80]	$\varphi_c = \dfrac{2.99 d/D}{1 + (2.99 d/D)}$
Rajagopal and Satyam	[112]	$\sigma = \sigma_f \dfrac{2r(\varphi 3D - 4d)(3D - 2d)}{D^2 d}$

Models based on the structure and direction of loads

Lewis and Nielsen	[113]	$\sigma = \sigma_m \dfrac{1 + AB\varphi}{1 - B\psi\varphi}$
		$\psi = 1 + \dfrac{(1 - \varphi_{max})\varphi}{\varphi_{max}^2}, \quad B = \dfrac{\sigma_f/\sigma_m - 1}{\sigma_f/\sigma_m + A}$

(Continued)

Model	Reference	Equation	Remarks
McCullough	[67]	$\dfrac{1}{\sigma} = \dfrac{1}{\sigma_m}\left(\dfrac{(1-\varphi)^2(1-l)}{V}\right)$ $+ \dfrac{1}{\sigma_f}\left(\dfrac{V^2 + l(1-\varphi)(1+V)}{V^2}\right)\varphi$ with $V = (1-\varphi)(1-l) + l\varphi$	shape and orientation of the fillers $A = \lambda_E - 1$ 2. λ_E: Einstein coefficient Using the length l of conductive chain, McCullough proposes a model to calculate the electrical conductivity of a composite containing particles without any orientation (case of spheres)

σ_m, σ_f: electrical conductivities of the polymeric matrix and the fillers, respectively; φ: filler volume fraction, φ_c: percolation threshold.

19.5.4 Models for Mechanical Properties

Model	Reference	Equation	Remarks
Linear elastic solid model Theret et al.	[114]	$\dfrac{L}{R_p} = \dfrac{\Phi_p \Delta P}{2\pi G}$ G is the shear modulus and is related to the Young's modulus E by $E = 2(1+v)G$ with v being the Poisson's ratio	1. L is the projection length 2. R_p is the pipette radius

Bilodeau	[115]	$F = \dfrac{1.4906 G}{(1-\nu)\tan\theta}\delta^2$
Harding and Sneddon	[116]	$F = \dfrac{4 R_1 G}{1-\nu}\delta^2$
Mijailovich et al.	[117]	$\dfrac{T}{\varphi} = k\alpha G$ $\dfrac{T}{d} = \dfrac{k\beta}{R} G$

3. G is the shear modulus
4. Φ_P is a function of the ratio of the pipette wall thickness to the pipette radius, $\Phi_P = 2.0$–2.1 when the ratio is equal to 0.2–1.0
5. ΔP is the aspiration pressure

1. F is the force of indentation
2. δ is the depth of indentation
3. θ is the inclination angle of the triangular faces

R_1 is the radius of the indenter

1. T (Pa) is the applied specific mechanical torque per unit bead volume
2. k is a shape factor ($k = 6$ for spherical beads)
3. φ is the measured bead rotation
4. d is the measured lateral bead translation
5. R is the radius of the bead
6. α and β are geometric coefficients depending on the degree of bead embedding and cell height for magnetic and optical MTC, respectively

(Continued)

Model	Reference	Equation	Remarks
Linear viscoelastic solid model			
Schmid-Schönbein et al.	[118]	$\tau_{ij} + \dfrac{\mu}{k_2}\dot{\tau}_{ij} = k_1\gamma_{ij} + \mu\left(1 + \dfrac{k_1}{k_2}\right)\dot{\gamma}_{ij}$	1. k_1 and k_2 are two elastic constants 2. γ is a viscous constant 3. γ_{ij} is defined as engineering strain 4. μ is a viscous constant
Standard linear solid (SLS) model	[119]	$J(t) = \dfrac{1}{k_1}\left[1 - \left(\dfrac{k_1}{k_1+k_2} - 1\right)e^{-t/\tau}\right]H(t)$	1. $J(t)$ is the creep compliance 2. $\tau = \mu(k_1+k_2)/(k_1 k_2)$ is the characteristic creep time 3. $H(t)$ is the Heaviside function
Sato et al.	[119]	$\dfrac{L(t)}{R_P} = \dfrac{\Phi_P \Delta P}{2\pi k_1}\left[1 + \left(\dfrac{k_1}{k_1+k_2} - 1\right)e^{-t/\tau}\right]H(t)$	Φ_P is defined as that in the elastic solution
Cheng et al.	[120]	$\delta(t) = \dfrac{F}{8R_1 k_1}\left[1 + \left(\dfrac{k_1}{k_1+k_2} - 1\right)e^{-t/\tau}\right]H(t)$	1. Similar to the micropipette aspiration problem, the flat-punch indentation of an SLS half-space can also be solved by applying the corresponding principle to the analogous elastic solution 2. The Poisson's ratio is taken to be 0.5 assuming incompressibility

Power-law structural damping model [121]

Alcaraz et al.

In the oscillatory AFM experiment, the oscillatory force is expressed as

$$F(t) - F_0 = R_1\left[A_F e^{i\omega t}\right]$$

On the other hand, the oscillatory indentation is expressed as

$$\delta(t) - \delta_0 = R_1\left[A_\delta e^{i(\omega t - \psi)}\right]$$

Applying Taylor expansion around the operating indentation depth δ_0 and the Fourier transform, Alcaraz et al. derived the equation for interpreting the complex modulus:

$$G^*(\omega) = G' + iG''$$
$$= \frac{(1-\nu)\tan\theta}{3\delta_0}\left[\frac{A_F}{A_\delta}e^{i\psi} - i\omega b(0)\right]$$

$$\begin{cases} G^*(\omega) = \frac{1}{ka}\frac{A_T}{A_\varphi}e^{i\psi} \\ G^*(\omega) = \frac{R}{k\beta}\frac{A_T}{A_d}e^{i\psi} \end{cases}$$

For magnetic MTC and optical MTC

1. F_0 is the operating force around which the indentation force F oscillates and A_F is the amplitude of this oscillation
2. δ_0 is the operating indentation depth around which the indentation δ oscillates, A_δ is the amplitude of this oscillation, ψ is the phase lag, and $R_1[\cdot]$ denotes the real part
3. G' and G'' are the dynamic storage modulus and loss modulus, respectively
4. $i\omega b(0)$ is a correction term that accounts for the hydrodynamic drag force due to the viscous friction imposed on the cantilever by the surrounding fluid

Mijailovich et al. [117]

1. $T(t) = R_1\left[A_T e^{i\omega t}\right]$

2. $\varphi(t) = R_1\left[A_\varphi e^{i(\omega t - \psi)}\right]$

3. $d(t) = R_1\left[A_d e^{i(\omega t - \psi)}\right]$

MTC: magnetic twisting cytometry, AFM: atomic force microscopy.

19.6
Conclusion

The use of metallic particle-filled composite in modern industries is increasing owing to the possibility to tailor their electrical and thermal conductivities within a quite wide range by adjusting the concentration shape and size of the filler, microstructure of the composite, and the processing and interfacial conditions. In addition, metal-filled polymers offer advantages of lightweight, chemical stability, processability, and cost-effectiveness compared to other conductive materials as metals. However, reaching high electrical or thermal conductivities requires a large amount of filler.

The design of specific properties of metal-filled polymer is a multidisciplinary task. This involves many fields, including material science, powder technology, polymer processing, and so on.

The modeling of the properties of metal-filled polymers is a very complex task and requires the measurement of several parameters that are sometimes difficult to obtain (parameters about the microstructure, thermal contact resistance, etc.). Further works in mathematical modeling, in experimental characterization at microscale, and in microstructure analysis could provide better prediction models for properties of metallic particle-filled polymers.

In this chapter, achieved properties using conductive nanoparticles in polymers are not mentioned since it was dedicated to composites with microsized particles. However, one may notice that higher electrical conductivity or mechanical performance of the composite can be obtained with nanoparticles using fewer amounts of filler than with microparticles. In addition, the increase in thermal conductivity is usually disappointing since very often only lower amount of nanoparticles can be added.

Future improvements of properties of polymers filled with microsized metallic particles could come from the manufacturing of new particles (spongelike particles, metalized particles with thicker coating) or from new ways to orient in the favorable direction particles with high aspect ratio (new design of injection gate, etc.) or to obtain more efficient microstructure of the composites.

References

1 Utracki, L.A. and Vu Khanh, T. (2002) Filled polymers, in *Multicomponent Polymer Systems*, Longman Scientific & Technical, London, pp. 207–268.
2 Sussman, V. (1987) Conductive fillers, in *Handbook of Fillers for Plastics*, Springer.
3 German, R.M. (1984) *Powder Metallurgy Science*, Metal Powder Industries Federation.
4 Bigg, D.M. (1979) Mechanical, thermal, and electrical properties of metal fiber-filled polymer composites. *Polymer Engineering & Science*, **19**, 1188–1192.
5 Boudenne, A., Ibos, L., Fois, M., Majesté, J.C., and Géhin, E. (2005) Electrical and thermal behavior of polypropylene filled with copper particles. *Composites Part A*, **36**, 1545–1554.

6 Danes, F., Garnier, B., and Dupuis, T. (2003) Predicting, measuring, and tailoring the transverse thermal conductivity of composites from polymer matrix and metal filler. *International Journal of Thermophysics*, **24**, 771–784.

7 Novák, I., Krupa, I., and Chodák, I. (2004) Electroconductive adhesives based on epoxy and polyurethane resins filled with silver-coated inorganic fillers. *Synthetic Metals*, **144**, 13–19.

8 Luyt, A.S., Molefi, J.A., and Krump, H. (2006) Thermal, mechanical and electrical properties of copper powder filled low-density and linear low-density polyethylene composites. *Polymer Degradation and Stability*, **91**, 1629–1636.

9 Mamunya, Y.P., Davydenko, V.V., Pissis, P., and Lebedev, E.V. (2002) Electrical and thermal conductivity of polymers filled with metal powders. *European Polymer Journal*, **38**, 1887–1897.

10 Ishigure, Y., Iijima, S., Ito, H., Ota, T., Unuma, H., Takahashi, M., Hikichi, Y., and Suzuki, H. (1999) Electrical and elastic properties of conductor–polymer composites. *Journal of Materials Science*, **34**, 2979–2985.

11 Dupuis, T. (2002) PhD thesis Ecole Polytechnique, University of Nantes.

12 Bigg, D.M. and Bradbury, E.J. (1981) Conductive polymeric composites from short conductive fibers. *Polymer Science and Technology*, **15**, 23–38.

13 Agari, Y. and Uno, T. (1986) Estimation on thermal conductivities of filled polymers. *Journal of Applied Polymer Science*, **32**, 5705–5712.

14 Bishay, I.K., Abd-El-Messieh, S.L., and Mansour, S.H. (2011) Electrical, mechanical and thermal properties of polyvinyl chloride composites filled with aluminum powder. *Materials & Design*, **32**, 62–68.

15 Kusy, R.P. and Corneliussen, R.D. (1975) The thermal conductivity of nickel and copper dispersed in poly(vinyl chloride). *Polymer Engineering & Science*, **15**, 107–112.

16 Malliaris, A. and Turner, D.T. (1971) Influence of particle size on the electrical resistivity of compacted mixtures of polymeric and metallic powders. *Journal of Applied Physics*, **42**, 614–618.

17 Boudenne, A., Ibos, L., Fois, M., Gehin, E., and Majesté, J.-C. (2004) *Thermophysical Properties of Polypropylene/Aluminum Composites*, vol. 42, John Wiley & Sons, Inc., New York.

18 Nakamura, S., Tommura, T., and Sawa, G. (1998) Electrical conduction mechanism of polymer–carbon black composites below and above the percolation threshold. Presented at Annual Report Conference on Electrical Insulation and Dielectric Phenomena, Atlanta, GA.

19 Roldughin, V.I. and Vysotskii, V.V. (2000) Percolation properties of metal-filled polymer films, structure and mechanisms of conductivity. *Progress in Organic Coatings*, **39**, 81–100.

20 Cecen, V., Boudenne, A., Ibos, L., Novak, I., Nogellova, Z., Prokes, J., and Krupa, I. (2008) *Electrical, Mechanical and Adhesive Properties of Ethylene-Vinylacetate Copolymer (EVA) Filled with Wollastonite Fibers Coated by Silver*, vol. 44, Elsevier, Kidlington, UK.

21 Agoudjil, B., Ibos, L., Majesté, J.C., Candau, Y., and Mamunya, Y.P. (2008) Correlation between transport properties of ethylene vinyl acetate/glass, silver-coated glass spheres composites. *Composites Part A*, **39**, 342–351.

22 Krupa, I., Mikova, G., Novak, I., Janigova, I., Nogellova, Z., Lednicky, F., and Prokes, J. (2007) *Electrically Conductive Composites of Polyethylene Filled with Polyamide Particles Coated with Silver*, vol. 43, Elsevier, Kidlington, UK.

23 Bigg, D.M. (1986) Thermally conductive polymer compositions. *Polymer Composites*, **7**, 125–140.

24 Tekce, H.S., Kumlutas, D., and Tavman, I.H. (2007) *Effect of Particle Shape on Thermal Conductivity of Copper Reinforced Polymer Composites*, vol. 26, Sage, London.

25 Tavman, I.H. (1996) Thermal and mechanical properties of aluminum powder-filled high-density polyethylene composites. *Journal of Applied Polymer Science*, **62**, 2161–2167.

26 Tavman, I.H. (1997) Thermal and mechanical properties of copper powder filled poly(ethylene) composites. *Powder Technology*, **91**, 63–67.

27 Molefi, J., Luyt, A., and Krupa, I. (2009) Comparison of the influence of Cu micro- and nano-particles on the thermal properties of polyethylene/Cu composites, *eXPRESS Polymer Letters*, **3**, 639–649.

28 Sofian, N.M., Rusu, M., Neagu, R., and Neagu, E. (2001) Metal powder-filled polyethylene composites. V. Thermal properties. *Journal of Thermoplastic Composite Materials*, **14**, 20–33.

29 Kumlutas, D. and Tavman, I.H. (2006) A numerical and experimental study on thermal conductivity of particle filled polymer composites. *Journal of Thermoplastic Composite Materials*, **19**, 441–455.

30 Maiti, S.N. and Mahapatro, P.K. (1990) Thermal properties of nickel powder filled polypropylene composites. *Polymer Composites*, **11**, 223–228.

31 Weidenfeller, B., Hofer, M., and Schilling, F.R. (2004) *Thermal Conductivity, Thermal Diffusivity, and Specific Heat Capacity of Particle Filled Polypropylene*, vol. **35**, Elsevier, Kidlington, UK.

32 Tlili, R., Cecen, V., Krupa, I., Boudenne, A., Ibos, L., Candau, Y., and Novak, I. (2011) Mechanical and thermophysical properties of EVA copolymer filled with nickel particles. *Polymer Composites*, **32**, 727–736.

33 De Araujo, F.F.T., Garrett, K.W., and Rosenberg, H.M. (1976) Presented at ICCM Proceedings of the International Conference on Composite Materials. vol. 2, p. 568.

34 Nieberlein, V. (1978) Thermal conductivity enhancement of epoxies by the use of fillers. *IEEE Transactions on Components, Hybrids, and Manufacturing Technology*, **1**, 172–176.

35 Katz, H.S. and Milewski, J.V. (1978) *Handbook of Fillers and Reinforcements for Plastics*, Van Nostrand Reinhold Company.

36 Chung, S.I., Im, Y.G., Jeong, H.D., and Nakagawa, T. (2003) The effects of metal filler on the characteristics of casting resin for semi-metallic soft tools. *Journal of Materials Processing Technology*, **134**, 26–34.

37 De-Araujo, F.F.T. and Rosenberg, H.M. (1976) The thermal conductivity of epoxy-resin/metal-powder composite materials from 1.7 to 300K. *Journal of Physics D*, **9**, 665–675.

38 Hansen, D. and Tomkiewicz, R. (1975) Heat conduction in metal-filled polymers: the role of particle size, shape, and orientation. *Polymer Engineering and Science*, **15**, 353–356.

39 Hamilton, R.L. and Crosser, O.K. (1962) Thermal conductivity of heterogeneous two-component systems. *Industrial & Engineering Chemistry Fundamentals*, **1**, 187–191.

40 Filip, C., Garnier, B., and Danes, F. (2007) *Effective Conductivity of a Composite in a Primitive Tetragonal Lattice of Highly Conducting Spheres in Resistive Thermal Contact with the Isolating Matrix*, vol. **129**, American Society of Mechanical Engineers, New York, NY.

41 Danes, F., Garnier, B., Dupuis, T., Lerendu, P., and Nguyen, T.P. (2005) Non-uniformity of the filler concentration and of the transverse thermal and electrical conductivities of filled polymer plates. *Composites Science and Technology*, **65**, 945–951.

42 Boutros, A., Garnier, B., Danes, F., Boudenne, A., Ibos, L., and Agoudjil, B. (2008) Effective thermal conductivity of composites with a polymer matrix in imperfect contact with embedded hollow metallic spheres. Presented at 18th European Conference on Thermophysical Properties, Pau, France.

43 Krupa, I., Cecen, V., Tlili, R., Boudenne, A., and Ibos, L. (2008) Thermophysical properties of ethylene–vinyl acetate copolymer (EVA) filled with wollastonite fibers coated by

silver. *European Polymer Journal*, **44**, 3817–3826.

44 Krupa, I., Boudenne, A., and Ibos, L. (2007) Thermophysical properties of polyethylene filled with metal coated polyamide particles. *European Polymer Journal*, **43**, 2443–2452.

45 Gungor, A. (2006) The physical and mechanical properties of polymer composites filled with Fe-powder. *Journal of Applied Polymer Science*, **99**, 2438–2442.

46 Rusu, M., Sofian, N., and Rusu, D. (2001) Mechanical and thermal properties of zinc powder filled high density polyethylene composites. *Polymer Testing*, **20**, 409–417.

47 Mansour, S.H., Gomaa, E., and Bishay, I.K. (2007) Effect of metal type and content on mechanical, electrical and free-volume properties of styrenated polyesters. *Journal of Materials Science*, **42**, 8473–8480.

48 Tan, S.T., Zhang, M.Q., Rong, M.Z., and Zeng, H.M. (1999) Effect of interfacial modification on metal fiber filled polypropylene composites and property balance. *Polymer Composites*, **20**, 406–412.

49 Martin, M., Hanagud, S., and Thadhani, N.N. (2007) Mechanical behavior of nickel + aluminum powder-reinforced epoxy composites. *Materials Science and Engineering A*, **443**, 209–218.

50 Molefi, J.A., Luyt, A.S., and Krupa, I. (2010) Comparison of LDPE, LLDPE and HDPE as matrices for phase change materials based on a soft Fischer–Tropsch paraffin wax. *Thermochimica Acta*, **500**, 88–92.

51 Koysuren, O., Yesil, S., and Bayram, G. (2006) Effect of composite preparation techniques on electrical and mechanical properties and morphology of nylon 6 based conductive polymer composites. *Journal of Applied Polymer Science*, **102**, 2520–2526.

52 Yung, K.C., Zhu, B.L., Yue, T.M., and Xie, C.S. (2010) Effect of the filler size and content on the thermomechanical properties of particulate aluminum nitride filled epoxy composites. *Journal of Applied Polymer Science*, **116**, 225–236.

53 Saini, G., Bhardwaj, R., Choudhary, V., and Narula, A.K. (2010) Poly(vinyl chloride)-*Acacia* bark flour composite: effect of particle size and filler content on mechanical, thermal, and morphological characteristics. *Journal of Applied Polymer Science*, **117**, 1309–1318.

54 Mottram, J.T. (1992) Design charts for the thermal conductivity of particulate composites. *Materials & Design*, **13**, 221–225.

55 Bigg, D. (1995) Thermal conductivity of heterophase polymer compositions, in *Thermal and Electrical Conductivity of Polymer Materials*, vol. **119**, Advances in Polymer Science, Springer, Berlin, pp. 1–30.

56 Zhou, W., Qi, S., Tu, C., Zhao, H., Wang, C., and Kou, J. (2007) Effect of the particle size of Al_2O_3 on the properties of filled heat-conductive silicone rubber. *Journal of Applied Polymer Science*, **104**, 1312–1318.

57 Huang, X., Birman, V., Nanni, A., and Tunis, G. (2005) Properties and potential for application of steel reinforced polymer and steel reinforced grout composites. *Composites Part B*, **36**, 73–82.

58 Radheshkumar, C. and Münstedt, H. (2005) Morphology and mechanical properties of antimicrobial polyamide/silver composites. *Materials Letters*, **59**, 1949–1953.

59 Visy, C., Bencsik, G., Németh, Z., and Vértes, A. (2008) Synthesis and characterization of chemically and electrochemically prepared conducting polymer/iron oxalate composites. *Electrochimica Acta*, **53**, 3942–3947.

60 Xia, X., Xie, C., Cai, S., Wen, F., Zhu, C., and Yang, X. (2006) Effect of the loading and size of copper particles on the mechanical properties of novel Cu/LDPE composites for use in intrauterine devices. *Materials Science and Engineering A*, **429**, 329–333.

61 Clingerman, M.L. (1998) Development and modelling of electrically conductive composite materials, in *Chemical*

62 Blaker, J.J., Nazhat, S.N., and Boccaccini, A.R. (2003) Development and characterisation of silver-doped bioactive glass-coated sutures for tissue engineering and wound healing applications. *Biomaterials*, **25**, 1319–1329.

63 Zhang, Q., Tian, M., Wu, Y., Lin, G., and Zhang, L. (2004) Effect of particle size on the properties of $Mg(OH)_2$-filled rubber composites. *Journal of Applied Polymer Science*, **94**, 2341–2346.

64 Xu, Y., Chung, D.D.L., and Mroz, C. (2001) Thermally conducting aluminum nitride polymer–matrix composites. *Composites Part A*, **32**, 1749–1757.

65 Nikzad, M., Masood, S.H., and Sbarski, I. (2011) Thermo-mechanical properties of a highly filled polymeric composites for fused deposition modelling. *Materials & Design*, **32**, 3448–3456.

66 Tang, Y., Xia, X., Wang, Y., and Xie, C. (2010) Study on the mechanical properties of Cu/LDPE composite IUDs. *Contraception*, **83**, 255–262.

67 McCullough, R.L. (1985) Generalized combining rules for predicting transport properties of composite materials. *Composites Science and Technology*, **22**, 3–21.

68 Sundstrom, D.W. and Lee, Y.-D. (1972) Thermal conductivity of polymers filled with particulate solids. *Journal of Applied Polymer Science*, **16**, 3159–3167.

69 Progelhof, R.C., Throne, J.L., and Ruetsch, R.R. (1976) Methods for predicting the thermal conductivity of composite systems: a review. *Polymer Engineering & Science*, **16**, 615–625.

70 Chen, C.R., Qin, S.Y., Li, S.X., and Wen, J.L. (1999) Finite element analysis about effects of particle morphology on mechanical response of composites. *Materials Science and Engineering A*, **278**, 96–105.

71 González, C. and Llorca, J. (1996) Prediction of the tensile stress–strain curve and ductility in Al/SiC composites. *Scripta Materialia*, **35**, 91–97.

72 Song, S.G., Shi, N., Gray, G.T., and Roberts, J.A. (1996) Reinforcement shape effects on the fracture behavior and ductility of particulate-reinforced 6061-Al matrix composites. *Metallurgical and Materials Transactions A*, **27**, 3739–3746.

73 Zeng, X., Fan, H., and Zhang, J. (2007) Prediction of the effects of particle and matrix morphologies on Al_2O_3 particle/polymer composites by finite element method. *Computational Materials Science*, **40**, 395–399.

74 Christman, T., Needleman, A., and Suresh, S. (1989) An experimental and numerical study of deformation in metal–ceramic composites. *Acta Metallurgica*, **37**, 3029–3050.

75 Xue, Q. (2004) The influence of particle shape and size on electric conductivity of metal–polymer composites. *European Polymer Journal*, **40**, 323–327.

76 Cho, J., Joshi, M.S., and Sun, C.T. (2006) Effect of inclusion size on mechanical properties of polymeric composites with micro and nano particles. *Composites Science and Technology*, **66**, 1941–1952.

77 Radford, K.C. (1971) The mechanical properties of an epoxy resin with a second phase dispersion, in *Journal of Materials Science*, vol. **6**, Springer, The Netherlands, pp. 1286–1291.

78 Mamunya, Y.P., Zois, H., Apekis, L., and Lebedev, E.V. (2004) Influence of pressure on the electrical conductivity of metal powders used as fillers in polymer composites. *Powder Technology*, **140**, 49–55.

79 Jiang, Y., Tohgo, K., and Yang, H. (2010) Study of the effect of particle size on the effective modulus of polymeric composites on the basis of the molecular chain network microstructure. *Computational Materials Science*, **49**, 439–443.

80 Bhattacharya, S.K. and Chaklader, A.C.D. (1982) Review on metal-filled plastics. Part 1. Electrical conductivity. *Polymer-Plastics Technology and Engineering*, **19**, 21–51.

81 Nogales, S. and Böhm, H.J. (2008) Modeling of the thermal conductivity and thermomechanical behavior of diamond reinforced composites. *International*

Journal of Engineering Science, **46**, 606–619.

82. Bayer, R.K., Ezquerra, T.A., Zachmann, H.G., Baltà Calleja, F.J., Martinez Salazar, J., Meins, W., Diekow, R.E., and Wiegel, P. (1988) Conductive PE–carbon composites by elongation flow injection moulding, *Journal of Materials Science*, **23**, 475–480.

83. Salazar, J.M., Bayer, R.K., Ezquerra, T.A., and Calleja, F.J.B. (1989) Conductive polyethylene–carbon black composites by elongational-flow injection molding Part 3. Study of the structure and morphology, *Colloid & Polymer Science*, **267**, 409–413.

84. Feller, J.F. (2004) Conductive polymer composites: influence of extrusion conditions on positive temperature coefficient effect of poly(butylene terephthalate)/poly(olefin)-carbon black blends. *Journal of Applied Polymer Science*, **91**, 2151–2157.

85. Weber, M. and Kamal, M.R. (1997) Estimation of the volume resistivity of electrically conductive composites. *Polymer Composites*, **18**, 711–725.

86. Coleman, J., Khan, U., and Gun'ko, Y. (2006) Mechanical reinforcement of polymers using carbon nanotubes. *Advanced Materials*, **18**, 689–706.

87. Al-Saleh, M.H. and Sundararaj, U. (2009) Electrically conductive carbon nanofiber/polyethylene composite: effect of melt mixing conditions. *Polymers for Advanced Technologies*, **22**, 246–253.

88. Torquato, S. (1984) Bulk properties of two-phase disordered media. I: Cluster expansion for the effective dielectric constant of dispersions of penetrable spheres. *The Journal of Chemical Physics*, **81**, 5079–5088.

89. McLachlan, D.S., Blaszkiewicz, M., and Newnham, R.E. (1990) Electrical resistivity of composites. *Journal of the American Ceramic Society*, **73**, 2187–2203.

90. Bruggemann, D.A.G. (1935) The calculation of various physical constants of heterogeneous substances. I. The dielectric constants and conductivities of mixtures composed of isotropic substances. *Annals of Physics*, **24**, 636–664.

91. Brown, W.F. (1955) Solid-mixture permittivities. *Journal of Chemical Physics*, **23**, 1514–1517.

92. Broadbent, S.R. and Hammersley, J.M. (1957) Percolation processes I. Crystals and mazes. *Proceedings of the Cambridge Philosophical Society*, **53**, 629–641.

93. Kirkpatrick, S. (1973) Percolation and conduction. *Reviews of Modern Physics*, **45**, 574–588.

94. McLachlan, D.S. (1985) Equation for the conductivity of metal-insulator mixture. *Journal of Physics C*, **18**, 1891–1897.

95. McLachlan, D.S. (1988) Measurement and analysis of a model dual-conductivity medium using a generalised effective-medium theory. *Journal of Physics C*, **21**, 188–191.

96. Deprez, N. and Maclachlan, D.S. (1988) The analysis of the electrical conductivity of graphite powders during compaction. *Journal of Physics D*, **21**, 101.

97. Deprez, N., McLachlan, D.S., and Sigalas, I. (1988) The measurement and comparative analysis of the electrical and thermal conductivities, permeability and Young's modulus of sintered nickel. *Solid State Communications*, **66**, 869–872.

98. Meredith, R.E., and Tobias, C.W. (1962) *Conduction in Heterogeneous Systems, Advances in Electrochemistry and Electrochemical Engineering* (ed. C.W. Tobias), Interscience, New York, pp. 15–47.

99. Lux, F. (1993) Models proposed to explain the electrical conductivity of mixtures made of conductive and insulating materials. *Journal of Materials Science*, **28**, 285–301.

100. Mottram, J.T. and Taylor, R. (1991) Thermal transport properties. *International Encyclopedia of Composite*, **5**, 476–496.

101. Rajinder, P. (2007) New models for thermal conductivity of particulate composites. *Journal of Reinforced Plastics and Composites*, **26**, 643–651.

102. Rajinder, P. (2008) Thermal conductivity of three-component composites of

core-shell particles. *Materials Science and Engineering A*, **498**, 135–141.
103 Hatta, H. and Taya, M. (1985) Effective thermal conductivity of a misoriented short fiber composite. *Journal of Applied Physics*, **58**, 2478–2486.
104 Torquato, S. (1997) Effective stiffness tensor of composite media – I. Exact series expansions. *Journal of the Mechanics and Physics of Solids*, **45**, 1421–1448.
105 Hashin, Z. and Shtrikman, S. (1962) A variational approach to the theory of the effective magnetic permeability of multiphase materials. *Journal of Applied Physics*, **33**, 3125–3131.
106 Torquato, S. (1985) Effective electrical conductivity of two-phase disordered composite media. *Journal of Applied Physics*, **58**, 3790–3797.
107 Nielsen, L.E. (1974) The thermal and electrical conductivity of two phase systems. *Industrial and Engineering Chemistry Fundamentals*, **13**, 17–18.
108 Sanada, K., Tada, Y., and Shindo, Y. (2009) Thermal conductivity of polymer composites with close-packed structure of nano and micro fillers. *Composites Part A*, **40**, 724–730.
109 Mamunya, E.P., Davidenko, V.V., and Lebedev, E.V. (1995) Percolation conductivity of polymer composites filled with dispersed conductive filler. *Polymer Composites*, **16**, 319–324.
110 Bueche, F. (1972) Electrical resistivity of conducting particles in an insulating matrix. *Journal of Applied Physics*, **43**, 4837–4838.
111 Slupkowski, T. (1984) Electrical conductivity of mixtures of conducting and insulating particles. *Physica Status Solidi A*, **83**, 329–333.
112 Rajagopal, C. and Satyam, M. (1978) Studies on electrical conductivity of insulator–conductor composites. *Journal of Applied Physics*, **49**, 5536–5542.
113 Lewis, T.B. and Nielsen, L.E. (1970) Dynamic mechanical properties of particulate-filled composites. *Journal of Applied Polymer Science*, **14**, 1449–1471.
114 Theret, D.P., Levesque, M.J., Sato, M., Nerem, R.M., and Wheeler, L.T. (1988) The application of a homogeneous half-space model in the analysis of endothelial cell micropipette measurements. *Journal of Biomechanical Engineering*, **110**, 190–199.
115 Bilodeau, G.G. (1992) Regular pyramid punch problem. *Journal of Applied Mechanics*, **59**, 519–523.
116 Harding, J.W. and Sneddon, I.N. (1945) The elastic stresses produced by the indentation of the plane surface of a semi-infinite elastic solid by a rigid punch. *Mathematical Proceedings of the Cambridge Philosophical Society*, **41**, 16–26.
117 Mijailovich, S.M., Kojic, M., Zivkovic, M., Fabry, B., and Fredberg, J.J. (2002) A finite element model of cell deformation during magnetic bead twisting. *Journal of Applied Physiology*, **93**, 1429–1436.
118 Schmid-Schönbein, G.W., Sung, K.L., Tözeren, H., Skalak, R., and Chien, S. (1981) Passive mechanical properties of human leukocytes. *Biophysical Journal*, **36**, 243–256.
119 Sato, M., Theret, D.P., Wheeler, L.T., Ohshima, N., and Nerem, R.M. (1990) Application of the micropipette technique to the measurement of cultured porcine aortic endothelial cell viscoelastic properties. *Journal of Biomechanical Engineering*, **112**, 263–268.
120 Cheng, L., Xia, X., Yu, W., Scriven, L.E., and Gerberich, W.W. (2000) Flat-punch indentation of viscoelastic material. *Journal of Polymer Science B*, **38**, 10–22.
121 Alcaraz, J., Buscemi, L., Grabulosa, M., Trepat, X., Fabry, B., Farré, R., and Navajas, D. (2003) Microrheology of human lung epithelial cells measured by atomic force microscopy. *Biophysical Journal*, **84**, 2071–2079.

20
Magnetic Particle-Filled Polymer Microcomposites
Natalie E. Kazantseva

20.1
Introduction

Magnetic composite materials produced by embedding ferro- and ferrimagnetic particles in a nonmagnetic matrix. This magnetic particle-filled composites have a significant advantage over bulk magnetic materials (metals, alloys, or ferrites) owing to their higher resistivity, lower density, chemical stability, and processability. Moreover, the fact is that the complex magnetic permeability and dielectric permittivity of these materials are structure sensitive that makes it possible to manipulate the electromagnetic properties of a material over a wide frequency range by changing the concentration, shape, and size of filler particles and the microstructure of the composite. Because of these friendly material handling properties, magnetic composites have many useful applications, such as magnetic substrates for microwave antennas, microwave inductors, radioabsorbing materials, and so on.

This chapter describes on the subject of magnetically soft composite materials in which a polymer serves as a dielectric binding component and micron-sized multi-domain magnetic particles serve as a filler material; the concentration of these particles does not exceed the critical value, which leads to the deterioration of the physical–mechanical properties of the composite. Depending on the type of the polymer, its mechanical properties, and the volume fraction in a composite, polymer magnetic composites (PMCs) can be fabricated as plastomagnets, which admit either cold or hot processing, and in the form of magnetoelastic materials, which endure considerable elastic deformation.

A field of application of a PMC determines the requirements on the initial permeability (μ_i), the level of magnetic losses, and the operating frequency range. Thus, the desirable properties for PMCs used in high-frequency devices are high initial permeability and low losses in a wide frequency range. In contrast, PMCs used as radioabsorbing materials should have high permeability and high magnetic losses.

The initial permeability of a PMC depends on the permeability of the original magnetic material (intrinsic permeability), on the loading factor, and on the

Polymer Composites: Volume 1, First Edition. Edited by Sabu Thomas, Kuruvilla Joseph,
Sant Kumar Malhotra, Koichi Goda, and Meyyarappallil Sadasivan Sreekala
© 2012 Wiley-VCH Verlag GmbH & Co. KGaA. Published 2012 by Wiley-VCH Verlag GmbH & Co. KGaA.

microstructure of the composite. In turn, the composite microstructure is determined by the morphology (the size and shape) of the magnetic particles and their microstructure and mutual arrangement in the bulk of the composite.

The most popular types of filler materials used in the large-scale production of PMCs are the powders of carbonyl iron, alsifer (triple eutectic alloys of Fe that contain from 7 to 8% of Al and from 9 to 11% of Si), permalloy (Fe–Ni alloy doped with Mo, Cu, and Cr), magnetically soft spinel-type ferrites, and hexagonal ferrites. At present, powders of magnetically soft amorphous alloys, amorphous microwire, as well as multicomponent fillers are also used in the design of PMCs.

This chapter is also devoted to the comparative analysis of the electromagnetic properties of PMCs produced on the basis of various magnetic fillers. The factors responsible for the difference between the magnetic properties of PMCs and of bulk magnetic materials are analyzed. It is shown how to control the values of complex permittivity and permeability and the position of the dispersion region on the frequency scale by choosing the type of magnetic filler, its concentration, morphology, as well as the microstructure of magnetic particles and their distribution in the polymer.

This chapter is organized in seven sections. Section 20.2 deals with the main characteristics of the original components of PMCs, namely polymers and magnetic fillers. In Section 20.3, comparative analysis of the methods for measuring the electromagnetic properties of PMCs over a wide range of frequencies is presented. In Section 20.4, by using example of various types of bulk magnetically soft materials, the important issues related to the magnetization of those materials in DC and AC magnetic fields is explained. In Section 20.5, the high-frequency permeability and permittivity of a series of PMCs in relation to the composition and the microstructure of the materials are discussed. In addition, a brief analysis of the effective medium theory (EMT) and its analogs (the so-called "mixing rules") is given from the viewpoint of their application to estimate the electromagnetic properties of PMCs on the basis of the material parameters of its components. In Section 20.6, the methods and techniques for increasing the high-frequency permeability of PMCs through the use of multicomponent fillers and multicomponent magnetic particles with core-shelllike structure are discussed. Finally, in Section 20.7, the features of high-frequency permeability of PMCs based on different types of magnetic fillers are summarized.

20.2
Basic Components of Polymer Magnetic Composites: Materials Selection

20.2.1
Introduction to Polymer Magnetic Composite Processing

The design of PMCs, especially highly filled ones, requires that one should achieve a particular balance between the processing ability and the properties of composites for a given range of applications. The primary purpose of the polymer matrix in magnetic composites is to bind the filler particles together owing to its cohesive and adhesive characteristics, while the filler particles are responsible for the electromagnetic properties of the composite. The physical, mechanical, and electromagnetic properties of a composite significantly depend on how uniformly the filler particles are distributed in a polymer. There are two main factors that influence the particles distribution in a polymer, namely, the filler–filler and the filler–matrix interactions. Thus, the selection of materials and their compatibility are crucial for the design of PMCs.

Polymers can generally be classed as thermoplastics and thermosets. In order to fabricate polymer-based composites, melted thermoplastics are mixed with filler and then processed into a final product, whereas filled thermosets must undergo further polymerization to complete cross-linking reactions of composite solidification. Many factors influence the processing operation, namely, viscosity, the orientation of heterogeneous phases, and the rate of the reactions. Among these factors, viscosity is the most crucial one for processing. Viscosity strongly depends on the chemical nature of the polymer (molecular weight and its distribution, molecular structure of the polymeric chains, degree of crystallinity, etc.) and based on the heterogeneity of materials. Moreover, viscosity also depends on the processing parameters, such as temperature and shear rate. Therefore, the study of rheological properties of polymers filled with magnetic particles is very important for choosing a proper condition for processing composites with the optimal balance between mechanical and physical properties.

The processing of PMCs utilizes the same techniques as polymer processing, which include injection molding, compression molding, and extrusion. Compression molding is the most extensively used method in the research study due to the simplicity of operation and flexibility regarding the polymer matrix selection. Injection molding is used for processing PMCs based on thermoplastics with the necessary precondition for thoroughly mixing raw materials. Extrusion method allows one to produce composite materials with a magnetic filler content of up to 80 vol.%, the value impossible with other technologies. To this end, high-oriented thermoplastics are used, for example, ultrahigh molecular weight polyethylene. In such a case, PMCs can be produced using the melts or solutions (gel-spinning technology) of thermoplastics.

20.2.2
Polymers

The characteristics of typical polymers used in the preparation of highly filled PMCs are described in Table 20.1, whereas Table 20.2 contains their advantages and limitations [1–11].

20.2.3
Magnetic Fillers

The most widespread magnetic fillers in PMCs are soft magnetic ferrites (NiZn, MnZn, Co_2Z, Co_2W, etc.), CI powders, and, more rarely, ferromagnetic alloys (alsifer,

Table 20.1 Physical and mechanical properties of polymers.

Type	Polymer	ϱ (g/cm)	T_g (°C)	ΔT (°C)	σ (MPa)	ε_b (%)	Hardness (shore A)	Dielectric constant	tan δ (1 kHz)
TP	Polyisoprene	0.91	−70	−54–90	4.5	650	75–90	2.37–2.45	0.001–0.003
	Chloroprene	1.25	−45	−40–100	25–38	500	70	6.5–8.1	0.03–0.086
	Butadiene–acrylonitrile	0.9–1.01	−60	−50–150	1.4–3.0	650	75–80	2.51	0.0009
	Poly(chlorotrifluoroethylene)	2.1–2.14	—	−240–150	30–40	80–250	75–80	2.2–2.8	0.02–0.03
	Polysiloxane	1.1–1.6	−70–95	−55–225	350	900	30–50	2.65–3.10	0.01–0.02
	Polyurethane	1.1–1.6	−70–140	−55–140	500	750	80–90	2.8–3.25	0.01–0.02
TS	Epoxy resin	1.1–1.4	−55	−35–140	22.7–72.3	3–4	60–76	2.8–5.6	0.008–0.09
LE	Dimethylsiloxane	1.05	−127	−60–370	6.2	100	60–70	2.2	0.001
	Polyurethane diisocyanate	1.0	−43–117	−70–140	500	3	80	2.8–3.25	0.01–0.02
HO	PE (ultrahigh molecular weight)	1.3–1.5	Melting T: 137	upper T: 55–95	20–40	500	50–70	2.3	0.0002–0.0005
	Ftorlon (copolymer of tetrafluoroethylene and vinylidene chloride)	1.8–2.0	Amorphous regions: −120; crystal regions: 327	−60–120	39–55	400–500	29–39	2	0.0002

The symbols and abbreviations used are as follows: ϱ: specific density; T: temperature; T_g: glass-transition temperature; ΔT: operating temperature range; σ: tensile strength; ε_b: elongation at break; tan δ: dissipation factor; TP: thermoplastic; TS: thermoset; LE: liquid elastomer; HO: high oriented.

Table 20.2 The advantages and limitations of polymers.

Polymer	Advantages	Limitations
Polyisoprene	Outstanding resilience; high tensile strength; superior resistance to tear and abrasion; good low-temperature flexibility	Fair resistance to heat, ozone, and sunlight; little resistance to oil, benzene, and hydrocarbon solvents
Chloroprene	Very good resistance to weather, ozone, and natural aging; good resistance to abrasion and flex cracking; moderate resistance to oil and benzene	Fair resistance to aromatic and oxygenated solvents; limited low-temperature flexibility
Butadiene–acrylonitrile	Excellent resistance to oil and benzene; superior to petroleum-based hydraulic fluids; good high-temperature performance; good resistance to sunlight and oxidation	Poor resistance to oxygenated solvents
Poly (chlorotrifluoroethylene)	Excellent resistance to oil, benzene, hydraulic fluids, and hydrocarbon solvents; very good heat resistance; very good resistance to weather, oxygen, ozone, and sunlight	Poor resistance to tear and cut growth
Polysiloxane	Resistance to oxidation, ozone, and weathering corona; high tensile strength; high tear strength; low cure shrinkage; properties constant with temperature; high dielectric strength; excellent resistance to heat aging	Poor adhesion to metals and other substrates
Polyurethane	Excellent resistance to oil, solvents, oxidation, and ozone; excellent wear, tear, and chemical resistance; fair electrical; excellent adhesion; reverts in humidity	Significant dependence of physical mechanical properties upon temperature gradient
Epoxy resin	Excellent adhesion; low shrinkage; high strength; shape stability; resistance to solvents; strong bases and hot water	Sacrifice of properties for high flexibility
Dimethylsiloxane	Outstanding heat resistance; excellent flexibility at low temperature; excellent resistance to weather, ozone, sunlight, and oxidation; the loss tangent is very low ($\tan \delta < 0.001$)	Fair resistance to oil benzene and solvents; poor resistance to abrasion, tear, and cut growth

(Continued)

Table 20.2 (Continued)

Polymer	Advantages	Limitations
Polyurethane diisocyanate	Outstanding resistance to tear and abrasion; high tensile strength and elongation; good weather resistance; good resistance to oil and benzene	Poor resistance to acids and alkalis; inferior resistance to hot water
PE (ultrahigh molecular weight)	High mechanical strength that allows one to apply, for example, the method of orientation drawing and rollmilling to the processing of highly filled composites	Poor contact wear resistance
Ftorlon	High mechanical strength, chemical and radiation resistance, resistance to atmospheric effects; very low water absorption (lower than 0.1% for 24 h), therefore, its dielectric properties weakly depend on the atmospheric humidity; high content of amorphous phase and the flexibility of the molecular chain make it possible to reach high values of the filling factor (up to 60 vol.%) in the composites	Poor wearing property, thus need of reinforcing agent

permalloy, etc.), which are characterized by high saturation magnetization, high initial magnetic permeability, wide range of electrical properties, and so on. Among all types of magnetic materials, only CI is produced in powder form, other fillers are obtained by high-energy milling. Thus, the magnetic properties of powders are a function of their chemical composition and melting practice also.

The main advantage of CI is that, by varying its type (chemical composition, particle shape and size, and microstructure), one can obtain a broad region of magnetic permeability dispersion in the RF and microwave bands [12–20]. The second advantage is the availability of CI, which is produced all over the world: BASF Corporation, Vogt (Germany), READE, GAF, and Amidon, Inc. (USA), INCO-MOND (UK), ONJA (France), Labdhi Chemical Industries (India), and SINTEZ (Russia), and so on. On the other hand, CI is not chemically stable due to the high content of α-iron (97–99 at.%); this especially applies to processed types of CI produced by chemical or mechanical treatment of primary CI with a view to improving their magnetic properties. Primary and processed CI powders differ radically in particle microstructure, which determines the electromagnetic properties of CI-filled composites, including their storage and thermal stability [12, 16, 17, 21]. The basic information about physical properties of various types of CI is given in Table 20.3.

Soft ferrites have significant advantages over metallic ferromagnets and CI in high resistivity, high chemical stability, and the capability of withstanding intense electromagnetic fields. However, polycrystalline ferrites with identical chemical composition and purity but different processing condition may exhibit nonidentical characteristics such as initial permeability μ_i, permeability dispersion $\mu(f)$, resonance frequency f_r (maximum absorption frequency), and so on [2, 9, 22–24]. Moreover, the high frequency μ of ferrites is limited by Snoek's law [25].

$$f_r(\mu_r-1)_r = \frac{2}{3}\gamma M_s \qquad (20.1)$$

where $\mu_r = \mu_a/\mu_0$ is the relative permeability (μ_a is the absolute permeability and μ_0 is the permeability of vacuum), f_r is the resonance frequency determined by the location of the magnetic loss peak, $\gamma = 2.8$ MHz/Oe is the gyromagnetic ratio, and M_s is the saturation magnetization.

Snoek's law predicts that no ferrite can have permeability higher than the Snoek's limit, as long as cubic magnetocrystalline anisotropy is present (Table 20.4).

For ferrites with a planar anisotropy, Snoek's limit can be overcome [33]. In such a case, the resonance frequency is given by

$$f_r(\mu_r-1) = \frac{1}{2}\gamma M_s \sqrt{H_\theta/H_\varphi} \qquad (20.2)$$

where H_θ is out-of plane and H_φ is in-plane magnetic anisotropy, belongs to angles defined in Figure 20.1. If H_φ is higher than H_θ, then this limit is higher than the Snoek's limit given by (20.1).

Table 20.5 lists various magnetic parameters for certain hexagonal ferrites for which the resonance frequency occurs at frequencies higher than the Snoek's limit.

Table 20.3 The main structural and electromagnetic characteristics of carbonyl iron powders.

Type			Hard grades (primary)			Soft grade processed	Special grade processed
		ES BASF	HQ BASF	R-20 SINTEZ		SL BASF	MCI SINTEZ
Chemical composition (at.%)	Iron	>97.7	>97.7	97.29		>99.5	>97.7
	Carbon	<0.9	<1.1	0.95		<0.05	<0.8
	Oxygen	<0.5	<0.4	0.9		<0.2	<0.8
	Nitrogen	<0.9	<1.1	0.86		<0.01	<0.7
Particle morphology		Spherical	Spherical	Spherical		Spherical	Flaky
Microstructure of particles		Multilayer, onionlike	Multilayer, onionlike	Multilayer, onionlike		Polycrystalline with magnetic texture	Disrupted onion
Particle size distribution (μm)	$d10$	1.4–1.9	0.5	—		—	—
	$d50$	3.0–4.0	1.1	2.5		~9	—
	$d90$	6.0–8.0	2.2	—		—	~1.0
Initial permeability[a]		~11	~8.5	~10		~21	~11.5

a) Data obtained by measuring highly filled CI-based polymer composites [16].

Table 20.4 The main structural and magnetic characteristics of spinel-type ferrites.

Chemical composition	Initial permeability	Magnetic dispersion region (Hz)	Resistance (Ohm × cm)	References
MnZn	750–15 000	10^6–10^9	10–10^2	[2, 26, 27]
NiZn	700–2000	5×10^7–10^8	10^6	[2, 27, 28]
Ni–Zn–Cu	1400	10^6–10^7	10^3–10^6	[29]
Mg–Zn	100–800	10^6–10^8	10^6–10^8	[2, 29]
Mg–Zn–Cu	150–600	10^6–10^9	10^3–10^6	[30]
Li–Ti	10–40	10^6–10^9	10^3–10^9	[31]
Li–Zn	100–800	10^7–10^9	10^3–10^5	[32]

Consequently, the dispersion region of permeability can be efficiently controlled by an appropriate choice of the CI particles microstructure and the ferrite composition and magnetocrystalline anisotropy.

20.3
Overview of Methods for the Characterization of Materials in the Radiofrequency and Microwave Bands

The interaction between a material and an electromagnetic field can be described in terms of Maxwell's equations

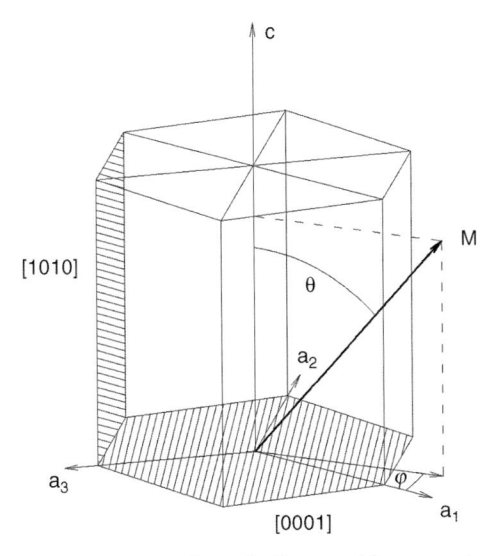

Figure 20.1 Crystalline cell of hexagonal ferrite together with an indication of preferential in-plane [0001] direction for a hexaferrites with easy-plane magnetic anisotropy.

Table 20.5 The main structural and magnetic characteristics of hexagonal ferrites.

Chemical formula	Description; anisotropy	μ_i	Magnetic field anisotropy (kOe)	Magnetic[a] dispersion region (GHz)	Resistance (Ohm × cm)	References
$Ba_3Co_2Fe_{24}O_{41}$	Co_2-Z planar	12.6	6	1–6	$10^5–10^9$	[34]
$Ba_3Co_2Ti_{0.8}Fe_{22.9}O_{41}$	Co_2-Z planar	12	11	4–10	10^8	[11]
$Ba_3Co_{2-x}Fe_{24+x-y}Cr_yO_{41}$	(Ba)-Z planar	16–20	4–7	0.4–1	10^8	[35]
$(Ba_{1-x}Sr_x)_3Co_2Fe_{24}O_{41}$ ($x = 0, 0.2, 0.4, 0.5, 0.8$)	(Ba–Sr,Co)-Z planar	15	6–10	1–10	$1.45 \times 10^3 – 6 \times 10^3$	[36]
$BaZn_{2-x}Co_xFe_{16}O_{27}$ ($x = 0, 0.5, 0.7, 1.0, 1.5, 2.0$)	CoZnW planar	8	8	5–7	10^8	[37, 38]
$BaZn_{1.3}Co_{0.7}Fe_{16}O_{27}$	BaW uniaxial	12	4.5	5–9	10^8	[37, 38]
$BaFe_{12-2x}A_xCo_xO_{19}$ (A – (Ti^{+4}, Ru^{+4})); Ti–Co ($x = 1.3$); Ru–Co ($x = 0.3; 0.5$)	BaM planar	6	7	3–6	10^8	[39]
$Ni_2BaSc_xFe_{16-x}O_{27}$ ($x = 0.5–0.8$)	Ni_2W uniaxial	5	6–11	14–26	10^8	[40]

a) Data obtained by measuring highly filled hexaferrite-based polymer composite.

20.3 Overview of Methods for the Characterization of Materials in the Radiofrequency

$$\text{div}\vec{D} = \varrho$$
$$\text{div}\vec{B} = 0$$
$$\text{rot}\vec{H} = \frac{\partial \vec{D}}{\partial t} + \vec{J} \qquad (20.3)$$
$$\text{rot}\vec{E} = -\frac{\partial \vec{B}}{\partial t}$$

supplemented with the following constitutive relations:

$$\begin{aligned}\vec{D} &= \varepsilon_0 \varepsilon \vec{E} = \varepsilon_0(\varepsilon' - j\varepsilon'')\vec{E} \\ \vec{B} &= \mu_0 \mu \vec{H} = \mu_0(\mu' - j\mu'')\vec{H} \\ \vec{J} &= \sigma \vec{E}\end{aligned} \qquad (20.4)$$

where $\vec{E}$ is the electric field strength vector; $\vec{H}$ is the magnetic field strength vector; $\vec{D}$ is the electric displacement vector; $\vec{B}$ is the magnetic flux density vector; $\vec{J}$ is the current density vector; ϱ is the charge density; σ is the conductivity of material; ε_0 is the permittivity of vacuum; μ_0 is the permeability of vacuum; $\varepsilon = \varepsilon' - j\varepsilon''$ and $\mu = \mu' - j\mu''$ are complex relative permittivity and permeability, respectively; the real parts of the permittivity (ε') and permeability (μ') characterize the electric and magnetic polarizability of the material, whereas the imaginary parts (ε'' and μ'') are related to the dissipation (loss) of electromagnetic field energy in the material.

Thus, the response of a material to an electromagnetic field is basically determined by three constitutive parameters, namely, ε, μ, and σ. However, depending on the conductivity of material, one can also use a different set of parameters. In a low-conductivity material, the whole volume of the material contributes to the response to an electromagnetic wave. Conversely, in a high-conductivity material, the field decays by a factor of e at a distance equal to the skin depth δ_s:

$$\delta_s = \frac{1}{\alpha} = \sqrt{\frac{2}{\omega \mu_0 \mu \sigma}} \qquad (20.5)$$

where α is the attenuation coefficient and ω is the angular frequency.

Consequently, the electromagnetic behavior of a low-conductivity material, as well as its skin depth, are mainly determined by two complex parameters, permittivity and permeability:

$$\delta_s = \frac{c}{\omega |\text{Im}\sqrt{\varepsilon \mu}|} \qquad (20.6)$$

where c is the velocity of light in vacuum.

Due to the skin effect, the behavior of high-conductivity materials at microwave frequencies is mainly determined by their surface impedance Z_s:

$$Z_s = (1+j)\sqrt{\frac{\mu_0 \mu \omega}{2\sigma}} \qquad (20.7)$$

Polymer magnetic composites, whose conductivity is usually much smaller than that of conductors, can be classified as low-conductivity materials.

A large number of methods have been developed over the years to measure the permittivity and permeability of monolithic and composite materials [41, 42]. Every method is limited to specific frequencies, materials, and applications. In the low-frequency range (from 1 MHz to 1 GHz), one usually applies impedance-measuring instruments, which determine the electromagnetic parameters of materials by measuring the inductance, capacitance, and resistance of an electrical circuit containing the sample under test [42]. Microwave methods for materials characterization can be basically categorized into *nonresonant* methods and *resonant* methods [41]. Nonresonant methods include reflection methods and transmission/reflection methods, where the properties of a material are calculated from measured reflection and transmission coefficients of an electromagnetic wave propagating in the material. In a resonant method, a sample is introduced into the resonator structure, and the properties of the sample are calculated from the change in the Q factor and the resonance frequency of the resonator loaded with the sample.

When using an impedance-measuring technique to measure the permeability, a sample in the form of a toroidal core is wrapped with a wire (Figure 20.2), and the relative permeability is derived from the results of inductance measurements.

To measure the permittivity by the same technique, one applies the parallel plate method. The parallel plate method involves sandwiching a thin sheet of a material between two electrodes to form a capacitor. The measured capacitance is then used to calculate the permittivity. An overview of the parallel plate method is shown in Figure 20.3.

As mentioned above, nonresonant methods include reflection methods and transmission/reflection methods. Here, any type of transmission media can be used to carry a wave, such as a coaxial line, hollow metallic waveguide, dielectric waveguide, a planar

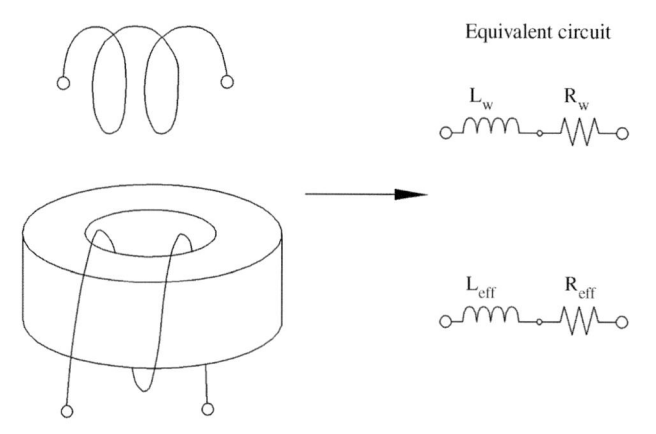

Figure 20.2 The inductance method for measuring permeability; L_w is the inductance of air-core coil, R_w is the resistance of wire, L_{eff} is the inductance of the toroidal core, R_{eff} is the equivalent resistance of the magnetic core loss including the wire resistance.

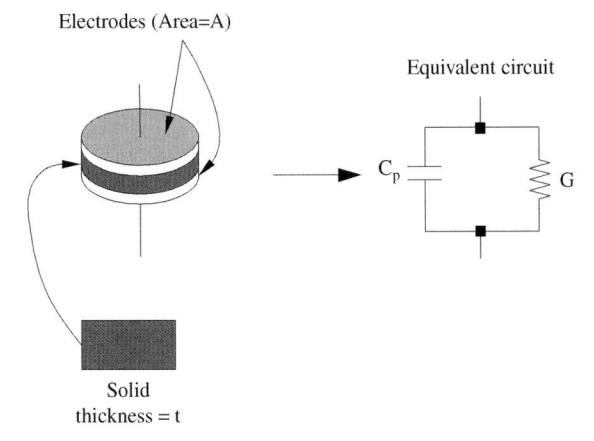

Figure 20.3 Parallel plate method for permittivity measurements.

transmission line, and free space. Figure 20.4 shows the configuration of the transmission/reflection method that involves waveguides and coaxial lines.

When the size of inclusions in a composite is comparable with the wavelength of the electromagnetic wave (e.g., fibers-based composites), one can use only the free-space reflection method to measure the ε and μ of such a composite. The free space methods use antennas to focus the electromagnetic-wave energy, and a flat sample of material under test (MUT) is placed between the transmitting and receiving antennas

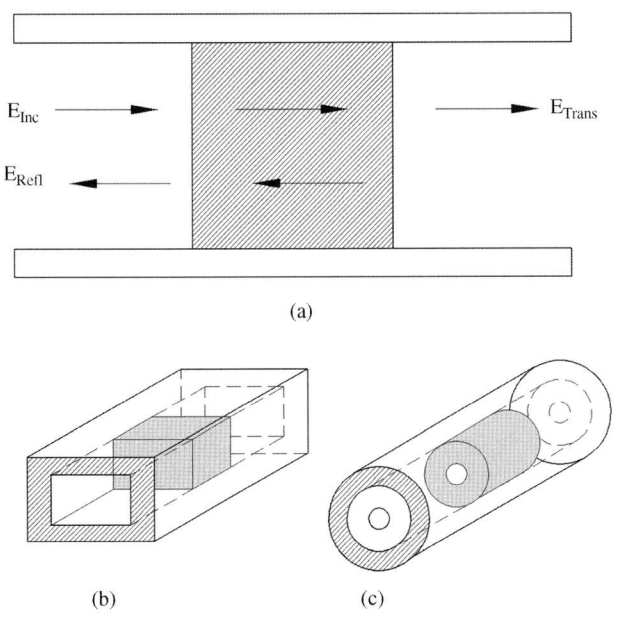

Figure 20.4 (a) Transmission/reflection method; (b) waveguide; and (c) coaxial line cases.

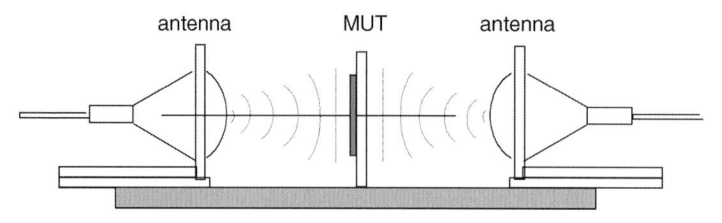

Figure 20.5 Setup for free-space measurements (MUT).

perpendicular to the incident wave. The focusing antennas are connected to the network analyzer as is shown in Figure 20.5.

In free-space measurements, one should take into consideration the sample size and the environment to satisfy the far-field criterion. To minimize the unwanted signals caused by reflections from the environment, one uses time-domain techniques; moreover, it is recommended to carry out free-space measurements in an anechoic chamber. The size of a sample should be at least twice as large as the wavelength of the electromagnetic wave and the antenna aperture. The distance between the antenna and the sample should satisfy the following far-field criterion:

$$d > \frac{2D^2}{\lambda} \tag{20.8}$$

where λ is the wavelength of the operating electromagnetic wave and D is the largest dimension of the antenna aperture.

Depending on the reflection coefficient of the materials under study, three kinds of free-space reflection methods are used: a short-circuited method, a movable metal-backing method, and a bistatic reflection method. The movable metal-backing method is used for measuring large values of reflection coefficient, and the short-circuited method is used in the case of small reflection coefficients. In the metal-backing method, the sample is placed tightly to the horn aperture. In the short-circuited method, the distance between the sample and the horn aperture is varied within half a wavelength. The bistatic reflection method (Figure 20.6) is important for the characterization of anisotropic materials; this method allows the measurement of the properties of materials in different directions. In this method,

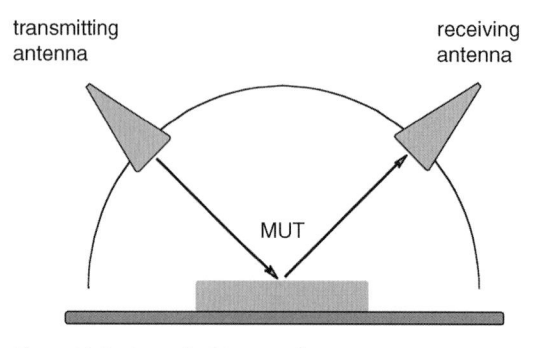

Figure 20.6 Setup for bistatic reflection measurements (MUT).

20.3 Overview of Methods for the Characterization of Materials in the Radiofrequency

Figure 20.7 Resonant cavity measurements: rectangular waveguide cavity (a), and the general view (b) and the cross section of a quasi-static cavity (c).

two antennas are used for transmitting and receiving signals, respectively, and the reflection can be measured at different angles of incidence.

Nonresonant methods are usually used to get general information about the electromagnetic properties of materials. Resonant methods can measure material properties only at a single or several discrete frequencies; however, these methods are widely used because of high accuracy and sensitivity. The main requirement of a resonant-perturbation method is that a sample forms a key part of the resonator, and the algorithm for calculating the material properties is related to the field distribution in the resonator and the sample. In principle, any kind of resonator can be used for material property characterization. A piece of a sample material inserted into the resonator changes the central frequency (f) and the quality factor (Q) of the cavity. For dielectric measurements, a sample is placed in the maximum of the electric field, and, for magnetic measurements, in the maximum of the magnetic field. The shift in f and the variation of Q allow one to calculate the complex permittivity or the complex permeability of the material at a fixed frequency. The test sensitivity of the method depends on the ratio of the sample volume to the resonator volume. The most widely used resonant-perturbation method uses a TE_{10n} mode rectangular waveguide with iris-coupled end plates (Figure 20.7a). Since the minimal dimensions of a rectangular measurement resonator are of the order of $\lambda/2$ and the cross section of a sample is small compared with λ, the sample-to-resonator volume ratio is small, and therefore the resonance frequency shift is small. To increase the accuracy of the resonant cavity method, the authors of Ref. [43] proposed the use of a quasi-static resonator (Figure 20.7b). In this resonator, the electric field is mainly concentrated in the gap (20.1), while the magnetic field is concentrated in the cylindrical volume (20.2) (Figure 20.7c). This volume and, naturally, the volume of the gap are much smaller than the volume of the rectangular resonator. Thus, the sample-to-resonator volume ratio becomes rather higher, and, hence, the sensitivity of the method is enhanced.

Table 20.6 provides a comparison between the measurement's methods discussed with emphasis placed on the features of the methods.

All the above-mentioned methods for materials characterization are often used in combination. As an example, Figure 20.8 shows the magnetic spectra of composites filled with different types of carbonyl iron (50 vol.%), measured by the impedance method and by a combination of the impedance and resonant methods.

Table 20.6 Comparison of the methods for characterization of materials.

Method	Operating frequency range	Method features
Impedance	1 MHz–1 GHz	Broadband; small samples; best for obtaining general information about electromagnetic properties of lossy and low-loss materials over an RF range; useful for low- and high-temperature characterization
Free-space (nonresonant)	2–20 GHz	Broadband; noncontacting; large flat samples; best for anisotropic materials; useful for high-temperature characterization
Transmission/reflection line (nonresonant)	2–20 GHz	Broadband; small machineable samples; best for lossy to low-loss materials
Resonant-perturbation (resonant)	Discrete frequencies 2–40 GHz	Accurate; small samples; best for low-loss materials

The impedance method overestimates the values of magnetic losses, whereas the combination of two methods allows one to get more accurate values of permeability and, moreover, to determine the resonance frequency.

20.4
Magnetization Processes in Bulk Magnetic Materials

20.4.1
Magnetostatic Magnetization Processes

The magnetic properties of a substance are determined by the spin and orbital magnetic moments of electrons and the magnetic moments of atomic nuclei. A characteristic feature of a ferromagnet is that, in the absence of external magnetic fields at temperatures below the Curie temperature, they exhibit a certain configuration of domains of spontaneous magnetization, which is called a domain structure. By the domain structure is meant a combination of the size, shape, and arrangement of domains, as well as the orientation of spontaneous magnetization vector ($\vec{M}_s$) in the domains and domain walls (Figure 20.9). The rearrangement of the domain structure in an external magnetic field determines the magnetization of a ferromagnet.

Figure 20.10 illustrates the magnetization process, the dependence of magnetization M (magnetic moment of a unit volume) of a sample as a function of the external magnetic field and, a magnetic hysteresis loop. The square insets on the top illustrate the magnetization distribution at various points on the magnetization curve.

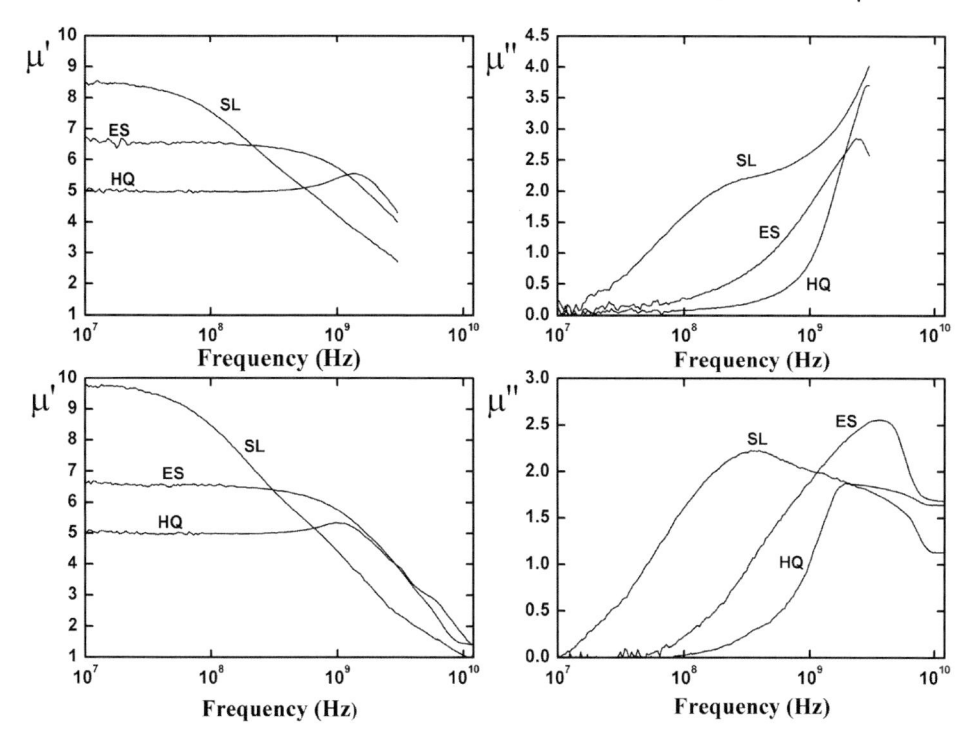

Figure 20.8 Magnetic spectra of polymer composites with different types of carbonyl iron (50 vol.%) measured by different methods: impedance method (top) and by a combination of impedance and resonant cavity methods (bottom).

In the absence of the magnetic field ($H=0$), a sample is demagnetized, and its magnetization M is zero. When exposed to a magnetic field, the domains with M aligned with the magnetic field grow, while the domains with M different from the direction of H decrease. When magnetized strongly enough (at saturation field H_s), the prevailing domain overruns all others and results in only one single domain, and the material is magnetized up to saturation M_s. If the saturation magnetization M_s

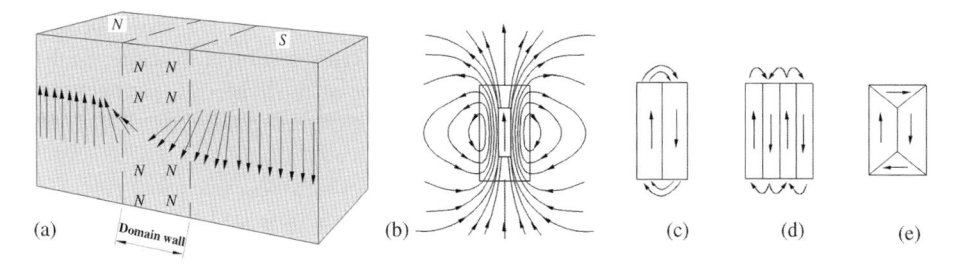

Figure 20.9 Schematic diagram of a 180° domain wall (a), and domains variety: (b) uniformly magnetized single domain; (c) two domains; (d) four domains in a lamellar pattern; and (e) two closure domains.

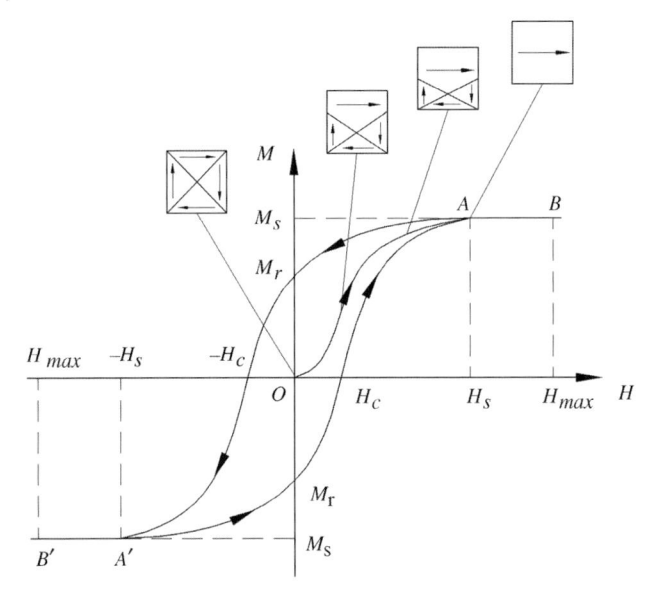

Figure 20.10 Initial magnetization curve and hysteresis loop for a typical ferromagnet.

does not exhibit any change under further increase in the magnetic field to a certain value H_{max}, it can be taken as the spontaneous magnetization. The curve OAB is the initial (main) magnetization curve. As the field decreases from H_s to H_0, the sample does not return to the state with $M = 0$, but has a remanent magnetization M_r at $H = 0$. The magnetization vanishes only under a certain opposite field of $-H_c$. The field H_c is called the coercive field, or the coercive force. Under a cyclic variation of the field $H_{max} \to 0 \to (-H_{max}) \to 0 \to H_{max}$, magnetization follows the closed loop $AM_rA'(-M_r)A$, called the limit hysteresis loop of magnetization. The area (S) of the magnetic hysteresis loop is proportional to the work of the external field on the remagnetization of the sample. The irreversible mechanism responsible for the magnetic hysteresis loop is domain-wall motion.

Depending on the shape of the hysteresis loop, magnetic materials are classified as magnetically soft (narrow loop) and magnetically hard (broad loop) materials. The physics behind this classification is related to the magnetic crystalline anisotropy.

Experiments on ferromagnetic single crystals have shown that the shape of the magnetization curve depends on the symmetry of the crystalline lattice – on the magnetic anisotropy axes. In one direction, along the easy magnetic axis (EMA), the magnetization of a crystal reaches saturation in a relatively lower magnetic field than in another direction, along the hard magnetic axis (HMA). This phenomenon is called a magnetic crystalline anisotropy, which is characterized by the magnetic energy necessary to turn the magnetization vector from the EMA to the HMA. The value of this energy is determined by the magnetic anisotropy constants K_A, which depend on the structural type and the chemical composition of a ferromagnet. Transition from cubic symmetry to tetragonal or hexagonal symmetry leads to a substantial increase in the K_A and to a change in the character of magnetic crystalline anisotropy. In addition to the EMA-type magnetic crystalline anisotropy, crystal lattices of ferromagnets may exhibit other types of anisotropy, such as "easy magnetic plane" and a "cone of easy magnetization." Since the energy of magnetic crystalline anisotropy is the magnetic energy of rotation of the vector of total magnetization, the phenomenon of anisotropy is characterized by the magnetic anisotropy field H_A:

$$H_A \cong \frac{2|K_A|}{\mu_0 M_s} \tag{20.9}$$

The field H_A is largely responsible for the magnetic properties of a material and, together with M_s, is one of the basic magnetostatic characteristics of ferromagnets. In real magnetic materials, in addition to H_A, one should also take into consideration the effects of inner and outer demagnetization factors. The anisotropy of inner demagnetization factors is associated with the structural inhomogeneity of the magnetic material. The anisotropy of outer demagnetization factors is associated with the anisotropy of the shape of a sample. Thus, in a real ferromagnet, H_A actually represents the effective magnetic field of anisotropy $H_A^{(\text{eff})}$, which provides a quantitative description of the effects of all types of anisotropy on the orientation of the magnetization vector in the material.

According to modern views, there exist three basic factors responsible for the magnetic hysteresis: (1) irreversible shift of domain boundaries; (2) irreversible rotation of spontaneous magnetization; and (3) delay in the formation and growth of magnetization nuclei [44, 45]. All the three factors are largely associated with the defects in a magnetic material, both microscopic (vacancies, foreign interstitial atoms, and dislocations) and macroscopic (shells, inclusions, cracks, etc.) ones. In the vicinity of a defect, the magnetic anisotropy field is reduced, magnetization vectors are distributed nonuniformly, and some of them deviate from easy magnetization axes. The defects hamper the motion of magnetic domain walls and are responsible for abrupt irreversible rotations of magnetization; it is the defects on which the magnetization nuclei – domains with opposite magnetization vectors – arise.

In addition to the main magnetization curve, the character of magnetization processes is also demonstrated by the magnetic field dependence of the magnetic permeability (Figure 20.11). There are several kinds of magnetic permeability. The magnetic permeability appearing in the relation $\vec{B} = \mu\mu_0 \vec{H}$ (see Eq. (20.4)) has two values, the initial μ_i and the maximal $\mu_{\max}$ permeabilities, which are frequently used in the description of the magnetic properties of ferromagnets. The initial magnetic permeability $\mu_i = (dB(H)/dH)|_{H=0}$ is determined by the slope of the magnetization curve at zero magnetic field. The initial permeability and the coercive force of a material are related by the empirical formula $(\mu_i - 1)H_c/M_s = 1$, which is obtained from the analysis of a large volume of experimental data [45]. Magnetic permeability attains its maximal value for a certain $H_{\mu\max}$, called the field of maximal magnetic permeability. The differential magnetic permeability $\mu_{\text{dif}} = (dB(H)/dH)$ characterizes the variation of the magnetic state of a ferromagnet under infinitesimal

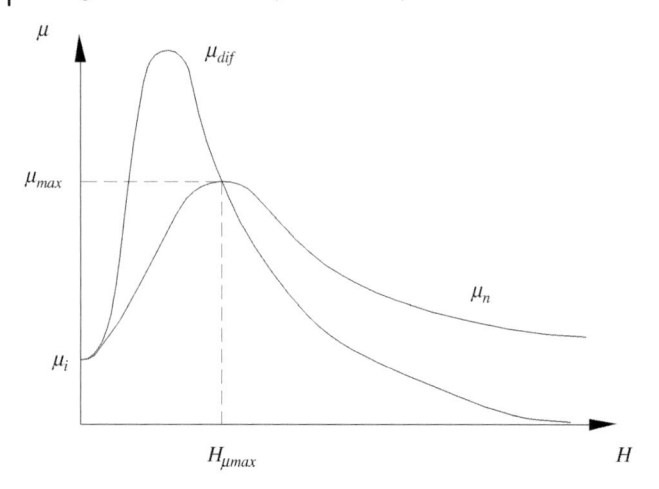

Figure 20.11 Relative magnetic permeability versus magnetic field strength.

variation in the magnetic field strength. The differential permeability exhibits strongest variations in the fields close to $H_{\mu max}$; in the fields from 0 to $H_{\mu max}$, μ_{dif} is greater than μ. In weak magnetic fields, all types of μ have identical values.

Thus, the basic macroscopic characteristics of a ferromagnet (extrinsic properties) are the parameters of the magnetic hysteresis loop (M_s, M_r, H_c, S) and μ, which, in contrast to the spontaneous magnetization, are structurally sensitive parameters and can be varied in wide limits by thermal, thermomagnetic, mechanical, and other types of processing. The spontaneous magnetization and the Curie temperature are determined by the quantum exchange interactions and are the fundamental characteristics of a ferromagnet (intrinsic properties).

20.4.2
Dynamic Magnetization Processes

In AC magnetic fields, ferromagnets exhibit irreversible magnetization processes resulting in power loss that may be expressed in terms of the phase lag between impressed field and induction change. If both $\vec{B}$ and $\vec{H}$ vary with time as $\exp(j\omega t)$, such a lag is described by a complex permeability $\mu = \mu' - j\mu''$ and magnetic loss tangent, $\tan \delta_\mu = \mu''/\mu'$, were δ_μ being the phase angle between $\vec{B}$ and $\vec{H}$.

At frequencies above 10^4 Hz, almost all ferromagnetic materials exhibit a decrease in μ with increasing frequency (permeability dispersion), which is associated with losses. The loss mechanisms in magnetic materials are divided into three traditional categories: hysteresis losses, eddy current or dielectric losses, and residual losses. The hysteresis losses result from irreversible mechanism of domain-wall motion. Eddy current losses are caused by ohmic losses through the finite conductivity of ferromagnet. This currents shield the interior of magnetic substance from the external field and the phase angle δ_μ increases with penetration depth. Residual losses are determined by various relaxation processes. Among the processes that contribute the residual losses are the resonance losses, namely domain-wall resonance and ferromagnetic resonance (FMR).

At present, extensive information has been obtained on the magnetic spectra (frequency dependence of μ) of materials with various compositions. In the magnetic spectra of ferrites, one distinguishes four regions of permeability dispersion: low-frequency (0–1 MHz), radiofrequency (1–1000 MHz), superhigh-frequency (10^3–10^5 MHz), and infrared (frequencies of about 10^7 MHz) regions [22, 24, 46]. In contrast to ferrites, the magnetic spectra of ferromagnetic metals and alloys are largely determined by eddy current losses and the skin effect [24, 45, 47].

Most of the publications on the magnetic spectra of ferrites are devoted to the study of radiofrequency and superhigh-frequency (microwave) regions of permeability dispersion, which is motivated by the application of these ferrites in radioelectronics. The radiofrequency dispersion of μ in ferrites is associated with the domain-wall motion, that is, the magnetic moments within the domain wall rotate as the wall moves to a new position. In this region, the dispersion curve may have either a relaxation or resonance character. The superhigh-frequency dispersion of μ is associated with a natural ferromagnetic resonance (or the gyromagnetic spin resonance). The physical origin and mechanisms of losses in magnetic materials are described below.

20.4.2.1 Natural Resonance

The dynamics of magnetization in a domain, when interacting with a magnetic field is described by the Landau–Lifshitz–Gilbert equation [48, 49]:

$$\frac{\partial \vec{M}}{\partial t} = -\gamma(\vec{M} \times \vec{H}_{\text{eff}}) + \frac{G}{\gamma M_s^2}\left[\vec{M} \times \frac{\partial \vec{M}}{\partial t}\right] \quad (20.10)$$

where $\vec{M}$ is the magnetization vector; $\gamma = 2.8$ MHz/Oe is the gyromagnetic ratio; M_s is the saturation magnetization; G is the Gilbert damping parameter; $\vec{H}_{\text{eff}}$ is total magnetic field to which the magnetic moment is exposed; and $\vec{H}_{\text{eff}}$ includes the external DC magnetic field $\vec{H}_0$, the demagnetizing fields, the fields associated with magnetic anisotropy, as well as the AC magnetic field $\vec{h}$.

The natural ferromagnetic resonance is the resonance absorption of electromagnetic field energy at a certain frequency called a natural ferromagnetic resonance frequency f_{NR}, when it coincides with the frequency of precession of the atomic moments about the direction of the internal magnetic field. A natural ferromagnetic resonance is a particular case of the induced FMR, which is associated with the precession of magnetization due to the simultaneous effect of the DC and AC magnetic fields [22–24, 46, 49].

The FMR frequency is described by

$$f_{\text{FMR}} = \gamma H_{\text{eff}} \quad (20.11)$$

In the case of a natural ferromagnetic resonance, $H_0 = 0$, and the role of the effective magnetic field is played by the magnetic anisotropy field H_A.

Equation (20.11) for materials with uniaxial anisotropy is rewritten as

$$f_{NR} = \gamma H_A \tag{20.12}$$

and, for materials with planar anisotropy, as

$$f_{NR} = \gamma (H_\theta \cdot H_\varphi)^{1/2} \tag{20.13}$$

Solving Eq. (20.10) for $H_0 = 0$ for a sample in the form of an ellipsoid (in particular, a sphere or a thin plate) [50], the expression for the frequency dependence of permeability is given by

$$\mu(f) = 1 + \frac{(\mu_s - 1) + j\beta f}{1 + j(f/f_d) - (f/f_r)^2} \tag{20.14}$$

where f is the operating frequency; μ_s is the static permeability; the parameters f_d and f_r are the Debye and the resonance characteristic frequencies, respectively, that determine the position of the magnetic loss peak and the shape of the dispersion curve; and β and f_d are functions of the Gilbert damping parameter.

Since Eq. (20.10) is real for $G \ll 1$, the term $(j\beta f)$ in Eq. (20.13) can be neglected; then $\mu(f)$ takes the form of a Lorentzian dispersion curve:

$$\mu(f) = 1 + \frac{\mu_s - 1}{1 + j(f/f_d) - (f/f_r)^2} \tag{20.15}$$

If $G \gg 1$, then the Lorentzian dispersion law for the permeability transforms into the Debye dispersion law:

$$\mu(f) = 1 + \mu_s \frac{1}{1 + j(f/f_d)} \tag{20.16}$$

20.4.2.2 Domain-Wall Resonance

The character of the radiofrequency dispersion of μ is determined by the domain structure in the material, by the properties of domain walls, and by quasi-elastic pinning forces.

Under the action of a low-energy alternating field, the domain walls experience reversible bulging and vibrate around their equilibrium positions. When the frequency of the AC magnetic field is equal to the frequency of the wall vibration, a resonance occurs. The permeability spectra associated with the domain-wall motion can be expressed by the equation [51]

$$\mu_{dw} = 1 + \frac{\chi_{dw}}{1 - (\omega/\omega_0)^2 + j\omega/\omega_r} \tag{20.17}$$

where $\omega = 2\pi f$ is the angular frequency of the applied magnetic field, ω_0 is the frequency at which the walls vibrate in resonance, ω_r is the relaxation frequency, and χ_{dw} is the initial susceptibility of the domain-wall displacement as $\omega \to \infty$.

The relaxation frequency of the above-mentioned magnetization mechanism can be expressed by the formula [22]

$$\omega_r = \frac{32 E_w}{\beta D_w^2} \tag{20.18}$$

where E_w is the domain-wall energy, β is the damping coefficient, and D_w is the width of the domain wall.

Therefore, the permeability and the characteristic frequencies associated with the domain-wall motion depend on the microstructure of the material. Generally, a large grain size leads to higher permeability and lower resonance frequency. In most cases, the wall resonance represents a Debye-type resonance.

For most bulk ferrites, the parameters of RF-permeability dispersion are related to each other by Snoek's law (Eqs. (20.1) and (20.2)), as discussed in Section 20.2.3. Snoek's law establishes a relation between the parameters of permeability dispersion and the magnetostatic parameters of ferrites with spinel and hexagonal structures; therefore, this law is useful for the analysis of experimental results and for choosing ferrites for different applications. However, Snoek's model assumes that the high-frequency permeability is related only to the natural ferromagnetic resonance [25], although the dominant process of magnetization in widely used polycrystalline ferrites is usually that of domain-wall motion [52]. Thus, the microstructure of ferrites may have considerable effect on the parameters of Snoek's law and must be taken into account.

In practice, most ferromagnetic materials are characterized by the Lorentzian form of the magnetic spectrum, but the region of magnetic dispersion in these materials is much broader. This can be attributed both to the inhomogeneities of the magnetic structure of a material and to the superposition of various resonance processes each of which is associated with one or other magnetization mechanism (domain-wall motion, natural resonance, etc.). In the case of a single-dispersion spectrum, such a superposition masks the superhigh-frequency absorption maximum. In this case, f_r is defined not as a characteristic resonance frequency that can be associated with one or other mechanism of magnetization, but as the mean resonance frequency that characterizes the position of the maximum of magnetic losses on the frequency dependence of the imaginary part of the complex permeability.

20.4.3
Experimental Magnetic Spectra of Bulk Ferromagnets

20.4.3.1 Bulk Ferrites

Figures 20.12–20.18 provide examples illustrating how various magnetization mechanisms manifest themselves as the frequency increases; these figures show the magnetic spectra of ferrites that differ in their composition and the types of their crystallographic structure and magnetic anisotropy field.

Figure 20.12 shows the frequency dependence of permeability (magnetic spectrum) of sintered polycrystalline MnZn, which is typical of ferrites with high initial permeability and cubic spinel crystal structure. In a low-frequency range

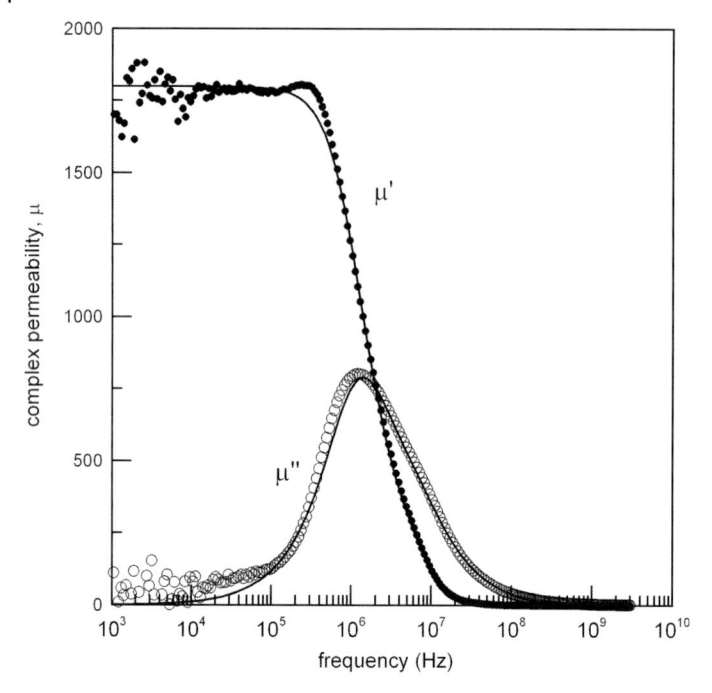

Figure 20.12 Magnetic spectrum of sintered polycrystalline MnZn ferrite approximated by Landau–Lifshitz equation.

($f < 10^4$ Hz), both real and imaginary parts of permeability remain constant. In the intermediate frequency region (10^4–10^6 Hz), the permeability shows a small change. The main change occurs in the high-frequency range, where the real part of permeability of MnZn decreases rapidly from the $\mu' \sim 1700$ at 1.2 GHz to $\mu' \sim 1$ at 1 GHz. At the same time, the imaginary part of permeability slowly increases with frequency and has a maximum of about 700–800 at about 2 MHz. The observed magnetic spectra are characterized by a single dispersion characteristic, which is resulted from the contributions of the domain-wall motion, magnetization rotation and natural ferromagnetic resonance.

The magnetic spectra of ferrites with hexagonal structure, which are characterized by an easy plane of magnetization, largely exhibit the same behavior as ferrites with spinel structure but have higher resonance frequencies. As an example, Figure 20.13 shows the magnetic spectra of polycrystalline Co_2Z magnetoplumbite and of $NiFe_2O_4$ spinel-type ferrite [22]. The comparison of these spectra shows that, although these ferrites have close permeabilities at low frequencies, the region of permeability dispersion of Co_2Z is much higher than that of $NiFe_2O_4$. For example, the resonance frequency of $NiFe_2O_4$ is $f_r \sim 200$ MHz, whereas that of Co_2Z is $f_r \sim 1.5$ GHz. This fact is attributed to the difference in the crystal structure of the ferrites and the associated value of magnetic anisotropy H_A. $NiFe_2O_4$ has a spinel structure, which is characterized by uniaxial magnetic anisotropy and low values of H_A, whereas Co_2Z magnetoplumbite has a

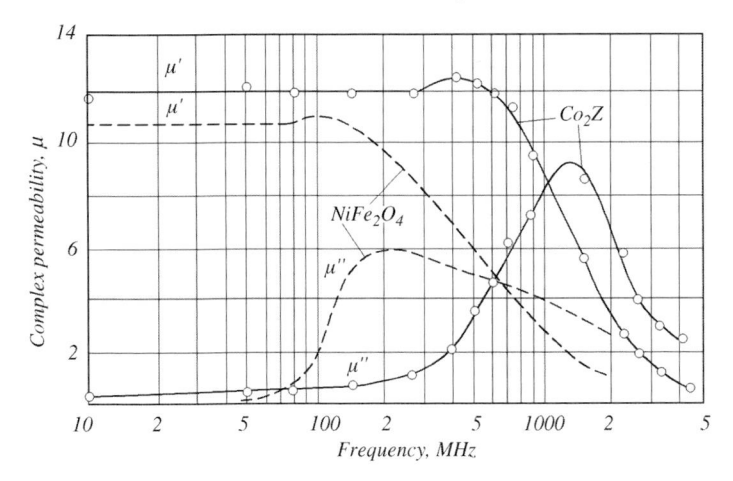

Figure 20.13 Magnetic spectra of sintered polycrystalline Co$_2$Z and NiFe$_2$O$_4$ ferrites.

hexagonal structure with the characteristic easy-plane magnetic anisotropy and high values of H_A (see Figure 20.1 and Table 20.3).

The magnetic spectra shown in Figures 20.12 and 20.13 are similar in that they both are characterized by single-stage permeability dispersion. However, this feature of dispersion in the magnetic spectra is associated with the superposition of the domain-wall motion with the natural ferromagnetic resonance (gyromagnetic spin rotation). Nakamura et al. [56] proposed a model that allows one to distinguish the contribution of each mechanism to the permeability dispersion. This model suggests considering the total magnetic permeability as a sum of the permeabilities (susceptibilities, χ) of two magnetization processes: $\mu(f) = 1 + \chi_{\text{spin}}(f) + \chi_{\text{dw}}(f)$, where χ_{spin} is the susceptibility of natural ferromagnetic resonance and χ_{dw} is the susceptibility of domain-wall motion resonance. It is assumed that the magnetic dispersion due to the natural ferromagnetic resonance has relaxation character, while the magnetic dispersion due to the domain-wall motion has a resonance character:

$$\chi_{\text{spin}}(\omega) = \frac{K_s}{[1 + j(\omega/\omega_{\text{res}})]} \quad (20.19)$$

$$\chi_{\text{dw}}(\omega) = \frac{K_{\text{dw}}\omega_{\text{dw}}^2}{[\omega_{\text{dw}}^2 - \omega^2 + j\beta\omega]} \quad (20.20)$$

where ω is the angular frequency of the external electromagnetic field; K_s and ω_{res} correspond to the static spin susceptibility and the spin resonance frequency; K_s, ω_{dw}, and β are the static spin susceptibility of domain-wall motion, the domain-wall motion resonance frequency, and the damping factor for the domain motion, respectively.

Figure 20.14 shows the results obtained by numerically fitting the measured complex permeability spectra of sintered polycrystalline NiZn and MnZn ferrites with those calculated by Eqs. (20.19) and (20.20) [27].

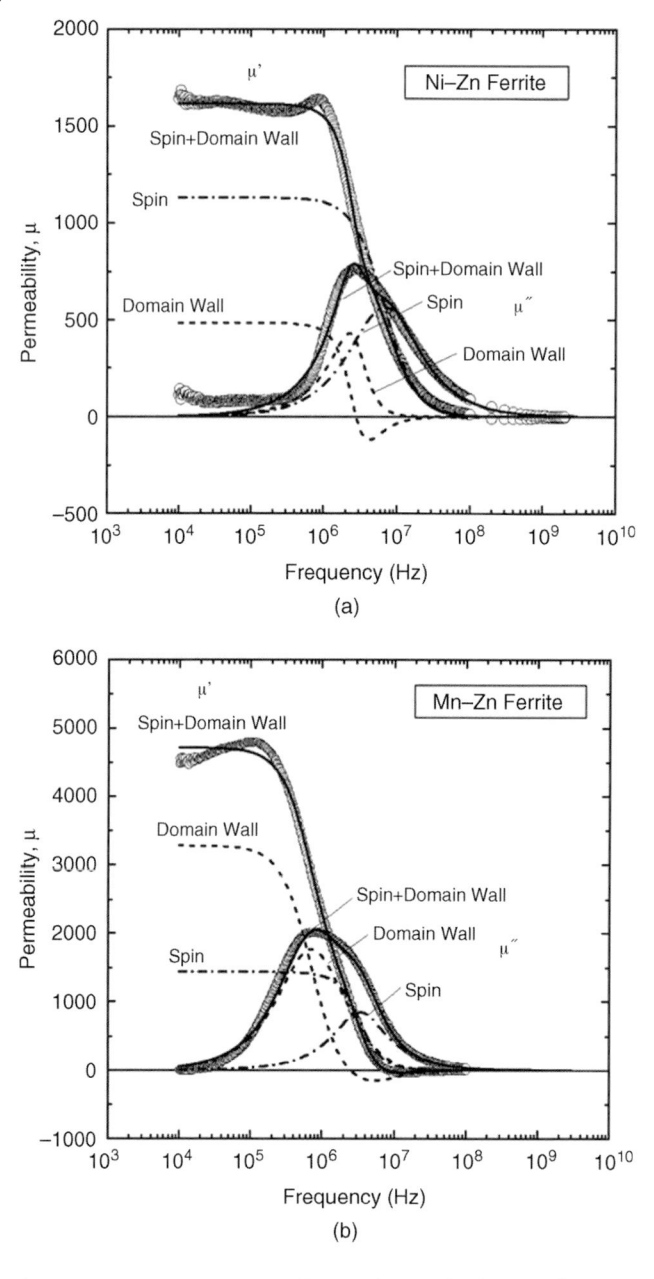

Figure 20.14 Magnetic spectra of sintered polycrystalline NiZn and MnZn ferrites. Open circles represent measured permeability. Solid, dotted, and dashed-dotted lines are the results of calculation for the domain wall and spin components. Reprinted with permission from Ref. [27]. Copyright 2003, American Institute of Physics.

20.4 Magnetization Processes in Bulk Magnetic Materials

These results show that the permeability spectra of polycrystalline NiZn and MnZn ferrites can be described by a gyromagnetic spin rotation component having the relaxation-type frequency dependence and a domain-wall motion component with resonance-type frequency dependence. One can see that the domain-wall contribution is dominant in the MnZn ferrite, while the gyromagnetic one is dominant in the NiZn ferrite.

20.4.3.2 Ferromagnetic Metals and Alloys

Until recently, it was assumed that the magnetic spectra of ferromagnetic metals are characterized by a smooth decay of μ' and μ'' in the radiofrequency region and do not exhibit any resonance features. However, the magnetic anisotropy field estimates of ferromagnetic metals, for example, iron and nickel, show that the natural ferromagnetic resonance in these metals should occur at frequencies from 100 MHz to 10 GHz. The absence of natural resonance in the magnetic spectra of these materials has been attributed to the shielding of internal regions of a metal by eddy currents due to the small skin depth [24, 57]. Indeed, at frequencies above 100 MHz, the penetration depth of an AC magnetic field into the bulk of a ferromagnetic metal is generally less than the equilibrium linear dimensions of domains (1–10 μm).

The development of measurement techniques for the electromagnetic parameters of materials [41, 47, 58] made it possible to carry out sufficiently detailed experimental investigations of the domain-wall motion and natural ferromagnetic resonance in ferromagnetic metals; usually, these studies were carried out on film samples of ferromagnetic metals and alloys (Figures 20.15–20.17) [53–55, 58].

According to the magnetic spectra shown in Figures 20.15–20.17, magnetic conductors exhibit permeability dispersion in the frequency range from hundreds of MHz to 10 GHz. At present, it is established that μ and the natural resonance frequency of both magnetic conductors and ferrites are mainly determined by their

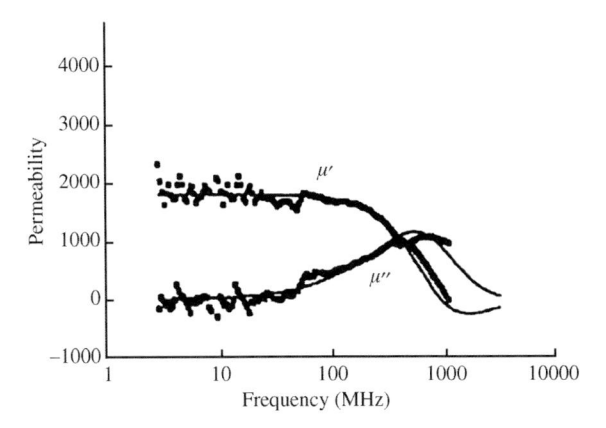

Figure 20.15 Complex permeability of a 0.63 μm thick CoZnFe film. Solid lines are $\mu'(f)$ and $\mu''(f)$ calculated by the Landau–Lifshitz equation with a damping parameter of 0.005. Filled squares are measured permeability. Reprinted with permission from Ref. [53]. Copyright 1999, American Institute of Physics.

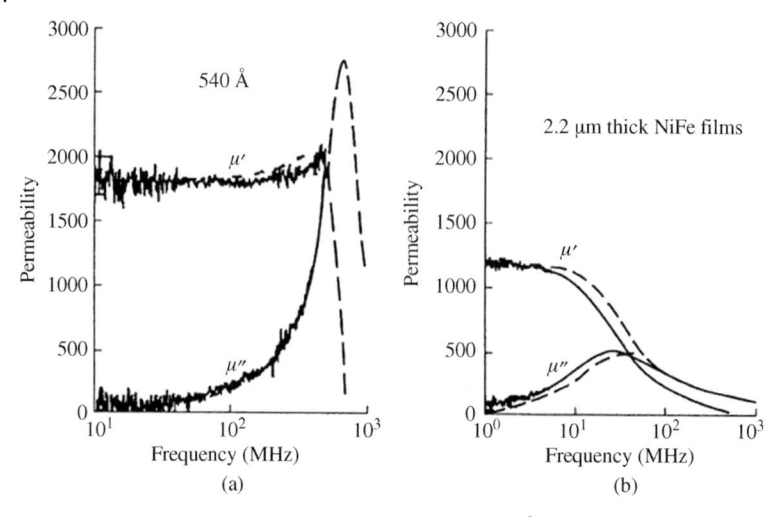

Figure 20.16 Measured and calculated permeability of 540 Å (a) and 2.2 μm (b) thick NiFe films. Solid lines are measured data. Dotted lines are μ′(f) and μ″(f) calculated by the Landau–Lifshitz equation. Reprinted with permission from Ref. [54]. Copyright 1998, IEEE.

composition and crystal structure. As regards the frequency dependence of μ of the films, it also depends on the morphology and thickness of a film, which determine the domain structure, magnetic anisotropy, and, as a consequence, the direction of the magnetization vector with respect to the film plane.

Figure 20.18 shows the magnetic spectrum of a monolithic sample of (SL-type) CI (99.5 at.% of α-Fe) prepared by pressing a powder under a high pressure of (1.5 GPa).

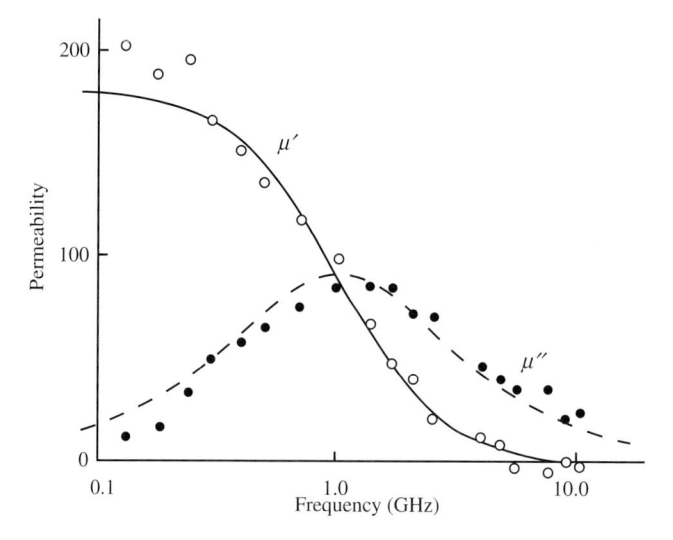

Figure 20.17 Complex permeability of a 0.7 μm-thick Fe film: dots are the results of measurements and lines are the results of fitting the experimental magnetic spectra by the Debye dispersion low. Reprinted with permission from Ref. [55]. Copyright 2007, with permission from Elsevier.

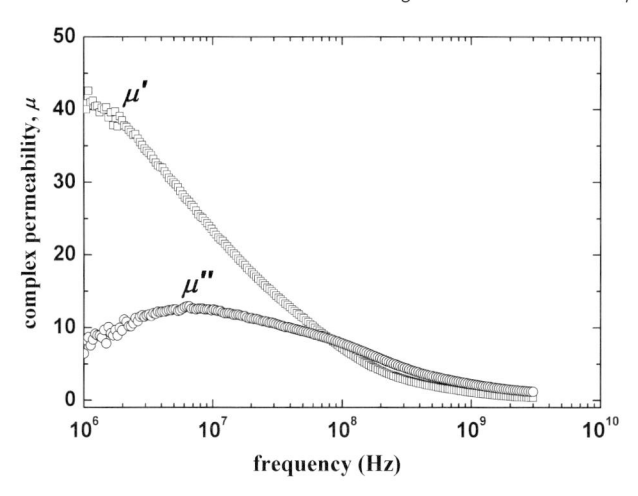

Figure 20.18 Complex permeability spectra of pressed sample of carbonyl iron (SL-type).

The comparison of the magnetic spectra of a pressed CI sample and of an iron sample with similar composition deposited as a film (Figure 20.17) shows that both materials are characterized by a broad resonance, which testifies to the inhomogeneous structure of the samples. However, in contrast to the film, which has permeability dispersion at frequencies of 0.1–10 GHz, the permeability dispersion region of the pressed CI sample lies in the range from 10^6 to 10^9 Hz. It is clear that, in the latter case, eddy current losses (skin effect) start to manifest themselves in the GHz range, which masks the magnetization processes associated with the natural ferromagnetic resonance.

20.5
Magnetization Processes in Polymer Magnetic Composites

20.5.1
Effect of Composition

Figure 20.19 shows the hysteresis loops of bulk MnZn ferrite (ceramic, $\mu_i \sim 1800$, Fig. 20.12) and of polymer composites filled with 40–60 μm polycrystalline particles of this ferrite. One can see that the magnetization of the net MnZn ferrite and its composites reaches saturated state at the same magnetic field of $H_s = 2.5$ kG. The value of saturation magnetization of polymer composites with different concentrations of ferrite ($M_s = 63$ emu/g for 40 vol.% and $M_s = 67$ emu/g for 50 vol.%) is less than that of the net ferrite ($M_s = 97$ emu/g).

Figure 20.20 shows the hysteresis loops for a pressed CI (SL-type) sample and for polymer composites filled with powders of this type of CI. In a monolithic sample, magnetization reaches a saturation value of $M_s = 232$ emu/g in fields of about 5–7 kG. In composites, the saturation state is achieved at higher fields (about 10 kG).

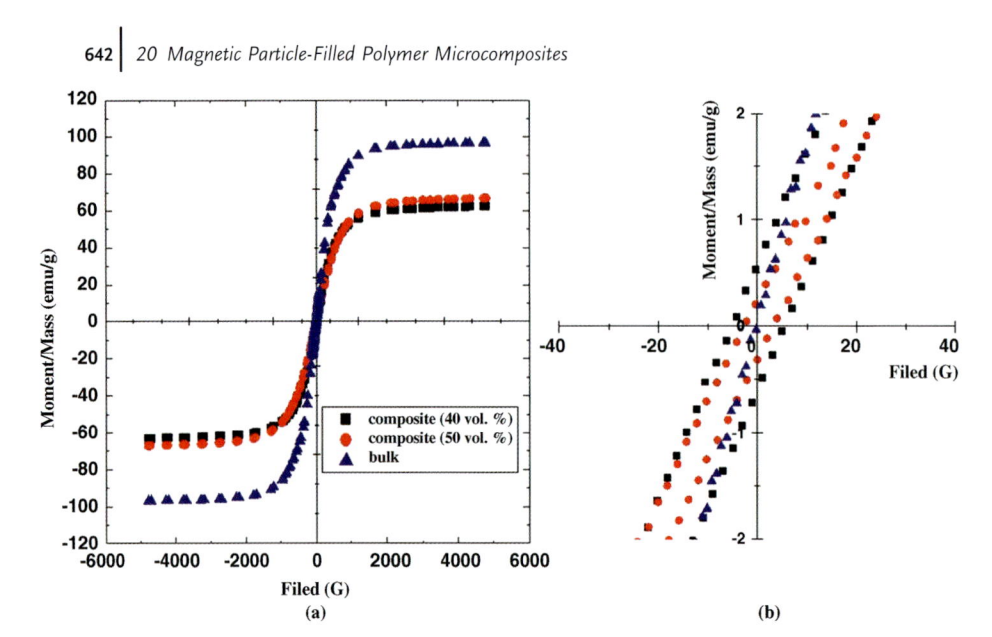

Figure 20.19 Hysteresis loops of bulk MnZn ferrite and its composites with polysiloxane (a); detail information about the coercitivity of bulk and composite materials (b).

The value of M_s increases logarithmically with the concentration of CI from $M_s = 96$ emu/g for a composite with 10 vol.% of CI to $M_s = 191$ emu/g for a composite with 50 vol.% of CI. However, the coercivities of the composites and of a bulk (pressed) sample differ little: $H_c = 13\,G$ for a monolithic sample, and $H_c \sim 10\,G$ for composites.

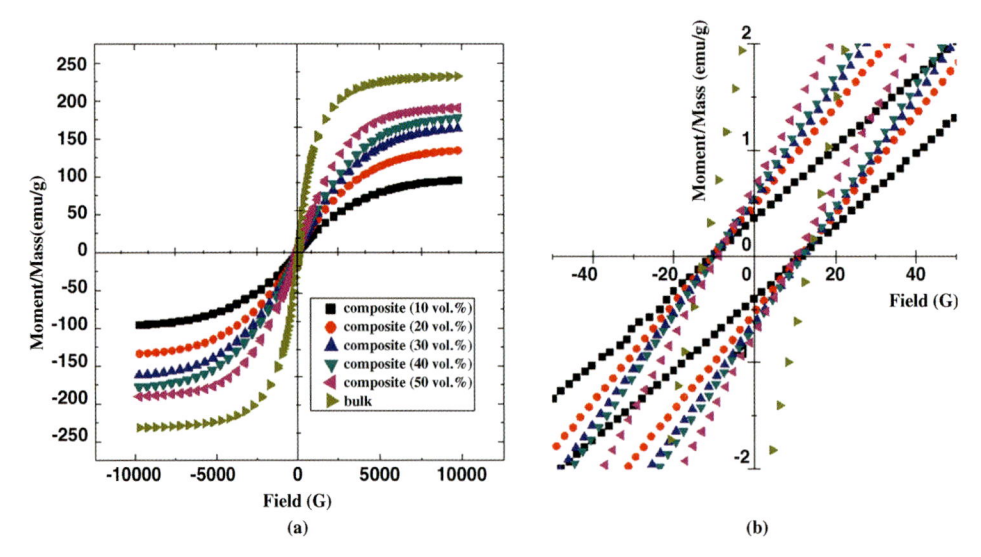

Figure 20.20 Hysteresis loops of bulk carbonyl iron (SL-type) and its composites with polysiloxane (a); detail information about the coercitivity of bulk and composite materials (b).

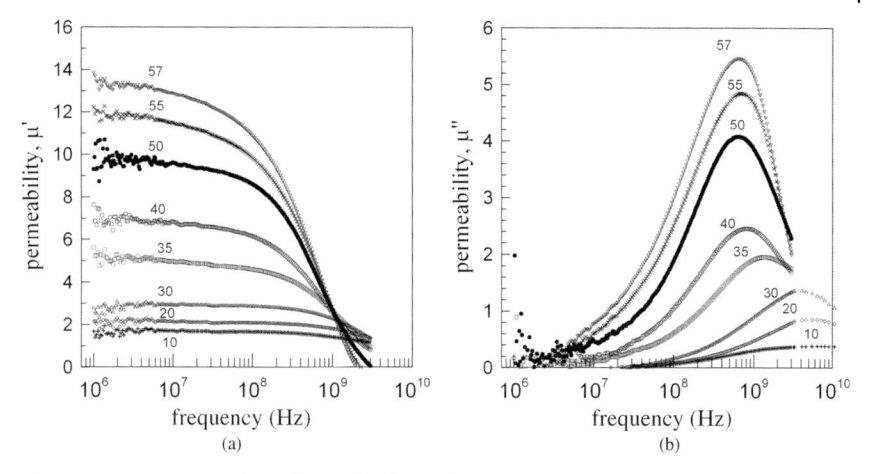

Figure 20.21 Frequency dependence of real (a) and imaginary (b) parts of complex permeability of MnZn ferrite in polyurethane. Ferrite particle size, 40–60 μm.

The principal feature of the magnetic spectra of composite materials is a significant decrease in μ at low frequencies and a shift of the dispersion region of μ to higher frequencies (Figure 20.21) compared with a bulk ferromagnet (Figure 20.12). Thus, in a high-frequency region, the permeability of polymer magnetic composites is greater than the permeability of bulk ferromagnets, as was reported in a number of works [27–29, 59–72].

The reduction of M_s and the change in the magnetic spectra of composites compared with those of bulk ferromagnets are associated with several physical mechanisms, first of all, with the magnetic polarization of magnetic particles isolated by a nonmagnetic polymer layer. The interphase polarization and the formation of effective magnetic charges/dipoles are responsible for the nonuniform distribution of magnetization over the bulk of the material. As a result, the effective magnetic field ($\vec{H}_i$) acting on a magnetic particle decreases by the value of the demagnetizing field ($\vec{H}_d$), which is proportional to the demagnetization factor [73]:

$$\vec{H}_i = \vec{H}_e + \vec{H}_d = \vec{H}_e - \hat{N}_M \vec{M} \qquad (20.21)$$

where $\vec{H}_e$ is the strength of the magnetic component of the applied electromagnetic field, $\vec{H}_i$ is the strength of the magnetic component of the internal field that acts on the magnetic composite, and $\hat{N}_M$ is the tensor of demagnetization factors, a second-rank material tensor that determines the relationship between the magnetization vector $\vec{M}$ and the demagnetization field vector $\vec{H}_d$.

$\hat{N}_M$ takes into account the inner and outer demagnetizing effects. As pointed out in Section 20.4.1, the inner demagnetizing effect is associated with the structural inhomogeneity of the magnetic substance, whereas the outer demagnetizing effect is associated with the anisotropy of the shape of a sample. The structural inhomogeneity of a magnetic composite is determined by the structural inhomogeneity of the magnetic (distortion of chemical composition, inhomogeneous elastic stresses, pores, cracks, surface roughness, etc.), as well as by the structural inhomogeneity of the composite due to the mutual arrangement of magnetic particles in the composite, that depends on the size, shape, and concentration of the particles. In the general form, the effective demagnetization factor of magnetic particles can be represented as [2]:

$$\hat{N}_M = \frac{L(\mu_f - 1) - (\mu_{eff} - 1)}{(\mu_f - 1)(\mu_{eff} - 1)} \qquad (20.22)$$

where L is the loading factor, μ_f is the permeability of ferromagnetic particles, and μ_{eff} is the effective permeability of the composite.

While the microstructural inhomogeneity of magnetic is ultimately responsible for the broadening of the ferromagnetic resonance linewidth in a bulk ferromagnet [74], the structural inhomogeneity of magnetic composites is responsible for the significant decrease in both components of μ compared with those in bulk magnetic materials and for the shift of the resonance frequencies to higher values due to the violation of the "magnetic coupling" of particles [56, 59, 63, 75]. The gaps/interphase boundaries (polymer layers) break the magnetic flux and give rise to local demagnetizing fields on the particle scale in the case of a low concentration (distributed magnetic charges on the particle surfaces) and to the external demagnetizing field on the sample scale when the concentration of the magnetic particles is greater than the critical value C_μ (which sometimes can be taken for the magnetic percolation threshold) [62–64]. For most polymer composites, the value of C_μ lies between 0.35 and 0.45 when the interaction between any two particles gives rise to a closed magnetic system, although the larger part of magnetic particles in the composite are isolated by a polymer layer even for the maximal loading (Figure 20.22) [16, 61].

Thus, the obstruction of the interparticle interaction is the main reason for the lower value of μ in polymer magnetic composites compared with that in bulk magnetic materials. The decrease of μ in a composite is the greater, the lower the

Figure 20.22 Morphology of polymer magnetic composites: (a) SEM microphotograph of MnZn (40 vol.%) filled polyurethane; SEM (b) and BSEM (c) microphotographs of the CI-filled polysiloxane at 50 vol.% fraction of primary CI (ES-type).

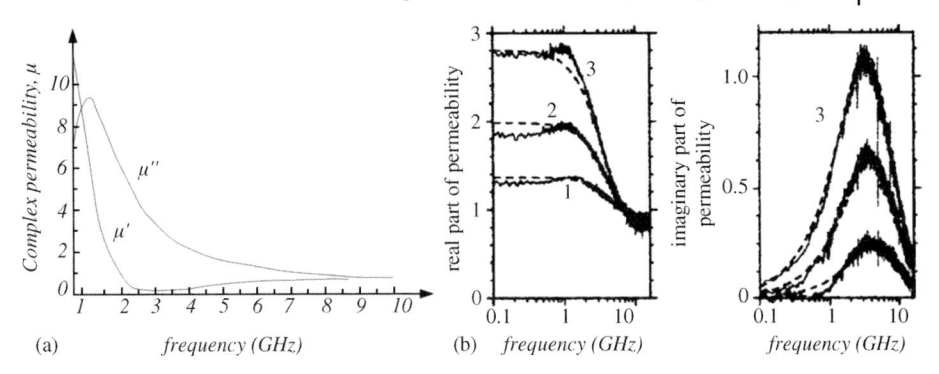

Figure 20.23 Frequency dependence of complex permeability of $Ba_3Co_2Fe_{24}O_{41}$ ferrite (a): 1 – real part of permeability; 2 – imaginary part of permeability; and polymer composites with different concentration of $Ba_3Co_2Fe_{24}O_{41}$ ferrite particles (b): 1–10 vol.%; 2–30 vol.%; 3–50 vol.%. Reprinted with permission from Ref. [34]. Copyright 2005, American Institute of Physics.

magnetic anisotropy field H_A of the magnetic component. For example, in composites filled with MnZn ferrite, which are characterized by very low values of H_A (units of Oe), the complex permeability is hundreds of times less than that in ferrite ceramics (Figure 20.21), whereas, in the composites filled with Co_2Z hexaferrite with high value of H_A (tens of kOe), this difference is insignificant (Figure 20.23) [34].

A slight decrease in the magnetic permeability compared with that of a bulk ferromagnet is also observed in the composites filled with carbonyl iron (Figure 20.24) [16]. In this case such a decrease is associated with the fact that the character of the magnetic spectra of a bulk sample of CI and of polymer

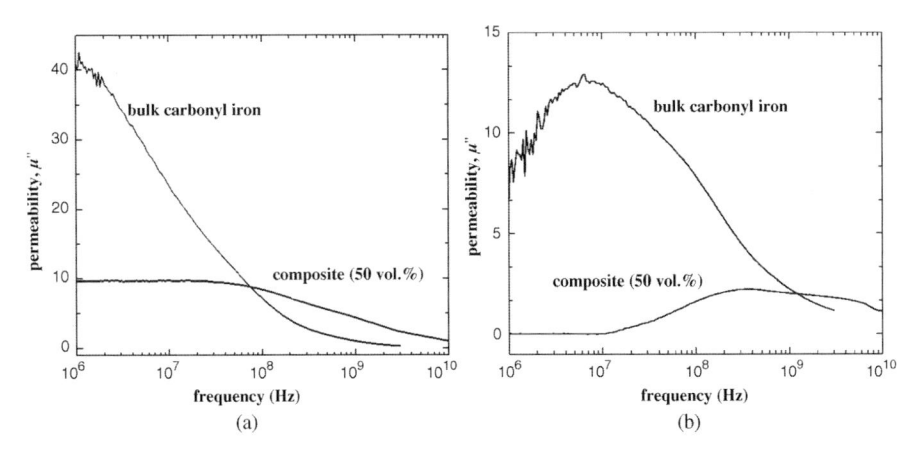

Figure 20.24 Frequency dependence of real (a) and imaginary (b) part of complex permeability of bulk CI (SL-type) and its composite with polysiloxane.

composites with CI is primarily determined by eddy current losses, which mask the processes associated with the domain-wall motion and a natural resonance (see Section 20.4.3).

20.5.2
Effect of Particle Size, Shape, and Microstructure

The effect of the particle size on the magnetic spectrum of a composite is sometimes similar to the effect of the loading of particles. For example, Figure 20.25 represents the magnetic spectra of MnZn-based composites with the same loading (40 vol.%) but different particle sizes [61]. One can see that a decrease in the average size of particles leads to a decrease in the permeability and shifts the resonance frequency to higher frequencies. This is caused by the internal demagnetization factor of particles, which is the higher, the smaller the magnetic particles, as well as by the effect of defective surface layer.

In contrast, the magnetic spectra (as well as the dielectric spectra, $\varepsilon = \varepsilon' - j\varepsilon''$) of composites filled with CI more strongly depend on the microstructure of CI particles than on their size (Figure 20.26) [16, 76].

This figure shows a significant difference in the high-frequency permeability of composites filled with primary and processed CI, although the chemical composition and the particle size distribution of these powders show small difference (Table 20.3). In the radiofrequency band, processed (SL-type) CI is characterized by the highest values of μ and ε. Moreover, a characteristic feature of the magnetic spectrum of SL-type composites is a very broad region of resonance on the magnetic loss curve. The composites filled with primary CI (of HQ, ES, and EA types) have higher permeability in the GHz band and lower values of ε over the whole radiofrequency band. The observed differences are attributed to the microstructure of particles;

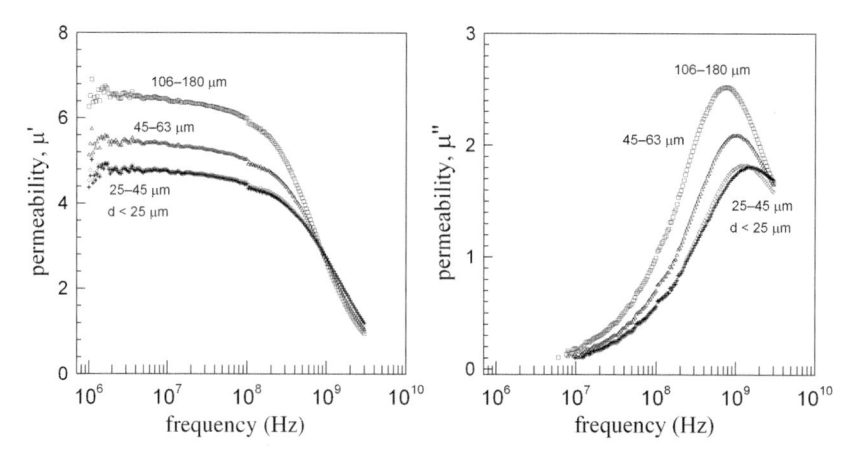

Figure 20.25 Effect of particle size on magnetic spectrum of MnZn-based composites with polyurethane.

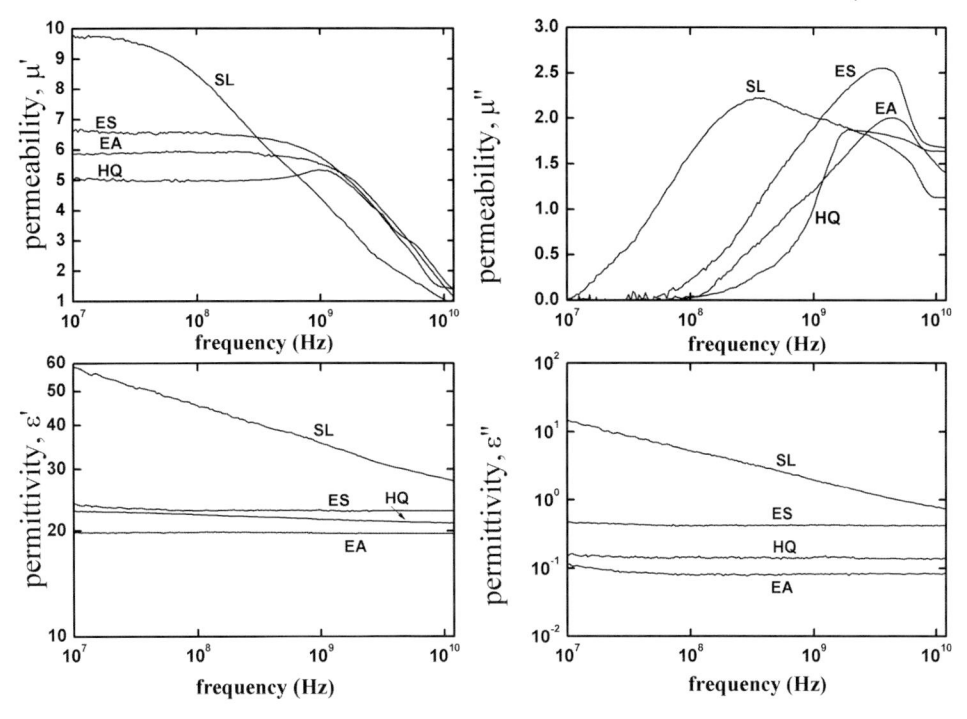

Figure 20.26 Effect of CI particles microstructure on magnetic and dielectric spectra of composites filled with primary and processed CI at the same content of CI 50 vol.%.

namely, these differences depend on whether or not the particles are characterized by "onionlike" multilayered morphology. The onionlike structure is typical of primary CI, while processed CI is characterized by polycrystalline structure (Figure 20.27) [12, 13, 16].

In contrast to primary CI (of HQ, ES, and EA types), processed CI has a distorted microstructure of particles due to the processing of primary CI powder in a hydrogen flow. This type of processing only slightly changes the elemental composition of CI (Table 20.3) but radically changes the microstructure of a particle [12]. The first and the most important change consists in the distortion of the "onionlike" structure of primary CI, in which αFe nanolayers alternate with nonmagnetic dielectric layers of cementite (Figure 20.27a, left). The hydrogen reduction of primary CI leads to the distortion of cementite layers in the particle structure and thus increases the content of αFe. In addition, this reduction is accompanied by the removal of nitrogen and carbon from the αFe crystalline lattice and by the growth of αFe crystallites. As a result, one obtains a particle with polycrystalline structure in which αFe crystallites are oriented in a specific order, so-called magnetic texture (Figure 20.27b, right). The presence of magnetic texture in (SL-type) CI is confirmed by the results of X-ray analysis [16]. The large size of αFe crystallites and the magnetic texture ensure high values of μ of (SL-type) CI composites in the low-frequency part of the radiofrequency

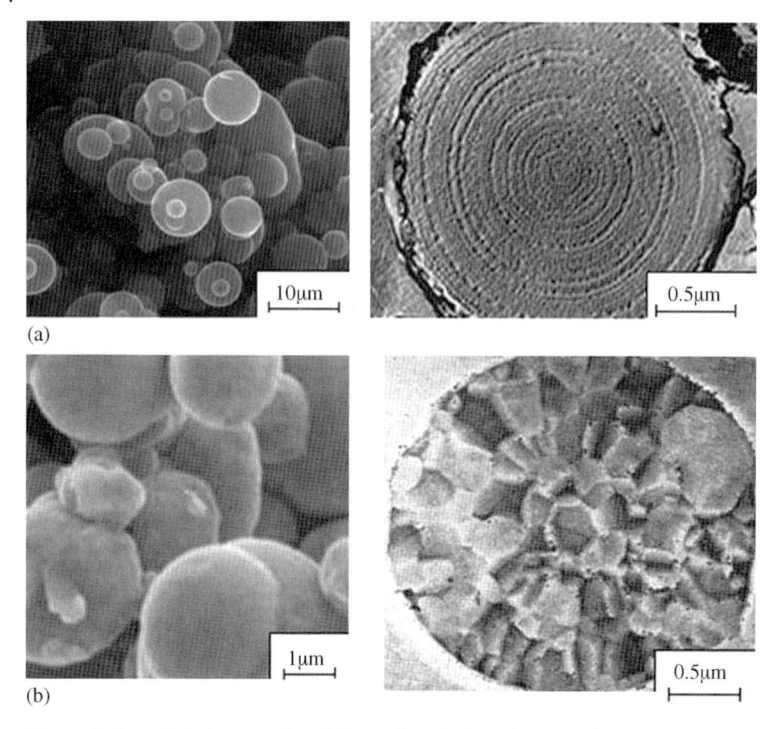

Figure 20.27 SEM photographs of CI powders (left) and views of CI particles microstructure (right): (a) primary CI with onionlike particle microstructure and (b) CI reduced with hydrogen with polycrystalline particle microstructure.

band due to the motion of domain walls, which is suppressed in the onionlike structure of primary CI. The large content of αFe and the polycrystalline structure of particles is also responsible for higher values of ε in composites based on (SL-type) CI compared with that in composites filled with primary CI.

The shape of particles can also produce an appreciable effect on the frequency dispersion of μ of the composites. As an example, Figure 20.28 demonstrates the magnetic spectra of composites filled with spherical and flaky CI particles [16].

Compared with the composites based on spherical CI particles, the composites with flakes have much higher values of μ' in the frequency range 10^7–10^9 Hz, as well as higher values of magnetic losses in the resonance region. The influence of particle shape on the μ is associated with the demagnetization form-factor of particles N_d (for spherical particles $N_x = N_y = N_z = 4\pi/3$; for flakes (normal in z-direction) $N_x = N_y = 0$ and $N_z = 4\pi$), as well as with packing density. Even the composites with spherical CI particles have higher packing density of particles (Figure 20.22b) than composites filled with flaky CI particles (Figure 20.29), the latter exhibits higher permeability due to the lower value of N_d.

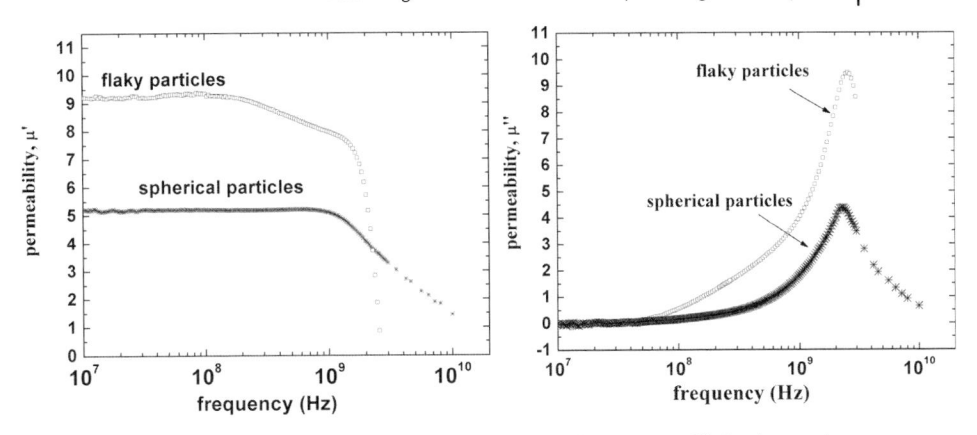

Figure 20.28 Effect of particles shape on the magnetic spectra of composites filled with CI at the same content of CI (50 vol.%) and different particles shape.

20.5.3
Estimation of the Effective Permeability of a PMC by Mixing Rules

The effective permeability of polymer magnetic composites μ_{eff} is a function of the intrinsic permeability of magnetic particles, the permeability of the matrix, the proportion between magnetic and dielectric components, and the shape and spatial distribution of the magnetic particles. Thus, to calculate μ_{eff}, one needs detailed information on the microstructure of the composite. In practice, the optimization of the composition of PMCs involves approximate methods, called "mixing rules" in the literature, to simulate the dependence of μ_{eff} on the permeabilities of the components

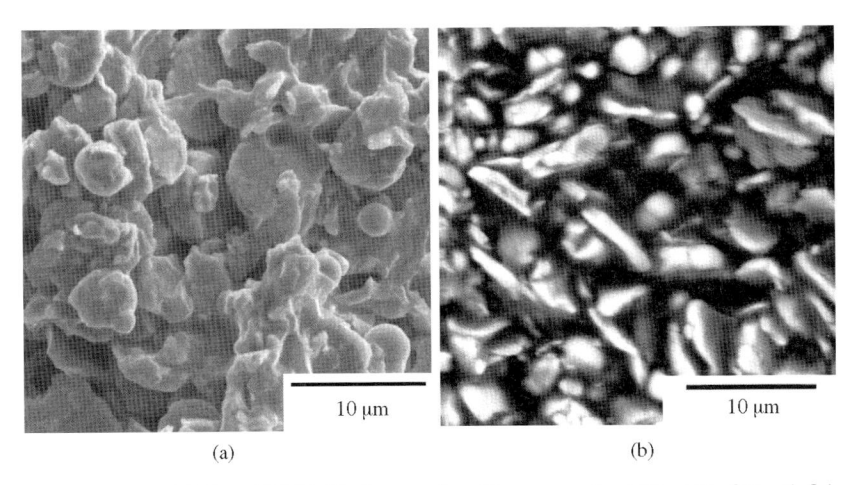

Figure 20.29 SEM (a) and BSEM (b) photographs of the composite at 50 vol.% of CI with flaky shape particles.

and theirs concentration. The most common mixing rules are the Maxwell–Garnett (MG) approximation and the Bruggeman EMT.

The Maxwell–Garnett mean-field model [77] considers a PMC as a homogeneous medium in which isolated particles of component a (inclusions) are embedded in component b (matrix). This assumption results in the following equation:

$$\frac{\mu_{\text{eff}} - \mu_b}{\mu_{\text{eff}} + 2\mu_b} = V_{f_a} \left(\frac{\mu_a - \mu_b}{\mu_a + 2\mu_b} \right) \quad (20.23)$$

where μ_{eff} is the effective macroscopic permeability of the medium; μ_b and μ_a are the intrinsic permeabilities of the host material (matrix) and the material of isolated particles, respectively; and V_{f_a} is the volume fraction of particles in the composite.

The MG model proved satisfactory in the case of dilute composites, when the interparticle interactions are negligible.

The Bruggeman effective medium theory [78] assumes that both components a and b have the same shape and are embedded in an effective medium with permeability equal to that of the composite μ_{eff}. This model is valid for not too high loading factors and is expressed as

$$0 = V_{f_a} \left(\frac{\mu_a - \mu_{\text{eff}}}{\mu_a + 2\mu_{\text{eff}}} \right) + V_{f_b} \left(\frac{\mu_b - \mu_{\text{eff}}}{\mu_b + 2\mu_{\text{eff}}} \right) \quad (20.24)$$

where μ_{eff} is the effective macroscopic permeability of the medium; μ_b and μ_a are the intrinsic permeabilities of the host material (matrix) and of the material of isolated particles, respectively; and V_{f_a} and V_{f_b} are the volume concentrations of particles and of the matrix, respectively.

Other mixing rules that are widely used for the interpretation of experimental results are the Landau–Lifshitz–Looyenga (LLL) mixing rule and the Musal (MU) and the Lichtenekker mixture equations.

The Landau–Lifshitz–Looyenga equation is given by [79, 80]

$$\mu_{\text{eff}} = \left(\mu_m + V_f \left(\mu_f^{1/3} - \mu_m^{1/3} \right) \right)^3 \quad (20.25)$$

where μ_{eff} is the effective macroscopic permeability of the medium, μ_m and μ_f are the intrinsic permeabilities of the matrix and of the ferromagnetic particles, and V_f is the volume concentration of ferromagnetic particles.

The LLL equation is rigorous when the difference between μ_f and μ_m is small.

According to Musal's approach [72], the Maxwell–Garnet expressions can be extended to the entire range of compositions $0 \leq V_{f_a} \leq 1$. Here, the medium is described as a combination of two different phases. The isolated phase consists of particles of a embedded in b (matrix), while in the agglomerated phase, the component b behaves as an inclusion in the extended medium a. The volume ratio of these two different phases changes as the volume fraction of the magnetic filler increases. The permeability of each phase can be calculated by the appropriate MG effective medium expression.

The original Musal mixture equation is as follows:

$$\mu_{eff} = \mu_m \left[1 + 3 \left(\frac{\mu_x + 2\mu_m}{V_x(\mu_x - \mu_m)} - 1 \right)^{-1} \right] \qquad (20.26)$$

Here, to calculate the permeability of an isolated ferromagnetic particle phase (μ_I), when it becomes the extended medium of a composite, μ_m is taken to be that of the matrix, μ_x is taken to be the permeability of ferromagnet μ_f, and V_x is taken to be the volume fraction of ferromagnet V_f. To calculate the properties of the agglomerated ferromagnetic particles phase (μ_A), $\mu_m = \mu_f$, $\mu_x = \mu_m$, and $V_x = (1 - V_f)$. Finally, to calculate the effective permeability of a composite (μ_{eff}), $\mu_m = \mu_I$, $\mu_x = \mu_A$, and V_x is the phase transition function that is simply taken to be V_f^n, were n is a fitting parameter.

The Lichtenekker empirical equation is often used in practice due to its simplicity [81]:

$$\log \mu_{eff} = V_f \cdot \log \mu_f + (1 - V_f) \log \mu_m \qquad (20.27)$$

where μ_{eff} is the effective macroscopic permeability of the medium, μ_m and μ_f are the intrinsic permeabilities of the matrix and of the ferromagnetic particles, and V_f is the volume concentration of ferromagnetic particles.

Presently it is recognized that the validity of mixing rules depends, first of all, on the difference between the permeabilities of the inclusions and of the host matrix. Thus, in the case of composites for which μ_f is close to μ_m, the MG and EMT mixing rules agree well with the measured effective permeability data. When the value of μ_f is high, which is the case of greatest practical interest, the Bruggeman expression and the Musal mixture equation agree reasonably well over a wide range of composition (Figure 20.30).

More information regarding the possibilities of "mixing rules" to predict the electromagnetic properties of PMC are available in the review article of Lagarkov and Rozanov [50].

20.6
Polymer Magnetic Composites with High Value of Permeability in the Radiofrequency and Microwave Bands

In Section 20.5, it has been shown that the effective magnetic permeability μ_{eff} of polymer magnetic composites is not a simple function of the magnetic permeabilities of individual components of the composite and their concentrations but it also depends on the size, shape, and the distribution of filler particles in the composite. The necessity to maintain the physical and mechanical properties of a PMC restricts the volume content of the magnetic phase, which cannot exceed 60 vol.% in the limiting case; therefore, one should seek new ways to further increase μ_{eff} in PMCs. There are a few techniques that allow one to obtain higher values of μ_{eff} in a PMC: the use of multicomponent filler [61, 82–88]; the use of multicomponent magnetic particles [54, 89–91]; formation of magnetic texture in a composite by orienting

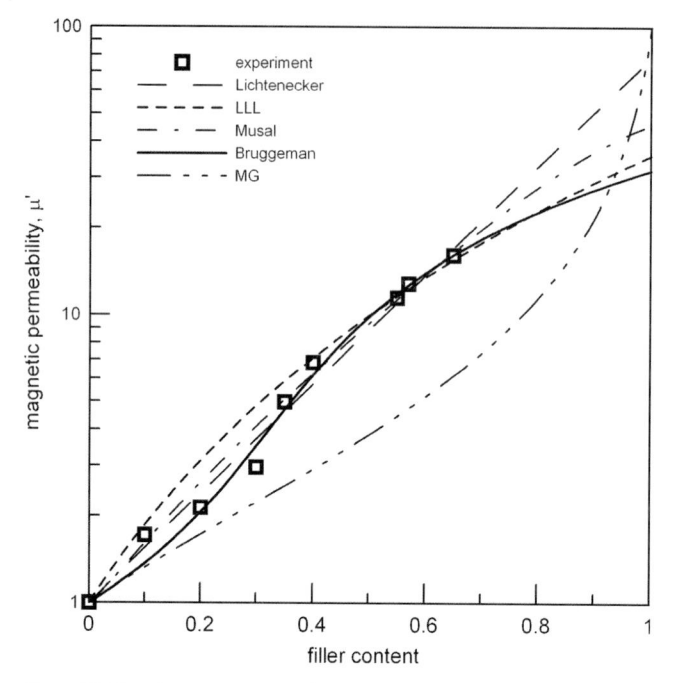

Figure 20.30 The concentration dependence of real part of complex permeability of MnZn-based polyurethane at 10 MHz. The experimental data is fitted by different mixing rules.

shape-anisotropic magnetic particles by external magnetic field or during forge rolling (magnetic elastomers) or by orientational drawing (magnetic fibers) [11, 92].

20.6.1
Polymer Magnetic Composites with Multicomponent Filler

In Ref. [82], the authors showed that the use of the mixture of NiZn ferrite and permalloy ($Fe_{55}Ni_{55}$) in the volume proportion of 30/20 allows one to significantly increase the μ_{eff} of a polymer magnetic composites based on polyphenylene sulfide (PPS) compared with the expected (calculated) values of μ_{eff} (Figure 20.31). Similar results were obtained for polymer composites filled with the mixture of NiZn ferrite and the powder of magnetically soft amorphous alloy (69% Co, 12% B, 4% Ni, and 2% Mo) [83]; with the mixture of NiZn ferrite and Ni [84]; with the mixture of Co_2Z hexaferrite with Ni [85]; with the mixture of magnetically soft iron and nanocrystalline alloy of $Fe_{73.5}Si_{15.5}B_7Cu_1Nb_3$ [86], and so on.

The authors of the above-listed studies attribute the increase in the μ_{eff} of the composites with multicomponent filler to a synergetic effect of the magnetic components of the filler, one of which is magnetic conductor (magnetic metal or alloy) and the other, magnetodielectric (as a rule, ferrite). However, the replacement of one of the components by nonmagnetic electrically conducting filler also results in an increased value of μ_{eff} of a composite [61, 88]. Figure 20.32 represents the

20.6 Polymer Magnetic Composites with High Value of Permeability | 653

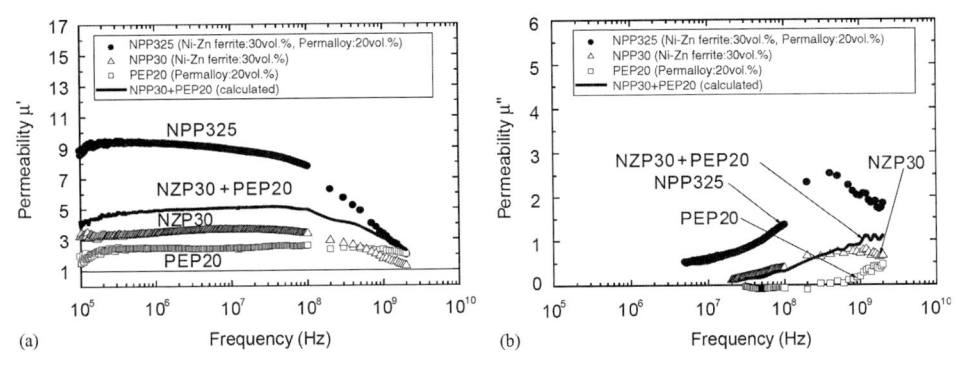

Figure 20.31 Magnetic spectra of composites filled with NiZn ferrite and permalloy and theirs mixture. Reprinted with permission from Ref. [82]. Copyright 2004, with permission from Elsevier.

magnetic spectra of composites filled with MnZn ferrite (particle size of 40–60 μm) and its mixtures with different electrically conducting fillers (carbon black, CB; carbon fiber, CF; Al powder, Al; polypyrrole, PPy).

The increase in the maximum of magnetic losses depends on the volume ratio between MnZn ferrite and conducting filler, which is in each case different. It was established that the increase in the magnetic losses in a composite occurs when the concentration of the conducting filler is higher than the electrical percolation threshold C_o in the two-component system polymer–conducting filler (Figure 20.33). For example, composites filled with the mixture of MnZn ferrite and carbon black show the increase in magnetic losses when the concentration of carbon black in a composite is 3–5 vol.% (Figure 20.34).

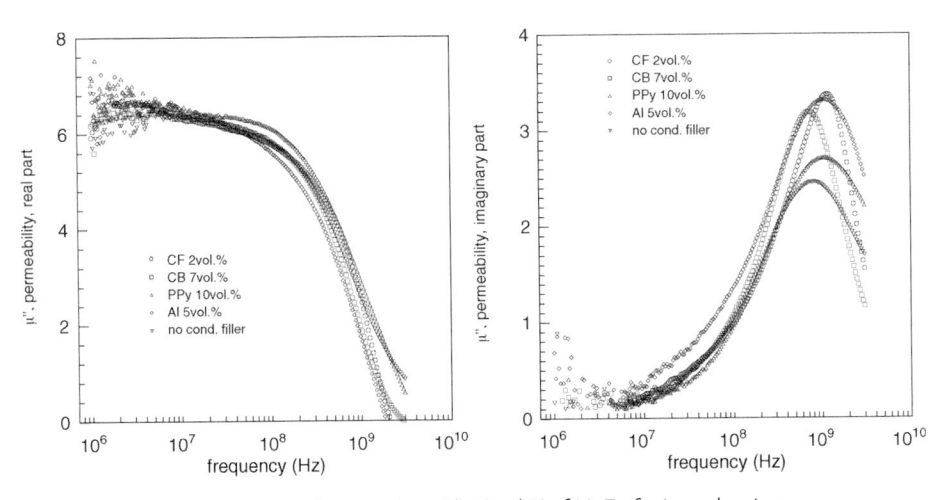

Figure 20.32 Magnetic spectra of composites with 40 vol.% of MnZn ferrite and various conducting fillers. The content of conducting filler is given for each curve (vol.%).

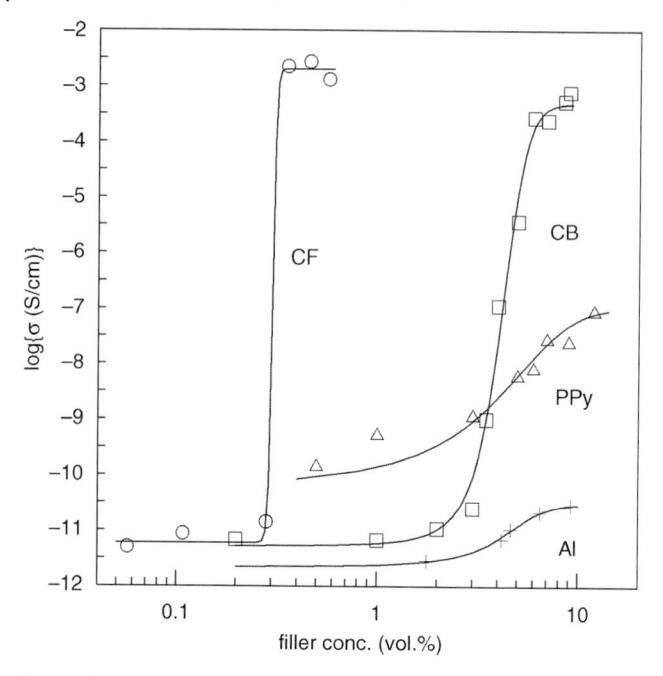

Figure 20.33 Concentration dependence of DC conductivity of polymer composites with different conductive fillers. Type of conductive filler is given for each curve.

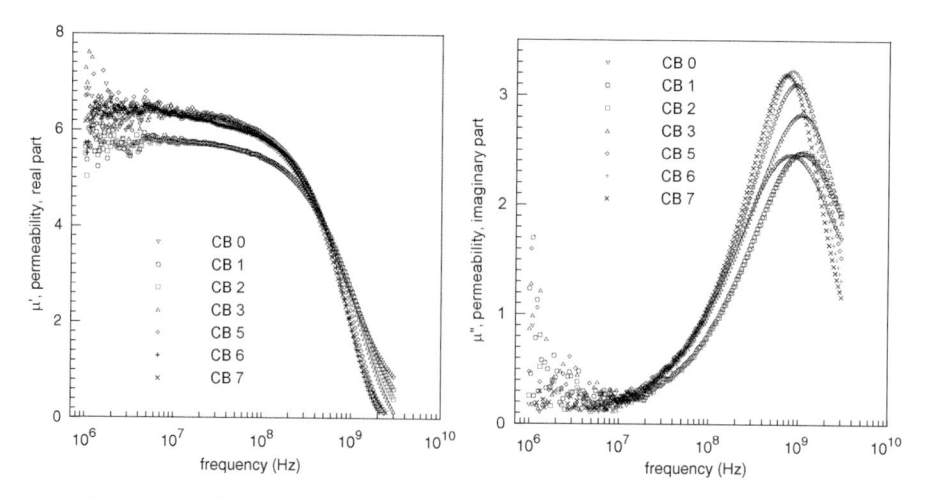

Figure 20.34 Magnetic spectra of composites with 40 vol.% of MnZn ferrite and different content of carbon black. The content of carbon black (in wt%) is given for each curve.

20.6 Polymer Magnetic Composites with High Value of Permeability

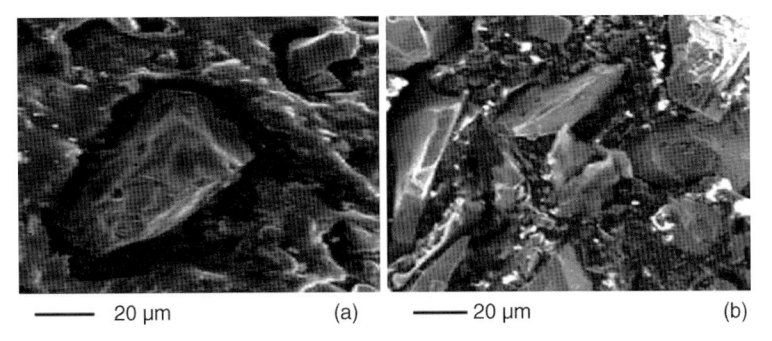

Figure 20.35 SEM photographs of polymer composites: (a) polyurethane with 40 vol.% of MnZn ferrite and (b) polyurethane with 40 vol.% of MnZn ferrite and 5 vol.% of Al.

The observed increase in magnetic losses is associated with the microstructure of the composites. Figure 20.35 shows the microphotographs of two types of composites: a two-component composite (polyurethane filled with 40 vol.% of MnZn ferrite) and a three-component composite (polyurethane filled with the mixture of MnZn ferrite and Al with the proportion of 40/5).

In both composites, the continuous phase is formed by a polymer, and ferrite particles are distributed in it as inclusions. However, in the case of a composite with multicomponent filler, ferrite particles are embedded in a polymer matrix filled with conducting particles. Each ferrite particle is surrounded by a conducting medium and can be considered as a "core–shell" structure, where the core is a ferrite particle and the shell represents a conducting polymer layer that separates the ferrites particles (Figure 20.36).

Core–shell structure in a polymer composite is formed under the following conditions:

- The concentration of ferrite particles in the composite is higher than the magnetic percolation threshold in the two-component composite polymer–ferrite.
- The size of ferrite particles is much greater than the size of the conducting filler particles, and the volume content of ferrite in the composite is much higher than that of electrically conducting filler. These two conditions ensure a "forced"

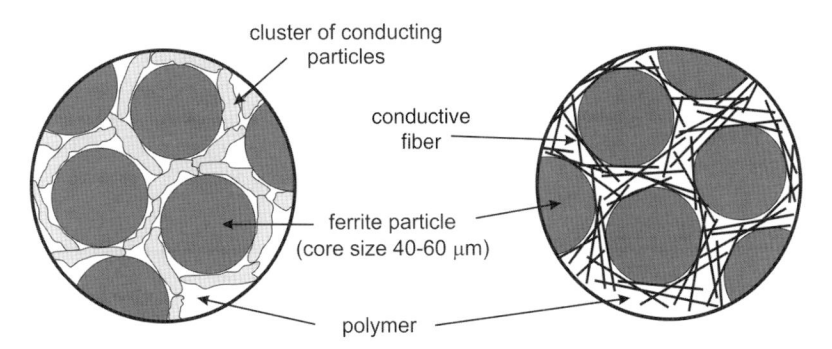

Figure 20.36 Schematic picture of composite with multicomponent filler.

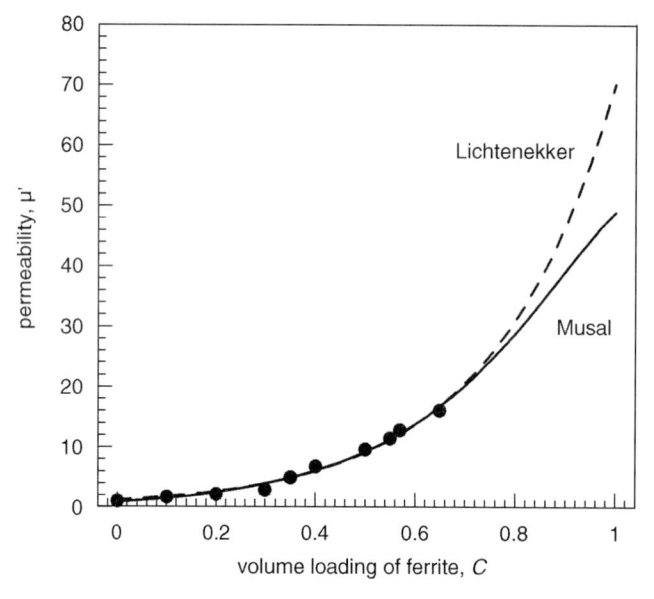

Figure 20.37 The concentration dependence of permeability of polyurethane filled with MnZn ferrite at 10 MHz. The experimental is fitted by Musal and Lichtenekker mixing equations.

arrangement (concentration) of conducting particles in the polymer interlayer between the ferrite particles.
- The volume content of conducting particles in the composite is higher than the electrical percolation threshold C_σ, so that they form an infinite conducting cluster in the polymer.

Figure 20.37 illustrates the concentration dependence of μ of composites based on polyurethane filled with MnZn ferrite. The experimental data are fitted by the Musal (Eq. (20.26)) and Lichteneker (Eq. (20.27)) equations. The concentration dependence of μ exhibits a percolation character with a magnetic percolation threshold of $C_\mu = 0.32$.

A core–shelllike structure is inherent in all polymer magnetic composites with multicomponent filler that exhibit an increase in μ_{eff} [61, 82–88]. This structure is formed roughly under the same conditions listed above as that in composites filled with MnZn ferrite mixed with electrically conducting fillers. As an example, Figures 20.38 and 20.39 present micrograph and magnetic spectra of poly(vinylidene fluoride) (PVDF) filled with a mixture of Co_2Z hexaferrite and Ni [85]. Figure 20.38 also illustrates the scheme of this three-phase composite, which more clearly demonstrates the core–shelllike structure, where the magnetic core is represented by large (75–150 μm) Co_2Z hexaferrite particles surrounded by smaller (1–3 μm) conducting particles of Ni embedded in a polymer. The proportion of components in the composite is as follows: PVDF (31–33 vol.%), Co_2Z (60 vol.%), Ni (7–9 vol.%).

Thus, the increase in the μ_{eff} of a three-phase composite is most likely to be attributed not to the mutual interaction of the magnetic components of fillers, as is

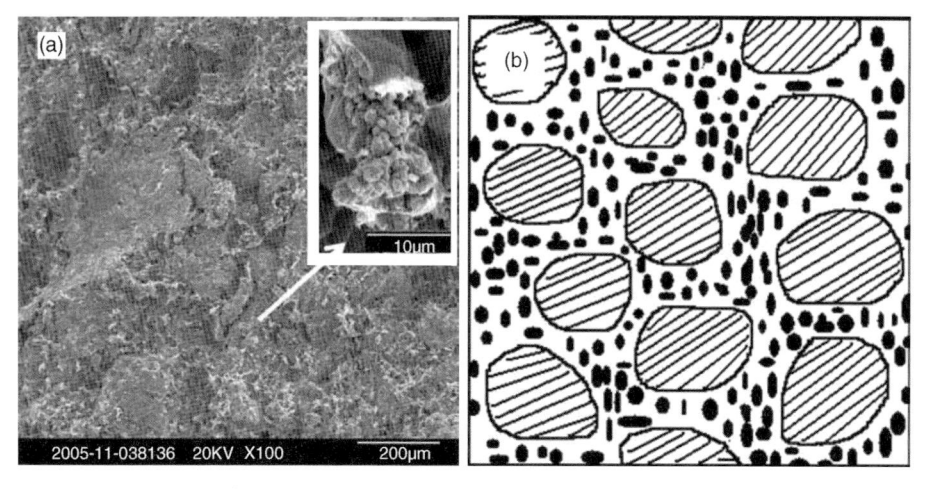

Figure 20.38 (a) SEM photograph of PVDF filled with Co_2Z (60 vol.%) and Ni (9 vol.%) and (b) schematic picture of composite with multicomponent filler. Reprinted with permission from Ref. [85]. Copyright 2006, American Institute of Physics.

shown in [82–86], but to the effect of the "conducting shell" on the magnetization processes in the composite under an AC magnetic field (see Section 20.6.2).

From the practical viewpoint, composites with multicomponent filler and core–shell structure have the following important property: the character of the frequency dispersion of the complex dielectric permittivity and the magnitude of the effective permittivity of a composite can be controlled by choosing the type and the concentration of electrically conducting filler in the mixture (Figure 20.40); this property may be crucial for the design of radioabsorbers (RA) on the basis of such composites. The reason is that these systems allow one to alter the electromagnetic properties of materials, namely to find frequency interval in which ε and μ have close values, which

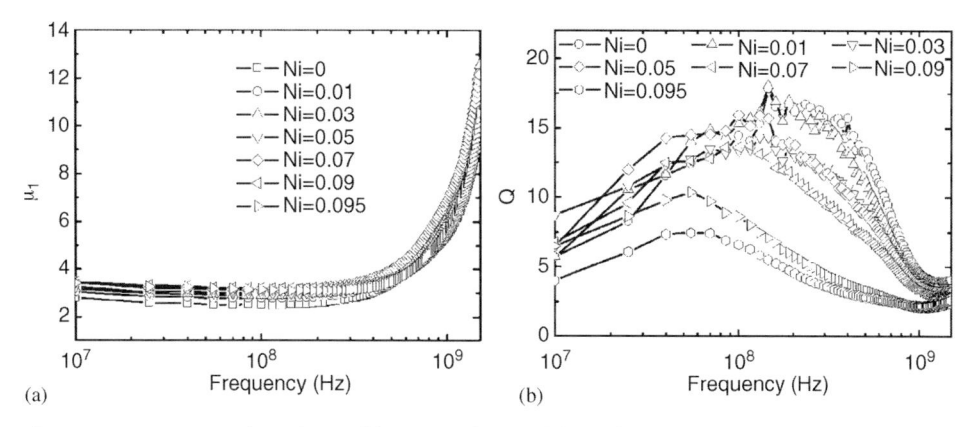

Figure 20.39 Frequency dependence of the (a) initial permeability and (b) quality factor Q for the composites based on PVDF filled with Co_2Z (60 vol.%) and different concentration of Ni. Reprinted with permission from Ref. [85]. Copyright 2006, American Institute of Physics.

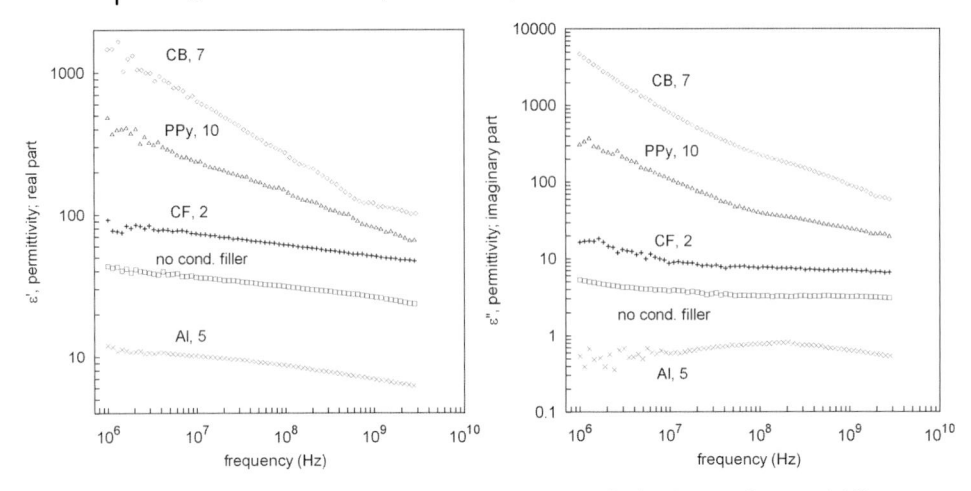

Figure 20.40 Dielectric spectra of polymer composites filled with MnZn ferrite and different types of conducting filler.

meet the desire requirements such as the matching condition in operating frequency range of RA. Thus, the matching frequency of RAs based on HCs can be effectively controlled in the radiofrequency range through an appropriate choice of the type of conducting filler (Figure 20.41) [87].

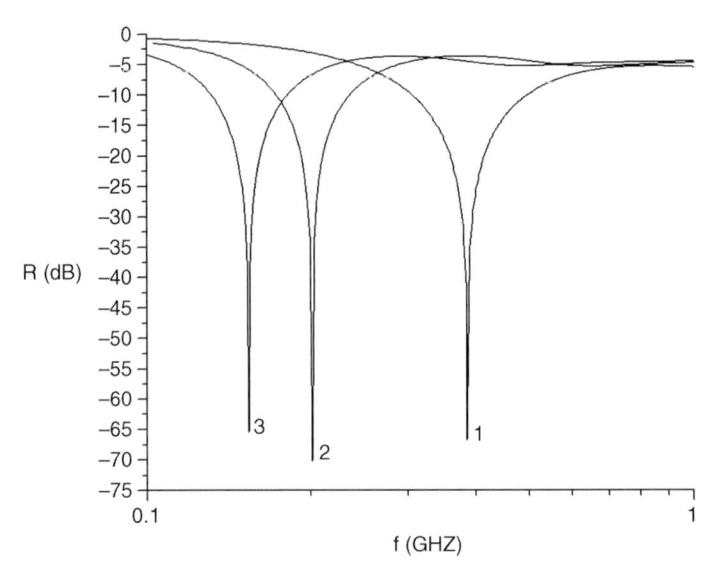

Figure 20.41 Frequency dependence of reflection coefficient, R, of metal-backed, single-layer radioabsorbers based on PMCs: 1—40 vol.% of MnZn ferrite; 2—40 vol.% MnZn ferrite and 2 wt% carbon fibers; 3—40 vol.% MnZn ferrite and 7 wt% Al.

20.6.2
Polymer Magnetic Composites with Multicomponent Magnetic Particles

Composites filled with multicomponent magnetic particles – MnZn ferrite particles coated with nanolayers of conducting polymers (polyaniline, PANI, and polypyrrole, PPy) possess the properties of corelike structure [59, 91], and the μ_{eff} of these composites exhibits behavior similar to that described in Section 20.6.1. The main parameters of these particles that determine the electromagnetic properties of the composites are the particle size and the thickness and the conductivity of the polymer shell. Furthermore, it was demonstrated that the thickness of the polymer shell and its conductivity depend on the reaction conditions and thus can be taken under control [59, 88, 91]. Therefore, before proceeding to the analysis of the magnetic properties of composites with corelike structure particles, one should dwell on the specific features of the manufacturing technology of particles with the required microstructure.

20.6.2.1 Preparation of Magnetic Particles with Corelike Structure

The manufacturing technology of magnetic particles with corelike structure is based on the deposition of electrically conducting polymers on the surface of particles by electrochemical and chemical *in situ* methods [59, 91, 93–96]. In order to form a polymer film on the surface of ferrite particles, ferrite powder is immersed in a reaction (polymerization) medium; the mass concentration of ferrite is chosen to be two or three times that of monomer (aniline or pyrrole). The formation of a polymer chains during the reaction is associated with the chemical or electrochemical oxidation of aniline (pyrrole). Figure 20.42 illustrates the oxidation polymerization of aniline in which the role of the oxidant is played by ammonium peroxydisulfate (standard polymerization [97]).

The polymerization reaction in acidic aqueous media consists of several stages: (1) induction period, associated with the formation of low molecular weight products; (2) exothermal process, associated with the rapid growth of polymer chains; and (3) postpolymerization period [97].

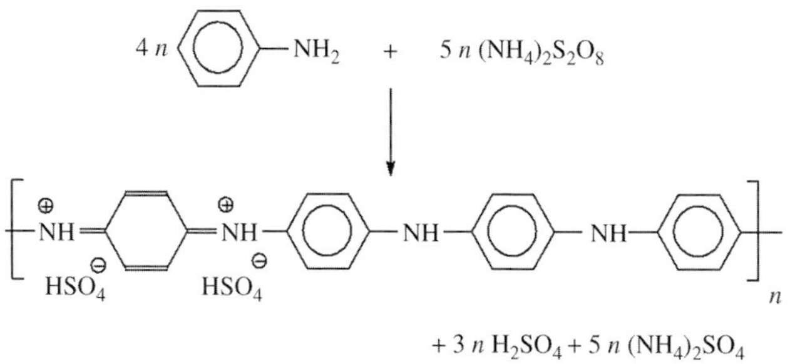

Figure 20.42 Stoichiometry of aniline oxidation with ammonium peroxydisulfate to polyaniline hydrogen sulfate in an acidic medium.

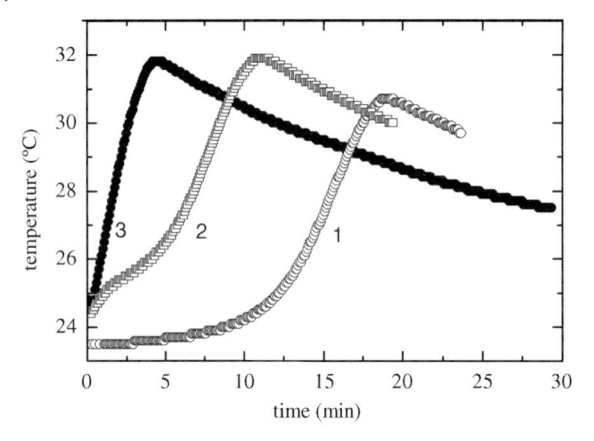

Figure 20.43 Temperature profile in the oxidative polymerization of aniline (0.2 mol/L) with ammonium peroxydisulfate (0.25 mol/L) in hydrochloric acid: (1) without MnZn ferrite; (2) and (3) in the presence of 200–250 μm and 40–60 μm MnZn ferrite particles, respectively.

The introduction of ferrite into the reaction medium changes the kinetic parameters of polymerization. Figure 20.43 demonstrates the temperature profile of polymerization of aniline in the presence of MnZn ferrite particles. The larger is the surface area of coated substrates, the faster is the oxidation process.

The sharp increase in the rate of oxidative polymerization of aniline in the presence of ferrite is associated with its heterophase character and indicates that the polymer chains grow on the interphase boundary between the reaction solution and the surface of ferrite particles; the latter fact is responsible for the formation of the core-shell structure of the particles [98]. The thickness of the polymer shell and its conductivity depend on the reaction conditions. These include, in particular, the chemical nature of the oxidants, the nature of the acids that protonate the aniline (pyrrole) and the reaction intermediates during the oxidation, the concentration of the reactants (aniline (pyrrole) and the oxidant) and their molar proportions, temperature, solvent components, the presence of additives (surfactants), and so on [99].

The shell thickness can be varied by changing of the monomer/oxidant ratio, as well as the concentration of ferrite in the reaction mixture. The most effective method, however, is the control of the reaction temperature. The film thickness on the surface of a particle is judged by the thickness of a film grown on spectrally transparent glass substrates placed in the reaction medium together with ferrite. The film thickness is calculated from the optical absorption. It was established that the film thickness increases as the reaction temperature decreases: the thickness of a film obtained at 50 °C is 50–60 nm, at 20 °C, 170–200 nm, and at 0 °C, 250–300 nm [100].

The conductivity of polymeric or oligomeric PANI (PPy) overlayer obtained during polymerization varies within $10^{-10}-10$ S/cm, and depends mostly on the pH of the reaction mixture and the strength of the selected acids (hydrochloric, sulfuric, phosphoric, picric, acetic, succinic, etc.) [101–103]. Alternatively, the conductivity

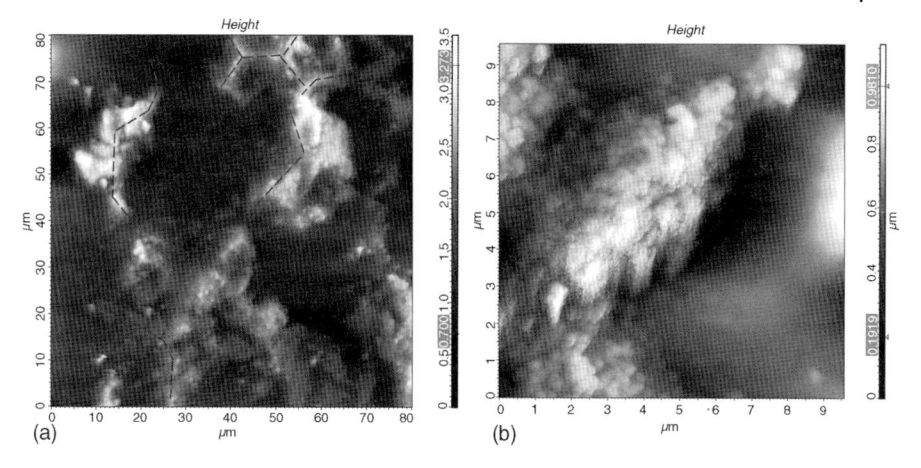

Figure 20.44 Atomic force micrographs of the surface of an islandlike structure of a PANI thin film on MnZn ferrite surface; scan: (a) 80 μm × 80 μm; (b) 9.5 μm × 9/5 μm. The dashed lines show the direction of PANI film growth.

of the shell can be monitored in broad range by deprotonating and followed steplike protonation of PANI (PPy) by acids of different strength [59, 101, 102], as well as by combination of PANI with conducting noble metals (silver or gold) [104, 105].

20.6.2.2 The Mechanism of PANI Film Formation

The morphology of *in situ* growth of PANI films on a single-crystal MnZn ferrite at the different stages of film growth (induction period and polymer chains development) was carried out by an atomic force microscopy (AFM). The internal structure of PANI films was studied in comparison with the surface relief and its phase contrast using amplitude modulation atomic force microscopy (AM AFM) technique [106, 107]. The results of AFM and AM AFM show that a discontinuous PANI film with unusual honeycomb structure is formed on the ferrite surface (Figure 20.44) that turns into a continuous polymer layer only at the final stage of synthesis (Figure 20.45).

The behavior of PANI film growth on any substrate is determined by molecular mechanism of aniline polymerization based on the formation of oligomers containing phenazine cyclic structures at the early stages of aniline oxidation (during the induction period) [99]. Phenazin-containing oligomers are hydrophobic because the nitrogen atoms that are associated within them are not protonated. Such oligomers are insoluble in water, and they separate from aqueous phase by absorbing and/or association at any available interface. The oligomers adsorbed on various types of surfaces immersed in the reaction mixture are present in their oxidized form, most likely as cation radicals. Such paramagnetic cation radicals serve as the initiation centers (nucleation centers) for the subsequent growth of PANI chains and give rise to thin films formation during the subsequent polymerization. Depending on the reaction conditions and the nature of template (hydrophobic or hydrophilic, magnetic, conducting, etc.), the sorption of oligomers can be random, selective or

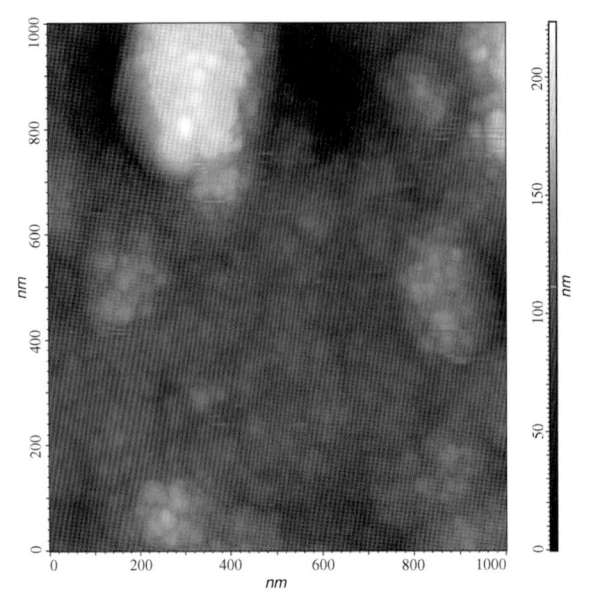

Figure 20.45 Atomic force micrographs of a globular structure of a PANI thick film on MnZn ferrite surface: scan 1 μm × 1 μm.

organized. As a result, PANI films in different morphology, density, and uniformity can be formed.

According to AFM results, discrete PANI film on the surface of MnZn ferrite is formed at the earlier stage of polymerization. It represents regular network of connected meshes with the size of tens microns (Figure 20.44). The analysis of this structure allow to mark out certain directions of film growth as well as to determine the angles between different directions: it is lines intersect at the 90° and 120° angles. Such morphology of thin film is not typical for PANI deposited on the other types of substrates, for example, polymers and glasses, when PANI islands with globular structure are spread randomly on the surface. The morphology of *in situ* grown PANI film on MnZn ferrite surface provide a means that ferrite have an impact on PANI film growth. In contrast to nonmagnetic substrates, ferrite surface absorbs nucleates selectively that manifest itself in formation of regular network from PANI islands. Therefore, the question arises which sites of the ferrite surface will be more favorable for absorption of nucleates.

MnZn ferrite has close stripe domain structure in the absence of magnetic field, which is directly visible in the Lorentz microscope image of the of MnZn ferrite (Figure 20.46) [108]. The white lines and black bands marked with arrowheads correspond to the grain boundary and domain walls, respectively. Figure 20.46 shows that MnZn ferrite has a mean grain size of approximately 10 μm and most of the domain walls exist on its grain boundaries.

If correlate data of Figures 20.44 and 20.46, it is obvious that adsorption of paramagnetic nucleates take place on the grain boundaries and/or domain walls and

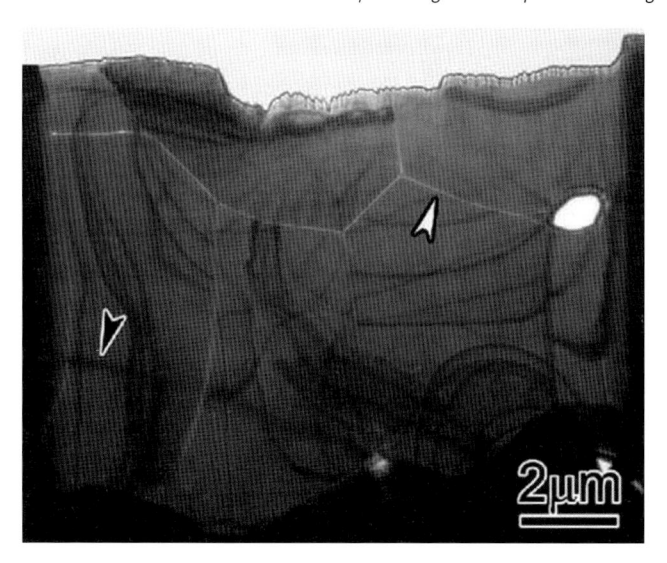

Figure 20.46 Lorentz microscope images of MnZn ferrite. The white lines and black bands marked with arrowheads correspond to the grain boundary and domain walls, respectively. Reprinted with permission from Ref. [108]. Copyright 2006, with permission from Elsevier.

then PANI chains start to grow. Moreover, if take into account that domains walls posses considerable magnetic inhomogeneity related to the rotation of magnetic moment when passing from one domain to another, the nucleates is predominantly anchored on the domain walls.

The analysis of the structure of a finite PANI film on different scales shows that it consists of closely packed spherically symmetric globules with a characteristic size of about 200 nm, each of which, in turn, consists of particles with a size of about 50 nm (Figure 20.47).

20.6.2.3 Electromagnetic Properties of Composites with Core-shell Structure Magnetic Particles

Figure 20.48 represents the magnetic spectra of composites based on polyurethane filled with MnZn ferrite particles (30–40 μm) coated with polypyrrole. According to thermogravimetric analysis, the concentration of PPy in the composites was 10 and 15 wt%.

The magnetic spectra show an increase in μ_{eff} at 15 wt% of PPy. According to SEM data, this concentration corresponds to the formation of a continuous (defect-free) PPy layer on the surface of particles and the formation of particles with core–shelllike structure (Figure 20.49).

In fact, the magnitude of μ_{eff} is determined by the thickness of the conducting polymer overlayer (PPy and PANI) rather than by the concentration of PPy or PANI. Figure 20.50 shows the magnetic spectra of MnZn–PANI (compressed samples) with different thickness of the PANI overlayer on MnZn ferrite particles. The PANI film

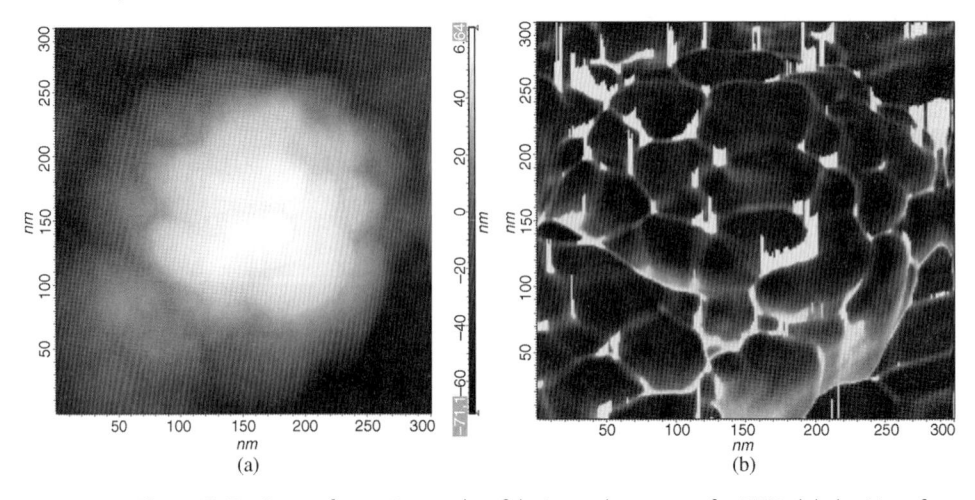

Figure 20.47 Atomic force micrographs of the internal structure of a PANI globule: (a) surface topography and (b) phase contrast.

thickness was varied during the synthesis of MnZn–PANI particles from 50 to 250 nm by changing the reaction temperature.

One can see that μ_{eff} (both μ' and μ'') shows a significant increase with the film thickness. A considerable increase in μ_{eff} occurs during the formation of a continuous film, whose thickness, according to AFM data amounts to 170–250 nm and has a globular structure [91]. The μ' for composites with a thin film (50 nm) are a half for thick films (170 and 250 nm). The maximum of magnetic losses have also been halved, the resonance region being appreciably broadened and resonance frequency is little

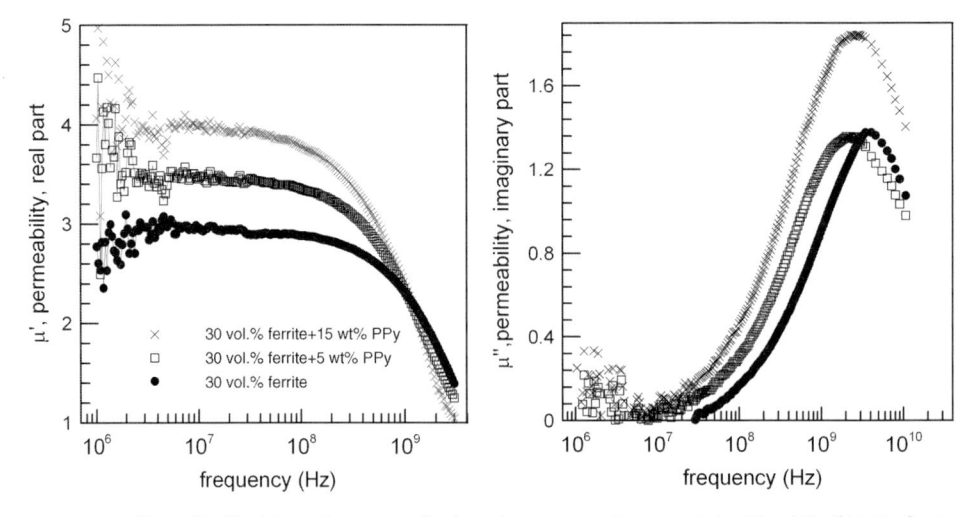

Figure 20.48 Magnetic spectra of polyurethane composites containing 30 vol.% of MnZn ferrite particles coated with polypyrrole. The content of PPy in composite is given for each individual curve.

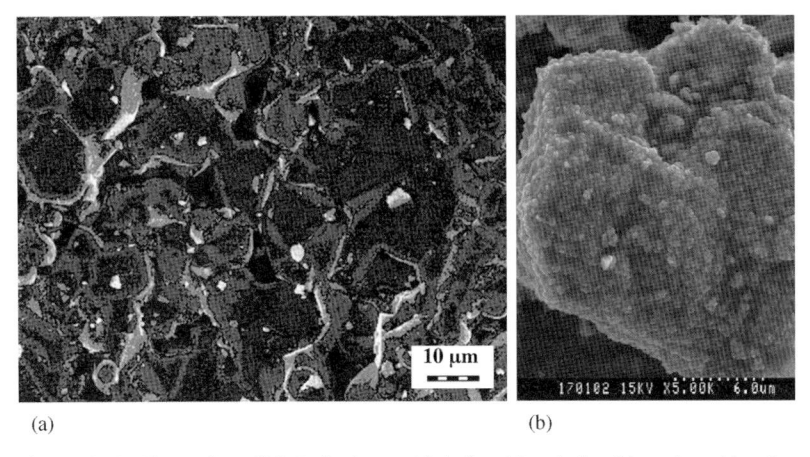

Figure 20.49 The surface of MnZn ferrite particle before (a) and after (b) coating with polypyrrole (15 wt%).

higher compared to composites with thick films. In the case of thin film, the decrease of μ_{eff} and the shift of permeability dispersion to high-frequency range are due to the effect of domain-wall pinning related to the adsorption of nucleates on the domain walls surface. Therefore, the discontinuous structure of a thin film (Figure 20.44) serves as a potential barrier against the motion of domain wall under the AC magnetic field. As a result, the low-frequency part of permeability associated with the motion of domain walls, decreases, the magnetization process associate mainly with magnetization rotation, thus, the resonance frequency (maximum of magnetic losses) has been observed at higher frequency. As the film thickness increases, individual islands merge together to form a continuous film (Figure 20.45). This smoothens the potential barrier, the characteristic geometrical defects on the surface of ferrite a particle (pores, cracks, surface roughness), and the domain walls move more freely.

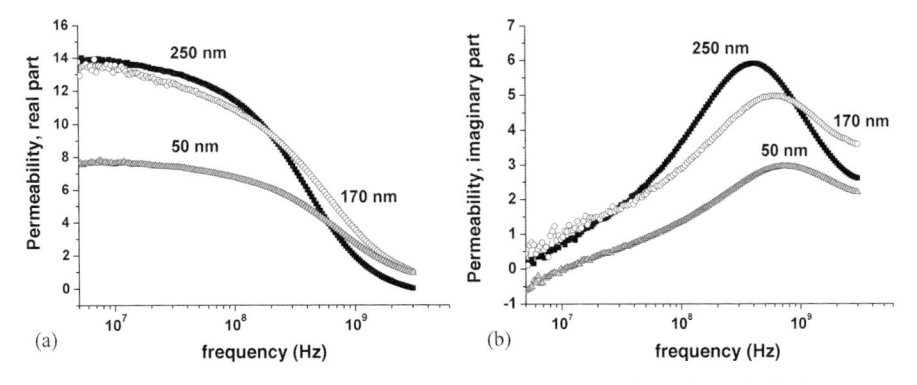

Figure 20.50 Magnetic spectra of MnZn–PANI composites (compressed samples). The thickness of the PANI conductive overlayer varies from 50 to 250 nm.

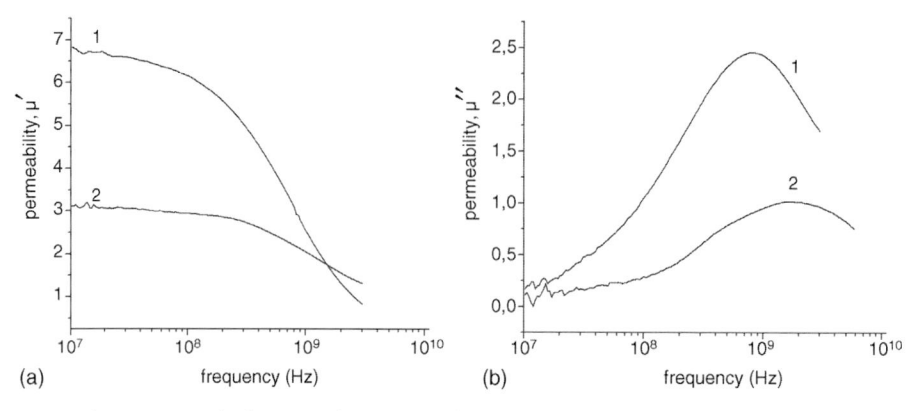

Figure 20.51 The frequency dependence of the real (a) and imaginary (b) parts of the complex permeability of a polyurethane composite containing 50 vol.% of 60 μm MnZn ferrite particles; (1) no coating and (2) coated with polyaniline with $\sigma = 4.4$ S/cm.

Another factor that determines the μ_{eff} of composites with multicomponent particles with core–shelllike structure is the conductivity of the shell. Figure 20.51 shows the magnetic spectra of composites based on polyurethane filled with MnZn ferrite particles (40–60 μm) coated with high-conductivity PANI. For comparison, the same figure represents the spectrum of a composite with pristine ferrite particles of the same size and concentration. The high-conductivity shell reduces the μ_{eff} of the composite and leads to a significant shift in the resonance frequency f_r from the MHz to GHz bands. For example, while the composites filled with pristine particles of MnZn ferrite have $\mu' \sim 7$ and the maximum of magnetic losses of $\mu'' \sim 2.5$ at $f_r \sim 900$ MHz, the composites filled with MnZn ferrite particles coated with PANI film with conductivity of 4.4 S/cm have $\mu' \sim 3$ and the maximum of magnetic losses of $\mu'' \sim 1$ at $f_r \sim 3$ GHz. The lower values of μ_{eff} in the composites filled with MnZn ferrite particles coated with a high-conductivity PANI film correlate with the lower values of the absolute value of the saturation magnetization M_s in these composites; however, the variation in M_s is small and amounts to about 10 G (Figure 20.52).

The lower value of μ_{eff} in a composite with MnZn ferrite particles coated with a high-conductivity PANI film is associated with the partial screening of the high-frequency electromagnetic field on the MnZn–PANI interphase. By the conductivity of a polyaniline overlayer one usually means the DC conductivity of pure PANI obtained under the same conditions as MnZn–PANI particles. However, the actual value of the conductivity of the PANI film on the interface of a ferrite particle can be much higher due to the perpendicular orientation grows of PANI chains on the surface of the particle according to the AFM results, while the conductivity of conjugated polymers along a polymer chain is, as it is known, much higher [109].

In addition, the MnZn–PANI interphase boundary is also responsible for the increase in the complex permittivity of composites due to the interphase polarization. The core–shell interfacial polarization manifests itself as the dielectric-loss peak at about 1 GHz on the frequency dependence of the permittivity of MnZn–PANI (compressed samples) (Figure 20.53).

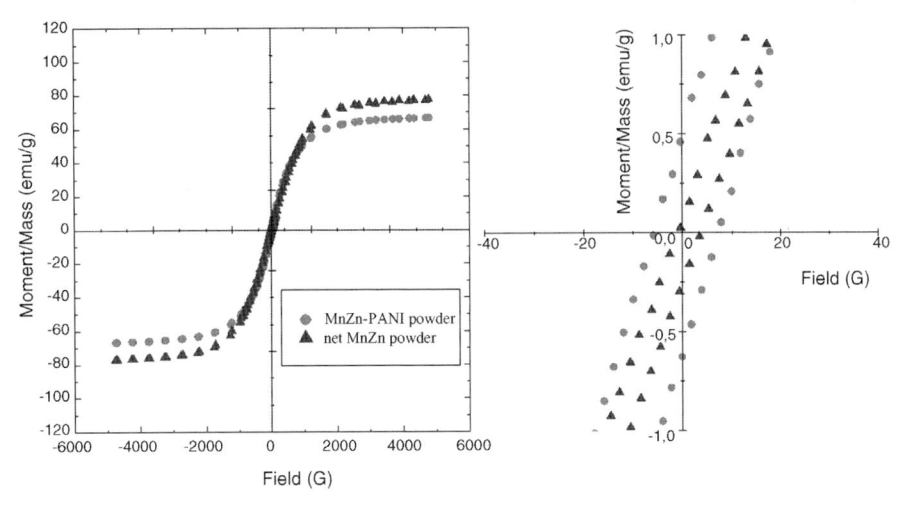

Figure 20.52 Magnetization curve of MnZn ferrite powders with and without polyaniline conducting coating.

In order to explain the observed behavior of the permeability and the resonance frequency in a polymer magnetic composite with a core–shell structure, in [88] was analyzed the effect of the conducting overlayer (shell) on the uniform ferromagnetic resonance in a spherical ferrite particle, and on the effective value of magnetic permeability μ_{eff} of an ensemble of such particles embedded into a nonmagnetic matrix.

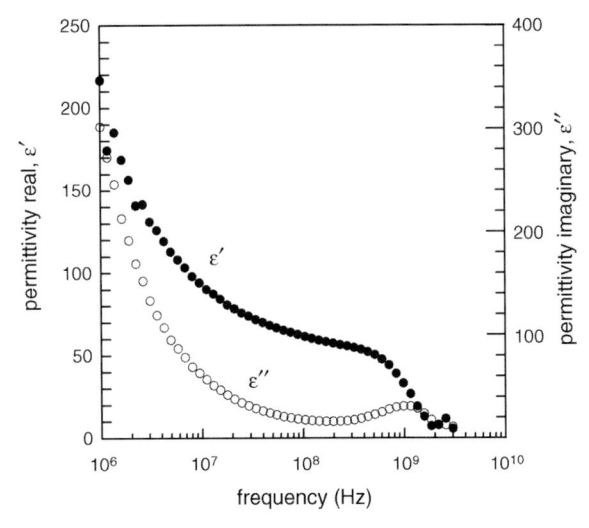

Figure 20.53 The frequency dependence of the complex permittivity of MnZn–PANI composites (compressed sample).

Theoretical analysis shows that the effect of a conducting overlayer on μ_{eff} is determined by a dimensionless quantity ξ,

$$\xi = \frac{d \cdot R_f}{\delta^2} \tag{20.28}$$

where d is the thickness of the conducting overlayer, R_f is the radius of a magnetic particle, $\delta_s = (2\pi\sigma|\omega|/c^2)^{-1/2}$ is the skin depth, σ is the conductivity of the conducting shell, $\omega = 2\pi f$ is the angular frequency, and c is the velocity of light.

Depending on conductivity, the modulus of ξ varies from 0 to ∞. A thin conducting overlayer partially screens the high-frequency field produced by the AC magnetization of a ferrite particle. This leads to the increase in the dipole field energy and, hence, in the frequency of uniform ferromagnetic resonance of a particle. The shift in the resonance frequency increases with the conductivity and the thickness of the overlayer; the latter factors, however, reduce the coupling between the magnetization oscillations and the external electromagnetic field, thus leading to the decrease in μ_{eff}. For the overlayer thickness of about the skin depth, the shift in the uniform ferromagnetic resonance frequency amounts to about $4\pi|\gamma|M_s$, where γ is the gyromagnetic ratio and M_s is the saturation magnetization of the ferrite. If the skin depth of the conducting shell is greater than the size of the magnetic particle, then its effect reduces to an increase in losses. Thus, depending on the relation between the parameters δ, R_f, and d, the conducting overlayer manifests itself in the shift of the ferromagnetic resonance frequency and (or) in the enhanced magnetic losses in the system.

20.7
Conclusions

Polymer magnetic composites represent a well-studied class of magnetic materials. They have a wide range of potential applications, because by choosing the type of magnetic filler and its concentration in a polymer and by varying the shape, size, and the microstructure of particles and their distribution in the polymer, one can design PMCs with high values of complex permeability in the microwave band. However, since the permeability of a PMC depends on many material parameters, a theory that would predict the frequency dependence of the permeability of a PMC on the basis of the material parameters of its components with a necessary degree of accuracy has not yet been developed. The only reliable method is the measurement of the electromagnetic parameters of PMCs over a wide range of frequencies by combining different methods (impedance, transmission/reflection, resonant, free-space methods, etc.).

The analysis of the electromagnetic properties of PMCs loaded with different types of magnetic fillers has shown as follows. The magnetic spectra of CI-filled composites are primarily determined by the microstructure of CI particles. For example, a PMC based on CI particles with polycrystalline structure has high values of permeability in the long-wavelength part of the microwave band, whereas a PMC with onionlike particles – in the centimeter-wave region of the microwave band. PMCs filled with ferrites with the spinel structure are characterized by the frequency dispersion of permeability in the long-wavelength part of the microwave band. The permeability dispersion region of PMCs based on easy-plane hexaferrites lies in the frequency range from 1 to 10 GHz. A remarkable advantage of PMCs filled with MnZn ferrite with high value of the initial permeability and low magnetic anisotropy is that they allow to control the permeability and make it possible to shift the region of permeability dispersion from the MHz to the GHz band by coating ferrite particles with conducting polymers (polyaniline or polypyrrole). One of promising directions in the design of PMCs with high values of permeability in the radiofrequency and microwave bands is the use of multicomponent fillers and the formation of a core–shelllike structure of composite.

Acknowledgment

The book chapter was written with the support of Operational Program Research and Development for Innovations co-funded by the European Regional Development Fund (ERDF) and national budget of Czech Republic, within the framework of project Centre of Polymer Systems (reg. number: CZ.1.05/2.1.00/03.0111). The author is also thankful to Prof. Petr Saha, Rector of Tomas Bata University in Zlin for providing excellent facilities to perform the research activities. She is also pleased to express her gratitude to her colleagues and PhD students for collaboration and support.

References

1 Brandrup, J., Immergut, E.H., and Grulke, E.A. (2004) *Polymer Handbook*, 4th edn, John Wiley & Sons, Inc., New York.

2 Letyuk, L.M., Balbashov, A.M., Krutogin, D.G., Gonchar, A.V., and Kudriashkin, I.G. (1994) *Production Engineering of Materials for Magneto-Electronics*, Metallurgy, Moscow.

3 Product information: Microwave absorbing materials, catalog, Laird Technologies, 2004.

4 Product information: Electromagnetic wave absorbers and ferrite sheets for EMI suppression, FE21051-1212-010, FDK corporation.

5 Product information: TDK Electromagnetic absorbers, TDK RF Solutions.

6 Hosoe, A. and Nitta, K. (2005) Electromagnetic wave absorber, US.

7 Kanda, K., Morimoto, M., Junichi, H., and Takumi, F. (2001) Electromagnetic wave absorbing material, US.

8 Toyodo, J., Iwashita, S., and Okayama, K. (2001) Radiowave absorbent and manufacturing method thereof, US.

9 Neelakanta, P.S. (1995) *Handbook of Electromagnetic Materials: Monolithic and Composite Versions and their Applications*, CRC Press, Boca Raton, London.

10 Suvajic, J. (2006) Absorber for mobile phones and production method.

11 Kazantseva, N.E., Ponomarenko, A.T., Shevchenko, V.G., and Klason, C. (2000) *Electromagnetics*, **20**, 453.

12 Volkov, V.S., Syrkin, V.G., and Tolmasskii, I.S. (1969) *Carbonyl Iron*, Metallurgy, Moscow.

13 Product information: Carbonyl iron powder, G-CAS/BS0106 CIP2, BASF, 2006.

14 Zhang, B.S., Feng, Y., Xiong, H., Yang, Y., and Lu, H.X. (2006) *IEEE Transactions on Magnetics*, **42**, 1778.

15 Wu, L.Z., Ding, J., Jiang, H.B., Chen, L.F., and Ong, C.K. (2005) *Journal of Magnetism and Magnetic Materials*, **285**, 233.

16 Abshinova, M.A., Lopatin, A.V., Kazantseva, N.E., Vilcakova, J., and Saha, P. (2007) *Composites Part A: Applied Science and Manufacturing*, **38**, 2471.

17 Abshinova, M.A., Kuritka, L., Kazantseva, N.E., Vilcakova, J., and Saha, P. (2009) *Materials Chemistry and Physics*, **114**, 78.

18 Kim, I., Bae, S., and Kim, J. (2008) *Materials Letters*, **62**, 3043.

19 Lin, G.Q., Li, Z.W., Chen, L., Wu, Y.P., and Ong, C.K. (2006) *Journal of Magnetism and Magnetic Materials*, **305**, 291.

20 Wu, L.Z., Ding, J., Jiang, H.B., Chen, L.F., and Ong, C.K. (2005) *Journal of Magnetism and Magnetic Materials*, **285**, 233.

21 Abshinova, M.A., Kazantseva, N.E., Saha, P., Sapurina, I., Kovarova, J., and Stejskal, J. (2008) *Polymer Degradation and Stability*, **93**, 1826.

22 Smit, J. and Wijn, H.P.J. (1959) *Ferrites: Physical Properties of Ferrimagnetic Oxides in Relation to Their Technical Applications*, Philips, Eindhoven.

23 Middelhoek, S. (1971) *Magnetic Properties of Materials*, McGraw Hill, New York.

24 Vonsovskii, S.V. (1971) *Magnetism*, Nauka, Moskow.

25 Snoek, J.L. (1948) *Physica*, **14**, 207.

26 Kim, D.Y., Chung, Y.C., Kang, T.W., and Kim, H.C. (1996) *IEEE Transactions on Magnetics*, **32**, 555.

27 Tsutaoka, T. (2003) *Journal of Applied Physics*, **93**, 2789.

28 Tsutaoka, T., Ueshima, M., Tokunaga, T., Nakamura, T., and Hatakeyama, K. (1995) *Journal of Applied Physics*, **78**, 3983.

29 Nakamura, T. (2000) *Journal of Applied Physics*, **88**, 348.

30 Haque, M.M., Huq, M., and Hakim, M.A. (2008) *Materials Chemistry and Physics*, **112**, 580.

31 Silvestrovich, M.I. (1970) Microwave ferrite in low fields, Moscow.

32 Nakamura, T., Miyamoto, T., and Yamada, Y. (2003) *Journal of Magnetism and Magnetic Materials*, **256**, 340.

33 Jonker, J.M., Wijn, H.P.J., and Brown, P.B. (1956–1957) *Philips Technical Review*, **18**, 145.

34 Rozanov, K.N., Li, Z.W., Chen, L.F., and Koledintseva, M.Y. (2005) *Journal of Applied Physics*, **97**, 7.

35 Tachibana, T., Nakagawa, T., Takada, Y., Shimada, T., and Yamamoto, T.A. (2004) *Journal of Magnetism and Magnetic Materials*, **284**, 369.

36 Nakamura, T. and Hankui, E. (2003) *Journal of Magnetism and Magnetic Materials*, **257**, 158.

37 Li, Z.W., Chen, L., Wu, Y., and Ong, C.K. (2004) *Journal of Applied Physics*, **96**, 534.

38 Li, Z.W., Chen, L.F., and Ong, C.K. (2003) *Journal of Applied Physics*, **94**, 5918.

39 Cho, H.S. and Kim, S.S. (1999) *IEEE Transactions on Magnetics*, **35**, 3151.

40 Nedkov, W., Cheparin, V.P., and Khanamirov, A. (1988) *Journal De Physique*, **49**.

41 Chen, L.F., Ong, C.K., Neo, C.P., Varadan, V.V., Varadan, V.K. (2004) *Microwave Electronics: Measurement and Materials Characterization*, John Wiley & Sons, Inc., New York.

42 Basics of measuring the dielectric properties of materials (application note), Agilent Technologies, 2005.

43 Apletalin, V.N., Kazantsev, Y.N., and Solosin, V.S. (2008) Cavity cell for measurements of electromagnetic characteristics of materials at 1-3GHz, in *Radiolocation and Communication*, Moscow.

44 Brown, W.F. (1962) *Magnetostatic Principles in Ferromagnetism*, North-Holland Pub. Co./Interscience Publishers, Amsterdam/New York.

45 Goodenough, J.B. (2002) *IEEE Transactions on Magnetics*, **38**, 3398.

46 Gurevich, A.G. and Melkov, G.A. (1996) *Magnetization Oscillations and Waves*, CRC, Boca Raton.

47 Lagarkov, A.N., Rozanov, K.N., Simonov, N.A., and Starostenko, S.N. (2005) *Handbook of Advanced Magnetic Materials*, Springer, Beijing.

48 Landau, L.D. and Lifshitz, E.M. (1935) *Phys Z Sowietunion*, **8**, 153.

49 Gilbert, T.L. (2004) *IEEE Transactions on Magnetics*, **50**, 3443.

50 Lagarkov, A.N. and Rozanov, K.N. (2009) *Journal of Magnetism and Magnetic Materials*, **321**, 2082.

51. Rado, G.T. (1953) *Reviews of Modern Physics*, **25**, 81.
52. Jankovskis, J. (2006) *Journal of Magnetism and Magnetic Materials*, **304**, e492.
53. Spenato, D., Fessant, A., Gieraltowski, J., Le Gall, H., and Tannous, C. (1999) *Journal of Applied Physics*, **85**, 6010.
54. Jayasekara, W.P., Bain, J.A., and Kryder, M.H. (1998) *IEEE Transactions on Magnetics*, **34**, 1438.
55. Lagarkov, A.N., Lakubov, I.T., Ryzhikov, I.A., Rozanov, K.N., Perov, N.S., Elsukov, E.P., Maklakov, S.A., Osipov, A.V., Sedova, M.V., Getman, A.M., and Ulyanov, A.L. (2007) *Physica B*, **394**, 159.
56. Nakamura, T., Tsutaoka, T., and Hatakeyama, K. (1994) *Journal of Magnetism and Magnetic Materials*, **138**, 319.
57. Kittel, C. (1946) *Physical Review*, **70**, 965.
58. Kourov, D.N., Kourov, N.I., and Tyulenev, L.N. (1998) *Physics of the Solid State*, **40**, 1723.
59. Kazantseva, N.E., Vilcakova, J., Kresalek, V., Saha, P., Sapurina, I., and Stejskal, J. (2004) *Journal of Magnetism and Magnetic Materials*, **269**, 30.
60. Slama, J., Vicen, R., Krivosik, P., Gruskova, A., and Dosoudil, R. (1999) *Journal of Magnetism and Magnetic Materials*, **197**, 359.
61. Moucka, R., Lopatin, A.V., Kazantseva, N.E., Vilcakova, J., and Saha, P. (2007) *Journal of Materials Science*, **42**, 9480.
62. Fiske, T.J., Gokturk, H.S., and Kalyon, D.M. (1997) *Journal of Materials Science*, **32**, 5551.
63. Mattei, J.L. and Le Floc'h, M. (2003) *Journal of Magnetism and Magnetic Materials*, **257**, 335.
64. Mattei, J.L. and Le Floc'h, M. (2003) *Journal of Magnetism and Magnetic Materials*, **264**, 86.
65. Chevalier, A. and Le Floc'h, M. (2001) *Journal of Applied Physics*, **90**, 3462.
66. Chevalier, A., Mattei, J.L., and Le Floc'h, M. (2000) *Journal of Magnetism and Magnetic Materials*, **215**, 66.
67. Laurent, P., Viau, G., Konn, A.M., Gelin, P., and LeFloc'h, M. (1996) *Journal of Magnetism and Magnetic Materials*, **160**, 63.
68. Paterson, J.H., Devine, R., and Phelps, A.D.R. (1999) *Journal of Magnetism and Magnetic Materials*, **197**, 394.
69. Lefloc'h, M., Mattei, J.L., Laurent, P., Minot, O., and Konn, A.M. (1995) *Journal of Magnetism and Magnetic Materials*, **140**, 2191.
70. Mattei, J.L., Laurent, P., Minot, O., and LeFloc'h, M. (1996) *Journal of Magnetism and Magnetic Materials*, **160**, 23.
71. Mattei, J.L., Minot, O., and Lefloc'h, M. (1995) *Journal of Magnetism and Magnetic Materials*, **140**, 2189.
72. Musal, H.M., Hahn, H.T., and Bush, G.G. (1988) *Journal of Applied Physics*, **63**, 3768.
73. Stratton, J.A. (1941) *Electromagnetic Theory*, McGraw-Hill, New York, London.
74. Rankis, G.Z. (1981) *Dynamics of Magnetisation of Polycrystalline Ferrites*, Zinatne, Riga.
75. Slama, J., Dosoudil, R., Vicen, R., Gruskova, A., Olah, V., Hudec, I., and Usak, E. (2003) *Journal of Magnetism and Magnetic Materials*, **254**, 195.
76. Kazantsev, Y.N., Lopatin, A.V., Kazantseva, N.E., Shatrov, A.D., Maltsev, V.P., Vilcakova, J., and Saha, P. (2010) *IEEE Transactions on Antennas and Propagation*, **58**, 1227–1235.
77. Garnett, J.C.M. (1904) *Philosophical Transactions of the Royal Society of London*, **203**, 385.
78. Bruggeman, D.A.G. (1935) *Annals of Physics*, **24**, 636.
79. Landau, L.D. and Lifsic, E.M. (1960) *Electrodynamics of Continuous Media*, Pergamon, Oxford.
80. Looyenga, H. (1965) *Physica*, **31**, 401.
81. Lichtenekker, K. (1926) *Phys Z S*, **27**, 118.
82. Kasagi, T., Tsutaoka, T., and Hatakeyama, K. (2004) *Journal of Magnetism and Magnetic Materials*, **272**, 2224.
83. Fiske, T.J., Gokturk, H., and Kalyon, D.M. (1997) *Journal of Applied Polymer Science*, **65**, 1371.
84. Shen, Y., Yue, Z.X., Li, M., and Nan, C.W. (2005) *Advanced Functional Materials*, **15**, 1100.
85. Li, B.W., Shen, Y., Yue, Z.X., and Nan, C.W. (2006) *Journal of Applied Physics*, **99**, 6.

86 Anhalt, M. and Weidenfeller, B. (2009) *Materials Science and Engineering: B. Advanced Functional Solid-State Materials*, **162**, 64.

87 Moucka, R., Vilcakova, J., Kazantseva, N.E., Lopatin, A.V., and Saha, P. (2008) *Journal of Applied Physics*, **104**, 11.

88 Bespyatykh, Y.I. and Kazantseva, N.E. (2008) *Journal of Communications Technology and Electronics*, **53**, 143.

89 Masanori, A. (2004) Composite magnetic material prepared by compression forming of ferrite-coated metal particles and method of preparation thereof, US.

90 Kathirgamanathan, P. (1993) *Journal of Materials Chemistry*, **3**, 259.

91 Kazantseva, N.E., Bespyatykh, Y.I., Sapurina, I., Stejskal, J., Vilcakova, J., and Saha, P. (2006) *Journal of Magnetism and Magnetic Materials*, **301**, 155.

92 Weidenfeller, B., Anhalt, M., and Riehemann, W. (2008) *Journal of Magnetism and Magnetic Materials*, **320**, E362.

93 Sapurina, I., Riede, A., and Stejskal, J. (2001) *Synthetic Metals*, **123**, 503.

94 Stejskal, J., Trchova, M., Brodinova, J., Kalenda, P., Fedorova, S.V., Prokes, J., and Zemek, J. (2006) *Journal of Colloid and Interface Science*, **298**, 87.

95 Jiang, J., Ai, L.H., and Li, L.C. (2009) *Journal of Materials Science*, **44**, 1024.

96 Yang, C.C., Gung, Y.J., Hung, W.C., Ting, T.H., and Wu, K.H. (2010) *Composites Science and Technology*, **70**, 466.

97 Stejskal, J. and Gilbert, R.G. (2002) *Pure and Applied Chemistry*, **74**, 857.

98 Stejskal, J. and Sapurina, I. (2004) *Journal of Colloid and Interface Science*, **274**, 489.

99 Sapurina, I. and Stejskal, J. (2008) *Polymer International*, **57**, 1295.

100 Stejskal, J., Sapurina, I., Prokes, J., and Zemek, J. (1999) *Synthetic Metals*, **105**, 195.

101 Stejskal, J., Hlavata, D., Holler, P., Trchova, M., Prokes, J., and Sapurina, I. (2004) *Polymer International*, **53**, 294.

102 Sedenkova, I., Trchova, M., Blinova, N.V., and Stejskal, J. (2006) *Thin Solid Films*, **515**, 1640.

103 Konyushenko, E.N., Trchova, M., Stejskal, J., and Sapurina, I. (2010) *Chemical Papers*, **64**, 56.

104 Stejskal, J., Prokes, J., and Sapurina, I. (2009) *Materials Letters*, **63**, 709.

105 Rapeski, T., Donten, M., and Stojek, Z. (2010) *Electrochemistry Communications*, **12**, 624.

106 Spivak, Yu.M. and Moshnikov, V.A. (2010) *Journal of Surface Investigation: X-ray, Synchrotron and Neutron Techniques*, **4**, 71.

107 Spivak, Yu.M., Moshnikov, V.A., Sapurina, I.Yu., and Kazantseva, N.E. (2010) Annual Proceedings of Technical University, Varna.

108 Kasahara, T., Park, H.S., Shindo, D., Yoshikawa, H., Sato, T., and Kondo, K. (2005) *Journal of Magnetism and Magnetic Materials*, **305**, 165.

109 Reghu, M., Cao, Y., Moses, D., and Heeger, A.J. (1993) *Physical Review B*, **47**, 1758.

21
Mica-Reinforced Polymer Composites
John Verbeek and Mark Christopher

21.1
Introduction

Particulate-reinforced polymer composites are often designed to have improved mechanical or physical properties or at the very least to have a lower overall cost. Mica is a platelike crystalline aluminosilicate and has been widely used as reinforcement in polymer composites. Commercially, mica is used in polymer composites of polyolefins, polyesters, polyamides, epoxies, and polyurethanes because of its excellent mechanical, electrical, and thermal properties.

Flakes or platelets represent a special class of reinforcing fillers for thermoplastics. Mica-filled polymer composites have become attractive because of their wide applications [1, 2]; mica improves various physical properties of the matrix such as mechanical strength, modulus, and heat distortion temperature. Mechanical properties typically depend on the size, shape, and distribution of mica particles in the matrix–polymer as well as interfacial adhesion. Mica can easily be cleaved into thin flakes during grinding or delaminated into very thin flakes. These ultrathin flakes have very high aspect ratios and if suitably aligned in a matrix, they may impart a high degree of reinforcement.

In this chapter, the use of mica in thermoplastic polymers is discussed in terms of its ability to improve mechanical properties, as well as other physical properties such as thermal stability, crystallinity, barrier properties, and processability. Factors affecting these are discussed and the chapter concludes with a discussion on methods to model the mechanical properties of mica-filled polymers.

The chapter mostly concerns thermoplastic microcomposites, although considerable work has also been done on mica-filled thermosetting polymers. Mica has also been used as nanoreinforcement in polymers, but this is beyond the scope of this work.

Polymer Composites: Volume 1, First Edition. Edited by Sabu Thomas, Kuruvilla Joseph, Sant Kumar Malhotra, Koichi Goda, and Meyyarappallil Sadasivan Sreekala
© 2012 Wiley-VCH Verlag GmbH & Co. KGaA. Published 2012 by Wiley-VCH Verlag GmbH & Co. KGaA.

Table 21.1 Mica properties [1, 3].

Property	Advantages
Chemical	Acid, alkali, and solvents resistant even at high temperatures
Mechanical	Good shape stability, high elasticity, high shear and compressive strength, perfect cleavability, incompressible
Electrical	High dielectric strength, high surface and volume resistivity, ability to resist sparking and corona effects
Thermal	High thermal stability, low thermal conductivity, nonflammable, and infusible
Optical	Translucent in thin films, absorbs, and stable to UV
Other	Damp proof, resists soiling

21.2
Structure and Properties of Mica

21.2.1
Chemical and Physical Properties

Silicate minerals make up approximately 90% of the earth's crust and, therefore, are an important class of rock-forming minerals. Phyllosilicates, such as clay, serpentine, and mica, are sheet silicates consisting of a parallel sheetlike structures made up of silicon and oxygen tetrahedra. Mica is a generic term for a series of hydrous potassium aluminum silicates with a highly laminated sheetlike structure. It is distinguished for being highly crystalline, having perfect cleavage between sheet layers, and being chemically inert.

The most common types of mica used commercially are muscovite (white mica), biotite (black mica), and phlogopite (amber mica). Muscovite tends to be aluminum rich, biotite iron rich, and phlogopite contains high levels of magnesium [1, 2].

Muscovite and phlogopite can be ground into platelike or flake particles that are thin, tough, and flexible. They are chemically inert except in hydrofluoric acid and concentrated sulfuric acid. Mica has the unique combination of high dielectric strength and a uniform dielectric constant. Muscovite is UV transparent, while phlogopite is opaque to UV radiation. Mica will not support a flame and has excellent stability to both high and low temperatures [2]. The most important properties of mica are listed in Table 21.1.

21.2.2
Structure

Mica consists of repeating tetrahedral, octahedral, tetrahedral (TOT), and interlayer cation layers (Figure 21.1). Each layer is approximately 9.94 Å or 0.1 nm thick. The general chemical formula for mica is $XY_2\text{-}3Z_4O_{10}(OH,F,Cl)_2$ [1, 3] where X is the cation interlayer, Y is the octahedral cation layer, and Z is the tetrahedral cation layer, as summarized in Table 21.2.

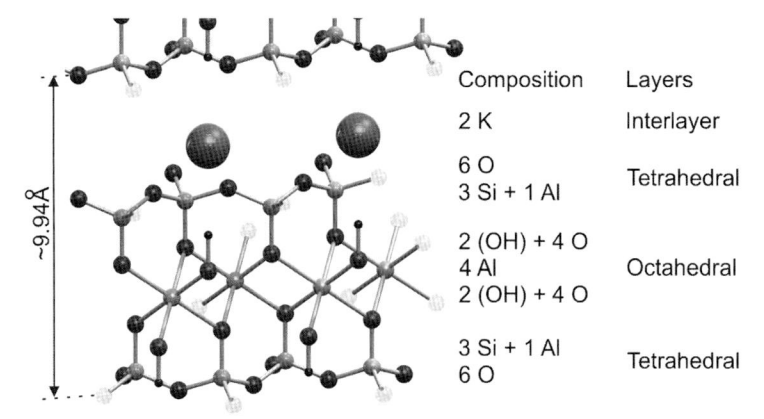

Figure 21.1 General structure of mica, muscovite [3].

Composition	Layers
2 K	Interlayer
6 O 3 Si + 1 Al	Tetrahedral
2 (OH) + 4 O 4 Al 2 (OH) + 4 O	Octahedral
3 Si + 1 Al 6 O	Tetrahedral

(~9.94 Å)

The cations in the octahedral layer are coordinated with six oxygen atoms, forming an octahedral shape (Figure 21.1). The octahedral layer may be occupied by two 3 + cations (e.g., Al^{3+}), making the mica dioctahedral (e.g., muscovite or margarite), or can contain three 2 + cations (e.g., Mg^{2+} or Fe^{2+}) making the mica trioctahedral (e.g., phlogopite or clintonite) [1].

The cations in the tetrahedral layers are coordinated with four oxygen atoms, forming a tetrahedral shape. The tetrahedrons are set out in a flat repeating hexagonal honeycomb lattice (Figure 21.2). This gives mica uniform strength in both axes parallel to the sheet plane. The tetrahedral layer generally consists of three Si^{4+} to one Al^{3+} that gives an overall negative charge to the TOT layer [3].

The negative charge of each TOT layer is balanced by an interlayer of cations such as K^+, Na^+, or Ca^{2+} (Figure 21.1 and Table 21.2). These are in 12-fold coordination with the TOT layers binding the structure together tightly [3].

Table 21.2 Mica types and composition [3].

Structure	Family	Name	X	Y	Z
Dioctahedral	True mica	Muscovite	K	Al_2	Si_3Al
		Paragonite	Na	Al_2	Si_3Al
		Glauconite	$(K, Na)_{0.6-1.0}$	$(Fe, Mg, Al)_2$	Si_3Al
	Brittle mica	Margarite	Ca	Al_2	Si_2Al_2
Trioctahedral	True mica	Phlogopite	K	$(Mg, Fe^{2+})_2$	Si_3Al
		Biotite	K	$(Fe^{2+}, Fe^{3+}, Mg, Al)_2$	$Si_{2.5-3}Al_{1-1.5}$
		Lepidolite	K	$(Li, Al)_{2.5-3}$	$Si_{2.5-3}Al_{1-1.5}$
	Brittle mica	Clintonite	Ca	$(Mg, Al)_3$	$Si_{1.25}Al_{2.75}$

X is the cation interlayer, Y is the octahedral layer, and Z is the tetrahedral layer.

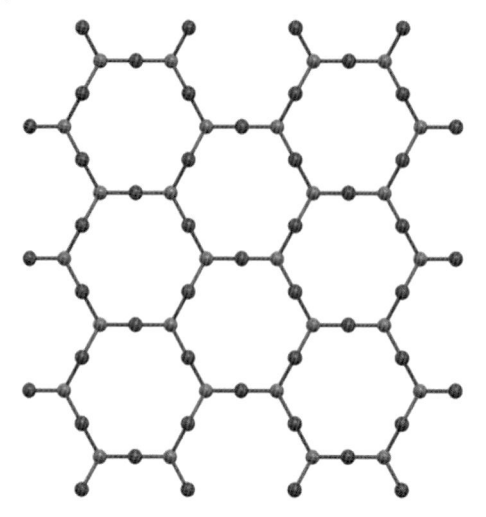

Figure 21.2 Hexagonal layout of the tetrahedral layer in mica.

21.2.3
Applications

India and Brazil are the main producers of muscovite mica, followed by the United States, Tanzania, Zimbabwe, and Argentina. The United States generally produces scrap mica. Malagasy Republic and Tanzania are the main producers of phlogopite, producing roughly 1000 and 300 tons annually, respectively. Smaller producers include Canada, Tanzania, and India [4].

Sheet mica is produced from large crystals and can be produced in thicknesses less than 0.03 mm to greater than 0.18 mm for use as insulation against heat and electricity [5]. It is stable at temperatures up to 700–900 °C, chemically resistant, and able to withstand 1500 V/mm without puncturing [4], making it ideal for the following:

- Viewing windows for stoves, petromax lamps, furnaces, and fireplaces; microscope slides; wave plates for optical instruments; and protective covering for sight glasses and liquid level gauges in high-pressure vessels [5].
- Wrapping for heating electrodes in soldering irons to protect the electrode from molten solder [5].
- Filters for metal foundries to remove impurities from molten metal [5].
- Mounting plates or tubes for resistance wire to be wrapped around for electric heating appliances and fuses [5].
- Mounting washers for diodes, heat sinks, rectifiers, and semiconductors; spacers in vacuum tubes; and backing plates in capacitors [5].

Scrap from producing sheet mica can be dry or wet ground or micronized [1], coated or dyed [6], and used to add sheen/glitter, color and texture to paints, coatings, polymers, printing inks, and rubber [1]. Flakes can be used as an additive in drilling mud in the oil industry to seal up pores in drill holes and prevent particulate settling [5].

Flakes can be reconstituted [1] using shellac, epoxy, alkyd, or silicone [7] to form heat-resistant mica sheets or papers for use as mounting blocks or spacers in low-temperature electrical applications, thermal insulation, and electrical insulation in electrical windings in generators, motors, commutators, electric iron elements, wound heater elements, variable resistors, and high-voltage cables [5, 7].

Mica powder is used as friction material for asbestos-free brake lining, brake pads, brake shoes, brake disks, and clutch facings; fillers in gypsum wallboards and asphalt and bituminous roofing felt and shingles [1]; antisticking and antiblocking compound in tire production for automobiles, motor bikes, and airplanes; coating for casting moulds in foundries; flux coating on welding electrodes to produce meltable slags and prevent weld cracking [1]; flow aid for powder fire extinguishers; and producing refractory bricks for furnaces [5].

Mica powder is widely used as mechanical reinforcement in the following:

- Paints, coatings, and cements to reduce sagging and running, shrinkage, and cracking [8], improve anticorrosive properties, and improve UV resistance [5].
- Rubber to improve resilience, hardness, tensile strength, and tear resistance [5].
- Plastic automobile components such as fascia and fenders, car seat backs, panels, and ignition systems to improve strength, rigidity, and temperature resistance and reduce warping, surface properties, and acoustic properties [5].

The technology behind tailoring physical and mechanical properties of these polymer microcomposites have been extensively researched and is discussed further in subsequent sections.

21.3
Mechanical Properties of Mica–Polymer Composites

Fillers are used as volume extenders or to improve physical and mechanical properties of polymers. Extenders are fillers used to increase the bulk volume of a polymer, while functional fillers improve mechanical or physical properties. These can be further classified as functional additives or reinforcements [9]. Additives are typically used to improve physical properties such as thermal or electrical conductivity, while reinforcements improve mechanical properties such as heat deflection temperature, stiffness, impact strength, and breaking strength [9, 10].

Mica can act as functional filler or as reinforcement. Mica is used as reinforcement because of its outstanding mechanical properties, most importantly its shape. It has a high area-to-volume (A/V) ratio because of its lamellar morphology that allows it to be ground into very thin platelets of high aspect ratio. It is well known that a large aspect ratio allows efficient stress transfer between the matrix and reinforcement, leading to improved mechanical properties. Along with its chemical and UV properties, mica as functional filler can impart some of the following properties:

- improved flame resistance [11],
- lower thermal degradation [12],
- improved UV stability [2],
- improved abrasion resistance [8],
- improved strength and stiffness,
- improved tear resistance [8],
- lower coefficient of thermal expansion (CTE), and
- improved fracture toughness.

It is not always possible to optimize all mechanical properties, such as impact strength, tensile strength, and stiffness independently; some may improve while others decrease [10]. The exact behavior depends on filler and polymer properties and any additives used, resulting in a broad range of mechanical properties (Table 21.3). Typically tensile or flexural strength can be improved only if adequate interfacial adhesion between the polymer and the mica is ensured. Stiffness either in tension or in bending is generally increased, regardless of the amount of filler used. Both strength and stiffness improvements, however, are influenced by factors such as particle size and distribution.

For polymer composites filled with rigid particles, such as mica, the most important mechanism for improving toughness is to induce large plastic deformation by interfacial debonding between the filler and the matrix, thereby ending crack propagation when under strain [20].

Adding unmodified mica can decrease the fracture toughness and crack resistance of polymer composites [28]. Mica may restrict plastic deformation [28], thereby reducing the total plastic work during the deformation process. When mica is coated with a modified polymer, crack propagation can be restricted increasing fracture toughness. A decrease in fatigue crack propagation resistance was also observed for mica/PE composites. The interface between the mica flakes and the polymer constitutes a weak region that helps initiate voids in front of the crack tip. It appeared that mica flakes in PE promoted the development of a plastic zone around the particle. Extending the damaged zone ahead of the crack reduces the local stress at the crack tip and slows down the crack growth process [29].

Compared to fillers such as calcium carbonate and silica, mica-filled polypropylene (PP) have a higher tensile strength and CTE [15]. It has been suggested that amorphous minerals have a lower CTE than the fully crystalline minerals [30]. Based on the low crystallinity of mica and its low CTE compared to those of $CaCO_3$ and silica, it can be expected that composites such as PP/mica would also show low CTE values [15].

21.3.1
Mechanism of Reinforcement

Generally, plastics consist of a combination of well-ordered crystalline regions where polymers form repeating structures and amorphous regions where the polymers are

Table 21.3 Summary of selected mechanical properties of mica/polymer composites.

Polymer	dp (μm)	% Mica	Coupling agent	Mechanical properties relative to unfilled polymer, expressed as a percentage (%)							Reference
				YS	TS	M	%E	FS	FM	IS	
cPP	20	10	Silane	105						71	[13]
		10	0.5 wt % iPP-MA	111						43	[14]
PP	80	35		116	109	340					[14]
		36.5	1.5 wt% aPP-pPBM	81	114	369					
		40	5 wt% aPP-pPBM	120	136	350					
	8	40			66	176					[15]
	40[a)]	30			93	254	55				[16]
	40[b)]	30			104	295	24				
	44	60		109			20		346	40	[17]
	58.5	1.5 wt% MAPP	123			24		346	37		
		60	Vinyl tri-methoxy silane	118			23		346	40		
	40	50	2 wt% silane		102			20	450	50	[10]
	250	40			78			92	340	150	[18]
		40	Sulfonyl azide, amino silane		109			133	460	142	
HDPE	20[c)]	20	Silicohydride		167	222					[19]
	20[d)]	20	Silicohydride		128	154					
	22	10	Silane	104	200	131	67			59	[20]
	21	10		105	99	240	3				[21]
LDPE		7.5		90	100	80	13				
Nylon6	75	5		74		614	5	121	122	180	[22]
		40		117		148	16	173	295	50	
		25		95		307		163	220	142	
		25	1 wt% tetra-isopropyl titanate			139		153	337	153	
		25	2.5 wt% e-caprolactum					165	287	130	

(Continued)

Table 21.3 (Continued)

Polymer	dp (μm)	% Mica	Coupling agent	\multicolumn{7}{c	}{Mechanical properties relative to unfilled polymer, expressed as a percentage (%)}	Reference
				YS	TS	M	%E	FS	FM	IS	
	75	25	3.5 wt% vinyl *tri*-methoxy silane			164		165	282	100	
	75	40	*tetra*-Isopropyl titanate	117		200	5	172	294	47	[23]
	37	40	*tetra*-Isopropyl titanate	108		631	5	155	256	94	
	75	30		96		183	5	167	237	126	
	37	30	1 wt% *tetra*-isopropyl titanate	111		581	6	149	347	153	
TPU	20	16.6			182						[24]
		25			100						
PAEK		40	PPTA		190	405	8				[25]
PEKK	50	40	Sulfonated PEKK		182	426	14				[25]
PDMS	F	14			310	1300					[26]
	86	14			330	600					
	37	14			400	380					
LPBD		40	Silane		309	168	131				[27]
SPBD		5			112	28	242				
SPBD		5	Silane		119	83	137				
SPBD		40			23	30	96				
SPBD		40	Silane		44	28	138				

dp: particle size, YS: yield strength, TS: tensile strength, M: Young's modulus, %E: percentage elongation, FS: flexural strength, FM: flexural modulus, IS: impact strength, cPP: block polypropylene-co-ethylene, iPP: isotactic polypropylene, aPP: atactic polypropylene, HDPE: high-density polyethylene, pPBM: *p*-phenylen-bis-maleamic acid, MA: maleic anhydride, TPU: thermoplastic polyurethane, SEBS: styrene–ethylene–butylene–styrene, EPDM: ethylene–propylene–diene monomer, TPU: thermoplastic polyurethane, PAEK: poly(aryl ether ketone), PPTA: poly(*p*-phenylene terephthalamide), PEKK: poly (ether ketone ketone), PVB: poly vinyl butyral, LDPE: linear low-density polyethylene, LPBD: linear poly(butadiene). SPBD: star-branched poly(butadiene), PDMS: poly(dimethylsiloxane).

a) Wet ground mica.
b) Dry ground mica.
c) Statically packed.
d) Dynamically packed.

more disordered. The polymer chains have a certain amount of flexibility and can move relative to each other when heated or placed under stress. Mica addition reduces chain movement by introducing physical obstructions and providing attachment sites for polymer chains to adhere to through primary and secondary valence bonds. This results in polymer chains adjacent to the surface of the filler particles being largely immobilized, resulting in a more ordered zone around the mica surface than the bulk polymer matrix. The filler particles effectively act as physical cross-links, preventing chain slippage, and as a consequence the polymer is stiffened [9].

In mica-reinforced polypropylene, for example, it was proposed that reinforcement particles disrupt the polymer bulk, leading to more order in some of the polymer segments. The amorphous regions at the interface, coating the particles, are then severely constrained. Only a small fraction of the amorphous phase would therefore become mobile upon heating as evident from a drop in glass transition temperature (T_g) [31].

In mica-reinforced linear and star-branched polybutadiene, stress at break increased with increasing mica loading. It was concluded that mica particles oriented along the direction of applied force, which is also evident from higher elongation values and the adsorption of polymer segments at several sites of the filler particle. This effectively introduced physical cross-links into the composite [27].

This theory has also been confirmed for polyamide reinforced with glass fiber and mica composites. It was proposed that the material consisted of two interpenetrating continuous phases: a polymer-adsorbed surface layer and the bulk and an unbound polymer phase. The phase adsorbed on the reinforcement surface was assumed to be mechanically stronger and to have a higher modulus of elasticity. The existence of two phases was confirmed using dielectric measurements, monitoring the dielectric loss angle (tan δ). Two maxima were observed: one between 40 and 80 °C and the other in the 120–140 °C region. It was thought that the first was associated with the glass transition of the bulk amorphous phase. However, the second maximum, also a glass transition, involved chain segments at the boundaries of crystalline regions. The number of possible chain conformations is limited in these regions leading to a rise in this glass transition temperature compared to the glass transition of the bulk [32].

The effect of time and temperature on the mechanical properties of polymers are significant enough to warrant consideration when evaluating mica-reinforced composites [31]. Rigid fillers decrease the damping of composites (as measured by tan δ) because their damping is much less than the polymer. This has been observed in polypropylene composites containing less than 30 wt% mica. However, an increase was observed when using more than 30 wt% mica, and was exacerbated by poor surface adhesion between the polymer and the filler. Increased damping is due to the introduction of new damping mechanisms such as mica–mica interactions in agglomerates and friction between mica and polymer in the absence of adhesion. Excess damping in the polymer at the interface due to changes in polymer morphology and conformation, as discussed above, also contributed to the overall increase in the composite's damping properties [33].

It has also been proposed that polymer chains may intercalate into mica galleries, restricting chain movement. Reduced chain movement has been confirmed by an increase in storage and loss modulus. The storage and loss modulus were also consistently higher for oriented flakes in polyethylene composites [19].

In general, the reinforcing action of functional fillers can be attributed to several factors [9, 21, 34, 35]:

- Large difference in stiffness and strength between filler and matrix.
- Filler concentration or volume that the filler particles occupy.
- Interfacial adhesion and slipping that may occur at the filler matrix interface.
- Particle size and its distribution.
- Shape and orientation of the reinforcing agent.

21.3.1.1 Shape and Orientation

Five commonly used filler shapes have been identified and was summarized in Table 21.4 [36]. They are grouped into three categories according to their dimensionality, that is, one, two, and three dimensional [27].

To maximize the reinforcement effect with respect to modulus, a filler should be used that provides the largest A/V ratio based on the chain confinement effect of fillers. The A/V ratio for spheres is fixed for any given volume; however, for fibers and flakes, it can still be changed by increasing the aspect ratio (length to diameter or depth ratio). Therefore, size and shape determine filler particle surface area and consequently composite properties [10, 27].

Mechanical properties in composites using spherical particles are almost isotropic. Reinforcement with spheres often increases compressive strength and flexural modulus, but mostly reduces impact strength [37]. Fiber reinforcement can improve tensile strength under certain conditions, while rigidity can be improved by sheetlike fillers. In both instances, improvement depends on the aspect ratio of the fillers. Impact strength usually cannot be modified by particulate fillers [10].

In fiber-reinforced polymers, fiber orientation is important as it allows anisotropic reinforcement or isotropic mechanical properties when oriented randomly. Fiber reinforcement increases tensile and flexural strength in the direction of alignment. Anisotropic fiber orientation typically leads to anisotropic shrinkage of parts during molding that leads to warping [37].

The reinforcement effect of platelike fillers is intermediate between that of spheres and fibers. These fillers modify polymer properties in two directions. As with fiber-reinforced composites, a high aspect ratio gives better reinforcement due to higher surface area and more efficient stress transfer from the matrix to the filler. Increased surface area results in stronger interaction between polymer chains and individual filler particle surface, increasing tensile and flexural strength [37]. But compared to fibers, the reinforcement effect is lower. Resin-rich regions, voids, or flake imperfections, as well as poor adhesion and inhomogeneous sites, will lower composite strength [38]. Shrinkage in platelet-reinforced composites is less compared to directional fiber composites and part warping is less likely.

The diameter and thickness of commercial mica are highly irregular and at low dosage levels, the alignment in the matrix is random, unless oriented by external means. Orientation is important in terms of mechanical properties. If mica content is

Table 21.4 Summary of commonly found filler shapes [10, 27].

	3D			2D	1D
Particle class	Sphere	Cube	Block	Flake	Fiber
Examples	Glass spheres Phenolic microballoons	Calcite Feldspar	Calcite Silica Barite Nephelite	Kaolin Mica Talc Graphite	Wollastonite Tremolite Wood Glass Carbon Hemp

increased or if shear is applied during the cooling phase of injection molding, mica particles can be oriented in the direction of flow, resulting in improved mechanical properties [19].

In polypropylene composites using $CaCO_3$, mica, and silica fillers, the tensile strength of $CaCO_3$ composites was shown to be lower than that of mica composites. Although $CaCO_3$ particles have a higher surface area than the larger mica and silica filler particles, it was thought that the effect of particle shape on the mechanical properties was dominant rather than the particle size. Mica/PP composites had the lowest void content, hence better mechanical properties, due to the increased wettability of the filler by the matrix because of mica's shape compared to $CaCO_3$ and silica composites [15].

21.3.1.2 Mica Concentration

The effect of filler content on composite mechanical properties has been extensively studied [39–42]. Two distinct regions of mechanical behavior have been identified. At relatively low filler content (<10% by mass), below the percolation threshold – where a continuous structure of solid particles is formed [43], reinforcement has little effect on mechanical properties. Above 10%, composite impact strength decreases, but rigidity and tensile strength improve with increasing filler content [10, 18].

Thermoplastic polymers are rarely filled beyond 30% by volume. Highly filled systems can be difficult to process and may lead to a loss of mechanical properties. Ultimate properties, such as tensile strength, are very sensitive to complete filler wet-out by the polymer and require coupling agents at high filler loadings.

Typically composite elastic modulus increases with increasing mica volume, with or without surface modification. This increase can be expected since polymer mobility is restricted by relatively high-modulus filler particles. More importantly, the interparticle distance between two neighboring particles decreases with increasing amount of filler and the number of percolation pathways formed around the particles increases, which in turn increases the elastic modulus [27].

Percentage elongation under stress typically decreases after addition of filler, which can also be expected due to reduced polymer chain mobility. At higher filler loading, there is insufficient polymer between filler particles to contribute to extension [22]. This has been confirmed by observations that a transition from ductile to brittle fracture occurs with increasing mica content in PP/mica composites [15].

Furthermore, when mica content is too high, the polymer matrix cannot fully wet the mica flakes, leading to an increase in void content and reduced interfacial adhesion. Although voids may toughen the polymer in some cases by allowing some matrix shear yielding to occur around the voids, composite mechanical properties are usually reduced [44, 45].

An increase in composite tensile strength with mica addition is not guaranteed and depends on the polymer type used. For example, the tensile strength of nylon6/mica composites initially decreased with increasing mica concentration and then increased at higher filler content [22]. The tensile strength of mica/PAEK composites reached a maximum at 20 wt% phlogopite, while the modulus increased significantly, leveling off at around 30 wt% [25]. In thermoplastic polyurethane/mica composites, 20 wt% mica caused a 75% increase in tensile strength [24], while the tensile strength of mica/PP composites were only marginally higher compared to unfilled PP [17].

In addition to the concept of polymer chain immobilization that occurs with increasing filler content, it has been suggested that filler agglomeration is also largely responsible for the observed changes in mechanical properties, especially modulus. At low filler loading, filler particles are encapsulated in the matrix and smaller particles occupy the space between them. On the other hand, at higher filler loading, agglomeration of filler particles can be observed [46]. The polymer fraction trapped in the agglomerates is less mobile and therefore less free to react to stress and strain compared to the bulk phase [47].

The composite's modulus strongly depends on the distribution of agglomerates or particles in the matrix, while its tensile strength is limited by particle agglomeration. Modulus is typically determined at low strain and the resulting stress could be too low to separate agglomerates, therefore the modulus is relatively unaffected by agglomeration [26]. If these agglomerates are distributed homogeneously, the modulus will increase with increasing filler loading. However, the composite modulus will plateau if the distribution is inhomogeneous. If the force used to determine strength is greater than what is required to break agglomerates, a crack is formed at that point, which lowers the composite's tensile strength [26].

In mica/nylon6 composites, flexural strength increased with increasing filler content regardless of particle size at low mica content. The rate of increase was lower for larger particles up to 20 wt% mica. Agglomeration of filler particles occurred at high mica content. However, the flexural modulus was higher for large particles, suggesting agglomeration was occurring in composites using small particles [46].

In another example, mica agglomeration occurred after increasing the amount of mica from 5 to 40% in polypropylene composites. However, particle distribution was good and the particles were spherical in shape. Coupling agents prevented reagglomeration of the particles and formed strong bonds between the matrix and the mica, leading to improved mechanical properties of the composite [48].

21.3.1.3 Particle Size and Size Distribution

In general, particles that are too large may act as points of discontinuity in a composite and adversely affect mechanical properties. The ideal filler would have a high aspect ratio and a size distribution where the larger fraction has been removed. It has been shown that the modulus of mica-filled polypropylene composites depends on the flake aspect ratio, whereas the strength is influenced more by the size, that is, the flake diameter [49]. Therefore, smaller particles with a large aspect ratio are preferred to maximize both modulus and strength [49]. However, smaller particles typically have a greater tendency to agglomerate [2, 50]. A decrease in PP composite tensile strength was observed at higher filler loadings, because the filler–matrix interaction was partly replaced by particle–particle contact and filler agglomeration [2, 50].

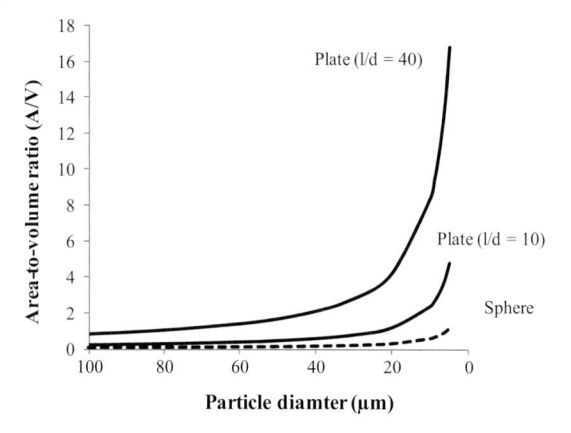

Figure 21.3 Relationship between particle diameter, area-to-volume ratio, and aspect ratio.

In mica/poly(dimethylsiloxane) composites, the modulus was increased substantially and also more significantly than in the glass spheres with equivalent diameter. It was shown that the modulus rose with increasing diameter and aspect ratio [26]. It is widely accepted that smaller particles is preferred when an increase in reinforcement effect is desired and is mostly true for spherical particles where available surface area is large. In case of mica, the size and aspect ratio varies considerably and the reverse effect is sometimes observed [26].

In earlier sections, it was pointed out that the mechanism of reinforcement highly depends on the restraining effect that mica has on polymer chain mobility and that more surface area leads to more sites where the polymer chains could be restrained. The overall effect is a stiffening of the polymer matrix.

The apparent reversed effect of the dependence of mechanical properties on particle size could therefore be explained as follows. For spheres, the A/V ratio increases with a reduction in particle size and the same is true for platelets, assuming the aspect ratio (l/d) is kept constant (Figure 21.3). Using smaller spheres as reinforcement is therefore desired. One of the favorable properties of mica is that when correctly ground, the mineral could delaminate, leading to an increase in aspect ratio; however, if this does not occur, reducing the particle size would actually reduce the aspect ratio. It is apparent from Figure 21.3 that reducing the particle size of mica, while at the same time reducing the aspect ratio, may then lead to a reduction in the overall surface-to-volume ratio. It therefore follows that larger particles may in fact have a larger aspect ratio with an accompanying high surface-to-volume ratio, and hence the reversed behavior is seen in platelet-reinforced composites.

However, in case of silicone/muscovite composites containing 20% unground and ground muscovite mica (average particle size 95 and 7 μm), it was shown that the modulus increased only 8% for coarse mica compared to 16% for fine mica. The tensile strength was reduced by 42% for coarse mica and 14% for fine mica. It is likely that unground mica would have a smaller aspect ratio and also, as can be inferred from Figure 21.3, that the reduction in particle size was enough to increase the surface area sufficiently, regardless of the effect of aspect ratio. However, the authors suggested that because mica was ground in ethanol, the ethanol adsorbed onto the hydroxyl groups on the surface of the particles, leaving the hydrophobic end of the ethanol group to adhere to the silicone polymer leading to improved surface adhesion [51].

Other work has shown that adding 20% mica, at any particle size, will increase the stiffness of mica/polypropylene composites. It was suggested that the finer the particle size, the smaller the overall size of the plate, implying that delamination occurred during particle size reduction. Assuming that it is the aspect ratio of the mica that provides the larger stiffening performance, the plates would usually result in higher stiffness [2]. The same trend was also observed by other researchers for mica/polypropylene composites. It was shown that flexural strength generally remained constant for untreated composites, but increased up to 26% at 40% mica for treated composites. Reducing mica particle size improved flexural strength due to the greater surface area available for adhesion to the polypropylene [18].

It was found that for mica/nylon6 composites, increasing filler content lead to an initial decrease in tensile strength before increasing again when using large particles (75 μm). A gradual increase has been observed for smaller particle sizes (37 μm). Considering the drastic reduction in percentage elongation, it was concluded that the addition of filler, in both cases, reduced the mobility polymer chains. For the same system, the Young's modulus increased with increase in filler concentration for both cases. However, when using larger particles, less reinforcement was required to achieve the same stiffness compared to using smaller particles [46]. These variations were explained in light of particle size distribution. At higher filler loadings, smaller particles are thought to occupy the volume between larger particles. These smaller particles tend to be agglomerated, but still show some continuity. Using mica with a smaller average particle size would typically have a narrower size distribution, and the continuity is not observed at higher loading [46].

One should therefore be careful in concluding on the effect of particle size on mechanical properties. It is assumed that when smaller particles are preferred rather than the larger particles, the aspect ratio either remained constant or was increased. The conclusion that larger particles are preferred over smaller particles implies a larger required surface area, which can only be achieved by having larger particles with high aspect ratio or very fine particles at any aspect ratio.

Particle size, shape, and particle size distribution are not the only factors affecting the mechanical properties of mica composites. In addition to these, reinforcement also depends on the effectiveness of the adhesion between the polymer and the filler [8, 52].

21.3.2
Interfacial Adhesion

It is well known that final properties of the composite materials depend not only on the intrinsic properties of each component (filler and matrix) but also on the interfacial adhesion between them, which is often the weakest point of the system [48]. Consequently, surface modification plays a major role in composite manufacture [8, 10, 30, 53]. If there is no adhesion, the filler particles simply represent flaws that may actually decrease the polymer load bearing ability. Adequate adhesion is essential for stress transfer between the reinforcing filler particles and the matrix [54].

There is a great difference between the polarity of mica and thermoplastics such as polypropylene. Mica surfaces are inert and therefore chemical bonding between the polymer chains and filler particle surface usually does not occur, resulting in poor particle dispersion and weak interfacial adhesion. The importance of relative polarity for mica can be illustrated by measuring mica sedimentation behavior and dispersion in a liquid. A small difference in surface polarity enables good wetting and dispersion – reflected by small sediment volume, whereas a large difference results in the opposite – poor wetting and dispersion and large sediment volume [28].

Mica flakes are generally highly irregular in both diameter and thickness. For optimal performance, the alignment, stacking, and distribution of the flakes should be random. Resin-rich regions, porosity, flake imperfections, poor adhesion, and uneven distribution will lower the composite strength. Several strategies are available for improving matrix/filler interaction: using coupling agents, surface active agents, or organic surface coatings on the filler or alternatively modifying the polymer polarity [28, 53].

An increase in mica content in PP composites lead to an increase in elastic modulus, but caused mica flakes to aggregate. Adding a coupling agent increased the affinity of PP for mica, improved mica distribution giving a more homogeneous composite, and improved the elastic modulus [33]. The same has been observed in calcium carbonate composites modified with stearic acid [38].

In particulate polymer composites, failure may start in the bulk polymer, at the polymer/filler interface, or within filler particle agglomerates [38]. Two failure modes are possible for mica-filled composites: flake fracture indicating strong interfacial adhesion and flake pull-out indicating weak adhesion. The second mode prevents full utilization of the filler's strength [18]. Modes of failure can be investigated by SEM analysis of composite fractured surfaces to qualitatively assess interfacial bonding [55].

Weak interfacial adhesion typically leads to debonded regions and cavities around the mica particles [28]. Including a surface modifier improves bonding and reduces these cavities significantly [28]. Furthermore, SEM photos of mica/epoxy composites revealed smooth filler surfaces when interfacial adhesion was poor (adhesive failure), while threadlike features were visible due to cohesive failure of the polymer when adhesion was adequate [54]. It has also been observed that unmodified particles appeared spherical in shape in the composite, which is representative of weak adhesion between the filler and the matrix. Mica surface modification with maleated polypropylene (MAPP) or silane coupling agents improves adhesion between the filler and the matrix leading to filler particles being separated from matrix with more difficulty during fracture. The resulting cavities' shape was more elliptical [17]. In untreated mica/nylon6 composites, voids formed near the mica surface due to insufficient matrix coating the particles, while sufficient polymer was seen after surface treatment [22].

A good surface modifier should comply with the following requirements [53]:

- Good adhesion between the filler surface and the modifier.
- Improved wettability of the modified surface with the matrix polymer.
- Shock absorbing and stress releasing properties of the interface between the filler and the matrix.
- High mechanical strength of the interfacial layer.

21.3.2.1 Coupling Agents

Many different coupling agents are available. Final composite properties depend on the type, concentration, and method and conditions used to incorporate the coupling agent [18]. Interfacial adhesion can be improved by chemically modifying mica particles or by adding a polymeric compatibilizer to the matrix. A coupling agent can be seen as a "bridge" at the interface, effectively coupling the matrix to the mica particle.

Conventional coupling agents include silanes, zirconates, titanates, and stearic acid [10, 49, 53, 56]. A coupling agent's basic role is to create a chemical linkage, exclude water, improve adhesion, and promote stress transfer. A coupling agent works on the principle that one end of the agent forms strong interactions, such as covalent or hydrogen bonding, with the filler while allowing the other end to entangle with the polymer chains, promoting physical interactions [56]. A silane coupling agent can also improve composite processability by modifying its rheological properties [33]. Its use has been shown to eliminate voids due to uneven filler distribution.

Improved adhesion has been reported for mica-filled polypropylene and acrylonitrile–butadiene–styrene copolymers after treatment with an aminimide-cured epoxy resin. Improvement of the filler/matrix interface was comparable to more expensive silane treatment. Treatment with conventional epoxy resins gave poor results in all aspects [53]. Adding silane-treated mica to styrene–butadiene rubber raised the modulus more than the untreated mica [8]. For linear poly(butadiene) and star-branched poly(butadiene), silane-treated mica substantially improved the ultimate tensile strength indicating enhanced matrix/filler interaction [27]. Using α-caprolactum as coupling agent also increased flexural modulus values for mica/nylon6 composites, giving results similar to that of silane-treated mica. Using a titanate coupling agent, however, plasticized the matrix and consequently increased composite flexibility [22].

The affinity of nonpolar polymers for mineral fillers can be improved by grafting polar groups on the chain, thereby improving adhesion to the filler. For example, LLDPE can be modified by graft copolymerization with polar monomers such as acrylic acid, maleic anhydride, or succinic anhydride [38]. The interfacial additives appear to be chemically similar to the polymer matrix [14], but the rigidity of these modified polymers may be different from that of the matrix and could affect the interfacial structure and consequently material properties [28]. However, in the absence of adhesion, only mechanical interlocking exists between the filler and the polymer and results in poor mechanical properties [30].

One of the most commonly used compatabilizers is maleated polypropylene (MAPP). The maleic acid-grafted groups on polypropylene give rise to ionic bonds between the filler and the polymeric coupling agent, which in turn increases physical interactions with the matrix. The result is an increase in adhesion between phases and better dispersion of mica particles, thereby improving mechanical properties of the composite [17].

An alternative to grafting is to modify the polymer's polarity by irradiation with an electron beam or with gamma rays. This process introduces polar groups onto the polymer chains by controlled oxidation, thereby increasing adhesion ability [57].

Other, less common grafted polymers have also been used in mica composites. Atactic polypropylene grafted with p-phenylene-bis-maleamic acid has been used as an effective interfacial modifier in mica/polypropylene composites [14]. Polystyrene-b-polyisoprene-b–polyvinyltriethoxysilane and polystyrene–polyvinyltriethoxysilane as block copolymer surface modifiers in mica/polystyrene and mica/polypropylene composites have been shown to improve composite mechanical properties. For mica/polystyrene composites, using a block copolymer was more effective than silane treatment [56]. The benefit of increased polarity of the polymer has also been shown for mica/poly(ether ketone ketone) (PEKK) composites using sulfonated poly (ether ketone ketone) as an interfacial modifier [25] as well as using chlorinated paraffin in phlogopite/polypropylene composites [18].

21.3.2.2 Mechanisms of Interfacial Modification

Macroscopically, polymers fail once a critical stress is reached. The ductile to brittle transition is known to be responsible for a composite yielding before failure. Yield and crazing are typically influenced by microscopic parameters such as chain entanglement and chain stiffness. These parameters can be increased by adding filler, and improved further by adding interfacial agents to increase interactions between the filler particles and polymer chains. Therefore, interfacial interaction is critical in governing polymer composite mechanical properties [31].

Most researchers seem to be in agreement that a heterogeneous composite system is comprised of the particle, an interface at the particle surface, an interphase, and the bulk polymer (Figure 21.4). The interphase is a dynamic region of finite thickness around the mica that consists of polymer chains attached to the mica surface that are intertwined or mixed with the bulk polymer.

The effect of the interphase depends on its volume and characteristics [14] such as polymer viscosity, rigidity, and diffusivity. Small amounts of interfacial modifiers are used to increase polymer interaction with the mica surface and polymer rigidity, thereby improving interphase properties but without introducing an additional phase. Their action depends on the modifier's ability to diffuse into the interphase, which is influenced by polymer viscosity, compatibility between the interfacial agent and the polymer, and processing conditions [28, 31].

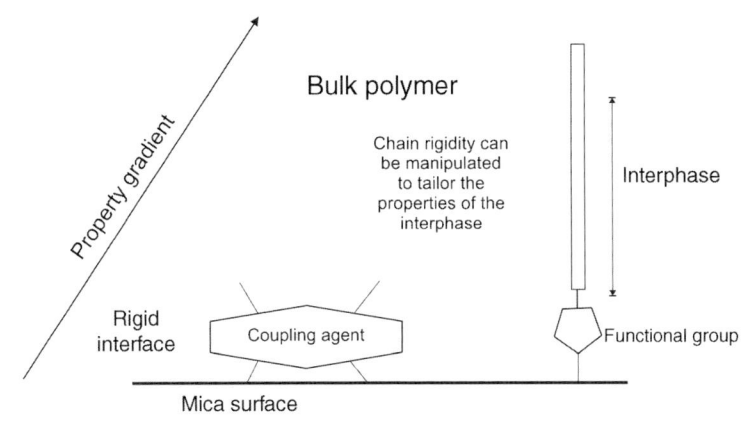

Figure 21.4 Morphology of the interfacial structure.

Block copolymer coupling agents have been shown to effectively improve the interfacial adhesion in mica/polystyrene and mica/polypropylene composites [56]. These block copolymers were designed to form chemical bonds with the mica interface via hydrolyzed silane groups. The polymeric chains are then free to entangle with the bulk polymer matrix. With this kind of coupling agent, the rigidity of the block copolymer can be manipulated by varying the chain length of the more rigid chains in the block copolymer. The coupling agent would therefore manipulate the interphase rather than the interface between the matrix and the filler material. This was successfully illustrated using the block copolymers polystyrene-b–polyisoprene-b–polyvinyltriethoxysilane and polystyrene–polyvinyltriethoxysilane. Manipulating the middle flexible polyisoprene block by increasing its length increased the impact strength of composites. Similarly, the tensile strength and modulus increased almost 10% with an increase in the polystyrene block length. Chain entanglements between polystyrene blocks and matrix polystyrene increased, leading to improved interfacial adhesion and consequently improved mechanical properties [56].

However, tensile and flexural strength and modulus decrease to some extent with increasing polystyrene block length. These segments increase the thickness of the interphase and also make it more ductile [56]. A ductile interphase can uniformly disperse the stress at the mica interface and can also absorb impact energy to prevent craze and crack propagation, thereby increasing impact strength and toughness. A ductile interphase could accelerate yielding and plastic deformation of the matrix near the filler surface, resulting in decreased tensile and flexural strength [22, 56]. To achieve the objective of toughening and reinforcing polymers simultaneously, it is necessary to form a thin ductile interphase to encapsulate the inclusions [20].

The rigidity of the interphase was also explored by using interfacial modifiers with different inherent rigidity. It was shown that because maleic anhydride-grafted polyolefin elastomer (POE-MA) is less rigid than maleic anhydride-grafted isotactic polypropylene (iPP-MA), it was less efficient in stress transfer and therefore resulted in lower yield strengths [28].

Assuming the interphase to be finite implies that any coupling agent must be constrained within it. It follows that a critical level of coupling agent can be expected. Levels close to this critical concentration are expected to improve interactions up to a certain point after which no further improvement or even a reduction in properties can be expected. It is thought that the interfacial region becomes saturated with coupling agent and any excess tends to diffuse into the matrix phase that can lead to a reduction in mechanical properties [14].

It was shown that the critical coupling agent level was between 2 and 8 wt% in mica/polypropylene composites, using silane as coupling agent [14]. The critical level was also shown to depend on the amount of mica used: lower quantities implied a greater amorphous phase in the matrix that can host the interfacial agent; therefore, more interfacial modifier was required for the same performance [14]. The same behavior was shown in mica/LLDPE composites [58, 59].

21.3.3
The Hybrid Effect in Systems Using More Than One Filler

Combining different filler types in a single polymer matrix offer hybrid structures with unique properties not achievable by individual fillers. This is often referred to as the hybrid effect. As a general rule, tensile strength can be improved using fibers providing sufficient adhesion between the matrix and reinforcement. Rigidity is mostly increased by sheetlike fillers such as mica, but depends on the aspect ratio. However, impact strength is seldom improved by mineral fillers [60].

The hybrid effect has been studied in polypropylene composites of mica and silica particles. It was found that when mica is used in combination with silica, the mechanical properties were improved over that of silica composites alone. Compared to using mica alone, the tensile strength was only marginally lower (at the same wt%), but elongation at break and tensile modulus were improved [15]. It has also been shown that combinations of fibers and platelets have synergistic effects on flexural strength in polypropylene [10].

The hybrid effect was also demonstrated for mica/nitrile–rubber composites. Mica was partially replaced by carbon black and it was shown that toughness increased, solvent swelling decreased, and vibrational damping increased [60].

The hybrid effect is most common in fiber-reinforced composites, where the best-known example is improving the low impact strength of carbon fiber-reinforced composites by adding glass or aramide fibers. Various hybrid systems using mineral fillers have been tested in polypropylene composites of glass beads, mica, wallosonite, and fly ash. It was found that all the single-component composites behaved as expected. However, using two or more filler types improved the mechanical properties in all cases [10].

Using mica in glass mat-reinforced thermoplastics has been illustrated for applications in the automotive industry. Tensile and flexural properties were improved only at low mica content, whereas impact strength showed a maximum at about 20 wt%. It was thought that mica influenced fiber–matrix adhesion positively at low mica content, but unfavorably at higher concentrations. Using maleic anhydride-grafted PP to enhance fiber–matrix adhesion at higher mica loading level increased the tensile properties, but impact strength was decreased. In this case, tensile and flexural properties decreased as the mica particle size was decreased. The opposite effect was observed for impact strength. It was thought that smaller mica particles penetrated the glass mat more efficiently, leading to more particles at the fiber/matrix interface and reducing adhesion. As a result, the impact strength increased, while the tensile and flexural properties decreased [44].

The properties of multicomponent compounds cannot be sufficiently described by conventional reinforced polymer theory. Different fillers tend to move and orient in different ways during processing and it is generally accepted that sheetlike and fibrous fillers have the strongest orientation tendency. Also, small particles are able to locate between the big particles and may facilitate melt flow [10]. It has been suggested that filler particles form a skeletonlike structure in the matrix characterized by a structural parameter such as the free distance in the polymer matrix. The free distance is seen as the size of the largest sphere that can fit between filler particles. It then follows that for longer free distances, fewer contacts exist between particles resulting in high stresses at these points, which in turn lowers the strength, modulus, and durability of the compound [10].

Orientation effects of mica in fumed silica-reinforced composites have been shown to improve the composite viscosity. The interaction between filler particles and matrix leads to polymer chain adsorption onto the particle surface, the degree of which can be manipulated by varying the nature of the polymer/filler interface. The increase in modulus is mainly attributed to an introduction of additional cross-links into the network by the filler. Reinforcement of polydimethylsiloxane (PDMS) by a hybrid filler system (mica and fumed silica) was not noticeable in stress–strain properties. However, swelling and orientation data revealed interaction between polymer chains and filler particles, leading to additional cross-links that increased the network chain density further. It was suggested that the effect of mica on these composites could be maximized by using thinner particles, that is, increasing aspect ratio of mica [61].

21.4
Thermal Properties

21.4.1
Crystallization

Semicrystalline polymers consist of a mixture of crystalline and amorphous regions where polymer chains are highly ordered and randomly packed, respectively. Semicrystalline polymers are generally very tough, as crystalline regions act as physical cross-links, preventing chain slippage during deformation.

Mica addition can improve crystallinity by providing nucleation points for crystal growth [62]. As with most fillers, polymer nucleation is heterogeneous [28]. Changes in crystallinity can be observed using differential scanning calorimetry. It has been shown that for polyethylene terephthalate containing more than 2 wt% muscovite, crystallinity was increased [62]. Also, the crystallization temperature can be increased with mica addition, that is, crystals start forming at higher temperatures or earlier in the cooling process after extrusion or injection molding, but the melting point and degree of crystallinity are increased only marginally [28]. Crystallinity can be further improved by increasing interfacial adhesion between mica and the matrix [17]. However, nucleation can be reduced slightly by precoating mica with a polymer. It has been found that crystallization and melting temperatures, as well as degree of crystallinity were lowered when mica was coated with MAPP [17].

Processing conditions can also influence crystallinity, for example, if the polymer is stressed during cooling. Mica flakes will orient parallel to the direction in which the stress is applied, with increased crystallization of the polymer on the mica surfaces [19].

Mica addition can also reduce crystallinity (e.g., mica/PP) [15] where mica at high wt% limits crystal growth by restricting polymer chain movement [15, 17]. It was found that the overall tensile strength of PP composites with 40 wt% filler was lower than those of the composites filled at 10 wt% and was concluded that this was due to the higher wt% mica reducing crystallinity [15].

Polymer chain movement restriction, while good for overall composite thermal stability [46], is problematic for mica composites that are going to be plastically welded. When mica was added as reinforcement to PP and ultrasonically welded, the weld strength was not as good as pure PP or $CaCO_3$/glass-filled polymers. It was concluded that this was due to the platelike nature of mica interfering with polymer diffusion across the weld interface [63].

21.4.2
Thermal Stability

A polymer's strength and rigidity depend on the interactions between polymer chains and between polymer and filler. These interactions are temperature dependent. At higher temperatures, but below the melting point of the polymer, bonds between polymer chains will break and reform, allowing rearrangement and reordering of the polymer chains and in some cases resulting in shrinkage, deformation, and warping.

Mica addition can improve thermal stability by restricting motion of the polymer chains [46] by steric hindrance (i.e., physical obstruction), by increasing crystallinity, or by providing attachment sites for polymer chains. In addition, mica itself has good thermal stability and insulation properties, reducing the material's heat diffusivity [64]. It has been shown that the heat distortion temperature of nylon can be improved by the addition of mica [46]. Treating mica with silane or adding MAPP to the polymer, both of which improve interfacial interaction between polymer and mica, increased the heat distortion temperature further [17, 22]. However, the thermal stability of sulfonated PEKK was not affected by mica addition [25].

Another mechanism by which a polymer's mechanical properties can deteriorate when heated is by thermal degradation or oxidation. The initial reaction in polymer thermal oxidation produces alkyl radicals. These react with oxygen to form hydroperoxides that can decompose to alkoxyl radicals. The alkoxyl radicals remove hydrogen from the polymer chain and other alkyl radicals to form water. Finally, various carbonyl species are formed [65].

Filler addition can also influence certain oxidation reactions, changing the yield of oxidation products [65]. For example, carbonyl, ester, and ketone species predominate in oxidation products of HDPE/mica, whereas in 100% HDPE, ketone species were about two times that of carbonyl, ester, and lactone. Lactone concentrations were too low to be detectable in HDPE/talc and HDPE/$CaCO_3$, while carbonyl species were small in HDPE/diatomite and HDPE/wollastonite [65]. The extent of oxidation in HDPE/mica was found to be less than that in HDPE/sericite–tridymite–cristobalite composites or pure HDPE, but greater than that in HDPE reinforced with kaolin, talc, $CaCO_3$, diatomite, or wollastonite [65].

Mica can limit thermal degradation by reducing overall diffusivity of the material to oxygen and volatiles formed, by forming regions largely impervious to diffusion, and by raising the temperature at which thermal degradation begins to occur (e.g., 5 wt % synthetic mica in PET fibers [64] and mica in thermoplastic polyurethane) [12].

While mica addition raises the temperature at which thermal degradation starts, it can increase the rate of degradation once it occurs [65]. For example, thermal degradation increased in poly(dimethylsiloxane) reinforced with large (85 µm) wet ground mica particles [26], while it was thought that mica addition could increase molecular degradation of polypropylene in the composite if kept in the melt for extended times [18].

While mica addition can improve thermal stability, it can reduce resistance to degradation by ultraviolet light. The effect of mica and black mica on photo-oxidation of HDPE was compared to HDPE filled with wollastonite, $CaCO_3$, kaolin, diatomite, and talc. It was found that mica, along with diatomite and kaolin, acted as catalysts for photo-oxidation, accelerating degradation. Black mica and talc showed less degradation than the previous three fillers, while wollastonite and $CaCO_3$ reflected nearly all the ultraviolet light protecting HDPE [66].

21.4.3
Flammability

Mica also has the potential to be used as fire retardant in composites. Polystyrene reinforced with sodium-fluorinated synthetic mica has shown reduced heat release rates of up to 25% compared to 100% polystyrene [67]. Mica addition decreased ignition time to two-thirds of that of 100% polystyrene, meaning that the composite caught fire quicker. It was thought that this was due to mica acting as a catalyst for thermal oxidation [67]. The ignition time and heat release rate of organically modified mica/polystyrene composites depended on the type of organic modifier used. Dimethyl, di(hydrogenated tallow) ammonium-modified mica reduced heat release rate by up to 60%, while ignition times were 10–20% shorter compared to that of 100% polystyrene. Triphenyl, n-hexadecyl phosphonium-modified mica/polystyrene composites had the same ignition time as 100% polystyrene, but up to 50% lower heat release rate. Once burning, mica composite continued to burn rather than self-extinguish [67]. XRD and TEM showed that unmodified mica had not intercalated, while dimethyl, di(hydrogenated tallow) ammonium-modified mica basal spacing had decreased and had formed layered stacks surrounded by polymer. Both had formed microcomposites, while the phosphonium-modified mica showed intercalation with the polymer with basal spacing being increased.

Mica has added benefits of being relatively inert compared to other fire retardants that can release toxic gases during thermal decomposition, such as compounds containing halogens, phosphorus, and heavy metals. Mica addition has been shown not to significantly affect the performance of other fire retardants, such as aluminum trihydrate, in composites, thereby not affecting the material's overall classification in terms of fire safety [68].

21.5
Other Properties

21.5.1
Processability

The amount of interfacial region able to host the coupling agent in particulate composites depends not only on the type of materials but also on the way in which it has been processed. Composites can be considered complex systems showing behavior distinctly different than expected from the characteristics of its constituents. By assuming this principle, processing is one of the main aspects influencing composite properties [14].

Unfortunately, mica addition tends to decrease processability, especially in blowing of polyethylene/mica composite films. Mica causes a reduction in blowup ratio and haul-off speed as well as an increase in film thickness. Apparent shear and elongational viscosity increased with increasing mica concentration and tension thinning played a major role in blown film extrusion [21]. Heat from the melted polymer during processing is transferred by both conduction and convection, which is influenced by mica addition. The heat transfer coefficient of PP/mica could be increased by about 20% when filled with mica [69]. This in itself can lead to different processing requirements.

The melt flow index (MFI) of mica-filled polymers typically decreases with increasing mica content as mica hinders plastic flow in the melt, but the MFI is higher than that of unfilled systems. Mica/PP MFI decreased with increasing mica loading, but showed a slightly higher MFI than that of $CaCO_3$/PP and silica/PP. Under shear, the platelike structure of mica enables particles to slide past each other allowing plastic flow [15, 20].

The melt viscosity of mica/PP composites is higher than that of unfilled systems and has been shown to decrease with increasing shear rate. At high shear rates, viscous stress exceeds particle interactions, resulting in a layered mica structure and lower melt viscosity [17]. Composites at low mica concentration showed classic viscoelastic behavior, similar to that of homogeneous materials. However, at higher mica concentration, classic viscoelastic behavior was not observed [17].

Modifying the mica surface affects composite rheology by changing particle dispersion. Using coupling agents may reduce processing time required to reach optimal performance. It was shown that adding 2 wt% chlorinated paraffin powder significantly reduced the time required to achieve maximal flexural strength in mica/polypropylene composites [18]. It was thought that using a coupling agent reduced thermal degradation during processing.

It has been discussed earlier that mica also nucleates crystallization that may lead to a lower melt viscosity and improve processability. This has been explained by the crystallization behavior of polyamide reinforced with glass fiber and mica. It was thought that the matrix was more flexible due to hindrance during crystallization due to the presence of mica, implying that the layer adjacent to the filler was denser and the polymer between the dense regions looser than before filling [32].

Mica can influence natural rubber processability. It was shown that the melt viscosity torque and specific mechanical energy showed only a slight increase with increasing filler content, implying weak interfacial adhesion. At low mica content (<20 wt%), the specific mechanical energy required for processing decreased, while at higher filler loading, the non-Newtonian behavior of mica-filled natural rubber was more apparent [70].

The benefit of using mica as a processing aid was also observed when aluminum trihydrate was used as flame retardant in thermoplastic polyurethane. At aluminum trihydrate levels required to be effective, the composite mechanical properties were severely compromised. However, processing is improved and cost is reduced without further loss of mechanical properties when using mica in the formulation [24].

During processing, particles may interact with each other or bind to the polymer chains, thereby changing polymer properties by restricting chain movement. Understanding the hydrodynamic influence of particles in the polymeric fluid forms the basis for describing the behavior of filled polymers [63]. During film blowing, mica tends to be oriented in a plane near the surface of the blown films, but further away from the surface they form less oriented overlapping and discontinuous layers [21]. Orientation can also be induced by applying shear during the cooling phase of injection-molded specimens in mica/HDPE composites [19].

Another consequence of processing, often overlooked, is that particle size does not necessarily remain unchanged during processing [16] and that breakdown of mica particles during mixing and processing causes reduction in aspect ratio [8]. Care must be taken to ensure mechanical breakage of mica or thermal breakage of the polymer does not occur [10] since these may lead to weak mechanical properties.

21.5.2
Barrier Properties

Mica can be used to reduce diffusivity of water, carbon dioxide, and oxygen in polymers. This is important in applications where the polymer protects a component or food from hydration or corrosion (e.g., aluminum in caustic solutions) [71].

Based on the hydrophobicity of synthetic polymers such as polypropylene, HDPE, PET, and so on, one would expect them to be excellent barriers against water vapor. Also, many natural polymers such as carbohydrate and pectins are hydroscopic and have high permeability to water vapor and CO_2. The barrier properties are further degraded when the polymer is modified by adding hydrophilic groups such kappa-carrageenan [72].

Because mica is largely impermeable, it forms regions of low diffusivity in polymers, resulting in a smaller cross-sectional area available in the polymer for diffusion, giving a lower overall permeability. Reduction in permeability is a function of the volume fraction Φ of mica in the polymer and its aspect ratio α. In polymers with low vol% mica addition, where the mica plates do not overlap, permeability reduction is proportional to $\Phi\alpha$ (where $\Phi < 1$ and $\Phi\alpha > 1$) [73]. It was found that permeability change was independent of mica flake size provided the aspect ratio of the different sized flakes was the same [71].

Permeability is further reduced if mica forms overlapping sheet layers that are perpendicular to the direction of solute flow (mica sheet orientation can happen if the polymer is under strain during cooling from the melt phase). This forces solute to take a tortuous route between the mica layers [21]. Therefore, in polymers with higher vol% mica and where the plates overlap, permeability reduction is proportional to $(\Phi\alpha)^2$ (where $\Phi > 1$ but $\Phi\alpha > 1$) [71, 73].

When using up to 10 wt% mica in carrageenan/pectin films, a 70% reduction in CO_2 permeability was achieved [72]; after 10 wt%, permeability increased. It was thought that this could be due to mica and polymer swelling when hydrated causing voids in the polymer matrix to open around the mica, increasing permeability [72].

21.5.3
Electrical Properties

Dielectric materials are poor conductors of electricity, typically having a high dielectric strength, but can support electrostatic fields, which can be used to temporarily store energy. This property is useful in capacitors, for example, in power supplies for electronic equipment that convert alternating current to direct current, in equipment for generating radio signals, or in coaxial radio frequency transmission lines.

Mica has a high dielectric strength of up to 70 MV/m that makes it useful as an insulating material in electrical applications. Polymers such as polystyrene or polyethylene have dielectric strengths of around 20 MV/m. Mica can withstand very high electric field strengths and voltages without the insulating material breaking down. When breakdown occurs in the insulating material, an electrically conductive path is formed, resulting in the loss of the material's insulative properties. Dielectric strength is proportional to the thickness of the insulating material and inversely proportional to operating temperature and frequency. Mica can be added to polymers to increase their overall dielectric strength, thereby increasing their insulative properties, either reducing the thickness of insulative material required for a given application or extending the range of electrical insulation applications the polymer can be applied to.

In mica/nylon6 composites, the dielectric strength increased with mica addition and plateaued at 20 wt% mica [22]. For fine mica particles, the dielectric strength peaked at 10 wt% and decreased thereafter. The initial increase was due to mica being uniformly distributed in the polymer at low concentrations, reducing cross-sectional area available for conduction. The decrease was thought to be due to the fine mica particles forming agglomerates at high concentrations, resulting in an increased cross-sectional area for conduction. For large mica particles, while agglomeration does occur, the mica flakes are sufficiently large to overlap each other, hence dielectric strength plateaus rather than decreases. Dielectric strength was reduced when mica was treated with coupling agents, compared to composites with untreated mica, possibly due to better encapsulation of the treated mica by the polymer or better mica dispersion through the polymer [22].

In dielectric materials, the dielectric constant is a measure of a material's ability to concentrate electrostatic lines of flux in an electrostatic field. If the dielectric constant is very low, lines of flux in the electrostatic field are uncompressed or unimpeded, meaning the field can store energy with little loss (losses are usually as heat). Vacuum and dry air have very low dielectric constants, while mica and some polymers have dielectric constants between 2 and 4. Metal oxides have very high dielectric constants, making them useful for small-volume high-value capacitors, but are generally not able to withstand intense electrostatic fields.

The effect of mica addition to polyimide films prepared using an ultrasonic and *in situ* polymerization process on the material's dielectric constant was examined. It was found that the dielectric constant decreased with increasing mica content at temperatures from -150 to $150\,°C$ and at frequency ranging from 1 kHz to 1 MHz. The resulting dielectric strength suggested the resulting films might be useful as cryogenic insulation in superconductive cables and magnets or as dielectric materials for microelectric applications [74].

Mica can also be coated with metals to increase the conductivity of the resulting composite. This is useful for electrical component casings that also serve as shielding against electromagnetic interference (EMI). Generally, a polymer's resistivity is 10^{13}–$10^{17}\,\Omega/cm$, making it transparent to radio waves. Including conductive fillers such as carbon, metal powders, or metal-coated mica flakes reduces the polymer's resistivity making it opaque to radio waves, therefore acting as a form of EMI shielding [75, 76].

Nickel-coated phlogopite was used in PP and ABS to increase the polymer's conductivity [75]. It was found that coating 40–50% of the mica surface was sufficient to make mica conductive and increasing coating improved the resulting polymer's conductivity further. It was also found that better conductivity was achieved when coated mica was confined to the amorphous regions of PP, and also when the resulting composite was processed into thin sheets due to the increased contact between mica flakes [75].

21.6
Modeling of Mechanical Properties

In practice, both the strength and stiffness of particulate composites are important. Ahmed and Jones presented an extensive review of the available theories for predicting these properties [54]. The theory for the strength of particulate composites is less developed than that for modulus. This is because it involves many more factors that need to be considered, that is, the complex interplay between the properties of the reinforcement, the polymer, and the interfacial layer [54]. Variables that are important include the shape, size, and distribution of the filler particles and the strength of the interfacial bond between the filler and the matrix. The fact that the reinforcements are discontinuous complicates the analysis. The reinforcing agents may have a nonuniform size and may even be nonuniformly distributed in the matrix [77].

Fibrous reinforcements allow properties to be maximized in the direction parallel to the fiber orientation [77]. Planar reinforcements allow mechanical properties to be developed in the plane of the reinforcement. Such planar reinforcements include flakes, ribbons, and continuous films [77]. Mica, the filler of interest in this study, approaches a flake structure.

21.6.1
Young's Modulus

The available models for predicting the Young's modulus vary from empirical to highly theoretical. One of the simplest models idealizes the composite as alternating layers of high-modulus reinforcement and a more compliant matrix. The elastic properties of these laminated composites depend on their orientation relative to the applied stress, as shown in Figure 21.5. The assumption is made that the layers are strongly bonded and this implies that the volume fraction rather than the thickness of

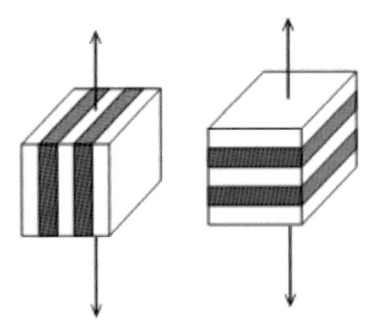

Figure 21.5 Mixing rule conditions for layered composites.

21.6 Modeling of Mechanical Properties

the individual layers determines the mechanical properties. The effective moduli when the layers are in parallel or in series yield the Voigt and Reuss average modulus, respectively [78]. Maximum stiffness is obtained when the stress is applied parallel to the layers. The assumption is that the strain will be the same in all the composite layers, that is, the iso-strain condition applies. The effective Young's modulus is given by Eq. (21.1) (Table 21.5).

When the layers are oriented transverse to the applied stress, the effective modulus is much lower. In this case, each layer is subjected to the same force. As it is assumed that the area remains constant through the stack, the stress is also the same in each layer, that is, the iso-stress condition applies. The effective modulus for this case is given by Eq. (21.2) (Table 21.5).

By analogy for these theoretically correct models, these equations are also used in the form of mixing rules for more general applications. In summary, the classical mixing rules are then as follows:

- Iso-strain, where the force acting on the composite is equal to the sum of the forces acting on each of the elements.
- Iso-stress, where the total strain is equal to the sum of the strains in each element [54, 79].

Padawer and Beecher [77] have shown that deviations from the iso-stress model do occur due to differences in the Poisson's ratios for the matrix and reinforcement. However, this effect is small when the reinforcement is significantly stiffer than the matrix [77].

Both forms of the mixing rule are applicable to unidirectional long fiber-reinforced composites [54]. For composites in general, the iso-strain and iso-stress mixing rules provide upper and lower bounds of the actual behavior. It is therefore of interest to consider modeling the modulus in terms of appropriate combinations of the mixing rules as in the Hirsch and Counto models. They are based on the geometric models for the composites shown in Figure 21.6. The Hirsch model (Eq. (21.3), Table 21.5) constitutes a linear combination of the two mixing rules, whereas the Counto model (Eq. (21.4), Table 21.5) is more complicated. Both are based on the assumption of perfect adhesion between the two phases.

Jacquet *et al.* have further modified the rule of mixture to account for isolated particulate inclusions, for example, microspheres [79]. In essence, their model is based on a three-dimensional array of Counto repeat units. Figure 21.7 shows how the composite is divided into columns to which the classical rules of mixture are applied to calculate the overall modulus. The result is a parameter-free model (Eq. (21.5), Table 21.5).

Halpin and Tsai [80] derived a model for the case of unidirectional plane-parallel ribbon or platelet-reinforced composites (Eq. (21.6), Table 21.5). E_c is the composite modulus in the plane of orientation (see Figure 21.8 for definition of symbols) [80]. This model reduces to the iso-strain mixing rule when $D \rightarrow \infty$. To calculate the transverse in-plane orientation modulus, D/t can be replaced by W/t. To calculate the transverse out-of-plane modulus, the inverse mixing rule can be used.

Table 21.5 Models for prediction of Young's modulus.

	Model	Symbols used	Reference
(21.1)	$E_C = E_m V_m + E_p(1 - V_m)$	E_c = composite modulus E_m = matrix modulus E_p = polymer modulus V_m = volume fraction matrix V_f = volume fraction filler	[78]
(21.2)	$\dfrac{1}{E_C} = \dfrac{V_m}{E_m} + \dfrac{(1 - V_m)}{E_p}$		[78]
(21.3)	$E_C = \gamma(E_p V_p + E_m V_m) + (1 - \gamma)\dfrac{E_p E_m}{E_p V_m + E_m V_p}$	γ = fraction distribution between parallel and series models	[54]
(21.4)	$\dfrac{1}{E_C} = \dfrac{1 - V_p^{1/2}}{E_m} + \dfrac{1}{\left(1 - V_p^{1/2}\right)/V_p^{1/2} E_m + E_p}$		[54]
(21.5)	$E_C = \dfrac{\alpha^2 E_m E_p}{\alpha E_p + (1 - \alpha) E_m} + (1 - \alpha^2) E_m, \quad \alpha = \sqrt[3]{V_m}$		[79]
(21.6)	$E_C = E_m \dfrac{1 + AB v_p}{1 - B v_p}$ $B = \dfrac{(E_p/E_m) - 1}{A + (E_p/E_m)}, \quad A = 2\left(\dfrac{W}{t}\right)$ $v_p = \left(\dfrac{W}{W + D_x}\right)\left(\dfrac{D}{D + D_x}\right)$	Symbols defined in Figure 21.8.	[80]

(21.7)

$$E_C = E_p V_p + \mu E_m (1 - V_p)$$

$$n = \frac{W}{t} \left(\frac{G_p (1 - V_p)}{E_m V_p} \right)^{1/2}$$

$\mu = 1 - (\tanh(n)/n)$ (A)

$\mu = 1 - (\ln(n+1)/n)$ (B)

G_p = shear modulus of polymer

[54, 77]

(21.8)

$$E_C = E_p E_1 E_2$$

$$E_1 = \frac{1 + AB(V_p - \phi_d)}{1 - B\psi(V_p - \phi_d)}; \quad E_2 = \frac{1 - \phi_d}{1 - B\psi \phi_d}$$

$$A = \frac{(7 - 5\nu_{ps})}{(8 - 10\nu_{ps})}; \quad B = -\frac{1}{A}; \quad \psi = 1 + \frac{V_p (1 - \phi_m)}{\phi_m^2}$$

ϕ_d = volume fraction debonded filler

ϕ_m = Maximum packing fraction

ν_{ps} = Poisson's ratio of polymer

[52]

(21.9)

$$E_C = E_m V_f (\text{MRF}) + E_p V_p$$

$$\text{MRF} = 1 - \frac{\tanh(\varphi)}{\varphi}; \quad \varphi = a \sqrt{\frac{(1-\chi)^3 G_p}{E_m} \frac{V_f}{V_p}}$$

$$\chi = \frac{\phi}{V_p (1 - \phi) + \phi}; \quad \phi = \frac{(1 - V_p)^2 \phi_m}{1 - V_p \phi_m}$$

a = Particle aspect ratio

[45]

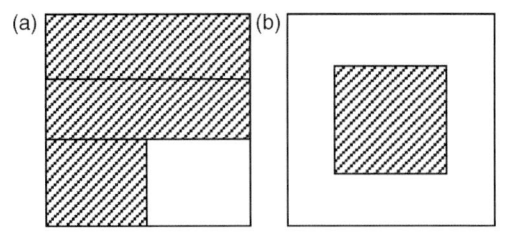

Figure 21.6 Schematic representation of the Hirsch (a) and Counto (b) models [54].

The models developed by Padawer and Beecher (Eq. (21.7A), Table 21.5) [77] and Lusis et al. (Eq. (21.7B) [54] assume a shear–stress transfer mechanism between the polymer and the filler. However, only the model by Lusis et al. [77] accounts for platelet interactions at high platelet concentrations [18, 80].

The Kerner and Lewis [52] model (Eq. (21.8), Table 21.5) takes into account that the mechanical behavior of composites depends on the degree of interfacial adhesion between the phases, the maximum packing fraction of the filler (φ_m), and the Poisson's ratio of the polymer (v_{ps}). The model assumes the following:

- Initially all filler particles are well bonded to the matrix ($\varphi_d = 0$),
- As the material is strained, the filler becomes progressively debonded from the matrix (φ_d = debonded filler volume fraction).
- The completely debonded composite behaves similar to a foamed matrix ($\varphi_d = V_p$)

By accounting for the void fraction in the composite due to debonding, the effective modulus of the composite is reduced. The Verbeek and Focke model [45] is a modification of the Padawer and Beecher model and is shown in Eq. (21.9). The model assumes perfect adhesion between the phases and that stress is transferred via a shear mechanism. The model also assumes that the porosity will decrease linearly with decreasing volume fraction polymeric binder and that the porosity is only in the polymeric phase. The composite modulus is therefore an average based on the individual component's modulus, but with the reinforcement's modulus reduced by a factor that depends on the aspect ratio of the reinforcement. The model shows very

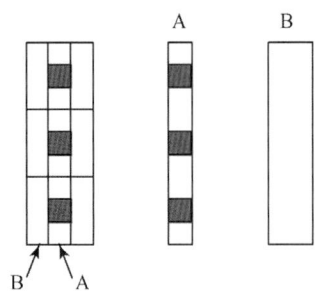

Figure 21.7 The Jacquet et al. model [79].

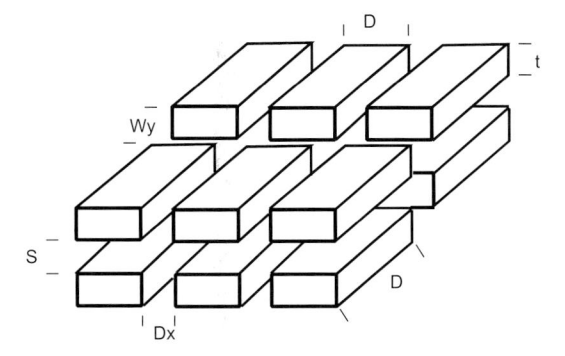

Figure 21.8 The Halpin and Tsai model [80] for regular platelet or ribbon reinforcement.

good agreement with experiment [45]. A summary of these models are presented in Table 21.5 and a selection of these have been graphed, as shown in Figure 21.9.

21.6.2
Tensile Strength

Models for predicting tensile strength are less well developed and generally give poor predictions [54, 77]. The particle–matrix bond is considered to act as an inherent flaw in the material when adhesion is poor. Poor adhesion basically precludes efficient stress transfer and the particle can therefore be seen as a void that weakens the composite [54]. A power-law model (Eq. (21.10), Table 21.6) may

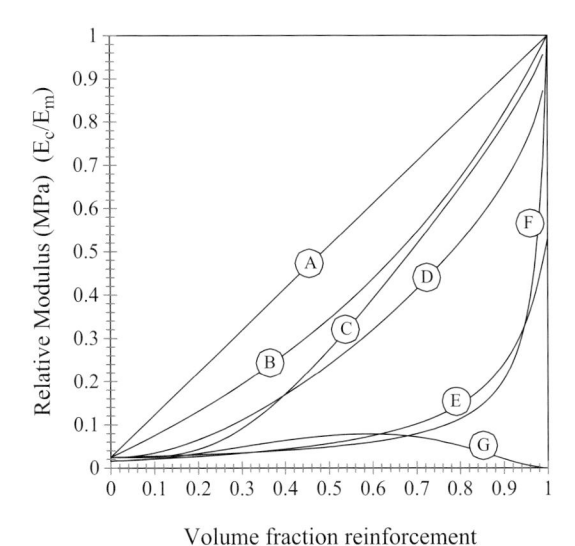

Figure 21.9 Graphical representation of models predicting Young's modulus. A: Eq. (21.1); B: Eq. (21.6); C: Eq. (21.7A); D: Eq. (21.7B); E: Eq. (21.8); F: Eq. (21.2); G: Eq. (21.9).

Table 21.6 Models predicting tensile strength of particulate composites.

	Model	Symbols used	Reference
(21.10)	$\sigma_C = \sigma_P(1-aV_f^b)$	σ_p = Polymer's ultimate strength σ_c = Composite's ultimate strength a, b = constants that depend on particle shape and orientation	[54]
(21.11)	$\sigma_C = \sigma_P(1-aV_f^b)$		[54, 81]
(21.12)	$\sigma_C = (\sigma_p + 0.83\tau_p) + \sigma_a K(1-V_p)$	K = stress concentration factor τ_p = Shear strength of matrix	[54]
(21.13)	$\sigma_C = 0.83\sigma_{th}a_{th}V_p + k\sigma_p(1-V_p)$	a_{th} = Coefficient of friction k = parameter that depends on particle size σ_{th} = Thermal compressive stress acting on particle	[54]
(21.14)	$\sigma_C = V_P\sigma_p + K_3'\tau_p\text{MPF}$ $\text{MPF} = V_m\left(\dfrac{\alpha}{\mu}\right)\left(\dfrac{1}{\tanh(\mu)} - \dfrac{1}{\mu}\right), \quad \mu = \alpha\left(\dfrac{G_pV_f}{E_m(1-V_f)}\right)^{1/2}$	α = Particle aspect ratio K_3' = correction factor to account for particle effects G_p = shear modulus of polymer	[77]
(21.15)	$\sigma_C = (V_p\sigma_p + K_3'\tau_p\text{MPF})(1-\chi)$ $\text{MPF} = V_m\left(\dfrac{\alpha}{\mu}\right)\left(\dfrac{1}{\tanh(\mu)} - \dfrac{1}{\mu}\right), \quad \mu = \alpha\left(\dfrac{G_pV_f}{E_m(1-V_f)}\right)^{1/2}$ $\chi = \dfrac{\phi}{V_p(1-\phi)+\phi}, \quad \phi = \dfrac{(1-V_p)^2\phi_m}{1-V_p\phi_m}$	ϕ_m = maximum packing fraction	[82]

be used to describe the strength of a composite with poor adhesion. It assumes that the strength of the composite is determined by the effective available area of load bearing matrix due to the presence of the filler [54]. These are the most important variables that determine the tensile strength of particular filled systems. The tensile strength generally increases with a decrease in particle size as smaller particles provide a greater interfacial area. The result is a more effective interfacial bond. Particle size is also related to the flaw size dependence of the material, and therefore the probability of finding a large flaw decreases with decreasing particle size [54].

Irregular fillers are expected to weaken the composite owing to high stress concentrations around the sharp edges. A stress concentration factor can also be introduced to take into account the reduction in strength due to stress concentrations caused by irregular shapes and has been illustrated by Nielson [54] and others [81] (Eq. (21.11), Table 21.6).

The Leider–Woodhams equation uses a simpler but more elaborate approach. In their model, spheres are represented by cylinders in order to determine the stress distribution at the bead, at breaking point. In case of nonbonded particles, stress transfer was assumed to be a result of particle–matrix fraction and residual compressive stress that act upon the particle surface. Stress is transferred via a shear mechanism for well-bonded particles and therefore depends on the shear strength of the matrix and the interfacial bond strength between the matrix and the particle [54]. Equation (21.12) (Table 21.6) shows the case for good adhesion between matrix and filler, while Eq. (21.13) (Table 21.6) is used for poor interfacial adhesion.

The model by Padawer and Beecher [77] for the Young's modulus was described earlier. They also postulated that the strength of the composite can be modeled in a likewise manner. If the maximum calculated stress in the particle is greater than the tensile strength of the particle, failure will be due to flake fracture. If the maximum calculated shear strength in the matrix exceeds the shear strength of the polymer, failure will be due to flake pull-out. It was found, however, that the latter case is predominant in most cases [77]. The above situation, for the tensile strength in the plane of orientation, can be described by Eq. (21.14). Verbeek modified this equation by introducing a correction for porosity (Eq. (21.15)) [82]. The various models are summarized in Table 21.6.

21.6.3
Limitations to Existing Models

Most models described in literature assume that the polymer forms a continuous phase with the filler suspended in it. Practically, composites may deviate from the idealized models if the filler particles are not completely separated from each other and the reinforcement element will therefore be an aggregate of smaller particles. The applied stress will therefore not be distributed evenly between the particles and the aggregates and the assumption of iso-stress or iso-strain will not be valid.

The most general assumptions are as follows:

- The reinforcement and polymer are linear elastic materials, with uniform modulus.
- The flakes have uniform width and thickness.
- The flakes are uniformly spaced and aligned in a plane-parallel fashion.
- The polymer adheres perfectly to the reinforcement.

It was said earlier that the mechanical properties of a composite depend on many factors:

- Volume fraction filler.
- Adhesion between the phases.
- Particle size and distribution (and therefore aspect ratio).
- Shape and orientation of the filler.
- Porosity in the composite.

Any model that predicts the mechanical properties of a composite should therefore account for these variables. However, most theoretical models predict that the mechanical properties of the composite depend only on the volume fraction of the filler. This represents a serious limitation for all these models.

The effect of particle shape, orientation, and size distribution will show up in the maximum possible packing of the filler. Polydispersed particle mixtures can pack more densely since smaller particles can fill the voids between larger particles [77]. Odd-shaped particles will also pack more loosely than highly regular particles such as spheres. One model that does recognize this phenomenon is the Kerner and Lewis equation [52]. Despite other limitations of this model [77], this is one of the few models that considers the influence of other factors on the mechanical properties of the composite.

A good model will provide accurate prediction over the whole concentration spectrum. Most of the models discussed in the previous section are based on various modifications to the mixing rule. Figure 21.9 shows the predicted effect of volume fraction on the relative modulus. All these models suffer the same drawback. They all predict that the composite modulus will approach the filler modulus when its volume fraction approaches one. Considering that the filler is present in the form of discrete particles, this is clearly incorrect. A bed of loose filler particles is expected to have no tensile properties unless the filler particles are bound together. This obviously implies the presence of a binder.

Any model used to describe the mechanical behavior of particulate composites should therefore be able to predict a decrease in mechanical properties, approaching zero, as the amount of binder present approaches zero volume fraction. Additional effects such as porosity should also be considered. The presence of voids will lower the modulus of the composite because they cannot carry any load.

Anderson and Farris [34] derived a model that takes into account the debonding process between the filler and the matrix at large values of strain. In the derivation, they used the Griffith energy balance approach for crack propagation. Despite the assumption that the individual components of the composite behave linear elastically, the model predicts highly nonlinear stress–strain curves.

Various other empirical models have also been proposed [35, 38, 54]. Some even use experimental data to predict optimum performance of a composite with respect to particle size distribution and filler content.

21.7
Conclusions

Mica-reinforced microcomposites are set apart from other particulate composites by the high aspect ratio attainable after appropriate grinding. It has been shown that by reducing mica particles size, an increase in surface area is attained, similar to other particulate fillers, which in turn leads to improved interaction between filler and matrix. However, if mica is not delaminated in the process, the aspect ratio increases, which could counteract the effectiveness of reinforcement.

It was found that the mechanism of reinforcement highly depends on the ability to manipulate interfacial adhesion. The morphology of a typical mica-filled system was proposed to comprise the mica interface, an interphase, and the bulk polymer. Various interfacial modification strategies have been identified that involved modifying the mica surface by relatively low molecular mass molecules, by introducing modified grafted polymers with an increased affinity for mica. The interphase can be manipulated by adjusting the properties of the coupling agent, such as the rigidity of the polymer. This allows careful tailoring of properties that can often not be adjusted independently.

References

1 Sims, C. (1997) *Mica: Building a Future on Dry Ground*, Industrial Minerals.
2 Usifer, D. and Fajardo, B. (1999) Mica and its effect in polyolefin compounds. International Conference on Polyolefins XI, Houston, TX, February 21–24, 1999, pp. 729–739.
3 Davis, L.L. (1994) Mica. *Minerals Review*, 73, 133–114.
4 http://www.mineralszone.com/minerals/mica.html (13/04/2010).
5 http://grmica.com/Products.php#aa (13/04/2010).
6 Ghannam, L., Garay, H., Shanahan, M.E.R., Francois, J., and Billon, L. (2005) A new pigment type: colored diblock copolymer–mica composites. *Chemistry of Materials*, 17 (15), 3837–3843.
7 http://www.icrmica.com/icrmica_micanite.html (13/04/2010).
8 Debnath, S., De, S.K., and Khastgir, D. (1987) Effect of silane coupling agent on vulcanization, network structure, polymer filler interaction, physical-properties and failure mode of mica-filled styrene butadiene rubber. *Journal of Materials Science*, 22 (12), 4453–4459.
9 Gachter, R. and Muller, H. (1987) *Plastic Additives Handbook: Stabilizers, Processing Aids, Plasticizers, Fillers, Reinforcements, Colorants for Thermoplastics*, Hanser Publishers, New York.
10 Jarvela, P.A. and Jarvela, P.K. (1996) Multicomponent compounding of polypropylene. *Journal of Materials Science*, 31, 3853.

11 Kozlowski, R., Lieeniak, B., Helwig, M., and Przepiera, A. (1999) Flame resistant lignocellulosic-mineral composite particle boards. *Polymer Degradation and Stability*, **64**, 528.

12 Baral, D., De, P.P., and Nando, G.B. (1999) Thermal characterization of mica-filled thermoplastic polyurethane composites. *Polymer Degradation and Stability*, **65** (1), 47–51.

13 Wang, L., Xie, B.H., Yang, W., and Yang, M.B. (2010) Grafted polyolefin-coated synthetic mica-filled polypropylene-*co*-ethylene composites: a study on the interfacial morphology and properties. *Journal of Macromolecular Science B*, **49** (1), 1–17.

14 Garcia-Martinez, J.M., Areso, S., and Collar, E.P. (2009) The role of a novel *p*-phenylen-bis-maleamic acid grafted atactic polypropylene interfacial modifier in polypropylene/mica composites as evidenced by tensile properties. *Journal of Applied Polymer Science*, **113** (6), 3929–3943.

15 Nurdina, A.K., Mariatti, M., and Samayamutthirian, R. (2009) Effect of single-mineral filler and hybrid-mineral filler additives on the properties of polypropylene composites. *Journal of Vinyl & Additive Technology*, **15** (1), 20–28.

16 Zhu, Y., Allen, G.C., Adams, J.M., Gittins, D., Heard, P.J., and Skuse, D.R. (2009) Characterization of size, aspect ratio and degree of dispersion of particles in filled polymeric composites using FIB. *Clay Minerals*, **44** (2), 195–205.

17 Khonakdar, H.A., Morshedian, J., and Yazdani, H. (2008) Investigation of thermal, rheological and mechanical properties of interfacially modified PP/mica composites. *E-Polymers*, **99**, 1–10.

18 Newman, S. and Meyer, F.J. (1980) Mica composites of improved strength. *Polymer Composites*, **1** (1), 37–43.

19 Xiang, Y.F., Hou, Z.C., Su, R., Wang, K., and Fu, Q. (2010) The effect of shear on mechanical properties and orientation of HDPE/mica composites obtained via dynamic packing injection molding (DPIM). *Polymers for Advanced Technologies*, **21** (1), 48–54.

20 Liang, J.Z. and Yang, Q.Q. (2007) Mechanical, thermal, and flow properties of HDPE–mica composites. *Journal of Thermoplastic Composite Materials*, **20** (2), 225–236.

21 Xanthos, M., Faridi, N., and Li, Y. (1998) Processing/structure relationships of mica-filled PE-films with low oxygen permeability. *International Polymer Processing*, **13** (1), 58–66.

22 Bose, S., Raghu, H., and Mahanwar, P.A. (2006) Mica reinforced nylon-6: effect of coupling agents on mechanical, thermal, and dielectric properties. *Journal of Applied Polymer Science*, **100** (5), 4074–4081.

23 Bose, S. and Mahanwar, P.A. (2005) Effects of titanate coupling agent on the properties of mica-reinforced nylon-6 composites. *Polymer Engineering and Science*, **45** (11), 1479–1486.

24 Pinto, U.A., Visconte, L.L.Y., and Nunes, R.C.R. (2001) Mechanical properties of thermoplastic polyurethane elastomers with mica and aluminum trihydrate. *European Polymer Journal*, **37** (9), 1935–1937.

25 Gan, D.J., Lu, S.Q., Song, C.S., and Wang, Z.J. (2001) Mechanical properties and frictional behavior of a mica-filled poly(aryl ether ketone) composite. *European Polymer Journal*, **37** (7), 1359–1365.

26 Osman, M.A., Atallah, A., Muller, M., and Suter, U.W. (2001) Reinforcement of poly(dimethylsiloxane) networks by mica flakes. *Polymer*, **42** (15), 6545–6556.

27 Nugay, N., Kusefoglu, S., and Erman, B. (1997) Swelling and static–dynamic mechanical behavior of mica-reinforced linear and star-branched polybutadiene composites. *Journal of Applied Polymer Science*, **66** (10), 1943–1952.

28 Rashid, E.S.A., Ariffin, K., Akil, H.M., and Kooi, C.C. (2008) Mechanical and thermal properties of polymer composites for electronic packaging application. *Journal of Reinforced Plastics and Composites*, **27** (15), 1573–1584.

29 Allard, R.C., Vukhanh, T., and Chalifoux, J.P. (1989) Fatigue crack-propagation in

mica-filled polyolefins. *Polymer Composites*, **10** (1), 62–68.

30 Ghosh, P., Chattopadhyay, B., and Sen, A.K. (1998) Modification of low density polyethylene (LDPE) by graft copolymerization with some acrylic monomers. *Polymer*, **39** (1), 193–201.

31 Garcia-Martinez, J.M., Laguna, O., Areso, S., and Collar, E.P. (2001) Polypropylene/mica composites modified by succinic anhydride-crafted atactic polypropylene: a thermal and mechanical study under dynamic conditions. *Journal of Applied Polymer Science*, **81** (3), 625–636.

32 Lushcheikin, G.A. (1998) Model representation of filled polymers. *Measurement Techniques*, **41** (8), 762–766.

33 Rochette, A., Choplin, L., and Tanguy, P.A. (1988) Rheological study of mica-filled polypropylene as influenced by a coupling agent. *Polymer Composites*, **9** (6), 419–425.

34 Anderson, L.L. and Farris, R.J. (1988) A predictive model for the mechanical-behavior of particulate composites. *Polymer Engineering and Science*, **28** (8), 522–528.

35 Sidess, A., Holdengraber, Y., and Buchman, A. (1993) A fundamental model for prediction of optimal particulate composite properties. *Composites*, **24** (4), 355–360.

36 Jarvela, P.A., Li, S.C., and Jarvela, P.K. (1997) Dynamic mechanical and mechanical properties of polypropylene poly(vinyl butyral) mica composites. *Journal of Applied Polymer Science*, **65** (10), 2003–2011.

37 Gachter, R. and Muller, H. (1987) *Plastic Additives Handbook: Stabilizers, Processing Aids, Plasticizers, Fillers, Reinforcements, Colorants for Thermoplastics*, Hanser Publishers.

38 Kovacevic, V., Lucic, S., and Cerovecki, Z. (1997) Influence of filler surface pre-treatment on the mechanical properties of composites. *International Journal of Adhesion and Adhesives*, **17** (3), 239–245.

39 Ismail, H. and Jaffri, R.M. (1999) Physico-mechanical properties of oil palm wood flour filled natural rubber composites. *Polymer Testing*, **18** (5), 381–388.

40 Kalita, D., Ghosh, S.R., and Saikia, C.N. (1999) Medium density particle board from weeds. *Journal of Scientific & Industrial Research*, **58** (9), 705–710.

41 Karr, G.S., Cheng, E.H., and Sun, X.S. (2000) Physical properties of strawboard as affected by processing parameters. *Industrial Crops and Products*, **12** (1), 19–24.

42 Khalil, H., Ismail, H., Rozman, H.D., and Ahmad, M.N. (2001) The effect of acetylation on interfacial shear strength between plant fibres and various matrices. *European Polymer Journal*, **37** (5), 1037–1045.

43 Verbeek, C.J.R. and Pickering, K.L. (2007) Recent developments in polymer consolidated composites. *Journal of Reinforced Plastics and Composites*, **26** (16), 1607–1624.

44 Zhao, R., Huang, J., Bin, S., and Dai, G. (2001) Study of the mechanical properties of mica-filled polypropylene-based GMT composite. *Journal of Applied Polymer Science*, **82** (11), 2719–2728.

45 Verbeek, C.J.R. and Focke, W.W. (2002) Modelling the Young's modulus of platelet reinforced thermoplastic sheet composites. *Composites Part A*, **33** (12), 1697–1704.

46 Bose, S. and Mahanwar, P.A. (2005) Influence of particle size and particle size distribution on mica filled nylon 6 composite. *Journal of Materials Science*, **40** (24), 6423–6428.

47 Mahanwar, P.A., Bose, S., and Tirumalai, A.V. (2006) The influence of interfacial adhesion on the predicted Young's modulus of mica-reinforced nylon-6. *Polymer-Plastics Technology and Engineering*, **45** (5), 597–600.

48 Yazdani, H., Morshedian, J., and Khonakdar, H.A. (2006) Effect of maleated polypropylene and impact modifiers on the morphology and mechanical properties of PP/mica composites. *Polymer Composites*, **27** (6), 614–620.

49 Busigin, C., Martinez, G.M., Woodhams, R.T., and Lahtinen, R. (1983) Factors affecting the mechanical properties of mica-filled polypropylenes. *Polymer Engineering & Science*, **23** (14), 766–770.

50 Pukanszky, B. and Moczo, J. (2004) Morphology and properties of particulate filled polymers. *Macromolecular Symposia*, **214**, 115–134.

51 Hanu, L.G., Simon, G.P., and Cheng, Y.B. (2005) Preferential orientation of muscovite in ceramifiable silicone composites. *Materials Science and Engineering A*, **398** (1–2), 180–187.

52 Meddad, A. and Fisa, B. (1997) Filler–matrix debonding in glass bead-filled polystyrene. *Journal of Materials Science*, **32** (5), 1177–1185.

53 Inubushi, S., Ikeda, T., Tazuke, S., Satoh, T., and Kumagai, Y. (1988) Aminimide-cured epoxy resins as surface modifiers for mica flakes in particle-reinforced thermoplastics. *Journal of Materials Science*, **23** (2), 535–540.

54 Ahmed, A. and Jones, F.R. (1990) A review of particulate reinforcement theories for polymer composites. *Journal of Materials Science*, **25**, 4933–4942.

55 Casenave, S., AitKadi, A., and Riedl, B. (1996) Mechanical behaviour of highly filled lignin/polyethylene composites made by catalytic grafting. *Canadian Journal of Chemical Engineering*, **74** (2), 308–315.

56 Zhou, X.D., Xiong, R.H., and Lin, Q.F. (2006) Effect of block copolymer coupling agents on properties of mica reinforced polymeric composites. *Journal of Materials Science*, **41** (23), 7879–7885.

57 Bhattacharya, A. (2000) Radiation and industrial polymers. *Progress in Polymer Science*, **25** (3), 371–401.

58 Verbeek, C.J.R. (2002) Highly filled polyethylene/phlogopite composites. *Materials Letters*, **52** (6), 453–457.

59 Verbeek, C.J.R. (2003) The influence of interfacial adhesion, particle size and size distribution on the predicted mechanical properties of particulate thermoplastic composites. *Materials Letters*, **57** (13–14), 1919–1924.

60 Nugay, N. and Erman, B. (2001) Hybrid reinforcement in nitrile rubber composites. *Macromolecular Symposia*, **169**, 269–274.

61 Bokobza, L. and Nugay, N. (2001) Orientational effect of mica in fumed silica reinforced composites. *Journal of Applied Polymer Science*, **81** (1), 215–222.

62 Sirelli, L., Prado, R.M.K., Tavares, M.I.B., Nunes, R.C.R., and Dias, M.L. (2008) Molecular dynamics of poly(ethylene terephthalate)/muscovite mica composite by low-field NMR. *International Journal of Polymer Analysis and Characterization*, **13** (3), 180–189.

63 Kuelpmann, A., Osman, M.A., Kocher, L., and Suter, U.W. (2005) Influence of platelet aspect ratio and orientation on the storage and loss moduli of HDPE–mica composites. *Polymer*, **46** (2), 523–530.

64 Chang, J.H. and Mun, M.K. (2007) Nanocomposite fibers of poly(ethylene terephthalate) with montmorillonite and mica: thermomechanical properties and morphology. *Polymer International*, **56** (1), 57–66.

65 Yang, R., Liu, Y., Yu, J., and Wang, K.H. (2006) Thermal oxidation products and kinetics of polyethylene composites. *Polymer Degradation and Stability*, **91** (8), 1651–1657.

66 Yang, R., Yu, J., Liu, Y., and Wang, K.H. (2005) Effects of inorganic fillers on the natural photo-oxidation of high-density polyethylene. *Polymer Degradation and Stability*, **88** (2), 333–340.

67 Morgan, A.B., Chu, L.L., and Harris, J.D. (2005) A flammability performance comparison between synthetic and natural clays in polystyrene nanocomposites. *Fire and Materials*, **29** (4), 213–229.

68 Pinto, U.A., Visconte, L.L.Y., Gallo, J., and Nunes, R.C.R. (2000) Flame retardancy in thermoplastic polyurethane elastomers (TPU) with mica and aluminum trihydrate (ATH). *Polymer Degradation and Stability*, **69** (3), 257–260.

69 Sato, S., Sakata, Y., Aoki, J., and Kubota, K. (2006) Effects of filler on heat transmission behavior of flowing melt polymer composites. *Polymer Engineering and Science*, **46** (10), 1387–1393.

70 Escocio, V.A., Visconte, L.L.Y., Nunes, R.C.R., and de Oliveira, M.G. (2008) Rheology and processability of natural rubber composites with mica.

International Journal of Polymeric Materials, **57** (4), 374–382.

71 Yang, C.F., Smyrl, W.H., and Cussler, E.L. (2004) Flake alignment in composite coatings. *Journal of Membrane Science*, **231** (1–2), 1–12.

72 Alves, V.D., Costa, N., and Coelhoso, I.M. (2011) Barrier properties of biodegradable composite films based on kappa-carrageenan/pectin blends and mica flakes. *Carbohydrate Polymers*, **79** (2), 269–276.

73 DeRocher, J.P., Gettelfinger, B.T., Wang, J.S., Nuxoll, E.E., and Cussler, E.L. (2005) Barrier membranes with different sizes of aligned flakes. *Journal of Membrane Science*, **254** (1–2), 21–30.

74 Zhang, Y.H., Dang, Z.M., Xin, J.H., Daoud, W.A., Ji, J.H., Liu, Y.Y., Fei, B., Li, Y.Q., Wu, J.T., Yang, S.Y., and Li, L.F. (2005) Dielectric properties of polyimide–mica hybrid films. *Macromolecular Rapid Communications*, **26** (18), 1473–1477.

75 Kandasubramanian, B. and Gilbert, M. (2005) An electroconductive filler for shielding plastics. *Macromolecular Symposia*, **221**, 185–195.

76 Du, J. and Kumagai, Y. (1994) Resistivity–temperature behaviours of epoxy nickel coated mica composites. *Journal of Materials Science Letters*, **13** (24), 1786–1788.

77 Padawer, G.E. and Beecher, N. (1970) On the strength and stiffness of planar reinforced plastic resins. *Polymer Engineering & Science*, **10** (3), 185–192.

78 Ward, I. and Hadley, D. (1993) *An Introduction to the Mechanical Properties of Solid Polymers*, John Wiley & Sons Ltd, Chichester.

79 Jacquet, E., Trivaudey, F., and Varchon, D. (2000) Calculation of the transverse modulus of a unidirectional composite material and of the modulus of an aggregate: application of the rule of mixtures. *Composites Science and Technology*, **60** (3), 345–350.

80 Akiva, U., Itzhak, E., and Wagner, H.D. (1997) Elastic constants of three-dimensional orthotropic composites with platelet/ribbon reinforcement. *Composites Science and Technology*, **57** (2), 173–184.

81 Firoozian, P., Akil, H.M., and Khalil, H.P.S.A. (2010) Prediction of mechanical properties of mica-filled epoxy composite using various mechanical models. *Journal of Reinforced Plastics and Composites*, **29**, 2368–2378.

82 Verbeek, C.J.R. (1991) The Young's modulus of compression moulded LLDPE-phlogopite composites. Thesis. University of Pretoria.

22
Viscoelastically Prestressed Polymeric Matrix Composites
Kevin S. Fancey

22.1
Introduction

22.1.1
Prestress in Composite Materials

Although the application of prestressing to structural materials such as concrete is a well-known concept, an awareness of possible benefits from making fiber-reinforced polymeric matrix composites with some residual level of (compressive) prestress seems to be comparatively recent. Prestressed concrete manufacturing principles may be used, so that applying tension to fibers (e.g., glass) as the matrix material cures, creates compressive stresses within the solidified matrix. Thus, matrix compression is balanced by residual tension within the fibers in the resulting elastically prestressed polymeric matrix composite (EPPMC). Focusing on composite laminates, Tuttle [1] was among the earliest investigators to evaluate this elastic prestressing principle; subsequently it has been found to reduce thermally induced warpage [2] and improve mechanical properties [3].

It seems that studies of (simpler) unidirectional glass fiber EPPMCs have been implemented only within the last decade. For beam-shaped geometries, Motahhari and Cameron [4, 5] found that elastic prestressing increased impact resistance, flexural stiffness, and strength by up to 33%, when compared with unstressed (control) counterparts. Results from a study by Hadi and Ashton [6] indicate that elastic prestressing can increase tensile strength and (tensile) elastic modulus by $\sim$25 and $\sim$50%, respectively. Explanations for these improvements have been based on matrix compressive stress and fiber tension effects, which can (i) impede and deflect crack propagation and (ii) reduce composite strain resulting from external tensile and bending loads [4–6].

Clearly, the elastic prestressing technique should offer opportunities for improved mechanical properties in fiber-reinforced polymeric matrix composites without the need to increase mass (e.g., through increased fiber content) or section dimensions. There are, however, two potential drawbacks [7]. First, the need to apply fiber tension during matrix curing can compromise fiber orientation, length, and spatial

Polymer Composites: Volume 1, First Edition. Edited by Sabu Thomas, Kuruvilla Joseph,
Sant Kumar Malhotra, Koichi Goda, and Meyyarappallil Sadasivan Sreekala
© 2012 Wiley-VCH Verlag GmbH & Co. KGaA. Published 2012 by Wiley-VCH Verlag GmbH & Co. KGaA.

distribution, which ultimately could lead to restrictions on product geometry. Second, the matrix material, being polymeric, may undergo creep in an attempt to counteract the compressive stresses it is subjected to. Specifically, localized matrix creep effects near the fiber–matrix interface would be expected to cause the prestress effect to deteriorate with time. To date, there appear to be no published investigations on possible changes to the long-term characteristics of EPPMCs.

This chapter covers the development of prestressed polymeric matrix composite (PPMC) materials based on an alternative prestressing principle [8], which avoids the potential drawbacks highlighted above. The resulting material is a viscoelastically prestressed polymeric matrix composite (VPPMC). Developing the techniques to investigate VPPMCs has enabled research to progress from simple embryonic proof-of-concept experiments to evaluation of key mechanical property improvements. The stage has now been reached where previously reported findings can be reviewed and integrated, enabling the potential for future applications to be discussed.

22.1.2
Principles of Viscoelastic Prestressing

Essentially, the principle involves the use of fibers to impart compressive stresses through viscoelastic recovery. Here, polymeric fibers are stretched under a load for a period of time to induce creep. The load is subsequently released and, although the fibers initially undergo (virtually instantaneous) elastic recovery, a significant proportion of the total fiber deformation is viscoelastic, that is, further recovery is time-dependent. The (now unrestrained) fibers are molded into a resin matrix. Thus, when the matrix has cured, compressive stresses are imparted by the viscoelastically strained fibers as they continue to attempt strain recovery against the surrounding (solid) matrix material. This viscoelastically generated compressive prestress can enhance the mechanical properties of the resulting composite material.

As a precursor to research with other fibers, investigations to date have focused on nylon 6,6, as this is a low cost fiber material with good mechanical properties. This was a continuous multifilament untwisted yarn (Goodfellow Cambridge Ltd, UK) with 140 filaments, 27.5 µm filament diameter. Room temperature curing polyester and epoxy resins were used as matrix materials, these being of sufficient optical transparency to facilitate internal inspection of composite samples, for example, to observe fiber lay and (following mechanical testing) fiber–matrix debonding characteristics.

22.2
Preliminary Investigations: Evidence of Viscoelastically Generated Prestress

22.2.1
Objectives

The initial need was to determine the feasibility of VPPMCs. Thus, evidence of viscoelastically generated prestress within a polymeric matrix was required and moreover, that it could be beneficial to mechanical properties. Therefore, early investigations were centered on the following objectives: (i) establish the fiber creep conditions for suitable viscoelastic recovery, (ii) demonstrate direct evidence of viscoelastically induced prestress, and (iii) evaluate by simple means a benefit to mechanical properties that VPPMCs may provide. Full details have been previously reported [9, 10] and are summarized below.

22.2.2
Creep Conditions for Viscoelastic Recovery

The principal aim was to establish the conditions necessary for fibers to retain a usable level of residual viscoelastic strain after releasing the applied creep load. Polyester resin with a room-temperature curing time of ∼2 h was to be used initially, hence viscoelastic recovery over this time would indicate the fiber strain available for creating prestress following molding and matrix curing of VPPMC samples. A stretching rig was used, in which 0.5 m lengths of yarn were loaded by applying weights to a counter-balanced platform. Creep and resulting recovery strain could be measured between two inked marks on the yarn by means of a digital cursor.

It was found that if the yarn was annealed prior to stretching, a significantly higher residual strain could be obtained, than from material in the as-received condition (Section 22.3.2). Referring to the work of others [11, 12], all annealing was performed at 150 °C for 30 min. A total strain of ∼12% could be obtained by applying a 24 h tensile creep stress of ∼350 MPa. On releasing the stress, the elastic strain was removed, leaving an initial viscoelastic recovery strain of ∼3%, dropping to ∼2.5% after 2 h, and ∼2% after 100 h; that is, the strain decreased only very slowly with time.

22.2.3
Visual Evidence of Viscoelastically Induced Prestress

By using an optically transparent matrix material, direct evidence of viscoelastically induced prestress could be achieved through photoelasticity principles. A 1.6 mm diameter nylon 6,6 monofilament was used in favor of yarn; this facilitated handling and ensured that any stress patterns would be attributed to a single line source. The annealed filament, being of relatively low strength, was only subjected to ∼20% of the creep stress normally applied to its fiber counterpart. The resulting test and control monofilaments were molded in a polyester resin with a peroxide-based hardener, using two identical open cast polished aluminum molds. The resin gelled in 10–15 min and, following curing (within 2 h at room temperature), the samples were demolded.

Figure 22.1 shows these samples mounted on a Sharples Polariscope under cross-polarized monochromatic (sodium) light. Quarter wave plates were fitted, so that only isochromatic fringes, representing lines of constant shear stress, would be observable. Although a result of low-level creep, Figure 22.1 shows clear evidence of residual stresses around the filament in the test sample. By physically distorting the test sample, the resulting fringe movement indicated that the residual stress was compressive [10].

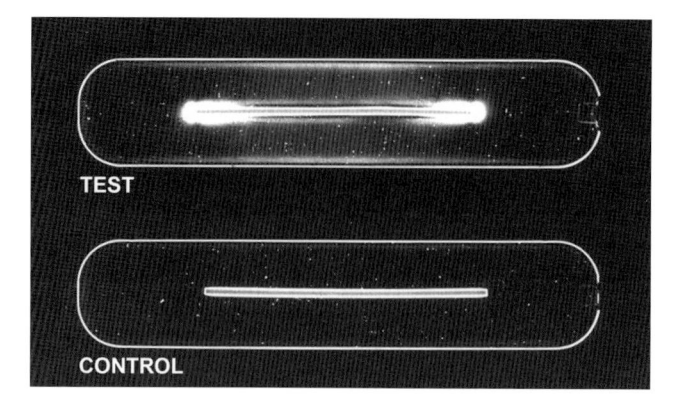

Figure 22.1 Polariscope image of nylon 6,6 monofilament molded in 150 × 30 × 2 mm polyester resin samples. Note the stress pattern resulting from viscoelastic recovery in the test sample, in contrast with the control sample. Photograph reprinted from Figure 4 in Ref. [10] with permission from Sage Publications.

22.2.4
Initial Mechanical Evaluation: Impact Tests

The most straightforward procedure for mechanical evaluation was to produce batches of composite samples for Charpy impact testing. For each batch, there was one set of VPPMC "test" samples and a corresponding set of "control" samples, the latter containing unstressed annealed fiber (but otherwise identical). Producing a batch required open-casting two strips of polyester resin (as used for the photo-elasticity study in Section 22.2.3) simultaneously from the same resin mix; two aluminum molds (with polished channels) were used. Each strip was embedded with a continuous length of either "test" (previously stretched) or "control" nylon yarn and in both cases, the yarn was combed out into a flat ribbon immediately prior to molding. Following sufficient curing, each strip was cut into five samples (80 × 10 × 3.2 mm), ready for impact testing. Owing to the limited fiber availability from the stretching rig (Section 22.2.2) for each batch and the tendency for fibers to sink to the bottom of the mold prior to the onset of curing, fibers were concentrated toward one side of the sample, occupying 50–60% of sample thickness. Averaged over the total sample, the fiber volume fraction, V_f, was 2–3%. With the objective being to evaluate the effects of fiber-induced prestress, samples were mounted flatwise for impact testing with the fiber-rich side facing away from the Charpy hammer, as this side would be put into tension during impact. Here, a Ceast Resil 25 Charpy machine with 7.5 J hammer was used, operating in accordance with the EN ISO 179 standard. Although sample dimensions concurred with EN ISO 179 Specimen Type 2 specifications, a 24 mm span (Specimen Type 3) was used instead of 60 mm, to prevent the possibility of some samples falling below minimum energy readings set by the standard.

Tests showed that the VPPMC samples absorbed, on average, 25% more energy than their control counterparts, with some batches approaching a 50% increase.

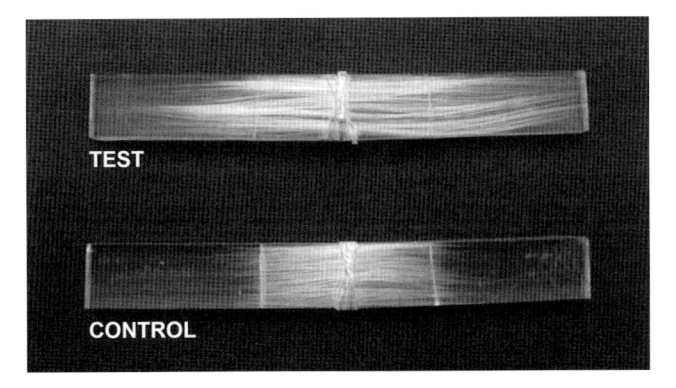

Figure 22.2 Typical appearance of test (VPPMC) and control samples following impact testing. Note the larger region of fiber–matrix debonding in the test sample.

Figure 22.2 shows a typical test and control sample following impact testing. The region of fiber–matrix debonding is larger in the test sample, and this was typical for all batches studied. Also, it was noted that fiber fracture was less frequent in the test samples. Comparable observations were made by Motahhari and Cameron, in that the Charpy impact damage area was much less localized in their EPPMC samples, compared with unstressed samples [4]. This concurrence thus provided further evidence of the capability for viscoelastic mechanisms to produce compressive stress.

22.3
Time–Temperature Aspects of VPPMC Technology

22.3.1
A Mechanical Model for Polymeric Deformation

To venture beyond the preliminary investigation stage required answers to two fundamental questions: (i) how long can viscoelastic prestressing last and (ii) how much force can it provide? Answering these questions was facilitated by a practical knowledge of time-dependent polymeric deformation. Thus, prior to addressing (i) and (ii), a mechanical model is outlined here.

Mechanical models are commonly used to represent polymeric deformation. The simplest are the Maxwell (spring and dashpot in series) and Voigt (spring and dashpot in parallel) models. More complex models use three or four elements, and these include the Zener and combined Maxwell–Voigt arrangements. Although these can be used to represent creep, recovery, and stress relaxation [13], it is said that real materials in general are not describable by models with a small number of springs and dashpots; they often lack the necessary accuracy for quantitative prediction [14, 15]. Essentially, these shortcomings arise from the restricted timescales that models with only a small number of elements can represent. To represent broad timescales, a model that provides a distribution of retardation or relaxation times is required, that is, one with many elements. Increasing the number of elements obviously improves the quantitative accuracy to which a model can represent the deformation response of a real polymeric material. Nevertheless, the mathematical complexity could reach a point where the number of parameters becomes impractically large and (potentially complex) methods of approximation are required [15, 16].

Irrespective of complexity, these spring and dashpot models all share a common factor, in that viscoelastic deformation is assumed to vary smoothly. This means that a material undergoing creep, recovery, or stress relaxation would do so continuously with time. An alternative approach has been presented, however, in which viscoelastic changes are said to occur through incremental jumps. On a molecular level, this could be considered as segments of molecules jumping between positions of relative stability within the polymer structure. These jumps are described by the action of time-dependent latch elements, and this concept was incorporated into a mechanical model, principally to explain viscoelastic recovery [17]. Subsequently, the mechanical latch approach was extended to describe the action of creep, recovery, and stress relaxation [18], and this is shown in Figure 22.3.

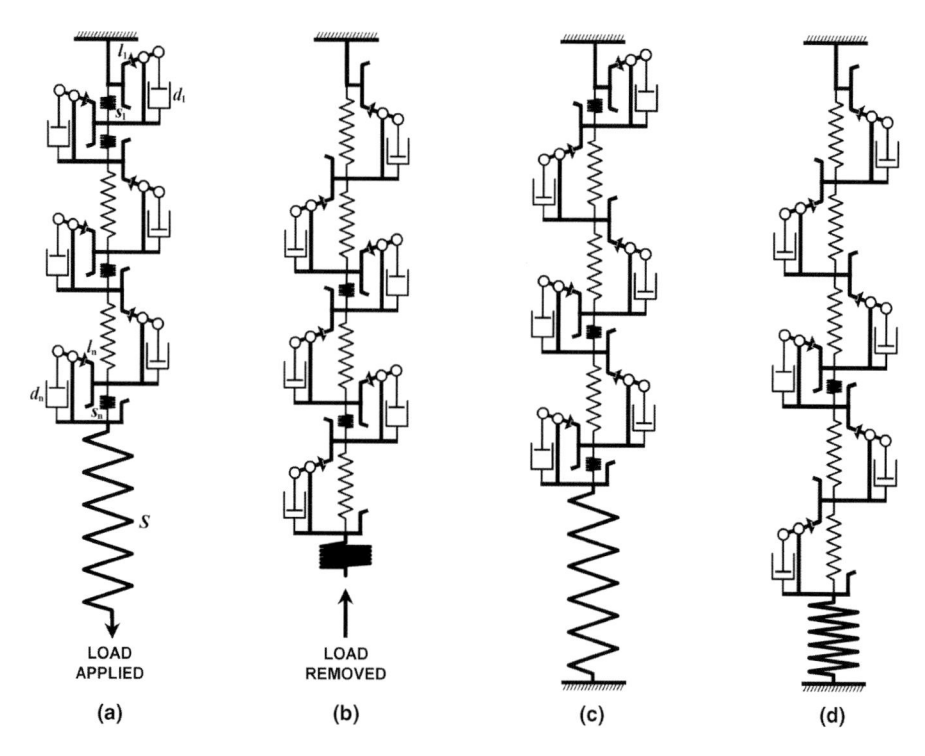

Figure 22.3 A latch-based spring and dashpot model to represent time-dependent polymeric deformation: (a) creep, (b) recovery, (c) stress relaxation at time t, and (d) stress relaxation at time $(t + \Delta t)$, showing an increase in the number of latch elements triggered into their extended positions. These deformation processes would need to be represented by many latch elements in a real polymeric material, to provide a broad distribution of trigger times. Reprinted from Figure 1 in Ref. [18] with permission from Springer Science and Business Media.

In contrast with conventional models, the springs and dashpots control latches in Figure 22.3. Thus once a latch is triggered, the structure extends (in creep) or contracts (in recovery) or extends to reduce the elastically stored stress (in stress relaxation). The triggering time of each latch ($l_1, l_2, \ldots l_n$) depends on the stiffness of the corresponding spring ($s_1, s_2, \ldots s_n$) and viscosity of the dashpot ($d_1, d_2, \ldots d_n$). Instantaneous (elastic) deformation is represented by the large spring, S.

Figure 22.3 enables the following comments to be made. Under creep or stress relaxation conditions in which only low loadings are applied, some latch units with very stiff springs may not be subjected to sufficient force for the latch to trigger over any timescale, that is, not all units are available. Increasing the creep or stress relaxation loading will thus increase availability of latch units and in general, reduce triggering times, thereby increasing (for creep) the strain rate. For recovery, however, these trigger times will be longer due to the absence of an externally applied load; hence recovery strain rates become lower than those observed under creep. During recovery, a latch coupled with a stiff spring and low viscosity dashpot will trigger earlier than one with a soft spring and high viscosity dashpot. If a proportion of latch units have triggering times that approach infinity on recovery, these can be considered to represent viscous flow effects.

Therefore, from a phenomenological viewpoint, the mechanical model shown in Figure 22.3 can be seen to represent all three forms of polymeric deformation. Mathematically, this incremental step argument is supported by stretched exponential functions for these three cases. Thus for example, when the applied load from a material undergoing creep is released, there will be some instantaneous (elastic) recovery, which is then followed by time-dependent recovery strain (Figure 22.3b). This time-dependent recovery strain, $\varepsilon_{rvis}(t)$, is given by the following equation [17]:

$$\varepsilon_{rvis}(t) = \varepsilon_r \left[\exp\left(-\left(\frac{t}{\eta_r}\right)^{\beta_r} \right) \right] + \varepsilon_f \quad (22.1)$$

Equation (22.1) is based on the Weibull or Kohlrausch–Williams–Watts (KWW) distribution function. Here, ε_f is the permanent strain from viscous flow effects, that is, it is the residual strain as recovery time t approaches ∞. The ε_r function, for viscoelastic strain recovery, depends on the Weibull shape parameter, β_r, and characteristic life, η_r.

Of key interest is that empirical data correlates very well with Eq. (22.1), over a very broad range of timescales, and also with corresponding Weibull-based equations for creep and stress relaxation [17, 18]. Although these equations are not, mathematically, direct representations of the mechanical model in Figure 22.3, they share common ground with the incremental jumps argument for two reasons. First, since the Weibull function is used in reliability engineering to represent the time-dependent failure of elements in a population [19], this is synonymous with polymeric deformation being represented by a population of time-dependent latches; thus each latch may be considered to have "failed" once it has been triggered. Second, the Eyring potential energy barrier relationship describes the motion of matter in terms of molecular jumps [20]. For stress relaxation, the KWW function, which is identical in form to the Weibull function, is considered to be an approximation to the Eyring relationship [21], and this is supported by empirical evidence [18].

22.3.2
Accelerated Aging: Viscoelastic Recovery

Localized matrix creep effects near the fiber–matrix interface in an EPPMC would be expected to cause the prestress effect to deteriorate with time (Section 22.1.1). For VPPMCs, however, long-term viscoelastic activity within the fibers should cause them to remain active in response to any changes in the matrix [7]. Thus, a key aspect to the likely success of VPPMCs is the length of time over which viscoelastic activity can occur. Figure 22.4 summarizes recovery strain data at 20 °C for annealed and as-received nylon 6,6 yarn after being subjected to identical creep conditions (342 MPa

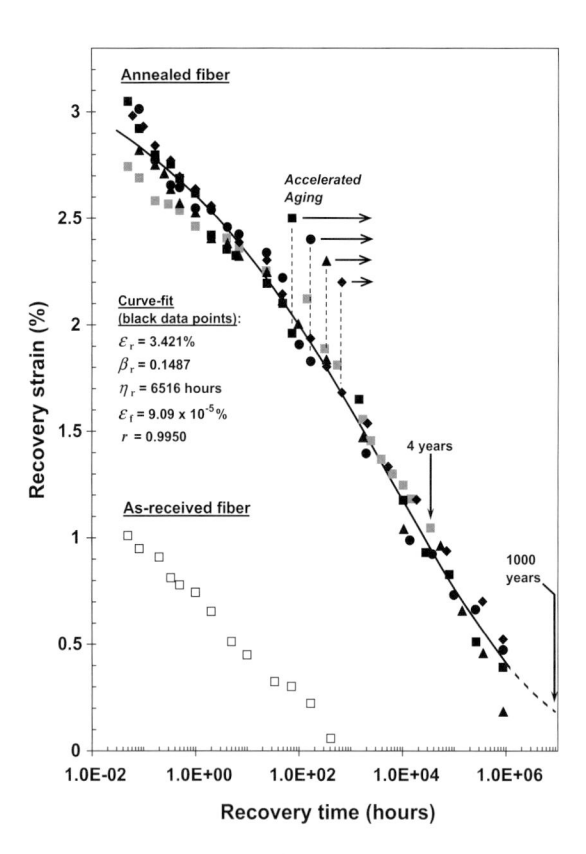

Figure 22.4 Recovery strain data [7, 10, 22] at 20 °C from nylon 6,6 yarn after 24 h creep at 342 MPa. For yarn annealed before creep, gray data points represent real-time measurements up to 3.5×10^4 h (4 years) and black data points are from four samples, each subjected to accelerated aging from different recovery times (as shown) up to 9×10^5 h (100 years). The curve shows Eq. (22.1) fitted to the black data, with listed parameters and correlation coefficient, r.

Curve-fit (black data points):
$\varepsilon_r = 3.421\%$
$\beta_r = 0.1487$
$\eta_r = 6516$ hours
$\varepsilon_f = 9.09 \times 10^{-5}\%$
$r = 0.9950$

for 24 h) [7, 10, 22]. Although recovery strain in the as-received yarn approaches zero within 1000 h of releasing the creep stress, data for the annealed yarn clearly shows viscoelastic activity over much longer timescales [10]. The gray data points in Figure 22.4 represent strain measurements from annealed yarn in real time and, at 4 years, the strain still exceeds 1% [7].

To investigate viscoelastic activity beyond a few years requires the use of accelerated aging techniques. Sets of creep or stress relaxation data measured over a limited period of time, taken from a polymeric material at different temperatures, can often be assembled into a single "master curve" by using an appropriate shift factor, a_T. Here, a_T equates a change in temperature to a shift on the time axis; hence, the master curve can represent a data plot spanning a vast range of timescales. This time–temperature superposition principle thus provides the basis for subjecting viscoelastically recovering samples to an elevated temperature, to accelerate the aging (recovery) process. Clearly, to exploit this principle, the a_T value must be known and, based on published data from nylon 6,6 fiber undergoing stress relaxation [11] and creep [23], the resulting linear relationship between log a_T and temperature enabled the former to be determined for a suitable temperature transition. The value for log a_T at 60 °C relative to 20 °C was found to be 3.577; therefore, stress relaxation or creep in fibers at 60 °C would be expected to occur 3776 times faster than at 20 °C under the same loading conditions [22]. With the log a_T–temperature relationship being common to both creep and stress relaxation, it was concluded in Ref. [22] that this log a_T value may also be applied to recovery strain data.

The black data points in Figure 22.4 represent recovery strain measurements from annealed yarn, using accelerated aging up to the equivalent of 100 years (9×10^5 h) at 20 °C. Here, four separate samples of yarn were subjected to the same anneal/creep conditions used for the gray data. These samples were then allowed to undergo recovery in real time (at ~20 °C), each over a different period of up to four weeks. Accelerated aging followed, in which samples were heated at 60 °C over successively increasing periods of time between recovery strain readings (taken at 20 °C). From the a_T value, 2.3215 h at 60 °C was equivalent to 1 year (8766 h) at 20 °C. For recovery times greater than 1 h, it is encouraging to note that Figure 22.4 shows excellent agreement between the black and gray data points. The slightly lower values in the gray data below 1 h are attributed to errors resulting from single readings with limited environmental control [22].

Figure 22.4 indicates that viscoelastic activity continues beyond the measured timescale but strain measurements below 0.5% become increasingly impractical. The curve in Figure 22.4, fitted to the black data points, represents the time-dependent recovery strain, $\varepsilon_{rvis}(t)$, from Eq. (22.1). This was fitted to the data with commercially available software (*CurveExpert 1.3*) and the resulting parameter values are listed in Figure 22.4. It is useful to note that ε_f is predicted to be close to zero ($<10^{-4}$%); thus virtually all available recovery is viscoelastic, indicating that the viscoelastic prestressing mechanism will not be limited by viscous flow effects. Extrapolating to 1000 years (9×10^6 h), Eq. (22.1) predicts that $\varepsilon_{rvis}(t)$ is ~0.2%, that is, recovery strain is still three orders of magnitude greater than ε_f. This suggests that viscoelastic activity, at least under these conditions, is a long-term phenomenon.

22.3.3
Force–Time Measurements

With long-term viscoelastic recovery strain decreasing to very low levels, the resulting prestress that may be generated in a VPPMC under these conditions must be questioned. Nevertheless, although the results in Section 22.3.2 demonstrate long-term viscoelastic activity, they provide no information on the associated force produced from fibers constrained within a solid matrix. Clearly, knowledge of the magnitude and time dependency of this force is vital to the development of VPPMCs. This aspect was addressed by subjecting annealed nylon 6,6 yarn to the creep-recovery test cycle represented by Figure 22.5. Here, a creep stress of 320 MPa was applied for 24 h (i.e., similar to conditions used for data in Figure 22.4) to a loop of annealed yarn. On removing the stress, the yarn was transferred to a bespoke recovery force measurement rig [24], attached in a loose state. Within a short time, Δt, the yarn was allowed to contract to a fixed strain to become taut. The force resulting from this state was then monitored with a transducer built into the rig, as a function of time.

Figure 22.6 shows the results, measured over 2700 h. The y-axis represents $\sigma(t)$, the stress from recovery force exerted by the yarn over its cross-sectional area, and the curve fit to the data points comes from the following Weibull/KWW-based equation [25]:

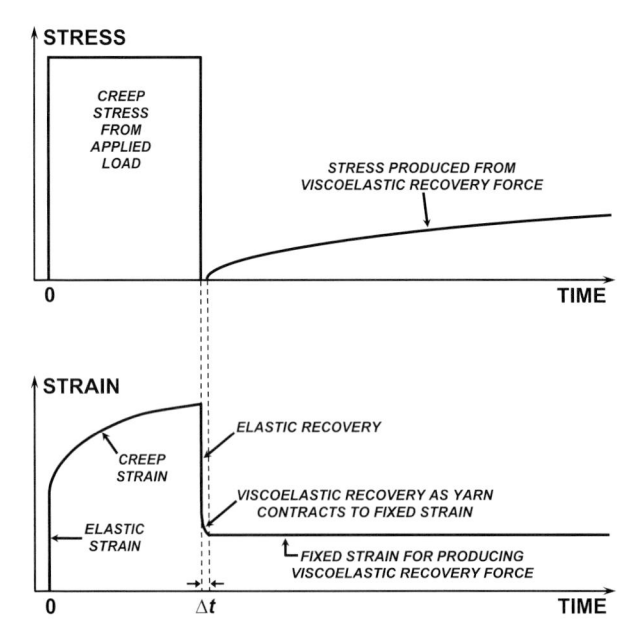

Figure 22.5 Schematic of the creep–recovery test cycle to investigate the recovery force–time characteristics from nylon yarn. After Figure 1 in Ref. [25].

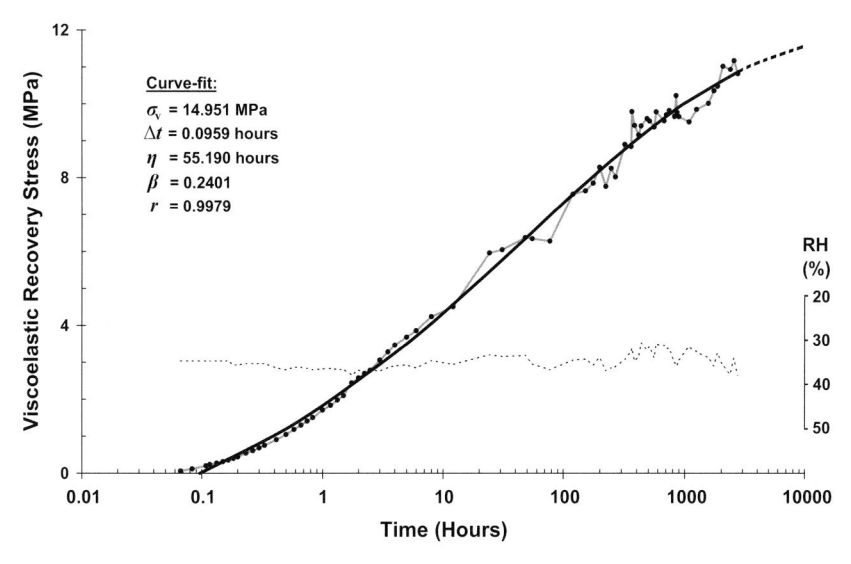

Figure 22.6 Viscoelastic recovery stress (force exerted across fibers) output from nylon 6,6 yarn, for readings recorded at 20–20.9 °C, 31–39% RH. The curve shows Eq. (22.2) fitted to the data, with listed parameters and correlation coefficient, r. Derived from data in Ref. [25].

Curve-fit:
σ_v = 14.951 MPa
Δt = 0.0959 hours
η = 55.190 hours
β = 0.2401
r = 0.9979

$$\sigma(t) = \sigma_v \left[\exp\left(-\left(\frac{\Delta t}{\eta}\right)^\beta\right) - \exp\left(-\left(\frac{t}{\eta}\right)^\beta\right) \right] \tag{22.2}$$

The σ_v function is the time-dependent viscoelastically generated stress, as determined by the characteristic life (η) and shape (β) parameters. Equation parameter values are shown in Figure 22.6. Despite the humidity-dependent fluctuations, the measured output fits well to Eq. (22.2), and $\sigma(t)$ is shown to reach 10.9 MPa at 2700 h and approach a limiting value of 12.0 MPa as t goes to infinity, that is, 3.8% of applied creep stress.

Since publication of the above findings in Ref. [25], the force output from this experiment has continued to be monitored and, with 25 000 h (2.8 years) of recovery time now being reached, $\sigma(t)$ has continued to progress in accordance with Eq. (22.2). This means, therefore, that viscoelastic recovery force does not deteriorate with time, at least under the simulated (ideal) matrix conditions represented by the experimental procedure outlined above. For real matrix conditions, the possibility of recovery force being reduced by localized matrix creep near the fiber–matrix interface should be negated by long-term viscoelastic activity (Section 22.3.2).

As mentioned in Section 22.3.1, viscoelastic changes may be considered to be molecular segments jumping between positions of relative stability within the polymer structure. For viscoelastic recovery strain, this will be a thermally activated process, that is, recovery rate will increase with ambient temperature as discussed in Section 22.3.2. Similar arguments can be proposed for the recovery force, in that thermally activated slippage of molecular segments control the release rate of elastic energy stored within the material acquired during creep. Nylon 6,6 is a semicrystalline polymer and although these arguments might relate more readily to amorphous regions, the jumping of line segments or kinks through the crystalline regions of nylon 6,6 in response to an applied stress has also been discussed [11]. As indicated by discussion in Ref. [25], however, the relative contributions to viscoelastic recovery force from the crystalline and amorphous regions of this material are unknown.

22.3.4
Accelerated Aging: Impact Tests

As discussed in Section 22.3.2, although strain data extrapolated to 1000 years (at 20 °C) indicates that viscoelastic recovery remains active (Figure 22.4), strain measurements below 0.5% become increasingly impractical. Moreover, strain data provide no information on the associated force produced from fibers, which actually grows with time to a limiting value (Section 22.3.3). Therefore, mechanical evaluation of VPPMCs becomes the preferred method for investigating long-term performance over timescales in excess of 100 years at 20 °C. Charpy impact testing of samples subjected to accelerated aging thus provides a simple means.

Sample processing followed the procedures described in Section 22.2.4, with annealed yarn for the test samples being subjected to a creep stress of 342 MPa for 24 h. Two polyester resins were used: one was a clear-casting resin (Reichhold Polylite 32 032-00); the other was a semiopaque general purpose lay-up resin, which contained an internal mold release agent (DSM Synolite 6061-P-1). The clear casting (designated as "CC") and general purpose (designated as "GP") resin gel times were ∼25 and ∼15 min, respectively, and were sufficiently cured for demolding after ∼2 h at room temperature. Each batch produced five test and five control samples (80 × 10 × 3.2 mm) from the molded strips of material. Batches were then stored at room temperature (18–22 °C) for at least 168 h (1 week) prior to accelerated aging. Samples were subjected to the accelerated aging principles covered in Section 22.3.2. A calibrated fan-assisted oven was used, in which both test and control samples were heated to 60 °C for up to 2321.5 h, that is, equivalent to 1000 years at 20 °C. Samples were subsequently stored at room temperature for at least 168 h prior to impact testing, the Charpy test set-up being identical to that described in Section 22.2.4.

Table 22.1 summarizes the data from impact tests on batches allowed to age in real time (to ∼1000 h) and those subjected to accelerated aging up to 100 years [22], together with the most recent data (500 and 1000 years) [26]. The data show that the test (VPPMC) samples absorb more impact energy than corresponding control samples and, within the range of batch-to-batch variation, impact performance does not seem to deteriorate, even with the oldest samples. Test samples from both resins exhibited larger regions of fiber–matrix debonding with less frequent fiber fracture when compared with the control samples; that is, these characteristics were consistent with observations from initial impact test studies (Section 22.2.4).

The average GP resin impact energy values (Table 22.1a) are higher for both test and control samples, but increases in energy from the test samples, below 100 years, tend to be lower than those from the CC resin (Table 22.1b). This can be attributed to the internal release agent within the GP resin, which may weaken fiber-matrix

Table 22.1 Charpy impact test results from composite sample batches, each batch comprising five (VPPMC) test and five control samples.

Actual age (h)	Exposure to 60 °C (h)	Age equiv at 20 °C (years)	Mean impact energy (kJ/m²)		Increase in energy (%)
			Test ± std. error	Control ± std. error	
(a) For the matrix: general purpose (GP) lay-up resin					
100	0	0.01	61.6 ± 2.7	50.9 ± 1.3	21.0
335	0	0.04	59.9 ± 4.6	49.2 ± 2.9	21.7
1008	0	0.11	54.9 ± 2.9	45.6 ± 4.4	20.4
—	69.6	30	60.2 ± 7.3	51.6 ± 3.0	16.7
—	232.2	100	66.0 ± 4.7	50.3 ± 3.6	31.2
—	1160.8	500	72.0 ± 3.5	46.4 ± 2.2	55.2
—	2321.5	1000	70.4 ± 2.7	48.7 ± 1.4	44.6
		Mean	63.6 ± 4.1	49.0 ± 2.7	30.1
(b) For the matrix: clear casting (CC) resin					
121	0	0.01	39.9 ± 2.3	27.4 ± 0.9	45.6
336	0	0.04	38.9 ± 3.0	26.5 ± 0.8	46.8
1022	0	0.12	41.4 ± 3.0	29.1 ± 2.3	42.3
—	69.6	30	42.8 ± 1.6	35.5 ± 1.5	20.6
—	232.2	100	42.8 ± 2.6	34.6 ± 2.1	23.7
—	1160.8	500	36.9 ± 2.8	26.1 ± 1.7	41.4
—	2321.5	1000	32.6 ± 1.8	27.1 ± 0.8	20.3
		Mean	39.3 ± 2.4	29.5 ± 1.4	34.4

For each age value, the energy and standard error averaged from two batches is shown. Data from Refs [22, 26].

adhesion. Thus irrespective of prestress effects, the release agent may allow greater absorption of impact energy; it is also clear, however, that this would adversely affect other parameters, such as tensile strength. Moreover, the GP resin data in Table 22.1a show that the greatest increase in impact energy occurs at 500 and 1000 years, that is, there is an apparent improvement in the energy absorbed with VPPMC sample age. Table 22.1a shows that although control sample energy values do not vary markedly, the average test sample energy rises from 100 years. Since there is no equivalent increase in the CC test sample energy values (Table 22.1b), the conclusion in Ref. [26] is that long-term exposure to 60 °C affected the prestress state in the GP samples. These samples were observed to exhibit some bending during heating, which must have occurred through creep of the GP resin at 60 °C. On reaching 500 years, sample deflection was 2.5–3 mm at the center, the bending direction being consistent with compressive stresses exerted by the fiber-rich side (Section 22.2.4). It is expected that this distortion, though small relative to sample length, would have increased impact resistance in the GP test samples.

No significant bending was observed in the equivalent CC samples. From Table 22.1b, the mean increase in energy from the youngest batches (121 and 336 h) is ~46% and this is in contrast with ~31% from the oldest batches (500 and

1000 years). Nevertheless, there is considerable variability and statistical analysis (from both one- and two-tailed hypothesis tests) indicates, to a significance level of 5%, there is no real difference between youngest and oldest batches [26].

22.3.5
The Long-Term Performance of VPPMCs

The information from Sections 22.3.1–22.3.4 suggests that viscoelastic activity, and the benefits associated with the resulting compressive stresses in a composite, will function over the long term. In this context, "long term" is quantified by ambient temperature and time, and Figure 22.7 shows the known time–temperature boundary, that is, 1000 years at 20 °C. Increasing the ambient temperature obviously reduces duration, as defined by the a_T-temperature relationship (Section 22.3.2). Thus, for example, the known duration of operation decreases to ~20 years at a constant temperature of 40 °C, but this would still make VPPMC technology a realistic option for many practical applications.

Based on current knowledge, resins requiring high temperature curing cycles over long periods of time may not be compatible with VPPMC processing using nylon 6,6 fibers. Nevertheless, since the known limit of duration from Figure 22.7 is, for example, ~40 h at 80 °C, a moderately elevated curing temperature over several hours should be possible, while still permitting (subsequently) an acceptable working life. It is worth noting here that low temperature curing resins are of interest for aerospace

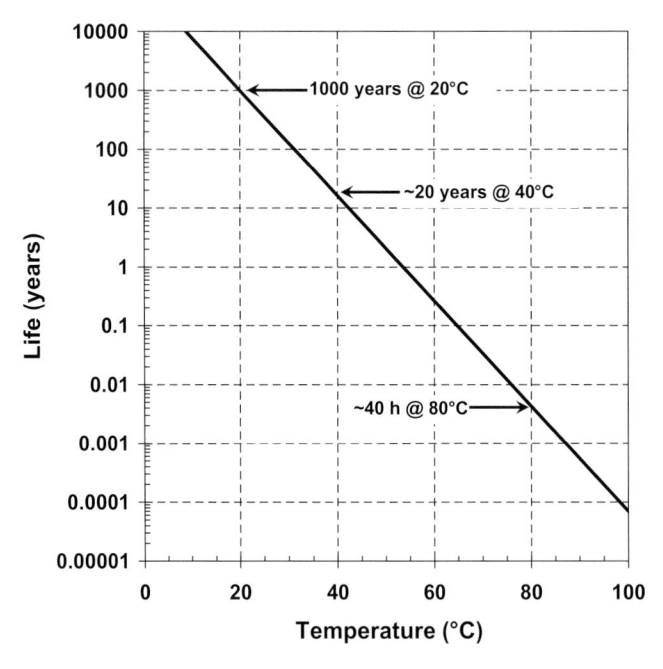

Figure 22.7 VPPMC life as a function of ambient temperature, based on the most recently established time–temperature boundary (1000 years at 20 °C). After Figure 4 in Ref. [26].

applications, since they would permit autoclave-free cures, cheaper tooling, and reduced spring-back of parts [27].

As a final comment on the long-term performance of VPPMCs, the following points should be noted. First, impact tests on samples subjected to accelerated aging (Table 22.1) verify that viscoelastic recovery mechanisms remain active, at least up to 1000 years (9×10^6 h) at 20 °C, and these support the extrapolated recovery strain curve in Figure 22.4. It is evident, however, that this curve could be extrapolated further; thus future investigations may demonstrate that VPPMCs can still perform successfully at elevated temperatures and durations that significantly exceed the currently defined boundary in Figure 22.7. Second, investigations to date have focused on nylon 6,6 fiber. Other viscoelastic fibers may offer superior long-term VPPMC performance characteristics.

22.4
VPPMCs with Higher Fiber Content: Mechanical Properties

22.4.1
Meeting the Objectives

Following the success achieved with Charpy tests, the next objectives were to investigate other basic mechanical characteristics of VPPMCs, that is, their flexural and tensile properties. Although the stretching rig described in Section 22.2.2 provided the means both to study viscoelastic strain and stretch yarn for Charpy impact tests covered in Sections 22.2 and 22.3, it could not provide sufficient yarn for batches of composite samples requiring V_f values in excess of 2–3%. Therefore, the next step was to scale up the fiber-processing techniques, to broaden the scope for mechanical evaluation. Essentially, this was achieved through the design and construction of a larger stretching rig, which enabled longer, multiple lengths of yarn to be subjected to tensile stress by means of a vertically mounted pulley wheel. This increased, by an order of magnitude, the yarn stretching capacity.

For tensile tests, an additional requirement was the production of composite samples only 1 mm in thickness. This could not be achieved with sufficient accuracy by the open casting method employed for Charpy tests, and a "leaky mold" method was adopted, based on principles from Ladizesky and Ward [28]. This consisted of an upper "T"-shaped section over a lower "U"-shaped section to form a closed channel. Once assembled, weights were positioned on top of the mold to force excess resin to flow from the (open) mold ends, with sample thickness being determined by spacers between the upper and lower sections.

22.4.2
Flexural Properties

Unidirectional fiber samples were produced by the same open casting techniques used for Charpy impact test samples (Section 22.2.4), with higher V_f values (8–16%).

A bisphenol-A epoxy resin (ABL-Stevens Resin & Glass, UK) with polyoxyalkyleneamine base catalyst was used. This resin, with good optical transparency, had a lower viscosity than the polyester resins previously adopted, thus facilitating the molding of higher V_f samples. In contrast with the polyester resins, however, the gel time was ~15 h at room temperature and molds had to be lined with release film for successful demolding. The resulting composite strips were cut to produce two test and two control samples per batch, each sample being 200 × 10 × 3.5 mm.

These samples were subjected to three-point bend tests using a freely suspended load. Sample test geometry was similar to ASTM D790M recommendations in terms of support pin dimensions and the span/thickness ratio (L/h) of ~30. The flexural modulus, $E(t)$, from deflections at $t = 5$ s (elastic deformation) and 900 s (short-term creep) was determined from:

$$E(t) = \frac{PL^3}{48\delta(t)I} \tag{22.3}$$

Here, $\delta(t)$ is the deflection at the center of the beam at time t; L is the span, P is the applied load, and I is the second moment of area. A low load value (4.15 N) minimized opportunities for specimen damage, but the resulting deflections (no more than a few millimeters) restricted measurement precision and accuracy.

Figure 22.8 summarizes the results. Two batches were evaluated at each V_f value, the large error bars resulting from batch-to-batch variability (from open casting) and errors in measurement (especially for the 5 s deflections). Nevertheless, since all test sample modulus values are higher than their corresponding control samples, it is evident that viscoelastic prestressing increases bending stiffness. For both deflection times, the modulus is increased by ~50%. There are no progressive changes in modulus with V_f and this is attributed to the narrow range of relatively low V_f values

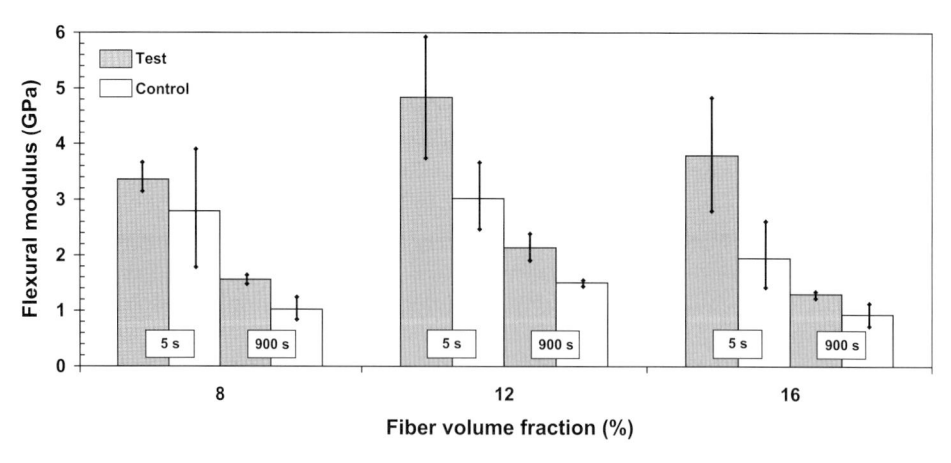

Figure 22.8 Flexural modulus values for test (VPPMC) and control samples from 5 and 900 s deflection. Error bars represent the standard error averaged from the two batches measured at each value. Derived from data in Ref. [29].

used, coupled with the nylon fiber modulus being close to that of the matrix, in contrast with more conventional fiber reinforcements (e.g., glass) [29].

The reduced modulus values at 900 s in Figure 22.8 indicate the contribution from flexural creep. Analysis of matrix-only specimens in Ref. [29] showed that this reduction was due to matrix creep. Following acquisition of data shown in Figure 22.8, the samples were aged to an equivalent of 100 years at 20 °C. Repeated measurements reported in Ref. [29] showed no significant change in increased modulus resulting from viscoelastic prestressing.

22.4.3
Tensile Properties

Using the leaky mold method (Section 22.4.1), unidirectional fiber composite samples with sufficiently high V_f values (16–53%) and appropriate dimensions (200 × 10 × 1 mm) were produced. The lower viscosity epoxy resin, as used for the flexural experiments (Section 22.4.2), further facilitated higher V_f sample production. Two identical molds were used, from which the resulting composite strips were cut to produce two test and two control samples per batch. Batches were produced with V_f values of 16, 28, 41, and 53%.

Tensile testing was performed with a Lloyd LR100K machine at a loading rate of 5 mm/min. For each V_f value (16, 28, 41, and 53%), typically three batches were tested, and Figure 22.9 shows representative stress–strain plots from a batch of these samples. At this V_f (28%), the test samples clearly show increased strength and stiffness. Although strength could be readily determined, curve shape resulted in stiffness being quantified by a modulus value obtained from the linear region as

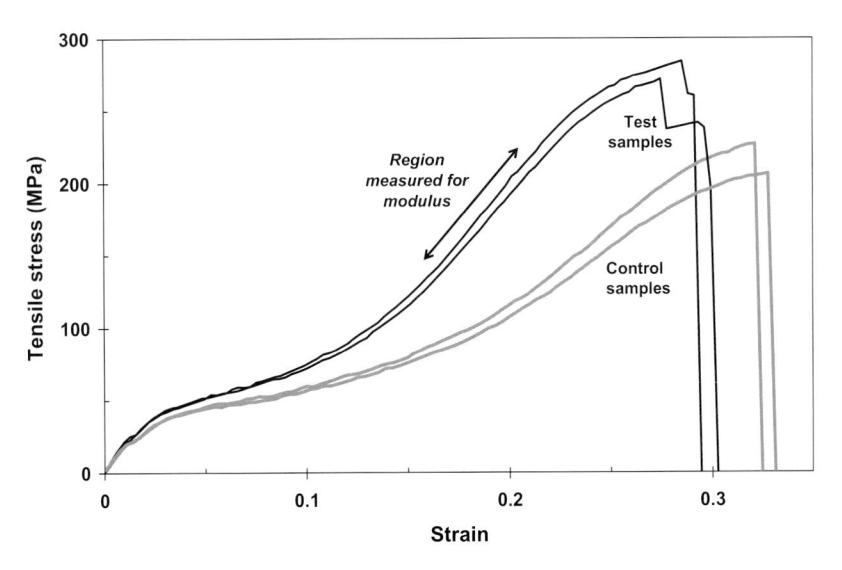

Figure 22.9 Typical tensile stress–strain plots for a batch of test (VPPMC) and control samples at 28% V_f. After Figure 4 in Ref. [30].

shown in Figure 22.9. In contrast with the elastic modulus (5 s) data in Figure 22.8, this modulus represented stiffness resulting from both elastic and viscoelastic contributions to tensile deformation [30]. In addition to strength and stiffness, the strain-to-failure (STF), tensile toughness (to fracture), and strain-limited toughness were determined. Here, the strain-limited toughness is the energy absorbed per unit volume to a fixed strain (0.25), as calculated from the area under the stress–strain curve from 0 to 0.25 strain. This may be considered as a more appropriate performance parameter than tensile toughness for deflection-limited design purposes.

Figure 22.10 summarizes the strength, modulus, and strain-limited toughness data and all three parameter values clearly increase with V_f. Although test and control strengths are similar for V_f at 16 and 53%, the prestressed samples show increased strength at the intermediate V_f values. The modulus and toughness data also show a comparable pattern, most clearly demonstrated by Figure 22.11. The curves in Figure 22.11 indicate an optimum V_f value (~35–40%) at which the benefits from

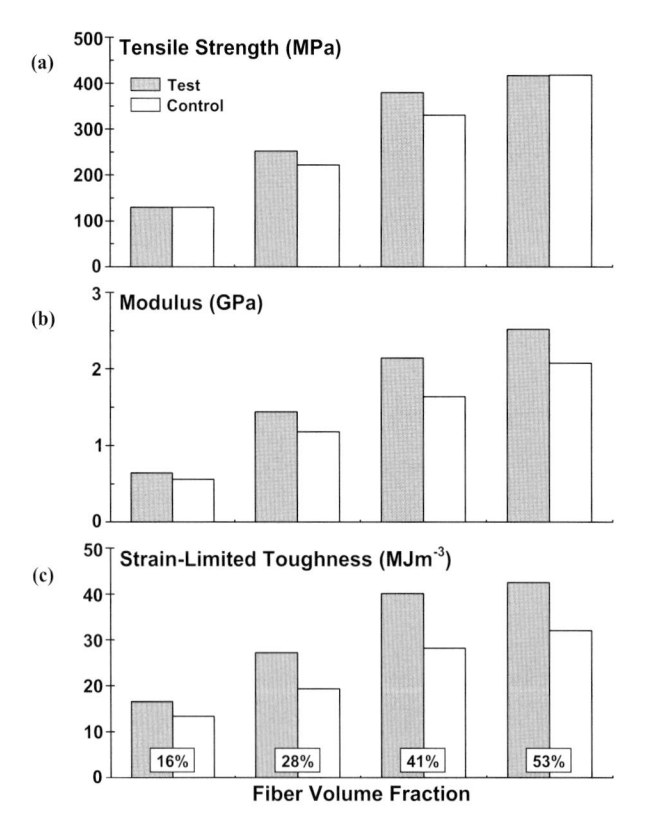

Figure 22.10 Effect of fiber volume fraction on the tensile properties of test (VPPMC) and control samples. Typical standard errors were (a) ±8 MPa; (b) ±0.05 GPa; and (c) ±0.5 MJ/m^3. Derived from data in Ref. [30].

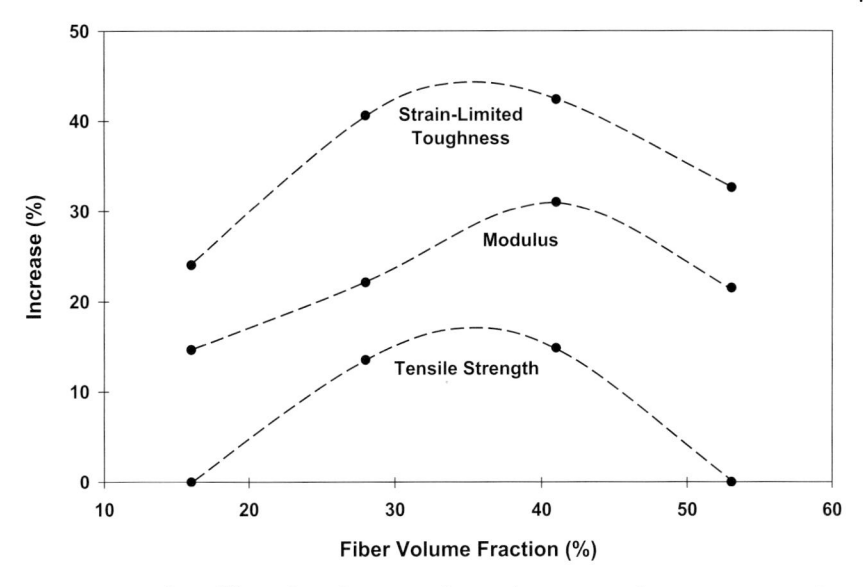

Figure 22.11 Effect of fiber volume fraction on the tensile properties of test (VPPMC) samples relative to their control sample counterparts, using data from Figure 22.10. After Figure 6 in Ref. [30].

prestressing are maximized, these increases exceeding 15, 30, and 40%, respectively for strength, modulus, and strain-limited toughness. The effect can be attributed to the competing roles between fibers and matrix: at lower V_f, less compressive stress will be produced due to too few fibers; at higher V_f, too many fibers will reduce the matrix cross-sectional area available for compression.

STF values for the test samples were consistently lower than their control counterparts by 10–20%, as exemplified by Figure 22.9. Thus for a test sample, while increased strength and stiffness contributed to a larger area under the stress–strain curve (toughness), the effect was reduced by a lower STF. Since tensile toughness (to failure) will be more sensitive to a lower STF than the strain-limited toughness, the observed increases in tensile toughness for test samples in Ref. [30] were found to be less than 10%.

22.5
Processing Aspects of VPPMCs

22.5.1
Background

It is evident from the preceding sections that the long-term viscoelastic recovery force from nylon fiber can be exploited to produce a prestressed composite material with improved mechanical properties compared to an unstressed counterpart. There are, however, processing-related aspects that have required investigation. Of particular importance, is whether the physical properties of the nylon fibers are changed during the stretching treatment, which in turn could (unknowingly) influence findings from subsequent evaluation of the composite samples. For similar reasons, the geometrical aspects of fibers in composite samples, that is, fiber spatial distribution and fiber length require consideration.

22.5.2
Fiber Properties

Concerns over whether the stretching process could change the size or surface characteristics of the fibers were initially addressed in early work [9]. Here, fiber samples from test and control yarns were studied by SEM. No differences in fiber topography were observed, that is, there was no evidence of surface damage, resulting from the stretching process, in the test yarn. Fiber diameters were also measured, to determine possible changes resulting from transverse (Poisson) contraction in the test yarn. For the test yarn at 4 and 220 h after releasing the stretching load, the mean diameters (from 10 readings) were 27.5 and 27.3 µm, respectively. The equivalent control yarn values were 28.5 and 27.7 µm and the distribution of fiber diameters was similar for both yarns (0.7 µm standard deviation). Thus, it was concluded in Ref. [9] that there were no significant changes in diameter, either between the test and control fibers or with age.

Further analysis of fibers was performed during tensile test studies [30]. Figure 22.12 shows the surface characteristics of test and control fiber samples, 385 h after releasing the stretching load (349 MPa for 24 h) on the test sample. Both fiber groups show longitudinal features, which had also been observed in earlier work [9], and these can be attributed to the original manufacturing process. Most importantly, Figure 22.12 verifies the findings from Ref. [9], in that there are no differences in surface features between the test and control fibers that could affect fiber–matrix bonding.

In addition to investigating fiber topography, fiber tensile properties were evaluated in Ref. [30], to determine whether the stretch treatment affected the mechanical behavior of the fibers. If such changes (e.g., work hardening) were present, then direct comparison between test and control composite samples would have led to erroneous inferences. By using a dynamic mechanical analyzer (DMA), individual fibers were tensile tested to evaluate strength and STF at four periods (24–1344 h) following release of the stretching load (349 MPa for 24 h) on the test yarn. The mean tensile strengths ($\pm$ standard error) were similar at 980 ± 37 MPa (test) and 975 ± 35 MPa (control). STF results were also similar at $31.9 \pm 1.8\%$ (test) and $33.4 \pm 2.3\%$ (control). Hypothesis tests on these data in Ref. [30] verified that there were no significant differences between the test and control groups.

The above findings demonstrate that the stretching treatment applied to test fibers cause no significant changes to their physical characteristics or tensile properties. Thus, it may be concluded that any improvements in mechanical properties observed from a VPPMC sample, compared with its control counterpart, must result from the prestress effects alone.

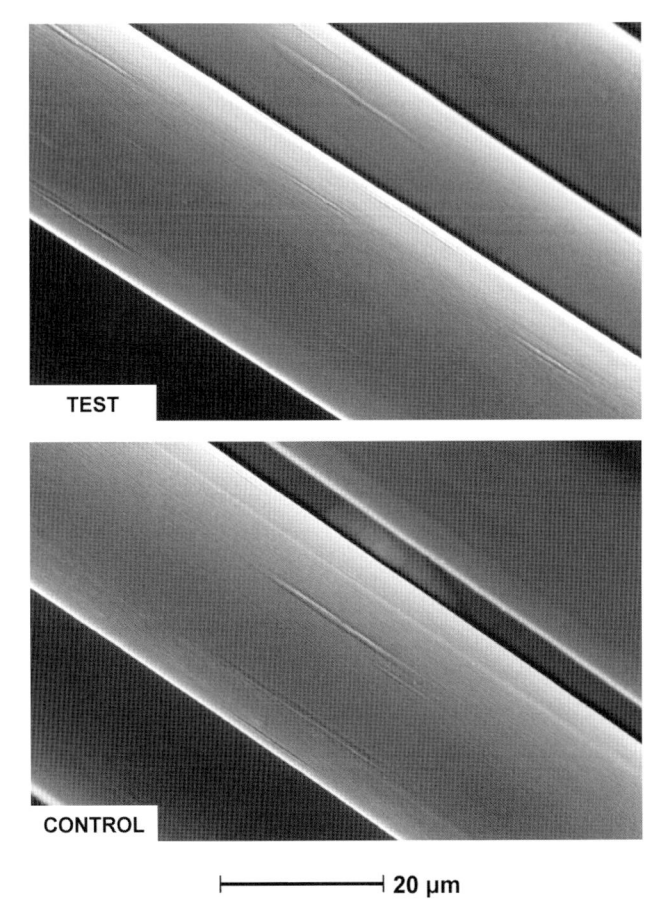

Figure 22.12 SEM micrographs of annealed test and control nylon 6,6 fibers, the test fibers being subjected to creep conditions used for VPPMC sample preparation. Reprinted from Figure 2 in Ref. [30] with permission from Elsevier.

22.5.3
Geometrical Aspects of Fibers in Composite Samples

Fiber spatial distribution, especially at low V_f values, may require careful consideration for some test procedures. It was mentioned in Section 22.2.4 that fibers tended to sink to the bottom of the mold during the production of composite samples by open casting with a polyester resin. Subsequent investigations, performed during flexural testing studies [29], suggested that the effect could be reduced by using an epoxy resin and evidence of this is shown in Figure 22.13. Here, cross-sections from open-cast composite samples (8% V_f) allow the fiber spatial distributions in epoxy resin (described in Section 22.4.2) and polyester resin (the CC resin described in Section 22.3.4) to be compared. Spatial variations with matrix-only regions can be seen in the epoxy resin, but on a macroscopic level, these are distributed randomly.

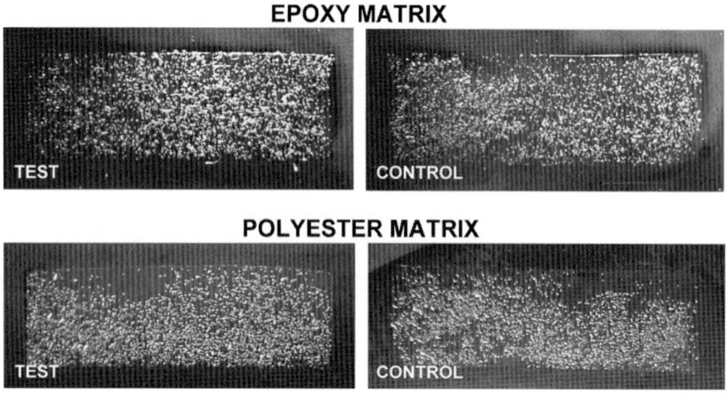

EPOXY MATRIX

POLYESTER MATRIX

Figure 22.13 Representative optical micrograph sections of test (VPPMC) and control samples (8% V_f) from open casting with epoxy and polyester resins; the polyester matrix shows greater variation in fiber spatial distribution. After Figure 2 in Ref. [29].

For the polyester matrix, however, there appears to be a greater fiber concentration in the lower 2/3 of the molding, that is, fewer matrix-only regions. Since the density of the fibers and both resins (in liquid state) were very similar, it was suggested in Ref. [29] that fiber settlement may have been affected by local variations in the relatively fast curing stages of the polyester resin.

It is encouraging to note that Figure 22.13 shows, for both resins, no substantial differences in fiber spatial distributions between corresponding test and control samples. Similarly, as shown in Figure 22.14, composite sample sections (same epoxy resin with 28% V_f) from tensile testing specimens [30] indicate no discernible differences between test and control samples, either in terms of fiber spatial distribution or matrix features.

Fiber length is another geometrical aspect. The effects of fiber length on load transfer in short fiber-reinforced composites are well known, and a short length sample of a composite with continuous fiber reinforcement can be similarly considered. For a conventional composite (no prestress), load transfer is characterized by the length of the fiber required for a tensile stress, transferred from the matrix, to fracture the fiber. Theoretically, this critical fiber length (l_c) may be very small, but practical aspects (differences in fiber–matrix mechanical properties and interfacial adhesion) can adversely affect load transfer. For maximum load transfer, l_c in practice could be 1000 fiber diameters [31]. Thus, for the nylon fiber diameter used in VPPMCs, l_c might exceed 25 mm; that is, this could represent the minimum length for maximum load transfer in a VPPMC sample.

In addition to the flexural stiffness studies on large (L/h) samples (Section 22.4.2), the elastic flexural modulus was also determined for small (L/h) samples produced by similar means, using a DMA in three-point bending mode [29]. Loading deflection rate was comparable to the 5 s deflection measurements performed on the long samples. With L at 20 mm and L/h of ~5, modulus values for both test and control samples were only ~30% of the long samples and the increase (from prestress) was substantially lower. It is well known that the contribution to beam deflection from

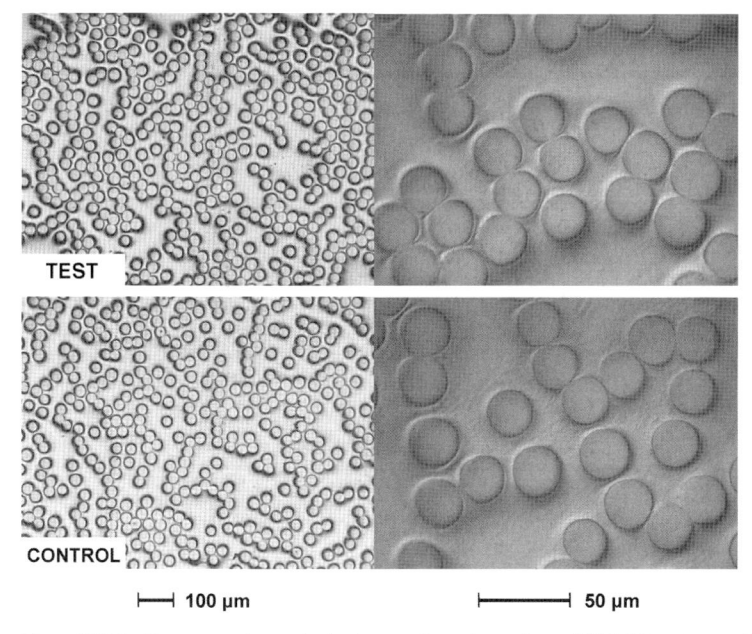

Figure 22.14 Representative optical micrograph sections of test (VPPMC) and control composite samples (28% V_f) with epoxy resin matrix. Reprinted from Figure 3 in Ref. [30] with permission from Elsevier.

shearing forces is increased as L/h decreases, and the observed reductions can be attributed primarily to this. Nevertheless, after considering published experimental work and modifications to Timoshenko bending theory, the magnitude could not be accounted for by shear effects alone, and it was suggested in Ref. [29] that fiber–matrix load transfer mechanisms may be less effective for fiber lengths of 20 mm. These findings thus provide some support to the suggestion that a minimum fiber length is required to maximize the viscoelastic prestressing effect, this length possibly being in excess of 20 mm.

22.6
Mechanisms for Improved Mechanical Properties in VPPMCs

22.6.1
Background

To date, research has been experimentally driven, determined by the need to demonstrate, by the simplest (and most reliable) means, the potential mechanical benefits that VPPMC technology may provide. Thus, mechanisms previously proposed to explain individually the observed improvements in flexural, tensile, and impact properties have been based on empirical evidence. Here, these mechanisms are reviewed and where possible, integrated, the aim being to provide a more coherent picture of VPPMC behavior.

22.6.2
Flexural Stiffness

Comparable increases in flexural stiffness have been observed in EPPMC samples, and the effect has been attributed to deflection-dependent forces opposing the applied bending load [5] and a more collective response to bending forces from the pretensioned fibers [32]. A further explanation has been proposed [29], in which the compressive prestress shifts the neutral axis in bending and this (i) reduces tensile forces so that bending stiffness is increased and (ii) increases the proportion of matrix in compression. Since matrix modulus can be greater in compression [33], (ii) may also contribute to increased bending stiffness.

It should be noted that the potential contributions from these proposed explanations for increased flexural stiffness remain unclear. Essentially, however, they may be considered to originate from Mechanisms (I) and (III) discussed in Section 22.6.3.

22.6.3
Tensile Strength

Of the tensile parameters outlined in Section 22.4.3, it is the strength results that have provided the most significant insight into the role of viscoelastically generated prestress. Clearly, the simplest explanation is that improved tensile strength can be attributed to the tensile load required to overcome compressive forces within the matrix. Matrix compression will impede crack formation within the matrix and fiber fracture and this is defined here as Mechanism (I).

Referring to Figure 22.10a, the increases in tensile strength at 28 and 41% V_f were found to be 30 and 49 MPa, respectively [30]. Combining this information with the curve in Figure 22.11, it can be inferred that the maximum increase in tensile strength could be ~50 MPa. Nevertheless, estimates from viscoelastic force measurement studies [25] suggested that the axial stress exerted by fibers on the matrix would not exceed ~10 MPa. Therefore, the maximum compressive stress transferred to matrix regions in close vicinity to the viscoelastically strained fibers could not have been more than ~10 MPa, that is, ~40 MPa less than the maximum increase in tensile strength. This (apparent) anomaly can also be seen in the tensile strength data [6] from EPPMCs using glass fibers: for prestress values up to 50 MPa, the tensile strengths of samples at 35 and 45% V_f (relative to unstressed counterparts) significantly exceed the originally applied prestress levels. This is most evident at 35% V_f (the lowest V_f value studied in Ref. [6]) which is close to the optimum value (35–40%) for prestress improvements reported in Section 22.4.3. From results in Ref. [6], increases in strength at 35% V_f are estimated to be 50–60 and 70–80 MPa for 25 and 50 MPa prestress, respectively. Since these increases are

~30 MPa higher than the values that could possibly be attributed to Mechanism (I), there must be other prestress-related effects occurring that contribute to increased tensile strength. It is evident that these effects may be applicable both to elastic and viscoelastic prestressing techniques and two further mechanisms have been suggested [30].

Mechanism (II) originates from considering the effects of dynamic overstress described by Manders and Chou [34]. When a fiber fractures within a composite of aligned fibers, a stress wave propagates outwards, subjecting neighboring fibers to a dynamic (oscillatory) overstress. This decays with time, resulting in a static stress concentration. The dynamic contribution generally exceeds the static effect, so that the probability of failure among neighboring fibers is increased, thereby causing the composite to be further weakened. Mechanism (II) represents the effects of matrix compression imparted by stress fields from the viscoelastically (or elastically) strained fibers; this attenuates the dynamic overstress effect, which in turn should reduce collective fiber failure.

Mechanism (III) originates from a study by Motahhari and Cameron [5] of EPPMCs subjected to flexural loads. Since the fibers are taut and straightened, their response to an applied load can be expected to be instantaneous and should occur more collectively. There will be fewer variations in the levels to which individual fibers are deformed as the load increases, so that the occurrence of eventual fiber fractures will be less progressive. Thus, when compared to a composite with no prestress, Mechanism (III) should increase tensile strength and composite displacement during fiber fracture should be reduced. For VPPMCs, some waviness in fiber lay has been observed; however, the fibers are taut, so that a faster and more collective response to tensile loads can be expected. STF results from VPPMC test samples were consistently 10–20% lower than control counterparts (Section 22.4.3), and this is seen in Ref. [30] as evidence of Mechanism (III).

22.6.4
Impact Toughness

The improved impact toughness observed in VPPMC samples from Charpy tests (Sections 22.2.4 and 22.3.4) can be attributed, in part, to matrix compression impeding crack propagation [9, 10], that is, Mechanism (I). A further mechanism, originally proposed for EPPMCs [4], has also been discussed [10]. During impact, crack propagation within the composite will occur through (i) transverse fracture of the fibers or (ii) fiber–matrix interfacial debonding. As mentioned in Section 22.2.4, there was clear evidence of increased fiber–matrix debonding in the test samples (Figure 22.2) with fewer fractured fibers. These effects can be explained by viscoelastically generated shear stresses at the fiber–matrix interface. The shear stresses, which are responsible for producing matrix compression, also reduce the forces required for initiating debonding. Thus in the test samples, crack propagation through fiber–matrix debonding tends to be promoted over transverse fracture. This is designated as Mechanism (IV).

22.7
Potential Applications

22.7.1
High Velocity Impact Protection

In contrast with the Charpy impact tests (Sections 22.2.4 and 22.3.4), high velocity impact (e.g., from blast fragments) is normally associated with low mass projectiles striking plate structures. Impact studies on fiber-reinforced polymeric composite plates, compressively preloaded (in-plane) by external means, have been reported [35–38], though investigations were restricted either to low velocity (drop weight) impact and/or uniaxial compression (thereby exacerbating impact damage in one direction). Nevertheless, in Ref. [35], biaxial compression was applied to glass fiber-reinforced polyester laminated plates, and an increase in absorbed (low velocity) impact energy was observed with little change in the damage or indentation region. Although encouraging, direct comparisons between preloaded composites and VPPMCs are not possible, even if the former were studied under suitable conditions (biaxial preload, high impact velocity impact). For preloaded composites, fibers and matrix are subjected to the same (externally applied) prestress but the situation is more complex for VPPMCs, as matrix compression occurs through shear at the fiber–matrix interface, the fibers themselves being in a state of tension.

Information from studies using compressively preloaded single-phase materials may at least provide some indication of matrix response within a VPPMC. Published evidence in this context is generally limited to ceramic (brittle) materials; for example, alumina tiles exhibited reduced damage from ballistic impact when subjected to biaxial compressive prestress [39]. Nevertheless, the compressive preload methodology has been applied to a brittle polymer [40]. Here, polymethyl methacrylate plates under biaxial compression showed decreased areas of damage at both low ($\sim$2 m/s) and high ($\sim$130 m/s) impact velocities. Although detailed studies were limited to low impact velocities in Ref. [40], impact energy was increased by $\sim$30% and the threshold stress level for projectile penetration was found to increase by an amount consistent with applied compression. Clearly, as these effects can be associated with Mechanism (I), a VPPMC matrix would be expected to exhibit similar characteristics.

Nylon 6,6 woven mesh PMCs are among the materials used for protection against ballistic threats [41, 42]. Kinetic energy from the cone of deformation surrounding the projectile is the dominant energy-absorbing effect, if the nylon fiber-reinforced PMC is perforated by ballistic impact [42]. Similar observations with woven glass fiber PMCs have been made [43], but for incomplete perforation, secondary yarn deformation energy was found to be the most significant effect. For woven nylon 6,6 PMCs impacted by fragment-simulating projectiles, delamination and tensile failure of fibers seem to be significant energy-absorbing processes [41]. Therefore, it appears that the dominant energy-absorbing effects (which may vary with impact conditions) depend principally on (i) fiber–matrix deformation (elastic and plastic), (ii) fiber tensile failure, and (iii) delamination. In a nylon fiber VPPMC, all four mechanisms discussed in Section 22.6 should enhance high velocity impact protection.

For (i), Mechanisms (I) and (III) would increase the energy absorption potential from fiber–matrix deformation, since external impact loads would need to work against matrix compressive stress and a more collective response from the fibers. For (ii), the dynamic overstress effect will be attenuated by Mechanism (II), thereby reducing collective fiber failure. This will result in fewer fractured fibers and these will be more confined to the actual area of impact. Therefore, the corresponding region of penetration damage to fibers is expected to be smaller. For (iii), Mechanism (IV) should promote fiber–matrix debonding over fiber transverse fracture, thus energy absorption through delamination will be increased.

From the above considerations, a VPPMC subjected to a low-mass high-velocity impact should be expected to possess a smaller penetration damage area within a larger region of (energy-absorbing) delamination, when compared with a conventional PMC. Therefore, VPPMC materials employed within a shielding system may enhance the effectiveness of lightweight protection from high velocity impacts. Fiber parameters (e.g., spatial density, weave architecture) may be optimized for specific applications; moreover, polymeric fibers (for exploiting VPPMC mechanisms) commingled with other (high strength) fibers or multilayered PMC/VPPMC/light alloy structures could be developed to maximize energy absorption for optimum post-impact structural integrity [26]. Applications could include body armor, especially where nonplanar geometries are required (e.g., helmets, footwear). Also ceramic-PMC systems, which employ a ceramic plate to spread the impact load over a (relatively ductile) PMC backing layer, have applications that include add-on armor for vehicle protection [44]. Thus, VPPMC technology may provide opportunities for enhancing the backing layer (load resistance) properties.

22.7.2
Crashworthiness

Vehicular structures require crashworthy structures with minimum mass for maximum fuel efficiency. For example, crashworthy automobile structures are designed so that net deceleration of the passenger cell is less than 20g (g being the acceleration due to gravity), for impact speeds up to 15.5 m/s, to avoid irreversible brain damage [45, 46]. The principal parameter for materials used in crashworthy structures is the specific energy absorption (SEA), which is the energy absorbed per unit mass. Experimentally, the SEA for PMCs is notably higher than that of steel or aluminum [45], but to prevent dangerously high deceleration rates occurring, especially during the initial stages of impact, the rate of energy absorption by a structure is also important [46]. Crashworthiness studies involving PMCs are dominated by axial compression studies using thin-walled tubes and to prevent unacceptably high-energy absorption rates, these must be triggered to fail progressively. This is often achieved by chamfering one of the tube ends. There are several progressive crushing modes, though ductile fiber-reinforced PMCs (e.g., Kevlar or Dyneema fibers in thermoset matrices) undergo progressive folding (or local buckling) [45–47]. Here, local buckles are formed by plastic deformation, with interlaminar cracks and delaminations occurring at the buckle sites [47]. VPPMCs have not been studied under axial crush-loading conditions, but since this crushing mode involves localized bending and matrix cracking, viscoelastic prestressing may enhance energy absorption. In particular, this could occur through the need to work against matrix compressive stress and a more collective response from the fibers, that is, Mechanisms (I) and (III).

Impacted structures can also often fail through crushing associated with bending [48]. Fiber-reinforced PMCs have, for example, been investigated for use in automotive applications, such as beams [49] and grid-stiffened panels [50] for car doors, to provide side-impact protection. For VPPMCs in bend-related impact failures, it may be expected that Mechanisms (I)–(IV) could make varying contributions toward enhancing energy absorption, at least in the tensile regions.

22.7.3
Enhanced Crack Resistance

In addition to the potential for high velocity (blast fragment) and crash protection, VPPMC technology may enable crack propagation resistance to be enhanced through Mechanisms (I) and (III). This could be particularly useful for rotor blade applications. Consider, for example, wind turbine blades for power generation. Here, carbon fiber PMCs offer better fatigue performance than glass fiber PMCs and they are also stiffer and lighter [51, 52]; however, the former are more brittle [51]. The addition of viscoelastically strained polymeric fibers to blade structures (e.g., by commingling in the most vulnerable regions) could either reduce brittleness in PMCs of carbon fiber or improve fatigue resistance in those utilizing glass fiber.

The development of fiber-reinforced concrete (FRC) has been progressing since the early 1960s [53] and this represents another potential application for enhancing crack resistance. FRC contains randomly oriented fibers to inhibit cracking and polymer (thermoplastic) fibers are routinely used [53–55]. The most common fiber is polypropylene, though FRC using nylon fiber has been found to sustain higher flexural stress levels [54]. Thus, through an extra processing step, VPPMC principles offer further opportunities for increasing crack resistance. The polymeric fibers could be treated (annealed, subjected to creep and then cut to size) and, if necessary, stored under refrigerated conditions, prior to being mixed on site.

22.8
Summary and Conclusions

A VPPMC is produced by applying tension to polymeric fibers, the tensile load being released prior to molding the fibers into a matrix. Following matrix solidification, compressive stresses are imparted by the viscoelastically strained fibers. To date, findings from research, using nylon 6,6 fiber, can be summarized as follows:

1) VPPMC samples (with unidirectional continuous nylon fibers) were produced for mechanical evaluation. Compared with their control (unstressed) counterparts:

22.8 Summary and Conclusions

a) Charpy tests showed that VPPMCs could absorb 25–30% more impact energy, with some tests demonstrating ~50% improvement.
b) Tensile tests demonstrated increases in strength, modulus, and energy absorbed (to 25% strain) exceeding 15, 30, and 40%, respectively.
c) Flexural tests, using a freely suspended load, showed that the elastic and short-term creep flexural moduli (from deflections at 5 and 900 s, respectively) were increased by ~50%.
d) Based on preliminary flexural experiments with short length samples, the minimum fiber length to maximize the viscoelastic prestressing effect may be in excess of 20 mm.

2) By using time–temperature superposition principles, results from fiber viscoelastic recovery studies and Charpy impact tests on composite samples suggest no significant deterioration in VPPMC performance over a duration equivalent to a constant 40 °C for ~20 years. These findings are also supported by observing that viscoelastically generated force grows to a limiting value with time. This long-term viscoelastic activity is expected to counteract any potential for deterioration in prestress resulting from localized matrix creep near the fiber–matrix interface regions.

3) It was found that the stretching treatment, as applied to the nylon fibers, had no significant effect on fiber diameter, topography, or tensile properties (e.g., possible work-hardening). Moreover, no differences in fiber spatial distributions could be observed between VPPMC and control samples.

4) Four principal mechanisms, which contribute toward the observed mechanical property improvements in VPPMCs, have been identified:
   (i) Matrix compression impedes crack propagation from external tensile loads.
   (ii) Matrix compression attenuates dynamic overstress effects, reducing the probability of fiber fracture outside the immediate area of impact.
   (iii) Residual fiber tension causes fibers to respond more collectively and thus more effectively to external loads.
   (iv) Residual shear stresses at the fiber–matrix interface regions promote (energy-absorbing) debonding over transverse fracture.

5) VPPMCs may be exploited for the following applications:
   a) Enhanced high velocity (blast fragment) impact protection. Mechanisms (I) and (III) would increase energy absorption from fiber–matrix deformation, Mechanism (II) may reduce collective fiber failure, and Mechanism (IV) should promote (energy-absorbing) delamination. Thus compared with a conventional PMC, a VPPMC is expected to possess a smaller penetration damage area within a larger delaminated region.
   b) Crashworthiness: Since progressive crushing of crashworthy structures can involve localized bending and matrix cracking, energy absorption may be enhanced through Mechanisms (I) and (III). For bend-related impact

failures, Mechanisms (I)–(IV) could make varying contributions toward increased energy absorption.

c) Enhanced crack resistance through Mechanisms (I) and (III). Applications include rotor blades (e.g., for wind turbines) and fiber-reinforced concrete.

VPPMC technology offers the means to produce composite materials with enhanced mechanical properties, without the need to increase mass or section dimensions. The fiber stretching and molding operations are decoupled, so that flexibility in composite part production and opportunities to produce complex component geometries should be comparable to conventional polymeric composite production routes. Moreover, prepreg principles may be exploited, so that following the stretching treatment, polymeric yarn may be stored under refrigerated conditions (to retard viscoelastic recovery) prior to subsequent use on another site.

To date, studies have focused on nylon 6,6 fiber; however, other viscoelastic fibers may offer superior characteristics for VPPMC applications. Moreover, future developments could include hybrid composites where polymeric fibers (for exploiting VPPMC mechanisms) are commingled or layered with other (high strength) fibers.

Acknowledgments

The author is grateful to a number of colleagues for their support over the past decade. Special thanks, however, go to Dr. Jody Pang, Garry Robinson, and Mechanical Workshop staff in the Department of Engineering for their research contributions and technical support. The financial support, initially from the University of Hull Research Support Fund and latterly from The Leverhulme Trust (Grant Ref. F/00181/K) is gratefully acknowledged.

References

1 Tuttle, M.E. (1988) A mechanical/thermal analysis of prestressed composite laminates. *Journal of Composite Materials*, **22** (8), 780–792.

2 Tuttle, M.E., Koehler, R.T., and Keren, D. (1996) Controlling thermal stresses in composites by means of fiber prestress. *Journal of Composite Materials*, **30** (4), 486–502.

3 Sui, G.X., Yao, G., and Zhou, B.L. (1995) Influence of artificial prestressing during the curing of VIRALL on its mechanical properties. *Composites Science and Technology*, **53** (4), 361–364.

4 Motahhari, S. and Cameron, J. (1998) Impact strength of fiber pre-stressed composites. *Journal of Reinforced Plastics and Composites*, **17** (2), 123–130.

5 Motahhari, S. and Cameron, J. (1999) Fiber prestressed composites: Improvement of flexural properties through fiber prestressing. *Journal of Reinforced Plastics and Composites*, **18** (3), 279–288.

6 Hadi, A.S. and Ashton, J.N. (1998) On the influence of pre-stress on the mechanical properties of a unidirectional GRE composite. *Composite Structures*, **40** (3–4), 305–311.

7 Fancey, K.S. (2005) Fiber-reinforced polymeric composites with

viscoelastically induced prestress. *Journal of Advanced Materials*, **37** (2), 21–29.

8 Fancey, K.S. (1997) Composite fibre-containing materials. UK Patent No. 2281299B.

9 Fancey, K.S. (2000) Investigation into the feasibility of viscoelastically generated pre-stress in polymeric matrix composites. *Materials Science and Engineering A*, **279**, 36–41.

10 Fancey, K.S. (2000) Prestressed polymeric composites produced by viscoelastically strained nylon 6,6 fibre reinforcement. *Journal of Reinforced Plastics and Composites*, **29** (15), 1251–1266.

11 Murayama, T., Dumbleton, J.H., and Williams, M.L. (1967) The viscoelastic properties of oriented nylon 6,6 fibers, Part III: Stress relaxation and dynamic mechanical properties. *Journal of Macromolecular Science – Physics*, **B1**, 1–14.

12 Babatope, B. and Isaac, D.H. (1992) Annealing of isotropic nylon 6,6. *Polymer*, **33**, 1664–1668.

13 Crawford, R.J. (1998) *Plastics Engineering*, 3rd edn, Butterworth-Heinemann, Oxford, p. 84.

14 Lakes, R.S. (1999) *Viscoelastic Solids, CRC Mechanical Engineering Series*, CRC Press, Boca Raton, p. 23.

15 Rosen, S.L. (1993) *Fundamental Principles of Polymeric Materials*, 2nd edn, John Wiley & Sons, Inc., New York, p. 299.

16 McCrum, N.G., Buckley, C.P., and Bucknall, C.B. (1988) *Principles of Polymer Engineering*, Oxford University Press, Oxford, p. 130.

17 Fancey, K.S. (2001) A latch-based Weibull model for polymeric creep and recovery. *Journal of Polymer Engineering*, **21**, 489–509.

18 Fancey, K.S. (2005) A mechanical model for creep, recovery and stress relaxation in polymeric materials. *Journal of Materials Science*, **40**, 4827–4831.

19 Carter ADS (1986) *Mechanical Reliability*, 2nd edn, Macmillan, London, p. 127.

20 Glasstone, S., Laidler, K.J., and Eyring, H. (1941) *The Theory of Rate Processes*, McGraw-Hill, New York, p. 480.

21 Dobreva, A., Gutzow, I., and Schmelzer, J. (1997) Stress and time dependence of relaxation and the Kohlrausch stretched exponent formula. *Journal of Non-Crystalline Solids*, **209**, 257–263.

22 Pang, J.W.C. and Fancey, K.S. (2006) An investigation into the long-term viscoelastic recovery of nylon 6,6 fibres through accelerated ageing. *Materials Science and Engineering A*, **431**, 100–105.

23 Howard, W.H. and Williams, M.L. (1963) The viscoelastic properties of oriented nylon 6,6 fibers. *Textile Research Journal*, **33**, 689.

24 Lamin, B.M. (2005) BE Thesis, Department of Engineering, University of Hull, UK.

25 Pang, J.W.C., Lamin, B.M., and Fancey, K.S. (2008) Force measurement from viscoelastically recovering nylon 6,6 fibres. *Materials Letters*, **62**, 1693–1696.

26 Fancey, K.S. (2010) Viscoelastically prestressed polymeric matrix composites – Potential for useful life and impact protection. *Composites Part B*, **41** (6), 454–461.

27 Soutis, C. (2009) Recent advances in building with composites. *Plastics, Rubber and Composites*, **38**, 359–366.

28 Ladizesky, N.H. and Ward, I.M. (1986) Ultra-high modulus polyethylene fibre composites: I – The preparation and properties of conventional epoxy resin composites. *Composites Science and Technology*, **26**, 129–164.

29 Pang, J.W.C. and Fancey, K.S. (2009) The flexural stiffness characteristics of viscoelastically prestressed polymeric matrix composites. *Composites Part A*, **40**, 784–790.

30 Pang, J.W.C. and Fancey, K.S. (2008) Analysis of the tensile behaviour of viscoelastically prestressed polymeric matrix composites. *Composites Science and Technology*, **68**, 1903–1910.

31 Nielsen, L.E. (1974) *Mechanical Properties of Polymers and Composites*, vol. **2**, Marcel Dekker, New York, pp. 469–470.

32 Cao, Y. and Cameron, J. (2006) Flexural and shear properties of silica particle modified glass fiber reinforced epoxy composite. *Journal of Reinforced Plastics and Composites*, **25** (4), 347–359.

33 Hine, P.J., Duckett, R.A., Kaddour, A.S., Hinton, M.J., and Wells, G.M. (2005)

The effect of hydrostatic pressure on the mechanical properties of glass fibre/epoxy unidirectional composites. *Composites Part A*, **36**, 279–289.

34 Manders, P.W. and Chou, T.W. (1983) Enhancement of strength in composites reinforced with previously stressed fibers. *Journal of Composite Materials*, **17**, 26–44.

35 Robb, M.D., Arnold, W.S., and Marshall, I.H. (1995) The damage tolerance of GRP laminates under biaxial prestress. *Composite Structures*, **32**, 141–149.

36 Zhang, X., Davies, G.A.O., and Hitchings, D. (1999) Impact damage with compressive preload and post-impact compression of carbon composite plates. *International Journal of Impact Engineering*, **22**, 485–509.

37 Herszberg, I. and Weller, T. (2006) Impact damage resistance of buckled carbon/epoxy panels. *Composite Structures*, **73**, 130–137.

38 Heimbs, S., Heller, S., Middendorf, P., Hahnel, F., and Weisse, J. (2009) Low velocity impact on GFRP plates with compressive preload: Test and modelling. *International Journal of Impact Engineering*, **36**, 1182–1193.

39 Sherman, D. and Ben-Shushan, T. (1998) Quasi-static impact damage in confined ceramic tiles. *International Journal of Impact Engineering*, **21**, 245–265.

40 Archer, J.S. and Lesser, A.J. (2009) Impact resistant polymeric glasses using compressive pre-stress. *Journal of Applied Polymer Science*, **114**, 3704–3715.

41 Iremonger, M.J. and Went, A.C. (1996) Ballistic impact of fibre composite armours by fragment-simulating projectiles. *Composites Part A*, **27A**, 575–581.

42 Morye, S.S., Hine, P.J., Duckett, R.A., Carr, D.J., and Ward, I.M. (2000) Modelling of the energy absorption by polymer composites upon ballistic impact. *Composites Science and Technology*, **60**, 2631–2642.

43 Naik, N.K., Shrirao, P., and Reddy, B.C.K. (2006) Ballistic impact behaviour of woven fabric composites: Formulation. *International Journal of Impact Engineering*, **32**, 1521–1552.

44 Sanchez Galvez, V. and Sanchez Paradela, L. (2009) Analysis of failure of add-on armour for vehicle protection against ballistic impact. *Engineering Failure Analysis*, **16**, 1837–1845.

45 Ramakrishna, S. (1997) Microstructural design of composite materials for crashworthy structural applications. *Materials & Design*, **18**, 167–173.

46 Jacob, G.C., Fellers, J.F., Simunovic, S., and Starbuck, J.M. (2002) Energy absorption in polymer composites for automotive crashworthiness. *Journal of Composite Materials*, **36**, 813–850.

47 Farley, G.L. and Jones, R.M. (1992) Crushing characteristics of continuous fiber-reinforced composite tubes. *Journal of Composite Materials*, **26**, 37–50.

48 Mamalis, A.G., Robinson, M., Manolakos, D.E., Demosthenous, G.A., Ioannidis, M.B., and Carruthers, J. (1997) Crashworthy capability of composite material structures. *Composite Structures*, **37**, 109–134.

49 Lim, T.S. and Lee, D.G. (2002) Mechanically fastened composite side-door impact beams for passenger cars designed for shear-out failure modes. *Composite Structures*, **56**, 211–221.

50 Jadhav, P., Mantena, P.R., and Gibson, R.F. (2006) Energy absorption and damage evaluation of grid stiffened composite panels under transverse loading. *Composites Part B*, **37**, 191–199.

51 Wilkes, S. (2004) (Jan 26–28: 2004) Going with the Wind. *Materials World*.

52 Kensche, C.W. (2006) Fatigue of composites for wind turbines. *International Journal of Fatigue*, **28**, 1363–1374.

53 Zollo, R.F. (1997) Fiber-reinforced concrete: An overview after 30 years of development. *Cement & Concrete Composites*, **19**, 107–122.

54 Kurtz, S. and Balaguru, P. (2000) Postcrack creep of polymeric fiber-reinforced concrete in flexure. *Cement and Concrete Research*, **30**, 183–190.

55 Carpinteri, A. and Brighenti, R. (2010) Fracture behaviour of plain and fiber-reinforced concrete with different water content under mixed mode loading. *Materials & Design*, **31**, 2032–2042.

Part Five
Applications

23
Applications of Macro- and Microfiller-Reinforced Polymer Composites
Hajnalka Hargitai and Ilona Rácz

23.1
Introduction

Fiber-reinforced polymer composite materials have been used for about 70 years, first in defense industry, particularly for use in aerospace and naval applications [1]. Fiberglass pipe (1948), solid rocket motor cases and tanks made of high performance composite materials can be also mentioned as the earliest applications.

Polymer composite materials own not only superior mechanical properties but also they are nonmagnetic and have a good resistance against the harsh marine environment. Because of the above-mentioned advantages the largest consumer of composite materials was the marine market in the 1960s. During the next decade the automotive market surpassed marine and has maintained that position. Application fields were expanded to electrical transmission (insulators, pool line hardware, and cross-arms, etc.), space applications, and in the construction sector, such as to repair or design of buildings and bridges, strengthen structures, and as stand-alone components [1, 2]. Further benefit of these materials is reducing the radar signature of the structures, such as a ship or aircraft.

The composite industry has shown considerable growth over the last decade and a growing market. In the last several years thermoplastic and thermosetting composites have been used in many sectors, such as aerospace (satellites and aircraft structures), automotive (molded parts, fuel, and gas tanks), rail, defense, construction, sports, medical (dental fixtures, prosthetic devices), electrical, oil/gas, energy and water.

Because of the superior and unique performances of polymer composite materials, and the wide applicability, they are still the focus of many research projects.

23.2
Some Features of Polymer Composites

The principal types of fibers in commercial use are glass, carbon (or graphite), and aramid. Other composite reinforcing materials include ultrahigh molecular weight polyethylene, polypropylene, polyester, nylon, and so on. During the last decade an

Polymer Composites: Volume 1, First Edition. Edited by Sabu Thomas, Kuruvilla Joseph,
Sant Kumar Malhotra, Koichi Goda, and Meyyarappallil Sadasivan Sreekala
© 2012 Wiley-VCH Verlag GmbH & Co. KGaA. Published 2012 by Wiley-VCH Verlag GmbH & Co. KGaA.

increasing environmental consciousness has developed, which has increased the interest to use natural fibers instead of man made fibers in composite materials [3].

The performance of any composites depends not only on the materials of which the composite is made, and the arrangement of the primary load-bearing portion of the composite (reinforcing fibers), but the interaction between the materials (fiber–matrix) also has a big influence on the reinforcing efficiency.

Beside fiber reinforcement a wide variety of additives can be used in composites to modify material properties and tailor the performances of the end products. They are used in a relatively low quantity related to the main constituents to perform critical functions, such as controlling shrinkage, fire resistance, emission control, viscosity control, mold release, coloration, and so on.

By addition of fillers (40–65% of the composite by weight) the cost of composites can be reduced, and they also frequently impart performance improvements that might not otherwise be achieved by reinforcement and resin ingredients alone, for example, resistance of weathering, surface smoothness, stiffness, dimensional stability, temperature stability, and so on. Commonly used fillers are calcium carbonate, carbon black (CB), some kind of metal and metal oxide, clay, alumina trihydrate, calcium sulfate, mica, silica, glass microspheres, flake glass.

23.3
Transportation

23.3.1
Land Transportation

Transportation is a dominant application sector for composites [4]. The predominant reason for use of composites in transportation perhaps ought to be the high specific stiffness and strength obtainable, so as to manufacture lighter and lighter vehicle that therefore can be equipped with a weaker drivetrain, suspension, and so on, to become even lighter (positive weight spiral) [4]. Such weight savings can be exploited to reduce fuel consumption or to increase payload. Beside specific stiffness and strength main advantages are reduced tooling costs, corrosive resistance, energy absorption, geometrical complexity, low manufacturing costs, and so on [4].

23.3.1.1 Shell/Body Parts
Body parts made of polymer-based composites can be found in all kinds of land transport vehicles. There are examples of full composite bodies – although it is not the general case. Weather a composite solution is chosen or not strongly depends on volume of production, partly due to the differences in mold costs and partly due to the fact that composite manufacturing in most cases is more time-consuming than sheet metal stamping [4].

23.3.1.1.1 **Carbon Fiber-Reinforced Composites** Although prohibitively high carbon fiber (CF) cost is often cited as the most difficult challenge that must be overcome

Figure 23.1 Mercedes SLR McLaren Roadster F1 style carbon fiber monocoque [6].

before carbon fiber-reinforced polymers (CFRP) can be widely used in the automotive sector, there are several other technical and market barriers that must be overcome [5] (Figure 23.1).

- supply chain maturity; long-term stable prices and supply;
- increased confidence and experience with CFRP design (design data, analytic tools);
- development of robust joining, testing, and nondestructive evaluation techniques;
- development of short-cycle time, high yield, molding technology;
- demonstration of cost-effective recycling/recovery and repair methods.

Carbon fiber is mostly used in premium category, high power sport cars [6]. Special editions, niche vehicles and customer demands to own "something different" are driving demands for more carbon fiber composite parts [7]. Combination of low weight with excellent safety characteristics can be achieved. In a collision, the fibers of the CFRP elements shred from front to rear, absorbing the energy of the impact with a constant rate of deceleration. Thanks to this uniform deformation behavior and the high-strength monocoque, the energy absorption of the CFRP side members can be precisely calibrated. The engineers achieve this, for example, by creating a constantly changing cross-sectional area for the components. This fine-tuning of the deceleration values results not only in predictable energy absorption behavior but also in a weight advantage, because this design uses only as much material as is actually needed [6].

23.3.1.1.2 **Glass Fiber-Reinforced Composites** In mass production, mostly glass fiber reinforcement is used. The underbody shields of the BMW 1 and 3 models are manufactured using the new SymaLite technology (an innovative, intelligent, light-weight-reinforced thermoplastic composite material, which is compression molded under application of low molding forces) [8]. SymaLite also enables components featuring different thickness to be produced: the degree of consolidation is smaller in the flat surface zones and greater in the edge zones and around attachment points, that is, where the greatest forces are at work. Weight savings of some 30% were also achieved compared with conventional solutions as a result of using the new SymaLite material concept, and the fully closed underfloor module makes the use of PVC for anticorrosion protection unnecessary. The high rigidity of the material combined with its low weight have enabled the aerodynamics to be improved significantly, thus reducing fuel consumption by 5%. Acoustics have also been optimized by the semiconsolidated structure, which generates less noise in rain or from stone chips, and – in contrast to fully consolidated structures – absorbs noise.

The advantages offered by the GMT/GMTex composite compared with other design/material combinations in the pedestrian beam include excellent performance for increased road safety and reduced pedestrian injury or death; integration into existing vehicle designs allowing for a small offset geometry/package space and lower spoiler geometry, while maintaining existing hood, grill, and headlamp position/angles; significant reductions in part and investment costs compared with steel designs – 50% weight reduction for better fuel economy and lower operating costs, 40% reduction in tooling costs, 50% lower materials costs compared with steel with ribs; and better recycling performance compared with metal/nylon–glass hybrids [9].

A glass fiber-reinforced, vinyl-ester resin laminate gives the Metro 45C composite bus a high strength-to-weight ratio providing significant weight reduction, simplicity of repair and absolute resistance to corrosion coupled with a rugged durability that is unmatched in the industry [10]. A further advantage of the solution is that a collision with an average weight car coming from the direction perpendicular to the body do not cause permanent deformation, reducing the maintenance from 2–3 weeks to 1–2 days after such an accident [11] (Figure 23.2).

High speed trains in Europe and Japan particularly, but also commuter trains contain increasing amounts of composites. There is even an example of entire railway cars being filament wound as square tubes [4].

A case study on coupled cost and life cycle assessment of metal, combined metal, and composite and full composite train bodies showed full composite variant was the optimum solution [13].

23.3.1.2 Compartment Parts

Plant fibers as fillers and reinforcements for polymers are currently the fastest-growing type of polymer additives [14]. The use of natural materials in automotive

Figure 23.2 NABI composite bus shell production by Seemann Composites Resin Infusion Molding Process (SCRIMP) [12].

applications is not a new idea. In the 1930s and 1940s, Henry Ford strongly advocated the use of natural materials, including hemp, producing reinforced soy resin composites in the manufacture of exterior body panels [15].

Daimler–Benz has been exploring the idea of replacing glass fibers with plant fibers in automotive components since 1991. Mercedes used jute-based door panels in its E-class vehicles since 1996 [14].

From a technical point of view, these bio-based composites will enhance mechanical strength and acoustic performance, reduce material weight, energy/fuel consumption and processing time, lower production cost, improve passenger safety and shatterproof performance under extreme temperature changes, and improve biodegradability for the auto interior parts [16].

Typical amounts of plant fibers used for different applications the automotive industry are [17]

- Front door linens: 1.2–1.8 kg.
- Rear door linens: 0.8–1.5 kg.
- Boot linens: 1.5–2.5 kg.
- Parcel shelves: up to 2.0 kg.
- Seat backs: 1.6–2.0 kg.
- Sunroof sliders: up to 0.4 kg.
- Headliners: average 2.5 kg.

Although the above-mentioned parts are still often made of glass-reinforced composites, both for car- and public transport (bus, train, etc.) vehicles. Interior composite components of mass transport vehicles also include major items like ceilings, floorings, bulkheads, vestibules, fire barriers, corridor adapter frames, handrails, and luggage bins or racks (Figure 23.3).

Figure 23.3 Natural fibers are used to make 50 Mercedes-Benz E-Class components [15].

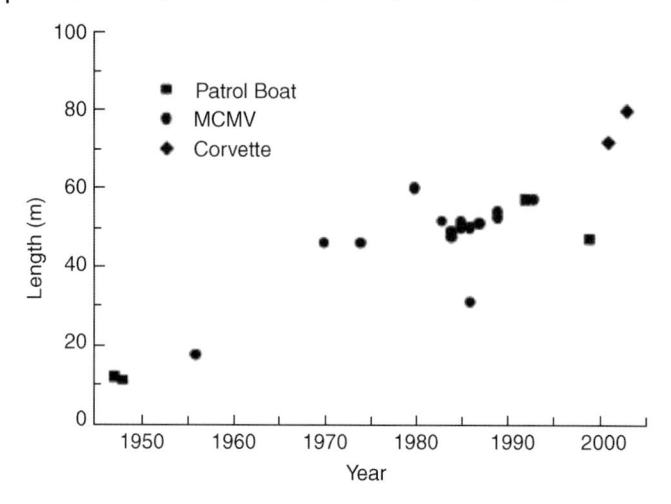

Figure 23.4 Plot of vessel length against year of construction for all-composite patrol boats, MCMV, and corvettes [18] (Source of data: [18, 19]).

23.3.2
Marine Applications

Application of composite materials in sport and recreational boats are detailed in Chapter 23.8.

Early uses of composite materials in naval applications were in the construction of small patrol boats and landing craft. The relatively poor fabrication quality and low stiffness of the hulls restricted these naval craft to less than about 15 m in length and 20 ton in displacement. In recent years the improved design, fabrication and mechanical performance of low-cost composites has led to an increase in the use of composites for large patrol boats, hovercraft, mine hunters, and corvettes [18].

Figure 23.4 presents the results of a survey on the length of naval vessels built entirely of composite between the years 1945 and 2000. Lengths have increased steadily with time, and currently there are all-composite naval ships up to 80–90 m long [18].

Advanced composite materials on large commercial ships have the potential to reduce fabrication and maintenance costs, enhance styling, reduce outfit weight, and increase reliability. George Wilhelmi, of the Navy's NSWC, Carderock Research Center in Annapolis summarized potential ship applications for composite materials as follows [20]:

Structural	Machinery	Functional
Topside superstructure	Piping	Shafting overwraps
Masts	Pumps	Life rails/lines
Stacks	Valves	Handrails
Foundations	Heat exchangers	Bunks/chairs/lockers

(Continued)

Structural	Machinery	Functional
Doors	Strainers	Tables/worktops
Hatches	Ventilation	Ducting insulation
Liferails fans	Blowers	Nonstructural partitions
Stanchions	Weather intakes	Seachest strainers
Fairings	Propulsion shafting	Deck grating
Bulkheads	Tanks	Stair treads
Propellers	Gear cases	Grid guards
Control surfaces	Diesel engines	Showers/urinals
Tanks	Electrical enclosures	Wash basins
Ladders	Motor housings	Water closets
Gratings	Condenser shells	Mast stays/lines

Glass, carbon, and aramid are the leading reinforcements, combined with different thermosetting resins. Honeycomb and other sandwich structures have crucial importance in marine applications.

23.3.3
Aviation

Use of composite materials can reduce the overall weight of an aircraft and improve fuel efficiency. Polymer composites resist fatigue and prevent corrosion. Because of the several advantages compared to aluminum they have gradually replaced metallic materials on parts of an aircraft's tail, wings, fuselage, engine cowlings, and landing gear doors. Composites are widely used in aircraft interiors to create luggage compartments, sidewalls, floors, ceilings, galleys, cargo liners, and bulkheads. Fiberglass with epoxy or phenolic resin utilizing honeycomb sandwich construction gives the designer freedom to create esthetically pleasing structures while meeting flammability and impact resistance requirements [20].

Over the past 20 years, military aircraft have driven the development of advanced materials, but today an increasing share of new commercial airplane development is devoted to composites.

23.3.3.1 Military Aircrafts
Composites used in military aircraft include high-strength fibers of glass, boron, plastic, or carbon that are embedded in an epoxy resin matrix [21]. Carbon–epoxy prepregs (or carbon with epoxy + BMI/cyanate-ester) can be used for structural components of fighter aircraft and helicopters, for example, wing skins, spars, fin, rudder, elevons, doors, and so on, or applied in frames, stiffeners, and rotor blades. E-glass fabric in epoxy resins has a wide range of application include fighter fairings, fin-radome, drop-tanks, or as a structural component in small transport aircraft, such as fuselage, wing, and so on [22].

23.3.3.2 Commercial Aircrafts

Prepregs composed of innovative resin systems based on phenolic, epoxy, cyanate ester and benzoxazine resin reinforced with glass, carbon, and aramid fiber fabrics, rovings, multiaxial complexes, or hybrids are used in both Airbus and Boeing commercial aircrafts, including the new Airbus A380 [23]. These composite materials are used for a variety of interior and structure components include floor panels, linings, ceiling panels, airducts, overhead compartments, seats, lavatories, galleys, bars, wardrobes, flap track, and belly fairings, winglet and fins, landing gear doors, trailing edges, and so on. The tail fin of the Airbus A310-300 consist of graphite/epoxy and aramid honeycomb and thus a 300 kg weight reduction could have been achieved and also reduced the number of parts from 2000 to 100. Polymer composites are also used as skins of aircraft engine cowls [24]. The Airbus 350 XWB, which is expected to enter into service in 2013, is projected to have more than 50% of its structural weight in composites [25, 26].

The rudders and elevators made of graphite/epoxy for the Boeing 767 and landing gear doors made of Kevlar (aramid)–graphite/epoxy composites. Composites are also used in panels and flooring of airplanes.

The Boeing 787 Dreamliner is unique in its utilization of composite materials, which is approximately 50% of the weight of the aircraft. The wings of the Dreamliner are composed of up to 50% composite material by weight (80% composite material by volume) and 20% aluminum. In comparison, the previous 777 model was composed of only 12% composite material and 50% aluminum by weight. The use of carbon/epoxy composites also offers significant mechanical properties [21, 25] (Figure 23.5).

Figure 23.5 The Boeing 787 Dreamliner [21].

23.3.3.3 Business Aviation

This segment started to use the composite materials only in the last few years, for example composite empennage in the Falcon7X aircraft. Bombardier's new C-Series aircrafts will have a composite content of approximately 45% [26].

23.3.3.4 Helicopter

Helicopter blades contain different materials including composites in its various versions. One example is where Nomex honeycomb is bonded to glass fiber/250 resin composite skin with low-temperature adhesives. The fibers in the composite blade skin are oriented in different directions [27]. When use graphite/epoxy and glass/epoxy rotor blades in helicopters are not only increase the life of blades by more than 100% over metals but also increase the top speeds [24].

23.3.3.5 Space

Polymer composites – primarily graphite/epoxy materials – have applications in space, because of their high specific modulus and strength and dimension stability during large changes in temperature in space. Kaw [24] reported some examples such as the graphite/epoxy–honeycomb payload of doors in the space shuttle. Also, for space shuttles, graphite/epoxy was chosen primarily for weight savings and for small mechanical and thermal deflections concerning the remote manipulator arm, which deploys and retrieves payloads. SpaceshipOne, the first privately manned vehicle was launched beyond the Earth's atmosphere in 2004. It is constructed from graphite/epoxy composite material as well.

23.4 Biomedical Applications

In the last decade polymers and polymer composites attracted a wider attention for use in medical especially in biomedical devices, primarily in hip replacement due to the coefficient of friction, low density, and good biocompatibility.

For orthopedic application PEEK, UHMWPE, and HDPE are the most preferable materials. UHMWPE has superior mechanical toughness, wear resistance and biocompatibility. However, the generation of large wear debris particles of UHMWPE likely cause adverse tissue reactions leading to osteolysis implant loosening. By helium plasma immersion ion implantation treatment Tóth *et al.* [28] could significantly modify the surface chemical and nanomechanical properties and increase the wear resistance of UHMWPE.

Because of monolithic polymers own poor mechanical strength, low hardness, low elastic modulus, and high wear rate strong efforts have been invested to improve the mechanical properties and wear resistance of polymers by incorporating ceramic fillers, or reinforcing by carbon or Kevlar (aramid) fibers. In biomaterials the biocompatibility of each constituent is required and have to be resist to body environment. There are several applications including dental filling composites, reinforced methyl methacrylate bone cement and UHMWPE, and orthopedic implants with porous surfaces [29]. Rubber used in catheters, rubber gloves, and so on, is usually reinforced with very fine particles of silica (SiO_2) to make the rubber stronger and tougher [29].

Ramakrishna *et al.* [30, 31] published a detailed survey on biomedical applications of polymer composite materials and discussed a great number of medical devices for both external prosthesis and implants used within the human body.

Plate and screw fixation is a rigid method for external fixation of the fractured bone. Intramedullary nails or rods are mainly used to fix the long bone fractures such as fracture of femoral neck or intertrochateric bone fracture. It is inserted into the intramedullary cavity of the bone and fixed in position using screws of friction fit approach. Table 23.1 summarizes the hard tissue applications [30, 31].

23.4.1
External Fixation

Polymer composites are often used for external fixation of bone fractures, glass or polyester fiber fabrics, and water-activated polyurethane are the most popular constituents. The biggest advantages of polymer composite materials over conventional steel include easy to handle, lightweight, comfortable to anatomical shape, strong, stiff, water proof, radiolucent, and easy to remove. CF/epoxy composites,

Table 23.1 Hard tissue application of polymer composites.

Application	Material
Dental implant	CF/C, SiC/C
Dental post	CF/C, CF/epoxy, GF/polyester
Dental bridges	UHMWPE/PMMA, CF/PMMA, GF/PMMA, KF/PMMA
Dental arch wire and brackets	GF/PC, GF/PP, GF/nylon, GF/PMMA
Vascular graft	Cells/PTFE, cells/PET, PET/collagen, PET/gelatin, PU/PU–PELA
Abdominal wall prosthesis	PET/PU, PET/collagen
Intramedullary nails	CF/LCP, CF/PEEK, GF/PEEK
Tendon/ligament	PET/PHEMA, KF/PMA, KF/PE, CF/PTFE, CF/PLLA, GF/PU
Cartilage replacement	PET/PU, PTFE/PU, CF/PTFE, CF/C
Bone plate, screws	CF/PEEK, CF/epoxy, CF/PMMA, CF/PP, CF/PS, CF/PLLA, CF/PLA, KF/PC, HA/PE, PLLA/PLDLA, PGA/PGA
Bone replacement material	HA/PHB, HA/PEG-PHB, CF/PTFE, PET/PU, HA/HDPE, HA/PE, bioglass/PE, bioglass/PHB, bioglass/PS, HA/PLA
Spine cage, plate, rods, screws, and disk	CF/PEEK, CF/epoxy, CF/PS, bioglass/PU, bioglass/PS, PET/SR, PET/hydrogel
Finger joint	PET/SR, CF/UHMWPE
Total hip replacement	CF/epoxy, CF/C, CF/PS, CF/PEEK, CF/PTFE, CF/UHMWPE, CF/PE, UHMWPE/UHMWPE
Bone cement	Bone particles/PMMA, titanium/PMMA, UHMWPE/PMMA, GF/PMMA, CF/PMMA, KF/PMMA, PMMA/PMMA, bioglass/bis-GMA
Total knee replacement	CF/UHMWPE, UHMWPE/UHMWPE
External fixation	CF/epoxy

Bis-GMA: bis-phenol A-glycidyl methacrylate, C: carbon, CF: carbon fibers, GF: glass fibers, HA: hydroxyapatite, KF: Kevlar fibers, LCP: liquid crystalline polymer, PC: polycarbonate, PEA: polyethylacrylate, PEEK: poly(etheretherketone), PEG: polyethyleneglycol, PELA: block copolymer of lactic acid and polyethylene glycol, PET: polyethylene-terephthalate, PGA: polyglycolic acid, PHB: polyhydroxybutyrate, PHEMA: poly(hydroxyethyl methacrylate) PLDLA: poly(L-DL-lactide), PLLA (L-lactic acid), PMA: polymethylacrylate, PMMA: polymethyl methacrylate, PP: polypropylene, PS: polysulfone, PTFE: polytetrafluoroethylene, PU: polyurethane, SR: silicon rubber, UHMWPE: ultrahigh molecular weight polyethylene.

which have sufficient strength and stiffness, is used in that typical external fixation system where wires or pins pierced through the bone and held under high tension by screw to the external frame [31, 32].

23.4.2
Dental Composites

These materials are usually filled with radiopaque silicate particles based on oxides of barium, strontium, zinc, aluminum, or zirconium. The resin matrix usually contains dimethacrylate monomers of which BisGMA is the most popular. Different types of reinforcing fibers such as glass, polyethylene and carbon are used in the composites for the laboratory and chairside fabrication of bridges [32].

Wear resistance is a prerequisite for a dental material to be accepted by both dentist and patient. Heintze et al. [33] listed more than 25 dental polymer composite materials, their manufacturer, main filler type and method of processing. The polymer matrix is consisted of a highly branched network of dimethylacrylates. The main fillers include fumed silica, Ba–Al–silicate, B-silicate, glass-ceramic, Ba–Al–F–B-silicate. Dental composites require materials that match tooth color and translucency so an optical index of 1.5 is required. Materials such as strontium glass, barium glass, quartz, borosilicate glass, ceramic, silica, prepolymerized resin, or the like are used [34]. Fillers are placed in dental composites to reduce shrinkage upon curing.

23.4.3
Bone Replacement

Among various biocompatible polymers, polyethylene-based materials have received wider attention because of its excellent stability in body fluid, inertness, and easy formability. Attempts have been made to improve their physical properties (modulus/strength) to enable them to be used as load-bearing hard tissue replacement applications [35]. Hydroxyapatite-reinforced high-density polyethylene (HA/HDPE) has been successfully used in human bodies as bone replacement. The production, structure, properties of HA/HDPE composites are described in details by Wang [36]. By varying the amount of HA in the composite, a range of mechanical properties of the material and biological responses to the material could be obtained. By increasing the HA content in HDPE the Young's modulus, shear modulus and tensile strength are increased with a corresponding decrease in the strain to fracture. The short-term creep resistance is also improved [36].

Implants made of HA/HDPE have been used for more than 20 years. A middle ear implant is commercially available since the mid-1990s. Subperiosteal orbital floor implants have been used in the correction of volume deficient sockets and in orbital floor reconstruction following trauma [36].

HAPEX™ being a bioactive composite of high-density polyethylene containing 40 vol. % hydroxyapatite has been used as a minor load-bearing bone replacement material [37]. It has already been used for total hip replacement acetabular cup and low load-bearing bone tissue replacement. Nath et al. [35] produced HDPE/20 vol.% HAp/20 vol.% Al_2O_3 composites and exhibit better mechanical and tribological properties in comparison with unfilled HDPE and the existing commercial composite, HAPEX. Subperisteal orbital floor implants made from HAPEX have been also used for patients [38].

23.4.4
Orthopedic Applications

Since fiber-reinforced polymers exhibit simultaneously low elastic modulus and high strength, they are proposed for several orthopedic applications. The properties and design of an implant can be varied and tailored to suit the mechanical and physiological conditions of the host tissues by controlling the volume fractions and local and global arrangement of the reinforcement phase in the polymer matrix [30]. Self-reinforced UHMWPE fiber/matrix composites for orthopedic bearing are described in several patents. For the same application, mineral and quasi-crystalline fillers, such as micron-sized quartz particles and Al–Cu–Fe powders have been used recently in UHMWPE, however these materials are still in the early phases of experimental investigation [39]. At a microscale untreated glass, aramid, and carbon-based fillers are all-insoluble in UHMWPE, and after blending will coat the UHMWPE powder surface [39].

Ultrahigh molecular weight polyethylene has excellent wear resistance but the wear particulate that is produced leads to the limited lifetime of the devices. Plumlee et al. [40] found that the inclusion of zirconium particles (3 µm, 20 wt%) into a UHMWPE matrix can effectively reduce the wear rate of the component without scarifying the impact toughness.

There are several polymer composite materials that are commercially available. One example is a carbon fiber-reinforced PEEK composite (ENDOLIGN™) with a high fiber content and a tailored fiber orientation distribution developed for use in implantable load-bearing medical device applications required for blood, bone or tissue contact of more than 30 days [41]. This material has the ability to closely match the modulus of natural bone while retaining high strength, good fatigue resistance, and compatibility with MRI, CT, and X-ray technologies and also resistant against sterilizing process without adversely affecting its mechanical properties or biocompatibility [42]. PEEK and 30% by weight short carbon fiber grades have been evaluated and approved for use in spinal applications, but unsuitable for high load-bearing applications such as hip implants. A PEEK-based preimpregnated carbon fiber composite tape material has been developed with a fiber content of approximately 62% by volume, providing high levels of mechanical performance and substantially bridging the gap between polymers and metals. The material offers strength and stiffness performance similar to those of commonly used metallic implant materials [42].

23.5
Civil Engineering, Construction

Polymer composites have been used heavily in the construction sector more than 20 years and today it is one of the largest segments in volume shipment. About 30% of all polymers produced each year are used in the civil engineering and building industries [2] and the largest market segment for fiberglass composites (44%, 2009) [43]. The dramatic increase in the use of both carbon- and glass fiber-reinforced PMCs for structures at rehabilitation over the past few years has been driven largely by the interest in seismic retrofitting of bridge columns and the strengthening of beams and slabs [44]. Typical application includes door, windows, rebar, utility pole, bridge deck, grating, cooling tower, bathtub, swimming pool, architectural applications, FRP panel, and others. According to Shi and Mo [45] these composites in the area of the construction can be conventionally classified as follows:

- all fiber-reinforced polymer (FRP) composite structures (building and civil structural systems),
- external reinforcement for metallic and RC structures (civil engineering structural systems) including Seismic retrofit of RC and masonry structures,
- FRP composite rebar reinforcement (civil engineering structural systems),
- replacement of degraded bridge deck systems.

Use of polymer composites in the construction sector has several benefits compared to steel or aluminum [46–49]. Unidirectional reinforced carbon fiber/epoxy laminates have 4–6 times greater specific tensile strengths and 3.5–5 times greater specific modulus than that of steel or aluminum, and decrease slowly over time. However FRP materials have a greater tendency than steel or concrete to undergo creep under sustained long-term loading. Corrosion properties of FRPs are superior to metals, but glass fiber-reinforced composite has higher coefficient of thermal expansion than steel and concrete and the resin matrix material of FRP composites absorb moisture, as do aramid fibers. The interaction of time, temperature, mechanical stress and other weathering conditions, such as moisture, UV, and freeze/thaw cycling affects the structural performance of FRP to a greater extent than any other building material [50].

23.5.1
External Strengthening

23.5.1.1 Repair and Retrofitting
FRP composite materials may be the most cost effective solution for repair, rehabilitation, and construction of portions of the highway infrastructure. A relatively widespread application of FRP-strengthened members is due to their additional flexural capacity in the postyielding stage related to the conventional reinforced concrete (RC) structures [51]. The most critical aspect of these materials involves long-term durability [52].

Externally used epoxy-bonded steel plates can be replaced by FRP strips or sheets. FRP laminates are available in forms of precured pultruded plates and uncured sheet systems utilizing carbon, glass and aramid. The most extensive applications include buildings and parking garages, although numerous bridge-related repairs have been conducted [47]. The composite plates and sheet systems used to externally bond to concrete elements such as girders and pier caps to increase flexural capacity, but the shear capacity of concrete members used to also increase. Retrofit applications aimed at controlling cracks and preventing spelling by using externally bonded FRP composites have also been conducted.

In 2000 a steel girder of a bridge in Newark, Delaware was strengthened using bonded carbon fiber plates [47]. Because FRP composites are corrosion resistant they can replace steel reinforcements. Glass fiber-reinforced polymer (GFRP) and CFRP rebars were found to be excellent replacements for steel rebars in reinforcing concrete bridge decks. FRP composite rebar and FRP composite strands (or tandons) have been developed and are available from various manufacturers, two- and three-dimensional composite grid reinforcements are also available. Several bridges have been constructed by using that composite rebar (for both flexural and shear reinforcement) and FRP strands for prestressing and for stay cables, of which one important advantage is the significant reduction in weight.

Another area that has received considerable attention is that of seismic retrofitting of concrete bridges using FRP composites. The primary application is column wrapping, where FRP sheets may be wrapped around RC elements, resulting in considerable increases in strength and ductility without and excessive stiffness change. Further advantage of this method that FRP wrapping may be tailored to fulfill the specific structural requirements by adjusting the placement of fibers in various directions [46]. CFRP fabric wrapping is widely used for repairing concrete piers, pier cap pedestals and also deteriorated or damaged precast prestressed and reinforced concrete beams [49]. FRP composite jacket applications have performed well and can be used as a retrofit technique for column strengthening [47].

23.5.1.2 Composite Decks

Within the field of highway structures bridge decks have received the greatest amount of attention in the past few years [46]. For rehabilitation of load-restricted bridges, without composite girders or stringers, FRP bridge decks represent a good potential replacement for existing concrete decks providing thus greater live-load capacity [47].

FRP decks commercially available at present time can be classified according to two types of construction [46]: sandwich and adhesively bonded pultruded shapes (EZSpan (Atlantic Research), Superdeck (Creative Pltrusions), DuraSpan (Martin Marietta Materials), square tube and plate deck (Strongwell)). Polyester resins are favored for their low cost, although vinylester resins are preferred in very moist environments. Woven and stitched fabrics are often employed (DuraSpan and Superdecks) for precise placement of multiaxial reinforcement for improved delamination resistance. EZSpan employs through-the-thickness braided performs as the reinforcement for the triangular tubes (Figure 23.6).

Most of the decks made from either the above-mentioned pultruded sections (e.g., honeycomb-shaped, trapezoidal, or double-web I-beams) or slabs made using a vacuum-assisted resin infusion process. Several have been made by hand with a wet lay-up process. Most of the bridges have a thin polymer concrete wearing surface, although sometimes asphalt is used [47] (Figure 23.7).

The Kings Stormwater Channel bridge on California State Route 86 (highly traveled NAFTA truck corridor, on the picture) is innovative in that it uses carbon fiber-reinforced epoxy tubes filled with concrete instead of traditional concrete and

Figure 23.6 Pultruded profiles from Strongwell (Bristol, Va.) have been used in structural applications for decades [53].

steel piers. It also has a carbon fiber deck, which is lighter and faster to construct than a concrete deck [55] (Figure 23.8).

In Salem Avenue Bridge in the city of Dayton (2000), originally built in 1952 the deteriorated concrete decks were replaced by lightweight FRP deck panels [47].

Figure 23.7 GRIDFORM™ FRP reinforcement of concrete bridge deck (prefabricated FRP double-layer grating, concrete-reinforcing system with integral stay-in-place form for vehicular bridge decks) can be seen in the figure [54].

A cable-stayed bridge in California (I-5/Gilman Advanced Technology Bridge in La Jolla) made entirely of FRP composite, will be 137 m long by 14.6 m wide, carry two 3.6-meter lanes of vehicular traffic and have two 2.4-meter bike lanes, a walkway and utility lines. Among many FRP composite components, the bridge will have girders and pylons made of CFRP composite tubes filled with concrete, an FRP deck and FRP stay cables [47].

Figure 23.8 The Kings Stormwater Channel bridge on California State Route 86.

23.5.1.3 Seismic Rehabilitation

Bousselham et al. [56] presented a comprehensive review and synthesis of experimental studies on the seismic rehabilitation of RC frame beam-column joints with FRP. The reported test results confirm the structural effectiveness of the FRP strengthening technique for the seismic retrofit of RC joints.

23.5.1.4 Unique Applications

FRP composites are also used for pedestrian bridges (Aberfeldy footbridge, 1993; Homestead Bridge, 1997; etc.) or cantilevered pedestrian/bike paths. Pedestrian bridges use standard pultruded shapes in both the cable tower and the decking system [46]. The Bonds Mill Bridge in England was manufactured from ten identical pultruded planks (automated building blocks), which produced mainly from unidirectional E-glass fiber/vinylester resin [45]. Other unique applications include piles, stay-in-place forms, glulams, signs, grates and drains, and guiderails/guardrails.

Odello et al. [57] describe Navy efforts to apply carbon fiber-reinforced epoxy composite materials to pier decking and support columns for the purpose of repairing deteriorating installations and strengthening structures not designed for current load requirements.

23.5.1.5 Structural Applications

Pultruded shapes, both standard and custom, have also been used in building and housing construction system. In nonbuilding structural product markets, custom structural shapes have been developed for latticed transmission towers, light poles, and highway luminaire supports and guardrails [46].

FRP composite rebar made from high strength glass fibers along with an extremely durable vinyl ester resin by combining the pultrusion process and an in-line winding and coating process for the outside sand surface has several advantages over steel rebars. A carbon/vinyl ester product is also available for structures that require more stiffness and enhanced mechanical properties. FRP rebar significantly improves the longevity of civil engineering structures where corrosion is a major factor [58] (Figure 23.9).

Figure 23.9 Transmission tower utility pole and composite rebar for reinforced concrete applications [59, 60].

Figure 23.10 Repair of corroded transmission pipeline by wrapping [61].

Mechanical properties, such as flexural, tensile, fatigue, creep, and so on, of different fiber-reinforced epoxy, vinyl ester, or polyester laminates are well studied in the scientific literature. According to Chin *et al.* [50], the resin type plays a minor role in static properties of the laminates but a major role in fatigue (Figure 23.10).

23.5.1.6 Pipe Applications
Growth of FRP pipes (chemical, oil and gas, offshore, sewage and retail fuel) used in last ten years has been steep because of their advantages over pipes made of traditional materials like iron, concrete and plastics. Its superior mechanical and anticorrosion properties, low conductivity, longer life cycle makes FRP pipes a natural

Figure 23.11 FRP vessels and pipes with a diameter of 4–15 m can be repaired by filament winding [63, 64].

choice both for general purpose as well as in specialty applications. With increasing cost of iron and steel, FRP pipe has also emerged competitive in large diameter pipe market, in high-pressure areas and in elaborate pipe networks extending over several thousands of kilometers [62] (Figure 23.11).

Chin et al. [50] describe several applications of FRP composites, including offshore applications (tension leg platforms, tether lines, risers, cables, tubing and drill pipes) by using carbon, aramid, glass fibers embedded in epoxy, vinyl ester resin.

The reinforcing fiber for composite pipes is limited mostly to glass at the moment due to its superior strength/cost ratio. Anhydride cured epoxies are utilized in flow lines and water injection systems, whereas oil gathering lines are fabricated from vinyl ester. Aromatic amine-cured epoxies have been showed to be capable of surviving 25 years in sour oil and brine. [50]. Glass-reinforced polyester pipes (2700 mm and 2900 mm) have been applied to a small hydroelectric plant in Brazil to replace conventional square galleries of framed concrete [65].

23.5.1.7 Wind Energy Applications

Wind energy is a rapidly growing market segment of the composite industry and the fastest growing energy sector. The wind energy market provides great opportunities to product manufacturers (gear box, tower, blade, generator, etc.) as well as material suppliers of the composites industry to expand their businesses. Wind turbines require the manufacture of large rotor blades, nacelles and other components using wet lay-up, VARTM, prepreg lay-up and other processes [66].

23.5.1.8 Others

CRFP is a novel approach for the repair and/or strengthening of smokestacks and chimneys. Repairs may be necessitated by corrosion of reinforcing steel and general deterioration of materials. Fiber-reinforced polymer composites are also used for strengthening of masonry parapets and to repair and prevent corrosion and/or leakage problems in metallic, concrete and fiber glass tanks and silos (by CFRP) [67]. CFRP is ideal solution for seismic repair or strengthening of tilt up walls. GFRP or CFRP can also be used for strengthening walls around door or window openings to create lintels or piers.

Autoclave-cured, three-ply, carbon fiber-epoxy laminate and knitted biaxial E-glass fabric, and later a Kevlar/glass FRP can be used to reinforce concrete walls. The externally reinforced walls suffered high displacements, but did not fail [52].

23.5.2
Internal Reinforcement for Concrete

Depending on the level of stress different types of constituent FRP materials are chosen for this application. Low-cost E-glass FRPs are generally chosen for non-prestressed applications, whereas high-strength carbon and aramid fiber FRPs are preferred for prestressed applications because of their capability of sustaining much higher stresses over the design life. Thermosetting resins such as vinyl ester and epoxy are the predominant polymers chosen for internal FRP reinforcements on account of their excellent environmental resistance, although affordable thermoplastic resins are recently gaining attention due to their potential for being heated and bent in the field [46].

FRC reinforcing bars have been used in magnetic resonance imaging facilities, an aircraft station compass calibration pad, tunnel boring operations, chemical plants, electrical substations, highway barriers, and a variety of seafront structures. Glass FRP dowel bars appear feasible for load transfer across highway pavement joints, provided the dowel diameter increases and/or the dowel spacing decrease relative to design with steel bars [47].

23.6
Electric and Electronic Applications

In recent years, conducting polymer composites (CPC) have been successfully used for many applications, for example, electrostatic discharge (ESD) protection, electromagnetic–radiofrequency interference (EMI/RFI) shielding, self-control heating, deformation-conductivity transducers and vapor sensing [68, 69].

Conducting polymer composites composed of insulating polymer and electrically conducting fillers carbon black and metal particles have to satisfy different characteristics of electrical properties (resistivity, percolation threshold) with temperature, pressure, or solvent solicitations, which give CPCs a sensitivity toward their environment [69]. For basic applications, to get cheap and easy processing, CPC it is useful to reduce the percolation threshold, that is, the quantity the filler beyond which the CPC becomes electrically conductive. This target can be achieved by using cocontinuous biphasic polymers, which is an immiscible blend, in which the fillers are dispersed in only one phase [69].

The expanding telecommunication industry has increased the demand for high dielectric constant/low loss microwave substrate. Filled polymers are potential electromagnetic (EM) crystal materials. Marett et al. [70] developed EM crystals for microwave applications from EPDM TPE/$BaTiO_2$ composite with a maximum filler level of 70% by weight. The EM crystal showed good reflection properties in the 12–15 GHz range.

Thermoplastic or thermosetting matrices filled generally with carbon black but more recently metal plated ceramic filler and metallic particles to support higher intensity and voltage working constraints. The perspectives of applications as smart materials are as temperature sensors and disruptor devices, materials for production of thermistors at self-regulating heaters, and so on [71]. Ramos et al. [72] prepared carbon black filled conductive PMMA composites for gas sensor application.

Ceramic–polymer composites consisting of ceramic particle filled in a polymer matrix are now widely used in the electronic industry as substrates for high frequency uses, since they combine the ceramics electrical properties and the mechanical flexibility, chemical stability and processing possibility of polymers [73]. Thomas et al. [73] reported $Sm_2Si_2O_7$ filled polyethylene and polystyrene composites for microwave packaging applications.

Hu et al. [74] presented a novel method for preparing conductive carbon black filled polymer composites with low percolation threshold for polyurethane emulsion. The insulator–conductor transition in the emulsion blended composites occurs at 0.8–1.4 vol.%, which is 12.3–13.3 vol.% for solution blended composites. It is demonstrated that the composite microstructure rather than chemical structure of the matrix polymer predominantly determines the electrical conduction performance of the composites.

The percolation limit depends on the shape of the conductive particle, the higher the aspect ratio (length to width ratio) of the particles the lower the concentration for percolation to take place [75].

The effects of electromagnetic interference are becoming more and more pronounced, caused by the demand for high-speed electronic devices operating at higher frequencies, the more intensive use of electronics in, for example, computers, communication equipment and cars, and the miniaturization of these electronics. These trends indicate the need to protect components against EMI in order to decrease the chances of these components adversely affecting each other or the outer world. The effects of electromagnetic interference can be reduced or diminished by positioning a shielding material between the source of the electromagnetic field and the sensitive component. This protection may be achieved by making the housing of electronic components electronically conducting. Electrical conductivity is a prerequisite of EMI shielding material [75].

For ESD protection surface conductivity is important, to allow a fast and controlled discharge of static charge. Filling a matrix of engineering plastic with an electrically conducting material combines the availability of a housing made of shielding material with the advantages of traditional compounding of these composite. Traditionally metal or carbon black particles have been used as electrically conductive filler materials [75]. Shielding and mechanical performance can be varied via filler loading or altered through wall thickness changes to satisfy demands associated with a particular device.

To create housings for asset tracking devices fire retardant polycarbonate with copper/nickel metallization can be also used for EMI shielding. There are several commercial polymer composites, which can increase the shielding effectiveness. LNP Faradex compound built on a resin, which offers exceptionally low temperature impact. The base polymers in the composite are PC, PC/ABS, ABS, PP, PA6, while the conducting fillers in the compound are stainless steel and carbon fiber (10–15 wt%).

Figure 23.12 EMI shielding gaskets made from metal particle loaded elastomeric polymers used in several applications.

The application fields include electronic housing, blood pressure monitoring device, which has a housing made of stainless steel fiber filled PC/ABS [76].

Premix Thermoplastics has introduced a full line of EMI shielding compounds for electronic enclosures called Pre-Elc EMI. The compounds using stainless-steel fiber, carbon fibers, or nickel coated carbon fibers in a large variety of thermoplastic polymer matrices, including PP, ABS, POM, PA6, PA66 PPS, PEEK, PES, PC, and so on. At 50% carbon fiber 50 dB shielding effectiveness can be obtained, while using nickel coated carbon fibers it can be 90 dB at higher loadings. SEBS elastomer filled with a variety of silver-coated and/or nickel-coated particles provide and ideal replacement for silicone gasketing materials because they are easier to process and are recyclable. Most compounds provide 90 dB or more shielding effectiveness [77].

Elastomeric polymers loaded with metal or metal-coated particles are commonly used for EMI shielding gaskets (see Figure 23.12 [78]).

Common filler materials include nickel-coated graphite, silver, and silver-coated glass. Nickel graphite is available in an average particle size ranging from 35 μm (75% Ni or greater) to 120 μm (60% Ni or greater) [79]. Acceptable physical and shielding properties are achieved by using about 60–65% conductive filler by weight of the gasket [80].

TechSIL 5000 conductive elastomer materials contain conductive fillers such as silver-plated copper, silver-plated aluminum, silver-plated glass, and nickel-coated graphite in silicon or fluorosilicon-based rubber. These materials are frequently used in military and aerospace applications [81].

23.7
Mechanical Engineering, Tribological Applications

The effect of wear on the reliability of industrial components is recognized widely and there are more and more technical applications in which friction and wear are critical issues.

Fiber-reinforced polymers are frequently used as tribo-materials because of their unique properties such as self-lubricity, high specific strength; resistance to wear, impact, corrosion, chemicals, solvents, nuclear radiation, contamination with oils, and so on, apart from easy processability in complex shapes and capacity to absorb vibrations leading to quiet operation. The FRP-based tribo components can be used in extreme conditions of temperatures (cryogenic to moderate up to 300 °C), pressures (vacuum to high), and so on, where liquid lubricants cannot be considered. Such tribo components fabricated from the polymer composites are used in typical situations where either hydrodynamic lubrication is not possible because of frequent starts and stops or low pressure–velocity conditions. They are unique solutions where no contamination is allowed, plain bearings or gears in industries, such as food, papers, pharmaceutical and textile, and so on; where maintenance is spasmodic or impossible (domestic appliances, toys, and instruments, etc.) or lubrication in sparse (aircraft linkage bearings) or as a safeguard in the event of failure of the lubricant systems (e.g., gears in train) [82].

23.7.1
Metal Forming Dies

Filled polymers are applied in producing prototype tools or in small batch production of sheet metal components. Sixty years ago, steel powder or sand filled epoxy, later polyurethane-based composites were applied, while nowadays the best polymeric materials commercially available consist of three or four component filler systems, aiming at improved wear resistance, mechanical strength, and self-lubrication properties [83].

Examples of typical tribological application can be cited as self-lubricant bearings, linear guides, mechanical seals, bushing, bearings cages, transporting belts, gears, and pulleys. Friedrich *et al.* presented how the different fillers, such as graphite, carbon fiber, bronze, and Al_2O_3 affect the sliding wear of PPS, PEEK, and PTFE polymers. As filler materials for tribo applications micrometer sized TiO_2, ZrO_2, SiC and copper compounds were incorporated into PEEK, PA, PPS POM, PTFE that led opposite trends in the wear resistance [84].

23.7.2
Seals

Internal lubricants such as polytetrafluoroethylene (PTFE), MoS_2 and graphite flakes are frequently incorporated to polymer matrix to reduce the adhesion to the counterpart material and to enhance their hardness, stiffness, and compressive strength [84]. Kawakame and Bressam [85] developed PTFE-based composite materials for applications involving dynamic seals in the area of electric motors with higher degree of protection. PTFE with 15% glass fibers type E and with 5% of solid lubricant molybdenum disulphide (MoS_2) and the PTFE with 15% graphite presented good characteristics for polymer wear resistance and they application is offered in lip seals in electrical motors. Short aramid, glass, or carbon fibers are used to increase the creep resistance and the compressive strength of the polymer matrix system used [84].

23.7.3
Bearings

Prehn et al. [86] produced highly filled (12% SiC and 7 vol.% graphite) epoxy composites for a journal bearing in pumps that showed excellent wear behavior under dry and water lubricated conditions, when operating against steel as the counterpart.

23.7.4
Brakes

Brake materials are usually made of composites. Commercial brake friction materials for an automotive break system normally contain more than 10 disparate ingredients, comprising polymers, metals, and ceramic constituents in various forms. The ingredients in a multiphase composite play roles as binders, reinforcements (fiberglass, inorganic fibers, carbon, metal fiber), friction modifiers (MoS_2, graphite, Sb_2S_3, $ZrSiO_4$, Al_2O_3), fillers (barium sulphite, aramid pulp, metal powder) and additives (calcium hydroxide, mica, vermiculite) [87]. Phenolic resin used widely as binder material (~30 vol.%), while rubber and thermosetting polymers are also applied as polymer binding substances. Rubber binders add up longer life to the abrasive tools and are preferred in making gringing wheels [88].

Kukutschova et al. [89] tested semimetallic polymer matrix composite brake material, which composition represents a typical semimetallic brake material available on United States, European, and Japanese market. Several constituents were blended in phenolic resin including steel fiber (10%), coke (15%), $BaSO_4$ (9%), synthetic graphite (5%), resilient graphitic carbon, vermiculite, nitrile rubber, copper (4%), iron powder (3%), Sb_2S_3, tin, MgO, $ZrSiO_4$ (2%), Twaron, and 1% of MoS_2, Al_2O_3 by mass.

23.7.5
Brake Pad

Mutlu [90] developed a new automotive friction material for the brake pads, used rice straw dust and rice husk dust among other fillers and additives including Cu particles, Al_2O_3, graphite, brass particles, steel fibers, cashew, and barite in phenolic resin.

23.7.6
Brake Lining

Polymer matrix composites are widely used in automotive, air, and railway transport systems as brake linings. In contrast to other tribological applications a relatively high friction coefficient in the range of 0.3–0.7 is normally desirable. A characteristic friction material is a multicomponent polymer matrix composite with a formulation that is often developed empirically. Copper metal, brass, Cu oxides, Cu sulphides have become popular additives, and the copper usage in brakes for new automobiles has increased by 40% in the last few years [91].

23.7.7
Gears

Du Pont offers a variety of internally lubricated polymer compositions as a base material of gear application, using acetal, nylon resin, polyamide, polyester, thermoplastic polyester as matrix materials reinforced by minerals, glass, aramid fibers, lubricated by PTFE, or silicon. Acetal resin reinforced with glass fiber can be used as garage door opener gears, door lock actuator gears, in windowlift motors [92].

Zhang and Friedrich [93] presented a good survey on tribological characteristics of micro- and nanoparticle filled polymer composites.

23.8
Recreation, Sport Equipments

Advanced materials were first introduced into high-level competitive sports. Weight savings have been a major driver, together with design optimization and performance improvements that enable sporting enthusiasts to compete at higher level for less effort [94]. The most important characteristics include strength, ductility, stiffness, temperature capability, and forgiveness (a collective term including fracture toughness, fatigue-crack growth rate, etc.) [95]. As the price of reinforcing materials decreased composite materials appeared in the fields of mass sport equipments and recreation.

There is almost no sport where composite materials are not used. Therefore – taking into account the limitations of the chapter – aims, application, and advantages of using composite materials in sport equipments are presented through selected examples.

23.8.1
Summersports

23.8.1.1 Bicycle
Polymer composites are now used in almost every part of a bicycle. Applications include the frame, handle bars, seat post, front fork, wheels and pedals. Even the chain can be replaced using rubber composites. A major advantage of polymer composites is that they facilitate fabrication of one-piece frames, eliminating the weight penalty associated with lugs and fittings that are commonly used to assemble the metal tubes. Continuous carbon, aramid, glass, boron, and ultrahigh molecular weight polyethylene fibers are used to reinforce different thermosetting matrices [96]. According to Hexel [94] carbon fiber prepregs are favored for the weight savings and high rigidity that is targeted in specific areas of the frame and wheels for maximum response when accelerating.

Reductions in weight and improvements in stiffness have significantly improved performance in different indoor and outdoor bicycle racing competition events [96]. The "Carbon Ultimate F10" racing bicycle frame developed by IVW and Canyon Bicycles GmbH is believed to first exceed the figure of 100 for its stiffness-to-weight ratio (STW coefficient) [97]. An example of the carbon/epoxy bicycle frame [98] only weights 1.36 kg, which is much less than the 5 kg weight of the corresponding steel frame.

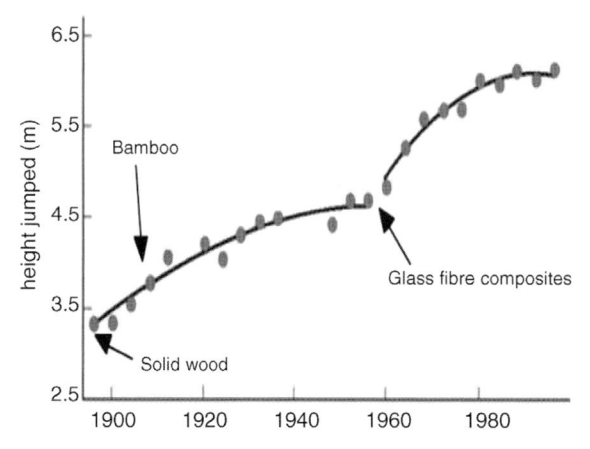

Figure 23.13 Change in Olympic record heights over last 100 years and associated vaulting pole material changes [100].

The phenomenal success of carbon fiber frames in the Tour de France Bicycle Race in 1989 and 1990 had a significant impact on the acceptance and popularity of such composites in the bicycle industry.

23.8.1.2 Athletics

Although as accessories one can find wide range of composites in athletics (bars, stands, hurdles, etc.) as sporting equipment pole (pole vaulting) has the greatest importance.

The sport has seen a huge change in the materials used for the pole, and in the world records achieved. The 1896 Olympics saw a height of 3.2 m achieved by bamboo pole. In the 1960s, after a pole vault record had rise upward for 60 years by centimeters, records began to fall as aluminum poles were introduced [95]. Glass fiber-reinforced polymer composite (GFRP) poles were then introduced, leading to a step change in performance. GFRP poles are flexible, allowing a different athletic style (feet first, vertically upside down approach) and a more energy-efficient vault [99]. Today, the world-class pole vaulter utilizes a highly sophisticated composite pole (Figure 23.13).

23.8.2
Wintersports

23.8.2.1 Skis, Snowboards

Skis were originally wooden planks made from a single piece of wood [101]. Later aluminum/wood bonded structures emerged, and were further developed with plastic base material and steel edges. In subsequent years, the once monolithic structure continued to evolve into a more complex design involving a base, core, side walls, top layer, reinforcement and damping layers. The design goal was to combine good strength and flexibility along the length of the ski with adequate torsion strength to sustain a platform for the skier. In addition, as skis were designed for ever-increasing speed and more demanding terrain, it became necessary to dampen the structures to suppress vibration over a wide range of frequency. Ski structures thus evolved as multicomponent systems incorporating a wide range of materials. The sandwich design continued to evolve with torsion boxes and tubes, and using glass and carbon fibers, wood and metal laminates [102].

There are attempts to develop eco-friendly sporting goods replacing high performance fibers by vegetable (i.e., hemp) fiber [103] although these are not yet commercial products.

23.8.2.2 Boots
To optimize stiffness, shell (or part of it) of high performance ski boots is made of composite materials [104]. Examples – without further details – can be found by the major producers.

23.8.2.3 Poles
Today's skiers can choose from two types of poles. Aluminum is sturdy, cost efficient, and similar in composition to the poles made in the mid-twentieth century. Recently, however, advanced and professional skiers have begun to prefer a new type of pole. Composite poles are self-described – manufacturers use various materials to create the pole. Examples of common materials are graphite, carbon, glass fiber, and even Kevlar. Combinations can be 90% carbon and 10% glass fiber, 60% carbon and 40% glass fiber, or any other construction the manufacturer thinks would be beneficial to the sport [105].

The benefits of composite poles are all intrinsic to a skier's performance. Not only are they generally sleeker than their old-fashioned cousins but they also are much lighter. Another advantage of composite poles is the strength and rigidity. Composite poles strength-to-weight ratio can be six times greater than aluminum poles. Aluminum poles, on the other hand, can give a slight bend under pressure and are not as shock-absorbent as their composite cousins.

23.8.2.4 Hockey
Nike Bauer's Supreme One95 carbon fiber composite stick is built for the strong and powerful features. With its low mid flex, and a torsionally stiff lower third, the overall design creates more energy transfer per shot, resulting in greater velocity [106]. Torsional stiffness in the lower third generates more stored energy, creating increased power in every shot – whether it is a booming slapshot or a blistering one-timer [106].

On the other hand, Stefanyshyn and Worobets [107] compared the energy return ability of 30 composite hockey sticks to 13 wood sticks. According to their pure mechanical tests, the composite sticks return significantly more energy (10%) than the wood sticks. However, shooting in hockey is not purely mechanical but depends on how an athlete transfers the energy from the stick to the puck. The average puck speeds generated by the hockey players were not significantly different between the wood and composite sticks, this was true for both the wrist and slap shots.

Protective equipments (also in other sports, e.g., baseball, American football, etc.) are also made of high performance composite materials.

23.8.3
Technical Sports

Composites play different roles in technical sport equipments. Beside applications where high strength and high stiffness-to-weight ratios are required polymer composites are also useful where energy is stored and released in a controlled manner, for example, a crossbow limb stores the energy required to propel the arrow [108–110]. The traditional materials for the crossbow limb are wood or spring steel. Kooi *et al.* [108] state that "The efficiency of the bow is affected by the relative mass of the arrow when compared to that of the limb, but for an arrow of constant mass the lighter the limb the better the efficiency." Composites offer resilience, low structural weight, long service life and are not adversely affected by the environment. Thus, a lightweight composite limb should outperform limbs of other materials [111]. According to Virk *et al.* [111] carbon fibers were considered too brittle for this application, aramid fibers combine low density and high stiffness but are weak in compression, E-glass fibers are relatively flexible but are of high density. The optimized design developed uses aramid fibers on the tension face with E-glass fibers on the compression side.

In case of sport shooting Carl Walther GmbH developed an air rifle with carbon fiber-reinforced barrel and carrier system. Carbon as a carrier system ensures an extremely light and highly stable structure allowing precise and individual adjusting of center of gravity by positioning externally moveable stock weights. The newly developed polymer core in the carbon carrier reduces vibrations considerably. Optimized system bedding and barrel fixation provide tension-free rest in the stock [112] (Figure 23.14).

One can consider Formula 1 as queen of technical sports using leading edge technologies and materials. Savage [113] published a paper on composite engineering for a new racing team that turned to be World Champion in their first season in 2009. The following are based on his work.

The introduction of fiber-reinforced composite chassis was one of the most significant developments in the history of Grand Prix motor racing. Technological advances gained from these advanced materials have produced cars that are lighter, faster and safer than ever before. The structural components of the car must be stiff, strong enough to satisfy the loading requirements, tolerant of and resistant to impact

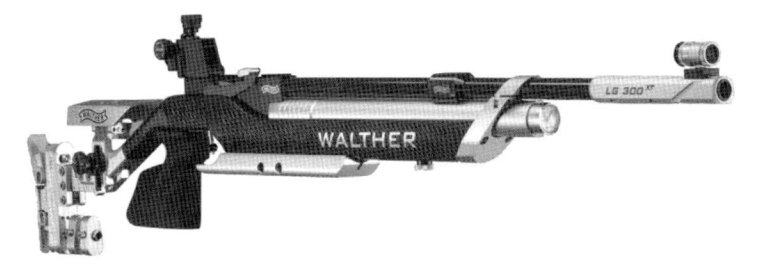

Figure 23.14 LG300 XT Carbontec air rifle of Carl Walther GmbH with carbon fiber-reinforced barrel and carrier system [112].

Figure 23.15 The "primary structure" of a Formula 1 car (chassis, engine, and gearbox) (left) and the complete car with secondary structures added (right) [113].

damage, and be of minimum weight. The solution to this problem is achieved by optimizing the geometry, the quality of construction and by using the most appropriate materials (Figure 23.15).

The primary mechanical properties of composites (strength, stiffness, and failure strain) are governed for the most part by the properties of the fibers, their volume fraction, orientation to the applied stress and their "architecture" within the structure.

Carbon fiber composites now make up almost 85% of the volume of a contemporary Formula 1 car while accounting for less than 30% of its mass. In addition to the chassis there is composite bodywork, cooling ducts for the radiators and brakes, front, rear and side crash structures, suspension, gearbox, and the steering wheel and column. In addition to the structural materials a number of "specialty" composites are also used. These include carbon–carbon brakes and clutches, and ablatives in and around the exhaust ports. In Formula 1, composites reinforced using PAN-based "standard modulus" carbon fibers are not generally used in structural applications. Rather they tend to be employed in bodywork, as "flat stock" for making inserts or as tooling prepregs. Intermediate modulus fibers are widely used throughout the F1 grid. "Ultrahigh strength" fibers are recently developed subdivision of the intermediate modulus grouping. The effect of using these fibers (having increased strength and ductility) manifest as improved impact properties, in particular resistance to and tolerance of damage. A car built in whole or in part from damage tolerant composites will be inherently safer since it is this property that contributes greatly to the integrity of the survival cell. High and ultrahigh modulus fibers tend to be used in lightly loaded, stiffness critical applications.

Three types of polymeric fiber have found use in racing car construction, these being aramids, "Zylon," and highly oriented polyethylene filaments.

Aramid fibers resemble inorganic fibers (carbon and glass) in terms of tensile properties but have much lower compressive strengths, lower density, and considerably greater toughness. They have been traditionally employed to exploit their impact performance (particularly foreign object damage from stones, etc.) and abrasion resistance. The use of aramid is, however, compulsory within the structure of front wing end plates and other aerodynamic appendages at the front of the car. This is done with the aim of reducing the probability of tire damage from sharp fragments of composite components damaged in impacts. Recent research has shown polypropylene fibers to be superior in this application so it is likely that they will replace aramid.

Zylon is an extremely strong fiber consisting of rigid-rod chain molecules of poly (p-phenylene-2,6-benzobisoxazole) (PBO). An epoxy/Zylon appliqué armor panel must be fitted to the survival cell of each monocoque in order to prevent penetration and protect the driver from sharp-force injuries.

Highly oriented polyethylene's high strength and very low density give it specific properties far superior to most other structural fibers. The fibers cannot, however, be used at high temperatures. These are used on F1 cars from time to time in the form of hybrids cowoven with carbon fibers for use in the statutory impact structures. The aim is to exploit their high tensile strength to maintain the components' integrity during the event and thus optimize the controlled disintegration of the carbon and honeycomb, which facilitates energy absorption. A carbon/Kevlar hybrid would perform a similar function but be less weight efficient.

It has been estimated that a mass of 20 kg above the weight limit (minimum weight for car plus driver is 605 kg according to current rules) equates to a loss of 0.4 s around a typical Grand Prix circuit, which during a full race distance to half a lap or several grid positions during a qualifying session. With modern materials it is relatively easy to build a car that satisfies all of the statutory requirements while still being well under the minimum weight limit. As a consequence the majority of the cars are required to carry ballast in order to make up the deficit. At first glance therefore it may seem fruitless to continually aim to reduce the mass of components only to increase the amount of ballast carried, but lowering the weight-center of the chassis is still very beneficial.

23.8.4
Racquets, Bats

A number of popular sports have benefited from extensive research on sporting equipment, including racquet sports (such as tennis and badminton) and club sports (such as golf, cricket, and baseball). Advanced composite materials are often used to reduce the weight and increase the durability of sports equipment. These materials may also allow the ball to be hit farther and with greater accuracy [114].

Wood tennis rockets were first replaced in the late 1960s by metal frames, generally fabricated from steel or aluminum [95]. In the early 1980s, "graphite" composites were introduced, and other materials were added to the composite, including ceramics, glass fiber, boron, and titanium [115]. Presently composite racquets rule the market. The improved stiffness of the frame allowed an increased head size. A larger head size generally means more power, and a larger "sweet-spot" [115] (central part of the racquet where no vibration occurs upon impact with the ball) that is more forgiving on off-center hits and reducing the possibility of forming tennis elbow [95]. A current composite racket can have a 40% larger head, be three times stiffer and 30% lighter than the most highly developed wooden version [116].

The next step in the evolution of the tennis racket may be the inclusion of piezoelectric materials that are capable of controlling the frame vibration. Advances in the technology used in skiing have already led to piezoelectric materials being attached to the surface of skis. At present these materials act in a passive way. The piezoelectric plates have a damping effect by converting the mechanical vibrations into electrical energy that is dissipated through a shunt circuit. A future possibility may be the conversion of the passive configuration to an active form in which the vibrations in the frame are sensed and then canceled by inverting the electrical signal applied to piezoelectric actuators sited in the handle. It could even be this technology that finally leads to the eradication of tennis elbow.

While use of composite materials made tennis sport safer (suppression of tennis elbow), in case of baseball their use is controlled by the sport federations due to safety reasons (increased number of injuries caused by too fast balls) [117].

Millions of golf clubs made of polymer composite have been made worldwide and have achieved phenomenal success in replacing the metallic versions [118]. Key design requirements are high bending and torsional stiffness and strength, lightweight, optimal weight distribution and damping effects. A high percentage of club weight is expected in the head. Polymer composites provide designers and manufacturers with great flexibility in design and fabrication techniques. The range of possibilities is almost unlimited. Carbon, Kevlar, glass, and ultrahigh molecular polyethylene fibers are used either individually or in hybrid format to satisfy different requirements. Virtually all golf clubs are currently manufactured using the filament winding process [118]. Filament wound clubs provide superior performance at low weight and cost compared to roll-wrapped clubs. Huntley *et al.* [119] have found significant increase in dynamic stiffness of filament wound golf shafts, compared with the static value, while in case of sheet laminated shafts there was no significant difference.

23.8.5
Watersports

Water sport equipments – kayak, canoe, sailing boats, power boats – were among the first civil applications of composite materials. According to US sources Ray Greene an Owens Corning employee produced a composite boat 1937, but did not proceed further at the time due to the brittle nature of the plastic used [117]. Beside strength and reduced weight composite materials provide the advantage of simpler maintenance since they do not rot or rust although data can be found on hydrolysis of polyester resin in seawater [120].

The applied materials and technologies are highly depending on the type and size of boats, if it is leisure craft or a competition sailing yacht. Smaller boats are generally sprayed up from discontinuous glass-reinforced unsaturated polyester or the hull is a single skin laminate [121]. Larger powerboats are generally hand laid up, sandwich structures are tend to be used throughout the hull, deck, and floorboards [121]. In competition powerboats and sailing yachts the hull and the deck is normally made if hand laid up sandwich structures of high performance continuous fiber-reinforced

Figure 23.16 A carbon racing sail boat [122].

laminates and cores. In extreme cases the low- or medium performance cores are replaced by high performance foams or honeycomb cores. More recently composites are becoming the material choice of masts, booms, and other parts too [121] (Figure 23.16).

In kayak and canoe sport the first reports on using carbon fiber were published in the late 1960s [123], still at the 1972 Munich Olympic Games only two rowing shell used carbon fiber beside glass reinforcement [124]. Nowadays competition kayaks and canoes are almost exclusively made of carbon fiber. Market leading manufacturers have used honeycomb for many years for weight savings and stiffness. Engineered fabrics and core material users have also moved from wet lamination to prepreg technology [94].

23.8.6
Leisure

Glass, carbon, and aramid-reinforced epoxies are being used for producing fishing poles all over the world. Their stiffness, strength, and foldable features and esthetic effects are their main advantages. Carbon fiber-reinforced epoxy fishing poles are lighter than bamboo and possess a better sensitivity and "feel." The poles are mainly made using a balanced wrap pattern. This technique puts a number of fibers in one direction and an equal number in the opposite direction so that the rod is balanced cylindrically and longitudinally. It is believed that such poles respond more quickly and steadily. Introduction of pultrusion technique allowed high degrees of automation and productivity improving both affordability and popularity of such polymer composites in leisure activity [96]. Carbon fiber prepregs are also popular options of rod productions as they can be used to "fine-tune" performance and provide an optimum combination of power and sensitivity, low weight, and durability [94].

Many other structural and nonstructural applications could be mentioned from step for step aerobics to cover panels of indoor bicycles. The list is endless (Figure 23.17).

Figure 23.17 A step aerobics group fitness class [125].

23.9
Other Applications

23.9.1
Fire Retardancy, High Temperature Applications

Because of the rigorous safety requirements fire protection has an increasing role in many applications, such as building industry, transportation, and so on. A diverse range of metal oxides and metal hydroxides are used as active flame retardant fillers, although the dominant ones are aluminum (most common is aluminum trihydroxide, Al(OH)$_3$ (ATH)) and magnesium hydroxides (MH), of which the former finds much greater commercial use. Other types of compounds containing aluminum, antimony (Sb_2O_3, Sb_2O_5), iron (e.g., ferrocene, FeOOH, FeOCl), molybdenum (MoO_3), magnesium ($Mg(OH)_2$), zinc, and tin are also used. They have different efficiency and changes by the filler ratio. Typically 20–60 wt% loading level is needed for the substantial reduction in flammability. Nitrogen-based compounds such as melamine are also highly effective flame retardants [126].

ATH is popular for several reasons but requires high loading levels (50 wt% or more), which resulted in the decrease of the mechanical and durability properties of most types of polymer composites. By combining it with other flame retardants, using lower concentration, this negative effect can be minimized. By using ATH in glass-reinforced epoxy or phenolic composites, the ignition time can be improved. The application of ATH is limited by its decomposition temperature of 220 °C. Magnesium hydroxide ($Mg(OH)_2$) and a hydromagnesite/hunite hybrid compound are active fillers suited to polymer composites that need to be processed above the decomposition temperature of ATH. Magnesium hydroxide is thermally stable up to 330–340 °C, and therefore can be used in most types of high temperature thermoplastics without decomposing during processing. Zinc oxide and borates are also used as fillers [126].

Various synergists can be used in combination with hydrated fillers to enhance flame retarding efficiency thereby lowering the overall levels of filler required to achieve a specific performance requirements.

A novel approach to improve flame resistance of composite materials is the addition of intumescent fillers to the polymer matrix. A large improvement was measured as in the fire resistance of glass-reinforced phenolic composites when filled with intumescent particles (ammonium polyphosphate/pentaerythritol compound). By heating the material a porous char layer develops, insulating and protecting the underlaying virgin composite material. The fibers strengthen the intumescent char and prevent it from spalling or flaking [126].

Inorganic or organic-based phosphorus fillers used to be added for improving the flammability resistance of polymers and polymer composites [126].

Toldy et al. [127] developed carbon fiber-reinforced epoxy composites for aerospace applications. By formation of multiplayer composite consisting of composite core and intumescent epoxy resin (by using phosphorus-containing reactive amine) coating layer a simultaneous increase could be achieved both in flame retardancy and mechanical performance.

Ethylene-vinyl acetate (EVA) copolymers with different acetate contents are widely used in many fields, particularly in the wire and cable industry as excellent jacket materials with the required physical and mechanical properties. To achieve the required flame retardant grade of low smoke and nontoxic character, metal hydroxides, for example, magnesium hydroxide or aluminum hydroxide, should be added in a great amount, up to 50–60% mass fraction, in general. That always leads to a deterioration of mechanical performance in the composites. By combining MH with other fillers the amount can be reduced by a synergetic effect. There are several publications on using zinc borate, MWNT, layered-double hydroxide, rubber, silicon talc, expandable graphite, ammonium-polyphosphate, and so on. Huang et al. [128] combined sepiolite (magnesium silicate) and MH in EVA polymer.

For temperatures up to approximately 320 °C certain advanced reinforced plastics offer a solution with thermosetting and thermoplastic-based composites. The most used matrix and fiber systems are phenol resins (max. 150–230 °C), special epoxy systems (phenol novolac, max. 230 °C), bismaleimide (BMI) resins (>230 °C, better mechanical habits than epoxy at high temperatures), polyimide resins (>315 °C), and thermoplastics include PEI and PEEK reinforced by glass fibers (E, S, and quartz glass), aramid (Twaron, Kevlar), boron fibers, carbon/graphite [129]. Examples of application are engine parts surrounding the hot area's of the engine, pipes for hot gasses/liquids, exhaust parts, insulations parts, and so on.

23.9.2
Self-Healing Polymer Composites

Self-healing, which is inspired by biological systems, is a novel alternative to damage tolerant design and removes the need to perform temporary repairs to damaged structures. To date, research into self-healing of polymeric materials has considered a number of approaches [130–132]. These include the use of either glass tubes or microcapsules, containing a healing agent, which is released into a damage side upon fracture. Dicyclopentadiene (DCPD) monomer has been used in numerous studies as healing agent, which possesses low viscosity and excellent shelf life when stabilized with 100–200 ppm *p-tert*-butylcatechol. Commercial microcapsules typically have a diameter of 3–800 μm, and consist of 10–90 wt% core materials. Several methods are used for their synthesis including the interfacial polymerization, *in situ* polymerization, extrusion, coacervation and sol–gel methods. But among of them, *in situ* polymerization is the easiest and best process to encapsulation, because it does not require high-level technology [131, 133].

Polymerization of the healing agent then restores the original mechanical properties. There has also been significant research into the use of such systems in fiber-reinforced polymers [131].

In the last few years several self-healing unidirectional glass fiber composites have been developed [132, 134–136]. Hucker *et al.* [137] embedded resin filled hollow glass fibers (30–100 μm diameter) within glass fiber-reinforced epoxy and infused with uncured resin to provide a healing functionality to the laminate. Williams *et al.* [135] wound HGF (healing glass fiber) directly onto uncured carbon fiber-reinforced epoxy plies prior to lamination, which host laminate is widely applied for aerospace application. Kessler *et al.* [138] also used woven graphite fiber-reinforced epoxy as host material and filled 20 wt% of microcapsules (DCPD) and 5 wt% catalyst.

These composite materials are in the research state now, but have good expectations for application in several fields including aerospace applications.

23.9.3
Shape-Memory Polymers

Shape-memory polymers (SMP) can be deformed and fixed into a temporary shape, and then recover their permanent shape by external stimulus [139]. The shape-memory effect (thermal shrinkage) of oriented polymers has been investigated intensively in the recent years because of their applications in microelectromechanical systems, actuators, for self-healing and health monitoring purposes and in biomedical devices. A very recent review of Ratna and Karger-Kocsis [140] highlights the recent progress in synthesis, characterization, evaluation, and proposed applications of SMPs and related composites.

SMPs in general have lower strength and stiffness and thus their use for many applications are limited. The effect of incorporation of reinforcing fillers has investigated by many researchers. Several types of reinforcing elements were applied, such as Kevlar fibers, SiC, glass fiber and carbon fibers. By blending carbon black conducting SMP composites could have been produced. Liu *et al.* [139] presented a review on electroactive SMP composites. In these composites carbon particles, conductive fiber and nickel zinc ferrite ferromagnetic particles, and so on, had been used.

There are some commercially available products, such as Veritex™, which is a fiber-reinforced composite that uses a shape-memory polymer, Veriflex®, as the matrix. This allows Veritex to easily change shape above its activation temperature. At lower temperatures, the material maintains high strength and high stiffness. When heated, Veritex will temporarily soften. It can then be reshaped and will harden in seconds, maintaining the new configuration. When reheated, it will return to its original cured shape. This versatility allows for structures to be stowed and then later deployed to the operational shape. Veritex can be manufactured with a variety of fiber types and shape-memory polymer formulations to fit unique applications requiring different properties [141].

23.9.4
Defense

Laminated ballistic composite materials may be used in protective helmets, or with ceramics and other materials for protective body armor. Standard hard plate protective body armor is made up of multiple layers, commonly including a ceramic plate to blunt and fracture projectiles and a laminated composite panel to stop the projectile while containing the ceramic particles. Armor may include an antitrauma layer to reduce potential injury caused by dynamic deformation of the armor into the wearer [142, 143]. High performance fibers used in ballistic products are glass fibers (S-and R-glass), aramid (Kevlar, Twaron), and high performance polyethylene (HPPE) fibers (Dyneema, Spectra). Recently ballistic products based on PBO (Zylon) have been introduced on the market [144]. Layers of woven or unidirectional fibers are bonded using a thermoplastic or thermosetting polymer matrix [143]. Based on studies of ballistic impact behavior of typical woven fabric E-glass/epoxy composites Naik et al. [145] determined that the possible damage and energy absorbing mechanisms are energy absorbed by moving cone, tension in the primary yarns, deformation of the secondary yarns, delamination, and matrix cracking.

Glass-reinforced polymer (GRP) composites are used in military platforms such as naval ships, patrol boats, submarines, and armored vehicles. Two benefits of using GRP in military platforms are good resistance to ballistic projectiles and explosive blasts. GRP can be highly resistant to perforation by small high-speed projectiles, such as bullets and shrapnel, because the composite rapidly absorbs the impact energy by various damage processes [146].

Compared to metal systems, composite armor systems made with S-2 Glass® fiber are lighter weight, have a lower radar and thermal signature, and resist degradation from corrosion, chemicals, biological agents, and nuclear radiation, while providing a comparable level of protection [147].

Glass fiber composite armor systems are design-engineered to meet specific requirements. Applications range from flat spall liners to contoured components that provide structure as well as nonspalling ballistic protection, from one integrated system [147].

Composite armor systems made with S-2 Glass fiber technology offer [147] (Figure 23.18)

- combination of structural as well as ballistic capabilities in a single system;
- combined weight savings of structure and armor of over 33% compared to a similar all-metallic vehicle with comparable protection;

Figure 23.18 Composite Armored Vehicle made of S-2 Glass Fiber Rovings by vacuum-assisted resin transfer molding (VARTM) technology [147].

- excellent energy absorption;
- signature management integrated into the composite armor system;
- improved survivability;
- improved durability;
- lease of repair; and
- costs comparable to current all-metallic hull systems.

23.10
Conclusion

As it was shown in the above chapter composite materials have growing impact on every fields of life from space to households. Although polymer-based nanocomposites receive a great attention in the recent years, the application possibilities of the "conventional" micro- and macrocomposite structures are continuously expanding. It is due to their superior characteristics over other materials, simple design opportunities, and relatively low cost preparation. The easily tailorable properties make easier the production of optimized solutions.

References

1 Busel, J.P., et al. (2007) ACI 440R-07. *Report on Fibre-Reinforced Polymer (FRP) Reinforcement for Concrete Structures*, American Concrete Institute.
2 Lechkov, M. (2009) Preface in *Encyclopedia of Polymer Composites: Properties, Performance and Application* (eds M. Lechkov and S. Prandzheva), Nova Science Publishers Inc. ISBN: 978-1-60741-717-0.
3 Hargitai, H., Rácz, I., and Anandjiwala, R. (2007) Development of hemp fibre reinforced polypropylene composites, Chapter 17, in *Textiles for Sustainable Development* (eds R. Anandjiwala,

L. Hunter, R. Kozlowsky, and G. Zaikov), ISBN: 1-60021-559-9, Nova Publishers, pp. 189–198.
4. Aström, B.T. (1997) *Manufacturing of Polymer Composites*, Chapman & Hall, London, Weinheim, New York, Tokyo, Melbourne, Madras, pp. 11–31.
5. Sullivan, R.A. (2006) Automotive carbon fibre: opportunities and challenges. *Journal of Materials Research JOM*, **58**, 77–79.
6. ZerCustoms (2007 Jul 02), http://www.zercustoms.com/news/Mercedes-SLR-McLaren-Roadster-F1-style-carbon-fibre-monocoque.html
7. Brosius D: Carbon fiber: The automotive materials of the twenty-first century starts fulfilling the promise, 3rd Annual SPE automotive Composites Conference (2003), http://www.speautomotive.com/SPEA_CD/SPEA2002/pdf/a13.pdf
8. Reinforced Plastics, 27 December 2009, http://www.reinforcedplastics.com/view/6117/thermoplastic-composite-specified-for-bmw-z4-underbody-shielding/
9. Source the Engineer, Glass-mat thermoplastic enables award-winning beam, 27 December 2005, http://source.theengineer.co.uk/materials-and-chemicals/general/materials-and-processing/glass-mat-thermoplastic-enables-award-winning-beam/377339.article
10. North American Bus Industries Inc, http://www.nabusind.com/NABI/metro-45c-bus.htm
11. Vezess.hu *online magazine* (in Hungarian), http://www.vezess.hu/magazin/amerikai_alom_magyar_autobusz/18488/
12. NABI Ltd, http://www.nabi.hu/oldal/compobus
13. Castella, P.S., Blanc, I., Ferrer, M.G., Ecabert, B., Wakeman, M., Manson, J.-A., Emery, D., Han, S.-H., Homg, J., and Jolliet, O. (2009) Integrating life cycle costs and environmental impacts of composite rail car-bodies for Korean train. *International Journal of Life Cycle Assessment*, **14**, 429–442.
14. Ashory, A. (2008) Wood–plastic composites as promising green-composites for automotive industries!. *Bioresource Technology*, **99**, 4661–4667.
15. Holbery, J. and Houston, D. (2006) Natural-fibre-reinforced polymer composites in automotive applications. *Journal of Materials Research JOM*, **58**, 80–86.
16. Chen, Y., Sun, L., Chiparus, O., Negulescu, I., Yachmenev, V., and Warnock, M. (2005) Kenaf/ramie composite for automotive headliner. *Journal of Polymers and the Environment*, **13**, 107–114.
17. Ellison, G.C. and McNaught, R. (2000) The use of natural fibres in nonwoven structures for applications as automotive component substrates. Ministry of Agriculture Fisheries and Food (now Department for Environment, Food and Rural Affairs), ⟨ http://www.defra.gov.uk/farm/acu/research/reports/Rdrep10.PDF ⟩ in Ref. [14].
18. Mouritz, A.P., Gellert, E., Burchill, P., and Challis, K. (2001) Review of advanced composite structures for naval ships and submarines. *Composite Structures*, **53**, 21–41.
19. Sharpe, R. (1999) *Jane's Fighting Ships 1999–2000*, Jane's Information Group Limited, Coulsdon, UK.
20. Eric Greene Associates, Inc.: Marine Composites, Ch1: Application, 2000, http://www.marinecomposites.com/
21. Lu, B., (2010) The Boeing 787 Dreamliner – Designing an Aircraft for the Future, *Journal Of Young Investigators*, **19** (24).
22. Mangalgiri, P.D. (1999) Composite materials for aerospace applications. *Bulletin of Material Science*, **22**, 657–664.
23. Gurit Aerospace: composite materials for aerospace, Brochure http://www.gurit.com/
24. Kaw, A.K. (2006) *Mechanics of Composite Materials*, CRC Press, ISBN: 978-0-8493-1343-1.
25. Sugita, Y., Winkelmann, C., and La Saponara, V. (2010) Environmental and chemical degradation of carbon/epoxy lap joints for aerospace applications, and effects on their mechanical performances. *Composites Science and Technology*, **70**, 829–839.
26. KPMG Corporate Finance Aerospace & Defense Advisory: The impact of composites on the Aerospace & Defense

industry, Summer 2008 http://www.kpmgcorporatefinance.com/engine/rad/files/library/Sector/Aerospace_Defense_Summer08.pdf

27 Ryerson, C.C., Dutta, P.K., and Pergantis, C.G. (2004) Response of polymer composite helicopter blades to thermal deicing. Proceedings of the Fourteenth International Offshore and Polar Engineering Conference, Toulon, France, May 23–28, p. 229.

28 Tóth, A., Mohai, M., Ujváry, T., and Bertóti, I. (2006) Advanced surface modification of ultra-high molecular weight poly(ethylene) by helium plasma immersion ion implantation. *Polymers for Advanced Technologies*, **17**, 898–901.

29 Lakes, R.S. (2003) Composite biomaterials, Chapter 4, in *Biomaterials Principles and Applications* (eds J.B. Park and J.D. Bronzino), CRC Press.

30 Ramakhrishna, S., Mayer, J., Wintermantel, E., and Leong, K.W. (2001) Biomedical applications of polymer-composite materials: a review. *Composites Science and Technology*, **61**, 1189–1224.

31 Huang, Z.M. and Ramakrishna, S. (2004) Composites in biomedical applications, Chapter 9, in *Materials for Biomedical Application* (ed. T.S. Hin), World Scientific Publishing Co. Pte. Ltd.

32 Yap, A.U.J. (2004) Biorestorative materials in dentistry, Chapter 5, in *Materials for Biomedical Application* (ed. T.S. Hin), World Scientific Publishing Co. Pte. Ltd.

33 Heintze, S.D., Zellweger, G., and Zappini, G. (2007) The relationship between physical parameters and wear of dental composites. *Wear*, **263**, 1138–1146.

34 DentalComposites.com, http://www.dentalcomposites.com/composite%20fillers.htm

35 Nath, S., Bodhak, S., and Basu, B. (2007) Tribological investigation of novel HDPE–Hap–Al_2O_3 hybrid biocomposites against steel under dry and simulated body fluid condition. *Journal of Biomedical Materials Research Part A*, **83**, 191–208.

36 Wang, M. (2004) Bioactive ceramic-polymer composites for tissue replacement, Chapter 8, in *Materials for Biomedical Application* (ed. T.S. Hin), World Scientific Publishing Co. Pte. Ltd.

37 McGregor, W.J., Tanner, K.E., and Bonfield, W. (2000) Fatigue properties of isotropic and hydrostatically extruded HAPEX™. *Journal of Materials Science Letters*, **19**, 1787–1788.

38 Shi, D. (2004) Bioactive composites, Chapter 1.5, in *Biomaterials and Tissue Engineering* (ed. D. Shi), Springer-Verlag, Berlin Heidelberg.

39 Kurtz, S.M. (2009) Composite UHMWPE biomaterials, Chapter 17, in *UHMWPE Biomaterials Handbook* (ed. S.M. Kurtz), Elsevier.

40 Plumlee, K. and Schwartz, C.J. (2009) Improved wear resistance of orthopedic UHMWPE by reinforcement with zirconium particles. *Wear*, **267**, 710–717.

41 Invibio Ltd, http://www.invibio.com/biocompatible-polymers/endolign-material-properties.php

42 Green, S., CFR PEEK composite for surgical applications. Medical Device Link, MDDI 2007 Jan http://www.devicelink.com/mddi/archive/07/01/006.html

43 Ridgeway, J. (2009) Minnesota's composite cluster, Minnesota economic trends, December, p. 2.

44 Karbhari, V.M. and Seible, F. (1998) Design considerations for the use of fibre reinforced polymeric composites in the rehabilitation of concrete structures. Proceedings of the NIST Workshop on Standards Development for the Use of Fibre Reinforced Polymers for the rehabilitation of Concrete and Masonry Structures, Jan 7–8, Tucson, Arizona.

45 Shi, C. and Mo, Y.L. (2008) Chapter 1, Introduction in engineering materials for technological needs – Vol. 1, in *High-Performance Construction Materials: Science and Applications* (eds C. Shi and Y.L. Mo), World Scientific Publishing Co.

46 Bakis, C.E., Bank, L.C., Asche, F., Brown, V.L., Asche, M., Cosenza, E., Davalos, J.F., Lesko, J.J., Machida, A., Rizkalla, S.H., and Triantafillou, T.C. (2002) Fibre-reinforced polymer composites for construction-State-of-the-art Review. *Journal of Composites for Construction*, **6**, 73–87.

47 Mertz, D.R., et al. (2003) Application of fibre reinforced polymer composites to the highway infrastructure. NCHRP Report 505, National Cooperative Highway Research Program,
48 Pendhari, S.S., Kant, T., and Desai, Y.M. (2008) Application of polymer composites in civil construction: a general review. *Composite Structures*, **84**, 114–124.
49 Harik, I.E. and Peiris, N.A. (2010) FRP composites applications in civil infrastructure. 14th European Conference on Composite Materials, 7–10 June 2010, Budapest, Hungary, Paper ID: 149-ECCM14.
50 Chin, J.W. (1996) Materials aspects of fibre reinforced polymer composites in infrastructure, NISTIR 5888.
51 Said, H. (2010) Deflection Prediction for FRP-strengthened concrete beams, *J. Compos. for Constr.* **14**, 244–248.
52 Busel, J.P., et al. (2007) ACI 440R-07. *Report on Fibre-Reinforced Polymer (FRP) Reinforcement for Concrete Structures*, American Concrete Institute.
53 Composites Technology October 2007, Lawrence Bank, Composites for Construction: The Design Basis for Pultruded FRP Members, http://www.compositesworld.com/articles/composites-for-construction-the-design-basis-for-pultruded-frp-members
54 http://www.strongwell.com/selected_markets/bridge_girders/
55 Bridging the gap, restorting and rebuilding the nation's bridges, American association os state highway and transportation officials, July 2008, http://www.compositesworld.com/articles/composites-for-construction-the-design-basis-for-pultruded-frp-members
56 Bousselham, A.J. (2010) State of research on seismic retrofit of RC beam-column joints with externally bonded FRP. *Journal of Composites for Construction*, **14**, 49–61.
57 Odello, R.J. (2002) Polymer composite retrofits strengthen concrete structures. *The AMPTIAC Quarterly*, **6**, 25–29.
58 Pultrall Inc, http://www.pultrall.com/Site2008/eng/V-ROD-1.htm
59 Ebert Composites corp, http://www.ebertcomposites.com/
60 http://combrae.org/htmlfolder/standaard.html
61 http://www.easervices.com/composite-wrap.htm
62 Opportunities of FRP Pipe Market in North America 2010–2015: Trends, Forecast and Opportunity Analysis, Lucintel, January 1, 2010.
63 Huaqiang FRP International Trading co, Ltd, http://www.frpfw.com/
64 http://www.lxfrp.com/english/cpjs.htm
65 Amitech Brail Ltd, www.amitech.com.br
66 Growth Opportunities in Wind Energy Market 2009–2014: Materials, Market and Technologies, Lucintel, June 2009.
67 QuakeWrap inc, www.quakewrap.com
68 Carotenuto, G., Nicolais, L., Pepe, G.P., Lanotte, L., Ausanio, G., and Barone, A. (2000) Low temperature anomalous magnetic properties of m-sized Ag/HDPE composites. *Journal of Materials Science Letters*, **19**, 425–427.
69 Droval, G., Glouannec, P., Salagnac, P., and Feller, J.F. (2008) Electrothermal behavior of conductive polymer composite heating elements filled with ceramic particles. *Journal of Thermoplastics and Heat Transfer*, **22**, 545–554.
70 Marrett, C., Moulart, A., Colton, A., and Tcharkhtchi, A. (2003) Flexible polymer composite electromagnetic crystals. *Polymer Engineering and Science*, **43**, 822–830.
71 Boiteux, G., Mamunya, Y.P., Lebedev, E.V., Adamczewski, A., Boullanger, C., Cassagnau, P., and Seytre, G. (2007) From conductive polymer composites with controlled morphology to smart materials. *Synthetic Metals*, **157**, 1071–1073.
72 Ramos, M.V., Al-Jumaily, A., and Puli, V.S. (2005) Conductive polymer composite sensor for gas detection, *Technology*, Nov 21–23, Palmerston North, New Zealand.
73 Sherin, T., Deepu, V., Uma, S., Mohaman, P., Philip, J., and Sebastian, M.T. (2009) Preparation, characterization and properties of $Sm_2Si_2O_7$ loaded polymer composites for microelectronic applications. *Materials Science and Engineering*, **163**, 67–75.

74 Hu, J.W., Li, M.W., Zhang, M.Q., Cheng, G.S., and Rong, M.Z. (2004) A novel method for preparing polyurethane based conductive composites with low percolation threshold. *Chinese Chemical Letters*, **15**, 1001–1004.

75 Brinkman, E., PolyCond: electromagnetic shielding with conducting polymers, http://www.polycond.eu

76 Sabic Innovative Plastics, http://www.sabic-ip.com/gep/Plastics/en/ProductsAndServices/ProductLine/lnpfaradex.html

77 Premix Oy, http://www.premixgroup.com/

78 W. l. Gore & Associates Inc, http://www.gore.com/en_xx/products/electronic/emi/gaskets/adhesive/goreshield_gs8000.html

79 Callen, B.W. and Mah, J. (2002) Practical considerations for loading conductive fillers into shielding elastomers, ITEM 130–137.

80 Callen, B.W. (2005) High performance composite powders for electronic devices. *Sulzer Technical Review*, **8**, 16–18.

81 http://www.leadertechnic.com/PDFs/conductiveSpecs.pdf

82 Bijwe, J., Hufenbach, W., Kunze, K., and Langkamp, A. (2008) Polymer composite bearings with engineered tribo-surfaces, Chapter 20, in *Tribology of Polymer Nanocomposites* (eds K. Friedrich and A.K. Schlarb), Elsevier, pp. 483–500.

83 de Souza F J.H.C.and Liewald, M. (2010) Analysis of the tribological behaviour of polymer composite tool materials for sheet metal forming. *Wear*, **268**, 241–248.

84 Friedrich, K., Zhang, Z., and Schlarb, A.K. (2005) Effects of various fillers on the sliding wear of polymer composites. *Composites Science and Technology*, **65**, 2329–2343.

85 Kawakame, M. and Bressan, J.D. (2006) Study of wear in self-lubricating composites for application in seals of electric motors. *Journal of Materials Processing Technology*, **179**, 74–80.

86 Prehn, R., Haupert, F., and Friedrich, K. (2005) Sliding wear performance of polymer composites under abrasive and water lubricated conditions for pump applications. *Wear*, **259**, 693–696.

87 Kim, S.J., Cho, M.H., Basch, R.H., Fash, J.W., and Jang, H. (2004) Tribological properties of polymer composites containing barite ($BaSO_4$) or potassium titanate ($K_2O·6(TiO_2)$). *Tribology Letters*, **17**, 655–661.

88 Koleva, M. and Boyiadjiski, Gr. (2008) Polymer composite material with abrasive features containing abrasive waste. *Journal of the University of Chemical Technology and Metallurgy*, **43**, 303–308.

89 Kukutschova, J., Roubicek, V., Malachova, K., Pavlickova, Z., Holusa, R., Kubackova, J., Micka, V., MacCrimmon, D., and Filip, P. (2009). Wear mechanism in automotive brake materials, wear debris and its potential environmental impact. *Wear*, **26** (5–8), 807–817.

90 Mutlu, I. (2009) Investigation of tribological properties of brake pads by using rice straw and rice husk dust. *Journal of Applied Science*, **9**, 377–381.

91 Roubicek, V., Raclavska, H., Juchelkova, D., and Filip, P. (2008) Wear and environmental aspects of composite materials for automotive braking industry. *Wear*, **265**, 167–175.

92 Du Pont catalogue: engineering polymers for high performance gears, http://plastics.dupont.com/plastics/pdflit/americas/markets/Gear_Brochure_11_06.pdf

93 Zhang, Z. and Friedrich, K. (2005) Tribological characteristics of micro- and nanoparticle filled polymer composites, Chapter 10, in *Polymer Composites: From Nano- to Macro-Scale*, Springer, US.

94 Prince, K. (2002) Composites win over sports market, *REINFORCEDplastics*, 48–51 September.

95 Froes, F.H. (1997) Is the use of advanced materials in sport equipment unethical? *Journal of the Minerals, Metals and Materials Society*, **49**, 15–19.

96 Overview of the Application of Composites to the Leisure Industry, at the website of National Composites Network http://www.ncn-uk.co.uk/DesktopDefault.aspx?tabid=433&tabindex=0

97 Grant, A. (2005) Sporting composites, *REINFORCEDplastics*, 46–49 May.

98 Daniel, I.M. and Ishai, O. (2006) *Engineering Mechanics of Composite*

Materials, Oxford University Press, New York;Liu, T.J.-C. and Wub, H.-C. (2010) Fibre direction and stacking sequence design for bicycle frame made of carbon/epoxy composite laminate. *Materials and Design*, **31**, 1971–1980.

99 Davis, C. (2007) Gaining a competitive edge. *Materials Today*, **10**, 60.

100 Davis, C.L., and Kukureka, S.N. (2004) Effect of materials and manufacturing on the bending stiffness of vaulting poles in M. Hubbard, R. D. Mehta, and J. M. Pallis (Eds.), The engineering of sport 5 (Vol. 2, pp. 245–252). Winfield, KS: Central Plain Books Manufacturing

101 http://en.wikipedia.org/wiki/Ski

102 Casey, H. (2001) Materials in ski design & development, in *Materials and Science in Sport* (eds F.H. Froes and S.J. Haake), TMS (The Minerals, Metals & Materials Society), pp. 11–17.

103 Santoro, T., Grant, C., and Franklin, C., The Eco-Core Snowboard, APD2006-02 at University of Michigan, http://designscience.umich.edu/pdf%20files/APD-2006-02.pdf

104 Sartor, M. and Fenato, P. (12. Nov. 2009) Ski boot, in particular for ski mountaineering, US patent Application, Appl. No. 2009/0277045.

105 Skiing And Snowboarding Equipment: The Benefits Of Composite Ski Poles, http://www.essortment.com/hobbies/skiingsnowboard_sjhv.htm

106 Nike Bauer Supreme One95 Composite Stick (2009 06. Jan) http://www.sportechblog.com/en/2009/01/nike-bauer-supreme-one95-composite-stick/comment-page-1/

107 Stefanyshyn, D.J. and Worobets, J.T. (2006) Energy return and puck speed of hockey sticks, Journal of Biomechanics, **39**, Supp. 1, S189.

108 Kooi, B.W. and Bergman, C.A. (1997) An approach to the study of ancient archery using mathematical modelling. *Antiquity*, **71**, 124–134.

109 Kooi, B.W. (1993) On the mechanics of some replica bows. *Journal of the Society of Archer-Antiquaries*, **36**, 14–18.

110 Kooi, B.W. and Sparenberg, J.A. (1980) On the static deformation of a bow. *Journal of Engineering Mathematics*, **14**, 27–45.

111 Virk, A.S., Summerscales, J., Hall, W., Grove, S.M., and Miles., M.E. (2009) Design, manufacture, mechanical testing and numerical modelling of an asymmetric composite crossbow limb. *Composites: Part B*, **40**, 249–257.

112 http://www.carl-walther.de/files/pdf/Sportkatalog2009.pdf

113 Savage, G. (2010) Formula 1 composites engineering. *Engineering Failure Analysis*, **17**, 92–115.

114 Shenoy, M.M., Smith, L.V., and Axtell, J.T. (2001) Performance assessment of wood, metal and composite baseball bats. *Composite Structures*, **52**, 397–404.

115 http://en.wikipedia.org/wiki/Racquet

116 http://www.itftennis.com/technical/equipment/rackets/history.asp

117 McDowell, M., Clocco, M.V., and Morreale, B. (2005) A composite softball revolution: Why the pitcher has little time to react to a batted-ball. *The Sport Journal*, **8**, http://www.thesportjournal.org/article/composite-softball-bat-revolution-why-pitcher-has-little-time-react-batted-ball

118 National Composite Network, Overview of the Application of Composites to the Leisure Industry, http://www.ncn-uk.co.uk/DesktopDefault.aspx?tabid=433&tabindex=0

119 Huntley, M.P., Davis, C.I., Strangwood, M., and Otto, S.R. (2006) Comparison of the static and dynamic behaviour of carbon fibre composite golf club shafts. *Journal of Materials Design and Applications*, **220**, 229–236.

120 Kootsookos, A. and Mouritz, A.P. (2004) Seawater durability of glass- and carbon–polymer composites. *Composites Science and Technology*, **64**, 1503–1511.

121 Aström, B.T. (1997) *Manufacturing of Polymer Composites*, Chapman & Hall, London, Weinheim, New York, Tokyo, Melbourne, Madras, pp. 19–22.

122 http://img.nauticexpo.com/images_ne/photo-g/sailboat-racing-sailing-yacht-reichel-pugh-design-in-carbon-91204.jpg

123 (1969) Carbon fibre reinforced racing canoes and kayaks. *Composites*, **1**, 112–113.

124 Marsh, G. (2006) 50 Years of reinforced plastic boats, *REINFORCEDplastics*, 16–19 Oct.

125 http://en.wikipedia.org/wiki/Step_aerobics
126 Mouritz, A., and Gibson, A. (2006) Flame retardant composites in (ed A. Mouritz and A. Gibson) *Fire Properties of Polymer Composite Materials*, Springer, The Netherlands, pp. 237–286.
127 Toldy, A., Szolnoki, B., and Marosi, Gy. (2011) Flame retardancy of fibre-reinforced epoxy resin composites for aerospace applications. *Polymer Degradation and Stability*. in press doi: 10.1016/j.polymdegradstab.2010.03.021.
128 Huang, N.H., Chen, Z.J., Yi, C.H., and Wag, J.Q. (2010) Synergistic flame retardant effects between sepiolite and magnesium hydroxide in ethylene-vinyl acetate (EEVA) matrix. *Express Polymer Letters*, **4**, 227–233.
129 United composites B.V. http://www.unitedcomposites.net/usapages/fireretardantcomposites2.htm
130 Sottos, N., White, S., and Bond, I. (2007) Introduction: self-healing polymers and composites. *Journal of the Royal Society Interface*, **4**, 347–348.
131 Yuan, Y.C., Yin, T., Rong, M.Z., and Zhang, M.Q. (2008) Self healing polymers and polymer composites. Concepts, realization and outlook: a review. *Express Polymer Letters*, **2**, 238–250.
132 Trask, R.S., Williams, H.R., and Bond, I.P. (2007) Self-healing polymer composites: mimicking nature to enhance performance. *Bioinspiration & Biomimetics*, **12**, 1–9.
133 Samadzadeh, M., Hatami Boura, S., Petikari, M., Kasiriha, S.M., and Ashraft, A. (2010) A review on self-healing coatings based on micro/nanocapsules. *Progress in Organic Coatings*, **68** (3), 159–164.
134 Williams, G., Bond, I., and Trask, R. (2009) Compression after impact assessment of self-healing CFRP. *Composite Part A*, **40**, 1399–1406.
135 Williams, G., Trask, R., and Bond, I. (2007) A self-healing carbon fibre reinforced polymer for aerospace applications. *Composite Part A*, **38**, 1525–1532.
136 Pang, J.W.C. and Bond, I.P. (2005) A hollow fibre reinforced polymer composite encompassing self-healing and enhanced damage visibility.
Composites Science and Technology, **65** (11–12), 1791–1799.
137 Hucker, MJ., Bond, IP., Bleay, S., and Haq, S. (2003) Experimental valuation of unidirectional hollow glass fibre (HGF)/epoxy composites under compressive loading. *Composite Part A*, **34**, 927–932.
138 Kessler, M.R., Sottos, N.R., and White, S.R. (2003) Self-healing structural composite materials. *Composite Part A*, **34** (8), 743–753.
139 Liu, Y., Haibao, Lv., Lan, X., Leng, J., and Du, S. (2009) Review of electro-active shape-memory polymer composite. *Composite Science and Technology*, **69**, 2064–2068.
140 Ratna, D. and Karger-Kocsis, J. (2008) Recent advances in shape memory polymers and composites a review. *Journal of Materials Science*, **43**, 254–269.
141 CRG Industries: Veritex^{TM} Shape Memory Polymer Composite, Product Data Sheet, http://www.crgrp.com/technology/materialsportfolio/veritex%20product%20data%20sheet.pdf
142 Cheeseman, B.A. and Bogetti, T.A. (2003) Ballistic impact into fabric and compliant composite laminates. *Composite Structures*, **61**, 161–173.
143 Gower, H.L., Cronin, D.S., and Plumtree, A. (2008) Ballistic impact response of laminated composite panels. *International Journal of Impact Engineering*, **35**, 1000–1008.
144 Mamivand, M. and Liaghat, G.H. (2010) A model for ballistic impact on multi-layer fabric targets. *International Journal of Impact Engineering*, **37**, 806–812.
145 Naik, N.K., Shrirao, P., and Reddy, B.C.K. (2006) Ballistic impact behaviour of woven fabric composites: formulation. *International Journal of Impact Engineering*, **32**, 1521–1552.
146 Mouritz, A.P. (2001) Ballistic impact and explosive blast resistance of stitched composites. *Composites Part B: Engineering*, **32**, 431–439.
147 ADVANCED GLASSFIBER YARNS LLC (2006): Defense Composite Armored Vehicle Advanced Technology Demonstrator (CAV-ATD), http://www.agy.com/markets/PDFs/NEW_AGY205nseCAVATD.pdf

Index

a

accelerated aging techniques 723
– principles 726
acidic aqueous media
– polymerization reaction stages 659
acrylonitrile butadiene rubber (NBR)-based composites 300
– NBR/butadiene rubber blends 558
acrylonitrile–butadiene–styrene copolymers 689
advanced functional fibers 444
adverse chemical reactions 85
aerobics group fitness class 780
aerospace industry 118, 125, 127, 130, 176, 179
– development of GFRP 203
aggregate model 412
A-glass 187
alkaline treatment 478
alkali treatment 12
aluminum
– flyer plates 19
– poles 774
aluminum trihydrate 697
aluminum trihydroxide, $Al(OH)_3$ (ATH) 780
– application of 780
3-aminopropyltriethoxysilane
– interactions with silica fiber surface, schematic presentation 345
ammonium peroxydisulfate
– stoichiometry, aniline oxidation with 659
amorphous silica
– equivalent loadings, vulcanizate based on 556
– filler, comparison 547
amorphous thermoplastics 315
– orientation of 315
aniline, oxidative polymerization
– rate 660
– temperature profile in 660
anterior cruciate ligament (ACL) 282
aramid fibers resemble inorganic fibers 776
area-to-volume (A/V) ratio 677
aspect ratio (AR) 577, 578, 686
ASTM cantilever bending device 372
asymptotic homogenization theory 87
atomic force microscopy (AFM) 661
– study of 229
atomization 576
autoclave consolidation cycle 125
automated extruder compression molding
– schematic presentation 116
automatic thermoplastic tape lay-up
– schematic presentation 116
Automotive Composites Consortium 177
automotive industry 9, 11, 112, 116, 121, 133, 176, 177, 361, 479, 480, 692, 753
average-field homogenization theory 87
aviation, transportation 755–757
– business aviation 756
– commercial aircrafts 756
– helicopter 757
– military aircraft 755
– space 757
axial modulus 410
axial ratio threshold 588

b

bag molding process 7, 8
banana fiber 12, 13
Beecher model 704
bending-induced damage 416
bending stiffness 384
– standard test method 373
beta-modification 323
beta-nucleated PP films 323

beta-polymorphs 323
BET technique 341
bias extension test
– three zones 371
biaxial tensile device 369
bidirectional weaving (BW) 399
– interlacing loops 401
– schematic drawing 401
bifunctional coupling agents 342, 549, 550
biocompatible polymers 759
biodegradable matrix 4
biofiber composite 4
biotite 674
bis-β-hydroxyterephthalate 276
bisphenol-A epoxy resin 730
bistatic reflection measurements
– setup for 626
bistatic reflection method 626
Boeing 787 dreamliner 756
Bombardier's, C-series aircrafts 756
π-bonding 154
bonds mill bridge 764
braided reinforcements 199
brake materials 771
Broutman test 86
Bruggeman effective medium theory 650
– approximation 594
Bruggeman expression 651
Bruggeman model 593
bulk carbonyl iron
– hysteresis loops 642
bulk ferrites 635–639
bulk ferromagnets
– bulk ferrites 635–639
– experimental magnetic spectra 635–641
– ferromagnetic metals and alloys 639–641
bulk magnetic materials, magnetization processes in 628–641
– bulk ferromagnets, experimental magnetic spectra 635–641
– dynamic magnetization processes 632–635
– magnetostatic magnetization processes 628–632
bulk MnZn ferrite
– frequency dependence 643
– hysteresis loops 642
bulk molding compound (BMC) 127, 191
butyl rubber 7

c
caprolactone 294
carbon black (CB) 516, 750
– aggregates, function of 535

– equivalent loadings, vulcanizate based on 556
– filled natural rubber composites 515
– filled sulfur-cured natural rubber (NR) 516
– loading
– – coefficient of thermal expansion (CTE) 520
– – dielectric relaxation, time-temperature superposition of 525–529
– – function of 535
– – tensile behavior 519, 520
– – viscoelastic properties 520–522
– – volume resistivity and conductivity 522–525
– nanofiller particles 515
– primary physical properties, comparison 556
carbon black (CB) aggregate network
– parameters, schematic diagram 534
– structure 535–538
– – dielectric properties, relationship 536
carbon black-filled polymer surface 506
carbon black (CB)-filled sulfur-cured NR vulcanizates 518, 519
– aggregate network structures 532
– – 3D network structures of 532
– – visualization of 533
– average minimum distance, dependence of 531
– d_p/d_g, definitions of 530
– 3D-TEM images of 529–538, 530
– loading 531
– – coefficient of thermal expansion (CTE) 520
– – dielectric relaxation, time-temperature superposition 525–529
– – tensile behavior 519, 520
– – viscoelastic properties 520–522
– – volume resistivity and conductivity 522–525
– nuclear magnetic resonance (NMR) 532
– standard deviations (STD) 529
carbon composite laminates 143
carbon/epoxy composites, uses 756
carbon–epoxy prepregs 755
carbon fiber (CF) 11, 137, 139, 140, 142, 143, 779
– characterization of 146, 147
– composite cored transmission conductors 178
– cost 750
– monocoque, Mercedes SLR McLaren Roadster F1 style 751
– properties of grades of 142

Index | 793

– reinforced epoxy fishing poles   779
– surface treatment   143
–– electrochemical oxidation   143
–– electroplating   143, 144
–– oxyfluorination   144, 145
–– plasma modification   145, 146
carbon fiber-reinforced polymer composites (CFRCs)   137, 138, 142, 161, 168–171, 176, 177, 180
– applications   176–179
– characterization   148
–– critical stress intensity factor   150
–– fracture behaviors   151–153
–– interlaminar shear strength   148–150
–– mode I interlaminar fracture toughness factor   151
– preparation   146
carbon fiber-reinforced polymers (CFRP)   751
– fiber glass tanks and silos   766
carbon-filled polymer composites
– surface features,   506
– surface topography,   506
–– PA12 and PA6.6,   506
– by TPM, mechanical property mapping,   501, 502
–– averaged function parameters over scanned area,   507
carbonization   139, 142
carbon nanofibers (CNFs)   286, 322
carbon racing sail boat   779
Carbon Ultimate F10   772
carbonyl iron (CI) powders   615, 645
– advantage   619
– chemical/mechanical treatment   619
– SEM photographs   648
– structural and electromagnetic characteristics   620
Carl Walther GmbH   775
– LG300 XT carbontec air rifle of   775
Cauchy–Green strain tensor   379, 380
Cauchy stress tensor   375, 381
C/C composite   405
– 3D, fracture in axial yarn   422
CC test sample energy values   727
cellulose fibers   4
central frequency   627
centrifugal casting   124, 125
ceramic matrix composites (CMCs)   4
CF/epoxy composites   758
C-glass   187
Charpy impact energy   476
Charpy impact testing   718, 726, 727, 729
chemical *in situ* methods   659
chemical methods   576

chemical resistance   189
chloroprene rubber   7
chopped strand composites   198
chopped strand mats (CSM)   190
closed magnetic system   644
coefficient of thermal expansion (CTE)   188, 479
cold plasma chemistry   12
Cole–Cole plots   536, 537
collagen   3, 13, 758
collar jackets   480
commercial glass fibers, properties   194
compatibilizers   209
compatibilizing effect of MA-g-PP   230
– DMTA analysis   233–235
– flow behavior   236, 237
– preparation of composites   230
– SEM study   237
– thermal properties   230–232
– X-ray study   233, 234
complementary test   386
complete particle debonding
– schematic diagram   95
complex dielectric permittivity   657
complex magnetic permeability   613
complex permeability   640
– frequency dependence   645
complex permeability spectra
– of carbonyl iron   641
composite
– modulus   685
– with multicomponent filler, schematic picture   655
composite fabrication process   347
composite laminates   315
composite materials   3, 15, 579
– characteristics   4
– classification   4, 5
– in current aeronautic structures   203
– dielectric spectra, CI particles microstructure effect   647
– electrical properties   592
– magnetic spectra   647, 653, 663
– particles shape effect on magnetic spectra   649
– poles, benefits   774
– predicted stress–strain responses   99
– properties, factors influencing   214
–– chemical stability   214
–– coupling agents, influence of   218
–– fiber length   217
–– fiber orientation   214
–– modulus   214
–– strength   214

– – voids, influence of   217
– – volume fraction   214–217
– properties, metallic particles effect   586
– SEM and BSEM photographs   649
– tensile strength   684
– textile reinforcements   373
– uses
– – helicopter blades   757
– – in naval applications   754
– varieties of polymers for   7
composite technology   176
composite transport properties theory   593–605
– effective medium theory   593, 594
– electrical conductivity models   598–605
– theory of percolation   594
– thermal conductivity models   595–597
compressed three-axis composite
– micrograph   421
compression-induced shear   421
compression molding   8, 112, 115–117, 121, 122, 302, 347, 476, 615
compressive stress   451, 457, 458
computer tomography   202
conducting filler, type   658
conducting polymer composites (CPC)   591, 767
conducting polymers, nanolayers   659
conductive compounds
– influence of injection rate on surface resistances,   491
– molding and electrical characterization,   489, 490
– percolation threshold,   490
– rheological characterization,   488, 489
– shear viscosity, filled with carbon black,   489
– surface resistances,   491
conductivity, filled polymers,   505
contact angle
– on carbon fibers as a function of current   154
– measurements   15
– method   147
– – dynamic, theoretical considerations   147, 148
– and surface free energy   153
continuous approach, advantage   382
continuous strand mats   190, 191
continuum mechanics approach   374
conventional circular fibers   192
conventional pull-out test   86
core–shelllike structure   655
correlation coefficient   458
corroded transmission pipeline, repair of   765

corrosion   194, 195
– properties   195
– resistance   3, 5, 10, 176, 187, 195, 203, 575
cotton fabric   473
coupling agent   343
covariant vectors   377
crack growth
– schematic drawing   426
creep-recovery test cycle   724
creep–recovery test cycle
– schematic presentation   724
creep resistance   197
critical amplitude   18, 46
critical fiber length   301
critical stress intensity factor (K_{IC})   155
cross-ply (CP)   319, 323
crystalline fibril model   316
crystalline lamellae   315
crystallites   315
crystallization   307
Cu/LDPE composites
– mechanical properties   585
Curie temperature   632
curing cycle   8, 34, 728
current density vector   623
curved neck specimen test   86
cycling loading   199
Cycom 4102 polyester resin matrix   20

d

damage/failure mechanisms   84
damage mechanisms   413
damping coefficient   635
damping mechanisms   681
data reduction methods   87
Debye dispersion law   634
deformation gradient tensor   376
deformation process   374
degree of adhesion   352
delamination   37, 61, 78, 174, 192, 305, 394, 687, 740, 741, 743, 762, 783
demagnetization factors   631, 643, 646
– magnetic particles   644
desized silica fiber
– DRIFT spectra   344
3D fabrics
– with complex shapes   404
– formation   405
2D fabrics continuous modeling   373–382
– geometrical approaches   373, 374
– hyperelastic model   378–382
– hypoelastic model   375–378
– mechanical approaches   374–382
2D fabrics discrete modeling   382–387

– global preform modeling   382–384
– longitudinal behavior   385
– mesoscale modeling use for permeability evaluation   386, 387
– transverse behavior   385, 386
– woven cell modeling   384–386
D-glass   187
diaminodiphenylmethane (DDM)   143
diaphragm–composite stack   119
diaphragm forming process   120
– schematic presentation   120
dicyclopentadiene (DCPD) monomer   782
dielectric constant   699
dielectric relaxation characteristics   535–538
dielectric strength   699
diffuse reflectance infrared Fourier transform (DRIFT) pectroscopy technique   344
diglycidylether of bisphenol-A (DGEBA)   143
di(hydrogenated tallow) ammonium-modified mica   695
dimethyl terephthalate (DMT)   276
direct processing long-fiber-reinforced thermoplastics (D-LFT)   113
– processing   117
dislocation-like model   93
dispersed (reinforcing) phase   3
distinction of image (DOI)   9
domain-wall
– resonance   634, 635
– schematic diagram   629
3D orthogonal weaving
– schematic drawing   397
double cantilever beam (DCB) test   151, 173–175
double strain amplitude (DSA)   564
dough molding   8
drainage process   437, 443
drainage system   456
dry reinforcement forming processes   382
2D textile composite reinforcement mechanical behavior   365
– continuous modeling, macroscopic scale   373–382
– discrete modeling, mesoscopic scale   382–388
– future trend   388
– and specific experimental tests   366–373
3D textile composites   14
3D woven composites design   429–431
– energy-absorption and damage resistance   430–431
– modulus   429
– strain-to-failure   430
– yield point   430

3D woven composites modeling   407–412
– bending fracture   413
– compressive damage   417–424
– failure behavior   412–428
– fiber volume fractions   407, 408
– fracture due to compression-induced bending   423, 424
– fracture due to transverse shear   424–426
– impact damage   427, 428
– microband   417–420
– miniband   420–423
– rule-of-mixtures   409
– surface loops influence   415–417
– tensile fracture   413
– weaker plane   414, 415
– yarn rotation   410–412
– yarns, elastic properties of   408
dynamic magnetization processes   632–635
– domain-wall resonance   634, 635
– natural resonance   633, 634
dynamic mechanical analyzer (DMA)   734
dynamic mechanical thermal analysis (DMTA)   226–228
dynamic wicking method   552

e

easy magnetic axis (EMA)   630
E-class vehicles   753
Eddy current   632
effective elastic modulus   89–92
– Mori–Tanaka method   90–92
– self-consistent method   89, 90
effective elastic stiffness tensor   92
effective medium approximation (EMA)   593
– type magnetic crystalline anisotropy   631
effective medium theory (EMT)   593, 614
effective opening size (EOS)   445
E-glass fibers   19, 33–35, 187, 193, 195, 199, 200, 204, 764, 775, 783
E-glass laminate   34
eigenstrain   88
eigenstress   88
elastically prestressed polymeric matrix composite (EPPMC)   715
– long-term characteristics   716
elastic compliances of GRP   34
elastic constants   19, 34, 62, 409, 410, 604
elastic–elastic bilaminates   23–25, 28, 29, 46
elastic modulus   412, 684
elastic precursor decay   23–25, 28, 31, 32
elastic prestressing technique   715
elastic stiffness tensor   91
elastic tensor   90
elastic–viscoelastic bilaminates   20, 23, 28

– late-time asymptotic solution 24–33
– solution at wave front 23, 24
– wave propagation in 21–23
elastomer composites 282, 283
elastomeric polymers 769
elastomers 7, 115, 282, 283, 299, 550, 553, 554, 557, 569, 769
electrical characterization, of conductive compounds, 489, 490
electrical conductivity models 586, 589, 598–605, 768
– of PP/Cu composites vs. filler content 578
electrical resistance, 496
electrical transmission 749
electric current density 155, 156
electric field strength vector 623
electrochemical in situ methods 659
electrochemical oxidations 141, 153
– of carbon fibers 155
electromagnetic (EM) crystal materials 767
electromagnetic field 637
electromagnetic interference (EMI) 699
– effects 768
– shielding 768
electromagnetic–radiofrequency interference (EMI/RFI) shielding 767
electron-beam-treated silica-filled compound 560
electron microscopy 15, 202, 257, 516
electroplating 143, 144, 156, 160
electroplating, of injection-molded components, 494
– microgear wheel made of nickel, 501
– sample, made of, 494, 495
– test results, transfer to standard material systems, 494–500
– – homogeneous deposition rates during, 499
– – injection rate, and optimization of parameter values, 499, 500
– – optical micrograph analysis, 498, 499
– – simulation of shear profile, and injection rate, 498
– Vestamid LR1-MHI specimen, 494–497
– effect of 156–163
electrospinning 302, 303
electrospun nanofibers 302
– for reinforcement 302
electrostatic discharge (ESD) protection 767
ensemble-volume average method 84
– micromechanical equation 94
EPDM-g-MA matrices 350
epichlorohydrin rubber 7
epoxy-based bonding agent 301

epoxy composite 4, 5, 200, 219, 221, 222, 286, 360, 522, 757, 783
epoxy resins (EP) 127, 137, 140, 143, 154, 319, 358, 689, 716, 735, 736, 755
– CTE of 319
equation of state (EOS) of KAST-V 19
erosion control process 439
Eshelby-based homogenization techniques 197
Eshelby's equivalent inclusion method 88, 89
– role in 90
Eshelby's tensor 88, 90, 91, 97
esterification 277
ethylene glycol (EG) 276
ethylene–octene copolymer 559
– master curves 565
ethylene propylene (EP) copolymer 222
– FTIR study 222, 223
– preparation of composites 222
– X-ray study 223–225
ethylene-propylene-diene terpolymer (EPDM) 349
– carbon black 330
– matrix 329
ethylene propylene rubber (EPR) compounds 7, 325, 551
– peroxide-cured 551
ethylene-vinyl acetate (EVA)
– copolymers 781
– resin 448
– thermal and electrical conductivities 590
Eulerian constitutive tensor 375
exponential function 22, 23, 721
extension ratio 563
extruders, role in 115
extrusion–compression molding 115

f

fabric induces reinforcing effects 472
fabrics
– thermal properties of 473
Falcon7X aircraft 756
fatigue
– crack propagation resistance 678
– properties 199, 200
– testing 281
FE model 383
ferrites 655, 659
– magnetic spectra 636
ferromagnetic resonance (FMR) 633, 667
– frequency 668
ferromagnet, initial magnetization curve and hysteresis loop 630
ferromagnetis, characteristic feature 628

fiber–fiber interaction 269
fiber-induced prestress 718
fiber–matrix adhesion 693
fiber–matrix debonding 726
fiber–matrix deformation 740, 741
fiber–matrix interface 739
fiber–matrix load transfer mechanisms 737
fiber orientation 113, 198, 214, 216, 257–259, 266–270, 377, 682, 700, 715, 760
– prediction, theoretical basis for 257–259
– simulation of 265
– tensor vs. normalized thickness 260
fiber-processing techniques 729
fiber-reinforced composites 14, 358, 475, 736
fiber-reinforced concrete (FRC) 742
fiber-reinforced plastics 220
fiber-reinforced polymers (FRP) 761, 762, 770
– based tribo components 770
– composite materials 749, 759
– composites 210
– 3-D composites 10, 11
– pipes growth 765
– vessels 766
fiber reinforcements 141–143, 682, 750
fiber–resin bonding 139, 141
fiber–resin interface 141, 474
fiber rotation tensor 375
fibers
– bundle 409
– bundle deformation 386
– fragmentation test 14
– mesh 9
– production 188, 189
– tension effects 715
– topography 734
– viscoelastic properties 281
– volume fraction 407
fibrous composites 4, 210, 214, 274, 296
fibrous fillers glass fiber 339
fibrous protein 3
fibrous reinforcements 700
filament winding process 7, 8, 129, 138, 176, 180, 189, 190, 198, 204, 326, 766, 778
– schematic presentation 129
filler–filler interactions 553, 554, 565, 568
fillers 339
– magnetic components 656
– particles 693
– shapes 683
filler surface modification, with bifunctional coupling agents 550–552
– pretreatment methods 552
– in situ method 551

film stacking 325
filtration process 437, 441
fisherman's net method 374
fishnet algorithm 374
fishnet method. See fisherman's net method
five-axis orthogonal fabric
– schematic drawing 398
five-axis weaving technology 13
floating platforms 178
Floquet's theory 20, 22
fluorinated Kevlar-reinforced composites 256
fluorination 209, 239, 319, 320
– as polymer surface modification tool 219, 220
fluoroelastomers 7
fluororubber release layer 516
Folgar–Tucker model 269
Formula 1 car
– primary structure of 776
fourth-order rank tensor. see Eshelby's tensor
Fractions of branched chains (F_{branch}) 535
Fractions of cross-linked chains (F_{cross}) 535
– F_{cir}, relationship 538
– relaxation strength on 537, 538
fracture behavior 75, 140, 151, 152, 221
fractured bone 758
– plate and screw fixation 758
fracture toughness 150, 151, 155, 168, 169, 174, 296, 478, 678, 772
free energy. See surface free energy
free sliding model 94
free-space measurements
– setup for 626
free surface
– particle velocity 19, 39, 40, 43–46, 45, 53, 59–63, 61, 77
– velocity profiles 45–48
FTIR spectroscopy 343
fully green composites 13
functional fillers
– reinforcing action 682

g

gel drawing 318
generalized effective medium (GEM) 593
geocomposites
– schematic presentation 436
geogrids
– specifications 447
– tensile strains 448
geomembranes 446, 461
– conducted soils interface, friction angles 462
geonet composites

– intrusion phenomena   454
geosynthetic clay liner (GCL)   435, 445
– schematic presentation   437
geosynthetics
– definition   435
– functions   437
– with polymeric materials   436
geotextile/geomembrane composites
– conducted soils interface
– – interface friction angles   465
– direct shear test results   463, 464
– photographs   462
geotextiles/geogrids composites   446
geotextile silt fences   439
geotextiles, polymer composites
– advanced functional fibers   444
– advanced trend   444–446
– application fields   440–443
– applications, examples   442
– definition   435–437
– developments   443–446
– for drainage and filtration   445, 446
– drainage function   439
– erosion control function   440
– filtration function   439
– function   437–440
– geosynthetic clay liners   445
– geotextiles/geonet, specification   450
– hybrid composite   447–462
– hydraulic properties   449
– in-plane permeability   461
– natural fibers   443
– nonwoven types   444, 445
– performance evaluation   462–467
– – required evaluation test items   465–467
– – test items   465
– raw materials   443, 444
– recycled fibers   444
– reinforced concrete by   446
– reinforced geomembrane by   446
– reinforcement function   440
– by role of function   441
– schematic presentation   436
– separation function   438
– specifications   447
– strength reinforcement   449
– synthetic fibers   444
– technology transition route   437
– transmissivity   461
– wide-width tensile strength-elongation curve   449
– woven types   445
Gilbert damping parameter   633, 634
glass, chemical structure   186

glass fiber composite armor systems   783
glass fiber-reinforced composites   19, 187, 196, 199, 200, 205, 751
– physical properties of   200, 202
– strain rate effect in   200
glass fiber-reinforced plastics   185, 199, 202–205, 762, 766, 773
glass fiber-reinforced polymer (GFRP)   33–35, 198, 202, 762
– composite   773
– composite tubes   763
– fabric wrapping   762
– fatigue resistance   199
– pipe systems   204
– properties of   201
glass fibers   11, 192, 198, 738
– composites, strength of   199
– fabrication of   188
– forms of   190
– in polymer composites   196
– production process   188, 189
– properties   192
– reinforcement   751
– technology   783
– types   187, 188
glass mat-reinforced thermoplastics (GMT)   116, 191, 692
glass/polypropylene plain weave   378
glass-reinforced polyester pipes   766
glass-reinforced polymer (GRP) composites   783
– plates   20, 40, 42, 59, 76
glass transition (T_g)   317
– temperature   7, 226, 239, 240, 244, 245, 250, 251, 255, 256, 279, 282, 305, 308, 582, 616, 681
glassy modulus   22
glassy silica, chemical network of   186
GNC-1, transmissivities   460
GP resin   726, 727
graft copolymerization   689
grafted silica   550
grand prix circuit   777
Grand Prix motor racing   775
graphite composite   4
graphitization   139, 142
green composites   4, 5, 13, 15
Green–Naghdi axes
– orientation   376
Green–Naghdi frame   377
Green's function   91
"Green-Tire" technology   550
GRIDFORM FRP reinforcement   763
Griffith energy balance approach   708

h

Halpin and Tsai model 705
hand lamination 123, 124
hand lay-up technique 7
hard magnetic axis (HMA) 630
healing glass fiber (HGF) 782
healing, polymerization 782
HEL. *See* Hugoniot elastic limit (HEL)
helix winding 118
Hertzian cone 427
hexadecyltrimethoxy silane (HDTMS) 550
hexagonal ferrite
– crystalline cell 621
– structural and magnetic characteristics 622
hexamethylenediamine 294
high density polyethylene (HDPE) 119
– films 325
– geomembranes 461
– photo-oxidation 695
– sericite–tridymite–cristobalite composites 695
high-modulus reinforcement 700
high-temperature epoxy resins 127
Hirsch and Counto models
– schematic representation 704
homogenization process 87
homogenized elastic operator 90
Hugoniot curve 68
Hugoniot elastic limit (HEL) 20, 55, 58, 69
Hugoniot strain 51–54, 71, 72, 77
Hugoniot stress 43, 52–54, 56, 57, 68, 71, 72, 77, 78
– *vs.* Hugoniot strain 51–54, 56
– – data of S2-glass GR 55, 56
– *vs.* particle velocity 54, 55, 57, 58
– – data of S2-glass GR 57
hybrid composite geotextiles 447–462
– for drainage 449–454
– – compressive stress and transmissivity 450–454
– – geotextiles/geonet composites manufacturing 449
– – nonwoven geotextiles/geonet composites 450
– – nonwoven/woven composite geotextiles 449
– for frictional stability 461, 462
– – frictional properties 461, 462
– – geotextiles/geomembranes composites manufacturing 461
– for protection and slope stability 454–461
– – thickness and compressive stress 457, 458
– – thickness and in-plane permeability 458–461

– – three-layered composite geotextiles manufacturing 454, 455
– – transmissivity 455–457
– for separation and reinforcement 447–449
– – geotextiles/geogrids composites manufacturing 447, 448
– – hydraulic properties 448, 449
– – wide-width tensile strength 448
hybrid composites 4, 133, 216, 217, 222, 327, 469, 744
– behavior of 469
– strength of 469
hybrid fabrics
– thermal properties of 473
hybrid filler system 693
hybridization 285
hybrid joints, tensile capacity 480
hybrid polymeric materials 444
hybrid reinforcement 327, 329
hybrid textile composites 478, 479
hybrid textile joints 479, 480
hybrid textile polymer composites 469–480
hydrodynamic effect 558
hydrogen bonding 154
– in Kevlar 212
hydrolyzed tetraethyl orthosilicate (TEOS) gels 360
hydrophobic–hydrophilic properties 154
hydroxyapatite-reinforced high-density polyethylene (HA/HDPE) 759
hyperelastic behavior law 379
hyperelastic constitutive model 379
hyperelastic models 375, 379
hypoelastic laws 375

i

impact stress 19, 44, 45, 49, 50, 62, 63, 68, 77, 78
impedance method 17, 24, 26, 29, 32, 40, 41, 60, 72, 76, 623, 624, 627, 628, 629
– measuring instruments 624
– measuring technique 624
imperfect interfaces 92
– characteristics 83
induced kink-bands 420
inductance method 624
infinite conductive cluster (IC) 586
inhomogeneity 89
initial in-plane permeability 459
injected PBT thermoplastic 581
injection-molded carbon-filled polymers, 486–488
injection-molded composite 592

injection-molded thermoplastic matrix composites
– modulus of elasticity   114
injection molding   8, 485
– advantages   8, 9
– cooling phase   684
in-plane shear   381
– moment   368, 371
– strain   372
– test   378
in situ method   343
INSTRON universal testing machine   272
intercrystalline bridge models   316
interface model   92–94, 212, 213
– characterization   14, 84
– dislocation-like model   93
– free sliding model   94
– interface stress model   93
– linear spring model   92, 93
– stress tensor   93
interfacial adhesion   6, 137, 138, 687–693
– coupling agents   689, 690
– hybrid effect in systems   692, 693
– interfacial modification mechanisms   690–692
interfacial damage modeling   94–100
– conventional Weibull's probabilistic approach   94–96
– cumulative damage model   98–100
– multilevel damage model   97, 98
interfacial debonding phenomenon   84
interfacial shear
– schemes   320
– strength   140
interlacing-induced deflection
– in axial yarns   406
interlacing loops
– covering weaker planes   428, 429
– holding axial yarns   429
– role   428, 429
interlaminar shear   192
– strength measurements   352
interleaving concept   322
interpenetrating network (IPN) structure
– phenolics of   328
intrusion phenomena
– of geonet composites   454
– schematic diagram   457
ionizing energy   305
Iosipescu method   424
iso-strain model   412
isotropic hardening law   94

j

Jacquet model   704
Jeffery model   269
J_2-type von Mises yield criterion   94

k

Kapok–cotton fiberreinforced polyester composites   473
KES-FB system   372
Kevlar composite   4, 221, 230, 238, 250, 271
Kevlar fibers   11, 209, 211, 218, 271, 757
– properties of   212, 213
– reinforced polymer composite   210
– structure   211, 212
Kevlar-reinforced thermosetting composites   270–272
Kevlar/vinyl ester composite   427
Kings stormwater channel bridge
– on California State Route   86 764
Kohlrausch–Williams–Watts (KWW) distribution function   721
Kraus equation   567

l

Lagrangian velocity   72, 73
laminate
– Al–PC laminate   29, 32
– ballistic composite materials   783
– composites   4
– Fe–Ti laminates   29, 30, 31
– made with 32 plies of prepregs   146
– Mo–Fe laminate   29
– Mo–Ti laminates   29, 31
– of unidirectional tapes   198
– used in analytical analysis   20, 21
– viscoelastic   18
– wave propagation in   21
Landau–Lifshitz equation   636
Landau–Lifshitz–Gilbert equation   633
Landau–Lifshitz–Looyenga equation   650
Landau–Lifshitz–Looyenga (LLL) mixing rule   650
land transportation   750
– compartment parts   752–754
– shell/body parts   750
– – carbon fiber-reinforced composites   750, 751
– – glass fiber-reinforced composites   751, 752
Lanthanum fluoride (LaF$_3$)   271
Laplace transform   20, 22, 76
latch-based spring and dashpot model   720
layered composites
– mixing rule conditions   700
L/D ratio   250

leaky mold method 729, 731
Leider–Woodhams equation 707
less-reactive filler 342
LFT processing 116
L-glass 187
Lichtenekker empirical equation 651
Lichtenekker mixture equations 650
life cycle analysis 15
LIGA process, for manufacturing high-quality microcomponents, 485
lignin 4, 13
lignocellulosic fabrics
– thermal properties of 474
linear spring model 93
liquid composite molding (LCM) processes 365, 385
liquid crystalline polymers (LCPs) 283, 444
– hybrid composites 250
liquid/gas barrier
– composite geotextiles 446
load–extension curves 152
localized matrix creep effects 722
London dispersive component 139, 148, 150, 154, 155, 160, 172, 173
Lorentzian dispersion curve 634
Lorentzian dispersion law 634

m

macro/microcomposites, interfaces in
– characterization 85–87
– composite materials, reinforcements surface treatments 85
– interfacial damage modeling 94–100
– micromechanics-based analysis 87–94
– microscale tests 85–87
macro/micropolymer composites
– future trends 130–133
– preparation and manufacturing techniques 111
– thermoplastic polymer composites 111–122
– thermosetting polymer composites 123–132
magnesium hydroxides 780
– decomposition temperature of 780
magnetically soft amorphous alloy 652
magnetic anisotropy 636
magnetic composite materials 613
magnetic conductors 639, 652
magnetic coupling 644
magnetic crystalline anisotropy 630
magnetic field 629
magnetic fillers 615–621
magnetic hysteresis 631

magnetic inhomogeneity 663
magnetic moments 628, 633
magnetic particle-filled polymer microcomposites 613
– bulk magnetic materials, magnetization processes in 628–641
– electromagnetic properties with corelike structure magnetic particles 663–668
– with high value of permeability in radiofrequency and microwave bands 651–668
– magnetic particles preparation with corelike structure 659–661
– with multicomponent filler 652–658
– with multicomponent magnetic particles 659–668
– PANI film formation mechanism 661–663
– polymer magnetic composites
– – basic components 614–621
– – magnetization processes 641–651
– radiofrequency and microwave bands, methods for characterization 621–628
magnetic particles
– demagnetization factor 644
– magnetic polarization 643
– manufacturing technology 659
magnetic permeability 631
– vs. magnetic field strength 632
magnetization mechanism 635
magnetization processes 628, 631, 641
– magnetostatic processes 628–632
– in polymer magnetic composites 641–651
MA-g-PP, as a compatibilizer 230
maleated polypropylene (MAPP) 688, 690
maleic anhydride-grafted isotactic polypropylene (iPP-MA) 691
maleic anhydride-grafted polyolefin elastomer (POE-MA) 691
manufacturing techniques, classification 130
marine applications, transportation 754, 755
mass transport vehicles
– interior composite components of 753
master curves
– regions 565
materials
– characterization methods, comparison for 628
material under test (MUT) 625
– free-space measurements, setup for 626
matrix compression 738, 743
matrix/filler interaction 688, 689
matrix materials 111
matrix polymer 347
matrix polystyrene 691

MAVSTAR project 178
maximum packing fraction 581
Maxwell–Garnet expressions 650
Maxwell–Garnett (MG) approximation 650
Maxwell–Garnett mean-field model 650
Maxwell–Garnett theory 588
Maxwell's equations 621
Maxwell–Voigt arrangements 719
mean stress 67, 69, 70
mechanical interfacial properties 155, 156
mechanical methods 576
melt blending 309
melt flow index (MFI) 465
meltmixing process 592
melt temperature 263
mesomodeling
– of unit cell 383
metal-coated particles 582
metal-filled polymers
– applications 575
– composites, rule for 585
– properties 576, 606
metallic conductive fillers 585
metallic particle-filled polymer microcomposites 575
– composite transport properties theory 593–605
– electrical conductivity 577–579
– main factors influencing properties 585–593
– mechanical properties 583–585
– metallic filled polymer, achieved properties 577–585
– metallic filler and production methods 576, 577
– particle shape, effect 587–589
– particle size, effect 589–591
– physical property prediction models 593–605
– preparation process, effect 591–593
– thermal conductivity 579–583
– volume fraction, effect 585–587
metallic particles 575
– filled composite use 606
– production methods 576
– suppliers for 576
– volumetric loading 581
metal matrix composites (MMCs) 4
metal–polymer composite 588, 591
methyl ethyl ketone (MEK) 143
M-glass 187
mica
– black (see biotite)
– heat-resistant sheets 677

– interfacial structure morphology 691
– powder, used as 677
– properties 674
– sheet 676
– tetrahedral layer, hexagonal layout 676
– tetrahedral, octahedral, tetrahedral (TOT) layers 674
– types 674, 675
– white (see muscovite)
mica-filled polymers
– melt flow index (MFI) 696
mica-filled polypropylene (PP) 678
mica-filled system 709
mica/nylon6 composites 685, 687, 699
mica/polymer composites
– mechanical properties 679, 680
mica/polypropylene composites 692
– melt viscosity 697
mica-reinforced microcomposites 709
mica-reinforced polymer composites 673
– applications 676, 677
– barrier properties 698
– chemical and physical properties 674
– crystallization 693, 694
– electrical properties 698–700
– flammability 695, 696
– interfacial adhesion 687–693
– limitations 707–709
– mechanical properties 677–693
– – modeling 700–709
– processability 696, 697
– reinforcement mechanism 678–687
– structure 674–676
– tensile strength 705–707
– thermal properties 693–696
– thermal stability 694, 695
– Young's modulus 700–705
mica-reinforced polypropylene 681
microair vehicles (MAVs) 178
microband(s) 417–420, 422
– connecting fiber kink-bands 418
– matrix cracks in transverse yarns 418–420
– microscopic kink-band 418
micro-debond test 14
microfabrication process
– based on 2C injection molding and electroplating, 486
microfibrillar composites 15, 285
microfibrillar-reinforced composite (MFC) 284
– PET composites 283–285
microfiller-reinforced polymer composites
– applications of 749
– biomedical applications 757–760

–– bone replacement   759
–– dental composites   759
–– external fixation   758, 759
–– orthopedic applications   760
– civil engineering, construction   760–767
–– composite decks   762, 763
–– pipe applications   765, 766
–– repair and retrofitting   761, 762
–– seismic rehabilitation   763
–– structural applications   764, 765
–– unique applications   764
–– wind energy applications   766
– electric and electronic applications   767–769
– features of   749, 750
– fire retardancy, high temperature applications   780–784
– internal reinforcement, for concrete   767
– mechanical engineering, tribological applications   769–772
–– bearings   , 771
–– brake lining   771
–– brake pads   771
–– brakes   771
–– gear   772
–– metal forming dies   770
–– seals   770
– recreation, sport equipments   772–780
–– leisure   779, 780
–– racquets, bats   777, 778
–– summersports   772, 773
–– technical sports   775–777
–– watersports   778, 779
–– wintersports   773, 774
– transportation   752–757
–– aviation   755–757
–– land transportation   752–754
–– marine applications   754, 755
– vessel length *vs.* year, plot of   754
microindentation test   14, 86
micromechanical damage models   94, 100
micromechanical formulation   197
micromechanical homogenization theory   87–89
– Eshelby's equivalent inclusion method   88, 89
– representative volume element   87, 88
micromechanical technique   14
micromechanics-based analysis   84, 87–94
– effective elastic modulus   89–92
– interface model   92–94
– micromechanical homogenization theory   87–89
micron-sized fibrous materials   361

microscale tests   85–87
– classification   86
microscopic fracture band. *See* microband
microscopic kink-bands   418, 430
microscopic techniques, for morphological changes   15
microstructure analysis   477
midpoint integration scheme   377
miniband   420–423, 422
– double-wave miniband   423
– through-thickness miniband   423
miniscopic fracture band. *See* miniband
mixing rules   614, 701
MnZn-based composites
– particle size effect on magnetic spectrum   646
MnZn-based polyurethane
– complex permeability   652
MnZn ferrite   653, 656, 662
– composites, magnetic spectra   654
– Lorentz microscope image   662, 663
– particles   666
– particle surface   665
– powders, magnetization curve   667
MnZn–PANI composites
– complex permittivity   667
– magnetic spectra   665
MnZn–PANI interphase boundary   666
MnZn–PANI particles
– synthesis   664
modulus of elasticity   584
molded filling simulations, tapered sample
– and calculations of maximum shear rates for,   490
– measured surface resistances, in correlation with calculated shear rate,   493
– simulation of maximum shear rate
–– occur when filling sample,   494
–– results of,   492
–– surface resistance of nozzle geometry,   494
– and surface resistances measured for different stages,   493
mold flow's fiber orientation model   259
mold flow technique   257
– simulation of fiber orientation   257–259
molding   7–10, 112, 115–117, 138, 190, 298, 682, 742. *See also* compression molding; reaction injection molding; thermoplastic injection molding
molecular jumps   721
monolithic polymers   757
monomer/oxidant ratio   660
Mori–Tanaka method   90–92, 471
multiaxis weave structures   13

multibeam VALYN VISAR system 36–38
multicomponent compounds
– properties 693
multicomponent magnetic particles 659
multidirectional reinforcing network 394
multifunction tribological probe
  microscope, 502–505
– displacement calibration, 505
– force-deformation curve, 504
– formulas used in calculation, 504
– friction sensor, calibration, 505
– measurement of functions of surface, 504
– multifunction mapping of, 507–512
– – cushion effect, 511
– – indentation curves, 510
– – mappings of topography, 508, 509
– – material distributions on specimen
    surfaces, 511
– – percolation of carbon particles, 511, 512
– – scanning rate, 510
– schematic diagram of, 503
– specifications of, 505
– specimen preparation for, 505–507
multilevel damage process
– schematic presentation 98
multilevel modeling techniques 197
multiple kink-bands 419
multiwall carbon nanotubes (MWCNTs) 287
Musal (MU) mixture equations 650, 651
muscovite 674, 676
– general structure 675

n

nanocatalysis 302
nanofiller-loaded vulcanized rubber
– 3D-TEM observation of 517, 518
nanolevel analysis 84
nanotechnology 302
natural aminosilane-treated silica fiber
    (NASF-AS)
– TG thermograms 346
natural amorphous silica fiber (NASF)
– scanning electron micrographs 341
– TG thermograms 346
natural ferromagnetic resonance 637
– frequency 633
natural fibers 443
– composites 11–13
– lower specific gravity 361
– Mercedes-Benz E-Class components 753
natural fibers *vs.* synthetic fibers 11
natural fiber textile-reinforced composites
– manufacture of 475–478
natural inorganic silica

– Silexil/Biogenic silica 340
natural resonance 633, 634, 639
natural rubber (NR) 7, 283, 477, 516, 553, 697
– composites 283
natural silica fiber 340
– reinforcement effect 348
– X-ray diffraction pattern 341
needle punching method 454
Ni-artifact made by UV-lithography, 501
nickel-coated phogopite 699
nickel electroplating, 489
nitrogen-based compounds 780
NiZn ferrite
– composites filled with, magnetic spectra 653
N_{NdNd}/TV
– dependence of number 534
nonbiodegradable matrix 4
noncrimp braided carbon fiber-reinforced
    plastics 13
noncrimp fabric (NCF) 365
non-E-coat polymers 9
nonlinear viscoelastic–viscoplastic
    model 281
nonresonant methods 624, 627
novolac-type phenolic composites
    reinforced 473
nucleates crystallization 697
numerical simulations 96, 97, 99, 100, 257,
    378, 382, 583
nylon 6 9, 294, 295, 302, 309, 348, 350, 351
– fiber-reinforced PMMA composites 309
nylon fiber-reinforced polymer
    composites 293
– polymer magnetic composites (PMCs) 740
nylon fibers 299
– as reinforcement 301
– used as reinforcements 294–299
nylon 6,6 fibers 728, 735, 742
– monofilament, polariscope image 718
– SEM micrographs 735
nylon-reinforced composites 305
– applications 310, 311
– basic steps to obtain microfibrils 309
– fracture morphology 309
– manufacturing of 305–311
nylon 6/silica fiber composite
– fracture surface, scanning electron
    micrograph 351
nylon 6,6 yarn
– recovery strain data 722

o

octadecyltrimethoxy silane (ODTMS) 550
one-shuttle scheme 402, 403

– weaving steps, schematic drawing 404
online monitoring of morphology of composites 15
ordinary differential equations (ODEs) 20
organic monomers
– *in situ* polymerization 557
overoxidation of fibers 140
oxidative chemicals 12, 219
oxidized carbon surfaces 139
oxidizing agents 85, 138
oxyfluorinated Kevlar fibers 228
oxyfluorination 144, 145, 163
– effect of 163–171
– as polymer surface modification tool 219, 220
– reactor 145

p

Padawer model 704
PANI film 666
PANI globule
– internal structure, atomic force micrographs 664
PANI thin film
– behavior 661
– globular structure, atomic force micrographs 662
– islandlike structure 661
– protonation 661
– *in situ* growth 661
parallel plate method
– for permittivity measurements 625
partially debonded damage model
– predicted elastoplastic responses, comparisons 96
partially oriented yarn (POY) 279
particle–matrix bond 705
particle–matrix fraction 707
particle velocity 22, 37, 46, 54, 67, 72, 73
particulate-filled composite
– chain formation in 587
particulate-reinforced polymer composites 673
Payne effect 563, 564
– role 555
PC/LCP/Kevlar composites 250
– dynamic mechanical analysis (DMA) 253–255
– SEM study 256
– thermal properties 251, 252
– X-ray diffraction pattern 253
pedestrian bridges 764
pelletized long-fiber-reinforced thermoplastics (P-LFT) 113

percolation theory 594
– mathematical model 594
percolation threshold 577, 586, 588, 591
permeability 619, 698
– dispersion region 621
– frequency dependence 635, 657
phase transition function 651
phenazin-containing oligomers 661
phenol formaldehyde 220
phenolic resin
– adsorption 354
– solution 352
picture frame test 370
– device 378
Piola–Kirchhoff second tensor 379
Piola–Kirchhoff stress tensor 380
pitch-based carbon fibers 142
plant fibers 752
– automotive industry 753
– glass fibers, replacing 753
plasma treatments 85, 138, 145, 171, 321
– effect of 171–176
plastic drain board (PDB) 445
plasticization effect 251
plate-impact experiments 19, 69
– on GRPs 33, 49
plate-impact shock compression experiments 36–38, 42–44
– Hugoniot stress
– – *vs.* Hugoniot strain 51–54
– – *vs.* particle velocity 54–58
– of S2-glass GRP 43
– shock waves in GRP, structure of 44–49
– state for S2-glass GRP, equation of 49–51
– stress–velocity diagram 37
– time *vs.* distance diagram 37
plate-impact shock-wave 20
plate-impact spall experiments 38–40, 58–66
– spall strength, determination of 59–61
– spall strength of GRP
– – following combined shock compression and shear loading 62–66
– – following normal shock compression 61, 62
– stress *vs.* velocity diagram 38
– time *vs.* distance diagram 38
platelet-reinforced composites 682, 701
platinum–rhodium alloy 188
Poisson's ratios 701
polar effect 154
polyacrylonitrile (PAN)
– based carbon fibers 137, 142
– fibers 142

polyamide (PA)   5, 13, 116, 119, 196, 294, 295, 296, 436, 581, 673, 681, 697, 772
polybutadiene   7, 557, 681
polybutylene oxide (PBO)   444
polybutylene terephthalate (PBT)   116
polycaprolactone (PCL)   302
– SEM micrographs   304
polycarbonate (PC)   7, 18, 29, 196, 251, 299, 758, 768
– composite   250
polycondensation   277
poly(dimethylsiloxane) (PDMS)
– cross-linking   355
– reinforcement   693
polydispersed particle mixtures   708
polyepoxides   9
polyester fiber-reinforced polymer composite (PFRPC)   281, 282–287
polyesters   5, 7, 9, 128, 130, 275, 282, 673
– physical properties   280
– properties   277, 279
– viscoelastic response   281
poly(ether ketone ketone) (PEKK)   5, 120, 137, 196, 210, 690, 757, 758
polyethylene (PE)   276, 315
– fibers
– – oxygen plasma-induced surface restructuring   321
– – reinforcement   326
poly(ethylene 2,6-naphthalate)   282
poly(ethylene naphthalate) (PEN)   275
– fibers   282
poly(ethylene terephthalate)   275, 276
– chemical structure   278
– fibres   275–277, 279, 281, 282, 286, 455, 695
– for industrial fibers   278
– market   276
– technologies for recycling   276
– viscoelastic properties   281
polyethylene terephthalate (PET)   119, 299
– fiber   276
– nanocomposites   286, 287
polyfluo wax (PFW)   271
polyimide   5, 7, 120, 196, 479, 699
polyisocyanurates   9
polymer chain immobilization concept   685
polymer chain movement restriction   694
polymer composite
– core–shelllike structure   655
polymer composites   3, 137, 360, 678, 749, 778
– applications of   471
– biomedical applications of   757
– characteristics   141, 577

– constituents   210
– DC conductivity, concentration dependence   654
– dielectric spectra   657
– electrical properties   589
– graphite/epoxy materials   757
– hard tissue application of   758
– magnetic spectra   629
– recent advances in   10–14
– saturation magnetization   641
– SEM photographs   655
– transportation   752–757
polymer–fiber adhesion   347, 355
polymer–fiber interactions   347
polymer fiber melt spinning production process   278
polymer–filler adhesion   353
polymer–filler interaction   549, 566, 567, 568
polymeric binders   189, 704
polymeric coupling agent   690
polymeric fibers   716
– types of   776
polymeric materials
– reinforcing filler   351
polymeric matrix   143, 359
polymeric/oligomeric PANI (PPy) overlayer
– conductivity   660
polymerization   277, 306, 307
– filling method   592
– of PET   278
polymer magnetic composites (PMCs)   656, 668
– advantage   669
– application   613
– ceramic systems   741
– components   614, 649
– electromagnetic parameters   668
– electromagnetic properties   614, 651
– fiber-reinforced   742
– magnetic fillers   615–621
– magnetization processes   641–651
– – effect of composition   641–646
– – effect of particle size, shape, and microstructure   646–651
– morphology   644
– permeability   613
– polymers in   615
– processing   614, 615
– production   614
polymer matrix composites   4–6, 177, 209, 342, 356, 358, 587, 613–615, 649–651, 741–743, 771
– dispersed phase inclusions, shape and orientation of   6

- fabrication of composites 7–10
- factors affecting properties 6
- interfacial adhesion 6
- for manufacturing 10
- properties of matrix 7
polymers 189, 575, 698
- advantages and limitations 617, 618
- filled with metal-coated particles
-- electrical conductivity 580
-- transverse thermal conductivity 583
- filled with metallic particles, electrical properties 578
- for glass fiber-reinforced composites 196
- low diffusivity in 698
- mechanical properties 681
- network 558
- physical and mechanical properties 616
- thermoplastics 615
- thermosets 615
polymer/silica fiber composites 349
polymethyl methacrylate (PMMA) 299, 326
- nylon composite, optical micrograph 307
polymorphs 322
polyolefin fiber-reinforced thermosets
- hybrid fiber-reinforced composites 328, 329
- polyolefin fiber-reinforced composites 327, 328
polyolefin fibers 319, 330
- family 315
- melting point of 327
- in rubbers
-- hybrid fiber-reinforced composites 329, 330
-- polyolefin fiber-reinforced composites 329
- and tapes 315–317
-- application of 321
-- basic characteristics of 317
-- gel drawing 318
-- hot drawing 317, 318
-- properties and applications 318–321
polyolefin/polyolefin combinations 327
polyolefin reinforced interface/interphase characteristics in 326
polyolefin-reinforced thermoplastics
- hybrid fiber-reinforced composites 327
- polyolefin fiber-reinforced composites 324
-- coextruded tapes, consolidation of 324, 325
-- film stacking 325
-- interphase 326, 327
-- matrix, *in situ* polymerization of 326
-- powder impregnation 326

-- solution impregnation 325, 326
- self-reinforced version 321
-- film stacking 322, 323
-- hot compaction 321, 322
-- wet impregnation 324
poly(phenylene ether) (PPE) 284
polyphenylene sulfide (PPS) 119, 652
polypropylene (PP)/mica 5, 7, 116, 119, 276, 310, 315, 447, 579, 678, 687, 692, 758, 777
- composites 684, 692
- fabric 323
- geotextiles 461
- heat transfer coefficient 696
- homopolymer
-- copolymer systems 324
-- fibers 325
-- tape 324
- tapes 318
polypropylene sulfide (PPS) 444
polystyrene (PS) domains 329
polysulfide rubber 7
polytetrafluoroethylene (PTFE) 771
polyureas 9
polyurethane (PU) composites 9, 128, 189, 196, 221, 349, 444, 616, 646, 673, 768
- fractured surface, scanning electron micrographs 349
- magnetic spectra 664
- permeability 666
-- concentration dependence 656
polyvinylacetate 189
polyvinylchloride (PVC) 359
- silica composite nanofiber 360
polyvinylidene fluoride (PVDF) 19, 586, 656, 657
- filled with Co_2Z, SEM photograph 657
Portland cement 4
powder technology 576
power-law model 705
prefabricated vertical drains (PVDs) 435
preformed molding compounds 8
premix thermoplastics 769
prepreg lay-up 125
prepreg sheets 306
prestressed polymeric matrix composite (PPMC) materials 716
prestressing
- application 715
properly shaped fibers 192
properties, determination of 197
pull-in of rods
- on specimen side 428
pull-out load *vs.* displacement curves 300
pull-out test preparation 14

pultruded glass fiber profiles   198
pultruded shapes   764
pultrusion process   8, 130, 138, 284
– schematic presentation   129
push-out test
– drawbacks   86
pyrolysis gas chromatography   552

q

quality factor   627
– frequency dependence   657
quartz   186
quasi-glass state   566
quasi-static resonator   627

r

radioabsorbers (RA)
– design   657
– matching frequency   658
Raman spectroscopy   343
Ramie-cotton fabrics
– configuration of   473
– textile reinforced polyester composites
–– tensile strength of   472
Rankine–Hugoniot conservation
    relations   49, 51, 52, 71, 72
ratio of modulus   562
Rayleigh line   69
rayon fibers   142
reaction injection molding (RIM)   7, 9, 35, 123, 127, 196, 326, 352, 472
recrystallization   305
recycled fibers   444
reflection coefficient   626
– frequency dependence   658
refraction index   193, 194, 202
reinforced concrete (RC) structures   761
reinforced reaction injection molding (RRIM)   9
reinforcement fibers
– advantages   446
– properties   294
reinforcement mechanism   339, 439, 441, 558–568, 678–687
– applications   568, 569
– dynamic mechanical properties   564, 565
– equilibrium swelling   566–568
– frequency dependence of storage modulus and loss tangent   565, 566
– mechanical properties   560–564
– mica concentration   684, 685
– particle size and size distribution   685–687
– rheological properties   559, 560
– shape and orientation   682–684

reinforcement of polymers   5
reinforcement's modulus   704
reinforcing agent   361
reinforcing fibers   750
– impregnation of   326
reinforcing fillers   567
reinforcing phase   3, 315, 325–327
reinforcing polyolefin films   315
relaxation function   22
remanent magnetization   630
representative unit cell (RUC)   367
representative volume element (RVE)   87
– definition   88
reshock–release loading   72
– calculation of off-Hugoniot states for   72–74
– stress *vs.* strain states   70
resin film infusion (RFI)   472
resin impregnation   393, 395
resin infusion techniques   199
resin matrix   354
resin solution   4
resin transfer molding (RTM) process   8, 9, 123, 126, 127, 326, 352, 472, 476, 477
– applications   125
– schematic presentation   125
resonance frequency   664, 665
resonant cavity measurements   627
resonant methods   624, 627
R-glass fibers   187, 199, 783
rheological characterization, of
    compounds,   488, 489
rheological parameters of composites   247
ribbons, surface modification of   327
roller transfer method   516
rotation tensor   376
rubber shell model   566
rule-of-mixtures   409, 410

s

sample-to-resonator volume ratio   627
saturation magnetization   629
sebacic acid   294
second-order identity tensor   93
seemann composites resin infusion molding
    process (SCRIMP)
– NABI composite bus shell production   752
selective free surface velocity profiles   47, 48
self-assembly techniques   360
self-consistent dynamic shear yield strength
    determination method   67–71
self-healing polymer composites   781, 782
self-reinforced polymer composites (SRPCs)
– development of   321

self-reinforced thermoplastic polymer composites 321
semicircular component, relaxation strength of 537
semicrystalline polymers 693
semicrystalline2010 polymers 307
SEM images 228, 229
SEM micrographs of tensile fractured surfaces 249
separation process 437
sequential probabilistic debonding analysis 97
S-glass 187
S2-glass 187
shaped glass fibers 192
shape-memory polymers (SMP) 782
sharples polariscope 717
shear angle 387
shear curve
– of glass plain weave 370
shear forces 546
shear modulus 378, 429
shear simulation 387
shear strain 62, 63
shear stresses 71, 113
– distribution 114
– types 424
shear–stress transfer mechanism 704
sheet molding compound (SMC) 7, 8, 123, 127, 191
shift factors 566
shock–release experiments 40–42, 66–75
shock–reshock experiments 40–42, 66–75
shock velocity vs. particle velocity
– data for S2-glass GRP 53
shock waves 17, 18, 41, 43, 44, 46, 68, 76, 78
silane coupling agents 342, 345, 346, 569
– chemical interaction 552
silane-treated silica fiber 355
silane-treated silica filler
– filler loading effect 564
silanization
– effect 353
– process 343, 344, 561, 568
silanized silica fiber-reinforced polymer composite 358
silanol groups 342
silica (SiO_2) 186, 340
silica fiber (SF) 339–347
– characteristics 340–342
– chemical modification 342
– chief component 340
– content effect 351
– effect 350

– plausible interaction, schematic presentation 359
– surface modification, types and methods 342, 343
– surface-pretreated silica fiber characterization 343–347
– surface treatment 342–347
– types 340
silica fiber-filled polymer composites
– composite fabrication 347
– composite properties, effect on 347–351
– reinforcement mechanism 358
– surface-modified 351–358
silica fiber-filled rubber 355
silica fiber/phenolics composites 353
– ILSS, experimental data 352
silica fiber-reinforced polymer composites 339
– applications 358–360
– general features 339–347
– new developments 360
– silica fiber-filled polymer composites 347–358
silica-filled polymer microcomposites 545
– as filler 545–552
– silica-filled rubbers 552–569
– thermoplastics and thermosets 569–571
silica-filled rubbers 552–569
– reinforcement mechanism 558–568
– silica filler effect on processability of unvulcanized rubber compounds 553, 554
– silica filler effect on vulcanizate properties of rubber compounds 554–557
– surface-modified silica-filled rubbers 557, 558
silica filler 515
– characteristics 546–548
– chemical modification 549
– filler surface modification
– – with bifunctional coupling agents 550–552
– – with monofunctional coupling agents 550
– physical modification 549
– silanol groups, types 551
– structure formation, schematic presentation 548
– surface modification, types and methods 549–552
– surface pretreated silica, characterization 552
– surface treatment of silica filler 548, 549
– TEM photographs 553
– types 546
silica reinforcement of polymers 571

silica–silane filler system   568
silica–silica interaction   554
silicate minerals   674
silicone/muscovite composites   686
silicone rubber   7
– composite, scanning electron micrographs   356
silicon monoxide
– oxidation   340
silicon nitride composite materials   360
silver-coated glass
– thermal and electrical conductivities   590
simulation
– of fiber orientation by mold flow technique   257–265
– – for PC/LCP/Kevlar composites   265–270
– – technique on s-PS/Kevlar Composites   259–265
– physical properties of material layers   29
single-fiber composites
– transverse tensile debonding stresses
– – schemes of   320
single-fiber compression test   86
single-fiber pull-out test   86
single-walled carbon nanotubes (SWNTs)   287
sintered polycrystalline
– ferrites, magnetic spectra   637, 638
– magnetic spectrum of   636
sisal/cotton composites   473
sisal fiber   12, 13, 248, 277, 477, 478
sisal fiber-reinforced vinyl ester composites
– cross-sections of   477
sisal textile-reinforced composites
– mechanical properties of   476
sizing agent   189, 190, 195
skew angles   20, 38, 58, 59, 62
smart composite geotextiles
– compressive stress   459, 460
– cross-sectional areas photographs   455
– in-plane permeability   459
– principal transmissivity mechanism   455
– specifications   455
– transmissivity   460
Snoek's law   619, 635
Snoek's limit   619
sodium-fluorinated synthetic mica   695
solid Kevlar rods
– three-axis fabrics combining   407
– vinyl ester rods   398
solid rods   406
– use   405–407
solvent species effects   354
SpaceshipOne   757

spall strength   59, 62
– determination of   59
– of E-glass GRP composite   77
– of GRP
– – following combined shock compression and shear loading   62–66
– – following normal shock compression   61
– of KAST-V   19
– in relationship with normal stress and shear strain   65
spatial orientation   410
specific energy absorption (SEA)   741
spectroscopic tests, for polymer surface   15
spinel-type ferrites
– main structural and magnetic characteristics   621
spinning. See gel drawing
split Hopkinson pressure bars (SHPBs)   19
spray gun   124, 125
spray-up   124, 131
s-PS/Kevlar composites (KSO)
– AFM study   245–246
– differential scanning calorimetric study   239–241
– dynamic mechanical thermal analysis   243–245
– FTIR study   238, 239
– mechanical properties   248
– preparation of   238
– properties of   238
– SEM study   245
– thermal properties   241, 242
– X-ray study   242, 243
square fabric blank
– pure shear on   384
steel   4
– properties of Kevlar fiber over other fibers   213
– properties of selected reinforcing fibers and tapes   317
stiffness matrix   409
Stokes/Brinkman flow simulation   387
storage modulus   566
– variation   357
– vs. frequency   560
– vs. loss modulus, log-log plot of   561
strain-limited toughness   732
strain–stress relationship   92
strain-to-failure (STF)   395, 732
stress   22
– vs. strain states   70
stress-free transformation strains. see eigenstrain
stress increment tensors   385

stress–strain curves   73, 74, 75, 100, 141, 560, 561, 584, 709
– for HDPZE/zinc composites   584
stress–strain relation   409, 411
stress–strain responses   96
stress tensor   380
stress-to-yield   395
stress transfer   322
– efficiency   84
Strongwell pultruded profiles   763
structural reaction injection molding (SRIM)   9, 123, 127, 128, 131, 132
styrene-butadiene rubber (SBR)   553, 557
styrene/butadiene/styrene (SBS)   329
styrene/ethylene–butylene/styrene (SEBS) structure   329
styrene maleic anhydride (SMA)   119
sulfonated poly(ether ketone ketone)
– thermal stability   694
sulfur cure systems   554
surface characteristics   153, 154, 171, 172, 219, 342, 549, 734
surface free energy   139, 148, 154, 155, 157, 159, 160–162, 165, 167, 172, 173, 213, 218, 219, 346, 352, 600
surface modification   256
– of fibers   218, 219
– of matrix polymers   219
surface oxidation on carbon fibers   155
surface swelling   324
surface tension   139
– of carbon fibers   139
surface-to-volume ratio   686
surface treatment   138
– of carbon fibers   139–141, 143
– dry methods   138
– of reinforcements for composite materials   85
– of silica fiber   342, 548, 549
– wet methods   138
swelling index   477
SymaLite technology   751
symmetric weaving (SW)   399
– interlacing loops   402
– schematic drawing   402
syndiotactic polystyrene composites. See s-PS/Kevlar composites (KSO)
synthetic fiber-reinforced composites   220–222
synthetic fibers   4, 210, 211, 220, 276, 444
synthetic polyisoprene   7
synthetic polymers
– hydrophobicity   698
synthetic silica   340, 545
– properties   546

t
tangential tensile stress   427
target assembly   42
TechSIL 5000 conductive elastomer materials   769
temperature curing cycles   728
tensile deformation   318
tensile force   371
tensile fracture   413
– fracture mechanisms, schematic drawing   414
tensile modulus   143
tensile properties
– of banana fiber   13
– effect of fiber volume fraction on   732, 733
– factors influencing   592
– of NR/SBR blends   557
– of PET fibers   280
– of PET/PP microfibrillar composites   285
– temperature-induced degradation and   222
tensile strain   305
tensile strength   12, 19, 187, 192, 194, 221, 228, 229, 279, 318, 356, 358, 439, 476, 557, 584, 585, 684, 685, 692, 705–707, 738, 739, 761, 777
– of $CaCO_3$ composites   684
– of glass fibers   194
– of mica/PP composites   685
– predicting models   706
– – graphical representation   705
tensile stress   39, 305, 428
tensile stress–strain curve   142
tensile surfaces   369
tensile tests   396, 731
tensile toughness   732
terephthalic acid (PTA)   276
testing methods, schemes of   320
tetrahydrofuran (THF)   343
tetrasulfide silane, bis-(3-triethoxysilylpropyl) tetrasulfide (TESPT)   550
– silane coupling agent   557
textile composites   4, 14, 370, 372, 388, 395, 412, 470–478
– reinforcements   366, 367, 388
– – mechanical behavior   365
textile fibers   329
textile reinforcements   365, 375
– geometry of   470
– mechanical behavior of   470
textile sheets
– bending stiffness   372

T-glass 187, 188
thermal conductivity 220, 575, 577, 579–583, 588
– models 595–597
– of polyamide/Cu composites 581
thermal decomposition 696
thermal diffusivity 473
thermal expansion coefficients 142, 583
thermal properties 195, 196, 216, 220, 225, 226, 230, 245, 252, 473, 474, 693–696
thermal protection systems 359
thermodynamic methods 15
thermoelastic waves 18
thermogravimetric analysis (TGA) 346
thermoplastic composites 209, 306, 324
– character 111
– manufacturing techniques
– – classification 121, 122
– – feasible series lengths 112, 123
thermoplastic elastomers 7, 329, 557
thermoplastic injection molding 128
thermoplastic matrix 286
– composites properties 115
– prepreg tapes 117
thermoplastic microcomposites 673
thermoplastic polymer composites 7, 111–120, 123, 196, 299, 673, 684
– compression molding 115–117
– diaphragm forming 119, 120
– extrusion 115
– injection molding 112–115
– mica, use 673
– thermoplastic composite manufacturing techniques, classification 120
– thermoplastic prepreg lay-up 117, 118
– thermoplastic tape winding 118, 119
thermoplastic polyurethane/mica composites 685
thermoplastics
– filled with metallic particles, transverse thermal conductivity 582
– pultrusion, scheme presentation 119
thermoplastic tape winding instruments
– general scheme 118
thermosets
– filled with metallic particles, transverse thermal conductivity of 583
thermosetting injection molding 128
thermosetting polymer composites 123–132
– bulk molding compound 127
– centrifugal casting 124, 125
– filament winding 128, 129
– hand lamination 123, 124

– manufacturing techniques, classification 130, 131, 132
– prepreg lay-up 125
– pultrusion 129, 130
– reaction injection molding (RIM) 127, 128
– resin transfer molding 126
– sheet molding compound 127
– spray-up 124
– structural reaction injection molding (SRIM) 127
– thermosetting injection molding 128
– vacuum-assisted resin transfer molding 126
thermosetting polymers 7, 9, 123, 472, 771
thermosetting resins 767
thermostability 209
Thornel-25 fibers 140
three-axis carbon/epoxy composite
– tensile fracture 413
three-axis orthogonal composite 398
– tensile fracture 413
– xy-section 399
three-axis orthogonal fabric 397
– schematic drawing 397
three-component composite 655
three dimensional transmission electron microscopy (3D-TEM) 516
three dimensional woven fabric composites 393
– design 429–431
– failure behavior 412–428
– formation 396–407
– general characteristics 394–396
– greater nonuniformity 394, 395
– higher fiber crimp 395
– interlacing loops role 428, 429
– lower fiber volume fraction 395
– lower stress-to-yield and higher strain-to-failure 395, 396
– material testing, difficult in 396
– modeling 407–412
– more complex in damage mechanisms 396
– multidirectional structural integrity 394
– near-net-shape design 394
– solid rods use 405–407
– three-axis orthogonal weaving 396–399
– weaving schemes design 399–403
– yarn distortion 403–405
three-layer structure geotextiles
– schematic diagram 456
three-point bending tests 415, 730
time-dependent latch elements 720
time–temperature superposition principles 723, 742

Timoshenko bending theory 737
titanium dioxide 295
TMCS-treated silica fiber 353
toluene
– swelling in, Kraus plots 567
torque 301
torsional stiffness 774
toughness 112, 115, 141, 150, 161, 197, 209, 218, 297, 299, 310, 678, 692, 760, 772, 776
transcrystalline layer
– formation of 323
transmission/reflection method 625
transmission tower utility pole 765
transmissivity 455–457, 459
– decrease ratio 452, 453
– disadvantage 450
– of nonwoven geotextiles with compressive stress 451, 452
– value 454
transverse matrix 415
– cracking 415
transverse shear test 424
– damaged specimens 425
– fixture 425
– fracture in 3D C/C composite 426
transverse yarns
– matrix cracks 420

u

ultrahigh modulus PE
– structural schemes for 316
ultrahigh molecular weight PE (UHMWPE) fibers 318, 319
– ballistic properties of 329
– bundles 325
– CF-reinforced layers 329
– coating of 326
– coefficient of thermal expansion (CTE) of 319
– dilute tetralin solution of 324
– ethylene/propylene/diene rubber (EPDM) matrix 329
– glass fiber (GF)
– – and CF 328
– – combination of 327
– oxygen plasma treatment for 328
– powder surface 760
– surface modification of 320, 327
uniaxial anisotropy 634
unidirectional (UD) composites
– fiber layers 319
– reinforced composite 197
– thermomechanical behavior of 470
– vinyl ester composites

– – short-beam shear test 320
unidirectional weaving (UW) 399
– interlacing loops 400
– schematic drawing 400
unit-cell 407
– schematic drawing 407, 408
unit woven cell
– loads on 367

v

vacuum-assisted resin transfer molding (VARTM) technology 126, 766, 784
– schematic presentation 126
vacuum bag 125
vacuum infusion (VI) 472
van der Waals attraction 137
van der Waals force 154
variation constant 457
veritex 782, 783
Vestamid LR1-MHI components. See also electroplating, of injection-molded components
– electrodeposition, deposition fail to occur in, 495
– surface resistance distributions, and screw feed rate, 496, 497
vinyl ester 5, 320, 328, 476, 766, 767
vinyltrimethoxysilane (VTMS) coupling agent 343
virtual work, principle 367–369
viscoelastically prestressed polymeric matrix composite (VPPMC) 716
– accelerated aging
– – impact tests 726–728
– – viscoelastic recovery 722, 723
– applications 743, 744
– batch of test, typical tensile stress–strain plots for 731
– in composite materials 715, 716
– feasibility 716
– flexural modulus values 730
– force–time measurements 724–726
– with higher fiber content 729–733
– – flexural properties 729–731
– – meeting the objectives 729
– – tensile properties 731–733
– impact toughness 739
– improved mechanical properties mechanisms 737–739
– – background 737, 738
– – flexural stiffness 738
– – impact toughness 739
– – tensile strength 738, 739
– long-term performance 728, 729

– polymeric deformation mechanical model 719–722
– potential applications 740–742
–– crashworthiness 741, 742
–– enhanced crack resistance 742
–– high velocity impact protection 740, 741
– preliminary investigations 716–719
–– initial mechanical evaluation, impact tests 718, 719
–– objectives 716, 717
–– viscoelastic recovery, creep conditions 717
–– visual evidence 717
– principles 716
– processing aspects 733–737
–– background 733, 734
–– fiber properties 734, 735
–– fibers in composite samples, geometrical aspects 735–737
– representative optical micrograph sections of test 736, 737
– tensile properties of test 732, 733
– time–temperature aspects 719–729
– typical appearance of test 719
viscoelastic bilaminates 18
viscoelastic deformation 720
viscoelastic modulus 22
viscoelastic recovery force 733
viscoelastic recovery stress 725
viscoelastic strain 717
viscosity 301
Voigt and Reuss average modulus 90, 701
volume-averaged strain tensor 91

w

water-activated polyurethane 758
water barrier 437, 443
wave propagation 18, 20
– in elastic–viscoelastic bilaminates 21–23
– time–distance diagram 39
wear resistance 759
weaving processes 393, 396, 402, 429
– fabrics from 403
weft yarns 478
Weibull distribution function 721
Weibull/KWW-based equation 724
Weibull's probabilistic approach 84, 94–96, 97
– probability distribution function 94, 99
Weibull's statistical approach 94
Weibull statistical function 99
wider test 197
wind energy 177, 178
wind turbine blades 204
wood 4
– tennis rockets 777
woven cell modeling 384–387
woven characteristics 475
woven fabrics 379, 470, 471, 474
– composite materials 470, 471
– reinforced composites 470
woven glass reinforcements 199
woven styles 475
woven unit cell
– bending behavior 372, 373
– biaxial tensile behavior 369, 370
– in-plane shear behavior 370–372
– load resultants 366, 367
– virtual work principle 367–369
woven yarns 366

x

X-ray photoelectron spectra (XPS) 320

y

yarn distortion
– types 403
yarn fracture
– schematic illustration 424
yarn split 420
yarn swelling 405
yield point 430
Young–Laplace equations 93
Young's modulus 7, 11, 142, 193, 279, 286, 287, 296, 322, 349, 356, 385, 429, 501, 504, 505, 507, 508, 509, 511, 584, 585, 590, 687, 700–705, 707, 759
– models for prediction 702, 703
YRR composite
– weaker planes 415

z

Z-glass 188
zinc oxide 780
zylon 777